This new, critical flora provides a definitive account of the native species, naturalised species, frequent garden escapes and casuals found in the British Isles. Planned in five volumes, its full keys and descriptions should enable the user to name all plants occurring in the wild, plus some ornamental trees and shrubs. For the first time full accounts of all the large apomictic genera are given and many infraspecific variants included.

Each species entry begins with the accepted Latin name, synonyms and the common English name. A detailed description follows, including information on flowering period, pollination and chromosome number. Separate descriptions are given for infraspecific taxa. Information on the status, ecology and distribution (including worldwide distribution) of the species and infraspecific taxa is also given.

Clear black and white line drawings illustrate an extensive glossary and illuminate the diagnostic features in a number of groups of plants.

FLORA OF GREAT BRITAIN AND IRELAND

FLORA
OF
GREAT BRITAIN
AND
IRELAND

VOLUME 5
BUTOMACEAE–ORCHIDACEAE

PETER SELL and GINA MURRELL
Herbarium, Department of Plant Sciences
University of Cambridge

CAMBRIDGE
UNIVERSITY PRESS

Published by the Press Syndicate of the University of Cambridge
The Pitt Building, Trumpington Street, Cambridge CB2 1RP
40 West 20th Street, New York, NY 10011-4211, USA
10 Stamford Road, Oakleigh, Melbourne 3166, Australia

First published 1996

Printed in Great Britain at the University Press, Cambridge

A catalogue record for this book is available from the British Library

Library of Congress cataloguing in publication data
Sell, P. D. (Peter D.)
Flora of Great Britain, Ireland, Isle of Man, and the Channel
Islands / Peter Sell and Gina Murrell.
p.   cm.
Includes bibliographical references and index.
Contents: – v. 5. Butomaceae–Orchidaceae.
ISBN 0 521 55339 3 (hardback: v. 5)
1. Botany–British Isles.   I. Murrell, Gina.   II. Title.
QK306.S44   1996
581.941–dc20   95-33535   CIP

ISBN 0 521 55339 3   hardback

To our Mentors

EDRED JOHN HENRY CORNER

JAMES EDGAR DANDY

HUMPHREY GILBERT CARTER

HARRY GODWIN

WILLIAM THOMAS STEARN

STUART MAX WALTERS

ALEXANDER STUART WATT

CYRIL WEST

Contents

Foreword

by S. M. Walters ScD, V.M.H.

It has been one of the continuing satisfactions of my academic career in Cambridge that the University Herbarium, of which I was Curator from 1948 to 1973, has provided an academic base for all my specialist interest in Angiosperm taxonomy to develop. Indeed, I count myself doubly fortunate that, twelve years after my retirement from academic life, the Herbarium, with its staff and visitors, still provides such a base where scholarship can be pursued for its own sake. With great pleasure I welcome this volume, the first of a set of five promised to us by Peter Sell and Gina Murrell. My association with Peter goes back more than half a century: though I was 'senior partner' in our happy collaboration in the post-war Herbarium, ours was a symbiotic relationship from which we both greatly benefited, and I was delighted when Gina, who had been part of the team in the 1960s and 1970s, returned to the fold as Herbarium Technician in 1991.

As explained in the Preface, this project to write an entirely new critical flora of Great Britain and Ireland comes to fruition some twenty years after an earlier scheme, in which the late Professor David Valentine took a leading part, had failed to find any financial support. Both Clive Stace, to whose *New Flora of the British Isles* (1991) Peter pays tribute in the Preface, and

Peter himself, were enthusiastic supporters of the Valentine project, and were prepared to play major parts in writing the Flora. It is fitting that both these eminent British taxonomists should separately carry on the tradition that David Valentine so enthusiastically advocated.

Two aspects of this new critical flora seem to be especially important. One concerns the acceptance, long overdue, of the 'alien element' in our flora as being equally worthy of taxonomic study: in this respect Stace's *Flora* represents a real change in attitude, which is to my mind unreservedly to be welcomed. The other, interestingly linked to the first by many examples, concerns the taxonomic recognition and treatment of hybrids and infraspecific variants. British botany lacks any single reference work from which the basic information about the variation of British vascular plants can be found, yet this information is increasingly needed by ecologists, conservationists, molecular biologists and biochemists, who will, as the century closes, determine the shape of much botanical study in Universities and specialised Institutions.

The authors of this impressive work have set themselves a colossal task. They have made an excellent start, and we can only wish them a successful conclusion.

Preface

For fifty years I have worked in the herbarium at Cambridge University on the British, Irish and European floras. I have collected some 25,000 specimens from most parts of Great Britain and Ireland and made many visits to Continental Europe. Particular attention has been given to most critical genera: *Cerastium*, *Crepis*, *Dactylorhiza*, *Euphrasia*, *Fumaria*, *Hieracium*, *Limonium*, *Pilosella*, *Prunus*, *Rhinanthus*, *Salicornia*, *Salix*, *Scleranthus*, *Sorbus* and *Ulmus*; and in helping friends in various ways I have considered the taxonomy of *Alchemilla*, Batrachian Ranunculi, *Chenopodium*, *Potamogeton*, *Rubus* and *Taraxacum*. I have also spent much time studying ecotypic and geographical variation, in particular a comparison of those variants which occur on the coasts in dunes, shingle and salt-marsh with those growing as arable weeds, and those in mountains. Special attention has also been given to trees and shrubs.

It has long been my wish to publish this information in a critical flora of Great Britain and Ireland. In the 1970s a group of us tried to get a grant to carry this out, but we were unsuccessful. Clive Stace then started on his *New Flora of the British Isles*, which is now published. In it he gives only abbreviated descriptions and omits the large apomictic genera and most of the infraspecific variants. Many introduced species are included by Stace in a British and Irish flora for the first time, detailed descriptions and specimens of many of which are difficult to find. Stace's flora is to my mind an excellent field guide, which it would be difficult to better, but it does not give the detailed descriptions which are needed to confirm the identification of a plant which is new to you. A good description in my opinion is one in which a picture of the plant unfolds before you as you read it.

I considered it was possible for me to write a flora in five volumes which gave a full description of all the species in Stace's flora and to add all the apomicts and many of the infraspecific variants, but it was too large a task to attempt to include all the biological information envisaged by the group in the 1970s. It was necessary, however, to have the help of another author, who lived in Cambridge, to deal with the large amount of work involved. My eye fell upon Gina Murrell who had worked with me in the 1960s and 1970s, when writing accounts for *Flora Europaea*, *Flora of Turkey* and *Flora of the Maltese Islands*. The work of one had complemented the work of the other and we were able to criticise one another without antagonism. We started field work on this flora on the 13 May 1987, by describing *Ceratocapnos claviculata*, which was flowering on Dunwich Heath in Suffolk, in a snowstorm. Since then we have as far as possible spent one day a week working in the field or at the Botanic Garden, Cambridge. We started writing this volume in 1992, and completed it by Easter 1994.

I have done most of the writing and made the taxonomic and nomenclatural decisions, while Gina has done most of the measuring, sometimes sitting at the microscope dictating the description while I, surrounded by a pile of books, wrote it down, and she has set out and put the whole onto a computer. At the start of the second volume we have where possible reversed roles and I sit at the microscope and dictate the description which she puts straight onto the computer, a much faster method of working. Gina has also done all the illustrations and organised our field work.

Peter Sell

Acknowledgements

We are greatly indebted to the following for reading through and commenting on our manuscript:
A. O. Chater (Cyperaceae, Juncaceae and Orchidaceae); J. Cullen (Iridaceae, Liliaceae and Orchidaceae); A. C. Leslie (Iridaceae and Liliaceae); D. C. McClintock (Bamboos); P. H. Oswald (*Allium*); C. D. Preston (water plants); R. H. Roberts (*Dactylorhiza*); P. J. O. Trist (Poaceae) and S. M. Walters (*Eleocharis*).

David Coombe and Max Walters have discussed many problems of taxonomy and translated much information, especially from German and Swedish. Arthur Chater has discussed many taxonomic and nomenclatural problems with us. Philip Oswald has dealt with questions of Greek and Latin grammar. Doug Kent has answered many nomenclatural questions for us, hunted out numerous obscure references, and read the proofs. Charlie Jarvis has helped us while working on the Hortus Cliffortianus at the British Museum, and Gina Douglas while working on the Linnaean collection at Burlington House. Barry Goddard has advised us on computer techniques. To Mrs J. E. Dandy we owe a special debt for giving us the second copy of her husband's manuscript of his detailed work on the nomenclature of the British flora. We are grateful to Harold Whitehouse for reading parts of the text and David Briggs for advice on breeding mechanisms.

At the Cambridge Botanic Garden we have received much help from Peter Kerley, Clive King, Caroline Lawes, D.-Z. Li, Peter Orriss, Donald Pigott, Norman Villis and Peter Yeo.

Chris Preston, as well as supplying us with almost verbatim accounts of the Potamogetonaceae and Ruppiaceae, has given us much information on other water plants, and has checked many records at the Monks Wood Records Centre.

To Clive Stace we owe a very special debt. Had he not written his *New Flora of the British Isles*, our task would have been insurmountable.

Introduction

HISTORICAL BACKGROUND

The first real flora of these islands was John Ray's *Catalogus Plantarum Angliae et Insularum Adjacentium* in 1670. The first flora to use the Linnaean binomial system of nomenclature was William Hudson's *Flora Anglica* nearly a hundred years later in 1762. This was followed by William Withering's *Botanical Arrangement of all the Vegetables naturally growing in Great Britain* in 1776–92, the first of many floras written primarily for the amateur.

James Sowerby's *English Botany*, whose text was written by J. E. Smith, was first published between 1790 and 1820. It presented for the first time a complete set of coloured illustrations of our plants, illustrations which are still unsurpassed for line and colour. The third edition published between 1863 and 1872 has inferior illustrations, but its text, rewritten by James Boswell Syme, is still important for its nomenclature and infraspecific taxa.

Three famous floras were produced in the nineteeth century. George Bentham's *Handbook of the British Flora* in 1858 was written as a before-breakfast relaxation. In it keys appeared for the first time in a British flora. It was revised by J. D. Hooker in 1886.

J. D. Hooker's *Students Flora of the British Islands*, first published in 1870 and finally revised in 1884, had beautifully clear and concise descriptions and was the main flora used by many generations of botanists up until the 1950s. It is also important in that Hooker was one of the first authors to use frequently the category of subspecies.

Charles Cardale Babington's *Manual of British Botany* first appeared in 1843 and the tenth edition, revised by A. J. Wilmott, was published in 1922. It contains many critical species and varieties not in other floras, but the descriptions are not clear and without keys it is difficult to use.

C. E. Moss's *Cambridge British Flora* (1914–20) was very detailed and would have supplied a much needed critical flora, but alas only two volumes were published.

The arrival of 'C. T. & W.', A. R. Clapham, T. G. Tutin and E. F. Warburg's *Flora of the British Isles*, in 1952, heralded the beginning of a new era in the study of British plants. It was the first up-to-date treatment this century. A much revised second edition appeared in 1962 and a third in 1987 when D. M. Moore replaced E. F. Warburg. This last edition included the information in Tutin et al., *Flora Europaea* **1–5** (1964–80). The nomenclature had been brought up to date by J. E. Dandy in his *List of British Vascular Plants* in 1958, and the work he did on this for *Flora Europaea*. Thus for the first time taxonomy and nomenclature had been brought in line with that of Continental Europe.

The Botanical Society of the British Isles's publication of the *Atlas of the British Flora* in 1962, edited by F. H. Perring and S. M. Walters, and the *Critical Supplement to the Atlas of the British Flora* in 1968, edited by F. H. Perring, gave us a much better idea of the distribution of our plants.

The arrival of Clive Stace's *New Flora of the British Isles* in 1991 and D. H. Kent's *List of Vascular Plants of the British Isles* in 1992 has brought about the end of the C. T. & W. era and given us a completely up-to-date account of our flora. The most important changes are the moving over of the main classification to A. Cronquist's *An integrated system of classification of flowering plants* (1981) and the inclusion of almost as many alien species as native ones.

The aim of our Flora is to supply full descriptions of all the species in Stace's flora, to include all the large apomictic genera and many infraspecific variants, and to add more information about hybrids, for which extensive use has been made of Stace's *Hybridization and the flora of the British Isles* (1975).

THE CONTENTS OF THE FLORA

The flora includes all the vascular plants, Lycopodiophyta (Clubmosses), Equisetophyta (Horsetails), Pteridophyta (Ferns), Pinophyta (Conifers) and Magnoliophyta (Flowering plants). The list of plants is made up of all our native species, including apomicts, and all the introduced plants given in Stace (1991), with a few more added, particularly planted trees. E. J. Clement and M. C. Foster's *Alien plants of the British Isles* arrived in 1994 after we had completed this volume, but we have been through it and added as much information as possible. It is a great pity that their account of the Poaceae is still unpublished. These alien taxa will be found to be more widespread when attention is given to them. In his coverage of alien taxa Stace considers inclusion is merited when an alien is either naturalised (i.e. permanent and competing with other vegetation, or self-perpetuating) or, if a casual, frequently recurrent so that it can be found in most years. These criteria were applied as much to garden escapes or throw-outs as to the unintentionally introduced plants, and rarity was not taken into consideration for any of them. Cultivated species were included if they are field crops or forestry crops, or, in the case of trees only, ornamentals grown on a large scale.

Stace's aim has been to include **all taxa that the botanist might reasonably be able to find in the wild in any one year**. To these we have added some ornamental trees and shrubs which are planted along streets and roadsides and which we consider to be part of the landscape. Usually plants in gardens are not mentioned at all, but some species, which seed freely and spread over areas of garden and lawn where they are not planted, are included. Most of the species which Stace has mentioned, but not numbered or included in the keys, are here included, while a few have been left out altogether. We have started with Volume 5 because the *The European Garden Flora* has already covered the Monocotyledons, which has made it easier for us to deal with the garden escapes. We will follow it with Volume 4, because it contains the large genera *Hieracium* and *Taraxacum* of which up-to-date accounts are badly needed.

GEOGRAPHICAL AREA

The flora includes England, Scotland and Wales, collectively known as Great Britain, Northern Ireland and Eire together forming Ireland, the Isle of Man, and the Channel Islands which include Jersey, Guernsey, Alderney, Sark, Herm and various small islands. In these respects it follows Stace (1991).

The smallest geographical area usually referred to is the county (which sometimes includes more than one vice-county), for which in Great Britain we have used the boundaries adopted by H. C. Watson in 1873 in *Topographical Botany* and in Ireland by R. L. Praeger in 1901 in *Irish Topographical Botany*. The vice-counties used by botanists have the benefit of not changing at regular intervals as do the political counties. With rare or local species the actual place or area may be given. The extra-limital distributions are those given in Clapham, Tutin and Moore (1987) with as much correcting as we can give them. Russia and Yugoslavia have been used in the sense of the old USSR and Yugoslavia before recent political disruptions.

CLASSIFICATION AND NOMENCLATURE

The classification follows that of Stace (1991) and Kent (1992) which is taken from A. Cronquist, *An integrated system of classification of flowering plants* (1981), with the exception that the main groups are called Divisions and the second groups Classes following H. C. Bold, C. Alexopoulos and T. Deleveryas, *Morphology of plants and fungi*, ed. 4 in 1980 and A. Cronquist, A. Takhtajan and W. Zimmermann, On the higher taxa of embryobionta in *Taxon* **15**: 129–134 in 1966, and set out by one of us, P. D. S., in *The Cambridge cyclopedia of life sciences* in 1985.

One of us (P. D. S.) has specialised in nomenclature for many years and it is here made as accurate as possible according to the latest *International code of botanical nomenclature*. The names of genera and species differ little from those in Stace (1991) and Kent (1992). Recent changes in the code of nomenclature have been used to get rid of some names which have been a long-standing source of confusion.

No rules have been made about the number of synonyms given, as many as possible being included, but an attempt has been made to include all names used in British and Irish floras. The abbreviation *auct.* following a name means only that the name has not been accepted for the plant; it does not mean the type has been checked and the name rejected. Only in the case of a later homonym, which has been checked, does the word *non* and an author follow the name and author.

The English name for the species follows Stace (1991) as far as possible, and where new ones were needed they have been created.

VARIATION

The recording of variation is most important for both ecology and conservation, and even more important for gardeners who go out of their way both to create and to conserve prominent variants.

Infraspecific variation is usually recorded by the recognition of subspecies, varieties, formas and cultivars. These taxa differ chiefly in ecology and distribution. A **forma** is a plant with a one or two gene difference which occurs with one or more other forms in a mixed population for most or all of its range. A **variety** is when one of these **formas** becomes more or less dominant in a particular ecological area, i.e. an *ecotype*. A **subspecies** is when one of these formas becomes dominant in a geographical area, i.e. a *race*. A **cultivar** is a forma which is selected by horticulturalists and perpetuated, usually vegetatively. Because ecotypes and races have become adapted morphologically to different conditions over a long period, it is likely that their physiology and biochemistry, and indeed their whole biology, is different and they probably differ by more than just one or two genes. Also, as they often flower at different periods their pollinators may be different and, if climatic conditions alter, one ecotype may be better able to survive than another. Variation thus becomes very important in conservation. It is unfortunate that many botanists tend to ignore variation completely, and they will certainly ignore it if it has no name at all; subspecies are usually more often recognised than varieties. Sometimes it is more important to conserve one variety rather than another, e.g. the Chilterns *Orchis militaris* var. *tenuifrons* is endemic, while the Suffolk var. *militaris* occurs in Continental Europe; *Liparis loeselii* var. *ovata* is rare in distribution, but frequent where it occurs, whereas var. *loeselii* is rare in Britain but occurs on the Continent. Sometimes the variant will tell us whether the plant is native or not; e.g. *Leucojum aestivum* subsp. *aestivum* is native, subsp. *pulchellum* is a naturalised garden escape. Escaped cultivars are named wherever

Shetland Islands

100 km

Orkney
Islands

Outer Hebrides

Inner Hebrides

GREAT BRITAIN

**Scottish
Highlands**

IRELAND

Lake District

River Tees

Yorkshire Dales

Isle of Man

Peak District

River Humber

Connemara

The Wash

**East
Anglia**

Snowdonia

Fens

The Broads

The
Burren

Brecon
Beacons

Breckland

River
Thames

River Severn

Lundy
Island

Isles of Scilly

**CHANNEL
ISLANDS**

Alderney

—Guernsey

FRANCE

Jersey

BRITISH ISLES

Vice–counties of the British Isles

(courtesy of *New Flora of the British Isles*, Stace, CUP 1991)

112

100km

111

108   109

107

110   105   106

95   93
94

96   92

104   91   Scotland

97

89   90

103   88

98   87   85   Great Britain

99   86   84   82
76   83   81
102   77   78   79   80   68   England
101   100   75   72   80
67

Ireland   35   74   73   70   66

34   40   39   69   65   62

36   64

33   37   38   60   61

27   28   29   32   71

30   59   63   54

26   25   24   31   57   56

16   17   23   22   52   58   53

15   18   19   21   49   50   51   28   27

9   14   20   48   47   40   55   31   29   26   25

10   13   39   38   32   30

8   11   12   46   43   37   33   23   24   20   19

2   7   45   44   42   36   18

6   41   35   33   34   21
Wales   46   7   22   17   16   15

4   5   6   8   12   13   14

3   4   5   11

9   10

CI

ENGLAND, WALES, SCOTLAND, ISLE OF MAN

1. West Cornwall
2. East Cornwall
3. South Devon
4. North Devon
5. South Somerset
6. North Somerset
7. North Wiltshire
8. South Wiltshire
9. Dorset
10. Isle of Wight
11. South Hampshire
12. North Hampshire
13. West Sussex
14. East Sussex
15. East Kent
16. West Kent
17. Surrey
18. South Essex
19. North Essex
20. Hertfordshire
21. Middlesex
22. Berkshire
23. Oxfordshire
24. Buckinghamshire
25. East Suffolk
26. West Suffolk
27. East Norfolk
28. West Norfolk
29. Cambridgeshire
30. Bedfordshire
31. Huntingdonshire
32. Northamptonshire
33. East Gloucestershire
34. West Gloucestershire
35. Monmouthshire
36. Herefordshire
37. Worcestershire
38. Warwickshire

39. Staffordshire
40. Shropshire
41. Glamorganshire
42. Breconshire
43. Radnorshire
44. Carmarthenshire
45. Pembrokeshire
46. Cardiganshire
47. Mongomeryshire
48. Merionethshire
49. Caernarvonshire
50. Denbighshire
51. Flintshire
52. Anglesey
53. South Lincolnshire
54. North Lincolnshire
55. Leicestershire
56. Nottinghamshire
57. Derbyshire
58. Cheshire
59. South Lancashire
60. West Lancashire
61. South-east Yorkshire
62. North-east Yorkshire
63. South-west Yorkshire
64. Middle-west Yorkshire
65. North-west Yorkshire
66. Co. Durham
67. South Northumberland
68. Cheviotland
69. Westmorland
70. Cumberland
71. Isle of Man
72. Dumfriesshire
73. Kirkcudbrightshire
74. Wigtownshire
75. Ayrshire
76. Renfrewshire

77. Lanarkshire
78. Peeblesshire
79. Selkirkshire
80. Roxburghshire
81. Berwickshire
82. East Lothian
83. Midlothian
84. West Lothian
85. Fifeshire
86. Stirlingshire
87. West Perthshire
88. Mid Perthshire
89. East Perthshire
90. Forfarshire
91. Kincardineshire
92. South Aberdeenshire
93. North Aberdeenshire
94. Banffshire
95. Morayshire
96. East Inverness-shire
97. West Inverness-shire
98. Main Argyllshire
99. Dunbartonshire
100. Clyde Islands
101. Kintyre
102. South Ebudes
103. Middle Ebudes
104. North Ebudes
105. West Ross-shire
106. East Ross-shire
107. East Sutherland
108. West Sutherland
109. Caithness
110. Outer Hebrides
111. Orkney Islands
112. Shetland Islands

IRELAND

H1. South Kerry
H2. North Kerry
H3. West Cork
H4. Mid Cork
H5. East Cork
H6. Co. Waterford
H7. South Tipperary
H8. Co. Limerick
H9. Co. Clare
H10. North Tipperary
H11. Co. Kilkenny
H12. Co. Wexford
H13. Co. Carlow
H14. Laois

H15. South-east Galway
H16. West Galway
H17 North-east Galway
H18. Offaly
H19. Co. Kildare
H20. Co. Wicklow
H21. Co. Dublin
H22. Meath
H23. Westmeath
H24. Co. Longford
H25. Co. Roscommon
H26. East Mayo
H27. West Mayo
H28. Co. Sligo

H29. Co. Leitrim
H30. Co. Cavan
H31. Co. Louth
H32. Co. Monaghan
H33. Fermanagh
H34. East Donegal
H35. West Donegal
H36. Tyrone
H37. Co. Armagh
H38. Co. Down
H39. Co. Antrim
H40. Co. Londonderry

they can be easily recognised and are considered important. All apomicts, where possible, are treated as species, long experience showing that any sort of lumping deprives them of having an interesting ecology or distribution. Hybrids are dealt with as fully as possible, especially those which spread vegetatively. No serious attempt has been made to decide on the correct infraspecific rank as taxa are often both ecological and geographical. Uniformity of infraspecific rank is often produced in a species or genus, but usually the only important thing considered is that a recognisable infraspecific taxon has a name.

HERBARIA AND LITERATURE CONSULTED

During the writing of the flora the following books were consulted for every species:

Clapham, A. R., Tutin, T. G. & Moore, D. M. *Flora of the British Isles*. Ed. 3. 1987. Cambridge.

Dandy, J. E. *List of British vascular plants*. 1958. London.

Hegi, G. *Illustrierte Flora von Mitteleuropa*. Ed. 1. 1906–31. Ed. 2. 1936–. Ed. 3. 1966–. Munich.

Kent, D. H. *List of vascular plants of the British Isles*. 1992. London.

Perring, F. H. & Walters, S. M. *Atlas of the British flora*. 1962. London & Edinburgh.

Perring, F. H. *Critical supplement to the atlas of the British flora*. 1968. London.

Stace, C. A. *Hybridization and the flora of the British Isles*. 1975. London, New York & San Francisco.

Stace, C. *New flora of the British Isles*. 1991. Cambridge.

Tutin, T. G., Heywood, V. H., Burges, N. A., Moore, D. M., Valentine, D. H., Walters, S. M. & Webb, D. A. *Flora Europaea*. **1–5**. 1964–80. Cambridge.

Walters, S. M., Brady, A., Brickell, C. D., Cullen, J., Green, P. S., Lewis, J., Matthews, V. A., Webb, D. A., Yeo, P. F. & Alexander, J. C. M. *The European garden flora*. **1–**. 1986–. Cambridge.

Many other books and journals were consulted, mainly in the Cambridge Department of Plant Sciences, including the N. D. Simpson collection of local floras, and the Cory Library at the Botanic Garden. Where these references were considered to be important for particular plants, we have cited them under the family or genus concerned.

The University herbaria at Cambridge, on which the flora is mainly based, are ideal for the study of the British flora for the following reasons:

1 The large British collection contains specimens from most of the main collectors of British plants from 1800 onwards, including sets of published exsiccatae and specimens sent through the Botanical Exchange Clubs. Most of the critical species have been named by experts.

2 The British herbarium contains 25,000 specimens collected by us in the last fifty years. The specimens are accompanied by detailed field notes and are often of critical species or infraspecific taxa.

3 There is a good herbarium of Continental European plants with which to compare the British plants.

4 The world collection in the Cambridge herbarium contains 50,000 sheets of John Lindley's herbarium made when he was secretary of the Royal Horticultural Society, when plants were coming into the country from all parts of the world; and the C. M. Leman collection, named by George Bentham, and put together at the same time. These collections are very important as regards the alien species when considered in conjunction with the Botanic Garden herbarium and recent gatherings of alien specimens.

5 The Botanic Garden herbarium contains a large collection of cultivated plants.

Thus, the libraries, herbaria, our own field notes and plants grown in the Botanic Garden have enabled us to do most of the work in Cambridge. Over many years books and specimens elsewhere have been consulted.

Kingdom PLANTAE

Volume 1.

Division 1. LYCOPODIOPHYTA

Order 1. LYCOPODIALES

1. LYCOPODIACEAE

Order 2. SELAGINELLALES

2. SELAGINELLACEAE

Order 3. ISOETALES

3. ISOETACEAE

Division 2. EQUISETOPHYTA

Order 1. EQUISETALES

4. EQUISETACEAE

Division 3. PTERIDIOPHYTA

Order 1. OPHIOGLOSSALES

5. OPHIOGLOSSACEAE

Order 2. OSMUNDALES

6. OSMUNDACEAE

Order 3. PTERIDALES

7. ADIANTACEAE   **8. PTERIDACEAE**

Order 4. MARSILEALES

9. MARSILEACEAE

Order 5. HYMENOPHYLLALES

10. HYMENOPHYLLACEAE

Order 6. POLYPODIALES

11. POLYPODIACEAE

Order 7. DICKSONIALES

12. CYATHEACEAE   **13. DICKSONIACEAE**

Order 8. DENNSTAEDTIALES

14. DENNSTAEDTIACEAE   **18. DAVALLIACEAE**

15. THELYPTERIDACEAE   **19. DRYOPTERIDACEAE**

16. ASPLENIACEAE   **20. BLECHNACEAE**

17. WOODSIACEAE (*ATHYRIACEAE*)

Order 9. SALVINIALES

21. AZOLLACEAE

Division **4. PINOPHYTA**

Class 1. PINOPSIDA

Order 1. PINALES

22. PINACEAE   **24. CUPRESSACEAE**

23. TAXODIACEAE   **25. ARAUCARIACEAE**

Class 2. TAXOPSIDA

Order 1. TAXALES

26. TAXACEAE

Division **5. MAGNOLIOPHYTA**

Class 1. MAGNOLIOPSIDA

Subclass 1. MAGNOLIIDAE (*DICOTYLEDONES*)

Order 1. MAGNOLIALES

27. MAGNOLIACEAE

Order 2. LAURALES

28. LAURACEAE

Order 3. ARISTOLOCHIALES

29. ARISTOLOCHIACEAE

Order 4. NYMPHAEALES

30. NYMPHACEAE **31. CERATOPHYLLACEAE**

Order 5. RANUNCULALES

32. RANUNCULACEAE **33. BERBERIDACEAE**

Order 6. PAPAVERALES

34. PAPAVERACEAE **35. FUMARIACEAE**

Subclass 2. HAMAMELIDAE

Order 1. HAMAMELIDALES

36. PLATANACEAE

Order 2. URTICALES

37. ULMACEAE **39. MORACEAE**

38. CANNABACEAE **40. URTICACEAE**

Order 3. JUGLANDALES

41. JUGLANDACEAE

Order 4. MYRICALES

42. MYRICACEAE

Order 5. FAGALES

43. FAGACEAE **45. CORYLACEAE**

44. BETULACEAE

Subclass 3. CARYOPHYLLIDAE

Order 1. CARYOPHYLLALES

46. PHYTOLACCACEAE

47. AIZOACEAE

48. CHENOPODIACEAE

49. AMARANTACEAE

50. PORTULACACEAE

51. BASELLACEAE

52. CARYOPHYLLACEAE
(*ILLECEBRACEAE*)

Order 2. POLYGONALES

53. POLYGONACEAE

Order 3. PLUMBAGINALES

54. PLUMBAGINACEAE

Subclass 4. DILLENIIDAE

Order 1. DILLENIALES

55. PAEONIACEAE

Order 2. THEALES

56. ELATINACEAE

57. CLUSIACEAE
(*GUTTIFERAE*; *HYPERICACEAE*)

Order 3. MALVALES

58. TILIACEAE

59. MALVACEAE

Order 4. NEPENTHALES

60. SARRACENIACEAE

61. DROSERACEAE

Order 5. VIOLALES

62. CISTACEAE

63. VIOLACEAE

64. TAMARICACEAE

65. FRANKENIACEAE

66. CUCURBITACEAE

Order 6. SALICALES

67. SALICACEAE

Volume 2.

Order 7. CAPPARALES

68. CAPPARACEAE

69. BRASSICACEAE (*CRUCIFERAE*)

70. RESEDACEAE

Order 8. ERICALES

71. CLETHRACEAE

72. EMPETRACEAE

73. ERICACEAE

74. PYROLACEAE

75. MONOTROPACEAE

Order 9. DIAPENSIALES

76. DIAPENSIACEAE

Order 10. PRIMULALES

77. MYRSINACEAE

78. PRIMULACEAE

Subclass 5. ROSIDAE

Order 1. ROSALES

79. PITTOSPORACEAE

80. HYDRANGEACEAE

81. ESCALLONIACEAE

82. GROSSULARIACEAE

83. CRASSULACEAE

84. SAXIFRAGACEAE

85. ROSACEAE

Volume 3.

Order 2. FABALES

86. MIMOSACEAE

87. CAESALPINIACEAE

88. FABACEAE

Order 3. PROTEALES

89. ELAEAGNACEAE

Order 4. HALORAGALES

90. HALORAGACEAE

91. GUNNERACEAE

Order 4. CYPERALES

166. CYPERACEAE   **167. POACEAE (*GRAMINEAE*)**

Order 5. TYPHALES

168. SPARGANIACEAE   **169. TYPHACEAE**

Subclass 4. ZINGIBERIDAE

Order 1. BROMELIALES

170. BROMELIACEAE   **171. PONTEDERIACEAE**

Subclass 5. LILIIDAE

Order 1. LILIALES

172. LILIACEAE (*ALLIACEAE*;   **174. AGAVACEAE**
***AMARYLLIDACEAE*; *TRILLIACEAE*)**

173. IRIDACEAE   **175. DIOSCOREACEAE**

Order 2. ORCHIDALES

176. ORCHIDACEAE

Division **MAGNOLIOPHYTA** Cronquist, Takht. & W. Zimm.

Class 2. LILIOPSIDA Cronquist, Takht. & W. Zimm. (*MONOCOTYLEDONIDAE*)

Mostly herbaceous plants, rarely trees or shrubs, frequently with bulbs, corms or rhizomes; rarely with secondary thickening and never from a permanent vascular cambium; vascular bundles usually scattered through the stem; primary roots usually short-lived, the mature root system wholly adventitious, fibrous. Cotyledon 1. Leaves usually alternate or all basal, usually with parallel venation, scarcely or not reticulate. Flower parts mostly in threes or multiples of three. Pollen grains mostly bilaterally symmetrical, commonly with one pore and/or furrow.

The class Liliopsida contains 5 subclasses, 19 orders, 65 families and nearly 50,000 species and occurs throughout the world.

Artificial key to families

1. Plant consisting of floating or submerged, more or less undifferentiated, pad-like fronds less than 10(–15) mm in diameter, sometimes with a narrow, stalk-like part at one end, with or without roots dangling in water, rarely stranded temporarily on mud   **162. LEMNACEAE**
1. If plants free-floating then with clearly differentiated stem and leaves   2.
2. Aquatic or mud plants with at least some leaves in whorls of more than 3, the leaves linear or more or less so, or divided into linear segments   3.
2. If aquatic or in mud then leaves not whorled and/or not linear or with linear segments   5.
3. Leaves with stipules free from the leaf-bases   **158. ZANNICHELLIACEAE**
3. Leaves without stipules, often with a sheathing leaf-base, or with 2 minute scales at the base in the axil   4.
4. Leaves wider near the base than near the apex, but narrowed at the extreme base, sometimes slightly clasping the stem, but not sheathing it.   **151. HYDROCHARITACEAE**
4. Leaves with a distinctly widened base shortly sheathing the stem   **157. NAJADACEAE**
5. Trees with an unbranched stem and terminal rosettes of huge pinnate or palmate leaves; or seedlings with the leaves ribbed alternately on each surface   **160. ARECACEAE**
5. If trees then not with a single terminal rosette of compound leaves; if seedlings then not with the leaves ribbed alternately on each surface   6.
6. Plant consisting of one to few rosettes of many, simple, linear leaves usually more than 1 m, either borne on the ground or at the tips of woody branches   **174. AGAVACEAE**
6. If leaves all in one or few rosettes then much shorter than 1 m and often not linear   7.

7. Plants consisting of a dense hemispherical mass often more than 1 m across, with narrow pineapple-like leaves which have strongly spiny margins   **170. BROMELIACEAE**
7. If plant is a dense hemispherical spiny mass, then a growth form of plant without pineapple-like leaves   8.
8. All inflorescences entirely replaced by vegetative propagules, i.e. bulbils or plantlets   9.
8. At least some inflorescences bearing flowers or fruits   11.
9. Inflorescences replaced by axillary, solitary or terminal clusters of small, solid bulbils   **172. LILIACEAE**
9. Inflorescences proliferating by producing small plantlets in place of flowers   10.
10. Stems solid; leaves flattened-cylindrical   **165. JUNCACEAE**
10. Stems hollow; leaves flat or inrolled or infolded   **167. POACEAE**
11. Perianth 0, or of 1 whorl, or of more than 2 whorls of, or a spiral of, similar segments   12.
11. Perianth of 2 (rarely more) distinct whorls or rarely a spiral, the inner and outer differing markedly in shape, size or colour   38.
12. Perianth more or less corolla-like, usually white or distinctly (often brightly) coloured   13.
12. Perianth more or less calyx-like or bract-like, green to brownish, or scarious, or much reduced, or absent   19.
13. Plants functionally dioecious   **172. LILIACEAE**
13. Plants monoecious, flowers unisexual or bisexual   14.
14. Ovary superior; flowers hypogynous to perigynous   15.
14. Ovary inferior; flowers epigynous   17.
15. Inflorescence an umbel   **149. BUTOMACEAE**
15. Inflorescence a spike   16.
16. Inflorescence a forked spike; perianth segments 1 or rarely 2, white   **152. APONOGETONACEAE**
16. Inflorescence a terminal spike; perianth segments 6, violet-blue   **171. PONTEDERIACEAE**
17. Stamens more than 6   **172. LILIACEAE**
17. Stamens less than 5   18.
18. Style obvious, with 3 obvious branches or 3 separate stigmas; stamens 3   **173. IRIDACEAE**
18. Styles absent; stamens 1 or 2   **176. ORCHIDACEAE**
19. Leaves at least partly opposite or whorled; aquatic or marsh plants with floating procumbent or very weakly ascending stems   20.
19. Leaves all alternate or all basal, or if some or all opposite, then plant not aquatic or if so then stems self-supporting   22.
20. Perianth segments 4–6 on at least some flowers; stamens 4–12, or flowers female   **155. POTAMOGETONACEAE**

20. Perianth segments absent; stamens 1–4, or flowers female   21.

21. Flowers bisexual; stamens 4; fruits on stalks more than 10 mm; only upper leaves opposite **156. RUPPIACEAE**

21. Flowers unisexual; male flowers with 1–2 stamens; fruits on stalks less than 10 mm; more or less all leaves opposite, or rarely in threes **158. ZANNICHELLIACEAE**

22. Leaves linear and grass-like, sheathing the stem proximally; flowers greatly reduced, arranged in units largely composed of leafy or membranous scaly bracts, with the perianth absent or represented by bristles or minute scales, aerial   23.

22. Leaves various; flowers with an obvious structure, mostly with a perianth, if greatly reduced without or with a greatly reduced perianth then not arranged in units as above and often subaquatic or on the surface of water   24.

23. Stems usually with solid internodes, often more or less triangular in section; leaf-sheaths usually cylindrical, with fused margins; flowers never with a bract above **166. CYPERACEAE**

23. Stems (culms) usually with hollow internodes, circular, or rarely compressed, or more or less quadrangular in section; leaf sheaths usually with free overlapping margins; flowers with a bract above as well as below, or if not stem hollow **167. POACEAE**

24. Aquatic or marsh plants with linear leaves   25.

24. If leaves linear then plants not in water or marshes; if aquatic then leaves not linear   34.

25. Leaves all basal; inflorescence a tight, capitate mass on a long, leafless stem **164. ERIOCAULACEAE**

25. If leaves all basal, the inflorescence not a single, terminal, tight, capitate mass   26.

26. Flowers very small with many tightly packed in a dense, spherical or elongated conspicuous cluster   27.

26. Flowers not many together in dense clusters   29.

27. Fresh leaves with a strong spicy scent when crushed; flowers bisexual **161. ARACEAE**

27. Leaves without a spicy scent; flowers unisexual, the male and female in neatly separated parts of the inflorescence   28.

28. Flowers in globose heads **168. SPARGANIACEAE**

28. Flowers in cylindrical spikes **169. TYPHACEAE**

29. Leaves very thin, ribbon-like or thread-like, mostly subaquatic   30.

29. Leaves thicker, not ribbon-like or thread-like   32.

30. Flowers bisexual, borne in stalked spikes **155. POTAMOGETONACEAE**

30. Plants dioecious or monoecious; flowers unisexual, borne in stalked or sessile spathes   31.

31. Flowers bisexual, in short- or long-stalked spathes; perianth segments 3; in fresh water **151. HYDROCHARITACEAE**

31. Flowers unisexual, in sessile spathes; perianth segments absent; marine **159. ZOSTERACEAE**

32. Flowers in branched cymes, sometimes compact **165. JUNCACEAE**

32. Flowers in a simple, terminal raceme   33.

33. Leaves several on a stem with a prominent pore at the apex; flowers less than 12, with bracts **153. SCHEUCHZERIACEAE**

33. Leaves on the stem absent, or with a few near the base, without a pore at the apex; flowers numerous, without bracts **154. JUNCAGINACEAE**

34. Flowers small, in a dense spike on an axis, the axis sometimes extended distally as a sterile projection, with a large spathe at the base often partly or wholly obscuring the flowers **161. ARACEAE**

34. If flowers in a single dense spike then without a bract at the base   35.

35. Plants aquatic or in wet bogs; leaves simple, entire or more or less so **155. POTAMOGETONACEAE**

35. Plants usually on dry ground; if in marshes or bogs leaves not simple or entire   36.

36. Leaves ovate **175. DIOSCOREACEAE**

36. Leaves or leaf-like organs linear, without basal lobes   37.

37. Leaves reduced to scales or not, but not replaced by a cluster of cladodes; flowers bisexual **165. JUNCACEAE**

37. Leaves reduced to scales, their normal function carried out by bunches of cladodes; plant dioecious and flowers unisexual **172. LILIACEAE**

38. Flowers functionally all male or all female, but vestigial parts of the other sex present   39.

38. At least some flowers bisexual, or both male and female flowers present   41.

39. Shrubs **172. LILIACEAE**

39. Herbs   40.

40. Aquatic or marsh plants with all the leaves in a basal rosette **151. HYDROCHARITACEAE**

40. Plants of dry ground, the leaves reduced to scales and replaced by more or less cylindrical cladodes **172. LILIACEAE**

41. Perianth segments fused at base for varying distances up to the apex   42.

41. Perianth segments free, or rarely more or less fused near the apex and free at the base   44.

42. Ovary superior (hypogynous or perigynous) or partly so **164. ERIOCAULACEAE**

42. Ovary inferior (epigynous) or partly so   43.

43. Stamens 3, with obvious filaments and anthers **173. IRIDACEAE**

43. Stamens 2, with sessile anthers **176. ORCHIDACEAE**

44. Ovary inferior (epigynous)   45.

44. Ovary superior (hypogynous or perigynous)   48.

45. Flowers zygomorphic; stamens 1–2 **176. ORCHIDACEAE**

45. Flowers actinomorphic; stamens 3–12   46.

46. Plant aquatic; outer whorl of perianth segments sepaloid; stamens 9–12, or absent in female flowers **151. HYDROCHARITACEAE**

46. Both whorls of perianth segments petaloid; stamens 3–6   47.

47. Stamens 6 **172. LILIACEAE**

47. Stamens 3 **173. IRIDACEAE**

48. Carpels and styles free, or carpels fused just at base   49.
48. Carpels and/or styles fused wholly or for the greater part   50.
49. Leaves linear; perianth segments purple with green tinge   **149. BUTOMACEAE**
49. Leaves ovate; outer perianth segments green
   150. ALISMATACEAE
50. Perennial herb   **163. COMMELINACEAE**
50. Woody shrub   **172. LILIACEAE**

Subclass 1. ALISMATIDAE Takht.

Herbs living in water or in wet places, sometimes marine. *Flowers* bisexual or unisexual. *Perianth segments* of 0 to 2 whorls. *Stamens* 1–6, rarely more. *Ovules* 1 to numerous; placentation basal, parietal, apical or scattered. *Seeds* with little or no endosperm.

This subclass consists of 4 orders, 16 families and about 500 species.

Order 1. ALISMATALES Lindl.

Herbs living in water or in wet places. *Flowers* hypogynous. *Perianth segments* generally divided into sepals and petals. *Carpels* distinct or only basally connate. *Seeds* without endosperm.

This order consists of 3 families and fewer than 100 species.

149. BUTOMACEAE Rich. nom. conserv.

Glabrous, rhizomatous, aquatic *perennials. Leaves* all basal, linear, erect. *Flowers* with long pedicels, in a terminal umbel, bisexual, hypogynous, actinomorphic. *Sepals* 3, petaloid, free, persistent. *Petals* 3, free. *Stamens* 9; filaments flattened; anthers basifixed, 2-celled, opening by lateral slits. *Ovary* superior. *Carpels* 6, connate at base, with numerous ovules all over the inner wall. *Seeds* small; endosperm 0; embryo straight.

One genus in temperate Eurasia.

1. Butomus L.

As above.

One species in temperate Eurasia.

Krahulcová, A. & Jarolimová, V. (1993). Ecology of two cytotypes of *Butomus umbellatus*. I. Karyology and breeding behaviour. *Folia Geobot. Phyt.-Tax.* **28**: 385–411.

1. B. umbellatus L.   Flowering Rush

Glabrous, rhizomatous *perennial*, rooted in mud and emergent through water. *Stems* up to 150 cm, erect, terete. *Leaves* all basal, shorter than to as long as the stem, up to 20 mm wide, bright but pale yellowish-green, linear, entire, acutely triquetrous, more or less spirally twisted especially at their acute withered tips, sheathing the stem with their broad, hollow, ribbed bases. *Inflorescence* a terminal umbel, subtended by bracts; bracts 3, 22–30 × 5–8 mm, lanceolate, entire, long acuminate at apex; pedicels 50–100 mm, unequal, brownish-purple. *Flowers* 28–32 mm in diameter, opening in succession. *Sepals* 3, 10–13 × 7–8 mm, brownish-pink in the centre with a diffusely pink to white margin, obovate, shallowly crenulate, rounded at apex. *Petals* 3, 13–15 × 13–15 mm, white, diffusely pink to pinkish-red, broadly obovate, shallowly crenulate, rounded at apex. *Stamens* 9; filaments 5–6 mm, white tinted pink at base, flattened; anthers pinkish-purple. *Style* 1, both it and stigma pinkish. *Fruit* a group of 6, dirty purplish follicles, connate at the base. *Seeds* numerous, small. *Flowers* 7–9. 2n = 26, 39.

Our plants need to be looked at to see if they are diploid or triploid, or both (cf. Krahulcová & Jarolimová, 1993).

Native. Ditches, ponds, canals and the margins of rivers. Scattered in Britain north to central Scotland and in Ireland, but often introduced. Europe and temperate Asia; naturalised in North America.

150. ALISMATACEAE Vent. nom. conserv.

Glabrous, aquatic, *annual* or *perennial herbs*, rooted in mud and often emergent through water. *Leaves* usually all basal, submerged ones often ribbon-like and very different from the aerial ones, simple, entire, sessile or petiolate, sometimes produced as tufts on rooting stolons; exstipulate. *Flowers* in simple or compound umbels or in whorls, sometimes solitary, with bracts at the base of the umbel or whorl, bisexual or unisexual, actinomorphic, hypogynous. *Sepals* 3, free, green. *Petals* 3, free, white to pink. *Stamens* 6 to numerous; anthers 2-celled. *Style* 0, or very short; stigma not lobed. *Ovary* superior. *Carpels* more than 6, free, with 1 to several ovules. *Fruit* a group of achenes or few-seeded follicles. *Seeds* small; endosperm 0.

Eleven genera and about 95 species, in temperate and tropical regions, mainly in the northern hemisphere.

Caldesia parnassifolia (L.) Parl., Parnassus-leaved Water Plantain, is established in woods at Kinfauns, East Perth.

1. Leaves often sagittate; flowers unisexual, male and
   female in same inflorescence; stamens more than 6
                          1. Sagittaria
1. Leaves never sagittate; flowers bisexual; stamens 6   2.
2. Stems procumbent or floating, rooting and producing
   tufts of leaves and inflorescences   3.
2. Stems erect, leafless, all leaves basal   4.
3. Floating or aerial leaves obtuse at apex; carpels in an
   irregular whorl or rather flat mass   **2. Luronium**
3. Floating or aerial leaves acute at apex; carpels spiral in
   a more or less globose head   **3. Baldellia**
4. Carpels spiral in a more or less globose head   **3. Baldellia**
4. Carpels in a single, often irregular whorl   5.
5. Leaves narrowed or subcordate at base; fruits curved
   inwards, more or less unbeaked, 1-seeded   **4. Alisma**
5. Leaves cordate; fruits divergent outwards, beaked,
   usually 2(–few)-seeded   **5. Damasonium**

1. Sagittaria L.

Perennial aquatic *herbs. Leaves* all basal, often sagittate. *Inflorescence* of several simple or rarely slightly branched whorls. *Flowers* unisexual. *Stamens* 7 to numerous. *Carpels* numerous, spirally arranged in a more or less globose head on a large receptacle, strongly compressed. In autumn stolons tipped with small bud-like turions are formed.

About 20 species, throughout the temperate and tropical regions.

S. graminea Michx, Grass-leaved Arrowhead, is recorded from a gravel pit near Sandhurst, Berkshire.

Brewis, A. (1975). *Sagittaria subulata* (L.) Buch. in the British Isles. *Watsonia* **10**: 411.

Hiern, W. P. (1908). *Sagittaria heterophylla* Pursh in Devon. *Jour. Bot. (London)* **46**: 273–277.

1. Emergent leaves mostly sagittate, with large basal lobes 2.
1. Emergent leaves absent, or if present narrowly to broadly elliptical, rarely with small basal lobes 3.
2. Petals white with a dark purple patch at the base; anthers purple; achenes 4–6 mm, the beak c. 1 mm **1. sagittifolia**
2. Petals white without a basal patch; anthers yellow; achenes 2.5–4.0 mm, the beak 1–2 mm **2. latifolia**
3. Flowers and leaves emergent; emergent leaves elliptical, rarely with short basal lobes; filaments with scale-like hairs **3. rigida**
3. Flowers and leaves floating or only submerged leaves; floating leaves elliptical or rarely with short basal lobes; filaments glabrous **4. subulata** var. **gracillima**

1. S. sagittifolia L. Arrowhead
S. aquatica Lam.; *Alisma sagittaria* Stokes nom. illegit.

Monoecious, glabrous, aquatic *herb*, perennating by means of turions borne at the ends of slender stolons. *Turions* up to 30 mm, bright blue with yellow spots, ovoid or subcylindrical. *Stems* up to 1 m, emergent, erect, simple, angular and furrowed. *Leaves* all basal, of two kinds, submerged and emergent, with intermediates; submerged leaves up to 2 m, yellowish-green, linear, entire, translucent; emergent leaves erect, the lamina 10–26 × 4–21 cm, yellowish-green to medium green, broadly to narrowly sagittate, entire, usually acute or acuminate at apex, the basal lobes acute, with a long petiole; intermediate leaves, often floating, like emergent leaves, but with rounded lobes and apex, or elliptical, lanceolate or ovate, with rounded apex and base. *Inflorescence* usually in whorls of 3–5, the male flowers in the upper and the female in the lower whorls, with 3 ovate bracts at the base of each whorl. *Flowers* 20–30 mm in diameter; pedicels of male up to 20 mm, of female up to 5 mm. *Sepals* 3, 3.5–5.5 × 3.0–3.5 mm, green, broadly ovate, rounded to a short point at apex. *Petals* 3, 10–15 × 12–17 mm, white with a dark violet patch at the base, rounded, broader than long. *Stamens* numerous; filaments about 1 mm, flattened; anthers violet. *Style* very short. *Achenes* in a globular head, 4–6 mm, obovate, compressed, surrounded by a broad, dilated, gibbous, membranous margin; beak about 1 mm, recurved. *Flowers* 7–8. *Fruits* 8–10. 2n = 22.

Native. Shallow water in ponds, canals and slow-flowing rivers on muddy substrate. Scattered throughout England, but rare in the north and south-west. Local in Wales and Ireland. Most of Europe, but rare in the north and south; Asia.

2. S. latifolia Willd. Duck Potato

Monoecious or *dioecious* rhizomatous aquatic *herb*, perennating by means of turions borne at the ends of stolons. *Stems* 20–50(–90) cm, erect, angled, glabrous. *Leaves* all basal, of 2 kinds, submerged and emergent with intermediates; submerged leaves yellowish-green,

linear, entire, translucent; emergent leaves erect, the lamina 10–26(–40) × 10–26(–40) cm, yellowish-green to medium green, broadly to narrowly sagittate or hastate, entire, acute or acuminate at apex, the basal lobes acute, with a long petiole; intermediate leaves occur without lobes. *Inflorescence* usually in whorls of 2–8, with one or more of the lower whorls female or unisexual, rhachis sometimes hairy, with 3, ovate or ovate-lanceolate, obtuse to acute, more or less connate bracts at the base of each whorl which are glabrous to densely hairy. *Flowers* 30–40 mm in diameter; pedicels up to 50 mm. *Sepals* 3, 5–10× 3–5 mm, green, broadly ovate, rounded to a short point at apex, glabrous to densely hairy. *Petals* 3, 10–15 × 12–16 mm, white, rounded at apex. *Stamens* numerous; filaments 1.5–2.0 mm, flattened; anthers yellow. *Style* very short. *Achenes* in a more or less globular head, 2.5 –4 mm, obovate, broadly winged especially near the apex; beak 1–2 mm, horizontal or slightly recurved. *Flowers* 7–9. *Fruits* 8–10. 2n = 22.

Garden escape. Naturalised in ponds and by streamsides in Surrey since 1941 and in a pond in Jersey since 1961. Native of North America.

3. S. rigida Pursh Canadian Arrowhead
S. heterophylla Pursh, non Schreb.

Monoecious aquatic *herb*, perennating by means of turions borne at the ends of stolons. *Stems* up to 80 cm, flexuous and weak, often shorter than the leaves, sometimes branched. *Leaves* all basal, of 3 kinds, submerged, floating and emergent; submerged leaves yellowish-green, linear and strap-like, entire; floating and emergent leaves similar, the lamina 5–20 × 0.5–7 cm, narrowly to broadly elliptical, obtuse to subacute at apex, entire, rarely with 2 short basal lobes, rounded or shortly cuneate at base, on long petioles. *Inflorescence* in 2–7 whorls, the male flowers in the upper, the female in the lower whorls, with 3, roundish, obtuse, connate bracts at the base of each whorl. *Flowers* 18–30 mm in diameter; pedicels of male up to 8 mm, of the female short or subsessile. *Sepals* 3, 6–8 mm, green, broadly ovate or oblong-ovate, obtuse at apex. *Petals* 3, 8–14 mm, white and slightly yellowish at base, rounded at apex. *Stamens* numerous; filaments short, flattened, with scale-like hairs; anthers yellow. *Style* very short. *Achenes* 2.5–4.0 mm, obovate-cuneate, rugose, winged; beak 1.0–1.5 mm, erect or arching. *Flowers* 7–10. *Fruits* 9–11. 2n = 22.

Introduced. Established in the River Exe and an adjacent canal near Exeter since 1898, and recorded near Camborne in Cornwall. Native of North America.

4. S. subulata (L.) Buchenau
Narrow-leaved Arrowhead

Monoecious, glabrous, rhizomatous, aquatic *herb*, perennating by means of turions borne at the ends of stolons. *Stems* up to 100 cm, rising to the surface of the water or just above it, flexuous, slender. *Leaves* all basal, of 2 kinds, submerged and floating; submerged leaves strap-like and much elongated according to the depth of water, yellowish-green, entire, obtuse at apex; floating leaves when present with lamina 15–50 mm, elliptical or ovate-oblong, rarely with rounded lobes, obtuse at apex, entire, with rounded base and long petioles. *Inflorescence* with 1–4, remote whorls, male in upper and female in lower whorls, with 3, scarious, acuminate, connate bracts at the base of each whorl. *Flowers* 20–30 mm in diameter opening in sequence and lying on the surface of the water; pedicels up to 60 mm, those of the male slender, those of the female shorter and thicker. *Sepals* 3, 3.0–4.5 mm, green, ovate, obtuse at apex. *Petals* 3, 10–15 mm, white. *Stamens* 7–9; filaments about 1.0 mm, flattened; anthers yellow. *Style* very short. *Achenes* in a globular head, 1.5–2.5 mm, crenate-winged; beak 0.1–0.3 mm, erect or curved upwards. *Flowers* 6–10. *Fruits* 8–10. $2n = $ c. 44.

Introduced. Naturalised in Shortheath Pond, Hampshire since 1962. Native of North America. Our plant is referable to var. **gracillima** (S. Watson) J. G. Sm. (*S. gracillima* S. Watson).

2. Luronium Raf.

Elisma Buchenau nom. illegit.

Slender, glabrous, *perennial herb*. *Stems* procumbent or floating, rooting at intervals and producing tufts of leaves and inflorescences. *Lower leaves* submerged and linear, floating ones elliptical and petiolate. *Flowers* solitary, or 2–5 in simple umbels, bisexual. *Stamens* 6. *Carpels* 6–15, in an irregular whorl or rather flat mass, oblong-ovoid.

One species in central and western Europe.

Stewart, A., Pearman, D. A. & Preston, C. D. (1994). *Scarce plants in Britain*. Peterborough.

1. L. natans (L.) Raf.　　　　Floating Water Plantain

Alisma natans L.; *Elisma natans* (L.) Buchenau

Slender, glabrous, *perennial herb*. *Stems* procumbent or floating, rooting at intervals and producing tufts of leaves and inflorescences. *Lower leaves* submerged, 20–100 × 2–4 mm, yellowish-green, long linear, obtuse at apex, entire; floating leaves 2–25(–40) × 2–12 mm, green, ovate or elliptical, rounded or obtuse at apex, entire, rounded or shortly cuneate at base; petioles very long, up to 30 cm. *Flowers* 12–18 mm in diameter, solitary or 2–5 in simple umbels; pedicels up to 7 cm; bracts small, ovate, acute at apex, scarious. *Sepals* 2.0–3.5 × 2.0–2.5 mm, green, subrotund or broadly ovate, rounded at apex. *Petals* 7–10 × 7–10 mm, more or less white with a yellow basal blotch, subrotund or broadly obovate, rounded at apex, often shallowly crenulate. *Stamens* 6; filaments about 1 mm, white; anthers yellow. *Achenes* 2.5–3.5 × 0.9–1.2 mm, oblong or oblong-ovoid, with a short beak, in an irregular whorl or rather flat mass. *Flowers* 7–8. Pollinated by insects; fruit dispersed by water. $2n = 42$.

Native. In lakes, tarns, ponds and canals which have acid water. Local in Wales and central England. West and central Europe to southern Norway and Bulgaria, rare and decreasing over much of its range.

3. Baldellia Parl.

Glabrous, sometimes stoloniferous, *perennial herbs*. *Leaves* all basal, narrowed at base. *Inflorescence* an umbel or in 2–3 simple whorls, sometimes solitary. *Flowers* bisexual. *Stamens* 6. *Carpels* numerous, spirally arranged, free, forming a crowded, more or less globose head, ovoid, narrowed to an acute apex.

Two species in western Europe and North Africa.

1. B. ranunculoides (L.) Parl.　　Lesser Water Plantain

Alisma ranunculoides L.; *Echinodorus ranunculoides* (L.) Engelm.

Glabrous, sometimes stoloniferous, *perennial herb*. *Stems* up to 30 cm, erect, or trailing, rooting and producing tufts of leaves. *Leaves* all basal; lamina 20–70 × 2–14 mm, yellowish-green, narrowly elliptical or linear, acute at apex, entire, attenuate at base; petioles up to 20 cm; sometimes long strap-like leaves produced in deep water. *Inflorescence* an umbel or in 2–3 simple whorls, sometimes solitary flowers from the shoots on the stolons. *Flowers* 10–17 mm in diameter, opening in succession; pedicels up to 10 cm, unequal; bracts small, scarious. *Sepals* 3, 5–6 mm, green, obovate, rounded at apex. *Petals* 6–8 mm, pale mauve or white, obovate, rounded at apex. *Stamens* 6; filaments 1–2 mm, pale yellowish-green; anthers pale yellow. *Achenes* in a more or less globose head, 2.0–3.5 mm, ovoid, acute at apex, curved, strongly 3-ribbed on the back with 2 ventral ribs. *Flowers* 5–8. *Fruits* 7–9. $2n = 16, 30$.

Native. In damp, often muddy places beside streams, ponds, lakes and fen ditches, sometimes in standing water. Scattered throughout the British Isles north to Ross-shire and the Outer Hebrides. Europe, north to south Norway and east to Lithuania and west Greece.

4. Alisma L.

Glabrous, *perennial herbs*. *Stems* erect. *Leaves* all basal, linear to elliptical-ovate, narrowed or subcordate at base. *Inflorescence* usually much branched. *Flowers* in whorls, bisexual. *Stamens* 6, in 3 pairs apparently in a single whorl. *Carpels* in a single whorl. *Fruits* more or less curved inwards, more or less unbeaked, usually 1-seeded.

Nine species in the northern hemisphere; introduced elsewhere.

Björkquist, I. (1967). Studies in *Alisma* L. I. Distribution, variation and germination. *Opera Bot.* **17**: 1–128.
Björkquist, I. (1968). Studies in *Alisma* L. II. Chromosome studies, crossing experiments and taxonomy. *Opera Bot.* **19**: 1–138.

1. At least some leaves usually elliptical to ovate-elliptical, rounded to subcordate at base; style arising about halfway up the fruit　　　　　**1. plantago-aquatica**
1. Leaves linear to elliptic-lanceolate or elliptical, cuneate at base; style arising in upper half of fruit　　　**2.**
2. Leaves elliptic-lanceolate or elliptical; fruits widest near the middle, with more or less straight, erect style　　　**2. lanceolatum**
2. Leaves linear or narrowly elliptical; fruits widest in upper half, with strongly coiled style　　　**3.**
3. Leaves 2–13 mm wide; inflorescence as long as or more or less overtopping the leaves; bracts 5–13 mm　　　**3(a). gramineum** subsp. **gramineum**

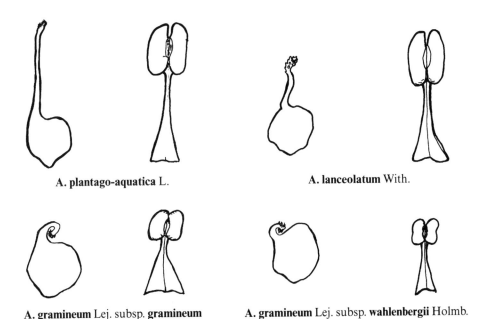

A. plantago-aquatica L. **A. lanceolatum** With.

A. gramineum Lej. subsp. **gramineum** **A. gramineum** Lej. subsp. **wahlenbergii** Holmb.

Styles and stamens of **Alisma** L. species

3. Leaves 1–3 mm wide; inflorescence not overtopping the leaves; bracts 2–5 mm
  3(b). gramineum subsp. **wahlenbergii**

1. A. plantago-aquatica L. Water Plantain
A. paniculatum Stokes nom. illegit.; *A. majus* Gray nom. illegit.; *A. gluckii* Druce

Glabrous, *perennial herb*, with well-developed rhizome. *Stems* 20–100 cm, erect, stout, usually unbranched in the lower half. *Leaves* variable according to water depth, entire; those submerged 3–6, 30–80 × 0.3–1.0 cm, greyish-green to deep green, linear, very thin, with 1–3 veins; those on surface 3–6, greyish-green to deep green, the lamina 2–8 × 1.0–2.5 cm, narrowly elliptical, acute at apex, truncate or obtuse at base and with 3–5 veins, the petiole 20–55 cm; those emersed 3–9(–12), greyish-green to deep green, the lamina 3–21 × 1–10 cm, ovate-elliptical or elliptical, acute to acuminate at apex, rounded to cordate at base and with (3–)5–9 veins, the petioles 10–35 cm; those terrestial similar to those emersed but petioles 6–16 cm. *Inflorescence* 20–100 cm, much branched; pedicels 10–36 mm; bracts at first branch 14–30 mm. *Flowers* 7–12 mm in diameter, open from 1 to 7 p.m. *Sepals* 1.7–3.2 × 1.4–2.3 mm, green with scarious margins, subrotund, acute-apiculate at apex. *Petals* 3.4–6.4 × 3.2–6.9 mm, white to pinkish- or purplish-white, subrotund, rounded at apex, subcrenulate. *Stamens* 6; filaments 1.0–1.5 mm; anthers green. *Style* 0.6–1.5 mm, straight, erect, arising about halfway up the fruit; stigma small. *Achenes* 15–28, 1.7–3.1 × 0.8–2.1 mm, more or less ellipsoid, in a more or less flat head. *Flowers* 6–8. 2n = 14.
  Native. In or by ponds, canals, ditches, lakes and slow-flowing rivers, usually some plants with emergent or ter-restial leaves. Common throughout most of the British

Isles, but rare in northern Scotland. Common in Europe; Asia except the northernmost parts; central and North Africa.

2. A. lanceolatum With. Narrow-leaved Water Plantain
A. major var. *lanceolata* (With.) Gray; *A. stenophyllum* (Asch. & Graebn.) Sam.

Glabrous, *perennial herb* with a thick rhizome. *Stems* 20–100 cm, erect, stout, usually unbranched in lower half. *Leaves* 4–10(–14) cm, more or less dark green, ter-restial or emersed; lamina 6–23 × 1.0–4.5 cm, narrowly to broadly elliptic-lanceolate or elliptical, more or less acute at apex, entire, narrowed at base, with (3–)5–7 veins. *Inflorescence* 20–70 cm, branched; pedicels 12–32 mm; bracts at first branch 7–17 mm. *Flowers* 7–12 mm in diameter, usually open from 9 a.m. to 2 p.m. *Sepals* 1.6–3.2 × 1.4–2.7 mm, green with scarious margins, more or less elliptical, acute at apex. *Petals* 4.4–6.5 × 4.5–6.9 mm, purplish-pink, broadly ovate, more or less acute or acuminate at apex, subcrenulate. *Stamens* 6; filaments 1.0–1.5 mm; anthers green. *Style* 0.5–0.9 mm, arising in the upper half, more or less straight, erect; stigma up to half the length of the style. *Achenes* 12–22, 2.0–2.9 × 1.2–1.7 mm, more or less obovate, in a more or less flat head. *Flowers* 6–8. 2n = 26, 28.
  Native. In or by ponds, canals, ditches, lakes and slow-flowing rivers. Scattered in Britain north to Yorkshire and in Ireland and the Isle of Man. Europe; North Africa; western Asia; Macaronesia.

  × **plantago-aquatica**
  = **A.** × **rhicnocarpum** Schotsman

Hybrids are often intermediate in all characters but sometimes nearer one parent than the other. They are usually highly sterile. Different times of flower opening may help to prevent frequent hybrids. $2n = 20, 21$.

Recorded from the London area, East Anglia, Isle of Man, central Scotland and Ireland. Also recorded for mainland Europe.

3. A. gramineum Lej.   Ribbon-leaved Water Plantain

Glabrous, usually submerged, *perennial herb* with a small rhizome. *Stems* (5–)15–30(–60) cm, stout and usually more or less curved, branched to below the middle. *Leaves* of two kinds; those terrestial or emersed 8–15(–20), dark green, 2–6 × 0.4–1.5 cm, narrowly elliptical-lanceolate, acute to acuminate at apex, entire, attenuate at base, the petioles 5–15 cm; those always submersed 5–20, dark green, 10–100 × 0.1 – 1.3 cm, linear, very thin, with 1–5 veins. *Inflorescence* 3–45 cm, little branched; pedicels 6–53 mm; branches at first branching 2–15 mm. *Flowers* 3.5–8.0 mm in diameter. *Sepals* 1.3–2.4 × 1.1–1.8 mm, subrotund, with an obtuse to acute apex. *Petals* 1.6–3.7 × 1.1–3.7 mm, faintly purplish-white, obovate, more or less obtuse at apex, entire. *Stamens* 6; filaments 1.0–1.5 mm; anthers dark. *Style* 0.3–0.6 mm, more or less coiled, arising in the upper quarter of the fruit; stigma up to half the length of the style. *Achenes* 11–20, 1.6–2.6 × 0.8–2.0 mm, more or less obovoid, in a more or less flat head. *Flowers* 6–8.

(a) Subsp. gramineum

A. graminea C. C. Gmel.; *A. loeselii* Gorski; *A. geyeri* Torr.; *A. arcuatum* Michalet; *A. validum* Greene
Leaves 2–13 mm wide. *Inflorescence* as long as to more or less overtopping the leaves. *Bracts* 5–13 mm. $2n = 14$.

(b) Subsp. wahlenbergii Holmb.

A. wahlenbergii (Holmb.) Juz.
Leaves 1–3 mm wide. *Inflorescence* not overtopping the leaves. *Bracts* 2–5 mm. $2n = 14$.

Possibly native, but more likely introduced sporadically. Shallow ponds and dykes. Known in one locality in Worcestershire since 1920 and for short periods in Lincolnshire, Norfolk and Cambridgeshire. Central and east Europe north to the St Petersburg region of Russia; North America. The Norfolk plants are said to be subsp. *wahlenbergii* and the remainder subsp. *gramineum*. Subsp. *gramineum* occurs throughout the range of the species. Subsp. *wahlenbergii* is thought to be native only in the Nyköping and Mälar region of Sweden and along the Finnish coast of the Gulf of Bothnia and the Gulf of Finland and west Russia.

5. Damasonium Mill.

Actinocarpus R. Br. nom. illegit.

Glabrous, *perennial herb*. *Stems* erect. *Leaves* all basal, submerged, floating or sometimes emergent, ovate-oblong, cordate at base. *Inflorescence* in one to several simple whorls. *Flowers* bisexual. *Stamens* 6. *Carpels* 6–10 in a single whorl. *Fruits* divergent, beaked, usually 2(–few)-seeded.

Four very disjunct species in Europe, North Africa, Asia, western North America and southern Australia.

1. D. alisma Mill.   Starfruit

Alisma damasonium L.; *D. stellatum* Thuill. nom. illegit.; *Actinocarpus damasonium* (L.) Sm.; *D. dalechampii* Gray nom. illegit.; *D. damasonium* (L.) Asch. & Graebn.

Glabrous, *perennial herb*. *Stems* up to 30(–60) cm, erect, rigid. *Leaves* all basal, submerged, floating or sometimes emergent, 1.5–7.0(–8.0) × 0.5–3.0 cm, yellowish-green, ovate, ovate-oblong or oblong, rounded at apex, entire, more or less cordate at base; petioles up to 30 cm. *Flowers* in several simple whorls, bisexual, 5–9 mm in diameter. *Sepals* 3, 2.5–3.5 × 1.5–2.0 mm, green with scarious margins, ovate-lanceolate, obtuse at apex. *Petals* 3–4 × 3–4 mm, white with a yellow spot at the base, subrotund, with a crinkly margin and broad apex. *Stamens* 6; filaments 1.0–1.5 mm; anthers pale yellow. *Style* terminal. *Fruits* 6–10 in a whorl, 5–14 mm, narrow and gradually tapered into a long beak, spreading stellately when ripe, more or less united at the base. *Flowers* 6–8. Entomophilus or autogamous. $2n = 28$.

Native. Muddy margins of acid ponds. Now found in only one place in Surrey and two in Buckinghamshire; formerly elsewhere in south and central England. West, south and south-east Europe; North Africa; Asia.

Order 2. HYDROCHARITALES Lindl.

Plants living in or by water. *Flowers* epigynous. *Perianth* generally divided into sepaloid and petaloid segments. *Ovary* compound and with parietal or modified laminar placentation. *Seeds* without endosperm.

Consists of a single family.

151. HYDROCHARITACEAE Juss. nom. conserv.

Glabrous, usually dioecious, aquatic *perennials*, free-floating or rooted in mud. *Leaves* submerged or floating, sometimes emergent, all basal or cauline, entire or serrate, sessile or petiolate, with or without stipules. *Inflorescence* axillary. *Flowers* 1–few, subtended by a spathe, epigynous, actinomorphic, conspicuous or inconspicuous. *Sepals* 3, free. *Petals* 3, or vestigial, white to purplish. *Stamens* 2–12. *Styles* 3 or 6; stigmas usually linear, often bifid. *Staminodes* sometimes present. *Ovary* 1-celled, with 3–6 parietal placentas bearing many ovules. *Fruit* an irregularly opening capsule.

Contains 17 genera and about 74 species, mostly in the warmer areas of the world, but some species in temperate regions.

Cook, C. D. K. (1990). *Aquatic plant book*. The Hague.

1. Leaves subrotund-reniform, cordate, with long petioles
                                        1. Hydrocharis
1. Leaves narrowed at apex or at both ends, sessile   2.
2. Leaves in a basal rosette   3.
2. Leaves cauline   4.
3. Leaves rigid, sharply serrate; petals conspicuous, white
                                        2. Stratiotes

3. Leaves flaccid, denticulate only near the apex; petals
   vestigial   **7. Vallisneria**
4. Leaves variously whorled to spiral, the lowest always
   spiral   **6. Lagarosiphon**
4. Leaves all whorled or opposite   5.
5. Middle and upper leaves in whorls of 3–6(–8), with 2
   minute fringed scales at the base   **5. Hydrilla**
5. Middle and upper leaves in whorls of 3–4(–5), with or
   without 2 entire scales less than 0.5 mm at the base   6.
6. Leaves in whorls of (3–)4–5; petals mostly more than
   5 mm, white   **3. Egeria**
6. Leaves in whorls of (2–)3–4(–5); petals mostly less than
   5 mm   **4. Elodea**

1. Hydrocharis L.

Floating *perennial herbs* with branched roots hanging in
water. *Leaves* in a basal rosette, subrotund-reniform,
cordate, long petiolate, with large stipules. *Flowers* unisexual, the male (1–)2–5 in a pedunculate spathe formed
by 2 bracts, the female solitary in a sessile spathe formed
by one bract. *Sepals* 3, greenish-white, free, narrower
and smaller than the petals. *Petals* 3, white, broad,
clawed. *Stamens* 12, in 4 whorls, the inner often sterile;
female flowers with 6 staminodes. *Styles* 6, bifid. *Fruit* a
globose, berry-like capsule.

Three species in temperate and subtropical Eurasia
and tropical Africa; naturalised in North America.

Cook, C. D. K. & Lüönd, R. (1982). A revision of the genus
  Hydrocharis (Hydrocharitaceae). *Aquatic Bot.* **14**: 177–204.
Stewart, A., Pearman, D. A. & Preston, C. D. (1994). *Scarce
  plants in Britain*. Peterborough.

1. H. morsus-ranae L.   Frogbit

H. rotundifolia Gilib. nom. invalid.; *H. asarifolia* Gray
nom. illegit.; *H. cordifolia* St-Lag., non Nutt.

Floating, glabrous *perennial herb*, rarely on mud; roots
up to 50 cm, in bunches at the nodes, green becoming
white, branched. *Stems* elongate-stoloniferous, up to
50(–100) cm, with seasonally dimorphic terminal buds,
those of spring and summer about 15 mm, narrowly
obclavate to fusiform, opening without delay, those of
autumn and winter 6–7(–9) mm, ellipsoidal, usually
with a dormancy mechanism and becoming detached
and sinking before germinating. *Leaves* all in a basal
rosette, floating on the surface of the water; lamina
1.2–6.0 × 1.3–6.3 cm, yellowish-green, tinged bronze,
subrotund-reniform to subrotund, broadly rounded at
apex, entire, cordate at base; primary veins 4, with a distinct midrib; petiole up to 14 cm, usually slender; stipules 2, up to 2.5 cm, free, translucent, ovate. *Flowers*
aerial, erect, unisexual, 20–30 mm in diameter. *Male
spathe* consisting of 2 bracts, each 10–12 mm, on peduncles 7–55 mm; flowers (1–)2–5; pedicels up to 40 mm;
sepals 4.0–5.5 mm, white or greenish-white, lanceolate
or elliptical, acute at apex, entire; petals 9–19 mm, white
with a greenish spot near the base, broadly obovate to
subrotund, broadly rounded at apex, entire, slightly narrowed at base to a claw; stamens 2.0–3.5 mm, in 4
whorls, the 2 outer whorls fully fertile, the first whorl
antesepalous with filaments united at their bases with

the filaments of the third whorl, the filaments of the
second whorl antepetalous, united with those of the
fourth whorl, filaments of the first and second whorls
glabrous and bearing fertile anthers; filaments of the
third whorl papillose and usually bearing fertile anthers,
but sometimes more or less sterile; filaments of the
fourth whorl papillose, usually sterile and appearing as
linear papillose appendages arising from the filaments of
the second whorl, but occasionally bearing reduced
anther thecae; anthers up to 1 mm, yellow. *Female
flowers* borne on pedicels up to 90 mm; sepals 4–5 mm,
greenish-white; petals 10–15 mm, white or with a faint
pinkish tinge, broadly obovate to subrotund; staminodes of antesepalous whorl linear, simple to bifid;
stigmas up to 5 mm, divided for one-quarter to two-
thirds of their length. *Fruit* a globose berry-like capsule
up to 10 mm, rarely if ever produced in the British Isles.
Seeds 1.0–1.3 mm, broadly ellipsoidal, densely covered
with blunt processes each with a spiral pattern on the
outer part. *Flowers* 7–8. Pollinated by small insects. The
plant normally spreads by buds which overwinter after
the plant has died down in autumn. Both seeds and buds
are distributed by water. $2n = 28$.

Native. Ponds, canals, ditches, lakes, slow-flowing
rivers and peat diggings. Locally frequent in England,
but declining in numbers; very scattered in Wales and
Ireland; formerly in the Channel Islands. Temperate
Eurasia, local and decreasing everywhere; naturalised in
North America.

2. Stratiotes L.

Submerged dioecious *perennial herb* rising to the surface
to flower; roots simple, rarely attached to the substrate.
Stems contracted, complexly branched, the branches
runner-like and bearing terminal buds. *Leaves* in a basal
rosette, sessile, linear or narrowly triangular and
spinous-serrate, tapering from the base; exstipulate.
Flowers unisexual, the male inflorescence with several in
a spathe, bracteolate, pedicellate; female inflorescence
with a solitary, sessile flower. *Sepals* herbaceous, smaller
than petals. *Petals* conspicuous, white. *Male androecium*
flowers up to 41 segments, the inner 5–17 fertile, the
female androecium with 15–30 staminodes. *Styles* 6,
bifid. *Fruit* a berry-like capsule.

One species in Europe and central Asia.

Cook, C. D. K. & Urmi-König, K. (1983). A revision of the
  genus *Stratiotes* (Hydrocharitaceae). *Aquatic Bot.* **16**:
  213–249.

1. S. aloides L.   Water Soldier

S. aquaticus Pall. nom. nud.; *S. ensiformis* Gilib. nom.
invalid.; *S. aculeatus* Stokes nom. illegit.; *S. generalis*
Krause nom. illegit.

Dioecious, stoloniferous *perennial herb*, submerged in
winter, rising to the surface of the water to flower; roots
simple, up to 180 cm, rarely attached to the substrate.
Stems contracted, depressed, conical, yellowish-green,
complexly branched, the branches runner-like, bearing
terminal buds and then rosettes which off-set and
produce new plants. *Leaves* forming large, basal rosettes,

linear or narrowly triangular, acute at apex, spinous-serrate on margin, sessile and semiamplexicaul, with numerous, longitudinal, parallel veins and cross-veins, those submerged up to 60(–110) × up to 1 cm, thin, flaccid but brittle, pale green turning bright green on drying and with weak spines, those emergent up to 40 × 4 cm, thick, rigid, dark green and with well-developed spines. *Flowers* 30–40 mm in diameter, unisexual, borne in spathes. *Male spathes* consisting of 2, free, overlapping bracts, 3- to 6-flowered, borne above the water on a flat-tened peduncle up to 30 cm; bracts elliptical, keeled, with or without spines, the outer enveloping the inner, (17–)26–36(–44) × 12–26 mm, somewhat hooded at the apex; flowers pedicellate, each subtended by a floral bract which resembles the spathe bract. *Male sepals* 3, (9–)11–13(–18) × (5–)6–8(–9) mm, white with a green central part and apex, translucent, ovate, obtuse at apex. *Male petals* 3, (15–)18–24(–29) × (14–)16–22(–24) mm, white, obovate or subrotund, often emarginate at apex. *Androecium* of 24–41 segments arranged in 5 whorls, the outer staminodal and the inner fertile; staminodes of 3 different kinds, the first 1.5–5.0 mm, intensely yellow spotted with darker yellow, nectar-secreting and in groups of 2–6 surrounding the fertile stamens, the second up to 8 mm, pale yellow at base and apex, dark yellow and nectar-secreting in the middle and linear, the third 1–2 mm, white, not nectar-secreting, capillary and interspersed with other staminodes or stamens; stamens 5–17, the filaments up to 2.5 mm, flattened, the anthers pale yellow. *Female spathes* like those of male but 1(–2)-flowered. *Female sepals* 3, (7–)9–12(–13) × (5–)6–9(–12) mm, green with purplish stripes, free, ovate, obtuse and rather hooded at apex. *Female petals* 3, (14–)16–21(–23) × (5–)6–9(–12) mm, white, free, obovate to obcordate and apex emarginate. *Androecium* of 15–20 staminodes, similar to those in the male flowers but the outer clavate; styles (3–)6, 7–13 mm, linear, rather fleshy, each bearing 2 stigmatic lobes. *Fruit* a berry-like capsule containing 24 seeds, ovoid or barrel-shaped, 12–34 × 9–15 mm, with 2–4 ridges, 2 of which bear spines. ·*Seeds* 5.8–10.6 × 2.3–3.0 mm, cylindrical, with more or less pronounced beak. *Flowers* 6–8. The seeds are separated from the ripe fruit and sink to the bottom in a gelatinous mass. $2n = 24$.

Native. Ponds, dykes, ditches and canals, usually cal-careous. In north-east Wales, north-west and north-east England and East Anglia. Introduced in scattered locali-ties in Britain north to central Scotland. Europe and central Asia. In the north of the range the plants are entirely female, in the south predominantly or entirely male; both sexes occur in an intermediate area. Only the female plant normally occurs in Britain, though plants with bisexual flowers have been recorded. The floating and submerging of the plant was said to be due to the amount of calcium carbonate on the leaves, but recent work shows it is more likely caused by the weight of the dead leaves taking it down and rising again when they break up.

3. Egeria Planch.

Dioecious submerged *perennial* rooted in mud. *Stems* elongated and regularly branched. *Leaves* all cauline, scale-like and opposite or elongate and in whorls of (3–)4–5; stipules absent. *Inflorescences* in the axils of elongate leaves; spathe of 2, united bracts, more or less sessile. *Male flowers* 2–4, held above the water surface on rigid pedicels; sepals 3, green; petals 3, white; stamens 9, or rarely more; filaments clavate; staminodes absent. *Female flowers* solitary, like males, but with the stamens replaced by 3 staminodes, with an elongate hypanthium bearing the flowers above the water surface. *Carpels* 3; styles 3. *Fruit* ellipsoidal, irregularly dehiscent. *Seeds* ellipsoidal.

Two species in subtropical and temperate South America; naturalised in almost all warm regions of the world.

Cook, C. D. K. & Urmi-König, K. (1984). A revision of the Genus *Egeria* (Hydrocharitaceae). *Aquatic Bot.* **19**: 73–96.

1. E. densa Planch.   Large-flowered Waterweed
Elodea densa (Planch.) Casp.; *Anacharis densa* (Planch.) Vict.; *Philotria densa* (Planch.) Small & St John

Dioecious submerged freshwater *perennial* rooted in mud. *Stems* 1–3 mm in diameter, elongated and regu-larly branched; internodes 2.5–24.0 mm. *Scale leaves* 1.4–2.5 × 1.6–2.8 mm, colourless to pale green, borne at thc bascs of stcms and branches, paired, more or less deltate with curved sides, some grading into elongate leaves. *Elongate leaves* 10–40 × 1.5–4.5 mm, usually densely clothing the stem, usually spreading, less fre-quently recurved, bright green, mostly in whorls of 4 at sterile nodes, linear to narrowly oblong, acute at apex and terminating in a single spine, serrulate; stipules absent. *Inflorescences* in the axils of elongate leaves. *Male spathes* 7.5–12.0 × 1.2–4.0 mm. *Male flowers* 2–4(–5) in each spathe; pedicels up to 80 mm or more long and 0.6–0.8 mm in diameter; sepals 3, 2.2–4.4 × 1.1–3.0(–4.0) mm, green, ovate, spreading; petals 3, 4.9–10.5 × 3.3–8.0 mm, white, obovate or subrotund; stamens 9 or rarely more, filaments 0.8–4.5 mm, yellow, clavate; staminodes absent. *Female spathes* 9–14 × 2–4 mm. *Female flowers* solitary; sepals 3–4 × 1.6–3.0 mm, green, ovate, spreading; petals 4.0–8.5 × 4–7 mm, white, obovate or subrotund; staminodes 0.9–2.4 mm, more or less clavate, yellow to orange; styles 2.4–3.8 mm, divided for at least two-thirds of their length. *Fruits* 11.5–14.5 × 4.0–5.5 mm, fusiform, sessile. *Seeds* 5.5–7.2 × about 2 mm; micropyle elongate and beak-like. Only male plants are known in Britain where it rarely flowers and only in well-warmed water. As long as a double node is present, short stem fragments will readily root and develop into new shoots. $2n = 48$.

Introduced. First recognised in Britain in 1953. In warm waters of canals and mill-lodges. Very local at Failsworth and Droylesden in Lancashire and in the canal and river system in the Huddersfield–Brighouse–Elland area in south-west Yorkshire. Also reported from Amberley in west Sussex and London area. Native in South America from south-east Brazil to Argentina. Widely grown as an aquarium plant and naturalised in many of the warmer areas of the world.

4. **Elodea** Michx

Anacharis Rich.; *Philotria* Raf. nom. illegit.; *Udora* Nutt. nom. illegit.

Dioecious submerged *perennials* rooted in mud. *Stems* long and branched. *Leaves* sessile, the lower opposite, the upper in whorls of (2–)3–4(–5), minutely serrulate, with 2, small, entire basal scales. *Inflorescence* in a sessile, axillary spathe. *Male flowers* solitary, remaining attached or becoming free-floating, usually opening on the surface of the water; sepals 3; petals 3; staminodes absent. *Female flowers* solitary, like male but smaller; staminodes 3; styles 3. *Fruit* ellipsoidal to ovoid, thin-walled, with an apical beak. *Seeds* fusiform, beaked.

Five species in temperate America; naturalised in many parts of the Old World and Australasia.

Cook, C. D. K. & Urmi-König, K. (1985). A revision of the genus *Elodea* (Hydrocharitaceae). *Aquatic Bot.* **21**: 111–156.
Marshall, W. (1852). Excessive and noxious increase of *Udora canadensis* (*Anacharis Alsinastrum*). *Phytologist* **4**: 705–715.
Simpson, D. A. (1984). A short history of the introduction and spread of *Elodea* Michx in the British Isles. *Watsonia* **15**: 1–9.
Simpson, D. A. (1986). Taxonomy of *Elodea* Michx in the British Isles. *Watsonia* **16**: 1–14.

1. Leaf apices obtuse to subacute, (0.7–)0.8–2.3 mm wide at 0.5 mm below the apex **1. canadensis**
1. Leaf apices acute to acuminate, 0.2–0.7(–0.8) mm wide at 0.5 mm below the apex **2.**
2. Usually some leaves strongly recurved and/or twisted; root-tips white to greyish when fresh; sepals 1.0–2.6 mm **2. nuttalii**
2. Usually no leaves strongly recurved or twisted; root-tips red when fresh; sepals 3.0–6.2 mm **3. callitrichoides**

1. **E. canadensis** Michx   Canadian Waterweed

Udora canadensis (Michx) Nutt.; *Serpicula canadensis* (Michx) Eaton; *Anacharis canadensis* (Michx) Planch.; *Philotria canadensis* (Michx) Britton; *Anacharis alsinastrum* Bab.; *E. planchonii* Casp.; *Philotria planchonii* (Casp.) Rydb.; *Anacharis planchonii* (Casp.) Rydb.; *E. latifolia* Casp.; *Philotria linearis* Rydb.; *Anacharis linearis* (Rydb.) Vict.; *E. linearis* (Rydb.) St John; *E. ioensis* Wylie; *E. brandegae* St John

Dioecious, glabrous, submerged *perennial* with its roots in mud. *Roots* simple, adventitious, developing only at nodes bearing lateral branches. *Stems* up to 3 m, pale green, slender, terete, branched. *Lower leaves* subopposite, 1.5–3.0 × 0.8–2.0 mm; upper leaves in whorls of (2–)3(–4), 4.5–17 × 1.4–5.6 mm, (0.7–)0.8–2.3 mm wide at 0.5 mm below apex, usually dark green, linear to oblong or ovate, obtuse, obtusely acuminate or widely acute at apex, serrulate in the distal two-thirds, often somewhat crisp, towards the stem apex often imbricate in regular rows and lying along the stem, those lower down the stem usually flat and spreading or recurved; basal scales 0.1–0.2 × 0.1–0.2 mm, elliptical. *Male spathes* at anthesis 6–10(–13) mm, with acuminate lobes; pedicels up to 15 cm. *Male flowers* solitary, becoming detached and floating on the surface of the water; sepals 3.0–4.5 × 1.4–2.0 mm, green with purple stripes; petals

1.4–3.0(–4.3) × 0.2–1.0 mm, translucent, white, linear, linear-triangular or narrowly elliptical; stamens (7–)9(–18); filaments united at the base into a column; inner 3 anthers slightly larger than outer. *Female spathes* 8.4–17.6 mm. *Female flowers* solitary; sepals 2.3–3.5 × 0.9–2.0 mm, petals 2–3 × 0.9–1.2 mm, translucent, white, elliptical; staminodes 1–2 mm; styles 2.6–4.0 mm, entire or bifid. *Ovary* 1.2–1.6 mm. *Fruit* 5.0–6.5 × 2.0–2.5 mm, obturbinate, with a beak 5–6 mm. *Seeds* 4.0–5.7 × about 1.0 mm, fusiform; shortly beaked, the beak with a collar of wart-like cells. *Flowers* 5–10. Male flowers rare in Britain. Perennating by means of winter buds. $2n = 24, 48$.

In Europe, where *E. canadensis* is introduced, very few genotypes seem to occur, and it shows little variation. In North America it is much more variable and sometimes difficult to distinguish from *E. nuttallii*.

Introduced. Naturalised in ponds, lakes, canals and slow-flowing rivers. Fairly common throughout the British Isles except in the north and north-west of Scotland. Probably first introduced into Britain in the 1820s and Ireland in 1836. It spread rapidly and attained great abundance so as to block many waterways, then diminished in quantity. Native in temperate North America where it is widespread, but absent from higher-lying land.

2. **E. nuttallii** (Planch.) St John   Nuttall's Waterweed

Anacharis nuttallii Planch.; *Philotria nuttallii* (Planch.) Rydb.; *E. columbiana* St John; *Serpicula verticillata* var. *angustifolia* Mühlenb. ex Asch. & Graebn.; *Philotria angustifolia* (Mühlenb. ex Asch. & Graebn.) Britton ex Rydb.; *Serpicula occidentalis* Pursh; *Udora occidentalis* (Pursh) Koch; *E. occidentalis* (Pursh) St John; *Philotria occidentalis* (Pursh) House; *Anacharis occidentalis* (Pursh) Vict.; *Udora verticillata* var. *minor* Engelm. ex Casp.; *Philotria minor* (Engelm. ex Casp.) Small; *E. minor* (Engelm. ex Casp.) Farw.; *Hydrilla lithuanica* auct.

Dioecious, glabrous, submerged *perennial*, with its roots in mud. *Roots* simple, with white to greyish tips when fresh, adventitious, developing only at nodes bearing lateral branches. *Stems* up to 3 m, pale green, slender, terete, branched. *Lower leaves* subopposite, 1.0–2.7 × 0.7–1.3 mm, transparent; upper leaves in whorls of (2–)3–4(–5), 5.5–35 × 0.8–3.0 mm, 0.2–0.7(–0.8) mm wide at 0.5 mm below the apex, usually pale green and flaccid, linear to linear-lanceolate, acuminate at apex, with undulate-serrulate margins, frequently strongly recurved and twisted, folded along the midrib; basal scales 0.1–0.4 × 0.1–0.3 mm, elliptical to subrotund. *Male spathe* at anthesis 2.2–4.0 × 1.5–3.0 mm, subglobose to subovoid, with apiculate lobes; pedicels up to 0.5 mm. *Male flowers* solitary, becoming detached and floating to the surface of the water; sepals 1.7–2.6 × 1.2–2.0 mm, green with purple stripes; petals 0.5–1.5 × 0.2–0.6 mm, translucent, white, triangular to narrowly triangular or rarely linear, mostly caudate; stamens (7–)9; inner 3 filaments united into a column, inner 3 anthers slightly larger than outer. *Female spathe* 8.5–14.5 × 0.8–1.8 mm.

Female flowers solitary; sepals 1.0–2.1 × 0.6–1.1 mm; petals 1.1–1.9(–2.5) mm, elliptical or obovate; staminodes 0.5–1.2 mm; styles 1.2–3.0 mm, linear, entire or bifid. *Ovary* about 0.8 mm in diameter. *Fruit* 5–10 × 1.5–2.0 mm, obclavate, with a beak 4.0–6.5 mm. *Seeds* 4.0–4.6 mm × about 1.0 mm, attenuate into a short beak at apex, the beak without a collar. $2n = 48$.

Introduced. Naturalised in ponds, lakes, canals and slow-flowing rivers. First recorded in 1966. Now locally common in England and very scattered in Wales, Scotland, Ireland and Jersey. Native of temperate North America, where it is largely sympatric with *E. canadensis*. Female plants are actively spreading in many parts of Europe and seem to be replacing *E. canadensis* in many localities. First discovered in Japan in the early 1960s where only male plants have been found.

3. E. callitrichoides (Rich) Casp.

South American Waterweed

Anacharis callitrichoides Rich.; *E. ernstae* St John; *E. richardii* St John

Dioecious, glabrous, submerged *perennial* with its roots in mud. *Roots* simple, with tips red when fresh, adventitious, developing only at nodes bearing lateral branches. *Stems* up to 3 m, pale green, slender, terete, branched. *Lower leaves* subopposite, (1–)1.5–2.0(–3) × 0.6–1.0(–1.5) mm; upper leaves in whorls of 3, 9–25 × 0.7–2.2 mm, 0.2–0.6 mm wide at 0.5 mm below the apex, linear to narrowly lanceolate or narrowly triangular, gradually narrowed to an acute apex, with serrulate margin, flat, spreading; basal scales about 0.2 × 0.4 mm, subrotund to transversely elliptical. *Male spathe* at anthesis (12–)16–20 × about 3 mm, cylindrical below and inflated apically, both lobes bearing apical teeth; pedicels up to 10 mm. *Male flowers* solitary; sepals 4.5–6.2 × 1.6–2.0 mm, green with purple stripes; petals linear and somewhat spathulate, longer and somewhat narrower than the sepals, very thin and flimsy, whitish-translucent, sometimes with a purple vein; stamens usually 9, sometimes 6; inner 3 filaments united into a column; anthers uniform. *Female spathe* at anthesis 11–16(–24) × 0.5–0.6 mm, cylindrical, not widened apically. *Female flower* solitary; sepals 3–4 × 0.7–1.0 mm; petals usually somewhat longer than the sepals, very thin and flimsy, translucent whitish to pale lilac, often with one purple vein; staminodes 0.8–1.2 mm; styles 5–8 mm, linear, bifid. *Fruit* about 11 mm, with a beak 4.5–6.0 mm. *Seeds* about 2.7 mm, with a beak about 1.5 mm, almost smooth. $2n = 48$.

Introduced. Slow-moving water, apparently only persisting in thermally polluted areas. First recorded in 1948. Very local in southern England and south Wales. Native of South America from Chaco and Corrientes southeastwards to Buenos Aires and extending to the Gulf San Matias. Also introduced in France and Germany.

5. Hydrilla Rich.

Serpicula L.

Gymnodioecious, *perennial* or occasionally annual, submerged water weed with its roots in mud. *Stems* long and branched. *Leaves* sessile, the lower opposite, the upper in whorls of 3–6(–8), minutely serrate, with 2, minute, fringed nodal scales. *Flowers* solitary, in sessile, axillary spathes, inconspicuous. *Petals* transparent, with red streaks, as large as sepals. *Stamens* 3. *Styles* 3(–5), simple. *Fruit* cylindrical, indehiscent. *Seeds* fusiform, smooth.

A single species in Eurasia, Africa and Australia.

Cook, C. D. K. & Lüönd, R. (1982). A revision of the genus *Hydrilla* (Hydrocharitaceae). *Aquatic Bot.* **13**: 485–504.
Scannel, M. J. P. & Webb, D. A. (1976). The identity of the Renvyle *Hydrilla*. *Irish Naturalists' Jour.* **18**: 327–331.

1. H. verticillata (L. fil.)Royle   EsthwaiteWaterweed

Elodea nuttallii auct.; *Serpicula verticillata* L. fil.; *Vallisneria verticillata* (L. fil.) Roxb.; *Udora verticillata* (L. fil.) Spreng.; *Elodea verticillata* (L. fil.) Mueller; *Hottonia serrata* Willd.; *Hydrilla lithuanica* (Besser ex Rchb.) Dandy; *Hydrospondylus submersus* Hassk.; *Udora pomeranica* Rchb.; *Anacharis pomeranica* (Rchb.) Peterm.; *H. najadifolia* Zoll. & Moritzi ex Moritzi; *H. angustifolia* Hassk.; *H. wightii* Planch.; *H. polysperma* Blatt.; *H. ovalifolia* Rich. nom. illegit.; *Udora occidentalis* auct.; *H. roxburghii* Steud. nom. illegit.; *Epigynanthus blumei* Hassk. nom. nud.; *H. dentata* Casp. nom. illegit.

Submerged *perennial* or occasionally annual, with unbranched roots in mud. *Stems* to 1 m, elongate, branched regularly but at distant intervals, horizontal and stoloniferous below, erect and spreading above, with bulbil-like turions developing either underground terminally on stolons, or terminally or axillary on erect stems or their branches. *Leaves* cauline, regularly spaced but contracted towards the apex, up to 20(–40) × 2–5 mm, scale-like and opposite or foliate and in whorls of 3–6(–8), linear to lanceolate or rarely ovate, acute at apex and terminating in a single spine, minutely serrate or dentate, the nodal scales about 5 mm, narrowly triangular to lanceolate and fringed with finger-like, orange-brown hairs. *Inflorescence* in the axils of normal foliage leaves; spathe sessile or subsessile, of 2 united bracts, tubular, 1-flowered. *Male flowers* small, abscising as buds and opening explosively on the water surface; sepals 3, about 3 × 2 mm, ovate and reflexed; petals 3, transparent with red streaks, linear and reflexed; stamens 3; staminodes 3, minute. *Female flowers* subsessile but with a long thread-like hypanthium carrying the flower to the surface of the water; sepals 3, 1.5–4.0 mm, oblong-ovate; petals 3, elongate-ovate, spreading and floating; staminodes 3, minute. *Carpels* 3; styles 3, up to 0.7 mm. *Fruit* cylindrical, indehiscent. *Seeds* rarely more than 5, about 2.5 mm, fusiform and smooth. Anemophilous, the pollen explosively liberated from the free-floating male flowers and caught by the floating female flowers. $2n = 16$.

Native. Lakes. Esthwaite Water in Westmorland from 1914 to about 1945, and Rusheenduff Lough in Galway from 1935. Very local in Europe and often impermanent; south and east Asia; east Africa; Australia; naturalised in North America.

Flowers have never been seen in the Esthwaite Water locality, and female flowers have been seen once only at Rusheenduff Lough. It closely resembles *Elodea nuttallii*, but is less robust and has more leaves per node which have minute teeth more or less to the base, whereas *E. nuttallii* has them only in the distal half.

6. Lagarosiphon Harv.

Dioecious *perennial* water plants, rooted in mud. *Stems* long, branched and submerged. *Leaves* variously whorled to spiral, the lowest always spiral, sessile, linear, subentire to minutely denticulate, with 2 minute nodal scales. *Flowers* inconspicuous, arising from a sessile, axillary spathe, the male spathe with many, the female with 1(–3). *Petals* reddish, as large as sepals. *Stamens* 3. *Styles* 3, bifid. *Fruit* a more or less ovoid capsule. *Seeds* ellipsoidal.

Nine species in Africa and Madagascar; naturalised in Europe and New Zealand.

Symoens, J. J. & Triest, L. (1983). Monograph of the African genus *Lagarosiphon* Harvey (Hydrocharitaceae). *Bull. Jard. Bot. Nat. Belg.* **53**: 441-488.

1. L. major (Ridl.) Moss Curly Waterweed
L. muscoides var. *major* Ridl.; *Elodea crispa* auct.

Dioecious, glabrous *perennial* water plant, rooted in mud. *Roots* unbranched. *Stem* elongate up to 3 m, branched regularly, but at distant intervals, horizontal below and erect above. *Leaves* 6.5–25.0 × 2.0–4.4 mm, cauline, variously whorled to spiral, the lowest always spiral, regularly spaced along the stem, but contracted towards the apex, dark green, linear, narrowly acute or acuminate at apex, subentire to minutely denticulate, recurved, sessile; nodal scales minute, entire. *Inflorescences* in the axils of normal foliage leaves; spathe of 2 united bracts, the male pedunculate, globose and containing numerous flowers, the female cylindrical and containing 1, sessile flower. *Male flowers* very small, becoming detached in bud and opening on the water surface; sepals 3, about 1.2 × 0.6 mm; petals 3, 1.0 × 0.2 mm, reflexed at anthesis; stamens 6, about 1.2 mm, 3 fertile and held parallel to the water surface and 3 papillose staminodes about 2 mm held vertically and acting as sails. *Female flowers* sessile, but with long, thread-like hypanthia which carry the flowers to the surface; sepals 3, about 1.2 mm; petals 3, spreading and floating; staminodes 3, linear. *Carpels* 3; styles 3, linear, each bifid almost to the base; stigmas purplish. *Fruit* about 4.5 × 1.0 mm, a more or less ovoid capsule. *Seeds* about 2 mm, ellipsoidal. Male flowers are mobile on the water surface where they are blown by wind and caught by the floating female flowers. Only female plants occur in the British Isles. 2*n* = 22.

Introduced. First recorded in 1944, now naturalised in ponds, lakes, canals and slow-flowing rivers, and much grown in aquaria. Native of South Africa.

7. Vallisneria L.

Dioecious, submerged, stoloniferous *perennial herb* rooted in mud. *Stems* with a complex branching system.

Leaves basal and ribbon-like, denticulate near the apex. *Inflorescences* axillary, in stalked spathes. *Male flowers* very small, numerous. *Female flowers* solitary, long-pedicelled. *Petals* 0, or vestigial. *Stamens* (1–)2(–3). *Styles* 3, bifid. *Fruit* a cylindrical capsule. *Seeds* numerous, ellipsoidal.

Two species in tropical or warm parts of the world.

Harris, S. & Lording, T. A. (1973). Distribution of *Vallisneria spiralis* L. in the River Lea Navigation Canal (Essex–Hertfordshire border). *Watsonia* **9**: 253–256.
Lowden, R. M. (1982). An approach to the taxonomy of *Vallisneria* L. (Hydrocharitaceae). *Aquatic Bot.* **13**: 269–298.

1. V. spiralis L. Tapegrass

Dioecious, glabrous, submerged *perennial* water plant rooted in mud. *Roots* fibrous, unbranched. *Stems* short, developing stolons. *Leaves* 2–80 cm × 1–10 mm, basal, green with reddish dots or short streaks, flaccid, linear and ribbon-like, obtuse or rounded at apex, denticulate towards the apex, submerged. *Inflorescence* axillary; spathe tubular, of 2 united bracts. *Male spathe* ovoid, shortly pedunculate. *Male flowers* numerous, very small, becoming detached in bud and rising to the surface of the water; sepals unequal, 2 large and one smaller, becoming reflexed at anthesis; petals rudimentary or absent; stamens (1–)2(–3). *Female spathe* with filiform peduncles which become spirally coiled after anthesis, floating to the surface at maturity; sepals 1.7–4.0 mm, pinkish-white; petals minute; styles 3, bifid, the lobes flattened and papillose. *Ovary* sessile within the spathe, cylindrical. *Fruit* breaking up irregularly. *Seeds* up to 2 mm, numerous, ellipsoidal. *Flowers* 6–10. 2*n* = 20.

Introduced. Naturalised in slow-flowing rivers and canals, often not permanently, usually where water is heated. Very local in southern England, particularly in the Lea Navigation Canal, in the Reddish Canal, Lancashire and in Yorkshire. Native of the warmer regions of the world, in Europe reaching as far north as north-central France and the central Ukraine. The British plant is probably var. **spiralis**.

Order 3. NAJADALES Nakai

Perennial herbs usually living in water or wet places, sometimes marine. *Leaves* linear, with scales (*squamulae intravaginales*) in their axils. *Flowers* bisexual or unisexual, hypogynous. *Perianth* absent or of one whorl, less often of 2 similar whorls. *Stamens* 1–6, rarely more. *Ovary* of (1–) few, free or more or less connate carpels; ovules 1, rarely more, placentation usually basal or apical. *Fruit* usually dry. *Seeds* with little or no endosperm.

This order contains 10 families and a little more than 200 species.

152. APONOGETONACEAE J. Agardh

Glabrous aquatic *perennials*. *Stems* elongated, rooted in mud at the tuberous base. *Leaves* all basal, floating, entire, long-petiolate with a sheathing base. *Flowers* in a forked spike on a long stalk at the surface of the water, with a deciduous spathe at the base of the spike, bisexual, hypogynous, actinomorphic except for the perianth, conspicuous. *Perianth segments* 1(–2), white or pinkish. *Stamens* 6–18. *Style* 1, short; stigmas linear. *Carpels* 2–6, free, each with several ovules. *Fruit* a group of follicles. *Seeds* without endosperm and with a straight embryo.

1. Aponogeton L. fil. nom. conserv.

As family.

The only genus, containing 43 species. Tropical to warm regions of the Old World.

Bruggen, H. W. E. van (1985). Monograph of the Genus *Aponogeton* (Aponogetonaceae). *Biblioth. Bot.* **137**: 1–76.

1. A. distachyos L. fil.   Cape Pondweed

Glabrous, aquatic *perennial* with an edible, tuberous stock rooted in mud. *Stems* long, green, spongy. *Leaves* all basal, floating, 6–25 × 1.5–7.5 cm, green, oblong-elliptical, rounded to acute at apex, entire, rounded or attenuate at base; veins 7 or 9; petioles up to 100 cm, sheathing at base. *Inflorescence* a forked dorsiventrel spike at the surface of the water on a stalk up to 80 cm, ebracteate, each fork of the spike with about 10 flowers. *Flowers* 20–40 mm in diameter, fragrant. *Perianth segments* 1(–2), 10–15 × 3.5–6.0 mm, up to 30 mm in fruit, white or pinkish, ovate, obtuse at apex. *Stamens* 6–18; filaments 3.0–4.5 mm, wider at base, white; anthers blackish-purple. *Styles* 1, short; stigma linear. *Carpels* 2–6, each with several ovules. *Fruit* a group of 3 follicles, up to 22 × 6 mm, with a curved or straight terminal beak. 2n = 16, 24.

Introduced. Planted in ponds and sometimes becoming more or less naturalised there and in canals. In scattered localities in Britain; known since 1924 in Keston Ponds in west Kent. Native of South Africa.

153. SCHEUCHZERIACEAE F. Rudolphi nom. conserv.

Glabrous, *perennial herbs* with creeping rhizomes clothed with old leaf-bases. *Stems* erect and leafy. *Leaves* alternate, linear, entire, with a sheathing base; ligule at top of sheath on adaxial side. *Flowers* few, in terminal racemes, with a bract at the base of each, bisexual, hypogynous, actinomorphic. *Perianth segments* 6, sepaloid, persistent. *Stamens* 6, with elongated filaments. *Ovary* superior; stigma sessile; carpels 3(–6), shortly connate at base, each with usually 2 ovules which are basal and erect. *Fruit* a group of follicles. *Seeds* without endosperm, straight.

One genus and one species in the colder parts of the northern hemisphere.

1. Scheuchzeria L.

As family.

1. S. palustris L.   Rannoch Rush

Glabrous, *perennial*, Juncus-like *herb*, with creeping rhizomes clothed with old leaf-bases. *Stems* 10–20(–40) cm, green, tinged red, erect, flexuous, striate, leafy. *Leaves* 20–170 × 1–2 mm, the upper often over-topping the inflorescence, bright medium green, linear, obtuse at apex, entire, with a sheathing base, slightly grooved, with a conspicuous pore near the tip; ligule triangular, acute at apex. *Inflorescence* a raceme, 3- to 10-flowered, very lax; bracts 3–7 × 1–2 mm, linear-lanceolate, gradually narrowed to an acute apex. *Perianth segments* 6, yellowish-green, 2–3 × 0.5–0.8 mm, oblinear-lanceolate, obtuse at apex, curved. *Stamens* 6; filaments 0.5–1.0 mm, greenish-yellow; anthers 2.0–3.5 mm, yellow. *Stigmas* sessile. *Carpels* 3(–6), 4–6 × 3–5 mm, broadly ellipsoid, obovoid or almost quadrangular; shortly beaked. *Seeds* 3.0–3.5 × 2.0–2.5 mm, pale brown, ellipsoid. *Flowers* 6–8. 2n = 22.

Native. Pools or wet *Sphagnum* bogs. Very rare in two localities in Perthshire. Formerly in Shropshire, Cheshire, Inverness-shire and Argyllshire and in one locality in Offaly, Ireland but a combination of drainage, peat cutting and afforestation has eliminated it. Scattered in the colder parts of the northern hemisphere.

154. JUNCAGINACEAE Rich. nom. conserv.

Glabrous *perennial herbs* with rhizomes. *Stems* erect, more or less leafless. *Leaves* mostly in a basal rosette, or a few alternate near the base. *Flowers* many, in a bractless, simple, terminal raceme, bisexual, hypogynous, inconspicuous. *Perianth segments* 6, sepaloid. *Stamens* 6. *Stigmas* papillate to long-fringed. *Ovary* superior, 6-celled, all cells with 1 ovule or alternate cells small and sterile. *Fruits* of 3 or 6, 1-seeded units breaking apart at maturity.

Juncus-like or *Plantago*-like in appearance, but distinguished by the 3 or 6, 1-seeded fruit segments.

Three genera with about 25 species, widely distributed, particularly in the temperate and cold regions of both hemispheres.

1. Triglochin L.

Rhizomatous herbs with fibrous roots and more or less tuberous stems. *Leaves* erect, linear, half-cylindrical. *Inflorescence* a raceme. *Perianth segments* deciduous. *Fruit* dehiscing by the carpels separating from the central axis.

About 15 species in temperate regions.

Davy, A. J. & Bishop, G. F. (1991). *Triglochin maritima* L. in Biological flora of the British Isles. *Jour. Ecol.* **79**: 531–555.

1. Leaves deeply furrowed on upper surface towards the base; fruit 7–10 × 0.9–1.5 mm, clavate, appressed to the stem   **1. palustris**
1. Leaves not furrowed; fruit 3–4 x 1.8–2.2 mm, oblong-ovoid, not appressed to the stem   **2. maritima**

1. T. palustris L.   Marsh Arrowgrass

Slender, glabrous *perennial herb* with long, slender rhizomes. *Stems* 15–70 cm, erect, pale green. *Leaves* mostly in a basal rosette, 6–40 cm × 0.5–2.0 mm, bright yellowish-green, half-cylindrical, deeply furrowed on the upper surface towards the base, obtuse at apex.

Inflorescence a terminal raceme, elongating after flowering, 1–25 cm, bractless. *Flowers* 2–3 mm in diameter; pedicels 1.0–2.5 mm, elongating after flowering. *Perianth segments* 6, 1–2 × 0.5–1.2 mm, green with purple edges, broadly ovate or elliptical, rounded at apex or shortly pointed. *Stamens* 6; filaments very short; anthers yellow. *Stigmas* long-fringed, white. *Fruit* 7–10 × 0.9–1.5 mm, clavate, appressed to the stem; with 3 carpels sterile and 3 fertile, remaining attached to the top of the triquetrous axis after dehiscence. *Flowers* 6–8. $2n = 24$.

Native. Marshy places and wet fields, sometimes at the back of salt-marshes. Throughout the British Isles but commoner in the north and very local in most southern and midland areas of Britain. Europe except for most of the Mediterranean area; North Africa; northern Asia; Greenland; North and South America.

2. T. maritima L.   Sea Arrowgrass

Rather stout *perennial herb* with short rhizomes. *Stems* 15–60 cm, erect, pale green. *Leaves* mostly in a basal rosette, 6–45 cm × 0.5–2.0 mm, bright green, half-cylindrical, not furrowed, obtuse at apex. *Inflorescence* a terminal raceme, scarcely elongating after flowering, 5–30 cm, bractless. *Flowers* 3–4 mm in diameter; pedicels 1–2 mm, elongating after flowering. *Perianth segments* 6, 1.0–1.5 × 0.8–1.2 mm, green with scarious margins, broadly ovate, broadly pointed at apex. *Stamens* 6; filaments very short; anthers yellow. *Stigmas* papillate, white. *Fruit* 3–4 × 1.8–2.2 mm, oblong-ovoid, not appressed to the stem; carpels 6, all fertile, separating completely at dehiscence. *Flowers* 6–8. $2n = 24$.

Native. Salt-marshes and salt-sprayed grassland. Round the coasts of the whole of the British Isles; rare in salty areas inland. Europe southwards to central Portugal and Bulgaria; North Africa; western and northern Asia; North America.

155. POTAMOGETONACEAE Dumort. nom. conserv.
By C. D. Preston, edited by P. D. Sell

Glabrous, aquatic *annual* or *perennial herbs* often with rhizomes. *Leaves* alternate or opposite, floating and/or submerged, or some more or less aerial, at least some with a sheathing base which is either free and stipule-like or fused to the leaf-base for most of its length, forming a sheath and a free distal ligule. *Flowers* in axillary or terminal, bractless spikes, bisexual, hypogynous, actinomorphic, usually aerial. *Perianth segments* 4, green. *Stamens* 4, sessile on the claws of the perianth segments. *Stigma* 1, more or less sessile, capitate. *Carpels* (1–)4(–7), free, each with 1 ovule. *Fruits* in a group of 1–4 achenes or drupes. *Seeds* non-endospermic; embryo with a massive hypocotyler foot.

Two genera and 80–90 species; cosmopolitan.

1. All or most leaves alternate, all with a membranous sheath or stipule   **1. Potamogeton**
1. All leaves opposite, only those subtending flowers on branches with membranous stipules   **2. Groenlandia**

1. Potamogeton L.

Annual or rhizomatous *perennial herbs*, overwintering by the rhizome or by specialised winter buds (*turions*) borne on the leafy stems. Some species are perennial and overwinter as the rhizome, by buds formed on the rhizome in autumn or as leafy shoots. Others have no rhizome, perennation being by the turions. *Leaves* all alternate or just those subtending the inflorescences opposite, all submerged and thin and translucent, or some floating which are usually more or less coriaceous and opaque; submerged leaves linear or more or less broadened into a sessile or stalked lamina; floating leaves usually narrowly to broadly elliptical-oblong; the leaf has in its axil a more or less delicate, membranous sheathing scale which may be free throughout (*stipule*), or may be adnate to the leaf-base in its lower part (*sheath*) and free above (*ligule*), in either case the basal part may be open and with overlapping (*convolute*) margins, or tubular. *Spikes* ovoid to cylindrical, dense, lax or interrupted, either submerged with pollen transferred on the surface of air bubbles, or on the surface of the water, or emergent and wind-pollinated. *Perianth* of 4, free, rounded, shortly-clawed segments, sometimes regarded as appendages of the connectives of the anthers. *Fruit* with a bony endocarp.

About 80–90 species. Cosmopolitan.

A taxonomically difficult genus because of frequent hybrids, which reproduce vegetatively, and the plasticity of the vegetative morphology. The leaves vary greatly in size and shape at different stages of development and in different conditions of light intensity, mineral nutrient supply and speed of water movement. Species which can form floating leaves in sufficiently shallow water may fail to do so in deep water, and some form leaves only of the floating type when growing subterrestrially. Some hybrids are widespread, others rare and they do not always occur with their parents. For this reason all hybrids have been dealt with fully in the text, and they have been placed as near as possible to those taxa to which they show the most resemblance. At the end of each species account the hybrids of that species are listed.

The most important diagnostic features are the range of leaf-shape, venation of leaves, the shape of the leaf-apex, the leaf margin, whether the stipules of the young leaves are open throughout or tubular in the lower part and the size of the fruits. Some of these characters are best examined when the plants are fresh. Hybrids, with the exception of *P.* × *zizii,* are more or less sterile, but reproduce vegetatively.

The following account is based on Preston (1995).

Dandy, J. E. & Taylor, G. Studies in British Potamogetons. *Jour. Bot. (London)* I in **76**: 89–92 (1938); II in **76**: 166–171 (1938); III in **76**: 239–241 (1938); IV in **77**: 56–62 (1939); V in **77**: 97–101 (1939); VI in **77**: 161–164 (1939); VII in **77**: 253–259 (1939); VIII in **77**: 277–282 (1939); IX in **77**: 304–311 (1939); X–XI in **77**: 342–343 (1939); XII in **78**: 1–11 (1940); XIII in **78**: 49–66 (1940); XIV in **78**: 139–147 (1940); XV in **79**: 97–101 (1941); XVI in **80**: 117–120 (1942): XVII–XVIII in **80**: 121–124 (1942).

Fryer, A. & Bennett, A. (1915). *The Potamogetons (Pond Weeds) of the British Isles*. London.

Preston, C. D. (1988). The *Potamogeton* L. taxa described by Alfred Fryer. *Watsonia* **17**: 23–35.

Preston, C. D. (1995). *Pondweeds of Great Britain and Ireland*. London.

Stewart, A., Pearman, D. A. & Preston, C.D. (1994). *Scarce plants in Britain*. Peterborough. [*P. coloratus, P. compressus, P. filiformis, P. friesii, P. praelongus* and *P. trichoides*.]

Key to fruiting material

1. Leaf margin serrate, with teeth which are easily seen with the naked eye; beak at least half as long as the rest of the fruit   **32. crispus**

1. Leaf margin entire or minutely denticulate, with teeth which are not or scarcely visible to the naked eye; beak much less than half as long as the rest of the fruit   2.

2. Some or all leaves narrowly elliptical to subrotund, with convex sides   3.

2. All leaves filiform to linear, with parallel sides   15.

3. All leaves floating or terrestrial, petiolate, usually coriaceous and opaque ('floating leaves') or submerged leaves, if present, reduced to opaque phyllodes without a distinct midrib and lamina   4.

3. Some or all leaves submerged, sessile or petiolate, with a distinct midrib and a thin, delicate and more or less translucent lamina ('submerged leaves')   6.

4 Floating leaves often with a discoloured junction 7–25 mm long between the petiole and the lamina; stipules 40–170 mm, the veins prominent when dry; fruits 3.8–5.0 mm   **1. natans**

4. Floating leaves without a discoloured junction between the petiole and the lamina; stipules 10–65 mm, the veins inconspicuous when dry; fruits 1.5–2.6 mm   5.

5. Floating leaves opaque, coriaceous, with inconspicuous secondary veins; fruits 1.9–2.6 mm, reddish-brown   **7. polygonifolius**

5. Floating leaves relatively translucent, not coriaceous, with conspicuous secondary veins; fruits 1.5–1.9 mm, olive green or greenish-brown, without a reddish tinge   **9. coloratus**

6. All submerged leaves opposite; inflorescence capitate, with 2 opposite flowers   **Groenlandia densa** (p. 37)

6. Submerged leaves (except for those immediately below an inflorescence) alternate; inflorescence cylindrical, with at least 12 flowers   7.

7. Submerged leaves semi-amplexicaul or amplexicaul, sessile, at least on the main stems   8.

7. Submerged leaves not amplexicaul, petiolate or sessile   9.

8. Submerged leaves semi-amplexicaul; stipules buff to dark green when dry, persistent or subpersistent; fruits 4.5–5.5 mm   **23. praelongus**

8. Submerged leaves amplexicaul; stipules hyaline when dry, fugacious; fruits 2.6–4.0 mm   **25. perfoliatus**

9. Submerged leaves linear   **33. epihydrus**

9. Submerged leaves broader   9a.

9a. Leaves at the base of the stem not reduced to phyllodes; submerged leaves obtuse to acute at apex, but not mucronate, the margins entire or (rarely) with fugacious teeth   10.

9a. Leaves at the base of the stem reduced to phyllodes;

submerged leaves mucronate or with the midrib excurrent, the margins minutely denticulate with persistent teeth   13.

10 Submerged leaves sessile; fruits 2.6–3.7 mm, pale brown, with a shiny appearance when mature   **20. alpinus**

10. Submerged leaves petiolate; fruits 1.5–2.6 mm, green or greenish-brown, with a matt appearance   11.

11. Floating leaves translucent or subcoriaceous, not markedly different in texture from the submerged leaves, truncate to cordate at base, the petioles (4–)8–45 mm, shorter than the lamina   **9. coloratus**

11. Floating leaves coriaceous, contrasting strongly in appearance with the translucent submerged leaves, cuneate to subcordate at base, the petioles (13–)30–210(–300) mm, longer or shorter than the lamina   12.

12. Submerged leaves 60–160 x 2.5–24 mm, 5–15(–30) times as long as wide, linear-elliptical to narrowly elliptical, the margin entire; fruits 1.9–2.6 mm   **7. polygonifolius**

12. Submerged leaves 160–280 × 22–38 mm, 6.0–7.5 times as long as wide, elliptical, the margin denticulate but the teeth minute and fugacious; fruits 2.7–4.1 mm   **11. nodosus**

13. Stems and main branches with petiolate submerged leaves 25–65 mm wide, the petiole 1–12(–25) mm; stipules on stems and main branches 35–80(–110) mm, with 2 parallel wings on the abaxial side extending from the base for at least half the length of the stipule; fruits 3.2–4.5 mm   **14. lucens**

13. Stems and main branches with sessile, or rarely petiolate, submerged leaves 4.5–25.0(–30.5) mm wide; stipules on stems and main branches 10–45(–55) mm, with 2 parallel ridges but no wings on the abaxial side, or narrowly winged at the base; fruits 2.4–3.4 mm   14.

14. Stems and main branches with submerged leaves 5–12 mm wide, with 3–4 lateral veins; stipules on these stems 10–25(–35) mm; fruits 2.4–3.1 mm   **12. gramineus**

14. Stems and main branches with submerged leaves 10–25(–30) mm wide, with 4–5(–6) lateral veins; stipules on these stems 20–45(–55) mm; fruits 2.7–3.4 mm   **16. × zizii**

15. Leaf lamina arising directly from the nodes, flat in section, with a conspicuous midrib, or leaves reduced to phyllodes   16.

15. Leaf lamina arising from the top of a sheath which surrounds the stem above the node (like the sheath and blade of a grass), elliptical to semicircular in section, with one large or several smaller air channels on each side of an inconspicuous midrib   26.

16. Leaves reduced to opaque phyllodes, without a distinct midrib and lateral veins; stipules 40–170 mm   **1. natans**

16. Leaves with a distinct midrib and lateral veins; stipules 4–55 mm   17.

17. Stems compressed, with a shallow groove running down one or both of the broader sides; leaf margin toothed at the apex   **32. crispus**

17. Stems terete to flattened, if compressed then without a groove running down the broader sides; leaf margin entire   18.

18. Stems strongly compressed to flattened; leaves with 1–2 lateral veins on each side of the midrib and many additional sclerenchymatous strands   19.

18 Stems terete to strongly compressed; leaves with 1–4 lateral veins on each side of the midrib but no additional sclerenchymatous strands   20.

19. Leaves with 2 lateral veins on each side of the midrib, the outer vein often faint; peduncles 28–95 mm; inflorescences with 10–20 flowers; most flowers with 2 carpels; fruits without a tooth on the ventral edge   **34. compressus**

19. Leaves with 1 lateral vein on each side of the midrib; peduncles 5–30 mm; inflorescences with 4–6 flowers; flowers with 1 carpel; fruits often with a tooth on the ventral edge   **35. acutifolius**

20. Plant rhizomatous; leaves 2.5–11 mm wide, with a broad band of lacunae which towards the base of the leaf extends at least as far as the inner lateral veins; lateral veins 2–4 on each side of the midrib   **33. epihydrus**

20. Rhizomes absent; leaves 0.5–3.5(–4) mm wide, without lacunae or with a band of lacunae on each side of the midrib which rarely reaches the inner lateral veins; lateral veins 1–2(–3) on each side of the midrib   21.

21 Stipules closed and tubular at the base when young   22.

21. Stipules open throughout their length   24.

22. Leaves 1.5–3.5(–4.0) mm wide; lateral veins (1–)2(–3) on each side of the midrib; leaves abruptly contracted to a mucronate apex; turions fan-shaped; fruits 2.4–3.0 mm   **37. friesii**

22. Leaves 0.5–1.4(–1.9) mm wide; lateral veins 1(–2) on each side of the midrib; leaves gradually tapering or rather abruptly narrowed to an acute but not mucronate apex; turions cylindrical; fruits 2.0–2.3 mm   23.

23. Leaf apex acute but not finely pointed; stipules translucent when dry, the veins not very prominent   **41. pusillus**

23. Leaf apex gradually tapering to a very fine point; stipules opaque when dry, the veins very prominent   **44. rutilus**

24. Leaves (1–)2.5–3.5 mm wide, often tinged pink or reddish-brown along the midrib or throughout; stipules with (8–)10–17 intracostal veins; turions 3–5 mm wide; inflorescences with 6–8 flowers   **40. obtusifolius**

24. Leaves 0.3–1.8(–2.3) mm wide, without any pink or reddish-brown tinge; stipules with 4–8(–9) intracostal veins; turions 0.6–1.7 mm wide; inflorescences with 2–5 flowers   25.

25. Leaves flaccid, with a midrib which occupies 10–20% of the leaf width near the base and in section has a shallowly convex lower side; flowers with (3–)4–5(–7) carpels; dorsal edge of fruits smooth   **39. berchtoldii**

25. Leaves relatively rigid, with a midrib which occupies 30–70% of the leaf width near the base and in section has a strongly convex lower side; flowers with 1(–2) carpels; dorsal edge of fruits muricate   **43. trichoides**

26. Ligules 5–15 mm long at the junction of the leaf-sheath and the lamina; leaf-apex entire; inflorescences with 4–14 flowers; mature fruits sessile   27.

26. Ligules absent; leaf-apex minutely denticulate; inflorescences with 2 flowers; mature fruits on stalks 3–35 mm long   28.

27. Leaf sheath closed and tubular at the base when young; mature fruits 2.2–3.2 mm   **45. filiformis**

27. Leaf sheath open and convolute along its entire length; mature fruits 3.3–4.7 mm   **47. pectinatus**

28. Peduncles 8–26 mm, 0.5–1.8 times as long as the longest fruit-stalk at maturity, rarely up to 3.3 times as long but only when peduncle is less than 10 mm long; fruits 2–2.8 mm, asymmetrical   **Ruppia maritima** (p. 38)

28. Peduncles 39–300(–770) mm, (1.6–)2–10(–30) times as long as the longest fruit-stalk at maturity; fruits 2.7–3.4 mm, symmetrical or slightly asymmetrical   **Ruppia cirrhosa** (p. 39)

Key to material without fruits

1. Some or all leaves linear-elliptical to subrotund, with convex sides   2.

1. All leaves filiform, linear or linear-oblong, with parallel sides   53.

2. All leaves floating or terrestrial, petiolate, usually coriaceous and opaque ('floating leaves') or submerged leaves, if present, reduced to opaque phyllodes without a distinct midrib and lamina   3.

2. Some or all leaves submerged, sessile or petiolate, with a distinct midrib and a thin, delicate and more or less translucent lamina ('submerged leaves')   5.

3. Floating leaves often with a djscoloured junction 7–25 mm between the petiole and the lamina; stipules 40–170 mm, the veins prominent when dry   **1. natans**

3. Floating leaves without a discoloured junction between the petiole and the lamina; stipules 10–63 mm, the veins inconspicuous when dry   4.

4. Floating leaves opaque, coriaceous, with inconspicuous secondary veins   **7. polygonifolius**

4. Floating leaves relatively translucent, with conspicuous secondary veins   **9. coloratus**

5. Floating leaves absent   6.

5. Floating leaves present in addition to the submerged leaves   31.

6. Leaves opposite; inflorescences with 2 opposite flowers   **Groenlandia densa** (p. 37)

6. Leaves (except for those immediately below an inflorescence) alternate; inflorescences with at least 3 flowers   7.

7. Stems terete or (very rarely) slightly compressed but not grooved along the broader sides; stipules rounded to acute at the apex   8.

7. Stems slightly compressed to compressed, with a shallow groove along one or both of the broader sides; stipules truncate to emarginate at the apex   26.

8. Plants with nodal glands; leaves linear or narrowly oblanceolate   **10. × lanceolatus**

8. Plants without nodal glands; leaves linear-elliptical to broadly ovate   9.

9. Most or all leaves sessile   10.

9. All leaves with petioles at least 1 mm long   21.

10. Leaves on the main stems amplexicaul or semi-amplexicaul   11.

10. Leaves on the main stems not amplexicaul   16.

11. Leaf margin entire; leaf apex markedly hooded
　　　　　　　　　　　　　　　　23. praelongus
11. Leaf margin denticulate; apex not or only slightly
　　hooded, rarely markedly hooded　　　　　　12.
12. Stipules fugacious, only present on the youngest
　　leaves　　　　　　　　　　　　**25. perfoliatus**
12. Stipules subpersistent or persistent, remaining for a
　　time on the mature leaves　　　　　　　　13.
13. Leaves with 7–12 lateral veins on each side of the
　　midrib; leaf apex markedly hooded　　**26. × cognatus**
13. Leaves with 3–8 lateral veins on each side of the
　　midrib; leaf apex not or only slightly hooded　14.
14. Stipules 20–55(–68) mm, often with 2 prominent ribs
　　which are narrowly winged towards the base on the
　　abaxial side　　　　　　　　　**15. × salicifolius**
14. Stipules (5.5–)11–28 mm, without 2 prominent ribs
　　or with 2 prominent ribs which are not winged on the
　　abaxial side　　　　　　　　　　　　15.
15. Leaf apex usually acute, sometimes apiculate or
　　mucronate, occasionally obtuse; leaf margin distinctly
　　denticulate; stipules flexible or rigid, often projecting
　　from the stems at an acute angle　　**19. × nitens**
15. Leaf apex obtuse or subacute, never apiculate or
　　mucronate; leaf margin obscurely denticulate;
　　stipules flexible, not projecting from the stems at an
　　acute angle　　　　　　　　　**22. × prussicus**
16. Leaf margin entire　　　　　　　　　　17.
16. Leaf margin denticulate　　　　　　　　18.
17. Stems unbranched; leaves with an obtuse, sometimes
　　slightly hooded but never markedly hooded apex
　　　　　　　　　　　　　　　　　20. alpinus
17. Stems branched towards the apex; leaves with an
　　obtuse or rounded and markedly hooded apex
　　　　　　　　　　　　　　　　21. × griffithii
18. Leaves on the main stems and branches 4.5–12 mm
　　wide, with 3–4 lateral veins on each side of the midrib
　　　　　　　　　　　　　　　　12. gramineus
18. Leaves on the main stems and branches 10–48 mm
　　wide, with 4–8 lateral veins on each side of the midrib　19.
19. Midrib bordered by a band of lacunae which extends
　　from the base to the apex; upper leaves copper-
　　coloured when dry　　　　　　**18. × nerviger**
19. Midrib only bordered by lacunae towards the base;
　　upper leaves without a copper tinge when dry　　20.
20. Leaves very rarely reduced to phyllodes at the base of
　　the stem; leaves linear-lanceolate to elliptical or
　　oblong, often inrolled but never recurved, with 4–8
　　lateral veins on each side of the midrib; stipules
　　rounded at the apex　　　　　　**15. × salicifolius**
20. Leaves reduced to phyllodes at the base of the stem;
　　leaves narrowly elliptical, not inrolled but often
　　recurved, with 4–5(–6) lateral veins on each side of
　　the lamina; stipules obtuse at the apex　**16. × zizii**
21. Leaves reduced to phyllodes at the base of the stem;
　　apex of mature leaves mucronate and sometimes with
　　an excurrent midrib; stipules with 2 strong ribs which
　　are often winged on the abaxial side　　　22.
21. Leaves not reduced to phyllodes at the base of the
　　stem; apex of mature leaves obtuse or acute but not
　　mucronate; stipules with 2 ribs which are not winged
　　on the abaxial side　　　　　　　　　24.

22. Petioles 25–70(–90) mm long　　**4. × fluitans**
22. Petioles 1–12(–25) mm long　　　　　23.
23. All leaves petiolate; leaves on stems and main
　　branches 25–65 mm wide, with stipules 35–80(–110)
　　mm long　　　　　　　　　　　　**14. lucens**
23. Some leaves sessile; leaves on stems and main
　　branches 11–25(–30.5) mm wide, with stipules
　　22–45(–55) mm　　　　　　　　**16. × zizii**
24. Margin of young leaves denticulate, with minute and
　　fugacious teeth　　　　　　　　　**11. nodosus**
24. Margin of young leaves entire　　　　　25.
25. Mature leaves linear-elliptical or narrowly elliptical,
　　2.5–24.0 mm wide, 5–15(–30) times as long as wide
　　　　　　　　　　　　　　　　7. polygonifolius
25. Mature leaves elliptical or broadly elliptical, 22–50
　　mm wide, 2.3–6.3 times as long as wide　**9. coloratus**
　　(N.B. *P. coloratus* and *P. polygonifolius* may not be
　　separable if only submerged leaves are available)
26. Leaf margin denticulate or serrate, especially towards
　　the apex　　　　　　　　　　　　27.
26. Leaf margin entire or very obscurely denticulate　30.
27. Leaves with 3–6 lateral veins on each side of the
　　midrib, semi-amplexicaul or more or less amplexicaul
　　at the base　　　　　　　　　**27. × cooperi**
27. Leaves with 1–3 lateral veins on each side of the
　　midrib, not or only slightly amplexicaul at the base　28.
28. Leaves with 1–2(–3) lateral veins on each side of the
　　midrib　　　　　　　　　　　**32. crispus**
28. Leaves with 2–3 lateral veins on each side of the
　　midrib　　　　　　　　　　　　29.
29. Stipules rigid, with 2 raised ridges towards the base
　　　　　　　　　　　　　　　28. × cadburyae
29. Stipules flexible, without ridges towards the base
　　　　　　　　　　　　　　　29. × olivaceus
30. Plant without a reddish-brown tinge when dry; leaves
　　abruptly narrowed to an auriculate or semi-
　　amplexicaul base　　　　　　　**24. × undulatus**
30. Plant often with a reddish-brown tinge when dry;
　　leaves gradually tapering to a slightly auriculate base
　　　　　　　　　　　　　　　29. × olivaceus
31. Submerged leaves linear, with parallel sides　32.
31. Submerged leaves linear-elliptical to broadly ovate,
　　with convex sides　　　　　　　　　36.
32. Submerged leaves sessile　　　　　　33.
32. Submerged leaves petiolate　　　　　35.
33. Nodal glands absent; stipules truncate or slightly
　　emarginate at apex; inflorescences 12–23 mm
　　　　　　　　　　　　　　　　33. epihydrus
33. Nodal glands present; stipules rounded to obtuse at
　　the apex but sometimes rolled so that they appear
　　acute; inflorescences 3.5–12 mm　　　34.
34. Submerged leaves 0.5–0.8 mm wide; floating leaves
　　sometimes with a discoloured junction between the
　　petiole and the lamina; petioles of floating leaves
　　26–154 mm　　　　　　　　　**6. × variifolius**
34. Submerged leaves 1.5–5.5(–7.5) mm wide; floating
　　leaves without a discoloured junction between the
　　petiole and the lamina; petioles of floating leaves
　　1–9 mm　　　　　　　　　　**10. × lanceolatus**

35. Stems unbranched; submerged leaves 1.0–3.9 mm wide; petioles of submerged leaves 45–175 mm long
    3. × gessnacensis

35. Stems sparingly to richly branched; submerged leaves 1.8–11.5 mm wide; petioles of submerged leaves not more than 55 mm long
    5. × sparganiifolius

36. All submerged leaves petiolate   37.

36. Some or all submerged leaves sessile   43.

37. Apex of submerged leaves mucronate, or with the midrib excurrent for up to 12 mm; stipules with 2 ridges which are often winged on the abaxial side
    4. × fluitans

37. Apex of submerged leaves obtuse or acute but never mucronate or with the midrib excurrent; stipules with 2 ridges which are never winged on the abaxial side   38.

38. Margin of young submerged leaves denticulate
    11. nodosus

38. Margin of submerged leaves entire   39.

39. Floating leaves relatively translucent, with conspicuous secondary veins not markedly different from the submerged leaves
    9. coloratus

39. Floating leaves opaque, coriaceous, with inconspicuous secondary veins, markedly different from the submerged leaves   40.

40. Submerged leaves without a prominent midrib; petioles 14–80(–165) mm   **7. polygonifolius**

40. Submerged leaves with a prominent midrib, appearing like phyllodes with a narrow lamina; petioles 45–355 mm   41.

41. Leaves at the base of the stem not reduced to phyllodes; submerged leaves 1.0–3.9 mm wide, with 1–2 lateral veins; midrib of submerged leaves without a band of lacunae along each side   **3. × gessnacensis**

41. Leaves at the base of the stem reduced or partially reduced to phyllodes; submerged leaves 2.5–14.5 mm wide, with 1–6 lateral veins; midrib of submerged leaves with a band of lacunae along each side   42.

42. Petioles of submerged leaves (60–)100–355 mm
    2. × schreberi

42. Petioles of submerged leaves less than 55 mm
    5. × sparganiifolius

43. Submerged leaves on main stems slightly amplexicaul to semi-amplexicaul   **19. × nitens**

43. Submerged leaves on main stems gradually or abruptly tapering to a sessile but not amplexicaul base   44.

44. Most submerged leaves more than 12 mm wide   45.

44. Most submerged leaves less than 12 mm wide   48.

45. Leaves at the base of the stem never reduced or partially reduced to phyllodes; margin of submerged leaves entire   46.

45. Leaves at the base of the stem usually reduced or partially reduced to phyllodes; margin of submerged leaves denticulate   47.

46. Stems unbranched; leaf apex narrowly obtuse and often shallowly hooded but not distinctly hooded
    20. alpinus

46. Stems branched; leaf apex obtuse or rounded and distinctly hooded   **21. × griffithii**

47. Apex of submerged leaves usually mucronate; floating leaves 57–103 × 22–40 mm   **16. × zizii**

47. Apex of submerged leaves obtuse and usually hooded; floating leaves 20–67 × 10–22 mm   **17. × billupsii**

48. Plants with nodal glands; leaves never reduced to phyllodes at base of stem; inflorescence with 6–11 flowers   **10. × lanceolatus**

48. Plants without nodal glands; leaves often reduced to phyllodes at base of stem; inflorescence with over 12 flowers   49.

49. Margin of submerged leaves denticulate, the teeth small but distinct   50.

49. Margin of submerged leaves entire or remotely and obscurely denticulate   51.

50. All submerged leaves sessile; apex of submerged leaves obtuse to acute, mucronate, never hooded
    12. gramineus

50. Some submerged leaves petiolate; apex of some submerged leaves obtuse and hooded   **17. × billupsii**

51. Submerged leaves linear or narrowly oblanceolate, over 16 times as long as wide   **5. × sparganiifolius**

51. Submerged leaves elliptical, oblong-elliptical or narrowly oblanceolate, less than 16 times as long as wide   52.

52. Some submerged leaves petiolate; submerged leaves 9–16 times as long as wide   **8. × lanceolatifolius**

52. All submerged leaves sessile; submerged leaves (4.7–)6.5–8.5(–10.5) times as long as wide on the main stems and branches   **13. × nericius**

53. Leaves arising directly from the nodes; leaf lamina reduced to opaque phyllodes or flat in section with a distinct midrib   54.

53. Leaves arising at the top of a sheath which surrounds the stem above the node (like the sheath and blade of a grass); leaf lamina elliptical to semicircular in section, with one large or several smaller air channels on each side of an inconspicuous midrib   72.

54. Leaves reduced to opaque phyllodes, without a distinct midrib and lateral veins; stipules 40–170 mm
    1. natans

54. Leaves with a distinct midrib and lateral veins; stipules 4–55 mm   55.

55. Rhizome absent; stipules rounded to obtuse at the apex   56.

55. Rhizome present; stipules truncate to slightly emarginate at the apex   66.

56. Stipules closed and tubular at the base when young   57.

56. Stipules open and convolute throughout their length   61.

57. Leaves with scattered sclerenchymatous strands in the lamina   **36. × pseudofriesii**

57. Leaves without scattered sclerenchymatous strands in the lamina   58.

58. Leaves 1.5–3.5(–4) mm wide; lateral veins (1–)2(–3) on each side of the midrib; leaves abruptly contracted to a mucronate apex; turions fan-shaped   **37. friesii**

58. Leaves 0.5–1.4(–1.9) mm wide; lateral veins 1(–2) on each side of the midrib; leaves gradually tapering to an acute but not mucronate apex; turions cylindrical   59.

59. Stipules white or buff-coloured and opaque when dry, the veins prominent   **44. rutilus**

59. Stipules hyaline and translucent when dry, the veins not very prominent   60.
60. Flowers with 4(–5) carpels   **41. pusillus**
60. Flowers with 1–3 carpels   **42. × grovesii**
61. Stems terete to compressed, rarely strongly compressed; leaves without sclerenchymatous strands in the lamina   62.
61. Stems strongly compressed to flattened; leaves with sclerenchymatous strands in the lamina   64.
62. Leaves (1.0–)2.5–3.6 mm wide, often tinged pink or reddish-brown along the midrib or throughout; stipules with (8–)10–17 intracostal veins; turions 3–5 mm wide; inflorescences with 6–8 flowers **40. obtusifolius**
62. Leaves 0.3–1.8(–2.3) mm wide, without any pink or reddish-brown tinge; stipules with 5–8(–9) intracostal veins; turions 0.6–1.7 mm wide; inflorescences with 2–5 flowers   63.
63. Leaves flaccid, with a shallow midrib which occupies 10–20 % of the leaf width near the base; flowers with (3–)4–5(–7) carpels   **39. berchtoldii**
63. Leaves relatively rigid, with a prominent midrib which occupies 30–70 % of the leaf width near the base; flowers with 1(–2) carpels   **43. trichoides**
64. Stems compressed, with nodal glands; leaves 1.1–3.0(–3.5) mm wide   **38. × sudermanicus**
64. Stems strongly compressed to flattened, without nodal glands; leaves 1.6–6.2 mm wide   65.
65. Leaves with 2 lateral veins on each side of the midrib, the outer vein often faint; peduncles 28–95 mm; inflorescences with 10–20 flowers; most flowers with 2 carpels   **34. compressus**
65. Leaves with 1 lateral vein on each side of the midrib; peduncles 5–30 mm; inflorescences with 4–6 flowers; most flowers with 1 carpel   **35. acutifolius**
66. Leaf margin denticulate, at least near the apex   67.
66. Leaf margin entire   70.
67. Stipules closed and tubular at the base when young   **30. × lintonii**
67. Stipules open and convolute throughout their length   68.
68. Leaves 2–5 mm wide; flowers with 2–3(–4) carpels   **31. × bennettii**
68. Leaves 5–12(–22) mm wide; flowers with (2–)4 carpels 69.
69. Leaves with 1–2(–3) lateral veins on each side of the midrib; teeth on the leaf margin usually easily seen with the naked eye; inflorescences with 3–8 flowers   **32. crispus**
69. Leaves with 2–4 lateral veins on each side of the midrib; teeth on leaf margin usually obscure, not visible to the naked eye; inflorescences with 10–12 flowers   71.
70. Stems terete to compressed, if compressed without a groove along one or both of the broader sides   **33. epihydrus**
70. Stems slightly compressed to compressed, with a groove along one or both of the broader sides 71.
71. Plant without a reddish-brown tinge when dry; leaves abruptly narrowed to an auriculate or semi-amplexicaul base   **24. × undulatus**
71. Plant often with a reddish-brown tinge when dry; leaves gradually tapering to a slightly auriculate base   **29. × olivaceus**

72. Leaf apex entire; ligule 5–15 mm, present at the junction of the leaf sheath and the lamina; inflorescences with 4–14 flowers   73.
72. Leaf apex minutely denticulate; ligule absent; inflorescences with 2 flowers   76.
73. All leaves open and convolute at the base when young; stigmas sessile   **47. pectinatus**
73. At least some sheaths closed and tubular at the base when young; stigmas sessile or borne on a distinct style   74.
74. Some leaf sheaths open and convolute when young, others on the same stem closed and tubular at the base   **46. × suecicus**
74. Leaf sheaths all closed and tubular at the base when young   75.
75. Stigmas sessile   **45. filiformis**
75. Stigmas borne on a distinct style   **46. × suecicus**
76. Peduncles 8–26 mm, 0.5–1.8 times as long as the longest fruit-stalk at maturity, rarely up to 3.3 times as long but only when peduncle is less than 10 mm
                                   Ruppia maritima (p. 38)
76. Peduncles 39–300(–770) mm, (1.6–)2–10(–30) times as long as the longest fruit-stalk at maturity
                                   Ruppia cirrhosa (p. 39)

Subgenus 1. Potamogeton

Leaves all submerged or some floating, variously shaped; stipules free from the leaf throughout, or adnate only at the very base. *Spike* not or hardly interrupted, its stalk and rhachis rigid. *Stigmas* with small papillae.

1. P. natans L.   Broad-leaved Pondweed
P. gessnacensis forma *hibernicus* Hagstr.; *P. hibernicus* (Hagstr.) Druce

Rhizome slender to very robust. *Stem* up to 1(–5.5) m, slender to very robust, terete, unbranched or very sparingly branched; nodal glands absent. *Submerged leaves* 120–450(–610) × 0.4–3.5 mm, 74–300 times as long as wide, reduced to entire, filiform or narrowly linear phyllodes, flat or shallowly concave on the upper side, convex on the lower side, opaque, mid-green, olive green or dark green, narrowly obtuse to acuminate at the apex which is often decayed or broken off. Transitional leaves to floating ones sometimes produced. *Floating leaves* with lamina (35–)50–100(–140) × (7–)20–45(–80) mm, (1.4–)1.6–3.5(–5.2) times as long as wide, opaque, coriaceous, pinkish-brown when young, mid-green, yellowish-green or olive green, often with a brownish tinge when mature, elliptical to ovate-oblong, acute or obtuse at apex and often apiculate, entire, cuneate, rounded or subcordate at base, shortly and narrowly decurrent on the petiole; lateral veins (6–)8–12(–16) on each side of the midrib, translucent in the living plant, the secondary veins numerous, transverse and rather obscure; petioles 50–150(–300) mm, shorter or longer than the lamina, with a flexible, pale, slightly swollen junction with the stem and usually with a flexible, pinkish-brown, discoloured section 7–25 mm long between the petiole and the lamina. *Stipules* 40–170 mm, rather rigid, enfolding the stem throughout their length or projecting from it at an acute angle, more or less translucent and colourless, slightly milky white or pale brown when fresh, opaque and green, brownish-green or buff when dry, narrowly obtuse to subacute, slightly cucullate and rolled so that they appear acuminate at apex, persistent although often eroded on the older leaves; veins prominent when dry, two stronger than the others and forming weak ridges along the back of the stipules. *Turions* absent. *Inflorescences* 20–60 × 4–9 mm; peduncles 40–100(–125) mm, slender to very robust, tapering slightly towards the apex, spongy. *Flowers* usually numerous, contiguous, with 4 carpels. *Fruits* 3.8–5.0 × 2.4–3.2 mm, olive-green or brownish-green; beak 0.3–0.8 mm, ventral, straight or recurved. *Flowers* 5–9. $2n = 52$.

The flexible junction between the petiole and the lamina of the floating leaves is diagnostic when present. In its absence, *P. natans* can be identified by the lack of laminar submerged leaves, the translucent veins of the floating leaves and the large fruits.

Native. Lakes, reservoirs, ponds, rivers, streams, canals and ditches, growing both in acidic and oligotrophic and in base-rich and eutrophic water. Common throughout the British Isles. Boreal and temperate regions of the northern hemisphere.

Hybrids occur between *P. natans* and *P. nodosus* (*P. × schreberi*, no. 2) *P. polygonifolius* (*P. × gessnacensis*, no. 3), *P. lucens* (*P. × fluitans*, no. 4), *P. gramineus* (*P. × sparganiifolius*, no. 5) and *P. berchtoldii* (*P. × variifolius*, no. 6).

2. P. × schreberi G. Fisch.   Schreber's Pondweed
= P. natans × nodosus

Rhizome robust to very robust. *Stem* up to 2 m, slender to robust, terete, unbranched; nodal glands absent. *Submerged leaves* reduced to phyllodes near the base of stem, otherwise with lamina 56–182 × 2.5–14.5 mm, (6–)10–20 (–30) times as long as wide, translucent, pale brown when young, green when mature, linear-elliptical, acute at the apex, entire, very gradually tapering to the petiole, midrib bordered by lacunae; the lateral veins 1–4 on each side of the midrib, more or less equally well developed; the secondary veins more or less transverse or ascending; petioles (60–)100–355 mm. *Floating leaves* with lamina 70–140 × 17–43 mm, 2.3–5.8 times as long as wide, opaque, coriaceous or subcoriaceous, brownish-green when young, rather dark olive green when mature, elliptical to oblong-elliptical, acute at apex, tapering to the base; lateral veins 5–10 each side of the midrib, paler than the lamina in the living plant, the secondary veins ascending near the midrib, transverse towards the margin, inconspicuous; petioles 89–330 mm, without a discoloured junction with the lamina. *Stipules* 60–141 mm, translucent, green, colourless with a greenish tinge or pale pink, rounded to obtuse and slightly cucullate at apex but rolled so that they appear acute, persistent, two veins more prominent than the others and forming ridges along the back of the stipule. *Turions* absent. *Inflorescences* 22–28 × about 6 mm; peduncles 46–79 mm, robust, of uniform diameter throughout their length, slightly compressed. *Flowers*

numerous, dense, with 4 carpels. *Fruits* do not develop. *Flowers* 7.

Like other hybrids of *P. natans*, *P. × schreberi* has submerged leaves which are expanded to a narrow lamina. The narrow submerged leaves with long petioles distinguish it from *P. nodosus*, *P. × fluitans* and *P. × sparganiifolius*.

Native. River Stour near Marnhull, Dorset, where it grows in the absence of both parents. Also occurs in Germany and Switzerland.

3. P. × gessnacensis G. Fisch.   German Pondweed
= P. natans × polygonifolius

Rhizome slender. *Stem* up to 0.9 m, slender to robust, terete, unbranched; nodal glands absent. *Submerged leaves* with lamina 15–120 × 1.0–3.9 mm, 8–175 times as long as wide, translucent or opaque, olive green, sometimes with a pinkish or reddish tinge, linear to narrowly elliptical, more or less acute at apex, entire at margin, very gradually tapering to the petiolate base; midrib not bordered by lacunae, the lateral veins 1–2 on each side of the midrib, the secondary veins occasional, transverse or ascending, or a few sometimes descending; petioles 45–175 mm. *Floating leaves* with lamina (22–)40–90 × (7–)12–32 mm, 2.6–6.8 times as long as wide, opaque, coriaceous, dark olive green with a reddish or pinkish-red tinge, elliptical, acute at apex, cuneate or rounded at base, the lamina shortly and narrowly decurrent; lateral veins (2–)4–10 on each side of the midrib, similar in colour to the lamina or appearing as very narrow translucent lines when the leaf is held up to the light, the secondary veins numerous, more or less ascending towards the midrib, more or less transverse towards the margin; petioles 80–270 mm, sometimes with a pale section 5–7 mm long between the petiole and the lamina. *Stipules* 38–85 mm, clasping the stems, translucent when fresh with a slightly milky tinge, pale brown, reddish-brown or brownish-green when dry, obtuse but rolled so that they appear acute at apex, persistent; veins prominent when dry, two stronger than the others and forming distinct ridges along the back of the stipules. *Turions* absent. *Inflorescences* 8–18 × 3.5–4.0 mm; peduncles 54–77 mm, slender, of uniform diameter throughout their length. *Flowers* 12–22, contiguous, with 4 carpels. *Fruits* do not develop. *Flowers* 6.

Native. Llyn Anafon, North Wales, and Hill of Nigg, Scotland. Germany.

4. P. × fluitans Roth   Thick-leaved Pondweed
= P. lucens × natans
P. crassifolius Fryer; *P. sterilis* Hagstr.

Rhizome robust to very robust. *Stem* up to 2.25 m, robust to very robust, terete, unbranched or sparingly branched; nodal glands absent. *Submerged leaves* with lamina 60–220 × 8–33 mm, 4.3–20 times as long as wide, translucent, dark green or olive green, usually reduced or partially reduced to phyllodes near the base of the stem, otherwise narrowly elliptical, acute, rarely acuminate and mucronate at apex, the midrib excurrent for up to 12 mm, entire or very obscurely denticulate on the margin, very gradually tapering to a petiolate base; midrib not bordered by lacunae, the lateral veins 3–7 on each side of the midrib, more or less equal or 1–2 more strongly developed than the others, the secondary veins ascending across the entire width of the leaf or more or less transverse near the margin; petioles 25–70(–90) mm. *Floating leaves* with lamina 70–155 × 11–53 mm, 2.0–4.5(–5.9) times as long as wide, more or less opaque, coriaceous or subcoriaceous, green with a strong brownish tinge when young, slightly yellowish-green or dark green when mature, narrowly elliptical to oblong-elliptical or ovate, acute to obtuse and often apiculate at apex, cuneate or rounded at base, the lamina narrowly decurrent; lateral veins translucent in the living plant, the secondary veins ascending across the entire width of the leaf or more or less transverse near the margin and inconspicuous; petioles 25–70(–90) mm, usually shorter than the lamina, with a pale, slightly swollen junction with the stem but without a flexible junction with the lamina. *Stipules* 35–100 mm, rigid, enfolding the stem or projecting from it at an acute angle, translucent, dark green or brownish green when fresh, obtuse and slightly hooded at apex, persistent, two veins stronger than the others and forming ridges or narrow wings along the back of the stipule in the basal part or for most of its length. *Turions* absent. *Inflorescences* 22–62 × 8.5–15 mm; peduncles 40–95 mm, robust to very robust, thickened and spongy towards the apex. *Flowers* numerous, densely packed or well spaced, with 4(–5) carpels. *Fruits* 3.6–4.5 × 2.5–3.3 mm, greenish-brown; beak 0.3–0.4 mm, straight or recurved. *Flowers* 6–8.

P. × *fluitans* has broader submerged leaves than the other hybrids of *P. natans*, and the narrow wings on the back of the stipules will also distinguish it from similar hybrids.

Native. Frequent in Moors River in Dorset and Hampshire; rare in ditches in East Anglia. Europe.

5. P. × sparganiifolius Laest. ex Fr.

Ribbon-leaved Pondweed

= **P. gramineus × natans**

P. longifolius auct.; *P. natans* subsp. *kirkii* Hook. fil.; *P. kirkii* (Hook. fil.) Syme ex Hook. fil.; *P. tiselii* K. Richt.

Rhizome slender to robust. *Stem* up to 1.4 m, slender to very robust, terete, sparingly to richly branched; nodal glands absent. *Submerged leaves* with lamina 60–520 × 1.8–11.5 mm, 16–72 times as long as wide, translucent or almost opaque, pale green or dark green, on the lowest part of the stem often reduced to linear phyllodes or with only a narrow lamina at the distal end, otherwise linear or narrowly oblanceolate, acute or acuminate at apex, entire or obscurely denticulate and plane at margin, very gradually tapering to a petiolate or occasionally sessile base; midrib bordered by a broad or narrow band of small lacunae, the lateral veins 1–6 on each side of the midrib, slightly stronger ones alternating with slightly weaker ones, the outermost sometimes faint, the secondary veins transverse or ascending; petioles up to 55 mm, sometimes with a slightly flexible,

slightly swollen junction with the stem. *Floating leaves* with lamina 38–115 × 9–34 mm, 1.9–8.3 times as long as wide, opaque, coriaceous, yellowish-green or dark green, narrowly elliptical or oblong-elliptical to ovate, obtuse to acute and often apiculate at apex, cuneate or rounded, rarely subcordate at base, the lamina often broadly or narrowly decurrent; lateral veins 6–12 on each side of the midrib, darker to lighter than the lamina in the living plant, the secondary veins numerous, more or less ascending towards the centre of the leaf, transverse towards the margin and inconspicuous; petioles 30–250 mm, sometimes with a discoloured section 5–18 mm long between the petiole and the lamina. *Stipules* 17–95 mm, enfolding the stem throughout their length or projecting from it at an acute angle, translucent and hyaline, sometimes with a green or milky tinge when fresh, more or less opaque and green, brown or buff when dry, broadly obtuse or round, and sometimes slightly cucullate and rolled at the apex so that they can appear acute, persistent; veins prominent when dry, two stronger than the others and sometimes forming ridges along the back of the stipules. *Turions* absent. *Inflorescences* 25–45 × 3.5–6.5 mm; peduncles 30–118 mm, of uniform diameter along their length or slightly broader towards the apex. *Flowers* numerous, contiguous, with 4 carpels. *Fruits* do not develop. *Flowers* 6–9.

The narrow submerged leaves with short petioles distinguish *P.* × *sparganiifolius* from its putative parents and from similar hybrids.

Native. Rivers, streams and fenland drains and pools. Local in Scotland and Ireland, rare in England and Wales. Europe, most frequent in Scandinavia.

6. P. × variifolius Thore　　Diverse-leaved Pondweed

= **P. berchtoldii × natans**

Rhizome very slender. *Stem* up to 0.4 m, very slender, terete to slightly compressed, sparingly branched; nodal glands poorly developed to well-developed. *Submerged leaves* with lamina 22–95 × 0.5–0.8 mm, (44–)80–145 times as long as wide, more or less translucent, pale green, linear, flat on the upper side, shallowly concave on the lower side, sessile, acute at apex, entire and plane at margin; midrib not bordered by lacunae, the lateral veins 3–5 on each side of the midrib and indistinct. *Floating leaves* with lamina 15–37 × 3.5–9.0 mm, 2.7–5.8 times as long as wide, opaque, olive green, often with a pinkish-brown tinge, narrowly elliptical to oblong-elliptical, acute and often apiculate at apex, cuneate or rounded at base; lateral veins 3–5 on each side of the midrib, translucent in the living plant, the secondary veins numerous and transverse; petioles 26–154 mm, sometimes with a pale brown, discoloured junction 3–6 mm long between the petiole and the lamina. *Stipules* 13–30 mm, open, convolute at the base, more or less translucent both when fresh and dry, obtuse but rolled at the apex so that they appear acute, persistent; veins prominent when dry, two stronger than the others but not forming distinct ribs. *Turions* not seen. *Inflorescences* 4.5–12.0 × 1.5–2.5 mm; peduncles 18–50 mm, slender, slightly compressed. *Flowers* densely packed. *Fruits* do not develop. *Flowers* 8.

A remarkable hybrid, differing from other taxa in its grass-like submerged leaves and small floating leaves.

Native. Glenamoy River, Co. Mayo. Also in France and Germany.

7. P. polygonifolius Pourr. Bog Pondweed
P. oblongus Viv.; *P. anglicus* Hagstr.; *P. natans* subsp. *polygonifolius* (Pourr.) Hook. fil.

Rhizome slender to very robust. *Stem* up to 0.65 m, slender to very robust, terete, unbranched; nodal glands absent. *Submerged leaves* with lamina 60–162 × 2.5–24 mm, 5–15(–30) times as long as wide, delicate, translucent, yellowish-green, olive green or pale reddish-green when dry, linear-elliptical or narrowly elliptical, gradually tapering to the base and to the narrow, obtuse apex, entire, petiolate; the midrib bordered by a narrow band of lacunae, the lateral veins 2–7 on each side of the midrib and all more or less equally well developed, the secondary veins ascending near the midrib and more or less transverse elsewhere; petioles 14–80(–165) mm. *Floating leaves* with lamina (15–)40–105 × (5.5–)15–68 mm, (1.1–)1.5–3.5(–4.3) times as long as wide, opaque, coriaceous, pinkish-brown when young, bright green when mature, often with a strong brownish or reddish tinge, elliptical, oblong-elliptical or ovate, acute to obtuse and often apiculate at apex, cuneate or rounded to subcordate at base, shortly and narrowly decurrent on the petiole; lateral veins (5–)6–9(–12) on each side of the midrib, darker than the rest of the lamina in the living plant, the secondary veins numerous, transverse or more or less ascending and becoming increasingly conspicuous as the leaf ages; petioles (13–)30–150(–300) mm, shorter or longer than the lamina. *Stipules* 10–50 mm, flexible, translucent, hyaline, often with a strong brownish tinge, very broadly obtuse to narrowly obtuse at apex, persistent; veins inconspicuous when dry, two more strongly developed than the others and forming distinct ridges on each side of the stipule. *Turions* absent. *Inflorescences* 11–42 × 3.5–5.5 mm; peduncles 25–100(–185) mm, slender to robust, of uniform diameter throughout their length, slightly compressed near the base, more or less terete towards the apex. *Flowers* numerous, contiguous, with 4 carpels. *Fruits* 1.9–2.6 × 1.3–1.9 mm, reddish-brown; beak 0.1–0.25 mm, more or less apical, straight. *Flowers* 5–10. $2n = 26, 28, 52$.

P. polygonifolius is closely related to *P. coloratus*, from which it differs in its larger, reddish fruits. Aquatic forms are often confused with *P. alpinus*, which also has entire submerged leaves, but the leaves of *P. polygonifolius* are petiolate, not sessile.

Native. Shallow acidic water in lakes, pools, rivers, streams, ditches and peat-cuttings, and as a subterrestrial plant in flushes and on boggy ground. Frequent in northern and western parts of Britain and Ireland; scarce and decreasing in the south and east. Europe, North Africa and North America (rare).

Hybrids of *P. polygonifolius* with *P. natans* (*P.* × *gessnacensis*, no. 3) and *P. gramineus* (*P.* × *lanceolatifolius*, no. 8) occur in the British Isles, but both are rare.

8. P. × lanceolatifolius (Tiselius) C. D. Preston
Spear-leaved Pondweed
= **P. gramineus × polygonifolius**
P. gramineus forma *lanceolatifolius* Tiselius

Rhizome very slender. *Stem* up to 0.55 m, very slender to slender, unbranched or sparingly branched; nodal glands absent. *Submerged leaves* with lamina 40–112 × 3.5–13 mm, 9–16 times as long as wide, translucent, olive green or yellowish-green when dry, sometimes with a slight reddish tinge, sometimes reduced or partially reduced to phyllodes towards the base of the stem, otherwise elliptical or narrowly oblanceolate, gradually tapering to a sessile or petiolate base and to an acute apex, obscurely and remotely denticulate on the margin; midrib bordered by a narrow band of lacunae, lateral veins 3–7 on each side of the midrib, more or less equally developed or two stronger than the others, the secondary veins frequent, ascending; petioles up to 14 mm. *Floating leaves* with lamina 25–75 × 7–17 mm, 3.5–5.1 times as long as wide, opaque, coriaceous, elliptical to ovate-elliptical, obtuse to acute and apiculate at apex, cuneate or rounded at base with a narrowly decurrent lamina; lateral veins 5–6 on each side of the midrib, the secondary veins numerous, transverse to ascending and obscure; petioles 52–98 mm. *Stipules* 16–40 mm, enfolding the stem or projecting from it at an angle, translucent or more or less opaque when dry, brown, obtuse at apex, persistent, with two veins which are stronger than the others and form low ridges on the abaxial side of the stipule towards the proximal end. *Turions* not seen. *Inflorescences* not seen in British material; in Swedish plants 12–17 × 2.5–3.0 mm; peduncles 43–65 mm, slender to robust. *Flowers* numerous, dense. *Fruits* do not develop.

Native. River Bladnoch, Spittal, Wigtownshire and a small loch near Nairn, East Inverness-shire; not collected from either locality since 1953. Also occurs in Sweden.

9. P. coloratus Hornem. Fen Pondweed
P. plantagineus Ducros ex Roem. & Schult.; *P. natans* subsp. *plantagineus* (Ducros ex Roem. & Schult.) Hook. fil.

Rhizome robust. *Stem* up to 0.72 m, slender to robust, terete, unbranched or very sparingly branched; nodal glands absent. *Submerged leaves* with lamina on short overwintering stems 70–175 × 10–30 mm, 3.0–8.6 times as long as wide, oblanceolate or narrowly elliptical, narrowly obtuse and hooded at apex, very gradually tapering to the base, petiolate or rarely sessile; on longer stems 75–146 × 22–50 mm, 2.3–6.3 times as long as wide, elliptical or broadly elliptical, tapering to an acute apex and gradually or more or less abruptly narrowed to base, petiolate; all translucent, bright green, sometimes with a pinkish or reddish tinge especially along the veins, entire; midrib bordered by a narrow band of lacunae, the lateral veins 4–8 on each side of the midrib, more or less equally developed or slightly stronger ones alternating with slightly weaker ones, the secondary veins numerous, more or less ascending in the centre of the leaf, and more or less transverse and rather wavy elsewhere, conspicuous; petioles (6–)20–65 mm. *Floating leaves* with lamina (15–)25–85 × (13.5–)16–55 mm, 1.05–2.1 times as long as wide, more or less translucent, only rarely subcoriaceous, bright green, older leaves often tinged reddish-brown, broadly elliptical, ovate or more or less subrotund, more or less acute or obtuse and sometimes apiculate at apex, more or less truncate or cordate at base; lateral veins 6–10 on each side of the midrib, opaque in the living plant, the secondary veins numerous, irregularly ascending near the midrib and more or less transverse elsewhere; petioles (4–)8–45 mm, shorter than the lamina. *Stipules* 20–63 mm, delicate, flexible, clasping the stems throughout their length or the distal portion projecting at an acute angle from the stem, translucent when fresh, translucent or more or less opaque when dry, obtuse at apex, persistent; veins inconspicuous when dry, two much stronger than the others and forming ridges along the abaxial side. *Turions* absent. *Inflorescences* 14–45 × 2.0–3.5 mm; peduncles (18–)30–180 mm, slender to robust, of uniform diameter throughout their length. *Flowers* numerous, contiguous, with 4 carpels. *Fruits* 1.5–1.9 × 1.0–1.3 mm, olive green or greenish-brown; beak 0.2–0.3 mm, more or less apical, straight. *Flowers* 6–7. $2n = 26$.

Native. Shallow calcareous water in pools, ditches, streams, flooded clay and marl pits and at the edges of lakes. It also grows subterrestrially in damp moss carpets. Local in Britain north to the Outer Hebrides, mainly in the east; widespread in Ireland. Europe, North Africa, south-west Asia.

Hybrids of *P. coloratus* with *P. berchtoldii* (*P.* × *lanceolatus*, no. 10) and *P. gramineus* (*P.* × *billupsii*, no. 17) are known in the British Isles. Both are rare.

10. P. × lanceolatus Sm.　　　　　Irish Pondweed
　= **P. berchtoldii × coloratus**
P. heterophyllus subsp. *lanceolatus* (Sm.) Hayw.;
P. perpygmaeus Hagstr. ex Druce; *P. hibernicus*
A. Benn. nom. nud.

Rhizome very slender. *Stem* up to 1.2 m, very slender to slender, terete to slightly compressed, unbranched or sparingly to richly branched; nodal glands poorly developed to well-developed. *Submerged leaves* with lamina (22–)30–75(–92) × 1.5–5.5(–7.5) mm, 10.5–24.0 times as long as wide, translucent, clear green, the upper leaves sometimes with a reddish tinge, linear or narrowly oblanceolate, often slightly twisted along their length or recurved, more or less acute or obtuse at apex, entire and plane or undulate at margin, usually sessile but occasionally shortly petiolate; midrib bordered by a band of rather large lacunae, the lateral veins (2–)3 on each side of the midrib, the outermost vein often faint, the secondary veins frequent and transverse or ascending; petioles up to 4 mm. *Floating leaves* with lamina 15–48 × 4.5–10.0 mm, 2.5–7.0 times as long as wide, more or less opaque, green or pale pinkish-green when young, dark green when mature, often with a pinkish-red tinge especially along the midrib, elliptical or obovate, acute at

apex, narrowly cuneate at base; midrib bordered by a band of relatively large lacunae, the lateral veins 3–4 on each side of the midrib; petioles 1–9 mm. *Stipules* 8–27 mm, open, convolute, translucent when fresh and dry, hyaline or sometimes with a slight greenish tinge, rounded to broadly obtuse at the apex, persistent; veins indistinct when dry, two stronger than the others and forming weak ribs along the back of the stipules. *Turions* axillary, 10–35 mm, with short, crowded leaves and stipules which conceal the stem. *Inflorescences* 3.5–6.5 × 2.0–3.5 mm; peduncles 9–69 mm, very slender to slender, of uniform diameter throughout their length or slightly broader towards the apex, terete or slightly compressed. *Flowers* 6–11, densely packed, with 4–5(–6) carpels. *Fruits* do not develop. *Flowers* 8.

Native. Well established in several more or less calcareous streams in western Ireland. Also known from Anglesey (perhaps extinct) and Cambridgeshire (extinct). Endemic.

11. P. nodosus Poir.　　　　　Loddon Pondweed
P. petiolatus Wolfg.; *P. drucei* Fryer

Rhizome robust. *Stems* up to 2.5 m, robust to very robust, terete, unbranched or sparingly branched. *Submerged leaves* with lamina 160–280 × 22–38 mm, 6.2–7.5 times as long as wide, translucent, pale green when fresh, green or brownish-green when dried, elliptical, gradually tapering to a petiolate base and an acute apex, denticulate at margin, the teeth minute and fugacious; midrib bordered by a band of lacunae, the lateral veins 5–10 on each side of the midrib, slightly stronger veins alternating with slightly weaker ones, the secondary veins numerous, more or less ascending, rather irregular, often wavy; petioles 70–210 mm. *Floating leaves* with lamina 70–130 × 34–49 mm, 1.9–4.3 times as long as wide, opaque, coriaceous, green, broadly elliptical, obtuse and more or less apiculate at the entire apex, rounded to cuneate at base; lateral veins 7–11 on each side of the midrib, opaque, darker or slightly lighter than the lamina on the living plant, the secondary veins numerous, obscure, ascending in the centre of the leaf and transverse towards the margin; petioles 30–210 mm, shorter or longer than the lamina. *Stipules* 45–125 mm, flexible, enfolding the stem, more or less translucent with a slight milky tinge when fresh, pale green to pale brown when dry, gradually tapering to a slightly cucullate apex, fairly persistent; veins inconspicuous when dry, two stronger than the others and forming ridges along the back of the stipules. *Turions* absent. *Inflorescences* 14–67 × 4–10 mm; peduncles 45–130 mm, robust to very robust, somewhat broader and slightly spongy at the apex. *Flowers* numerous, more or less contiguous, with (2–)4(–5) carpels. *Fruits* not seen in British material; in European plants 2.7–4.1 × 2–3 mm, reddish-brown or brown; beak 0.3–0.8 mm, ventral, straight or slightly recurved. *Flowers* 5–9. $2n = 52$.

The denticulate submerged leaves with long petioles are distinctive, especially when combined with the large floating leaves.

Native. Calcareous, lowland rivers. Southern England, rare. Europe, Africa, Asia, North and South America.

P. nodosus × *natans* (*P.* × *schreberi*, no. 2) has recently been recorded in Britain for the first time.

12. P. gramineus L. Various-leaved Pondweed
P. heterophyllus Schreb.; *P. palustris* Teesd.; *P. gracilis* Wolfg.; *P. lonchites* Tuck.; *P. varians* Morong ex Fryer; *P. falcatus* Fryer; *P. graminifolius* H. & J. Groves; *P. heterophyllus* subsp. *lonchites* (Tuck.) Hook. fil.

Rhizome very slender to very robust. *Stem* up to 0.8(–2.5) m, very slender to robust, terete, usually richly branched. *Submerged leaves* 40–90(–140) × 4.5–12 mm on the main stems and branches, as small as 22 × 3 mm on short branches, 4.6–12(–19.3) times as long as wide, translucent, green without any pinkish tinge, reduced or partially reduced to phyllodes at base of stem, otherwise narrowly elliptical or narrowly oblong-elliptical, often recurved, acute to obtuse and usually mucronate at apex, denticulate, gradually tapering or abruptly narrowed to the base, sessile; midrib bordered by a narrow band of lacunae, the lateral veins 3–4 on each side of the midrib, equally well developed or the outermost fainter than the others, the secondary veins frequent and ascending. *Floating leaves* with lamina 19–70(–92) × 7.5–34 mm, 1.9–2.7 times as long as wide, opaque, coriaceous, yellowish-green to dark green, narrowly to broadly elliptical or oblong-elliptical, obtuse to acute and apiculate at apex, gradually tapering or abruptly narrowed to a cuneate, truncate or subcordate base; lateral veins slightly lighter to slightly darker than the rest of the lamina in the living plant, the secondary veins numerous, ascending near the midrib, transverse towards the margins and rather obscure to relatively conspicuous; petioles 18–60 mm, shorter or longer than the lamina. *Stipules* 10–25(–35) mm on the stems and main branches, as short as 6.5 mm on the short branches, rigid, sometimes (in sparingly branched forms) enfolding the stem but usually rolled and projecting from it at an acute angle, pale green to blackish-green when dry, broadly obtuse and slightly hooded at apex, persistent; veins inconspicuous when dry, two more prominent than the others and forming ribs towards the base of the stipule on the abaxial side. *Turions* absent. *Inflorescences* 12–41 × 4–7 mm; peduncles (20–)35–100(–280) mm, slender to very robust, terete, broader and spongy towards the apex. *Flowers* usually numerous, contiguous, with (3–)4(–5) carpels. *Fruits* 2.4–3.1 × 1.6–2.1 mm, dark olive green, rugose; beak 0.2–0.4 mm, usually more or less ventral, occasionally subapical, straight. *Flowers* 6–9. 2*n* = 52.

Although it is a variable species, *P. gramineus* can usually be identified by its narrow, sessile and denticulate leaves which often have a mucronate apex.

Native. Mesotrophic, acidic or base-rich moderately shallow water in lakes, reservoirs, rivers, streams, canals and ditches. Widespread in Scotland and Ireland, less frequent and declining in England and Wales. Boreal and temperate regions throughout the northern hemisphere.

Hybrids of *P. gramineus* with *P. lucens* (*P.* × *zizii*, no. 16) and *P. perfoliatus* (*P.* × *nitens*, no. 19) are frequent and can be difficult to distinguish from it. Those with *P. alpinus* (*P.* × *nericius*, no. 13), *P. coloratus* (*P.* × *billupsii*, no. 17), *P. natans* (*P.* × *sparganiifolius*, no. 5) and *P. polygonifolius* (*P.* × *lanceolatifolius*, no. 8) are scarcer.

13. P. × nericius Hagstr. Swedish Pondweed
= P. alpinus × gramineus

Rhizome very slender to slender. *Stem* up to 1.1 m, very slender to robust, terete, richly branched at the base, the longer shoots simple or with a few axillary branches. *Submerged leaves* with lamina 43.5–80.0 × 6.5–13.0 mm and (4.7–)6.5–8.5(–10.5) times as long as wide on the main stems and branches, as small as 26 × 2.5 mm and 7.5–11.4 times as long as wide on short branches, translucent, green or brown when dry, sometimes reduced or partially reduced to phyllodes at the base of the stem, otherwise elliptical or oblong-elliptical, acute to obtuse and sometimes mucronate at apex; entire or obscurely and remotely denticulate and undulate at the margin, gradually tapering or abruptly narrowed to the sessile base; midrib bordered by a band of lacunae which is broad at the base of the leaf and extends almost as far as the apex, the lateral veins 3–5 on each side of the midrib, more or less equally well developed or the outermost fainter than the others, the secondary veins frequent and ascending. *Floating leaves* with lamina 38–70 × 9.5–20.5 mm, 3.2–5.0 times as long as wide, opaque, subcoriaceous, green or slightly reddish-brown when dry, elliptical to oblong-elliptical, obtuse and often apiculate at apex, tapering to a cuneate base; secondary veins numerous, ascending, obscure; petioles 9–62 mm, usually shorter than the lamina. *Stipules* 16–26 mm on the main stems and branches, as short as 6.5 mm on short branches, rigid, enfolding the stem or projecting from it at an acute angle, translucent when fresh, brownish-green when dry, obtuse at apex, persistent; veins prominent when dry, two stronger than the others and forming ribs towards the base of the stipule on the abaxial side. *Turions* absent. *Inflorescences* 13.5–23.0 × 3.5–4.5 mm; peduncles 132–210 mm, robust, of uniform diameter throughout their length. *Flowers* 13–24, contiguous, with 4 carpels. *Fruits* do not develop. *Flowers* 7–8.

Native. River Don, Bridge of Alford, South Aberdeenshire. Also in Iceland, Norway and Sweden.

14. P. lucens L. Shining Pondweed
P. lucidus Salisb. nom. illegit.; *P. acuminatus* Schumach.; *P. cornutus* J. & C. Presl

Rhizome robust to very robust. *Stems* up to 2.5(–6) m, robust to very robust, terete, sparingly to richly branched. *Submerged leaves* 75–200(–260) × 25–65 mm on the stems, as small as 55 × 17 mm on short branches, (2.25–)3–6(–10) times as long as wide, translucent, usually rather glossy yellowish-green, occasionally with a reddish tinge, often reduced or partially reduced to phyllodes on the lower stems, otherwise narrowly elliptical to oblong-elliptical, rounded to acuminate at apex which is always at least mucronate and frequently with the midrib excurrent, denticulate and minutely undulate, gradually tapering or rather abruptly narrowed to the petiolate base; midrib prominent, not bordered by lacunae, the lateral veins 4–5(–6) on each side of the midrib, the inner veins equally well developed, the 1–2 outer veins fainter than the others, the secondary veins frequent, ascending across the entire width of the leaf and almost as conspicuous as the laterals; petioles 1–12(–25) mm, with the lamina extending down the petiole as a wing along each side so that the leaf can appear subsessile. *Floating leaves* absent. *Stipules* 35–80(–110) mm on the stems, as short as 20 mm on short branches, rigid, translucent, pale green, rounded, obtuse or rarely acute at the apex, very persistent; veins inconspicuous when dry, two much stronger than the others, forming conspicuous green ribs, the ribs winged on the abaxial side, usually for at least half of the length of the stipule. *Turions* absent. *Inflorescences* 22–70 × 7–13 mm; peduncles 50–200(–270) mm, very robust, broader and spongy towards the apex, slightly compressed especially towards the base. *Flowers* numerous, dense, with 4–5(–6) carpels. *Fruits* 3.2–4.5 × 2.4–3.0 mm, brown; beak 0.5–0.8 mm, more or less ventral, straight. *Flowers* 6–9. $2n = 52$.

A robust plant, easily distinguished from other species by its denticulate and shortly petiolate submerged leaves and winged stipules.

Native. In rather deep, calcareous water in lakes, sluggish rivers, canals and major fenland drains. Widespread and locally frequent in eastern England and central Ireland; rare elsewhere. Europe; western Asia; North Africa; Uganda.

P. lucens x *gramineus* (*P.* × *zizii*, no. 16) and *P. lucens* × *perfoliatus* (*P.* × *salicifolius*, no. 15) are two widespread hybrids, both of which can resemble *P. lucens* very closely. Hybrids between *P. lucens* and *P. alpinus* (*P.* × *nerviger*, no. 18), *P. crispus* (*P.* × *cadburyae*, no. 28) and *P. natans* (*P.* ×*fluitans*, no. 4) are much rarer.

15. P. × salicifolius Wolfg.   Willow-leaved Pondweed
= P. lucens × perfoliatus

P. decipiens Nolte ex Koch; *P. burtonii* Hopkins ex Syme nom. nud.; *P. upsaliensis* Tiselius; *P. salignus* Fryer; *P. brotherstonii* A. Benn.; *P. kupfferi* A. Benn.; *P. lucens* subsp. *decipiens* (Nolte ex Koch) Hook. fil.

Rhizome robust. *Stems* up to 3 m, robust to very robust, terete, sparingly to richly branched. *Submerged leaves* with lamina 60–120(–215) × 14–40 mm, 2.7–9.8 times as long as wide, translucent, yellowish-green or dark green, occasionally tinged with pink especially along the veins, very rarely reduced or partially reduced to a phyllode at base of stem, otherwise linear-lanceolate to elliptical or oblong, acute or rounded and mucronate at apex, denticulate and undulate at margin, abruptly or gradually narrowed to a sessile, sometimes semi-amplexicaul, or rarely very shortly petiolate base; midrib prominent, sometimes bordered at least below by a narrow band of lacunae, the lateral veins 4–8 on each side of the midrib, 1–2 sometimes bordered by a narrow band of lacunae

and hence more prominent than the rest, the secondary veins frequent, more or less ascending towards the centre of the leaf and transverse towards the margins; petioles up to 0.5 mm. *Floating leaves* absent. *Stipules* 20–55 (–68) mm, flexible or rigid, translucent, hyaline, sometimes with a pinkish tinge, rounded at apex, fairly persistent; veins inconspicuous when dry, two slightly stronger than the others and forming very weak ribs or much stronger than the others and forming distinct green ribs, the ribs sometimes narrowly winged on the abaxial side for up to 75% of the length of the stipule. *Turions* absent. *Inflorescences* 12–41 × 6.5–9.5 mm; peduncles 30–100 mm, robust to very robust, broader and spongy towards the apex, slightly compressed especially towards the base. *Flowers* numerous, contiguous, with (3–) 4 carpels. *Fruits* do not develop. Flowers 7–8. $2n = 52$.

This hybrid is easily mistaken for *P. lucens*, but it differs in its sessile rather than shortly petiolate leaves.

Native. Lakes, rivers, canals and major fenland drains. Local in those areas of England, southern Scotland and Ireland where *P. lucens* occurs. Europe; south-west Asia.

16. P. × zizii Koch ex Roth   Ziz's Pondweed
= P. gramineus × lucens

P. lucens var. *coriaceus* Mert. & Koch; *P. longifolius* auct.; *P. coriaceus* (Mert. & Koch) A. Benn.; *P. babingtonii* A. Benn.; *P. lucens* subsp. *zizii* (Koch ex Roth) Nyman

Rhizome robust to very robust. *Stem* up to 1.2(–2) m, slender to robust, terete, sparingly to richly branched. *Submerged leaves* with lamina 50–129 × 11–25(–30.5) mm on the stems and main branches, as small as 30 × 8.5 mm on short branches, 3.8–7.1 times as long as wide, translucent, glossy yellowish-green, sometimes with a pinkish tinge, green or dark olive green, reduced to phyllodes at base of stem, otherwise narrowly elliptical and often recurved, acute to obtuse and usually mucronate at apex, denticulate and undulate at margins, gradually tapering or rather abruptly narrowed to a sessile or occasionally shortly petiolate base; midrib bordered by a narrow band of lacunae, the lateral veins 4–5(–6) on each side of the midrib, the inner veins more or less equally strong, the outer 1–2 veins usually fainter than the others, the secondary veins frequent, regularly or irregularly ascending across the entire width of the leaf; petioles up to 3(–7) mm. *Floating leaves* with lamina 57–103 × 22–40 mm, 1.8–3.3 times as long as wide, opaque, coriaceous, dark green, elliptical to broadly elliptical or oblong-elliptical, obtuse to subacute and apiculate at apex, tapering or abruptly narrowed to a cuneate or subcordate base, the lamina shortly and narrowly decurrent; lateral veins lighter or darker than the lamina in the living plant, the secondary veins relatively conspicuous; petioles (10–)20–65 mm, usually shorter than the lamina. *Stipules* 22–45(–55) mm on the main stems, as short as 11 mm on short branches, rigid, clasping the stem or projecting from it at an acute angle, green when fresh, olive green or brownish-green when dry, obtuse and slightly hooded at the apex, persistent; veins inconspicuous when dry, two much stronger than the others, forming strong ribs in the proximal half of the stipule, the ribs sometimes narrowly winged on the abaxial side towards the base of the stipule. *Turions* absent. *Inflorescences* 21–49 × 8–10 mm; peduncles 30–150(–290) mm, robust to very robust, broader and spongy towards the apex, terete. *Flowers* numerous, contiguous, with (3–)4 carpels. *Fruits* 2.7–3.4 × 1.9–2.35 mm, green or pale brownish-green; beak 0.2–0.5 mm, ventral, straight or recurved at the apex. *Flowers* 6–8.

Native. Mesotrophic lakes and rivers, reservoirs, fenland drains and ditches. Scattered throughout much of Britain and Ireland, although rare and decreasing or absent in the south and west. North and central Europe.

17. P. × billupsii Fryer   Billups's Pondweed
= **P. coloratus × gramineus**

Rhizome slender to robust. *Stem* up to 0.35 m, slender to robust, terete, unbranched or with a few short axillary branches. *Submerged leaves* with lamina 22–76 × 4–17 mm, 2.9–10.7 times as long as wide, translucent, bright green, olive green or yellowish-green, often with a pinkish tinge along the veins or throughout, sometimes reduced or partially reduced to phyllodes at base of stem, otherwise elliptical, oblong-elliptical or oblanceolate, obtuse and usually hooded, rarely acute and apiculate at apex, minutely denticulate at margin, gradually tapering to a sessile or petiolate base; midrib bordered by a narrow band of lacunae; lateral veins 3–6 on each side of the midrib, more or less equally developed or slightly stronger veins alternating with weaker ones, the secondary veins frequent, ascending in the centre of the leaf, ascending or more or less transverse towards the margin and conspicuous; petioles up to 27 mm. *Floating leaves* with lamina 20–67 × 10–22 mm, 1.7–3.75 times as long as wide, translucent or more or less opaque, subcoriaceous, bright green, broadly elliptical or ovate, obtuse and often apiculate at apex, cuneate or rounded at base; lateral veins 6–8 on each side of the midrib, opaque, darker than the lamina in the living plant, the secondary veins numerous, transverse or ascending; petioles 10–95 mm, shorter than or longer than the lamina. *Stipules* (7.5–)12–34 mm, clasping the stem or rolled and projecting from it at an acute angle, translucent when fresh, obtuse at apex, persistent; veins inconspicuous when dry, two stronger than the others and forming ridges towards the base of the stipule on the abaxial side. *Turions* absent. *Inflorescences* 10–14 × 2.5–3 mm; peduncles 20–125 mm, slender to robust, slightly broader towards the apex. *Flowers* densely packed. *Fruits* do not develop. *Flowers* 8–9.

Native. Formerly known from a ditch in the Isle of Ely, and still present at Loch na Liana Moire, Benbecula, Outer Hebrides. Sweden.

18. P. × nerviger Wolfg.   Copper-leaved Pondweed
= **P. alpinus × lucens**

Rhizomes very robust. *Stem* up to 2 m, robust, terete, richly branched. *Submerged leaves* with lamina 125–230 × 15–48 mm, 4.0–8.2 times as long as wide, translucent, olive green, with a pinkish-brown tinge when fresh especially in the upper leaves and along the midrib and often with the upper leaves copper-coloured when dry, occasionally reduced to phyllodes, otherwise elliptical or narrowly elliptical, acute, hooded and recurved at the extreme tip, minutely denticulate and undulate at margin, gradually tapering to a sessile base; midrib bordered by a band of lacunae which narrows about the base but extends to the apex, the lateral veins 4–6 on each side of the midrib, one or two slightly more strongly developed than the others, the secondary veins ascending across the whole width of the leaf. *Floating leaves* not seen. *Stipules* 52–78 mm, rigid, translucent, pale green with a slightly milky tinge, obtuse at apex, very persistent; veins inconspicuous when dry, two much stronger than the others and forming conspicuous green ribs, the ribs winged on the abaxial side for 20–42% of their length. *Turions* absent. *Inflorescences* 12–20 × 6–8 mm; peduncles 9–38 mm. *Flowers* dense. *Fruits* do not develop.

Native. River Fergus, Co. Clare. Lithuania.

19. P. × nitens G. Weber   Bright-leaved Pondweed
= **P. gramineus × perfoliatus**

P. curvifolius Hartm.; *P. heterophyllus* subsp. *nitens* (G. Weber) Hook. fil.; *P. pseudonitens* Sturrock ex F. B. White nom. nud.; *P. lundii* K. Richt.; *P. nitens* forma *involutus* Fryer; *P. involutus* (Fryer) H. & J. Groves

Rhizome very slender to robust. *Stem* up to 2.45 m, very slender to robust, terete, sparingly to richly branched. *Submerged leaves* with lamina 42–112 × 9–22.5 mm on the main stems and branches, as small as 23.5 × 5 mm on short branches, 3.6–8.5 times as long as wide, translucent, yellowish- to brownish-green, sometimes tinged pink along the midrib and principal lateral veins, sometimes reduced to phyllodes near base of stem, otherwise broadly elliptical to ovate-oblong, often recurved, usually acute and occasionally obtuse and hooded and sometimes apiculate or mucronate at apex, denticulate on margin, the teeth often lost on the older leaves, abruptly narrowed to a sessile, slightly amplexicaul to semi-amplexicaul base; midrib bordered by a narrow band of lacunae, the lateral veins 3–8 on each side of the midrib, 1–2 of which are more strongly developed than the others, the secondary veins frequent, transverse or ascending and often irregular. *Floating leaves* with lamina 35–65 × 9–23 mm, 2.4–3.7 times as long as wide, semi-opaque or opaque, subcoriaceous, pale green when young, becoming brownish-green with age, elliptical to broadly elliptical, acute and apiculate at apex, tapering to a cuneate base; lateral veins slightly lighter to slightly darker than the lamina in the living plant, the secondary veins inconspicuous; petioles 12–40 mm, shorter than the lamina. *Stipules* 11–28 mm, as short as 5.5 mm on short branches, flexible or rigid, clasping the stem or rolled and projecting from it at an acute angle, buff and translucent to dark green and opaque when dried, obtuse to rounded and slightly hooded at the apex, subpersistent, often lost on the older leaves; two veins often stronger than the others but not forming ridges or wings. *Turions* absent. *Inflorescences* 5–25 × 3.5–7.0 mm; peduncles 19–80(–175) mm, slender to very robust, broader towards apex but not or only slightly spongy, terete. *Flowers* usually numerous, contiguous and often densely packed, with 4–5(–7) carpels. *Fruits* do not develop. *Flowers* 6–9. 2n = 52.

Native. Lakes, reservoirs, rivers, streams, canals, fenland lodes and ditches. Rare and decreasing in England and Wales, frequent in Scotland and Ireland where it is the commonest hybrid in the genus and often occurs in the absence of at least one of its parents. Northern and central Europe, temperate Asia, North America.

20. P. alpinus Balb.   Red Pondweed
P. rufescens Schrad.; *P. palmeri* Druce nom. nud.; *P. alpinus* var. *palmeri* Druce

Rhizome slender to robust. *Stem* up to 2.75 m, usually robust, rarely slender, terete, unbranched. *Submerged leaves* with lamina 70–180(–220) × (6.5–)10–25(–33) mm, (4–)5–10 (–17) times as long as wide, translucent, clear green, sometimes with a reddish or brownish tinge when fresh and usually developing a strong reddish tinge when dried, especially in the upper leaves and along the veins, narrowly elliptical to oblong-elliptical, tapering to a narrow, obtuse and often slightly hooded apex, entire and shallowly undulate at margin, gradually tapering to base, sessile; midrib bordered by a broad band of lacunae which occupies over half the width of the leaf base, the lateral veins 4–7 on each side of the midrib, one usually more prominent than the others, the secondary veins ascending in the centre of the leaf and more or less transverse towards the edge. *Floating leaves* with lamina 45–90 × 9–25 mm, 2.0–6.5 times as long as wide, opaque, subcoriaceous, green or yellowish-green when fresh with a pinkish tinge especially along the veins, green with a strong pinkish-red tinge to reddish-brown when dry, narrowly elliptical to oblong-elliptical or obovate, obtuse at apex, gradually tapering to base; lateral veins 4–9, similar in colour or darker than the rest of the lamina in the living plants, the secondary veins numerous, ascending near the midrib, transverse elsewhere and rather obscure; petioles 10–35 mm, shorter than the lamina. *Stipules* 20–44 mm, delicate, flexible, clasping the stem along their entire length, translucent, hyaline, often with a pinkish tinge, obtuse or rounded at apex, often lost on the older leaves or only a basal, non-fibrous portion remaining; veins inconspicuous when dry, two somewhat more prominent than the others, but not forming distinct ridges. *Turions* absent. *Inflorescences* 15–32 × 4–8 mm; peduncles 30–150(–310) mm, robust, of more or less uniform diameter throughout their length, terete or slightly compressed. *Flowers* numerous, contiguous, with 4 carpels. *Fruits* 2.6–3.7 × 1.6–2.3 mm, pale brown, often with a reddish tinge, smooth; beak about 0.5 mm, more or less apical, often hooked at tip. *Flowers* 6–9. 2n = 26, 52.

The entire and sessile leaves of *P. alpinus* with a broad band of lacunae towards the base are distinctive.

Native. Neutral or mildly acidic, still or slowly flowing water in lakes, rivers, canals, ditches and flooded mineral workings. Frequent in northern and western parts of Britain and Ireland; rare and perhaps declining in the south and east. Arctic and boreal regions throughout the northern hemisphere.

P. alpinus × *crispus* (*P.* × *olivaceus*, no. 29) is a scarce hybrid which might be confused with *P. alpinus*. The other 4 recorded hybrids, with *P. gramineus* (*P.* × *nericius*, no. 13), *P. lucens* (*P.* × *nerviger*, no. 18), *P. perfoliatus* (*P.* × *prussicus*, no. 22), and *P. praelongus* (*P.* × *griffithii*, no. 21), are rare.

21. P. × griffithii A. Benn.   Griffith's Pondweed
     = **P. alpinus × praelongus**
P. macvicarii A. Benn.

Rhizomes slender. *Stem* up to 1.7 m, robust, terete, unbranched below, branched towards the apex. *Submerged leaves* with lamina (72–)120–240(–330) × 11–25 mm, 7.7–15.5 times as long as wide, translucent, mid-green to dark green with an oily sheen, sometimes with a pinkish tinge especially along the midrib when fresh and a reddish tinge when dry, ligulate, oblong-elliptical or oblong-lanceolate, slightly recurved, obtuse or rounded and hooded at apex, sometimes splitting when pressed, entire on margin and sometimes minutely undulate for up to 20 mm above the base, otherwise plane or shallowly undulate, gradually tapering or more abruptly contracted to a shortly petiolate or sessile, sometimes slightly amplexicaul base; midrib bordered by a broad band of lacunae especially towards the base, the lateral veins 4–9 on each side of the midrib, more or less similar or 1–2 more strongly developed than the others, the secondary veins frequent, rather conspicuous, ascending or more or less transverse; petioles up to 3.5 mm. *Floating leaves* with lamina 85–105 × 13–19 mm, 6–7 times as long as wide, opaque, coriaceous, dark green, elliptical, gradually tapering to the petiole; lateral veins opaque in the living plant, the secondary veins conspicuous; petioles 50–90 mm. *Stipules* 16–103 mm, colourless with a slight milky tinge when fresh, sometimes also with a pinkish tinge, more or less opaque when dry, broadly obtuse to rounded at apex, persistent, becoming eroded or detached on the oldest stems but not splitting into fibrous remnants; veins prominent when dry, two slightly stronger than the others but not forming distinct ribs. *Turions* absent. *Inflorescences* 10–21 × 3.5–7.5 mm; peduncles 45–190 mm, robust, of uniform diameter throughout their length. *Flowers* contiguous. *Fruits* do not develop. *Flowers* 6–8.

Native. Lakes in North Wales, Scotland and Ireland, very rare. Endemic.

22. P. × prussicus Hagstr.   Hebridean Pondweed
     = **P. alpinus × perfoliatus**
P. johannis Hesl.-Harr.

Rhizome not seen. *Stem* up to 0.53 m, slender, unbranched. *Submerged leaves* with lamina 45–80 × 8–14 mm, 5.0–7.9 times as long as wide, olive green when dry, the upper leaves sometimes with a strong orange-brown tinge, oblong-lanceolate to oblong-elliptical, obtuse or subacute and sometimes slightly hooded at apex, minutely and rather obscurely denticulate at margin, semi-amplexicaul or amplexicaul at base; midrib bordered by a narrow band of lacunae, the lateral veins 4–7 on each side of the midrib, one of which is usually more strongly developed than the others, the secondary veins frequent, more or less ascending in the centre of the leaf and more or less transverse towards the margin. *Floating leaves* not seen. *Stipules* 13–17 mm, translucent, obtuse at apex, subpersistent, becoming eroded to a basal portion on older leaves. *Turions* absent. *Inflorescences* not seen.

Native. Lakes in Colonsay and Benbecula in the Hebrides; also tentatively identified from a canal in Co. Kildare, Ireland. Germany, Norway and Sweden.

23. P. praelongus Wulfen   Long-stalked Pondweed
P. undulatus auct.

Rhizome robust to very robust. *Stem* up to 3 m, robust to very robust, terete, unbranched or sparingly to richly branched. *Submerged leaves* with lamina (45–)60–150(–220) × 14–40 mm, (2.5–)3.0–6.5(–15.0) times as long as wide, translucent, clear green with an oily sheen when young, becoming dark green with age, sometimes tinged pink along the midrib and principal lateral veins, very rarely with a brownish-red tinge throughout when dry, lanceolate or oblong-lanceolate, rarely linear-lanceolate or oblong, obtuse at apex and markedly hooded, often splitting when pressed, entire, plane or shallowly undulate on margin, sessile and semi-amplexicaul at base; midrib bordered by a narrow band of lacunae which occupies less than a third of the width of the leaf base, the lateral veins 5–9 on each side of the midrib, 1–2 of which are usually more strongly developed than the others, the secondary veins numerous and more or less transverse. *Floating leaves* absent. *Stipules* 10–80 mm, flexible, the entire stipule or the proximal half enfolding the stem, milky-white with a slight pinkish tinge and more or less translucent when fresh, buff-coloured, opaque and conspicuous when dry, obtuse at apex, persistent but becoming eroded to fibrous remnants with age; veins prominent when dry, two slightly more prominent than the others but not forming distinct ribs. *Turions* absent. *Inflorescences* 25–55 × 5–8 mm; peduncles (50–)80–200(–350) mm, robust, of uniform diameter throughout their length or becoming broader towards the apex, slightly compressed towards the base, more or less terete towards the apex. *Flowers* 15–20, more or less contiguous but rather widely spaced, with (2–) 4 carpels. *Fruits* 4.5–5.5 × 2.5–3.6 mm, dark brownish-green; beak 0.6–0.9 mm, ventral, straight. *Flowers* 5–8. 2n = 52.

The entire, sessile and semi-amplexicaul leaves of *P. praelongus* with a markedly hooded apex are diagnostic.

Native. Deep, mesotrophic and slightly to strongly calcareous water in lakes, rivers, canals and major fenland drains. Locally frequent in northern and western parts of Britain and Ireland, but scarce and decreasing in the south and east.

Hybrids with *P. alpinus* (*P. × griffithii*, no. 21), *P. crispus* (*P. × undulatus*, no. 24) and *P. perfoliatus* (*P. × cognatus*, no. 26) are recorded, but only rarely. All might be confused with *P. praelongus*.

24. P. × undulatus Wolfg.   Lagan Pondweed
  = P. crispus × praelongus

Rhizome robust. *Stem* up to 1.4 m, robust, slightly compressed, with a shallow groove running down one or both of the broader sides, contracted at the nodes, unbranched below, sparingly or richly branched above. *Submerged leaves* with lamina 45–150 × 8–22 mm, (4.8–)6.0–12.2 times as long as wide, translucent, glossy bright green, olive green or dark green when fresh, sometimes with a pink tinge along the midrib, linear-oblong to oblong-lanceolate, obtuse or subacute at apex, not hooded or hooded but not splitting when pressed, the margin entire or very obscurely toothed towards the apex and undulate, sessile, with an auriculate or semi-amplexicaul base; midrib bordered by a band of lacunae which occupies 44–80% of the width of the leaf-base but narrows rapidly above, the lateral veins 2–4 on each side of the midrib, the central vein sometimes more strongly developed than the others, the outermost vein often faint, the inner secondary veins ascending, those towards the margin of the leaf more or less ascending, transverse or wavy. *Floating leaves* absent. *Stipules* 6–25 mm, translucent and hyaline when fresh, translucent or opaque and buff-coloured when dry, truncate or slightly emarginate at apex, persisting as entire stipules or as fibrous remnants, two veins slightly stronger than the others but not forming distinct ridges. *Turions* axillary, 15–70 mm, slender, with 2–5 free leaves, resembling the slender turions produced by *P. crispus*. *Inflorescences* not seen.

Native. Llynheilyn, Radnorshire (perhaps extinct) and River Lagan, Six Mile Water and Lough Neagh, Northern Ireland. Denmark, Germany, Poland and Russia.

25. P. perfoliatus L.   Perfoliate Pondweed

Rhizome robust to very robust. *Stem* up to 3 m, exceptionally up to 8 m, robust to very robust, terete, unbranched or sparingly to richly branched. *Submerged leaves* with lamina 20–115 × 7–42 mm, 1.3–10.0 times as long as wide, translucent, bright green, dark green, yellowish-green, olive green or brownish-green, sometimes tinged pink along the midrib and numerous principal lateral veins, narrowly lanceolate to broadly ovate, rounded to obtuse or more or less acute and often slightly hooded at apex; denticulate and sometimes minutely undulate on the margin, amplexicaul at base, the edges sometimes overlapping on the far side of the stem; midrib bordered by a narrow band of lacunae, the lateral veins 5–12 on each side of the midrib, 1–3 of which are usually more prominent than the others, the secondary veins numerous and more or less transverse. *Floating leaves* absent. *Stipules* 2.5–22.0 mm, delicate, translucent and hyaline when fresh and dry, rounded at the apex, usually fugacious, rarely persisting on mature leaves; veins inconspicuous when dry, two slightly stronger than the others but not forming distinct ribs. *Turions* absent. *Inflorescences* 13–25 × 3–8 mm; peduncles 20–110 mm, robust, slightly broader towards the apex and slightly compressed near the base. *Flowers* 12–20, contiguous. *Fruits* 2.6–3.5(–4.0) × 1.7–2.9 mm, brownish-green, rugose; beak 0.4–0.7 mm, apical, straight. *Flowers* 6–9. 2*n* = 52.

One of the most distinctive *Potamogeton* species, easily recognised by its broadly amplexicaul leaves.

Native. Moderately deep water in lakes, reservoirs, rivers, streams, canals, fenland lodes and flooded sand and gravel pits. Widespread and locally frequent in both Britain and Ireland. Europe, Asia, North America, North and central Africa and Australia.

Three hybrids of *P. perfoliatus* are relatively frequent. *P. crispus × perfoliatus* (*P. × cooperi*, no. 27) could easily be confused with *P. perfoliatus* but *P. gramineus × perfoliatus* (*P. × nitens*, no. 19) and *P. lucens × perfoliatus* (*P. × salicifolius*, no. 15) usually resemble the other parent. Two hybrids are rare: *P. alpinus × perfoliatus* (*P. × prussicus*, no. 22) and *P. perfoliatus × praelongus* (*P. × cognatus*, no. 26).

26. P. × cognatus Asch. & Graebn.

Small-toothed Pondweed
= P. perfoliatus × praelongus

Rhizome robust. *Stem* up to 1.25 m, robust, terete, lower stem unbranched or sparingly branched, upper stem sometimes richly branched. *Submerged leaves* with lamina (20–)45–110 × (11–)14–39 mm, 2.1–3.8 times as long as wide, translucent, dark green, brownish-green or reddish-brown when dry, ovate-lanceolate to ovate-oblong, obtuse and hooded at apex and often splitting when pressed, denticulate at margin, sessile with an amplexicaul base, with auricles which do not overlap; midrib bordered by a narrow band of lacunae, the lateral veins 6–12 on each side of the midrib, 1–3 of which are more strongly developed than the others, the secondary veins more or less transverse. *Floating leaves* absent. *Stipules* 16–64 mm, buff-coloured and somewhat opaque when dried, obtuse at apex, persisting for a while on the younger stems; veins inconspicuous when dry and not persisting as fibrous remnants when the stipule decays, two slightly stronger than the others but not forming distinct ribs. *Turions* absent. *Inflorescence* 8–24 × 3–6.5 mm; peduncles 33–200(–255) mm, robust, of uniform diameter throughout their length. *Flowers* numerous, densely packed, with 4 carpels. *Fruits* do not develop. *Flowers* 6–8.

Native. Fenland drains in North Lincolnshire, where it has not been seen since 1944, and a highly calcareous lake, Loch Borralie, West Sutherland. Northern Europe.

27. P. × cooperi (Fryer) Fryer   Cooper's Pondweed
= P. crispus × perfoliatus
P. perfoliatus var. *jacksonii* F. Lees; *P. undulatus* var. *cooperi* Fryer; *P. undulatus* var. *jacksonii* (F. Lees) Fryer; *P. cymatodes* Asch. & Graebn.; *P. jacksonii* H. Wilson nom. in syn.

Rhizome slender to robust. *Stem* up to 1.5 m, rarely to 4 m, robust, slightly compressed to compressed, with a

shallow groove running down the broader sides of the mature stems, unbranched or with numerous axillary branches. *Submerged leaves* with lamina 26–85 × 8–25 mm, 2.3–6.0 times as long as wide, translucent, bright green, dark green or brownish-green, sometimes tinged red along the midrib and principal lateral veins, linear-oblong to ovate, sessile, obtuse or subacute and sometimes slightly but distinctly hooded at apex, denticulate or serrulate on margin especially towards the apex of the leaf, and undulate, the base semi-amplexicaul or more or less amplexicaul; midrib bordered by a band of lacunae which occupies over half the width of the leaf base but rapidly narrows above the base, the lateral veins 3–5(–6) on each side of the midrib, the secondary veins transverse or ascending between the midrib and the innermost lateral veins, otherwise more or less transverse. *Floating leaves* absent. *Stipules* 5–21 mm, clasping the stem along their entire length, delicate, translucent, hyaline, truncate or shallowly emarginate at the apex, fugacious, the bases sometimes persisting as fibrous remnants, sometimes with two veins very slightly stronger than the others. *Turions* axillary, 7–55 mm, slender, with 5–10 short, distant, erecto-patent to slightly recurved leaves, resembling the slender turions produced by *P. crispus*. *Inflorescences* 4–13 × 2–6 mm; peduncles (8–)40–96 mm, slender to robust, of uniform diameter throughout their length, slightly compressed to compressed at base, terete to slightly compressed towards the apex. *Flowers* 8–15, contiguous, with (3–)4 carpels. *Fruits* do not develop. *Flowers* 7–8.

The compressed stem of *P. × cooperi* and the fewer lateral veins distinguish this hybrid from the frequent and variable *P. perfoliatus*.

Native. Canals, rivers, lakes, reservoirs, flooded brick pits and major fenland drains. Scattered localities in Britain and Ireland north to Edinburgh. Europe.

28. P. × cadburyae Dandy & G. Taylor
Cadbury's Pondweed
= P. crispus × lucens

Rhizome slender. *Stem* up to 0.12 m, slender, unbranched. *Submerged leaves* with lamina 52–95 × 7–15 mm, 4.3–8.1 times as long as wide, yellowish-green, sometimes with a reddish tinge especially along the midrib, oblong-elliptical, usually obtuse and sometimes mucronate, rarely acute at apex, the margin denticulate or serrulate towards the base, serrulate towards the apex and minutely undulate, gradually or rather abruptly tapering to a sessile or very shortly petiolate base; midrib bordered by a band of lacunae, the lateral veins 2–3 on each side of the midrib, the outermost usually the faintest, the secondary veins frequent, ascending across the entire width of the leaf or more or less transverse towards the margin; petioles to 1.5 mm. *Floating leaves* absent. *Stipules* 12–22 mm, apparently rigid, translucent, obtuse at apex, persistent; veins prominent, two stronger than the others and towards the base of the stipules forming raised ridges on the abaxial side. *Turions* unknown. *Inflorescences* unknown.

Native. Seeswood Pool, Warwickshire, where it was collected once, in 1948. Endemic.

29. P. × olivaceus Baagøe ex G. Fisch.
Olive-leaved Pondweed
= **P. alpinus × crispus**
P. venustus Baagøe ex A. Benn.; *P. semipellucidus* auct.

Rhizome slender. *Stem* up to 0.9 m, slender to robust, slightly compressed to compressed, with a shallow groove running down one or both of the broader sides, usually unbranched, occasionally with short axillary branches. *Submerged leaves* with lamina 47–121 × 6–15 mm, (5.2–)8.5–15.0(–19.6) times as long as wide, translucent, bright green, olive green or dark green, with a pink tinge along the midrib and sometimes along the lateral veins when fresh, often with a reddish-brown tinge when dry, obtuse or subacute at apex; more or less entire, denticulate or bluntly serrulate at margin, especially towards the apex and undulate, tapered to a sessile, slightly auriculate base; midrib bordered by a band of lacunae which is broad at the base of the leaf but narrows rapidly above, extending almost to the leaf apex, the lateral veins 2–3 on each side of the midrib, the outermost vein often faint, the secondary veins frequent, ascending in the centre of the leaf, ascending, transverse or wavy towards the margin. *Floating leaves* not seen, described as tapering to a short petiole. *Stipules* 8–23 mm, delicate, translucent, hyaline, sometimes with a pinkish tinge, truncate or slightly emarginate at the apex, frequently persisting on the lower leaves although often with the distal portion eroded away; veins indistinct when dry, two slightly stronger than the others but not forming distinct ribs. *Turions* not seen. *Inflorescences* 5.0–9.5 × 2.0–6.5 mm; peduncles 7–60 (–105) mm, slender to robust, of uniform diameter throughout their length or slightly broader at the base, slightly compressed to compressed at the base, more or less terete or slightly compressed towards the apex. *Flowers* (7–)10–12, contiguous, with (3–)4 carpels. *Fruits* do not develop. *Flowers* 7.

Native. Rivers in Wales, northern England and Scotland; very rarely in nearby pools. Denmark.

30. P. × lintonii Fryer Linton's Pondweed
= **P. crispus × friesii**

Rhizome very slender to slender. *Stem* up to 0.9 m, slender, rarely very slender, compressed, the most robust stems with a shallow groove running down one or both of the broader sides, simple or much branched; nodal glands absent. *Submerged leaves* with lamina 25–59 × 1.75–5.0 mm, 7–17 times as long as wide, dark green or brownish-green, tinged pinkish-red along the midrib, linear to linear-oblong, often slightly twisted, rounded, obtuse or acute at apex, the margin entire in proximal part, usually denticulate towards apex, rarely serrulate or more or less entire, plane or slightly undulate, sessile; midrib bordered by a band of lacunae which is broad at base but rapidly contracts to a narrow band which extends or almost extends to the apex, the lateral veins 1–2 on each side of the midrib, the secondary veins few, ascending in the centre of the leaf, ascending or

transverse towards the margin. *Floating leaves* absent. *Stipules* 6–12 mm, tubular at base for 0.5–2.2 mm, translucent and pale buff when fresh and dry, truncate or shallowly concave in outline at apex, persistent although soon eroding to fibrous strands at the apex, with a rather weak green rib along each side; veins rather prominent when dry, two slightly stronger than the others but not forming raised ribs. *Turions* usually axillary, rarely terminal, 12–75 mm, very slender, with 3–12 well-spaced short, narrow, erect, erecto-patent or slightly recurved leaves which do not conceal the axis. *Inflorescences* 3–11 × 2–5 mm; peduncles 8–43 mm, very slender to slender, slightly compressed. *Flowers* 2–4, dense, with (3–)4 carpels. *Fruits* do not develop. *Flowers* 8.

The stipules, which are tubular towards the base, provide the critical distinction between *P. × lintonii* and forms of *P. crispus* with plane, inconspicuously toothed leaves.

Native. Canals, lakes, rivers, streams and gravel pits. Scattered localities in England, southern Scotland and eastern Ireland. Belgium, Holland, Czechoslovakia.

31. P. × bennettii Fryer Bennett's Pondweed
= **P. crispus × trichoides**

Rhizome slender. *Stem* up to 1.5 m, slender, compressed, with a shallow groove running down one or both of the broader sides, sparingly to much branched; nodal glands absent or rudimentary. *Submerged leaves* with lamina 22–78 × 2–5 mm, 10–20 times as long as wide, olive green, with a reddish tinge along the midrib of the older leaves, linear-oblong, often slightly twisted, subacute or acute at apex, the margin entire in proximal part, usually denticulate towards the apex, rarely more or less entire throughout, sessile with the base sometimes slightly auriculate; midrib bordered by a band of lacunae which almost extends to the apex, the lateral veins 1–2 on each side of the midrib, the secondary veins few and more or less transverse or ascending. *Floating leaves* absent. *Stipules* 4.0–13.5 mm, open, convolute at base, translucent, hyaline, truncate or shallowly concave in outline at apex, soon becoming eroded at the apex but the basal part persistent; veins prominent when dry, two stronger than the others but not forming distinct ribs. *Turions* axillary, 12–33 mm, slender, with 3–7 well-spaced, short, narrow, erecto-patent or slightly recurved leaves which do not conceal the axis, resembling the slender turions produced by *P. crispus*. *Inflorescences* 3–7 × 2.5–3.5 mm; peduncles 8–43 mm, slender, slightly compressed. *Flowers* 3–5, dense, with 2–3(–4) carpels. *Fruits* do not develop. *Flowers* 8–9.

P. × bennettii is very similar to *P. × lintonii*, but has open stipules and flowers with fewer carpels.

Native. Forth and Clyde Canal, central Scotland. Endemic.

32. P. crispus L. Curled Pondweed
P. serratus auct.; *P. lucens* var. *angustifolius* Gray, non Hornm.

Rhizome very slender to robust. *Stem* up to 1.5 m, robust to very robust, compressed, with a shallow groove running down one or both the broader sides, contracted at the nodes, unbranched or with short branches with crowded leaves or, more rarely, long branches with distant leaves; nodal glands absent. *Submerged leaves* 25–95 × 5–12(–18) mm, 5–9(–13) times as long as wide, bright green, olive green or brownish-green, often with a reddish tinge especially along the midrib, linear-oblong to oblong, obtuse or acute at apex, serrate on the margin, the lower leaves often plane, the upper leaves usually strongly undulate, sessile, the base of the broader leaves auriculate; midrib bordered by a band of lacunae which is broad at the base of the leaf but rapidly narrows above, and extends almost to the leaf apex, the lateral veins 1–2(–3) on each side of the midrib, the secondary veins few, transverse to ascending. *Floating leaves* absent. *Stipules* 4–17 mm, open, delicate, translucent, hyaline with a pinkish or brownish tinge, truncate or emarginate at apex, soon becoming torn or reduced to wispy remains, without distinct ribs; veins indistinct. *Turions* axillary, 7–25(–47) mm, usually robust, with crowded, short, broad leaves, the hardened leaf-bases imbricate and completely concealing the axis, occasionally slender, with the leaves less crowded, narrow and not concealing the axis. *Inflorescences* 5–16 × 4–6 mm; peduncles 14–65(–124) mm, slender to robust, of uniform diameter throughout their length, slightly compressed to compressed. *Flowers* 3–8, more or less contiguous, with (2–)4 carpels. *Fruits* 4.0–6.2 × 2.0–2.5 mm, the ventral edge with a single tooth at the base; beak 1.5–2.4 mm, at least half and up to four-fifths the length of the rest of the fruit, tapering from a broad base to a slender apex, more or less falcate. *Flowers* 5–10. 2n = 52.

Although it is a variable plant, *P. crispus* can usually be easily distinguished by its undulate and conspicuously toothed leaves. Forms with plane leaves differ from *P. friesii* and *P. obtusifolius* in having grooved stems and teeth which may be inconspicuous but are always present, at least towards the leaf apex.

Native. Mesotrophic or eutrophic, standing or flowing water in lakes, reservoirs, ponds, rivers, streams, canals, fenland lodes and disused mineral workings. Widespread and frequent in much of Britain and Ireland, but rarer in south-west England, Wales and northern Scotland. Europe, Asia, Africa, Australia; introduced in North America.

Hybrids with the linear-leaved species *P. friesii* (*P. × lintonii*, no. 30) and *P. trichoides* (*P. × bennettii*, no. 31) could be confused with *P. crispus*. Those with *P. alpinus* (*P. × olivaceus*, no. 29), *P. lucens* (*P. × cadburyae*, no. 28), *P. perfoliatus* (*P. × cooperi*, no. 27) and *P. praelongus* (*P. × undulatus*, no. 24) all bear a closer resemblance to the broad-leaved parent.

33. P. epihydrus Raf.   American Pondweed
P. nuttallii Cham. & Schltdl.; *P. pensylvanicus* Willd. ex Cham. & Schltdl.

Rhizome slender or robust. *Stem* up to 1.85 m, very slender to robust, stems slightly compressed to compressed or (in flowering stems) more or less terete towards the apex, unbranched or sparingly branched;

nodal glands absent. *Submerged leaves* sometimes markedly distichous on the vegetative shoots, 65–240 × 2.5–10.5 mm, 18–30(–60) times as long as wide, delicate, translucent, pale brown, pale green or olive green, sometimes with a pinkish tinge, linear, narrowly obtuse to more or less acute at apex, entire and plane at the base of leaf and plane or undulate on margin towards the apex, sessile; midrib bordered by a broad band of lacunae which towards the base of the leaf extends at least as far as the inner lateral veins and sometimes occupies the entire leaf width, narrowing above until it ceases below or reaches the apex, the lateral veins 2–4 on each side of the midrib, the innermost the strongest and the outermost often very faint, the secondary veins occasional, rather irregular, ascending, more or less transverse or descending. *Floating leaves* with lamina 35–77 × 7–22 mm, 2.6–5.0 times as long as wide, opaque, coriaceous, pinkish-brown when young, dark olive green when mature, elliptical to oblong-elliptical, obtuse at apex, cuneate at base; lateral veins opaque in the living plant, the secondary veins numerous, ascending and relatively conspicuous towards the centre of the leaf, more or less transverse and inconspicuous towards the margin; petioles 20–60(–90) mm, usually shorter than the lamina. *Stipules* 11–45 mm, open, convolute at the base, delicate, translucent, truncate and sometimes slightly emarginate at apex, soon decaying; veins inconspicuous when dry, two more prominent than the others but not forming distinct ribs. *Turions* absent. *Inflorescences* 12–23 × 3–6 mm; peduncles 23–90 mm, slender or robust, of uniform diameter throughout their length or somewhat broader towards the apex, terete or slightly compressed. *Flowers* numerous, contiguous, with 4 carpels. *Fruits* 2.5–3.1 × 1.9–2.75 mm, olive green or brownish-green; beak 0.1–0.35 mm, ventral. *Flowers* 6–8. 2n = 26.

P. epihydrus has been divided into two varieties. Var. *ramosus* (Peck) House differs from var. *epihydrus* in having narrower submerged and floating leaves, the submerged leaves markedly distichous on the young shoots, and in its smaller fruit. In America the larger var. *epihydrus* tends to occur in richer waters. The plants from the oligotrophic waters of the Outer Hebrides are clearly referable to var. *ramosus*, whereas those from the more nutrient-rich canals of northern England are intermediate between the two varieties.

Native. Shallow, oligotrophic lochans in South Uist; introduced in mesotrophic canals in Lancashire and Yorkshire. Unknown elsewhere in Europe, but widespread in North America.

No hybrids of *P. epihydrus* are recorded in the British Isles.

34. P. compressus L.   Grass-wrack Pondweed
P. zosterifolius Schumach.; *P. cuspidatus* Schrad.; *P. gramineus* subsp. *zosterifolius* (Schumach.) Hayw.

Rhizome absent. *Stem* to 0.85 m, robust or very robust, flattened, sometimes narrowly winged at the edges on one or both sides, slightly swollen just below the nodes, richly branched; nodal glands absent. *Submerged leaves* 87–240(–272) × 3.0–6.2 mm, 20–60(–90) times as long as wide, olive green or dark green, the upper leaves sometimes with a reddish tinge especially along the midrib, linear, bordered by a strong marginal vein, truncate or rounded to acute and mucronate at apex, strongly thickened, entire and plane on margin, sessile; midrib bordered by a band of lacunae which is variable in width at the base of the leaf and sometimes fills the space between the inner lateral veins (occupying 40–56% of leaf width) but narrows rapidly until it ceases in the distal half of the leaf, sometimes well below the apex but sometimes almost reaching it, the lateral veins 2 on each side of the midrib, the inner stronger than the outer, the secondary veins absent; numerous and conspicuous sclerenchymatous strands present in the lamina. *Floating leaves* absent. *Stipules* 21–55 mm, open, convolute at the base, translucent and hyaline with a brownish tinge when fresh, opaque and buff-coloured when dry, obtuse at apex, persistent but soon eroding to fibrous strands at the apex, with a strong green rib along each side which remains after the intervening tissue has decayed, the stipules eventually splitting into a V-shaped remnant; veins prominent when dry. *Turions* terminal at the ends of short axillary branches, 25–45 × 3.5–8.0 mm, composed of appressed, short leaves with more or less truncate apices surrounded by conspicuous stipules, with 2–4 erecto-patent to erect leaves arising near the base. *Inflorescences* 11–25 × 2.5–4.5 mm; peduncles 28–95 mm, slender or robust, compressed. *Flowers* 10–20, more or less contiguous, with (1–)2 carpels. *Fruits* 3.4–4.0 × 2.1–3.0 mm, brown or brownish-green, the dorsal edge entire or muricate; beak 0.4–0.7 mm, ventral, recurved. *Flowers* 6–9. 2n = 26.

P. compressus differs from the closely related *P. acutifolius* in its longer peduncles, longer inflorescences, predominantly 2-carpellate flowers and fruits which lack a tooth on the ventral edge.

Native. Lakes and pools, the backwaters of rivers, canals and species-rich drainage ditches in grazing marshes. Scarce and declining in the English Midlands and East Anglia; very rare in Scotland where it is now restricted to a single locality. Widespread in the boreal and temperate zones of Europe and Asia.

No hybrids involving *P. compressus* are known in Britain.

35. P. acutifolius Link   Sharp-leaved Pondweed
P. compressus subsp. *acutifolius* (Link) Hook. fil.

Rhizome absent. *Stem* up to 1.0 m, slender to robust, strongly compressed to flattened, slightly swollen just below the nodes, unbranched below, richly branched above; nodal glands absent. *Submerged leaves* 34–100(–135) × 1.6–5.4 mm, 13–30(–40) times as long as wide, dark green, often with a strong brownish or reddish-brown tinge, linear, bordered by a strong marginal vein, acute or more or less abruptly contracted to a distinctly mucronate tip, entire and plane on margin, sessile; midrib bordered by a band of lacunae which is broad towards the base of the leaf and usually occupies the space between the lateral veins (occupying 50–60% of leaf width) but narrows more or less rapidly until it

ceases below the apex, the lateral veins one on each side of the midrib, the secondary veins absent; rather inconspicuous sclerenchymatous strands present in the lamina. *Floating leaves* absent. *Stipules* 13–26(–38) mm, open, often convolute at the base, translucent and hyaline with a milky or brownish tinge when fresh, opaque and buff-coloured when dry, obtuse at apex, persistent but soon eroding to fibrous strands at the apex, with a strong green rib along each side which remains after the intervening tissue has decayed, the stipules eventually splitting into a V-shaped remnant; veins prominent when dry. *Turions* terminal on the axillary branches, 22–36 × 3.0–3.5 mm, composed of appressed leaves with more or less mucronate apices, the surrounding stipules not especially conspicuous, with 2–3 erecto-patent leaves arising near the base. *Inflorescences* 2.5–5.0 × 1.5–2.0 mm; peduncles 5–20(–29) mm, slender, compressed. *Flowers* 4–6, contiguous, with 1 carpel. *Fruits* 3.0–4.0 × 2.3–3.0 mm, green, the ventral edge with a single rather blunt tooth; beak 0.4–1.1 mm, ventral, straight or recurved. *Flowers* 6–7. 2n = 26.

Native. Shallow, species-rich drainage ditches in grazing marshes. Southern England, formerly occurring north to Yorkshire; rare and declining. Temperate Europe.

Its hybrids with *P. berchtoldii* (*P. × sudermanicus*, no. 38) and *P. friesii* (*P. × pseudofriesii*, no. 36) have each been recorded from a single locality.

36. P. × pseudofriesii Dandy & G. Taylor
                                    Buckenham Pondweed

  = P. acutifolius × friesii

Rhizome absent. *Stem* up to 0.5 m, slender, probably strongly compressed, with long axillary branches; nodal glands well-developed. *Submerged leaves* with lamina 25–50 × 1.3–2.1 mm, 21–32 times as long as wide, linear, abruptly contracted to a mucronate tip, entire and plane at margin, sessile; midrib bordered by a narrow band of lacunae towards the base of the leaf; lateral veins 2 on each side of the midrib, the secondary veins occasional, sclerenchymatous strands scattered along the length of the lamina. *Floating leaves* absent. *Stipules* 11–19 mm, tubular at the base, opaque and buff-coloured when dry, obtuse at apex, persistent but soon becoming eroded into fibrous strands at the apex and splitting longitudinally into two with age. *Turions* unknown. *Inflorescences* unknown.

Native. Ditch near Buckenham Ferry, Norfolk; not seen since the original collection in 1952. Endemic.

37. P. friesii Rupr.   Flat-stalked Pondweed
P. pusillus var. *major* Hook., non Mert. & Koch; *P. mucronatus* Schrad. ex Sond.; *P. pusillus* var. *mucronatus* (Schrad. ex Sond.) Hook. fil.; *P. pusillus* subsp. *friesii* (Rupr.) Hook. fil.

Rhizome absent. *Stem* up to 1.5 m, slender to robust, compressed to strongly compressed, usually with numerous, short, axillary branches; nodal glands well developed, green, semi-globose. *Submerged leaves*

35–85(–100) × 1.5–3.5(–4.0) mm, 15–45 times as long as wide, light green, linear, usually obtuse, rarely subacute and distinctly mucronate at apex, entire, plane at margin, sessile; midrib bordered by a narrow band of lacunae which is restricted to the base of the leaf or extends almost to the apex, the lateral veins (1–)2(–3) on each side of the midrib and distinct, the secondary veins infrequent. *Floating leaves* absent. Stipules 9.5–25.0 mm, tubular in the basal 2–9 mm when young but splitting with age, opaque and buff-coloured when dry, tapering to an obtuse apex, persistent but soon eroding to fibrous strands at the apex, with a strong green rib along each side which remains after the intervening tissue has decayed, the stipules eventually splitting into a V-shaped remnant; veins prominent when dry. *Turions* usually sessile in the leaf axils or terminal at the end of short axillary branches, occasionally terminal on the main stem or long axillary branches, 12–22 × 2.5–8.5 mm, fan-shaped, composed of appressed dark green leaves enveloped in conspicuous, prominently veined, buff-coloured stipules, with 2–4, erect to erecto-patent, free leaves which exceed the turions in length and are often recurved at the apex. *Inflorescences* (5–)8–13 × 2.0–3.5 mm; peduncles 15–30(–70) mm, very slender to slender, compressed. *Flowers* 4–8, contiguous or more or less distant, with 4 carpels. *Fruits* 2.4–3.0 × 1.3–1.9 mm, olive green; beak 0.3–0.55 mm, more or less apical, straight or recurved. *Flowers* 6–8. 2n = 26.

Native. Base-rich lakes, rivers, canals, ditches and flooded mineral workings. Locally frequent in eastern England, north to Yorkshire, local in Ireland, rare in Wales and Scotland. Boreal and temperate zones throughout the northern hemisphere.

Hybrids occur with *P. crispus* (*P.* × *lintonii*, no. 30) and *P. acutifolius* (*P.* × *pseudofriesii*, no. 36).

38. P. × sudermanicus Hagstr.   Dorset Pondweed
= P. acutifolius × berchtoldii

Rhizome usually absent, if present short and very slender. *Stem* up to 1.05 m, very slender to robust, compressed, unbranched or sparingly branched with long axillary branches below, richly branched above; nodal glands poorly developed to well-developed, semiglobose to more or less globose. *Submerged leaves* with lamina 27–105 × 1.1–3.0(–3.5) mm, 16–40 times as long as wide, green or dark green, sometimes with a strong brown or reddish-brown tinge, linear, gradually tapering or more frequently abruptly contracted to a distinctly mucronate tip, entire and plane at margin, sessile; midrib occupying 6–14% of leaf width near base, bordered by a band of lacunae which is very broad towards the base of the leaf (occupying 32–67% of leaf width) but narrows gradually until reaching the apex, the lateral veins 1–2 on each side of the midrib, the outer vein, if present, faint, the secondary veins very few; sclerenchymatous strands scattered along the length of the lamina, inconspicuous in the living plant. *Floating leaves* absent. *Stipules* 12–25 mm, open, convolute at the base, translucent, hyaline, rounded to obtuse at apex, persistent but eroding to fibrous strands at the apex,

with a green or colourless rib along each side; veins fairly prominent when dry. *Turions* terminal on the main stems and the axillary branches, 18–40 × 1.8–4.5 mm, composed of tightly appressed, short, dark green leaves surrounded by conspicuous or inconspicuous stipules, with 2–4 erecto-patent leaves at the base. *Inflorescences* 3.5–8.0 × 2.0–3.0 mm; peduncles 10–44 mm, very slender to slender, slightly compressed to compressed. *Flowers* 2–4(–5), contiguous, with 1–3(–4) carpels. *Fruits* 2.0–3.1 × 1.7–2.7 mm, green, the dorsal edge very slightly muricate; beak 0.6–0.9 mm, ventral, straight.

Native. Ditches in grazing marshes near Wareham, Dorset. Holland, Sweden.

39. P. berchtoldii Fieber   Small Pondweed
P. pusillus auct.; *P. pusillus* var. *tenuissimus* Mert. & Koch; *P. gracilis* auct.; *P. sturrockii* auct.; *P. pusilliformis* Hagstr.; *P. dualis* Hagstr.; *P. pusillus* subsp. *lacustris* Pearsall & Pearsall fil.; *P. lacustris* (Pearsall & Pearsall fil.) Druce; *P. millardii* Hesl.-Harr.

Rhizome absent. *Stem* up to 0.6(–1.0) m, very slender, terete to slightly compressed, sparingly or richly branched; nodal glands well-developed, translucent. *Submerged leaves* (17–)25–50(–75) × (0.5–)0.8–1.8(–2.3) mm, (16–)20–45(–55) times as long as wide, flaccid, pale green, olive green or brownish-green, linear, more or less abruptly narrowed to an acute or obtuse, often slightly asymmetrical apex which is not bordered by a marginal vein, entire and plane at margin, sessile; midrib occupying 10–20% of leaf width near base, bordered on each side by a band of lacunae which is sometimes narrow and confined to the proximal part of the leaf but is often well developed and extends to the apex, the lateral veins one on each side of the midrib, distinct, rarely with a faint second vein in the proximal half, the secondary veins absent. *Floating leaves* absent. *Stipules* 5–15 mm, open, convolute at the base at least when young, hyaline or greenish and translucent when fresh and dry, rounded to obtuse at apex, often lost on the oldest leaves, with very weak ribs along each side which are similar in colour to the other veins and 5–8(–9) intercostal veins, the veins indistinct when dry. *Turions* terminal on the main shoots and on axillary branches, 8.5–24 × (0.65–)0.8–1.6 mm, cylindrical, dark green, with 2–4 erect to erecto-patent free leaves. *Inflorescences* 4.0–7.5 × 2.0–3.5 mm; peduncles (7–)10–30(–45) mm, very slender, slightly compressed to compressed. *Flowers* 2–4, more or less contiguous, with (3–)4–5(–7) carpels. *Fruits* 1.8–2.7(–3.0) × 1.1–1.75 mm, olive green; beak 0.2–0.7 mm, ventral to apical, recurved to straight. *Flowers* 6–9. 2n = 26.

Native. Lakes, reservoirs, ponds, rivers, canals, streams, ditches and disused mineral workings throughout the British Isles. Widespread in the northern hemisphere.

Hybrids between *P. berchtoldii* and the broad-leaved species *P. natans* (*P.* × *variifolius*, no. 6) and *P. coloratus* (*P.* × *lanceolatus*, no. 10) and with the narrow-leaved *P. acutifolius* (*P.* × *sudermanicus*, no. 38) are recorded in the British Isles, but all are rare.

40. P. obtusifolius Mert. & Koch

Blunt-leaved Pondweed

P. gramineus subsp. *obtusifolius* (Mert. & Koch) Hayw.; *P. perthensis* Sturrock ex F. B. White nom. nud.; *P. pusillus* subsp. *sturrockii* A. Benn.; *P. sturrockii* (A. Benn.) A. Benn.; *P. foliosus* var. *diffusus* A. Benn.

Rhizome absent. *Stem* up to 1.9 m, very slender to slender, compressed, richly branched throughout or sparingly branched below, richly branched above; nodal glands well-developed, rather irregular in shape. *Submerged leaves* 48–85(–100) × (1.0–)2.5–3.6 mm, 16–30(–35) times as long as wide, flaccid, pale green, often tinged pink along the midrib or pinkish- or reddish-brown throughout and with an oily sheen, linear, usually rounded to obtuse, rarely subacute and often very shortly and rather obscurely mucronate at apex, entire and plane at margin, sessile; midrib occupying 8–20% of leaf width near the base, bordered by a band of lacunae which is broad towards the base but narrower above and ceases just below the apex, the lateral veins 1–2 on each side of the midrib and rather indistinct, the secondary veins infrequent. *Floating leaves* absent. *Stipules* 11–29 mm, open, convolute at the base, translucent when fresh, buff-coloured and more or less opaque when dry, rounded to broadly obtuse at apex, persistent, with a distinct green or an indistinct pale brown rib on each side and (8–)10–17 intercostal veins, the veins indistinct when dry. *Turions* terminal on short or long axillary branches, 23–40 × 3–5 mm, composed of appressed, dark green leaves enveloped in conspicuous stipules, with 3–5, erect to erecto-patent, free leaves. *Inflorescences* 4–9 × 1.5–3.3 mm; peduncles (6–)8–20(–35) mm, slender, slightly compressed or compressed. *Flowers* 6–8, contiguous, with (3–)4(–5) carpels. *Fruits* 2.6–3.2 × 1.7–2.1 mm, brownish-green, the dorsal edge entire or muricate; beak 0.2–0.5 mm, more or less apical or closer to the ventral than the dorsal edge, straight. *Flowers* 6–9. 2n = 26.

Native. Still or slowly flowing water in lakes, ponds, rivers, canals, ditches and flooded mineral workings. Widespread in both Britain and Ireland, but rare or absent from many western areas. Europe, eastern North America and scattered localities elsewhere in the northern hemisphere.

No hybrids of *P. obtusifolius* are recorded in the British Isles.

41. P. pusillus L.   Lesser Pondweed

P. panormitanus Biv.; *P. trinervius* G. Fisch.

Rhizome absent. *Stem* up to 0.7(–1.0) m, very slender to slender, slightly compressed to compressed, branching sparingly near the base in deep water but richly branched in shallow water; nodal glands usually absent or poorly developed, rarely well-developed. *Submerged leaves* (15–)20–50(–100) × (0.5–)0.8–1.4(–1.9) mm, 22–50(–70) times as long as wide, flaccid or firm, mid-green or olive green, linear, tapering or rather abruptly narrowed to an acute apex, bordered by a marginal vein, entire and plane at margin, sessile; midrib occupying 17–36% of leaf width at base, not bordered by lacunae or the lacunae poorly developed and restricted to the

proximal half of the leaf, rarely bordered by well-developed lacunae, the lateral veins 1(–2) on each side of the midrib and distinct, the secondary veins absent. *Floating leaves* absent. *Stipules* 5–17 mm, tubular for most of their length when young but splitting with age, hyaline and translucent when fresh and dry, obtuse at apex, persistent, usually with a green rib along each side; veins indistinct when dry. *Turions* usually sessile and axillary but in autumn also terminal on axillary branches, 10–23 × 0.65–1.0 mm, narrowly cylindrical, with 1–3 erecto-patent to recurved free leaves. *Inflorescences* 6–13 × 2.0–4.5 mm; peduncles 10–40(–80) mm, very slender to slender, slightly compressed to compressed. *Flowers* 3–6, more or less contiguous, with 4(–5) carpels. *Fruits* 1.8–2.3 × 1.2–1.5 mm, pale olive green; beak 0.2–0.4 mm, apical, straight. *Flowers* 6–9. 2n = 26.

P. pusillus appears to be related to *P. friesii*, but can usually be distinguished by its less well developed nodal glands and by its narrower, usually 3-veined leaves with an acute but not mucronate apex. There is a closer but probably more superficial resemblance to *P. berchtoldii*, which has open rather than tubular stipules.

Native. Lakes, reservoirs, ponds, rivers, canals, streams, ditches and disused mineral workings; particularly frequent in eutrophic and brackish water. Throughout the British Isles but more frequent in England than elsewhere. Europe; Africa; Asia; North and central America.

A hybrid with *P. trichoides* (*P. × grovesii*, no. 42) has been recorded from the British Isles, but it is apparently extinct.

42. P. × grovesii Dandy & G. Taylor

Groves's Pondweed

= P. pusillus × trichoides

Rhizome not seen, probably absent. *Stem* up to 0.35 m, very slender, much branched especially towards the apex; nodal glands absent or poorly developed. *Submerged leaves* with lamina 14–30 × 0.35–0.7 mm, 32–50(–60) times as long as wide, linear, gradually tapering to an acuminate apex, entire and plane at margin, sessile; midrib prominent, occupying 20–40% of leaf width near the base, not bordered by lacunae, the lateral veins one on each side of the midrib, indistinct, the secondary veins absent. *Floating leaves* absent. *Stipules* 5.5–9.0 mm, tubular for 0.5–2.5 mm at base when young but splitting with age, hyaline with a greenish tinge, rather abruptly tapering to an obtuse apex, persistent, with a green rib along each side; veins indistinct when dry. *Turions* terminal on axillary branches, 8–11 × 0.5–0.8 mm, cylindrical, with (1–)2, erecto-patent, free leaves. *Inflorescences* 4.0–5.5 × 1.5–2.5 mm; peduncles 12–55 mm, very slender. *Flowers* 3–4, contiguous, with 1–3 carpels. *Fruits* do not develop.

Native. Ditches in East Norfolk; not seen since 1900. Endemic.

43. P. trichoides Cham. & Schltdl.   Hairlike Pondweed

Rhizome absent. *Stem* up to 0.9(–2.0) m, very slender, more or less terete to slightly compressed, sparingly to richly branched especially towards the apex; nodal glands absent or poorly developed. *Submerged leaves* 16–80(–130) × 0.3–1.0(–1.75) mm, (28–)40–80(–107) times as long as wide, rigid, green, becoming dark green or brownish-green with age, linear, gradually tapering to an acute or acuminate apex, not bordered by a marginal vein, entire and plane at margin, sessile; midrib prominent, occupying 30–70% of the leaf width near the base; lacunae absent or restricted to a very narrow band at base of leaf, very rarely well-developed and extending almost to apex, with one indistinct lateral vein on each side of the midrib, the secondary veins absent. *Floating leaves* absent. *Stipules* 5–30 mm, open, convolute and tightly inrolled, translucent when fresh and dry, green or hyaline with a greenish tinge, obtuse at apex, persistent; veins indistinct when dry, two stronger than the others and sometimes forming faint ribs, the intracostal veins 4–7. *Turions* terminal on the main stems and axillary branches, 9–22 × 0.8–1.7 mm, fusiform, dark green or brownish-green, with 2–3, erect to erecto-patent, free leaves. *Inflorescences* 5–10 × 1.5–4 mm; peduncles 10–75 mm, very slender, terete to slightly compressed. *Flowers* 3–5, contiguous, with 1(–2) carpels. *Fruits* 2.5–3.2 × 1.8–2.1 mm, the ventral edge often with a single tooth towards the proximal end, the dorsal edge more or less muricate; beak 0.3–0.5 mm, more or less apical, straight. *Flowers* 6–9. $2n = 26$.

Native. Lakes, reservoirs, ponds, rivers, canals, ditches and flooded mineral workings, often in recently disturbed sites. Rather local in Britain, north to central Scotland. Europe; Africa; south-west Asia.

Two hybrids of *P. trichoides* are known from the British Isles: that with *P. crispus* (*P.* × *bennettii*, no. 31) is most unlikely to be confused with *P. trichoides*; that with *P. pusillus* (*P.* × *grovesii*, no. 42) is similar to both parents but is apparently extinct. A hybrid with *P. compressus* is described from Denmark.

44. P. rutilus Wolfg.   Shetland Pondweed

Rhizome absent. *Stem* up to 0.45 m, very slender, compressed, simple or sparingly branched near the base, sometimes also with short axillary branches; nodal glands inconspicuous, sometimes absent. *Submerged leaves* 32–75 × 0.5–1.1 mm, 37–78 times as long as wide, rather rigid, bright green or brownish-green, linear, gradually tapering to a very finely pointed apex, bordered by a strong marginal vein, entire and plane at margin, sessile; midrib prominent, occupying 17–27% of leaf width near base, not bordered by lacunae, the lateral veins one on each side of the midrib or occasionally two towards the base of the leaf and distinct, the secondary veins absent. *Floating leaves* absent. *Stipules* 15–19 mm, tubular in the basal 2–3 mm when young but splitting with age, milky white and translucent when fresh, white or buff-coloured and opaque when dry, tapering to an obtuse apex, persistent but eroding to fibrous strands at the apex, with a colourless or green rib along each side; veins prominent when dry. *Turions* axillary, 35–75 mm, slender, cylindrical, the outermost leaves curling away from the axis, the remaining leaves

appressed, the whitish prominently veined stipules conspicuous when dry. *Inflorescences* 3–7 × 2–3 mm; peduncles (3–)10–17 mm, very slender, slightly compressed to compressed. *Flowers* 6, contiguous, with (2–)4 carpels. *Fruits* 2.0–2.1 × 1.1–1.3 mm, brown or olive brown; beak 0.3–0.4 mm, more or less apical. *Flowers* 8.

Native. Moderately base-rich, lowland lochs in northern and western Scotland. Northern Europe south to France, Germany and Poland.

No hybrids of *P. rutilus* are recorded in the British Isles.

Subgenus 2. **Coleogeton** (Rchb.) Raunk.

Leaves all submerged, alternate with only the involucral ones opposite, narrowly linear, entire, sheathing at the base, with two longitudinal, air-filled canals, one on each side of the midrib, occupying the greater part of their interior; stipules adnate below to the leaf-base, forming the basal sheath, but free above as a ligule. *Spike* few-flowered, interrupted, its stalk and rachis flexible. *Stigmas* with large papillae.

P. filiformis and *P. pectinatus* are the two British species in subgenus *Coleogeton*. *Eleogiton fluitans* and *Ruppia* species, aquatic monocots with a similar vegetative structure, lack a ligule at the top of the leaf sheath. The aquatic form of *Juncus bulbosus* has auricles at the top of the sheath which can easily be mistaken for a ligule, but differs from *Potamogeton* in its growth habit (it lacks a rhizome but the stem nodes bear untidy clusters of leaves and roots) and the very slender leaves are transversely septate.

45. P. filiformis Pers.   Slender-leaved Pondweed
P. setaceus Schumach., non L.; *P. marinus* auct.; *P. pectinatus* subsp. *filiformis* (Pers.) Hook. fil.

Rhizome very slender to robust. *Stem* up to 0.3(–0.5) m, very slender to slender, terete, unbranched or sparingly branched; nodal glands absent. *Submerged leaves* 30–176 × 0.25–1.2 mm, (50–)100–300 times as long as wide, green, filiform, semicircular or slightly canaliculate in cross-section, obtuse or more or less acute at apex, entire and plane on margin, sessile; midrib bordered on each side by one large and several smaller air channels, the lateral veins one on each side of the midrib and inconspicuous. *Floating leaves* absent. *Leaf-sheaths* 8–27 mm long, the basal part closed and tubular when young, eventually splitting with age, the side from which the leaf arises green, the opposite side hyaline; ligules 5–15 mm, hyaline, obtuse to rounded or truncate and often asymmetrical at the apex. *Turions* absent. *Inflorescences* 17–75 × 2–5 mm; peduncles (25–)50–150(–220) mm, very slender, terete, flexuous. *Flowers* 5–9, in 3–5 groups of 1–2(–3) flowers, all groups distant at anthesis or the upper two more or less contiguous; carpels 4–6; stigmas sessile. *Fruits* 2.2–2.8(–3.2) × 1.4–2.0 mm, greenish-brown, becoming brown with age; beak 0.2–0.3 mm, apical or subapical, straight. *Flowers* 5–8. $2n = 78$.

Native. In open vegetation in shallow water at the edge of lowland lakes and reservoirs, and less frequently in pools, rivers, streams, ditches and flooded quarries. Scotland, northern England (rare), Wales (extinct), northern and western Ireland. Widespread in the northern hemisphere.

The hybrid with *P. pectinatus* (*P.* × *suecicus*, no. 46) is more likely to be confused with *P. pectinatus* than *P. filiformis*.

46. P. × suecicus K. Richt.   Western Pondweed
= P. filiformis × pectinatus

Rhizome very slender to very robust. *Stem* up to 1.3 m, very slender to robust, terete, branched near the base and otherwise simple or sparingly or richly branched; nodal glands absent. *Submerged leaves* with lamina (24–)50–156 × 0.3–2.3 mm, (20–)50–169 times as long as wide, mid-green or dark green, linear to filiform, semicircular to canaliculate in cross-section, broad leaves on pioneer shoots obtuse at apex, the other leaves subacute to acuminate, entire and plane at margin, sessile; midrib bordered on each side by one large or 0–4 large and several smaller air channels, the lateral veins one on each side of the midrib, inconspicuous. *Floating leaves* absent. *Leaf-sheaths* 10–40 mm, some open and convolute throughout their length, others on the same plant closed and tubular for up to 5(–12) mm at the base, the side from which the leaf arises green, the opposite side hyaline; ligules 7.5–24.0 mm, hyaline, rounded, asymmetrically obtuse, emarginate or irregularly truncate at the apex. *Turions* absent. *Inflorescences* 14–26 × 3.5–4.5 mm; peduncles 30–225 mm, very slender, terete, flexuous. *Flowers* (4–)8–10, in 3–5 groups of 1–3 flowers, all groups usually distant; carpels 4; stigmas sessile or borne on a distinct style 0.15–0.3 mm long. *Fruits* do not develop.

Native. Rivers, streams, lakes and pools. Scattered localities in Scotland and Ireland within the range of the rarer parent, *P. filiformis*, and in the Rivers Ure and Wharfe in Yorkshire, south of the current limit of that species. Northern Europe; Siberia; North America.

47. P. pectinatus L.   Fennel Pondweed
P. marinus L.; *P. interruptus* Kit.; *P. vaillantii* Roem. & Schult.; *P. zosteraceus* Fr.; *P. flabellatus* Bab.; *P. pectinatus* subsp. *flabellatus* (Bab.) Syme; *P. junceus* A. Kern. ex Druce nom. illegit.

Rhizome very slender to very robust. *Stem* up to 2.25 m, very slender to robust, terete, richly branched; nodal glands absent. *Submerged leaves* 27–115 × 0.2–4.0 mm, 24–160(–200) times as long as wide, bright green or olive green, linear to filiform, circular, semicircular, elliptical or canaliculate in cross-section, broader leaves with rounded to narrowly obtuse and mucronate apex, the narrower leaves more or less acute to finely acuminate at apex, entire and plane at margin, sessile; midrib bordered on each side by one large or 0–5 larger and several smaller air channels, the lateral veins 1–2 on each side of the midrib and inconspicuous. *Floating leaves* absent. *Leaf-sheaths* 10–70 mm long, open and convolute, the side from which the leaf arises green, the free edges hyaline, often lustrous; ligules 5–15 mm, hyaline or pale

brown, rounded or truncate at the apex. *Turions* absent. *Inflorescences* 13–45(–60) × 3.5–7.0 mm; peduncles 20–90(–285) mm, very slender, rarely slender, terete, flexuous. *Flowers* (4–)8–14, in 2–7 groups of 1–2 flowers, all groups distant at anthesis or the upper groups more or less contiguous; carpels 4; stigmas borne on a distinct style about 0.2 mm long. *Fruits* 3.3–4.7 × 2.6–3.6 mm, pale brown, becoming brown with age; beak 0.2–0.45 mm, ventral, rarely subventral. *Flowers* 5–9. 2*n* = c. 78.

Native. Eutrophic and brackish waters in lakes, ponds, rivers, streams, canals, ditches and flooded mineral workings; often abundant. Throughout the British Isles. Almost cosmopolitan.

P. filiformis and *P. pectinatus* hybridise; their hybrid, *P.* × *suecicus* (no. 46), can easily be overlooked as *P. pectinatus*.

2. Groenlandia Gay

Aquatic *perennial herbs* with all leaves submerged and flowering spike emergent. *Leaves* in opposite pairs, sessile, amplexicaul; stipules 0 except for the leaves subtending branches and spikes, where they are adnate to the leaf-base and form 2 lateral auricles. *Spike* 2-flowered. *Fruit* with a thin pericarp, but no bony endocarp.

One species occuring in Europe, western Asia and North Africa.

1. G. densa (L.) Fourr.   Opposite-leaved Pondweed
Potamogeton densus L.; *P. serratus* L.; *P. setaceus* L.

Rhizome slender to robust. *Stem* up to 0.65 m, slender to robust, terete, unbranched or sparingly to richly branched, often rooting at the lower nodes. *Submerged leaves* opposite, with lamina 6–42 × 1.5–12.5 mm, 1.8–9.0 times as long as wide, translucent, bright green, the upper leaves sometimes tinged reddish-brown, lanceolate to ovate, often recurved, more or less hyaline, usually narrowly obtuse, sometimes broadly obtuse at apex, denticulate at margin, sessile with base amplexicaul; midrib bordered by a band of lacunae which is broad towards the base of the leaf, narrows rapidly above and ceases just below or in the apex, the lateral veins 1–3 on each side of the midrib, the secondary veins few, transverse or ascending. *Floating leaves* absent. *Stipules* present only on leaves which subtend branches or peduncles, 2.0–4.6 × 1.0–1.5 mm, oblong to ovate-oblong or triangular-oblong, delicate, hyaline, obtuse at apex, forming lateral auricles on each side of the leaf base with the lower part adnate to the leaf; veins very faint. *Turions* absent. *Inflorescences* 2.0–4.5 × 2.5–4.0 mm; peduncles 4–14 mm, erect in flower, becoming strongly recurved in fruit, very slender to slender, of uniform diameter throughout their length, slightly compressed to compressed. *Flowers* 2, opposite, with 4 carpels. *Fruits* 3.0–4.0 × 2.0–2.8 mm, brown; beak 0.5–0.9 mm, more or less apical, broad at the base, often recurved at the tip. *Flowers* 5–9. 2*n* = 30.

Native. Shallow, base-rich water in rivers, streams, ponds and ditches. England, but rare in the south-west, rare in Wales and Ireland, extinct as a native species in Scotland. Europe; North Africa, western Asia.

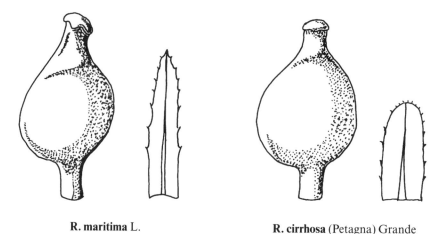

R. maritima L.   **R. cirrhosa** (Petagna) Grande

Fruits and leaf tips of **Ruppia** L.

156. RUPPIACEAE Hutch. nom. conserv.
By C. D. Preston, edited by P. D. Sell

Glabrous, aquatic *perennials* with monopodial rhizomes. *Stems* elongate, submerged, leafy. *Leaves* mostly alternate, but those subtending the inflorescence opposite, linear, sessile and minutely denticulate near the apex, sheathing at base; ligule absent. *Inflorescence* a short terminal spike at anthesis, appearing umbel-like in fruit. *Flowers* in pairs on very slender short peduncles which lengthen before and may then coil after pollination, bisexual, hypogynous, submerged or floating; perianth absent; stamens 2, filaments short and broad, anther with 2 widely separated lobes, so appearing as 4; stigma more or less sessile, peltate; carpels 2–8, each with 1 ovule. *Fruits* in a group of up to 4 drupelets, each druplet becoming very long-stalked.

One genus and from 5 to 10 species, cosmopolitan in temperate and warm regions of the world.

1. Ruppia L.

As for the family.

Ruppia is a taxonomically difficult genus, as the differences between species are slight and they show considerable environmental variation. Identification should be based on plants with mature fruits; vegetative characters do not seem to be reliable.

Reese, G. (1962). Zur intragenerischen Taxonomie der Gattung *Ruppia* L. *Zeitsch. für Bot.* **50**: 237–264.

Reese, G. (1963). Über die deutschen *Ruppia*- und *Zannichellia*-Kategorien und ihre Verbreitung in Schleswig-Holstein. *Schriften Naturwiss. Vereins für Schleswig-Holstein*, **34**: 44–70.

Stewart, A., Pearman, D. A. & Preston, C.D. (1994). *Scarce plants in Britain*. Peterborough. [*R. cirrhosa.*]

Van Vierssen, W., van Wijk, R. J. & van der Zee, J. R. (1981). Some additional notes on the cytotaxonomy of *Ruppia* taxa in western Europe. *Aquatic Bot.* **11**: 297–301.

Verhoeven, J. T. A. (1979). The ecology of *Ruppia*-dominated communities in western Europe. I. Distribution of *Ruppia* representatives in relation to their autecology. *Aquatic Bot.* **6**: 197–268.

1. Peduncle (8–)11–26 mm, 0.5–1.8(–3.3) times as long as the longest carpel stalk when the fruit is mature; fruits 2.0–2.5 × 1.3–1.8 mm, asymmetrical   **1. maritima**
1. Peduncles 39–300(–770) mm, (1.6–)2–10(–29) times as long as the longest carpel stalk when the fruit is mature; fruits 2.8–3.4 × (1.2–)1.4–1.9 mm, symmetrical or slightly asymmetrical   **2. cirrhosa**

1. R. maritima L.   Beaked Tasselweed
R. rostellata Koch

Rhizome very slender to slender. *Stem* to 0.4 m, very slender, terete, sparingly to richly branched; nodal glands absent. *Leaves* submerged, (20–)36–115 × 0.3–0.9 mm, 50–210 times as long as wide, bright green, filiform, elliptical in cross-section, acute at apex, denticulate at the leaf apex, otherwise entire, plane, sessile; midrib bordered on each side by an air channel; lateral veins absent. *Leaf-sheaths* 5–22 mm, open and convolute, the side from which the leaf arises green, the opposite side hyaline; ligule absent; sheaths of involucral leaves slightly dilated, hyaline when young, becoming brown with age. *Inflorescence* with 2 flowers about 1.5 mm apart. *Flowers* small, perianth absent; stamens 2; carpels 3–5, initially with a very short stalk so that they appear sessile at anthesis, the stalk then elongating after fertilisation and becoming on mature fruits 3–35 mm (the longest stalk on a peduncle (3–)10–35 mm), very slender, terete, straight; peduncles (8–)12–26 mm, 0.5–1.8 times as long as the longest carpel stalk when the fruit is mature, rarely as much as 3.3 times as long as the longest carpel stalk in dwarf forms, very slender, terete, straight, slightly recurved, arcuate-recurved or coiled with a single turn. *Fruits* 2.0–2.8 × 1.3–1.8 mm, pyriform, asymmetrical about the longitudinal axis, dark brown, with slightly raised, more or less elongated, wine-red tubercles on the surface; beak 0.4–0.6 mm, apical, straight. *Flowers* 7–9(–1). $2n = 20, 40$.

Native. Coastal lakes and ditches, rocky cliff-top pools, creeks and pools in salt-marshes, slow-flowing streams and tidal estuaries. Local round most coasts of

Britain and Ireland, formerly in Guernsey. Most of Europe; North Africa; western Asia; North America; Australia.

2. R. cirrhosa (Petagna) Grande   Spiral Tasselweed
R. maritima auct.; *R. spiralis* L. ex Dumort.

Rhizome very slender to slender. *Stem* to 0.6 m, very slender, terete, richly branched; nodal glands absent. *Leaves* submerged, 45–120(–175) × (0.2–)0.4–1.4 mm, 70–250(–500) times as long as wide, bright green or dark green, filiform to linear, elliptical in cross-section, obtuse to acute at apex, denticulate at apex, otherwise entire, plane, sessile; midrib bordered by one large or several smaller air channels, the air channels sometimes absent from the younger leaves; lateral veins absent. *Leaf-sheaths* 10–25 mm, open and convolute, the side from which the leaf arises green, the opposite side hyaline; ligule absent; sheaths of involucral leaves dilated, hyaline when young, becoming brown with age. *Inflorescences* with 2 flowers about 1.5 mm apart. *Flowers* small; perianth absent; stamens 2; with carpels 2–8, initially with a very short stalk so that they appear sessile at anthesis, the stalks then elongating after fertilisation and becoming on mature fruits 4–32 mm (the longest stalk on a peduncle 14–32 mm), very slender, terete, straight; peduncles 39–300(–770) mm, (1.6–)2–10(–29) times as long as the longest carpel stalk when the fruit is mature, very slender, terete, sinuous or spirally coiled. *Fruits* 2.7–3.4 × 1.2–1.9 mm, pyriform, symmetrical or slightly asymmetrical about the longitudinal axis, brown or greyish-brown, with raised, more or less elongated, wine-red tubercles on the surface; beak 0.5–0.95 mm, usually subapical or apical, occasionally ventral, straight. *Flowers* 7–9. $2n = 40, 60$.

Native. Ditches, ponds, coastal lagoons, tidal inlets and in lakes near the sea. Round the coasts of Britain and Ireland, but frequent only in east and south-east England. Temperate and subtropical regions of the world.

157. NAJADACEAE Juss. nom. conserv.

Glabrous, aquatic monoecious or dioecious *annuals* or *perennials* rooted in mud. *Stems* leafy, submerged. *Leaves* opposite or more or less whorled, linear, sessile with a sheathing base, minutely denticulate to conspicuously dentate or more or less lobed. *Flowers* inconspicuous, 1–3 sessile in each leaf-axil, unisexual, hypogynous. *Male flowers* surrounded by 2 scales, with 1 sessile anther. *Female flowers* without scales, with 2 elongated stigmas, 1 carpel, with 1 ovule. *Fruit* a sessile drupe.

One genus and about 50 species in temperate and tropical regions.

1. Najas L.

As family.

Stewart, A., Pearman, D. A. & Preston, C. D. (1994). *Scarce plants in Britain*. Peterborough. [*N. flexilis*.]
Triest, L. (1988). A revision of the genus *Najas* L. (Najadaceae) in the Old World. *Mém. Acad. Roy. Sci. Outre-Mer., Sci. Nat.* nov. ser. **22**: 1–172.

1. Leaves with many large spiny teeth   **1. marina**
1. Leaves minutely denticulate or nearly entire   2.
2. Leaf-sheaths rounded   **2. flexilis**
2. Leaf-sheaths with long, narrow auricles   **3. graminea**

Section 1. Najas

Plants mostly dioecious, mostly robust. *Stems* usually with spines on the internodes and on the abaxial side of the midrib of the leaf. *Seed* asymmetrically ovate, areoles unequal in shape and size, irregularly arranged.

1. N. marina L.   Holly-leaved Naiad
N. major All.

Plants submerged, dioecious, mostly robust. *Stems* up to 150 cm × 2 mm in diameter, mostly roughened with spines. *Leaves* 10–45(–60) × 1–6 mm, fleshy, acute at apex, serrulate with conspicuous spiny teeth on broad, triangular bases, the teeth up to 2 mm, sometimes also on the back of the midrib; sheath rounded, entire or serrulate. *Inflorescence* axillary, solitary. *Male flowers* enclosed by 2 scales 2–5 × 0.8–3 mm, tapering at the top; inner envelope protruding 0.2–0.7 mm above the anther; anther 1.2 × 0.7–2.5 mm, tetrasporangiate. *Female flowers* naked, 2.5 mm long; ovary 1.0–3.5 × 0.3–1.9 mm; style and stigma 0.3–1.6 mm; stigma (2–)3 (–4) lobed. *Fruit* (3–)4–6(–8) × 1.5–3.0 mm, with persistent, thin, membranous pericarp and the remaining parts of the style. *Seeds* (2.8–)3–4(–5.1) × (0.9–)1.2–2.0(–2.8), ovoid, slightly asymmetrical; testa pitted with areoles which are irregular in shape and dimensions. *Flowers* 7–8. $2n = 12$.

Native. In slightly brackish water. In a few of the Norfolk Broads. It appears to be declining where open to boat traffic and has disappeared in some localities where it used to be abundant. Cosmopolitan except for the colder parts of the temperate regions, in Europe extending northwards to about 63° N in Finland. The British plant is subsp. **intermedia** (Wolfg. ex Gorski) Casper. It is widespread in Continental Europe.

Section 2. Hyas Dumort.
Caulinia Willd.

Plant monoecious, more rarely dioecious, mostly slender. *Stems* and midrib of leaf unarmed. *Seed* elliptical-oblong; areoles all about the same shape and size, except near the raphe, mostly regularly arranged.

2. N. flexilis (Willd.) Rostk. & W.L.E. Schmidt
                                              Slender Naiad
Caulinia flexilis Willd.; *Fluvialis flexilis* (Willd.) Pers.

Plants submerged, annual, monoecious, slender. *Stems* unarmed, up to 30 cm, around 0.5 mm in diameter. *Leaves* 10–25(–40) × about 1 mm, flat, linear-lanceolate, acute at apex, serrulate with inconspicuous spiny teeth up to 0.7 mm, midrib without spines; sheaths sloping, rounded. 1.3–2.1 × 1.5–1.8 mm. *Inflorescence* axillary, male and female flowers at different nodes, but the male ones more to the top of the plant. *Male flowers* enclosed in two scales 1.2–1.5 mm, which tapers at the top; inner envelope protruding about 0.1 mm above the anther; anther 1.0–1.3 × 0.4–0.6 mm, unisporangiate. *Female flowers* naked; ovary 1.6–2.1 × 0.6–0.9 mm; style and stigma 0.8–2.0 mm; stigma 4-lobed. *Fruit* 2.5–3.5 mm, with persistent, thin, membranous pericarp and the remaining part of the style. *Seeds* straight, elliptical-oblong 2–3(–3.5) × 0.6–0.9 mm, with testa smooth, with areoles squarish to hexagonal. *Flowers* 8–9. Pollination occurs under water. 2*n* = 24.

Native. In clear water in mesotrophic lowland lakes over silty substrates. Lancashire (Esthwaite Water), Islay, Outer Hebrides, Kerry, Galway, Leitrim, Mayo and Donegal. Rare in north and central Europe south to Switzerland; northern Asia; North America.

3. N. graminea Delile   Grass-leaved Naiad
N. alagnansis Pollini

Plants submerged, monoecious, slender or robust. *Stem* up to 60 cm, unarmed, 0.4–1.5(–2.3) mm in diameter, often plumose above because of the closely packed leaves. *Leaves* (7.5–)14–25(–60) × 0.5–0.9(–4) mm wide, flat, linear-lanceolate, acute at apex, minutely serrulate with inconspicuous spiny teeth, the teeth up to 0.07 mm; midrib without spiny teeth; sheath (1.4–)2–4(–10.5) mm, deeply auriculate, the auricle (0.4–)0.8–1.2(–2.6) × (0.1–)0.2–0.3(–0.5) mm, with an acute apex. *Inflorescence* axillary, male and female flowers solitary or 2–4 together at the same node, but the male ones toward the top of the plant. *Male flowers* naked; inner envelope protruding 0.05–0.13 above the anther; anther (0.7–)1.3(–2.7) mm, tetrasporangiate. *Female flowers* naked, 1.6–3.7 mm; ovary 0.7–1.8 × 0.2–0.8 mm; style and stigma 0.5–2.6 mm; stigma 2(–3) lobed. *Fruit* with a persistent, thin, membranous pericarp and remaining parts of the style. *Seed* (1.2–)1.5–2.4 × 0.5–0.7(–0.9) mm, elliptical-oblong, sometimes slightly recurved; testa pitted with areoles regularly arranged in rows; areoles squarish to hexagonal or rectangular 0.6–0.8 mm. *Flowers* 8–9. 2*n* = 24, 36.

Introduced. Occurred in the Reddish Canal near Manchester, with warmed water, between 1883 and 1947. Native of the Old World Tropics.

158. ZANNICHELLIACEAE Dumort. nom.
conserv.

Glabrous, aquatic, rhizomatous, monoecious *annuals* or *perennials* rooted in mud. *Stems* leafy and submerged. *Leaves* mostly opposite, but sometimes alternate on sterile shoots, linear, sessile, entire, with sheathing base more or less free from the leaf. *Flowers* inconspicuous, solitary in the leaf-axils, unisexual, hypogynous, submerged. *Male flowers* with 1–2 stamens, naked, on long stalks; pollen grains spherical. *Female flowers* (2–)4–8 with one ovule; stigma peltate or lingulate on a distinct style in a cup-shaped scale. *Fruits* 3–6 mm, in a group of one to several, often 4, achenes, each with a persistent style.

Four genera, each with one or more polymorphic species.

1. Zannichellia L.

As family. Can be distinguished from *Potamogeton* particularly *P. pectinatus* in brackish water, by its linear,

mostly opposite leaves and unisexual axillary flowers without a perianth. From *Najas* it differs in its entire leaves and several carpels.

One or several species. Almost cosmopolitan.

Vierssen, W. Van (1982). Reproductive strategies of *Zannichellia* taxa in western Europe in Symoens, J. J., Hooper, S. S. & Compère, P. *Studies on Aquatic Vascular Plants.* Brussels.

1. Habit more robust; stigmas peltate; style 0.5–1.5 mm;
   achenes 2–6, subsessile or with a short stalk up to 1
   mm, dorsal margins more or less muricate
                                 (a) palustris subsp. **palustris**

1. Habit more slender; stigma lingulate; style 2.0–2.5 mm;
   achenes 2–5, stalks 1.5–2.5 mm and often on a
   common peduncle, dorsal and ventral margins often
   muricate   **(b) palustris** subsp. **pedicellata**

1. Z. palustris L.   Horned Pondweed

Rhizome not clearly differentiated from the stem. *Stems* up to 0.4 m, very slender to slender, terete, rooting at the lower nodes or throughout, sparingly to richly branched. *Submerged leaves* opposite, sometimes alternate on sterile shoots, 15–95 × 0.2–1.3 mm, pale green, bright green or dark green, linear, acute or acuminate at apex, entire and plane at margin, sessile; midrib bordered on each side by an air channel, lateral veins absent; enclosed at the base in a hyaline, truncate sheath 2.5–15.0 mm long, which is tubular when young but splits with age, filiform, circular or elliptical in cross-section. *Inflorescences* consisting of a single male or a single female flower, usually borne at the same node. *Male flowers* consisting of a single stamen; filaments 1.4–6.6 mm; anther 0.7–1.4 mm; stalks when mature 0.5–2.5 mm, often borne on a common peduncle. *Female flowers* with 2–5, shortly stalked carpels, surrounded by a translucent, tubular perianth; stigmas peltate or lingulate; style 0.5–2.5 mm. *Fruits* 2–6, each 2.6–5.0 × 0.6–1.3 mm, the dorsal edge entire or more or less muricate, the ventral edge usually entire or rarely more or less muricate; beak 0.4–2.6 mm, ventral, straight; stalks of mature fruits 0.5–2.5 mm. *Flowers* 5–8.

(a) Subsp. palustris
Z. brachystemon Gay ex Reut.; *Z. palustris* subsp. *brachystemon* (Gay ex Reut.) Hook. fil.; *Z. polycarpa* Nolte ex Rchb.; *Z. palustris* subsp. *polycarpa* (Nolte ex Rchb.) Hook. fil.; *Z. macrostemon* Gay; *Z. palustris* subsp. *macrostemon* (Gay) Hook. fil.
Habit more robust. *Stigma* peltate; style 0.5–1.5 mm. *Achenes* 2–6, subsessile or with a short stalk up to 1 mm, dorsal margin more or less muricate. 2*n* = 24.

(b) Subsp. pedicellata (Fr.) Syme
Z. pedicellata Fr.; *Z. maritima* Nolte ex G. Mey. nom. illegit.; *Z. gibberosa* Rchb.; *Z. pedunculata* Rchb.; *Z. palustris* subsp. *pedunculata* (Rchb.) Hook. fil.
Habit more slender and more variable than in subsp. *palustris*. *Stigma* lingulate; style 2.0–2.5 mm. *Achenes* 2–5; stalks 1.5–2.5 mm and often on a common peduncle, dorsal and ventral margins often muricate. 2*n* = 36.

Native. Rivers, streams, ditches and pools of fresh or brackish waters, up to 213 m in Ireland. Throughout most of the British Isles, though more frequent in England and eastern Ireland. Cosmopolitan. The distribution of the subspecies in the British Isles is not known, but subsp. *pedicellata* is thought to have a preference for brackish water. In Continental Europe, where they are often regarded as species, subsp. *palustris* is commoner in the north and can be in fresh or brackish water, and subsp. *pedicellata* is mostly in brackish water and is more common in the south.

159. ZOSTERACEAE Dumort. nom. conserv.

Perennial, glabrous, submerged, marine, monoecious *herbs* rooted in substratum, with a creeping rhizome, often exposed at low tide, bearing at each node 2 or more unbranched roots and a leaf with a short shoot in its axil. *Short shoots* with several, distichous, linear leaves with sheathing base and ligule. *Flowering stems* lateral or terminal. *Inflorescence* (spadix) enclosed in the sheathing base of a spathe, with a ligule at the top, the flowers in congested, compound cymes. *Flowers* unisexual, hypogynous. *Male flowers* with 1 dorsifixed, sessile stamen. *Female flowers* with one, 1-celled ovary, one ovule and 2, filiform stigmas on one common style. *Pollen* grains filiform. *Fruit* a sessile drupe.

Contains 3 genera and 18 species. Mainly found in temperate seas of both the northern and southern hemispheres.

1. Zostera L.

Rhizome monopodial, bearing alternate, distichous leaves. *Flowering shoots* annual, simple or branched. *Pollination* by water. *Reproduction* mostly vegetative by the breaking up of the rhizome.

Hartog, O. den (1970). The Sea-grasses of the world. *Verh. Kon. Nederl. Akad-Wetens. Natuurk.* **59** (1).
Stewart, A., Pearman, D. A. & Preston, C. D. (1994). *Scarce plants in Britain.* Peterborough.
Tutin, T. G. (1936). New species of *Zostera* from Britain. *Jour. Bot. (London)* **74**: 227–230.
Tutin, T. G. (1942). *Zostera* L. in Biological flora of the British Isles. *Jour. Ecol.* **30**: 217–226.

1. Flowering stems lateral, simple or sparingly branched; leaf-sheaths open **2. noltii**
1. Flowering stems terminal, freely branched; leaf-sheaths tubular, splitting when old 2.
2. Leaves (2–)4–12 mm wide; stigma about twice as long as style; fruit 3.0–3.5 mm **1(i). marina** var. **marina**
2. Leaves 1–2(–3) mm wide; stigmas about as long as style; fruit 2.5–3.0 mm **1(ii). marina** var. **stenophylla**

Subgenus **1. Zostera**

Leaf-sheaths tubular. *Connective of anthers* without an appendage.

1. Z. marina L. Eelgrass
Z. trinervis Stokes nom. illegit.; *Alga marina* Lam.; *Z. maritima* Gaertn.; *Z. stenophylla* Raf.; *Z. oregana* S. Watson; *Z. pacifica* S. Watson; *Z. latifolia* Morong

Perennial with creeping rhizome 2–5 mm thick, and numerous roots with a leaf at each node. *Rhizome leaves* 25–40 × 3.5–4.0 mm, tubular, without a blade, membranous, transparent, obtuse at apex, ventral apical margin depressed; veins 3, joining the midrib below the apex; midrib not reaching the apex but extending beyond the junction with the outer veins. *Internodes* 10–35 mm. *Cortex* with 2 bundles of fibres in its outer layer. *Short branches* with 3–8 leaves arising from the axils of the rhizomatous leaves. *Leaf-lamina* up to 120 cm × 1–12 mm, dark green, linear, rounded and often mucronate at apex; veins 5–11, the midrib slightly widening at the apex, the intermediate and outer veins joining each other in arches below the apex, the outer veins marginal in leaves with 5–7 veins, intermarginal in leaves with 7–11 veins, between every 2 veins 4–7 accessory bundles, the cross-veins more or less perpendicular or ascending, at intervals of 1.5–5.0 mm; leaf-sheath 5–20 cm, wider than the lamina, tubular, membranous, becoming irregularly torn with age, auriculae up to 1 mm, ligule about 0.5 mm, veins 3–7(–9). *Flowering stem* terminal, 60–150 cm, repeatedly branched, with numerous spathes. *Spathal sheath* 40–85 × 2–4 mm, amplexicaul, dorsal side green with 3–5 veins, flaps green, with a membranous, uncoloured margin broadly overlapping, with 2, obtuse or truncate auricles and a very short ligule; lamina of spathe 5–20 cm, narrower than that of a leaf, distinctly constricted at the base, obtuse at apex, with 5–7 veins; peduncle 20–30 mm. *Spadix* linear, with up to 20 female and 20 male flowers. *Stamen* 1, sessile, dorsifixed. *Style* 1, 1.5-2.5 mm; stigmas 2, 2.0–3.5 mm, usually deciduous. *Fruit* 2.5–4.0 mm, ellipsoid to ovoid, often beaked. *Seed* ellipsoid or ovoid; testa dark brown to straw-coloured, with 16–25 ribs. *Flowers* 6-9. *Fruits* 8–10. Germination in autumn. $2n = 12$.

(i) Var. marina
Leaves (2–)4–12 mm wide. *Stigmas* about twice as long as style. *Fruit* 3.0–4.0 mm.

(ii) Var. stenophylla Asch. & Graebn.
Z. hornemanniana Tutin; *Z. hagstromii* Druce
Leaves 1–2(–3) mm wide. *Stigmas* about as long as style. *Fruit* 2.5–3.0 mm.

Native. On fine gravel, sand or mud in the sea or in estuaries. Round the coasts of the British Isles, but becoming rarer northwards. Formerly covered large areas in suitable localities, but decreased markedly in abundance about 1933 and now local. Europe from about 71° N in Norway to the Mediterranean; west Greenland; Atlantic and Pacific coasts of America from North Carolina to Hudson Bay and California to Unalaska Bay.

The two varieties are often accepted as species and seem to be ecologically distinct. Var. *marina* occurs from low-water spring tides down to 4 m, rarely in estuaries and throughout the range of the species. Var. *stenophylla* occurs on mud-flats in estuaries and in shallow water, from half-tide mark to low tide mark, rarely down to 4 m, and is scattered round the coasts of the British Isles and in Denmark and Sweden.

Subgenus 2. Zosterella (Ascherson) Ostenf.

Leaf-sheaths open. *Connective of anthers* with an appendage.

2. Z. noltei Hornem.   Dwarf Eelgrass

Z. nana auct.; *Z. marina* var. *angustifolia* Hornem.; *Z. angustifolia* (Hornem.) Rchb.

Perennial with creeping rhizome 0.5–2.0 mm thick, and 1–4 roots with a leaf at each node. *Rhizome leaves* up to 15 × 2 mm, consisting of a sheath only, flaps overlapping or almost so, amplexicaul, membranous, transparent, truncate at apex; veins 3; midrib reaching the apex; outer veins parallel, joining the midrib just below the apex. *Internodes* 4–35 mm. *Cortex* with bundles of fibres in its innermost layers. *Short branches* arising from the axils of the rhizomatous leaves, with 2–5 leaves. *Leaf-lamina* 6–22 cm × (0.5–)0.7–1.0(–1.5) mm, green, linear, emarginate and often asymmetrical at apex, the outer veins marginal or just intramarginal, becoming distinctly intramarginal in the apical region and joining the midrib below the apex, the cross-veins more or less perpendicular at intervals of 2–4 mm; leaf-sheath 0.5–4 cm, wider than the leaf-lamina, open, with overlapping membranous flaps, auriculae 0.3–0.5 mm, obtuse, ligule very short, veins 3. *Flowering stem* lateral, unbranched, 2–25 cm, with 1–3(–6) spathes. *Spathal sheath* 12 20 × 1.5 2.0 mm, amplexicaul, dorsal side green, with 3 veins, the flaps scarious, transparent, not coloured and broadly overlapping, with 2, obtuse auricles and a very short ligule; lamina of spathe 30–80 mm, not narrowed at the base, deciduous after flowering, veins as in the leaf-lamina. *Spadix* lanceolate, apiculate, with 4–5 female flowers and 4–5 male ones. *Stamen* 1, sessile, dorsifixed. *Style* 1.2–1.7 mm; stigmas 2, 1.5–2.0 mm, usually deciduous. *Fruit* 1.5–2.0 mm, ellipsoid, beaked. *Seed* 1.5–2.0 mm, ellipsoid; testa reddish-brown, smooth and shiny, with fine longitudinal and transverse striations. *Flowers* 6–10. *Fruits* 7–11. Germination in autumn. $2n = 12$.

Native. On mud-banks in creeks and estuaries from half-tide mark to low-tide mark. Locally common in suitable habitats round the coasts of the British Isles, north to Sutherland; east coast of Ireland. Coasts of Europe from the Mediterranean to south-west Norway and Sweden.

Subclass 2. ARECIDAE Takht.

Herbs or trees with syncarpous flowers, often with a spadix subtended by a spathe, rarely aquatic. *Endosperm* present.

Order 1. ARECALES Nakai

This order consists of a single family.

160. ARECACEAE Schultz Sch. nom. conserv.
PALMAE Juss. nom. altern.

Shrubs or trees. *Stem* usually unbranched. *Leaves* large, persistent, petiolate, with a conspicuous sheath.

Inflorescence a simple or branched spadix, enveloped by one or several spathes. *Flowers* unisexual, hypogynous. *Perianth* usually of 2 whorls of 3. *Anthers* dehiscing longitudinally. *Ovary* of 1–3, free or united carpels; ovules 1 in each carpel. *Fruit* a berry or drupe.

About 202 genera with about 2,780 species. Chiefly tropical, with some subtropical and a few temperate species.

1. Leaves pinnately divided   **1. Phoenix**
1. Leaves palmately divided   **2. Trachycarpus**

1. Phoenix L.

Dioecious *trees*. *Leaves* pinnatisect, with numerous, induplicate, linear-lanceolate, long-acuminate segments, the proximal shorter, spinescent; petiole much shorter than lamina, unarmed. *Spadix* pedunculate, with few, simple branches; spathe 1. *Flowers* yellowish. *Stamens* with subulate filaments. *Ovary* of 3 free carpels. *Fruit* a berry, developing usually from 1 carpel only. *Seeds* deeply grooved along the ventral side.

About 17 species in the drier parts of the Old World tropics and subtropics.

P. dactylifera L., the Date Palm, is frequent on rubbish tips as seedlings, but is killed off by the first frosts. The one to few, leathery, narrowly elliptical leaves arise from the ground and have several parallel veins raised alternately on either surface. It is native of south-west Asia and North Africa.

1. P. canariensis Chab.   Canary Palm

Dioecious *tree*. *Trunk* solitary, up to 20 m, about 80–90 cm in diameter, when old with leaf-stalk scars broader than high. *Leaves* very numerous, up to 6 m long, the outer arching, sometimes twisted at an angle from the horizontal, with very many, crowded, more or less regularly arranged or paired, dark green leaflets 40–50 cm, in one plane. *Inflorescence* elongate. *Female flowers* with calyx nearly as long as petals. *Fruit* up to 1.5–2.3 × about 1.5 cm, cylindrical to ellipsoid, yellow to reddish- or dark purple. $2n = 36$.

Introduced. Planted by the sea in small numbers in south-west England. Native of the Canary Islands.

2. Trachycarpus H. Wendl.

Dioecious *trees*. *Leaves* palmate, with numerous induplicate, bifid segments; petiole slender, with sharp teeth. *Spadix* pedunculate, branched. *Flowers* yellow. *Stamens* with fleshy filaments. *Ovary* of 3, distinct carpels. *Fruit* a berry, developing from 1 carpel.

Six species in the Himalayas of northern India to northern Thailand and China.

1. T. fortunei (Hook.) H. Wendl.   Chusan Palm
Chamaerops fortunei Hook.; *T. excelsus* auct.

Dioecious *tree* up to 5 m. *Trunk* erect, cylindrical, clothed with coarse, dark, stiff fibres which are the disintegrating sheathing bases of the leaves. *Leaves* up to 75 × 120 cm, palmately divided into deep palmate, acute segments up to 5 cm wide. *Inflorescence* a decurved panicle. *Male flowers* more ornamental than female. *Fruit* suborbicular, bluish-black. $2n = 36$.

Introduced. Planted in south and south-west England and self-sown near Helford and at Trelowarren in Cornwall, and at Abbotsbury in Dorset. Native of China.

Order 2. ARALES Lindl.

Perennial herbs, or sometimes floating aquatics. *Flowers* very small, bisexual or unisexual, hypogynous, densely crowded on a spadix or rather few together, the inflorescence usually more or less enclosed in a large bract, the spathe. *Perianth* present and small or absent. *Ovary* 1- to many-celled, placentation various. *Fruit* usually a berry. *Seeds* with endosperm.

This order consists of two very well defined families, the Araceae with about 2,000 species and the Lemnaceae with 28 species.

161. ARACEAE Juss. nom. conserv.

Glabrous *perennial herbs* often with tuberous or elongated rhizomes; raphids often present. *Stems* with 0–few leaves. *Leaves* alternate, simple, linear to ovate, entire or rarely deeply lobed, often cordate to sagittate at base, sessile or petiolate, but usually sheathing at base; exstipulate. *Flowers* closely packed on a sterile axis (*spadix*) which often extends distally as a succulent appendix, usually subtended or partially enclosed by the leaf-like but often coloured *spathe*, bisexual or unisexual, if the latter the upper usually male and lower female, hypogynous, actinomorphic. *Perianth segments* absent or 4–6, free or connate into a truncate cup. *Stamens* 1–6, free or connate; anthers opening by pores or slits. *Staminodes* sometimes present. *Stigmas* capitate, more or less sessile. *Ovary* 1- to 3-celled, with one to many ovules. *Fruit* a berry with one to several seeds.

About 115 genera and about 2,000 species. Widely distributed throughout the world, but the greatest number of species in the tropics.

Species in this family have phallic and vaginal connotations, hence the vernacular names of Lords-and-Ladies, Cuckoo-pint and Jack-in-the-Pulpit for *Arum maculatum*.

1. Leaves linear; spadix apparently lateral and lacking a spathe **1. Acorus**
1. Leaves lanceolate or ovate, narrowed at base and usually petiolate; spadix terminal, with a spathe 2.
2. Flowers covering spadix to its apex 3.
2. Flowers overtopped by a succulent sterile appendix 5.
3. Spathe more or less flat, not enclosing the spadix even at its extreme base **3. Calla**
3. Spathe wrapped round the basal part of the spadix 4.
4. Leaves truncate to cuneate at base, with petiole shorter than lamina; perianth segments 4 **2. Lysichiton**
4. Leaves cordate at base, with petiole longer than lamina; perianth segments absent **4. Zantedeschia**
5. Leaves palmately divided **6. Dracunculus**

5. Leaves simple 6.
6. Spathe overlapping at base, not fused into a tube, no more than acuminate at apex **5. Arum**
6. Spathe fused into a tube proximally, with a filiform projection more than 5 cm at apex **7. Arisarum**

1. Acorus L.

Perennial herbs with rhizomes. *Scape* flattened. *Leaves* linear, entire, sessile, with a spicy scent when bruised. *Spadix* apparently lateral, without appendix; spathe apparently lacking. *Flowers* bisexual. *Perianth segments* 6, in two ranks, free, membranous. *Stamens* 6. *Stigmas* small. *Ovary* 2- to 3-celled. *Fruit* a berry. *Seeds* few, oblong, with a fleshy endosperm.

Two species in Europe, temperate Asia and America.

1. Leaves 50–125 cm × 7–25 mm, with a well-defined midrib; spadix 5–9 cm × 6–12 mm **1. calamus**
1. Leaves 8–50 cm × 28 mm, without an obvious midrib; spadix 5–10 cm × 3–5 mm **2. gramineus**

1. A. calamus L. Sweet Flag

Glabrous *perennial herb* with aromatic rhizomes. *Stems* up to 1 m, erect, stout, flattened, reddish at base. *Leaves* 50–125 cm × 7–25 mm, crowded, distichous, bright shining green, ensiform, acute at apex, entire but wavy, rather stiff, with a well-defined midrib, sheathing at base, usually transversely wrinkled in places, smelling of tangerines when bruised. *Spathe* apparently lacking, but the long leaf-like extension of the stem beyond the spadix could be regarded as such. *Spadix* 5–9 × 6–12 mm, sessile, making an angle of about 45° with the stem, gradually tapering upwards to an obtuse apex. *Flowers* yellowish-green, tightly packed and completely covering the spadix. *Perianth segments* 6, free, obovate, obtuse at apex, membranous. *Stamens* 6; filaments pale green; anthers yellow. *Fruit* a reddish berry, not formed in the sterile triploid in the British Isles. *Flowers* 5–7. $2n = 27$.

Introduced. In shallow water at the edges of rivers, canals, ponds and lakes. Scattered over most of the British Isles, but frequent only in England. Native of Asia and North America. Our plant is the sterile triploid, var. **calamus**.

2. A. gramineus Aiton Slender Sweet Flag

Glabrous *perennial herb* with much-branched, creeping, aromatic rhizomes. *Stems* up to 70 cm, erect, stout, flattened. *Leaves* 8–50 cm × 2–8 mm, deep green, tufted, linear-ensiform, gradually narrowed to an acute apex, entire, without an obvious midrib, soft, smooth, sheathing at base. *Spathe* apparently lacking, the long leaf-like extension of the stem beyond the spadix could be regarded as such. *Spadix* 5–10 cm × 3–5 mm, sessile, ascending to nearly erect, narrowly cylindrical, obtuse at apex. *Flowers* yellow, tightly packed and completely covering the spadix. *Perianth segments* 6, free, obovate, truncate at apex, thick, incurved. *Stamens* 6; filaments linear. *Stigma* small. *Fruit* a reddish berry, but not formed in the British Isles. *Flowers* 5–7. $2n = 24$.

Introduced. Naturalised by Mytchett Lake in Surrey since 1986. Native of eastern Asia.

2. Lysichiton Schott

Perennial herbs with rhizomes. *Stems* absent. *Leaves* ovate-oblong, entire, truncate to cuneate at base, shortly petiolate, without a sheath and appearing at about the same time as the flowers. *Spathe* arising from ground level, the lower part consisting of a narrow sheath enclosing the very long stipe of the spadix, but with free margins, withering before the fruit is ripe. *Spadix* terminal, cylindrical, without appendix. *Flowers* bisexual, crowded. *Perianth segments* 4, often unequal. *Stamens* 4. *Stigmas* sessile. *Ovary* (1–)2-celled. *Fruit* a green berry with 2 seeds, partly embedded in the spadix.

Two species in the North Pacific area. Hybrids (which may have been overlooked when naturalised) occur between the species in gardens and they may be better treated as subspecies.

Hultén, E. & St John, H. (1932). The American species of *Lysichitum*. *Svensk Bot. Tidskr.* **25**: 453–464.

1. Flowers foul-smelling; spathe yellow; perianth
   segments 3–4 mm; anthers 0.9–2.0 mm   **1. americanus**
1. Flowers scentless; spathe white; perianth segments 2–3
   mm;; anthers 0.6–0.8 mm   **2. camtschatcensis**

1. L. americanus Hultén & St John
American Skunk-cabbage

Glabrous, robust *perennial herb* with dark, erect rhizome. *Stipe* 10–40 cm, elongating after anthesis, erect. *Leaves* with lamina 30–150 × 25–70 cm, elongating after anthesis, oblanceolate or elliptical, more or less obtuse at apex, entire, sessile or subpetiolate, the petiole broadly channelled. *Spathe* tightly covering the stem at base, 10–35 cm, yellow, elliptical, obtuse or shortly acuminate at apex, withered in fruit. *Spadix* 3.5–12 cm, cylindrical or narrowly fusiform-cylindrical, with many dense flowers, without an appendix. *Flowers* foul smelling. *Perianth segments* 4, 3–4 mm, narrowly oblong, rounded at apex, incurved below the apex. *Stamens* 4; filaments rather flat; anthers 0.9–2.0 mm. *Fruit* a green berry. 2*n* = 28.

Introduced. Grown for ornament and spreading into swampy ground where it becomes established. Reproducing freely from seed. Scattered through south and west Britain and Ireland. Native of western North America.

2. L. camtschatcensis (L.) Schott   Asian Skunk-cabbage
Dracontium camtschatcense L.; *Pothos camtschaticus* Spreng.; *Arctiodracon japonicum* A. Gray; *Arctiodracon camtschaticum* (Spreng.) A. Gray

Glabrous *perennial herb* with thick, white, short, erect rhizomes. *Stipe* 10–30 cm, elongating after anthesis, erect. *Leaves* with lamina 40–80 × 15–30 cm, elongating after anthesis, elliptical to narrowly oblong, more or less obtuse at apex, entire, gradually narrowed at base, with a thick midrib, the lateral veins slender, curved inwards at the tip and not reaching the margin; petiole flat and spongy. *Spathe* tightly covering the stem at base, 8–12 cm, white, elliptical to ovate, abruptly acute at apex, withered in fruit. *Spadix* 4–8 cm in flower, to 12 × 5 cm in fruit, oblong to cylindrical, with many dense flowers,

without an appendix. *Flowers* scentless. *Perianth segments* 4, 2–3 mm, narrowly oblong, rounded at apex, incurved below the apex. *Stamens* 4; filaments rather flat; anthers 0.6–0.9 mm, yellow, ovoid. *Fruit* a green berry. *Flowers* 5–7. 2*n* = 28.

Introduced. Grown for ornament and spreading into swampy ground. Distribution uncertain due to confusion with *L. americanus*. Native of eastern Asia.

3. Calla L.

Aquatic *perennial herbs* with rhizomes. *Leaves* all basal and mixed with scale-leaves, ovate to reniform, entire, cordate, long-petiolate, the sheathing bases of the petioles persisting on the rhizome. *Spathe* open, more or less flat, not concealing spadix, greenish outside, white inside, persistent in fruit. *Spadix* terminal, green, shortly cylindrical, without an appendix. *Flowers* mostly bisexual, but uppermost usually male, without a perianth. *Stamens* 6. *Stigma* sessile. *Ovary* 1-celled. *Fruit* a red berry with several seeds; with abundant endosperm.

A single species of cool temperate and Subarctic regions of North America and Eurasia.

1. C. palustris L.   Bog Arum

Rather soft *perennial herb* with long-creeping rhizomes and numerous, pale, brown roots. *Leaves* all basal and mixed with lanceolate scale-leaves; lamina 5–12 × 4–10 cm, broadly ovate to reniform, cuspidate at apex, entire; veins numerous, slender, inconspicuous, curved inwards at apex; petiole 10–25 cm, terete, long-sheathing. *Spathe* 3–8 × 3–6 cm, greenish outside, white inside, ovate to ovate-oblong, cuspidate at apex, more or less flat and not covering spadix, persisting in fruit. *Spadix* 1–3 × 0.7–2 cm, green, densely flowered, shortly cylindrical; peduncle up to 30 cm above water. *Stamens* 6; filaments broad and flat. *Stigma* sessile. *Berry* globose, red when ripe. *Seeds* several, oblong. *Flowers* 6–7. 2*n* = 36, 72.

Introduced. Grown for ornament. Persistent and spreading in marshy ground and shallow ponds, often in shade. Very scattered in Britain from south-eastern England to central Scotland. North, central and east Europe; northern Asia; North America.

4. Zantedeschia Spreng. nom. conserv.

Perennial herbs with short, tuberous rhizomes. *Leaves* ovate, entire, cordate, long-petiolate. *Spathe* wrapped around and concealing the base of the spadix, ivory white inside. *Spadix* shorter than spathe, terminal, without an appendix, yellowish. *Flowers* unisexual, without a perianth. *Stamens* 2–3. *Style* short; stigma entire. *Ovary* (1–)3-celled, each with 1–4 ovules. *Fruit* an orange berry with several seeds, but very rarely produced in the British Isles.

Six species from southern Africa.

Letty, C. (1973). The genus *Zantedeschia. Bothalia* **11**: 5–26.
Traub, H. P. (1949). The genus *Zantedeschia. Plant Life* **4**: 8–32.

1. Z. aethiopica (L.) Spreng.   Altar Lily

Glabrous *perennial herb* up to 60(–250) cm, with short tuberous rhizomes. *Leaves* evergreen, leathery and more or less spreading; lamina 15–20 × 10–15 mm, ovate or broadly so, obtuse to acute at apex, entire, cordate or hastate at base; petiole long, green. *Spathe* up to 15 × 12 cm, ivory white inside, bright green at base outside, merging into white upwards, longitudinally veined, folded from slightly below the insertion of the spadix into a wide-mouthed funnel, the limb obliquely spreading and ending in a green recurving apiculus about 2 cm. *Spadix* sessile, cylindrical. *Stamens* 2–3; anthers 1.5–2.0 × 1.0–1.5 mm, bright yellow. *Style* short, persistent. *Ovary* about 4 mm, globose, grooved, pale yellowish-green grading to whitish at the tip. *Fruits* 10–12 × 10–12 mm, green at first becoming soft and orange-coloured at the base and greenish at the apex, tapering to a triangular base. *Flowers* 8–10. $2n = 24$.

Introduced. Grown for ornament, escaping and persistent and spreading in ditches, damp hedgerows and scrub, and in neglected fields. Channel Islands, southwest England, Kent, Glamorgan and Kerry. Native of South Africa.

5. **Arum** L.

Perennial herbs with short, tuberous rhizomes. *Leaves* all basal, triangular-ovate, entire, hastate or sagittate, long-petioled with a shortly sheathing base. *Spathe* wrapped around and concealing the base of the spadix, pale greenish-yellow. *Spadix* terminal, with a long appendix, with zones of male and female flowers usually separated by some sterile ones which are also often present above the male. *Flowers* unisexual, without a perianth. *Stamens* 3–4. *Stigma* sessile. *Ovary* 1-celled. *Fruit* a berry, with one to several seeds.

Twelve species ranging from the Canary Islands to Iran and northwards to Sweden.

Prime, C. T. (1960). *Lords and Ladies.* London.

1. Leaves appearing in spring, 7–20 cm, with concolorous midrib; spathe 10–25 cm; spadix appendage purple or yellow, usually reaching about halfway up the expanded part of the spathe, when fruiting 3–5 cm
   1. maculatum

1. Leaves appearing in early winter, 15–35 cm, with pale midrib; spathe 15–40 cm; spadix appendage always yellow, usually reaching about one-third of the way up the extended part of the spathe, when fruiting 10–15 cm   **2.**

2. Leaves with whitish veins and basal lobes divergent; fruits with 2–4 seeds   **2(a). italicum** subsp. **italicum**

2. Leaves with paler green veins and basal lobes somewhat convergent and sometimes overlapping; fruits with 1–2 seeds   **2(b). italicum** subsp. **neglectum**

1. **A. maculatum** L.   Lords-and-Ladies

Glabrous *perennial herb* with acrid juice, 30–50 cm, with rather round, horizontal, tuberous rhizome, a fresh tuber being produced at the base of the stem each year. *Leaves* few, all basal, appearing in spring; lamina 7–20 × 3–19 cm, dark green with a concolorous midrib, sometimes with blackish-purple spots and blotches, occasionally with whitish veins, triangular-hastate, rounded to acute at apex, with an entire margin, cordate at base with obtuse or acute basal lobes; petiole up to 30 cm, sheath-

ing at base. *Spathe* 10–25 cm, erect, acute at apex, pale yellowish-green, edged and sometimes spotted with brownish-purple, wrapped round the base of and concealing the spadix. *Spadix* about half as long as the spathe, cylindrical, obtuse at apex and stalked; appendage purple or yellow; fruiting spike 3–5 cm. *Stamens* 3–4; anthers purple, sessile. *Stigmas* sessile. *Berry* 8–9 × 7–8 mm, orange-red to red when ripe, almost globose. *Seeds* brownish-white, almost globose, reticulate. *Flowers* 4–5. *Fruits* 7–8. Protogynous and cross-pollinated by small flies, particularly midges. $2n = 56$.

Very variable in size, leaf shape, folding of the spathe round the spadix, colour of spadix appendage and marking of the leaves and spathe. The type plant has leaf spots and is var. *maculatum*. Plants without spots or blotches on the leaves are var. *immaculatum* Mutel. Plants with a yellow spadix appendage are var. *latrelii* Corb. These characters, however, together with clockwise or anti-clockwise folding of the spathe, occur in different combinations, none of which seem to have any ecological or geographical significance. *A. maculatum* does reproduce by seed, but the main means of spread is by the budding off of daughters from the corm, thus forming large clones, plants of which look exactly alike.

Native. Woods, hedgerows and shady banks, usually on base-rich soils, occasionally becoming a persistent weed in gardens and very shade tolerant. Generally distributed throughout England, Wales and Ireland, less frequent in Scotland and not native in the north. West, central and south Europe, northwards to Scotland and eastwards to the western Ukraine; North Africa. As far as is known all British plants are referable to subsp. **maculatum** with $2n = 56$. Subsp. *danicum* Prime with $2n = 28$ occurs in Denmark. This subspecies was later transferred to *A. orientale* M. Bieb., which only certainly differs from *A. maculatum* in the vertical, not horizontal tuberous rhizome.

2. **A. italicum** Miller   Italian Lords-and-Ladies

Glabrous *perennial herb* 20–55 cm, with tubers about 5 cm. *Leaves* few, all basal, well-developed by November; lamina 15–35 × 8–25 cm, dark green with a pale midrib, sometimes with blackish-purple spots or whitish veins, triangular-hastate, rounded to acute at apex, with an entire margin, cordate at base with obtuse to acute, divergent or somewhat congested and overlapping, curved basal lobes; petiole 10–40 cm, sheathing at base. *Spathe* 15–40 cm, erect, obtuse or shortly acute at apex, pale yellowish-green, never spotted, wrapped round the base of and concealing the spadix. *Spadix* about one-third as long as the expanded part of the spathe; appendage yellow; fruiting spike 10–15 cm. *Stamens* 3–4; anthers yellow. *Stigmas* sessile. *Berry* 10–11 × 8–9 mm, red when ripe, subglobose or broadly ellipsoid. *Flowers* 4–5. *Fruits* 8–9.

(a) Subsp. **italicum**

Leaves never dark-spotted, with whitish veins, the basal lobes divergent. *Berry* with 2–4 seeds. $2n = 84$.

(b) Subsp. **neglectum** (F. Towns.) Prime
A. italicum var. *neglectum* F. Towns.
Leaves sometimes dark-spotted, the veins slightly paler than the rest of the leaf, the basal lobes somewhat convergent and sometimes overlapping. *Berry* with 1–2 seeds. $2n = 84$.

Native and introduced. Subsp. *neglectum* is native in hedgerows, scrub and stony field borders in the extreme south and south-west of England from Kent to Cornwall, in Glamorgan in south Wales and in the Channel Islands and is very rarely naturalised elsewhere. It is common in the Channel Islands and the only taxon in the Isles of Scilly. It is in western Europe and the west Mediterranean region. Subsp. *italicum* is an introduced garden throw-out which is naturalised in hedgerows, scrub and waste places and is found in the Channel Islands, where it is possibly native, and scattered over Britain to the Isle of Man, Dunbarton and Lincolnshire and in eastern Ireland. It occurs almost throughout the range of the species which extends through southern Europe. Other subspecies occur in the eastern Balkans and the Crimea.

× maculatum

This hybrid is intermediate between the species, its leaves appear in early winter and are often spotted, and it is sterile.

It occurs rarely through the range of *A. italicum*. Both subspecies of *A. italicum* are involved.

6. Dracunculus Mill.

Perennial herbs with short, tuberous rhizomes. *Leaves* basal, deeply and more or less palmately lobed, cordate, long-petiolate. *Spathe* wrapped round and concealing spadix at base, purplish-brown with mottling inside. *Spadix* terminal, with sterile florets above and below the male zone, with long, dark, blackish-red appendage. *Flowers* unisexual, without a perianth. *Stamens* 2–3. *Ovary* 1-celled. *Fruit* a red berry with several seeds.

Three species in the Mediterranean region and the Canary Islands.

1. D. vulgaris Schott   Dragon Arum
Arum dracunculus L.

Perennial herb 70–180 cm, with short, tuberous rhizome. *Leaves* all basal; lamina 15–20 × 25–35 cm, green, often spotted or striped with white, more or less reniform in outline, deeply and more or less palmately lobed, the 9–15 segments up to 20 cm, lanceolate-oblong, and acute at apex; petiole with long, wide, sheathing base, spotted with dark purple, concealing base of scape so as to make some leaves appear cauline. *Spathe* 20–40(–50) cm, outer surface greenish, inner surface dark brownish-purple; upper part lanceolate and erect; margins undulate. *Spadix* nearly as long as the spathe, with stout, blackish-red appendage. *Stamens* 2–3. *Berry* orange-red. *Flowers* 5–7. $2n = 28$.

Introduced. A garden throw-out, naturalised in hedges, rough waste ground and old gardens. Scattered records in south and south-east England and the

Channel Islands. Native of the east and central Mediterranean region.

7. Arisarum Mill.

Perennial herbs with a rootstock and slender rhizomes. *Leaves* triangular-ovate, entire, sagittate, long-petiolate. *Spathe* forming a tuber round the spadix and concealing most of it. *Spadix* terminal, with a long whitish appendix. *Flowers* unisexual, without a perianth. *Stamens* 1. *Ovary* 1-celled. *Fruit* a greenish berry with several seeds.

Two species from the Mediterranean region and the Atlantic Islands.

1. A. proboscoideum (L.) Savi   Mousetail Plant
Arum proboscoideum L.

Perennial herb with a slender, creeping rhizome. *Leaves* solitary or few; lamina 6–15 × 3–8 cm, yellowish-green, hastate, the middle lobe ovate or oblong, obtuse or apiculate at apex, and entire and recurved at the margin, the lateral lobes as long as or shorter than the middle one and obtuse and rather recurved; petioles 10–15 cm, cylindrical. *Spathe* 2–4 cm, dark brown or brownish-green above, paler below, sometimes striped with dull purple, the apical part strongly hooded, curved forwards and terminating in a filiform process 5–15 cm. *Spadix* with a white appendage, included in the spathe except for the uppermost part. *Stamen* 1; filaments very short; anthers broadly reniform. *Style* very short; stigmas capitate. *Ovaries* very few at the base of the spadix in front, subglobose. *Berry* greenish. *Flowers* 3–5. $2n = 18$.

The upper part of the spathe, which is all that protrudes from beneath the leaves, bears a striking resemblance to the tail and hind quarters of a mouse.

Introduced. A garden throw-out, naturalised in hedges. Very widely scattered in southern England. Native of Spain and Italy.

162. LEMNACEAE Gray nom. conserv.

Small floating or submerged, *aquatic monoecious herbs*, sometimes stranded on mud, not or variously adhering together. *Roots* 0–21, simple. *Fronds* up to 15 mm. *Flowers* rarely produced, borne in (1–)2 hollows on the frond, naked or at first enclosed in a sheath. *Perianth* absent. *Male flowers* with 1–2 stamens and 1- to 2-celled anthers. *Female flowers* with funnel-shaped stigma; ovary 1, with 1–2 ovules. *Fruits* of 1–2 seeds in a sac.

Four genera with 28 species, cosmopolitan in fresh water. Mainly vegetatively produced, thus often forming clones.

Spring and summer fronds are best used for identification, as those produced in autumn and overwintering are often atypical in vein and root number.

Daubs, E. H. (1965). A monograph of Lemnaceae. *Illinois Biol. Mongr.* **34**.

1. Fronds rootless and veinless, spherical to ellipsoid
                 1. Wolffia
1. Fronds with (0–)1–16(–21) veins   2.
2. Fronds with (0–)1 root and 1–5(–7) veins   **2. Lemna**
2. Fronds with 7–16(–21) roots and veins   **3. Spirodela**

1. **Wolffia** Horkel ex Schleid. nom. conserv.

Tiny plants, usually floating and forming a cover on the water surface. *Fronds* usually not exceeding 1.2 mm, without roots or veins, strongly swollen on both sides. *Inflorescence* in hollows on the upper surface of the frond, without a sheath, with one male and one female flower, the latter nearest the base of the frond. *Stamen* 1; anthers unilocular, opening by a pigmental line of dehiscence across the top. *Fruit* spherical, with persistent style. *Seed* globose or slightly compressed.

Eight species distributed throughout the world.

Stewart, A., Pearman, D. A. & Preston, C. D. (1994). *Scarce plants in Britain.* Peterborough.

1. **W. arrhiza** (L.) Horkel ex Wimm.

Rootless Duckweed

Lemna arrhiza L.; *Lenticula arrhiza* (L.) Lam.; *Lemna globosa* Roxb.; *W. michelii* Schleid.; *Grantia globosa* (Roxb.) Griff.; *Bruniera vivipara* Franch.; *W. delilii* Kurz, non Schleid.; *W. cylindracea* Hegelm.

Fronds 0.5–1.2 × 0.4–1.0 mm, floating on water and usually forming a cover, green, without pigmentation, spherical to ellipsoid, the upper surface weakly arched or convex, with a strongly delineated, rounded margin, the lower surface much swollen, producing daughter fronds by budding from one end. *Flowers* unknown in Britain. No special resting fronds are produced, but the ordinary ones sink in winter. $2n = 50$.

Native. Ponds and ditches, very local. Southern England, from Somerset to Kent. Europe from Lithuania to west-central Portugal, Sicily and Bulgaria, local; Africa; Asia; America; Australia.

2. **Lemna** L.

Lenticula Hill; *Hydrophace* Haller; *Telmatophace* Schleid.; *Staurogeton* Rchb.

Plants floating upon or beneath the surface of the water, solitary, or many remaining attached, with a single root peltately attached, with one to several veins. *Inflorescence* or young fronds, or both, borne in hollows on the upper surface of the frond, enclosed in a membranous sheath, with 2 male and one female flowers. *Fruit* 1- to 6-seeded. *Seeds* typically ribbed.

Nine species distributed throughout the world.

Leslie, A. C. & Walters, S. M. (1983). The occurrence of *Lemna minuscula* Herter in the British Isles. *Watsonia* **14**: 243–248.
Reveal, J. L. (1990). The neotypification of *Lemna minuta* Humb., Bonpl. & Kunth, an earlier name for *Lemna minuscula* Herter (Lemnaceae). *Taxon* **39**: 328–330.

1. Fronds narrowed to a stalk-like portion at one end, usually submerged, usually connected in branched chains of 3–50 **4. trisulca**
1. Fronds without a stalk-like portion at one end, usually on the surface of the water, cohering in small groups, not chains **2.**
2. Fronds usually strongly swollen on underside, with (3–)4–5(–7) veins originating from one point **3. gibba**
2. Fronds more or less flattened on both surfaces, with 1–3(–5) veins, if 4 or 5 then 3 of them originating from one point **3.**

3. Fronds 0.8–3.0(–4.0) mm, usually elliptical, rarely oblong or obovate, with (0–)1 vein **1. minuta**
3. Fronds (1–)2–5(–8) mm, usually ovate, with 3(–5) veins **2. minor**

1. **L. minuta** Kunth

Least Duckweed

L. minuscula Herter nom. illegit.; *L. minima* Phil.; *L. valdiviana* auct.

Fronds 0.8–4.0 × 0.5–2.5 mm, floating on the surface of the water, usually pale green, more or less translucent, solitary or cohering in twos or threes, more or less flattened on both surfaces, usually elliptical, sometimes oblong or obovate, entire, often symmetrical along the long axis, usually with one short vein, sometimes without veins; upper surface often with a longitudinal ridge which forms a slight point at the obtuse apex and appears as a pale line to the naked eye; with a single root. *Flowers* as genus, but not recorded in the British Isles.

Introduced. First recorded in 1977 in Cambridgeshire and now known in ponds, canals, ditches and streams in southern Britain in scattered localities north to Norfolk and Flintshire and in the Channel Islands; probably much overlooked. Native of temperate North and South America.

2. **L. minor** L.

Common Duckweed

Lenticula minor (L.) Scop.; *Lenticula vulgaris* Lam.; *L. palustris* Haenke ex Mert. & Koch; *L. obcordata* Bojer; *L. ovata* A. Br. ex Krauss; *Lenticula cyclostasa* Kurz; *Lenticula minima* Kurz; *L. monorhiza* Montandon

Fronds 1–8 × 0.6–5.0 mm, floating on the surface of the water and forming a mat, deep green on upper surface, tinged purple or red beneath, commonly cohering in groups of 2–5, more or less flattened on both surfaces with air-spaces on the upper surface, visible as a reticulum, less than 0.3 mm across, subrotund to elliptic-obovate, entire, very nearly symmetrical; veins 3(–5); with a single root up to 15 cm. *Inflorescence* in hollows on the upper surface of the frond, the sac-like sheath open only at the top, with 2 male and 1 female flowers. *Fruit* broadly ovoid, wingless, projecting about one-third beyond the frond margin. *Seed* about 0.6 mm, obovate, flattened, smooth. *Flowers* 6–7, not uncommonly, usually in shallow ditches fully exposed to the sun. $2n = 40$.

Native. Ponds, lakes, ditches, canals and slow-flowing streams and rivers. Common throughout the British Isles except for north Scotland where it is rare. The most widely distributed of all duckweeds, found in almost all the tropical and temperate regions of the world and occurring in Alaska nearly to the Arctic Circle.

3. **L. gibba** L.

Fat Duckweed

Lenticula gibba (L.) Moench; *Lenticula gibbosa* Renault; *Telmatophace gibba* (L.) Schleid.; *L. tricorrhiza* Thuill. ex Schleid.; *Telmatophace gibbosa* (Renault) Montandon

Fronds 1–8 × 0.8–0.6 mm, floating on the surface of the water, mottled yellowish-green on upper surface, frequently strongly reddish-purple pigmented on both surfaces, solitary or commonly cohering in groups of 2–4, flat to convex on upper surface with air spaces, visible as a reticulum, more than 0.3 mm across, flat to convex or strongly inflated on lower surface, obovate then orbicular, entire; veins (3–)4–5(–7); with a single root. *Inflorescence* in hollows on the upper surface of the frond, enclosed in a sheath, with 2 male and one female flowers. *Fruit* ovate-cordate to obovate, winged, 1- to 3(–6)-seeded. *Seed* ribbed. *Flowers* 6–7, less frequently than in *L. minor*. $2n = 64$.

Winter fronds are not swollen beneath and are chiefly formed after flowering.

Native. Ponds, lakes, canals and ditches, usually in rich, often brackish waters. Frequent in central and southern Britain, scattered in the Channel Islands and Ireland. Widely distributed throughout the temperate and tropical regions of the world.

4. L. trisulca L.   Ivy-leaved Duckweed
Lenticula trisulca (L.) Scop.; *L. cruciata* Roxb.; *L. intermedia* Ruthe ex Schleid.; *Staurogeton trisulcus* (L.) Schur; *L. bisulca* Veesenm.

Fronds 3–15 (plus stalk 2–20) × 1–5 mm, floating beneath the surface of the water except when flowering, yellowish-green, usually connected in branched chains of 3–50, more or less flattened on both surfaces, air-spaces in centre of upper surface only, elongate-lanceolate, more or less acute and serrulate at apex, narrowed abruptly to the stalk, symmetrical; one vein prominent, 2 lateral ones indistinct; roots often lacking, but solitary when present; 2 young fronds arise on opposite sides of, at right angles to, and in the same plane as each old one; fertile fronds smaller than the sterile and producing stomata on the upper surface, not branching, usually 2–3 cohering. *Inflorescence* in hollows on the upper surface of the fronds, enclosed in a sheath, with 2 male and one female flowers. *Fruit* 1-seeded. *Seed* about 1 mm, 12- to 15-ribbed. *Flowers* 6–7, not very frequently. $2n = 40, 44$.

Native. Ponds, ditches, canals, lakes and slow-flowing rivers. Frequent in most of the British Isles, but scattered in Scotland. Common in temperate areas of the northern hemisphere; rare in the tropics.

3. Spirodela Schleid.

Plants floating on the water surface, 3–5 or more often remaining connected, often by elongated strips, with 2–16(–21) roots. *Fronds* 1.5–10.0 mm. *Inflorescence* borne in one of the two hollows on the upper surface of the frond, enclosed in a sheath, with 2(–3) male and one female flowers. *Fruit* broadly elliptical, slightly winged.

Five species, cosmopolitan except in South America.

1. S. polyrhiza (L.) Schleid.   Greater Duckweed
Lemna polyrhiza L.; *Lemna orbiculata* Roxb.; *Lemna orbicularis* Kit. ex Schult.; *Lemna thermalis* P. Beauv. ex Nutt.; *Lemna bannatica* Waldst. & Kit. ex Schleid.; *Telmatophace polyrhiza* (L.) Godr.; *Lemna major* Griff.; *Telmatophaca orbicularis* (Kit. ex Schult.) Schur; *Lemna transsylvanica* Schur; *Lemna umbonata* A. Br. ex Hegelm.; *S. atropurpurea* Montandon; *Lemna maxima* Blatt. & Halib.; *S. maxima* (Blatt. & Halib.) McCann

Fronds 3.0–10.0 × 2.5–8.0 mm, floating on the surface of the water, usually 2–5 remaining connected by elongated, peltately attached strips, flat and yellowish-green above, flat to slightly convex and usually purplish to red beneath, subrotund to obovate, slightly asymmetrical, with 5–15 prominent veins, the roots 5–12, fascicled, the fronds producing towards the end of summer purplish-brown, reniform buds 2–4 mm in diameter, which become detached and often sink, rising to the surface in spring. *Flowers* in July, but only recorded in Somerset. $2n = 40$.

Native. In still waters of ditches and ponds, local. Central and south Britain, very scattered in northern Britain and Ireland, formerly in Channel Islands. Europe, except the extreme north and south-west; Madeira; Africa; Asia; America; Australia; rather uncommon throughout its range.

Subclass 3. COMMELINIDAE Takht.

Herbs, often with rather large flowers never aggregated into a spadix.

Order 1. COMMELINALES Lindl.

Perennial herbs. Leaves entire with a sheathing base. *Flowers* bisexual, hypogynous, actinomorphic, few in terminal paired cymes. *Perianth segments* 6, the 3 outer sepaloid and free, the 3 inner petaloid and free. *Stamens* 6. *Ovary* 3-celled with 2 ovules in each cell. *Fruit* a capsule.

This order consists of 4 families totalling about 1,000 species of which 700 are in the Commelinaceae.

163. COMMELINACEAE R. Br. nom. cons.

Perennial herbs. Leaves alternate, entire, sheathing at base, exstipulate. *Flowers* few, in terminal paired cymes with a large leaf-like bract at the base of each, bisexual, hypogynous, actinomorphic. *Sepals* 3, green, free or united at base. Petals 3, white or coloured, free or united at base. *Stamens* 6; filaments long-hairy. *Style* 1; stigma capitate. *Ovary* 3-celled, with 2 ovules in each cell. *Fruit* a capsule.

Contains 38 genera and about 600 species. A mainly tropical, subtropical and warm temperate family represented in the British Isles only by ornamental species escaped from gardens.

Commelina coelestis Willd. and **C. diffusa** Burm. fil. occasionally occur as casuals or garden escapes.

1. Tradescantia L.
Zebrina Schnizl.

As family.

About 65 species found in North and South America.

1. Plant tufted or shortly creeping and ascending from a rhizomatous base; leaves linear-lanceolate, rarely variegated   **3. virginiana**
1. Plant creeping and rooting at the nodes; leaves ovate-oblong to broadly ovate or elliptical, often variegated   **2.**

2. Leaves typically green, silver-striped above, purplish beneath; petals pinkish-purple, united at base into a slender white tube   **1. zebrina**
2. Leaves not silver-striped but sometimes variegated white; petals white, free   **2. fluminensis**

1. T. zebrina Loudon   Striped Wandering Jew
Zebrina pendula Schnizl

Perennial herb. Stems up to 1 m or more, decumbent, branched, often purplish, creeping and rooting at the nodes. *Leaves* 2.5–10.0 × 1.5–3.5 cm, usually bluish-green on upper surface and often with silver stripes, more or less purplish below, ovate-oblong to broadly ovate or elliptical, acute to acuminate at apex, entire but sometimes wavy, rounded at the sessile, sheathing base, somewhat succculent, glabrous or with long, scattered simple eglandular hairs on the sheath. *Flowers* 10–15 mm in diameter, clustered between 2 unequal bracts. *Sepals* united into a whitish tube at base, with irregular green segments 5–8 mm. *Petals* united into a slender white tube at the base up to 10 mm; segments 5–12 × 3–7 mm, pinkish-purple. *Stamens* 6, equal, inserted at the throat of the corolla tube. *Stigma* capitate, 3-lobed. *Fruit* a capsule. Possibly *flowers* throughout the year. $2n = 24$.

Introduced. Much cultivated and found as a throw-out on dumps and in waste places where it may persist for a short time. Native of Mexico and widely naturalised in the tropics.

2. T. fluminensis Vell.   Wandering Jew
T. albiflora Kunth; *T. tricolor* auct.; *T. viridis* auct.

Perennial herb. Stems up to 1 m or more, procumbent or decumbent, rooting at the nodes. *Leaves* 2–10 × 1.0–3.5 cm, all green, or purplish beneath, or variegated with several cream or white longitudinal stripes, thin to fleshy, elliptical, ovate or ovate-oblong, acute at apex, asymmetrical and sometimes petiolate at base, sheathing, glabrous except for the sheath and margin. *Flowers* few, 10–15 mm in diameter, in paired cymes subtended by bracts; pedicels 10–20 mm, slender. *Sepals* up to 9 mm, white with a green midrib, lanceolate, obtuse at apex, glabrous except for the hairy midrib, free. *Petals* up to 12 mm, white, free, narrowly ovate, obtuse at apex. *Stamens* 6; filaments hairy; anthers orange. *Fruit* a capsule. *Flowers* throughout the year. $2n = 40, 50, 60, 108, 132, 140, 144$.

Introduced. Much grown as a pot plant, sometimes persisting in shrubberies and frost-free tips and waste ground where thrown out. Rare in southern England and the Channel Islands. Native of South America.

3. T. virginiana L.   Spiderwort
Incl. *T.* × *andersoniana* W. Ludw. & Rohwer nom. nud.

Tufted *perennial herb* with short rhizomes. *Stems* 30–60 cm, erect, glabrous or nearly so. *Leaves* 15–35 × 0.5–2.5 cm, green, slightly fleshy, linear-lanceolate, gradually narrowed to a more or less acute apex, entire, sheathing at the base, with a variable amount of short, simple eglandular hairs. *Flowers* numerous, in cymes subtended by bracts similar to the leaves but smaller; pedicels 20–30

mm. *Sepals* free, 10–15 mm, green, broadly ovate, obtuse to acute at apex, often with long simple eglandular hairs. *Petals* free, 15–20 × 10–15 mm, pale to deep violet, purplish or rarely white, broadly ovate, obtuse to acute at apex. *Stamens* 6; filaments pale to deep violet, hairy; anthers yellow. *Style* more or less violet, with a greenish-yellow stigma. *Fruit* a capsule. *Flowers* 7–9. $2n = 16$.

Included here are the common garden plants, many of which appear to be hybrids between *T. virginiana* and other species. The name *T.* × *andersoniana*, which is invalid, has been proposed for complex hybrids involving *T. ohiensis* Raf., *T. subaspera* Ker Gawl. and *T. virginiana*.

Introduced. Grown in gardens and occasionally persisting on tips and waste ground when thrown out. Rare in south-east England. Native of North America.

Order 2. ERIOCAULALES Nakai

Aquatic *perennial herbs. Leaves* appearing to be all basal, simple, subulate, entire. *Flowers* unisexual with the male flowers in the centre of the inflorescence and female flowers around them, hypogynous, actinomorphic, with bracteoles below each, in a terminal, whitish, capitate mass on a long leafless stem. *Perianth segments* 4, the 2 outer fused (male) or more or less free (female), the 2 inner free. *Stamens* 4. *Ovary* 2-celled, each cell with one ovule. *Fruit* a capsule. *Seeds* 2.

This order consists of a single family.

164. ERIOCAULACEAE P. Beauv. ex Desv. nom. conserv.

Perennial, monoecious, more or less aquatic *herbs* rooted in mud. *Stems* long and leafless. *Leaves* appearing to be all basal, subulate, entire, sessile, exstipulate. *Flowers* in terminal, bracteate, whitish heads, unisexual with male flowers in the centre of the inflorescence and female around them, hypogynous, actinomorphic, with bracteoles below each flower. *Perianth segments* 4, the 2 outer fused in the male flower or more or less free in the female flower and the 2 inner free, membranous and hairy or fringed. *Stamens* 4. *Style* 1; stigmas 2, filiform. *Ovary* 2-celled, each cell with 1 ovule. *Fruit* a capsule with 2 seeds.

About 13 genera and around 1,150 species. Distributed throughout most of the world except the mainland of Europe, few in temperate regions, frequent in swampy habitats, abundant in South America.

1. Eriocaulon L.

As family.

About 400 species with the distribution of the family.

1. E. aquaticum (Hill) Druce   Pipewort
E. septangulare With.; *Cespa aquatica* Hill

Slender, glabrous *perennial herb* with a creeping stock; roots white, soft and worm-like, with conspicuous striations. *Stems* up to 7–30(–150) cm according to depth of water, usually emergent, erect, twisted, yellowish-green, 6- to 8-furrowed, smooth, without leaves. *Leaves* all basal, 10–100 × 1.0–2.5 mm, deep green, narrowly linear-lanceolate, gradually narrowed to a finely acute apex, entire, clearly transversely septate, translucent. *Flowers* in terminal heads 5–12(–20) mm in diameter; bracts lead-coloured, obovate, obtuse at apex; bracteoles black, with scarious margins, oblanceolate, obtuse at apex. *Perianth segments* 4, the 2 male outer fused or the female more or less free, the 2 inner free, ovate, obtuse at apex, membranous, white and hairy or fringed at top, the male dull cream, the female greyish-black. *Stamens* 4; anthers purplish-black. *Stigmas* filiform. *Fruit* a membranous, brown capsule. *Flowers* 7–9. $2n = 64$.

Native. In shallow lakes and pools or in bare, wet, peaty ground sometimes forming dense mats. Very local in west Scotland and local in the west of Ireland from Cork to Donegal. Widely distributed in eastern North America.

Order 3. JUNCALES Lindl.

Annual or *perennial herbs* often more or less aquatic. *Leaves* grass-like to rush-like with a sheathing base. *Flowers* bisexual, hypogynous, actinomorphic, in simple to complex, often congested cymes which are terminal, but often appear lateral. *Perianth* of 3 inner and 3 outer segments, greenish, brownish or membranous. *Stamens* (3–)6. *Ovary* 1-celled with 3 ovules, 1-celled with many ovules on parietal placentas, or 3-celled with many ovules per cell on axile placentas. *Fruit* a capsule. *Seeds* 3–numerous.

This order consists of two families, the Juncaceae with about 400 species and the Thurniaceae with only three.

165. JUNCACEAE Juss. nom. conserv.

Annual or *perennial herbs*, often more or less aquatic, frequently tufted or with creeping, sympodial rhizomes. *Leaves* alternate, or all basal, simple, entire, grass-like (*bifacial*) to rush-like (*cylindrical* to flattened but *unifacial*), with sheathing base, sometimes reduced to scales, exstipulate; ligule membranous, at top of sheath. *Flowers* in various simple to complex, often congested cymes, which are terminal but often appear lateral, bisexual, hypogynous, actinomorphic, wind-pollinated. *Perianth segments* 6, 3 inner and 3 outer, greenish, brownish or membranous, free. *Stamens* (3–)6, free. *Style* 0 or 1; stigmas 3, linear. *Ovary* 1-celled with 3 ovules, 1-celled with many ovules on parietal placentas, or 3-celled with many ovules per cell on axile placentas. *Fruit* a loculicidal capsule. *Seeds* 3 to many, often with appendages; endosperm starchy.

Nine genera and about 400 species. Cosmopolitan, but mainly in temperate or cold climates or at high altitudes in the tropics.

Buchenau, F. (1906). Juncaceae in Engler, A. *Das Pflanzenreich* IV. 36. Leipzig.

1. Leaves bifacial or unifacial, glabrous; ovary with many ovules; capsule with many seeds   **1. Juncus**
1. Leaves bifacial, sparsely hairy at least near the base when young; ovary with 3 ovules; capsule with 3 seeds
    2. Luzula

1. Juncus L.

Juncastrum Heist.; *Juncinella* Fourr.; *Phylloschoenus* Fourr.; *Tenageia* (Ehrh.) Rchb.

Glabrous *annual* to *perennial herbs*, more or less tufted and often with rhizomes. *Stems* usually erect. *Leaves* various, bifacial to unifacial, the cauline usually with a slit, sheathing base which is often produced above into auricles; lamina channelled, compressed or more or less terete, sometimes wanting. *Inflorescence* a cluster of cymes, sometimes condensed into a head, terminal but in some species seeming lateral because the lowest bract appears to be a continuation of the stem. *Flowers* rarely few or solitary. *Ovary* 1- to 3-celled, with many ovules. *Capsule* many-seeded. *Seeds* sometimes with appendages; testa usually finally sculptured, often becoming mucilaginous.

About 300 species. Cosmopolitan.

Blackstock, T. H. (1986). Observations on the morphology and fertility of *Juncus × surrejanus* Druce ex Stace & Lambinon in north-western Wales. *Watsonia* 16: 55–63.
Cope, T. A. & Stace, C. A. (1978). The *Juncus bufonius* L. aggregate in western Europe. *Watsonia* 12: 113–128.
Cope, T. A. & Stace, C. A. (1983). Variation in the *Juncus bufonius* L. aggregate in western Europe. *Watsonia* 14: 263–272.
Grime, J. P. et al. (1988). *Comparative plant ecology*. London. [Accounts of *Juncus bufonius, bulbosus, effusus* and *squarrosus*.]
Jones V. & Richards, P. W. (1953). *Juncus acutus* L. in Biological flora of the British Isles. *Jour. Ecol.* 42: 639–650.
Richards, P. W. & Clapham, A. R. (1941). *Juncus conglomeratus* L. in Biological flora of the British Isles. *Jour. Ecol.* 29: 381–384.
Richards, P. W. & Clapham, A. R. (1941). *Juncus effusus* L. in Biological flora of the British Isles. *Jour.Ecol.* 29: 375–380.
Richards, P. W. & Clapham, A. R. (1941). *Juncus inflexus* L. in Biological flora of the British Isles. *Jour. Ecol.* 29: 369–374.
Richards, P. W. & Clapham, A. R. (1941). *Juncus* L. in Biological flora of the British Isles. *Jour. Ecol.* 29: 362–374.
Richards, P. W. & Clapham, A. R. (1941). *Juncus subnodulosus* Schrank in Biological flora of the British Isles. *Jour. Ecol.* 29: 385–391.
Richards, P. W. (1943). *Juncus filiformis* L. in Biological flora of the British Isles. *Jour. Ecol.* 31: 60–65.
Richards, P. W. (1943). *Juncus macer* S. F. Gray in Biological flora of the British Isles. *Jour. Ecol.* 31: 51–59.
Stace, C. A. (1972). The history and occurrence in Britain of hybrids in *Juncus* subgenus *Genuini*. *Watsonia* 9: 1–11.
Stewart, A. Pearman, D. A. & Preston, C. D. (1994). *Scarce plants in Britain*. Peterborough. [Accounts of *J. acutus, alpinoarticulatus, balticus, biglumis, castaneus* and *filiformis*.]
Welch, D. (1966). *Juncus squarrosus* L. in Biological flora of the British Isles. *Jour. Ecol.* 54: 535–548.
Willis, A. J. & Davies, E. W. (1960). *Juncus subulatus* Forsk. in the British Isles. *Watsonia* 4: 211–217.

1. Leaves with 2 faces, flat and more or less grass-like with 2 opposite surfaces but sometimes inrolled, or subcylindrical and more or less rush-like but with a

distinct channel on the upperside for most or all of
their length   2.

1. Leaves with one face or more or less so, or apparently
absent, cylindrical to flattened-cylindrical and rush-
like, not deeply channelled or with a deep channel
only near the ligule and extending less than halfway
to the leaf apex, sometimes with shallow grooves   12.

2. Easily uprooted annual, with simple, fibrous root
system   3.

2. Perennial with rhizomes, usually firmly rooted   6.

3. Stems unbranched, with basal leaves and leaf-like
bracts at the top but bare in between   **10. capitatus**

3. Stems branched, leafy but leaves often short and very
narrow   4.

4. Leaves usually more than 1.5 mm wide; perianth
segments usually with a dark line on either side of
the midrib; anthers 1.2–5.0 times as long as filaments;
seeds with rib-like longitudinal ridges   **6. foliosus**

4. Leaves rarely more than 1.5 mm wide; perianth
segments rarely with a dark line on either side of the
midrib; anthers usually 0.3–1.1 times as long as
filaments; seeds without longitudinal ridges   5.

5. Inner perianth segments more or less acute at apex;
capsule obtuse to acute at apex, rarely more or less
truncate, usually shorter than the inner perianth
segments   **7. bufonius**

5. Inner perianth segments more or less rounded to
emarginate and mucronate at apex; capsule truncate
at apex, at least as long as the inner perianth
segments   **8. ambiguus**

6. Outer perianth segments obtuse to rounded at apex   7.

6. Outer perianth segments acute to acuminate at apex   8.

7. Anthers 0.5–1.0 mm, 1–2 times as long as filaments;
seeds 0.3–0.5 mm   **3. compressus**

7. Anthers 1.0–2.2 mm, 2–3 times as long as filaments;
seeds 0.5–0.7 mm   **4. gerardii**

8. Stem with (1–)2–4 (–5) well-developed leaves, usually
near the base; lowest 2 bracts of inflorescence leaf-
like, usually far exceeding the inflorescence   9.

8. Stem with 0(–1) well-developed leaves; all bracts of
inflorescence mostly scarious, usually much shorter
than the inflorescence   11.

9. Inflorescence 1–3(–4) flowers in a tight cluster;
anthers longer than filaments; seeds 0.9–1.0 mm, with
a long appendage at each end   **5. trifidus**

9. Inflorescence of 5–40 flowers, usually more or less
diffuse; anthers shorter than filaments; seeds 0.3–0.4
mm, with short appendages   10.

10. Leaves about as long as stem; auricles long, obtuse,
thin and whitish   **2(i). tenuis** var. **tenuis**

10. Leaves about one-third as long as stem; auricles short,
cartilaginous and yellowish   **2(ii). tenuis** var. **dudleyi**

11. Leaves rounded on lower side, with a deep channel
on upper side; stamens 6.   **1. squarrosus**

11. Leaves flat or more or less inrolled; stamens 3
   9. planifolius

12. Leaves on stem represented near the base only by
blade-less scarious sheaths without a lamina   13.

12. At least one stem-leaf with a well-developed green
lamina   21.

13. Stems strongly glaucous, with pith conspicuously and
regularly interrupted at least in the region just below
the inflorescence   **25. inflexus**

13. Stems not glaucous, with pith well-formed and
conspicuous or ill-formed and irregular   14.

14. Rhizomes extended, forming straight lines or diffuse
patches of aerial stems; stems rarely more than 50 cm;
inflorescence usually less than 20-flowered   15.

14. Rhizomes short, forming dense clumps of aerial
stems or large dense patches in very old plants; stems
usually more than 50 cm; inflorescence usually more
than 20-flowered   16.

15. Stem without subepidermal sclerenchyma girders,
not ridged when dry; inflorescence in upper quarter
of apparent stem, usually elongated   **23. balticus**

15. Stem with subepidermal sclerenchyma girders, with
fine longitudinal ridges when dry; inflorescence in
lower two-thirds to three-quarters of apparent stem,
more or less globose   **24. filiformis**

16. Fresh stem dull, ridged, with usually less than 35
ridges; main stem-like bract opened out and more or
less flat, adjacent to inflorescence and causing it to
hinge over backwards at the end of the season   17.

16. Fresh stems glossy, smooth, becoming finely ridged,
with usually more than 35 ridges when dry; main
bract scarcely opened out, adjacent to inflorescence
and not hinging over at end of season   18.

17. Inflorescence a compact, rounded head
   27(i). conglomeratus var. **conglomeratus**

17. Inflorescence of several stalked heads
   27(ii). conglomeratus var. **subuliflorus**

18. Larger stems usually more than 1.2 m; stamens 6;
capsule as long as or longer than perianth segments,
2.5–3.5 mm when fully fertile   **28. pallidus**

18. Stems usually less than 1.2 m; stamens 3(–6); capsule
shorter than perianth segments, 2.0–2.5 mm when
fully fertile   19.

19. Stems spiral and more or less spreading or flattened
   26(iii). effusus var. **spiralis**

19. Stems straight and erect   20.

20. Inflorescence lax, with suberect to widely divergent
branches   **26(i). effusus** var. **effusus**

20. Inflorescence with a single, rounded, compact head
   26(ii). effusus var. **subglomeratus**

21. Leaves and main bract with a very sharp apex, with
subepidermal sclerenchyma girders and vascular
bundles scattered through the pith   22.

21. Leaves and main bract with a soft apex, without
subepidermal sclerenchyma girders, without vascular
bundles in pith   26.

22. Inner perianth segments obtuse, the extreme apex
not exceeded by membranous margins; capsule
2.5–3.5 mm, not or slightly longer than the perianth
segments   23.

22. Inner perianth segments retuse, with membranous
margins extended into lobes on each side of the
extreme apex and exceeding it; capsule 4–6 mm,
much longer than the perianth segments   25.

23. Panicle a compact, orbicular head
   20(i). maritimus var. **congestus**

23. Panicle lax   24.

24. Plant robust; panicle exceeded by lowest bract
    20(ii). maritimus var. maritimus
24. Plant tall and weak; lowest bract rarely more than
quarter the length of the panicle
    20(iii). maritimus var. atlanticus
25. Inflorescence a dense, more or less rounded head
    21(i). acutus var. acutus
25. Inflorescence more elongated and open
    21(ii). acutus var. effusus
26. Leaves cylindrical, with continuous pith within the
vascular cylinder; each flower with 2 small bracteoles
immediately beneath the perianth segments **22. subulatus**
26. Leaves cylindrical to flattened-cylindrical, with the
pith usually interrupted by transverse septa and often
with large cavities; perianth segments without
bracteoles immediately beneath them   27.
27. Easily uprooted annual, with a simple fibrous root
system; found only in Cornwall   **16. pygmaeus**
27. Perennial, usually firmly rooted, either with rhizomes
or swollen stem-bases   28.
28. Anthers less than one-third as long as filaments;
seeds with conspicuous whitish appendages at each
end, each about as long as actual seed; alpine
                                                29.
28. Anthers more than one-third and up to twice as long
as filaments; seeds with at most minute points at each
end; lowland or alpine   31.
29. Stems usually solitary from the rhizome system;
outer perianth segments acute at apex; capsule
usually more than 6 mm   **19. castaneus**
29. Stems usually in small tufts; outer perianth segments
obtuse to rounded at apex; capsule less than 6 mm   30.
30. Lowest bract usually exceeding inflorescence; flowers
mostly 2 per inflorescence; capsule 3–4 mm, retuse at
apex   **17. biglumis**
30. Lowest bract usually shorter than inflorescence:
flowers mostly 3 per inflorescence; capsule 4–6 mm,
obtuse at apex   **18. triglumis**
31. Leaves with more than 2 empty or pith-filled
longitudinal cavities separated by thin walls bearing a
few vascular bundles   32.
31. Leaves with one empty or loosely pith-filled
longitudinal cavity   33.
32. Plant rhizomatous; flowers very rarely vegetatively
proliferating; outer perianth segments obtuse,
incurved at apex   **11. subnodulosus**
32. Plant not rhizomatous; stem base usually swollen;
flowers commonly vegetatively proliferating; outer
perianth segments acute, not incurved   **15. bulbosus**
33. Perianth segments acuminate at apex, the outer with
recurved apical points   **14. acutiflorus**
33. Outer perianth segments obtuse to shortly acute at
apex with erect apical points; inner perianth segments
acute to rounded at apex   34.
34. Outer perianth segments subacute to obtuse at apex;
inner perianth segments obtuse to rounded at apex;
capsule obtuse (ignore beak)   35.
34. Outer perianth segments acute at apex; inner
perianth segments acute to subacute at apex; capsule
acute (ignore beak)   36.
35. Flowers subsessile
    12(i). alpinoarticulatus var. alpinoarticulatus

35. Flowers partly sessile, partly on pedicels up to 6 mm
    12(ii). alpinoarticulatus var. marshallii
36. Stems up to 80 cm, thick and rigid; inflorescence
usually large with numerous heads
    13(iii). articulatus var. articulatus
36. Stems less than 30 cm, thick or slender; inflorescence
small and contracted with few heads   37.
37. Stems spreading or ascending, in dense fasciculate
clumps, sometimes rooting at the nodes, thick; leaves
attenuate at apex (coastal sands)
    13(i). articulatus var. littoralis
37. Stems erect, rather slender; leaves shortly acute at
apex (mountains)   **13(ii). articulatus var. nigritellus**

Subgenus 1. Poiophylli Buchenau

Annual or *short-lived perennial herbs. Leaves* all cauline,
those at the base often withered by flowering time, flat or
inrolled, not sharply pointed, with one subepidermal
sclerenchyma girder at each margin, with cavities developing between the vascular bundles but without pith.
Inflorescence terminal, usually very diffuse and occupying most of the plant, interspersed with more or less
leaf-like bracts. *Flowers* with 2, small bracteoles. *Seeds*
without appendages. Cleistogamous or wind-pollinated.
Seeds mucilaginous and easily carried on the feet of
animals.

1. J. squarrosus L.   Heath Rush
J. squamosus Link; *J. sprengelii* Willd.; *Tenageia
squarrosa* (L.) Fourr.

Glabrous *perennial herb*, densely tufted to mat-forming,
with a short, suberect, branched rhizome. *Stems* 15–50
cm, erect, rigid, rarely with a single leaf or a leafless
sheath. *Leaves* mostly basal, 7–30 cm × 1–2 mm, yellowish-green, subcoriaceous, long and narrow, acute but not
sharply pointed at apex, channelled on upper side, with
auricles of varying length. *Inflorescence* with 10–30(–40)
flowers; branches more or less erect; lowest bract much
shorter than inflorescence. *Perianth segments* subequal,
4–7 mm, dark brown with a whitish hyaline margin; outer
narrowly ovate, more or less acute at apex; inner lanceolate, more or less obtuse at apex. *Stamens* 6, usually about
half as long as the perianth segments; anthers 1.5–2.0
mm, 2.5–6.0 times as long as the filaments. *Capsule* more
or less equalling the perianth segments, ovoid to ellipsoid,
obtuse–mucronate at apex. *Seeds* 0.6–0.8 mm, obliquely
ovoid, with conspicuous, verrucose striae; appendages
absent. *Flowers* 6–7. $2n = 42$.

Native. Bogs, wet moors and heaths on acid soils.
Common throughout the British Isles where there are
acid soils, but absent from much of central Ireland and
central England. West-central and northern Europe,
from the mountains of south Spain and northern Italy
to 68° 15' N in Norway, east to the Ukraine and 41° E
in north-west Russia; Morocco; Iceland; southern
Greenland.

2. J. tenuis Willd.   Slender Rush
J. gracilis Sm., non Roth; *J. macer* Gray; *J. gesneri* Sm.
nom. illegit. *J. bicornis* Michx; *J. aristatus* Link; *J.
chloroticus* Schult.; *J. smithii* Kunth; *J. lucidus* Hochst.

Glabrous *perennial herb* with stems in dense tufts. *Stems* 10–80 cm, erect, rigid to weak, slightly compresssed, with a few basal sheaths and 2–3 basal or lower cauline leaves. *Leaves* from one-third as long to about as long as stems, 0.5–2.0 mm wide, pale green, long and narrow, acute but not sharply pointed at apex, flat, often conspicuously convolute, often curved; auricles long, obtuse, thin and whitish or short, cartilaginous and yellowish. *Inflorescence* usually lax, rarely with the flowers in a few clusters, with 5–40 flowers; lowest 1(–2) bracts usually much longer than inflorescence; branches more or less erect. *Perianth segments* subequal, 3–4 mm, greenish, becoming straw-coloured, narrowly ovate, long-acute at apex. *Stamens* 6, half as long as the perianth segments; anthers 0.7–0.8 mm, half to two-thirds as long as the filaments. *Capsule* shorter than the perianth segments, broadly ovoid, obtuse to truncate and shortly mucronate at apex. *Seeds* 0.3–0.4 mm. *Flowers* 6–9. Germination in spring.

(i) Var. **tenuis**
Leaves about as long as the stem; auricles long, obtuse, thin and whitish. $2n = 84$.

(ii) Var. **dudleyi** (Wiegmann) F.J. Herm.
J. dudleyi Wiegmann
Leaves about one-third as long as the stem; auricles short, cartilaginous and yellowish. $2n = 84$.

Introduced. Damp, rather bare ground on roadsides, tracks and paths and in waste ground. Var. *tenuis* was first recorded in 1795/6 and is now locally frequent throughout most of the British Isles. It is native of North and South America. Var. *dudleyi* is naturalised in two localities in Perthshire and is native of temperate North America.

3. J. compressus Jacq.　　　　Round-fruited Rush
J. bulbosus subsp. *compressus* (Jacq.) Syme

Glabrous *perennial herb*, loosely tufted, with a creeping horizontal rhizome and sometimes forming extensive patches. *Stems* 10–50 cm, erect, usually slightly compressed, often curved, with 0–3 basal sheaths and 1–4 basal and 1–2 upper cauline leaves. *Leaves* 5–25 cm × 0.8–2.0 mm, glaucous, long and narrow, acute but not sharply pointed at apex, flat or inrolled, rarely channelled, with obtuse auricles. *Inflorescence* usually lax, with 10–60 flowers; lowest bract usually exceeding inflorescence; branches more or less erect. *Perianth segments* equal, 1.5–3 mm, chestnut-brown with a broad, dull green midrib and a hyaline margin, ovate, rounded-obtuse at apex. *Stamens* 6, half to two-thirds as long as the perianth segments; anthers 0.5–1.0 mm, 1–2 times as long as the filaments. *Style* shorter than the ovary. *Capsule* up to 1.5 times as long as the perianth segments, globose to ovoid, often slightly trigonous at apex, mucronate. *Seeds* 0.3–0.5 mm, with about 20 striae; without appendages. *Flowers* 6–7. $2n = 44$.
　　Native. Marshes, alluvial meadows and grassy places where the vegetation is kept low by mowing or grazing, chiefly on non-acid soils and often near the sea.

Scattered in England and Wales, mostly in central and east England. Eurasia except the Arctic, and North America.

4. J. gerardii Loisel.　　　　Saltmarsh Rush
J. bottnicus Wahlb.; *J. consanguineus* Ziz; *J. coenosus* Bicheno; *J. attenuatus* Viv.; *J. nitidiflorus* Dufour; *Tenageia gerardi* (Loisel.) Fourr.

Glabrous *perennial herb*, tufted or with a far-creeping rhizome. *Stems* 5–50 cm, usually slender, sometimes compressed, stiffly erect, triquetrous above, with 0–2 basal sheaths, 4–5 basal and 0–2 upper cauline leaves. *Leaves* 2–30 cm × 0.5–2.5 mm, long and narrow, acute but not sharply pointed at apex, flat to subterete; auricles obtuse. *Inflorescence* lax, with 20–50 flowers; lowest bract usually shorter than inflorescence; branches more or less erect. *Perianth segments* equal, 2.5–4.0 mm, usually dark brown, ovate, obtuse at apex. *Stamens* two-thirds as long as to almost as long as the perianth segments; anthers 1.0–2.2 mm, 2–3 times as long as the filaments. *Styles* equalling or slightly longer than the ovary. *Capsule* about equalling the perianth segments, ovoid to broadly ellipsoid, usually trigonous at the obtuse, mucronate apex. *Seeds* 0.5–0.7 mm, striate; without appendages. *Flowers* 6–7. $2n = 84$.
　　Native. In salt-marshes, from just above the high-water mark of spring tides upwards, abundant and locally dominant; occasionally in saline sites inland. All round the coasts of the British Isles. Coast and inland salt areas of Eurasia; North Africa; North America. The plant of the British Isles is subsp. **gerardii**, which occurs almost throughout the range of the species. Other subspecies occur in northern, southern and eastern Europe.

5. J. trifidus L.　　　　Three-leaved Rush
Glabrous *perennial herb*, densely tufted to mat-forming, with stems in very dense rows from a branching rhizome. *Stems* 5–40 cm, slender, rigid, erect, terete, finely striate, with 4–6 basal sheaths, rarely with one lower cauline leaf, and 2–4 upper leaves. *Leaves* 2–10 cm, filiform, finely serrulate; the upper usually appearing as bracts of the inflorescence; auricles lacerate. *Inflorescence* of one terminal and 0–3 lateral, sessile or shortly stalked flowers, much exceeded by the lowest bract and 1–2 upper cauline leaves. *Perianth segments* 2–5 mm, equal or subequal, dark brown, lanceolate to ovate, acute to apiculate and the outer sometimes aristate at apex. *Stamens* 6, two-thirds to three-quarters as long as the perianth segments; anthers 1.2–1.5 mm, about twice as long as the filaments. *Capsule* exceeding the perianth segments, trigonous-ovoid, dark brown and mucronate at apex. *Seeds* 0.9–1.3 mm, irregular and variable in form, with unequal appendages. *Flowers* 6–8. $2n = 30$.
　　Native. Detritus, rock ledges and other rather bare places on high mountains, where it is sometimes the most abundant plant over large areas. Locally frequent in central and west Scotland and at a lower altitude in Shetland. High mountains of Europe from central Spain to Caucasus and eastern Siberia, and in the

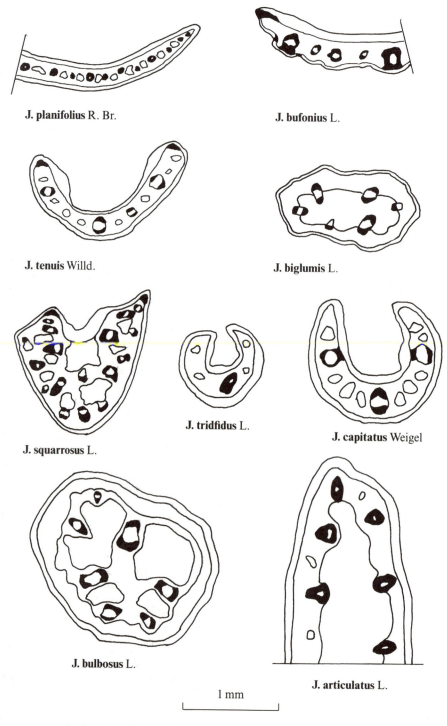

J. planifolius R. Br.   **J. bufonius** L.

J. tenuis Willd.   **J. biglumis** L.

J. tridfidus L.

J. squarrosus L.   **J. capitatus** Weigel

J. bulbosus L.

J. articulatus L.

1 mm

Leaf-sections of **Juncus** L. Sclerenchyma in black. (After C. A. Stace, 1991.)

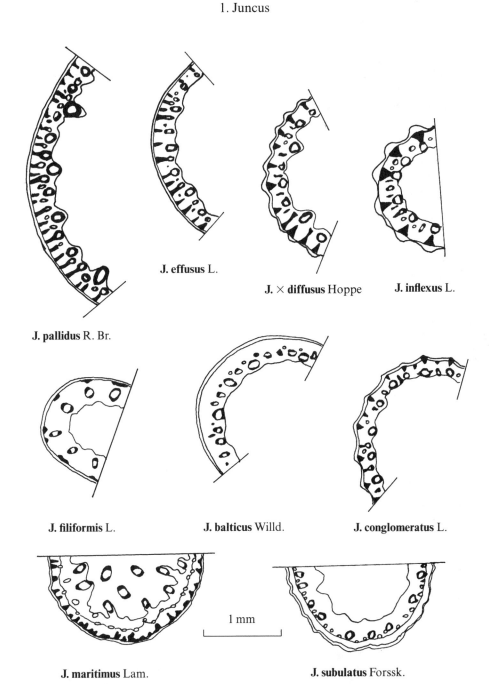

J. effusus L.

J. × **diffusus** Hoppe **J. inflexus** L.

J. pallidus R. Br.

J. filiformis L. **J. balticus** Willd. **J. conglomeratus** L.

1 mm

J. maritimus Lam. **J. subulatus** Forssk.

Leaf-sections of **Juncus** L. (main bract in cases of **J. pallidus, effusus,** × **diffusus, inflexus, filiformis, balticus** and **conglomeratus**). Sclerenchyma in black. (After C. A. Stace, 1991.)

Subarctic and Arctic areas to 71° 30' N; North America. Our plant is subsp. **trifidus** which occurs throughout most of the range of the species. Another subspecies occurs in the Alps and Apennines.

Subgenus 2. **Pseudotenageia** V. I. Krecz. & Gontsch.

Perennial herbs with rhizomes, sometimes densely tufted. *Leaves* basal and cauline, flat or strongly channelled, not sharply pointed, with one subepidermal sclerenchyma girder at each margin, with cavities developing between the vascular bundles but without pith. *Inflorescence* terminal, more or less compact, exceeded or not by 1–2, leaf-like or scale-like bracts. *Flowers* with 2 small bracteoles. *Seeds* with or without terminal appendages.

6. J. foliosus Desf.   Leafy Rush
J. bufonius var. *major* Boiss.; *J. bufonius* var. *foliosus* (Desf.) Buchenau; *J. riphaenus* Pau & Font Quer; *J. sphaerocarpus* var. *riphaenus* (Pau & Font Quer) Maire; *J. bufonius* subsp. *foliosus* (Desf.) Maire & Weiller

Glabrous *annual* or short-lived *perennial herb*. *Stems* 10–50 cm, densely tufted, erect or ascending from a decumbent base, with 1–2 large leaves below the inflorescence, the lower bracts leaf-like. *Leaves* all cauline, those at the base often withered at the time of flowering, up to 20 cm × 1.5–5 mm, pale green, long and narrow, acute but not sharply pointed at apex, flat; stomata 31–45 μm. *Inflorescence* often occupying most of the stem, many-flowered, lax; branches more or less straight, often widely diverging or almost horizontal; flowers 1–3(–5) per ultimate branch. *Perianth segments* unequal, usually with dark brown lines between the herbaceous centre and the scarious margin; outer 4.6–6.8 mm, narrowly ovate, apiculate at apex; inner 3.6–5.4 mm, lanceolate to narrowly ovate, subacute to apiculate at apex. *Stamens* 6, half to two-thirds as long as the perianth segments; anthers 1.2–2.0 mm, usually 1.2–5.0 times as long as the filaments. *Capsule* about equalling the inner perianth segments, obtusely angled, obtuse at apex, with a raised style-base. *Seeds* 430–600 × 270–400 μm, obovoid, often truncate at one end and tapered at the other, interstices of testa large, about 60 × 20 μm, with 20–30 pronounced ribs and weaker transverse striae. *Flowers* 5–9. 2n = 26.

Native. Muddy margins of pools, ponds, lakes, streams and rivers, in wet fields and marshes, in roadside ditches and on waste land. South and west Britain and widespread in Ireland. South and west Europe; western North Africa; Madeira.

7. J. bufonius L.   Toad Rush
J. divaricatus Gilib.; *J. prolifer* Humb., Bonpl. & Kunth; *J. bufonius* var. *congestus* Wahlb.; *J. bufonius* var. *gracilis* St Amans; *J. inaequalis* Willd.; *J. bufonius* var. *grandiflorus* Schult. & Schult. fil.; *J. bufonius* var. *fasciculatus* Koch; *J. dregeanus* C. Presl; *J. bufonius* var. *parvulus* Hartm.; *J. bufonius* var. *longiflorus* Kit.; *J. bufonius* var. *alpinus* Schur; *J. bufonius* var. *compactus* Čelak.; *J. bufonius* var. *laxus* Čelak.; *Tenageia bufonius* (L.) Fourr.; *J. bufonius* var. *jadarensis* Bryhn; *J.*

bufonius var. *pumilio* Griseb.; *J. bufonius* var. *leucanthus* Asch. & Graebn.; *J. bufonius* var. *subauriculatus* Buchenau; *J. bufonius* forma *minutulus* Albert & Jahand.; *J. minutulus* V. I. Krecz. & Gontch.; *J. bufonius* subsp. *minutulus* (V. I. Krecz. & Gontch.) Soó

Glabrous *annual herb*. *Stems* 5–50 cm, tufted or solitary, erect or ascending from a decumbent base, with 1–5 leaves below the inflorescence, the lower bracts leaf-like. *Leaves* all cauline, those at the base often withering by flowering time, 0.5–2.0 cm × less than 1 mm, dark green, long and narrow, acute but not sharply pointed at apex, flat or subterete; stomata 29–47 μm. *Inflorescence* lax, rarely with some flowers in clusters; branches usually straight, diverging at less than 90°; flowers 1–5 per ultimate branch. *Perianth segments* usually without dark lines; outer 4.1–7.3 mm, usually herbaceous with a narrow, scarious margin, narrowly ovate, acute or shortly acuminate at apex; inner 3.4–5.8 mm, narrowly ovate, more or less acute at apex. *Stamens* 6, one-third to half as long as the perianth segments; anthers 0.5–1.0 mm, two-fifths as long as the filaments. *Capsule* 3.1–4.9 mm, usually shorter than the inner perianth segments, ovoid to almost ellipsoid, acute to obtuse at apex, very variable in colour, the style usually persisting as a short beak. *Seeds* 340–520 × 210–350 μm, obliquely obovoid, rarely ellipsoid or ovoid, interstices of testa small, about 15 × 5 μm or perfectly smooth. *Flowers* 8–9. 2n = 108 (54, 60, 70, 72, 80, 104–110).

Extremely variable, but although numerous infraspecific taxa have been named they seem to have no geographical or ecological significance.

Native. Wherever the water-table is high, at least seasonally, on muddy, sandy or gravelly margins of ponds, lakes, streams and rivers, in marshes and less frequently in acid bogs. It is frequent in brackish conditions such as estuarine mud- and sand-flats, dune-slacks and the margins of saline or brackish lakes. It also grows in bare patches among crops, on or by paths and tracks, in wheel-ruts and drainage ditches and on waste ground. Common throughout the British Isles. Cosmopolitan, but probably native only in Eurasia, North Africa and North America.

8. J. ambiguus Guss.   Frog Rush
J. ranarius Songeon & Perrier; *Tenageia ranaria* (Songeon & Perrier) Fourr.; *J. bufonius* var. *halophilus* Fernald & Buchenau; *J. bufonius* var. *ambiguus* (Guss.) Husn.; *J. bufonius* subsp. *ambiguus* (Guss.) Schinz & Thell.; *J. bufonius* subsp. *ranarius* (Songeon & Perrier) Hiitonen.; *J. juzepczkii* V. I. Krecz. & Gontsch.

Glabrous *annual herb*. *Stems* usually many, 3–20 cm, erect or ascending from a decumbent base, with 1–3 leaves below the inflorescence, the lower bracts long and leaf-like. *Leaves* all cauline, those at the base often withering by flowering time, 1–10 cm × 0.5–1.0 mm, dark green, long and narrow, acute but not sharply pointed at apex, flat with convolute margins or subterete; stomata 24–36 μm. *Inflorescence* lax, or usually with some flowers in capitate clusters; branches scorpioid with the ultimate 2 or 3 flowers on each close together; flowers

2–4(–5) per ultimate branch. *Perianth segments* without dark lines; outer 4.0–6.8 mm, narrowly ovate, acute at apex; inner 3.3–5.3 mm, mostly scarious, lanceolate, more or less obtuse or rounded and sometimes emarginate at apex. *Stamens* 6, one-third to half as long as the perianth segments; anthers 0.4–0.8 mm, around half as long as the filaments. *Capsule* 3.3–5.3 mm, equalling or slightly shorter than inner perianth segments, obtusely angled, obtuse to more or less truncate at apex, the style usually deciduous. *Seeds* 330–440 × 250–350 μm, broadly ovoid to ellipsoid, interstices of testa small, about 15 × 5 μm, smooth. *Flowers* 8–9. Flowers often cleistogamous. $2n = 34$ (30, 32, 60).

Native. Typically found on the coast on sand- and mud-flats above the high-water mark and on the margins of saline and brackish lakes. It also occurs on bare mud and waste ground associated with inland salt-flushes and salt-workings and on the highly basic substrate provided by lime-waste tips. Scattered in suitable habitats throughout the British Isles. Over much of Europe and parts of North Africa, Asia and North America.

Subgenus 3. Graminifolii Buchenau

Tufted *perennial herbs* with short rhizomes. *Leaves* basal, not sharply pointed, without subepidermal sclereanchyma girders, with cavities developing between vascular bundles but without pith. *Inflorescence* terminal, rather compact, with a short main bract. *Flowers* without bracteoles. *Seeds* without appendages.

9. J. planifolius R. Br.   Broad-leaved Rush
J. homalophyllus Steud.

Tufted *perennial herb*, often flowering in the first year. *Stems* 10–30(–60) cm, striate, covered with minute, sessile glands, usually with about 10 basal leaves. *Leaves* 4–20 cm × 2–8 (–10) mm, much shorter than stems, pale green, long-linear, thin, translucent, flat becoming slightly involute above and gradually narrowed to the minutely mucronate apex, the sheathing base suffused with pink or red, without auricles. *Inflorescence* with several, unequal main branches and (1–)4–12(–50), dense, rounded heads of 8–10 flowers, much longer than the narrow, leaf-like lowest bract. *Perianth segments* 1.5–3.0 mm, subequal, dark brown with hyaline margins, ovate to lanceolate, the outer acuminate at apex, the inner obtuse. *Stamens* 3, almost as long as the perianth segments; anthers one-third as long as the filaments. *Capsule* exceeding the perianth segments, dark brown, trigonous-obovoid, with a mucro about 0.5 mm at the apex. *Seeds* 0.3–0.4 mm, broadly ellipsoid, without appendages. *Flowers* 7–8.

Introduced. By lakes and streams, in wet meadows and along paths. First discovered in 1971, it is locally abundant over an area of about 40 sq km near Carna and Cashel in Galway. Native of Australia, New Zealand and South America.

Subgenus 4. Juncinella V.I. Krecz. & Gontsch.

Annual herbs. *Leaves* all basal, flat to strongly channelled, not sharply pointed, without subepidermal sclerenchyma girders, with cavities developing between vascular bundles but without pith. *Inflorescence* terminal, very compact, exceeded or not by 1–2 main bracts. *Flowers* without bracteoles. *Seeds* more or less without appendages.

10. J. capitatus Weigel   Dwarf Rush
J. ericetorum Pollich; *Schoenus minimus* T. F. Forst.; *Scirpus michelianus* Gouan; *J. gracilis* Roth; *J. tenellus* Geuns; *J. triandrus* Gouan; *J. stellatus* Roem. & Schult.; *Juncinella capitata* (Weigel) Fourr.; *J. supinus* Bicheno nom. illegit.

Small, tufted, glabrous *annual herb*. *Stems* solitary or several, 1–5(–20) cm, stiffly erect, becoming reddish in fruit, with several basal leaves. *Leaves* 0.5–5 cm × 0.5–1.0 mm, green, setaceous, flat or channelled, with a short, brown, sheathing base; auricles absent. *Inflorescence* of 2–10 sessile flowers in a terminal head, rarely with 1–2 lateral heads, with 2 leaf-like bracts, the longer much exceeding the inflorescence. *Perianth segments* unequal, the outer (2–)3–5 mm, greenish, becoming reddish-brown, ovate or ovate-lanceolate and with curved, acuminate apex, the inner shorter, almost entirely scarious and acute to acuminate at apex. *Stamens* 3, half to two-thirds as long as the perianth segments; anthers 0.2–0.7 mm, about half as long as the filaments. *Style* short or long; stigmas short, directed downwards. *Capsule* (1–) 1.5–2.0 mm, usually much shorter than the perianth segments, pale brown to reddish-brown, ovoid to globose, obtuse at the sometimes shortly mucronate apex. *Seeds* about 0.3 mm, ovoid or obovoid, about twice as long as wide. *Flowers* 3–6. Flowers of 2 kinds, short-styled and cleistogamous or long-styled and wind-pollinated. $2n = 18$.

Native. Rather bare ground on heaths especially where water has stood during the winter. Very local in Cornwall and the Channel Islands; formerly in Anglesey. South and west Europe and sparingly through central Europe to south Sweden, Finland and northwest Russia, east to Lower Don; Africa; Newfoundland; South America; Australia.

Subgenus 5. Septati Buchenau

Annual or *perennial herbs* with or without rhizomes. *Leaves* all cauline or some basal, terete or flattened-terete, not sharply pointed, with 1–several central cavities, divided by transverse septa, often visible externally, without subepidermal sclerenchyma girders. *Inflorescence* terminal, very compact to very diffuse, with a more or less leaf-like bract usually shorter than it. *Flowers* without bracteoles. *Seeds* without appendages.

11. J. subnodulosus Schrank   Blunt-flowered Rush
J. obtusiflorus Ehrh. ex Hoffm.; *J. articulatus* var. *sylvaticus* L.; *J. sylvaticus* (L.) Reichardt, non Huds.; *J. divaricatus* Davies nom. in syn.; *J. retroflexus* Rafn; *J. divergens* Koch; *J. neesii* F. Heller; *J. aquaticus* Seb. & Mauri; *J. obtusatus* Kit.; *Phylloschoenus obtusiflorus* (Ehrh. ex Hoffm.) Fourr.

Glabrous *perennial herb* with thick, creeping rhizome, growing in extensive patches but not densely tufted.

Stems 40–130 cm, bright green, terete, smooth, hollow, with 3–4 basal sheaths and 1–2 cauline leaves; non-flowering shoots short, with one long leaf. *Leaves* 20–120 cm × 2–4 mm, bright green, pluritubulose with a wider central tube, a central vascular bundle, and very distinct transverse septa; auricles short, wide and firm. *Inflorescence* with 10–50(–100) heads, diffuse, with very widely spreading to more or less reflexed branchlets; heads hemispherical to subglobose with 5–10(–30) flowers. *Perianth segments* 1.8–2.5 mm, pale brown with paler margins, equal, elliptical, obtuse at apex, the outer boat-shaped. *Stamens* 6, two-thirds to three-quarters as long as the perianth segments; anthers 0.8–1.1 mm, 1.5–2.0 times as long as the filaments. *Capsule* slightly exceeding the perianth segments, pale brown, trigonous-ovoid, compressed laterally, attenuate and mucronate at the apex, 3-locular. *Seeds* 0.5–0.6 mm, pyriform, markedly reticulate with about 30 longitudinal ridges. *Flowers* 7–9. Germinating in spring. $2n = 40$.

Native. Fens, marshes and dune-slacks on peaty base-rich soils. Locally frequent in England, Wales and Ireland, in some fens dominant over a large area, very local in Jersey and south-west and central-east Scotland. West, central and south Europe, northwards to southern Sweden and eastwards to the Black Sea; Kurdistan; North Africa.

12. J. alpinoarticulatus Chaix   Alpine Rush

Glabrous *perennial herb* with a horizontally creeping rhizome. *Stems* 20–40(–60) cm, smooth, usually terete and rather weak, erect to ascending, often more or less tufted, with 0–1(–2) basal sheaths and (2–)3(–5) cauline leaves. *Leaves* 4–35 cm × 0.5–1.0 mm, very slender, nearly terete but channelled on the upper side, much shorter than the inflorescence, with very distinct transverse septa. *Inflorescence* very variable in form and size, narrow to wide, with more or less erect, straight or curved branches, with 5–25 heads. *Perianth segments* 1.8–2.5 mm, dark brown to blackish, equal or subequal, ovate or narrowly oblong, the inner broadly obtuse to rounded at apex, the outer obtuse to subacute and often mucronate. *Stamens* 6, half to two-thirds as long as perianth segments; anthers 0.4–0.7 mm, half to two-thirds as long as the filaments. *Capsule* equalling or exceeding the perianth segments, ovoid, obtuse and shortly mucronulate at apex. *Seeds* 0.5–0.6 mm, ovoid. *Flowers* 7–9. $2n = 40$.

(i) Var. alpinoarticulatus
J. alpinus Vill. nom. illegit.
Stems slightly tufted. *Inflorescence* branches straight. *Flowers* subsessile. *Perianth segments* ovate. *Capsule* blackish, ovoid, slightly exceeding the perianth.

(ii) Var. marshallii (Pugsley) P. D. Sell
J. marshallii Pugsley; *J. alpinus* var. *marshallii* (Pugsley) Lindq.; *J. nodulosus* var. *marshallii* (Pugsley) P. W. Richards
Stems more tufted than var. *alpinoarticulatus*. *Inflorescence* branches often curved. *Flowers* partly sessile, partly on pedicels up to 6 mm. *Perianth segments*

oblong. *Capsule* dark reddish-brown, obovoid, about equalling the perianth.

Native. Marshes, flushes and gravelly stream-beds in mountains. Local in mainland Britain north from Yorkshire. Northern Eurasia, extending southwards to the mountains of southern Europe and the Caucasus; Greenland; North America. Var. *alpinoarticulatus*, to which most British material belongs, is part of subsp. *alpinoarticulatus* which occurs throughout much of the range of the species. Var. *marshallii* is a connecting link between subsp. *alpinoarticulatus* and subsp. *nodulosus* (Wahlenb.) which occurs in Fennoscandia, Iceland and northern Russia. It is endemic to the shores of Loch Ussie, near Conan, Ross-shire and near Braemar in Aberdeenshire.

× articulatus = × buchenaui Dörfl.
This hybrid is intermediate between the parents in habit and shape of the perianth segments and has low, if any, fertility. $2n = 60$.

Recorded from Upper Teesdale, where var. *alpinoarticulatus* is involved, and by Loch Ussie where var. *marshallii* in one parent. The hybrid is widespread in mainland Europe, where both parents show wide variation.

13. J. articulatus L.   Jointed Rush
J. compressus Relhan, non Jacq.; *J. lampocarpus* Ehrh. ex Hoffm.; *J. articulatus* subsp. *lampocarpus* (Ehrh. ex Hoffm.) Hook. fil.

Glabrous *perennial herb*, tufted or with a creeping rhizome, rarely with underground stems rooting and branching at the nodes. *Stems* 5–80 cm, erect to decumbent, rarely floating, terete, with 0–2 basal sheaths and 3–6 cauline leaves. *Leaves* 3–30 × 0.5–3.0 mm, deep green, unitubulose, terete or somewhat laterally compressed, usually shortly acute, sometimes attenuate at apex, with very distinct transverse septa, long-sheathing at base; auricles short and obtuse. *Inflorescence* diffuse, much branched, the branches erect or erecto-patent, with (1–)5–20(–80) well-spaced heads, each head with 5–15(–30) well-spaced flowers; lowermost bracts leafy. *Perianth segments* 2.5–3.5 mm, dark brown to blackish, the inner with broad, colourless margins, equal or unequal, ovate to lanceolate, the outer acute or rarely obtuse and mucronate at apex, the inner acute to obtuse and often mucronate. *Stamens* 6, half to three-quarters as long as the perianth segments; anthers 0.7–1.0 mm, equalling or longer than the filaments. *Capsule* 2.5–3.5(–4) mm, usually exceeding the perianth segments, trigonous-ovoid or rarely ellipsoid, acute or rarely obtuse at apex, mucronate. *Seeds* 0.5–0.6 mm, ovoid, reticulate. *Flowers* 6–9. Germinates in spring. $2n = 80$.

A very variable species especially in size and habit. The three following variants seem to represent ecotypes, but do not cover all the variation.

(i) Var. littoralis Patze
J. lampocarpus var. *littoralis* (Patze) Buchenau
Rhizome long. *Stems* up to 30 cm, spreading or ascending, in dense, fasciculate clumps, sometimes rooting at

the nodes, thick. *Leaves* rather fleshy, abruptly attenuate at apex. *Inflorescence* small and contracted. *Heads* few with few flowers.

(ii) Var. **nigritellus** (D. Don) Druce

J. nigritellus D. Don; *J. lampocarpus* var. *nigritellus* (D. Don) Buchenau; *J. polycephalus* D. Don, non Michx
Rhizome short to rather long. *Stems* up to 25 cm, erect, rather slender. *Leaves* slender, often curved, shortly acute at apex. *Inflorescence* small; branches suberect. *Heads* few.

(iii) Var. **articulatus**

Rhizome short to long. *Stems* up to 80 cm, erect or ascending, rather thick and rigid. *Leaves* stiff, sometimes curved, shortly acute at apex. *Inflorescence* large; branches erecto-patent. *Heads* numerous.

Native. Damp grassland, heaths, moors, marshes, dune-slacks and margins of rivers and ponds. Common throughout the British Isles. Europe and Asia except the Arctic; Iceland; North Africa; North America. Introduced in South Africa, Australia and New Zealand. Var. *littoralis* occurs on coastal sands. Var. *nigritellus* is a plant of the mountains. Var. *articulatus* is the lowland and upland plant of heaths, moors, marshes and river margins and ponds.

14. J. acutiflorus Ehrh. ex Hoffm.
Sharp-flowered Rush
J. nemorosus Sibth., non Pollich; *J. sylvaticus* auct.; *J. spadiceus* Schreb.; *J. micranthus* Desv.; *Phylloschoenus acutiflorus* (Ehrh. ex Hoffm.) Fourr.

Glabrous *perennial herb* with a thick, creeping, sparingly branched rhizome, the internodes usually 0.5–1.5 cm. *Stems* 30–160 cm, erect, with 2–3 basal sheaths and 2–4 cauline leaves. *Leaves* 5–50 cm × 0.5–2.0 mm, deep green, unitubulose, with very distinct transverse septa, straight, subterete. *Inflorescence* richly branched, the branches erecto-patent or diverging, with (10–)50–80 (–250) heads, the heads with (3–)5–8(–20) flowers. *Perianth segments* 1.5–2.7 mm, the inner longer than the outer, mid- to dark brown, ovate to narrowly ovate, acuminate and apiculate to cuspidate and often deflexed at apex. *Stamens* 6, half to two-thirds as long as perianth segments; anthers 0.8–1.0 mm, half to twice as long as the filaments. *Capsule* usually exceeding the perianth segments, chestnut-brown, trigonous-ovoid to narrowly pyramidal, usually gradually narrowed into a beak 0.5–1.0 mm. *Seeds* about 0.5 mm, ovoid to ellipsoid, reticulate. *Flowers* 7–9. Germinates in spring. $2n = 40$.

Native. Wet meadows, moorlands, marshes, bogs, swampy woodlands and margins of ponds and rivers. Common throughout the British Isles. West, central and south Europe, north to Denmark and east to Moscow; Newfoundland.

× **articulatus**
= × **surrejanus** Druce ex Stace & Lambinon
This hybrid is variable in habit, and is intermediate between the parents during flowering time in the branching of the inflorescence and shape and size of the perianth segments. The inflorescence sometimes proliferates. The capsules normally do not swell and produce no seed but viable seed has been obtained from this hybrid in Wales. Because of the variability of *J. articulatus*, hybrids are most easily recognised in the field where they can be compared with the variant of *J. articulatus* occupying that area. An erect, fertile rush with the chromosome number of *J. articulatus* ($2n = 80$), but characters of the hybrid have been recorded. The hybrid usually has $2n = 60$.

Throughout the British Isles with the parents and in some places is commoner than either. It is also recorded for central Europe.

15. J. bulbosus L.   Bulbous Rush
J. supinus Moench; *J. viviparus* Relhan; *J. uliginosus* Roth; *J. fluitans* Lam.; *J. subverticillatus* Wulfen; *J. kochii* F.W. Schultz; *J. articulatus* subsp. *supinus* (Moench) Hook. fil.; *Phylloschoenus supinus* (Moench) Fourr.

Glabrous, tufted *perennial herb* without a rhizome, the new shoots produced from the basal internodes of the stem. *Stems* 1–20 cm, erect and grass-like or procumbent and rooting at the nodes in land forms, up to 30 cm and much branched in floating forms and up to 4 m and flexible in submerged forms, slender and often slightly swollen at base. *Leaves* mostly more or less basal and up to 10 cm in land forms, much longer and often cauline in aquatic forms, setaceous, terete and slightly flattened to furrowed dorsally, pluritubulose, with very indistinct transverse septa, with sheathing base. *Inflorescence* diffuse, with 3–20 heads, each with 2–15 flowers, the flowers often partly replaced by adventitious shoots; branches suberect to patent. *Perianth segments* 1.5–3.5 mm, green to dark brown with a scarious margin, equal or the inner slightly longer than the outer, ovate to lanceolate or the inner oblong, the outer acute, the inner obtuse at apex. *Stamens* 3–6, half to two-thirds as long as perianth; anthers 0.3–1.0 mm, one-third as long to as long as the filaments. *Capsule* 2.2–3.5 mm, equalling or exceeding the perianth segments, trigonous, obtuse to truncate at apex, unilocular. *Seeds* 0.5–0.6 mm, turbinate, reticulate. *Flowers* 6–9. $2n = 40$.

Very variable. The land forms (var. *bulbosus*) are short, erect and usually with 3 stamens. Taller land forms bear a superficial resemblance to *J. articulatus*. The mud or floating forms (var. *uliginosus* (Roth) Fr.) have procumbent stems often rooting at the nodes and usually 3 stamens. The submerged forms (var. *fluitans* (Lam.) Fr.) have long drawn out stems and leaves and usually 3 stamens. These variants all grade into one another and are probably habitat states. Rather robust plants with 6 stamens, oblong anthers shorter than the filaments and dark brown perianth and capsule have been called *J. kochii*, but the characters are said to occur in different combinations with *J. bulbosus* and no distinct taxon seems to be recognisable. They are common in western and northern Britain and elsewhere in western Europe.

Native. Moist heaths, bogs, wheel-ruts, rides in woods

and all kinds of wet and damp places. Abundant throughout the British Isles. Europe, except the southeast, northwards to 69° 15' N in Norway, eastwards to western Russia; North Africa; Macaronesia; eastern North America.

16. J. pygmaeus Rich.   Pigmy Rush
J. mutabilis auct.; *J. nanus* Dubois; *Juncinella pygmaea* (Rich.) Fourr.

Glabrous, tufted *annual herb* usually with few stems. *Stems* 1–10 cm, erect or rarely ascending, often becoming suffused with purple, with a few basal sheaths, 0–1 cauline leaves and 1–2 leaf-like bracts. *Leaves* 0.5–3.0 cm × up to 0.5 mm, mostly basal, subulate from a sheathing base, unitubulose, with rather indistinct transverse septa rarely visible externally on fresh material; auricles 2, pointed, scarious. *Inflorescence* of 1–5 heads each with 2–15 flowers; lowest bract much exceeding the inflorescence. *Perianth segments* 4–7 × 0.5–1.0 mm, subequal, greenish or purplish, narrowly lanceolate, the outer obtuse to apiculate, the inner acuminate at the sometimes incurved apex. *Stamens* 3–6, variable and often unequal, the outer usually a quarter to half as long as the perianth segments, the inner often smaller or absent; anthers 0.3–1.0 mm, not more than half as long as the filaments. *Capsule* 2.5–3.5(–4) mm, straw-coloured, shorter than perianth segments, narrowly pyramidal to narrowly ovoid, sometimes shortly mucronate, thin-walled, unilocular. *Seeds* 0.4–0.5 mm, turbinate to ellipsoid, reticulate. *Flowers* 5–6. Flowers usually cleistogamous. $2n = 40$.

Native. Damp hollows and rutted tracks on heathland. Confined to the Lizard area of Cornwall. Western and southern Europe, northwards to Denmark and eastwards to Yugoslavia; North Africa; Turkey.

Subgenus 6. **Alpini** Buchenau

Dwarf, alpine *perennial herbs* with rhizomes. *Leaves* all basal or some on stems, terete or flattened-terete, not sharply pointed, with 1–several central cavities, divided by transverse septa, often not visible externally, without subepidermal sclerenchyma girders. *Inflorescence* terminal, very compact, exceeded or not by leaf-like or scale-like main bracts. *Flowers* without bracteoles. *Seeds* with terminal appendages.

17. J. biglumis L.   Two-flowered Rush

Small, glabrous, neatly tufted *perennial herb*, the rhizomes with short internodes and producing leafy non-flowering shoots and often long subterranean runners. *Stems* 2–12(–20) cm, erect, channelled along one side, with several basal leaves. *Leaves* 1–8 cm × 0.5–1.5 mm, all basal, subulate, curved, bluntly pointed at apex, sheathing at base, 1- to 3-tubulose at base, unitubulose above, transverse septa usually present but not visible externally; auricles very small. *Inflorescence* with a single (1–)2(–4) -flowered head; flowers in a vertical row; the first bract erect, exceeding the flowers and often with a small lamina. *Perianth segments* 2.5–3.0 mm, equal, mostly dark brown to blackish, elliptical to ovate, obovate or oblong, obtuse at apex. *Stamens* 6, usually

slightly exceeding or equalling the perianth segments; 0.3–0.4 mm, one-eighth to one-sixth as long as the filaments. *Capsule* 3–4 mm, 1.5–2.0 times as long as the perianth segments, obtusely angled, emarginate and mucronate at apex. *Seeds* 0.7–0.9 mm, with 2 subequal appendages 0.3–0.6 mm. *Flowers* 6–7. $2n = 60$.

Native. Wet stony places and rock ledges in base-rich soil on mountains between 600 and 1000 m. Very local in central, west and north-west Scotland. Throughout the Arctic to 83° 6' N in Greenland, extending southwards to the Alps (one locality), south Norway, Altai, Kamchatka and to 40° in the Rocky Mountains.

18. J. triglumis L.   Three-flowered Rush

Small, glabrous, tufted *perennial herb* with larger and taller tufts than *J. biglumis* and a shortly creeping rhizome. *Stems* 10–20 cm, slender but stiffly erect, terete, channelled along one side, with several basal leaves and sometimes one upper cauline leaf. *Leaves* 1–10 cm × 0.5–1.5 mm, subulate, curved, bluntly pointed at apex, sheathing at base, 2-tubulose at middle, often up to 5-tubulose at base, transverse septa usually present but not visible externally; auricles large. *Inflorescence* with a single (2–)3(–5)-flowered head; flowers in a horizontal row; bracts wide, erecto-patent, usually shorter than the flowers. *Perianth segments* 3.5–4.5 mm, pale to dark reddish-brown, equal or subequal, ovate-lanceolate, obtuse at apex. *Stamens* 6, almost equalling the perianth segments; anthers 0.7–0.9 mm, one-quarter to one-third as long as the filaments. *Capsule* 4–6 mm, trigonous-ellipsoid, obtuse to rounded and mucronate at apex. *Seeds* 1.7–2.5 mm, with 2 subequal appendages 0.3–0.6 mm. *Flowers* 6–7. $2n = 50$, c. 134.

Native. Bogs and wet rock-ledges on high mountains, decending to 60 m in Shetland, preferring non-calcareous rocks. Very local in North Wales, the Lake District and northern Pennines, local in central, west and north-west Scotland. Arctic and Sub-arctic to 80° N, extending southwards to the Pyrenees, Alps, Apennines, Balkan Peninsula and Caucasus; Himalayas; Colorado, Labrador.

19. J. castaneus Sm.   Chestnut Rush
J. triceps Rostk.; *J. jacquini* auct.

Glabrous *perennial herb* with a creeping rhizome forming long subterranean runners between the usually solitary stems. *Stems* 8–32 cm, erect, smooth, terete, with a few basal sheaths, 3–5 basal and sometimes one upper cauline leaf. *Leaves* 5–20 cm × 0.5–4.0 mm, subulate, bluntly pointed at apex, soft, flat and more or less convolute, with a long sheathing base, pluritubulose, septa present, sometimes visible externally at apex; auricles absent. *Inflorescence* of 1–3 heads, each with 3–10 flowers; flowers in a horizontal row; lowest bract not exceeding the inflorescence. *Perianth segments* 4.5–5.5 mm, equal or unequal, blackish-brown, but paler at apex, narrowly ovate, the outer acute, the inner obtuse at apex. *Stamens* 6, equalling or slightly shorter than the perianth segments; anthers 1.0–1.5 mm, one-quarter to one-third as long as the filaments, with a shortly acute apex. *Capsule* (5–)6.0–7.5(–9) mm, about 1.5 times as long as the perianth, ovoid to ellipsoid, obtuse-mucronate at apex, shining and reddish-brown at apex with a paler base. *Flowers* 6–7. 2*n* = 40.

Native. Boggy places and flushes on high mountains. Very local in central, west and north-west Scotland, particularly Clova and the Breadalbanes. Circumpolar, reaching 78° 30' N in Spitsbergen, extending south along the mountain ranges to the Alps, Urals, Altai and Shensi; in North America southwards to Quebec and New Mexico.

Subgenus 7. **Juncus**
Subgenus *Thalassii* Buchenau

Robust, maritime *perennial herbs* with rhizomes, but sometimes tufted. *Leaves* basal and cauline, cylindrical, very sharply pointed, with central compact pith bearing scattered vascular bundles, not transversely septate, with very numerous, subepidermal sclerenchyma girders. *Inflorescence* terminal, with sharply pointed, leaf-like main bract shorter than or longer than it, more or less compact. *Flowers* without bracteoles. *Seeds* with terminal appendages.

20. **J. maritimus** Lam. Sea Rush
J. spinosus auct.; *Juncastrum maritimum* (Lam.) Fourr.

Densely to scarcely tufted, usually very tough *perennial herb* with a creeping rhizome; intravaginal shoots absent. *Stems* 30–100(–150) cm, pale green, smooth, erect, with 2–4 leaves. *Leaves* up to 70 cm, basal and on stems, cylindrical, very sharply pointed at apex, non-septate; auricles absent. *Inflorescence* many-flowered, usually lax, rarely compact, forming an interrupted, irregularly compound panicle with ascending branches, the first bract usually long and forming an apparent prolongation of the stem, the second usually short. *Perianth segments* 3.0–4.5 mm, unequal, pale yellowish-brown, the outer ovate, more or less boat-shaped, acute and shortly mucronate at apex, the inner shorter, narrowly elliptical and obtuse at apex, without auricles. *Stamens* about two-thirds as long as perianth segments; anthers about twice as long as filaments. *Capsule* 2.5–3.5 mm, trigonous-ovoid, obtuse to subacute and mucronate at apex, equalling or slightly exceeding the perianth segments. *Seeds* 0.8–1.2 mm, obliquely ovoid, with a large appendage. *Flowers* 7–8. 2*n* = 48.

(i) Var. **congestus** L.B. Hall
Plant robust. *Panicle* congested into compact, orbicular heads, exceeded by lowest bract.

(ii) Var. **maritimus**
Plant robust. *Panicle* lax, exceeded by lowest bract.

(iii) Var. **atlanticus** J. W. White
Plant tall and weak. *Panicle* very large and diffuse; lowest bract rarely more than a quarter of its length.

Native. Salt-marshes above high-water mark of spring tides, often dominant over large areas. Common round the coasts of the British Isles except the extreme north of Scotland. Atlantic and Mediterranean coasts of Europe, extending into the Baltic and Black Sea and east to Sind; North and South Africa; North and South America; Australasia; inland in salt areas in south Europe and central Asia. The usual plant of the British Isles is var. *maritimus*. Var. *congestus* is recorded only from Studland Heath in Dorset. Var. *atlanticus* is confined to the Isles of Scilly.

21. **J. acutus** L. Sharp Rush
Juncastrum acutum (L.) Fourr.

Glabrous, robust, densely tufted *perennial herb* forming prickly tussocks with intervaginal shoots present. *Stems* numerous, (15–)25–150 cm, usually 2–4 mm in diameter, rigid, terete, smooth when fresh, finely striate when dry, the barren ones much more numerous than the fertile and terminating in a very acute, pungent point. *Leaves* mostly as chestnut sheaths, a few of the upper ones terminating in a lamina which is extremely similar to the barren stems, not transversely septate; auricles absent. *Inflorescence* usually a dense, more or less rounded head with erect to reflexed branches, rarely lax and elongated, lowest bract well developed, pungent and usually exceeding inflorescence. *Perianth segments* 2.5–4.0 mm, equal or subequal, reddish-brown, with a wide scarious margin, becoming almost woody, ovate-lanceolate to oblong, the outer obtuse or acute, the inner wide and retuse at apex with membranous margins extended into lobes on each side. *Stamens* equalling or shorter than perianth segments; anthers 1.2–2.0 mm, several times as long as filaments. *Capsule* 4–6 mm, much exceeding the perianth segments, obovoid to ovoid, with a broadly conical to obtuse, shortly mucronate apex. *Seeds* 1.5–2.5 mm, obliquely ovoid, with equal appendages. *Flowers* 6. 2*n* = 48.

(i) Var. **acutus**
Inflorescence a dense, more or less rounded head.

(ii) Var. **effusus** Buchenau
Inflorescence more elongated and open.

Native. Sandy seashores and dune-slacks and the drier part of salt-marshes. South and east coasts of England and in Wales north to Caernarvonshire and Norfolk, extinct in Yorkshire; south-east coast of Ireland from Cork to Dublin. Mediterranean region of Europe and North Africa, extending up the Atlantic coast to the British Isles and east to Transcaucasia; Macaronesia; California; South America; South Africa. Our plant is subsp. **acutus**. Our commonest plant is var. *acutus*. Var. *effusus* occurs in the Channel Islands, but its total distribution is unknown.

Subgenus 8. **Subulati** Buchenau

Maritime *perennial herbs* with rhizomes. *Leaves* basal and cauline, cylindrical, not sharply pointed, with central soft pith, not transversely septate, without subepidermal sclerenchyma girders. *Inflorescence* terminal, with short, much reduced main bract, diffuse. Each *flower* with 2 bracteoles. *Seeds* with short, terminal appendages.

22. J. subulatus Forssk.   Somerset Rush
J. multiflorus Desf., non Retz.; *Tenageia multiflora*
Fourr.

Glabrous *perennial herb* with a strong creeping rhizome.
Stems 30–120 cm, greyish-green, erect, channelled, with
basal sheaths and 2–4 cauline leaves, in autumn turning
vinous-red at the base. *Leaves* basal and on stems,
bluish-green, cylindrical, hollow, somewhat channelled,
bluntly pointed at apex, sheathing at base. *Inflorescence*
terminal, lax, interrupted, many-flowered, with suberect
branches and a short, much-reduced main bract; each
flower with 2 small bracteoles immediately beneath the
perianth segments. *Perianth segments* 2.0–3.5 mm, pale
green to pale yellow; equal or subequal, the outer lanceolate and apiculate to acute at apex, the inner wider and
obtuse at apex with whitish, membranous margins.
Stamens 6, about half as long as the perianth segments;
anthers 1.0–1.2 mm, 4–5 times as long as the filaments.
Style about 1 mm, with 3 pink, papillose stigmas.
Capsule about equalling the perianth segments, brown
and shiny, trigonous-ovoid, obtuse and mucronulate at
apex. *Seeds* 0.6–1.0 mm, reticulate, with short
appendages. *Flowers* 7–9. Only producing fruit in some
years. $2n = 42$.

Introduced. Known only from two localities. First
discovered in a salt-marsh at Berrow in Somerset in
1957. In 1983 found on wet, reclaimed land by docks at
Grangemouth, Stirling. It is native of the
Mediterranean region and the coast of northern Spain.

Subgenus 9. **Genuini** Buchenau

Perennial herbs with rhizomes, but often densely tufted.
Leaves reduced to brown sheaths at the base of the stem
and the lowest bract, which is stem-like and much
exceeds the inflorescence which appears lateral; main
bract cylindrical, not or sharply pointed at apex, with
central soft pith sometimes regularly interrupted, not
transversely septate, with or sometimes without subepidermal sclerenchyma girders. *Inflorescence* very
compact to rather diffuse; each flower with 2 small
bracteoles. *Seeds* more or less without appendage.

23. J. balticus Willd.   Baltic Rush
J. helodes Link nom. illegit.; *J. arcticus* auct.; *J. arcticus*
subsp. *balticus* (Willd.) Hyl.

Glabrous, bluish-green *perennial herb* with a strong, far-
creeping rhizome. *Stems* (10–)25–100 cm, deep green,
erect, rigid, terete, smooth and glossy when fresh, with
continuous pith, without subepidermal sclerenchyma
girders, non-flowering ones few or absent. *Leaves*
reduced to brown sheaths at base of stem. *Inflorescence*
rather lax, with (5–)25–60(–80) flowers; branches
suberect; lowest bract stem-like and much exceeding the
inflorescence which appears lateral. *Perianth segments*
3.2–5.0 mm, usually subequal, dark brown, ovate-lanceolate to ovate, obtuse and mucronate or apiculate at
apex. *Stamens* 6; anthers 0.8–1.5 mm, 1.5–2.0 times as
long as filaments. *Capsule* 3.0–4.5 mm, equalling or
exceeding perianth segments, usually dark brown, trigonous-ovoid, shortly mucronate at apex. *Seeds* 0.8–1.0

mm, ovoid, with inconspicuous appendages. *Flowers*
6–8. $2n = ?80, 84$.

Native. Maritime dune-slacks and rarely on upland
river terraces on bare or grassy ground. East and north
coasts of Scotland from Fifeshire to Sutherland and in
the Hebrides; rarely inland up to 278 m in Inverness-
shire; still in South Lancashire and formerly in North
Lancashire. Northern Europe south to Holland; Faeroes;
Iceland; North America; Patagonia; New Caledonia.

× **effusus = J.** × **obotritorum** Rothm.

This hybrid has rather weak, slender stems, but spreads
rapidly by extensive rhizomes. It is intermediate in stem
anatomy and inflorescence characters, but the anthers
are larger than in either parent and apparently produce
perfect pollen.

Occurred as three patches, one large, of completely
sterile clones in Lancashire from 1933 to 1980. It is also
reported from Denmark, Germany, Norway and Russia,
but at least the first and last areas for those records are
erroneous.

× **inflexus**

This hybrid occurs as three large patches of strongly rhizomatous, completely sterile clones, two patches of
which are very tall (up to 2 m) and have interrupted pith
as in *J. inflexus*, while the other is close to *J. inflexus* in
height, but has a continuous pith.

Known only from Lancashire. Endemic.

24. J. filiformis L.   Thread Rush

Glabrous, slender *perennial herb* with a horizontal
rhizome. *Stems* 10–60 cm × 0.7–1.0 mm, stiffly erect,
very faintly ridged when fresh, filiform, with continuous
pith, with subepidermal sclerenchyma girders, bearing
several brownish leaf-sheaths of which the uppermost
often has a short green lamina; non-flowering shoots
usually few or absent. *Inflorescence* of 4–10 flowers
forming a more or less compact head; lowest bract
(0.3–)1.0–1.5 times as long as stem, second bract sometimes leaf-like, though short. *Perianth segments* 2.5–3.5
mm, unequal, pale brown, becoming straw-coloured,
ovate to lanceolate or the inner oblong, the outer aristate, the inner obtuse at apex. *Stamens* 6; anthers about
0.5 mm, one-third to half as long as filaments. *Capsule*
(2.5–)3.0–3.5(–4.5) mm, pale brown, about equalling the
perianth segments, trigonous-ovoid to globose, usually
shortly mucronate at apex. *Seeds* about 0.5 mm,
obliquely ovoid, faintly reticulate, with one inconspicuous appendage. *Flowers* 6–9. $2n = 84$.

Native. On stony, silty shores of lakes and reservoirs.
Local in Britain from Leicestershire to Inverness-shire.
Apparently increasing, but easily overlooked. Northern
and Arctic Eurasia, south to the mountains of Portugal,
northern Apennines and Caucasus; Iceland; North
America; Patagonia.

25. J. inflexus L.   Hard Rush
J. glaucus Sibth.; *J. tenax* Poir.

Glabrous, densely tufted *perennial herb* with horizontal
matted rhizomes with very short internodes. *Stems*

20–120 cm, bluish-green, slender, stiffly erect, with 10–20 prominent ridges when fresh, with interrupted pith and very strong subepidermal sclerenchyma girders, with glossy, dark brown or blackish leaf-sheaths. *Inflorescence* rather lax, with suberect branches, many-flowered; lowest bract long. *Perianth segments* 2.5–4.0 mm, dark brown, unequal, lanceolate, subulate at apex. *Stamens* 6, about half as long as perianth segments; anthers 0.8–1.0 mm, 1.0–1.5 times as long as filaments. *Capsule* equalling or exceeding perianth segments, trigonous-ovoid to trigonous-ellipsoid, acute to obtuse and mucronate at apex. *Seeds* about 0.5 mm, obliquely ovoid, reticulate. *Flowers* 6–8. Germinating in spring. $2n$ = 40, 42.

Native. Marshes, dune-slacks, wet meadows, ditches and by lakes and rivers, usually on neutral or base-rich soils. Common throughout most of the British Isles north to central Scotland. Europe north to south Sweden and central Russia, east to India and Mongolia; North Africa; Macaronesia; introduced in North America and New Zealand. This range is that of subsp. **inflexus**. Another subspecies occurs in South Africa and east Java.

× **pallidus**

This hybrid was recorded with the parents in Middlesex and Bedfordshire in the 1950s, but see under *J. pallidus*.

26. **J. effusus** L. Soft Rush
J. communis E. Mey. nom. illegit.; *J. communis* subsp. *effusus* (L.) Syme

Glabrous, densely tufted *perennial herb* sometimes forming large patches. *Stems* 30–150 cm × 1.5–3.0 mm, erect, spreading or flattened, sometimes spiral, smooth and glossy when fresh, bright to yellowish-green, with 30–90 weak striae, with continuous pith and subepidermal sclerenchyma girders, with reddish- to dark brown leaf-sheaths. *Inflorescence* lax with suberect to widely divergent branches or compact, many-flowered; with long basal bract well beyond the inflorescence. *Perianth segments* 1.5–3.0 mm, subequal, pale brown, lanceolate, finely pointed at apex. *Stamens* 3(–6), half to three-quarters as long as perianth segments; anthers 0.6–0.8 mm, as along as the filaments. *Capsule* usually shorter than perianth segments, yellowish- to chestnut-brown, broadly ovoid to globose, obtuse and sometimes slightly emarginate at apex. *Seeds* about 0.5 mm, obliquely ovoid, reticulate. *Flowers* 6–8. Occasionally cleistogamous. Germinating in spring. $2n$ = 40, 42.

(i) Var. **effusus**
Stem straight and erect. *Inflorescence* lax, with suberect to widely divergent branches.

(ii) Var. **subglomeratus** DC.
J. effusus var. *compactus* Lej. & Courtois
Stem straight and erect. *Inflorescence* condensed into a single, rounded, compact head.

(iii) Var. **spiralis** J. McNab
J. effusus var. *suberectus* D. M. Hend.; *J. effusus* forma *spiralis* (J. McNab) Hegi

Stems spiral and more or less ascending or flattened. *Inflorescence* more or less congested in a rounded head.

Native. Marshes, ditches, bogs, wet meadows, by rivers and lakes and in damp woods, mostly on acid soils. Abundant and locally dominant throughout the British Isles. Europe to 65° 25' N in Norway. North and temperate zones and on mountains in the tropics. The distribution of the varieties is not fully known, but they do not seem to form ecological or geographical races although var. *spiralis* seems to be recorded only from western and northern Scotland, Carmarthenshire, Surrey and western Ireland, usually with one of the other varieties.

× **inflexus** = **J. × diffusus** Hoppe
This hybrid is densely tufted and its inflorescence resembles *J. inflexus*, but the floral characters and stem texture are intermediate. The pith is continuous or nearly so. The pollen appears perfect but seed production is much reduced, but variable.

Widespread, but nowhere common and is recorded from most areas where both parents occur. It is widespread on the mainland of Europe.

× **pallidus**
This hybrid has been recorded with the parents in Middlesex and Bedfordshire in the 1950s, but see under *J. pallidus*.

27. **J. conglomeratus** L. Compact Rush
J. leersii T. Marsson; *J. communis* subsp. *conglomeratus* (L.) Syme

Glabrous, densely tufted *perennial herb*. *Stems* 40–100 cm, dull greyish-green when fresh, erect, with 12–30 ridges which are especially prominent below the inflorescence, with continuous pith and subepidermal sclerenchyma girders, with brown leaf-sheaths. *Inflorescence* usually very compact, sometimes of several stalked heads; lowest bract with a wide sheath and hinging over backward at end of season. *Perianth segments* 2–4 mm, pale brown, lanceolate, finely pointed at apex. *Stamens* 3(–6), half to two-thirds as long as perianth segments; anthers 0.4–0.7 mm, usually shorter than filaments. *Capsule* about equalling perianth segments, yellowish- to chestnut-brown, broadly ovoid to globose, usually emarginate and mucronate at apex. *Flowers* 5–7, about a month earlier than *J. effusus*. Germinates in spring. $2n$ = 42.

(i) Var. **conglomeratus**
Inflorescence a compact, rounded head.

(ii) Var. **subuliflorus** (Drejer) Asch. & Graebn.
J. subuliflorus Drejer
Inflorescence of several stalked heads.

Native. Ditches, marshes, wet meadows, bogs, by lakes and rivers and in damp woods, but with a narrower ecological range than *J. effusus* and a more marked preference for acid soils. Common throughout the British Isles. Europe north to Faeroes and to 68° 55' N in Norway, extending east to Syria, Kurdistan and

Dzungaria in western China; North Africa; Macaronesia; eastern North America. The distribution of the varieties is not known, but they do not seem to form ecological or geographic races.

× effusus

= **J. × kernreichgeltii** Jansen & Wachter ex Reichg. This hybrid is difficult to identify because of the closeness of the parents and their variability. The best characters are the intermediate ridging of the stems and degree of inrolling shown by the base of the main bract. These hybrids are apparently fertile.

Occurs sporadically with the parents in north and west Britain. It is also recorded in central and northern Europe and Newfoundland.

28. J. pallidus R. Br.　　　　　　Great Soft Rush
J. macrostigma Colenso

Large, densely tufted *perennial herb*. *Stems* 50–200 cm × 3–8 mm, pale green, robust, smooth, more or less shining, rather soft, with more than 35 ridges, with continuous pith and subepidermal sclerenchyma girders, with pale brown leaf-sheaths. *Inflorescence* many-flowered, rather lax with long branches or more condensed, with a long lower bract exceeding the inflorescence. *Perianth segments* 2.3–3.0 mm, pale brown, ovate-lanceolate, more or less acute at apex, coriaceous. *Stamens* 6, one-third as long as perianth segments; anthers about equalling the filaments. *Capsule* 2.8–3.5 mm, as long as or longer than the perianth segments, very pale greenish-brown, ovoid-trigonous. *Seeds* about 0.5 mm, reddish-brown, oblong, reticulate. *Flowers* 8–10.

Introduced. Wool and bird-seed alien formerly naturalised in gravel pits in Middlesex and Bedfordshire, but still present in gravel-pit ditches near Evesham, Worcestershire. Native of Australia and New Zealand.

In the Bedfordshire localities and orchards at Barming, Kent, as well as the hybrids mentioned under *J. effusus* and *J. inflexus*, other Australian and New Zealand wool aliens occurred which L. A. S. Johnson named *J. aridicola* L. A. S. Johnson, *J. australis* Hook. fil., *J. procerus* E. Meyer; *J. vaginatus* R. Br.; *J. gregiflorus* L. A. S. Johnson; *J. sarophorus* L. A. S. Johnson; *J. subsecundus* N. A. Wakef.; *J. radula* Buchenau; *J. ochrocoleus* L. A. S. Johnson; *J. flavidus* L. A. S. Johnson; *J. distegus* Edgar; *J. continuus* L. A. S. Johnson and *J. usitatus* L. A. S. Johnson. These do not seem to have persisted, except for some of them in the orchards at Barming. *J. ensifolius* Wikström and *J. imbricatus* Leharpe from America have been recorded as casuals.

2. Luzula DC. nom. conserv.
Gymnodes (Griseb.) Fourr.; *Juncoides* Ség.; *Luciola* Sm.; *Nemorinia* Fourr.; *Pterodes* (Griseb.) Börner

Tufted *perennial herbs*, sometimes stoloniferous. *Leaves* mostly basal, grass-like, bifacial, variously hairy, but rarely without hairs near the base of the leaves on the margins, with a closed sheathing base, without auricles. *Inflorescence* of few- or many-flowered cymes, sometimes condensed into a head. *Ovary* 1-celled, with 3 ovules. *Capsule* with 3 seeds, which are smooth and shiny and usually with a conspicuous appendage (*aril*).

About 80 species. Cosmopolitan, but chiefly in the cold and temperate regions of the northern hemisphere.

Chrtek, J. & Krisa, B. (1962). A taxonomical study of the species *Luzula spicata* (L.) DC. sensu lato in Europe. *Bot. Not.* **115**: 293–310.

Kay, G. M. (1993). Novel concept of *Luzula multiflora* and *L. congesta*. *B.S.B.I. News* **63**: 14.

Kirschner, J. (1990). *Luzula multiflora* and allied species (Juncaceae): a nomenclatural study. *Taxon* **39**: 106–114.

Nordenskiöld, H. (1951 & 1956). Cyto-taxonomical studies in the genus *Luzula*. *Hereditas* **37**: 325–355 (1951); **42**: 7–73 (1956).

Grime, J. P. et al. (1988). *Comparative plant ecology*. London. [Accounts of *L. campestris* and *L. pilosa*.]

Stewart, A., Pearman, D. A. & Preston, C. D. (1994). *Scarce plants in Britain*. Peterborough. [*L. arcuata*.]

1.　Flowers mostly borne singly in the inflorescence, each on a distinct pedicel more than 3 mm, rarely some in pairs　　　　　　　　　　　**2.**

1.　Flowers mostly borne in groups of 2 or more, each one in a group which are sessile or with pedicels up to 2 mm, often a few solitary　　　　　**3.**

2.　Basal leaves rarely more than 4 mm wide; inflorescence branches erect to widely erecto-patent in fruit; seeds with terminal appendage less than half as long as the rest of the seed, more or less straight　　　　　　　　　　**1. forsteri**

2.　Some basal leaves usually more than 4 mm wide; lower inflorescence branches reflexed in fruit; seeds with terminal appendage more than half as long as (often longer than) the rest of the seed, often curved or hooked.　　　　　　**2. pilosa**

3.　Perianth segments pale straw to whitish or suffused red.　　　　　　　　　　　**4.**

3.　Perianth segments yellowish- to dark brown　　**6.**

4.　Perianth segments pure white, 4.5–5.5 mm, about twice as long as the capsule　　**5. nivea**

4.　Perianth segments whitish or suffused red, 2.5–3.5 mm, about as long as the capsule　　**5.**

5.　Perianth segments whitish
　　　　　　4(a). luzuloides subsp. **luzuloides**

5.　Perianth segments flushed red
　　　　　　4(b). luzuloides subsp. **cuprina**

6.　All or most basal leaves more than 8 mm wide
　　　　　　　　　　　　3. sylvatica

6.　All leaves less than 8 mm wide　　　　　**7.**

7.　Inflorescence drooping, spike-like, with flower-groups subsessile along the main axis, or the lower, themselves forming lateral spikes　**10. spicata**

7.　Inflorescence without a single main axis; either all flower-clusters congested in a dense head or some or all with distinct stalks arising from a short main axis near the base of the inflorescence　　**8.**

8.　Leaves deeply channelled, glabrous or sparsely hairy just near the base; seed with inconspicuous appendage; Scottish mountains　　**9. arcuata**

8.　Leaves more or less flat, conspicuously hairy; seeds with conspicuous whitish terminal appendage one-quarter to half as long as rest of the seed; widespread　**9.**

9. Plant with rhizomes or stolons; anthers more than 1.5 times as long as filaments **6. campestris**

9. Plant without rhizomes or stolons; anthers less than 1.5 times as long as filaments 10.

10. Perianth segments 1.5–2.5 mm, pale yellowish-brown **8. pallidula**

10. Perianth segments 2.5–3.5 mm, reddish- to dark brown 11.

11. Inflorescence with all flower clusters subsessile; lower bract usually longer than inflorescence; perianth segments longer than capsule; seed 0.9–1.0 mm wide **7(a). multiflora** subsp. **congesta**

11. Inflorescence of several stalked corymbose clusters of 8–18 flowers; lower bract not longer than inflorescence; perianth segments about as long as or slightly shorter than the capsule; seed about 0.8 mm wide 12.

12. Capsule valves ovate to elliptical, dark brown to black, basal appendage of seeds 0.2–0.4 mm **7(b). multiflora** subsp. **frigida**

12. Capsule valves obovate, green to brown; basal appendage of seeds (0.3–)0.4–0.7 mm **7(c). multiflora** subsp. **multiflora**

Subgenus 1. Pterodes (Griseb.) Buchenau

Perennial herbs. Inflorescence a lax cyme. *Flowers* borne singly. *Seeds* with a basal appendage.

1. L. forsteri (Sm.) DC. Southern Woodrush

Juncus forsteri Sm.; *Juncoides forsteri* (Sm.) Kuntze; *Pterodes forsteri* (Sm.) Börner; *L. caspica* Rupr. ex Bordza; *Luciola forsteri* (Sm.) Sm.; *Nemorinia forsteri* (Sm.) Fourr.; *Juncoides forsteri* (Sm.) Kuntze

Tufted *perennial herb*, with short, slender stolons. *Stems* (15–)20–35(–40) cm, more or less erect, pale greyish-green, striate, glabrous. *Leaves* mostly basal, 2–15 cm × 2.5–3.5(–5.0) mm, yellowish-green, long-linear, narrowed to an obtuse but apiculate apex, flat, fringed with long, pale simple eglandular hairs. *Inflorescence* a lax cyme, usually slightly one-sided, with erect to widely erecto-patent branches bearing flowers singly or rarely in pairs. *Perianth segments* 3.5–5.0 mm, usually chestnut brown, with a scarious margin, subequal, lanceolate, acute at apex, usually longer than the capsule. *Stamens* 6; anthers 1–4 times as long as filaments. *Style* as long as ovary. *Capsule* 2.8–4.5 mm, ovoid, acuminate-mucronate at apex. *Seeds* 1.3–1.6 mm, with a short, straight basal appendage. *Flowers* 4–6. $2n = 24$.

Native. Woods and hedgebanks. Locally common in the Channel Islands and Britain north to Bedfordshire and Herefordshire. Southern and western Europe north to Belgium and east to Crimea, Syria and Iran; North Africa; Macaronesia. The British Plants are referable to subsp. **forsteri**.

× **pilosa = L. × borreri** Bromf. ex Bab.

This hybrid is intermediate between the parents in inflorescence shape and leaf width and has very low fertility. $2n = 45$.

Frequent in Britain inside the range of *L. forsteri* and formerly outside it in Co. Wicklow. Also recorded for France and Germany.

2. L. pilosa (L.) Willd. Hairy Woodrush

Juncus pilosus L.; *Juncus vernalis* Reichard; *L. vernalis* (Reichard) DC.; *Juncoides pilosim* (L.) Kuntze; *Juncus nemorosus* Lam., *Luciola pilosa* (L.) Sm.; *Nemorinia vernalis* (Reichard) Fourr.

Tufted *perennial herb* with a short upright stock and short, slender stolons. *Stems* (15–)20–35(–40) cm, more or less erect, pale greyish-green, striate, glabrous. *Leaves* mostly basal, 2–20 cm × (3.5–)5–10 mm, yellowish-green, long-linear, narrowed to a small, truncate swelling at apex, flat, sparsely to densely fringed with long, pale simple eglandular hairs. *Inflorescence* a lax cyme, with unequal, spreading capillary branches deflexed in fruit and bearing flowers singly or rarely in pairs. *Perianth segments* 3.0–4.5 mm, usually dark brown with a wide hyaline margin, subequal, ovate-lanceolate, acute at apex, shorter than or almost equalling the capsule. *Stamens* 6; anthers longer than filaments. *Style* about as long as ovary. *Capsule* 3.5–4.5 mm, ovoid, very wide below, suddenly contracted above the middle to a truncate, conical top. *Seeds* 1.2–1.8 mm, pale brown, with a long, hooked basal appendage. *Flowers* 4–6 (sometimes later). $2n = 66$.

Native. Woods, hedgebanks and amongst heather on moors. Common throughout most of Britain, scattered over Ireland. Europe except the Mediterranean, east to the Caucasus and Siberia.

Subgenus 2. Anthelaea (Griseb.) Buchenau

Annual or *perennial herbs. Flowers* usually in pairs, sometimes borne singly or in larger groups. *Seeds* with short to inconspicuous basal appendage or appendage absent.

3. L. sylvatica (Huds.) Gaudin Great Woodrush

Juncus sylvaticus Huds.; *Juncus maximus* Reichard; *L. maxima* (Reichard) DC.; *Juncoides sylvatica* (Huds.) Kuntze; *Luciola sylvatica* (Huds.) Sm.

Robust, laxly tufted *perennial herb* with numerous, short, ascending stolons sometimes forming bright green mats. *Stems* (30–)40–80(–100) cm, erect, bright green, striate, glabrous. *Leaves* mostly basal, 10–50 cm × 5–12(–30) mm, bright green, long-linear, gradually tapering to a very acute apex, flat, fringed with sparse to dense, long, pale simple eglandular hairs. *Inflorescence* a lax terminal cyme, with erect to reflexed subumbellate branches bearing flowers in groups of (2–)3–5. *Perianth segments* 3–4.5 mm, brown, lanceolate, acuminate at apex, about equalling the capsule. *Stamens* 6; anthers up to 6 times as long as filaments. *Style* longer than ovary. *Capsule* (3.5–)4.0–4.5 mm, ovoid, finely beaked. *Seeds* 1.5–2.0(–2.2) mm, slightly shiny, with a small basal appendage. *Flowers* 5–6. $2n = 12$.

This species varies considerably in size. Plants of upland moorland in Shetland with shorter, narrower leaves and reduced inflorescences have been called var. *gracilis* Rostr. Very large plants from Jersey have been called var. *latifolia* Gérard. Most plants fall between these extremes. Var. *gracilis* may be a distinct ecotype, but more study of the great variation which occurs in mainland Europe needs to be carried out before it can be recognised.

Native. Woods, especially oak, on acid soil, on peat and open moorland, and on rocky ground near streams. Locally common throughout the British Isles, but rare in East Anglia and most abundant in the west and north where it ascends to 1000 m or more. West, central and south Europe to 68° 20' N in Norway; Caucasus; Turkey. Doubtfully native in South America. All plants from the British Isles are referable to subsp. **sylvatica**.

4. L. luzuloides (Lam.) Dandy & Wilmott
                                    White Woodrush
Juncus nemorosus Pollich; *Juncus luzuloides* Lam.; *Juncus albidus* Hoffm.; *L. albida* (Hoffm.) DC.; *L. nemorosa* (Pollich) E. Mey., non Hornem.; *Juncoides nemorosa* (Pollich) Kuntze

Loosely tufted *perennial herb* with short stolons. *Stems* (30–)40–65(–75) cm, slender, more or less erect, greyish-green, striate, glabrous. *Leaves* mostly basal, 5–30 cm × 3–6 mm, greyish-green, long-linear, gradually narrowed to a fine point, flat, fringed with long, pale simple eglandular hairs. *Inflorescence* corymbose, of many erect to erecto-patent branches bearing flowers in groups of (1–)2–5(–8). *Perianth segments* (2–)2.5–3.5 mm, whitish, sometimes suffused with red, the outer shorter than the inner, lanceolate, acute at apex, about as long as the capsule. *Stamens* 6; anthers about 3 times as long as filaments. *Style* longer than ovary. *Capsule* about as long as the perianth segments, ovoid, beaked. *Seeds* 1.1–1.2 mm, brown to black. *Flowers* 6–7. 2n = 12.

(a) Subsp. **luzuloides**
Inflorescence lax. *Perianth segments* whitish.

(b) Subsp. **cuprina** (Rochel ex Asch. & Graebn.) Chrtek & Krisa
L. albida var. *erythranthema* Wallr.; *L. nemorosa* var. *cuprina* Rochel ex Asch. & Graebn.; *L. rubella* Hoppe; *L. intermedia* Baumg.; *L. fuscata* Schur; *L. alpigena* Schur
Inflorescence often more or less condensed. *Perianth segments* suffused with red.

Introduced. Grown for ornament and naturalised in woods and by shady streams. Scattered through much of Britain and established in some localities for over a hundred years. Native of central and west Europe to north Germany, south to the Pyrenees and east to Moscow. Introduced in Scandinavia and North America. Both subspecies occur in Britain. Both occur through much of the native range of the species, but subsp. *cuprina* is mainly in the mountains.

5. L. nivea (L.) DC.   Snow-white Woodrush
Juncus niveus L.; *Juncoides nivea* (L.) Kuntze

Tufted *perennial herb* with long stolons. *Stems* 40–60(–80) cm, erect, greyish-green, striate, glabrous. *Leaves* mostly basal, 5–25 cm × 3–4(–5) mm, greenish-grey, long-linear, gradually narrowed to a fine point, flat, fringed with long, pale simple eglandular hairs. *Inflorescence* more or less corymbose, with erect

branches bearing groups of 6–20 flowers. *Perianth segments* 4.0–5.5 mm, whitish, unequal, lanceolate, acute at apex. *Stamens* 6; anthers about as long as or shorter than filaments. *Style* longer than ovary. *Capsule* 2.0–2.5 mm, about half as long as the perianth, globose. *Seeds* 1.3–1.5 mm, reddish-brown, with short basal appendage. *Flowers* 6–7. 2n = 12.

Introduced. Grown in gardens for ornament from which it sometimes escapes, but does not seem to persist for long. Native of Europe from north-central France southwards to north-east Spain, central Italy and Slovenia.

Subgenus 3. Luzula
Subgenus *Gymnodes* (Griseb.) Buchenau

Perennial herbs. Inflorescence spike-like or a subumbellate panicle. *Flowers* in dense clusters. *Seeds* with a conspicuous basal appendage.

Section 1. Luzula

Inflorescence subumbellate, with pedunculate or subsessile clusters on straight, erect branches at anthesis.

6. L. campestris (L.) DC.   Field Woodrush
Juncus campestris L.; *Juncoides campestris* (L.) Kuntze; *L. subpilosa* (Gilib.) V. I. Krecz.; *Gymnodes campestris* (L.) Fourr.

Loosely tufted *perennial herb* with short stolons. *Stems* 10–30(–40) cm, erect, greyish-green, striate, glabrous. *Leaves* mostly basal, 3–20 cm × 2–4 mm, bright green, long-linear, gradually narrowed to a small, truncate swelling at apex, flat, fringed with long, pale simple eglandular hairs. *Inflorescence* of one sessile and 2–several, stalked, corymbose clusters of 3–12 flowers, the branches more or less deflexed in fruit. *Perianth segments* 3–4 mm, dark reddish- to pale brown with scarious margins, subequal, lanceolate, long-acute at apex, longer than the capsule. *Stamens* 6; anthers 2–6 times as long as filaments. *Style* longer than ovary. *Capsule* 2.5–3.0 mm, obovoid, obtuse-apiculate at apex. *Seeds* 1.1–1.3 mm, with a white basal appendage up to half their length. *Flowers* 3–6. 2n = 12.

Native. Short grassland of various kinds. Common throughout the British Isles. Europe to 63° 55' N in Norway; North Africa and mountains of tropical Africa; North America; introduced in various other parts of the world. Plants from the British Isles are all referable to subsp. **campestris**.

7. L. multiflora (Ehrh.) Lej.   Heath Woodrush
Juncus campestris var. *multiflorus* Ehrh.; *Juncus multiflorus* (Ehrh.) Ehrh.; *Juncus erectus* Pers. nom. illegit.; *L. erecta* Desv. nom. illegit; *Juncoides multiflora* (Ehrh.) Druce; *L. campestris* subsp. *multiflora* (Ehrh.) Buchenau; *Juncus pallescens* Wahlb. nom. illegit.; *L. pallescens* Sw. nom. illegit.; *Juncoides pallescens* Druce nom. illegit.; *Juncus intermedius* Thuill.; *L. intermedia* (Thuill.) Spenn., non Nocc. & Balb.; *L. nemorosa* Hornem.

Densely tufted *perennial herb* with few or no stolons. *Stems* 10–60 cm, erect, pale green, striate, glabrous.

Leaves mostly basal, 2–30 cm × 3–4(–6) mm, bright green, long-linear, narrowed to an obtuse apex, flat, fringed with rather sparse, long, pale simple eglandular hairs. *Inflorescence* of several, stalked corymbose clusters of 8–18 flowers or flower clusters subsessile in a compact, lobed head. *Perianth segments* 2.5–3.5 mm, subequal, reddish- to pale brown with hyaline margins, broadly lanceolate, acute-aristate at apex, shorter than to longer than capsule. *Stamens* 6; anthers as long as to slightly longer than filaments. *Style* about as long as ovary. *Capsule* 2.0–2.8(–3) mm, ovoid to globose, apiculate at apex, shorter than to about as long as the perianth segments. *Seeds* 1.1–1.3 mm, about twice as long as wide; basal appendage up to half as long as their length. *Flowers* 4–6.

It is difficult to define precisely the following three subspecies. Subsp. *frigida* is a handsome alpine ecotype. Subsp. *multiflora* and subsp. *congesta* are generally separable, but plants occur which show combinations of characters which suggest hybridisation.

(a) Subsp. **congesta** (Thuill.) Arcang.
Juncus congestus Thuill.; *L. congesta* (Thuill.) Lej.; *Juncus liniger* With.
Flower-clusters subsessile. *Lower bract* usually longer than inflorescence. *Perianth segments* distinctly longer than capsule. *Capsule valves* broadly ovate, dark brown. *Seeds* 0.9–1.0 mm wide, with basal appendage 0.3–0.5 mm. 2*n* = 48.

(b) Subsp. **frigida** (Buchenau) V. I. Krecz.
L. campestris subsp. *multiflora* var. *frigida* Buchenau
Flower-clusters stalked. *Lower bract* not longer than inflorescence. *Perianth segments* as long as capsule. *Capsule valves* ovate to elliptical, dark brown to black. *Seeds* about 0.8 mm wide, with a basal appendage 0.2–0.4 mm. 2*n* = 36.

(c) Subsp. **multiflora**
Flower-clusters stalked. *Lower bract* not longer than inflorescence. *Perianth segments* as long as capsule. *Capsule valves* obovate, green to brown. *Seeds* about 0.8 mm wide, with a basal appendage (0.3–)0.4–0.7 mm. 2*n* = 24, 36.

Native. Grassland, heaths, moors and woods on acid soil. Common throughout the British Isles except in parts of central and eastern England. Throughout Europe but restricted to mountains in the south; North Africa; Asia; North America; introduced in New Zealand. Subsp. *multiflora* and subsp. *congesta* are both widespread, especially in Scotland and partially fertile intermediates with 2*n* = 42 occur. On the mainland of Europe subsp. *congesta* is more strongly calcifuge and more oceanic in its distribution. Subsp. *frigida* occurs on Scottish mountains and is also found through northernmost Europe.

8. L. pallidula Kirschner   Fen Woodrush
L. pallescens auct.; *Juncoides pallescens* auct.

Tufted *perennial herb* with few or no stolons. *Stems* 10–35(–45) cm, erect, pale green, striate, glabrous.

Leaves 2–20 cm × (1–)2–5 mm, mostly basal, yellowish-green, long-linear, narrowed to an obtuse apex, flat, fringed with rather sparse, long, pale simple eglandular hairs. *Inflorescence* subumbellate, of 5–10, globose-oblong clusters of 12–20(–25) flowers, the central cluster subsessile, the others on straight, erect branches. *Perianth segments* 1.5–2.5 mm, pale yellowish-brown, unequal, broadly lanceolate, long acuminate-aristate at apex, the outer longer than the capsule. *Stamens* 6; anthers slightly shorter than to as long as the filaments. *Style* shorter than ovary. *Capsule* 1.5–2.1 mm, obovoid, usually shorter than the perianth segments. *Seeds* 0.8–1.1 mm, with a basal appendage 0.2–0.4 mm, less than half as long as the seed. *Flowers* 5–6. 2*n* = 12.

Native. Open grassy places in dry parts of fens. Woodwalton Fen and Holme Fen in Huntingdonshire and Antrim in Ireland. Introduced in Surrey. Central and northern Europe east to Japan; eastern North America.

Section **2. Alpinae** Chrtek & Krisa

Inflorescence dense, often interrupted, drooping, spike-like.

9. L. arcuata Sw.   Curved Woodrush
Juncus arcuatus Wahlenb. nom. illegit.; *Juncoides arcuata* (Sw.) Kuntze; *Luciola arcuata* (Sw.) Sm.

Tufted *perennial herb*, with short stolons. *Stems* 5–20(–30) cm, erect, pale green, striate, glabrous. *Leaves* mostly basal, 2–7 cm × 1–3 mm, green, rigid, recurved, narrowly linear, narrowed to an acute apex, channelled, nearly glabrous. *Inflorescence* of several, more or less stalked clusters of 2–5 flowers, the longer stalks curved downwards, the others erect. *Perianth segments* 2.0–2.3 mm, equal, dark brown, broadly lanceolate, acute at apex, much longer than the capsule. *Stamens* 6; anthers as long as or shorter than the filaments. *Capsule* about 2 mm, broadly ovoid, shorter than the perianth segments. *Seeds* (0.8–)1.0–1.2 mm, oblong, with a short dark brown basal appendage. *Flowers* 6–7. 2*n* = 36, 42.

Native. Open, stony ground on high mountains, chiefly over 1000 m. Very local in central and northern Scotland. Arctic and Subarctic Europe; rarer in Asia and North America; Greenland.

Section **3. Nivales** Chrtek & Krisa

Inflorescence a subumbellate panicle of several clusters on erect or recurved stalks.

10. L. spicata (L.) DC.   Spiked Woodrush
Juncus spicatus L.; *Juncoides spicata* (L.) Kuntze; *Luciola spicata* (L.) Sm.; *Gymnodes spicata* (L.) Fourr.

Tufted *perennial herb* with short stolons. *Stems* (5–)7–30(–45) cm, pale green, striate, erect but nodding at apex, glabrous. Leaves mostly basal, 2–12 cm × 1–2(–4) mm, green, recurved, slightly channelled, linear, narrowed to an acute apex, fringed with a few, long, pale simple eglandular hairs. *Inflorescence* with a main axis and many, subsessile, many-flowered clusters. *Perianth segments* (2.0–)2.5–3.0 mm, subequal, dark brown, lanceolate, acute-aristate at apex, equalling or slightly longer than capsule. *Stamens* 6; anthers about as long as the filaments. *Style* as long as the ovary. *Capsule* (1.9–)2.1–2.5(–2.6) mm, dark brown to blackish, broadly ellipsoid, equalling the perianth segments. *Seeds* (1.0–)1.1–1.5(–1.6) mm, pale brown, with a short appendage at the base. *Flowers* 6–7. $2n = 24$.

Native. Rocks, screes, open stony ground and heaths on non-calcareous substrate. Local in central and northern Scotland; formerly in the Lake District. Arctic and Subarctic extending south into the mountains of southern Europe, Corsica, the Atlas Mountains, Himalayas; North America south to Arizona and New England. British populations belong to subsp. **spicata**.

Order 4. CYPERALES Burnett

Annual or *perennial herbs*, often with rhizomes. *Leaves* linear and sheathing, sometimes reduced to sheaths. *Flowers* bisexual or unisexual, hypogynous, small, in heads, spikes, racemes or panicles, either enclosed between 2 bracts or subtended by one bract. *Perianth* absent, or of scales. *Stamens* usually 3; anthers basifixed or versatile. *Ovary* 1-celled; ovule solitary. *Fruit* dry. *Seeds* with endosperm.

This order consists of the two large families, the Cyperaceae with nearly 4,000 species and the Poaceae with about 8,000 species.

166. CYPERACEAE Juss. nom. conserv.

Usually *perennial*, often rhizomatous, monoecious or sometimes dioecious *herbs*, most often growing in wet places. *Stems* usually solid, often trigonous. *Leaves* alternate, usually linear, flat to subcylindrical, entire, with a sheathing base, some or all often reduced to sheaths; ligules at the top of the sheath on the adaxial side, membranous; exstipulate. *Flowers* bisexual or unisexual, hypogynous, more or less actinomorphic, one in each of the axils of the bract-like glumes, one to many in spikelets, the spikelets terminal and solitary or in terminal spikes, racemes or panicles, often with extra sterile glumes. *Perianth* absent, or represented by one to many bristles or scales which sometimes elongate in fruit. *Stamens* (1–)2–3; anthers basifixed. *Style* absent or very short; stigmas 2 or 3, elongated. *Ovary* unilocular; ovule solitary, erect. *Fruit* an indehiscent nut, globose or trigonous in plants with 3 stigmas, biconvex in plants with 2 stigmas. *Seed* erect; embryo small; endosperm abundant. *Flowers* wind-pollinated.

Sedges differ from grasses in the stems often being trigonous, the leaf-sheath never split, the ligule partly fused to the upper surface of the lamina, the bracts subtending the spikes often leaf-like, the flowers spirally arranged and the anthers basifixed.

About 100 genera and 4,000 species in all parts of the world.

Bulbostylis humilis (Kunth) C. B. Clarke and **B. striatella** C. B. Clarke have been recorded as casual wool aliens.

1. Stems hollow; leaves with saw-edged margins and lower side of midrib   **14. Cladium**
1. Stems solid, the centre often occupied by very soft pith; leaves not saw-edged or very softly so   2.
2. Perianth of bristles which elongate and greatly exceed the glumes when fruiting, forming a whitish, cottony head   3.
2. Perianth absent or of inconspicuous bristles shorter than the glumes   4.
3. Flowers with numerous perianth-bristles; spikelets 1–several, more than 10 mm (excluding bristles)
                                         1. Eriophorum
3. Flowers with 4–6 perianth-bristles; spikelet 1, terminal, less than 10 mm (excluding bristles)
                                        2. Trichophorum
4. Flowers all unisexual, the male and female in different spikes or different parts of the same spike, or rarely on different plants; ovary and fruit enclosed or closely enfolded in a membranous innermost glume   5.
4. Flowers all bisexual; ovary and fruit not enclosed or closely enfolded in an innermost glume   6.
5. Ovary and fruit closely enfolded in the innermost glume the margins of which are not fused, leaving the fruit exposed at the top; male and female flowers in the same spikes, the spikes crowded and more or less sessile; stigmas and stamens 3   **15. Kobresia**
5. Ovary and fruit entirely enclosed in an extra fused membranous glume, the utricle, usually ending in a short or long beak; male and female flowers in same or different spikes or on different plants, the spikes variously crowded or distant, and stalked or sessile; stigma and stamens 2 or 3   **16. Carex**
6. Inflorescence a solitary terminal spikelet; lowest bract not leaf-like or stem-like, shorter than spikelet   7.
6. Inflorescence of more than 2 spikelets, or sometimes of 1, but then the lowest bract leaf-like or stem-like and exceeding the spikelet   9.
7. Most or all leaf-sheaths on stems with a leafy lamina
                                          9. Eleogiton
7. Most or all leaf-sheaths on stems without a lamina   8.
8. Uppermost leaf-sheath on stem with a short lamina
                                        2. Trichophorum
8. Uppermost leaf-sheath on stem, and all or most below it, without a lamina   **3. Eleocharis**
9. Inflorescence with more than 2 bifacial leaf-like bracts very close together at base   10.
9. Inflorescence with basal bracts stem-like or leaf-like, if leaf-like then either 1 or more than 2 and well spaced out   12.
10. Spikelets flattened, with glumes on 2 opposite sides of an axis   **11. Cyperus**
10. Spikelets more or less terete, with glumes spirally arranged   11.
11. Inflorescence dense; spikelets more than 8 mm
                                       4. Bolboschoenus
11. Inflorescence diffuse; spikelets less than 5 mm   **5. Scirpus**
12. Inflorescence a flattened, compact terminal head, with spikelets only on 2 opposite sides of the main axis   **10. Blysmus**

12. Inflorescence various, if a compact terminal head then spikelets not confined to 2 opposite sides of the axis 13.
13. Spikelets flattened, with glumes on 2 opposite sides of the axis **12. Schoenus**
13. Spikelets terete, with glumes spirally arranged 14.
14. Inflorescence obviously terminal with a leaf-like main bract; sheaths with several well-developed laminas **13. Rhynchospora**
14. Inflorescence usually apparently lateral, with the main bract more or less stem-like and continuing apically as a stem sheath with 0–1(–2) reduced laminas 15.
15. Stems very slender, less than 1 mm wide, rarely more than 20 cm high **8. Isolepis**
15. Stems stouter, more than 1.5 mm wide, rarely less than 30 cm high 16.
16. Inflorescences composed of (1–) several, sessile to stalked, globose apparent spikelets **6. Scirpoides**
16. Inflorescence composed of (1–) several, sessile to stalked, ovoid spikelets **7. Schoenoplectus**

Subfamily 2. Cyperoideae

Flowers bisexual in many-flowered spikelets or solitary and unisexual.

1. Eriophorum L.

Perennial rhizomatous *herbs*, either tufted or with a far-creeping rhizome. *Stems* terete to more or less triquetrous, leafy. *Leaves* remaining green through the winter, but nearly or quite dead at flowering time, variously shaped in section. *Inflorescence* of one to several large spikelets in a terminal umbel. *Lowest bract* leaf-like or glume-like. *Flowers* bisexual. *Perianth* of numerous bristles elongating to form conspicuous, white, cottony heads in fruit. *Stamens* 3. *Stigmas* 3; style-base not swollen. *Ovary* not enfolded or enclosed by a glume. *Nut* compressed-trigonous.

About 20 species. Mainly in Arctic and north temperate regions.

Grime J. P. et al. (1988). *Comparative plant ecology.* London. [*E. angustifolium* Roth and *E. vaginatum* L.]
Phillips, M. E. (1954). *Eriophorum angustifolium* Roth in Biological flora of the British Isles. *Jour. Ecol.* **42**: 612–622.
Wein, R. W. (1973). *Eriophorum vaginatum* L. in Biological flora of the British Isles. *Jour. Ecol.* **61**: 601–615.

1. Spikelet solitary, erect, without a leaf-like bract at base; leaf-lamina more or less triangular in section, absent or very reduced on uppermost stem leaf-sheath **4. vaginatum**
1. Spikelets several, more or less nodding in fruit, with 1–3 more or less leaf-like bracts at base; leaf-lamina flat to V-shaped in section, well developed on uppermost stem leaf-sheath 2.
2. Stems more or less terete to very bluntly 3-angled; peduncles smooth; anthers 2.5–5.0 mm **1. angustifolium**
2. Stems distinctly 3-angled; peduncles with numerous, very small, forwardly directed, rigid hairs; anthers 1.5–2.0 mm 3.
3. Plant loosely tufted; lamina of leaf 3–8 mm wide; glumes 1-veined **2. latifolium**

3. Plant with long rhizomes and solitary stems; lamina of leaf 0.5–2.0 mm wide; glumes with several veins **3. gracile**

1. E. angustifolium Honck. Common Cottongrass
E. polystachion L. nom. rej.; *E. polystachion* subsp. *angustifolium* (Honck.) Hook. fil.; *E. capitatum* auct.; *E. gracile* auct.; *E. scheuchzeri* auct.

Glabrous, extensively creeping, rhizomatous *perennial herb.* *Stems* 15–75 cm, subterete, trigonous only at apex, smooth, leafy, scattered, erect. *Leaves* 2–60 cm × (2–)3–5(–7) mm, yellowish- or bluish-green, linear, channelled, narrowed at apex into a long triquetous point; lamina of uppermost cauline leaf usually at least 1.5 times as long as the more or less inflated sheath; ligule very short, rounded. *Inflorescence* of (1–)3–7, pendulous spikelets; peduncles up to 8 cm, slender, smooth. *Bracts* 1–2, shortly sheathing, about equalling the spikelet. *Lower glumes* 4–10 × 3–4 mm, reddish- or greyish-brown, with a broad hyaline margin, ovate-lanceolate, narrowed to a more or less obtuse apex, 1-veined; upper glumes similar but narrower and lanceolate. *Perianth bristles* 40–50 mm, white, entire at apex. *Anthers* 2.5–5.0 mm. *Nut* 1–3 × 0.7–1.0 mm, dark brown, obovoid-trigonous. *Flowers* 5–6. *Fruits* 6–7. $2n = 58$.

Very variable in size and number of spikelets. Var. *alpinum* Gaudin (var. *minus* Koch), a dwarf plant with few sessile spikelets, may be an ecotype of alpine bogs worth recognising.

Native. Wet bogs, shallow bog pools and acid fens. Throughout the British Isles, locally abundant in the north and west, local and declining in the south and east. Europe except the most southerly Mediterranean region; Arctic regions; Siberia; North America; Greenland.

2. E. latifolium Hoppe Broad-leaved Cottongrass
E. pubescens Sm.; *E. polystachion* subsp. *latifolium* (Hoppe) Hook. fil.; *E. polystachion* auct.; *E. paniculatum* auct.

Nearly glabrous, tufted, rhizomatous, *perennial herb.* *Stems* 20–70 cm, triquetous throughout, leafy, smooth, grouped together, erect. *Leaves* 2–60 cm × 3–8 mm, yellowish- or bluish-green, linear, flat except for the short triquetous point; lamina of uppermost cauline leaf about equalling the close-fitting sheath; ligule absent. *Inflorescence* of 2–12, pendulous spikelets; peduncles up to 5 cm, slender, rough with very short, rigid, forwardly directed hairs. *Bracts* 2–3, very shortly sheathing, black at least below, about equalling the spikelet. *Lower glumes* 5–6 × 1.5–2.0 mm, reddish-brown below, greyish-brown or sometimes black above, with very narrow hyaline margins, ovate-lanceolate, narrowed to a more or less acute apex, 1-veined; upper glumes similar but narrower and lanceolate. *Perianth bristles* 20–25 mm, white, papillose at apex. *Anthers* 1.5–2.0 mm. *Nut* 3.0–3.5 × 1.0–1.4 mm, reddish-brown, narrowly obovoid-trigonous. *Flowers* 5–6. *Fruits* 6–7. $2n = 54, 72$.

Native. In wet places on base-rich soils. Scattered throughout the British Isles, but local and much less common than *E. angustifolium*. Throughout most of Europe; Turkey; Caucasus; Siberia; North America.

3. E. gracile Koch ex Roth   Slender Cottongrass
E. triquetrum Hoppe

Nearly glabrous, extensively creeping, rhizomatous *perennial herb*. *Stems* 20–60 cm, slender, triquetous, leafy only in lower half, scattered, erect. *Leaves* 2–40 cm × 0.5–2 mm, yellowish- or bluish-green, linear, obtuse at apex, 3-angled and slightly channelled; lamina of cauline leaf rarely more than half as long as the close-fitting sheath; ligule short. *Inflorescence* of 3–6, drooping spikelets; peduncles up to 3 cm, triquetous, slender, rough with very short, rigid, forwardly-directed hairs. *Bracts* 1–2, shorter than spike, brownish. *Glumes* 4–5 × 1.5–2.0 mm, yellowish- or greenish-brown tinged with grey, without hyaline margin, ovate, subacute at apex, with several veins. *Perianth bristles* 15–20 mm, white, entire at apex. *Anthers* 1.5–2.0 mm. *Nut* 2.5–3.0 mm, yellowish-brown, ellipsoid-cylindrical. *Flowers* 6. *Fruits* 7–8. 2n = 60, 76.

Native. In wet, more or less neutral to acid fen. Very local in southern Britain from Surrey to Caernarvonshire, and in central and western Ireland; formerly in Yorkshire, Northamptonshire and Norfolk. Central and north Europe, very local in the south and absent from the Mediterranean region and the Hungarian plain.

4. E. vaginatum L.   Hare's-tail Cottongrass
E. opacum auct.; *E. brachyantherum* auct.

Glabrous, shortly rhizomatous *perennial herb* forming large, compact tussocks. *Stems* (15–)30–60(–80) cm, terete below, triquetous above, leafy in lower half, erect. *Leaves* up to 50 cm × about 1 mm, yellowish- or bluish-green, linear, or more or less setaceous, triquetous, pointed but obtuse at apex, with brown fibrous sheaths; cauline leaves 2–3, lamina short, that of the uppermost leaf almost absent or a dark membranous apex to the strongly inflated sheath; ligule absent. *Inflorescence* a single terminal spikelet. *Bracts* absent. *Glumes* 6–7 mm, grey and translucent, ovate-lanceolate, acuminate at apex, 1-veined, the lower 10–15 darker and sterile, deflexed after anthesis. *Perianth bristles* 20–30 mm, white, entire at apex. *Anthers* 2.5–3.0 mm. *Nut* 2–3 mm, yellowish-brown, ovoid-trigonous. *Flowers* 4–5. *Fruits* 5–6. 2n = 58.

Native. Damp peaty places, especially on moorland bogs. Common in Ireland and west, central and north Britain; very local in central, east and south England. North and central Europe, southwards in the mountains to south Spain and Macedonia.

2. Trichophorum Pers. nom. conserv.

Tufted *perennial herbs*. *Stems* terete to trigonous, smooth, transverse section showing air canals bounded by very thick-walled cells. Only uppermost *leaf-sheath* with a lamina, thickly crescent-shaped in section. *Inflorescence* a solitary terminal spikelet, the lowest glume generally fertile, though usually larger than the others. *Flowers* bisexual. *Perianth* of 4–6 bristles, shorter or rarely longer than the glumes. *Stamens* 3. *Stigmas* 3. Ovary not enfolded or enclosed by a glume. *Nut* ovoid- or obovoid-trigonous.

1. Plant shortly creeping; stems trigonous; bristles longer than the glumes   **1. alpinum**
1. Plant densely tufted; stems terete; bristles shorter than the glumes   **2.**
2. Uppermost sheath with an opening of about 1 mm; spikelet with 3–10 flowers; glumes brown with a yellowish-brown midrib   **2(a). cespitosum** subsp. **cespitosum**
2. Uppermost sheath with an opening of 2–3 mm; spikelets with 8–20 flowers; glumes brown with a green midrib   **2(b). cespitosum** subsp. **germanicum**

1. T. alpinum (L.) Pers.   Cotton Deergrass
Eriophorum alpinum L.; *Eriophorum hudsonianum* Michx; *Scirpus hudsonianus* (Michx) Fernald

Diffusely tufted, shortly creeping, rhizomatous *perennial herb*. *Stems* 10–40 cm, yellowish-green, erect, very slender, trigonous, rough with minute, forwardly-projecting hairs. *Lower leaf-sheaths* without a lamina, the upper with a lamina up to 1(–3) cm, pale yellowish-green, setaceous, channelled and keeled, obtuse at apex, minutely hairy on margin. *Inflorescence* of a solitary, terminal spikelet. *Spikelet* 5–7 × about 3 mm, ellipsoid to lanceolate, with 8–12 flowers. *Glumes* 3.5–4.5 × 1.5–2.0 mm, yellowish-brown with a green midrib, ovate-oblong, obtuse at apex, the lowest with an apical projection. *Perianth bristles* 4–6, up to 25 mm in fruit, white, smooth and forming a cottony head. *Nut* 1.0–1.5 mm, shining dark brown, obovoid. *Flowers* 4–5. *Fruits* 6. 2n = 58.

T. alpinum closely resembles *T. cespitosum* when in flower, but looks like a very small *Eriophorum* when fruiting.

Native, but extinct. In a bog in Angus from 1791 to about 1813. Europe southwards to the mountains of central Spain and south Urals; Siberia; North America.

2. T. cespitosum (L.) Hartm.   Deergrass
Scirpus cespitosus L.; *Eleocharis cespitosus* (L.) Link

Densely tufted or tussock-forming *perennial herb* without rhizomes or stolons. *Stems* 5–35(–60) cm, slender, erect, more or less terete, smooth. *Lower leaf-sheaths* without a lamina, the upper with a lamina 3–10 mm. *Inflorescence* of a solitary terminal spikelet. *Spikelet* 3–6(–8) × about 3 mm, obovoid to linear, with 3–20 flowers. *Glumes* 2.5–4.0 × 1.5–2.0 mm, yellowish- to reddish-brown with a greenish midrib, linear-lanceolate, acute at apex, the 2 lowest with obtuse apical projections. *Perianth bristles* 5–6, about 1.5 times as long as nut, brownish, smooth or papillose at apex. *Nut* 1.5–2.0 mm, dull greyish- to yellowish-brown, obovoid-trigonous. *Flowers* 5–6. *Fruits* 7–8. 2n = 104.

(a) Subsp. cespitosum
T. austriacum Palla; *T. cespitosum* subsp. *austriacum* (Palla) Hegi

Basal sheaths shining; uppermost sheath with a subrotund opening about 1 mm and with a narrow, scarious margin clasping the stem. *Spikelets* with 3–10 flowers. *Perianth bristles* usually smooth.

(b) Subsp. **germanicum** (Palla) Hegi

T. germanicum Palla; *Scirpus germanicus* (Palla) Lindm.; *Scirpus cespitosus* subsp. *germanicus* (Palla) Brodd.

Basal sheaths dull; uppermost sheath with an oblanceolate opening 2–3 mm and with a reddish, scarious margin loosely surrounding the stem. *Spikelets* with 8–20 flowers. *Perianth bristles* usually papillose at apex.

Native. Bogs, wet moors and heaths on acid soils where it is locally dominant. Common throughout the British Isles in suitable habitats, but absent from most of central and eastern England. Europe, southwards to central Spain and south Bulgaria; Himalayas; North America; Greenland. Subsp. *cespitosum* is said to occur throughout most of the range of the species and subsp. *germanicum* is in western Europe extending eastwards to Bornholm. All our plants seem to be subsp. *germanicum*, for although there are records of subsp. *cespitosum* they have not been substantiated, but plants intermediate between the subspecies have been recorded in widely scattered localities.

3. Eleocharis R. Br.

Limnochloa R. Br.; *Baeothyron* A. Dietr.

Perennial herbs with long or short, stout or slender rhizomes. *Stems* terete to ridged, in transverse section with numerous approximately equal air canals without vascular bundles at the intersections of the strips of tissue separating the canals. *Leaf-sheaths* without lamina, although some spikeless stems may resemble basal leaves. *Inflorescence* of a solitary terminal spikelet. *Lowest bract* leaf-like or glume-like. *Flowers* bisexual. *Perianth bristles* 0–6, not elongating in fruit, shorter than or not much exceeding nut. *Stamens* 3. *Stigmas* 2 or 3. *Ovary* not enfolded or enclosed by a glume.

Contains about 200 species. Cosmopolitan.

E. nodula (Roth) Schult. from South America has been recorded as a wool casual.

Grime, J. P. et al. (1988). *Comparative plant ecology.* London. [*Eleocharis palustris* subsp. *vulgaris* Walters.]
Stewart, A., Pearman, D. A. & Preston, C. D. (1994). *Scarce plants in Britain.* Peterborough. [*E. acicularis.*]
Svenson, H. K. (1929). Monographic studies in the genus *Eleocharis.* I. *Rhodora* **31**: 121–129.
Svenson, H. K. (1939). Monographic studies in the genus *Eleocharis.* V. *Rhodora* **41**: 1–19; 43–77; 90–110.
Walters, S. M. (1948). *Eleocharis* R. Br. in Biological flora of the British Isles. *Jour. Ecol.* **37**: 192–206.
Walters, S. M. (1963). *Eleocharis austriaca* Hayek: A species new to the British Isles. *Watsonia* **5**: 329–335.

1. Lowest glume at least half as long as spikelet; spikelets usually 3- to 12-flowered — 2.
1. Lowest glume less than half as long as spikelet; spikelets 10- to many-flowered — 4.
2. Rhizomes very slender, pale, ending in small, pale tubers; upper sheath very delicate and inconspicuous; glumes greenish; style-base not enlarged — **6. parvula**
2. Rhizomes brownish, not bearing tubers; upper sheath conspicuous, often brownish; glumes brown, often with a green midrib; style-base enlarged, persistent — 3.
3. Stems more than 0.5 mm wide, more or less terete; spikelets 4–10 mm; lowest glume 2.5–6.0 mm — **5. quinqueflora**
3. Stems less than 0.5 mm wide, usually 4-ridged; spikelets 2–5(–7) mm; lowest glume 1.5–2.5(–3) mm — **7. acicularis**
4. Plant densely tufted; upper sheath oblique at apex; stigmas 3; nuts triquetrous — **4. multicaulis**
4. Plant not densely tufted; upper sheath almost transversely truncate; stigmas 2; nuts biconvex — 5.
5. Lowest glume empty, more or less completely encircling base of spikelet — **3. uniglumis**
5. Two lowest glumes empty, not more than half encircling base of spikelet — 6.
6. Perianth bristles (4–)5(–6); style-base long and narrow, hardly constricted at junction with nut — **2. austriaca**
6. Perianth bristles 4, very rarely 0; style-base broader, obviously constricted at junction with nut — 7.
7. Spikelets usually with 40–70 flowers; glumes from middle of spikelet 2.7–3.5 mm; nut 1.3–1.4(–1.5) mm; stomatal length 35–36 μm — **1(a). palustris** subsp. **palustris**
7. Spikelets usually with 20–40 flowers; glumes from middle of spikelets 3.5–4.5 mm; nut (1.3–)1.5–2.0 mm; stomatal length 50–77 μm — **1(b). palustris** subsp. **vulgaris**

Series 1. Palustriformes Svenson

Style-base spongy, beak-like, rarely depressed. *Stigmas* 2. *Nuts* bright yellow to brown or olivaceous, biconvex or trigonous, smooth to punctate.

1. E. palustris (L.) Roem. & Schult.

Common Spike-rush

Scirpus palustris L.; *E. boissieri* Podp.; *E. crassa* Fisch. & C. A. Mey. ex A. K. Becker

Glabrous, *perennial herb* with well-developed, far-creeping rhizome. *Stems* 10–75 cm × 1–4 mm in diameter, loosely to rather densely tufted, slender to stout, pliable and not easily cracked, with more than 20 vascular bundles not especially obvious as ridges when dry. *Sheaths* all without lamina, yellowish-brown, the uppermost more or less truncate at apex. *Spikelet* 5–30 mm, ovoid or lanceolate-ovoid. *Glumes* 1.5–3.5 mm, reddish-brown, with a pale midrib and a hyaline margin, ovate, obtuse to acute at apex, the 2 basal subequal, empty, and each half encircling the base. *Stamens* 3. *Stigmas* 2. *Style-base* swollen, constricted at base. *Perianth bristles* (0–)4, shorter than to longer than nut. *Nut* 1.2–2.0 mm, yellow to deep brown, biconvex, very finely punctate. *Flowers* 5–7. *Fruit* 6–8.

(a) Subsp. **palustris**

E. palustris subsp. *microcarpa* Walters; *E. intersita* Zinserl.

Spikelet usually with 40–70 flowers. *Glumes* from middle of spikelet 2.7–3.5 mm. *Nut* 1.2–1.4(–1.5) mm. *Stomatal length* 35–56 μm. 2n = 16.

(b) Subsp. **vulgaris** Walters

Spikelet usually with 20–40 flowers. *Glumes* from middle of spikelet 3.5–4.5 mm. *Nut* (1.3–)1.5–2.0 mm. *Stomatal length* 50–77 μm. 2n = (37–)38 (39, 40).

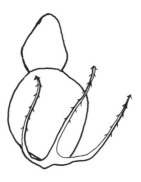

E. palustris (L.) Roem. & Schult.
subsp. **palustris**

E. palustris (L.) Roem. & Schult.
subsp. **vulgaris** S. M. Walters

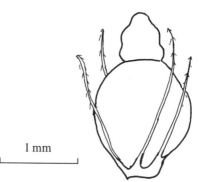

1 mm

E. austriaca Hayek

E. uniglumis (Link) Schult.

Fruits of **Eleocharis** R. Br.

Native. In or by ponds, marshes, ditches and riversides. Throughout the British Isles. North temperate regions. Subsp. *palustris* occurs in south and central England from Kent to Hampshire, Worcestershire and Nottingham-shire, but even there it is much rarer than subsp. *vulgaris*. It occurs in much of mainland Europe. Subsp. *vulgaris* is frequent throughout the British Isles and is widespread on mainland Europe. Intermediate, partially fertile plants (2*n* = 27), have been found in Oxfordshire.

× **uniglumis**
Morphologically intermediate plants between the two parents are regarded as putative hybrids. They do not show any significant sterility. They are found mainly on the west coast of Britain and are also recorded from Sweden.

2. E. austriaca Hayek   Northern Spike-rush
E. benedicta Beauverd; *E. leptostylopodiata* Zinserl.; *E. palustris* subsp. *austriaca* (Hayek) Podp.

Glabrous, *perennial herb* usually with a well-developed rhizome. *Stems* up to 60 cm, weak and easily cracked, with 10–16 vascular bundles visible especially in dried material as distinct ridges. *Sheaths* all without lamina, yellowish-brown with a little reddish colouring, upper more or less truncate at apex. *Spikelet* (5–)8–20 mm, ovoid or lanceolate-ovoid, often becoming conical in fruit, dense-flowered. *Glumes* 2–3 mm, deep reddish-brown with narrow paler midrib and narrow hyaline margin, ovate, obtuse at apex, caducous in fruit, with the 2 lower subequal, empty and each half-encircling its base. *Stamens* 3. *Stigmas* 2. *Style-base* typically very narrow, not more than 0.5 mm wide, rather high and gradually attenuate into a narrow neck. *Perianth bristles* (4–)5(–6), long and usually much exceeding the nut but easily broken, with rather long, thin teeth. *Nut* 1.0–1.5 mm, yellowish-brown when ripe, broadly elliptical-obovoid, very finely punctate. *Flowers* 5–7. *Fruits* 6–8. 2*n* = 16.

Native. Wet, marshy and flushed areas in or by rivers and lakes. Local in northern England and south Scotland. Scattered localities from north Norway to north Spain, central Yugoslavia and southern Urals.

3. E. uniglumis (Link) Schult.　　　Slender Spike-rush
Scirpus tetragonus Walker, non Poir.; *Scirpus uniglumis* Link; *E. watsonii* Bab.; *E. palustris* subsp. *uniglumis* (Link) Hook. fil.; *Scirpus palustris* subsp. *uniglumis* (Link) Syme; *E. klingei* (Meinsh.) B. Fedtsch.; *E. korshinskyana* Zinserl.

Glabrous *perennial herb* usually with a well-developed rhizome. *Stems* up to 60 cm and up to 1.5 mm in diameter, slender but firm, smooth, often shiny, with 15–20 vascular bundles (not especially obvious as ridges when dry). *Sheaths* without lamina, more or less truncate at apex. Base of stems, rhizome and leaf-sheaths reddish-purple. *Spikelet* 5–12 mm, lanceolate-ovoid, with 10–30 flowers more widely spaced than in allied species. *Glumes* 3.5–5.0 mm, dark reddish-brown with a narrow hyaline margin, ovate or ovate-lanceolate, obtuse at apex, the lowest empty and more or less encircling the base of the spikelet. *Stamens* 3. *Stigmas* 2. *Style-base* broad, spongy and beak-like. *Perianth bristles* 0–4(–5), often poorly developed and rarely much longer than the nut. *Nut* 1.4–2.2 mm, yellow to yellowish-brown, obovoid, coarsely punctate. *Flowers 5–7. Fruits 6–8. 2n =* (40–44) 46 (54–92).

In the British Isles *E. uniglumis* is a much less variable plant than *E. palustris*, but both cytology and morphology are very variable on the mainland of Europe.

Native. Estuarine flats, dune-slacks, brackish grazing marshes, lowland fen, grazing pasture and upland calcareous marsh. Scattered throughout the British Isles but mainly coastal, particularly in the south and west. Most of Europe; west Asia; North Africa.

Series 2. **Multicaules** Svenson

Style-base sharply trigonous with rounded basal lobes. *Stigmas* 3. *Nuts* yellowish- or olive brown, triquetrous, smooth, very faintly striolate.

4. E. multicaulis (Sm.) Desv.　Many-stalked Spike-rush
Scirpus multicaulis Sm.; *Baeothryon multicaule* (Sm.) Börner

Densely tufted, glabrous *perennial herb* with thickened, yellowish roots. *Stems* 15–40(–50) cm × 1.0–1.5 mm in diameter, slender but rigid, often recurved, markedly striate. *Sheaths* without lamina, pale yellow to purplish, acutely oblique at apex. *Spikelet* 5–15 mm, lanceolate-ovoid, acute at apex, 10- to 30-flowered, often proliferating vegetatively. *Glumes* 2.5–5.0 mm, reddish-brown, sometimes with a green midrib, ovate-lanceolate, obtuse to acute at apex, the lowest sterile, about quarter as long as the spikelet and not fully encircling its base. *Stamens* 3. *Stigmas* 3. *Style-base* acuminate, sharply trigonous with rounded basal lobes. *Perianth bristles* 4–6, longer than nut. *Nut* 1.2–1.5 mm, yellowish- or olive brown, narrowly obovoid, triquetrous, smooth, very faintly striolate. *Flowers 7–8. Fruits 8–10. 2n = 20.*

Native. Bogs, wet peaty places and wet sandy heaths,

usually on acid soils. Throughout the British Isles, common in the west, sparse in the east. Western Europe north to 61° 50' N in Norway and east to south-east Sweden, south-west Poland and north-west Yugoslavia; western North Africa; Azores.

Series 3. **Pauciflorae** Svenson

Style-base more or less confluent with the apex of the nut. *Stigmas* 3. *Nuts* blackish, greyish or pale yellow, trigonous, usually reticulate under high magnification.

5. E. quinqueflora (Hartmann) Schwarz
　　　　　　　　　　Few-flowered Spike-rush
Scirpus quinqueflorus Hartmann; *Scirpus pauciflorus* Lightf.; *Scirpus baeothyron* L. fil.; *Scirpus campestris* Roth; *E. pauciflora* (Lightf.) Link; *Limnochloa baeothryon* (L. fil.) Rchb.; *Baeothryon pauciflorum* (Lightf.) A. Dietr.; *Scirpus sepium* Honck.; *Scirpus halleri* Vill.; *Scirpus campestris* Roth; *E. baeothryon* (L. fil.) C. Presl; *Clavula baeothryon* (L. fil.) Dumort.; *Baeothryon halleri* (Vill.) T. Nees; *E. atacamensis* Philippi; *Scirpus andinus* Philippi; *Cyperus pauciflorus* (Lightf.) Missback & Krause

Tufted, glabrous *perennial herb*, with short, stout, brownish rhizome and slender underground stolons which often have conspicuously thickened terminal buds. *Stems* up to 30(–40) cm × 0.5–1.0 mm in diameter, more or less terete, greenish, slender, erect, striate. *Sheaths* without lamina, the upper conspicuous, pale yellow or brownish, obliquely truncate at apex. *Spikelet* 4–10 mm, ovoid, 2- to 7-flowered. *Glumes* all flower-bearing, 2.5–6.0 mm, brown with a green midrib and pale scarious margin and apex, lanceolate to oblong-lanceolate, obtuse to acute at apex, the lowest more than half as long as the spikelet and encircling its base. *Stamens* 3, the filaments often whitened and elongated. *Stigmas* 3. *Style-base* elongate, very slightly constricted at junction of and more or less confluent with the nut. *Perianth bristles* 4–6, slender, irregularly toothed, equalling or exceeding the nut, but sometimes poorly developed. *Nut* 2–3 mm, black when fresh, turning grey when dry, obovoid or fusiform, trigonous, prominently reticulate with small rectangular cells. *Flowers 6–7. Fruits 7–8. 2n = 132 (134).*

Dwarf plants from dune-slacks have been called var. *campestris* (Roth) and those from mountains var. *minor* (Sonder). They may be distinct ecotypes, but require further investigation.

Native. Wet places in moorland, fens and dune-slacks, usually in base-rich soils. Scattered throughout the British Isles and commonest in the north-west. Most of Europe, but rare in the south; temperate Asia; North America; Greenland.

6. E. parvula (Roem. & Schult.) Link ex Bluff, Nees & Schauer　　　　　　　　　　Dwarf Spike-rush
Scirpus pollicaris Del. nom. nud.; *Scirpus nanus* Spreng., non Poir.; *Scirpus parvulus* Roem. & Schult.; *Scirpus pusillus* Vahl; *Eleogiton parvula* (Roem. & Schult.) Link; *Limnochloa parvula* (Roem. & Schult.) Rchb.; *Baeothryon nanum* A. Dietr.; *Baeothryon*

pusillum A. Dietr.; *E. pygmaea* Torr.; *Scirpus translucens* Legall; *Chaetocyperus pygmaeus* (Torr.) Walp.; *Cyperus parvulus* (Roem. & Schult.) Missback & Krause

Slender, glabrous *perennial herb*, sparsely tufted, the tufts annual, with capillary stolons terminating in easily detached, whitish tubers. *Stems* 2–8 cm, greenish or straw-coloured, often spongy and translucent, terete, becoming somewhat striate on drying, some basal without spikelets and sometimes regarded as leaves. *Uppermost stem sheath* very thin, pale brownish, obtusely oblique at apex, without a lamina. *Spikelets* 2–4 mm, broadly ovoid, 3- to 5(–9)-flowered. *Glumes* 1.8–2.0 mm, green to yellowish, often dull brown on the sides, ovate, obtuse to acute at apex, scarcely keeled, striate and chartaceous, the lowest empty, at least half as long as the spikelet and more or less fully encircling the base of the spikelet. *Stamens* 3. *Stigmas* 3. *Style-base* very small, triangular, greenish and more or less confluent with nut. *Perianth bristles* straw-coloured, equalling or exceeding the nut. *Nut* 1.0–1.4 mm, pale yellow, obovoid, equilaterally triangular with prominent angles, smooth and shining, under high magnification usually slightly striate-reticulate. *Flowers* 8–9. $2n = 10$.

Native. Muddy places by the sea and in estuaries. Very local in south-west Britain from Hampshire and Devon to Merionethshire and Caernarvonshire. Recorded from Kerry, Wicklow and Londonderry in Ireland, but may be extinct. Central Scandinavia southwards to Portugal and south-east Russia; North and South Africa; Japan; North and South America.

Series 4. Aciculares (C. B. Clarke) Svenson

Style-base not confluent with apex of the nut. *Nuts* obscurely trigonous or terete, elongated, with longitudinal ridges separated by numerous latitudinal lines. *Lowest glume* infertile. *Styles* 3.

7. E. acicularis (L.) Roem. & Schult.
Needle Spike-rush

Scirpus acicularis L.; *Cyperus acicularis* (L.) With.; *Mariscus acicularis* (L.) Moench; *E. costata* C. Presl; *Isolepis acicularis* (L.) Schltr.; *Scirpus chaeta* Schult.; *Clavula acicularis* (L.) Dumort.; *Clavula comosa* Dumort.; *Scirpidium aciculare* (L.) Nees; *Chaetocyperus acicularis* (L.) Nees

Slender, glabrous, rhizomatous *perennial herb*, the rhizome brown and creeping, the roots firm and white. *Stems* in non-flowering submerged state up to 50 cm, but in flowering terrestial state on mud not more than 10 cm, less than 0.5 mm in diameter, deep green, usually 4-ridged and sulcate. *Sheaths* without lamina, loose, thin, reddish-striate at base, the uppermost conspicuous, often brownish, acutely to obtusely oblique at apex. *Spikelet* 2–5(–7) mm, ovoid to linear, 3- to 11(–15)-flowered. *Glumes* 1.5–2.5(–3.0) mm, green with reddish-brown sides and scarious margins, ovate-lanceolate, acute at apex, the lowest usually empty, obtuse at apex, about half as long as the spikelet and fully encircling its base. *Stamens* 3. *Stigmas* 3. *Style-base* narrow, some-

what compressed, conical-triangular, separated from the top of the nut by a deep groove. *Perianth bristles* 0–1(–4), caducous, never more than half as long as nut. *Nut* 0.7–1.0 mm, pale brown or whitish, obovate-oblong, obscurely trigonous, with many longitudinal ribs and close latitudinal lines. *Flowers* 8–10. $2n = 20$.

Often occurring in a totally submerged and persistently vegetative state. Tall flowering plants in deep water have been called var. *longicaulis* H. C. Watson.

Native. In wet sandy and muddy places at the margins of lakes and pools, or totally submerged. Scattered throughout the British Isles, but rather local. Most of Europe; northern Asia; Australia; North and South America; Greenland.

4. Bolboschoenus (Asch.) Palla
Schoenus taxon *Bolboschoenus* Asch.

Rhizomatous *perennial herbs* with fleshy roots. *Stems* solitary, solid, triquetrous, leafy, with swollen tuberous bases. *Leaves* flat, widely V-shaped in section. *Inflorescence* of (1–) 3–many spikelets either sessile or variously clustered on one to several stalks. *Lowest bract* leaf-like. *Flowers* bisexual. *Glumes* spirally arranged. *Perianth bristles* 1–6, not elongating in fruit. *Stamens* 3. *Stigmas* 2–3. *Ovary* not enfolded or enclosed by a glume. *Nut* obovoid, trigonous below, smooth.

Contains about 8 species. Europe, temperate Asia and eastern North America.

1. B. maritimus (L.) Palla   Sea Club-rush
Scirpus maritimus L.; *Schoenoplectus maritimus* (L.) Lye

Nearly glabrous, rhizomatous *perennial herb*, with fleshy roots; rhizome about 2 mm in diameter, often far-creeping, branching, sometimes tuberiferous, covered with sheath-like scales, becoming hard and black; tubers up to 15 mm in diameter, subglobose or ellipsoid. *Stems* (10–)25–100 cm, solitary, solid, sometimes curved, leafy in the lower half, triquetrous, with the angles sharp or narrowly winged and the sides concave above, flat below and ribbed. *Leaves* 10–35 cm × 5–7 mm, often exceeding the stems, long-linear, gradually long-acuminate at apex, flat, many-nerved, scaberulous and thickened on the entire margins, midrib channelled above and prominent beneath; sheaths often long, tight, membranous and red-dotted on the ventral side below the shallowly convex mouth; ligule absent. *Inflorescence* capitate or subumbellate, of 1–10 solitary or glomerate, sessile or rayed spikelets; bracts scabrid, the lower leaf-like, with the lowermost generally about twice as long as the rays; rays flattened, up to 40 mm, or almost wanting. *Spikelets* 10–20(–40) × 5–6 mm, terete, ovoid becoming oblong, acute at apex. *Glumes* 3.5–4.0 × about 2.5 mm, ovate or elliptical, membranous, stramineous with a dark border, or dark reddish-brown, sparsely hispid, apically retuse or lacerate with a narrow, raised midrib excurrent in a long, recurved, scabrid awn. *Perianth bristles* (0–)3–6, white, minutely retrorsely scabrid, unequal, shorter than, to slightly exceeding the nut. *Stamens* 3; anthers about 3 mm, with a scabrid, conical connective-tip. *Stigmas* (2–)3. *Nut* about 2.2 × 1.3 mm, obovoid, trigonous below, planoconvex above, or occasionally lenticular, shortly beaked, brown, smooth. *Flowers* 7–8. *Fruits* 8–9. $2n = 80$, c. 104.

Native. Wet muddy places in estuaries or by the sea. Common round the coasts of the British Isles except in the extreme north of Scotland, rarely inland. Most of Europe and temperate Asia.

5. Scirpus L.

Perennial herbs with creeping rhizomes. *Stems* solid, trigonous, leafy. *Leaves* flat, with well-developed lamina, minutely denticulate. *Inflorescence* of very numerous, terete spikelets, one to several at the ends of a diffusely branching panicle. *Lowest bract* leaf-like. *Glumes* spirally arranged. *Flowers* bisexual. *Perianth bristles* 6, not elongating in fruit. *Stamens* 3. *Stigmas* 3. *Ovary* not enfolded or enclosed by a glume. *Nut* ovoid or subglobose, compressed-trigonous.

S. nodosus Rottb. has been recorded as a casual wool alien.

1. S. sylvaticus L.　　　　　　Wood Club-rush

Stout *perennial herb* with a creeping rhizome. *Stems* 30–120 cm, glabrous, solid, erect, solitary, trigonous, smooth, leafy, striate. *Leaves* up to 40 cm × 5–20 mm, yellowish-green, long linear, entire, gradually tapered to a narrow apex, flat, margins minutely denticulate. *Bracts* 2–4, leaf-like or the shorter one setaceous. *Inflorescence* of numerous spikelets, 1–several at the ends of a diffusely branching panicle; main rays up to 10 cm, shortly hairy; secondary rays up to 4 cm, shortly hairy. *Spikelets* 3–4 mm, ovoid, terete, obtuse at apex. *Glumes* 1.5–2.9 mm, green or greenish-brown, ovate, obtuse at the mucronulate apex. *Perianth bristles* 6, not elongating in fruit, as long as or slightly longer than the nut, straight, retrorsely barbed. *Stamens* 3; filaments flattened. *Stigmas* 3. *Nut* about 1 mm, dull yellowish-brown, obovoid or subglobose, compressed-trigonous. *Flowers* 6–7. *Fruits* 7–8. $2n = 62, 64$.

Native. By streams and in marshes and other damp areas in woods and shady places. Scattered and locally frequent throughout the British Isles north to central Scotland. Most of Europe; Caucasus; Siberia.

6. Scirpoides Ség.

Holoschoenus Link

Tall, tufted *perennial herbs*, strongly rhizomatous. *Stems* densely clustered, solid, terete. *Leaf-sheaths* mostly without a lamina, but uppermost 1(–2) with a well-developed lamina which is semicircular in section. *Inflorescence* of (1–)5 to many, variously stalked or sessile, globular heads, each composed of numerous, terete, tightly packed spikelets. *Lowest bract* more or less stem-like making the inflorescence appear lateral. *Flowers* bisexual. *Glumes* spirally arranged. *Perianth bristles* absent. *Stamens* 3. *Stigmas* 3. *Ovary* not enfolded or enclosed by a glume. *Nut* broadly ellipsoid, unequally trigonous.

A single species or species complex, widely distributed in tropical and warm temperate zones of the Old World.

1. S. holoschoenus (L.) Soják Round-headed Club-rush
Scirpus romanus L.; *Isolepis holoschoenus* (L.) Roem. & Schult.; *Scirpus holoschoenus* L.; *Holoschoenus vulgaris* Link; *Isolepis paniculatus* Gray nom. illegit.

Densely tufted, nearly glabrous, glaucous *perennial herb*, with sand-binding roots and a short, woody, creeping rhizome. *Stems* 30–90 cm × 2–3 mm, densely clustered, solid, much thickened at the base by sheaths, soft, flattened below, smooth or ridged, harsh above, branched intervaginally. *Leaves* basal, mostly reduced to sheaths, short, hard, terete, channelled, becoming flat at the apex, scabrid on the margins at least below, with an obtuse, scabrid, often brown, flattened apex; sheaths long, dorsally herbaceous, becoming brown and sometimes hard, ventrally membranous, pale, red-spotted, prolonged into a rounded antiligule, soon splitting into fibres; ligule absent. *Inflorescence* pseudolateral, shortly once or twice branched; bracts leaf-like, the upper 10–40 cm, erect and much exceeding the inflorescence, the lower spreading or deflexed, much shorter than the upper but very variable in length, usually flat; rays usually less than 50 mm, firm, flattened, sometimes marginally scabrid. *Spikes* 1–4(–10), 2–12 mm in diameter, sessile or pedunculate, spherical, composed of many spikelets. *Spikelets* 2.0–3.5 × 1–2 mm, glomerate, apparently confluent when young, ovoid, rather obtuse at apex. *Glumes* about 1.2 × 0.8–1.0 mm, obovate, obtuse at apex, rounded on the back with a green, acute keel, prominent above and excurrent to a short, obtuse mucro, the sides membranous, nerved, red-tinged and scabrid with white apiculae in the upper half. *Stamens* 3; anthers about 0.8 mm, connective-tip small, rounded. *Style* about 0.3 mm; stigmas 3, about 0.7 mm, thick, crystalline-papillose. *Nut* about 1.0 mm × 0.6 mm, greyish, broadly ellipsoid, shortly beaked, unequally trigonous, thickened on the angles with a minutely reticulate surface. *Flowers* 8–9. *Fruits* 9–10. $2n = 164$.

Native. Damp, sandy flats near the sea. Very rare in Devon and Somerset; occasional elsewhere in southern Britain as a introduction. Europe north to south-west Britain and White Russia; north-west Africa; Siberia; Canaries.

7. Schoenoplectus (Rchb.) Palla nom. conserv.
Scirpus subgen. *Schoenoplectus* Rchb.

Tall, stout *perennial herbs* with creeping rhizomes. *Stems* terete to triquetrous, solid, usually nearly or quite leaf-less, but upper 1(–3) sheaths each with a small lamina; in transverse section crescent-shaped with numerous, approximately equal air canals with vascular bundles at the intersections of the strips of tissue separating the canals. *Inflorescence* of (1–)few to numerous, variously stalked or sessile, ovoid spikelets 5–8 mm. *Lowest bract* more or less stem-like making the inflorescence appear lateral. *Flowers* bisexual. *Glumes* spirally arranged. *Perianth bristles* 0–6, not elongating in fruit, rough with short downwardly directed hairs. *Stamens* 3. *Stigmas* 2 or 3. *Ovary* not enfolded or enclosed by a glume. *Nut* ovoid, compressed.

About 25 species in the north temperate region.

1. Stems terete   2.
1. Stems triquetrous   3.
2. Glumes more or less smooth; stigmas usually 3; nut 2.5–3.0 mm   **1(a). lacustris** subsp. **lacustris**
2. Glumes with numerous reddish papillae; stigmas usually 2; nut 2.0–2.5 mm   **1(b). lacustris** subsp. **tabernaemontani**
3. Uppermost leaf-sheath with a short lamina; glumes with obtuse lateral lobes; perianth bristles 6, more than half as long as nut   **2. triqueter**
3. Uppermost leaf-sheath with the lamina up to 30 cm; glumes with acute lateral lobes; perianth bristles 0–6, less than half as long as nut   **3. pungens**

1. S. lacustris (L.) Palla   Common Club-rush
Scirpus lacustris L.; *Scirpus glaucus* Sm., non Lam.; *Scirpus lacustris* subsp. *glaucus* Hartm. fil. (type intermediate)

Robust, erect, nearly glabrous *perennial herb*, with a thick, creeping rhizome. *Stems* up to 350 cm × 15 mm, solid, thickened at base by sheaths, terete, spongy, green or glaucous. *Leaves* usually reduced, but sometimes well-developed when submerged, when submerged ribbon like, when emergent up to 15 cm, linear-subulate and tapering to a blunt apex; sheaths long, membranous, splitting early, brownish or purplish; ligule brown, membranous. *Inflorescence* of numerous spikelets, lax and once or twice branched or congested; rays up to 10 cm, flattened, scabridulous; bracts subulate, the lowermost erect and shorter to longer than inflorescence, the upper with a membranous base and subulate, scabridulous tip up to 10 mm. *Spikelets* 6–15 × 3–5 mm, solitary or in small clusters, narrowly ovoid, acute or subacute at apex. *Glumes* 2.5–4.0 × 1.5–3.0 mm, broadly ovate, scarious, tightly appressed, dorsally rounded, with a thin, raised midrib sometimes excurrent in a narrow, scabrid apiculus, the sides reddish-brown, nerveless, smooth or glandular and red-spotted, transversely wrinkled, margins broadly hyaline-membranous, often ciliate and becoming lacerate apically. *Perianth bristles* 4–6, about as long as the nut, conspicuously retrorse-strigose, brownish. *Stamens* 3; filaments strap-shaped; anthers 2.0–2.5 mm, connective-tip blunt, hairy or glabrous. *Stigmas* 2 or 3. *Nut* 2.0–3.0 × 1.5–2.5 mm, broadly obovoid, lenticular or bluntly trigonous, grey or fuscous, smooth or minutely punctulate, shining. *Flowers* 6–7. *Fruits* 8–9.

(a) Subsp. **lacustris**
Stems up to 350 cm, green. *Glumes* more or less smooth. *Stigmas* usually 3. *Nut* 2.5–3.0 mm, compressed-trigonous. $2n = 42$.

(b) Subsp. **tabernaemontani** (C. C. Gmel.) Å. & D. Löve
Scirpus tabernaemontani C. C. Gmel.; *Scirpus medius* Gray nom. illegit.; *Schoenoplectus tabernaemontani* (C. C. Gmel.) Palla; *Scirpus lacustris* subsp. *tabernaemontani* (C. C. Gmel.) Syme

Stems up to 150 cm, more or less glaucous. *Glumes* with numerous reddish papillae, especially near the apex. *Stigmas* usually 2. *Nut* 2.0–2.5 mm, planoconvex or biconvex. $2n = 42, 44$.

Native. The species occurs throughout the British Isles in rivers, lakes and ponds, usually where there is abundant silt. It occurs almost throughout Europe and the same or closely related taxa occur in Asia, Africa, North and Central America and Australia. The subspecies are usually ecologically separated, but are sometimes found in each others' habitats and intermediates sometimes occur. Subsp. *lacustris* is scattered throughout the British Isles in fresh water, and usually inland. Subsp. *tabernaemontani* is usually in brackish water near the coast or where there is a tidal influence inland.

× **triqueter** = S. × **carinatus** (Sm.) Palla
Scirpus duvalii Hoppe; *Scirpus carinatus* Sm.; *Scirpus lacustris* subsp. *carinatus* (Sm.) Syme
Stems triquetrous above, terete below.

(a) Nothosubsp. **carinatus**
Glumes smooth.

(b) Nothosubsp. **kuekenthalianus** (Junge) P. D. Sell
Scirpus kuekenthalianus Junge; *Scirpus arunensis* Druce; *Scirpus scheuchzeri* Brugger
Glumes with reddish papillae.

Nothosubsp. *carinatus* has been recorded from by the River Tamar in Cornwall and Devon, and by the River Thames in Kent, Surrey and Middlesex. Nothosubsp. *kuekenthalianus* occurs by the River Tamar in Cornwall and Devon, the River Arun in Sussex and the River Medway in Kent. Both also occur in central Europe.

2. S. triqueter (L.) Palla   Triangular Club-rush
Scirpus triqueter L.; *Scirpus mucronatus* auct.

Stout, glabrous *perennial herb* with a creeping rhizome. *Stems* 50–150 cm × up to 8 mm, solid, thickened towards the base, solitary, green or rarely glaucous, erect, triquetrous. *Leaf-sheaths* without a lamina except for the upper which usually has a short one. *Inflorescence* mostly in fascicles of few to numerous, sessile spikelets with some pedunculate ones, variously clustered; lower bract up to twice as long as the mature inflorescence. *Spikelets* 5–10 × 3–4 mm, ovoid, subacute at apex. *Glumes* 3.5–4.0 × 2.0–2.5 mm, reddish-brown with a green mid-vein and hyaline margins, broadly obovate, emarginate at apex and mucronate with a rounded lobe on the back side of the projection, fimbriate-ciliate. *Perianth bristles* 6, linear, not dilated at apex, about equalling the nut, retrorsely scabrid. *Stamens* 3. *Stigmas* 2. *Nut* 2.5–3.0 × 1.5–2.5 mm, reddish-brown, ellipsoid-obovoid, plano- or biconvex, smooth, shiny. *Flowering* 8–9. *Fruits* 9–10. $2n = 40$.

Native. Muddy banks of tidal rivers, very local. By the Rivers Tamar, Arun, Thames and Medway in southern England and by the River Shannon in Ireland. West, central and southern Europe, but local in the north of its range; western Asia.

3. S. pungens (Vahl) Palla   Sharp Club-rush
Scirpus mucronatus auct.; *Scirpus americanus* auct.; *S. americanus* auct.

Glabrous *perennial herb* with a creeping rhizome. *Stems* 30–60 cm × 2–5 mm, solid, thickened towards the base, green, solitary, erect, triquetrous, smooth. *Leaf-sheaths* with 2–3 uppermost each with a lamina, the lamina up to 30 cm, linear, incurved, with a prominent mid-vein beneath and more or less acute at apex. *Inflorescence* with few, sessile spikelets in a tight cluster. *Spikelets* 1–6, ovoid, subacute at apex. *Glumes* 3.5–4.0 × 2.5–3.0 mm, reddish-brown with a paler mid-vein, broadly ovate, emarginate at apex with acute lateral lobes, mucronate, sometimes fringed. *Perianth bristles* absent or up to 6 rudimentary ones which are less than half as long as the nut. *Stamens* 3. *Stigmas* 2. *Nut* 2.0–2.5 × 1.5–2.0 mm, greyish-brown, obovoid, biconvex, smooth, shiny. *Flowering* 6–7. *Fruiting* 8–9. 2n = c. 80.

Native. Margins of ponds near the sea and wet dune-slacks. In a pond-margin in Jersey where it has not been seen since the 1970s. In wet dune-slacks in Lancashire it was discovered in 1928 and lost about 1980, but was planted back from the same stock. West and central Europe extending to central Italy.

8. Isolepis R. Br.

Small, slender, tufted *annual* or *perennial herbs*. *Stems* solid, terete. *Leaf-sheaths* confined to near the base of the stem, the upper 1(–2) with a short lamina. *Leaves* crescent-shaped in section. *Inflorescence* of 1–4, sessile, terete spikelets 1.5–5.0 mm. *Lowest bract* usually more or less stem-like, making the inflorescence appear lateral, sometimes very short and more or less glume-like. *Flowers* bisexual. *Glumes* spirally arranged. *Perianth bristles* absent. *Stamens* 1–2(–3). *Stigmas* (2–)3. *Ovary* not enfolded or enclosed by a glume. *Nut* broadly obovoid or suborbicular.

About 30 species. Cosmopolitan.

1. Bract usually longer than inflorescence; nut longitudinally ridged, shiny   **1. setacea**
1. Bract usually shorter than inflorescence; nut smooth, not shiny   **2. cernua**

1. I. setacea (L.) R. Br.   Bristle Club-rush
Scirpus setaceus L.

Small, tufted *annual* or *perennial herb* with fine, fibrous roots and sometimes a short, knotted rhizome. *Stems* 2–15(–30) cm × 0.3–0.5 mm, terete, ridged. *Leaves* capillary, shorter than stems, crescentiform in section, obtuse at apex, the lowermost reduced to setaceous points; sheaths, loose, membranous, purple-tinged. *Inflorescence* of 1–4(–10), clustered, sessile spikelets, subtended by an erect, stem-like bract, variable in length up to 3 cm, but often much shorter though usually longer than the inflorescence. *Spikelets* 1.5–5.0 × 1–3 mm, ovoid. *Glumes* 1.0–2.1 × 0.6–1.2 mm, ovate, rounded at apex, with the green, thickened, keeled midrib exserted as a blunt mucro, the sides membranous, dark purplish-brown, clearly nerved. *Stamens*

1–2(–3); anthers about 0.3 mm, connective-tip blunt. *Stigmas* (2–)3. *Nut* 0.5–1.2 × 0.4–0.8 mm, broadly obovoid, dark brown, shiny, angles obtuse, rounded-apiculate at apex, conspicuously longitudinally ridged by the short ends of transversely elongate, rectangular cells in vertical rows. *Flowers* 5–7. *Fruits* 6–9. 2n = 28.

Native. Wet, more or less open ground in ditches, fens, marshes and dune-slacks, by ponds and lakes and on heaths. Throughout the British Isles. Most of Europe except the north-east; North Africa; temperate Asia; Madeira; Azores.

2. I. cernua (Vahl) Roem. & Schult.   Slender Club-rush
Scirpus cernuus Vahl; *Scirpus filiformis* Savi, non Burm. fil.; *Scirpus pygmaeus* (Vahl) A. Gray, non Lam.; *Scirpus savii* Sebast. & Mauri; *Fimbristylis pygmaea* Vahl; *I. pygmaea* (Vahl) Kunth

Densely tufted, slender *annual herb*, with fine fibrous roots. *Stems* up to 20 cm × about 0.5 mm, setaceous, terete. *Leaves* inconspicuous, much shorter than stems, usually reduced to a short, blunt apiculus, or the uppermost up to 20 mm, crescentiform in section, ridged; sheaths membranous, loose, tinged purple. *Inflorescence* of 1 or 2, sessile spikelets; bracts up to 10 mm, but usually not exceeding the spikelets, erect when young, but becoming deflexed. *Spikelets* 2–4 × 1.5–2.5 mm, ovoid, acute or subacute at apex, often fuscous. *Glumes* 1.5–2.0 × 1.0–1.5 mm, ovate, distinctly keeled, with a thickened green midrib ending in a minute apiculus, the sides slightly veined. *Stamens* 1–2; anthers about 0.3 mm, connective tip blunt or subacute. *Stigmas* (2–)3. *Nut* about 1.0 × 0.7–0.9 mm, broadly obovoid or suborbicular, unequally trigonous with one rounded angle, apically blunt with a minute apiculus, black, smooth, not shiny, minutely white-papillose. *Flowers* 6–8. *Fruits* 7–9. 2n = 48.

Native. Wet more or less open peaty or sandy ground in ditches, marshes and dune-slacks, mostly near the sea. Frequent in Ireland and extreme western Britain, east to Hampshire and very local in Norfolk. South and west Europe; North Africa; Macaronesia.

9. Eleogiton Link

Floating or submerged, rarely terrestrial *perennial herbs*. *Stems* elongated, compressed, solid, branched, leafy and rooting at the nodes. Usually devoid of leaf-sheaths without a lamina. *Leaves* more or less flat. *Inflorescence* a solitary, terminal spikelet. *Lowest bract* more or less glume-like. *Flowers* bisexual. *Perianth bristles* absent. *Stamens* 3. *Stigmas* 2–3. *Ovary* not enfolded or enclosed by a glume. *Nut* obovoid, compressed-trigonous.

Several species. Cosmopolitan.

1. E. fluitans (L.) Link   Floating Club-rush
Scirpus fluitans L.; *Eleocharis fluitans* (L.) Hook.

Floating or submerged, aquatic, rarely terrestrial, glabrous *perennial herb*, with slender rhizomes. *Stems* 5–50(–130) cm, yellowish-green, solid, compressed, branched, leafy, striate, rooting at the nodes. *Leaves* up to 100 × 2 mm, narrowly linear, more or less acute at apex. *Inflorescence* a solitary, terminal spikelet; peduncles up to 10 cm, smooth, terete below, triquetrous above; lowest bract glume-like. *Spikelets* 2–5 mm, narrowly ovoid, obtuse at apex, with 3–5 flowers. *Glumes* 2.0–2.5 mm, pale greenish-hyaline, sometimes edged with brownish-red, ovate, obtuse at apex. *Stamens* 3. *Stigmas* 2–3. *Nut* 1.2–1.5 mm, whitish or yellowish, obovoid, compressed-trigonous, crowned by the persistent style-base. *Flowers* 6–9. *Fruits* 7–10. 2n = 60.

Native. In or by peaty ditches, streams, lakes and ponds. Fairly widespread but local throughout the British Isles, but commoner in the west. West and west-central Europe extending to south Sweden and central Italy; Africa; Asia; Australia.

10. Blysmus Panz. ex Schult.

Perennial herbs with long rhizomes. *Stems* solid, more or less terete, sometimes trigonous above, leafy mainly below. *Leaves* flat to strongly inrolled. *Inflorescence* a flattened, more or less compact terminal head with spikelets on 2 opposite sides of an axis. *Lowest bract* leaf-like to more or less glume-like. *Flowers* bisexual. *Perianth bristles* 0–6. *Stamens* 3. *Stigmas* 2. *Ovary* not enfolded or enclosed by a glume. *Nut* plano-convex or biconvex.

Four species in temperate Europe and Asia.

1. Leaves flat, keeled, rough; spikelets usually 10–25 per spike; nut 1.5–2.0 mm **1. compressus**
1. Leaves involute, more or less rush-like, smooth; spikelets 3–8 per spike; nut 3–4 mm **2. rufus**

1. B. compressus (L.) Panz. ex Link   Flat Sedge
Schoenus compressus L.; *Carex uliginosa* L.; *Scirpus caricis* Retz. nom. illegit.; *Cyperus horizontalis* Salisb.; *Scirpus compressus* (L.) Pers., non Moench; *Chaetospora compressa* (L.) Gray

Glabrous *perennial herb* with a far-creeping rhizome. *Stems* 3–45 cm, yellowish-green, solid, solitary or caespitose, smooth, striate, more or less terete, sometimes trigonous above. *Leaves* 40–300 × 1–4 mm, bluish-green, linear, gradually narrowed towards the obtuse apex, flat, keeled, antrorsely barbed near the apex, sheathing below. *Inflorescence* a flattened, more or less compact terminal head with (3–)10–25 spikelets on 2 opposite sides of an axis; lowest bract green, longer or shorter than inflorescence, other bracts several-veined and shorter than spikelet. *Spikelets* 4–10 × 2–3 mm, ovate-lanceolate, obtuse at apex. *Glumes* 3–5 × 2–3 mm, yellowish- or reddish-brown with a pale midrib and a narrow hyaline margin, ovate to lanceolate, obtuse to acute at apex, keeled, 2- to 9-veined. *Perianth bristles* 3–6, at least twice as long as the nut, brown, retrorsely barbed. *Stamens* 3. *Stigmas* 2. *Nut* 1.5–2.0 mm, dark brown, shining, smooth, obovoid, plano-convex. *Flowers* 6–7. *Fruits* 8–9. 2n = 44.

Native. Marshy places in rather open communities. Locally frequent in England and Wales, but has disappeared in many localities; very local in Scotland. Europe; Atlas Mountains; temperate Asia.

2. B. rufus (Huds.) Link   Saltmarsh Flat Sedge
Schoenus rufus Huds.; *Scirpus rufus* (Huds.) Schrad.; *Chaetospora rufa* (Huds.) Gray

Glabrous *perennial herb* with a far-creeping rhizome. *Stems* 3–45 cm, yellowish-green, solid, solitary or tufted, smooth, striate, more or less terete, sometimes trigonous above. *Leaves* 40–300 × 1.0–2.5 mm, bluish-green, linear, gradually narrowed towards the obtuse apex, involute and more or less rush-like, smooth, sheathing below. *Inflorescence* a flattened, more or less compact terminal head with 3–8 spikelets on opposite sides of an axis; bracts, except the lowest, 1- to 3-ribbed, equalling the spikelet. *Spikelets* 5–10 × 2–4 mm, ovate-lanceolate, obtuse at apex. *Glumes* 4–6 × 2–3 mm, dark brown with a pale midrib, ovate to lanceolate, obtuse to acute and mucronate at apex, keeled. *Perianth bristles* absent, or 3–6 and much shorter than the nut, white and antrorsely barbed. *Stamens* 3. *Stigmas* 2. *Nut* 3–4 mm, yellow or yellowish-brown, obovoid, plano-convex. *Flowers* 6–7. *Fruit* 8–9. 2n = 40.

Native. In short turf in salt-marshes and dune-slacks. Locally frequent on the coasts of Britain and Ireland south to Lincolnshire and Pembrokeshire; common in western Scotland. Coasts of northern Europe and temperate Asia; inland in north Germany.

11. Cyperus L.

Annual or *perennial herbs*, tufted or rhizomatous. *Stems* solid, triquetrous, leafy. *Leaves* flat to keeled. *Inflorescence* a simple or more often compound umbel or umbel-like raceme with the flattened spikelets clustered on the ultimate branches or all clustered in a more or less dense head. Lowest 2–10 bracts leaf-like, often much exceeding the inflorescence. *Flowers* bisexual. *Glumes* on 2 opposite sides of an axis. *Perianth bristles* absent. *Stamens* 1–3. *Stigmas* (2–)3. *Ovary* not enfolded or enclosed by a glume. *Nut* trigonous.

About 550 species in all the warmer parts of the world, rare in colder regions.

C. brevifolius (Rottb.) Endl. ex Hassk., **C. clarus** S. T. Blake, **C. congestus** Vahl, **C. cyperinus** (Retz.) Sur, **C. dactylotes** Benth., **C. erectus** (Schum.) Mattf. & Kük., **C. esculentus** L., **C. flavus** (Vahl) Boeck., **C. globulosus** Aubl., **C. gunnii** Hook. fil., **C. luzulae** (L.) Retz., **C. ovularis** (Michx) Torr., **C. reflexus** Vahl, **C. rigidifolius** Steud., **C. rotundus** L., **C. rutilans** (Torr.) Maiden & Betche, **C. sesquiflorus** (Torr.) Mattf. & Kük., **C. sporobolus** R. Br., **C. tenuis** Sw., **C. ustulatus** Rich. and **C. vaginatus** R. Br. have been recorded as casual wool aliens. **C. involucratus** Rottb. has been recorded as a garden escape in Essex and Middlesex.

1. Tufted annual; leaves 1.5–5.0 mm wide; glumes 1.0–1.5 mm **3. fuscus**
1. Tufted or rhizomatous perennial; leaves 3–10 mm wide; glumes 2–3 mm **2.**
2. Rhizomes long; inflorescence more or less diffuse; glumes reddish-brown to dark brown **1. longus**
2. Rhizomes short; inflorescence more or less compact; glumes greenish to yellowish-brown **2. eragrostis**

1. C. longus L. Galingale

Loosely tufted *perennial* herb; roots pectinate with many rootlets; rhizomes robust, horizontal, flexuous, thickened at the stem bases, covered with purple-nerved scales which become fibrous with age. *Stems* 20–150 cm × 3–5 mm, erect, acutely triquetrous, smooth. *Leaves* both basal and cauline up to 60 cm × 3–10 mm, linear, with long-acuminate apex, scabrid especially towards the tip; sheaths long, herbaceous and distinctly ribbed dorsally, hyaline above on the ventral side, the outermost dingy reddish-purple; ligule absent. *Inflorescence* a more or less diffuse, simple or compound umbel; bracts leafy, 2–6, up to 60 cm, often exceeding the inflorescence, gradually acuminate; rays 2–10, 8–12(–35) cm, flattened, smooth, secondary rays very slender. *Spikes* few to numerous, broadly ovate, cuneate at base, with loose or dense spikelets. *Spikelets* (2–)5–20 × 1.0–2.5 mm, digitately arranged, erecto-patent to erect, narrowly oblong to oblong-ellipsoid, flattened, acute at apex; rhachila broadly winged. *Glumes* 2.0–2.5 × 1.0–1.7 mm, regularly but rather loosely imbricate on 2 opposite sides of an axis, ovate, obtuse to subacute at apex, sides reddish-brown to dark brown, indistinctly 2-nerved, margin hyaline and decurrent into a persistent wing, the keel acute, green, 3-nerved and excurrent as a minute apiculus. *Stamens* 3; anthers 1.2–2.0 mm, connective-tip conical. *Style* 1; stigmas 3. *Nuts* 0.9–1.0 × 0.3–0.4 mm, oblong-ellipsoid, trigonous, very shortly apiculate, almost smooth, shining dark brown to iridescent-black. *Flowers* 8–9. $2n$ = c. 60.

Native. Marshes, pond-sides and ditches near the coast. Very local, Channel Islands and south-west Britain east to Kent and Suffolk and north to Caernarvon; introduced in scattered localities in southern and central England. South, west and central Europe; west and central Asia; North Africa.

2. C. eragrostis Lam. Pale Galingale
C. vegetus Willd.

Glabrous *perennial herb* with a short, thick rhizome. *Stems* 25–60(–90) cm, solitary, rather greyish-green, striate, triquetrous, smooth, leafy. *Leaves* shorter than to equalling the stems, 4–10 mm wide, greyish-green, linear, gradually narrowed to the acute apex, minutely antrorsely barbed on the margin in the upper part; sheath herbaceous and distinctly ribbed; ligule absent. *Inflorescence* a simple or compound umbel of dense heads; bracts 5–11, exceeding the inflorescence, gradually narrowed to an acute apex, minutely antrorsely barbed on margin; rays 8–10, up to 12 cm, flattened, smooth. *Spikes* few to numerous. *Spikelets* 8–13 × 1.8–3.0 mm, flattened, narrowly lanceolate, acute at apex, patent, with 14–30 flowers; rhachilla not winged. *Glumes* on 2 opposite sides of an axis, 2–3 × 1.0–1.5 mm, greenish or yellowish-brown with a green centre, lanceolate to ovate-lanceolate, trigonous, acuminate at apex. *Stamens* 1. *Stigmas* 3. *Nuts* about half as long as the glumes, grey. *Flowers* 8–10. $2n$ = 42.

Introduced. Grown for ornament and escaping on roadsides, rough ground and by water; also a frequent wool and grass-seed alien. Scattered in southern Britain; in the Channel Islands where it is well naturalised in Guernsey. Native of tropical America.

3. C. fuscus L. Brown Galingale
C. haworthii Gray

Glabrous, tufted, *annual herb*, with reddish, fibrous roots. *Stems* 10–45 cm × 1–3 mm, clustered, soft, triquetrous with concave sides. *Leaves* equalling or exceeding stems, 1.5–5.0 mm wide, flat, linear, acuminate at trigonous tip, slender, very rough; sheaths loose, membranous, red-veined and spotted; ligule about 1 mm, membranous, obtuse at apex. *Inflorescence* a simple or compound umbel, often rather compact; bracts 3, leaf-like, unequal, at least the lowest much exceeding the rays; rays (1–)3–8(–15), up to 4 cm, somewhat flattened, smooth. *Spikes* dense, subspherical; rhachis short or almost none; branches if present short, patent. *Spikelets* 2.5–4.5 × about 1.5 mm, lanceolate to oblong, subcompressed, obtuse at apex, generally dark-coloured; rhachilla not winged, elongating in fruit. *Glumes* on 2 opposite sides of an axis, 1.0–1.5 × 0.8–1.0 mm, ovate, obtuse at apex, closely imbricate when young, spreading later; keel broad and flattened below, more or less sharp above, green, 3-nerved and minutely excurrent in a recurved mucro; sides nerveless, thin, generally tinged with dark purple; margin hyaline, very shortly decurrent. *Stamens* 2; anthers 0.2–0.3 mm, connective-tip minute. *Style* 1, very short; stigmas (2–)3. *Nut* about 0.7 × 0.3–0.4 mm, narrowly obovoid, sharply trigonous or occasionally lenticular, distinctly apiculate, shortly stipitate, greenish to golden-brown with pale angles, microscopically reticulated by the quadrate surface-cells. *Flowers* 7–9. *Fruits* 8–10. $2n$ = 72.

Native. Damp, rather bare ground by ponds and ditches. Very rare in Hampshire, Somerset and Middlesex and in Jersey; formerly elsewhere in southern England. Most of Europe; Asia; North Africa; Madeira, Tenerife.

Subfamily 2. **Rhyncosporoideae** Asch. & Graebn.

Flowers bisexual or unisexual in few-flowered spike-like racemes of spikes or heads.

12. **Schoenus** L.

Densely tufted *perennial herbs*. *Stems* solid, terete. *Leaf-sheaths* only near the base of the stem and bearing long or short laminas which are very thickly crescent-shaped in section to subterete. *Inflorescence* a compact head of flattened spikelets. *Lowest bract* leaf-like to more or less glume-like. *Flowers* bisexual. *Glumes* on 2 opposite sides of the axis. *Perianth bristles* 0–6. *Stamens* 3. *Stigmas* 3. *Ovary* not enfolded or enclosed by glumes. *Nuts* trigonous.

About 100 species, mainly in Australia and New Zealand, a few in Europe, Asia and America.

Sparling, J. H. *Schoenus nigricans* L. in Biological flora of the British Isles. *Jour. Ecol.* **56**: 883–899.

1. Leaves at least half as long as stems; inflorescence of (2–)5–10 spikelets, with the lowest bract usually conspicuously exceeding it   **1. nigricans**
1. Leaves not more than one-third as long as the stems; inflorescence of 1–3(–4) spikelets, with the lowest bract shorter than to as long as it   **2. ferrugineus**

1. S. nigricans L.   Black Bog-rush
Cyperus nigricans (L.) With.; *Chaetospora nigricans* (L.) Kunth

Densely tufted, glabrous *perennial herb*, with fleshy, often dark purple roots which often grow up through the sheaths. *Stems* 15–80 cm × 1.0–1.5 mm, smooth, wiry, usually erect, rarely prostrate, solid, terete below, somewhat flattened above, ribbed. *Leaves* all basal, half as long as the stem or more, linear, wiry, subterete or laterally compressed, adaxially channelled, apically trigonous; sheaths open, with overlapping membranous edges, stiff, ribbed, keeled at least above, all except the innermost dark reddish-brown and shining inside and out; ligules very short, minutely ciliate. *Inflorescence* 10–18 × 8–18 mm, capitate, ovoid or subglobose, made up of crowded bracteate fascicles, each including about (2–)5–10 spikelets, lowest 2 bracts with ovate, sheathing bases and leaf-like tips, the lowermost 2–5 times the length of the inflorescence. *Spikelets* 10–12 × 2–4 mm, lanceolate, laterally flattened, often curved; rhachilla robust, zigzag, winged. *Two lowermost glumes* about 5 mm, lanceolate, apiculate at apex and strongly ciliate on the keel, the 2 lower fertile glumes about 6 mm, subacute at keel and the keel smooth or apically rough, the upper 3–4 fertile glumes ciliate on the keel. *Perianth bristles* 3–4, very short, lanceolate, antrorsely ciliate to apiculate. *Stamens* 3; filaments 10–12 mm; anthers about 4 mm, connective-tip yellow, conical or bifurcate. *Style* 3–5 mm; stigmas 3, long-papillose. *Nut* 1.5–1.9 × about 1.2 mm, ovoid-ellipsoid, obtusely trigonous with convex sides, ivory to grey, smooth and shining. *Flowers* 5–6. *Fruits* 7–8. 2n = 44, 54.

Dwarf variants from the serpentine of Unst, Shetland have been called var. *nanus* Lange. They require further investigation. A prostrate variant occurs on exposed cliff-tops at the Lizard, Cornwall. It is said to retain its habit in cultivation.

Native. Damp peaty places, serpentine heathland, bogs, salt-marshes, fens and flushes. Locally frequent in the British Isles especially near the west coasts and in East Anglia, but absent from much of England, Wales and east Scotland. Most of Europe eastwards to Estonia and the Crimea; North Africa; west Asia.

2. S. ferrugineus L.   Brown Bog-rush
Tufted, glabrous *perennial herb*, with fleshy, dark roots. *Stems* 10–40 cm, pale green, slender but rigid, striate, solid, terete. *Leaves* all basal, up to 20 cm, not more than one-third as long as the stems, pale green, linear, wiry, obtuse to acute at apex, adaxially channelled; sheaths open, with overlapping membranous edges, rigid, striate, keeled at least above, shining reddish-brown to nearly black; ligules very short. *Inflorescence* 7–12 × 2–8 mm, capitate, ovoid or obovoid, each made up of

1–3(–4) spikelets; lower bract shorter than to about equalling the inflorescence, its lamina 3–7 mm and shorter than or equalling the sheathing base. *Spikelets* 7–10 × 1.5–2.5 mm, lanceolate, laterally flattened, acute at apex; rhachilla robust, angled. *Glumes* 5–7 × 1–2 mm, dark reddish-brown with a narrow hyaline margin, lanceolate, acute at apex, keels smooth. *Perianth bristles* (3–)6, exceeding the nut. *Stamens* 3; filaments 10–12 mm; anthers about 3 mm. *Stigmas* 3. *Nut* about 1.5 mm, whitish, narrowly ovoid, trigonous. *Flowers* 6–7. 2n = 76.

Native. Semi-open ground in base-rich flushes and lakesides. Two localities and formerly in a third in Perthshire. From Fennoscandia and north-west Russia to south-east France and south-east Russia.

13. Rhynchospora Vahl nom. conserv.

Perennial herbs, more or less tufted or with far-creeping rhizomes. *Stems* solid, terete to trigonous, leafy. *Leaves* channelled. *Inflorescence* of one to few, rather compact heads, each of 1- to 3-flowered, terete spikelets. *Lowest bract* leaf-like to more or less glume-like. *Flowers* bisexual. *Glumes* spirally arranged. *Perianth bristles* 5–13. *Stamens* 2–3. *Stigmas* 2, the style-base persistent and forming a beak to the fruit. *Ovary* not enfolded or enclosed by a glume. *Nuts* biconvex or trigonous.

About 200 species, cosmopolitan but mainly tropical.

Stewart, A., Pearman, D. A. & Preston, C. D. (1994). *Scarce plants in Britain.* Peterborough. [*R. fusca*.]

1. Plant tufted, with a short rhizome; bracts not or only a little longer than the terminal head; glumes whitish in flower turning pale reddish-brown; perianth bristles 9–13   **1. alba**
1. Plant with a far-creeping rhizome; bracts 2–4 times as long as the terminal head; glumes dark reddish-brown; perianth bristles 5–6   **2. fusca**

1. R. alba (L.) Vahl   White Beak-sedge
Schoenus albus L.; *Juncus stygius* auct.

Slender, glabrous, more or less tufted *perennial herb*, with a short rhizome. *Stems* 10–40(–60) cm, pale green, erect, solid, terete or trigonous above, striate, leafy. *Leaves* 10–15 cm × 1–2 mm, pale greyish-green, narrowly linear, obtuse at apex, channelled; lower sheaths without a lamina, often bearing bulbil-like buds in their axils. *Inflorescence* of one to few rather compact heads, each of several spikelets; lowest bract leaf-like or more or less glume-like, not or slightly exceeding the terminal head. *Spikelets* 4–5 mm, fusiform, acute at apex, usually 2-flowered. *Glumes* usually 4–5, 1–2 of lowest sterile and 2–3 mm, the next fertile, the next sterile and the upper fertile and 4–5 mm, whitish at anthesis, becoming pale reddish-brown, ovate-lanceolate, acute-aristate at apex, keeled. *Stamens* 2–3. *Stigmas* 2, the persistent style-base forming a beak to the nut. *Perianth bristles* 9–13, shorter than or equalling the nut, minutely retrorse-hairy. *Nut* 1.5–2.0 mm, obovoid, biconvex or trigonous, smooth, beaked. *Flowers* 7–8. *Fruits* 8–9. 2n = 26, 42.

Native. In wet, usually peaty places on acid soils. Scattered throughout much of the British Isles and locally common in some places in the north and west, but almost absent from much of England, east and south Wales and eastern Scotland. Europe, except the Mediterranean region and the south-east; Siberia; North America.

2. R. fusca (L.) W. T. Aiton   Brown Beak-sedge
Schoenus fuscus L.

Slender, glabrous *perennial herb,* with a far-creeping rhizome. *Stems* 10–40 cm, pale greyish-green, erect, solid, terete or trigonous, leafy. *Leaves* up to 15 cm × 1 mm, usually much shorter than stems, pale greyish-green, narrowly linear, obtuse at apex, channelled; lower sheaths usually with a short lamina, not bearing bulbils. *Inflorescence* of one to few, rather compact heads, each of several spikelets; bracts 2–4 times as long as terminal head. *Spikelets* 5–6 mm, oblong-ovoid, pointed at apex, 1- to 3-flowered. *Glumes* spirally arranged, usually 4–5, 1–2 of lowest sterile and 2–3 mm, the next fertile, the next sterile and the upper fertile and 4.5–5.5 mm, dark reddish-brown, lanceolate to ovate-lanceolate, acute-aristate at apex, keeled. *Stamens* 2–3. *Stigmas* 2, the persistent style-base forming a beak to the nut. *Perianth bristles* 5–6, longer than nut, minutely antrorse-hairy. *Nut* 1.5–2.0 mm, shining brown, obovoid, biconvex or trigonous, smooth; beak minutely pubescent. *Flowers* 5–6. *Fruits* 8–9. $2n = 32$.

Native. Damp peaty soils on heaths and at margins of bogs. Rare and local, Devon, Dorset, Hampshire and Surrey; Cardiganshire; western Scotland; western Ireland from Cork to Sligo and east to Kildare and Cavan. North-west and central Europe extending eastwards to northern Italy.

14. Cladium P. Browne

Mariscus Scop., non Gaertn.

Rhizomatous *perennial herbs. Stems* more or less terete or trigonous, usually leafy. *Leaves* channelled, with fiercely serrated cutting edges and keel. *Inflorescence* much branched, with many rather compact heads each of several 1- to 3-flowered spikelets. *Lowest bract* of each head leaf-like or glume-like, with long leaf-like bracts at the base of the branches. *Flowers* bisexual. *Perianth bristles* absent. *Stamens* 2(–3). *Stigmas* (2–)3; style-base small, persistent. *Ovary* not enfolded or enclosed by a glume. *Nuts* terete, but flattened basally.

About 60 species in tropical and temperate regions.

Conway, V. M. (1942). *Cladium mariscus* (L.) R. Br. in Biological flora of the British Isles. *Jour. Ecol.* **30**: 211–216.

1. C. mariscus (L.) Pohl   Great Fen Sedge
Schoenus mariscus L.; *C. germanicum* Schrad. nom. illegit.; *Mariscus mariscus* (L.) Borbás nom. illegit.

Harsh, densely tufted, stout, somewhat glaucous *perennial herb,* with thick, fleshy roots; rhizome more or less elongated and branched, about 8 mm in diameter, covered with dark, purplish-brown, triangular to lanceolate scales. *Stems* up to 2 m × 4–8 mm, clustered or remote, thickened basally by sheaths, terete, finely ridged, with hollow internodes. *Leaves* several, mostly basal, but 2–3 cauline, up to 2 m × 10–15 mm, erect,

linear, gradually narrowed to a solid, triangular tip, rounded on the back, thick, hard, finely nerved with transverse anastomoses, sharply serrate on the margin and midrib, especially towards the tip; sheaths of outermost soon splitting, those of inner terete-ridged; ligule absent. *Inflorescence* an elongate, leafy panicle 30–70 × 5–12 cm, with 1–2 lateral, compound, bracteose, more or less corymbose partial panicles at each node; bracts long-sheathing, the lower leaf-like, the upper with a reduced lamina; peduncles unequal, flattened, with acutely serrulate margins, twice or thrice terminally branched; branches subtended by slender, non-sheathing bracts, unequal, those of the last order bearing subglobose, subdigitate clusters of few spikelets. *Spikelets* 2–4 × 0.7–2.0 mm, ovoid, terete, 2- to 4-flowered, pale reddish-brown. *Glumes* papery, acute at apex, with one thin medium nerve, the lowermost 2(–4) empty; small fertile glumes, about 4 × 3 mm, ovate. *Stamens* 2; anthers about 3 mm, with a conical, red, densely papillose connective-tip. *Stigmas* 3. *Perianth bristles* absent. *Nut* about 3.0 × 1.5 mm, dark brown, ellipsoid, shining, brittle, terete but flattened basally, with a conical beak frequently shed with an obscurely 3-lobed, saucer-like basal disc; exocarp spongy; endocarp stony. *Flowers* 7–8. *Fruits* 8–9. $2n = 36$, c. 60.

Native. Wet, base-rich areas in fens and by streams and ponds. Scattered over the British Isles and locally common, especially in Norfolk and west and central Ireland. Where it grows over a large dominant area it has in the past been cut for thatching, and is still used in some areas. Europe, north to central Finland; North Africa; Asia.

Subfamily 3. Caricoideae Pax

Flowers unisexual, naked, usually in many-flowered spikes, females surrounded by a utricle.

15. Kobresia Willd.

Densely tufted *perennial herbs. Stems* trigonous, solid, leafy at base. *Leaves* mostly basal, channelled. *Inflorescence* of 1-flowered spikelets arranged in a terminal cluster of 3–10 spikes. *Lowest bract* a sheath with a short, usually brown blade. *Flowers* unisexual, the upper spikelets male and lower female in each spike. *Perianth bristles* absent. *Stamens* 3. *Stigmas* 3. *Female flowers* with an extra glume folded, but not fused around the ovary. *Nut* trigonous.

About 50 species in north temperate regions.

1. K. simpliciuscula (Wahlenb.) Mack.   False Sedge
Carex simpliciuscula Wahlenb.; *Schoenus monoicus* Sm.; *K. caricina* Willd.; *K. bipartita* auct.

Densely tufted, glabrous *perennial herb. Stems* 5–20 cm, greyish-green, erect, trigonous, solid, stiff, striate, rough, leafy at extreme base. *Leaves* shorter than stems, 0.5–1.5(–2) mm wide, greyish-green, narrowly linear, more or less obtuse at apex, channelled, minutely antrorsely barbed on the margin near the apex; basal sheaths pale orange-brown, dull. *Inflorescence* of 1-flowered spikelets arranged in a terminal cluster of 3–10 spikes; lowest bract a sheath with a short, usually brown lamina. *Spikes* with male spikelet at top and female below. Spikelets 4–6 mm, lanceolate, obtuse at apex, 1-flowered. *Glumes* 4–5 mm, reddish-brown with a green midrib and a scarious margin, elliptic-ovate with an acuminate-aristate apex. *Stamens* 3. *Stigmas* 3. *Perianth bristles* absent. *Nut* 2–3 mm, pale dull brown, trigonous. *Flowers* 6–7. *Fruits* 7–8. $2n = 72$.

Native. Flushed grassy areas and on damp banks on moors and mountains. Very local in Teesdale and Perthshire; formerly in adjacent counties. Northern Europe; Pyrenees, Alps, Carpathians; Caucasus, Altai; Greenland; North America.

16. Carex L.

Vignea P. Beauv. ex Lestib.; *Trasus* Gray; *Edritria* Raf.; *Facolos* Raf.; *Diemisa* Raf.; *Onkerma* Raf.; *Physiglochis* Raf.; *Psyllophora* Heuff.; *Leucoglochin* Heuff.; *Vignantha* Schur; *Caricina* St-Lag.; *Caricinella* St-Lag.; *Proteocarpus* Börner; *Manochlaenia* Börner; *Limivasculum* Börner; *Rhaptocalymma* Börner; *Rhynchopera* Börner, non Klotzsch; *Desmigrestis* Börner

Extensively rhizomatous to densely tufted *perennial herbs*. *Stems* simple, triquetous, trigonous or terete, usually leafy, up to halfway or only at the extreme base. *Leaves* flat to channelled or inrolled, usually sheathing, with a ligule at the junction of the lamina and sheath. *Inflorescence* of 1-flowered spikelets grouped in variously arranged spikes all except the terminal subtended by a bract. *Lowest bract* without a sheath but with a short, green or brown lamina. *Flowers* unisexual, each in the axis of one glume, the sexes variously arranged from mixed in a single spike to separate plants, but often with the upper spikes entirely male and the lower entirely female. *Perianth bristles* absent. *Stamens* 2 or 3. *Stigmas* 2 or 3. *Female flowers* with an extra inner glume, the utricle, which is completely fused around the ovary forming a false fruit enclosing the nut and usually with a long or short beak. *Nut* biconvex or trigonous.

Probably about 2,000 species world-wide, often in wet places and abundant in the Arctic.

This account, with the exception of *C. flava*, is based on Jermy, Chater and David (1982).

There are four kinds of rhizome and shoot systems. (i) A sympodial rhizome system which produces shoots and branches every few nodes, the general direction of growth being upwards, thus forming a tussock. (ii) A sympodial rhizome system in which the internodes between the shoots are few and usually short, radiating from a centre, the shoots either few or many forming a tufted plant. (iii) A sympodial system in which at least some of the rhizomes are either short or far-creeping, the shoots in dense or loose tufts joined by rhizomes. (iv) A monopodial rhizome system in which a single or rarely two shoots are produced at regular intervals along a creeping rhizome. Thus the creeping, tufted or tussock habit is diagnostic but to some extent can be affected by habitat. The only aerial stem is that bearing a terminal inflorescence. It is usually trigonous, although in a few British species it is terete and ridged.

The leaves are on the whole grass-like and vary in width and shape of the upper part and apex. In some species the flat surface continues to the tip, the midrib gradually petering out. In others the tip becomes subulate and trigonous with the midrib channel ending abruptly. Leaves of sterile shoots may not show this so clearly. Most species have obvious laminas to the leaves, but they may be keeled, channelled, inrolled or folded. A few species have setaceous leaves which are round or triangular in section. The leaf-sheath, that is the lower part of the leaf which is tubular around the shoot or stem, shows important colour and texture characteristics. The most important part of the leaf-sheath is the thin, colourless, hyaline inner face, of which the texture and spotting and the shape of its apex should be noted. The ligule, situated between the leaf-sheath and the lamina, differs from that of the grasses in being fused for most of its length to the upper surface of the lamina.

The inflorescence is basically a spike or a panicle whose flowers are unisexual, but both kinds are borne in the same inflorescence except in two species. The most common arrangements of the flowers is where the terminal and upper spikes are entirely male and the lower are all female, or occasionally the higher female spikes have male flowers at the top. A variation of this is where the inflorescence is a single spike with male flowers at the top and female at the base. The other arrangement, rare in the British Isles, is where the terminal spike has female flowers at the top and male at the base; the lower spikes may repeat this pattern, or more likely be entirely female.

In Subgenus *Carex* all the spikes except the terminal one are subtended by a bract, which may not be green, but like a large glume. It may, however, have the green midrib of the glume exaggerated and the remainder reduced, and it is then called setaceous. On most lower spikes the green midrib is exaggerated still further to form a leaf-like bract.

In the Subgenus *Vignea* the inflorescence is often more compound and the spikes or spikelets, as the secondary spikes are often called, are small in comparison with those of Subgenus *Carex*, but the arrangement of the male and female flowers is basically the same. Bracts in this subgenus tend to be glumaceous or setaceous and some secondary spikes lack bracts altogether.

The male flowers consist of three stamens arising from a receptacle and subtended by a glume. As the anthers mature the filaments elongate and the stamens hang from the sheathing glume. The anther falls but the white filaments are left, thus showing the position of the male flowers in fruiting plants.

The female flowers consist of a bottle-shaped utricle containing a single ovary and subtended by a glume. The colour, shape and size of the glumes are important in identification. The utricles of some species become abnormally swollen and shiny when infected by gall flies. The apex of the utricle is drawn out into a beak and the orifice through which the style protrudes may be split, bifid, notched or obliquely truncate.

The ovary has either two stigmas and is a flat circular disc which matures into an ellipsoid biconvex nut, or has three stigmas and is usually cylindrical and matures into a trigonous, ellipsoid or obovoid nut.

As far as is known all species of *Carex* are outbreeding and there is no record of apomixis or vivipary. Wind is the main pollinating agent, although pollen-eating beetles may transport pollen effectively.

If the genus *Carex* was treated in the same way as *Scirpus* it would be split into a number of genera and numerous names are available for such genera. No modern treatment, however, has accepted such a classification.

C. capitata L., **C. bicolor** All. and **C. glacialis** Mackenzie were recorded from the Western Isles of Scotland in the 1940s, but have not been seen recently and were probably planted. **C. appressa** R. Br., **C. devia** Cheeseman, **C. deweyana** Schweinf., **C. flagellifa** Colenso, **C. hubbardii** Nelmes, **C. inversa** R. Br., **C. longebrachiata** Boeck., **C. secta** Boott, **C. solandri** Boott, **C. tereticaulis** F. Muell. and **C. virgata** Boott have been recorded as casual wool aliens.

David, R. W. (1977). The distribution of *Carex montana* L. in Britain. *Watsonia* **11**: 377–378.

David, R. W. (1978a). The distribution of *Carex digitata* L. in Britain. *Watsonia* **12**: 47–49.

David, R. W. (1978b). The distribution of *Carex elongata* L. in the British Isles. *Watsonia* **12**: 158–160.

David, R. W. (1979a). The distribution of *Carex humilis* Leyss. in the British Isles. *Watsonia* **12**: 257–258.

David, R. W. (1979b). The distribution of *Carex rupestris* All. in Britain. *Watsonia* **12**: 335–337.

David, R. W. (1981a). The distribution of *Carex ericetorum* Poll. in Britain. *Watsonia* **13**: 225–226.

David, R. W. (1981b). The distribution of *Carex punctata* Gaud. in Britain, Ireland and Isle of Man. *Watsonia* **13**: 318–321.

David, R. W. (1982a). The distribution of uncommon *Carices*: addenda and corrigenda. *Watsonia* **14**: 68–70.

David, R. W. (1982b). The distribution of *Carex maritima* Gunn. in Britain. *Watsonia* **14**: 178–180.

David, R. W. (1983). The distribution of *Carex tomentosa* L. (*C. filiformis* auct.) in Britain. *Watsonia* **14**: 412–414.

David, R. W. (1990). The distribution of *Carex appropinquata* Schumacher (*C. paradoxa* Willd.) in Great Britain and Ireland. *Watsonia* **18**: 201–204.

David, R. W. & Kelcey, J. G. (1985). *Carex divulsa* Stokes, *C. muricata* L. and *C. spicata* Huds. in Biological flora of the British Isles. *Jour. Ecol.* **73**: 1021–1039.

Davies, E. W. (1953). Notes on *Carex flava* and its allies. I. A sedge new to the British Isles. *Watsonia* **3**: 66–69.

Davies, E. W. (1953). Notes on *Carex flava* and its allies. II. *Carex lepidocarpa* in the British Isles. *Watsonia* **3**: 70–73.

Davies, E. W. (1953). Notes on *Carex flava* and its allies. III. The taxonomy and morphology of the British representatives. *Watsonia* **3**: 74–84.

Faulkner, J. S. (1972). Chromosome studies on *Carex* section *Acutae* in north-west Europe. *Bot. Jour. Linn. Soc.* **65**: 271–301.

Faulkner, J. S. (1973). Experimental hybridization of north-west European species in *Carex* section *Acutae* (Cyperaceae). *Bot. Jour. Linn. Soc.* **67**: 233–253.

Grime, J. P. et al. (1988). Comparative plant ecology. London. [*Carex acutiformis, caryophyllea, flacca, nigra, panicea* and *pilulifera*.]

Jermy, A. C., Chater, A. O. & David, R. W. (1982). *Sedges of the British Isles*. B.S.B.I. Handbook No. 1. Ed. 2.

Kükenthal, G. (1909). *Das Pflanzenreich* **4**(20). Cyperaceae: Caricoideae. Leipzig.

Noble, J. C. (1982). *Carex arenaria* L. in Biological flora of the British Isles. *Jour. Ecol.* **70**: 867–886.

Schmid, B. W. (1983). Notes on the nomenclature and taxonomy of the *Carex flava* group in Europe. *Watsonia* **14**: 309–319.

Stewart, A., Pearman, D. A. & Preston, C. D. (1994). *Scarce plants in Britain*. Peterborough. [*C. appropinquata, aquatilis, atrata, capillaris, digitata, divisa, elata, elongata, ericetorum, humilis, magellanica, maritima, montana, punctata, rupestris, saxatilis, vaginata* and *vulpina*.]

Taylor, F. J. (1956). *Carex flacca* Schreb. in Biological flora of the British Isles. *Jour. Ecol.* **44**: 281–290.

1. Spike solitary   2.
1. Spikes 2 or more   7.
2. Stigmas 2; nut biconvex   3.
2. Stigmas 3; nut trigonous   5.
3. Plant monoecious   **74. pulicaris**
3. Plant dioecious   4.
4. Plant rhizomatous; stems usually smooth and terete
   19. dioica
4. Plant densely tufted; stems usually scabrid and more
   or less trigonous above   **20. davalliana**
5. Leaves more or less curled at apex; female glumes
   persistent; utricles erecto-patent to erect at maturity
   73. rupestris
5. Leaves more or less straight; female glumes caducous;
   utricles deflexed at maturity   6.
6. Utricles 3.5–4.5(–6) mm, with a bristle arising from
   the base of the nut and protruding from the top of
   the beak along with the style-base   **71. microglochin**
6. Utricles 5–7 mm, without a bristle   **72. pauciflora**
7. Spikes all more or less similar in appearance, the
   terminal usually at least partly female   8.
7. Spikes dissimilar in appearance, the terminal or
   upper usually entirely male, the lower usually entirely
   female   35.
8. Stigmas 2; nut biconvex   9.
8. Stigmas 3; nut trigonous   32.
9. All spikes with female flowers at apex   10.
9. At least one spike with male flowers at apex   16.
10. Body of utricles distinctly winged for at least part of
    its length   11.
10. Utricles unwinged, except sometimes narrowly so on
    the beak   12.
11. Leaves usually much shorter than stems; spikes 2–9;
    female glumes 3.0–4.5 mm   **16. ovalis**
11. Leaves often more or less equalling stems; spikes
    3–15; female glumes 2.5–3.0 mm   **17. crawfordii**
12. Lowest bract leaf-like, exceeding the inflorescence
    15. remota
12. Lowest bract usually not leaf-like, shorter than the
    inflorescence   13.
13. Spikes subglobose; utricles usually very divaricate at
    maturity, not more than 10   **18. echinata**
13. Spikes ovoid to oblong; utricles erect or erecto-
    patent, more than 10   14.

14. Spikes whitish, greenish or pale brown   **23. curta**
14. Spikes dark reddish-brown   15.
15. Spikes 5–18; utricles without a slit down the back of the beak; lowland wet places   **21. elongata**
15. Spikes 2–5; utricles with a slit down the back of the beak; high mountains   **22. lachenalii**
16. Plant densely caespitose, without creeping rhizomes   17.
16. Plant not or laxly caespitose, with creeping rhizomes   27.
17. Utricles weakly to strongly convex on adaxial side, strongly convex on abaxial side   18.
17. Utricles plane on adaxial side, weakly convex on abaxial side   21.
18. Utricles 2.0–2.6 mm   **6. vulpinoidea**
18. Utricles (2.7–)3–4 mm   19.
19. Utricles broadly winged in upper half   **1. paniculata**
19. Utricles not or only very narrowly winged   20.
20. Usually tussock-forming; basal sheaths fibrous; lower clusters of spikes more or less pedunculate   **2. appropinquata**
20. Not tussock-forming; basal sheaths entire; lower spikes or clusters of spikes sessile   **3. diandra**
21. Stems more than 2 mm wide; leaves 4–10 mm wide; utricles prominently veined more or less throughout   22.
21. Stems less than 2 mm wide; leaves 2–5 mm wide; utricles without veins or with faint ones at base   23.
22. Ligule much wider than long, overlapping edges of leaf; utricles dull, papillose, with isodiametric, thick-walled epidermal cells   **4. vulpina**
22. Ligules longer than wide, not overlapping edges of leaf; utricles shining, smooth, with oblong, thin-walled, epidermal cells   **5. otrubae**
23. Roots and often basal sheaths and base of stems, ligules and occasionally glumes purplish-tinged; ligule distinctly longer than wide; utricles corky and thickened at base   **7. spicata**
23. Roots, basal-sheaths and base of stems not purplish-tinged; ligule not or only slightly longer than wide; utricles not corky and thickened at base   24.
24. Lowest 3–4 spikes or branches separated from each other by a gap of much more than their own length; ripe utricles not divaricate   **9(a). divulsa subsp. divulsa**
24. Lowest spikes overlapping, or separated from each other by a gap of not more than their own length; ripe utricles divaricate   25.
25. Ligules usually wider than long; inflorescence 3–5(–8) cm; utricles 4.5–5.0(–5.5) mm, equally narrowed at both ends   **9(b). divulsa subsp. leersii**
25. Ligules about as wide as long; inflorescence 1–4 cm; utricles (2.6–)3.0–4.5 mm, truncate or rounded at base   26.
26. Spikes globose; female glumes blackish or dark reddish-brown, much darker and shorter than the greenish or brownish utricles; utricles (3.5–)4.0–4.5 mm, strongly divaricate   **8(a). muricata subsp. muricata**
26. Spikes ovoid; female glumes pale brown, similar in colour to or paler than, and almost as long as the utricles; utricles 2.6–3.5(–4) mm, erecto-patent   **8(b). muricata subsp. lamprocarpa**
27. Body of utricle distinctly winged for at least part of its length   28.

27. Utricle unwinged except sometimes narrowly so on the beak   29.
28. Middle spikes male at top, female below; terminal spike entirely male   **10. arenaria**
28. Middle spikes entirely male; terminal or upper spikes entirely female   **11. disticha**
29. Stems smooth   30.
29. Stems rough at least towards the top   31.
30. Stems 15–40 cm; utricles abruptly contracted into the beak; very wet bogs   **12. chordorrhiza**
30. Stems 1–18 cm; utricles gradually narrowed into the beak; coastal sands or rocks   **14. maritima**
31. Plant laxly caespitose, with short, ascending rhizomes; beak comprising more than one-third of the length of the utricle; lowest bract shorter than its spike   **3. diandra**
31. Rhizomes far-creeping; beak comprising less than one-third of the length of the utricle; lowest bract longer than its spike and usually longer than the whole inflorescence   **13. divisa**
32. Lowest spike erect   33.
32. Lowest spike drooping or pendulous   34.
33. Spikes more or less remote or the lowest arising at least 10 mm from the one above; utricles pale green; lakeside meres   **62. buxbaumii**
33. Spikes in a compact cluster; utricles greenish-brown; high alpine ledges and flushes   **63. norvegica**
34. Lowest bract with a sheath at least 5 mm; terminal spike male at top, female at base   **57. atrofusca**
34. Lowest bract not sheathing or with a sheath less than 3 mm; terminal spike female at top, male at base   **61. atrata**
35. Stigmas 2; utricles usually plano-convex or biconvex; nuts biconvex   36.
35. Stigmas 3; utricles usually trigonous or inflated; nuts trigonous   45.
36. Utricles inflated, patent   37.
36. Utricles not inflated, erect or erecto-patent   38.
37. Utricles 4–5 mm, ovoid-ellipsoid, sterile   **31. × grahamii**
37. Utricles 3.0–3.5 mm, ovoid, fertile   **32. saxatilis**
38. Densely caespitose, sometimes tussock-forming plants, without creeping rhizomes   39.
38. Not or laxly caespitose plants, with creeping rhizomes   40.
39. Basal sheaths brown or reddish-brown, not fibrous; margins of leaves rolling inwards on drying, the stomata mostly confined to the upper surface; lowest bract usually exceeding spike but not inflorescence   **68. nigra**
39. Basal sheaths yellowish-brown, becoming conspicuously reticulately fibrous; margins of leaves rolling outwards on drying; the stomata confined to the lower surface; lowest bract usually shorter than spike   **69. elata**
40. Utricles without veins   41.
40. Utricles with veins, sometimes obscurely so   42.

41. Stems bluntly trigonous, brittle; lowest bract
    exceeding inflorescence **65. aquatilis**
41. Stems sharply trigonous, not brittle; lowest bract not
    exceeding inflorescence **70. bigelowii**
42. Female glumes, at least in lower part of spikes,
    aristate, up to 5 times as long as utricles **64. recta**
42. Female glumes obtuse or acute, not more than 1.5
    times as long as utricles 43.
43. Leaves 3–10 mm wide, the margins rolling outwards
    on drying, the stomata confined to the lower surface;
    lowest bract exceeding and often concealing the
    inflorescence; male spikes 2–4 **66. acuta**
43. Leaves 1–3(–5) mm wide, the margins rolling inwards
    on drying, the stomata mostly confined to the upper
    surface; lowest bract narrow, not exceeding or
    concealing inflorescence; male spike usually solitary 44.
44. Leaves greyish- or glaucous-green, channelled; lower
    bract channelled and exceeding the inflorescence;
    female glumes 3-veined; utricles 3.5–5.0 mm **67. trinervis**
44. Leaves glaucous, more or less flat; lower bract more
    or less equalling inflorescence; female glumes with a
    pale midrib; utricles 2.5–3.5 mm **68. nigra**
45. Utricles hairy on at least part of the surface of the
    body 46.
45. Utricles glabrous on surface of body, though
    sometimes ciliate or hispid-denticulate on margin or
    on beak 56.
46. Utricles with a prominent, bifid beak usually more
    than 0.5 mm 47.
46. Utricles without or with a short, usually conical,
    entire or notched beak not more than 0.5 mm 48.
47. Leaves 2–5 mm wide, hairy at least on the sheaths;
    utricles (4.5–)5.0–7.0 mm **24. hirta**
47. Leaves 1–2 mm wide, glabrous; utricles 3–5 mm
    25. lasiocarpa
48. Plant not or only laxly caespitose, with creeping
    rhizomes 49.
48. Plant more or less densely caespitose, without
    creeping rhizomes 52.
49. Basal sheaths remaining entire; male spikes usually 2
    or more; utricles papillose **37. flacca**
49. Basal sheaths becoming fibrous; male spike solitary;
    utricles strongly hairy 50.
50. Stems usually more than 20 cm; leaves erect; basal
    sheaths red, shiny; lowest bract exceeding spike
    53. filiformis
50. Stems usually less than 20 cm; leaves more or less
    recurved; basal sheaths brown; lowest bract usually
    shorter than spike 51.
51. Lowest bract with a sheath 3–5 mm; female glumes
    equalling utricles, green or brownish, without or with
    a narrow, scarious margin, acute at apex
    52. caryophyllea
51. Lowest bract not sheathing or with a sheath less than
    2 mm; female glumes usually shorter than utricles,
    purplish-black, with a wide scarious and often ciliate
    margin **54. ericetorum**
52. Inflorescence comprising more or less all of stem;
    female spikes with 2–4 flowers **51. humilis**
52. Inflorescence comprising not more than two-thirds of
    stem; female spikes with usually more than 4 flowers 53.

53. Flowering stems lateral and leafless; female spikes
    not more than 3 mm wide, lax 54.
53. Flowering stems terminal, leafy at base; female
    spikes 4–6 mm wide, dense 55.
54. Female spikes separated; utricles 3.0–4.5 mm, more
    or less equalling the purplish glumes **49. digitata**
54. Female spikes all arising from more or less the same
    point; utricles 2–3 mm, 1.5–2.0 times as long as the
    straw-coloured glumes **50. ornithopoda**
55. Leaves soft, pale green; basal sheaths reddish-brown;
    utricles 3.0–4.5 mm, pyriform **55. montana**
55. Leaves more or less rigid, mid- or greyish-green;
    basal sheaths brown; utricles 2.0–3.5 mm, obovoid
    56. pilulifera
56. At least the lowest spike pendulous 57.
56. Spikes not pendulous 73.
57. Male spikes 2 or more 58.
57. Male spike solitary 63.
58. Utricle without or with an entire or weakly notched
    beak less than 0.5 mm 59.
58. Utricle with a usually strongly bifid beak more than
    0.5 mm 60.
59. Leaves more than 7 mm wide; lowest bract with a
    sheath (30–)50–100 mm **33. pendula**
59. Leaves less than 7 mm wide; lowest bract not
    sheathing or with a sheath not more than 3(–10) mm
    37. flacca
60. Plants with creeping rhizomes and stout erect stems 61.
60. Plants caespitose, without or with short creeping
    rhizomes 62.
61. Female glumes exceeding utricles **27. riparia**
61. Female glumes shorter than utricles **30. vesicaria**
62. Female spikes 3–5 mm wide **34. sylvatica**
62. Female spikes 6–8 mm wide **41. laevigata**
63. Utricles without or with a truncate or obliquely
    truncate beak 64.
63. Utricles with a distinct, bifid or prominently notched
    beak 69.
64. Plants densely caespitose, without creeping rhizomes 65.
64. Plants not or laxly caespitose, with distinct creeping
    rhizomes 66.
65. Leaves more than 7 mm wide; female spikes 70–160
    mm **33. pendula**
65. Leaves less than 3 mm wide; female spikes 5–25 mm
    35. capillaris
66. Utricles without veins, minutely papillose **37. flacca**
66. Utricles more or less veined, smooth 67.
67. Female spikes 3–4 mm wide, with 5–8 flowers; not
    occurring below 600 m **59. rariflora**
67. Female spikes 5–7 mm wide, with 7–20 flowers;
    usually not occurring above 600 m 68.
68. Leaves 1–2 mm wide; lowest spikes entirely female;
    female glumes less than 1.5 times as long as and at
    least as wide as utricles **58. limosa**
68. Leaves (1.5–)2–4 mm wide; lowest spike with male
    flowers at base; female glumes more than 1.5 times as
    long as and only two-thirds as wide as utricles
    60. magellanica

69. Female spikes ovoid, of shaggy, blackish appearance
57. atrofusca
69. Female spikes cylindrical, green or pale to purplish-brown in appearance   70.
70. All spikes overlapping; lowest bract at least twice as long as inflorescence   **28. pseudocyperus**
70. At least the lower spikes distant; lowest bract less than twice as long as (usually even shorter than) the inflorescence   71.
71. Female spikes 3–5 mm wide; utricles with a smooth beak   **34. sylvatica**
71. Female spikes 5–8 mm wide; utricles with a scabrid beak   72.
72. Leaves 4–12 mm wide; ligules 7–15 mm   **41. laevigata**
72. Leaves 3–6 mm wide; ligules 1–2 mm   **42. binervis**
73. Sheaths and usually lower surface of leaves pubescent   **48. pallescens**
73. Plant glabrous   74.
74. Lowest bract not sheathing   75.
74. Lowest bract with a cylindrical sheath at least 1 mm   80.
75. Utricles papillose   **37. flacca**
75. Utricles smooth   76.
76. Utricles with a notched beak about 0.5 mm
26. acutiformis
76. Utricles with a bifid beak at least 0.7 mm   77.
77. Female glumes exceeding utricles   **27. riparia**
77. Female glumes shorter than utricles   78.
78. Utricles sterile; alpine flushes on mainly inorganic soils   **31. × grahamii**
78. Utricles fertile; meres and reedswamps, if on mountains then in deep peat flushes   79.
79. Leaves usually glaucous; utricles 3.5–6.5 mm, patent, more or less abruptly contracted into a beak **29. rostrata**
79. Leaves usually yellowish-green; utricles (4–)5–8 mm, erecto-patent, gradually narrowed into the beak
30. vesicaria
80. Plant not or only laxly caespitose, with long creeping rhizomes   81.
80. Plant more or less densely caespitose, without or with very short creeping rhizomes   85.
81. Utricles papillose   **37. flacca**
81. Utricles smooth   82.
82. Utricles 7–9 mm   **40. depauperata**
82. Utricles less than 7 mm   83.
83. Utricles with a prominent, scabrid, bifid beak
46. hostiana
83. Utricles without or with a smooth, entire or weakly notched beak   84.
84. Leaves glaucous; sheath of lowest bract not inflated, tight; utricle abruptly contracted into a very short beak less than 0.5 mm   **38. panicea**
84. Leaves yellowish to bronzy green; sheath of lowest bract inflated, loose; utricles gradually narrowed into a beak 0.5–1.0 mm   **39. vaginata**
85. Flowering stems lateral, leafless   **50. ornithopoda**
85. Flowering stems terminal in the middle of leaf rosettes, usually leafy at base   86.
86. Utricles 7–9 mm   **40. depauperata**

86. Utricles less than 7 mm   87.
87. At least the lower and middle utricles in each spike patent or deflexed   88.
87. All utricles erecto-patent or appressed 97.
88. Apex of sheath of stem leaves with a projection opposite the ligule; female glumes with a prominent, wide silvery, scarious margin   **46. hostiana**
88. Apex of sheath of stem leaves without a projection; female glumes without a prominent scarious margin   89.
89. Female spikes oblong to cylindrical, more or less distant   **44. punctata**
89. Female spikes ovoid, on some stems at least the 2 upper approximate   90.
90. Beak of utricle deflexed or curved, usually at least half as long as the usually more or less curved body   91.
90. Beak of utricle neither deflexed nor curved, usually less than half as long as the more or less straight body   94.
91. Utricles 5–7 mm; male spike usually subsessile; leaves up to 7 mm wide   **47(a). flava** subsp. **flava**
91. Utricles (3–)3.5–5.0 mm; male spike usually pedunculate; leaves not more than 4 mm wide   92.
92. Leaves from two-thirds as long to equalling stems; utricles gradually narrowed into the beak
47(b). flava subsp. **jemtlandica**
92. Leaves about half as long as stems; utricles abruptly contracted into the beak   93.
93. Leaves 1.5–2.5(–3.5) mm wide; female glumes pale brownish-yellow   **47(c). flava** subsp. **brachyrrhyncha**
93. Leaves 2.5–4.5 mm wide; female glumes dark chestnut brown   **47(d). flava** subsp. **scotica**
94. Male spike usually sessile; utricles abruptly contracted into the beak   **47(g). flava** subsp. **serotina**
94. Male spike usually pedunculate; utricles gradually narrowed into the beak   95.
95. Utricles 1.7–2.0 mm   **47(h). flava** subsp. **pulchella**
95. Utricles 3–4 mm   96.
96. Stems erect; leaves 1.0–3.5 mm wide; utricles usually not more than 3.5 mm   **47(e). flava** subsp. **bergrothii**
96. Stems ascending; leaves usually more than (3–)3.5 mm wide; utricles usually more than 3.5 mm
47(f). flava subsp. **oedocarpa**
97. Apex of sheath of stem-leaves truncate opposite the ligule; female spikes 2–3 mm wide   **36. strigosa**
97. Apex of sheath of stem-leaves with a projection opposite the ligules; female spikes at least 4 mm wide, dense   98.
98. Leaves channelled or with inrolled margins; lowest bract equalling or exceeding inflorescence   99.
98. Leaves flat or plicate; lowest bract shorter than inflorescence   100.
99. Stems up to 40 cm; leaves usually more or less straight; utricles 3–4 mm   **45(i). extensa** var. **extensa**
99. Stems 4–12 cm; leaves often curved; utricles 2–3 mm
45(ii). extensa var. **minor**
100. At least some leaves on the plant more than 6 mm wide; ligules more than 5 mm   **41. laevigata**
100. Leaves not more than 6 mm wide; ligule less than 5 mm   101.

101. Leaves abruptly contracted below the linear, veinless apex; female glumes with a prominent, wide, silvery, scarious margin **46. hostiana**
101. Leaves gradually narrowed to the apex; female glumes without a prominent scarious margin 102.
102. Basal sheaths orange-brown; female glumes dark reddish- or purplish-brown **42. binervis**
102. Basal sheaths pale to dark brown, not orange; female glumes pale brown or pale reddish-brown **43. distans**

Subgenus 1. Vignea (P. Beauv. ex Lestib.) Kük.
Vignea P. Beauv. ex Lestib.

Monoecious, with 2 or more spikes all similar in appearance, hermaphrodite or unisexual, in a simple or paniculately branched inflorescence, or with a solitary spike with male flowers at base, or dioecious with a solitary spike. *Lateral spikes* without a scale between the bract and the lowest glume. *Stigmas* usually 2.

Section 1. Heleoglochin Dumort.

Tufted herbs. Stems trigonous or triquetrous, rough above. *Inflorescence* usually branched; spikes numerous, male above and female below (or the lower entirely female); lowest bract shorter than branch, usually setaceous or glumaceous. *Female glumes* equalling or slightly shorter than utricles. *Utricles* ovoid or suborbicular to ovoid-pyriform, erect or erecto-patent, weakly to strongly convex on abaxial side, with a serrulate or scabrid, bifid beak. *Stigmas* 2.

1. C. paniculata L. Greater Tussock Sedge
Vignea paniculata (L.) Rchb.; *Caricina paniculata* (L.) St-Lag.

Glabrous *perennial herb*, forming dense tussocks up to 1.5 m high and 1 m in diameter; rhizomes very short; roots thick, dark brown, felty; scales dark brown, ribbed, persistent. *Stems* 60–150 cm, dark green, trigonous, spreading, rough. *Leaves* overwintering, 20–120 cm × 4–7 mm, dark green, shiny beneath, long-linear, stiff, channelled or involute, tapering abruptly to a short, trigonous tip, margins serrulate; lowermost sheaths brown with shorter lamina, inner face hyaline, brown at apex, concave; ligules 2–5 mm, more or less rounded at apex. *Inflorescence* 5–15 cm, a compact panicle. *Spikes* numerous, mostly compound, the lower clusters pedunculate; spikelets 5–8 mm, male at top, female below, some lower spikelets all female; bracts setaceous or glumaceous, with a broad hyaline margin and excurrent midrib. *Male glumes* 3–4 mm, greyish- or orange-brown, hyaline, with a pale midrib, ovate-lanceolate, with an acute apex. *Female glumes* 3–4 mm, greyish- or orange-brown, with a wide hyaline margin, appressed to utricle, triangular-ovate, with an acute or acuminate apex. *Utricles* 3–4 mm, green to dark brown, ovoid, more or less trigonous, ventricose at base, ribbed; beak 1.0–1.5 mm, winged, serrulate, deeply split on adaxial side; stigmas 2; nut about 1.5 mm, ovoid, biconvex. *Flowers* 5–6. *Fruits* 7. 2*n* = 60, 62, 64.

Native. On peaty, medium base-rich soils where water levels are at least seasonally high, in fens and beside slow-flowing streams. Common throughout Ireland and

Britain, but becoming less frequent in Scotland. Europe to 63° N in Russia, rare in the south; Caucasus; northwest Africa. Our plant is subsp. **paniculata** which occurs throughout much of the range of the species.

× **remota** = **C. × boenninghauseniana** Weihe
C. hailstonii S. Gibson
Sterile, and variable in appearance between the two parents, but can always be recognised by its long, lax, densely tufted stems and long, narrow, dull green leaves. Spikes numerous, with a few, remote ones below and more congested ones above.

Wet alder and willow swamps, often among the tussocks of *C. paniculata*, or at the edge of runnels in the swamp. Scattered over the British Isles where it is one of the most frequent sedge hybrids. Central and west Europe.

2. C. appropinquata Schumach. Fibrous Tussock Sedge
C. paradoxa Willd., non J. F. Gmel.; *C. fulva* Thuill., non Gooden.; *C. teretiuscula* subsp. *paradoxa* Hook. fil.; *C. paniculata* subsp. *paradoxa* (Hook. fil.) Hook. fil.; *Vignea paradoxa* Rchb.; *Caricina paradoxa* St-Lag.

Glabrous *perennial herb*, forming dense tussocks up to 1 m high and 80 cm in diameter; rhizomes very short; roots dark brown, thick, felted; scales black, soon breaking up into wiry fibres. *Stems* 40–100 cm, rough, trigonous, faces flat. *Leaves* 20–80 cm × 1–3 mm, yellowish-green, long-linear, stiff, keeled, more or less flat; lower sheaths leafless, becoming fibrous when old, blackish-brown; ligule 2–3 mm, rounded at apex. *Inflorescence* 4–8 cm, a compact panicle. *Spikes* 4–6, more or less sessile; spikelets 4–7 mm, male at top, female below, some spikelets all female and more or less pedunculate; bracts setaceous, shorter than spikes. *Male glumes* 3–4 mm, red-brown-hyaline, ovate, acute at apex. *Female glumes* 3–4 mm, red-brown-hyaline, ovate, acuminate at apex. *Utricles* 2.7–4.0 mm, broadly ovoid to subglobose, abruptly contracted into the beak, planoconvex, distinctly 3–7-veined in the lower half, some of the veins reaching the base of the beak; beak 0.7–1.5 mm, notched, serrulate, not winged; stigmas 2; nut about 1.7 mm, ovoid, biconvex. *Flowers* 5–6. *Fruits* 6–7. Reproduces by seed, but many populations show no young plants. 2*n* = 64.

Native. In similar habitats to *C. paniculata* on medium base-rich soils where seasonal water-levels are high, but perhaps requiring less water movement and occasionally in more basic fen. It is a relic in fen carr of when conditions were more open. Predominantly a lowland plant, but found up to 380 m in Malham Tarn. Frequent in East Anglia with its headquarters in Broadland and formerly on the Middlesex–Hertfordshire–Buckinghamshire borders. It is very local in south-east and mid-west Yorkshire and in Roxburghshire and Selkirkshire. Widespread but local in central Ireland (cf. R. W. David, 1990.) Europe, except the south and temperate Asia.

× **paniculata** = **C. × rotae** De Not.
C. × solstitialis Figert
Sterile hybrid differing from *C. appropinquata* in its tufted habit, broader leaves, stouter stems and denser

1. **C. paniculata** L.

2. **C. appropinquata** Schumach.

3. **C. diandra** Schrank

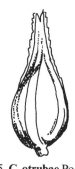

4. **C. vulpina** L.

5. **C. otrubae** Podp.

6. **C. vulpinoidea** Michx

7. **C. spicata** Huds.

8a. **C. muricata** L. subsp. **muricata**

8b. **C. muricata** L. subsp. **lampro-carpa** Čelak.

9a. **C. divulsa** Stokes subsp. **divulsa**

9b. **C. divulsa** Stokes subsp. **leersii** (Kneucker) W. Koch

10. **C. arenaria** L.

11. **C. disticha** Huds.

12. **chordorrhiza** L. fil.   13. **C. divisa** Huds.

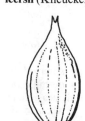

14. **C. maritima** Gunnerus

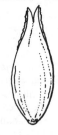

15. **C. remota** L.

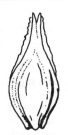

16. **C. ovalis** Gooden.

17. **C. crawfordii** Fern.

18. **C. echinata** Murray

19. **C. dioica** L.

Utricles of **Carex** L. species

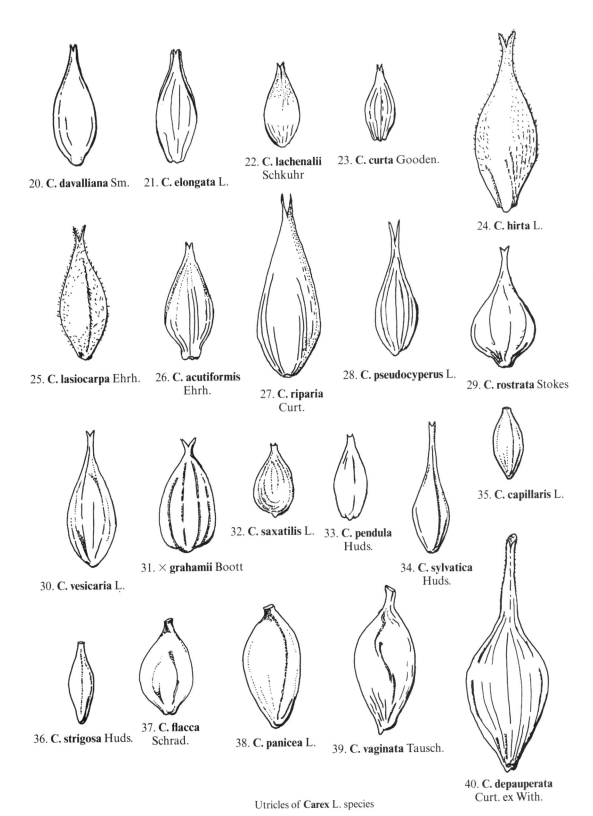

20. **C. davalliana** Sm.

21. **C. elongata** L.

22. **C. lachenalii** Schkuhr

23. **C. curta** Gooden.

24. **C. hirta** L.

25. **C. lasiocarpa** Ehrh.

26. **C. acutiformis** Ehrh.

27. **C. riparia** Curt.

28. **C. pseudocyperus** L.

29. **C. rostrata** Stokes

30. **C. vesicaria** L.

31. × **grahamii** Boott

32. **C. saxatilis** L.

33. **C. pendula** Huds.

34. **C. sylvatica** Huds.

35. **C. capillaris** L.

36. **C. strigosa** Huds.

37. **C. flacca** Schrad.

38. **C. panicea** L.

39. **C. vaginata** Tausch.

40. **C. depauperata** Curt. ex With.

Utricles of **Carex** L. species

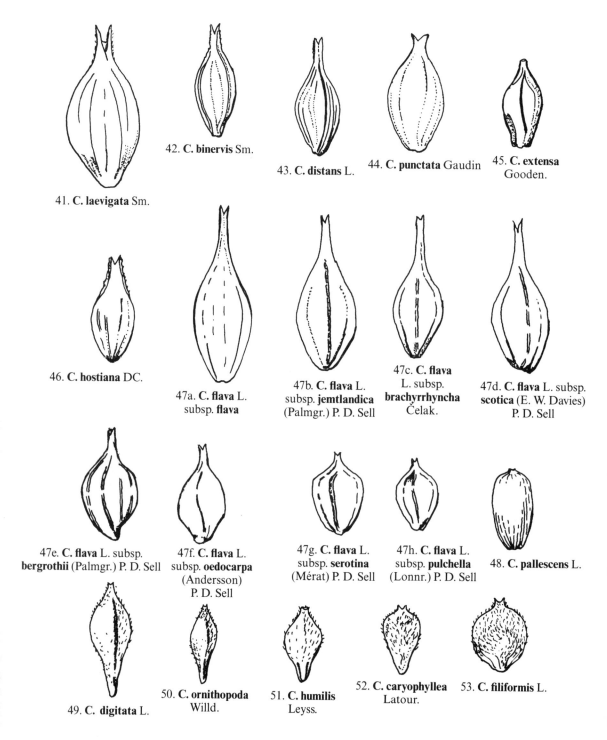

41. **C. laevigata** Sm.

42. **C. binervis** Sm.

43. **C. distans** L.

44. **C. punctata** Gaudin

45. **C. extensa** Gooden.

46. **C. hostiana** DC.

47a. **C. flava** L. subsp. **flava**

47b. **C. flava** L. subsp. **jemtlandica** (Palmgr.) P. D. Sell

47c. **C. flava** L. subsp. **brachyrrhyncha** Čelak.

47d. **C. flava** L. subsp. **scotica** (E. W. Davies) P. D. Sell

47e. **C. flava** L. subsp. **bergrothii** (Palmgr.) P. D. Sell

47f. **C. flava** L. subsp. **oedocarpa** (Andersson) P. D. Sell

47g. **C. flava** L. subsp. **serotina** (Mérat) P. D. Sell

47h. **C. flava** L. subsp. **pulchella** (Lonnr.) P. D. Sell

48. **C. pallescens** L.

49. **C. digitata** L.

50. **C. ornithopoda** Willd.

51. **C. humilis** Leyss.

52. **C. caryophyllea** Latour.

53. **C. filiformis** L.

Utricles of **Carex** L. species

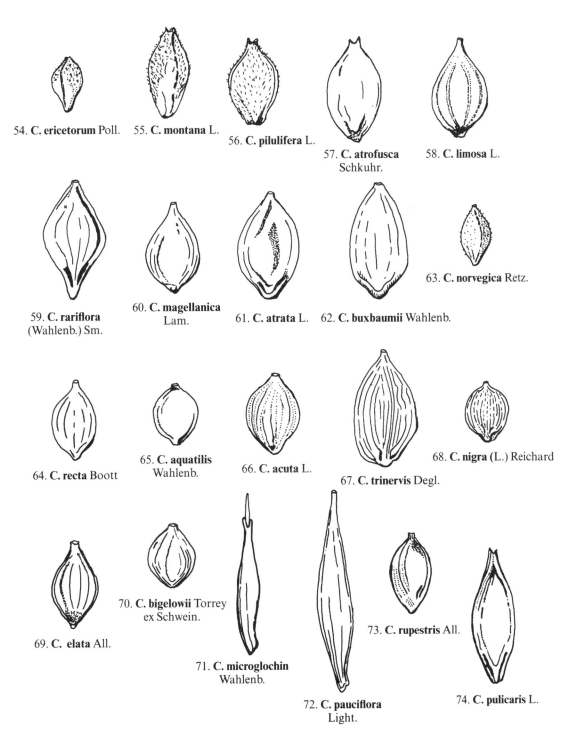

54. **C. ericetorum** Poll. 55. **C. montana** L.

56. **C. pilulifera** L.

57. **C. atrofusca** Schkuhr.

58. **C. limosa** L.

59. **C. rariflora** (Wahlenb.) Sm.

60. **C. magellanica** Lam.

61. **C. atrata** L.

62. **C. buxbaumii** Wahlenb.

63. **C. norvegica** Retz.

64. **C. recta** Boott

65. **C. aquatilis** Wahlenb.

66. **C. acuta** L.

67. **C. trinervis** Degl.

68. **C. nigra** (L.) Reichard

69. **C. elata** All.

70. **C. bigelowii** Torrey ex Schwein.

71. **C. microglochin** Wahlenb.

72. **C. pauciflora** Light.

73. **C. rupestris** All.

74. **C. pulicaris** L.

Utricles of **Carex** L. species

spikes. From *C. paniculata* it is distinguished by its more slender stems, less rough leaves and smaller inflorescence of the dark fuscous colour of *C. appropinquata*. Occurs locally in Norfolk and Suffolk. Recorded from central and west Europe.

3. C. diandra Schrank   Lesser Tussock Sedge
C. teretiuscula Gooden.; *Vignea teretiuscula* (Gooden.) Rchb.; *Physiglochis teretiuscula* (Gooden.) Raf.; *C. pseudoparadoxa* S. Gibson; *Caricina teretiuscula* (Gooden.) St-Lag.; *Rhynchopera teretiuscula* (Gooden.) Börner

Glabrous *perennial herb*, the shoots forming a loose tussock or appearing at more or less regular intervals; rhizomes shortly creeping; roots grey or orange-brown; scales black or dark greyish-brown, persistent. *Stems* 25–60 cm, greyish-green, slender, more or less triquetrous, rough on the angles. *Leaves* 20–40 cm × 1–2 mm, often overwintering, greyish-green, long-linear, more or less flat or slightly keeled, gradually tapered to a fine, more or less trigonous point at apex, rough at least on the upper surface; lower sheaths greyish- or dark brown, persistent, inner face hyaline, more or less straight at apex, not becoming fibrous when old; ligules 1–2 mm, obtuse at apex. *Inflorescence* 1–5 cm, a more or less compact, cylindrical head. *Spikes* 6–10, 5–8 mm, sessile, male at the top, female below, sessile; bracts glumaceous, lowest rarely setaceous. *Male glumes* 3–4 mm, pale brown-hyaline, elliptic-lanceolate, acute at apex. *Female glumes* about 3 mm, pale purplish-brown with a short green midrib and broad hyaline margin, elliptic-lanceolate, acute or mucronate at apex. *Utricles* 2.7–4.0 mm, shining dark brown, broadly ovoid or suborbicular, not winged, plano-convex, distinctly 2- to 5-ribbed in lower half; beak 1–2 mm, broad, serrulate, bifid; stigmas 2; nut about 2 mm, turbinate, plano-convex, stipitate. *Flowers* 5–6. *Fruits* 6–7. $2n = 60$.

Native. In very wet, peaty meadows, alder-sallow carr and more acid swamps, often in overgrown ditches or old peat cuttings. Scattered throughout the British Isles north to Caithness, but absent from the south-west peninsula and in Ireland mainly in the centre. Mainly lowland, infrequent in calcareous districts and becoming scarcer as sites are drained. Scattered in Europe, except for much of the south; temperate Asia; North America; Japan.

× **paniculata = C. × beckmannii** Keck ex F. W. Schultz
C. × beckmanniana Figert; *C. × germanica* K. Richt. This sterile hybrid is intermediate in leaf width, panicle and bract length and in the shape and venation of the empty utricles. The leaves are serrulate as in *C. paniculata* and come to a fine point as in *C. diandra*.

Has been recorded from Yorkshire, Ayrshire, Argyllshire and Co. Cork. Widely recorded in central Europe.

Section **2. Vulpinae** (J. Carey) H. Christ
Tufted herbs, without creeping rhizomes, sometimes forming tussocks. *Stems* triquetrous with sharp or winged angles, scabrid above. *Inflorescence* branched, with numerous spikes in 5–10 (–14), ovoid, contiguous or overlapping clusters (or rarely the lowest slightly separate); spikes male above and female below; lowest bract setaceous to leaf-like. *Utricles* biconvex, distinctly veined, patent, gradually narrowed into a scabrid, almost winged, bifid beak. *Stigmas* 2.

4. C. vulpina L.   True Fox Sedge
Vignea vulpina (L.) Rchb.

Glabrous *perennial herb* forming dense tufts; rhizomes short; roots greyish-brown, thick; scales dark brown, remaining as black fibres. *Stems* 30–100 cm, stout, smooth below, rough above, triquetrous, their faces more or less concave and the rough angles more or less winged. *Leaves* up to 80 cm × 4–10 mm, dark green even when dry, more or less erect, long-linear, keeled, more or less abruptly tapering to a flat, sharp point, not auriculate at base; sheath glandular and transversely wrinkled on inner face; ligule 2–6 mm, truncate at apex, overlapping the edges of the leaf-lamina. *Inflorescence* a stout, dense panicle. *Spikes* numerous, compound, 8–14 mm, the upper male, the lower female; bracts short, with more or less prominent auricles. *Male glumes* 3.5–4.0 mm, dark or rusty brown with a green midrib, oblanceolate elliptical, acute at apex. *Female glumes* 4–5 mm, dark or rusty brown, ovate, acuminate at apex. *Utricles* 4–5 mm, ovoid-ellipsoid, not shiny, prominently veined, minutely papillose, with epidermal cells more or less square or roundish and thick-walled, readily dropping at maturity; beak 1.0–1.5 mm, split on the back; stigmas 2; nut about 2.5 mm, oblong-obovoid, biconvex. *Flowers* 5–6. *Fruits* 6–7. Spreads by abundant seed. $2n = 68$.

Native. In damp places, often in standing water in ditches, usually on chalk or limestone. Local in south-east and south-central England, extending northwards to Yorkshire. Europe, but very local in the west and in southern Siberia.

5. C. otrubae Podp.   False Fox Sedge

Glabrous *perennial herb* forming dense tufts; rhizomes short; roots greyish-brown, thick; scales brown, remaining as dark fibres. *Stems* 30–100 cm, smooth below, rough above, triquetrous, their faces more or less flat, hardly winged on the angles. *Leaves* up to 60 cm × 4–10 mm, bright green, becoming greyish-green when dry and pale orange-brown on decay, long linear-lanceolate, more or less erect, keeled, more or less abruptly tapered to a flat, sharp point, more or less auriculate at base, margins rough; sheaths white with green veins, becoming brown, soon decaying, inner face hyaline and not wrinkled, with a straight apex; ligule 5–10 mm, tubular, acute at apex, contained within the width of the leaf. *Inflorescence* an elongate panicle, becoming dense when in fruit, the lower bracts setaceus with leaf-like bases, sometimes as long as the inflorescence, the upper glumaceous. *Spikes* numerous, compound, 8–14 mm, all or most with male flowers above and female below. *Male glumes* 3.5–4.0 mm, pale orange-brown with a green midrib, oblanceolate-elliptical, acute at apex. *Female glumes* 4–5 mm, pale reddish-brown or orange-brown, with a green midrib, ovate, acuminate at apex. *Utricles* (4.5–)5–6 mm, green turning dark brown when mature, shiny, smooth, ovoid, plano-convex, prominently veined, epidermal cells oblong and thin-walled, not readily dropping; beak 1–2 mm, bifid, rough at apex, more or less not split at back; stigmas 2; nut oblong-obovoid, biconvex. *Flowers* 6–7. *Fruits* 7–9. $2n = 58, 60$.

Native. Usually in damp places on heavy soils such as roadside ditches and by dykes. Uncommon on the sandy soils of Norfolk and Suffolk and mostly absent from the peat uplands of Wales, northern England and Scotland where it is mostly a coastal plant. Scattered throughout Ireland, but most common near the coast. The world distribution of *C. otubrae* is still imperfectly known. It occurs in south Scandinavia, west, central and south Europe (much commoner than *C. vulpina* in Mediterranean region), southern Russia; Macaronesia; North Africa and the Near East to the Himalayas and central Asia.

× **remota = C. × pseudoaxillaris** K. Richt.
C. axillaris Gooden., non L.; *C. kneuckeriana* Zahn; *Vignea axillaris* Rchb.
Intermediate between the parents and usually sterile. It differs from *C. otrubae* in its longer, laxer, narrower leaves and distant spikes of which the lower are often compound, and from *C. remota* by its stouter stems, longer, broader leaves and more robust, pale brown spikes.

Locally frequent where both parents occur throughout much of Britain. Also occurs in the Channel Islands and is of scattered occurrence in Ireland. It is recorded from Germany and Holland.

Section **3. Phleoideae** (Meinsh.) T.V. Egorova
Carex Phleoideae Meinsh.

Tufted; without creeping rhizomes. *Stems* triquetrous, scabrid above. *Inflorescence* branched, the lower branches usually remote, with a terminal cluster of short spikes, male at base, female above; bracts long. *Utricles* veinless or with very faint veins, abruptly contracted into an unwinged beak. *Stigmas* 2.

6. C. vulpinoidea Michx   American Fox Sedge
Glabrous *perennial herb* forming dense tufts; roots brownish; scales brown, eventually becoming fibrous. *Stems* 30–100 cm, slender, triquetrous, scabrid above. *Leaves* up to 180 cm × 2–4 mm, mid-green, long-linear, flat or nearly so, gradually tapering; sheaths sparsely red-dotted; ligule 6–8 mm, acute at apex. Inflorescence 5–10 cm, spike-like, narrow, with numerous short branches. *Spikes* clustered on numerous branches, crowded, ovoid-oblong, the lower male, the upper female; bracts filiform or leaf-like, the lower exceeding the spikes. *Male glumes* 3–5 mm, pale brown, ovate-oblong, aristate at apex. *Female glumes* 3–5 mm, pale reddish-brown with a hyaline margin, ovate-oblong, aristate at apex. *Utricles* 2.0–2.6 mm, pale reddish-brown, ovoid-orbicular, plano-convex, abruptly contracted into a short beak 0.7–1.2 mm; stigmas 2; nut

1.3–1.6 mm, ovoid-orbicular, plano-convex. *Flowers* 5–6. *Fruits* 6–7. $2n = 52, 54$.

Possibly introduced with fodder or other seed, and established in Kent and Surrey for many years, where it seems now to have disappeared. Also a wool alien, scattered over Britain north to central Scotland. Native of North America.

Section **4. Phaestoglochin** Dumort.

Tufted herbs, without creeping rhizomes. *Stems* triquetrous or trigonous. *Inflorescence* usually simple, with 3–8 spikes; spikes usually male above, female below; lowest bract setaceous. *Utricles* veinless or with faint veins at base, sometimes erect, sometimes abruptly contracted into the always unwinged beak. *Stigmas* 2.

7. C. spicata Huds.   Spiked Sedge
C. contigua Hoppe; *C. muricata* auct.

Glabrous, densely tufted *perennial herb*; rhizomes short; roots purple; scales brown or reddish-brown, with black veins, eventually becoming fibrous. *Stems* 10–85 cm, often stout, triquetrous. *Leaves* up to 45 cm × 2–4 mm, mid-green, long-linear, keeled, gradually tapering to a flat tip; sheaths sometimes stained wine-red, with a straight or concave apex; ligule 4–8 mm, more or less acute at apex, with much soft, white, loose tissue. Inflorescence 1–4 cm, spikes contiguous or the lowest slightly distant from the next. *Spikes* 3–8, 5–10 mm, sessile, the upper male, the lower female; bracts glumaceous, with a setaceous point. *Male glumes* 3–4 mm, lanceolate, acute at apex. *Female glumes* 4.0–4.5 mm, reddish-brown with a green midrib, lanceolate, acuminate at apex. *Utricles* 4–5 mm, bronze-green, eventually turning black, irregularly bulbous and corky at base, narrowly attenuated above; beak 1–2 mm, scabrid; stigmas 2; nut rounded-cubical. *Flowers* 6–7. *Fruits* 7–8. $2n = 58, 60$.

Native. On roadsides, in waste ground and in meadows, often on fairly heavy and damp soils, though also found on chalk. Frequent throughout southern and midland England, but absent from Cornwall and becoming rare towards the east coast; possibly only casual north of the Rivers Ribble and Tees, in southern and eastern Scotland, in west Wales, and in Ireland. Europe to 62° N in Norway, but absent from much of the south; North Africa; Macaronesia; introduced in North America.

8. C. muricata L.   Prickly Sedge
Vignea muricata (L.) Rchb.; *Caricina muricata* (L.) St-Lag.

Glabrous, densely tufted *perennial herb*; rhizomes short; roots brown; scales brown with black veins, soon becoming fibrous. *Stems* 10–85 cm, often stout, pentagonal to trigonous, more or less scabrid. *Leaves* up to 5 cm × 2–4 mm, mid-green, sometimes bronzing on drying, long-linear, keeled, gradually or abruptly tapering to a flat tip; sheaths with apex straight or more or less concave; ligule 2.0–3.5 mm, ovate. *Inflorescence* 1–4 cm, sometimes interrupted. *Spikes* 3–8, contiguous or crowded, the lowest sometimes separated by up to 10 mm, ovoid or globose, sessile, the upper male the lower female, or the lower spike entirely female; bracts glumaceous with a setaceous point. *Male glumes* 3–4 mm, reddish-brown or brown with a green midrib, elliptical or lanceolate-elliptical, with an acute apex. *Female glumes* 2.5–4.5 mm, dark brown, reddish-brown or pale brown turning whitish, ovate, acute or apiculate at apex. *Utricles* 2.6–4.0 mm, yellowish- or greyish-green becoming dark or reddish-brown, shiny; beak about 0.7–1.3 mm, minutely toothed; stigmas 2; nut orbicular-compressed. *Flowers* 5–7. *Fruits* 7–8. $2n = $ c. 56.

(a) Subsp. muricata
C. pairaei subsp. *borealis* Hyl.
Stems usually strongly scabrid above. *Spikes* globose. *Female glumes* 2.5–3.5 mm, dark reddish-brown or almost blackish, much darker and shorter than the greenish or brownish utricles. *Utricles* (3.5–)4.0–4.5 mm, rounded at base, prominently flanged at the margins, strongly patent. *Flowers* May and June.

(b) Subsp. lamprocarpa Čelak.
C. pairii F. W. Schultz; *C. bullockiana* Nelmes
Stems usually weakly scabrid. *Spikes* ovoid. *Female glumes* 3.0–4.5 mm, pale brown, similar in colour to or paler than, and almost as long as the utricles. *Utricles* 2.6–3.5(–4.5) mm, truncate at base, scarcely flanged at the margin, erecto-patent. *Flowers* June and July.

Native. Scattered over England and Wales, less frequent in Scotland and Ireland, and absent from the north. Most of Europe; west Asia; North Africa; Macaronesia; introduced in North America. The two subspecies are ecologically and to a certain extent geographically vicarious. Subsp. *muricata* occurs in Britain on steep, dry, limestone slopes near Woodchester in Gloucestershire, near Wrexham in Denbighshire, Gordale and Ribblesdale in Yorkshire and Lauder in Berwickshire. In Scandinavia and eastern Europe it is the commoner subspecies. Subsp. *lamprocarpa* is a plant of banks and heaths in open situations on drier, lighter and more acid soils. It is frequent in southern England and East Anglia and common in the south-west and in west Wales where its favourite habitat is on walls. It occurs in quantity in Galloway and on the Cheviots where it frequently occupies the talus below basalt cliffs. It is uncommon in Ireland. In Europe it is mainly in the south and west.

9. C. divulsa Stokes   Grey Sedge
Glabrous, more or less densely tufted *perennial herb*; rhizomes short; roots thick, greyish-brown; scales dark brown with black veins, soon becoming fibrous. *Stems* 25–90 cm, sometimes striate, slender to robust, trigonous. *Leaves* up to 75 cm × 2–5 mm, bright yellowish-green to greyish-green, bronzy or dark green, long-linear, flaccid or stiff, flat or channelled, gradually narrowed to a slender, flat apex; lower sheaths pale brown and ribbed with a hyaline inner face, straight or more or less concave at apex; ligule up to 2 mm, obtuse or truncate at apex. *Inflorescence* 3–18(–20) cm. *Spikes* 4–8, upper 5–8 mm, contiguous and all female, the lower

distant by up to 20 mm, sometimes branched, male above, female below; bracts 20–30 mm, setaceous, upper glumaceous. *Male glumes* 3.5–5.0 mm, more or less hyaline, sometimes brownish, lanceolate, acute at apex. *Female glumes* 3.0–4.5 mm, golden yellow or pale brown, sometimes more or less hyaline with a green midrib, ovate-elliptical, acute at apex and often attenuate. *Utricles* 3.0–5.5 mm, pale yellow or whitish-green, turning dark reddish-brown or dull black, more or less erect to stongly patent, ovoid, tapered above and below, faintly nerved at base, shiny; beak about 1 mm, rough, split; stigmas 2; nut obovoid, biconvex. *Flowers* 5–10. *Fruits* 7–11. $2n = 56, 58$.

(a) Subsp. divulsa
C. virens Lam.; *C. nemorosa* Lumn., non Schrank; *Vignea divulsa* (Stokes) Rchb.; *C. muricata* subsp. *divulsa* (Stokes) Syme; *Caricina divulsa* (Stokes) St-Lag.; *Desmiograstis divulsa* (Stokes) Börner
Stems slender. *Leaves* 2–3 mm wide; ligule about as wide as long. *Inflorescence* 5–10(–20) cm, lowest 3–4 spikes or branches separated from one another by a gap of much more than their own length. *Utricles* 3.0–4.5 mm, pale to yellowish-brown, appressed to erecto-patent. *Flowers* June to October. *Fruits* July to November.

(b) Subsp. leersii (Kncuck.) Walo Koch
C. polyphylla Kar. & Kir.; *C. leersii* F. W. Schultz, non Willd.
Stems robust. *Leaves* 3–4 mm wide; ligule usually wider than long. *Inflorescence* 3–5(–8) cm, lowest 2 spikes separated from each other by a gap of not more than their own length. *Utricles* 4.5–5.0(–5.5) mm, dark brown, patent. *Flowers* May to July. *Fruits* July and August.

Native. Common in England and Wales, extending northwards to south-east Scotland and mainly in south and east Ireland. Much of Europe, north to 59° N in Sweden; North Africa; Asia; introduced in North America.

The extreme forms of the two subspecies are very distinct, especially if you compare subsp. *divulsa* from western Europe with subsp. *leersii* from central Asia, but there are many intermediates. Subsp. *divulsa* in a plant of hedgerows, wood-borders and waste ground where competition is not too great. It is common throughout southern England from Kent to Cornwall and from the Channel to the Forest of Dean, the Chilterns and East Anglia, becoming rare in the Midlands, but with strong colonies in north-eastern Wales and Yorkshire east of the Pennines. There are colonies in south and east Ireland. Subsp. *leersii* occurs in similar habitats, but is strongly calcicole. The most extreme forms are found in Yorkshire, Derbyshire, Rutland, Denbighshire and the Cotswolds. It is scattered over much of the rest of southern England and eastern Britain north to Edinburgh, but there are many populations with intermediate characters between the two subspecies.

× **muricata**
Recorded from East Cork in 1987. ?Endemic.

× **otrubae**

Recorded in Sussex and Worcestershire in the last century and in Kent and Oxfordshire in the 1980s.

× **remota**   = C. × **emmae** L. Gross

Recorded from Waldron, Heathfield in Sussex and in Co. Cork.

Section **5. Ammoglochin** Dumort.

Rhizomes far-creeping. *Stems* triquetrous, scabrid above. *Inflorescence* simple, lobed; spikes 3–30, with male above and female below, or female above and male below, or entirely male or entirely female, linear-oblong to ovoid; lowest bract setaceous to leaf-like. *Female glumes* usually equalling or slightly shorter than utricles, ovate to ovate-lanceolate, subacute to acuminate at apex, with narrow, hyaline margin. *Utricles* ovoid, plano-convex, with a scabrid, bifid beak. *Stigmas* 2.

10. C. arenaria L.   Sand Sedge
C. witheringii Gray; *Vignea arenaria* (L.) Rchb.

Glabrous *perennial herb*; rhizomes extensively creeping, with a single shoot at about every fourth node; roots pale brown, much branched; scales dark brown, soon becoming fibrous. *Stems* 10–90 cm, varying in thickness, triquetrous, rough towards the top, often curved. *Leaves* up to 60 cm × 1.5–3.5 mm, dark green, shiny, often dark brown on dying, thick, rigid, rough, often recurved and keeled or channelled in open habitats, tapering gradually to a fine trigonous point; lower sheaths pale or grey-brown, persistent, inner face hyaline, becoming brown, membranous, apex straight; ligule 3–5 mm, tubular, obtuse at apex. *Inflorescence* a dense untidy, pyramidal head up to 8 cm. *Spikes* 5–15, 8–14 mm, the terminal entirely male, the upper part of middle male, the lower part of the middle ones and the lower entirely female; bracts glumaceous, the lower with setaceous points. *Male glumes* 5–7 mm, pale reddish-brown with hyaline margins, elliptic-lanceolate, with an acute apex. *Female glumes* 5–6 mm, pale reddish-brown, with a pale or green midrib and hyaline margins, ovate, with an acute or acuminate apex. *Utricles* 4.0–5.5 mm, pale greenish-brown, ovoid, plano-convex, many-ribbed, broadly winged, serrate in upper half; beak 1–2 mm, bifid, scabrid, winged below; stigmas 2; nut oblong-ellipsoid. *Flowers* 6–7. *Fruits* 7–8. 2n = 58, 64.

Native. A dominant plant on fixed dunes and wind-blown sand, usually with a low lime content. Common round the coasts of the British Isles and locally inland, especially the Breckland of East Anglia and Lincolnshire heaths. Europe; North America.

11. C. disticha Huds.   Brown Sedge
C. spicata Poll., non Huds.; *C. intermedia* Gooden., non Retz.; *Vignea intermedia* Rchb. nom. illegit.; *Vignea disticha* (Huds.) Peterm.; *Caricina disticha* (Huds.) St-Lag.

Glabrous *perennial herb*; rhizomes extensively creeping, with shoots single or in pairs; roots greyish-brown, much branched; scales brown, becoming dark and fibrous. *Stems* 20–100 cm, rough, triquetrous. *Leaves* 15–60(–90) cm × 2–4 mm, mid-green, long-linear, thick, more or less flat but with a keeled midrib, rough on the veins beneath, gradually tapering to a flat, rough tip; sheaths with their inner face hyaline only around the concave apex, otherwise herbaceous, the lower sheaths with very short, brown, persistent blades; ligule 3–7 mm, tubular, obtuse at apex. *Inflorescence* a dense untidy panicle 2–7 cm, forming a broadly lanceolate to ellipsoid head narrowed in the middle. *Spikes* numerous, contiguous, sessile, the terminal female (sometimes overtopped by male spikelets), the intermediate male, the lower all female or sometimes male at base; bracts glumaceous, brown-hyaline or the lowest leaf-like and exceeding the inflorescence. *Male glumes* 4–5 mm, pale reddish-brown with hyaline margins, lanceolate, acute at apex. *Female glumes* 3.5–4.5 mm, pale reddish-brown with hyaline margins, ovate-lanceolate, with an acute apex. *Utricles* 4.0–5.5 mm, reddish-brown, ovoid, plano-convex, with very narrow, more or less serrate lateral wings; beak about 1.0–1.5 mm, rough, bifid; stigmas 2; nut ovoid, biconvex, shortly stalked. *Flowers* 6–7. *Fruits* 7–8. 2n = 62.

Native. Mixed herb-sedge fens and wet meadows, prefering areas with a somewhat fluctuating water-table. Rather local throughout the British Isles north to Caithness and most frequent in the calcareous fens of eastern England and south Scotland. Central and northern Europe (except the Arctic), rare in south Europe; temperate Asia; introduced in North America.

Section **6. Divisae** H. Christ ex Kük.

Rhizomes long. *Stems* trigonous or subterete. *Inflorescence* simple, ovoid to oblong; spikes 2–8, ovoid, crowded, male above, female below, the upper sometimes entirely male, the lower sometimes entirely female; lowest bract glumaceous to leaf-like. *Female glumes* slightly shorter to slightly longer than utricles, ovate to ovate-oblong, acute or aristate at apex, usually with a narrow, scarious margin. *Utricles* ovoid or ellipsoid, plano- or concavo-convex, with a bifid beak. *Stigmas* 2.

12. C. chordorrhiza L. fil.   String Sedge
Vignea chordorrhiza (L. fil.) Rchb.

Glabrous *perennial herb*; rhizomes extensively creeping, often obliquely ascending; shoots solitary, arising from the elongated, decumbent base of the flowering stem; roots reddish-brown; scales pale yellowish-brown, chaffy, soon decaying. *Stems* 15–40 cm, stout, more or less terete, striate, with a few short leaves at the base. *Leaves* up to 30 cm × 1–2 mm, mid-green, long-linear, stiff, erect, flat or more or less involute, gradually narrowed to a fine more or less trigonous point; sheaths with inner face hyaline-brown with a concave apex, lower sheaths pale brown-hyaline and often darkening; ligules 1–2 mm, rounded at apex. *Inflorescence* 7–15 mm, a compact, more or less ovoid head. *Spikes* 2–4, 4–8 mm, male at top, female below, lower almost entirely female; bracts glumaceous. *Male glumes* about 3.5 mm, pale reddish-brown, oblanceolate-elliptical, acute at apex. *Female glumes* 3–4 mm, pale reddish-brown, hyaline towards margin, broadly ovate-oblong, more or less acute at apex. *Utricles* 3.5–4.5 mm, yellowish-brown or dark brown, shiny, ovoid, more or less compressed, faintly ribbed; beak about 0.5–1.0 mm, bifid; stigmas 2; nut oblong-obovoid, truncate at apex, stalked. *Flowers* 6–7. *Fruits* 7–8.

Native. A plant of very wet, base-poor mires. First found in 1897 at the head of Loch Naver, Altnarhara, Sutherland, and since found in several more places within a three mile radius of it. In 1978 it was found in quantity in Easterness. Scattered throughout central and north Europe south to the Pyrenees and east to central Ukraine; northern Asia and North America.

13. C. divisa Huds.   Divided Sedge
Vignea divisa (Huds.) Rchb.; *Caricina divisa* (Huds.) St-Lag.

Glabrous *perennial herb*; rhizomes thick, woody, sometimes far-creeping and branched; shoots more or less densely clustered, arising from short, lateral branches; roots black or greyish-brown; scales pale brown, ribbed, very quickly becoming dark and fibrous. *Stems* 15–80 cm, rough at top, wiry, trigonous. *Leaves* 15–60 cm × 1.5–3.0 mm, mid- to greyish-green, long-linear, stiff, more or less flat, channelled or inrolled, tapering to a slender, more or less flat tip, overwintering; lower sheaths brown and soon decaying, hyaline on inner face, straight or concave at apex; ligules 2–3 mm, more or less tubular, obtuse at apex. *Inflorescence* 1–3 cm, compact, ellipsoid or pyramidal. *Spikes* 3–8, contiguous, the lower sometimes remote, 3–13 mm, the upper male at the top and female below, the lower often all female; bracts leaf-like or setaceous, the lowest usually much exceeding the inflorescence. *Male glumes* 3.5–4.5 mm, pale reddish-brown, lanceolate-elliptical, acute at apex. *Female glumes* 3.5–5.0 mm, purplish-brown, with a pale midrib and hyaline margin, ovate, aristate at apex. *Utricles* 3.2–4.0 mm, pale brown, ovoid or broadly ellipsoid, plano-convex, faintly nerved, not winged; beak 0.5–0.8 mm, parallel-sided, bifid; stigmas 2; nut suborbicular, biconvex. *Flowers* 5–6. *Fruits* 7–8. Much of its spread is by rhizomes, but dispersal by seed is frequent. $2n = 60, 62$.

Native. Damp, lightly grazed depressions in pasture or marshy ground, mostly in inorganic soils; also beside ditches and rarely in the water. It flourishes in brackish conditions and is rarely found far from the coast, but is absent from saltmarshes themselves. Mainly in south and east England, reaching the south Wales coast and north to the River Humber and Holy Island. It is declining in most of its British range. The two Scottish records may be introductions, and it has not been seen recently in Ireland. South, south-central and west Europe, mainly coastal in the north; western Asia; North Africa.

Section 7. Foetidae (L. H. Bailey) Kük.
Carex Foetidae L. H. Bailey

Rhizome creeping, not tufted. *Stems* trigonous. *Inflorescence* dense, ovoid or subglobose; spikes 4–8, the terminal or upper entirely male and often hidden, the lower female with occasional male at top; bracts gluma-ceous or absent. *Female glumes* about two-thirds as long as utricles. *Utricles* ovoid to subglobose, biconvex, with faint veins, gradually narrowed into a bifid beak. *Stigmas* 2.

14. C. maritima Gunnerus   Curved Sedge
C. incurva Lightf.; *C. banata* Sm. nom. illegit.; *Vignea incurva* (Lightf.) Rchb.; *Caricina incurva* (Lightf.) St-Lag.; *Rhaptocalymma incurva* (Lightf.) Börner

Glabrous, much-branched *perennial herb*; rhizomes far-creeping; shoots loosely tufted or solitary and terminal on short branches of the rhizome; roots pale reddish-brown with numerous branched laterals; scales dark brown, soon becoming fibrous. *Stems* 1–18 cm, often curved, trigonous, striate, solid. *Leaves* 3–15 cm × 0.5–2.0 mm, mid-green, often overwintering, long-linear, thick, stiff, channelled, often inrolled, tapering gradually to an almost trigonous point; sheaths hyaline or brown, thin, ribbed, persistent, inner face with a thin, hyaline, membranous strip, straight at apex; ligules 0.5–1.0 mm, rounded. *Inflorescence* 5–15 mm, a compact, ovoid or subglobose head. *Spikes* 4–8, clustered, 3–6 mm, few-flowered, the upper male and often hidden, the lower female with occasional male at top; bracts glumaceous. *Male glumes* about 4 mm, dark orange-brown, elliptical, more or less acute at apex. *Female glumes* 3–4 mm, reddish-brown with a paler midrib and narrowly hyaline margins, broadly ovate, with an acute or obtuse, mucronate apex. *Utricles* 3.5–4.5 mm, brown to almost black at maturity, ovoid to subglobose, faintly ribbed; beak 0.5–1.0 mm, bifid; stigmas 2, often persistent; nut orbicular, biconvex. *Flowers* 6. *Fruits* 7. $2n = 60$.

Native. A rare species of wet sand by the sea, able to withstand salt spray and silt accretion to a limited extent. Often at the mouth of a stream where it terminates on a beach, in wet dune-slacks or turf beside rock-pools. Scattered round the north and east coast of Scotland formerly reaching as far south as Holy Island. Also in the Outer Hebrides and the Northern Isles, with an isolated station at Humphrey Head in north Lancashire where it is now apparently extinct (cf. David, 1982b). The largest known colonies are on the fairways of golf courses at St Andrews. Coasts of Arctic and north-west Europe; Alps; higher mountains of North Africa, Asia and North and South America.

Section 8. Remotae (Asch.) C. B. Clarke
Carex Remotae Asch.

Densely tufted plants usually forming tussocks. *Stems* trigonous, with 2 serrulate angles near the apex. *Inflorescence* very lax; spikes 4–9, oblong-ovoid, male at the top, female at the base, or the lower entirely female; lowest bract leaf-like, exceeding inflorescence. *Female glumes* slightly shorter than utricles. *Utricles* ovoid-ellipsoid, plano-convex, suberect, gradually narrowed into a short, broad beak. *Stigmas* 2.

15. C. remota L.   Remote Sedge
C. axillaris L. nom. illegit.; *C. tenella* auct.; *Vignea remota* (L.) Rchb.; *Caricina remota* (L.) St-Lag.

Glabrous *perennial herb*; rhizomes short; shoots densely tufted forming a stool up to 30 cm high; roots pale purplish-brown; scales brown, persistent or more rarely becoming fibrous. *Stems* 30–75 cm, spreading, trigonous or with 2 serrulate angles near the apex. *Leaves* 25–60 cm × 1.5–2.0 mm, mid-green, overwintering, long-linear, channelled, gradually tapering to a long, slender, pendulous point; sheaths pale yellowish-brown, persistent, the inner face narrow and hyaline, concave at apex; ligule 1–2 mm, rounded or obtuse at apex. *Inflorescence* one-quarter to one-third the length of the stem. Spikes 4–9, 3–10 mm, oblong-ovoid, sessile, the upper more or less contiguous, male at the top, female at the base, the lower remote and entirely female; lower bracts leaf-like, exceeding inflorescence, the upper glumaceous. *Male glumes* 2.5–3.0 mm, pale brown hyaline with a green midrib, ovate-elliptical, acute at apex. *Female glumes* about 2.5 mm, pale brown or hyaline with a green midrib, lanceolate to ovate, acute at apex. *Utricles* 2.5–3.8 mm, green, more or less shiny, ovoid-ellipsoid, plano-convex, suberect; beak about 0.5–0.8 mm, broad, split; stigmas 2; nut ovoid, biconvex. *Flowers* 6. *Fruits* 7. $2n = 62$.

Native. Shady situations on peaty or siliceous soils with a high water level for at least part of the year. Common in alder or wet birch carr. Mainly a lowland plant and common throughout England and Wales except in the winter-cold Fenland–Breckland basin. In lowland and west Scotland, but absent from the extreme north and Outer Hebrides. Throughout Ireland, sometimes on calcareous soils, but rarer in the west. Europe to 63° N in Norway, but rare in the south; Caucasus, north-west Africa.

Section 9. Ovales (Kunth) H. Christ
Carex Ovatae Kunth

Tufted plants with short *rhizomes*. *Stems* triquetrous or trigonous, rough at top. *Inflorescence* a compact, ovoid head; spikes 2–9, ovoid, female at the top, male at the base; lowest bract setaceous and equalling the spike. *Utricles* ellipsoid-ovoid, plano-convex, suberect, gradually narrowed into a distinct, winged, rough, bifid beak. *Stigmas* 2.

16. C. ovalis Gooden.   Oval Sedge
C. leporina auct.; *C. malvernensis* S. Gibson; *Caricina ovalis* (Gooden.) St-Lag.

Glabrous *perennial herb*; rhizomes short; shoots densely tufted, often more or less prostrate; roots pale or purplish-brown; scales dark brown, becoming fibrous. *Stems* 10–90 cm, rough at top, stiff, often curved, trigonous, solid. *Leaves* up to 50 cm × 1–3 mm, mid- to dark green, often overwintering, long-linear, thin, more or less soft, with rough margins, more or less flat, gradually tapering to a more or less fine trigonous point; sheaths becoming pink- or greyish-brown, persistent, the inner face narrow, hyaline and more or less straight at apex; ligule about 1 mm, tubular, obtuse at apex. *Inflorescence* a compact, ovoid head. *Spikes* 2–9, ovoid, contiguous or overlapping, 5–15 mm, sessile, the upper with female at the top

and male at the base, the lower all female; lower bracts often setaceous and equalling the spike, the upper glumaceous. *Male glumes* 4–5 mm, pale orange-brown, broadly lanceolate, with a keeled midrib and broadly hyaline margin, acute at apex. *Female glumes* 3.0–4.5 mm, dark or reddish-brown with a green or paler midrib and a hyaline margin, lanceolate-elliptical, acute at apex. *Utricles* 3.8–5.0 mm, pale brown, suberect, ellipsoid-ovoid, plano-convex, narrowly winged at the top, distinctly nerved; beak 1.0–1.5 mm, winged, rough, bifid; stigmas 2; nut obovoid or oblong-ellipsoid, biconvex, shortly stalked. *Flowers* 7–8. *Fruits* 8–9. $2n = 64, 66, 68$.

Native. Wet meadows, woodland rides, rough heathland where water accumulates and upland *Festuca–Agrostis–Nardus* grasslands. Common throughout the British Isles, but local in the East Anglia fens and less frequent in south-central Ireland. Europe except the Arctic and only on mountains in the south; mountains of North Africa; temperate Asia; introduced in North America.

17. C. crawfordii Fernald   Crawford's Sedge

Glabrous *perennial herb*; shoots densely tufted; roots pale brown; scales dark brown, becoming fibrous. *Stems* 10–60 cm, numerous, triquetrous, slender but stiff, roughened on the angles near the inflorescence, brownish-tinged at base. *Leaves* 7.5–60 cm × 1–3 mm, yellowish-green, flat or channelled, slender and pointed at apex; sheaths white-hyaline; ligules about 1 mm, rounded at apex. *Inflorescence* linear-oblong or oblong, often slightly flexuous. *Spikes* 3–15, closely aggregated, 3–9 mm, tapering at apex, the upper female, the lower male; lower bracts setaceous, the upper not developed. *Male glumes* 3–4 mm, dull light brown with a 3-nerved, green centre, ovate, acute or shortly acuminate at apex. *Female glumes* 2.5–3.0 mm, dull light brown with a green centre, ovate, acute or shortly acuminate at apex. *Utricles* 3.5–4.5 mm, pale brown, ellipsoid-ovoid, narrowly winged nearly to the base; beak about 1 mm, winged, serrulate, bifid at apex; stigmas 2; nut about 1.5 mm, oblong-ellipsoid, shortly stalked. *Flowers* 7–8. *Fruits* 8–9. $2n = 68$.

Introduced. A fodder or agricultural seed alien formerly more or less naturalised in Kent and Surrey; now a rare wool alien. Native of North America.

Section 10. Stellulatae (Kunth) H. Christ
Carex Stellulatae Kunth

Densely or laxly tufted plants. *Stems* trigonous to subterete. *Inflorescence* simple and lax; spikes 2–5, the terminal female at top and male below, the lower all female; lowest bract usually glumaceous. *Female glumes* about two-thirds as long as utricles, ovate, acute at apex. *Utricles* ovoid, plano-convex, divaricate at maturity; abruptly contracted into a serrate, bifid beak. *Stigmas* 2.

18. C. echinata Murray   Star Sedge
C. stellulata Gooden.; *C. leersii* Willd.; *C. grypos* Schkuhr; *Vignea stellulata* (Gooden.) Rchb.; *Caricina stellulata* (Gooden.) St-Lag.; *Desmiograstis stellata* (Gooden.) Börner

Glabrous *perennial herb*; rhizomes very short; shoots densely tufted; roots whitish; scales pale brown, persistent. *Stems* 10–40 cm, slender, trigonous and more or less striate to subterete. *Leaves* up to 30 cm × 1.0–2.5 mm, mid- to dark green, long-linear, thick, keeled or becoming flat, gradually tapered to a rough, trigonous tip; sheaths often white with green veins, becoming pale brown and soon decaying, inner face hyaline-green, more or less straight at apex; ligules about 1 mm, tubular, rounded at apex. *Inflorescence* 1–3 cm. Spikes 2–5, more or less distant, 3–6 mm, characteristically star-like in fruit, sessile, the terminal female at the top and the male below, the lower all female; bracts glumaceous or rarely setaceous and equalling the inflorescence. *Male glumes* 2.5–3.0 mm, pale brown with a broad hyaline margin, broadly lanceolate, obtuse at apex. *Female glumes* 2.0–2.5 mm, pale reddish-brown with a green midrib and broad hyaline margin, broadly ovate, embracing lower part of utricle, acute at apex. *Utricles* 2.8–4.0 mm, green becoming yellowish-brown, ovoid, plano-convex, faintly ribbed, divaricate at maturity; beak about 1.0–1.5 mm, broad, serrate, bifid; stigmas 2; nut obovoid or orbicular-compressed. *Flowers* 5–6. *Fruits* 6–8. $2n = 56, 58$.

Native. On mesotrophic soils which are seasonally or permanently waterlogged, such as heath reverting to bog. Also in oligotrophic mires usually within the pH range of 4.5–5.7, or in more eutrophic mires with silty, not peaty soils with a high base status such as those over limestone. Abundant in the north and west of the British Isles and on the wet, sandy heaths of the south, south-east and east; rarer in the Midlands due to drainage. Most of Europe, temperate Asia, Siberia and Japan; mountains of North Africa; North America.

Section **11.** Physoglochin Dumort.
Section *Dioicae* (Tuck.) Pax

Plant sometimes tufted; usually dioecious with a solitary spike. *Stems* more or less terete, smooth or scabrid above. *Male spike* rarely with female flowers at the base; glumes ovate-oblong, obtuse to acute at apex. *Female spike* subglobose to cylindrical; glumes ovate, obtuse to acute at apex. *Utricles* ovoid or linear-lanceolate, compressed, with a distinct beak. *Stigmas* 2.

19. C. dioica L.　　　　　　　　　　Dioecious Sedge
C. linnaeana Host; *Vignea dioica* (L.) Rchb.;
Maukschia laevis Heuff.; *Caricinella laevis* (Heuff.) St-Lag.; *Psyllophora dioica* (L.) Schur

Dioecious, glabrous *perennial herb*; rhizomes shortly creeping; shoots loosely tufted, often decumbent; roots often as thick as the rhizome, pale purplish-brown; scales pale purple or orange-brown, soon decaying or becoming fibrous. *Stems* 5–30 cm, erect, terete, striate, usually smooth. Leaves 5–20 cm × 0.3–1.0 mm, with 3 veins, dark green, dying to a dull reddish-brown, long-linear, rigid, more or less erect, channelled, incurved, rounded at apex; sheaths becoming pale orange-brown, persistent, the inner face hyaline, straight at apex, but soon split; ligules about 0.5 mm, tubular, rounded at

apex. *Inflorescence* a single terminal, usually unisexual spike. *Male spike* 8–20 mm, rarely with female flowers at the base. *Female spike* 5–20 mm, subglobose to cylindrical. *Bracts* glumaceous or absent. *Male glumes* 3–4 mm, reddish-brown with a hyaline margin, ovate-oblong, obtuse or acute at apex. *Female glumes* 2.5–3.5 mm, reddish- to purplish-brown with a pale midrib, dark nerve and more or less hyaline margin, ovate, with obtuse or acute apex, that of the lowest flowers often enlarged with an acuminate apex. *Utricles* 2.5–3.5 mm, pale red- to purplish-brown with dark ribs, patent or rarely deflexed when ripe, broadly ovoid, compressed; beak 0.5–1.3 mm, serrulate, notched; stigmas 2; nut subglobose, compressed. *Flowers* 5. *Fruits* 6–7. $2n = 52$.

C. dioica is variable in the size of the female spike, utricle and whole plant. *C. parallela* (Laest.) Sommerf., a north Scandinavian species, has been confused with *C. dioica*, but can be distinguished by its narrower utricles with a smooth beak. It has not been confirmed for Britain, but could possibly occur.

Native. Eutrophic mires in wet, silty muds, rarely in pure peat in the pH range 5.5–6.5. Also found in calcareous flushes and springs at various altitudes. Common on the mica-schists and limestone of north and west Scotland and likewise in north Wales, the north Pennines and the Lake District. Scattered in lowland Britain and Ireland, but rapidly disappearing due to drainage of its habitats. Europe; Siberia.

× **echinata** = C. × **gaudiniana** Guthnick
This hybrid is intermediate between the parents and is sterile. It is a tall plant with very narrow leaves and usually has one male spike with two female spikes below it.

It has been recorded from Denbighshire and Co. Mayo. There are also records from central Europe.

20. C. davalliana Sm.　　　　　　　　Davall's Sedge
Vignea davalliana (Sm.) Rchb.; *Maukschia scabra* Heuff.; *Caricinella scabra* (Heuff.) St-Lag.; *Psyllophora davalliana* (Sm.) Schur

Dioecious, glabrous, densely tufted *perennial herb*, without creeping rhizomes; roots brown; scales dark brown. *Stem* (1–)10–40(–50) cm, erect, terete, striate, usually scabrid above. *Leaves* up to 30 cm × 0.5–0.7 mm, shorter than stems, greyish-green, setaceous, acute at apex, channelled. *Inflorescence* a single, terminal, usually unisexual spike. *Male spike* 15–20 × 1.5–2.5 mm, oblong. *Female spike* oblong. *Bracts* glumaceous or absent. *Male glumes* 2.5–3.5 mm, brown with a hyaline margin, lanceolate, acute at apex. *Female glumes* 2–3 mm, brown, ovate, acute at apex. *Utricles* 3.5–4.5 mm, dark reddish- or blackish-brown, linear-lanceolate, with weak veins, patent or recurved when ripe, gradually narrowed into a long, more or less smooth beak 1.0–1.7 mm; stigmas 2; nut subglobose. *Flowers* 5–6. *Fruits* 6–7. $2n = 46$.

Possibly formerly native, now extinct. Known only from a calcareous fen in Lansdown near Bath, Somerset until about 1845. Central Europe eastwards to Macedonia.

Section **12.** Elongatae (Kunth) Kük.
Carex Elongatae Kunth

Densely tufted plants. *Stems* trigonous, rough. *Inflorescence* simple, lax; spikes 5–18, oblong to oblong-ovoid, more or less contiguous, the upper female at the top and male at the base, the lower entirely female; lowest bract usually setaceous and shorter than spike. *Female glumes* half to two-thirds as long as utricles, ovate-elliptical, acute to obtuse at apex. *Utricles* lanceolate-ellipsoid, plano-convex, often recurved, gradually narrowed into a smooth or minutely serrulate, entire beak. *Stigmas* 2.

21. C. elongata L.   Elongated Sedge
C. multiculmis Ehrh.; *C. divergens* Thuill.; *C. multiceps* Gaudin; *Vignea elongata* (L.) Rchb.; *Caricina elongata* (L.) St-Lag.

Glabrous *perennial herb*; rhizomes short; shoots densely tufted; roots pale brown; scales greyish-brown, usually persistent. *Stems* 30–80 cm, rough with upwards pointing teeth, trigonous. *Leaves* 25–90 cm × 2–5 mm, mid-green, often reddish-brown and persistent on dying, long-linear, thin, rough beneath, more or less flat or slightly keeled, tapering gradually to a very fine, flat tip; lower sheaths pale or pinkish-brown, shiny and persistent, the inner face hyaline, concave at apex; ligules 4–8 mm, acute at apex, with little or no free margin. *Inflorescence* 3–7 cm, lax. *Spikes* 5–18, more or less contiguous, 5–15 mm, oblong to oblong-ovoid, more or less erect, becoming divaricate on ripening, sessile, the upper female at the top and male at the base, the lower entirely female; lower bracts setaceous, upper glumaceous. *Male glumes* 2.5–3.0 mm, pale reddish-brown with a green midrib and broadly hyaline margin, ovate-oblong, rounded or obtuse and more or less mucronate at apex. *Female glumes* about 2 mm, reddish-brown with a green midrib, ovate-elliptical, acute or obtuse at apex. *Utricles* 2.5–4.0 mm, green becoming dark brown, lanceolate-ellipsoid, plano-convex, often curved, distinctly ribbed; beak 0.5–0.8 mm, often minutely serrulate, truncate; stigmas 2; nut compressed-cylindrical, stalked. *Flowers* 5. *Fruits* 6–7. Seems to seldom set viable seed in some localities. $2n = 56$.

Native. Local plant of damp soil in water meadows, beside ditches, ponds, canal sides and lakes and often in boggy woodland. It cannot tolerate continuous swamp conditions and benefits from winter flooding and drying out in summer. In south-east Britain as far west as the Hampshire–Dorset border and in Warwickshire; north-east Wales scattered in Denbigh and Cheshire to Yorkshire and Cumberland north to Loch Lomond; formerly frequent by canals in the Manchester area (cf. David, 1978b). In Ireland around Lough Neagh, and in Fermanagh, Cavan and Roscommon. Central and north Europe (except the Arctic), rare in south Europe; Caucasus.

Section **13. Canescentes** (Fr.) H. Christ
Carex Canescentes Fr.

More or less tufted plants. *Stems* trigonous or triquetrous, smooth or rough above. *Inflorescence* simple; spikes 2–8, ellipsoid or ovoid-oblong, female at the top, male below; lowest bract glumaceous or setaceous. *Female glumes* ovate or ovate-oblong. *Utricles* ovoid or ovoid-ellipsoid, with a short beak. *Stigmas* 2.

22. C. lachenalii Schkuhr   Hare's-foot Sedge
C. lagopina Wahlenb.; *C. approximata* Hoppe, non All.; *C. leporina* L. nom. rej.; *Vignea lagopina* (Wahlenb.) Rchb.; *Caricina approximata* St-Lag.

Glabrous *perennial herb*; rhizomes shortly creeping; shoots closely tufted; roots brown; scales brown, some fibrous. *Stems* up to 20(–30) cm, trigonous, striate, smooth or scabrid at apex. *Leaves* shorter than stems, 1–2 mm wide, dark green, long-linear, flat; basal sheaths brown; ligules about 1.0 mm, rounded at apex. *Inflorescence* 5–20 mm, dense, ovoid to oblong. *Spikes* 2–5, closely contiguous, the upper at the apex of a triangle completed by the next two, 5–10 mm, ellipsoid, female at the top, male below; lower bracts glumaceous. *Male glumes* 2–3 mm, reddish-brown, elliptical, obtuse at apex. *Female glumes* about 2.5 mm, reddish-brown with broad hyaline margins, ovate, obtuse at apex. *Utricles* 2.3–4.0 mm, usually green below and brown above, ovoid or obovoid, markedly narrowed at base, erect; beak 0.5–1.0 mm, smooth, split at the back, the halves overlapping; stigmas 2; nut obovoid, biconvex. *Flowers* 6–7. *Fruits* 7–9. $2n = 58, 62, 74$.

Native. Very local, but sometimes abundant, on wet slopes and rock-ledges between 750 and 1140 m on acid mountains in the Cairngorms. Also known in one station in the Ben Nevis range and one in Glencoe. Circumpolar mountains of central and south-west Europe and North America.

23. C. curta Gooden.   White Sedge
Facolos curtus (Gooden.) Raf.; *C. brunnescens* auct.; *C. persoonii* auct.; *C. vitilis* auct.; *C. canescens* auct.

Glabrous *perennial herb*; rhizomes very shortly creeping; shoots loosely tufted; roots pale brown or whitish; scales pale or pinkish-brown, usually persistent. *Stems* 10–50 cm, rough above, slender, triquetrous. *Leaves* 15–55 cm × 2–3 mm, pale green, soft, thin, long-linear, flat or more or less keeled, tapering more or less gradually to a fine, rough, more or less flat tip; sheaths thin, the lower pinkish-brown, soon decaying, the inner face hyaline, more or less straight at apex; ligule 2–3 mm, acute at apex. *Inflorescence* 3–5 cm. *Spikes* 4–8, contiguous or more or less distant, 5–8 mm, ovoid-oblong, female at top, male at base; bracts glumaceous, lower sometimes setaceous. *Male glumes* 2.0–2.5 mm, hyaline with a green midrib, ovate or broadly elliptical, obtuse at apex. *Female glumes* about 2 mm, hyaline with a green midrib, ovate-oblong, acute or apiculate at apex. *Utricles* 2–3 mm, pale or bluish-green to yellow, with yellowish ribs; ovoid-ellipsoid, plano-convex; beak 0.5–0.7 mm, minutely rough, notched; stigmas 2; nut obovoid or ellipsoid, biconvex. *Flowers* 6–8. *Fruits* 8–9. $2n = 56$.

Native. A component of oligotophic alpine mires around 720 m, and in the lowland more mesotrophic mires and wet places. On wet sandy heaths in south-east

England, becoming frequent in north-west Wales, northern England and Scotland. In Ireland it is rare in the south and west. Europe south to the Pyrenees and central Ukraine; temperate Asia; Japan; North and South America.

× echinata = C. × biharica Simonk.

C. × *tetrastachya* Traunst. ex Saut., non Scheele
This hybrid is intermediate between the parents and sterile. The spikes are stouter and with fewer flowers than *C. curta*, with slightly larger and more spreading utricles which have serrated beaks, while it has narrower, longer spikes and smaller utricles than *C. echinata.*

It occurs on the higher mountains of central Scotland. It is also recorded from central Europe.

× lachenalii = C. × helvola Blytt ex Fr.

This hybrid is sterile and intermediate between its parents. It differs from *C. lachenalii* in its more numerous, rather longer spikes which are more slender and less congested, and from *C. curta* in its more robust, darker-coloured spikes. It is somewhat variable and has been confused with *C. curta* × *echinata.*

It has been recorded from two localities in Aberdeenshire. It is also recorded from northern Europe.

× paniculata = C. × ludibunda Gay

Has been found with the parents in Sussex, Pembrokeshire, Caernarvonshire and Breconshire. Has also been recorded from western Europe.

Subgenus 2. Carex

Monoecious plants. *Inflorescence* usually simple. *Spikes* 2 or more, dissimilar in appearance and at least the terminal one male and the basal one female. Flowers mostly with 3 stigmas, or spikes similar in appearance and at least the terminal one with both male and female flowers, with 2 stigmas; rarely the spike solitary with both male and female flowers, and the flowers with 2 stigmas. *Lateral spikes* usually with a scale between the bract and the lowest glume.

Section 14. Carex

Laxly tufted plants with long rhizomes. *Stems* trigonous; basal sheaths without a lamina. *Male spikes* 1–3, linear-lanceolate, shortly pedunculate. *Female spikes* oblong to cylindrical, erect; lowest bract sheathing. *Female glumes* slightly shorter to slightly longer than utricles. *Utricles* ovoid, prominently ribbed, hairy, gradually narrowed into a bifid beak. *Stigmas* 3.

24. C. hirta L.   Hairy Sedge

C. villosa Stokes nom. illegit.; *Trasus hirtus* (L.) Gray
Perennial herb; rhizomes often far-creeping; shoots tufted; roots pale brown, often much branched; scales brown or reddish-brown, eventually becoming fibrous. *Stems* 15–70 cm, trigonous, glabrous and shiny. *Leaves* 10–50 cm × 2–5 mm, mid-green, often overwintering, long-linear, flat or more or less keeled, gradually tapering to a fine point, more or less hairy on both surfaces; sheaths hairy, those of the sterile shoots forming a rigid,

false stem, the inner face hyaline and often densely hairy, with a more or less straight apex; ligule 1–2 mm, obtuse at apex, the free margin fringed with hairs. *Inflorescence* up to three-quarters the length of the stem, male spikes above, female below; lower bracts leaf-like, longer than spike but not exceeding the inflorescence, the upper setaceous. *Male spikes* 2–3, 10–25 mm; male glumes 4–5 mm, reddish-brown with a pale midrib, often more or less hyaline throughout, obovate-oblanceolate, mucronate at apex, more or less hairy. *Female spikes* 2–3, contiguous or more or less distant, 10–45 mm, cylindrical, erect; peduncles smooth, up to twice as long as the spike, half ensheathed; female glumes 6–8 mm, green-hyaline, ovate-oblong, midrib excurrent as a green ciliate awn. *Utricles* 4.5–7.0 mm, green, ovoid, many-ribbed, hairy; beak 1.5–2.5 mm, rough, hairy, deeply bifid; stigmas 3; nut obovoid, trigonous, stalked. *Flowers* 5–6. *Fruits* 6–9. 2n = 112.

Native. Usually in grass associations on hedgebanks, sand-dunes and in meadows on a variety of soils. Usually in hollows or channels where moisture accumulates and occasionally in damp woods. Throughout Britain, becoming less common in Scotland and very rare north of the Great Glen. Markedly absent from areas of heavy rainfall. Scattered throughout Ireland except in blanket bog areas. Europe, except the north; mountains of North Africa; Caucasus; introduced in North America.

× vesicaria = C. × grossii Fiek

This hybrid is intermediate between the parents and the fruits do not develop. The utricles are much less hairy and with shorter hairs than *C. hirta*, but there is a well-developed rhizome system.

The only accepted record is from a dune-slack in Co. Wicklow. It is also widely recorded in central Europe.

25. C. lasiocarpa Ehrh.   Slender Sedge

C. filiformis auct.; *C. tomentosa* auct.; *Trasus filiformis* auct.; *C. splendida* Willd.

Nearly glabrous *perennial herb*; rhizomes far-creeping; shoots loosely tufted, slender; roots pale yellowish-brown; scales pale or yellowish-brown, occasionally wine-red. *Stems* 45–120 cm, slender, stiff, smooth or slightly rough above, trigonous, striate. *Leaves* 30–100 cm × 1–2 mm, greyish-green, long-linear, stiff, flat but usually inrolled, long attenuate into a fine acicular, whip-like point; sheaths dark purplish-brown to pale reddish-brown, soon decaying, the inner face membranous and purplish or pale brown, persistent or often fibrillose, straight at the dark purplish apex; ligule 2–3 mm, obtuse at apex. *Inflorescence* one-eighth to one-sixth of the stem, the male spike above, female below; bracts leaf-like, very slender, the lower often exceeding the inflorescence. *Male spikes* 1–3, 20–70 mm; male glumes purplish-brown with a green or pale midrib, lanceolate, acute at apex. *Female spikes* 1–3, contiguous or more or less distant, 10–30 mm, cylindrical-oblong, erect, more or less sessile; female glumes 3.5–5.0 mm, chestnut-brown with a pale midrib, lanceolate, acute or acuminate at apex. *Utricles* 3–5 mm, greyish-green, ovoid, tomentose; beak 0.5–1.0 mm, more or less deeply bifid; stigmas 3; nut ovoid, trigonous. *Flowers* 6–7. *Fruits* 7–9. $2n = $ c. 56.

Native. Mesotrophic to eutrophic mires and reed swamp on substrata ranging from sedge peats to raw sandy exposed lake shores up to 600 m. Scattered throughout the British Isles and locally common. North and central Europe southwards to the Pyrenees, central Italy and south Ukraine; north Asia; North America.

× **riparia** = **C.** × **evoluta** Hartm.
This hybrid is intermediate between its parents and is sterile. It differs from *C. lasiocarpa* by its broader, flatter leaves, more robust spikes and less hairy utricles, and from *C. riparia* by its narrower leaves, smaller spikes and hairy utricles.

Recorded only from Somerset and Cambridgeshire. It is also recorded from mainland Europe.

Section 15. Paludosae (Fr.) H. Christ
Carex Paludosae Fr.

Herbs with long, stout rhizomes. *Stems* triquetrous; basal sheaths without a lamina. *Male spikes* 2–6, oblong or oblong-lanceolate, subsessile or shortly pedunculate. *Female spikes* cylindrical or oblong, dense, erect. *Female glumes* oblong-lanceolate, acuminate or acute at apex, purplish-brown. *Utricles* ellipsoid-ovoid and plano-convex or ovoid and inflated, more or less beaked. *Stigmas* (2–)3.

26. C. acutiformis Ehrh.   Lesser Pond Sedge
C. paludosa Gooden.; *Trasus paludosus* (Gooden.) Gray; *Edritria paludosa* (Gooden.) Raf.

Glabrous *perennial herb;* rhizomes widely creeping; shoots tufted; roots brown, often dark; scales greyish-brown, soon becoming fibrous. *Stems* 60–150 cm, rough, often smooth below, triquetrous, solid. *Leaves* up to 160 cm × 7–10(–12) mm, glaucous at first, becoming dull, often reddish-green at apex, reddish-brown on dying, long-linear, thin, keeled or plicate, arcuate, gradually tapering to the apex; sheaths brown, usually red-streaked, the persistent inner sheaths not translucent, the inner face hyaline-brown, persistent, usually fibrillose on splitting, concave at apex; ligules 5–15 mm, usually shorter than width of leaf, acute at apex. *Inflorescence* about one-third the length of the stem, the male spikes above, female below; bracts leaf-like, exceeding the inflorescence. *Male spikes* 2–3, clustered, 10–40 mm, the lower with setaceous bracts; male glumes 5–6 mm, purplish-brown with a pale midrib, oblong to oblanceolate, obtuse or subacute at apex. *Female spikes* 3–4, more or less contiguous, 20–50 mm, cylindrical or oblong, erect, the upper sessile and often male at top, the lowest only shortly pedunculate; female glumes 4–5 mm oblong-lanceolate, reddish- or purplish-brown with a paler midrib, acute or acuminate at apex. *Utricles* 3.5–5.0 mm, greyish-green, ellipsoid-ovoid, somewhat flattened, strongly ribbed; beak about 0.5 mm, emarginate or weakly bifid; stigmas 3, rarely 2; nut obovoid, trigonous, flat at apex. *Flowers* 6–7. *Fruits* 7–9. $2n = $ c. 38, 78.

Native. Swamps, semi-swamp carr, wet meadows and woods, and beside slow-flowing rivers, canals and ponds. Scattered throughout the British Isles north to Banff, locally abundant, but rare in Scotland. Europe, south to central Spain, Sicily and south Bulgaria; North Africa; temperate Asia; introduced in North America.

× **vesicaria** = **C.** × **ducellieri** Beauverd
Known only from Hampshire where it was discovered in 1986. It is also recorded for Switzerland.

27. C. riparia Curtis   Greater Pond Sedge
Trasus riparius (Curtis) Gray

Glabrous *perennial herb*; rhizomes far-creeping; shoots tufted; roots thick, pale brown; scales greyish-brown, soon becoming fibrous. *Stems* 60–130 cm, rough, triquetrous, solid. *Leaves* up to 160 cm × 6–15 mm, glaucous, persisting as pale brown litter, long-linear, rigid, usually erect, thin, sharply keeled or plicate, rather abruptly attenuate to a short trigonous apex; sheaths persistent, greyish-brown, often red-tinged, white hyaline at base with distinct transverse septa, the inner face hyaline, becoming brown and persisting, rarely forming fibrillae or splitting, concave at apex; ligules 5–10 mm, usually longer than width of leaf, obtuse or rounded at apex. *Inflorescence* about one-third the length of the stem, male spikes above, female below; lower bracts leaf-like, exceeding the inflorescence, the upper setaceous. *Male spikes* 3–6, contiguous, 2–6 cm; male glumes 7–9 mm, dull dark brown with the midrib and margins pale, oblong-lanceolate, acuminate at apex. *Female spikes* 1–5, more or less contiguous, 3–10 mm, cylindrical-fusiform, the upper erect, more or less sessile and often male at the top, the lower pedunculate; peduncles rough, often as long as the spike, shortly ensheathed; female glumes 7–10 mm, dark often purplish-brown with a paler or green midrib, oblong-lanceolate or narrowly ovate, acuminate at apex. *Utricles* 5–8 mm, ovoid, somewhat inflated, brown or yellowish-brown, obscurely ribbed, tapered at apex; beak 1–2 mm, strongly bifid; stigmas 3; nut oblong-ovoid, trigonous, stalked. *Flowers* 5–6. *Fruits* 6–9. $2n = 72$.

Native. Often forming large stands by slow-flowing rivers, in ditches, around ponds and in other wet places. Mainly a lowland plant and thus most common in south and east England, in Scotland not found north of Argyll and Aberdeenshire and in Ireland almost confined to the south and east. Europe north to 62° N in Finland; North Africa; Caucasus; west Asia.

× **vesicaria** = **C.** × **csomadensis** Simonk.
This hybrid is sterile. It differs from *C. riparia* by its much narrower leaves, more slender male spikes, longer and more slender peduncles to the female spikes, glumes with a paler margin and a more slender beak to the utricle, and from *C. vesicaria* by its glaucous leaves and darker and smaller utricles.

It has been found in Co. Wicklow and there are some doubtful records from southern England. It is also recorded from central Europe.

Section **16. Pseudocyperae** (L. H. Bailey) Kük.
Carex Pseudocyperae L. H. Bailey

Laxly tufted herbs with short rhizomes. *Stems* triquetrous, the angles rough. *Male spike* solitary, oblong-lanceolate, shortly pedunculate. *Female spikes* oblong to cylindrical, all at the same level, equalling or exceeding the male spike, pendulous; peduncles long and slender. *Female glumes* ovate, long-aristate. *Utricles* ovoid-ellipsoid, patent, prominently ribbed, gradually narrowed into a smooth, deeply bifid beak. *Stigmas* 3.

28. C. pseudocyperus L.   Cyperus Sedge
Trasus chlorostachyos Gray nom. illegit.

Nearly glabrous *perennial herb*; rhizomes short; shoots loosely tufted; roots thick, orange-brown, felted; scales dark or greyish-brown, persistent. *Stems* 40–90 cm, solid, triquetrous, angles rough. *Leaves* up to 120 cm × 5–12 mm, bright yellowish-green, passing through yellow to greyish-brown on dying, long-linear, rigid, erect, thin, plicate, very rough on the keel and margins, tapered gradually to a fine point, persistent; sheaths becoming pink- or greyish-brown, persistent, the inner face hyaline, forming fibrillae on splitting, straight or concave at apex; ligules 10–15 mm, obtuse at apex. *Inflorescence* one-sixth to one-quarter the length of the stem; bracts leaf-like, the lowest 3–4 times longer than inflorescence. *Male spikes* 1, 20–60 mm; male glumes 5–7 mm, brown with a green or paler midrib, oblong-lanceolate, long-acuminate and ciliate at apex, shortly pedunculate. *Female spikes* 3–5, clustered at top of the stem and often exceeding the male spike, 20–100 mm, oblong to cylindrical, pendulous; peduncles long and slender, rough, the lowest only shortly ensheathed; female glumes 5–10 mm, brownish-hyaline with a green midrib, ovate, drawn out into a long, fine, ciliate arista at apex. *Utricles* 4–5 mm, broader than glumes, ovoid-ellipsoid, green, ribbed, patent, soon falling when ripe; beak 1.5–2.5 mm, smooth, deeply bifid; stigmas 3; nut obovoid, trigonous. *Flowers* 5–6. *Fruits* 7–8. $2n = 66$.

Native. Eutrophic and mesotrophic open-water swamps, often along sides of slow-flowing dykes and in ponds, oxbows and derelict canals. Predominantly a lowland plant, mainly in the south-east and Midlands extending as far north as north Lancashire; rare in Wales and in one isolated locality in Moray; scattered but less frequent than formerly in Ireland. Europe from south-central Finland to central Spain and Macedonia; Algeria; temperate Asia to north Japan; North America.

× **rostrata** = **C.** × **justischmidtii** Junge
This hybrid differs from *C. pseudocyperus* by its partly glaucous leaves, less pendulous and aggregated spikes which are longer and laxer, and by its more inflated and divergent utricles, and from *C. rostrata* by its flatter, wider and greener leaves, longer and more pendulous spikes and longer utricles tapering gradually into the beak.

Known only from peat in carr at Cranberry Rough, Hockham in Norfolk in 1955. It is also recorded in northern Europe.

Section **17. Vesicariae** (O. Lang) H. Christ
Carex Vesicariae O. Lang

Creeping, rhizomatous *herbs*. *Stems* trigonous. *Male spikes* linear, shortly pedunculate or more or less sessile. *Female spikes* subglobose to cylindrical, more or less erect; lowest bract not sheathing. *Female glumes* shorter than utricles, lanceolate or oblong-lanceolate to ovate, more or less acute at apex. *Utricles* more or less inflated, patent to erecto-patent, with a smooth, more or less bifid beak.

29. C. rostrata Stokes   Bottle Sedge
C. ampullacea Gooden. nom. illegit.; *C. inflata* auct.; *Trasus ampullaceus* Gray nom. illegit.; *C. catteyensis* A. Benn.

Glabrous *perennial herb*; rhizomes far-creeping; shoots few in each tuft; roots thick, purple- or orange-brown; scales pale greyish-brown, rarely red-tinged, soon decaying. *Stems* 20–100 cm, terete and smooth below, trigonous and rough above. *Leaves* 30–120 cm × 2–7 mm, glaucous with dense stomata on upper surface, dark green and shiny without or with few lines of stomata beneath, long-linear, rough, rigid, flat, keeled or plicate, or in some habitats inrolled, tapering to a long (2–6 cm) acicular point, overwintering; sheaths herbaceous, thick and more or less spongy, dark brown, often streaked with red, the inner sheaths pink, the inner face hyaline, becoming brown and often fibrillose on splitting, straight at apex; ligules 2–3 mm, shorter than width of leaf, rounded at apex. *Inflorescence* up to half the length of the stem, the male spikes above, the female below; bracts usually leaf-like, equalling or exceeding the inflorescence. *Male spikes* 2–4, 20–70 mm, linear, lower ones with setaceous bracts; male glumes 5–6 mm, brown with a paler midrib, elliptic-lanceolate, acute or obtuse at apex. *Female spikes* 2–5, contiguous or the lowermost distant, 30–80 mm, cylindrical, suberect, subsessile or the lowest shortly stalked; female glumes 3.0–5.5 mm, purplish-brown with a pale midrib, narrower than utricle, oblong-lanceolate, acute at apex. *Utricles* 3.5–6.5 mm, yellowish-green, ovoid and abruptly contracted above, inflated, faintly ribbed, more or less patent; beak 1.0–1.5 mm, very slender and parallel-sided, smooth, bifid; stigmas 3; nut subglobose-trigonous. *Flowers* 6–7. *Fruits* 7–9. $2n = $ c. 60, 76.

Native. Swamps, lake margins and peaty areas with high water level where the pH is between 4.5 and 6.5. Also occurs as a flush plant where the base status is not too high. Common and important component of upland mesotrophic wetlands throughout Scotland, Ireland and northern England; in the south it is scattered in the more acid fens. Most of Europe, but rare in the south; west Asia to Altai; Japan; North America.

× **vesicaria** = **C.** × **involuta** (Bab.) Syme
C. vesicaria var. *involuta* Bab.; *C. pannewitziana* Figert; *C. inflata* var. *involuta* (Bab.) Druce
A variable hybrid, some plants being nearer one parent than the other. It differs from *C. rostrata* in its flatter,

greener leaves, longer, less inflated utricles and more conspicuous female glumes, and from *C. vesicaria* in its narrower leaves and smaller utricles which are rather more crowded and patent. It is highly sterile.

Scattered over the British Isles in wet places and by water. Also recorded from much of central and northern Europe.

30. C. vesicaria L.   Bladder Sedge
C. inflata Huds.; *Trasus vesicarius* (L.) Gray

Glabrous *perennial herb*; rhizomes shortly creeping; shoots slender, markedly trigonous, 2–3 in each tuft; roots thick, pale yellowish-brown; scales brownish- or purplish-red, usually persistent. *Stems* 30–120 cm, trigonous, rough on the angles above, smooth below. *Leaves* up to 150 cm × 4–8 mm, mid- or yellowish-green, without stomata above and dense stomata beneath, long-linear, serrulate for the entire length, thin, rigid, flat or plicate, gradually tapered to a fine point; sheaths becoming purplish-red, persistent, the inner face hyaline, fibrillose on splitting, straight or concave at apex; ligule 5–8 mm, as long as or longer than the leaf, acute at apex. *Inflorescence* one-quarter to one-third the length of the stem, the male spikes above, female below; bracts leaf-like, the lower exceeding the inflorescence. *Male spikes* 2–4, 10–40 mm, linear, the lower often with setaceous bracts; male glumes 4–6 mm, purplish-brown with a green or pale midrib and hyaline margins, elliptical or oblanceolate, more or less acute at apex. *Female spikes* 2–3, more or less contiguous but distant from the male spikes by 2–4 cm, oblong-cylindrical, erect, subsessile or the lowest with a peduncle as long as the spike; female glumes 4–6 mm, purplish-brown with a pale or green midrib, narrowly lanceolate, hyaline, acute or acuminate at apex. *Utricles* (4–)5–8 mm, more or less shiny olive green, ovoid-ellipsoid and gradually narrowed above, inflated, ribbed, ascending; beak 1.2–2.5 mm, smooth, bifid; stigmas 3; nut obovoid, trigonous. *Flowers* 6. *Fruits* 7–8. $2n = 70, 74, 82, 86, 88$.

Native. Wet peatland, often forming a characteristic community round Scottish lochs. Also in more open situations and on inorganic soils at the edges of streams, dykes and canals. · Scattered throughout Britain and Ireland to Shetland, although uncommon north of the Great Glen. Europe south to north-east Spain and south Bulgaria; temperate Asia; North America.

31. C. × grahamii Boott   Mountain Bladder Sedge
C. × ewingii E. S. Marshall; *C. stenolepis* auct.; *C. vesicaria* subsp. *grahamii* (Boott) Hook. fil.

Glabrous *perennial herb*; rhizomes far-creeping; shoots 2–3 in tufts at frequent, more or less regular intervals; roots pale yellowish-brown, often darkening on drying; scales purplish or wine-red, soon decaying. *Stems* 25–50 cm, more or less rough above, trigonous, often curved. *Leaves* 20–40 cm × 2–4 mm, dull mid-green becoming straw-coloured or greyish-brown, long-linear, thinner and more sharply keeled than in *C. saxatilis*, tapering to a trigonous point; sheaths often dark purplish- or wine-red, persistent; upper face more or less hyaline, straight

at apex; ligules 3–6 mm, acute at apex. *Inflorescence* one-quarter to one- fifth the length of the stem, male spikes above, female below; lowermost bract leaf-like, often with its subtending spike abortive, more or less equalling the inflorescence, the upper setaceous or glume-like. *Male spikes* 1–2, rarely 3, 10–25 mm, linear; male glumes 4–5 mm, purplish- or orange-brown, oblanceolate, subacute at apex. *Female spikes* 1–3, more or less contiguous, 10–30 mm, cylindrical or ovoid, erect, the lower pedunculate; female glumes 3–4 mm, reddish-brown with a paler midrib and hyaline margin, ovate, subacute and broadly hyaline at apex. *Utricles* 4–5 mm, ovoid-ellipsoid, more or less tapered to the apex; beak 0.5–0.9 mm, translucent greenish-brown, distinctly ribbed; stigmas 2–3; nut not forming. *Flowers* 6–8.

Native. High mountain flushes between 750 and 900 m in inorganic mires on steep slopes or broad ledges, mainly on the schistose soils of the Breadalbanes, with an outlier in Clova. Colonies are compact, spreading vegetatively, widely separated from each other and often show morphological differences. The utricles are empty suggesting a hybrid origin. One parent is almost certainly *C. saxatilis*. The morphology suggests the other parent is *C. vesicaria*, but that species does not grow with it. Another putative parent is *C. rostrata* which does grow with it. It is similar to *C. stenolepis* Less., a fertile species possibly also of hybrid origin.

32. C. saxatilis L.   Russet Sedge
C. pulla Gooden.; *C. saxatilis* subsp. *pulla* (Gooden.) Syme; *C. vesicaria* subsp. *saxatilis* (L.) Hook. fil.

Glabrous *perennial herb*; rhizomes far-creeping; shoots in tufts of 1–3 at frequent, more or less regular intervals; roots pale yellowish-brown; scales greyish-brown, often tinged wine-red, soon becoming fibrous. *Stems* 15–40 cm, more or less rough above, trigonous, often curved. *Leaves* 12–40 cm × 2–4 mm, more or less shiny midgreen, becoming straw-coloured, long-linear, often curved, thick, bluntly keeled or channelled, gradually tapering to a trigonous point up to 5 cm; sheaths thick, white, tinged with wine-red, persistent, inner face hyaline, straight at apex; ligule 2–4 mm, rounded at apex. *Inflorescence* one-fifth to one-quarter the length of the stem, male spike above, female below; lower bracts leaf-like, shorter than or equalling the inflorescence, upper usually glume-like. *Male spike* 1, rarely 2, 10–15 mm, linear; male glumes 3–4 mm, purplish-black with a hyaline margin, oblanceolate, more or less acute at apex. *Female spikes* 1–3, more or less contiguous, 5–20 mm, ovoid or subglobose, erect, only the lowest rarely with a short peduncle; female glumes 2–3 mm, purplish- to dark reddish-brown with a paler midrib and hyaline margins, ovate, more or less acute at apex. *Utricles* 3.0–3.5 mm, often dark purplish-green on exposed faces, shiny, ovoid, more or less inflated; beak about 0.5 mm, more or less notched; stigmas 2–3; nut subglobose. *Flowers* 7. *Fruits* 8–9. $2n = 80$.

Native. Mires where the water movement is not great especially on saddles or slightly sloping hillsides, usually between 460 and 1125 m. Tolerant of a wide range of pH. Confined to the higher Scottish mountains, mainly in the high rainfall areas of the west and usually in places where snow lies late; very local in the Cairngorms and Clova mountains with scattered outposts north of the Great Glen. North Europe, south to 56° N in Scotland; circumpolar.

Section 18. Rhynchocystis Dumort.
Section *Maximae* (Asch.) Kük.

Densely tufted herbs. *Stems* tetraquetrous. *Male spike* cylindrical. *Female spikes* cylindrical; lowest bract about equalling the inflorescence, sheathing. *Female glumes* ovate, acute or acuminate at apex. *Utricles* broadly ellipsoid or ovoid, gradually or abruptly narrowed into a short beak. Stigmas 3.

33. C. pendula Huds.   Pendulous Sedge
Trasus pendulus (Huds.) Gray; *Manochlaenia pendula* (Huds.) Börner; *C. maxima* Scop.

Glabrous *perennial herb*; rhizomes short; shoots in tufts often up to 70 cm across; roots up to 3 mm thick, reddish-brown; scales reddish-brown, persistent or becoming fibrous. *Stems* 60–180 cm, tetraquetrous. *Leaves* 20–100 cm × 15–20 mm, yellowish-green above, more or less glaucous beneath, reddish-brown on dying, often overwintering, long-linear, rigid, thin, keeled, more or less flat, abruptly tapered to a blunt point, rough on margins; sheaths reddish-brown, persistent, the inner face hyaline only at the concave apex; ligules 30–60 mm, acute at apex. *Inflorescence* about one-third the length of the stem, male spikes at top, female below; bracts leaf-like, more or less equalling or slightly shorter than the inflorescence. *Male spikes* 1–2, 6–10 cm, cylindrical; male glumes 6–8 mm, brownish- hyaline, lanceolate, acuminate at apex. *Female spikes* 4–5, more or less contiguous, lowest often distant, 70–160 mm, cylindrical, erect at first, becoming pendulous; peduncles rough, ensheathed, the lowest half as long as the spike; female glumes 2.0–2.5 mm, reddish-brown with a pale midrib, ovate, acute or acuminate at apex. *Utricles* 3.0–3.5 mm, more or less glaucous-green turning brown, broadly ellipsoid or ovoid, trigonous; beak 0.2–0.6 mm, truncate at apex; stigmas 3; nut obovoid, trigonous. *Flowers* 5–6. *Fruits* 6–7. $2n = 58, 60$.

    Native. Prefers acid, but base-rich, heavy soils, and is frequent in hazel–ash–oak woods and scrubland on the Wealden and Oxford clays. Also found on less clayey soils where there is a constant water supply as in runnels and by wet ditches. Mainly a lowland species in south and east England, scattered in Wales and Ireland and scarce and mainly coastal in Scotland. Frequently introduced. West, central and south Europe; Caucasus; North Africa; Madeira and Azores.

Section 19. Strigosae (Fr.) H. Christ
Carex Strigosae Fr.

Tufted herbs with or without short, creeping rhizomes. *Stems* more or less trigonous or subterete, smooth. *Male spikes* oblong-lanceolate to cylindrical, more or less pedunculate. *Female spikes* oblong to cylindrical, lax, usually pendulous; lowest bract sheathing. *Female*

glumes shorter than utricles, ovate or ovate-lanceolate. *Utricles* ovoid to ellipsoid-ovoid or ellipsoid, with a more or less distinct bifid or truncate beak. *Stigmas* 3.

34. C. sylvatica Huds.   Wood Sedge
Trasus sylvaticus (Huds.) Gray; *Edritria sylvatica* (Huds.) Raf.; *Proteocarpus sylvaticus* (Huds.) Börner

Glabrous *perennial herb*; rhizomes very short; shoots often densely tufted; roots greyish-brown; scales brown, soon fibrous. *Stems* 15–60 cm, spreading or nodding, slender, trigonous. *Leaves* 5–60 cm × 3–6(–11) mm, mid- to yellowish-green, becoming brown then bleached on drying, overwintering, long-linear, soft, slightly keeled or plicate, abruptly tapered to a sharp point; sheaths hyaline, becoming brown, the inner face splitting and persisting as a brown membrane, concave at apex; ligules about 2 mm, obtuse at apex. *Inflorescence* one-third to half the length of the stem, male spikes above, female below; lowermost bracts leaf-like, sometimes longer than inflorescence, sheathing, upper setaceous and shorter than spike. *Male spike* usually 1, 10–40 mm, very slender; male glumes 4–5 mm, brownhyaline, oblong-lanceolate to cylindrical, acute or obtuse and mucronate at apex. *Female spikes* 3–5, more or less distant, 20–65 mm, lax-flowered, oblong to cylindrical, pendulous; peduncles rough, filiform, up to three times the length of the spike, base ensheathed; female glumes 3–5 mm, straw-coloured or brown with a green midrib, ovate-lanceolate, hyaline, acute or acuminate at apex. *Utricles* 3–5 mm, green, ellipsoid- or obovoidtrigonous, with two prominent lateral nerves; beak 1.0–2.5 mm, bifid; stigmas 3; nut ellipsoid, more or less trigonous. *Flowers* 4–7. *Fruits* 6–9. $2n = 58$.

    Native. On heavy, often wet soils in woods, sometimes on chalky soils with a little clay, occasionally in open scrub and grassland which is probably relict woodland. Frequent throughout Britain and Ireland, but uncommon in central and north Scotland. Most of Europe north to 65° N in Norway; North Africa; temperate Asia; introduced in North America. Our plants belong to subsp. **sylvatica**.

35. C. capillaris L.   Hair Sedge
Trasus capillaris (L.) Gray; *C. plena* Clairv.

Glabrous *perennial herb*; rhizomes short; shoots in more or less open tufts; roots slender, purplish-brown; scales dark or reddish-brown, soon becoming fibrous. *Stems* 10–40 cm, trigonous to subterete. *Leaves* 5–10 cm × 0.5–2.5 mm, greyish-green, changing to reddish- then greyish-brown on dying, usually overwintering, narrowly long-linear, stiff, sometimes arcuate, flat or more or less channelled, gradually tapered to a short subulate tip; sheaths reddish-brown, soon becoming fibrous, the inner face hyaline, straight at apex; ligule about 1 mm, rounded at apex. *Inflorescence* up to half the length of the stem; bracts leaf-like, usually a single one subtending a cluster of spikes, more or less exceeding inflorescence, that of the distant spike shorter. *Male spike* 1.5–10.0 mm, few-flowered, overtopped by the female spikes; male glumes 2–3 mm, hyaline with a brown midrib, obtuse to rounded at apex. *Female spikes* 2–4, 5–25 mm, rarely more than 10-flowered, clustered and appearing to arise from a single node, the lowest sometimes distant, oblong; peduncles up to 4 cm, hair-like, half-ensheathed; female glumes 2–3 mm, hyaline or straw-coloured, caducous, broadly ovate, acute or mucronate at apex. *Utricles* 2.5–3.0 mm, olive to dark brown, narrowly ovoid-ellipsoid, smooth, shiny; beak 0.5–1.0 mm, truncate at apex; stigmas 3; nut ellipsoid, trigonous. *Flowers* 7. *Fruits* 7–8. Probably spreads mostly by rhizomes. $2n = 54$.

Native. Dry to permanently wet herb-rich, grassy hillsides, mostly on base-rich soils or areas flushed by base-rich water especially limestone and calcareous mica-schists, also in mineral-rich bogs. Rare in Snowdonia, local in the north Pennines and southern uplands, and frequent in the Grampians up to at least 1025 m and in limestone areas of north Scotland where it descends to sea level. North Europe, mountains of central and south Europe and North Africa; northern Asia and North America. Our plants belong to subsp. **capillaris**.

36. C. strigosa Huds.   Thin-spiked Wood Sedge
Trasus strigosus (Huds.) Gray; *Manochlaenia strigosa* (Huds.) Börner

Glabrous *perennial herb*; rhizomes usually short; shoots tufted; roots pale brown; scales orange- or reddish-brown, becoming fibrous. *Stems* 35–75 cm, trigonous-subterete, often spreading. *Leaves* 15–40 cm × 6–10 mm, mid-green, pinkish-brown on dying, long-linear, thin, more or less arcuate, plicate, with two lateral veins prominent on the upper surface, abruptly tapered to a sharp point; sheaths thin, brown, occasionally red-tinged, persistent, the inner face hyaline, concave at apex; ligules 5–8 mm, acute at apex. *Inflorescence* half to three-quarters the length of the stem, male spike above, female below; bracts leaf-like, longer than spikes, not exceeding inflorescence, sheathing. *Male spike* 1, 30–40 mm, cylindrical; male glumes 4.5–5.5 mm, brown with a green midrib, narrowly obovate, acuminate at apex. *Female spikes* 3–6, distant, lowest often remote, 25–80 mm, lax–flowered, oblong to cylindrical, more or less erect; peduncles smooth, half-ensheathed; female glumes about 2.5 mm, green becoming brown, ovate or ovate-lanceolate, acute at apex. *Utricles* 3–4 mm, green, oblong- or narrowly ellipsoid, often curved; beak 0.3–0.5 mm, truncate at apex; stigmas 3; nut subglobose, trigonous, shortly stalked. *Flowers* 5–6. *Fruits* 6–9. $2n = 66$.

Native. On base-rich loamy soils, or less frequently heavy soils especially near streams or in damp hollows; in woodland usually in open glades. Most frequent in the lowland wooded areas of the Weald, Somerset and the Severn Valley, and scattered throughout England, becoming rare in the east and north as far as east Yorkshire. In Ireland in scattered localities in the north and east. South, central and west Europe from Denmark southwards and eastwards to Bulgaria and the Caucasus.

Section **20. Glaucae** (Asch.) H. Christ
Carex Glaucae Asch.

Not or laxly tufted herbs with long creeping rhizomes. *Stems* trigonous-subterete, usually smooth. *Male spikes* (1–)2–3, cylindrical, more or less pedunculate. *Female spikes* cylindrical, erect or pendulous; lowest bract leaf-like, usually at least as long as the inflorescence, often sheathing. *Female glumes* oblong-ovate. *Utricles* broadly ellipsoid to obovoid, minutely papillose; abruptly contracted into a short beak.

37. C. flacca Schreb.   Glaucous Sedge
C. glauca Scop.; *C. diversicolor* auct.; *C. recurva* Huds.; *C. micheliana* Sm.; *C. ambleocarpa* Willd. nom. illegit.; *Trasus glaucus* (Scop.) Gray; *C. strictocarpa* Sm.

Glabrous *perennial herb*; rhizomes often far-creeping; shoots loosely tufted; roots pale greyish-brown; scales dark brown or reddish-brown, persistent. *Stems* 10–60 cm, rigid, trigonous-subterete, usually smooth. *Leaves* up to 50 cm × 1.5–4.0 mm, glaucous beneath, dark dull green above, persisting as dark or rich-brown litter, long-linear, rigid, often arcuate, flat, with the midrib channel extending to the extreme tip, gradually tapering to a fine point; sheaths becoming dark brown, often wine-red, persistent, entire, the inner face hyaline-brown, straight or concave at apex; ligules 2–3 mm, shortly tubular, rounded at apex. *Inflorescence* one-fifth to one-third the length of the stem, male spikes above, female below; bracts leaf-like, the lowest slightly exceeding the inflorescence and with a sheath 0–3(–10) mm. *Male spikes* (1–)2–3, 10–35 mm, cylindrical, more or less pedunculate; male glumes 3–4 mm, purplish-brown with a pale midrib and hyaline margin, rounded or subacute at apex. *Female spikes* 1–5, 15–55 mm, contiguous, cylindrical, the upper erect, subsessile, often male at the top, the lower more or less nodding; peduncles rough, as long as the spike, half-ensheathed; female glumes 2–3 mm, purplish-black, pruinose, with a wide pale midrib and hyaline margin, oblong-ovate, obtuse-mucronate at apex. *Utricle* 2–3 mm, yellowish-green, often turning deep purplish-black, broadly ellipsoid to obovoid, often inflated on the adaxial side, minutely papillose; beak 0.2–0.3 mm, truncate at apex; stigmas 3; nut ellipsoid, trigonous. *Flowers* 5–6. *Fruits* 6–9. $2n = 76, 90$.

Very variable in structure, length of spikes and coloration of utricles and glumes. Forms with dark glumes are sometimes mistaken for *C. nigra*, but that species has 2 stigmas and a flattened utricle. All plants from the British Isles are included in subsp. **flacca**.

Native. The commonest sedge of calcareous grassland and also on sand-dunes and boulder clay. It can withstand some salinity and is often found in estuarine marshes, and can be a component of eutrophic flushes. Europe, except the north-east; North Africa; introduced in North America.

Section **21. Paniceae** (O. Lang) H. Christ
Carex Paniceae O. Lang

Herbs with short, slender, creeping rhizomes. *Stems* trigonous-subterete, nearly smooth. *Male spike* solitary, more or less cylindrical or clavate, with long peduncles. *Female spikes* cylindrical, lax, erect; lowest bract leaf-like, sheathing. *Female glumes* shorter than utricles, ovate, acute at apex. *Utricles* broadly obovoid, usually with a short, truncate beak. *Stigmas* 3.

38. C. panicea L.   Carnation Sedge
Trasus paniceus (L.) Gray

Glabrous *perennial herb*, rhizomes shortly creeping; shoots tufted; roots pale greyish- or purplish-brown; scales greyish-brown, soon decaying. *Stems* 10–60 cm, often curved above, trigonous-subterete, striate. *Leaves* up to 60 cm × 1.5–5.0 mm, glaucous, becoming pale straw on drying, rough at top, long-linear, more or less flat, with midrib channel ceasing before tip, tapering to a trigonous point; sheaths not loose, white or pale pinkish-brown, ribbed, persistent, but fibrous on decay, the inner face hyaline, straight at apex; ligules 1.2–2.0 mm, obtuse at apex. *Inflorescence* one-sixth to one-quarter the length of the stem, male spike above, female below; bracts leaf-like, not loose-sheathed, 1–2 times length of spike, the lowest with a sheath 10–15 mm. *Male spike* 1, 10–20 mm, more or less cylindrical, with a long peduncle; male glumes 3.0–4.5 mm, purplish-brown with a pale midrib and hyaline margin, ovate-oblong to elliptical, more or less obtuse at apex. *Female spikes* 1–3, 10–15 mm, cylindrical, erect, more or less distant, few and lax-flowered; female glumes 3–4 mm, purple or reddish-brown with a pale midrib and hyaline margin, ovate, acute at apex. *Utricles* 3–4 mm, olive green or more or less purplish-tinged, broadly obovoid, inflated on the adaxial side so that the apex points outwards; beak 0.2–0.5 mm, truncate at apex; stigmas 3; nut oblong-obovoid, trigonous. *Flowers* 5–6. *Fruits* 6–9. 2n = 32.

Native. In mountain grassland, dwarf shrub heaths on deep peat, oligotrophic mires, meso- and eutrophic mires and alpine flushes with a pH from 4 to 7 of varying base content and less frequent where there is more or less continuous irrigation. Throughout the British Isles up to 1215 m in Scotland. Europe, rare in the south; temperate Asia; North Africa; introduced in North America.

39. C. vaginata Tausch   Sheathed Sedge
C. sparsiflora (Wahlenb.) Steud.; *C. mielichhoferi* auct.; *C. panicea* var. *sparsiflora* Wahlenb.; *C. smithii* Tausch; *Trasus erectus* Gray; *C. scotica* Spreng.; *C. phaeostachya* Sm.

Glabrous *perennial herb*; rhizomes far-creeping; shoots tufted; roots brown; scales greyish-brown, fibrous. *Stems* 15–40 cm, trigonous-subterete, striate. *Leaves* up to 60 cm × 3–6 mm, yellowish- or bronze green, keeled, long-linear, parallel-sided for much of their length and rather abruptly tapered to a flat point, the basal decumbent, those on the stem with loose sheaths and subulate laminas 1–3 cm long; ligules very short, truncate at apex. *Inflorescence* one-quarter the length of the stem, male spikes above, female below; lower bracts loosely sheathing and shorter than spike. *Male spike* 1, 10–15

mm, more or less clavate, with long peduncles; male glumes more than 4.5 mm, orange-brown with a paler midrib. *Female spikes* 1–2(–3), 10–20 mm, cylindrical, more or less distant, erect, few- and lax-flowered; female glumes about 3 mm, reddish-brown with a paler midrib, broadly ovate, acute or mucronate at apex. *Utricles* 3.5–5.0 mm, broadly obovoid, hardly inflated, ribbed; beak 0.5–1.0 mm, obliquely truncate; stigmas 3; nut ovoid-trigonous. *Flowers* 7. *Fruits* 7–9. Only a few plants flower in any one year. 2n = 32.

Native. Often abundant, but usually shy-flowering on wet, rocky sills between 600 and 1150 m. A characteristic constituent of flushed grassland on neutral or slightly acid soils in the Breadalbanes and Cairngorms; very local in Dumfriesshire and Roxburghshire with single stations in Ross-shire and Sutherland. North Europe, locally in the mountains of central and south-west Europe; Siberia; North America.

Section 22. Rhomboidales Kük.

Laxly tufted herbs. *Stems* trigonous-subterete. *Male spike* solitary, cylindrical, pedunculate. *Female spikes* obovoid, erect, distant; lowest bract leaf-like, sheathing. *Female glumes* broadly lanceolate to obovate, acute at apex, with a scarious margin. *Utricles* rhomboid-obovoid, abruptly contracted into a beak. *Stigmas* 3.

40. C. depauperata Curtis ex With.
                                          Starved Wood Sedge
C. ventricosa Curtis; *Trasus depauperata* (Curtis ex With.) Gray; *Proteocarpus depauperatus* (Curtis ex With.) Börner

Glabrous *perennial herb*; rhizomes shortly creeping; shoots few, loosely tufted; roots pale purplish-brown; scales purplish- to reddish-brown, shiny, persistent. *Stems* 30–100 cm, stout, trigonous-subterete. *Leaves* 20–60 cm × 2–4 mm, mid- to yellowish-green, dying to a pale straw, thin, flat, narrowly linear-lanceolate, tapering from a broad base to a fine point; sheaths purplish- or reddish-brown, shiny, persistent, the inner face hyaline, flecked with red and becoming brown, straight or concave at apex; ligules 2–3 mm, obtuse at apex. *Inflorescence* one-quarter the length of the stem, male spike at top, female below; bracts leaf-like, longer than the spikes, the upper sometimes exceeding the inflorescence. *Male spike* 1, 18–30 mm, cylindrical, pedunculate; male glumes 5.0–6.5 mm, pale or reddish-brown with a paler or green midrib and hyaline margins, elliptical, more or less acute at apex. *Female spikes* 2–4, distant, 2- to 6-flowered, 10–20 mm, obovoid, erect; peduncles rough, half-ensheathed; female glumes 4.5–6.0 mm, brown with a green midrib and broad hyaline margin, broadly lanceolate to obovate, acute or mucronate at apex. *Utricles* 7–9 mm, shiny brownish-green, rhomboid-obovoid, narrowed into a solid base, distinctly ribbed; beak 2.5–3.0 mm, smooth or scabrid, obliquely truncate at apex, split in front; stigmas 3; nut obovoid, trigonous. *Flowers* 5. *Fruits* 6–7. 2n = 44.

Native. Dry woods and hedgebanks on chalky or limestone soils, very rare. Somerset, Surrey and Co.

Cork, probably extinct in Anglesey and Kent. West and south Europe; Caucasus.

Section 23. Elatae Kük.

Tufted herbs with short, creeping rhizomes. *Stems* trigonous, usually slightly rough above. *Male spikes* 1 or 2, cylindrical. *Female spikes* ovoid-cylindrical; lowest bract leaf-like, sheathing. *Female glumes* ovate or ovate-lanceolate, acuminate at apex. *Utricles* ovoid or subglobose, with a bifid beak. *Stigmas* 3.

41. C. laevigata Sm.   Smooth-stalked Sedge
C. helodes Link; *Trasus laevigatus* (Sm.) Gray; *Edritria laevigata* (Sm.) Raf.; *Proteocarpus laevigatus* (Sm.) Börner

Glabrous *perennial herb*; rhizomes short; shoots forming dense tufts up to 30 cm across; roots yellowish-brown, felty; scales brown or reddish, rarely becoming fibrous. *Stems* 30–120 cm, stout, trigonous with slightly rounded faces. *Leaves* 15–60 cm × 4–12 mm, bright or yellowish-green, brown and persisting when dead, long-linear, smooth, thin, shallowly keeled or plicate, abruptly tapered to the apex; sheaths persistent, brown, the inner face hyaline, convex or lingulate at apex; ligule 7–15 mm, acute or obtuse at apex. *Inflorescence* one-quarter to two-thirds the length of the stem, male spikes above, female below; bracts leaf-like, longer than spike, but not exceeding the inflorescence. *Male spikes* 1 or 2, 20–60 mm, cylindrical; male glumes 5–6 mm, pale orange-brown with hyaline margins and base, oblong-oblanceolate, obtuse and often mucronate at apex, less often long-acute. *Female spikes* 2–4, distant, 20–50 mm, ovoid-cylindrical, more or less erect, but lowest usually pendulous; peduncles up to 80 mm, ensheathed; female glumes 3–5 mm, pale reddish-brown with a green midrib which is often rough at the tip, ovate or ovate-lanceolate, acuminate at apex. *Utricles* 4–6 mm, ascending at an angle of 45–60° to the axis, green with fine reddish dots, ovoid or subglobose, partially inflated, more or less strongly ribbed; beak 1–2 mm, more or less scabrid, deeply bifid; stigmas 3; nut trigonous-globose, shortly stalked. *Flowers* 6. *Fruits* 7–8. 2n = 72.

Native. Mostly in shady, moist situations, more rarely in open habitats. Most common in woodlands on acid, but base-rich clay soils. Frequent in Britain and Ireland where annual rainfall is over 750 mm, as in the Weald, south-west and west Britain and south-east and west Ireland. West Europe and north-west Africa.

× pallescens
This hybrid has the broad leaves and long-beaked utricle of *C. laevigata* and the few, long, white hairs on the leaf-sheath of *C. pallescens*.

It has been found only once in Inverness-shire in 1973, where it was growing with both parents and other Carices and was sterile.

Section 24. Spirostachyae (Drejer) L. H. Bailey
Carex Spirostachyae Drejer

Tufted herbs, sometimes with short, creeping rhizomes. *Stems* trigonous or trigonous-terete. *Male spike* usually solitary, subcylindrical, sessile or pedunculate. *Female spikes* ovoid to cylindrical or subglobose; lowest bract leaf-like. *Female glumes* shorter than utricles, usually obtuse and mucronate at apex, but sometimes more or less acute, erect or pendulous. *Utricles* speckled, ellipsoid or ovoid-ellipsoid, with an emarginate to bifid beak. *Stigmas* 3.

42. C. binervis Sm.   Green-ribbed Sedge
C. sadleri E. F. Linton; *Trasus binervis* (Sm.) Gray; *Proteocarpus binervis* (Sm.) Börner

Glabrous *perennial herb*; rhizomes shortly creeping; shoots forming dense clumps in wet habitats, less tufted in closed grasslands; roots thick, greyish-brown; scales orange-brown, persistent. *Stems* 15–150 cm, trigonous-terete, often with a single furrow. *Leaves* 7–30 cm × 3–6 mm, dull dark green with wine-red blotches on dying, persisting as pinkish- or orange-brown litter, frequently overwintering, long-linear, rigid, often arcuate in dwarf plants, keeled or more or less flat, more or less abruptly tapering to a fine point; sheaths dull reddish-brown, persistent, the inner face hyaline, lingulate at apex at least on stem leaves, convex or straight on lower leaves; ligules 1–2 mm, rounded at apex. *Inflorescence* up to half the length of the stem; lower bracts leaf-like, 2–4 times as long as the spike, the upper glume-like. *Male spike* 1, 20–45 mm, subcylindrical; male glumes 4.0–4.5 mm, purplish with a paler midrib, oblong-obovate, obtuse or rounded and mucronate at the erose, scarious apex. *Female spikes* 2–4, distant, 15–45 mm, cylindrical, erect, the lowermost usually nodding; peduncles half-ensheathed, the lowest up to 10 cm; female glumes 3–4 mm, black or dark purplish-brown with a green or pale brown midrib, ovate, obtuse-mucronate at apex. *Utricles* 3.5–4.5 mm, ascending to 45–60° to axis of spike, purplish-brown or rarely green, broadly ellipsoid, with 2, prominent, green lateral ribs; beak 1.0–1.5 mm, rough, bifid; stigmas 3; nut obovoid, trigonous, olive brown. *Flowers* 6. *Fruits* 6–8. 2n = 74.

Native. Mainly on acid, siliceous soils especially in mountain grasslands, but occurs also on peaty soil and wet alpine cliff-ledges. In lowland areas it is found on sandy heaths, rough pastures and rocky cliffs. Frequent throughout Britain and Ireland, but less so in the drier areas of the south Midlands and East Anglia. West Europe northwards to about 64° N in Norway; north-west Africa.

× flava = C. × corstorphinei Druce
This hybrid has the habit of *C. binervis*, but the female spikes are fuscous in colour and the utricles are shorter, more shortly beaked and sometimes inflated.

Recorded only from Glen Phee, Forfarshire in 1915. The *C. flava* segregate was subsp. *oedocarpa*.

× laevigata = C. × deserta Merino
This hybrid is intermediate between the parents and sterile. The utricles are 4.0–5.5 mm and lack conspicuous stripes or dots, the female glumes are 4.0–4.5 mm, obovate-acuminate and with an excurrent, pale brown midrib and a very narrow scarious margin.

Found only in Caernarvonshire in 1961. Described from Spain in 1909.

× punctata

This sterile hybrid resembles *C. binervis* in its robust habit, very distant female spikes, purplish-brown female glumes and spreading mature utricles, but the more slender, paler male spike and slender, shortly bifid beak of the utricle show the presence of *C. punctata*.

Known only from near Barmouth, Merionethshire between 1954 and 1960. It has also been reported from France.

43. C. distans L.   Distant Sedge
Trasus distans (L.) Gray

Glabrous *perennial herb*; rhizomes short; shoots more or less densely tufted; roots reddish-brown; scales dark brown or black, rarely wine-red. *Stems* 15–100 cm, smooth, trigonous-terete. *Leaves* 10–15 cm × 2–6 mm, greyish-green rapidly becoming brown, then ashy-grey and persisting on dying, overwintering, long-linear, rigid, more or less erect, flat, tapered to a fine point; sheaths persisting, eventually becoming fibrous, dark to mid-brown, the younger ones orange-brown, the inner face herbaceous, hyaline towards the top with a brown margin, convex or straight at apex, or in the upper leaves protruding; ligules 2–3 mm, obtuse at apex. *Inflorescence* compact at flowering, elongating on fruiting to at least two-thirds the length of the stem, male spike above, female below; bracts leaf-like, the lower shorter than the adjacent internode, the upper longer, but not exceeding the stem. *Male spike* usually 1, 15–30 mm, subcylindrical; male glumes 3–4 mm, pale to purplish-brown, obovate, subacute to obtuse-mucronate at apex. *Female spikes* 2–4, 10–20 mm, oblong-cylindrical, erect; peduncles up to 40 mm, ensheathed; female glumes 2.5–3.5 mm, pale brown or rarely chestnut with a greenish midrib and hyaline margin, ovate-oblong, obtuse to acute and mucronate at apex. *Utricle* 3.0–4.5 mm, greenish or rarely dark brown, trigonous-ellipsoid, rounded at base, tapered at apex, distinctly nerved, ascending at an angle of 45–60° on the stem-axis; beak 0.7–1.0 mm, more or less rough, bifid; stigmas 3; nut trigonous-ellipsoid, yellowish-brown. *Flowers* 5–6. *Fruits* 6–7. $2n = 72, 74$.

Native. Mostly a coastal plant of rocky or sandy places within the spray zone or on beaches in reach of the higher spring tides, also in brackish marshes. All round the coasts of the British Isles. In the south and east also in inland mineral-rich marshes. Most of Europe except the north-east; west Asia; mountains of North Africa.

× extensa = C. × tornabenii Chiov.

Hybrids between these parents are intermediate in the characters of the leaves, bracts and spikes, and are sterile.

This hybrid persisted for over twenty years on salt-marshes at Mochras and at Morfa Harlech in Merionethshire. It is also recorded for Cornwall. On the mainland of Europe it is known only from Sweden.

× **flava** = **C.** × **luteola** Sendtn.
C. × *binderi* Podp.
Sterile plants more or less intermediate between these species were found in Kintyre in 1971 and there are more doubtful records from elsewhere. It is also recorded from central Europe.

× hostiana = C. × muelleriana F.W. Schultz

This sterile hybrid has the female spikes cylindrical and fuscous when mature and the male and female glumes have a silvery margin. If *C. flava* segregates are present in the area this hybrid becomes more difficult to recognise.

It has been found in Hampshire and Co. Dublin in base-rich damp meadows and a valley bog. It is also recorded from central Europe.

44. C. punctata Gaudin   Dotted Sedge

Glabrous *perennial herb*; rhizomes shortly creeping; shoots tufted; roots orange-brown to black, not felty; scales brown, rarely red, becoming fibrous. *Stems* 15–100 cm, trigonous. *Leaves* 10–50 cm × 2–5(–7) mm, usually as long as the stem, but variable, pale or yellowish-green, persisting on dying as a greyish-brown litter, doubtfully overwintering, long-linear, flat or shallowly keeled, abruptly tapered to a fine tip; sheaths persistent, orange- or pinkish-brown, the inner face hyaline, dark brown and concave at top; ligules around 3 mm, tubular at least on stem, obtuse at apex. *Inflorescence* about half the length of the stem, male spike above, female below; bracts leaf-like, at least one usually exceeding the inflorescence. *Male spike* 1, 10–30 mm, subcylindrical; male glumes 3–4 mm, orange-brown, oblong-obovate, mucronate and often fimbriate at apex. *Female spikes* 2–4, the upper more or less contiguous, the lower distant, 5–25 mm, ovoid-cylindrical; peduncles ensheathed; female glumes 2.5–3.5 mm, yellowish or pale brown with a green midrib and hyaline margin, obovate, acuminate or obtuse and mucronate at apex. *Utricles* 3–4 mm, shiny pale green, minutely dotted with reddish-brown, obovoid-ellipsoid, more or less inflated, prominently ribbed when dry, with lateral nerves prominent, ascending at an angle of 75–90° to the stem axis and therefore strongly patent; beak about 0.7 mm, widely bifid; stigmas 3; nut obovoid, trigonous, dark brown, shortly stalked. *Flowers* 6–7. *Fruits* 7–8. $2n = 68$.

Native. Sandy soils, wet cliffs and raised beaches, usually within reach of salt spray; occasionally in non-brackish marshes with a high base status. Its main stations are in south-west Ireland, but it reaches as far north as north Donegal and the Solway Firth, and as far east as Hampshire (cf. David, 1981b). Old records from the east coast may be *C. distans*, but it has been confirmed from Berwickshire. West and south Europe locally northwards to south-west Sweden and north Poland; North Africa; Asia Minor.

45. C. extensa Gooden.   Long-bracted Sedge
Trasus extensus (Gooden.) Gray; *Proteocarpus extensus* (Gooden.) Börner

Glabrous *perennial herb*; rhizomes short; shoots often forming large tufts; roots reddish-brown, often stained black; scales dark greyish-brown, becoming more or less fibrous. *Stems* 4–40 cm, rigid, bluntly trigonous, sometimes arcuate, solid. *Leaves* 5–35 cm × 2–3 mm, greyish-green or glaucous, becoming reddish-brown, then grey on dying, overwintering, rigid, thick, more or less keeled, often inrolled, gradually tapered to a blunt apex; sheaths orange-brown, occasionally red-tinged, darkening and often blackish and fibrous on decay, the inner face narrowly hyaline, concave at apex; ligules around 2 mm, rounded at apex. *Inflorescence* one-third to half the length of the stem, male spike above, female below; bracts leaf-like, usually reflexed, far exceeding the inflorescence. *Male spike* usually 1, rarely 2–3, 5–25 mm, subcylindrical; male glumes 3–4 mm, reddish-brown with a paler midrib, obovate-elliptical, obtuse at apex. *Female spikes* 2–4, erect, contiguous or the lowest sometimes distant, 5–20 mm, subglobose to cylindrical; peduncles ensheathed, the lowest more or less exserted; female glumes 1.5–2.0 mm, reddish-brown with a pale midrib and more or less hyaline margin, broadly ovate, mucronate at apex. *Utricles* 2–4 mm, greyish-green or brownish with purple blotches, ovoid or ellipsoid, weakly ribbed; beak 0.5–1.0 mm, smooth, notched; stigmas 3; nut broadly ovoid, trigonous. *Flowers* 6–7. *Fruits* 7–8. 2n = 60.

(i) Var. **extensa**
Stems up to 40 cm. *Leaves* usually more or less straight. *Utricles* 3–4 mm.

(ii) Var. **minor** Syme
C. extensa forma *pumila* Andersson
Stems 4–12 cm. *Leaves* often curved. *Utricles* 2–3 mm.

Native. A coastal species usually within reach of salt water or sea spray on both muddy and sandy estuarine or littoral flats. Around the coasts of Britain and Ireland where there is a suitable habitat. Europe to about 61° N in Sweden; western Asia; North Africa; introduced in North America. The usual plant throughout most of its range is var. *extensa*. Var. *minor* occupies a very narrow belt along the drift-line on western coasts of Britain particularly in western Scotland. It also occurs on the shores of the Baltic.

Section 25. Ceratocystis Dumort.
Section *Flavae* (O. Lang) H. Christ

Tufted herbs, sometimes with short, creeping rhizomes. *Stems* trigonous. *Male spike* usually solitary, subcylindrical, sessile or pedunculate. *Female spikes* ovoid to oblong, obtuse at apex, erect; lowest bracts leaf-like or glumaceous. *Female glumes* ovate, acute or subobtuse at apex, not mucronate. *Utricles* not speckled, ovoid-ellipsoid or ovoid, sometimes plano-convex, with an emarginate to bifid beak. *Stigmas* 3.

46. C. hostiana DC. Tawny Sedge
C. fulva auct.; *C. hornschuchiana* Hoppe; *Trasus hostianus* (DC.) Gray; *C. speirostachya* Swartz ex Sm.; *Trasus fulvus* auct.

Glabrous *perennial herb*; rhizomes shortly creeping; shoots hardly tufted; roots of varying colours; scales

pale brown, soon decaying and leaving robust fibres. *Stems* 15–60 cm, slightly rough, trigonous. *Leaves* 5–30 cm × 2–5 mm, pale green to yellowish-green, greyish-brown when dead, rarely overwintering, long-linear, more or less flat or shallowly keeled, abruptly contracted to a trigonous linear point at apex; sheaths dark greyish-brown becoming fibrous, the inner sheaths pale, the inner face hyaline, with a convex or ligulate apex; ligules about 1 mm, rounded at apex. *Inflorescence* one-quarter to half the length of the stem, male spikes above, female below; bracts leaf-like, longer than spike, but not exceeding inflorescence, upper bract sometimes very short. *Male spikes* 1 or 2, 10–20 mm; male glumes 3.5–4.5 mm, brown with a broad hyaline margin, obovate-elliptical, obtuse at apex. *Female spikes* 1–3, more or less distant, 8–20 mm, ovoid-cylindrical, erect; peduncles ensheathed for half their length; female glumes 2.5–3.5 mm, dark chestnut brown with a broad conspicuous silvery-hyaline margin and pale, often green midrib, broadly ovate, acute at apex. *Utricles* 4–5 mm, ascending at an angle of 45–60° to the axis, yellowish-green, obovoid, ribbed; beak 0.8–1.2 mm, more or less deeply bifid, serrulate; stigmas 3; style often shortly exserted; nut obovoid, trigonous, shortly stalked. *Flowers* 6. *Fruits* 7–8. 2n = 56.

Native. Wet flushes and marshy ground where the water is fairly base-rich and has a pH of 5.5–6.5, for example on schistose and igneous rock flushes. Common in the hilly areas of north England and Scotland, more local along spring lines and in valley bogs in the south and east and in Wales. In Ireland probably in more acid and less base-rich mires. Europe, except the north-east and the Mediterranean region; North America.

47. C. flava L. Yellow Sedge
Trasus flavus (L.) Gray

Glabrous *perennial herb*; rhizomes short; shoots 2 to many, more or less tufted; roots white to pale or yellowish-brown; scales greyish-brown, sometimes pinkish, soon decaying. *Stems* 5–75 cm, trigonous-terete. *Leaves* 5–70 cm × 1.7–7.0 mm, yellowish- to greyish-green, usually becoming bleached, erect or spreading, straight or curved, long-linear, flat, channelled or inrolled, often keeled, narrowed to a usually blunt tip; sheaths whitish or hyaline becoming brownish, with a more or less straight apex; ligules 1–5 mm, rounded, sometimes notched at apex. *Inflorescence* up to three-quarters the length of the stem, male spikes above, female below; bracts leaf-like, much exceeding the spike, the upper sometimes glumaceous, often patent or reflexed. *Male spikes* 10–20 mm; male glumes 3–4 mm, orange- or reddish-brown with a pale green midrib, lanceolate-elliptical or oblong-lanceolate, obtuse to acute at apex. *Female spikes* 2–5, the lowest sometimes more or less distant, the upper clustered, the lower often more or less distant, 5–15 mm, ovoid, erect; sessile or lower with partly ensheathed peduncle; female glumes 2.0–4.5 mm, orange- or reddish-brown or greenish-yellow, with a green midrib and more or less hyaline margin, lanceolate-elliptical to oblong-lanceolate, more or less acute at apex. *Utricles* 1.7–7.0 mm, yellowish-green to golden, broadly ellipsoid to obovoid, more or less ribbed, more or less deflexed when ripe; beak 0.2–2.5 mm, more or less split or bifid; stigmas 3; nut obovoid, trigonous. *Flowers* 5–8. *Fruits* 7–9. $2n = 30, 33, 58, 60, 64, 68, 70$.

The following subspecies are to a considerable extent ecological. They have sometimes been regarded as species and sometimes *C. flava* subsp. *flava* has been separated as a species from the rest, but there are intermediates, apparently fertile, between all of them.

(a) Subsp. **flava**
C. flavella V. I. Krecz.
Stems (10–)20–50(–90) cm, erect. *Leaves* equalling or slightly shorter than stems, 3–7 mm wide, flat, yellowish-green. *Male spikes* usually subsessile, sometimes with a peduncle up to 20 mm. *Female spikes* 10–15 × 10–12 mm, crowded, the lowest usually somewhat remote; lowest bract greatly exceeding inflorescence, patent or deflexed. *Utricles* 5–7 mm, broadly ellipsoid, prominently veined, the central ones patent, the lower deflexed, all curved and gradually narrowed into a beak 1.5–2.5 mm usually comprising at least one-third of the length of the utricle. $2n = 30, 33, 60, 64$.

(b) Subsp. **jemtlandica** (Palmgr.) P. D. Sell
C. jemtlandica (Palmgr.) Palmgr.; *C. lepidocarpa* subsp. *septentrionalis* forma *jemtlandica* Palmgr.; *C. lepidocarpa* subsp. *jemtlandica* Palmgr.
Stems (15–)20–50 cm, erect. *Leaves* from one-third as long as to equalling stem, 2–4 mm wide, dark green, flat. *Male spike* almost absent or up to 10 mm. *Female spikes* 7–12 × 7–10 mm, the upper crowded or at least overlapping, the lowest sometimes remote; lowest bract exceeding inflorescence, patent or deflexed. *Utricles* 3.5–5.0 mm, broadly ellipsoid, prominently veined, the central one patent, the lower deflexed, curved, gradually narrowed into a curved beak 1–2 mm and usually comprising at least one-third of the length of the utricle.

(c) Subsp. **brachyrrhyncha** Čelak.
C. lepidocarpa Tausch; *C. flava* subsp. *lepidocarpa* (Tausch) Nyman; *C. viridula* subsp. *brachyrrhyncha* (Čelak.) B. Schmid
Stems (20–)30–50(–75) cm, slender, erect and often curved. *Leaves* about half as long as the stems, 1.5–2.5 (–4.0) mm wide, green or yellowish-green, flat. *Male spike* with peduncle 5–30 mm. *Female spikes* 6–13 × 5–8 mm, the upper often overlapping, the lower usually somewhat remote; lowest bract equalling or exceeding the inflorescence, patent or deflexed. *Female glumes* pale brownish-yellow, hyaline. *Utricles* (3.5–)4.2–4.5(–4.7) mm, yellowish-green, obovoid-ellipsoid, prominently veined, the central ones patent, the lower deflexed, all curved and rather abruptly contracted into a recurved or deflexed beak 1.5–2.0 mm and usually comprising at least one-third of the length of the utricle. $2n = 58, 68$.

(d) Subsp. **scotica** (E.W. Davies) P. D. Sell
C. lepidocarpa subsp. *scotica* E.W. Davies; *C. viridula* subsp. *brachyrrhyncha* var. *scotica* (E.W. Davies) B. Schmid

Stems (8–)15–20(–30) cm, stout, rigid and erect. *Leaves* about one-third as long as the stems, 2.5–4.5 mm wide, green or yellowish-green, flat. *Male spikes* with peduncle 5–30 mm. *Female spikes* 6–13 × 5–8 mm, the upper often overlapping, the lower usually somewhat remote; patent or deflexed. *Female glumes* dark chestnut-brown, seldom caducous. *Utricles* (4.0–)4.2–5.0 mm, dark green, obovoid-ellipsoid, prominently veined, the central ones patent, the lower deflexed, all curved and rather abruptly contracted into a recurved or deflexed beak 1.5–2.0 mm and usually comprising at least one-third of the length of the utricle.

(e) Subsp. **bergrothii** (Palmgr.) P. D. Sell
C. bergrothii Palmgr.; *C. viridula* var. *bergrothii* (Palmgr.) B. Schmidt
Stems 12–30 cm, erect. *Leaves* about equalling stems, 1.0–3.5 mm wide, green or yellowish-green, flat. *Male spike* sessile or with peduncle up to 10 mm. *Female spikes* 5–12 × 6–8 mm, the upper overlapping, the lowest sometimes remote; lowest bract usually exceeding inflorescence and suberect. *Utricles* 3.0–3.5(–4.0) mm, greenish or brownish-yellow, broadly ellipsoid and distinctly inflated, prominently veined, the central ones patent, the lower sometimes deflexed, straight, abruptly narrowed into a straight beak about 1 mm, and comprising less than one-third the length of the utricle.

(f) Subsp. **oedocarpa** (Andersson) P. D. Sell
C. oederi subsp. *oedocarpa* (Andersson) Lange; *C. tumidocarpa* Andersson; *C. demissa* Vahl ex Hartm.; *C. viridula* subsp. *oedocarpa* (Andersson) B. Schmid
Stems 5–20(–40) cm, ascending. *Leaves* at least as long as stems, 2–5 mm wide, green or yellowish-green, flat. *Male spike* with a peduncle 3–25 mm. *Female spikes* 7–13 × 6–8 mm, the upper usually overlapping, the lowest often very remote; lowest bract usually greatly exceeding the inflorescence, erect to deflexed. *Utricles* (3–)3.5–4.0 mm, usually green, prominently veined, all patent or ascending, straight, gradually narrowed into a straight beak about 1 mm usually comprising less than one-third of the length of the utricle. $2n = 70$.

(g) Subsp. **serotina** (Mérat) P. D. Sell
C. oederi auct.; *C. viridula* auct.; *C. serotina* Mérat; *Trasus oederi* auct.; *C. flava* subsp. *oederi* auct.
Stems (3–)8–20(–40) cm, erect or ascending. *Leaves* about equalling or exceeding stems, (1–)1.5–3.0(–4) mm wide, yellowish-green, flat. *Male spike* usually sessile. *Female spikes* 5–8 × 3–6 mm, crowded, the lowest sometimes remote; lowest bract greatly exceeding inflorescence, patent. *Utricles* (2–)2.5–3.0(–3.5) mm, greenish-yellow, abruptly contracted into a beak 0.5–1.0 mm; nut not filling the utricle. $2n = 35, 68, 70$.

(h) Subsp. **pulchella** (Lönnr.) P. D. Sell
C. pulchella (Lönnr.) Lindm., non S. Berggr.; *C. oederi* subsp. *pulchella* (Lönnr.) Palmgr.; *C. serotina* subsp. *pulchella* (Lönnr.) Ooststr.; *C. viridula* var. *pulchella* (Lönnr.) B. Schmid; *C. scandinavica* E.W. Davies

Stems (2–)5–8(–18) cm. *Leaves* 1.0–2.5 mm wide, about equalling or exceeding stems, greyish-green, flat or channelled. *Male spike* usually sessile. *Female spikes* 3–5 × 2–4 mm, crowded, the lowest sometimes remote; lowest bract greatly exceeding inflorescence. *Utricles* 1.7–2.0 mm, greyish-green, gradually narrowed into a beak less than 0.5 mm; nut filling the utricle. $2n = 70$.

Native. Throughout the British Isles. Most of Europe; North America. Subsp. *flava* is now known in the British Isles only from Roudsea Wood in Westmorland. It formerly occurred in Cumberland. Most of Europe, but absent from much of the Mediterranean region; North America. Subsp. *jemtlandica* is known from Yorkshire, Hampshire and several sites in Ireland. It also occurs in Fennoscandia and the Baltic region. It is fertile and intermediate between subsp. *flava* and subsp. *brachyrrhyncha* and may have arisen through hybridisation. Subsp. *brachyrrhyncha* is found in base-rich fens especially in areas with seasonal flooding or flushing and peaty habitats where there is a concentration of calcium. It is scattered throughout the British Isles in base-rich strata, sodium perhaps replacing calcium in coastal habitats. Europe, except for most of the south; North America. Subsp. *bergrothii* is found in a number of sites in Ireland. It is also found in Fennoscandia and northwest Russia. It is fertile and intermediate between subsp. *oedocarpa* and subsp. *serotina* and may have arisen through hybridisation. Subsp. *oedocarpa* occurs in similar habitats to subsp. *brachyrrhyncha* but is less tolerant of a high base content, less calcium concentration and can withstand a higher aluminium concentration. It is widespread in the British Isles, especially in the north and west, and local in the Midlands and East Anglia. Europe southwards to north-west Portugal and eastwards to south Finland; North America. Subsp. *serotina* is a plant of wet habitats, tolerant of a wide range of pH, but most frequent in acid soils in valley bogs and stony lake shores. Scattered throughout the British Isles mainly in lowland habitats. Most of Europe eastwards to about 40° E in central Russia. The North American *C. viridula* Michx is considered to be a distinct subspecies. Subsp. *pulchella* occurs on maritime sands, salt-marshes and lake shores in north and west Scotland. It also occurs elsewhere in north-west Europe, the Baltic region and inland in scattered localities in central Europe and north-west Russia.

× **hostiana = C. × fulva** Gooden.
C. × *appeliana* Zahn; *C.* × *leutzii* Kneuck.; *C.* × *alsatica* Zahn; *C.* × *pieperiana* Junge; *C.* × *ruedtii* Kneuck.; *C.* × *xanthocarpa* Degl.; *Trasus fulvus* (Gooden.) Gray; *C. distans* subsp. *fulva* (Gooden.) Hook. fil.
These hybrids are easily recognised by their intermediate characters and complete sterility.

They occur throughout the British Isles wherever *C. hostiana* meets any of the subspecies of *C. flava*. They are also widespread in mainland Europe.

× **laevigata**
This hybrid is intermediate between the parents and is highly sterile. It differs from *C. flava* in its tall stems, the lowest female spike exserted on a long peduncle and larger female glumes and utricles, and from *C. laevigata* by its curved stem-base, narrower leaves, less acuminate female glumes and utricles with a less deeply notched beak. It resembles the common *C. flava* × *hostiana* but is taller and has larger female glumes and utricles and more distant female spikes.

Known only from a single plant found in 1970 near Egryn Abbey in Merionethshire. The *C. flava* segregate was subsp. *oedocarpa*.

Section **26. Porocystis** Dumort.
Tufted herbs, without creeping rhizomes. *Stems* tetraquetrous, rough above. *Male spike* solitary, cylindrical. *Female spikes* subglobose to ovoid, dense, suberect, usually all about the same level, or only the lowest distant; lowest bract leaf-like, usually exceeding inflorescence, not or scarcely sheathing. *Female glumes* shorter than utricles, ovate to lanceolate, acuminate at apex. *Utricles* ovoid-oblong, not beaked. *Stigmas* 3.

48. C. pallescens L.   Pale Sedge
C. pallida Salisb. nom. illegit.; *Trasus pallescens* (L.) Gray; *C. undulata* Kuntze; *C. microstoma* French.

Perennial herb; rhizomes very short; shoots tufted; roots usually dark reddish-brown; scales brown, often red-tinged. *Stems* 20–60 cm, tetraquetrous, rough on the sharp angles. *Leaves* 15–50 cm × 2–5 mm, mid-green, turning greyish-brown on dying, soft, long-linear, flat or often keeled, gradually tapering to a fine point, more or less hairy beneath; sheaths brown, hairy, persistent, the inner face hyaline and hairy, concave at apex; ligules about 5 mm, more or less obtuse to acute at apex. *Inflorescence* up to one-quarter the length of the stem, but usually much shorter, male spike above, female below; bracts leaf-like, the lower exceeding the inflorescence, crimped at base, the uppermost setaceous. *Male spike* 1, 8–12 mm, cylindrical, often concealed by the female spikes; male glumes 3–4 mm, pale brown with the midrib often darker, obovate-oblong, mucronate at apex. *Female spikes* 2–3, clustered or the lower remote, 5–20 mm, subglobose to ovoid, suberect or the lower nodding; peduncles smooth, the lower often longer than the spike; female glumes 3–4 mm, pale brown or hyaline with a broad midrib, ovate to lanceolate, acuminate at apex. *Utricles* 2.5–4.0 mm, shiny mid-green, ovoid-oblong, faintly nerved, rounded at apex and without a beak; stigmas 3; nut ellipsoid, trigonous, stalked. *Flowers* 5–6. *Fruits* 6–7. $2n = 62, 64, 66$.

Native. Open woodland, either on heavy clays where it can form large clumps, or on better drained soils where water is always available. On open wet ledges and stream banks in hilly districts where it is possibly a woodland relict, but it does occur in Scotland above the forest limit. In England more frequent on the heavier soils of the south-east and south Midlands, common in rough grassland of the north Pennines and south-west

Scotland and common in west and Highland Scotland in acid grassland vegetation of rocky hillsides. In Ireland mainly in the mountainous districts in damp heaths and pasture. Most of Europe, but absent from parts of the south and north-east; temperate Asia; North America.

Section 27. **Digitatae** (Fr.) H. Christ
Carex Digitatae Fr.

Tufted herbs, with short creeping rhizomes. *Stems* trigonous-subterete. *Male spike* solitary, linear. *Female spikes* linear or fusiform, lax; bract glumaceous. *Female glumes* obovate to broadly elliptical, with a scarious margin. *Utricles* obovoid or pyriform, usually more or less stipitate, with a very small or obscure beak. *Stigmas* 3.

49. **C. digitata** L.   Fingered Sedge
Tragus digitatus (L.) Gray

Almost glabrous *perennial herb*; rhizomes short, branched; shoots tufted, of 2 kinds, the apical leafy shoots with overwintering terminal buds and the lateral short shoots ending in the flowering stem; roots deep reddish-brown, wiry, fibrous; scales purplish-crimson. *Stems* 5–25 cm, slender, trigonous-subterete with 2–4, short (up to 20 mm), often setaceous leaves at base. *Leaves* 5–25 cm × 1.5–5.0 mm, pale to bronze-green, often overwintering, long-linear, soft, more or less flat or slightly keeled, tapered abruptly to a blunt point, usually sparsely hairy on the upper surface and the margins rough; sheaths bright crimson, even in very young shoots, inner face herbaceous, concave at the apex and soon splitting; ligules 0.5–1.5 mm, rounded at apex. *Inflorescence* one-fifth to one-quarter the length of the stem, the lowest branch 1 cm or more away from the next above, male spike above, female below; bracts glumaceous. *Male spike* 1, 8–15.0 mm, linear, few-flowered, overtopped by the upper 1–2 female spikes; male glumes about 5 mm, reddish-brown with a pale midrib and hyaline margins, oblong, rounded or emarginate at apex. *Female spikes* 1–3, rather distant, 10–15 mm, linear, very lax, 5- to 10-flowered; peduncles slender, rather short; female glumes 3.0–4.5 mm, purplish-crimson, obovate, obtuse or emarginate at apex. *Utricles* 3.0–4.5 mm, greenish-brown, obovoid, hairy; beak 0.2–0.5 mm, truncate at apex; stigmas 3; nut obovoid, trigonous, stalked. *Flowers* 4–5. *Fruits* 4–6. Sets abundant seed if not too shaded. $2n = 48, 50, 52, 54$.

Native. Locally abundant in open woodland or scrub, and amongst rocks and on stabilised screes of hard chalk and limestone. It requires a pH of 7.3–8.0 and good drainage. Very local from Somerset and Dorset to Westmorland and north-east Yorkshire (cf. David 1978a). Most of Europe, but absent from parts of the south; temperate Asia.

50. **C. ornithopoda** Willd.   Bird's-foot Sedge
C. digitata subsp. *ornithopoda* (Willd.) Hook. fil.

Nearly glabrous *perennial herb*; rhizomes short, much branched; shoots tufted, of 2 kinds, apical leafy shoots with overwintering terminal buds and lateral short shoots ending in a flowering stem; roots deep brown, wiry, fibrous; scales deep reddish-brown, becoming fibrous. *Stems* 5–20 cm, slender, trigonous-subterete with 2–4, short, often setaceous leaves at the base. *Leaves* 5–20 cm × 1–3 mm, bright or mid-green, often overwintering, soft, flat or slightly keeled, long-linear, tapering abruptly to a blunt point; sheaths orange- to dark crimson-brown, becoming fibrous, inner face herbaceous, concave and soon splitting at apex; ligule 0.5–1.0 mm, rounded at apex. *Inflorescence* one-tenth to one-eighth the length of the stem, compact, more or less digitate, male spike above, female beneath; bracts glumaceous. *Male spike* 1, 5–8 mm, linear, few-flowered, overtopped by the lower female spikes; male glumes about 2.5 mm, reddish-brown with a pale midrib and hyaline margin, obovate, acute or mucronate at apex. *Female spikes* 2–3, 5–10 mm, linear, 2- to 4-flowered, more or less acute and often erose at apex. *Utricles* 2–3 mm, obovoid or pyriform, yellowish-green to brown, hairy; beak 0.1–0.3 mm, truncate at apex; stigmas 3; nut obovoid, trigonous, stalked. *Flowers* 5. *Fruits* 5–6. $2n =$ c. 46, 52. 54.

Native. Dry, well-drained limestone grassland or in crevices in limestone pavement, rarely in the shade. Very local in Derbyshire, Yorkshire, Westmorland and Cumberland. Most of Europe, but absent from the Mediterranean region and the south-east; Turkey. Our plant is subsp. **ornithopoda** which occurs throughout most of the range of the species.

51. **C. humilis** Leyss.   Dwarf Sedge
C. clandestina Gooden. nom. illegit.; *Trasus clandestina* Gray nom. illegit.; *C. prostrata* All.

Nearly glabrous, long-lived *perennial herb* forming mats; rhizomes shortly creeping, often much branched; shoots densely fasciculate at the branch tips; roots purplish-brown, woody, much branched; scales reddish-brown, persisting as fibres. *Stems* 2–10(–15) cm, slender, often arcuate or flexuous, subterete, solid, often hidden by the leaves. *Leaves* up to 20 cm × 1.0–1.5 mm, dark green, pale purplish-brown on decay, overwintering, rough, stiff, arcuate, at first flat, later becoming channelled, long-linear, tapering from base to a fine trigonous point; sheaths white with green veins, becoming orange- and reddish-brown, persistent, eventually becoming fibrous and clothing rhizome, inner face hyaline, concave at apex; ligules 0.5–1.0 mm, rounded at apex. *Inflorescence* up to three-quarters the length of the stem, male spike above, female below; bracts glumaceous, hyaline or pale brown, almost enclosing the female spike and with a sheath 3–8 mm. *Male spike* 1, 10–15 mm, linear, male glumes 5–7 mm, reddish- or purplish-brown with a very broad hyaline margin and a pale midrib, elliptic-lanceolate, obtuse or subacute at apex. *Female spikes* 2–4, distant, 4–10 mm, fusiform, with 2–4 flowers only; peduncles 1–2 mm, ensheathed; female glumes 2–3 mm, reddish-brown, hyaline at edges and base, obovate to broadly elliptical, clasping utricle and appearing narrower, obtuse and often mucronate at apex. *Utricle* 2–3 mm, obovoid, pyriform, trigonous, shortly hairy; beak about 0.1 mm, truncate at apex; stigmas 3; nut more or less ellipsoid, trigonous, stalked. *Flowers* 3–5. *Fruits* 4–7.

Gradually spreads vegetatively and ripe seed is distributed by ants. $2n = 36, 38$.

Native. Locally abundant in short turf and limestone and chalk grassland, often on south, south-west or west facing slopes out of reach of ploughing. Most frequent on the Dorset and south Wiltshire downs; also in Hampshire, Wiltshire, Somerset, Gloucestershire and Herefordshire (cf. David, 1979a). Central and south Europe; Caucasus; west Siberia east to Manchuria.

Section 28. Mitratae Kük.

Tufted herbs with short creeping rhizomes. *Stems* trigonous, rough above. *Male spike* solitary, sessile or shortly pedunculate. *Female spikes* ovoid; lowest bracts often leaf-like, sheathing. *Female glumes* ovate, more or less obtuse to acute at apex. *Utricles* obovoid-ellipsoid, tomentose, with a short, emarginate beak. *Stigmas* 3.

52. C. caryophyllea Latourr.　　　　Spring Sedge
C. verna Chaix, non Lam.; *C. praecox* Jacq., non Schreb.; *Trasus praecox* Gray nom. illegit.

Nearly glabrous *perennial herb*; rhizomes shortly creeping; shoots loosely tufted; roots dark or purplish-brown; scales rich or blackish-brown, often shiny, soon becoming fibrous. *Stems* 2–30 cm, trigonous, erect, rigid, rough above, leafy below. *Leaves* up to 20 cm × 1.5–3.0 mm, mid- or dark green, sometimes overwintering, recurved, rough on upper surface, shiny, more or less flat, long-linear, tapered more or less abruptly to a short, trigonous point; sheaths herbaceous, becoming brown or blackish-brown and fibrous, inner face hyaline, soon decaying, straight at apex; ligules 1–2 mm, obtuse at apex, free margin very narrow, entire. *Inflorescence* 2–4 cm, male spikes above, female below; lower bracts often leaf-like, the lowest with a sheath 3–5 mm, the upper glumaceous. *Male spike* 1, 10–15 mm, clavate; male glumes 4–5 mm, greenish- or reddish-brown, hyaline towards the base with a darker midrib, oblanceolate-elliptical, more or less acute or mucronate at apex. *Female spikes* 1–3, more or less contiguous, 5–12 mm, ovoid, erect, sessile or with peduncles ensheathed; female glumes 2.0–2.5 mm, reddish-brown or greenish with an excurrent green midrib, broadly ovate, often attenuate but more or less acute to acuminate at apex. *Utricles* 2–3 mm, green, obovoid-ellipsoid, more or less trigonous, with two lateral ribs, tomentose; beak 0.2–0.3 mm, notched; stigmas 3; nut obovoid-ellipsoid, trigonous. *Flowers* 4–5. *Fruits* 5–7. $2n = 62, 64, 66, 68$.

Native. Frequent in calcareous grasslands and on acid soils on mountains where flushed occasionally by base-rich water. Throughout the British Isles, but local north of the Great Glen. Europe north to about 63° N in Sweden; introduced in North America.

Section 29. Acrocystis Dumort.

Herbs tufted or not. *Stems* trigonous or tetraquetrous. *Male spike* often solitary, occasionally 2, cylindrical to subclavate, more or less sessile. *Female spikes* ovoid, ovoid-oblong or subglobose; lowest bracts glumaceous to leaf-like, sometimes sheathing. *Female glumes* usually shorter than utricles, ovate, obovate or subrotund.

Utricles ovoid to subglobose, usually puberulous or tomentose; beak very small. *Stigmas* 3.

53. C. filiformis L.　　　　Downy-fruited Sedge
C. tomentosa L.; *Trasus tomentosus* (L.) Gray

Nearly glabrous *perennial herb*; rhizomes long creeping, often very slender; shoots 2–3 per tuft; roots pale; scales reddish-brown, shiny, with sharp points, persistent. *Stems* 20–50 cm, rough above, trigonous, slender. *Leaves* 15–40 cm × 1.5–2.0 mm, more or less glaucous, greyish-brown on decay, often overwintering, rough, flat, long-linear, gradually tapering to a fine point; sheaths red or reddish-purple, inner face hyaline, becoming brown, often persistent and fibrillose on splitting, concave at apex; ligules 1–2 mm, tubular, obtuse at apex. *Inflorescence* one-sixth or less of the stem length, male spikes above, female beneath; upper bracts setaceous, lower leaf-like, more or less equalling inflorescence, the lowest not or shortly sheathing. *Male spikes* 1–2, 12–25 mm, subclavate; male glumes 4–5 mm, reddish-brown with a paler midrib and more or less hyaline margin, more or less acute and often apiculate at apex. *Female spikes* 1–2, more or less contiguous, 5–14 mm, oblong-ovoid to subglobose; female glumes 2–3 mm, purplish- or reddish-brown with a pale midrib, ovate to subrotund, acute at apex or the lowermost in the spike mucronate. *Utricles* 2–3 mm, green, subglobose or pyriform, trigonous, tomentose; beak 0.1–0.3 mm, notched; stigmas 3; nuts obovoid or pyriform, trigonous. *Flowers* 5–6. *Fruits* 6–7. $2n = 48$.

Native. Fairly rich pastures, damp meadows, roadsides and rough ground. Very local in Wiltshire, Oxfordshire, Surrey and Gloucestershire, formerly in Middlesex. Europe from Estonia to east Spain and central Greece; Caucasus; Siberia.

54. C. ericetorum Pollich　　　　Rare Spring Sedge
C. approximata All.

Nearly glabrous *perennial herb;* rhizomes shortly creeping; shoots tufted, forming a close mat; roots dark or purplish-brown; scales rich or blackish-brown, becoming fibrous. Stems 2–20 cm, erect, rigid, trigonous, more or less leafless or with 3 very short leaves at base. *Leaves* up to 15 cm × 1.5–4.0 mm, mid- or dark shiny green with usually a broad scarious margin, sometimes overwintering, long-linear, rigid, often recurved, rough and sometimes almost papillose on upper surface, more or less flat, tapered more or less abruptly to a short trigonous point; sheaths herbaceous, becoming dark brown and fibrous; ligules less than 1 mm, rounded at apex, more or less entire. *Inflorescence* 2–3 cm, male spikelet above, female beneath; bracts glumaceous, the lowest scarcely sheathing. *Male spike* 1, 10–15 mm, narrowly cylindrical; male glumes 2–3 mm, purplish-brown with scarious margins, oblong, fringed, rounded at apex. *Female spikes* 1–3, more or less contiguous, 5–12 mm, ovoid, erect, sessile; female glumes 2.0–2.5 mm, purplish-brown becoming black, rounded, with a fringed, scarious margin. *Utricles* 2–3 mm, dark green, obovoid-subglobose, trigonous, tomentose; beak up to 0.3 mm;

stigmas 3; nut subglobose. *Flowers* 4–5. *Fruits* 4–6. Spreads slowly by rhizomes as well as seed. 2n = 30.

Native. Dry, short calcareous grassland up to 400 m in the Lakes. Local on the East Anglian chalk heaths, and on the limestone in Lincolnshire, Derbyshire, Yorkshire, Durham and Westmorland (cf. David, 1981a). Most of Europe, but absent from the Mediterranean region; Caucasus; Siberia (Urals).

55. C. montana L.   Soft-leaved Sedge
C. collina Willd.; *C. emarginata* Willd.

Nearly glabrous *perennial herb* with a very distinctive habit, making widening tufts that often die out in the middle like fairy rings; rhizomes creeping, thick, often much branched; shoots tufted at apex; roots purplish-brown, woody; scales reddish-brown, becoming fibrous. *Stems* 10–40 cm, drooping or prostrate, slender, rough at top, flaccid, trigonous or often with 6 angles, more or less solid, leafless or with few short leaves at base. *Leaves* 10–35 cm × 1.5–2.0 mm, usually erect, pale to mid-green, greyish-brown and persisting on dying, soft, flat, long-linear, gradually drawn out to a slender point, sparsely hairy on upper surface, becoming glabrous; sheaths dark reddish-brown to almost bright red, ribbed, becoming fibrous and densely clothing the rhizome, inner face hyaline, soon decaying, concave at apex; ligules about 1 mm, obtuse at apex, fimbriate. *Inflorescence* 1–2 cm, very congested, male spikelets above, female beneath; bracts glumaceous or the lowest setaceous, the lowest scarcely sheathing. *Male spike* 1, 10–20 mm, cylindrical; male glumes 4–5 mm, reddish-brown with a pale midrib, oblanceolate or broadly elliptical, acute or more or less mucronate at apex. *Female spikes* 1–4, 6–10 mm, ovoid, with a few lax flowers clustered beneath the male spike, erect, sessile; female glumes 3–5 mm, reddish-black with a pale midrib, broadly ovate or obovate, obtuse or retuse-mucronate at apex. *Utricles* 3.0–4.5 mm, brown or blackish on the exposed face, ovoid-pyriform, bluntly trigonous, tapered to a stout stalk, densely minute hairy, lightly ribbed; beak up to 0.3 mm, notched; stigmas 3; nut obovoid, trigonous, stalked. *Flowers* 5. *Fruits* 5–6. 2n = 38.

Native. Neutral to acidic grasslands and in light shade in woodlands. The underlying rock is usually basic, but is covered with a layer of non-calcareous drift. Very local across southern England and Wales north to Derbyshire; not recorded for Ireland. Europe from southern Sweden and central Russia to north-west Spain and south-west Bulgaria; temperate Asia.

56. C. pilulifera L.   Pill Sedge
C. montana auct.; *Trasus pilulifer* (L.) Gray

Nearly glabrous *perennial herb*; rhizomes short; shoots densely tufted, often decumbent; roots purplish-brown; scales brown, soon becoming fibrous. *Stems* 10–30 cm, more or less rough above, wiry, tetraquetrous, often arcuate and prostrate. *Leaves* 5–20 cm × 1.5–2.0 mm, yellowish- or mid-green, or bronze-green, pinkish-brown and persistent on decay, overwintering, rough

above, more or less flat, papillose on upper surface, long-linear, more or less abruptly tapered to a short trigonous point; sheaths pale reddish-brown or wine-red, becoming fibrous, inner face hyaline, straight at apex; ligule 0.5–1 mm, rounded at apex, minutely fimbriate. *Inflorescence* 2–4 cm, clustered at apex of stem, male spikelet above, female beneath; bracts leaf-like or the upper setaceous, rarely exceeding inflorescence, the lowest not sheathing. *Male spike* 1, 8–15 mm, cylindrical; male glumes 3.5–4.0 mm, brown or chestnut, hyaline towards the margin with a pale midrib, oblanceolate-elliptical, with an acute apex. *Female spikes* 2–4, 5–8 mm, ovoid or subglobose, erect, sessile; female glumes 3.0–3.5 mm, reddish-brown, hyaline towards the margin with a green midrib, broadly ovate, acute or acuminate at apex. *Utricles* 2.0–3.5 mm, green, obovoid-subglobose, more or less minutely downy; beak 0.3–0.5 mm, notched; stigmas 3; nut obovoid, trigonous. *Flowers* 5–6. *Fruits* 6–7. 2n = 18.

Native. Leached, sandy, peaty and sometimes more loamy soils with a low base content and a pH usually between 4.5 and 6.0. Frequent in hill pasture and also in open heath. Throughout Britain, but more frequent in sandy regions and common in Scotland up to 750 m; scattered, but less frequent, in Ireland. Europe eastwards to St Petersburg and the east Carpathians. Our plants are all subsp. **pilulifera** which occurs throughout the range of the species except for the Azores.

Section **30. Aulocystis** Dumort.
Section *Frigidae* H. Christ

Tufted herbs with shortly creeping rhizomes. *Stems* trigonous. *Male spike* solitary. *Female spikes* 2–4, usually clustered, ovoid-globose; lowest bracts leaf-like, more or less sheathing. *Female glumes* shorter than utricles, oblanceolate, acuminate at apex. *Utricles* narrowly obovoid or ellipsoid, trigonous; beak notched. *Stigmas* 3.

57. C. atrofusca Schkuhr   Scorched Alpine Sedge
C. ustulata Wahlenb. nom. illegit.; *Trasus ustulatus* Gray nom. illegit.

Glabrous *perennial herb*; rhizomes shortly creeping; shoots in loose tufts; roots yellowish; scales pale brown, soon decaying. *Stems* 5–35 cm, trigonous. *Leaves* 2–12 cm × 2–5 mm, mid-green or very rarely glaucous, soft, more or less flat or slightly keeled, long-linear, those of the stem abruptly tapered to a short trigonous point, those of the sterile shoots gradually tapered to a fine trigonous point 1–3 cm long; sheaths pale with green veins, becoming pale yellowish-brown, soon decaying, inner face hyaline, concave at apex; ligules 1.0–2.5 mm, obtuse at apex. *Inflorescence* one-sixth to one-quarter the length of the stem, male spikelet above, female beneath; lower bracts narrowly leaf-like, shorter than the spike, the lowest with a sheath 5–15(–25) mm, upper glumaceous. *Male spike* 1, 5–10 mm, broadly ellipsoid; male glumes about 3.5 mm, dark reddish-brown, rarely pale, with pale or green midrib, acute to acuminate at apex. *Female spikes* 2–4, clustered, lowest rarely distant,

5–12 mm, ovoid-globose, nodding; peduncles smooth, up to 3 times as long as the spike, half ensheathed; female glumes about 3 mm, purplish or reddish-black, with pale, thin or often distinct midrib, oblanceolate, acuminate at apex. *Utricles* 4.0–4.5 mm, purplish-black, narrowly obovoid or ellipsoid; beak 0.3–0.7 mm, notched; stigmas 3; nut more or less ellipsoid, trigonous, stalked. *Flowers* 7. *Fruits* 7–9. $2n = 38, 40$.

Native. Micaceous stony flushes between 540 and 1050 m. Rare in mid-Perthshire, Inverness-shire and Argyllshire. Alps, north and west Scandinavia; north Urals; Greenland.

Section **31. Limosae** (O. Lang) H. Christ
Carex Limosae O. Lang

Tufted herbs with creeping rhizomes. *Stems* trigonous. *Male spike* solitary, linear or narrowly lanceolate-elliptical. *Female spikes* sometimes with a few male flowers at base, ovoid, pendulous; lowest bract leaf-like, sometimes sheathing. *Female glumes* lanceolate, ovate or ovate-oblong. *Utricles* ellipsoid to obovoid-globose; beak obscure. *Stigmas* 3.

The taxonomy of the species in this section can be difficult, especially on mainland Europe where *C. limosa* and *C. rariflora* can be difficult to separate.

58. C. limosa L.   Bog Sedge
Trasus limosus (L.) Gray

Glabrous *perennial herb*; rhizomes often far-creeping, partly ascending; shoots loosely tufted, initially decumbent, slender; roots yellow, felted; scales brown, persistent. *Stems* 10–40 cm, usually rough, slender, rigid, trigonous, striate, often decumbent at the base. *Leaves* 15–40 × 1–2 mm, folded or inrolled, pale to bluish-green, often rich brown on decay, more or less rough, thin, keeled, long-linear, gradually tapered to a fine rough point; sheaths red-flushed, becoming brown or reddish-brown, persistent, inner face hyaline, persistent, concave at apex; ligules 1–2 mm, obtuse at apex. *Inflorescence* about one-sixth the length of the stem, male spikelet above, female beneath; bracts leaf-like, lowest about as long as the spike, not sheathing or with a sheath less than 5 mm. *Male spike* 1, 10–25 mm; male glumes 3–4 mm, reddish-brown, often with a green midrib, broadly lanceolate, more or less acute and apiculate at apex. *Female spikes* 1–3, 7–20 mm, more or less ovoid, 7–20 flowered, usually nodding; peduncles up to 40 mm, slender, smooth; female glumes 3.5–4.5 mm, slightly wider than utricle, brown or reddish-purple with a green midrib, ovate, acute at apex. *Utricles* 3–4 mm, bluish-green, obovoid to broadly ellipsoid, compressed, strongly ribbed; beak up to 0.2 mm, truncate at apex; stigmas 3; nut obovoid, trigonous. *Flowers* 5–6. *Fruits* 6–9. $2n = 56, 64$.

Native. Edges of pools, in very wet blanket or marginal valley mires and mesotrophic mires, usually below 450 m, but ascending to 817 m in the Breadalbanes. Mainly in north Wales, northern England and Scotland where it is frequent in the blanket mires of the west. Also in the south Dorset–Hampshire valley mires and in east

Norfolk where it is decreasing due to drainage. In Ireland it is mainly in the north and west. North and central Europe, south to the Pyrenees and south-east Russia; north Asia; North America.

59. C. rariflora (Wahlenb.) Sm.   Mountain Bog Sedge
C. limosa var. *rariflora* Wahlenb.; *Trasus rariflorus* (Wahlenb.) Gray

Glabrous *perennial herb*; rhizomes shortly creeping, often producing a close carpet of single shoots; roots yellow, felted; scales reddish-brown. *Stems* up to 20 cm, usually smooth, trigonous, more or less solid. *Leaves* up to 15 cm × 1–2 mm, shorter than stems, flat, with about 9 veins, long-linear, incurved at the tips; sheaths broadly membranous in upper part, soon becoming fibrous; ligules about 3 mm, acute at apex. *Inflorescence* one-third to one-fifth the length of the stem, male spikelet above, female beneath; lower bracts narrow, shorter than inflorescence, with a sheath up to 5 mm. *Male spike* 1, 8–12 mm, erect; male glumes about 4 mm, dark brown or purple with a pale midrib, oblong-ovate, obtuse-mucronate at apex. *Female spikes* usually 2, rarely 3, about 10 × 4 mm, with up to 8 flowers, nodding, lax; female glumes 3–4 mm, dark purple or chocolate brown with a pale midrib, oblong-ovate, obtuse-mucronate at apex. *Utricles* 3.0–4.5 mm, narrower and shorter than female glumes, ellipsoid, strongly ribbed, markedly tapered at both ends; beak up to 0.2 mm, stigmas 3; nut ellipsoid, trigonous, stalked. *Flowers* 6. *Fruits* 7–8. $2n = 50, 54$.

Native. Rare and local on wet slopes of oligotrophic peat at between 750 and 1050 m. East central Highlands of Scotland, very rare in the Breadalbanes. Circumpolar. A shy-flowering species, which often conceals its presence.

60. C. magellanica Lam.   Tall Bog Sedge
C. limosa var. *irrigua* Wahlenb.; *C. paupercula* Michx; *C. irrigua* (Wahlenb.) Sm. ex Hoppe; *C. magellanica* subsp. *planitiei* (Asch. & Graebn.) Schultze-Motel; *C. limosa* subsp. *irrigua* (Wahlenb.) Hook. fil.

Glabrous *perennial herb*; rhizomes shortly creeping, often ascending; shoots loosely tufted; roots yellow, felted; scales reddish-brown, persistent, shiny. *Stems* 12–40 cm, trigonous, usually more or less solid. *Leaves* up to 25 cm × 1.5–4.0 mm, pale or bright green, soft, thin, with up to 15 veins, rough above, more or less flat, long-linear, abruptly tapered to a rough tip; sheaths becoming reddish-brown, persistent, inner face hyaline, concave at apex; ligules 4–5 mm, tubular, acute at apex. *Inflorescence* about one-fifth the length of the stem; male spikelet on top, female beneath; bracts leaf-like, the lower usually exceeding the inflorescence, the lowest not or scarcely sheathing. *Male spike* 1, 10–20 mm; male glumes 5–6 mm, pale reddish-brown, hyaline at margin, lanceolate-elliptical, acuminate at apex. *Female spikes* 2–4, 5–18 mm, ovoid, lax and up to 10-flowered, upper more or less erect, lower nodding; peduncles smooth, 5–20 mm; female glumes 5.0–6.5 mm, narrower than utricle, reddish- or purplish-brown with a paler midrib, lanceolate, acuminate or aristate at apex, caducous.

Utricles 3.0–3.5 mm, bluish-green, ovoid-globose, more or less compressed, faintly ribbed; beak up to 0.1 mm; stigmas 3; nut ellipsoid, trigonous. *Flowers* 5–6. *Fruits* 6–7. Spreading largely by seed, but shy-flowering in some seasons. $2n = 58$.

Native. *Sphagnum* upland bogs and valley mires, but does not tolerate standing water, between 170 and 658 m. Often on a wet level ledge, near the crest of a moor, close to the watershed, being neither inundated nor subject to much movement of water. Scattered from the Lake District to North Uist and Sutherland; very local in Co. Antrim, Co. Tyrone and north mid-Wales. Circumpolar south to Mount Cenis in France, Utah and Pennsylvania; South America. The plants of the northern hemisphere are subsp. **irrigua** (Wahlenb.) Hiitonen. Subsp. *magellanica* occurs in South America southwards from 40° S.

Section **32. Atratae** H. Christ

Tufted herbs with a creeping rhizome. *Stems* trigonous or tetraquetrous. *Terminal spikes* female above and male at base, lateral spikes female; lowest bract usually leaf-like, not or very shortly sheathing. *Female glumes* ovate or ovate-elliptical, usually blackish. *Utricles* obovoid-ellipsoid or oblong-ovoid, with a short truncate or notched beak. *Stigmas* 3.

61. C. atrata L.   Black Alpine Sedge
Trasus atratus (L.) Gray

Glabrous *perennial herb*; rhizomes short; shoots tufted; roots pale brown; scales brownish- or reddish-purple, persistent. *Stems* 30–55 cm, thick, rigid, trigonous, solid. *Leaves* 15–30 cm × 2–6 mm, glaucous, more or less flat or slightly keeled, long-linear, abruptly tapered to a rough apex; sheaths dark brown, often wine-red, persistent, inner face hyaline-brown, persistent, straight at apex; ligules 2–3 mm, obtuse at apex. *Inflorescence* one-eighth to one- sixth the length of the stem; bracts leaf-like, the lower exceeding the inflorescence, the lowest not or scarcely sheathing. *Spikes* 3–5, 8–20 mm, ovoid, clustered in a more or less nodding head, terminal with female at top, male at base, lower female only; lower peduncles rough, as long as the spike. *Male glumes* 4.5–6.0 mm, reddish-black with a pale midrib, ovate-elliptical, acute or rounded at apex. *Female glumes* 3.5–4.5 mm, purple or reddish-black with a pale midrib, ovate-elliptical, more or less acute at apex. *Utricles* 3–4 mm, green, obovoid-ellipsoid, compressed at edges; beak 0.3–0.5 mm, slightly notched; stigmas 3; nut obovoid, trigonous. *Flowers* 6–7. *Fruits* 7–9. Probably maintained by vegetative growth. $2n = 54, 56$.

Native. Wet, ungrazed ledges, usually where there is calcareous veining, from 550 to 1060 m. Wales (Snowdon), northern England (rare) and Scotland (frequent in the Breadalbanes). North Europe and on higher mountains in the south; northern Asia and the mountains south to the Caucasus, Turkestan and Baikal; North America from the Rocky Mountains south to Utah. Our plant is subsp. **atrata** which occurs throughout much of the range of the species.

62. C. buxbaumii Wahlenb.   Club Sedge
C. polygama Schkuhr, non J. F. Gmel.; *C. canescens* L. nom. rej.

Glabrous *perennial herb*; rhizomes shortly creeping; shoots single or in small tufts; roots pale yellowish-brown, much branched; scales reddish-brown or sometimes blackish, shiny, persistent. *Stems* 30–70 cm, rigid, tetraquetrous. *Leaves* 25–60 cm × 1.5–2.0 mm, glaucous, more or less flat or keeled, long-linear, gradually tapered to a fine rough point; sheaths reddish- or orange-brown, persistent, inner face hyaline-brown, fibrillose on splitting; ligules about 3 mm, acute at apex. *Inflorescence* about one-quarter as long as the stem; upper bracts glumaceous or setaceous, lowest leaf-like and more or less equalling the inflorescence, not or scarcely sheathing. *Spikes* 2–5, contiguous or more or less remote, 7–15 mm, erect, the terminal with female at the top, male at base, the lowest spikes female; peduncles 2–12 mm, rough. *Male glumes* about 4 mm, reddish-black with a pale midrib, lanceolate-elliptical, acute at apex. *Female glumes* 3–5 mm, dark reddish-brown with a pale midrib, ovate or ovate-elliptical, acuminate or aristate at apex. *Utricles* 3.0–4.5 mm, pale green, indistinctly ribbed, oblong-ovoid, more or less inflated, conical and truncate at apex; beak 0–0.2 mm, stigmas 3; nut ellipsoid, trigonous. *Flowers* 6–7. *Fruits* 7–8. $2n = $ c. 74.

Native. Rare in mesotrophic fens. Known only from four localities in Inverness-shire. Formerly by Lough Neagh, Co. Antrim, where it has disappeared since drainage. North and central Europe, local; Siberia; Japan; North America. Our plant is subsp. **buxbaumii** which occurs throughout much of Europe, but is less widespread in the north.

63. C. norvegica Retz.   Close-headed Alpine Sedge
C. halleri auct.; *C. vahlii* Schkuhr; *C. alpina* Lilj., non Schrank

Glabrous *perennial herb*; rhizomes short; shoots tufted; roots pale yellowish-brown; scales brown, reddish-brown or blackish, becoming fibrous on decay. *Stems* 6–30 cm, trigonous. *Leaves* 5–20 cm × 1.5–3.0 mm, mid-green, not glaucous, more or less flat or keeled, long-linear, gradually tapered to a short, fine trigonous point; sheaths white, green-veined, becoming red, reddish-brown or brown, persistent, inner face hyaline, straight at apex; ligules 0.5–1.0 mm, obtuse at apex. *Inflorescence* a compact, terminal cluster of 1–4 subglobose, erect spikes the terminal one of which is female at the top and male at the base (often only to be detected by the presence of white filaments) the rest being female; upper bracts glumaceous, lower bracts leaf-like, exceeding inflorescence, the lowest not or scarcely sheathing. *Male glumes* 2.0–2.5 mm, dark reddish-brown with a pale midrib, lanceolate-elliptical, acute at apex. *Female glumes* 1.5–2.0 mm, reddish-black with narrow, pale midrib, ovate, acute at apex. *Utricles* 1.8–2.5 mm, greenish-brown, obovoid-ellipsoid, minutely papillose; beak 0.1–0.3 mm, slightly notched; stigmas 3; nut ellipsoid, trigonous. *Flowers* 6–7. *Fruits* 7–8. $2n = 56$.

Native. Rare on wet ledges and rocky slopes in mountains between 690 and 990 m in north-facing crevices, where the snow lies late. Perthshire, Angus and Aberdeenshire. Northern Europe and Alps; Arctic Siberia; North America. Our plant is subsp. **norvegica** which occurs throughout much of the species range in Europe.

Section 33. Phacocystis Dumort.
Section *Acutae* (Fr.) H. Christ; *Carex Acutae* Fr.

Herbs tufted or not. *Stems* trigonous, tetraquetrous or subterete. *Male spikes* 1–several, more or less cylindrical, occasionally female below. *Lateral spikes* female, cylindrical or cylindrical-ovoid; lowest bract leaf-like, not sheathing. *Female glumes* ovate, ovate-lanceolate, elliptical, oblong-obovate or ovate-elliptical. *Utricles* obovoid, suborbicular, ellipsoid, ovoid-ellipsoid or suborbicular-ellipsoid, more or less abruptly contracted into a very short, truncate or notched beak. *Stigmas* 2.

Partially fertile hybrids have been synthesized between many of the species in this section and have been found in the wild. Much information about them can be found in J. S. Faulkner (1972, 1973).

64. C. recta Boott   Estuarine Sedge
C. kattegatensis Fr. ex Krecz.; *C. salina* auct.

Glabrous *perennial herb*; rhizomes often far-creeping; shoots tufted; roots pale purplish-brown; scales reddish-brown, often blackened, persistent. *Stems* 30–100 cm, stiff, trigonous. *Leaves* up to 110 cm × 3–6 mm, mid- to yellowish-green, smooth beneath, rough above and on the margins, more or less flat or weakly keeled, long-linear, more or less abruptly tapering to a flat, stiff point; sheaths herbaceous becoming brown or tinged wine-red, persistent, transverse septa distinct, inner face hyaline, more or less persistent; ligule 2–4 mm, tubular, obtuse at apex. *Inflorescence* one-sixth to one-fifth the length of the stem; bracts leaf-like with purplish auricles, at least the lower exceeding the inflorescence. *Male spikes* 1–4, 10–40 mm, the lower pedunculate; male glumes 4–5 mm, purplish-brown with pale midrib and hyaline margins, oblanceolate-elliptical, obtuse at apex. *Female spikes* 2–4, contiguous or more or less overlapping, 30–80 mm, lax-flowered at base, cylindrical, the upper often male at top; peduncles 1–3 cm, rough; female glumes 4–5 mm, dark brown or purple with a pale midrib, ovate-lanceolate, acute to acuminate at apex, aristate in lower flowers. *Utricles* 2.5–3.0 mm, green, obovoid or suborbicular, faintly nerved; beak up to 0.3 mm, truncate at apex; stigmas 2; nut orbicular or obovate, biconvex. *Flowers* 7. *Fruits* 8–9. 2n = 84.

Native. In stiff peaty alluvium or in more sandy situations, but usually where silt is periodically deposited or the water-table fluctuates seasonally. A very local species of estuarine or lower riverine situations in Inverness-shire, Ross-shire and Sutherland where it often forms extensive colonies. Atlantic coast of North America from Labrador to Massachusetts, apparently always local. Doubtfully recorded from Norway and Faeroes.

65. C. aquatilis Wahlenb.   Water Sedge

Glabrous *perennial herb*; rhizomes far-creeping; shoots tufted; roots up to 2 mm thick, much branched, reddish-brown, persistent. *Stems* 20–110 cm, smooth, bluntly trigonous to subterete, brittle. *Leaves* 15–100 cm × 3–5 mm, dark or yellowish-green without stomata beneath, more or less glaucous with dense stomata above, smooth, shiny, flat or more or less channelled, long-linear, gradually tapered to a flat point, the margins rolling inwards on drying; sheaths herbaceous, wine-red or reddish-brown, persistent, inner face hyaline, soon decaying, straight or concave at apex; ligules 4–15 mm, tubular, obtuse or acute at apex. *Inflorescence* one-fifth to one-quarter the length of the stem; bracts leaf-like with purplish auricles, the lowest exceeding the inflorescence. *Male spikes* 2–4, 5–50 mm; male glumes 3–4 mm, brown with a pale midrib and hyaline margin, obovate to oblanceolate, obtuse at apex. *Female spikes* 2–5, contiguous or the lowest more or less distant, 20–60 mm, cylindrical, slender, dense-flowered, becoming lax below, upper more or less sessile, often male at top; peduncles 10–20 mm, smooth; female glumes 2–3 mm, dark purplish-brown with a pale midrib and hyaline tip, ovate or broadly elliptical, subacute or obtuse at apex. *Utricles* 2.0–2.5 mm, green, ellipsoid to obovoid, nerveless; beak up to 0.2 mm, truncate at apex; stigmas 2; nut ellipsoid to obovoid, biconvex. *Flowers* 7. *Fruits* 7–9. Spreads by rhizomes and often sterile over large areas. 2n = 76, 77, 84.

A dwarf alpine variant occurring in the Scottish mountains which has been called var. *sphagnophila* Fr. (var. *minor* Boott) may be an ecotype worth recognising.

Native. A species of mires and swamps by lakes, along rivers and on gently sloping deep peat. It has a wide altitudinal range from sea level to 750 m. Mainly a northern species, from Shetland to the Scottish border though extinct in many former localities. In the Lake District and west Wales it is local, and is scattered over Ireland. North Europe southwards to Wales, northern Germany and White Russia; Siberia; North America south to Arizona and California.

× **bigelowii = C.** × **limula** Fr.
C. × *epigejos* auct.
The stems of this hybrid are taller than in *C. bigelowii*, but curved as in that species, the spikes are more slender than in *C. bigelowii* and the fruits sterile.

The hybrid occurs between 800 and 900 m in open, stony ground in the Scottish mountains between Dalwhinnie and Lochnagar where *C. aquatilis* is the small variant of alpine bogs. It also occurs in northern Europe.

× **nigra = C.** × **hibernica** A. Benn.
This hybrid is intermediate between the parents and sterile. The stem leaves and bracts are longer and the spikes longer than in *C. nigra*, the leaves narrower and more glaucous and the spikes more slender and more tapering at the base than in *C. aquatilis*.

Occurs on river banks, loch shores and mountain

mires in central and northern Scotland and the west of Ireland. It is also recorded from northern Europe.

× recta = C. × grantii A. Benn.
This hybrid is very variable even in the same locality, some colonies approaching one parent and some the other. It has shorter bracts and spikes than *C. aquatilis* and the midrib of the female glumes is shorter.

Known only from the banks of the Wick River in Caithness where there is a hybrid swarm. Seed-set is higher when *C. recta* is the male parent.

66. C. acuta L.   Slender Tufted Sedge
C. gracilis Curtis; *C. tricostata* Fr.; *Vignantha acuta* (L.) Schur; *Limivasculum gracile* (Curtis) Börner

Glabrous *perennial herb*; rhizomes far-creeping; shoots tufted; roots brown or reddish-brown; scales reddish-brown, often soon decaying. *Stems* 30–120 cm, rough and tetraquetrous above, subterete at base. *Leaves* 30–140 cm × 3–10 mm, glaucous, without stomata on upper surface, with dense stomata beneath, rough on the edges, thin, plicate, long-linear, gradually narrowed to a pendulous tip, the margins rolling outwards on drying; sheaths brown or reddish-brown, persistent, transverse septa prominent, inner face hyaline, persisting as a brown membranous strip, straight or concave at apex; ligules 4–6 mm, obtuse or truncate at apex. *Inflorescence* one-sixth to one-quarter the length of the stem; bracts leaf-like, the lowest exceeding the inflorescence. *Male spikes* usually 2–4, 20–60 mm; male glumes 4.5–5.5 mm, purplish with a pale midrib and black at tip, elliptical to obovate-oblong, obtuse to acute at apex. *Female spikes* 2–4, usually contiguous, 30–100 mm, cylindrical, often lax-flowered at base, erect, the upper sessile, often male at top, the lower shortly pedunculate; female glumes 2.5–4.0 mm, purplish-black with a pale midrib in lower half, oblong-obovate, obtuse at apex or the margin inrolled to form a cusp, the midrib sometimes excurrent. *Utricles* 2.0–3.5 mm, green, ellipsoid-obovoid to subglobose, either longer or shorter than glume, faintly ribbed; beak up to 0.2 mm, truncate at apex; stigmas 2; nut obovoid, biconvex, shortly stalked. *Flowers* 5–6. *Fruits* 6–7. $2n = $ c. 74–76, 78, 82–85.

Native. Ponds, dykes, riversides and marshy places where there is more or less a constantly high water level. Mainly in the south and west, becoming rare in Scotland; local in the north and east of Ireland. Most of Europe; Siberia.

× acutiformis = C. × subgracilis Druce
This hybrid is intermediate between the parents and flowers with 2 and 3 stigmas occur in the same spike.

Recorded from scattered localities in England and Wales.

× elata = C. × prolixa Fr.
C. × curtisii Druce nom. nud.
Putative hybrids are intermediate between the parents and show partial sterility, and may be more common than records show.

It has been recorded in a few places in southern

England and south Wales. It also occurs in central Europe.

× nigra
C. ? × elytroides Fr.
This hybrid has rather glaucous, semicylindrical leaves, pale green utricles and brown, orbicular, punctulate, flat fruits which are apparently fertile, and is said to be very distinct in appearance when fresh.

It occurs in scattered localities in northern Britain, including some sites without *C. acuta*. It is also recorded from central and northern Europe.

67. C. trinervis Degl.   Three-nerved Sedge
Glabrous *perennial herb*; rhizomes long and slender; shoots tufted; roots reddish-brown, fibrous. *Stems* 10–30(–60) cm, trigonous, smooth or rough at apex. *Leaves* up to 40 cm × 1.5–2.0(–3.0) mm, equalling or exceeding stems, greyish- or glaucous-green, channelled, narrowed to a point at apex, minutely serrulate near apex; sheaths pale brown, entire, rounded at apex; ligule 1–2 mm, rounded at apex. *Inflorescence* one-sixth to one-quarter as long as the stem; bracts leaf-like, lowest exceeding the inflorescence. *Male spikes* 1–4, 20–40 mm; male glumes 3–5 mm, reddish-brown with a pale midrib and a hyaline margin, ovate-lanceolate, rounded at apex. *Female spikes* 2–4, 10–40 mm, dense, crowded; female glumes 2.5–3.0 mm, equalling or slightly shorter than utricles, brownish, oblong-ovate, subobtuse at apex, with 2 lateral veins. *Utricles* 3.5–5.0 mm, yellowish- to greyish-green, often purplish-spotted, ellipsoid to broadly ovoid, prominently veined; beak up to 0.2 mm, truncate or slightly emarginate at apex; stigmas 2; nut ellipsoid. *Flowers* 5–7. *Fruits* 6–8.

Possibly formerly native, now extinct. Found in an inland area of Norfolk at Ormesby Common in 1869, but has never been refound. Damp maritime sands and heaths of western Europe from Denmark to central Portugal.

68. C. nigra (L.) Reichard   Common Sedge
C. acuta var. *nigra* L.; *C. fusca* All.; *C. caespitosa* auct.; *C. angustifolia* Sm.; *C. goodenowii* Gay nom. illegit.; *C. vulgaris* Fr. nom illegit.; *C. gibsonii* Bab.; *C. juncella* auct.; *Vignantha vulgaris* Schur nom. illegit.; *C. eboracensis* Nelmes; *Limivasculum goodenoughii* Börner nom. illegit.

Glabrous *perennial herb*; rhizomes far-creeping with scales which often become fibrous; shoots tufted; roots pale or reddish-brown, more or less shiny, persistent or fibrous. *Stems* 7–70 cm, rough above, slender, trigonous, solid. *Leaves* up to 90 cm × 1–3(–5) mm, glaucous, thin, more or less flat, long-linear, gradually tapering to a fine point, the margins rolling inwards on drying; sheaths brown, black or rarely red, more or less fibrous, inner face hyaline, straight at apex; ligules 1–3 mm, rounded at apex. *Inflorescence* one-sixth to one-quarter as long as the stem; bracts leaf-like, lowest more or less equalling inflorescence. *Male spikes* 1–2, 5–30 mm; male glumes 3–5 mm, purplish with a pale midrib or rarely brown, obovate-oblong, obtuse or subacute at apex. *Female spikes* 1–4, more or less contiguous, 7–50 mm, cylindrical, erect, upper often male at top, lower sometimes distant, with short peduncle; female glumes 2.5–3.5 mm, black sometimes with a pale midrib and narrow hyaline margin, rarely brown, lanceolate-oblong, obtuse to acute at apex. *Utricles* 2.5–3.5 mm, broader than glumes, green, often tinged dark purple, ovoid-ellipsoid, faintly but distinctly ribbed; beak up to 0.2 mm, truncate at apex; stigmas 2; nut ellipsoid, biconvex. *Flowers* 5–7. *Fruits* 6–8. $2n = 82$–85.

A very variable species especially in habit and leaf form. Plants from dune-slacks and growing around hill sheep shelters have short, wide, often arcuate leaves and have been called var. *stolonifera* (Hoppe). Plants from mineral-poor mires with slender, channelled leaves have been called var. *strictiformis* (L. H. Bailey). A tufted plant with wide rigid leaves and a dense spike is referable to var. *tornata* (Fr.). Var. *recta* (Fleisch.) Hyl. with long bracts and red leaf-sheaths seems to show characters towards *C. aquatilis*. A tussocky form from stagnant mires, var. *subcaespitosa* (Kük.), has been mistaken for var. *juncea* (Fr.) Hyl. which is a north Scandinavian plant forming a tussock and without long rhizomes and orange-brown basal sheaths. The taxonomic significance of this variation, which is even more complicated on the mainland of Europe, is not well understood and needs much experimental work before any useful formal classification can be attempted.

Native. In mires and bogs with some degree of water movement or mineral enrichment, flushes in upland areas, lowland marshes, dune-slacks and by streams. Throughout the British Isles in suitable habitats up to 1010 m. Europe; northern Asia; mountains of north-west Africa; North America.

69. C. elata All. Tufted Sedge

C. stricta Gooden., non Lam.; *C. hudsonii* A. Benn.; *C. cespitosa* auct.; *Vignea stricta* Rchb.; *Diemisa stricta* (Rchb.) Raf.; *C. reticulosa* Peterm.; *Vignantha stricta* (Rchb.) Schur; *Linivasculum strictum* (Rchb.) Börner

Glabrous *perennial herb*; shoots densely tufted forming tussocks up to 40 cm; rhizomes very short, erect; roots up to 2 mm thick, purplish-brown; scales light brown, shiny, persistent. *Stems* 25–100 cm, rough, tetraquetrous, solid. *Leaves* 40–100 cm × 3–6 mm, glaucous, rough, thin, plicate, long-linear, gradually tapering to a flat apex, the margins rolled outwards on drying; sheaths becoming yellowish-brown, persistent, inner face hyaline, fibrillose on splitting, concave at apex; ligules 5–10 mm, more or less tubular, acute at apex. *Inflorescence* about one-seventh the length of the stem; bracts leaf-like to setaceous, the lowest not half the length of the inflorescence. *Male spikes* 1–3, 15–50 mm, lowermost occasionally female at base; male glumes about 5 mm, dark purplish-brown with a pale midrib and hyaline margin, oblanceolate, obtuse at apex. *Female spikes* 2–3, usually contiguous, 15–40 mm, cylindrical, erect, more or less sessile, often male at top; female glumes 3–4 mm, purplish-brown with paler

midrib and hyaline margins, ovate-elliptical, obtuse or subacute at apex. *Utricles* 2.5–4.0 mm, green, broadly ovoid-ellipsoid, ribbed; beak up to 0.3 mm, truncate at apex; nut obovoid, biconvex, stalked. *Flowers* 5–6. *Fruits* 6–7. Spreads by seed, but often fails to flower. $2n = 74$, 78, 80.

Native. Eutrophic mires where there is at least seasonal flooding and therefore common by fen ditches, rivers and lakes where it can form extensive stands; also in upland fens. Most frequent in eastern England and central Ireland, scattered elsewhere in England; rare in Scotland. Europe north to central Finland, but absent from much of the south; west Siberia; north-west Africa; North America. Our plant is subsp. **elata**. Eastern populations have been called subsp. *omskiana* (Meinsh.) Jalas.

× **nigra = C. × turfosa** Fr.

This hybrid forms some viable fruits. It differs from *C. elata* in its non-tufted, more slender habit, narrower leaves, more slender spikes and shorter glumes and utricles, and from *C. nigra* by its taller stems, broader leaves and less slender spikes.

It is found in scattered localities in central and south Britain and in Ireland. It also occurs in central and northern Europe.

70. C. bigelowii Torr. ex Schwein. Stiff Sedge

C. rigida Gooden., non Schrank; *C. hyperborea* Drejer; *Onkerma rigida* Raf.; *Vignantha rigida* (Raf.) Schur; *C. concolor* auct.; *Limivasculum rigidum* (Raf.) Börner; *C. bigelowii* subsp. *nardeticola* Holub

Glabrous *perennial herb*; rhizomes shortly or long-creeping; shoots solitary or in pairs, more or less close together; roots greyish-brown, often as thick as rhizomes; scales reddish- or purplish-brown, shiny, coriaceous, entire, persistent. *Stems* 4–30 cm, rough towards the top, rigid, thick, tetraquetrous. *Leaves* up to 25 cm × 2–7 mm, glaucous, dying to a reddish-brown, stiff or arcuate, more or less rough, keeled, long-linear, more or less abruptly tapered to a flat apex, margins often revolute and rolling outwards on drying; sheaths brown or reddish-brown and shiny, persistent, inner face hyaline, soon decaying, more or less straight at apex; ligules 1–2 mm, shortly tubular, acute at apex. *Inflorescence* one-eighth to one-fifth the length of the stem; bracts leaf-like, with blackish auricles, the lowest shorter than inflorescence. *Male spikes* 1, rarely 2, 5–20 mm; male glumes 3–4 mm, purplish or almost black with pale base and midrib, obovate-elliptical, acute or rounded at apex. *Female spikes* 2–3, contiguous or the lower more or less distant, 5–15 mm, cylindrical-ovoid, erect, sessile or the lowest only sometimes pedunculate; female glumes 2.5–3.5 mm, purple or blackish with a pale or sometimes inconspicuous midrib, ovate-elliptical, obtuse at apex. *Utricle* 2–3. mm, broader than glumes, green, suborbicular-ellipsoid, veinless; beak up to 0.2 mm, truncate or slightly notched; nut ovoid-ellipsoid, biconvex. *Flowers* 6–7. *Fruits* 7–8. $2n = 68$–71.

Native. Lichen-rich heaths, areas of deep snow cover

and flushed gullies above 600 m. North Wales, north Pennines, Lake District, southern Uplands and western Highlands of Scotland and mountain summits and ridges in Ireland. Circumpolar and on mountains southwards to southern Urals and Japan. Our plant is subsp. **bigelowii** which is confined to northern Europe.

× **nigra = C.** × **decolorans** Wimm.

This hybrid is intermediate between the parents and sterile.

It occurs in the Scottish mountains from near Dalwhinnie across to Lochnagar with the parents and often *C. aquatilis* × *bigelowii*. It is also recorded from central and northern Europe.

Subgenus **3. Primocarex** Kük.

Dioecious herbs, or monoecious ones with male flowers at top of spike and female ones at base. *Spikes* solitary. *Stigmas* 2–3.

A variable assemblage of sections whose affinities are a matter of much conjecture.

Section **34. Leucoglochin** Dumort.

Monoecious herbs. Stems subterate or trigonous. *Spike* lax. *Female glumes* caducous. *Utricles* narrowly conical or fusiform with faint veins, glabrous, patent or deflexed, gradually narrowed into a long, conical, smooth beak. *Stigmas* 3.

71. C. microglochin Wahlenb.   Brittle Sedge
Uncinia microglochin (Wahlenb.) Spreng.; *Caricinella microglochin* (Wahlenb.) St-Lag.

Glabrous *perennial herb*; rhizomes slender, shortly creeping; shoots often single; roots pale yellowish-brown; scales pale brown, persistent. *Stems* 5–12 cm, stiff, erect, long-linear, solid, subterete, striate. *Leaves* 1–5 cm × 0.5–1.0 mm, mid-green, thick, stiff, erect, channelled, more or less truncate and rounded at apex; sheaths soon becoming brown and decaying, rarely fibrous, inner face hyaline becoming brown, straight at apex, serrate; ligules about 0.5 mm, tubular, rounded at apex. *Inflorescence* a single, few-flowered terminal spike 3–5 mm, male above, female below; bracts absent. *Male glumes* 2.5–3.0 mm, reddish-brown with a pale midrib, ovate-lanceolate, more or less acute at apex. *Female glumes* about 2 mm, reddish-brown with a pale margin, ovate-lanceolate, acute and hyaline at apex, caducous. *Utricles* 3.5–4.5(–6) mm, pale yellowish-green, becoming straw-coloured, narrowly conical, rounded at base, faintly ribbed, tapered at apex; beak 1.0–1.5 mm, deflexed at maturity; a stiff bristle, arising from the base of the nut, protrudes from the top of the beak; stigmas 3; nut cylindrical-trigonous. *Flowers* 7–8. *Fruits* 7–9. *2n* = 48, 58.

Native. In gently sloping, stony, micaceous flushes between 600 and 900 m where the total plant cover is usually less than 50%. First found in 1923 in the Breadalbanes, Perthshire, its only British station. Circumpolar, southwards to the Alps in Europe; Caucasus, Altai, Himalayas, Tibet; southernmost South America.

72. C. pauciflora Lightf.   Few-flowered Sedge
C. patula Huds., non Scop.; *Trasus pauciflorus* (Lightf.) Gray; *Leucoglochin pauciflora* (Lightf.) Heuff.; *Psyllophora pauciflora* (Lightf.) Schur; *Caricinella pauciflora* (Lightf.) St-Lag.

Glabrous *perennial herb*; rhizomes slender, shortly creeping, often much branched, forming an open mat; shoots loosely tufted; roots cream or pale yellowish-brown; scales pale brown (rarely dark), striate, persistent. *Stems* 7–27 cm, stiff, solid, trigonous, often curved. *Leaves* up to 20 cm × 1–2 mm, with about 9 veins, mid-green, stiff, thick, more or less channelled, long-linear, gradually narrowed to a wide, rounded apex, those of sterile shoots often narrower and setaceous; sheaths pink- (rarely reddish-) brown, persistent, some with very short subulate green laminas only, inner face hyaline, straight at apex; ligules about 0.5 mm, tubular, rounded at apex. *Inflorescence* a single, few-flowered terminal spike 3–8 mm, male above, female below; bracts absent. *Male glumes* 3.5–5.0 mm, pale reddish-brown with hyaline margins, lanceolate, more or less acute at apex. *Female glumes* 3.5–4.5 mm, clasping the utricle, pale reddish-brown with hyaline margins at apex, broadly lanceolate, more or less acute at apex, caducous. *Utricles* 5–7 mm, pale yellow or rarely reddish-brown, sub-fusiform, tapered abruptly below, faintly nerved, tapered above; beak 1–2 mm; stigmas 3; style persistent in fruit, protruding from apex of utricle; nut oblong-cylindrical, trigonous. *Flowers* 5–6. *Fruits* 6–7. *2n* = 46, 76.

Native. Oligotrophic bogs. Throughout the Highlands and west Scotland becoming less frequent in the east and scattered in the Outer Isles; not recently recorded in Orkney and absent from Shetland. In England in the Lake District and north Pennines with one isolated locality in north-east Yorkshire and another in Snowdonia; in Ireland only in Co. Antrim. Circumpolar.

Section **35. Petraea** (O. F. Lang) Kük.
Carex Petraceae O. F. Lang

Monoecious herbs. Stems trigonous. *Spike* lax or dense. *Female glumes* persistent. *Utricles* obovoid, with faint veins, glabrous, erecto-patent to erect, abruptly contracted into a short, smooth beak. *Stigmas* 3.

73. C. rupestis All.   Rock Sedge
C. petraea Wahlenb.; *Caricinella rupestris* (All.) St-Lag.

Glabrous *perennial herb*; rhizomes shortly creeping, often much branched; shoots more or less tufted; roots purplish-brown, slender; scales shiny, reddish-brown, persistent. *Stems* 7–20 cm, stiff, solid, trigonous. *Leaves* 5–20 cm × 1.0–1.5 mm, dull, dark green, often curled or twisted, flat below, long-linear, becoming keeled towards the gradually tapered, trigonous apex, persistent on dying; sheaths reddish- or orange-brown, ribbed, persistent, inner face hyaline, becoming brown and persisting, straight and brown at apex; ligules 0.5–2.0 mm, more or less tubular, rounded at apex. *Inflorescence* a single, terminal spike 7–15 mm, male above, female below; bracts glumaceous with setaceous points, caducous. *Male glumes* 2.5–3.0 mm, reddish- or purplish-brown with an obscure midrib, ovate, acute at apex. *Female glumes* 2.5–3.5 mm, dark reddish- or purplish-brown with an obscure midrib and narrowly hyaline margin, elliptical to broadly ovate, obtuse and sometimes mucronate at apex. *Utricles* 2.0–3.5 mm, greyish-green to brown, obovoid, trigonous, faintly ribbed below, erecto-patent to erect at maturity; beak 0.2–0.3 mm, slightly notched; stigmas 3; nut broadly ellipsoid, trigonous. *Flowers* 6–7. *Fruits* 7–8. Sometimes flowers sparingly, sometimes sets abundant seed. $2n = 50, 52$.

Native. Local on limestone cliff-ledges or on siliceous rock-ledges where influenced by calcareous flushing; often on unstable slopes. Confined to Scotland, usually above 600 m on the calcareous mica-schists of the Grampians and Cairngorms, but much lower on dolomitic limestones in Skye and in Ross-shire and Sutherland where it decends to sea level. North Europe and mountains south to the Alps, Pyrenees, Carpathians and southern Urals; Siberia; Caucasus; Greenland; North America.

Section 36. Unciniiformes Kük.

Monoecious herbs. Stems terete. Spike lax. Female glumes caducous. Utricles ellipsoid or oblanceolate, veinless, glabrous, deflexed; beak short and emarginate. Stigmas 2.

74. C. pulicaris L.   Flea Sedge
Psyllophora vulgaris Heuff.; *Psyllophora pulicaris* (L.) Schur; *Caricinella pulicaris* (L.) St-Lag.

Glabrous *perennial herb*; rhizomes shortly creeping; shoots often densely tufted; roots reddish-brown, very fine; scales pale or purplish-brown, often becoming fibrous. *Stems* 10–30 cm, slender, stiff, terete. *Leaves* 5–25 cm × 0.5–1.0 mm, with about 9 veins, dark green, more or less stiff, keeled, long-linear, obtuse at apex; sheaths becoming brown, ribbed, lower fibrous or soon decaying, inner face hyaline, more or less straight at apex; ligules usually less than 0.5 mm, rounded at apex. *Inflorescence* a single, few-flowered terminal spike 10–25 mm, male above, female below; bracts glumaceous. *Male glumes* 4.5–5.0 mm, purplish- or rarely reddish-brown, with paler margins, oblong-elliptical, obtuse or more or less acute at apex. *Female glumes* 3.5–4.0 mm, reddish- or purplish-brown the margins sometimes narrowly hyaline, broadly lanceolate, midrib keeled, acute or more or less obtuse at apex. *Utricles* 3.5–6.0 mm, dark brown and shiny, ellipsoid or oblanceolate, shortly and stoutly pedicelled, deflexed on maturity; beak 0.2–0.5 mm, slightly notched; stigmas 2; nut narrowly obovoid, biconvex, tapered to a thick stalk. *Flowers* 5–6. *Fruits* 6–7. $2n = 58, 60$.

Native. On meso- to eutrophic silty soils with impeded drainage or on wet slopes where drainage water comes from calcareous rocks. In central and north-west Scotland on mica-schists and similar calcareous rocks, in the southern uplands, northern England and north

Wales more often in species-rich grassland, in the east and south of England in valley bogs and calcareous mires. Scattered throughout Ireland in a variety of habitats. North, central and western Europe, southwards to north Spain and eastwards to Estonia.

167. **POACEAE** Barnhart
GRAMINEAE Juss. nom. altern.

Annual or *perennial herbs*, rarely woody, often with rhizomes below ground or with stolons above ground, sometimes with sterile leafy shoots, *tillers. Stems* (*culms*) usually hollow, cylindrical, rarely flattened or other shapes but not 3-angled. *Leaves* alternate, with long usually linear, entire lamina, with long stem-sheathing, often cylindrical lower part (*sheath*), usually with a membrane, a fringe of hairs, or a membrane with a distal fringe of hairs at the top of the sheath on the adaxial side (*ligule*), sometimes with a small wing-like extension on either side of the top of the sheath (*auricle*). *Flowers* much reduced, 1–many in discrete units (*spikelets*), variously arranged in terminal inflorescences, mostly bisexual, but often unisexual and bisexual mixed in same spikelet, rarely male and female in different spikelets or parts of plants, hypogynous. *Perianth* of 2 minute scales (*lodicules*) at the base of the ovary (rarely fused or 0, 1 or 3). *Stamens* usually 3, rarely 2, 4 or 6. *Ovary* 1-celled, with 1 ovule. *Styles* 2, rarely 1 or 3; stigma elongated, feathery. *Fruit* a typical *caryopsis*, rarely the wall not fused to the seed inside. *Spikelet* consisting of a series of bracts; usually 2 sterile, the lower and upper *glumes*, rarely 1 or 0, with empty axils. *Flowers* above consisting of the bisexual or unisexual flower proper, plus 2 fertile glumes on either side, the *lemma* on the abaxial side and the *palea* on the adaxial side, but sometimes both absent; the flowers borne on a slender axis, the *rhachilla*, often one or more sterile or even reduced to vestigial scales; lemmas often with a horny region, the *callus*, at the base, which is often vestigial, but sometimes well developed; lemmas and glumes often with short to long, dorsal to terminal bristles, the *awns*.

This account follows the subfamilial, tribal and generic arrangements and descriptions of Stace (1991), which in turn follows the arrangement of Clayton and Renvoize (1986). Most of the descriptions of the native species are based on those of Hubbard (1984). The descriptions of the alien species are taken from a great variety of sources. The whole has been put together while consulting the large collection of specimens in the Cambridge University herbarium (**CGE**) all of which have been recently determined by P. J. O. Trist. Trist has not only read the whole of this text but has added all his field notes made over many years.

Many species of grass have hairy and glabrous forms. As no serious work seems to have been done on the distribution and ecology of these, they have on the whole, not been dealt with, although the variation is covered in the descriptions.

Bean, W. J. (1970–1988). *Trees and shrubs hardy in the British Isles*. Ed. 8. London. [For bamboos.]

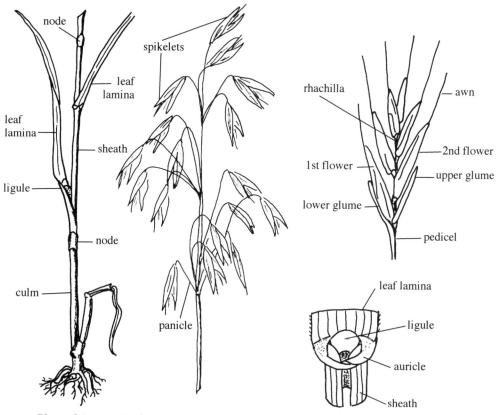

Plant of **Avena sativa** L.

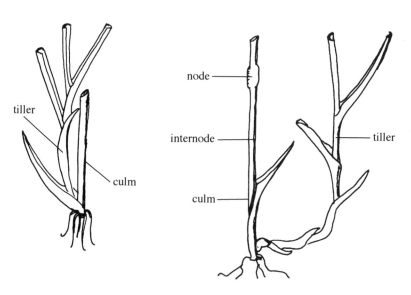

Grass morphology

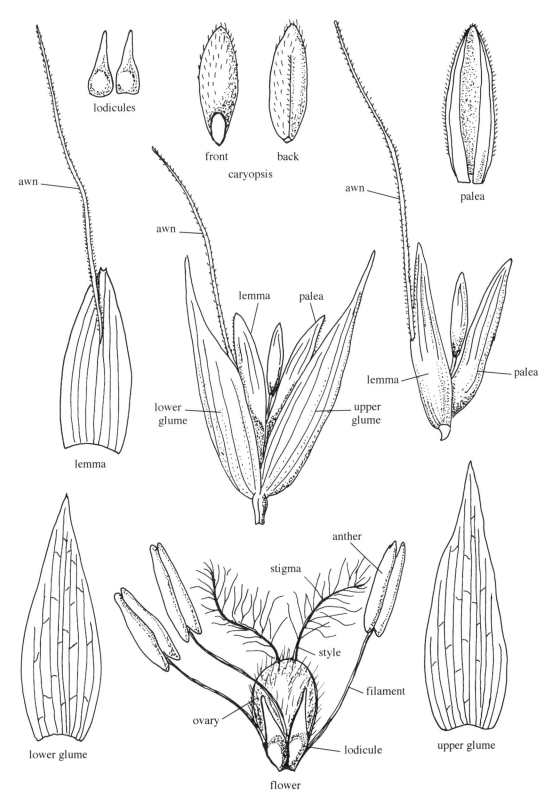

Spikelet of **Avena sativa** L. (After Hubbard, 1984).

Bor, N. L. (1968) in Townsend et al., *Flora of Iraq* 9. Baghdad.

Chao, C. S. (1989). *A guide to Bamboos grown in Britain*. Kew.

Clayton, W. D. & Renvoize, S. A. (1986). *Genera Gramineum : Grasses of the world*. London.

Dajun, W. & Shap-Jin, S. (1987). *Bamboos of China*. London.

Fernald, M.L. (1950). *Gray's Manual of botany*. Ed. 8. New York et al.

Hitchcock, A. S. (1959). *Manual of the grasses of the United States*. Ed. 2. Washington.

Hubbard, C. E. (1954). *British Grasses*. Ed. 1; ed. 2 (1968); ed. 3 (1984).

Lawson, A. H. (1968). *Bamboos. A gardener's guide to their cultivation in temperate climates*. London.

McClintock, D. (1980). Descriptive key to Bamboos naturalised in the British Isles. *Watsonia* 13: 59–61.

McClintock, D. (1992). The shifting sands of Bamboo genera. *Plantsman* 14: 169–177.

Meredith, D. (Edit.) (1955). *The grasses and pastures of South Africa*.

Rehder, A. (1956). *Manual of the trees and shrubs hardy in North America*. Ed 2. New York. [For bamboos.]

Stanley, T. D. & Ross, E. M. (1989). *Flora of south-eastern Queensland* 3. Brisbane.

Suzuki, S. (1978). *Index to Japanese Bambusaceae*. Tokyo. [In Japanese and English.]

Watson, L. & Dallwitz, M. J. (1992). *The grass genera of the world*. Cambridge.

1.   Ligules a dense fringe of hairs, or membranous but breaking into a dense fringe of hairs at the apex   2.

1.   Ligules membranous, sometimes jagged or shortly hairy but not densely fringed with hairs at the apex; sometimes ligules absent   28.

2.   Female spikelets in a simple raceme (cob) low down on the plant; male spikelets in a terminal panicle or umbel of racemes (tassel)   **92. Zea**

2.   Male and female spikelets not in separate inflorescences   3.

3.   Spikelets arising in pairs, one fatter and bisexual, sessile and often awned, the other thinner, male or sterile, often stalkless and awnless   **91. Sorghum**

3.   Spikelets all bisexual and similar   4.

4.   Spikelets with stout, hooked spines on the back of the glume   **81. Tragus**

4.   Spikelets without hooked spines, but sometimes with barbed awns   5.

5.   Inflorescence a spike or a spike-like panicle; spikelets or groups of spikelets with 1– several, sometimes fused at apex, barbed bristles at the base   6.

5.   Inflorescence of several spikes or racemes, or obviously a panicle; spikelets without barbed bristles at the base, though often with soft hairs   7.

6.   Bristles remaining on the axis when the spikelets or flowers fall, not fused   **88. Setaria**

6.   Bristle fused proximally to form a small capsule around the spikelets, falling with spikelets to form a bur   **90. Cenchrus**

7.   Inflorescence an umbel or raceme of spikes, or of racemes whose spikelets have pedicels less than 2 mm   8.

7.   Inflorescence a panicle   17.

8.   Spikes all arising from the same point at the tip of the culm   9.

8.   At least some spikes arising from a different point

along the apical part of the culm, the points often close together   12.

9.   Spikes terminating in bare prolongations of the axis   **76. Dactyloctenium**

9.   Spikes terminating in a spikelet   10.

10. Plant markedly stoloniferous; spikelets with one bisexual flower only   **79. Cynodon**

10. Plant tufted, not stoloniferous; spikelets with several bisexual flowers, if only one then with sterile flowers or scales distal to it   11.

11. Spikelets with more than 3 bisexual flowers; lemmas not awned   **75. Eleusine**

11. Spikelets with 1 bisexual flower plus 1–few sterile ones more distal; lemmas awned   **78. Chloris**

12. Spikelets with more than 3 flowers   **75. Eleusine**

12. Spikelets with 1–2 flowers   13.

13. Spikelets more than 8 mm   **80. Spartina**

13. Spikelets less than 6 mm   14.

14. Spikelets with a small, globose swelling at the base, immediately below the glumes   **86. Eriochloa**

14. Spikelets without a globose swelling at the base   15.

15. Spikelet with 2 scales (upper and lower glumes) less than two-thirds as long as the spikelet, with 1 bisexual flower   **77. Sporobolus**

15. Spikelets without or with 1 scale (lower glume) at base, less than two-thirds as long as spikelet   16.

16. Upper lemma awnless; lower lemma and upper glume obtuse at apex   **84. Brachiaria**

16. Upper lemma with awn 0.3–1.0 mm (hidden between upper glume and lower lemma); lower lemma and upper glume shortly acuminate at apex   **85. Urochloa**

17. Plants dioecious; most leaves more than 1 m, with very rough cutting edges   **70. Cortaderia**

17. Plants monoecious; leaves less than 1 m, without a cutting edge   18.

18. Glumes minute, distant from the lowest lemmas and separated from them by a conspicuously short-hairy callus   **9. Ehrharta**

18. Glumes overlapping the lowest lemma   19.

19. Spikelets with a basal tuft of long silky hairs becoming very conspicuous in fruit   **72. Phragmites**

19. Spikelets glabrous to shortly hairy, but not with basal tufts of long silky hairs   20.

20. Spikelets with 1 flower only   21.

20. Spikelets with more than 2 flowers, sometimes only 1 bisexual and more than 1 male or sterile   23.

21. Lemma with a conspicuous, bent awn   **11. Stipa**

21. Lemmas awnless   22.

22. Spikelets all with one flower; lemmas with 1 vein   **77. Sporobolus**

22. Usually many spikelets with more than 1 flower; lemmas with 3–5 veins   **71. Molinia**

23. Spikelets with 2 flowers, the distal bisexual, the proximal male or sterile   **82. Panicum**

23. Spikelets with more than 2 flowers, at least the lowest bisexual   24.

24. Glumes much shorter than the rest of the spikelet; lemmas entire at apex, not awned, with (1–)3 –5) -veins   25.

24. Glumes as long as the spikelets excluding awns, or nearly so; lemma notched or 2- to 3-lobed at apex, sometimes awned with 5–9 veins   26.

25. Perennial of wet, peaty areas; lemmas more than 3 mm   **71. Molinia**

25. Alien annuals or sometimes perennials; lemmas less than 3 mm   **74. Eragrostis**

26. Lemmas with a bent awn more than 5 mm   **68. Rytidosperma**

26. Lemmas awnless or with a straight awn less than 1 mm   27.

27. Perennial; flowers falling separately from glumes; lemmas more than 4 mm, with a very narrow hyaline margin, minutely 2- 3-toothed at apex   **67. Danthonia**

27. Annual; spikelets falling whole; lemmas less than 3 mm, with a very broad hyaline margin, deeply 2-lobed at apex   **69. Schismus**

28. Bamboos with woody culms; leaves with a distinct, short petiole between the blade and the sheath   29.

28. Culms not woody; leaves without a petiole between the blade and the leaf   37.

29. Main culms more or less square in section   **7. Chimonobambusa**

29. Main culms more or less cylindrical throughout, except sometimes just above each node   30.

30. Nodes of mid-region of main culm mostly with 1 (–2) lateral branches   31.

30. Nodes of mid-region of main culm mostly with more than 3 lateral branches   34.

31. Leaves with short hairs on the lower side   **5. Sasaella**

31. Leaves glabrous, except sometimes on the margins   32.

32. Leaves 4–5 times as long as wide, with 5–14 veins on either side of the midrib   **4. Sasa**

32. Leaves more than 6 times as long as wide, with 2–9 veins on either side of the midrib   33.

33. Culms 0.2–2.0 m; leaves 2.5–20.0 × 0.3–2.5 cm, with 2–7 veins on either side of the midrib   **3. Pleioblastus**

33. Culms 2.5–5.0 m; leaves 15–30 × 2–4 cm, with 5–9 veins on either side of the midrib   **6. Pseudosasa**

34. Culms less than 2.5 m   **3. Pleioblastus**

34. Culms more than 2.5 m   35.

35. Leaves 15–25 mm wide, mostly with 4–7 veins on either side of the midrib   **3. Pleioblastus**

35. Leaves less than 12(–13) mm wide, mostly with 2–4 veins on either side of the midrib   36.

36. Rhizome long-necked up to 1 m, but terminating in a clump of culms; leaves with 2–3 veins either side of the midrib; inflorescence simple, exserted   **1. Yushania**

36. Rhizome long-creeping with culms arising at intervals; leaves with 3–4 veins either side of the midrib; inflorescence of fascicules partly enclosed in a spathe   **2. Fargesia**

37. Female spikelets in a simple raceme (cob) low down on the plant; male spikelets in a terminal panicle (tassel)   **92. Zea**

37. Male and female spikelets not in separate inflorescences   38.

38. Spikelets arising in groups of 2–7, one fatter and bisexual, the other 1–6 thinner and sterile and male   39.

38. Spikelets all bisexual and similar   43.

39. Inflorescence a spike with 3 spikelets per node, each with 1 flower, the middle one bisexual, the 2 laterals male or sterile   **64. Hordeum**

39. Inflorescence a panicle, sometimes strongly contracted and spike-like, not with regularly 3 spikelets per node and not all spikelets with 1 flower   40.

40. Sterile or male spikelets with fewer than 3 flowers, with glumes at least nearly as long as the spikelets   41.

40. Sterile or male spikelets clearly with more than 5 flowers, with glumes much shorter than the spikelets   42.

41. Spikelets in groups of 3–7, one bisexual, the rest male or sessile, all falling as a unit   **43. Phalaris**

41. Spikelets in pairs, one bisexual, the other male or sterile   **91. Sorghum**

42. Bisexual spikelets with (1–) 2–5 flowers accompanied by 1–few sterile spikelets, the latter and the glumes of the fertile spikelets not falling   **17. Cynosurus**

42. Bisexual spikelets with 1 flower, plus a sterile vestige, accompanied by 2–4 sterile spikelets all falling as a unit   **18. Lamarckia**

43. Inflorescence a simple spike, or a simple raceme whose spikelets have pedicels less than 2 mm   44.

43. Inflorescences more complex than a simple spike or raceme of spikelets, but often condensed   69.

44. Spikelets with stout, hooked spines on the dorsal surface of the glumes   **81. Tragus**

44. Spikelets without hooked spines, but sometimes with barbed awns   45.

45. Spikelets or groups of spikelets with 1–several (sometimes proximally fused) stiff bristles at the base (do not confuse with bristle-like glumes)   46.

45. Spikelets without stiff bristles at the base   47.

46. Bristles remaining on the axis when spikelets or seeds fall, not fused   **88. Setaria**

46. Bristles fused proximally to form a small capsule around the spikelets, falling with the spikelets to form a bur   **90. Cenchrus**

47. One, 1-flowered spikelet at each node   48.

47. Each node with more than 1 spikelet, or with one spikelet with more than 1 flower   51.

48. Densely tufted perennial; lemma awned; glumes 1–2, much shorter than the flowers   **10. Nardus**

48. Annual; lemma not awned; glumes 1–2, at least as long as the flower   49.

49. Spikelets not sunk in hollows in the axis; glumes acute to acuminate at apex; lemma shortly hairy   **50. Mibora**

49. Spikelets sunk in hollows in the axis; lemma glabrous   50.

50. Glumes 2 on all spikelets   **26. Parapholis**

50. Glumes 2 on terminal spikelets; 1 on lateral spikelets   **27. Hainardia**

51. Spikelets 2–3 at each node   52.

51. Spikelets one at each node   55.

52. Spikelets with 1(–2) flowers; lemmas long-awned   53.

52. Spikelets with (2–)3–6 flowers; lemmas awned or not   54.

53. Two glumes of each spikelet fused at base   **63. Hordelymus**

53. Two glumes of each spikelet free at base   **64. Hordeum**

54. Leaves not glaucous; upper nodes with 1 spikelet; lowest lemma less than 14 mm   **61 × 64. × Elytrordeum**

54. Leaves very glaucous; all or more or less all nodes with 2 spikelets; lowest lemma more than 14 mm   **62. Leymus**

55. Lower glume, except in terminal spikelet, absent or less than three-quarters as long as upper   56.

55. Lower glume more than three-quarters as long as upper   59.

56. Lemma with a bent awn from the dorsal surface   **33. Gaudinia**

56. Lemma awnless or with a straight to curved terminal or subterminal awn   57.

57. Upper glume with 1–3 veins; lemma very gradually narrowed to a long terminal awn   **16. Vulpia**

57. Upper glume with 5(–9) veins; lemma awnless or rather abruptly narrowed to a long or short, terminal or subterminal awn   58.

58. Lower glume present on all or most spikelets   **14 × 15. × Festulolium**

58. Lower glumes absent except in the terminal spikelet   **15. Lolium**

59. Perennial; sterile shoots and often rhizomes present   60.

59. Annuals; sterile shoots and rhizomes absent   64.

60. Spikelets scarcely flattened, on distinct pedicels 0.2–2.0 mm   **59. Brachypodium**

60. Spikelets flattened, sessile or with vestigial pedicels less than 0.5 mm   61.

61. Inflorescence axis not breaking up at maturity   62.

61. Inflorescence breaking up at maturity, one segment falling off with each spikelet   63.

62. Plant densely tufted, without rhizomes; spikelets breaking up below each lemma at maturity, leaving the glumes on the rhachis; lemmas usually with awns more than 7 mm; anthers less than 3 mm   **60. Elymus**

62. Plant with long rhizomes, not densely tufted; spikelets eventually falling whole, not leaving 2 glumes alone on the rhachis; lemmas rarely with awns more than 7 mm; anthers more than 3.5 mm   **61. Elytrigia**

63. Lemma awned   **61 × 64. × Elytrordeum**

63. Lemmas not awned   **61. Elytrigia**

64. Glumes truncate at apex   65.

64. Glumes acuminate to obtuse at apex, sometimes shouldered or notched at apex, awned or not   66.

65. Glumes rounded on the back, except near the apex and sometimes awned   **(Aegilops**, p. 220)

65. Glumes 1- to 2-keeled, not awned   **66. Triticum**

66. Glumes and lemmas less than 4 mm, not awned   **24. Catapodium**

66. Glumes and lemmas more than 4 mm, the lemmas long-awned   67.

67. Spikelets with (2–)3 flowers, all bisexual; glumes linear-lanceolate   **65. Secale**

67. Spikelets with 2–3 bisexual flowers plus 1–few distal sterile ones or with more than 4 bisexual ones   68.

68. Spikelets with more than 4 bisexual flowers, usually flattened with the narrow side against the inflorescence axis   **59. Brachypodium**

68. Spikelets with 2–3 bisexual flowers plus 1–few, distal, sterile ones; flattened with the broad side against the inflorescence axis   **65 × 66. × Triticosecale**

69. Inflorescence an umbel or raceme of spikes or of racemes whose spikelets have pedicels less than 2 mm 70.

69. Inflorescence a panicle or raceme whose spikelets have pedicels more than 3 mm   91.

70. Spikes all arising from the same point at the tip of the culm   71.

70. At least some spikes arising at different (though often very close) points along the apical part of the culm   76.

71. Lemmas conspicuously awned   **78. Chloris**

71. Lemmas not awned   72.

72. Spikelets with 2–10 flowers, if with 2 then the lower one bisexual   73.

72. Spikelets with 1–2 flowers, if with 2 then the lower one male or sterile   74.

73. Spikes ending in a spikelet   **75. Eleusine**

73. Spikes ending in a bare prolongation of the axis   **76. Dactyloctenium**

74. Annual; fertile flower with 3–4 scales (1–2 glumes, lemma and palea of lower sterile flower) below it   **89. Digitaria**

74. Stoloniferous perennial; fertile flower with 2 scales (2 glumes, or upper glume and lemmas representing lower sterile flower) below it   75.

75. Spikelet with 2 lower scales (glumes) 1-veined, shorter than the lemmas of the fertile flower   **79. Cynodon**

75. Spikelet with 2 lower scales (glume and lemma of sterile flower) 3-veined, longer than the lemma of the fertile flower   **87. Paspalum**

76. Spikelets with 2 flowers, the upper bisexual, the lower male or sterile   77.

76. Spikelets 1-many flowered, at least the lower bisexual 83.

77. Ligule absent   **83. Echinochloa**

77. Ligules present, membranous or as a ring of hairs   78.

78. Ligules membranous   79.

78. Ligules a fringe of hairs or membranous with a fringe of hairs at the apex   81.

79. Annual; fertile flower with 3–4 scales (1–2 glumes, lemmas and palea of the sterile flower) below it   **89. Digitaria**

79. Perennial; fertile flower with 2 scales (2 glumes, or upper glume and lemma representing the lower sterile flower below it   80.

80. Spikelet with 2 lower scales (glumes) 1-veined, shorter than the lemma of the fertile flower   **79. Cynodon**

80. Spikelet with 2 lower scales (glume and lemma of sterile flower), 3-veined, longer than the lemma of the fertile flower   **87. Paspalum**

81. Spikelet with a small bead-like swelling at the base; lower glume more or less absent   **86. Eriochloa**

81. Spikelet without a bead-like swelling at the base; lower glume present   82.

82. Upper (and lower) lemma awnless   **84. Brachiaria**

82. Upper lemma with an awn 0.3–1.0 mm (hidden between the upper glume and the lower lemma)   **85. Urochloa**

83. Spikelets with 1–2 flowers   84.

83. Spikelets with more than 3 flowers — 88.
84. Spikelets more than 8 mm — **80. Spartina**
84. Spikelets less than 6 mm — 85.
85. Ligule membranous — 86.
85. Ligule a fringe of hairs — 87.
86. Spikes usually more than 4, 0.5–2.0 cm; spikelets glabrous on both sides — **53. Beckmannia**
86. Spikes 2(–4), 2–7 cm; spikelets appressed hairy on one side, glabrous on the other — **87. Paspalum**
87. Two basal scales of spikelet much shorter than rest of spikelet, glabrous — **77. Sporobolus**
87. Two dorsal scales of spikelet longer than rest of spikelets, densely silky-hairy — **86. Eriochloa**
88. All except the terminal spikelet in each branch with one glume — **15. Lolium**
88. All spikelets with 2 glumes — 89.
89. Lemma 5-veined, more than 5.5 mm — **14 × 15. × Festulolium**
89. Lemma 1- to 3-veined (sometimes with 1–3 extra veins very close to the midrib, forming a thickened keel), more than 5.5 mm, keeled — 90.
90. Axis of inflorescence much longer than the longest spike; tip of lemma bifid (often awned from the notch) — **73. Leptochloa**
90. Axis of inflorescence usually shorter (rarely slightly longer) than the longest spike; tip of lemma acute to obtuse (often awned from tip) — **75. Eleusine**
91. Spikelets regularly proliferating to form small leafy plantlets (sexual spikelets present or not) — 92.
91. Spikelet not or only irregularly proliferating (if so diseased, or because of sterility, or very late in season) — 94.
92. Lemma, whether proliferating or not, distinctly keeled on the back along the midrib — **21. Poa**
92. Lemma, whether proliferating or not, rounded on the back — 93.
93. Lemma with the awn arising from the dorsal surface with a tuft of hairs arising from just below the base — **37. Deschampsia**
93. Lemmas with a terminal awn or the awn absent, without a tuft of hairs arising from just below the base — **14. Festuca**
94. Spikelets with only 1 flower, bisexual and not accompanied by vestigial flowers or scales — 95.
94. Spikelets with more than 2 bisexual flowers, or with only 1 fertile flower but also more than 1 male or sterile flowers or scales — 119.
95. Spikelets with a small, bead-like swelling at the base — **86. Eriochloa**
95. Spikelets without a small bead-like swelling at the base — 96.
96. Glumes more or less absent, reduced to a rim below the flower — **8. Leersia**
96. Both glumes well developed — 97.
97. Panicle a soft, ovoid, dense head; lemmas tapered to 2 apical bristles, with a longer awn — **48. Lagurus**
97. Panicle rarely a soft, woolly, ovoid dense head, if so then lemmas blunt and awnless; lemmas awned or not but with 2 long, apical bristles — 98.

98. Both glumes notched at the apex and with an awn from the sinus; lemma with a dorsal awn — 99.
98. Glumes usually not awned, often tapered to a fine apex, but not notched at apex, if awned then the lemmas awnless — 100.
99. Awns of glumes (not lemma) more than 3 mm; lemma awned from the apex; fertile annual with deciduous spikelets — **51. Polypogon**
99. Awns of glumes less than 3 mm; lemma awned from below the apex; sterile perennial, with more or less persistent spikelets — **44 × 51. × Agropogon**
100. Lemmas with a long terminal awn with a much twisted, stout proximal part — **11. Stipa**
100. Lemmas with terminal or dorsal awns or without awns — 101.
101. Flower with a tuft of white hairs at the base, the hairs more than a quarter as long as the lemma — 102.
101. Flowers variously hairy or glabrous, but without a basal tuft of white hairs more than a quarter as long as the lemma — 106.
102. Spikelets less than 9(–10) mm; anthers less than 3 mm — 103.
102. Spikelets more than 9(–10) mm: anthers more than 3 mm — 105.
103. Lemma keeled, acute to obtuse at apex, awnless; palea almost as long as lemma — **21. Poa**
103. Lemma rounded on its back, truncate or variously toothed at apex, often awned; palea about two-thirds to three-quarters as long as the lemma — 104.
104. Tuft of hairs at base of lemma about 0.3–0.6 mm — **44. Agrostis**
104. Tuft of hairs at base of lemma more than 1 mm — **45. Calamagrostis**
105. Panicle very pale, spike-like; spikelets 10–16 mm; lemmas with basal hairs less than half as long as the lemma, with an awn up to 1.0 mm; anthers 4–7 mm, shedding pollen — **46. Ammophila**
105. Panicle usually purplish-green, usually more or less lobed; spikelets 9–12 mm; lemmas with basal hairs more than half as long as the lemma, with an awn 1–2 mm; anthers 3.0–4.5 mm, not opening — **45 × 46. × Calammophila**
106. Both glumes less than three-quarters as long as the spikelet, usually one or both about half as long — 107.
106. Both glumes more than three-quarters as long as spikelet, usually one or both as long or longer — 110.
107. Annual; lemmas with awns more than 3 mm — **49. Apera**
107. Normally perennial; lemmas awnless — 108.
108. Ligule membranous; lemmas more or less truncate at apex — **23. Catabrosa**
108. Ligule a fringe of hairs; lemmas obtuse to acute at apex — 109.
109. Usually some spikelets with more than 1 flower; lemmas with 3–5 veins, more than 3 mm — **71. Molinia**
109. Spikelets all with one flower; lemmas with 1 vein, less than 3 mm — **77. Sporobolus**
110. Glumes with a swollen, more or less hemispherical base and long very tapering distal part — **47. Gastridium**
110. Glumes without a swollen base — 111.

111. Lemmas shiny and much harder and tougher than the glumes when mature   112.

111. Lemmas about the same texture as the glumes or more flimsy and delicate   113.

112. Ligules about 1 mm; lemma with a deciduous awn 2–5 mm   **12. Oryzopsis**

112. Ligules 2–10 mm; lemma awnless   **13. Milium**

113. Lemma with a subterminal awn 4–10 mm   **49. Apera**

113. Lemma with an awn less than 4 mm or absent, or with the awn more than 4mm and arising from the lower half of the back of the lemma   114.

114. Panicle very compact, ovoid or oblong to cylindrical, spike-like, with very short branches   115.

114. Panicle diffuse or contracted but with obvious branches   116.

115. Spikelets falling as a whole at maturity; lemmas usually awned (often shortly); palea absent
52. Alopecurus

115. Flowers falling at maturity leaving the glumes on the panicle; lemmas awnless; palea present   **54. Phleum**

116. Lemmas acute to obtuse at apex, strongly keeled **21. Poa**

116. Lemmas truncate or variously toothed at apex, rounded on back   117.

117. Glumes with short pricklets or at least rough all over the back; lemma about half as long as the glumes; palea nearly as long as the lemma   **51. Polypogon**

117. Glumes often with short pricklets on the midrib but otherwise more or less smooth; lemma about two-thirds to three-quarters as long as the glumes   118.

118. Panicle branches with a clear, bare region at the base, disarticulating (sometimes not) at the base of the lemmas; very common   **44. Agrostis**

118. Panicle branches bearing spikelets more or less to base, disarticulating (if at all) near the base of the pedicels; very rare   **44 × 51. × Agropogon**

119. Spikelets with 1 bisexual, terminal flower and 1 or more male or sterile flowers or scales below it   120.

119. Spikelets with more than 2 bisexual flowers or with 1 fertile flower, but more than 1 male or sterile flowers or scales above it   129.

120. Spikelets or groups of spikelets with 1–several (proximally fused or free) stiff bristles at the base   121.

120. Spikelets without stiff bristles at the base   122.

121. Bristles not fused, retained on the axis when the spikelets or flowers fall   **88. Setaria**

121. Bristles fused proximally to form a small capsule around the spikelets, falling with the spikelets to form a bur   **90. Cenchrus**

122. Glumes minute, well separated from the lowest lemma by a conspicuously hairy callus   **9. Ehrharta**

122. Glumes overlapping the lowest lemma   123.

123. Ligule a fringe of hairs, or membranous with a fringe of hairs round the apex   **82. Panicum**

123. Ligules membranous or absent   124.

124. Bisexual flower with 1 male or sterile flower below it 125.

124. Bisexual flower with 2 male or sterile flowers below it 127.

125. Perennial; lower lemma awned from the upper half of its back   **31. Arrhenatherum**

125. Annual; lower lemma awnless or with a terminal awn   126.

126. Ligules membranous; sterile flower a minute scale; glumes equal and keeled   **43. Phalaris**

126. Ligules absent; sterile flower as large as fertile one; glumes very unequal and not keeled   **83. Echinochloa**

127. Lower glume about half as long as the upper; lemmas of lower 2 flowers with an awn more than 1.5 mm   **42. Anthoxanthemum**

127. Glumes equal or subequal; lemmas of lower 2 flowers awnless or with an awn less than 1 mm   128.

128. Panicle more or less diffuse; lower 2 flowers male, longer than the bisexual flower   **41. Hierochloe**

128. Panicle compact; lower 2 flowers sterile, shorter than bisexual flower   **43. Phalaris**

129. Ovary with a hairy terminal appendage extending beyond the base of the styles   130.

129. Ovary glabrous or hairy, but style-bases on the apex of the ovary (sometimes wide apart) not exceeded by the ovary appendage   137.

130. At least the lowest lemma with a conspicuously bent awn arising dorsally from the lower two-thirds of its length, or awnless with the glumes longer than the rest of the spikelet   131.

130. Lemmas with a straight or slightly bent awn arising from the apex or near the apex, or awnless with the glumes much shorter than the rest of the spikelet   133.

131. Easily uprooted annual without non flowering shoots; upper glume more than (15–)20 mm   **32. Avena**

131. Firmly rooted perennial with non-flowering shoots; upper glume less than 15(–20) mm   132.

132. Spikelets usually more than 11 mm (excluding awns); all lemmas with an awn arising from near the middle or above it; upper glume 10–20 mm; lower flower bisexual   **30. Helictotrichon**

132. Spikelets usually less than 11 mm (excluding awn); awn of the lowest lemma arising from about one-third up from the base; upper glume 7–11 mm; lowest flower usually male (sometimes bisexual)
31. Arrhenatherum

133. Lemmas strongly keeled on the back   **58. Ceratochloa**

133. Lemmas rounded on the back or keeled near the apex   134.

134. Perennial with sterile shoots at flowering time, often with rhizomes   135.

134. Annual (or biennial), without sterile shoots at flowering time, without rhizomes   136.

135. Spikelets less than 1 mm (excluding awns), narrowed to apex; lemmas less than 8 mm (excluding awns)   **14. Festuca**

135. Spikelets more than 15 mm (excluding awns) more or less parallel-sided almost to apex; lemmas more than 8 mm (excluding awn)   **56. Bromopsis**

136. Spikelets ovate to lanceolate, slightly compressed, markedly narrowed towards the top; lower glume 3- to 7-veined, upper glume 3- to 9-veined; lemmas 5–11 mm, with an awn about as long as or shorter   **55. Bromus**

136. Spikelets more or less straight-sided, widening towards the top; lower glume 1(–3)-veined, upper glume 3(–5)-veined; lemmas 9–36 mm, with awn about as long as or longer   **57. Anisantha**

137. Lemmas with a dorsal or subterminal, usually bent

137. awn, often bifid at apex, sometimes awnless and then clearly bifid at apex 138.

137. Lemmas awnless and entire at apex, or with a terminal, straight or curved awn and then sometimes with a bifid or several-toothed apex 146.

138. Lemmas more than 6 mm, usually with an awn more than 10 mm 139.

138. Lemmas less than 6 mm, never with an awn more than 10 mm 141.

139. Easily uprooted annual without non-flowering shoots; upper glume more than (15–) 20 mm **32. Avena**

139. Firmly rooted perennial with non-flowering shoots; upper glume less than 15 (–20) mm 140.

140. Spikelets usually more than 11 mm (excluding awns); upper glume 10–20 mm; all lemmas with awns arising from near the middle or above it **30. Helictotrichon**

140. Spikelets usually less than 11 mm (excluding awns); upper glume 7–11 mm; awn of lowest lemma arising from about one-third up from the base; upper lemma(s) often more or less unawned **31. Arrhenatherum**

141. Easily uprooted annual without non-flowering shoots 142.

141. Firmly rooted (unless in sand) perennial with non-flowering shoots 143.

142. Spikelets with 3–5 flowers; lemmas with a usually more or less straight, subterminal awn **36. Rostraria**

142. Spikelets with 2 flowers; lemmas with a bent awn arising from below halfway **40. Aira**

143. Awns slightly but distinctly widened towards the apex (club-shaped) **39. Corynephorus**

143. Awns parallel-sided or tapered to the apex 144.

144. Glumes hairy at least along the midrib, falling with the flowers at maturity; lower lemma unawned; upper flower usually male **38. Holcus**

144. Glumes glabrous, remaining on the plant when the flowers fall; lowest lemma awned; all flowers usually bisexual 145.

145. Spikelets with 2–4 flowers; upper glume distinctly shorter than the spikelet; lemmas acute and finely bifid at apex **34. Trisetum**

145. Spikelets with 2 flowers; upper glume as long as the spikelet or almost so lemmas truncate to very blunt and usually jagged at apex **37. Deschampsia**

146. Leaf-sheaths fused almost to apex to form a tube round the stem 147.

146. Leaf-sheaths with free, usually overlapping margins 149.

147. Lemmas finely pointed or awned, 3-veined **14. Festuca**

147. Lemmas acute to rounded, or more or less 3-lobed at apex, 7(–9)-veined 148.

148. Fertile flowers 4–16 or more, with 1 reduced sterile flower beyond them, or the sterile flower absent **28. Glyceria**

148. Fertile flowers 1–3, with a club-shaped group of sterile flowers beyond them **29. Melica**

149. Lemmas with 3–5, short, very pointed or shortly awned teeth at the more or less truncate apex 150.

149. Lemmas 1-pointed to rounded or 2-lobed at apex, awned or not 151.

150. Ligules membranous; glumes 1-veined, the veins running into the apical points **25. Sesleria**

150. Ligule a fringe of hairs; glumes 3- to 5-veined (lateral veins often short); lemmas 7- to 9-veined, the veins ending short of the apical points **67. Danthonia**

151. Lemmas 2-lobed or toothed at apex, awned or not from the sinus 152.

151. Lemmas 1-pointed (sometimes very minutely notched) to rounded at apex, awned or not from the tip 154.

152. Lemmas with a bent awn more than 5 mm **68. Rytidosperma**

152. Lemmas awnless or with a straight awn less than 1 mm 153.

153. Perennial; lemma minutely 2-toothed; flowers falling separately from glumes **67. Danthonia**

153. Annual; lemma deeply 2-lobed; spikelets falling whole **69. Schismus**

154. Ligule a fringe of hairs 155.

154. Ligule membranous 156.

155. Perennial of wet peaty areas; lemmas more than 3 mm **71. Molinia**

155. Alien annuals or sometimes perennials; lemmas less than 3 mm **74. Eragrostis**

156. Lemma strongly keeled throughout its length 157.

156. Lemma rounded on back or keeled only distally 159.

157. Lemmas 3-veined, with wide, membranous, more or less shiny margins rendering the panicle silvery **35. Koeleria**

157. Lemmas 5-veined, often with membranous margins but scarcely shiny and panicle not silvery 158.

158. Spikelets not aggregated into dense, 1-sided clusters; lemmas obtuse to acute at apex **21. Poa**

158. Spikelets borne in dense, 1-sided clusters; lemmas long-acuminate to shortly awned **22. Dactylis**

159. Lemmas awned 160.

159. Lemmas not awned 163.

160. Easily uprooted annual without non-flowering shoots **16. Vulpia**

160. Firmly rooted perennials with non-flowering shoots 161.

161. Lower glume mostly more than three-quarters as long as upper; anthers dehiscent **14. Festuca**

161. Lower glumes mostly less than three-quarters as long as upper; anthers indehiscent 162.

162. Pointed auricles present at the junction of the leaf-sheath and blade; lemmas rather abruptly narrowed to the awn; upper glume with 3–9 veins **14 × 15. × Festulolium**

162. Auricles absent; lemmas very gradually narrowed to the awn; upper glume with 1–3 veins **14 × 16. × Festulpia**

163. Lemmas acute to acuminate at apex 164.

163. Lemmas subacute to rounded at apex 165.

164. At least the lower glume, and usually both glumes less than half as long as spikelets; lemma 5-veined **14. Festuca**

164. Both glumes more than half as long as the spikelet; lemma 3-veined **35. Koeleria**

165. Lemmas strongly cordate at base, wider than long;
spikelets pendulous at maturity   **20. Briza**

165. Lemmas not or scarcely cordate at base, longer than
wide; spikelets not pendulous   166.

166. Leaf-sheaths compressed, keeled on back; spikelets
with 2–3 flowers; lemmas 3-veined   **23. Catabrosa**

166. Leaf-sheaths rounded on back; spikelets usually with
more than 4 flowers; lemmas 5-veined   167.

167. Pointed auricles present at the junction of the leaf-
sheath and blade; lower glume mostly less than three-
fifths as long as upper   **14 × 15. × Festulolium**

167. Auricles absent; lower glume more than three-fifths
as long as upper   168.

168. Usually perennial; lemmas minutely hairy at base,
with conspicuous membranous tips and margins
19. Puccinellia

168. Annual; lemmas glabrous, with very narrow
membranous margins   **24. Catapodium**

Subfamily 1. **Bambusoideae** Asch. & Graebn.

Woody bamboos or *herbaceous perennials. Leaves* often
with a short false petiole separating them from the
sheath, often with conspicuous cross-veins as well as
longitudinal veins; ligule membranous, rarely a membrane fringed with hairs; sheaths not fused, the lower
ones often without or with very reduced leaves.
Inflorescence a panicle. *Spikelets* with more than 3
flowers, with all flowers (except often the distal one)
bisexual, or with one bisexual flower with or without 2
large sterile ones below it. *Glumes* (0–)2(–3). *Lemmas*
firm, 5- to many-veined, awnless or with a terminal awn.
Paleas with 1–many veins, one on the mid-line. *Stamens*
3–6. *Stigmas* 2 or 3. *Lodicules* 2 or 3. *Embryos* bambu-
soid or oryzoid; photosynthetic C3-type alone, with
non-Kranz leaf anatomy; fusoid cells often present;
micro-hairs present. *Chromosome base-number* 12.

Tribe 1. **Bambuseae** Nees
Arundinarieae Asch. & Graebn.

Woody bamboos with clumped culms (*pachymorph*) or
with culms rising at intervals (*leptomorph*). *Leaves* with
a short false petiole and cross-veins as well as longitudinal ones; ligule membranous; lower sheaths usually
without or with very reduced leaves. *Spikelets* with more
than 3 flowers, with all flowers except often the distal
ones bisexual. *Glumes* (0–)2(–3). *Lemmas* 5- to many-
veined, awnless. *Paleas* with more than 4 veins. *Stamens*
3–6. *Stigmas* 2 or 3. *Lodicules* 2 or 3.

Bamboos are grown in parks, gardens and estates
where they persist after neglect, some spreading by a
strong rhizome system. Flowering is very spasmodic and
a few species have never been known to flower, while
others have only recently flowered for the first time.

1. **Yushania** Keng fil.
Sinarundinaria auct.

Woody bamboos with a pachymorph rhizome which is
long-necked up to 1 m, but terminating in another
clump of culms. *Culms* mostly over 3 m, clumped but

with a distance between clumps, terete except for just
above the nodes, the nodes with mostly more than 3
lateral branches. *Leaves* 5–12 mm wide, mostly with 2–3
veins on either side of the midrib, often with a false
petiole separating them from the sheath, often with
cross-veins as well as longitudinal ones; ligule membranous. *Inflorescence* simple, exserted, supported by
narrow sheaths. *Stamens* 3. *Stigmas* 2–3.

About 30 species mostly in tropical Asia and
Madagascar, but one species on tropical African mountains and two in central America.

1. Y. anceps (Mitford) Lin   Indian Fountain Bamboo
Arundinaria anceps Mitford; *Arundinaria jaunsarensis*
Gamble; *Yushania jaunsarensis* (Gamble) Yi;
Sinarundinaria anceps (Mitford) C. S. Chao &
Renvoize; *Chimonobambusa jaunsarensis* (Gamble)
Bahadur & Naithani

Woody bamboo with a pachymorph rhizome which is
long-necked up to 1 m and terminating in another
clump of culms. *Culms* arising in clumps at intervals,
erect then arching, mostly 3–6 m by up to 20 mm thick,
purplish-green at first changing to brownish-green,
glabrous, terete except for just above the nodes, the
nodes mostly with more than 3 lateral branches, the
purple branches forming dense clusters on the older
culms; sheaths mottled within, truncate at apex,
hairy on the margin, with a narrow, falcate, bristly
auricle on each side of the short, subulate lamina.
Leaves 6–12 cm × 5–12 mm, brilliant green above,
slightly glaucous beneath, thin, linear-lanceolate to
narrowly oblong, gradually narrowed at apex, mostly
with 2–3 veins on either side of the midrib, glabrous;
sheath pale or purplish-brown, glabrous, but fringed
with bristles and short hairs where it joins the lamina;
ligule 0.5–1.0 mm. *Inflorescence* simple, exserted, supported by narrow sheaths. *Spikelets* with more than 3
flowers, all (except often the distal ones) bisexual.
Glumes (0–)2(–3). *Lemmas* with 5–many veins and
awnless. *Palea* with more than 4 veins. *Stamens* 3.
Stigmas 2–3. Rarely flowering, but did so in 1910–11
and 1980–1.

Introduced. Naturalised in parks, estates and waste
places. West and south Britain, Ireland and the Channel
Islands. Native of the north-western Himalayas.

2. **Fargesia** Franch.
Sinarundinaria Nakai; *Thamnocalamus* auct.

Woody bamboos with a leptomorph rhizome. *Culms*
terete, arising at intervals; stem sheaths caducous.
Leaves with 3–4 veins either side of the midrib.
Inflorescence of fascicles partly enclosed in a spathe.
Palea long-acuminate at apex as long as or longer than
the lemma. *Stamens* 3. *Stigmas* 3, sessile. *Lodicules* 3.

Fifty species in China, Sri Lanka and the Himalayas.

Stapleton, C. (1995). Muriel Wilson's Bamboo. *The Bamboo
Society Newsletter* **21**: 10–20.

1. Leaves drawn out into a long fine point at apex **1. murieliae**
1. Leaves pointed at apex, but not drawn out   **2. nitida**

1. F. murieliae (Gamble) Yi   Umbrella Bamboo
F. spathacea auct.; *Thamnocalamus spathaceus*
(Gamble) Soderstrom; *Arundinaria spathacea*
(Gamble) D. C. McClint.; *Arundinaria murieliae*
Gamble; *Sinarundinaria murieliae* (Gamble) Nakai;
Thamnocalamus murieliae (Gamble) Demoly

Woody tufted bamboo with a leptomorph rhizome.
Culms 3–4(–12) m, terete, arising at intervals, erect, yellowish-brown, much branched at the nodes, where at the
base of each is a pale glaucous ring; culm-sheaths very
smooth and glossy inside, rounded at the apex except for
an awl-like prolongation up to 5 cm, glabrous. *Leaves*
6–12 cm × 8–13 mm, oblong, finely drawn out in a long
fine point at apex, tapered to a short stalk at base, of thin
texture, with 3–5 veins either side of the midrib with
faint cross-veins giving a tasselated appearance, one
margin of leaf minutely toothed; sheath ribbed and
fringed at apex, without auricles or oral setae; ligule very
short. *Inflorescence* a large panicle, with 1–4 spikelets in
the axil of each branch. *Spikelets* with 2–4 flowers.
Glumes 2. *Stamens* 3. *Stigmas* 2–3, sessile. This species
flowered from 1978 to 1980.

Introduced. Sometimes found in neglected parks
where it can survive. Native of western China.

2. F. nitida (Mitford) Keng fil.
Chinese Fountain Bamboo
Arundinaria nitida (Mitford) Stapf; *Sinarundinaria
nitida* (Mitford) Nakai

Woody bamboo with a leptomorph scaly rhizome. *Culms*
4–6 m × about 10–15 mm thick, but often much less,
arising at intervals, brownish-green or purple, terete,
hollow, very crowded, branching from the second season
and becoming heavily laden with foliage from the top,
arching gracefully; sheaths purplish, more or less hairy.
Leaves 3.5–8.0 cm × 6–18 mm, bright green above,
rather glaucous beneath, linear or oblong, pointed at
apex, rounded at base, glabrous with 3–4 veins on either
side of the midrib, the margin minutely bristly on one
side; sheaths hairy, with very short oral setae. The flowering of this species has only recently become known
and no detailed account of it is available, but it does have
the *Fargesia*-type inflorescence.

Introduced. Naturalised in parks, estates and waste
places. Scattered records in southern Britain. Native of
China.

3. Pleioblastus Nakai

Woody bamboos with slender, elongate leptomorph rhizomes. *Culms* 0.2–5.0 m, erect, terete, hollow but thick-
walled; nodes with one to many lateral branches, or 1–2
branches in the lower part of the culm. *Leaves* flat,
glabrous, mostly with 2–7 veins on either side of the
midrib and tesselated; sheaths persistent, bearing
smooth, whitish bristles which are occasionally absent.
Inflorescence a spike or raceme. *Spikelets* with 5–13
flowers. *Glumes* 2. *Stamens* 3. *Stigmas* 3. *Lodicules* 3,
ciliate.

About 20 species in China and Japan.

1. Upper margin of leaf-sheath oblique   **4. simonii**
1. Upper margin of leaf-sheath horizontal   2.
2. Culms 3–4 m; leaves the same colour on both surfaces;
   branches numerous   **1. chino**
2. Culms up to 2 m; leaves paler on one side of lower
   surface; branches often fewer than 3   3.
3. Culms up to 2 m; leaves 8–20 cm × 10–25 mm   **2. humilis**
3. Culms to 50 cm or less; leaves 2.5–8.0 cm × 3–15 mm
   3. pygmaeus

1. P. chino (Franch. & Sav.) Nakai   Chino Bamboo
Arundinaria chino (Franch. & Sav.) Makino; *P.
maximowiczii* (Rivière) Nakai, non Munro; *Bambusa
chino* Franch. & Sav.; *Nipponocalamus chino* (Franch.
& Sav.) Nakai

Woody bamboo with far-running leptomorph rhizomes.
Culms 2–4 m × 0.6–2 cm, purple when young, with a
white, waxy bloom below the nodes when mature, terete,
hollow but thick-walled, erect, branches numerous; new
shoots arising in spring. *Leaves* variable, 5–25 cm ×
15–22 mm, green on both sides, sometimes striped with
cream, chartaceous, narrowly to broadly lanceolate,
long-tapered to an acute apex, rounded at base, glabrous
or hairy on one side beneath; sheath slightly downy
when young, glabrous on the surface but ciliate on the
margin when mature, upper margin horizontal; without
auricles; ligule very short. *Inflorescence* racemose-fasciculate, at the nodes of branches and culms. *Spikelets* few,
3–7 cm, with numerous flowers, lanceolate, purplish or
green, flattened. *Glumes* 2, small and chartaceous, the
lower 3–4 mm, lanceolate, without a vein, the upper
8–10 mm, lanceolate to ovate, 5- to 7-veined. *Lemmas*
10–19 mm, green to purplish, chartaceous, broadly
lanceolate, acuminate at apex, 11- to 13-veined. *Palea*
11–13 mm, 2-keeled, the keels ciliate. *Stamens* 3,
exserted. *Stigmas* 3, feathery. *Ovary* short and cylindrical.

A very variable species in leaf shape, size and coloration. It has flowered sporadically since 1974.

Introduced. Widely grown in gardens and sometimes
found in neglected parks and estates. Native of Japan.

2. P. humilis (Mitford) Nakai   Dwarf Bamboo
Arundinaria humilis Mitford; *Sasa humilis* (Mitford)
Camus; *Arundinaria nagashima* Mitford;
Nipponocalamus humilis (Mitford) Nakai

Woody bamboo with far-running leptomorph rhizome.
Culms 0.6–2.0 m × 3–7 mm, very slender, green or purplish, the nodes usually glabrous with a white, waxy
bloom below, hollow but thick-walled, erect, with 1–3
long branches from low down; culm sheaths densely
hairy with long yellowish-grey hairs at base; new shoots
arising in spring. *Leaves* 8–25 cm × 10–25 mm, bright
green, lanceolate or linear-lanceolate, narrowly pointed
at apex, rounded at the base, usually glabrous but sometimes downy beneath towards the base; sheaths purplish
at first, slightly downy, ciliate; upper margin horizontal,
with small auricles and a few oral setae; ligule minute,
downy. *Inflorescence* racemose-fasciculate, at the tips of
stems. *Spikelets* few. *Glumes* 2, chartaceous. *Lemmas*

chartaceous. *Palea* as long as lemma, 2-keeled. *Stamens* 3. *Stigma* 3.

This species has flowered sporadically since 1964. It varies in hairiness.

Introduced. Grown in gardens and sometimes found in neglected parks and estates. Native of Japan in southern Hokkaido and northern and central Honshu.

3. P. pygmaeus (Miq.) Nakai   Pygmy Bamboo
Bambusa pygmaea Miq.; *Arundinaria pygmaea* (Miq.) Mitford; *Bambusa nana* auct.

Woody bamboo with far-running leptomorph rhizome. *Culms* (20–)40–50(–120) cm × 3–6 mm, erect, solid and flattened above, or hollow, with a white, waxy bloom, glabrous except for the nodes which have a dense fringe of brownish bristles at least at first, with 1–2 long branches at the lower culm nodes. *Leaves* 2.5–8.0 cm × 3–8(–15) mm, chartaceous-coriaceous, pale green above, slightly glaucous beneath, narrowly to broadly lanceolate, abruptly acuminate at apex, rounded at base, glabrous or hairy beneath; sheaths hairy on the margins and sometimes the veins, with few oral setae. *Inflorescence* racemose-fasciculate. *Spikelets* few. *Glumes* 2, small and chartaceous. *Lemmas* chartaceous. *Palea* as long as lemma, 2-keeled. *Stamens* 3. *Stigmas* 3.

This species flowered between 1967 and 1970.

Introduced. Naturalised rarely but widely in the British Isles. Origin unknown. Our usual plant is var. **distichus** (Mitford) Nakai (*Bambusa disticha* Mitford; *Arundinaria disticha* (Mitford) Pffizer; *Sasa disticha* (Mitford) Camus; *P. distichus* (Mitford) Muroi & Okamura).

4. P. simonii (Carrière) Nakai   Simon's Bamboo
Bambusa simonii Carrière; *Nipponocalamus simonii* (Carrière) Nakai; *Arundinaria simonii* (Carrière) Rivière & M. Rivière

Woody bamboo with a shortly-running leptomorph rhizome. *Culms* 2.5–8.0 m × 20–35 mm, erect, dark green and glabrous, hollow, the outer often arching, with a white, waxy bloom below the hairy or glabrous nodes, with 5–10 spreading branches at each node; sheath green, glabrous or puberulous at base. *Leaves* 15–25(–30) cm × 10–35 mm, green or half green and half glaucous, sometimes variegated, narrowly lanceolate to broadly linear, sharply acute at apex, rounded at base, puberulent only on the upper surface of the petiole; sheaths conspicuous, long-persistent, bearded at the base when young, with few, white oral setae, upper margin oblique; ligule about 1 mm, downy, truncate or concave at apex. *Inflorescence* racemose-fasciculate in the leaf axils. *Spikelets* rather numerous, 4- to 10-flowered, linear to narrowly lanceolate. *Glumes* 2, small and chartaceous, equal, broadly lanceolate, the lower 15–16 mm, 11-veined, the upper a little longer and 15- to 17-veined. *Lemma* 10–20 mm, sometimes purplish, ovate, acuminate at apex to a sharp point. *Palea* about as long as lemma, 2-keeled, ciliate on the keels, densely hairy. *Stamens* 3, exserted; anthers about 7 mm. *Stigmas* 3, feathery.

This species has flowered irregularly since 1965.

Introduced. Commonly cultivated in gardens and sometimes found in neglected parks and estates. Native of Japan.

4. Sasa Makino & Shibata
Woody bamboos with leptomorph rhizomes. *Culms* up to 3 m, terete except just above the nodes; nodes mostly with one lateral branch. *Leaves* 2.5–9.0 cm wide, mostly with 5–13 veins either side of the midrib, glabrous or sometimes with marginal hairs. *Inflorescence* a panicle. *Stamens* 6. *Stigmas* 3. *Lodicules* 3.

About 50 species, mainly in Japan, and extending to Korea and adjacent parts of China.

1. Leaves 12–40 cm, green with a yellow midrib; false petioles greenish-yellow   **1. palmata**
1. Leaves 10–25 cm, green with a broad white edge; false petioles purplish   **2. veitchii**

1. S. palmata (Burb.) Camus   Broad-leaved Bamboo
Bambusa palmata Burb.; *Arundinaria palmata* (Burb.) Bean

Woody bamboo with extensive leptomorph rhizomes. *Culms* 2–3 m × 7–10 mm thick, slightly ascending from base, green, usually soon purple-streaked, nodes rather prominent with a white, waxy bloom below, often glabrous, terete, hollow, unbranched below, one branch at each node above. *Leaves* 12–30(–40) × 3.5–9.0 cm, bright shining green, paler beneath, the midrib yellow, narrowly oblong, acuminate at apex, the margin sometimes decaying in hard winters, usually glabrous, sometimes puberulent beneath; sheath glabrous, without auricles or oral bristles, permanent; ligule about 2 mm, downy; false petioles greenish-yellow. *Inflorescence* a panicle. *Spikelets* many and loosely arranged, 2.5–4.0 cm, with 4–10 flowers. *Glumes* 2, membranous, lanceolate, the lower 0.5–3.0 mm, the upper 2–4 mm. *Lemma* 10–18 mm, sometimes purplish, rather small, ovate, acute at apex with 9 veins. *Palea* about equal to the lemma, truncate at apex, 2-keeled. *Stamens* 6, exserted; anthers about 4 mm. *Stigmas* 3; plumose. *Ovary* ovoid.

The widest-leaved of all our bamboos, which flowered profusely in the 1960s.

Introduced. Cultivated and widely naturalised in the British Isles. Native of Japan.

2. S. veitchii (Carrière) Rehder   Veitch's Bamboo
Bambusa veitchii Carrière; *Arundinaria veitchii* (Carrière) N. E. Br.

Woody bamboo with extensive leptomorph rhizomes. *Culms* 0.5–2.0 m × 5–7 mm thick, glaucous, finely purplish, hairy at first but hairs soon deciduous, with a white, waxy bloom below the nodes, erect, terete except just above the nodes, nodes mostly with one lateral branch; culm sheaths hairy, oral setae weak or absent. *Leaves* 10–25 × 2.5–6.0 cm, rather dull green above, glaucous beneath, soon with a broad, white margin, narrowly oblong, abruptly acuminate at apex, rounded at base, with 5–9 veins either side of the midrib; shortly and prominently hairy beneath; sheath white-hairy when young and with auricles and short oral bristles which soon fall and the sheath becomes glabrous; ligules up to 1.6 mm, dark in colour; false petioles purplish. *Inflorescence* a panicle. *Spikelets* many, 2.0–3.5 cm, linear, with few to many flowers. *Glumes* 2, membranous, narrowly lanceolate, the lower about 1 mm, the upper about 1.5 mm and 1-veined. *Lemma* small, ovate, acuminate at apex, 9-veined, ciliate on the margin. *Palea* about equal to lemma, obtuse at apex, 2-keeled. *Stamens* 6, exserted; anthers about 4 mm. *Stigmas* 3, plumose.

As far as is known this species has not flowered outside of Japan.

Introduced. Widely grown, but rarely naturalised in the British Isles. Native of Japan.

5. Sasaella Makino

Woody bamboos with leptomorph rhizomes. *Culms* erect, terete except just above the nodes; nodes usually with one lateral branch. *Leaves* 1–3 cm wide, mostly with 3–5 veins either side of the midrib. *Inflorescence* a panicle. *Glumes* 1–2, small and narrow. *Lemma* chartaceous. *Palea* about as long as lemma, 2-keeled. *Stamens* 6. *Stigmas* 3. *Lodicules* 3.

About 50 species in Asia, mostly in Japan.

McClintock, D. (1983). On the nomenclature and flowering in Europe of the Bamboo, *Sasaella ramosa* (*Arundinaria vagans*). *Kew Bull.* **38**: 191–195.

1. S. ramosa (Makino) Makino & Shibata

Hairy Bamboo

Bambuso ramosa Makino; *Arundinaria ramosa* (Makino) Nakai; *Sasa ramosa* (Makino) Makino; *Arundinaria vagans* Gamble

Woody bamboo with far-running, leptomorph rhizomes. *Culms* 0.5–1.5 m × 3–8 mm thick, terete except for just above the nodes, purplish, glabrous, nodes mostly with one lateral branch; sheaths glabrous, pale with purple stripes. *Leaves* 8–25 × 1.0–3.5 cm, linear-lanceolate, very shortly acute at apex, rounded or truncate at base, glabrous above, downy beneath, mostly with 3–5 veins either side of the midrib; sheaths glabrous and with few oral bristles. *Inflorescence* a panicle. *Spikelets* few, purplish, linear to narrowly lanceolate, 5- to 10-flowered. *Glumes* 1–2, small, narrow lanceolate, sharply acute at apex, membranous. *Lemma* 12–17 mm, chartaceous, ovate, acute to a sharp point at apex. *Palea* about as long as the lemma, 2-keeled. *Stamens* 6, exserted; anthers about 5 mm. *Stigmas* 3, long plumose. *Ovary* ovoid.

A few clumps flowered in Britain for the first time in 1981.

Introduced. Grown in gardens and widely but rarely naturalised in the British Isles, especially in woods. Native of Japan.

6. Pseudosasa Makino ex Nakai

Woody bamboos with rhizomes, but usually clump-forming with an occasional runner. *Culms* 2–5 m, terete except just above the nodes; nodes mostly with one lateral branch. *Leaves* 2–5 cm wide, glabrous, mostly with 5–9 veins either side of the midrib. *Inflorescence* a panicle. *Stamens* 3. *Stigmas* 3. *Lodicules* 3.

Four species in Japan and Taiwan.

1. P. japonica (Siebold & Zucc. ex Steud.) Makino ex Nakai
Arrow Bamboo

Arundinaria japonica Siebold & Zucc. ex Steud.; *Arundinaria metake* Nichols; *Yadakeya japonica* (Siebold & Zucc. ex Steud.) Makino

Woody bamboo with rhizomes, but usually clump-forming with an occasional runner. *Culms* 2–5 m × 1–2 cm thick, terete except for just above the nodes, erect, thin-walled, unbranched in first year then richly branched above, arching when mature, nodes often oblique with some white waxy bloom below. *Leaves* 15–35 × 2–5 cm, deep green above, narrowly lanceolate, long-tapered to an acute apex, narrowed at base, with 5–9 veins either side of the midrib, glabrous; sheaths purplish at tip, densely and roughly hairy at first, persistent, usually without oral setae; ligule 1.5–3.0 mm, minutely downy. *Inflorescence* a panicle. *Spikelets* 3–5 cm, linear, flattened, purplish, with 3–8 flowers. *Glumes* 2, the lower 10–11 mm, lanceolate and 7-veined, the upper 12–13 mm, ovate and 13- to 15-veined. *Lemma* 10–15 mm, narrowly ovate, acute at apex, 17-veined, scaberulus above. *Palea* 8–12 mm, with 2 ciliate keels. *Stamens* 3, rarely 4, exserted. *Stigmas* 3, feathery. *Ovary* narrowly ovoid.

Introduced. Frequently naturalised throughout the British Isles and easily the commonest bamboo grown in gardens, and flowers in some years. Native of Japan and Korea.

7. Chimonobambusa Makino

Woody bamboos with leptomorph rhizomes. *Culms* 3–8 m, square in section with rough edges and faces; nodes mostly with 3 lateral branches. *Leaves* 1–3 cm wide, minutely pubescent when young, becoming glabrous, mostly with 8–14 veins on either side of the midrib. *Inflorescence* a panicle. *Spikelets* solitary or few at the ends of branchlets. *Glumes* absent. *Lemma* membranous. *Stamens* 3. *Stigmas* 2. *Lodicules* 3.

About 15 species in China and Japan.

1. C. quadrangularis (Fenzl) Makino
Square-stemmed Bamboo

Bambusa quadrangularis Fenzl; *Arundinaria quadrangularis* (Fenzl) Makino; *Tetragonocalamus quadrangularis* (Fenzl) Nakai; *Phyllostachys quadrangularis* (Fenzl) Rendle

Woody bamboo with often far-running, leptomorph rhizomes. *Culms* 3–8 m, 1–4 cm thick, square in section with rough edges and faces, green; nodes very prominent and densely fringed with yellowish-brown bristles, mostly with 3 lateral branches, sometimes with a purple band below; internodes 8–22 cm. *Leaves* 8–29 × 1–3 cm, narrowly lanceolate, acute at apex, narrowed at base, minutely pubescent when young becoming glabrous, papery-thin, mostly with 8–14 veins on either side of the midrib; sheaths coriaceous, glabrous, thin, open and soon falling, without auricles and with erect oral bristles; ligules short, truncate at apex, finely ciliate and hairy on the abaxial surface. *Inflorescence* a panicle, primary branches 3, becoming multi-branched and clustered after accessory buds develop; branches smooth with nodes prominently swollen. *Spikelets* solitary or a few terminal on branchlets, many-flowered, subtended by bracts. *Glumes* absent. *Lemma* membranous, lanceolate, acute at apex, smooth, glabrous. *Palea* nearly as long as lemma, 2-keeled, glabrous. *Stamens* 3. *Stigmas* 2.

Introduced. Naturalised in south-west England and south-west Ireland where it forms thickets. Native of eastern, southern and southwestern China.

Tribe **2. Oryzeae** Dumort.

Perennial herbs with rhizomes. *Leaves* without a false petiole and cross-veins; ligule membranous. *Spikelets* with one flower. *Glumes* vestigial or absent. *Lemmas* 5-veined, awnless. *Palea* 3-veined. *Stamens* 3. *Lodicules* 2.

8. Leersia Sw. nom. conserv.

Perennial herbs with rhizomes. *Leaves* without false petioles and cross-veins; ligules membranous. *Spikelets* with one flower. *Glumes* vestigial or absent. *Sterile lemmas* absent.

Eighteen species in tropical and warm temperate regions.

1. L. oryzoides (L.) Sw.   Cut Grass
Oryza oryzoides (L.) Brand; *Phalaris oryzoides* L.

Perennial herb; rhizomes long and slender; shoots in loose tufts or patches. *Culms* 30–120 cm, erect or ascending, slender to somewhat stout, often branched, few- to many-noded, hairy at the nodes, rough towards the upper nodes or smooth. *Leaves* 8–30 cm × 5–10 mm, yellowish-green, flat, linear, finely pointed at apex, rough, with bristles on the margin, the abaxial surface with a prominent midrib; sheaths rounded on the back, finely ribbed, the upper rough with short, stiff, reflexed hairs between the ribs; ligules 0.5–1.5 mm. *Inflorescence* a panicle enclosed in the leaf-sheath or partially or wholly exserted from it, contracted to very loose, 10–22 cm × up to 14 cm; branches spreading, wavy, very fine, flexuous, bare in the lower part; pedicels 0.3–1.0 mm. *Spikelets* overlapping on one side and towards the tips of the branches, (3.5–)4–5 mm, pale green, semi-elliptic-oblong, flattened, 1-flowered, falling at maturity. *Glumes* reduced to a narrow rim at the tip of the pedicels. *Lemmas* semi-elliptic-oblong, abruptly acute at apex, firm, fringed on the keel with stiff, patent hairs and with much shorter ones on the sides, 5-veined, the outermost veins marginal. *Palea* as long as or slightly longer than the lemma, narrow, 3-veined, stiffly hairy on the back. *Anthers* 0.4–0.7 mm in closed spikelets, 1.5–3.0 mm in open spikelets. *Caryopsis* enclosed by lemma and palea. *Flowers* 8–10. Chasmogamous, or cleistogamous with the panicle partially or wholly enclosed in the uppermost leaf-sheath, depending on climatic conditions. 2*n* = 48.

Native. Lake and pond sides, part-filled ditches and by canals and rivers. Very locally frequent and decreasing, from Somerset to Surrey and Sussex. South and central Europe, north to south Sweden and Finland; temperate Asia; North America.

Tribe **3. Ehrharteae** Nevski

Perennial herbs with rhizomes. *Leaves* without false petioles and cross-veins; ligule a membrane fringed with hairs. *Spikelets* with one small bisexual flower with awnless, 3-veined lemma and 1-veined palea, and 2 much larger sterile flowers below each consisting of one long-awned, 5-veined lemma only, the former obscured by the two latter. *Glumes* 2, very small and separated from the flowers by a stalk-like callus. *Stamens* 4. *Lodicules* 2.

9. Ehrharta Thunb. nom. conserv.

Perennial herbs, tufted or with few to numerous stems from a scaly rhizome. *Culms* up to 70 cm, erect to procumbent. *Leaves* without false petioles or cross-veins; ligule a membrane, often fringed with hairs. *Inflorescence* a loose panicle or simple raceme. *Spikelets* with one bisexual flower and 2 much larger sterile ones. *Glumes* 2, very small, separated from the flowers by a stalk-like callus. *Stamens* 4. *Styles* 2. *Lodicules* 2.

About 35 species, 25 in South Africa, one of them extending to Ethiopia, the rest from Indonesia to Australia and New Zealand.

Connor, H. E. & Matthews, B. A. (1977). Breeding systems in New Zealand grasses: VII Cleistogamy in Microlaena. *New Zealand Jour. Bot.* **15**: 531–534.

1. E. stipoides Labill.   Weeping Grass
Microlaena stipoides (Labill.) R. Br.

Wiry *perennial herb*, sometimes tufted, or with few to numerous culms from a scaly rhizome. *Culms* 30–70 cm, erect to procumbent, slender, more or less smooth, sparsely branched. *Leaves* 7–17 cm × 2–5 mm, bright to dark green, narrowly subulate, acute at apex, flat or convolute, rough with upwardly directed bristles; sheaths usually not fused; ligule about 0.5 mm, the hyaline rim with or without sparse, silky caducous hairs. *Inflorescence* a contracted loose panicle or simple raceme. *Spikelets* solitary, shortly pedicellate, flowers 3, the lower 2 reduced to lemmas. *Glumes* 2, unequal, 0.5–1.0 mm, membranous, ovate, the upper nearly twice as long as the lower. *Lemmas* unequal, laterally compressed, scabrid along the keel, those of the fertile flower 3-veined and awnless, the sterile flowers 6–12 mm, with an awn up to 20 mm. *Palea* shorter than lemma, thinly membranous, 1-veined. *Stamens* 4. *Styles* 2, free. *Caryopsis* linear-obloid, compressed, free. *Flowers* 8–10. 2*n* = 40.

There is much variation in glume length, lemma awns and palea veins, and as cleistogamy occurs the characters may be perpetuated.

We have followed Clayton & Renvoize (1986) and Stace (1991) in keeping this species in the genus *Ehrharta*, but Australian and New Zealand botanists put it in a separate genus, *Microlaena* R. Br.

Introduced. A rather infrequent wool alien, with scattered records in England. Native of Australia and New Zealand.

Subfamily 2. **Pooideae** Macfarlane & Watson
Subfamily *Festucoideae* Rouy

Annual to *perennial herbs. Leaves* without false-petioles or cross-veins; ligule membranous, sometimes a membrane fringed with hairs; sheaths usually not fused. *Inflorescence* a spike, raceme or panicle. *Spikelets* with 1 to many flowers, with all flowers (except often the distal ones) bisexual or some variously male or sterile. *Glumes* (0–)2. *Lemmas* firm to membranous, 3- to 9-veined, awnless or with a terminal or dorsal awn. *Palea* usually 2(–4)-veined, rarely with a midrib or with the midrib absent. *Stamens* 1–3. *Stigmas* 1–2. *Lodicules* absent or 2(–3), or 2 fused laterally. *Embryo* pooid; photosythesis C3-type alone with non-Kranz leaf anatomy; fusoid cells absent; microhairs usually absent; *chromosome base-number* usually 7, rarely higher or lower.

Tribe 4. Nardeae Koch

Perennial herbs. Leaves with a membranous ligule; sheaths not fused. *Inflorescence* a more or less 1-sided spike with one spikelet at each node. *Spikelets* with one bisexual flower. *Glumes* very short, the upper often absent. *Lemma* with 2–3 keels, 3-veined, with a terminal awn. *Stamens* 3. *Stigma* 1. *Lodicules* absent. *Ovary* glabrous.

10. **Nardus** L.

Perennial tufted herbs. Leaves tightly inrolled. *Inflorescence* a more or less 1-sided spike with spiklets at each node. *Spikelets* with one bisexual flower.

One species in Europe, temperate Asia, Caucasus and Greenland.

Chadwick, M. J. (1960). *Nardus stricta* L. in Biological flora of the British Isles. *Jour. Ecol.* **48**: 255–267.
Grime, J. P. et al. (1983). *Nardus stricta* L. in *Comparative plant ecology* 418–419.

1. **N. stricta** L.   Mat Grass

Wiry, densely tufted *perennial herb*; rhizomes short; shoots close together; roots coarse. *Culms* 10–40(–60) cm, erect, slender, simple, 1-noded towards the base. *Leaves* 4–30 cm × about 0.5 mm, greyish-green or green, bristle-like, tightly inrolled, sharp-pointed at apex, hard, stiff, grooved, minutely hairy in the grooves or glabrous and smooth; sheaths smooth, the basal shoot crowded, persistent, pale, tough, shining; ligules 0.5–2.0 mm, membranous, obtuse at apex. *Inflorescence* a spike 3–8 cm, green or purplish, erect, very slender, one-sided; rhachis rough on the margins, with a bristle up to 10 mm at the apex. *Spikelets* 5–9 mm, narrow, finely pointed, sessile, loosely to closely overlapping, in 2 rows along one side of the rhachis, 1-flowered. *Glumes* persistent, the lower very small, the upper usually absent. *Lemmas* 6–9 mm, narrowly lanceolate or lanceolate-oblong, 2- to 3-keeled, rough on the keels, 3-veined, firm; awn 1–3 mm at tip. *Palea* slightly shorter than the lemma, 2-veined. *Anthers* 3.5–4.0 mm. *Caryopsis* 3–4 mm, narrow, tightly enclosed by and falling with the hardened lemma and palea. *Flowers* 6–8. Apomictic. $2n = 26$.

Native. Abundant on the poorer siliceous and peaty soils where it covers great areas on moors and mountains. Rejected by sheep and thus common in overgrazed areas. Throughout the British Isles, but absent from some lowland and calcareous districts, though occasionally common on sandy heaths. Throughout Europe; temperate Asia; Caucasus; Greenland; only on mountains in the southern part of its range and always calcifuge; doubtfully native in North America.

Tribe 5. Stipeae Dumort.
Tribe *Milieae* Dumort.

Annual or *perennial herbs. Leaves* with a membranous ligule or sometimes a membrane with a fringe of hairs; sheaths not fused. *Inflorescence* a more or less diffuse panicle. *Spikelets* with one bisexual flower. *Glumes* more or less equal, longer than the body of the lemma. *Lemma* rounded on its back, 3- to 5-veined, awnless or with a long terminal awn, becoming hard and tightly wrapped around the caryopsis. *Stamens* 3. *Stigmas* 2. *Lodicules* 2–3. *Ovary* glabrous.

11. **Stipa** L.

Annual or *perennial herbs,* often tufted. *Leaves* usually plicate or convolute, at least when dry; ligule membranous with a fringe of hairs; sheaths not fused. *Inflorescence* a diffuse panicle. *Spikelets* somewhat laterally compressed, with one bisexual flower. *Glumes* usually subequal, hyaline or membranous, much longer than the lemma, 1- to 3-veined. *Lemma* usually coriaceous, convolute, terete, entire or shortly bifid at the apex, hairy; awned from the apex or sinus, with a long basal callus which is bearded and sharply pointed, the awn usually with a thick, twisted lower part (*column*) and a thinner, straight upper part (*seta*), usually 2-geniculate. *Palea* hyaline, 2-veined, usually enclosed by the lemma. Rhachilla not prolonged.

About 300 species in temperate regions of the world.

1. Awns 5–9 cm   2.
1. Awns less than 4 cm   3.
2. Lemma including callus 9–11 mm, glabrous except for a dense tuft of hair at the base   **1. neesiana**
2. Lemma including callus 6–8 mm, hairy with a ring of longer hairs at the apex   **2. capensis**
3. Lemma including callus 5.5–6.5 mm, silky appressed-hairy with an awn 2.5–4.0 cm   **3. aristiglumis**
3. Lemma including callus about 5 mm, glabrous except for a dense tuft of hairs at the base and awns 2–4 cm   **4. formicarum**

1. **S. neesiana** Trin. & Rupr.   American Needle Grass

Tufted *perennial herb. Culms* up to 60 cm, striate, erect, glabrous. *Leaves* very variable in length, 6–20 cm × 0.7–2.5 mm, bluish-green, long-linear, gradually narrowed to an acute apex, convolute or flat, glabrous or sparsely hairy, not fused; sheath glabrous or sparsely hairy; ligule very short, membranous, ciliate. *Inflorescence* a dense, narrow but diffuse panicle 5–15 cm. *Spikelets* compressed, with one bisexual flower. *Glumes* equal or nearly so, up to 20 mm including the long, fine, 3-

pronged apex, purplish with a hyaline margin and apex, linear. *Lemmas* including the callus 9–11 mm, elliptical-linear, acute at apex, glabrous except for a dense tuft of hairs at the base and on the callus, with a distinct ring of hairs at the apex and an awn 5–9 cm. *Palea* about as long as and enclosed by the lemma, narrowly elliptical-linear, 2-veined. *Caryopsis* fusiform, tightly enclosed in the lemma and palea. *Flowers* 5–7. $2n = 28$.

Introduced. Wool alien. Sometimes more or less naturalised in south-east England, and scattered records from other parts of England. Native of South America.

2. S. capensis Thunb.   Mediterranean Needle Grass
S. tortilis Desf.

Annual or *biennial herb* with fibrous roots. *Culms* 10–30(–80) cm, glabrous. *Leaves* 6–15 cm × 0.2–1.5 mm, bluish-green, narrowly long-linear, gradually narrowed to an acute apex, convolute or flat, glabrous or sparsely hairy, smooth; sheath bearded at mouth; ligule very short, truncate at apex, ciliate. *Inflorescence* a dense panicle 3–10(–15) cm. *Spikelets* compressed, with one bisexual flower. *Glumes* 15–20 mm including the awn, linear, long-attenuate at the apex to an awn 4–5 mm, hyaline at the margin. *Lemma* 6–8 mm, narrowly elliptical-linear, acute at apex, with appressssed, rigid hairs and a ring of longer hairs below the base of the awn, the awn 7–10 cm, the column with hairs about 0.8 mm, the seta with more or less appressed hairs about 0.1 mm. *Palea* enclosed by the lemma, hyaline, 2-veined. *Caryopsis* tightly enclosed in the hardened lemma. *Flowers* 5–7. $2n = 36$.

Introduced. Wool alien. Scattered records in Britain. Native of the Mediterranean region.

3. S. aristiglumis F. Muell.   Australian Needle Grass
S. fusiformis Hughes

Robust, tufted *perennial herb* with a very contracted, scaly rhizome. *Culms* 0.3–1.2(–2) m, erect, glabrous. *Leaves* very variable in length, 6–15 cm × 0.2–1.5 mm, green, narrowly long-linear, acute at apex, convolute, glabrous; ligules 0.5–1.0 mm, ciliate on the outer edge. *Inflorescence* a loose panicle up to 45 cm; branches filiform, clustered at nodes, clearly separated in the lower part of the axis. *Spikelets* compressed, with one bisexual flower, pedicellate, not falling entire. *Glumes* 6–14 mm, subequal, persistent, longer than flower, hyaline, 3- to 5-veined, linear, acute to attenuate at apex, glabrous. *Lemma* 5.5–6.5 mm, narrowly lanceolate, acute at apex, silky-pubescent, with a terminal, geniculate, slightly curved, scabrid awn 2.5–4.0 cm. *Palea* enclosed by the lemma. *Caryopsis* tightly enclosed in the hardened lemma. *Flowers* 5–7.

Introduced. Wool alien. Scattered records in Britain. Native of Australia.

4. S. formicarum Del.   Ant Needle Grass

Perennial herb with a strong rhizome. *Culms* 40–80 cm, erect, slender, glabrous. *Leaves* 10–30 cm × 1–2 mm, yellowish-green, flat or convolute, long-linear, narrowed at the apex, glabrous; ligule about 1 mm, obtuse at apex, membranous. *Inflorescence* a panicle 10–20 cm, with

many flowers. *Spikelets* compressed, with 1 bisexual flower. *Glumes* subequal, 8–11 mm, purplish with a hyaline margin, lanceolate, narrowed at apex, 3-veined. *Lemma* including callus about 5 mm, glabrous apart from the basal hair-tuft; awns 2–4 cm. *Palea* about 1 mm, hyaline, truncate at apex, glabrous, without a vein. *Caryopsis* 2.5–2.8 mm, fusiform, tightly enclosed in the hardened lemma and palea. *Flowers* 5–7.

Introduced. Wool alien. In scattered localities in England. Native of South America.

12. Oryzopsis Michx
Piptatherum P. Beauv.

Perennial herbs with a short rhizome. *Leaves* flat; ligule membranous. *Inflorescence* a panicle with a long central axis bearing at each node many slender branches with spikelets clustered at the ends. *Glumes* 3–4 mm. *Lemma* shiny, with a short, smooth basal callus and a long, straight, deciduous terminal awn. *Palea* similar to lemma. *Anthers* bearded. *Caryopsis* compressed.

Thirty-five species in temperate and subtropical regions of the northern hemisphere.

1. O. miliacea (L.) Benth. & Hook. fil. ex Asch. & Schweinf.   Smilo Grass
Piptatherum miliaceum (L.) Cosson; *Agrostis miliacea* L.

Perennial herb with a knotty, woody base or short rhizome. *Culms* up to 1.5 m, erect or geniculately ascending, slender, branched at the base or at the nodes, more or less scabrid, glabrous. *Leaves* up to 30 cm × 2–5 mm, long-linear, acuminate at apex, flat, finely involute or rolled, scabrid on the upper surface and on the margins, hairy on the upper surface, smooth and glabrous beneath; sheaths much shorter than the internodes, rounded, clasping, smooth, glabrous; ligules 1.0–1.5 mm, truncate at apex, minutely pubescent. *Inflorescence* a panicle up to 40 cm, lax or dense, erect or nodding; rhachis smooth, glabrous; branches fascicled, often very numerous, ascending, unequal, slender, bare for half their length, the lower sometimes without spikelets, the upper carrying numerous spikelets. *Spikelets* 2–4 mm, green or flushed with violet, narrowly elliptical, acute or acuminate at apex. *Glumes* 3–4 mm, more or less equal, membranous, elliptical, acute at apex, the lower a little longer than the upper, 3-veined. *Lemma* about 2.5 mm, chartaceous, smooth, glabrous, shining, 3-veined, awned. *Palea* similar to lemma, with a scabrid awn up to 3 mm. *Anthers* about 1.5 mm, bearded. *Caryopsis* elliptical, somewhat compressed. *Flowers* 7–8. $2n = 24$.

Introduced. Ornamental escape and wool alien, casual in waste places. Scattered in England and south Wales, naturalised in Jersey since 1931 and in west Kent. Mediterranean region eastwards to Iran; Macaronesia. Introduced elsewhere.

13. Milium L.

Glabrous *annual* or *perennial herbs*. *Leaves* with a membranous ligule. *Spikelets* with one flower, dorsally compressed. *Glumes* subequal, membranous, exceeding the flower. *Lemma* shiny, with a minute callus, 5-veined. *Palea* similar to lemma, but smaller and 2-veined, awnless. *Lodicules* 2.

Four species in the north temperate zone of the Old World and in eastern North America.

Grime, J. P. et al. (1988). *Milium effusum* L. in *Comparative plant ecology* 400–401.
Tutin, T. G. (1950). *Milium scabrum* Merlet. *Watsonia* 1: 345–348.

1. Perennial 45–180 cm; sheaths smooth; panicle
   spreading   **1. effusum**
1. Annual 2.5–15 cm; sheaths scabrid; panicle contracted
                                  2. vernale subsp. **sarniense**

1. M. effusum L.   Wood Millet

Loosely tufted *perennial herb. Culms* 45–180 cm, erect or bent at the base, usually very slender, sometimes stout, smooth, with 3–5 nodes below the middle. *Leaves* 10–30 cm × 5–15 mm, dull green, glabrous, linear, pointed at apex, flat, rough on the margins, smooth or only slightly rough on the veins; sheaths rounded on the back, smooth; ligules 3–10 mm, membranous. *Inflorescence* a panicle 10–40 × up to 20 cm, lanceolate to ovate or oblong, very loose, nodding; rhachis smooth; branches in clusters, fine, flexuous, spreading or deflexed, rough in the upper part; pedicels 1–3 mm. *Spikelets* 3–4 mm, pale green, rarely purple, narrowly elliptical to ovate, pointed or at length obtuse at apex, slightly compressed from the back, 1-flowered, breaking up above the glumes at maturity. *Glumes* persistent, ovate to elliptic-ovate, as long as the spikelet, greenish, membranous except for the whitish margins, minutely rough, 3-veined, equal, the upper narrower than the lower. *Lemma* lanceolate to elliptical in back view, pointed at apex, as long as the glumes or slightly shorter, rounded on the back, becoming hard and tough; very smooth and shining, finely 5-veined. *Palea* as long as the lemma and similar in texture. *Anthers* 2–3 mm. *Caryopsis* tightly enclosed by the hard, brown lemma and palea. *Flowers* 5–7. 2n = 28.

Native. Locally abundant in oak and beech woods, especially on damp, heavy, calcareous soils with humus and frequently in considerable shade. Locally frequent throughout England, scattered in Wales, Ireland and lowland Scotland, but absent from the Outer Isles. Most of Europe, temperate Asia; Japan; North America.

2. M. vernale M. Bieb.   Early Millet
M. scabrum Rich.

Glabrous *annual herb. Culms* up to 10(–15) cm, procumbent to decumbent, with small nodes in the lower third, scabrid. *Lower leaves* 10–20 × 1.5–3.0 mm, the uppermost 0.5–5.0 cm × 1–3 mm, scabrid; sheaths often purplish, with a wide scarious margin, scabrid; ligule 2–4 mm, acute at apex. *Inflorescence* a panicle 1–4 cm, with appressed branches. *Spikelets* 2.5–3.5 mm, olive to pale green or sometimes purplish-tinged. *Glumes* 2–3 mm, usually green with hyaline margins, ovate, acute at apex, scaberulous. *Lemma* 1.5–2.0 mm, coriaceous and

shining, ovate, obtuse at apex. *Anthers* 1.5–1.8 mm. *Flowers* 4–5. 2n = 8.

Native. Short turf on fixed sand-dunes and cliffs by the sea. Three small sites on L'Ancresse Common and small groups on L'Hommet Headland, both in Guernsey, where it was first found in 1899. West and south Europe; temperate Asia; North Africa. The Guernsey plant is subsp. **sarniense** D. C. McClint. which is endemic to the island. Other subspecies occur elsewhere in its range.

Tribe 6. Poeae
Tribe *Festuceae* Dumort.; Tribe *Seslerieae* Koch

Annual or *perennial herbs. Leaves* with a membranous ligule; sheaths not or sometimes fused. *Inflorescence* a panicle, or a raceme, or a spike with normally one spikelet at each node. *Spikelets* with (1–)3–many flowers, all bisexual or a group of the apical male or sterile flowers, sometimes the spikelets in pairs, one of each pair entirely sterile. *Glumes* equal to very unequal, sometimes the lower vestigial or absent. *Lemma* rounded or keeled on the back, 3- to 9-veined, awnless or with a terminal awn. *Stamens* 1–3. *Stigmas* 2. *Lodicules* 2. *Ovary* glabrous or pubescent at apex.

14. Festuca L.

Perennial herbs with or without rhizomes, without stolons. *Leaves* flat or folded; sheaths fused or not; ligule membranous. *Inflorescence* a panicle. *Spikelets* with (2–)3 to many flowers, all except the apical bisexual, sometimes proliferating. *Glumes* subequal. *Lemmas* rounded on the back, 3- to 5-veined, more or less acute at apex; awned or not. *Stamens* 3. *Stigmas* 2. *Lodicules* 2.

About 300 species in temperate regions and on higher mountains in the tropics.

The *F. rubra* aggregate, species 6–7, and the *F. ovina* aggregate, species 8–16, are extremely critical groups. The *F. rubra* aggregate follows the work done by A.-K. K. A. Al-Bermani and the *F. ovina* aggregate that of M. J. Wilkinson. The following character measurements apply to those groups.

Leaf, sheath and ligule characters refer to leaves on the tillers unless otherwise stated. Spikelet length is not the total length of the spikelet, but the length from the base of the lower glume to the apex (excluding awn) of the fourth lemma. The length of the spikelet when there are only 3 flowers is obtained by addition to the total length of an increment equal to the distance between the tips of the second and third lemmas. Lemma lengths exclude the awn, and refer to only the lower 2 per spikelet. Awn lengths are calculated by averaging the lengths of all the awns of one spikelet, and then finding the mean of this value for 5–10 spikelets. All measurements given are a range of means of 5–10 measurements per plant and not a range of extremes. This may be mathematically all right, but these are not practical characters to use. They are, however, retained here because they have been used by the two authors who have done the most work on these two critical groups in the British Isles.

The distinction between extravaginal and intravaginal tillers is fundamental. *Intravaginal tillers* arise parallel to the parent shoot and remain enclosed within the parental leaf sheath for some distance. *Extravaginal tillers* arise more or less at right angles to the parent shoot and break through the parental leaf-sheath at its base. Rhizomes always start off intravaginally; hence the presence of rhizomes indicates the existence of intravaginal shoots but the absence of rhizomes does not necessarily mean that all the tillers are intravaginal. Leaf-sheath fusion should be observed on the next to the most apical sheath of a tiller by stripping off all the more mature sheaths below it.

Al-Bermani, A.-K. K. A. & Stace, C. A. (1991). A new sub-species of *Festuca rubra* L. *Watsonia* 18: 315–316.

Auquier, P. (1970). Typification et taxonomie de *Festuca tenuifolia* Sibth. *Lejeunia* nouvelle série 53: 1–7.

Auquier, P. (1973). Une féstuque nouvelle de Bretagne, *Festuca huonii*. *Candollea* 28: 15–19.

Auquier, P. (1973). Qu'est-ce que le *Festuca caesia* Sm. (Poaceae)? *Lejeunia* nouvelle série 70: 1–12.

Auquier, P. (1977). Taxonomie et nomenclature de quelques *Festuca* tetraploides du groupe de *F. ovina* L. s. l. (Poaceae) en Europe moyenne. *Bull. Jard. Bot. Nat. Belg.* 47: 99–116.

Auquier, P. & Kerguélen, M. (1977). Un groupe embrouillé de *Festuca* (Poaceae): les taxons désignèr par l'epithète "*glauca*" en Europe occidentale et dans les régions voisines. *Lejeunia* nouvelle série 89: 1–82.

Auquier, P. & Rammeloo, J. (1973). Nombres chromosomiques dans le genre *Festuca* en Belgique et dans les régions limotrophes. *Bull. Soc. Roy. Bot. Belg.* 106: 317–328.

Howarth, W. O. (1925). On the occurrence and distribution of *Festuca ovina* L. sensu ampliss., in Britain. *Jour. Linn. Soc. Lond. Bot.* 47: 29–39.

Howarth, W. O. (1948). A synopsis of the British Fescues. *Bot. Soc. Exch. Cl. Brit. Isles* 13: 338–346.

Jarvis, C. E., Stace, C. A. & Wilkinson, M. J. (1987). Typification of *Festuca rubra* L., *F. ovina* L. and *F. ovina* var. *vivipara* L. *Watsonia* 16: 299–302.

Jenkin, T. J. (1922). Notes on vivipary in *Festuca ovina*. *Bot. Soc. Exch. Cl. Brit. Isles* 6: 418–432.

Kerguélen, M. (1975). Les Gramineae (Poaceae) de la flore française. Essai de mise au point taxonomique et nomenclaturale. *Lejeunia* nouvelle série 75: 1–343.

Kerguélen, M. (1983). Les Graminées de France au travers de 'Flora Europaea' et de la 'Flore' du C.N.R.S. *Lejeunia* nouvelle série 110: 1–79.

Kerguélen, M. & Plonka, F. (1988). Le genre *Festuca* dans le flore française. Taxons nouveaux, observations nomenclaturales et taxonomiques. *Bull. Soc. Bot. Centre-Ouest*, Nouv. Sér. 19: 15–30.

Onder, A. & Jong, K. (1977). The occurrence of B. chromosomes in *Festuca ovina* L. sensu lato from Scotland. *Watsonia* 11: 327–330.

Stace, C.A., Al-Bermani, A.-K. K. A. & Wilkinson, M.J. (1992). The distinction between the *Festuca ovina* L. and *Festuca rubra* L. aggregates in the British Isles. *Watsonia* 19: 107–112.

Stewart, A., Pearman, D. A. & Preston, C. D. (1994). *Scarce plants in Britain*. Peterborough [*F. altissima* and *F. arenaria*.]

Trist, P. J. O. (1973). *Festuca glauca* Lam. and its var. *caesia* (Sm.) K. Richter. *Watsonia* 9: 257–262.

Watson, P. J. (1958). The distribution in Britain of diploid and tetraploid races within the *Festuca ovina* group. *New Phytol.* 57: 11–18.

Wilkinson, M. J. & Stace, C. A. (1987). Typification and status of the mysterious *Festuca guestfalica* Boenn. ex Reichb. *Watsonia* 16: 303–309.

Wilkinson, M. J. & Stace, C. A. (1985). The status of *Festuca ophiolitocola* Kerguélen and related taxa. *Soc. l'Echange P. Vasc. l'Europe Basin Médit. Bull.* 20: 69–73.

Wilkinson, M. J. & Stace, C. A. (1989). The taxonomic relationships and typification of *Festuca brevipila* Tracey and *F. lemanii* Bastard (Poaceae). *Watsonia* 17: 289–299.

Wilkinson, M. J. & Stace, C. A. (1991). A new taxonomic treatment of the *Festuca ovina* L. aggregate (Poaceae) in the British Isles. *Bot. Jour. Linn. Soc.* 106: 347–397.

Wycherley, P. R. (1953). Proliferation of spikelets in British grasses. I. The taxonomy of   the viviparous races. *Watsonia* 3: 41–56.

Wycherley, P. R. (1953). The distribution of the viviparous grasses in Great Britain. *Jour. Ecol.* 41: 275–289.

1.   Base of leaf on each side extended into pointed auricles clasping the stem at the level of and on the opposite side to the ligule   2.

1.   Leaves without auricles, or with short, rounded auricles not clasping the stem   4.

2.   Exposed nodes of culms deep violet-purple; lemmas with awns 10–18 mm   **3. gigantea**

2.   Exposed nodes of culm green, sometimes tinged purplish; lemmas with awns absent or up to 4 mm   3.

3.   Leaf-auricles glabrous; lowest 2 panicle nodes with 2 unequal branches, the shorter with 1–2(–3) spikelets   **1. pratensis**

3.   Leaf-auricles usually fringed with minute hairs (often only a few) tearing off with age; lowest 2 panicle nodes each with 2 subequal branches, the shorter with (3–)4–many spikelets   **2. arundinacea**

4.   Leaves of culms and tillers flat (or folded longitudinally when dry), more than 4 mm wide   5.

4.   Leaves of tillers and usually culms folded longitudinally or the edges also inrolled, less than 4 mm wide   6.

5.   Rhizomes absent; leaves 4–14 mm wide; ligules more than 1 mm; lemma 3-veined, awnless; caryopsis with a hairy apex   **4. altissima**

5.   Rhizomes present; leaves less than 5 mm wide; ligules less than 1 mm; lemma 5-veined and awned; caryopsis with a glabrous apex
                             7(g). rubra subsp. **megastachya**

6.   Some or all tillers extravaginal; young leaves with sheaths fused almost up to the top   7.

6.   All tillers intravaginal; young leaves with sheaths not fused, but with overlapping margins   14.

7.   Leaves with 3(–5) veins; leaves of culms and tillers markedly different, the former flat and 2–4 mm wide, the latter folded and less than 0.6 mm from midrib to edge; ovary and caryopsis with hairy apex
                                    5. heterophylla

7.   Leaves with 5–9(–11) veins; leaves of culms and tillers similar to obviously different; ovary and caryopsis glabrous   8.

8.   Leaves with densely hairy adaxial ribs, rounded on midrib abaxially, with abaxial sclerenchyma usually continuous or semi-continuous, with distinct

sclerenchyma bundles in adaxial ribs; always coastal **6. arenaria**

8. Leaves with scabrid or sparsely hairy adaxial ribs, obtuse to keeled on the midrib abaxially, with abaxial sclerenchyma in discrete islets, often without or with very sparse sclerenchyma in adaxial ribs; coastal or inland 9.

9. Rhizomes absent or very few and very short **7(d). rubra** subsp. **commutata**

9. Rhizomes well-developed (plants densely tufted or not) 10.

10. Leaves 0.8–1.4(–2.5) mm from midrib to edge, with distinct islets of sclerenchyma in adaxial ribs; lemma usually 6–8 mm **7(f). rubra** subsp. **scotica**

10. Leaves 0.5–1.2 mm from midrib to edge, usually without or with very sparse sclerenchyma except in subsp. *juncea*; lemma 4.2–6.5 mm 11.

11. Lemmas 4.2–6.0 mm, with awns (0–)0.1–2.2 mm, glaucous, usually with dense white hairs, sometimes glabrous, but usually some hairy plants nearby; Scotland and Caernarvon only **7(e). rubra** subsp. **arctica**

11. Lemma 4.4–8.0 mm, with awns 0.5–3.0 mm, glaucous or not, the hairs if present not dense and white; widespread 12.

12. Spikelets 8.7–11.2 mm; lemmas 5.7–8.0 × more than 2 mm, with awns 1.1–2.8 mm; saline sand or mud, often forming dense mats **7(c). rubra** subsp. **litoralis**

12. Spikelets 6.8–10.2 mm; lemmas 4.4–6.7 × 1.2–2.4, with awns 0.5–2.2 mm; rarely in saline soil 13.

13. Rhizomes medium to long, forming loose patches; plants rarely glaucous **7(a). rubra** subsp. **rubra**

13. Rhizomes short, forming dense tufts, often some plants in a population with very glaucous leaves **7(b). rubra** subsp. **juncea**

14. Leaves with 5–9 veins; with 4(–6) adaxial grooves; lemma with awns usually more than 1.2 mm and often more than 1.6 mm 15.

14. Leaves with 5–7 veins, with 2(–4) adaxial grooves; lemmas with awns less than 1.6 mm and often less than 1.2 mm 16.

15. Leaves usually very glaucous, with abaxial sclerenchyma in more than 5 main islets at midrib, edges and variously in between; pedicels 0.5–1.8 mm; sheaths glabrous, spikelets 5.4–7.0 mm; lemmas with awns 0.5–1.5 mm **14. longifolia**

15. Leaves not or slightly pruinose with abaxial sclerenchyma in 3 main islets at midrib and edges; sheaths often sparsely hairy; pedicels 1.2–2.8 mm; spikelets 6.1–8.5 mm; lemmas with awns 1.2–2.6 mm **16. brevipila**

16. Leaves usually less than 0.6 mm from midrib to edge; panicles mostly less than 8 cm, with fewer than 26 spikelets and lowest 2 nodes less than 2 cm apart; spikelets 4.7–7.0 mm; lemmas with awns usually 0–1 mm 17.

16. Leaves more than 0.6 mm from midrib to edge; panicles less than 13 cm, with fewer than 40 spikelets and lowest 2 nodes less than 3.7 cm apart; spikelets often more than 7 mm, rarely less than 6 mm; lemmas with awns usually more than 1 mm 22.

17. Spikelets all or mostly proliferating; sexual flowers (if present) with lemmas 3.4–4.2 mm and awns 0–0.2 mm **9. vivipara**

17. Spikelets not or rarely some proliferating; sexual flowers with lemmas 2.5–4.9 mm and awns 0–1 mm 18.

18. Leaves glabrous, 0.3–0.5 mm from midrib to edge; spikelets less than 5.5 mm; lemmas less than 3.5 mm with awns 0–0.6 mm **10. filiformis**

18. Leaves often hairy at base, 0.3–0.9 mm from midrib to edge; spikelets more than 5.2 mm; lemma more than 3.2 mm, with awns 0–1.7 mm 19.

19. Leaves with (5–)7 veins; spikelets 5.5–7.5 mm; lemmas 3.6–4.9 mm 20.

19. Leaves with 5–7 veins; spikelets 5.3–6.3 mm; lemmas 3.1–4.2 mm 21.

20. Leaves (0.4–)0.5–0.7 mm from midrib to edge; awns (0–) or 0.5–1.0(–1.6) mm **8(c,i). ovina** subsp. **ophioliticola** var. **ophioliticola**

20. Leaves 0.3–0.4 mm from midrib to edge; awns 0–0.3(0.4) mm **8(c,ii). ovina** subsp. **ophioliticola** var. **hibernica**

21. Leaves and lemmas usually scabrid; stomata mostly less than 31.5 μm; awns 0.0–1.2 mm **8(a). ovina** subsp. **ovina**

21. Leaves usually pubescent at base; stomata mostly more than 31.5 μm; awns 0.0–0.8 mm **8(b). ovina** subsp. **hirtula**

22. Culms 19–66 cm; panicle 3.7–8.6 cm; pedicels 0.6–2.5 mm; widespread **13. lemanii**

22. Culms 11–35 cm; panicles 2.3–5.9 cm; pedicels 0.3–1.5 mm; coastal, in the Channel Islands only, or a garden escape 23.

23. Leaves up to 14 cm, strongly glaucous, with 4(–6) grooves on the adaxial surface; garden escape **15. glauca**

23. Leaves up to 10.2(–12) cm, green or slightly glaucous; grooves on adaxial surface 2 (–4); Channel Islands 24.

24. Culms erect; leaves usually not glaucous; panicles well exserted from sheath at anthesis; usually on dunes **11. armoricana**

24. Culms erect to procumbent; leaves often slightly glaucous; panicles not completely or only just exserted from the sheath at anthesis; usually on cliffs **12. huonii**

1. F. pratensis Huds. Meadow Fescue
F. elatior auct.

Loosely tufted *perennial herb* forming large tussocks when growing alone. Culms up to 80(–120) cm, erect or spreading, moderately slender to stout, unbranched, with 2–4 nodes, smooth. *Leaves* up to 45 cm × 3–8 mm, bright green, long-linear, glabrous, tapering to a fine tip, flat, glossy beneath, rough on the margins and sometimes above; sheaths not fused, rounded on the back, smooth, bearing narrow, spreading, glabrous, pointed auricles at the apex; ligules about 1 mm, membranous. *Inflorescence* a panicle 10–35 cm, loose, erect or more often nodding, lanceolate to ovate, more or less 1-sided, green or purplish; rhachis rough in the upper part; the lowest 2 panicle nodes with branches usually in pairs, unequal, slender, angular, rough, the shorter bearing 1–2(–3) spikelets, the longer several; pedicels up to 5 mm. *Spikelets* 10–20 mm, cylindrical, becoming lanceolate or narrowly oblong, 5- to 14-flowered, breaking up at maturity beneath each lemma. *Glumes* persistent, narrowly lanceolate to oblong, slightly unequal, firm except for the membranous tips and margins, the lower 2–4 mm, 1-veined, the upper 3–5 mm and 1- to 3-veined. *Lemmas* 6–7 mm, overlapping, narrowly oblong or lanceolate-oblong in side view, pointed at apex, usually awnless, when awned the awn 0.4–1.2 mm, rounded on the back, firm except for the membranous margins and tip, 5-veined, smooth or minutely rough near the tip. *Paleas* as long as the lemmas, with rough keels. *Anthers* 3–4 mm. *Caryopsis* tightly enclosed by the hardened lemma and palea. *Flowers* 6–8. 2n = 14.

Native. Meadows, waysides, hedgerows and by ditches and rivers, often on rich, moist soil, and a valuable constituent of grazing and hay meadows. Frequent throughout most of the British Isles, but rare in northern Scotland. Most of Europe; North Africa; temperate Asia; introduced in North America.

2. F. arundinacea Schreb.   Tall Fescue
F. elatior L. nom. rej.

Tufted *perennial herb*, sometimes forming dense tussocks. *Culms* 45–200 cm, mostly erect, usually stout to robust, unbranched, with 2–5 nodes, smooth, or rough towards the panicle. *Leaves* 10–60 cm × (1–) 3–12 mm, green, linear, long-tapering to a fine tip, stiff, rough or smooth beneath, flat; sheaths not fused, rounded on the back, smooth or rough, not shredding, with small, narrow, spreading, pointed auricles at the apex, minutely hairy on the auricles and at the junction with the lamina; ligules up to 2 mm, membranous. *Inflorescence* a panicle (6–)10–50 cm, green or purplish, erect or nodding, lanceolate to ovate, loose and open or contracted; rhachis rough; branches rough, later angular, spreading, at the lowest 2 nodes usually in pairs, with the shorter one bearing (3–)4 or more spikelets; pedicels up to 8 mm. *Spikelets* 10–18 mm, elliptical to oblong, closely 3- to 10- flowered, breaking up beneath each lemma at maturity. *Glumes* persistent, often purple tinged, slightly unequal to equal, pointed at apex, the lower 3–6 mm, narrowly lanceolate, 1-veined, the upper 4.5–7.0 mm, lanceolate to lanceolate-oblong, 3-veined. *Lemmas* 6–10 mm, overlapping or later with their margins incurved, often purple tinged, lanceolate or oblong-lanceolate in side view, pointed to obtuse at apex, broadly rounded on the back, awnless, or with the middle vein continued as a fine, rough awn 0.5–3.5(–6) mm, firm except for the membranous upper margins, 5-veined, rough especially on the veins. *Paleas* as long as the lemmas, with rough keels. *Anthers* 3–4 mm. *Caryopsis* tightly enclosed by the lemma and palea. *Flowers* 6–8. 2n = 42.

Native. Grassy places by sides of rivers and streams, pastures and rough hills and downs. Not as useful as *F. pratensis* as a grazing or hay grass. Very variable and robustly represented by a number of ecotypes in different habitats. Common throughout the British Isles. Most of Europe; western Siberia; North Africa.

× **gigantea = F. × fleischeri** Rohlena
F. × gigas O. Holmb.

This hybrid is a sterile intermediate between the parents with a more dwarf appearance and shorter awns than *F. gigantea* and the minute hairs on the leaf auricles are like *F. arundinacea*. 2n = 42.

Scattered in Britain from Huntingdonshire northwards and probably elsewhere. It also occurs in central and northern Europe.

× **pratensis = F. × aschersoniana** Dörfl.

This hybrid is sterile, but is difficult to distinguish morphologically from *F. arundinacea*. 2n = 28.

It has been recorded from Argyllshire and may be widespread. It has been widely recorded in Continental Europe.

3. F. gigantea (L.) Vill.   Giant Fescue
Bromus gigantea L.

Loosely tufted, glabrous *perennial herb*. *Culms* 45–150 cm, erect or spreading, usually stout, unbranched, with 2–5, deep violet-purple nodes, smooth. *Leaves* up to 60 cm × 6–18 mm, dark green, long-linear, long-tapering to a fine tip, flat, usually drooping, smooth and glossy beneath, rough on the margins and sometimes also above; sheaths not fused, rounded on back, smooth, or the lower rough, with prominent, spreading, glabrous, narrow, pointed, deep violet auricles at the apex; ligules up to 2.5 mm, membranous. *Inflorescence* a panicle 10–50 cm, green, nodding, lanceolate to ovate, loose; rhachis rough; branches angular, rough, spreading, flexuous, usually in pairs, bare for some distance at the base, unequal, the shorter one with several spikelets; pedicels 1.5–6.0 mm. *Spikelets* 8–20 mm, lanceolate to narrowly oblong, loosely 3- to 10-flowered, readily breaking up at maturity beneath each lemma. *Glumes* persistent, slightly unequal, finely pointed, firm except for the broad membranous margins, the lower 4–7 mm, narrowly lanceolate, 1- to 3-veined, the upper 5–8 mm, lanceolate, 3-veined. *Lemmas* 6–9 mm, at first overlapping, later with the margins incurved, broadly rounded on the back, lanceolate in side view, narrowed at the tip into a straight or flexuous, hair-like, wavy, rough awn 10–18 mm, firm except for the membranous upper margins, minutely rough, 5-veined. *Paleas* as long as the lemmas with minutely rough keels. *Anthers* 2.5–3.0 mm. *Caryopsis* tightly enclosed by the lemma and palea. *Flowers* 7–8. 2n = 42.

Native. Woods, hedgerows and other shady places. Common throughout the British Isles, except in northern Scotland. Most of Europe except the Arctic and rarely south of the Alps; Asia.

× **pratensis = F. × schlickumii** Grantzow

This hybrid resembles *F. gigantea* in appearance, but is male-sterile with non-dehiscent anthers and very few viable caryopses.

In Britain it has only been recorded from Norfolk. It has also occurred in central and northern Europe.

4. F. altissima All.   Wood Fescue
F. sylvatica (Pollich) Vill., non Huds.

Compactly tufted, long-lived *perennial herb* without rhizomes. *Culms* 50–120(–150) cm, erect, slender to moderately stout, unbranched, with 3–4 nodes, smooth.

Leaves up to 60 cm × 4–14 mm, green, glabrous, long-linear, gradually narrowed to a fine point, flat, thin to firm, finely veined, minutely rough on both sides or only on the margins; sheaths not fused, smooth, or rough upwards, rounded on the back, without auricles; ligules up to 5 mm, thinly membranous, becoming torn. *Inflorescence* a panicle 10–18 × up to 12 cm, loose, open, nodding, green; branches usually in pairs, fine, spreading, smooth or slightly rough; pedicels very unequal, 1.5–15.0 mm. *Spikelets* 5–8 mm, oblong or wedge-shaped, 2- to 5-flowered, breaking up at maturity above the glumes and between the lemmas. *Glumes* persistent, very narrow, shorter than the lowest lemma, slightly unequal, finely pointed at apex, 1-veined, smooth, the lower 2–3 mm and narrowly lanceolate, the upper 3.0–4.5 mm and narrowly oblong. *Lemmas* 4–6 mm, lanceolate, finely pointed at apex, rounded on the back below, keeled in the upper part, firmly membranous, 3-veined, minutely rough, awnless. *Paleas* about as long as the lemmas, with 2 rough keels. *Anthers* 2.5–3.0 mm. *Caryopsis* hairy at the top, enclosed by the hardened lemma and palea. *Flowers* 5–7. 2*n* = 14, 42.

Native. Moist, wooded valleys, rocky slopes, wood margins and rocks by streamsides. Scattered throughout the British Isles, particularly in the north and west and absent from East Anglia and much of southern England. North and central Europe southwards to north Spain and south-central Russia; west Asia.

5. F. heterophylla Lam.   Various-leaved Fescue

Densely tufted *perennial herb* without rhizomes; at least some tillers extravaginal. *Culms* 60–100 (–120) cm, erect, or slightly bent at the base, moderately slender, 2- to 3-noded near the base, smooth; young shoots growing up within the old leaf-sheaths. *Leaves* green, those of the tillers up to 60 cm × 0.3–0.6 mm from midrib to edge, fine and thread-like, infolded, 3(–5)-veined and 3-angled, rough on the margins or smooth, rather weak, with 3 small abaxial sclerenchyma islets, the culm leaves conspicuously broader, up to 25 cm × 2–4 mm, flat, shortly hairy on the veins above; sheaths entire when young, fused more or less to apex, smooth, rounded on the back, without or slightly auricled; ligules 0.3–0.5 mm. *Inflorescence* a loose, open or contracted, nodding panicle 6–18 cm, green, one-sided, the axis and branches angular and antrorsely scabrid, the branches paired or solitary. *Spikelets* 7–14 mm, lanceolate to oblong, loosely 3- to 9-flowered, breaking up at maturity beneath the lemmas. *Glumes* persistent, slightly unequal, finely pointed, rough upwards on the keels, the lower 3.0–5.5 mm, narrowly lanceolate and 3-veined, the upper 4.0–6.5 mm, and 3-veined. *Lemmas* (4.7–)5.0–6.5(–8) mm, at first overlapping, finally loose, with the margins incurved, rounded on the back, lanceolate to narrowly oblong-lanceolate in side view, narrowed into a fine straight rough awn 1.5–4.5(–6.0) mm, firm except for the narrow, membranous margin, finely 5-veined, minutely rough in the upper part. *Paleas* as long as the lemmas, minutely rough upwards on the keels. *Anthers* 2.5–4.5 mm. *Caryopsis* minutely hairy at

the top, enclosed by the hardened lemma and palea. *Flowers* 6–7. 2*n* = 28, 42.

Introduced. Grown for ornament and appearing as a contaminent of grass seed. Naturalised in woods and wood-borders on light soils. Scattered records over Britain, but mainly the south. Thinly distributed through central and southern Europe and in south-west Asia.

6. F. arenaria Osbeck   Rush-leaved Fescue
F. juncifolia Chaub.; *F. rubra* subsp. *arenaria* (Osbeck) F. Aresch.

Perennial herb with slender, extensively creeping rhizomes with many fibrous roots and scattered shoots and culms; tillers extravaginal. *Culms* 20–90 cm, erect or spreading, rather stout, about 2-noded towards the base, ribbed near the panicle, smooth. *Leaves* up to 30 cm × 0.5–1.9 mm from midrib to edge or opening up to 5 mm, glaucous, very narrowly linear, with a hard, sharply pointed tip, tough, smooth beneath, prominently 5- to 9-veined with the veins densely and minutely hairy; usually with continuous or subcontinuous abaxial sclerenchyma; sheaths fused almost to apex, dark purplish-brown, rounded on the back, smooth and glabrous, the basal purplish, without auricles; ligules membranous, those of the basal leaves very short, those of the culm leaves 0.5–4.0 mm. *Inflorescence* an erect or inclined, lanceolate panicle 8–20 cm, contracted and rather dense before and after flowering, greyish-green or tinged with purple, the branches angular, tough, scabrid and minutely hairy, in pairs or solitary, the longer bare towards the base. *Spikelets* 10–18 mm, compressed, elliptical or oblong, 4- to 12-flowered, breaking up at maturity beneath each lemma; pedicels 2–4 mm, stout, scabrid. *Glumes* persistent, narrowly lanceolate, finely pointed, slightly unequal, firm, rough near the tip, the lower (4.5–)6–8 mm and 1- to 3-veined, the upper 3.5–10 mm, with a hairy tip and 3-veined. *Lemmas* 6–10 mm, overlapping, rounded on the back, narrowly lanceolate in side view, very finely pointed and scabrid at apex or narrowed into a rough awn up to 2.6 mm, usually softly and densely long-hairy, rarely glabrous (var. *glabrata* Lebel), firm, finely 5-veined. *Paleas* shorter than the lemma, with minutely rough keels. *Anthers* 4–5 mm. *Caryopsis* glabrous at tip. *Flowers* 6–7. 2*n* = 56.

Native. Mobile sand-dunes and sandy shingle by the sea. Local on the coasts of Britain north to Ross. Coasts of western Europe and the Baltic.

7. F. rubra L.   Red Fescue

Perennial herb forming dense tufts or loose or dense patches, without rhizomes or with slender, creeping rhizomes; at least some tillers extravaginal. *Culms* 20–100 cm, erect, spreading or slightly bent at the base, slender to fairly stout, 1- to 3-noded, smooth, young shoots growing up within or outside the old leaf-sheaths. *Leaves* 5–40 cm × 0.5–1.4(–2.5) mm from midrib to edge, flat or folded, green or greyish-green, narrowly long-linear, acute or obtuse at apex, bluntly keeled, stiff to rigid, smooth or rough near the tip; with abaxial sclerenchyma in 5–9 discrete islets, sometimes shortly hairy on the veins above, with 9–11 veins; sheath fused more or less to apex, without auricles, smooth; ligules very short. *Inflorescence* 3–12 cm, an erect or slightly nodding panicle, loose, open or contracted, purplish, reddish, yellowish or green, axis and branches angular and rough; pedicels 1–5 mm. *Spikelets* 5–14 mm, lanceolate to oblong, 3- to 12-flowered, breaking up at maturity beneath each lemma. *Glumes* persistent, narrowly lanceolate to oblong-lanceolate, pointed at apex, lower 2–4 mm and 1-veined, the upper 3–5 mm and 3-veined. *Lemmas* 4–6 mm, overlapping, finally incurved, narrowly lanceolate or oblong-lanceolate, finely pointed, rounded on the back, 5-veined, tipped with a fine rough awn 0.5–3.0 mm. *Palea* shorter than to as long as the lemma, with 2 round keels. *Anthers* 2–3 mm. *Caryopsis* glabrous at tip, tightly enclosed by the hardened lemma and palea. *Flowers* 6–8.

This very variable species is divided into seven subspecies by Stace (1991), based on the work of A.-K. K. A. Al-Bermani. This treatment is followed here.

(a) Subsp. **rubra**
Rhizomes well developed, plants usually forming loose patches. *Culms* to 75 cm. *Leaves* 0.6–1.3 mm from midrib to edge, folded, rarely flat. *Spikelets* 6.9–10.2 mm; pedicels 1–5 mm. *Upper glume* 2.6–5.3 mm. *Lemmas* 4.6–6.7 mm with awns 0.8–2.1 mm. $2n = 42$.

(b) Subsp. **juncea** (Hack.) K. Richt.
F. rubra subvar. *juncea* Hack.; *F. rubra* subsp. *pruinosa* (Hack.) Piper
Rhizomes short, plants forming dense tufts. *Culms* up to 75 cm. *Leaves* folded, 0.5–1.2 mm from midrib to the edge. *Spikelets* 6.8–9.9 mm. *Upper glume* 3.2–5.3 mm. *Lemmas* 4.4–6.5 mm, with awns 0.5–2.2 mm. $2n = 42$.

(c) Subsp. **litoralis** (G. Mey.) Auq.
F. rubra var. *litoralis* G. Mey.
Rhizomes rather short, plants often forming dense mats. *Culms* up to 55 cm. *Leaves* 0.6–1.0 mm from midrib to edge, folded. *Spikelets* 8.7–11.2 mm. *Upper glume* 4.3–6.2 mm. *Lemmas* 5.7–8.0 mm, with awns 1.1–2.8 mm. $2n = 42$.

(d) Subsp. **commutata** Gaudin
F. nigrescens Lam.; *F. rubra* subsp. *caespitosa* Hack.; *F. rubra* var. *fallax* auct.
Rhizomes absent or very few and short, plants forming dense tufts. *Culms* up to 75 cm. *Leaves* 0.6–1.1 mm from midrib to edge, folded; sheaths pink, generally smooth, but sometimes shortly hairy, shredding at maturity; ligule 0.1–0.2 mm *Spikelets* 7.2–9.1 mm; pedicels 1–2 mm. *Upper glume* 3.2–5.0 mm. *Lemmas* 4.4–6.3 mm, with awns 1.1–2.8 mm. $2n = 42$.

(e) Subsp. **arctica** (Hack.) Govoruchin
F. rubra forma *arctica* Hack.; *F. richardsonii* Hook.
Rhizomes well developed, plants sometimes forming loose patches, but more often single-culmed plants. *Culms* up to 50 cm. *Leaves* 0.5–1.2 mm from midrib to edge, folded; sheaths pink, often shortly hairy. *Panicle*

dense. *Spikelets* 6.6–8.8 mm, with dense, white hairs many curled or hooked at tips. *Upper glumes* 3.1–4.3(–5.5) mm. *Lemmas* 4.2–6.0 mm, with awns (0.0–)0.1–2.2 mm. $2n = 42$.

A proliferate form with culms 2–10 cm, leaves 0.5–1.2 mm wide and folded, panicles 1–3 cm, spikelets 6.0–7.5 mm, and lemmas almost absent occurs above 1000 m.

(f) Subsp. **scotica** S. Cunn. ex Al-Berm.
Rhizomes well developed, plants forming loose patches. *Culms* 30–70 cm. *Leaves* 0.8–1.4 mm from midrib to edge, folded. *Panicle* dense. *Spikelets* 8.8–11.8 mm. *Upper glume* 4.3–6.5 mm. *Lemmas* 6–8 mm, with awns 0.5–2.5 mm. $2n = 56, 70$.

(g) Subsp. **megastachya** Gaudin
F. diffusa Dumort.; *F. fallax* Thuill.; *F. heteromalla* Pourr.; *F. rubra* subsp. *fallax* (Thuill.) Nyman; *F. rubra* subsp. *multiflora* Piper; *F. rubra* var. *fallax* (Thuill.) Hack.
Rhizomes well developed, forming diffuse patches. *Culms* up to 100 cm. *Leaves* 1.0–1.4(–2.5) mm from midrib to the edge. *Spikelets* 7.0–11.2(–13.0) mm. *Upper glume* 3.5–6.5 mm. *Lemma* 4.7–7.9 mm, with hairy margins, with awns 0.5–3.0 mm. $2n = 56$.

The species as a whole is mostly native, but some populations are probably introduced. All kinds of grassy places, flushes and rocky areas. Throughout the British Isles. Almost throughout Europe and in North America. Subsp. *rubra* occurs throughout the range of the species. It is the most variable of the subspecies and populations often contain plants approaching the other subspecies. Subsp. *juncea* occurs on maritime cliffs and inland grassy, rocky places. All round the coast of the British Isles and in hilly areas in north Britain, rarely inland elsewhere. Shorter coastal plants with markedly pruinose leaves have been separated as subsp. *pruinosa*, but even on the coast non-pruinose populations occur. Subsp. *litoralis* is native in salt-marshes and sandy saline areas. Probably in suitable places all round the coasts of the British Isles. It also occurs on the coast of western Europe and the Baltic. Subsp. *commutata* is native in grassy places and rough ground, usually in well-drained soils. Probably throughout Britain due to extensive use as grass-seed (Chewing's Fescue). Subsp. *arctica* is native on wet mountain slopes and in gulleys, rock crevices and flushes down to sea level, often on serpentine. Scattered from central Scotland to Shetland and in Caernarvonshire. Also in Arctic and Subarctic Europe. Subsp. *scotica* is native in grassy and rocky places from sea level to over 800 m. It is confined to Scotland from Argyll and Outer Hebrides to Shetland. Subsp. *megastachys* is probably introduced in grassy places, especially waysides. It is scattered throughout Britain and Continental Europe.

8. F. ovina L.   Sheep's Fescue

Densely tufted *perennial herb*, but without rhizomes; vegetative shoots all intravaginal, usually shedding old leaves, but sometimes retaining them. *Culms* 5–60 cm, erect or spreading, very slender, stiff, 1- to 2-noded, angular, smooth to rough just below the panicle, young shoots growing up within the leaf-sheath. *Leaves* 3–19(–25) cm × 0.3–0.7 mm from midrib to edge, green or greyish-green, filiform or bristle-like, obtuse at apex, tightly infolded, with 5–7 veins and 2(–4) adaxial grooves, with abaxial sclerenchyma in thin broken or sometimes a continuous band, glabrous or hairy below, rough all over or only near the tip; sheaths (1.0–)1.3–4.0(–5.0) cm, not fused, rounded on the back, smooth, tipped with rounded auricles; ligules extremely short. *Inflorescence* an erect, lanceolate or narrowly oblong panicle 1.5–7.8(–8.1) cm, open in flower, later rather dense, somewhat one-sided, green or purplish; axis angular and rough; branches erect or slightly spreading; pedicels 0.8–3.6 mm. *Spikelets* (4.6–)5.3–7.2(–7.5) mm, with 2–9 flowers, elliptical to oblong, breaking up at maturity beneath each lemma. *Glumes* persistent, slightly unequal, pointed at apex, the lower 1.7–2.9(–3.4) × 0.5–0.9 mm, narrowly lanceolate and 1-veined, the upper (2.4–)2.7–4.2(–4.9) × 0.8–1.4 mm, more or less lanceolate and 3-veined. *Lemmas* (3.1–)3.3–4.9 × 1.3–2.0 mm, lanceolate or narrowly oblong-lanceolate, finely pointed at apex, rounded on back, at first overlapping, later loose, tipped with an awn (0–)0.2–1.6 mm. *Paleas* 3.2–4.7 × 0.5–0.9 mm, linear-lanceolate, with 2 keels, rough upwards. *Anthers* 1.6–2.5 (–3.0) mm, yellow or purple. *Caryopsis* 1.9–3.7 × 0.5–0.7 mm, oblong, enclosed by the hardened lemma and palea. *Flowers* 5–7.

(a) Subsp. ovina
Culms 10–35 cm. *Leaves* 0.3–0.7 mm from midrib to edge, with 5–7 veins. *Panicles* 2.2–7.3 cm; pedicels 0.8–2.2 mm. *Lemmas* 3.1–4.2 mm, with awns 0–1.2 mm. $2n = 14$.

(b) Subsp. hirtula (Hack. ex Travis) M. J. Wilk.
F. ovina subvar. *hirtula* Hack. ex Travis; *F. filiformis* subsp. *hirtula* (Hack. ex Travis) Kerguélen; *F. hirtula* (Hack. ex Travis) Kerguélen; *F. ophioliticola* subsp. *hirtula* (Hack. ex Travis) Auq.; *F. tenuifolia* var. *hirtula* (Hack. ex Travis) Howarth
Culms 5–45 cm. *Leaves* 0.3–0.6 mm from midrib to edge, usually scabrid above, generally hairy below; sheaths usually hairy; lemmas frequently hairy. *Panicles* 1.5–6.6 cm; pedicels 0.9–2.7 mm. *Lemmas* with awns 0.0–0.8 mm. $2n = 28$.

(c) Subsp. ophioliticola (Kerguélen) M. J. Wilk.
F. ophioliticola Kerguélen; *F. opioliticola* subsp. *calaminaria* Auq.; ?*F. ophioliticola* subsp. *guestfalica* (Boenn. ex Rchb.) K. Richt.; *F. guestfalica* Boenn. ex Rchb.
Culms 20–52(–60) cm. *Leaves* 0.3–0.7 mm from midrib to edge, with (5–)7 veins. *Panicles* 2.8–8.0 cm; pedicels 1.3–3.6 mm. *Spikelets* 5.5–7.5 mm. *Lemmas* 3.6–4.9 mm with awns 0–1.6 mm. $2n = 28$.

(i) Var. ophioliticola (Kerguélen) M. J. Wilk.
Leaves (0.4–)0.5–0.7 mm from midrib to edge. *Awns* (0–)0.5–1.0(–1.6) mm.

(ii) Var. hibernica (Markgr.-Dann.) M. J. Wilk.
F. ovina subsp. *ovina* var. *vulgaris* subvar. *hibernica* Markgr.-Dann.
Leaves 0.3–0.4 mm from midrib to edge. *Awns* 0–0.3(–0.4) mm.

Native. Grassy places. Common throughout the British Isles. North and central Europe. Subsp. *ovina* occurs in grassy places on well-drained, usually acid soils. It is common in north, central and south-west Britain, very sparse in south-central and south-east England and questionably in Ireland. It occurs also in north and central Europe. Subsp. *hirtula* is found in grassy places on well-drained, usually acid soils. It is common throughout the range of the species. Subsp. *ophioliticola* occurs in well-drained, often calcareous or serpentine soils. It is locally common throughout Britain and Ireland. It also occurs in Continental Europe. Var. *ophioliticola* occurs throughout the range of the sub-species. Var. *hibernica* occurs in scattered localities in northern Ireland, southern Britain and Wales.

9. F. vivipara (L.) Sm.   Viviparous Sheep's Fescue
F. ovina var. *vivipara* L.; *F. supina* subsp. *vivipara* (L.) K. Richt.; *F. vivipara* var. *hirsuta* Schol.; *F. vivipara* subsp. *hirsuta* (Schol.) Fred.

Densely to laxly tufted *perennial herbs* without rhizomes; vegetative shoots all intravaginal, usually retaining old leaves, but occasionally shedding them. Culms (8–)12–44(–50) cm, erect, smooth or scabridulous above, glabrous or weakly hairy, sometimes grooved, with 2–3 nodes, the uppermost node green and not visible or visible beyond the subtending sheath. *Leaves* 3–14(–20) cm × 0.3–0.6 mm from midrib to edge, green, filiform, obtuse at apex, flaccid or rigid, curved at tip, scabrid to smooth, usually glabrous but occasionally weakly hairy at base, 5- to 7(–9)-veined, sclerenchyma in a thin, broken or sometimes unbroken ring; sheaths (0.5–)1.5–3.6(–4.7) cm, not fused, usually smooth but rarely scabridulous distally, usually glabrous but occasionally hairy, the auricles short and minutely ciliate; ligules less than 0.5 mm, minutely ciliate. *Inflorescence* a panicle (1.5–)2.5–5.0(–8.0) cm, usually erect but occasionally shortly nodding, proliferous or partially sexual, with (5–)8–26(–42) bulbils and/or spikelets; branches green, not narrowed below bulbils/spikelets, scabrid to smooth, covered in hairs or prickles. *Bulbils* 0.8–6.0 cm; pedicels (1.0–)1.3–2.4(–3.3) mm. *Spikelets* when present green, (5.5–)5.8(–6.2) mm with 5–10 flowers (plus one sterile flower); pedicels (1.1–)1.4(–1.8) mm. *Glumes* unequal, the lower (2.2–)2.5(–3.0) × (0.7–)0.8(–1.0) mm, narrowly triangular to narrowly lanceolate, glabrous to weakly hairy, 1-veined and with a ciliate margin, the upper (3.2–)3.4(–3.8) × (1.1–)1.3(–1.4) mm, lanceolate, scabrous in the distal third, or hairy, 3-veined and ciliate on the margin. *Lemmas* (3.4–)3.9(–4.2) (excluding awn) × (1.4–)1.6(–1.8) mm, lanceolate, 5-veined, hairy or scabrid in the distal half; awns 0–0.1(–0.2) mm. *Paleas* (3.5–)3.9(–4.2) × 0.7–0.8 mm, linear-lanceolate. *Flowers* 5–7. $2n = 28$.

F. arenaria Osbeck   **F. heterophylla** Lam.

F. arenaria Osbeck

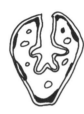

F. rubra L.
subsp **rubra**

F. rubra L.
subsp. **litoralis** (G. Mey.) Auq.

F. rubra L.
subsp. **arctica** (Hack.) Govoruchin

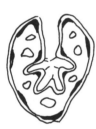

F. rubra L. subsp. **juncea** (Hack) K. Richt.   **F. rubra** L. subsp. **scotica** S. Cunn. ex Al-Berm.

F. rubra L. subsp. **megastachys** Gaudin   **F. rubra** L. subsp. **commutata** Gaudin

1 mm

Transverse sections of innovation leaves of **Festuca rubra** aggr. Sclerenchyma in black.
(After C. A. Stace, 1991.)

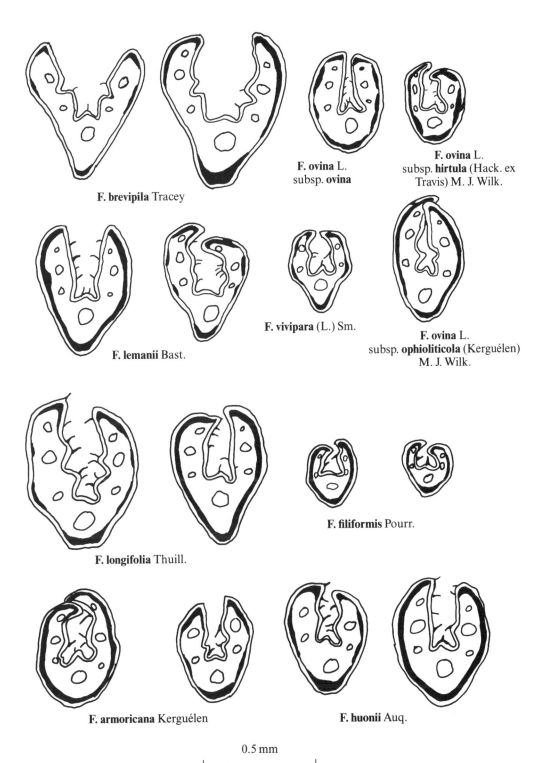

F. brevipila Tracey

F. ovina L.
subsp. **ovina**

F. ovina L.
subsp. **hirtula** (Hack. ex
Travis) M. J. Wilk.

F. lemanii Bast.

F. vivipara (L.) Sm.

F. ovina L.
subsp. **ophioliticola** (Kerguélen)
M. J. Wilk.

F. longifolia Thuill.

F. filiformis Pourr.

F. armoricana Kerguélen

F. huonii Auq.

0.5 mm

Transverse sections of innovation leaves of **Festuca ovina** agg. Sclerenchyma in black.
(After C. A. Stace, 1991.)

Spikelets of some individuals occasionally possess seminiferous flowers. These may be slightly or greatly deformed, or may be normal in appearance. They are borne singly or in pairs on otherwise proliferous spikelets or they may form wholly seminiferous spikelets. Various compositions occur of seminiferous and proliferous spikelets in an inflorescence and occasionally an entirely seminiferous inflorescence occurs. The species is very variable and some populations which are vegetative apomicts should perhaps be recognised. Proliferating triploids (2*n* = 21) are said to be hybrids between *F. ovina*, *F. filiformis* and *F. vivipara* in various combinations. In the British Isles the leaves of *F. vivipara* tend to be more like *F. filiformis*, while elsewhere they are more like *F. ovina*.

Native. Grassy and rocky places in hilly districts decending to almost sea level in parts of western Scotland. Common in central and north Scotland, local in south Scotland, north England, Wales and Ireland. Arctic and north-east Europe, south to south-west Ireland.

10. F. filiformis Pourr.   Fine-leaved Sheep's Fescue
F. tenuifolia Sibth.; *F. ovina* subsp. *tenuifolia* (Sibth.) Dumort.; *F. capillata* auct.; *F. paludosa* Gaudin; *Poa capillata* auct.; *F. mutica* Wulf.

Densely tufted *perennial herb* without rhizomes; vegetative shoots all intravaginal, retaining or shedding old leaves. *Culms* (8.5–)14–35(–55) cm, erect, scabrid or smooth just below the panicle, glabrous, weakly grooved or angular above, with 2–3 nodes, uppermost node visible or not beyond the subtending sheath. *Leaves* (1.6–)3.7–10.0(–25.0) cm × 0.3–0.6 mm from midrib to edge, green, filiform, obtuse or finely pointed and curved at the apex, usually scabrid at least in the upper half, but occasionally smooth, glabrous or very rarely very weakly hairy at base, with (4–)5(–7) veins and 2–3 adaxial grooves, with abaxial sclerenchyma in thin broken or sometimes continuous bands; sheath (1.1–)1.3–2.6(–4.8) cm, not fused, smooth or rarely scabridulous, distally glabrous or very rarely weakly hairy, the auricles short and minutely ciliate; ligules less than 0.3 mm, minutely ciliate, membranous. *Inflorescence* an erect panicle (2.1–)2.3–6.0(–7.2) cm, with (8–)10–19(–23) spikelets; branches rarely subpruinose, not narrowed below spikelets, scabrid or scabridulous, covered in prickles. *Spikelets* (4.5–)4.7–5.2(–5.5) mm, with 2–8 flowers (plus one sterile flower); pedicels (0.6–)0.7–1.4(–2.3) mm. *Glumes* unequal, the lower (1.4–)1.5–2.0(–2.2) × (0.3–)0.4–0.6(–0.8) mm, subulate to lanceolate, usually glabrous or occasionally scabrid in the distal third, 1-veined, and with a ciliate or serrate margin, the upper (2.0–)2.2–2.9(–3.2) × (0.6–)0.7–1.0 mm, oblong-lanceolate to narrowly lanceolate, glabrous or scabrid at tip, 3-veined and with ciliate or serrate margin. *Lemmas* (2.5–)2.7–3.2(–3.5) mm (excluding awn) × 1.1–1.3 mm, lanceolate to oblong-lanceolate, 3-veined, glabrous to scabrid at tip; awns 0–0.3(–0.6) mm. *Palea* (2.7–)2.8–3.3(–3.5) × 0.5–0.7 mm, minutely rough upwards on the 2 keels. *Anthers* 1.5–1.9 mm, yellow or

purple. *Caryopsis* 1.6–2.2 × 0.5–0.7 mm, oblong, tightly enclosed by the hardened lemma and palea. *Flowers 5–6*. 2*n* = 14.

Native. Heaths, moorland, parkland, open woodland and other grassy places, usually on acid, sandy soil. Frequent throughout the British Isles. West and central Europe.

11. F. armoricana Kerguélen   Breton Fescue
F. ophioliticola subsp. *armoricana* (Kerguélen) Auq.

Densely tufted *perennial herb* without rhizomes; vegetative shoots all intravaginal, retaining old leaves. *Culms* 9–40 cm, erect, 2- to 3-noded, smooth, more or less grooved, glabrous to shortly hairy above. *Leaves* 3.1–10.2 cm × 0.5–0.8 mm from midrib to edge, green or rarely glaucous, filiform, wavy, acute at apex, scabrid above, with abaxial sclerenchyma in thin broken or sometimes continuous bands, with 7 veins and 2–4 adaxial grooves; sheaths 1.3–2.3 cm, not fused, but usually closed in the lower one-third to half, glabrous or sparsely hairy, auricles short, minutely ciliate; ligules less than 0.5 mm, minutely ciliate. *Inflorescence* an erect panicle 3.3–4.0(–5.9) cm, rather dense, with (8–)11–15(–27) spikelets; branches not pruinose, not narrowed below spikelets, usually scabridulous or smooth, usually covered with hairs, sometimes with prickles. *Spikelets* 6.0–6.6(–8.0) mm, with 3–6 flowers (plus one sterile flower); pedicels 0.8–1.5 mm. *Glumes* unequal, the lower 2.2–2.7 × 0.6–0.8 mm, subulate to narrowly triangular, 1-veined, glabrous or scabrid in the upper third and with a ciliate margin, the upper 3.1–3.7(–4.1) × 0.9–1.3 mm, narrowly to broadly lanceolate, 3-veined, scabrid to weakly hairy at tip and ciliate on the margin. *Lemmas* 3.9–4.4(–5.1) (excluding awn) × 1.2–2.1 mm, lanceolate, 5-veined, scabrid or hairy in upper half; awns (0.8–)1.2–1.6 mm. *Palea* 3.9–4.3(–5.1) × 0.7–1.8 mm, linear-lanceolate. *Anthers* 1.8–2.4 mm, yellow or purple. *Caryopsis* 2.3–3.2 × 0.6–0.7 mm, oblong.

Native. Covering large fixed dunes. St Ouen's and St Brelade's Bays, Jersey in the Channel Islands. Also in Brittany in north-western France.

12. F. huonii Auq.   Huon's Fescue

Densely tufted *perennial herb* without rhizomes; vegetative shoots all intravaginal, old leaves retained. *Culms* 7–18(–25) cm, green, prostrate or erect, 2- to 3-noded, usually smooth but occasionally scabridulous above, usually glabrous but sometimes weakly hairy, more or less grooved above. *Leaves* (3–)5–10(–12) cm × 0.4–0.7 mm from midrib to edge, green or slightly bluish, filiform, curved at tip, smooth, glabrous or rarely weakly hairy at base, with abaxial sclerenchyma in a thin broken or sometimes continuous band; sheaths (0.9–)1.4–2.9(–3.2) cm, not fused, smooth, glabrous or rarely weakly hairy with short, minutely ciliate auricles; ligules less than 0.5 mm, minutely ciliate. *Inflorescence* a dense, erect panicle 2.3–4.3(–5.4) cm, with (8–)9–19(–26) spikelets; branches green or slightly pruinose, not narrowed below spikelets, usually smooth or scabridulous, occasionally scabrid, usually covered in hairs. *Spikelets* (5.5–)6.0–7.0(–7.5) mm, with (2–)3–5(–6) flowers (plus one sterile flower); pedicels (0.7–)0.8–1.1 (–1.8) mm. *Glumes* unequal, the lower (2.2–)2.3–3.0(–3.1) × (0.4–)0.6–0.8(–1.0) mm, subulate or narrowly triangular, glabrous or scabrid at tip, 1-veined, occasionally weakly hairy and with a ciliate margin, the upper (3.3–)3.5–3.9(–4.3) × 1.1–1.4 mm, lanceolate to oblong-lanceolate, scabrid at tip, hairy or glabrous, 3-veined and ciliate on the margins. *Lemmas* (3.5–)3.7–4.3(–4.9) mm (excluding awn) × 1.6–1.9 mm, 5-veined, scabrid above or hairy; awns (0.9–)1.2–1.5(–1.8) mm. *Paleas* (3.6–)4.1–4.4(–4.9) × 0.6–0.9 mm, linear-lanceolate. *Anthers* (1.5–)1.9–2.4 mm, usually yellow, occasionally purple. *Flowers* 6–8. 2n = 42.

Native. Characteristic of acid, maritime cliff-tops. Guernsey, Jersey and smaller islands in the Channel Islands. Southern and western coasts of the Brittany Peninsula, France.

13. **F. lemanii** Touss. Bast.   Confused Fescue
F. bastardii Kerguélen & Plonka; *F. duriuscula* auct.; *F. glauca* auct.; *F. hirsuta* auct.; *F. longifolia* auct.

Perennial herb fairly densely to loosely tufted and without rhizomes; vegetative shoots all intravaginal, retaining or shedding old leaves. *Culms* (19–)28–66 cm, generally smooth or scabridulous just below the panicle, glabrous or hairy, more or less grooved, 2- to 3-noded. *Leaves* (4–)6–11(–17) cm × 0.4–0.9 mm from midrib to edge, green or rarely slightly pruinose, narrowly linear, obtuse at apex and curved, usually scabrid at least in the distal part, glabrous or hairy near the base, with 5–7 veins and 2–4 adaxial grooves, with abaxial sclerenchyma in a continuous or sometimes broken band; sheaths (1.3–)2.1–4.4(–4.7) cm, not fused, smooth or scabridulous above, glabrous or hairy, the auricles short and usually minutely ciliate; ligules short, usually minutely ciliate, rarely glabrous. *Inflorescence* a panicle, 3.0–7.8(–8.6) cm, usually erect, but sometimes slightly nodding; branches sometimes more or less pruinose, not narrowed below the spikelets, smooth to scabrid, covered in hairs or prickles, rarely nearly glabrous. *Spikelets* (6.1–)6.5–7.5(–8.5) cm, sometimes subpruinose, with (2–)3–6(–8) fertile flowers (plus one sterile flower); pedicels (0.6–)1.4–2.4(–2.5) mm. *Glumes* unequal, the lower (2.1–)2.4–3.2(–3.5) mm × 0.5–0.8 mm, subulate to narrowly lanceolate, 1-veined, usually glabrous, but sometimes with a scabrid tip or rarely laxly hairy, the upper 3.4–4.6(–4.8) × (0.8–)1.0–1.2(1.3) mm, narrowly lanceolate to lanceolate, 3-veined, usually glabrous or with a scabrid distal third, rarely laxly hairy, sometimes with a ciliate margin. *Lemmas* 4.0–4.9(–5.4) × 1.5–1.8(–2) mm, green, but often purple at apex, lanceolate with an awn (0.3–)0.6–1.6(–1.9) mm, 5-veined, scabrid in the distal half or hairy. *Paleas* (3.9–)4.1–4.6(–5.2) × 0.6–0.9 mm. *Anthers* 1.8–2.5 mm, yellow or purple. *Caryopsis* 2.3–3.1 × 0.6–1.0 mm, oblong, enclosed between the hardened lemma and palea. *Flowers* 6–8. 2n = 21, 42, sometimes with 2B chromosomes.

Probably native. Grassy places on well-drained, acid or calcareous soils, often with *F. ovina*. Very scattered in Britain, but probably much under-recorded. Western Europe, in clefts of limestone banks in partial shade.

14. **F. longifolia** Thuill.   Blue Fescue
F. caesia Sm.; *F. involuta* Moench ex Roem. & Schult.; *F. ovina* var. *caesia* (Sm.) Sm.; *F. glauca* auct.; *F. duriuscula* var. *longifolia* (Thuill.) Holandr.; *F. duriuscula* var. *caesia* (Sm.) Husnot; *F. glauca* var. *caesia* (Sm.) Howarth

Densely tufted, usually glabrous *perennial herb* without rhizomes; vegetative shoots all intravaginal, retaining old leaves. *Culms* 7–30(–45) cm, erect, usually glabrous, but sometimes with a few prickles under the panicle, grooved at least in the upper half, 2 (–3)-noded near the base, uppermost node pruinose, usually not visible beyond the subtending sheath. *Leaves* 2–12 cm × 0.5–0.9 mm from midrib to edge, glaucous, filiform, curved at tip, with 7–9 veins and 4(–6) adaxial grooves, with abaxial sclerenchyma in a thin broken or sometimes continuous band, smooth and glabrous; sheaths 0.9–4.2(–5.1) cm, fused for up to one-third of their length, smooth and glabrous, the auricles short, glabrous or occasionally minutely ciliate; ligules less than 0.5 mm, glabrous or ciliate. *Inflorescence* an erect, more or less dense panicle 2.5–8.0 cm, with 11–27 spikelets; branches glaucous, sometimes narrowing below the spikelets, usually smooth but occasionally scabridulous. *Spikelets* 5.4–7.0 mm, with (2–)3–6 flowers (plus one sterile flower); pedicels 0.5–1.8 mm. *Glumes* unequal, the lower 1.7–2.5 × 0.4–0.7 mm, narrowly triangular, usually glabrous but occasionally scabridulous towards the apex, 1-veined and with a glabrous or ciliate margin, the upper (2.6–)2.8–3.5(–3.8) × 0.4–1.2(–1.5) mm, narrowly lanceolate to oblong-lanceolate, 3-veined, glabrous and with a glabrous or ciliate margin. *Lemmas* (3.5–)3.6–4.4(–5.5) (excluding awn) × 1.4–1.9 mm, 5-veined, glabrous or scabrid at the tip; mucronate or with a fine awn 0.3–2.5 mm. *Paleas* (3.5–)3.7–4.5(–4.8) × 0.6–0.8 mm, linear-lanceolate. *Anthers* 1.8–2.6 mm, yellow or purple. *Caryopsis* tightly enclosed between the hardened lemma and palea. *Flowers* 5–6. 2n = 14.

Native. Dry acid heaths and maritime cliff-tops. On roadside sandbanks and on a Breckland heath in East Anglia, two localities in Lincolnshire, and Sark and Guernsey in the Channel Islands. It is extinct in Jersey. Elsewhere it occurs on the maritime cliffs of Normandy and the acid heaths of central and north-western France.

15. **F. glauca** Vill.   Garden Fescue
F. glauca Lam.; *Schedonorus glaucus* (Vill.) P. Beauv.; *F. duriuscula* var. *glauca* (Vill.) Bréb.; *F.*; *ovina* var. *glauca* (Vill.) Koch; *F. duriuscula* subsp. *glauca* (Vill.) Mutel; *F. ovina* subsp. *glauca* (Vill.) Hack. ex Hegi; *F. ovina* subsp. *indigesta* var. *durissima* forma *pruinosa* St-Yves; *F. ovina* subsp. *indigesta* forma *villiflora* Litard; *F. cinerea* var. *glauca* (Vill.) Stohr

Densely tufted *perennial herb* without rhizomes; vegetative shoots all intravaginal, generally retaining old leaves. *Culms* (15–)22–35 cm, erect, smooth, glabrous, grooved, with 2–3 nodes, the uppermost node pruinose and sometimes visible beyond the subtending sheath. *Leaves* 5–14 cm × 0.5–0.9 mm from midrib to edge, glaucous-pruinose, filiform, curved at tip, smooth, glabrous or hairy at base, with abaxial sclerenchyma usually forming a thin broken, rarely unbroken ring; sheaths 2.9–4.5 cm, not fused, smooth and hairy, the auricles short, rounded and ciliate; ligules less than 0.4 mm, minutely ciliate. *Inflorescence* an erect, dense panicle 2.7–5.0 cm, with 17–33 spikelets; branches more or less pruinose, narrowed below the spikelets, smooth or scabridulous, covered in hairs. *Spikelets* (5.7–)6.1–6.5(–6.7) mm, with 2–5 flowers (plus one sterile flower); pedicels 0.3–0.5(–0.8) mm. *Glumes* unequal, the lower (1.8–)2.1–2.5(–2.9) × 0.5–0.6 mm, narrowly triangular, 1-veined, glabrous on the surface and with ciliate margins, the upper 2.8–3.7 × 1.0–1.4 mm, oblong-lanceolate, 3-veined, glabrous on the surface and with ciliate margins. *Lemmas* 3.5–4.6 (excluding awn) × 1.6–1.9 mm, lanceolate, 5-veined, glabrous; awns (0.6–)1.0–1.5(–1.7) mm. *Palea* 3.7–4.0 × 0.7–0.9 mm, linear-lanceolate. *Anthers* yellow or purple. *Caryopsis* tightly enclosed between the hardened lemma and palea. *Flowers* 5–6. $2n = 42$.

Introduced. Widely grown in parks and gardens throughout the British Isles for ornament and occasionally occurring as a throw-out, but it is not known to be naturalised. Native of Roussillon, southern France.

16. F. brevipila Tracey   Hard Fescue
F. ovina subsp. *euovina* var. *duriuscula* subvar. *trachyphylla* Hack.; *F. ovina* var. *glaucescens* Hack.; *F. duriuscula* subsp. *trachyphylla* (Hack.) K. Richt.; *F. ovina* var. *trachyphylla* (Hack.) Druce; *F. longifolia* var. *trachyphylla* (Hack.) Howarth; *F. trachyphylla* (Hack.) Kraj., non Hack. ex Druce; *F. ovina* subsp. *ovina* var. *duriuscula* subvar. *trachyphylla* (Hack.) Maire; *F. longifolia* auct.; *F. cinerea* var. *trachyphylla* (Hack.) Stohr; *F. stricta* subsp. *trachyphylla* (Hack.) Patzke nom. inval.; *F. glauca* auct.; *F. guestfalica* auct.

Laxly or densely tufted *perennial herb* without rhizomes; vegetative shoots all intravaginal, retaining or shedding old leaves. *Culms* (9–)30–47(72) cm, erect, smooth to scabrid above, grooved, with 2(–3) nodes, uppermost node not pruinose, sometimes visible beyond the subtending sheath. *Leaves* (3.5–)5.5–15.0(–19) cm × 0.6–1.0 mm from midrib to edge, green but slightly pruinose, filiform, curved at apex, scabrid at least in the upper one-third, usually hairy at base, occasionally glabrous with (5–)7–9 veins and 4–6 adaxial grooves, with abaxial sclerenchyma usually as 3 islets with their tails at the medial or the edge, often with smaller islets opposite other veins, rarely subcontinuous; sheaths (1.3–)1.7–4.2(–4.9) cm, not fused, smooth or scabridulous in the upper one-third, usually laxly hairy but sometimes glabrous, the auricles short and minutely ciliate; ligules less than 0.5 mm and minutely ciliate.

Inflorescence an erect to nodding panicle (3.5–)4.2–8.8(–9.5) cm, with (10–)14–28(–39) spikelets; branches sometimes slightly pruinose, not narrowing below the spikelets, scabrid to smooth, covered in hairs or prickles. *Spikelets* (5.0–)6.4–8.0(–8.5) mm, with (2–)3–6(–7) flowers (plus one sterile flower), sometimes slightly pruinose; pedicels (1.2–)1.5–2.6(–2.8) mm. *Glumes* unequal, the lower glume (2.0–)2.5–3.1(–4.3) × (0.4–)0.6–0.8(–0.9) mm, subulate to narrowly lanceolate, usually glabrous or scabrid at tip, rarely weakly hairy, 1-veined and with ciliate or serrate margins, the upper (3.3–)3.6–4.5(–4.7) × (1–)1.2–1.4(–1.6) mm, narrowly lanceolate to lanceolate, usually scabrid at tip, occasionally hairy or glabrous, 3-veined and with a ciliate margin. *Lemmas* (3.9–)4.2–5.1(–5.5) (excluding awn) × (1.5–)1.6–2.1(–2.5) mm, lanceolate, 5-veined, usually hairy or scabrid in upper part, occasionally glabrous; awn (0.2–)1.4 × 2.5 (–3.2) mm. *Palea* (3.8–)4.1–5.0(–5.3) × (0.7–)0.8–1.0 mm, linear-lanceolate. *Anthers* (1.9–)2.1–2.9 mm, yellow or purple. *Caryopsis* tightly enclosed between the hardened lemma and palea. *Flower* 5–6. $2n = 42$.

Introduced. Formerly included in grass-seed mixtures called *Festuca duriuscula* and naturalised on roadsides, commons and rough ground especially on acid, well-drained soils. Frequent in south-east England and East Anglia, scattered north to Yorkshire; Jersey. Probably much under-recorded. Native of central Europe and introduced in France, Scandinavia and Canada.

14 × 15. × Festulolium Asch. & Graeb.
Festuca × Lolium

The *inflorescence* is variously intermediate between the two parent genera. At one extreme it is a simple raceme, but rarely a spike as in *Lolium*. At the other extreme it is a panicle, often a raceme with a few racemose branches near the base. *Glumes* usually 2, the lower much shorter than the upper. *Anthers* more or less indehiscent with more or less empty pollen grains. Some degree of fertility exists and some back-crossing may occur.

Beddows, A. R. (1964). *Lolium multiflorum* Lam. × *Festuca arundinacea* Schreb. Natural and artificial hybrids. *Jour. Linn. Soc. London Bot.* **59**: 89–98.
Gymer, P. T. & Whittington, W. J. (1973). Hybrids between *Lolium perenne* L. and *Festuca pratensis* Huds. 1. Crossing and incompatibility. *New Phytol.* **72**: 411–424.
Gymer, P. T. & Whittington, W. J. (1973). Hybrids between *Lolium perenne* L. and *Festuca pratensis* Huds. 2. Comparative morphology. *New Phytol.* **72**: 861–865.
Jenkin, T. J. (1955). Interspecific and intergeneric hybrids in herbage grasses. 17. Further crosses involving *Lolium perenne*. *Jour. Genet.* **53**: 442–466.

14/2 × 15/2. F. arundinacea × L. multiflorum

This hybrid differs from all other × *Festulolium* hybrids in having both ciliate leaf auricles and awned lemmas. It resembles *F. arundinacea* in general habit, but has a more compact appearance due to reduced branching and shorter internodes. It has non-dehiscent anthers. $2n = 28$.

Native and probably introduced with grass-seed. Grassy places, rough ground and waysides. Very scattered in southern England north to Warwickshire. It has also been recorded in Austria and Denmark.

14/2 × 15/1. × F. holmbergii (Dörfl.) P. Fourn.
F. arundinacea × L. perenne
Festuca × holmbergii Dörfl.

This hybrid resembles *F. arundinacea* in its gross vegetative and reproductive morphology, but the panicle may be less branched and the spikelets almost sessile giving it a more compact appearance. The hybrids have ciliate auricles like *F. arundinacea*. They are male sterile.

Native. In pastures, meadows, riversides and roadsides often on damp, rich soils. Very scattered in southern England north to Warwickshire. The parents are often sympatric. It has also been recorded in Sweden.

14/3 × 15/2. × F. nilssonii Cugnac & A. Camus
F. gigantea × L. multiflorum

This hybrid almost certainly occurs in the British Isles, but it is morphologically indistinguishable from × *F. brinkmannii*. Even if growing wih the parents in the absence of *L. perenne*, the origin would not be certain owing to the presence of an awn in *F. gigantea*. It is recorded from Sweden.

14/3 × 15/1. × F. brinkmannii (A. Br.) Asch. & Graeb.
F. gigantea × L. perenne
× *Festuca brinkmannii* A. Br.

This hybrid has *Lolium*-like racemes, weakly awned lemmas and non-dehiscent anthers. It sometimes resembles *F. gigantea* in habit, the stems stout with deep mauve nodes and the leaves broad and thin and the inflorescences large, lax and branched, but it can have weak stems and narrow leaves. The spikelets are often 12–15 mm, the glumes 5.3–7.0 mm, and the lemmas 5.5–6.5 mm with awns 5.2–7.5 mm.

Native. Grassy places with its parents, usually in partial shade of a hedge or scrub. Cambridgeshire, Pembrokeshire and Merionethshire. Reported elsewhere from Denmark, France, Germany and Sweden. The habitat preferences of the parents are usually quite distinct.

14/1 × 15/2. × F. braunii (K. Richt.) A. Camus
F. pratensis × L. multiflorum
Festuca × braunii K. Richt.

This hybrid is variable in morphology between the parents. Usually the inflorescence resembles *L. multiflorum* in structure, but the spikelets tend to be more widely spaced on the rhachis and some branching may occur. The spikelets are subtended by 2 glumes, the lower much reduced and occasionally absent. It differs from × *F. loliaceum* by its short, awned lemmas ranging from 0 to 6 mm. The hybrids are functionally male sterile. 2n = 14.

Native and probably introduced in grass-seed. Grassy places, rough ground and waysides. Scattered in England and in Denbighshire. Also recorded in central Europe, Portugal and Sweden.

14/1 × 15/1. × F. loliaceum (Huds.) P. Fourn.
Hybrid Fescue
F. pratensis × L. perenne
Festuca × loliacea Huds.

This hybrid is variable in its degree of resemblance to either parent. In general it resembles *L. perenne* in having the spikelets arranged alternately in two rows on opposite sides of the axis which is more frequently unbranched than branched. The spikelets may be sessile or have short pedicels, and resemble those of *F. pratensis*, but are more compressed and have very unequal glumes, the lower one much reduced or occasionally absent. The anthers are non-dehiscent. The leaf-auricles are glabrous and the lemmas awnless. 2n = 14, 21.

Native. Pastures, meadows, low places in marshes by riversides and roadsides, often on damp, rich soils. Throughout most of the British Isles, but commonest in southern England. It is widespread in Continental Europe.

14/7 × 15/1. × F. fredericii Cugnac & A. Camus
F. rubra × L. perenne

This hybrid agrees with *F. rubra* in all respects except that its rhizomes are less well developed. The parents grow together in old pasture and it is probable that it is under-recorded because of the difficulty of recognition. 2n = 28.

Native. A single record from a meadow in Borrowdale, Cumberland in 1956. Elsewhere there is a single record from Sweden.

14 × 16. × Festulpia Melderis ex Stace & R. Cotton
Festuca × Vulpia

These hybrids are perennials and vegetatively close to the *Festuca* parents, but have fewer and shorter rhizomes and some overlapping sheaths, the panicles are narrower and less branched and they have markedly longer awns. The lower glume is about half as long as the upper and the anthers are indehiscent.

Benoit, P. M. (1958). A new hybrid grass. *Proc. B.S.B.I.* **3**: 85–86.
Benoit, P. M. (1960). *Festuca rubra × Vulpia membranacea* at Harlech and Newborough. *Nature Wales* **6**: 59–60.
Melderis, A. (1955). A hybrid between *Festuca rubra* and *Vulpia membranacea*. *Proc. B.S.B.I.* **1**: 390–391.
Melderis, A. (1957). *Festuca rubra* var. *arenaria × Vulpia membranacea*. *Proc. B.S.B.I.* **2**: 243.
Melderis, A. (1965). *Festuca rubra × Vulpia bromoides*, a new hybrid in Britain. *Proc. B.S.B.I.* **6**: 172–173.
Stace, C. A. & Cotton, R. (1974). Hybrids between *Festuca rubra* L. sensu lato and *Vulpia membranacea* (L.) Dum. *Watsonia* **10**: 119–138.
Trist, P. J. O. (1971). *Festuca rubra* L. × *Vulpia bromoides* (L.) Gray. *Watsonia* **8**: 311.

14/6 × 16/1. × F. melderisii Stace & R. Cotton
F. arenaria × V. fasciculata

This sterile hybrid has the lower glume 5.2–8.0 mm, the upper glume 8.0–11.5 (including awn), the lemmas 9.5–10.5 mm with the awn 3.5–5.0 mm and the anthers 3, 1.5–2.0 mm. 2n = 42.

Native. On open sand-dunes with the parents. Very local in southern England and south Wales. ?Endemic.

14/7 × 16/2. F. rubra × V. bromoides

This hybrid is a rhizomatous perennial and has the lower glume 2.0–3.4 mm, the upper glume 3.4–5.9, the lemmas 4.3–7.0 mm, the awn 3.2–6.0 mm and the 3 anthers 0.8–1.7 mm. $2n = 28$. It differs from *F. × hubbardii* in not having markedly thickened pedicels at the apex, one, not 2–several ovary-less flowers at the apex of the spikelet and in the awnless upper glumes. It is likely both *F. rubra* subsp. *rubra* and subsp. *commutata* have produced this hybrid. It is highly male sterile.

Native. With the parents on waste and rough ground and coastal shingle and sand. Scattered in Britain north to Yorkshire. ?Endemic.

14/7 × 16/1. × F. hubbardii Stace & R. Cotton
F. rubra × V. fasciculata

This hybrid is a tufted or shortly rhizomatous perennial and has the lower glume 2.4–4.4 mm, the upper glume including the awn 3.5–7.2 mm, the lemmas 6.0–9.5 mm with an awn 2.0–5.5 mm and the 3 anthers 1.5–2.0 mm. $2n = 35$.

Native. Occurs with the parents on open sand-dunes. Probably frequent in the Channel Islands and in Britain north to Lancashire. There is an unconfirmed record for Spain.

14/7 × 16/4. F. rubra × V. myuros

This hybrid is a densely tufted perennial without creeping rhizomes and has the lower glume 1.5–3.3 mm, the upper glume 3.2–5.0 mm, the lemmas 4.5–6.2 mm with an awn 3–6 mm and the 3 anthers 0.6–1.5 mm. $2n = 42$. *Festuca rubra* subsp. *rubra*, subsp. *commutata* and subsp. *litoralis* have probably all produced this hybrid.

Native. On waste and rough ground with the parents. Scattered records in Britain north to Lancashire.

15. Lolium L.

Annual to *perennial herbs* without rhizomes or stolons. *Leaves* with sheaths not fused; ligules obtuse at apex. *Inflorescence* usually a simple spike, very rarely branched, with laterally compressed solitary spikelets lying edgeways to the concavities of the rhachis. *Spikelets* with 2 to many flowers, all except the most apical bisexual. *Glumes* 2 in the terminal spikelet, one in the lateral spikelets, membranous, 3- to 7-veined. *Lemma* membranous or chartaceous, rounded on the back, sometimes indurate, sometimes turgid in fruit, obtuse to subacute or sometimes minutely bifid at apex, with or without a subterminal awn. *Palea* like the lemma, narrowly keeled, usually ciliate. *Stamens* 3.

Abnormal plants with branched inflorescences occur. All the species are diploid and produce fertile hybrids. All species may have awned or awnless lemmas. *L. multiflorum × temulentum*, *L. multiflorum × rigidum* and *L. rigidum × temulentum* have been recorded as rare casuals in grain, wool and distillery refuse.

Eight species in temperate Eurasia; introduced elsewhere.

Beddows, A. R. (1967). *Lolium perenne* L. in Biological flora of the British Isles. *Jour. Ecol.* **55**: 567–587.

Beddows, A. R. (1973). *Lolium multiflorum* Lam. in Biological flora of the British Isles. *Jour. Ecol.* **61**: 587–600.

Grime, J. P. et al. (1988). *Lolium perenne* ssp. *perenne* in *Comparative plant ecology*.

Jenkins, T. J. (1935). Interspecific and intergeneric hybrids in herbage grasses. II. *Lolium perenne × L. temulentum. Jour. Genetics* **31**: 379–411.

Jenkins, T. J. (1954). Interspecific and intergeneric hybrids in herbage grasses. IV. *Lolium rigidum* et alia. *Jour. Genetics* **52**: 239–251.

Jenkins, T. J. (1954). Interspecific and intergeneric hybrids in herbage grasses. VII. *Lolium perenne* with other *Lolium* species. *Jour. Genetics* **52**: 300–317.

Terrell, E. E. (1966). Taxonomic implications of genetics in Ryegrasses (*Lolium*). *Bot. Rev.* **32**: 138–164.

1. Lemmas elliptical to ovate, less than 3 times as long as wide, becoming thick and hard in fruit; seed less than 3 times as long as wide   **2.**

1. Lemmas narrowly oblong-ovate, more than 3 times as long as wide, not becoming thick and hard; seed more than 3 times as long as wide   **3.**

2. Lowest 2 lemmas (4.6–)5.0–8.5 mm; glumes (7–)10–30 mm   **4. temulentum**

2. Lowest 2 lemmas 3.5–5.0(–5.5) mm; glumes 5–12(–15) mm   **5. remotum**

3. Perennial, with tillers at flowering and fruiting times; leaves folded along their midribs when young; lemma usually awnless   **1. perenne**

3. Annual or biennial, without tillers at flowering or fruiting time; leaves rolled along their long axis when young; lemmas awned or awnless   **4.**

4. Spikelets usually with more than 11 flowers; lemmas nearly always awned   **2. multiflorum**

4. Spikelets usually with less than 11 flowers; lemmas usually awnless   **3. rigidum**

1. L. perenne L.   Perennial Rye Grass

Loosely to densely tufted *perennial herb*, with tillers at flowering and fruiting time. *Culms* 10–90 cm, erect or spreading, slender, smooth, with 2–4 nodes. *Leaves* 3–20 cm × 2–6 mm, green, glabrous, linear, pointed or obtuse at apex, folded when young, with small, narrow projections at the base, smooth and glossy beneath, smooth or rather rough above; sheaths smooth, the basal usually pinkish when young; ligules up to 2 mm, membranous. *Inflorescence* a spike 2–30 cm, straight or slightly curved, stiff, slender to somewhat stout, flattened, green or purplish, sometimes branched; rhachis smooth. *Spikelets* 7–20 mm, sessile, alternating on opposite sides of the rhachis, spaced, or less than their own length apart, their edges fitting into hollows in the rhachis, oblong to elliptical, 4- to 14-flowered, breaking up at maturity beneath the lemma. *Upper glume* persistent, usually shorter than the spikelet, narrowly lanceolate to oblong-lanceolate, obtuse at apex, rounded on the back, 5- to 7-veined, smooth; lower glume present only in the terminal spikelet and similar to the upper. *Lemmas* 5.0–7.5 mm, overlapping, oblong or ovate-oblong, obtuse or slightly pointed at apex, rounded on the back, awnless, firm below, thin at the tips, 5-veined, smooth. *Paleas* as long as the lemmas, the 2 keels minutely rough. *Anthers* 3–4 mm. *Caryopsis* tightly enclosed by the hardened lemmas and palea. *Flowers* 5–8. $2n = 14$.

Some striking forms of this species have been given names. Forma *sphaerostachyum* Masters has the spikelets short and 5–15 mm apart, and the glumes 5–8 mm and more or less equalling the spikelets. Forma *ramosum* Schumach. is normal except for the inflorescence branches up to 7 cm. Var. *longiglume* Grantzow also has branched inflorescences but the glumes are 10–16 mm and exceed the spikelet by 1–3 mm. Forma *aristata* (Pes.) Döll has the spikes 2–10 cm, the spikelets 14–18 mm with all spikelets touching or the lower 2 spaced. The glumes are about 12.5 mm and the lemmas 5.5–7.5 mm. It maintains its status when grown from seed in cultivation.

Native. A valuable grazing and hay grass, common in old pastures and meadows, especially on rich heavy soils of the lowlands. Also in waste and rough grounds, and waysides and lawns. Abundant throughout the British Isles. Europe, except the Arctic; temperate Asia; North Africa; introduced in North America and Australia.

2. L. multiflorum Lam. Italian Rye Grass
L. perenne subsp. *multiflorum* (Lam.) Husn.

Annual or *biennial herbs* without tillers at flowering and fruiting time. *Culms* 30–100 cm, tufted or solitary, erect or spreading, slender to somewhat stout, unbranched or branched towards the base, with 2–5 nodes, smooth, or rough towards the spike. *Leaves* 6–25 cm × up to 10 mm, green, glabrous, linear, finely pointed at apex, rolled along their long axis in young shoots, glossy and smooth beneath, smooth or rough above, with narrow, spreading auricles at the base; sheaths rounded on the back, smooth or rough; ligules 1–2 mm, membranous. *Inflorescence* a spike 10–30 cm, slender to rather stout, erect or nodding, compressed, green or purplish; rhachis mostly rough. *Spikelets* 8–25 mm, sessile, alternate and singly in rows on opposite sides of the rachis, with their edges in hollows, oblong, compressed, awned, overlapping on their own length or more apart, 5- to 15-flowered, breaking up at maturity beneath each lemma. *Upper glume* persistent, varying in length but much shorter than the spikelet, narrowly oblong or lanceolate-oblong, obtuse or pointed at apex, 4- to 7-veined, smooth, lower glume present only in the terminal spikelet and similar to upper. *Lemmas* 5–8 mm, overlapping, oblong or lanceolate-oblong, obtuse or minutely 2-toothed at apex, rounded on the back, firm except for the thin margins and tip, 5-veined, smooth or minutely rough, with a fine straight awn up to 10 mm from near the tip, rarely awnless. *Paleas* as long as the lemmas, with 2 minutely rough keels. *Anthers* 3.0–4.5 mm. *Caryopsis* tightly enclosed by the hardened lemma and palea. *Flowers* 6–8. $2n = 14$.

Introduced. A valuable fodder plant introduced into Britain about 1830. Rough and waste ground, field margins and waysides. Scattered throughout the British Isles, common in lowland Britain. Cultivated and natu-

ralised in temperate regions. Perhaps native of southern Europe.

× **perenne = L.** × **baucheanum** Kunth
L. × *hybridum* Haussk.
This hybrid may be annual or perennial, it is mostly awned, has rolled young leaves and intermediate spikelet structure. The spikelets are often 8–20 mm, the lemmas 4–9 mm and awnless or with an awn 0.7–6.0 mm and the anthers 3–4 mm. The parents are completely interfertile and have a sympatric distribution. *Flowers* 7–8. $2n = 14$.

Occurs occasionally throughout Britain and Ireland as a natural product. It is a valuable pasture or meadow grass and often escapes from cultivation in lowland Britain. It also occurs in Continental Europe.

3. L. rigidum Gaudin Mediterranean Rye Grass

Annual herb. Culms up to 45(–70) cm, usually tufted, rarely solitary, the outer shortly decumbent but eventually erect, smooth and glabrous. *Leaves* 7–22 cm × 6–8 mm, linear, acuminate at apex, flat, glabrous, very scabrid on the nerves of the upper surface, scaberulous to almost smooth on the margins; sheaths rounded on the back, markedly striate, smooth and glabrous, rarely slightly scaberulous; ligule 1.0–2.5 mm, membranous, truncate or rounded at apex. *Inflorescence* a spike 4–18 cm, rigid, straight or rarely curved. *Spikelets* 10–13 mm, 6- to 8-flowered, oblong, finally cuneate. *Glume* 8–13 mm, lanceolate, acute or obtuse at apex, firm, 5- to 7-veined, smooth and glabrous, scarious on the margins. *Lemma* 5–9 mm, oblong, obtuse at apex, 5-veined, smooth and glabrous, or slightly rough, scarious on the margins, rarely with awns 0.7–3.5 mm, mainly near the apex of the spike. *Palea* as long as the lemma, 2-keeled, rough on the keels. Anthers 1.5–3.5(–4.8) mm. *Caryopsis* tightly enclosed by the hardened lemma and palea. *Flowers* 3–6. $2n = 14$.

Introduced. Casual on waste ground and tips. Scattered over Britain. Native of south Europe and the Mediterranean region east to central Asia.

4. L. temulentum L. Darnel

Rigid *annual herb.* Culms 30–90 cm, tufted or solitary, erect or slightly spreading, slender to stout, rough towards the spike, or smooth, with 2–4 nodes. *Leaves* 6–40 cm × 3–13 mm, green, glabrous, with narrow spreading auricles at the base, linear, narrowed to a fine hard point at the apex, flat, firm, rough, or smooth except for the margins; sheaths rounded on the back, smooth or rough; ligules up to 2 mm, obtuse at apex. *Inflorescence* a spike 10–30 cm × 5–12 mm, erect, green, rigid; rhachis stout, rough or smooth on the back. *Spikelets* 12–26 × 4–6 mm, about their own length or more apart, with one edge against the rhachis, oblong, 4- to 10-flowered, breaking up at maturity beneath each flower. *Upper glume* (7–)10–30 mm, usually extending to the tip of or exceeding the uppermost lemma, narrow, obtuse at apex, rigid, flat, smooth or rough, 7- to 9-veined, lower glume usually suppressed, except in the terminal spikelet. *Lemmas* (4.6–)5.0–8.5 mm, elliptical to ovate, obtuse at apex, rounded on back, slightly hairy on the keel distally, becoming tumid and hard, smooth, 5- to 9-veined, awned from near the apex, the awn straight, rough and up to 20 mm, or awnless. *Paleas* as long as the lemmas, 2-keeled. *Anthers* about 2.5 mm. *Caryopsis* tightly enclosed by and more or less adhering to the hardened lemma and palea. *Flowers* 6–8. $2n = 14$.

Introduced. Formerly common in cornfields, now casual on tips and in waste places. Scattered throughout the British Isles. Probably native to south-east Europe and temperate Asia.

5. L. remotum Schrank   Flaxfield Rye Grass
L. arvense Schrad., non With.; *L. linicola* A. Br.

Rigid *annual herb.* *Culms* up to 75 cm, tufted or solitary, erect, slender to rather stout, smooth, with 2–4 nodes. *Leaves* 6–30 cm × 2–7 mm, green, long-linear, gradually narrowed at the apex to a hard point, flat, slightly rough or smooth, with 2 narrow, pointed auricles at base; sheaths rounded on the back, smooth; ligules 0.5–2.0 mm, obtuse at apex. *Inflorescence* a spike 4–22 cm × 5–7 mm, erect, green, rigid; rhachis slender, rough. *Spikelets* 12–15 × 4–5 mm, up to 20 mm apart, with one edge against the rhachis, oblong, 4- to 10-flowered, breaking up at the maturity beneath each flower. *Glume* 5-16 mm, from two-thirds to one and a half times as long as the rest of the spikelet, narrow, acute at apex. *Lemma* of the lowest 2 flowers 3.5–5.0(–5.5) mm, more or less ovate, obtuse at apex, usually unawned but when awn present less than 10mm. *Paleas* as long as the lemma, 2-keeled. *Anthers* about 2.5 mm. *Caryopsis* 3.2–4.5 mm, tightly enclosed by and more or less adhering to the hardened lemma and palea. Flowers 6–8. $2n = 14 + 1B$.

Introduced. Formerly a typical flax-field alien, now a very occasional alien from grain and other sources. In scattered, sporadic localities in England. East Europe and south-west Asia, widely introduced elsewhere.

16. Vulpia C. C. Gmelin
Nardurus (Bluff, Nees & Schauer) Rchb.

Glabrous *annual herbs.* *Leaves* flat, convolute when dry; sheaths not fused. *Inflorescence* a sparsely branched narrow panicle or a raceme, usually more or less secund. *Spikelets* with 2–many flowers, all except the most apical bisexual, or with a group of sterile flowers at the apex. *Glumes* 2, very unequal, the upper 1- to 3-veined, acute to acuminate at apex, awned or not, the lower at most three-quarters as long as the upper, veinless or 1-veined. *Lemmas* 3- to 5-veined, rounded on back, chartaceous with hyaline margins, gradually tapered upwards into a long, straight awn. *Palea* about equalling lemmas, 2-veined, bifid. *Stamens* 1–3. *Caryopsis* narrowly ellipsoid, enclosed by the palea; hilum linear, at least a quarter as long as the seed.

Twenty-two species in temperate and subtropical regions of the northern hemisphere; introduced to the southern hemisphere.

Cotton, R. & Stace, C. A. (1976). Taxonomy of the genus *Vulpia* (Gramineae). 1. Chromosome numbers and geographical distribution of the Old World species. *Genetica* **46**: 235–255.

Cotton, R. & Stace, C. A. (1977). Morphological and anatomical variation of *Vulpia* (Gramineae) *Bot. Not.* **130**: 173–187.
Stace, C. A. & Cotton, R. (1976). Notes on alien *Vulpias* in Britain. *Watsonia* **11**: 72–73.
Stace, C. A. & Cotton, R. (1976). Nomenclature, comparison and distribution of *Vulpia membranacea* (L.) Dumort., and *V. fasciculata* (Forskål) Samp. *Watsonia* **11**: 117–123.
Stace, C. A. & Cotton, R. (1976). *Vulpia hybrida* (Brot.) Pau, nomen ambiguum. *Watsonia* **11**: 237.
Stewart, A., Pearman, D. A. & Preston, C. D. (1994). *Scarce plants in Britain.* Peterborough. [*V. ciliata* subsp. *ambigua, V. fasciculata* and *V. unilateralis.*]

1. Lemma with a pointed, minutely scabrid basal callus; ovary and seed with a minute, hairy apical appendage; lemmas 8–18 mm; upper glume 10–20 mm with awn   **1. fasciculata**
1. Lemma with a rounded basal callus; ovary and seed glabrous; lemmas 3.0–7.5 mm, excluding awn; upper glume 1.5–9.0 mm including awn if present   2.
2. Inflorescence more or less always a raceme; lemma 3–5 mm excluding awn; anthers 3, 0.7–1.3(–1.9) mm, well exserted at anthesis   **6. unilateralis**
2. Inflorescence a panicle except in starved plants; lemmas 4–7 mm; anthers 1–3, 0.4–0.8(–1.8) mm, usually not exserted at anthesis   3.
3. Spikelets with 1–3 bisexual and 3–7 distal sterile flowers; lemma of sterile flowers, 3(–5)-veined   4.
3. Spikelets with 2–3 bisexual and 1–2 distal, much reduced sterile flowers; lemma of fertile flowers 5-veined   5.
4. Spikelets mostly 7.0–10.5 mm excluding awn; fertile lemmas 5.0–6.5 mm excluding awn, pubescent on dorsal midline and margins; sterile lemmas less than 8 mm excluding awn, densely ciliate.   **5(a). ciliata** subsp. **ciliata**
4. Spikelets mostly 5–7 mm excluding awn; fertile lemmas 4–5 mm excluding awn, minutely scabrid; sterile lemmas less than 6 mm excluding awn, minutely scabrid.   **5(b). ciliata** subsp. **ambigua**
5. Lemma 1.3–1.9 mm wide; lower glume 2.5–5.0 mm, half to three-quarters as long as upper   **2. bromoides**
5. Lemmas 0.8–1.3 mm wide; lower glume 0.5–3.0 mm, usually less than half as long as upper   6.
6. Inflorescence usually well exserted from the uppermost leaf-sheath; lower glume one-quarter to half as long as upper   **3. muralis**
6. Inflorescence usually not fully exserted from the uppermost leaf-sheath; lower glume one-tenth to two-fifths as long as upper   7.
7. Lemma scabrid   **4(1). myuros** forma **myuros**
7. Lemma ciliate or pubescent dorsally   8.
8. Lemma ciliate   **4(2). myuros** forma **megalura**
8. Lemma pubescent dorsally   **4(3). myuros** forma **hirsuta**

Section **1. Monachne** Dumort.

Annual herbs. Inflorescence a panicle. *Flowers* cleistogamous or the anthers just exserted at anthesis. *Spikelets* disarticulating below each fertile flower or additionally at the base of the pedicel with a distant apical group of small, sterile flowers; pedicels dilated distally. *Anthers* 1–3, 0.5–2.0 mm.

1. V. fasciculata (Forssk.) Fritsch   Dune Fescue
Festuca uniglumis Aiton; *V. uniglumis* (Aiton) Dumort.;
V. membranacea auct.; *Festuca fasciculata* Forssk.

Annual herb. Culms 6–20 cm, loosely tufted or solitary,
erect or spreading, slender, usually unbranched,
smooth, with 2–3 nodes. *Leaves* 1–10 cm, inrolled or
opening out up to 3 mm, green, obtuse at apex, stiff to
rather weak, smooth beneath, minutely hairy above on
the prominent ribs; sheaths mostly overlapping,
smooth, rounded on the back; ligules very short, membranous. *Inflorescence* a panicle 2–12 cm, erect, stiff,
narrowly oblong, dense, secund, green or purplish,
branched in the lower part or raceme-like throughout;
rhachis rough; branches erect; pedicels 3–7 mm, erect,
thickened upwards, rough. *Spikelets* 12–16 mm (excluding awn), closely overlapping, narrowly oblong or
wedge-shaped, loosely 2- to 3-flowered, and with 3–4
sterile lemmas at the apex, breaking up at maturity, or
falling with the pedicel attached. *Glumes* very unequal;
the lower 0.2–1.6 mm; the upper 10–14 mm, narrowly
lanceolate, keeled, tipped with an awn 4–6 mm, rough,
firm except for the membranous margins, 3-veined.
Fertile lemmas 8–18 mm, narrowly lanceolate in side
view, keeled, narrowed into a fine, straight, rough awn
up to 25 mm, firm , rough, finely 3- to 5-veined, with a
pointed basal callus. *Sterile lemmas* much smaller and
narrower than the fertile ones. *Paleas* with 2 rough keels.
Anthers 1–3, 0.8–2.0 mm. *Caryopsis* hairy at the tip,
tightly enclosed by the lemma and palea. *Flowers* 5–6. *2n*
= 28.

Native. Sand and small shingle between foreshore
beaches and on open parts of sand dunes. Locally frequent on the coast of the British Isles north to Norfolk,
Cumberland and Co. Louth. West Europe and the
Mediterranean region.

Section **2. Vulpia**

Annual herbs. Inflorescence a panicle. *Flowers* cleistogamous or the anthers just exserted at anthesis. *Spikelets*
disarticulating below each fertile flower; most flowers
fertile or with an apical group of sterile but not smaller
flowers. *Pedicels* not or scarcely dilated. *Anthers* 1–3,
0.3–0.8 (–1.8) mm.

2. V. bromoides (L.) Gray   Squirrel-tail Fescue
V. sciuroides (Roth) C. C. Gmel.; *Festuca sciuroides*
Roth; *Festuca bromoides* L.

Annual herbs. Culms 5–60 cm, loosely tufted or solitary,
erect, or ascending from a bent or prostrate base, very
slender, rather stiff, often branched in the lower part,
smooth, with 2–4 nodes. *Leaves* 1–14 cm × 0.5–3.0 mm,
green, narrowly linear, finely pointed at apex, flat or
rolled, flaccid to rather stiff, rough near the tip and on
the margins, minutely hairy above; sheaths smooth,
rounded on the back; ligules up to 0.5 mm, membranous. *Inflorescence* a panicle 1–10 cm, green or purplish,
long-exserted from the uppermost sheath, erect or
slightly nodding, lanceolate to narrowly oblong, rather
loose to compact, one-sided, sometimes reduced to a
single spikelet; rhachis angular, scaberulous; branches

erect or slightly spreading; pedicels 1–4 mm, thickened.
Spikelets 7–14 mm, oblong or wedge-shaped, 5- to 10-
flowered, breaking up between the lemmas. *Glumes* persistent, finely pointed at apex, the lower 2.5–5.0 mm,
half to three-quarters the length of the upper, subulate-
lanceolate, 1-veined, the upper 6–10 mm, lanceolate or
oblong-lanceolate, 3-veined. *Lemmas* 5–7 × 1.4–1.9
mm, at first overlapping, rounded on the back, with the
margins finally incurved, linear-lanceolate in side view,
narrowed into a fine, rough awn up to 13 mm, firm,
finely 5-veined, rough. *Paleas* about as long as the
lemmas, rough on the 2 keels. *Anthers* usually 1, 0.4–0.6
mm. *Caryopsis* tightly enclosed by the lemma and palea.
Flowers 5–7. *2n* = 14.

Native. Heaths, waysides, waste ground and open
grassy places on well-drained soils. Frequent over most
of the British Isles and common in south Britain. South,
west and central Europe.

3. V. muralis (Kunth) Nees   Wall Fescue
V. broteri Boiss. & Reut.; *V. dertonensis* var. *broteri*
(Boiss. & Reut.) Hegi; *V. sciuroides* var. *longearistata*
Willk.; *V. dertonensis* var. *longearistata* (Willk.) Aznav.;
V. australis auct.

Annual herb. Culms 10–60 cm, usually erect, smooth.
Leaves 1–14 cm × 0.5–3.0 mm, green, narrowly linear,
flat or rolled, pointed at apex, flaccid to rather stiff.
Inflorescence a usually erect panicle, rarely a raceme
3–15 cm, usually well exserted from the uppermost leaf-
sheath; pedicels 0.5–3.5 mm. *Spikelets* 5–10 mm, disarticulating below each fertile flower, most flowers fertile;
distal 1–2(–3) flowers gradually reduced and male or
sterile. *Lower glume* 0.5–2.5 mm, one-quarter to half as
long as the upper, the upper 2.5–6.5 mm. *Lemma* 4.5–7.5
× 0.8–1.3 mm, with an awn usually 1–2 times as long,
finely 5-veined; callus about 0.5 mm, rounded, glabrous.
Palea about equalling the lemma. *Anthers* 1(–3), 0.4–0.7
mm, usually included at anthesis. *Ovary* glabrous.
Caryopsis tightly enclosed by the lemma and palea.
Flowers 5–7. *2n* = 14.

Introduced. A rare casual, especially as a wool alien.
Native of southern Europe.

4. V. myuros (L.) C. C. Gmel.   Rat's-tail Fescue
Festuca myuros L.

Annual herb. Culms up to 50, 10–70 cm, densely tufted
or solitary, erect, ascending or spreading from a bent
base, slender to very slender, often branched towards the
base, smooth, usually sheathed up to the base of the
panicle, with 2–3 nodes. *Leaves* 2–15 cm × 0.5–3.0 mm,
green, narrowly linear, finely pointed at apex, inrolled or
opening out, flaccid to firm, smooth beneath, rough on
the margins, shortly hairy on the upper surface; sheaths
smooth, rounded on the back; ligules up to 1 mm, membranous. *Inflorescence* a panicle 5–30 cm, not fully
exserted from the uppermost leaf-sheath, green or purplish, linear, contracted, lax to rather dense, usually
curved or nodding, sometimes raceme-like in the upper
part; rhachis angular, rough; branches mostly appressed
to the rhachis; pedicels 1–3 mm, thickened. *Spikelets*

(5.5–)7–10 mm, oblong or wedge-shaped, 3- to 7-flowered, readily breaking up at maturity beneath each lemma. *Glumes* persistent, very unequal, shorter than the adjacent lemmas, finely pointed at apex, the lower 1.0–3.5 mm, one-tenth to two-fifths the length of the upper, linear-lanceolate, 1-veined. *Lemmas* 5–7 mm, at first overlapping, at length with the margins incurved, rounded on the back, linear-lanceolate in side view, narrowed into a fine, straight, rough awn up to 15 mm, firm, finely 5-veined, scabrid, ciliate or pubescent dorsally. *Paleas* about as long as the lemmas, with 2 rough keels. *Anthers* 1–2, 0.4–0.6 mm. *Caryopsis* tightly enclosed by the hardened lemma and palea. *Flowers* 5–7. 2n = 42.

(1) Forma myuros
Lemma scabrid.

(2) Forma megalura (Nutt.) Stace & R. Cotton
V. megalura (Nutt.) Rydb.; *Festuca megalura* Nutt.
Lemma ciliate.

(3) Forma hirsuta (Hack.) Blom
Lemma pubescent dorsally.

Probably native. Forma *myuros* occurs in open ground, rough and waste land, waysides and along railways and on walls. Throughout most of the British Isles except central and north Scotland and most of Northern Ireland; common in southern Britain. West, central and south Europe; south-west Asia; North Africa; Macaronesia; introduced in South Africa, North and South America and Australia. Forma *megaleura* and forma *hirsuta* both appear to be introductions.

5. V. ciliata Dumort.   Bearded Fescue
Loosely tufted *annual herb*. *Culms* 5–30 cm, erect or bent at the base and spreading, very slender, sometimes branched in the lower part, sheathed up to the panicle, with 1–2 nodes. *Leaves* 1–10 cm × up to 2 mm, green or reddish, narrowly linear, with a fine blunt or pointed tip, inrolled or opening out, rough on the margins, minutely hairy on the upper surface; sheaths smooth, rounded on the back; ligules extremely short, membranous. *Inflorescence* a panicle 3–13 cm, green, purplish or reddish, erect, very narrow, linear or lanceolate, contracted, one-sided, sparingly branched, raceme-like in the upper part or throughout; rhachis angular; branches short, erect; pedicels swollen, up to 1 mm. *Spikelets* 5.0–10.5 mm, overlapping, narrowly oblong or wedge-shaped, loosely 3- to 7-flowered, breaking up at maturity beneath each fertile lemma. *Glumes* persistent, very unequal, much shorter than the lowest lemma, membranous, the lower 0.2–1.0 mm, ovate or oblong and veinless, the upper 2–3 mm, linear-lanceolate and 1-veined. *Lemmas* 4.0–8.0 mm, at first overlapping, later with the margins incurved, narrowly lanceolate in side view, firm, finely 5-veined, rough, minutely scabrid or densely ciliate, narrowed upwards into a fine, straight, rough awn up to 10 mm. *Paleas* nearly as long as the lemmas, 2-keeled, with the keels rough. *Anthers* 1–3, 0.4–0.6 mm. *Caryopsis* tightly enclosed by the hardened lemmas and palea. *Flowers* 5–6.

(a) Subsp. **ciliata**
Spikelets mostly 7.0–10.5 mm excluding awn. *Fertile lemmas* 5.0–6.5 mm, pubescent on the dorsal mid-line and margins; sterile lemmas up to 8 mm, densely ciliate. 2n = 28.

(b) Subsp. **ambigua** (Le Gall) Stace & Auq.
Festuca ambigua Le Gall; *V. ambigua* (Le Gall) More
Spikelets mostly 5–7 mm. *Fertile lemmas* 4–5 mm, minutely scabrid; sterile lemmas up to 6 mm, minutely scabrid. 2n = 28.

Native. Subsp. *ambigua* occurs on maritime or sub-maritime sand or shingle and by roads and tracks in sandy areas inland. It is local in southern Britain north to mid-Wales and Yorkshire and was formerly in Derbyshire. A few localities in Belgium and north-west France. Subsp. *ciliata* is an occasional casual from wool and grain in southern England and has been naturalised in railway sidings at Ardingly in Sussex since 1967. West and south Europe; North Africa; south-west and central Asia.

Section 3. Nardurus (Bluff, Nees & Schauer) Stace
Nardurus (Bluff, Nees & Schauer) Rchb.

Annual herbs. Inflorescence racemose or with a few branches near the base. *Spikelets* disarticulating below each fertile flower; pedicels not or scarcely dilated; most flowers fertile. Anthers 3, 0.7–1.3(1.9) mm.

6. V. unilateralis (L.) Stace   Mat-grass Fescue
Triticum unilaterale L; *Nardurus maritimus* (L.) Murb.; *V. hispanica* (Reichard) Kerguélen nom. invalid.

Annual herb, green or suffused with purple. *Culms* 5–40 cm, erect or spreading, solitary or loosely tufted, very slender, sometimes branched in the lower part, with a few nodes towards the base. *Leaves* 5–50(–80) × 0.2–0.5 mm, narrowly linear, obtuse or pointed at apex, flat or rolled, minutely hairy above, glabrous or nearly so beneath, flaccid; sheaths very narrow, minutely hairy or glabrous; ligules up to 1 mm, obtuse at apex, membranous. *Inflorescence* a solitary, erect or slightly nodding spike-like raceme, 1–5(–10) cm × 2–3 mm, slender, secund, green or purplish; rhachis very slender, minutely rough; pedicels 0.8–1.5 mm. *Spikelets* 4–7 mm, lanceolate-oblong to oblong, compressed, 2- to 6-flowered, alternating on one side of the main axis and slightly overlapping, breaking up at maturity beneath the lemmas. *Glumes* lanceolate, pointed, unequal, shorter than the lemmas, keeled, firm, the lower 1–2 mm, 1-veined, the upper 2.5–4.0 mm, 3-veined. *Lemmas* 2.5–5.0 mm, lanceolate in side view, pointed at apex, rounded on the back, 5-veined, becoming hardened, minutely and stiffly hairy in the upper part or all over, or glabrous, tipped with a fine, straight, rough awn 1–6 mm. *Paleas* as long as the lemmas, narrowly oblong, minutely rough on the keels. *Anthers* 3, 0.7–1.3(–1.9) mm, well exserted at anthesis. *Caryopsis* 2.5–3.0 mm, tightly embraced by the firm lemma and palea. *Flowers* 5–7. 2n = 14.

Native. Open, grassy places on chalk, and waste

places and waysides. Scattered in southern Britain north to Norfolk and Gloucestershire, formerly to Derbyshire. South and west Europe; North Africa; south-west Asia to the Himalayas.

17. Cynosurus L.

Annual or *perennial herbs* without rhizomes or stolons. *Leaves* with sheaths not fused. Inflorescence a compact, spike-like panicle. *Spikelets* of two kinds, fertile ones with (1–)2–5 flowers and all except the most apical bisexual, and sterile ones which consist of numerous, very narrow lemmas with an acuminate apex in a herringbone arrangement, normally one fertile and one sterile together. *Glumes* 2, subequal. *Fertile lemmas* rounded on the back, acute to obtuse or minutely bifid at apex, 5-veined, awnless or with a long terminal or subterminal awn. *Stamens* 3.

Eight species in temperate regions of the Old World.

Grime, J. P. et al. (1988). *Cynosurus cristatus* L. in *Comparative plant ecology* 210–211.
Lodge, R. W. (1959). *Cynosurus cristatus* L. in Biological flora of the British Isles. *Jour. Ecol.* **47**: 511–518.

1. Perennial herb; leaves 1–4 mm wide; uppermost sheath hardly inflated; panicle narrowly oblong   **1. cristatus**
1. Annual herb; leaves 3–10 mm wide, uppermost sheath obviously inflated; panicle ovoid, squarrose   **2. echinatus**

1. C. cristatus L.   Crested Dog's-tail

Compactly tufted *perennial herb*. *Culms* 5–75 cm, erect or slightly spreading, stiff, unbranched, smooth, with 1–3 nodes. *Leaves* up to 15 cm × 1–4 mm, green, long-linear, with a fine tip, flat, rather rough, minutely hairy on the upper surface or glabrous; sheaths rounded on the back, glabrous, smooth; ligules 0.5–1.5 mm, very obtuse at apex, membranous. *Inflorescence* a panicle 1–14 cm × 4–10 mm, green or tinged with purple, spike-like, erect, or slightly curved, dense, one-sided, narrowly oblong, stiff; rhachis rough or minutely hairy; branches very short. *Spikelets* in a dense cluster, of 2 kinds, fertile and sterile mixed, the sterile almost concealing the fertile. *Fertile spikelets* 3–6 mm, oblong or wedge-shaped, 2- to 5-flowered, breaking up at maturity beneath each lemma. *Glumes* 3–5 mm, persistent, keeled, narrow, pointed at apex, thin, 1-veined, nearly equal. *Lemmas* 3–4 mm, rounded on the back, their tips exceeding the glumes, oblong-ovate and obtuse at apex when opened out, usually tipped with an awn up to 1 mm, firm except for the membranous margins, rough upwards, finely 5-veined. *Paleas* slightly shorter than the lemmas, 2-keeled. *Anthers* about 2 mm. *Caryopsis* about 2 mm, oblong, tightly enclosed by the hardened yellowish or brownish lemma and palea. *Sterile spikelets* 4–6 mm, persistent, ovate, becoming obovate, flattened, composed of up to 18 very narrow, finely pointed, 1-veined bracts 3–5 mm. *Flowers* 6–8. 2*n* = 14.

Native. Grasslands, both in the lowlands and the hills, on a wide variety of soils and in dry or damp situations. Common throughout the British Isles. Europe except the extreme north; Caucasus; Madeira; Azores; introduced in North America.

2. C. echinatus L.   Rough Dog's-tail

Annual herb. *Culms* 10–100 cm, tufted or solitary, erect or spreading, slender to somewhat stout, often branched, smooth, with 2–6 nodes. *Leaves* 5–20 cm × 3–10 mm, green, glabrous, long-linear, tapering to a fine point at apex, flat, rough above, smooth beneath; sheaths rounded on the back, smooth, the upper slightly inflated; ligules up to 10 mm, obtuse at apex, membranous. *Inflorescence* a panicle 1–8 × 1–2 cm, green or purplish, shining, spike-like, dense, ovate to oblong or almost globose, one-sided, bristly; branches short. *Spikelets* densely clustered, of two kinds, fertile and sterile, the latter more or less concealing the former. *Fertile spikelets* 8–14 mm, wedge-shaped, 1- to 5-flowered, breaking up at maturity beneath each lemma. *Glumes* 7–12 mm, whitish, persistent, nearly equal, narrowly lanceolate, very finely pointed at apex, keeled, very thin, 1-veined. *Lemmas* 5–7 mm, rounded on the back, ovate when opened, 2-toothed or entire at tip, becoming firm except for the membranous margins, rough upwards, 5-veined, awned from near the tip, the awn 6–16 mm, straight and rough. *Paleas* about as long as the lemmas, 2-keeled. *Anthers* 2.5–4.0 mm. *Caryopsis* 3–4 mm, oblong, tightly enclosed by the hardened lemma and palea. *Sterile spikelets* 7–13 mm, persistent, broadly obovate, flattened, bearing up to 18 bracts, which are narrowly lanceolate and tipped with a fine awn, bract and awn together 4–8 mm, firm to rigid, at first overlapping, finely supported and spreading. *Flowers* 6–7. 2*n* = 14.

Introduced. Naturalised in sunny places on sandy or rocky ground on the coasts of southern England and the Channel Islands. Elsewhere a casual on waste and rough ground north to central Scotland. Native of south Europe; North Africa; south-west Asia; Macaronesia.

18. Lamarckia Moench nom. conserv.

Annual herbs. *Leaves* with sheaths not fused. *Inflorescence* a rather compact panicle. *Spikelets* of two kinds, fertile ones with one bisexual and one vestigial flower, and sterile ones which consist of numerous flat, overlapping, obtuse lemmas, normally 3 sterile and 2 fertile together, but one of the two fertile often reduced and not producing seed. *Glumes* 2, subequal. *Lemmas* rounded on back, 5-veined, minutely bifid with a long awn protruding from the sinus. *Stamens* 3.

One species in the Mediterranean area and the Middle East.

1. L. aurea (L.) Moench   Golden Dog's-tail
Cynosurus aureus L.

Annual herb. *Culms* up to 30 cm, erect or shortly decumbent at the base, usually loosely or densely fasciculate, rarely solitary, smooth and glabrous, but rough just beneath the panicle. Leaves up to 10 cm × 2–6 mm, green, acuminate at apex, flaccid, flat, scabrid on the margins, smooth and glabrous on the surfaces, or the upper scaberulous; sheaths rather loose, markedly striate, smooth and glabrous, the lower slipping from the stems; ligules up to 10 mm, membranous, rounded or pointed at apex. *Inflorescence* a panicle 3–9 × 2–3 cm, erect, dense, one-sided. *Spikelets* of two kinds, the outer ones sterile and shielding the fertile inner ones. *Sterile spikelet* fascicled, the 3–4 sterile ones surrounded by 1 fertile one, each on a hairy pedicel, 6–7(–8) mm, linear, the glumes 2, 3–4 mm and lanceolate, followed by 6–8, empty, distichous, obovate, emarginate, pale often tinged purple scales 1.5–2.0 mm. *Fertile spikelets* 1- to 2-flowered; glumes about 3 mm, equal, narrow, acuminate at apex, 1-veined, membranous; lower lemma about 2.5 mm, broadly elliptical, truncate at apex, smooth on the back below, scabrid towards the tip, awned below the tip; palea 2-veined; rhachis crowned by a rudimentary lemma which is awned, the awns 10–12 mm, straight, scaberulous. *Anthers* about 0.5 mm. *Flowers* 4. $2n = 14$.

Introduced. Casual, sometimes persisting for a few years on tips or in rough or waste ground. Very scattered over Britain, mainly in south and central England and south Wales. Native of the Mediterranean region and extending into central Asia.

19. Puccinellia Parl. nom. conserv.
Atropis (Trin.) Griseb.

Annual or *perennial herbs* without rhizomes, with or without stolons. *Leaves* with sheaths not fused. *Inflorescence* a panicle. *Spikelets* with 2 to many flowers, all except the most apical bisexual. *Glumes* 2, slightly unequal. *Lemmas* rounded on back, 5-veined, subacute to rounded at apex, without awns. *Stamens* 3.

Gilbert, O. L. & Holligan, P. M. (1979). *Puccinellia capillaris* (Liljebl.) Jans. × *P. maritima* (Huds.) Parl. in North Rona, Outer Hebrides. *Watsonia* 12: 338–339.

Gray, A. J. & Scott, R. (1977). *Puccinellia maritima* (Huds.) Parl. in *Biological flora of the British Isles. Jour. Ecol.* 65: 699–716.

Jones, B. M. G. & Newton, L. E. (1970). The status of *Puccinellia pseudodistans* (Crép.) Jansen & Wachter in Great Britain. *Watsonia* 8: 17–26.

Stewart, A., Pearman, D. A. & Preston, C. D. (1994). *Scarce plants in Britain*. Peterborough. [*P. fasciculata* and *P. rupestris*.]

Trist, P. J. O. & Butler, J. K. (1995). *Puccinellia distans* (Jacq.) Parl. subsp. *borealis* (O. Holmb.) W. E. Hughes (Poaceae) in Scotland and the Outer Islands. *Watsonia*. 20: 391–396.

1. Lemmas 2.8–4.6 mm   2.
1. Lemmas 1.8–2.5(–2.8) mm   3.
2. Perennial herb with many tillers and usually rooting stolons at anthesis; anthers 1.3–2.5 mm   **1. maritima**
2. Annual or biennial herbs with no or few tillers and without stolons at flowering; anthers 0.7–1.0 mm
                                                 4. rupestris
3. Panicle branches at the lower nodes more or less all with a conspicuous basal region bare of spikelets; lemma broadly obtuse to rounded at apex with the midrib falling short of the apex; anthers mostly more than 0.7 mm   4.
3. At least some of the panicle branches at the lower nodes bearing spikelets more or less to the base; lemmas subacute to obtuse at apex and with the midrib reaching the apex; anthers mostly less than 0.7 mm   5.

4. Leaves usually flat; lower panicle branches strongly reflexed at maturity; lemmas 2.0–2.5 mm
                                      2(a). distans subsp. **distans**
4. Leaves usually folded along the midrib; panicle branches usually patent to suberect at maturity; lemmas 1.8–2.8 mm   **2(b). distans** subsp. **borealis**
5. Panicle symmetrical; anthers 0.6–1.0 mm
                                      3(i). fasciculata var. **fasciculata**
5. Panicle asymmetrical; anthers 0.4–0.5 mm
                                      3(ii). fasciculata var. **pseudodistans**

1. P. maritima (Huds.) Parl. Common Saltmarsh Grass
Glyceria maritima (Huds.) Thunb.; *Poa maritima* Huds.; *Atropis maritima* (Huds.) Griseb.; *Sclerochloa maritima* (Huds.) Lindl. ex Bab.; *Panicularia maritima* (Huds.) Scribn; *Sclerochloa festuciformis* auct.; *Glyceria foucaudii* Hack. ex E. S. Marshall nom. nud.; *Glyceria burdoni* Druce; *Phippsia maritima* (Huds.) Å. Löve

Perennial herb, densely tufted, with creeping stolons up to 40 cm, rooting at the nodes and forming a compact tuft. *Culms* 10–80 cm, erect, spreading or prostrate, slender to moderately stout, smooth, with 2–4 nodes. *Leaves* 2–20 cm × 1–3 mm, greyish-green, glabrous, narrowly long-linear, with an obtuse or abruptly pointed, slender, hooded tip, folded or inrolled or opening out, smooth beneath, rough on the veins above; sheaths rounded on the back, smooth; ligules 1–3 mm, obtuse at apex, membranous. *Inflorescence* a panicle 3–25 × 0.4–2.5 cm opening up to 8 cm wide, erect, stiff, linear to ovate, usually contracted and rather dense, but sometimes opening up branches usually erect after flowering, sometimes permanently spreading, stiff, rough; pedicels very short. *Spikelets* 5–13 mm, 3- to 10-flowered, breaking up at maturity beneath each lemma. *Glumes* persistent, lanceolate to ovate, unequal, the lower 1.5–3.5 mm and 1- to 3-veined, the upper 2.0–4.2 mm and 3-veined. *Lemmas* 2.8–5.0 mm, much exceeding the glumes, overlapping, rounded on the back, elliptical to broadly oblong, rather obtuse or slightly pointed, firm except for the white membranous tips and margins, silky hairy towards the base, 5-veined. *Paleas* as long as the lemmas, narrowly elliptical, with the 2 keels minutely hairy. *Anthers* 1.3–2.5 mm. *Caryopsis* 2–3 mm, ellipsoid, enclosed between the hardened lemma and palea. *Flowers* 6–7. $2n = 14, 49, 56, 60, 63, 70, 77$.

Native. Bare or semi-bare mud in salt-marshes and estuaries where it is often dominant, rarely saline areas and salted roadsides inland. Common round the coasts of the British Isles and in a few places in central England. Coasts of western Europe; North America.

× **rupestris** = **P.** × **krusemaniana** Jansen & Wachter
This hybrid is said to be intermediate between the parents and sterile.

It was recorded from Sussex in 1920 and Hampshire in 1977. Also recorded from Holland.

2. P. distans (Jacq.) Parl.   Reflexed Saltmarsh Grass
Glyceria distans (Jacq.) Wahlenb.; *Atropis distans* (L.) Griseb.; *Poa distans* Jacq.; *Poa retroflexa* Curt.;

Sclerochloa distans (Jacq.) Bab.; *Sclerochloa*
multiculmis Syme nom. illegit.; *Sclerochloa multiculmis*
subsp. *distans* (Jacq.) Syme; *Glyceria pulvinata* auct.

Loosely to densely tufted *perennial herb*. *Culms* 6–60 cm,
erect, spreading or prostrate, slender, smooth, with 2–4
nodes. *Leaves* 2–12 cm × 1–4 mm, greyish- or whitish-
green, glabrous, long-linear, with an abruptly pointed or
blunt, hooded tip, flat or rolled, rough above.
Inflorescence a panicle 2–18 × 0.5–14 cm, linear, lanceolate, narrowly to broadly ovate or triangular, loose and
open or contracted and dense, greenish; branches erect,
spreading or finely deflexed, stiff, rough; pedicels short,
rough. *Spikelets* 3–9 mm, narrowly oblong, 2- to 9-flow-
ered, greenish or purplish, variegated with white or
yellow, breaking up at maturity beneath each lemma.
Glumes persistent, blunt or pointed at apex, ovate, elliptical or oblong, unequal, the lower 0.7–2.0 mm and 1-
veined, the upper 1.5–2.8 mm, 3-veined. *Lemmas*
1.5–2.8 mm, overlapping, rounded on the back, oblong
or oblong-elliptical, very blunt at apex, minutely hairy at
the base, 4-veined, the veins not reaching the white, yellowish, purplish or green tips. *Paleas* as long as or
slightly shorter than the lemmas, narrowly oblong,
minutely hairy on the 2 keels. *Anthers* 0.5–1.2 mm.
Caryopsis narrowly ellipsoid, enclosed between the
hardened lemma and palea. *Flowers* 6–8.

(a) Subsp. **distans**
Culms up to 60 cm. *Leaves* usually flat. *Lower panicle*
branches strongly reflexed at maturity. *Lemmas* 2.0–2.5
mm. $2n = 42$.

(b) Subsp. **borealis** (O. Holmb.) W. E. Hughes
P. retroflexa subsp. *borealis* O. Holmb.; *P. capillaris*
(Liljeblad) Jansen
Culms 3–15(–40) cm. *Leaves* usually folded along the
midrib. *Panicle* 2–12 cm, dense, lower branches some-
times spreading. *Lemmas* 1.8–2.8 mm.

Native. Subsp. *distans* occurs on semi-bare mud,
rough and waste ground in estuaries, ditches in coastal
marshes, inland saline areas and by salted main roads. It
occurs round the coasts of the British Isles north to
central-east Scotland, commonly in the east, rarely in
the west; common inland in central and eastern
England. Europe; west Siberia; North Africa; introduced in North America. Subsp. *borealis* is found in
stony or rocky places and on sea-walls. It occurs on the
north and east coasts of Scotland south to Fife and in
the Outer Hebrides, Orkney and Shetland. North and
west Europe, southwards to the Netherlands.

× **fasciculata**
This hybrid is intermediate between the parents in
panicle shape and lemma shape and is sterile. $2n = 35$.
It occurs rarely on the coasts of England from Sussex
to Norfolk. ?Endemic.

× **maritima**   = P. × **hybrida** O. Holmb.
Glyceria × *jansenii* P. Fourn.; *Glyceria* × *salina* Druce
nom. nud.; *P.* × *kattegatensis* (Neum.) O. Holmb.;
Glyceria × *kattegatensis* Neum.

This hybrid often has stolons, is intermediate between
the parents in panicle shape and lemma size, and is
sterile. The lower glume is 1.5–2.0 mm, the upper glume
2–3 mm, the lemma 2.5–3.7 mm and the anthers 0.5–1.0
mm in nothosubsp. *hybrida*, while in nothosubsp. *mixta*
the lower lemma is 3.3–3.5 mm and the anthers 1.0–1.8
mm. $2n = 49, 51$.

Nothosubsp. *hybrida* occurs rarely on the coast of
England from Sussex to Durham and has been found by
a salted road in Northumberland. In the Outer Hebrides
where it involves subsp. *borealis* it is nothosubsp. **mixta**
(O. Holmb.) P.D. Sell (*P. mixta* O. Holmb.) The hybrid
has also been recorded from France, Holland and
Scandinavia.

× **rupestris** = **P.** × **pannonica** (Hack.) O. Holmb.
Atropis × *pannonica* Hack.; *Glyceria distans* var.
pseudoprocumbens W.-Dod.; *Glyceria*
pseudoprocumbens (W.-Dod.) Druce
This hybrid is intermediate between the parents in
panicle shape and length of lemma and is sterile. $2n =$
42.

It occurs rarely by tracks and in damp places on the
coasts of England from Devon to Norfolk. It is also
recorded in Holland and Norway.

3. P. fasciculata (Torrey) E. P. Bicknell
                    Borrer's Saltmarsh Grass
Poa fasciculata Torrey; *Glyceria borreri* (Bab.) Bab.;
Festuca borreri Bab.; *Poa borreri* (Bab.) Hook.;
Glyceria conferta Fr.; *Sclerochloa borreri* (Bab.) Bab.;
Sclerochloa multiculmis subsp. *borreri* (Bab.) Syme;
Atropis borreri (Bab.) K. Richt.; *Glyceria distans* subsp.
conferta (Fries) Hook. fil.; *Glyceria distans* subsp.
borreri (Bab.) Hook. fil.

Loosely to densely tufted *perennial herb*. *Culms* 6–60 cm,
erect or spreading, slender to relatively stout,
unbranched, smooth, with 1–3 nodes. *Leaves* 2–16 cm ×
1.5–5.0 mm, greyish-green, glabrous, long-linear,
hooded at the blunt or slightly pointed apex, flat or
folded upwards, firm, rough above, smooth beneath;
sheaths rounded on the back, smooth; ligules 1.0–2.5
mm, very obtuse at apex, membranous. *Inflorescence* a
panicle 1.5–18 cm, erect, lanceolate to narrowly oblong
or ovate, contracted and dense or loose and open, unilateral or asymmetrical, greyish-green or tinged with
purple; branches spreading, stiff, rough, crowded with
spikelets to the base, or the longer rarely bare; pedicels
extremely short. *Spikelets* 3–6 mm, densely clustered,
oblong, 3- to 8-flowered, breaking up at maturity
beneath each lemma. *Glumes* persistent, rounded on the
back, green and firm except for the white, membranous
margins and tips, unequal, ovate to elliptical, mostly
obtuse at apex, the lower 1.0–1.5 mm and 1-veined, the
upper 1.3–2.2 mm and 3-veined. *Lemmas* 1.7–2.3 mm,
exceeding the glumes, overlapping, elliptical, obtuse to
subacute at apex, green and firm except for the white
membranous margins and tips, minutely hairy at the
base, 5-veined, the middle vein usually minutely projecting at the tip. *Paleas* as long as the lemmas, oblong, 2-

keeled, the keel minutely hairy. *Anthers* 0.4–0.7(–1.0) mm. *Caryopsis* about 1.5 mm, ellipsoid, enclosed between the hardened lemma and palea. *Flowers* 6–9. 2*n* = 28.

(i) Var. fasciculata

Panicle unilateral; branches obliquely erect. *Anthers* 0.6–1.0 mm. 2*n* = 28.

(ii) Var. pseudodistans (Crépin) P. D. Sell

P. pseudodistans (Crépin) Jansen & Wachter; *Glyceria pseudodistans* Crépin; *P. fasciculata* forma *pseudodistans* (Crépin) L. E. Newton & B. M. G. Jones *Panicle* symmetrical; branches spreading. *Anthers* 0.4–0.5 mm. 2*n* = 28.

Native. Rather bare mud and shingle on sea-walls and banks, and by dykes and in open spaces in saline meadows. Coasts of southern Britain north to Norfolk and Carmarthen, and by salted roads in Kent. Both varieties have a similar distribution, but var. *fasciculata* usually occurs in a drier habitat. Var. *fasciculata* is a plant of the coasts of western Europe. Var. *pseudodistans* is found in the western Mediterranean. All our plants belong to subsp. **fasciculata** which is found through most of the range of the species.

4. P. rupestris (With.) Fernald & Weatherly
Stiff Saltmarsh Grass

Glyceria rupestris (With.) E. S. Marshall; *Glyceria procumbens* (Curtis) Dumort.; *Sclerochloa procumbens* (Curtis) P. Beauv.; *Poa rupestris* With.; *Poa procumbens* Curtis; *Festuca procumbens* (Curtis) Kunth, non Muhl.; *Scleropoa procumbens* (Curtis) Parl.; *Atropis procumbens* (Curtis) Thurb.; *Panicularia procumbens* (Curtis) Kunze; *Sclerochloa rupestris* (With.) Britten & Rendle

Loosely to densely tufted *annual* or *biennial herb*. *Culms* 4–40 cm, spreading or prostrate, slender to relatively stout, unbranched, smooth, with 1–3 nodes below the middle. *Leaves* 1–10 cm × 2–6 mm, greyish-green, glabrous, long-linear, hooded at the obtuse or slightly pointed tip, firm, rough above, smooth beneath; sheaths rounded on the back, smooth; ligules 1.0–2.5 mm, very blunt to pointed, membranous. *Inflorescence* a panicle 2–8 × 1.0–4.5 cm, ovate to oblong, stiff, dense to somewhat loose, one-sided; branches stiff, spreading, slightly rough; pedicels extremely short. *Spikelets* 5–9 mm, close together on one side of the branches, narrowly oblong, 3- to 5-flowered, breaking up at maturity beneath each lemma. *Glumes* persistent, green and firm except for the membranous tips and margins, ovate to elliptical, obtuse at apex, unequal, the lower 1.5–2.5 mm and 1- to 3- veined, the upper 2.0–3.0 mm, and 3-veined. *Lemmas* 2.8–4.0 mm, exceeding the glumes, overlapping, rounded on the back, broadly elliptical, obtuse at apex, firm and green except for the white membranous tips and margins, minutely hairy at the base, 5-veined, the middle vein sometimes minutely projecting at the tip. *Paleas* as long as the lemmas, narrowly oblong, 2-keeled, the keels minutely hairy. *Anthers* 0.7–1.0 mm. *Caryopsis*

about 2 mm, narrowly ellipsoid, enclosed between the hardened lemma and palea. *Flowers* 5–8. 2*n* = 42.

Native. On mud and clay and amongst rocks and stones, open areas in marshes occasionally subject to saline tides, and consolidated areas of shingle and tracks. Coasts of Britain from Pembrokeshire to Yorkshire, and by salted roads in Kent. Western Europe from Holland to central Spain; also recorded from Syria.

20. Briza L.

Annual or *perennial* herbs. *Leaves* with sheaths not fused. *Inflorescence* a panicle, or sometimes a raceme with pedicels over 5 mm. *Spikelets* with 4–many flowers, all except the most apical bisexual, characteristically flattened, broadly ovate and pendulous. *Glumes* 2, subequal. *Lemmas* rounded on the back, cordate at base, rounded to very obtuse at apex, without awns. *Stamens* 3.

Twenty species in temperate Eurasia and South America.

Grime, J. P. et al. (1988). *Briza media* L. in *Comparative plant ecology* 132–133.
Stewart, A., Pearman, D. A. & Preston, C. D. (1994). *Scarce plants in Britain*. Peterborough. [*B. minor.*]

1. Perennial herb with tillers at the time of flowering; leaves 2–4 mm wide; ligules 0.5–1.5 mm   **1. media**
1. Annual herb without tillers at the time of flowering; leaves 2–10 mm wide; ligules 2–6 mm   **2.**
2. Spikelets 2.5–5 mm, over 20 per panicle   **2. minor**
2. Spikelets 8–25 mm, less than 15 per panicle   **3. maxima**

1. B. media L.
Quaking Grass

Perennial herb forming loose tufts, with short rhizomes bearing leafy vegetative shoots (tillers). *Culms* 15–75 cm, mostly erect, slender, stiff, smooth, with 2–3 nodes. *Leaves* 4–15 cm × 2–4 mm, green, glabrous, linear or linear-lanceolate, with a slender, obtuse tip, minutely rough on the margins; sheaths entire, soon splitting, smooth; ligules 0.5–1.5 mm. *Inflorescence* a panicle 4–18 × 4–18 cm, loose, more or less pyramidal; branches spreading, sparingly divided; pedicels 5–20 mm, hair-like, curved. *Spikelets* 4–7 × 4–7 mm, drooping, loosely scattered, very broadly elliptical to broadly ovate, laterally compressed, 4- to 12-flowered, shining, usually purplish. *Glumes* 2.5–3.5 mm, horizontally spreading, persistent, deeply concave, hooded at the apex, firmly membranous, 3- to 5-veined. *Lemmas* about 3.5–4 mm, similar to the glumes, closely overlapping, cordate at the base, rounded on the back, 7- to 9-veined, smooth, variegated with purple and green, and with whitish margins. *Paleas* slightly shorter than the lemmas, elliptical, flat, the 2 keels narrowly winged. *Anthers* 2.0–2.5 mm. *Caryopsis* ellipsoid, enclosed by the papery lemma and palea, pale brown, rounded on the back, flattened in front. *Flowers* 6–8. 2*n* = 14.

Native. On light to heavy, dry to damp, acid to calcareous grasslands. Locally common throughout the British Isles, except the north of Scotland. Most of

Europe except the Arctic and parts of the south; temperate Asia; introduced in North America. Our plant is subsp. **media** which occurs throughout the range of the species.

2. B. minor L.   Lesser Quaking Grass

Loosely tufted *annual herb*. *Culms* 10–60 cm, erect, or slightly bent at the base, slender, round, smooth, with 2–4 nodes. *Leaves* 3–14 cm × 2–10 mm, green, glabrous, linear or narrowly linear-lanceolate, finely pointed at apex, flat, finely veined, minutely rough above and on the margins; sheaths round, smooth; ligules 3–6 mm, obtuse at apex, membranous. *Inflorescence* a panicle 4–20 × 2–10 cm, loose, obovate; branches finely divided, minutely rough, with curved hair-like pedicels 4–12 mm. *Spikelets* 2.5–5.0 × 3–5 mm, nodding, orbicular to triangular-ovoid, 4- to 8-flowered, shining, green or tinged with purple. *Glumes* 2.0–3.5 mm, persistent, horizontally spreading, hooded at the apex, firmly membranous, 3- to 5-veined. *Lemmas* closely overlapping, similar to the glumes, very broad, cordate at the base, rounded at the top and on the back, deeply concave, becoming hardened and shining in the centre, but with broad membranous margins, glabrous, 7- to 9- veined. *Paleas* shorter than the lemmas flat, with the 2 keels very narrowly winged. *Anthers* about 0.6 mm. *Caryopsis* broadly ellipsoid, enclosed by the lemma and palea, flat in front, rounded on the back, pale brown. *Flowers* 3–9. $2n = 10$.

Probably introduced, but perhaps native. Arable fields, bulb-fields and waste places. Locally frequent in south-west and south-central England and the Channel Islands, a rare casual elsewhere. South and west Europe; North Africa; south-west Asia; Macaronesia.

3. B. maxima L.   Greater Quaking Grass

Annual herb. *Culms* 10–60 cm, loosely tufted or solitary, erect, or bent below, slender, round, smooth, with 2–4 nodes. *Leaves* 5–20 cm × 3–8 mm, green, glabrous, linear, finely pointed at apex, flat, finely veined, minutely rough only on the margins; *sheaths* rounded, smooth; ligules 2–5 mm, membranous. *Inflorescence* a panicle 3–10 cm, loose, nodding, sparingly branched, with 1–12 spikelets; pedicels hair-like, 6–20 mm. *Spikelets* (8–)14–25 × 8–15 mm, nodding, ovate to oblong, 7- to 20-flowered, glabrous or minutely hairy, pale green, silvery, or often suffused with reddish-brown or purple. *Glumes* 5–7 mm, persistent, horizontally spreading, broadly rounded, deeply concave, firmly membranous, 5- to 9-veined. *Lemmas* 6–8 mm, closely overlapping, rounded on the back, cordate at the base, very broad, glabrous or with minute gland-tipped hairs on the hardened centre of the back and fine appressed hairs on the broad, firmly membranous margins, 7- to 9-veined. *Paleas* broadly elliptical, up to two-thirds as long as the lemmas, the 2 keels narrowly winged and minutely hairy. *Anthers* up to 2 mm. *Caryopsis* 2.5 mm, pale brown, narrowly ellipsoid, enclosed by the lemma and palea, flat in front, rounded on the back. *Flowers* 5–7. $2n = 14$.

Introduced. Grown in gardens for ornament, escaping and becoming naturalised on banks, field margins

and other dry, open places. Local in south-western and central-southern England and in the Channel Islands. Native of the Mediterranean region; Macaronesia.

21. Poa L.
Parodiochloa C. E. Hubb.

Annual or *perennial herbs* with or without stolons or rhizomes. *Leaf-sheaths* not fused. *Inflorescence* a panicle. *Spikelets* laterally compressed, with 1–many flowers all of which except for the most apical bisexual and sometimes proliferating. *Glumes* 2, subequal, keeled, membranous, usually 3-veined, awnless or rarely with a short, terminal awn, its callus often with a tuft of cottony hairs. *Stamens* 3. *Caryopsis* ellipsoid or narrowly ovoid; hilum basal.

About 500 species in cool temperate regions throughout the world, extending through the tropics on mountain tops.

Although a fairly uniform genus it lacks useful discriminating characters and is complicated by apomixis and introgression.

Barling, D. M. (1962). Studies in the biology of *Poa subcaerulea* Sm. *Watsonia* **5**: 163–173.

Barling, D. M. (1967). *Poa pratensis* L., *P. angustifolia* L. and *P. subcaerulea* Sm. *Proc. B.S.B.I.* **6**: 363–364.

Grime, J. P. et al. (1988). *Poa annua* L., *Poa pratensis* L. and *Poa trivialis* L. in *Comparative plant ecology* 442–447.

Hutchinson, C. S. & Seymour, G. B. (1982). *Poa annua* L. in Biological flora of the British Isles. *Jour. Ecol.* **70**: 887–901.

Nannfeldt, J. A. (1937). On *Poa jemtlandica* (Almqu.) Richt., its distribution and possible origin. *Bot. Not.* **1937**: 1–27.

Stewart, A., Pearman, D. A. & Preston, C. D. (1994). *Scarce plants in Britain*. Peterborough. [*P. alpina, P. bulbosa, P. glauca* and *P. palustris*.]

Trist, P. J. O. (1986). A reconsideration of the taxonomic status of *Poa balfouri* Parnell (Gramineae). *Watsonia* **16**: 37–42.

Trist, P. J. O. (1989). Spreading meadow grass, *Poa subcaerulea* Sm. *Nature Cambridgeshire* **31**: 57–60.

Tutin, T. G. (1957). A contribution to the experimental taxonomy of *Poa annua* L. *Watsonia* **4**: 1–10.

1.	At least some leaves more than 6 mm wide	2.
1.	Leaves 6 mm or less wide	3.
2.	Ligules 0.5–2.0 mm; panicle open and loose; lemma without an awn	**7. chaixii**
2.	Ligules 8–12 mm; panicle dense and spike-like; lemma with a terminal awn up to 2 mm	**16. flabellata**
3.	At least some spikelets proliferating	4.
3.	Spikelets not proliferating	6.
4.	Base of culms swollen and bulb-like	**14(ii). bulbosa** var. **vivipara**
4.	Base of culms not swollen	5.
5.	Leaves 1.0–2.5 mm wide when flattened, gradually tapered to apex, the uppermost usually arising above half way up the culm	**9. × jemtlandica**
5.	Leaves 2.0–4.5 mm wide when flattened, parallel-sided and more or less abruptly narrowed to the apex, the uppermost usually arising below halfway up the culm	**15(ii). alpina** var. **vivipara**
6.	Plants with distinct, often far-creeping rhizomes	7.
6.	Plants without rhizomes, sometimes with stolons	10.

7. Culms strongly compressed, 4–6(–9)-noded, usually slightly bent at each node · · · **10. compressa**

7. Culms terete to somewhat compressed, 2- to 4-noded, usually straight except near the base · · · 8.

8. Culms usually all solitary; glumes subequal, both usually with 3 veins and distinctly acuminate at the apex · · · **4. humilis**

8. Culms usually in small or dense clusters; glumes distinctly unequal, the lower often 1-veined, acute at the apex · · · 9.

9. Culms usually in small clusters; tiller leaves 2–4(–5) mm wide; lowest lemma 3–4 mm · · · **5. pratensis**

9. Culms usually in dense clusters; tiller leaves 0.5–2.0 mm wide; lemmas 2.0–3.0 mm · · · **6. angustifolia**

10. Base of culms swollen and bulb-like · · · **14(i). bulbosa var. bulbosa**

10. Base of culms not swollen · · · 11.

11. Base of culms surrounded by a dense mass of dead leaf-sheaths; mountains · · · 12.

11. Base of culms with few or no persistent dead leaf-sheaths; widespread · · · 13.

12. Leaves 2.0–4.5 mm wide when flattened, parallel-sided and more or less abruptly narrowed to the apex, the uppermost usually arising below halfway up the culm · · · **15(i). alpina var. alpina**

12. Leaves 1 2 mm wide when flattened, gradually tapered to the apex, the uppermost usually arising above halfway up the culm · · · **8. flexuosa**

13. Annual, or perennial due to procumbent stems rooting; culms usually procumbent to ascending; anthers less than 1.0(–1.3) mm, usually some leaves transversely wrinkled · · · 14.

13. Perennial; culms usually erect; anthers usually more than 1.3 mm, leaves not transversely wrinkled · · · 15.

14. Panicle branches spreading and usually becoming erect to suberect by fruiting time; anthers 0.2–0.5 mm, 1.0 to 1.5 times as long as wide · · · **1. infirma**

14. Panicle branches usually patent to reflexed when fruiting; anthers 0.6–0.8(–1.3) mm, 2 to 3 times as long as wide · · · **2. annua**

15. Ligule of uppermost culm-leaf 4–10 mm, acute at apex; sheaths rough · · · **3. trivialis**

15. Ligule of uppermost culm-leaf less than 5 mm, obtuse to rounded at apex; sheaths smooth · · · 16.

16. Ligule of uppermost culm-leaf 0.2–0.5 mm, usually truncate at apex · · · **13. nemoralis**

16. Ligule of uppermost culm-leaf 0.8–4.0(–5) mm, usually obtuse at apex · · · 17.

17. Lowland plant about 30–150 cm; lowest panicle node with (3–)4–6(–8) branches; ligule of uppermost culm-leaf 2–4(–5) mm · · · **11. palustris**

17. Mountain plants 10–40 cm; lowest panicle node with 1–2(–4) branches; ligule of uppermost culm-leaf 1.0–2.5(–3.0) mm · · · **12. glauca**

1. P. infirma Kunth   Early Meadow Grass

Loosely tufted yellowish-green *annual herb. Culms* 1–10(–25) cm, erect, spreading or prostrate, very slender, usually branched near the base, smooth, with 1–3 nodes. *Leaves* 2–30 × 1–4 mm, green, glabrous, linear, abruptly pointed or obtuse at apex, folded or opening out, thin,

minutely rough on the margins, usually some transversly wrinkled; sheaths keeled, smooth, thin; ligules up to 3 mm, membranous. *Inflorescence* a panicle 0.5–1.5 cm, pale green, lanceolate to ovate, loose; branches in pairs or solitary, spreading, but becoming more or less erect by fruiting time, very fine, smooth, bare and undivided in the lower part; pedicels 0.3–3.0 mm. *Spikelets* 2–4 mm, ovate or oblong, compressed, 2- to 4-flowered, readily breaking up at maturity beneath each lemma. *Glumes* persistent, keeled, obtuse, thinly membranous, the lower 1.0–1.5 mm, ovate and 1-veined, the upper 1.3–2.5 mm, elliptical or oblong and 1- to 3-veined. *Lemmas* 2.0–2.5 mm, oblong and obtuse at apex in side view, keeled, membranous, with very thin tips and margins, 5-veined, densely hairy, with short, silky hairs up to or beyond the middle. *Paleas* narrowly elliptical, as long as the lemmas with the two keels densely hairy. *Anthers* 0.2–0.5 mm. *Caryopsis* ellipsoid, tightly embraced by the thin lemma and palea. *Flowers* 3–5. $2n = 14$.

Native. Rough, sandy ground, waysides and paths, usually near the sea. Channel Islands, Isles of Scilly, Cornwall, Devon and Dorset. South-west Europe; North Africa; south-west Asia.

2. P. annua L.   Annual Meadow Grass

Loosely to compactly tufted *annual* or short-lived *perennial herb. Culms* 3–30 cm, erect, spreading or prostrate, sometimes with a creeping base and rooting at the nodes, very slender, weak, unbranched, or branched towards the base, smooth, with 2–4 nodes. *Leaves* 1–14 cm × 1–5 mm, green, glabrous, linear, abruptly pointed or obtuse and hooded at apex, folded then opening out, weak, often crinkled when young, minutely rough only on the margins; sheaths compressed, keeled, smooth; ligules 1–5 mm, thinly membranous. *Inflorescence* a panicle 1–12 cm, pale to bright green, reddish or purplish, ovate or triangular, open and loose, or somewhat dense; branches mostly paired or solitary, spreading, smooth, patent to reflexed when fruiting, bare and undivided in the lower part; pedicels 0.3–4.0 mm. *Spikelets* 3–10 mm, ovate or oblong, 3- to 10-flowered, readily breaking up beneath each lemma at maturity. *Glumes* persistent, pointed at apex, keeled, the lower 1.5–3.0 mm, lanceolate to ovate and 1-veined, the upper 2–4 mm, elliptical or oblong and 3-veined. *Lemmas* 2.5–4.0 mm, overlapping, semi-elliptical or oblong and rather obtuse in side view, keeled, 5-veined, membranous and with broad delicate tips and margins, sparsely to densely hairy on the nerves below the middle or glabrous. *Paleas* slightly shorter than the lemmas, narrowly linear-elliptical, with hairy or rarely glabrous keel. *Anthers* 0.6–0.8 (–1.3) mm. *Caryopsis* narrowly ovoid, enclosed by the lemma and palea. *Flowers* throughout the year. $2n = 28$.

Very variable and the variants tending to remain distinct because of self-pollination.

Native. Cultivated and waste land, paths, waysides, lawns and close-cut turf, in damp as well as dry places, in the open or in shade, on a wide range of soils from sea level to high in mountains. Abundant throughout the

British Isles. Throughout nearly the whole world, but in the tropics mainly on mountains.

3. P. trivialis L.   Rough Meadow Grass

Loosely tufted *perennial herb*, with creeping leafy stolons. *Culms* 20–100 cm, erect, or usually spreading from a decumbent base, slender to rather stout, smooth, with 3–5 nodes. *Leaves* 3–20 cm × 1.5–6.0 mm, green or purplish, glabrous, linear, abruptly and sharply pointed at the apex, folded at first, afterwards flat, flaccid or firm, minutely rough or nearly smooth, glossy beneath; sheaths usually rough, rarely smooth, keeled; ligules 4–10 mm, pointed at the apex, membranous. *Inflorescence* a panicle 3–20 × up to 15 cm, purplish, reddish or green, ovate to oblong, erect or nodding, open and very loose, or contracted and rather dense; branches mostly in clusters of 3–7, fine, spreading, bare and undivided in the lower part, minutely scabrid; pedicels 0.3–2.0 mm. *Spikelets* 3–4(–6) mm, ovate to elliptical or oblong, compressed, 2- to 4-flowered, breaking up at maturity beneath each lemma. *Glumes* persistent, finely pointed, slightly unequal, with minutely rough keels, the lower 2.0–3.8 mm, lanceolate and 1-veined, the upper 2.5–4.3 mm, ovate and 3-veined. *Lemmas* 2.5–3.8 mm, overlapping at first, later with incurved margins, narrowly oblong and pointed at apex in side view, keeled, with short hairs on the keel up to about the middle and long crinkled hairs at the base up to 3.8 mm, the rest usually glabrous, distinctly 5-veined, with membranous tips and margins. *Paleas* nearly as long as the lemmas, linear-lanceolate, with 2 minutely rough keels. *Anthers* 1.5–2.0 mm. *Caryopsis* ellipsoid, tightly enclosed by the hardened lemma and palea. *Flowers* 6–7. $2n = 14$.

Native. Meadows and pastures in the lowlands, waste and cultivated land, ponds, streams, ditches and river margins, and marshes. At one time included in seed for permanent pasture. Very common throughout the British Isles. Europe; temperate Asia; North Africa; Macaronesia; North America.

4. P. humilis Ehrh. ex Hoffm.
                                       Spreading Meadow Grass
P. subcaerulea Sm.; *P. irrigata* Lindm.; *P. pratensis* subsp. *irrigata* (Lindm.) Lindb. fil.; *P. pratensis* subsp. *alpigena* auct.; *P. pratensis* subsp. *subcaerulea* (Sm.) Tutin

Perennial herb, with scattered, solitary vegetative shoots and stems, or small tufts, arising from slender, extensively creeping rhizomes. *Culms* 3.5–25.0(–40) cm, usually all solitary, more or less terete, erect or ascending from a bent base, slender, with 1–2 nodes, smooth. *Leaves* 3–15(–18) cm × (0.7–)1.5–4.0 mm, green or greenish-white, linear, obtuse or with a slightly pointed, hooded tip, smooth or slightly rough at the tip, folded or opening out; sheaths smooth, the basal compressed and keeled, often fringed with minute hairs at the junction with the lamina; ligules 0.5–2 mm, membranous. *Inflorescence* a panicle 2–8(–11) × 2–6 cm, green, purplish or whitish, ovate, loose and open, erect or nodding; branches mostly in pairs or threes, sometimes up to 5,

spreading, bare near the base, smooth, or rough near the tips; pedicels 0.5–3.5 mm. *Spikelets* 4–7 mm, ovate or elliptical, 2- to 4-flowered, compressed, breaking up at maturity beneath each lemma. *Glumes* persistent, tapering to a finely pointed tip, slightly unequal, rough on the keels, usually 3-veined, with membranous margins, the lower 3.0–4.5 mm and narrowly ovate, the upper 3–5 mm and ovate. *Lemmas* 3.0–5.0 mm, overlapping, oblong and abruptly pointed at apex in side view, the keel and marginal veins fringed up to about the middle with soft, short hairs, with longer crinkled hairs at the base, firm except for the membranous margins, finely 5-veined. *Paleas* nearly as long as the lemmas, narrowly elliptical with rough keels. *Anthers* 1.5–3.0 mm. *Caryopsis* narrowly ovoid, tightly enclosed by the hardened lemma and palea. *Flowers* 6–7. $2n = 38$–117.

Native. Grassland, waysides, old walls, stream and river sides, damp coastal sands and mountain slopes up to 2000 m or more. Probably throughout the British Isles, but much under-recorded and perhaps more frequent in the north and west. North and central Europe.

5. P. pratensis L.   Smooth Meadow Grass

Very variable *perennial herb*, with strong creeping rhizomes, forming loose to compact tufts. *Culms* 10–190 cm, in small clusters, erect or bent below, slender to relatively stout, cylindrical, smooth, with 2–4 nodes. *Leaves* up to 30 cm × 2–4(–6) mm, green or greyish-green, linear, abruptly pointed or bluntly hooded at apex, folded then opening out, rough to almost smooth; sheaths smooth, the lower compressed and keeled, glabrous or minutely hairy; ligules membranous, the lower very short, the upper 1–3 mm. *Inflorescence* a panicle 2–20 × 1–12 cm, purplish, green or greyish, ovate to pyramidal or oblong, erect or nodding, loose and open to contracted and rather dense; branches mostly in clusters of 3–5, spreading, unequal, hair-like, flexuous, minutely rough; pedicels 0.2–2.0 mm. *Spikelets* 4–6 mm, ovate to oblong, compressed, 2- to 5-flowered, breaking up at maturity beneath each lemma. *Glumes* persistent, pointed at apex, unequal, rough on the keels, the lower 2.0–3.5 mm, ovate and 1- to 3-veined, the upper 2.5–4.0 mm and 3-veined. *Lemmas* 3–4 mm, overlapping, oblong or ovate-oblong in side view, obtuse or slightly pointed at apex, thinly to densely hairy on the keel and marginal veins up to the middle, with long, fine crinkled hairs at the base, finely 5-veined, with thin tips and margins. *Paleas* narrowly elliptical, about as long as the lemmas, with 2 rough keels. *Anthers* 1.5–2.0 mm. *Caryopsis* narrowly ovoid, tightly enclosed by the hardened lemma and palea. *Flowers* 5–7. Sexual or apomictic. $2n = 42, 50$–78, 91, 98.

A very variable species with many strains.

Native. Meadows, pastures, waysides, waste places and walls mainly on well-drained, sandy, gravelly or loamy soils. It was an important pasture grass, but not now used in seed mixtures. Very common throughout the British Isles. Europe; south-west Asia; North America.

6. P. angustifolia L.   Narrow-leaved Meadow Grass
P. pratensis subsp. *angustifolia* (L.) Dumort.

Perennial herb, forming small compact tufts and spreading by slender, wiry rhizomes. *Culms* 20–70 cm, in dense clusters, erect, slender, stiff, smooth, with about 2 nodes below the middle. *Leaves* 3–30 cm × 1.5–2.0(–3.0) mm, green, long-linear, abruptly pointed or obtuse at apex, the basal bristle-like, folded about the midrib and bluntly keeled or flat, smooth or slightly rough near the tip, glabrous or minutely hairy above; sheaths smooth or the lower minutely rough, the basal keeled; ligules membranous, the lower extremely short, the upper up to 2 mm. *Inflorescence* a panicle 3–14 × up to 9 cm, purplish or green, erect or nodding, lanceolate to ovate, loose or contracted; branches hair-like, spreading, mostly in clusters of 3–5, unequal, minutely rough, bare in the lower part; pedicels 0.3–2.0 mm. *Spikelets* 2.5–5.0 mm, ovate to oblong, 2- to 5-flowered, compressed, breaking up at maturity beneath each lemma. *Glumes* persistent, pointed at apex, minutely rough on the keels, the lower 1.5–2.5 mm, narrowly ovate and 1-veined, the upper 2–3 mm, ovate or elliptical and 3-veined. *Lemmas* 2.0–3.5 mm, overlapping, narrowly oblong in side view, obtuse or slightly pointed at apex, keeled, with a fringe of short, white hairs on the keel and marginal veins below the middle, with longer crinkled hairs at the base, membranous at tip and margins, finely 5-veined. *Paleas* as long as the lemmas, narrowly elliptical, with minutely rough keels. *Anthers* 1.5–2.0 mm. *Caryopsis* narrowly ovoid, tightly enclosed by the hardened lemma and palea. *Flowers* 4–7. 2*n* = 46–63.

Native. Grassy places, rough ground, on walls and banks on chalky, sandy and gravelly soils and in partial shade, particularly in beech hangers. Probably frequent throughout the British Isles. Much of Europe and south-west Asia.

7. P. chaixii Vill.   Broad-leaved Meadow Grass

Compactly tufted *perennial herb*. *Culms* 60–130 cm, erect or bent at the base, moderately stout, often compressed, rough near the panicle or smooth, with 2–3 nodes, vegetative shoots much flattened. *Leaves* up to 45 cm × 4.5–12.0 mm, bright green, linear, abruptly and sharply pointed at apex, at first folded and hooded at the tip, afterwards flat, firm, rough beneath on the projecting midrib and on the margins or only on the latter, minutely hairy on the upper surface, finely veined and also with numerous lateral veins; sheaths compressed and sharply keeled, prominently veined, usually minutely rough on the veins; ligules 0.5–2.0 mm, membranous. *Inflorescence* a panicle 10–25 × 5–12 cm, green, erect or slightly nodding, open and loose, ovate to ovate-oblong; rhachis minutely rough, or smooth below; branches in clusters of up to 7, very fine, spreading or the lower drooping, flexuous, bare and undivided towards the base, rough; pedicels 0.5–3.0 mm. *Spikelets* 4–7 mm, crowded, ovate or oblong, compressed, loosely 3- to 4-flowered, breaking up at maturity beneath each lemma. *Glumes* persistent, unequal, pointed at apex, firm, rough upwards on the keels, the lower 2.0–3.5 mm, lanceolate and 1- to 3-

veined, the upper 3–4 mm, ovate-elliptical and 3-veined. *Lemmas* 3.5–4.0 mm, at first overlapping, lanceolate-oblong in side view, abruptly pointed at apex, keeled, minutely rough, glabrous, prominently 5-veined, firm. *Paleas* as long as the lemmas, narrowly elliptical, with 2 rough keels. *Anthers* 2.0–2.5 mm. *Caryopsis* narrowly ellipsoid, tightly enclosed by the hardened lemma and palea. *Flowers* 5–7. 2*n* = 14.

Introduced. Grown in gardens for ornament and escaping and becoming naturalised in woods and copses. Sparse in England and Wales, more frequent in the policies of large houses in Scotland and in Co. Down in Ireland. Native of central and south Europe, mainly in mountainous districts.

8. P. flexuosa Sm.   Wavy Meadow Grass
P. laxa var. *scotica* Druce; *P. laxa* subsp. *flexuosa* (Sm.) Hyl.

Tufted *perennial herb* without rhizomes. *Culms* 6–15 cm, erect or slightly spreading, very slender, unbranched, smooth, with 1–2 nodes towards the base. *Leaves* 2–6 cm × 1–2 mm, the uppermost usually arising less than halfway up the culm, green, glabrous, linear, with a slender, slightly hooded, pointed tip, folded or opening out, minutely rough on the margins and on the upper surface towards the tip, smooth beneath; sheaths smooth, narrow; ligules 1–3(–4) mm, whitish-membranous. *Inflorescence* a panicle 1–5 × 1–2 cm, erect, loose, lanceolate to narrowly ovate; rhachis smooth; branches usually in pairs, very fine, smooth, bearing few spikelets, flexuous; pedicels 1–5 mm. *Spikelets* 3.0–5.5 mm, ovate to elliptical, compressed, 2- to 4-flowered, variegated with purple, green, gold and white, breaking up beneath each lemma at maturity. *Glumes* persistent, slightly unequal, keeled, minutely rough on the keels towards the tips, pointed at apex, with broad, membranous margins, the lower 2.5–3.5 mm, ovate and 1- to 3-veined, the upper 2.8–4.0 mm, oblong-elliptical and 3-veined. *Lemmas* 3–4 mm, closely overlapping, oblong and abruptly pointed or blunt in side view, keeled, with a fringe of short hairs along the keel and marginal veins in the lower third, and with a few crinkled hairs at the base, rough on the keel, finely 5-veined, membranous in the upper part. *Palea* about three-quarters the length of the lemma, elliptical, with minutely rough keels. *Anthers* 0.8–1.2 mm. *Caryopsis* narrowly ovoid, enclosed by the lemma and palea. *Flowers* 7–8. 2*n* = 42.

Native. Mountain screes and ledges between 800 and 1100 m. Very rare on Ben Nevis, Cairntoul and Cairn Gorm and formerly on Lochnagar in central Scotland. Mountains of north-west Europe.

9. P. × jemtlandica (S. Almq.) K. Richt.
                                          Swedish Meadow Grass
    P. alpina × flexuosa
P. alpina subsp. *jemtlandica* S. Almq.

Densely tufted, pruinose, slender *perennial herb* with both intravaginal and extravaginal shoots. *Culms* 10–20 cm, with the remains of leaf-sheaths at the base. Leaves 2–6 cm × 1.0–2.5 mm, the uppermost usually arising above halfway up the culm, flat or folded, linear, tapering gradually to an acute apex; ligule 1.0–3.5 mm, *Inflorescence* a panicle 2–5 cm, lax, irregular, purplish, proliferous, branches more or less nodding and spreading in flower; rhachis and branches not flexuous. *Spikelets* rather few towards the apex of the panicle branches, with 2–4 proliferating flowers. *Glumes* subequal, long-pointed, the lower 2.5–4.5 mm, the upper 3.2–5.0(–6.0) mm. *Lemma* hairy on keel and marginal veins at base. *Flowers* 7–8. $2n = 36, 42$.

This hybrid is remarkably uniform throughout its range and is always sterile, reproducing by bulbils (proliferous). Its basal leaves are less persistent than in *P. alpina* and it has narrower leaves than *P. flexuosa*.

Native. It is known from rock-ledges and damp stony slopes. It tolerates a fair amount of shade under and between rocks and survives in water-fed gullies. Ben Nevis in Inverness-shire and Lochnagar and Cairntoul in Aberdeenshire. It also occurs on mountains in southern and central Norway and central Sweden.

10. P. compressa L.   Flattened Meadow Grass

Stiff *perennial herb*, spreading by wiry rhizomes. *Culms* 10–60 cm, in loose tufts or scattered, erect or bent and ascending, slender, flattened but not always in the upper part, wiry, smooth, with 4–6(–9) nodes in lower half. *Leaves* 2–12 cm × 1–4 mm, bluish- or greyish-green, glabrous, linear, abruptly pointed or obtuse at apex, folded or opening out, usually stiff, more or less rough, or smooth beneath; sheaths compressed, keeled, smooth; ligules 0.5–3.0 mm, obtuse at apex, membranous. *Inflorescence* a panicle 1.5–10.0 × 0.5–3.0 cm, green, yellowish-green or purplish, stiff, narrowly oblong to ovate, often contracted and dense or open and moderately loose; branches paired or clustered, angular, slightly rough, with spikelets to the base, or the longer ones bare; pedicels 0.3–2.0 mm. *Spikelets* 3–8 mm, densely clustered, ovate, elliptical or oblong, compressed, 3- to 10-flowered, breaking up at maturity beneath each lemma. *Glumes* 1.6–3.0 mm, persistent, ovate, oblong, or elliptical, pointed at apex, equal or slightly unequal, firm except for the membranous margins, 3-veined, rough on the keels. *Lemmas* 2.5–3.3 mm, at first overlapping, later with inflexed margins, narrowly oblong and blunt in side view, firm except for the membranous tips, obscurely 5-veined, softly hairy on the keel and marginal veins below the middle or only near the base, sometimes also with long crinkled hairs at the base, or glabrous. *Paleas* nearly as long as the lemmas, narrowly elliptical, with 2 rough keels. *Anthers* 1.3–1.5 mm. *Caryopsis* ellipsoid, tightly enclosed by the hardened lemma and palea. *Flowers* 6–8. $2n = 35, 42, 49, 56$.

Native. Paths, walls, waysides, dry banks, waste ground and poor grassland. Scattered throughout the British Isles except in north Scotland. Most of Europe; Siberia; introduced in North America.

11. P. palustris L.   Swamp Meadow Grass

Short-lived, loosely tufted *perennial* herb without rhizomes. *Culms* 30–150 cm, erect or spreading, sometimes bent and rooting at the basal nodes, weak and slender to relatively stout, usually unbranched, smooth, with 3–4 nodes. *Leaves* up to 20 cm × 2–4 mm, green, glabrous, thin, linear, pointed at apex, usually flaccid, rough; sheaths smooth, the lower slightly keeled; ligules 2–5 mm, oblong, membranous becoming torn. *Inflorescence* a panicle 10–30 cm × up to 15 cm, yellowish-green or purplish, ovate to oblong, open and loose, erect or mostly nodding; branches mostly in distant clusters of 3–8, spreading, fine, flexuous, rough, bare and undivided in the lower part, loosely divided above; pedicels 1–5 mm. *Spikelets* 2.5–5.2 mm, ovate to oblong, compressed, 2- to 5-flowered, breaking up at maturity beneath each lemma. *Glumes* persistent, equal or slightly unequal, finely pointed at apex, keeled, rough on the keel, the lower 2–3 mm, lanceolate and 1- to 3-veined, the upper 2.5–3.0 mm, narrowly ovate or elliptical and 3-veined. *Lemmas* 2.5–3.3 mm, overlapping, narrowly oblong and rather obtuse in side view, keeled, usually with golden or bronze tips, finely 5-veined, firm except for the membranous tip and margins, the keels and marginal veins fringed below the middle with short white hairs, also with longer crinkled hairs at the base. *Paleas* about as long as the lemmas, narrowly elliptical, with 2 rough keels. *Anthers* 1.0–1.5 mm. *Caryopsis* ellipsoid, tightly enclosed by the hardened lemma and palea. *Flowers* 6–7. $2n = 28$.

Introduced about 1814 as a fodder grass. Marsh and fen, ditches, shallow water-filled gravel workings, damp grassland and loch margins in northern England and Scotland, rarely by railways and on waste ground. Scattered through the British Isles north to central Scotland, but mainly in south and central England. Most of Europe except much of the west and part of the south; temperate Asia; North America.

12. P. glauca Vahl   Glaucous Meadow Grass
P. balfourii Parn.; *P. caesia* Sm.

Tufted *perennial* herb, without rhizomes. *Culms* (10–)15–25(–40) cm, mostly erect, stiff, slender, smooth, with up to 3 nodes below the middle. *Leaves* 2–8 cm × 2–3(–4) mm, usually bluish-grey from a covering of whitish wax, sometimes green, glabrous, linear, with an abruptly pointed hooded tip, folded or opening out, stiff, spreading, smooth or nearly so; sheath mostly overlapping, rounded on the back or slightly keeled, smooth; ligules 1.0–2.5(–3.0) mm, membranous, blunt at apex. *Inflorescence* a panicle (1.2–)2.0–6.0(–7.0) cm, usually stiffly erect, sometimes lax and nodding, lanceolate to ovate, open to contracted, usually variegated with mauve; branches mostly in pairs or threes, angular, stiff, rough, more or less spreading, bare and undivided in the lower part; pedicels 0.5–4 mm. *Spikelets* (2.5–)3.0–4.5(–6.3) mm, ovate to oblong, compressed, 3- to 6-flowered, breaking up at maturity beneath each lemma. *Glumes* persistent, equal or nearly so, (2.5–)2.8–4.5(–4.7) mm, ovate to elliptical, pointed at apex, keeled, 3-veined, minutely rough on the keels in the upper part. *Lemmas* 3.0–4.0(–4.6) mm, overlapping, oblong and rather blunt in side view, keeled, densely silky-hairy on the keels and marginal veins up to the middle, sometimes with shorter hairs on the intermediate veins, finely 5-veined, firm and with broad membranous tip and margins. *Palea* about as long as the lemmas, narrowly elliptical, with the keels minutely hairy or rough. *Anthers* 1.5–2.0 mm. *Caryopsis* ellipsoid, enclosed by the lemma and palea. *Flowers* 7–8. 2*n* = 42, 44, 49, 56, 63.

P. balfourii is a shade variant of *P. glauca* with laxer habit and an absence of a deep mauve colour in the leaf and spikelets.

Native. Damp mountain rock-ledges and crevices and rocky slopes between 550 and 1100 m especially on Dalradian schist. Very local in central and north-west Scotland, the Lake District and Caernarvonshire. North Europe south to the Pyrenees and north Greece; north Asia; Japan; North and South America.

13. P. nemoralis L.   Wood Meadow Grass

Loosely tufted *perennial herb*, without rhizomes. *Culms* (15–)75–90 cm, erect or spreading, slender, smooth, with 3–5 nodes. *Leaves* 5–12 cm × 1–3 mm, green, glabrous, linear, finely to abruptly pointed, flat, usually weak, minutely rough or nearly smooth; sheaths smooth; ligules 0.2–0.5 mm, membranous. *Inflorescence* a panicle 3–20 cm, greenish or purplish, usually nodding, lanceolate to ovate or oblong, very lax and open or sometimes contracted; branches 3–6, clustered, hair-like, flexuous, spreading, bare and undivided in the lower part, minutely rough; pedicels 0.5–6.0 mm. *Spikelets* 3–6 mm, lanceolate to ovate or oblong, compressed, 1- to 5-flowered, breaking up at maturity beneath each lemma. *Glumes* persistent, equal or slightly unequal, finely pointed at apex, 3-veined, membranous, rough on the keels, the lower 2–3 mm and lanceolate, the upper 2.5–3.6 mm and lanceolate to ovate or oblong. *Lemmas* 2.6–3.6 mm, overlapping, narrowly oblong to lanceolate-oblong in side view, obtuse or slightly pointed at apex, keeled, finely 5-veined, with delicate tips and margins, the keel and marginal veins fringed with fine hairs up to the middle, with or without a few long hairs at the base. *Paleas* about as long as the lemmas, narrowly elliptical, with minutely rough keels. *Anthers* 1.3–2.0 mm. *Caryopsis* ellipsoid or narrowly ovoid, enclosed in the hardened lemma and palea. *Flowers* 6–7. 2*n* = 28–33, 42, 56.

Native. Rides and open areas in woodlands, hedgerows, walls and other shady places. Frequent to common in most of the British Isles, but probably introduced in much of Ireland and north-west Britain. Europe, but only on mountains in the south; temperate Asia; North Africa; Japan; North America.

14. P. bulbosa L.   Bulbous Meadow Grass

Tufted, bulbous-based *perennial herb*. *Culms* 5–40 cm, erect or spreading, very slender, unbranched, smooth, with 2–4 nodes and a swollen bulb-like base. *Leaves* 1–10 cm × 0.5–2 mm, green or greyish-green, glabrous, linear, abruptly pointed at apex, folded or opening out, firm to soft, minutely rough on the margins, otherwise smooth,

soon dying and breaking up; sheaths smooth, purplish or the upper green, the inner basal ones enlarged and fleshy, forming a bulbous, pear-shaped thickening at the base of the vegetative shoot, the outer basal sheaths membranous; ligules up to 4 mm, whitish-membranous. *Inflorescence* a panicle 2–6 × 1.0–2.5 cm, erect, ovate or oblong, contracted and rather dense; branches ascending, hair-like and often wavy, minutely rough; pedicels 0.3–3.0 mm. *Spikelets* 3–5 mm, sometimes proliferous, variegated with green, purple, gold or white, ovate to broadly oblong, compressed, 3- to 6-flowered, breaking up at maturity beneath each lemma. *Glumes* 2–3 mm, persistent, finely pointed at apex, equal, rough on the keels, with broad, membranous margins, the lower ovate and 1- to 3-veined, the upper broader and 3-veined. *Lemmas* 2.5–3.5 mm, closely overlapping, lanceolate to lanceolate-oblong in side view, pointed at apex, keeled, membranous, finely 5-veined, with a dense fringe of short, white hairs on the keel and marginal veins, and with fine, long crinkled hairs at the base. *Paleas* as long as the lemmas, narrowly elliptical, minutely hairy on the keels. Anthers 1.0–1.5 mm. *Caryopsis* ellipsoid, tightly enclosed between the lemmas and palea. *Flowers* 3–5.

(i) Var. bulbosa

Spikelets not proliferous. The above-ground plants wither early leaving the bulbous bases of the stem lying on the surface. These 'bulbils' can be dispersed by wind and re-grow in the autumn. 2*n* = 28, 45.

(ii) Var. vivipara Koel.

Spikelets proliferous. The plantlets lie on or near the surface until autumn 2*n* = 35.

Native. Rather bare places in short grassland and open ground on sandy soil, shingle or limestone near the sea, very rare inland. Coasts of southern Britain from Glamorgan to Lincolnshire. Europe, except the extreme west or parts of the north; temperate Asia; North Africa; Macaronesia; introduced in North America. Most plants are var. *bulbosa*. Var. *vivipara* occurs in Glamorgan (all plants) and Surrey and Jersey (with var. *bulbosa*).

15. P. alpina L.   Alpine Meadow Grass

Tufted *perennial herb*, thickened at the base with old leaf-sheaths, without rhizomes. *Culms* 5–40 cm, erect or spreading from a bent base, slender, unbranched, smooth, with 1–2 nodes below the middle or towards the base. *Leaves* 2–12 cm × 2–5 mm, green, mostly basal, glabrous, linear, obtuse or abruptly pointed at apex, folded about the middle vein or flat, minutely rough on the margins or nearly smooth; sheaths smooth, rounded on the back, the lower with membranous margins; ligules 2–6 mm, obtuse or pointed at apex, membranous. *Inflorescence* an ovate, erect or nodding, open, moderately loose to dense panicle 3–7 cm × up to 7 cm, purplish or green; axis smooth; branches usually in pairs and spreading. *Spikelets* usually proliferating, the upper part of each spikelet being replaced by a miniature new plant 4–7 mm, crowded towards the ends of branches, ovate to oblong, compressed, 2- to 5-flowered, breaking up at maturity beneath each lemma. *Glumes* persistent, equal or slightly unequal, keeled, ovate to elliptical, curved, pointed at apex, rough on the keels towards the tips, with broad, whitish-membranous margins, the lower 2.5–4.0 mm and 1- to 3-veined, the upper 3.0–4.5 mm and 3-veined. *Lemmas* 3.5–5.0 mm, closely overlapping, oblong or narrowly elliptical in side view, pointed at apex, keeled, finely 5-veined, with a dense fringe of fine, white hairs along the keel and marginal veins below the middle and with shorter hairs on the inner veins, with a broad membranous apex. *Paleas* as long as the lemmas, elliptical, with minutely hairy keels. *Anthers* 1.5–2.5 mm. *Flowers* 7–8. $2n = 28, 32$–$34, 35, 38, 42$.

(i) Var. alpina
Spikelets not proliferous, sexual. *Caryopsis* ovoid, tightly enclosed between lemma and palea.

(ii) Var. vivipara L.
Spikelets proliferous, the panicles becoming heavy and bending until they touch the ground, take root and eventually become separate plants.

Native. Damp mountain rock-ledges and crevices and rocky slopes between 300 and 1200 m. Local in central and north Scotland, north-west England, Caernarvonshire, south Kerry and Co. Sligo. The most frequent plant is var. *vivipara*. The sexual var. *alpina* occurs in a few places in northern England and Scotland. The species is found in Arctic Europe; Asia; America; North Africa; on mountains except in the north.

16. P. flabellata (Lam.) Rasp.   Tussock Grass
Festuca flabellata Lam.; *Paradiochloa flabellata* (Lam.) C. E. Hubb.
Densely tufted *perennial herb* forming large tussocks, the shoots extravaginal. *Culms* 45–55(–150) cm, stout, erect. *Leaves* 30–40(–70) cm × 4–12(–15) mm, linear, flat to channelled; ligules 8–12 mm, acute or obtuse at apex, usually lacerate. *Inflorescence* a panicle 8–21 × 1–4 cm, narrow, spike-like, dense, nodes with 2–4 branches. *Spikelets* with 1–4 flowers. *Glumes* subequal, the lower 1-veined, the upper 3-veined. *Lemmas* scabrid on the veins, with a scabrid terminal awn up to 2 mm. *Palea* keels aculeolate. *Anthers* 2.8–4.0 mm. *Flowers* 5–7.
Introduced. Very persistent where planted in yards and on walls in Shetland. Native of the extreme south of South America.

22. Dactylis L.

Perennial herbs without stolons and rhizomes, with strongly compressed tillers. *Leaf sheaths* not fused. *Inflorescence* a more or less one-sided panicle, simply lobed or formed of stalked, dense clusters of spikelets. *Spikelets* with 2–5 flowers, all except the most apical bisexual. *Glumes* 2, unequal. *Lemma* keeled on back, 5-veined, without or with a short terminal awn. *Stamens* 3.

Often regarded as one variable species, but consisting of a large and extremely variable tetraploid complex,

with a dozen or so little enclaves of diploids particularly around the Mediterranean region, but many segregates have been described. It is found throughout Eurasia and introduced elsewhere.

Beddows, A. R. (1959). *Dactylis glomerata* L. in Biological flora of the British Isles. *Jour. Ecol.* **47**: 223–239.
Domin, K. (1943). Monograficka studie o rodu *Dactylis* L. *Acta Bot. Bohem.* **14**: 3–147.
Grime, J. P. et al. (1988). *Dactylis glomerata* L. in *Comparative plant ecology* 214–215.

1. Lemma divided into 2 lobes
                        1(b). glomerata subsp. **hispanica**
1. Lemma not divided at apex   2.
2. Loosely tufted perennial; leaves usually green, not or only slightly rough; panicle slender with small clusters of spikelets; awns less than 0.7 mm
                1(a). glomerata subsp. **aschersoniana**
2. Densely tufted perennial herb; leaves more or less bluish- or greyish-green, very rough; panicles stout with large spikelets; awns 1–2 mm
                1(c). glomerata subsp. **glomerata**

1. D. glomerata L.   Cock's-foot
Densely to loosely tufted *perennial* herb, with compressed tillers. *Culms* 15–140 cm, erect or spreading, slender to stout, rough or smooth, more or less hairy, with 3–5 nodes. *Leaves* 10–45 cm × 2–14 mm, green, bluish-green or greyish-green, linear, sharply pointed at apex, at first folded, then opening and flat, firm, rough; sheaths keeled, at first entire, rough, glabrous or rarely shortly hairy; ligules 2–12 mm, membranous. *Panicles* 2–30 cm, green, purplish or yellowish, one-sided, erect, oblong to ovate, with the branches close together and spike-like, or usually with the lower distant; branches up to 18 cm, erect, spreading, or sometimes deflexed, stiff, angular, rough or hairy, rarely almost smooth, the longer bare at the base. *Spikelets* 5–9 mm, in dense one-sided masses at the ends of the branches, compressed, oblong or wedge-shaped, 2- to 5-flowered, nearly sessile, breaking up above the glumes. *Glumes* 4.0–6.5 mm, persistent, lanceolate to ovate, finely pointed at apex, membranous, 1- to 3-veined, rough or hairy on the keel. *Lemmas* 4–7 mm, closely overlapping, exceeding the glumes, lanceolate to oblong in side view, pointed or nearly obtuse at apex, sometimes divided at apex, tipped with a rigid awn up to 2 mm, firm except for the membranous margins, 5-veined, keels fringed with hairs or rough. *Paleas* lanceolate, shorter than or as long as the lemmas, the 2 keels minutely hairy or rough. *Anthers* 3–4 mm. *Caryopsis* narrowly ovoid, tightly enclosed by the hardened lemma and palea. *Flowers* 6–9.

(a) Subsp. aschersoniana (Graebn.) Thell.
D. glomerata subsp. *lobata* (Drejer) Lindb. fil.; *D. polgama* Horv.; *D. aschersoniana* Graebn.
Loosely tufted *perennial herb*. *Culms* up to 70 cm. *Leaves* green, glabrous to slightly prickly on the keel. *Panicle* slender with small clusters of spikelets. *Lemma* not split into 2 lobes, glabrous or obscurely prickly on the keel; awn less than 0.7 mm. $2n = 14$.

(b) Subsp. **hispanica** (Roth) Nyman
D. hispanica Roth

Densely tillering *perennial herb*. *Culms* up to 70 cm. *Leaves* green, glabrous to slightly prickly on the keel. *Panicle* compact and lobed. *Lemma* split into 2 lobes between which arises a short awn. $2n = 28$.

(c) Subsp. **glomerata**

Densely tufted *perennial herb*. *Culms* up to 140 cm. *Leaves* more or less glaucous, very rough. *Panicles* stout with large spikelets. *Lemma* not split into 2 lobes, with hairs or prickles on the keel; awns 1–2 mm. $2n = 28$.

Native or sometimes introduced. Grassland, open woodland, waysides and waste and cultivated ground. Much grown for hay and pasture, the more robust plants probably of introduced stock, often cultivars from the Plant Breeding Station at Aberystwyth. Common throughout the British Isles. Europe; temperate Asia; North Africa; Macaronesia; introduced in other temperate areas. Subsp. *glomerata* occurs throughout the range of the species. Subsp. *aschersoniana* is grown for ornament and is naturalised in woods in scattered localities in southern England from Buckinghamshire and Surrey to Dorset. It is also in western and central Europe. Subsp. *hispanica* has been recorded once from Co. Cork where it may be native or have been introduced. It is native in coastal habitats in the Mediterranean region.

23. Catabrosa P. Beauv.

Perennial herb with stolons, but without rhizomes. *Leaf-sheaths* not fused. *Inflorescence* a diffuse panicle. *Spikelets* with 1–3 flowers, all except the most apical bisexual. *Glumes* 2, unequal. *Lemma* rounded on back, 3-veined, truncate at apex, awnless. *Stamens* 3.

Two species in the north temperate zone and in Chile.

Perring, F. H. & Sell, P. D. (1967). *Catabrosa aquatica* (L.) P. Beauv. subsp. *minor* (Bab.) Perring & Sell stat. nov. *Watsonia* **6**: 317–318.

1. C. aquatica (L.) P. Beauv.   Whorl Grass

Creeping *perennial herb*, spreading by stolons and rooting at the nodes. *Culms* 5–75 cm, erect or ascending from a bent base, slender to somewhat stout, unbranched, smooth. *Leaves* 2–14 cm × 2–10 mm, bright green, glabrous, smooth, linear, equally wide throughout, obtuse at apex, folded when young, afterwards flat, rather thin; sheaths compressed, with free margins, the lower overlapping, the basal often purplish; ligules 2–8 mm, whitish, membranous. *Inflorescence* a panicle 5–30 × 2.5–10.0 cm, ovate to oblong, loose, erect; branches clustered, spreading, very slender, minutely rough; pedicels short. *Spikelets* 3–5 mm, green, yellow or brown, ovate to oblong, loosely 1- to 3-flowered, breaking up at maturity beneath each lemma. *Glumes* persistent, thinly membranous, smooth, blunt at apex, unequal, purple or white, the lower 1.0–1.5 mm and ovate to elliptical, the upper 1.5–2.5 mm and broader. *Lemmas* 2.5–3.5 mm, elliptic-oblong to oblong, rounded on the back, truncate at apex, firmly

membranous except for the whitish tips, prominently 3-veined, smooth, or with the veins minutely hairy. *Paleas* as long as the lemmas, narrowly elliptical, 2-keeled, smooth or minutely hairy on the keels. *Anthers* about 1.5 mm. *Caryopsis* ellipsoid, loosely enclosed between the lemmas and the palea. *Flowers* 5–7. $2n = 20$.

(a) Subsp. **aquatica**

Culms up to 75 cm, usually erect or ascending. *Leaves* up to 14 cm. *Panicles* up to 30 cm. *Spikelets* 1- to 3-flowered.

(b) Subsp. **minor** (Bab.) F. H. Perring & P. D. Sell
C. aquatica var. *uniflora* Gray; *C. aquatica* var. *littoralis* Parn.; *C. aquatica* var. *minor* Bab.

Culms 8–30 cm, often flopping. *Leaves* up to 5(–9) cm. *Panicles* up to 8 cm. *Spikelets* 1-flowered.

Native. Subsp. *aquatica* occurs on the muddy margins of ponds, slow-running streams and in swampy places, particularly where cattle have trodden, and sometimes floats in shallow water. Scattered throughout much of lowland British Isles, but becoming rare and decreasing in the south. Most of Europe; north and west Asia; North Africa. Subsp. *minor* occurs where fresh water runs through sand and muddy sand on the coast. It occurs in west Britain, particularly the Western Isles of Scotland, and in scattered localities in Ireland. Not recorded outside the British Isles.

24. Catapodium Link

Annual herbs. Leaf-sheaths not fused. *Inflorescence* a stiff raceme or a little-branched panicle. *Spikelets* with (3–)5–14 flowers, all except the apical bisexual. *Glumes* 2, subequal. *Lemma* rounded on the back or keeled distally, 5-veined, acute to obtuse or emarginate at apex, awnless. *Stamens* 3.

Clark, S. C. (1974). *Catapodium rigidum* (L.) C. E. Hubbard in Biological flora of the British Isles. *Jour. Ecol.* **62**: 937–958.
Perring, F. H. & Sell, P. D. (1967). *Catapodium rigidum* (L.) C. E. Hubbard subsp. *majus* (C. Presl) Perring and Sell comb. nov. *Watsonia* **6**: 317.

1. Inflorescence unbranched and spikelets sessile, sometimes shortly stalked, rhachis flattened   **1. marinum**
1. Inflorescence usually somewhat branched and at least some of the spikelets distinctly stalked, rhachis slender and angled   **2.**
2. Culms up to 15(–20) cm; leaves 0.8–1.5 mm wide; panicle linear in outline, 1-sided; lemmas 2.0–2.5 mm   **2(a). rigidum** subsp. **rigidum**
2. Culms up to 18–38 cm; leaves up to 3.5 mm wide; panicle loose, ovate in outline, branches spreading in all directions; lemmas 1.8–3.0 mm   **2(b). rigidum** subsp. **majus**

1. C. marinum (L.) C. E. Hubb.   Sea Fern Grass
Festuca marina L.; *Festuca rottboellioides* Kunth; *Poa loliacea* Huds.; *C. loliaceum* Link; *Desmazeria loliacea* Nyman; *Desmazeria marina* (L.) Druce

Annual herb. Culms 3–20 cm, tufted or solitary, erect, spreading or prostrate, slender, rigid, often branched towards the base, with few nodes, smooth. *Leaves* 1–10 cm × 1.0–3.5 mm, dark green, glabrous, linear, narrowed to a fine but obtuse apex, flat or rolled, minutely rough above, smooth beneath; sheaths smooth; ligules 0.5–3.0 mm, obtuse at apex, membranous. *Inflorescence* a panicle 0.5–7 cm × 4–12 mm, green or purplish, spike-like, narrow, rigid, branched in the lower part, with the branches up to 10 mm, erect, and bearing up to 4 spikelets, or unbranched and forming a raceme; rhachis flattened on the back, angular in front; pedicels extremely short. *Spikelets* 4–9 mm, in two rows on one side of the rhachis, touching or overlapping, lanceolate-oblong to oblong, slightly compressed, 4- to 12-flowered, slowly breaking up at maturity beneath each lemma. *Glumes* 2.0–3.5 mm, persistent, equal or nearly so, the lower lanceolate and 1- to 3-veined, the upper ovate or oblong and 3-veined. *Lemmas* 2.5–3.8 mm, exceeding the glumes, overlapping, narrowly elliptical or oblong-elliptical in side view, obtuse at apex, rounded on the back below, keeled above, nearly smooth, tough except for the narrow membranous margins, 5-veined. *Paleas* a little shorter than the lemmas, elliptical, with the 2 keels minutely hairy. *Anthers* 0.5–1.0 mm. *Caryopsis* shorter than the palea, ellipsoid, tightly enclosed between it and the hardened lemma. *Flowers* 5–7. $2n = 14$.

Native. Rather bare places on banks, sand, shingle and walls by the sea. Locally common round the coasts of the British Isles except for parts of the east and north coasts of Britain. South and west Europe; North Africa; south-west Asia; Macaronesia.

× rigidum
This hybrid is intermediate between the parents in habit, length of spikelet stalks, shape of spikelets, size of glumes and size, shape and imbrication of flowers. $2n = 14$.

One sterile plant on railway ballast at Aberdovey Station, Merioneth in 1960 is the only time this hybrid has ever been found.

2. C. rigidum (L.) C. E. Hubb.   Fern Grass
Poa rigida L.; *Festuca rigida* (L.) Raspail, non Roth; *Sclerochloa rigida* (L.) Link; *Desmazeria rigida* (L.) Tutin

Glabrous *annual herb*. *Culms* 2–38 cm, tufted or solitary, erect or spreading, very slender, often rigid, smooth, with 2–5 nodes. *Leaves* 1–10 cm × 0.5–3.5 mm, green or purplish, linear, finely pointed at apex, inrolled or flat, finely veined, minutely rough on the veins; sheaths smooth; ligules 1–3 mm, obtuse at apex, membranous. *Inflorescence* a panicle 1–8 × up to 2.5 cm, linear to ovate in outline, one-sided or spreading in all directions, stiff, rather dense to somewhat loose, branched in the lower part and unbranched above, or in weak plants with the branches reduced to solitary spikelets, green or purplish; branches up to 22 cm, rigid, 3-angled, smooth; pedicels up to 1.5 mm. *Spikelets* 4–7 × 1.0–1.5 mm, overlapping, appressed to one side of the rhachis and branches, narrowly oblong, 3- to 10-flowered, slowly

breaking up at maturity beneath each lemma. *Glumes* up to 2.3 mm, persistent, slightly unequal, pointed at apex, the lower lanceolate and 1- to 3-veined, the upper elliptical and 3-veined. *Lemmas* 1.8–3.0 mm, overlapping at first, becoming spaced, much exceeding the glumes, narrowly oblong in side view, obtuse at apex, rounded on the back, nearly smooth, 5-veined, tough except for the thin narrow margins. *Paleas* almost as long as the lemmas, narrowly elliptical, with the keels minutely rough. *Anthers* about 0.3 mm. *Caryopsis* narrowly oblong, tightly enclosed between the broadened lemma and palea. *Flowers* 5–7. $2n = 14$.

(a) Subsp. **rigidum**
Culms up to 15(–20) cm. *Leaves* up to 0.8–1.5 mm wide. *Panicles* linear in outline, 1-sided. *Lemmas* 2.0–2.5 mm.

(b) Subsp. **majus** (C. Presl) F. H. Perring & P. D. Sell
C. rigidum var. *majus* (C. Presl) Lainz; *Sclerochloa patens* C. Presl; *Sclerochloa rigida* var. *major* C. Presl
Culms up to 18–38 cm. *Leaves* up to 3.5 mm wide. *Panicles* loose, ovate in outline, branches spreading in all directions. *Lemmas* 1.8–3.0 mm.

Native. Subsp. *rigidum* occurs on dry banks, walls, stony, rocky, chalky and sandy places, and sometimes in short grassland. Locally common in the British Isles north to central Scotland. South-west Europe; North Africa; south-west Asia; Macaronesia. Subsp. *majus* occurs on the coasts of south-west Britain, Ireland and the Channel Islands. It rarely occurs as an introduction inland. It is also found on the coasts of south-west Europe and the Mediterranean.

25. Sesleria Scop.

Tufted *perennial herbs* with short rhizomes, but without stolons. *Leaf-sheaths* not fused. *Inflorescence* a small, very compact panicle. *Spikelets* with 2(–3) flowers, all bisexual. *Glumes* 2, subequal. *Lemma* rounded on back, 3- to 5-veined, 3- to 5-toothed at apex, each tooth with an awn 0–1.5 mm. *Stamens* 3.

Twenty-seven species in Europe, especially the Balkans. A difficult genus with closely related integrating species, perhaps better treated as subspecies.

1. S. caerulea (L.) Ard.   Blue Moor Grass
S. albicans Kit. ex Schultes; *Cynosurus caerulea* L.

Tufted *perennial herb*, with short, slender rhizomes. *Culms* 10–45 cm, erect, slender, wiry, smooth, with nodes only near the base. *Leaves* up to 20 cm × 2–6 mm, mostly basal, glabrous, bluish-green above, linear, equally wide up to the abruptly pointed, hooded tip, the uppermost very short, folded or flat, keeled beneath, firm, rough on the margins and below on the middle vein near the tip, otherwise smooth. *Inflorescence* a panicle 1–3 cm × 5–10 mm, spike-like, dense, ovate or oblong, cylindrical, usually bluish-grey or purplish-grey, glistening, with short, broad scales at the base; pedicels very short. Spikelets 4.5–7.0 mm, oblong, 2- to 3-flowered, breaking up at maturity beneath each lemma. *Glumes* 3–6 mm, persistent, equal or slightly unequal, ovate, mostly finely pointed, translucent, 1-veined. *Lemmas* 4–5 mm, with their tips equalling or projecting from the glumes, rounded on the back, keeled upwards, broadly oblong or elliptical, 3- to 5-veined, with a broad 3- to 5-toothed tip, the veins running into awn-points, the middle awn 0.5–1.0 mm, firm except for the membranous tips and margins, minutely rough or minutely hairy, especially on the nerves and margins. *Paleas* as long as or slightly longer than the lemmas, narrowly elliptical, densely short hairy on the 2 keels. *Anthers* 2.0–2.5 mm. *Caryopsis* narrowly ellipsoid, nearly 2 mm, hairy at the tip, loosely enclosed by the firm lemma and palea. *Flowers* 4–6. $2n = 28$.

Native. Abundant on calcareous grassland, rock-crevices, screes and limestone pavement in northern England and central and north-west Ireland, rare on micaceous schists in central Scotland and very rare in north Wales. The species occur throughout Europe. Our plant is subsp. **calcarea** (Čelak.) Hegi which is found in west and central Europe and Iceland.

Tribe 7. **Hainardieae** Greuter
Monermeae auct.

Annual herbs. Leaves with a membranous ligule; sheaths not fused. *Inflorescence* a very slender, cylindrical spike with alternating spikelets partly sunk in cavities in the rhachis which breaks up into 1-spikeleted segments at fruiting. *Spikelets* with one bisexual flower. *Glumes* 1–2, strongly veined and more or less horny. *Lemma* delicate, 3-veined with very short lateral veins, acute at apex, awnless. *Stamens* 3. *Stigmas* 2. *Ovary* with a rounded, glabrous appendage protruding beyond the style-bases, glabrous. *Lodicules* 2.

26. **Parapholis** C. E. Hubb.

Annual herbs. Culms up to 40 cm, erect to prostrate. *Leaves* up to 2.5 mm wide, flat or rolled. *Inflorescence* a spike. *Glumes* 2, inserted side by side and together covering the rhachis cavity except at anthesis. *Spikelets* 1-flowered. *Lemma* with its side towards the rhachis, membranous, 3-veined, the lateral veins very short. *Rhachis* disarticulating at maturity below each spikelet.

Six species in the Middle East and Mediterranean, northwards along the Atlantic coast of Europe to the Baltic.

Runiemark, H. (1962). A revision of *Parapholis* and *Monerma* in the Mediterranean. *Bot. Not.* **115**: 1–17.
Stewart, A., Pearman, D. A. & Preston, C. D. (1994). *Scarce plants in Britain.* Peterborough. [*P. incurva.*]

1. Spike straight or slightly curved; anthers mostly more than 2 mm   **1. strigosa**
1. Spike often strongly curved and tardily straight; anthers 0.5–1.0(–1.5) mm   **2. incurva**

1. **P. strigosa** (Dumort.) C. E. Hubb. Hard Grass
Lepturus strigosus Dumort.; *Lepturus filiformis* auct.

Slender *annual herb. Culms* up to 40 cm, loosely tufted or solitary, erect or bent, or curved and spreading, very slender, usually loosely branched, with few to many

nodes, smooth. *Leaves* 10–60 × 1.0–2.5 mm, greyish-green or green, linear, pointed at apex, flat or rolled, smooth beneath, rough on the veins above and on the margins, glabrous; sheaths rounded on the back, smooth, uppermost not inflated; ligules 0.3–1.0 mm, membranous. *Inflorescence* a spike 2–20 cm × 1.0–1.5 mm, usually straight and erect, sometimes curved, cylindrical, very slender, stiff, green or purplish; rhachis smooth, jointed, the joints deeply hollowed out on one side and breaking horizontally at maturity beneath each spikelet. *Spikelets* embedded in the hollows in the rhachis, solitary and alternating on opposite sides of the rhachis, 3–7 mm, closely pressed to the rhachis, falling with the joints of the rhachis at maturity, 1-flowered. *Glumes* 3–7 mm, equal or nearly so, as long as the spikelet, narrowly lanceolate, pointed at apex, rigid, hardened and thickened, 3- to 5-veined, glabrous, placed side by side in front of and covering the cavity in the rhachis. *Lemma* as long as or nearly as long as the glumes, ovate-oblong to oblong, obtuse at apex, membranous, finely 3-veined. *Palea* about as long as the lemma, lanceolate. *Anthers* 1.5–3.0(–4.0) mm. *Caryopsis* ellipsoid, tightly enclosed between the glumes and the joint of the spike rhachis. *Flowers* 6–8. $2n = 28$.

Native. Margins of and in salt-marshes, on sea defence walls and dyke sides and on waste ground near the sea on damp, heavy soils and sandy and gravelly muds. Along the coasts of the British Isles north to Lothian and the Isle of Mull. West Europe from Denmark southwards and eastwards to north Yugoslavia, but rare in the Mediterranean region.

2. **P. incurva** (L.) C. E. Hubb. Curved Hard Grass
Lepturus incurvus (L.) Druce; *Pholiurus incurvatus* A. S. Hitchc.; *Pholiurus incurvus* (L.) Schinz & Thell.; *Lepturus incurvatus* Trin.

Annual herb. Culms 2–20 cm, loosely to densely tufted or solitary, prostrate, curved and ascending, or erect, very slender, rigid, usually much branched in the lower part, smooth, with few to many nodes. *Leaves* up to 3(–8) cm × 1–2 mm, green, linear, flat or rolled, smooth beneath, rough on the veins above and on the margins, glabrous; sheaths rounded on the back, smooth, the uppermost inflated. *Ligules* 0.5–1.0 mm, membranous. *Inflorescence* a spike 1–8 cm × 1–2 mm, green or purplish, cylindrical, rigid, curved, or rarely straight, slender; rhachis smooth, jointed, the joints shorter than the spikelets, deeply hollowed out on one side, breaking horizontally beneath each spikelet at maturity. *Spikelets* embedded in the hollows of the rhachis, solitary and alternating on opposite sides of the rhachis and closely pressed to it, 4–6 mm, 1-flowered, oblong, falling with the joint of the rhachis at maturity. *Glumes* 4–6 mm, as long as the spikelet, placed side by side and closing the cavity of the axis, narrowly oblong-subulate, acute at apex, thick and rigid, 3 to 4-veined, glabrous. *Lemmas* slightly shorter the glumes, lanceolate to narrowly ovate, thinly membranous, finely 3-veined. *Palea* nearly as long as the lemma, lanceolate. *Anthers* 0.5–1.0(–1.5) mm.

Caryopsis narrowly ellipsoid, enclosed between the hardened glumes and the joints of the rhachis. *Flowers* 6–7. Cleistogamous. $2n = 38$.

Native. In bare places on the more elevated parts of salt-marshes and on shingle, wooden mooring stays, gravelly mud banks, sea walls, cliff-tops and ledges. It is occasionally introduced inland. Locally by the sea from Somerset and Dorset to Merionethshire, Norfolk and Northumberland; and in the Channel Islands. Very local in south-east Ireland. South and west Europe; south-west Asia; North Africa; Macaronesia.

27. **Hainardia** Greuter
Monerma auct.

Annual herbs. Culms erect to ascending. *Leaves* flat or rolled; ligule membranous. *Spikelets* sunk in cavities in the rhachis, dorsally compressed, with one flower. *Glumes* 2 in the terminal spikelet, one (the upper) in all the others and inserted so as to cover the rhachis-cavity except at anthesis. *Lemma* with its back towards the rhachis, membranous, 3-veined, the lateral veins very short. *Rhachis* disarticulating at maturity below each spikelet.

A single species in the Mediterranean area. Formerly known as *Monerma* which is actually an illegitimate name for *Lepturus*.

1. H. cylindrica (Willd.) Greuter
One-glumed Hard Grass
Rottboellia cylindrica Willd.; *Monermia cylindrica* (Willd.) Cosson & Durieu; *Lepturus cylindricus* (Willd.) Trin.

Annual herb. Culms up to 30(–45) cm, straight or somewhat curved, erect or spreading, solitary or loosely fascicled, smooth and glabrous. *Leaves* up to 6 cm × 1.0–2.5 mm, flat at first, folded or rolled in older plants, linear, acuminate at apex, scabrid on the upper surface and margins, smooth below, glabrous; sheaths rather loose below, tighter above, markedly striate, smooth and glabrous; ligules 1.0–1.5 mm, membranous, lacerate. *Inflorescence* a spike up to 10 cm, usually partly enveloped at the base by the uppermost leaf-sheath, consisting of a noded, fragile axis at each node of which, in alternate distichous excavations, are seated the solitary spikelets. *Spikelets* 6–7 mm, fitting the excavations in the axis but longer than the internodes. *Glumes* 2 in the terminal spikelet, solitary in the lateral, as long as the spikelet, coriaceous, 5- to 7-veined, the central vein particularly strong, acute to acuminate at apex, smooth and glabrous. *Lemmas* about 5 mm, hyaline, lanceolate, acute at apex, with the dorsal surface appressed against the excavation wall. *Palea* 2-veined, hyaline. *Anthers* about 4 mm. *Caryopsis* oblong, somewhat dorsally compressed with an appendage at the apex. *Flowers* 5–6. $2n = 26$.

Introduction. Fairly frequent casual from bird-seed on tips and waste ground. Scattered in southern England. Native of the Mediterranean region eastwards to Iran and in Macaronesia; introduced in Australia, South Africa and North and South America.

Tribe **8. Meliceae** Rchb.
Tribe *Glycerieae* Endl.

Perennial herbs with rhizomes or stolons. *Leaves* with a membranous ligule; sheath fused into a tube, often splitting later. *Inflorescence* a little or much-branched panicle or a raceme with pedicels more than 3 mm. *Spikelets* with 1–many bisexual flowers, if with less than 4 flowers then with a group of sterile ones beyond. *Glumes* 2, subequal or equal. *Lemmas* rounded on back, 7- to 9-veined, subacute to rounded at apex, awnless. *Stamens* 3. *Stigmas* 2. *Lodicules* fused laterally into a single scale shorter than wide. *Ovary* glabrous.

28. **Glyceria** R. Br. nom. conserv.

Glabrous aquatic or marsh *perennial herbs* with rhizomes and/or stolons. *Leaves* flat; sheaths with connate margins; ligules 2–10(–15) mm, acute or acuminate at apex. *Panicle* simple to much branched. *Spikelets* with 4–16 flowers, all except the most apical bisexual, and each falling separately as the seed ripens. *Glumes* unequal, hyaline, shorter than the first flower, 1(–3)-veined. *Lemmas* membranous with a hyaline apex, with 7 prominent veins. *Palea* nearly or quite equalling the lemma, tough, 2-veined, bifid. *Lodicules* 2, more or less connate, truncate, fleshy. *Stamens* 3. *Styles* terminal, short.

About 40 species in temperate regions throughout the world.

Lambert, J. M. (1947). *Glyceria maxima* (Hartm.) Holmb. in Biological flora of the British Isles. *Jour. Ecol.* **34**: 310–344.

1. Culms erect and self-supporting, usually more than 1 m; spikelets 5–12 mm, with 4–10 flowers; paleas not winged on keels .... **1. maxima**
1. Culms decumbent to ascending, if erect not self-supporting, rarely more than 1 m; spikelets 10–35 mm, with 6–17 flowers; paleas winged on keels .... **2.**
2. Spikelets remaining intact after not flowering or forming seeds; anthers remaining indehiscent; pollen grains all or most empty and shrunken .... **3. × pedicellata**
2. Spikelets breaking up between the flowers when the seed is ripe; anthers dehiscent; pollen grains full and turgid .... **3.**
3. Lemmas 5.5–6.5(–7.5); anthers 1.5–2.5(–3.0) mm **2. fluitans**
3. Lemmas 3.5–5.5 mm; anthers 0.6–1.5 mm .... **4.**
4. Lemmas distinctly 3(–5)-toothed at apex, exceeded by the 2 sharply pointed teeth of the palea .... **4. declinata**
4. Lemmas not or scarcely toothed at apex; not exceeded by the 2 very short teeth of the palea .... **5. notata**

Section **1. Hydropoa** Dumort.

Stock long and thick. *Culms* robust, erect. *Leaf-sheaths* terete. *Panicle* wide, much branched, with many spikelets. *Spikelets* 5–12 mm, moderately compressed before anthesis. *Palea* unwinged on the keels. *Stamens* 3.

1. G. maxima (Hartm.) O. Holmb.   Reed Sweet Grass
G. aquatica (L.) Wahlenb., non J. & C. Presl; *Molinia maxima* Hartm.

Stout *perennial herb*, with numerous vegetative shoots, spreading by stout rhizomes and covering large areas.

Culms 90–250 cm, erect, stout to robust, smooth or rough towards the panicle. *Leaves* 30–60 cm × 7–20 mm, green, glabrous, with lateral nerves, linear, abruptly pointed at apex, rough on the margins and sometimes beneath; sheaths entire, later splitting, keeled upwards, rough towards the lamina or smooth; ligules 3–6 mm, obtuse at apex, but generally with a central point, firmly membranous. *Inflorescence* a panicle 15–45 cm, open and loose or becoming contracted and rather dense, broadly ovate to oblong, much branched; branches clustered, very slender, rough, the lower up to 20 cm or more; pedicels 1–10 mm. *Spikelets* 5–12 × 2.0–3.5 mm, narrowly oblong or oblong, slightly compressed, closely 4- to 10-flowered, green or tinged with yellow or purple, slowly breaking up at maturity beneath each lemma. *Glumes* persistent, broadly ovate to oblong or elliptical, membranous, 1-veined, the lower 2–3 mm, the upper 3–4 mm. *Lemmas* 3–4 mm, overlapping, rounded on the back, elliptical to ovate-elliptical, very obtuse at apex, firm except for the membranous apex, prominently 7-veined, minutely rough on the veins. *Paleas* about as long as the lemmas, oblong, with 2 rough keels. *Anthers* (1.0–)1.5–2.0 mm. *Caryopsis* 1.5–2.0 mm, dark brown, ellipsoid, enclosed by the hardened lemma and palea. *Flowers* 6–8. 2n = 60.

Native. Ditches, by rivers, canals, ponds and lakes and marshy areas subject to flooding. Common in most of England except the north, scattered in Wales, Scotland and Ireland, but absent from north Scotland. Most of Europe; temperate Asia; Canada, where it is probably introduced.

Section 2. Glyceria

Stock short. *Culms* slender, often decumbent and ascending. *Leaf-sheath* compressed. *Panicle* wide to narrow, sometimes with a few spikelets only. *Spikelets* 8–35 mm, terete before anthesis. *Palea* winged on the keels. *Stamens* 3.

2. G. fluitans (L.) R. Br.   Floating Sweet Grass
Festuca fluitans L.

Loosely tufted *perennial herb* or forming loose masses in shallow water. *Culms* up to 1 m, erect or spreading, sometimes with a prostrate or floating base, with few nodes, slender to rather stout, smooth. *Leaves* 5–25 cm × 3–10 mm, green, glabrous, linear, pointed at apex, folded or flat, smooth except for the rough margins; sheaths sometimes purplish, glabrous, tubular, smooth; ligules 5–15 mm, lanceolate-oblong, membranous. *Inflorescence* a panicle 10–54 cm, open in flower, afterwards contracted and narrow, erect or curved and nodding, sparingly branched in the lower part, rhachis smooth; branches usually in pairs or solitary, the longer of a pair bearing 1–4 spikelets, the shorter with one spikelet, appressed to the rhachis after flowering; pedicels 1–4 mm. *Spikelets* 18–35 × 2.0–3.5 mm, narrowly oblong, green or purplish, 8- to 16-flowered, breaking up at maturity beneath the lemmas. *Glumes* persistent, elliptic-oblong or oblong, obtuse, 1- to 3-veined, thin, the lower 2–3 mm, the upper 3–5 mm.

Lemmas 5.5–6.5(–7.5) mm, rounded on the back, at first overlapping, later with incurved margins, elliptic-oblong or oblong, rather obtuse at apex, entire, 7-veined, firm except for the thin, whitish apex, minutely rough. *Paleas* oblanceolate, sharply 2-toothed, the teeth reaching the tip of the lemmas or usually shortly projecting. *Anthers* 1.5–2.5(–3.0) mm, violet. *Caryopsis* 2–3 mm, dark brown, elliptical, enclosed by the hardened lemma and palea. *Flowers* 5–8. 2n = 40.

Native. On mud or in shallow water by ditches, rivers, canals, ponds, lakes and marshy places, often in deeper water than other species of the genus. Common throughout the British Isles. Most of Europe; west Siberia; North Africa; North America.

3. G. × pedicellata F. Towns.   Hybrid Sweet Grass
G. fluitans × notata

Perennial herb, sometimes in large patches and with long floating stolons. *Culms* up to 1 m, ascending from an extensively creeping, branched base, slender to rather stout, fleshy, smooth. *Leaves* up to 35 cm × 5–12 mm, green, glabrous, linear, abruptly pointed or rather obtuse at apex, folded or flat, rough on the veins beneath and sometimes above, or smooth except for the rough margins; sheaths often minutely rough towards the lamina, or quite smooth; ligules up to 10 mm, whitish, membranous. *Inflorescence* a panicle 10–50 cm, lanceolate or oblong, loose; branches erect or finally spreading, slender, mostly in pairs or threes in the lower part of the panicle, single above or sometimes throughout, unequal, the longer up to 11 cm and bearing up to 9 spikelets; the shorter branches with 1 or 2 spikelets, smooth; pedicels 1–6 mm. *Spikelets* 10–35 mm, green or rarely purplish, linear-oblong, becoming slightly compressed, more or less persistent. *Glumes* broadly oblong to broadly elliptical, obtuse at apex, very thin, whitish, 1-veined, the lower 2–3 mm, the upper 3.0–4.5 mm. *Lemmas* 4.0–6.0 mm, overlapping, rounded on the back, elliptic-oblong, very obtuse at apex, firm except for the whitish membranous apex, prominently 7-veined, minutely rough. *Paleas* as long as the lemmas, oblong, shortly 2-toothed, with the keels narrowly winged in the upper part. *Anthers* 1.0–1.8 mm, pale yellow, remaining closed with imperfect pollen. *Caryopsis* not formed. *Flowers* 6–8. 2n = 40.

Native. Shallow ponds, streams, ditches and swampy depressions in pastures. A male-sterile hybrid which spreads vegetatively and may grow with one or both parents, or with neither. Scattered throughout the British Isles and frequent in England. West Europe.

4. G. declinata Bréb.   Small Sweet Grass

Perennial herb, usually shortly tufted. *Culms* 10–45(–60) cm, erect, or ascending from a curved or bent base, or prostrate, with 1–3 nodes, smooth. *Leaves* 3–18 cm × 1.5–8.0 mm, greyish-green or tinged with purple, glabrous, linear, equally wide throughout, abruptly pointed or obtuse, at first folded, becoming flat, smooth except for the rough margins, often rather stiff; sheaths keeled, entire, usually smooth; ligules 4–9 mm, long-pointed, membranous. *Inflorescence* a panicle 4–30 cm, linear to lanceolate, straight or curved, often secund, sparingly branched, rhachis smooth; branches solitary or in pairs or threes, appressed to or spreading on one side of the rhachis, smooth; pedicels 1.5–4.0 mm. *Spikelets* 13–25 × 1.5–2.0 mm, green or purplish, narrowly oblong, slightly compressed, 8- to 15-flowered, breaking up at maturity beneath each lemma. *Glumes* persistent, ovate to oblong, obtuse, membranous, usually 1-veined, smooth, the lower 1.5–2.5 mm, the upper 2.5–3.0 mm. *Lemmas* 3.5–5.5 mm, overlapping, much exceeding the glumes, broadly elliptic-oblong, usually with a broad 3-lobed or 3- to 5-toothed apex, becoming firm except for the thin whitish apex, 7-veined, minutely rough. *Paleas* narrowly elliptical, narrowed into a sharply toothed apex, which usually slightly projects beyond the tip of the lemma, with the 2 keels narrowly winged. *Anthers* 0.6–1.3 mm, purple or yellow. *Caryopsis* 1.5–2.3 mm, chestnut brown, ellipsoid, enclosed by the hardened lemma and palea. *Flower* 6–9. 2n = 20.

Native. On mud or in shallow water by ditches, rivers, canals, ponds and lakes. Scattered throughout the British Isles and probably under-recorded. West and central Europe; Macaronesia.

× fluitans
This is a sterile hybrid differing from *G.* × *pedicellata* in its obscurely 3-toothed lemmas. It has persistent spikelets, blunt lemmas 5.0–5.5 mm and sterile anthers 0.5–1.8 mm. 2n = 20.

Occurs rarely with the parents. Very rare in England. It has also been recorded from Sweden.

5. G. notata Chevall.   Plicate Sweet Grass
G. plicata (Fr.) Fr.; *G. fluitans* var. *plicata* Fr.

Perennial herb forming tufts or loose patches. *Culms* 30–75 cm, ascending from a prostrate base, rooting at the nodes, branched in the basal part, unbranched above, slender to relatively stout, spongy, smooth. *Leaves* 5–30 cm × 3–14 mm, green or greyish-green, linear, pointed at apex, folded or flat, rough on both sides, or nearly smooth above; sheaths entire, keeled, rough or minutely hairy; ligules 2–8 mm, oblong, membranous, whitish. *Inflorescence* a panicle usually rather broad, 10–45 cm, lanceolate to oblong, or broadly ovate, loose; branches finely and widely spreading, the lower in clusters of 2–5 with one branch longer than the rest and up to 12 cm, the others shorter and with one to few spikelets, slender, rough; pedicels 1–6 mm. *Spikelets* 10–25 × 1.5–2.0 mm, linear-oblong, at first cylindrical, later slightly compressed, 7- to 16-flowered, green or purplish, breaking up at maturity beneath each lemma. *Glumes* persistent, oblong to broadly elliptical, very obtuse at apex, membranous, 1-veined, the lower 1.5–2.5 mm, the upper 2.5–4.0 mm. *Lemmas* 3.5–5.0 mm, overlapping, later with incurved margins, rounded on the back, broadly elliptical to broadly obovate-oblong, very obtuse or very slightly 3-lobed at apex, prominently 7-veined, firm except for the broad, thin,

whitish tip, minutely rough. *Paleas* oblong, very obtuse at apex, as long as or more often shorter than the lemmas, narrowly winged on the 2 keels. *Anthers* 0.7–1.0(–1.5) mm. *Caryopsis* about 2 mm, ellipsoid, enclosed by the hardened lemma and palea. *Flowers* 6–8. 2n = 40.

Native. Ponds, ditches, streams and swampy places. Frequent throughout most of the British Isles except the north of Scotland. Most of Europe; west Asia.

29. Melica L.
Perennial herbs of woodland and mountains with short rhizomes. *Culms* slender. *Leaf-sheath* connate; ligules less than 2 mm, truncate at apex. *Inflorescence* a raceme or a sparsely branched panicle. *Spikelets* with 1–3, bisexual flowers plus a distal more or less clavate cluster of sterile vestiges, all the flowers falling as a unit when the seed is ripe. *Glumes* subequal, thin, unawned, 3- to 5-veined, nearly as long as the flowers. *Lemmas* obtuse at apex, rounded on the back, 5- to 9-veined. *Palea* membranous, 2-veined. *Lodicules* very short, truncate and connate. *Stamens* 3. *Styles* short and spreading.

About 80 species in temperate regions throughout the world, except Australia. Both our species belong to the Subgenus **Melica**.

Hempel, W. (1970). *Feddes Repert.* **81**: 131–145.

1. Leaf-sheaths without apical bristles; spikelets eventually nodding, with 2–3 flowers   **1. nutans**
1. Leaf-sheaths with a long bristle at apex; spikelets erect, with 1 fertile flower   **2. uniflora**

1. M. nutans L.   Mountain Melick
Perennial herb, with slender rhizomes. *Culms* 20–60 cm, loosely clustered or solitary, erect or spreading, slender, angular, minutely rough near the raceme. *Leaves* 4–20 cm × 2–6 mm, bright green, rolled when young, linear, thin and slender, with a fine but obtuse tip, flat, with short hairs above, minutely rough; sheaths tubular, 4-angled, minutely rough, the lower purplish; ligules very short, obtuse at apex, membranous. *Inflorescence* a raceme 3–15 cm, loose, nodding, one-sided, sometimes branched; pedicels hair-like, 3–15 mm, curved, the tips minutely hairy. *Spikelets* 6–8 mm, solitary or paired, nodding, elliptical-oblong, obtuse at apex, purplish or reddish-purple, glabrous, with 2–3 fertile flowers and with the rhachis terminating in a club-shaped mass of smaller sterile lemmas, the flowers falling together at maturity. *Glumes* 4–6 mm, persistent, similar, slightly unequal, elliptic-ovate to elliptical, obtuse at apex, rounded on the back, papery, membranous, finely 5-veined. *Fertile lemmas* 5–7 mm, closely overlapping, with their tips slightly exceeding the glumes, elliptical or elliptical-oblong, obtuse at apex, rounded on the back, 7- to 9-veined, becoming hardened, minutely rough. *Paleas* as long as or shorter than the lemmas, elliptical, with the 2 keels thickened, narrowly winged and minutely hairy. *Anthers* 1.5–2.0 mm. *Caryopsis* about 3 mm, ellipsoid, tightly enclosed by the tough lemma and palea, brown. *Flowers* 5–7. 2n = 18.

Native. Limestone outcrops in woods and scrub and shady rock-crevices on limestone, frequently near running water. Scattered through north and west Britain south to Monmouthshire and Northamptonshire. Most of Europe, eastwards to central Asia and Japan.

2. M. uniflora Retz.   Wood Melick

Perennial herb, forming loose, leafy patches, with slender, creeping, whitish rhizomes. *Culms* 20–60 cm, erect or spreading, slender, smooth. *Leaves* 5–20 cm × 3–7 mm, bright green, linear, narrowed to a fine point at apex, rather thin, shortly hairy above, minutely rough beneath and on the margins; sheaths tubular, tight, loosely hairy with short reflexed hairs or glabrous, on the side opposite the lamina with a slender bristle 1–4 mm at the apex, the basal overlapping, usually purplish; ligules short, membranous. *Inflorescence* a panicle 6–22 × 1–12 cm, very loose, sparingly branched, erect or nodding; branches spreading naked below, bearing 1–6 spikelets towards the tips, fine, rough; pedicels 2–5 mm. *Spikelets* 4–7 mm, erect, elliptic-oblong, with one fertile flower and 2–3 sterile lemmas in a club-shaped mass, the whole falling together at maturity. *Glumes* persistent, rounded on the back, firmly membranous, smooth, purple or brownish, equal or slightly unequal, with the upper 4.5–4.7 mm, as long as the spikelet, elliptical and 5-veined, the lower 3.8–4.4 mm, narrowly elliptical, 3-veined. *Fertile lemma* 4.0–5.5 mm, green, broadly rounded on the back, boat-shaped, elliptical, obtuse at apex, eventually tough and rigid, 7-veined, smooth. *Palea* as long as the lemma, elliptical, with 2, tough keels, narrowly winged and minutely hairy. *Sterile lemmas* up to 3 mm. *Anthers* 1.5–2.3 mm. *Caryopsis* about 3.5 mm, ellipsoid, tightly enclosed between the hardened lemma and palea. *Flowers* 5–7. $2n = 18$.

Native. Woods and shady hedgebanks on light and heavy soils; often abundant in open beech woods. Scattered and locally common throughout the British Isles except for north Scotland and the Channel Islands. Most of Europe; Caucasus; North Africa.

Tribe 9. Aveneae Dumort.

Annual or *perennial herbs*, with or without rhizomes and without stolons. *Leaves* with a membranous ligule; sheaths not or rarely fused to form a tube. *Inflorescence* a panicle, sometimes contracted, rarely a spike. *Spikelets* with 2–6(–11) flowers, sometimes one or more flowers sometimes proliferating. *Glumes* 2, more or less equal to unequal, often as long as or longer than the rest of the spikelet, often with wide, more or less shiny, hyaline margins. *Lemmas* 5- to 9-veined, rounded on the back, acute or obtuse to bifid or toothed at apex, awnless or with a short terminal or dorsal awn, conspicuously bent. *Stamens* 3. *Stigmas* 2. *Lodicules* 2. *Caryopsis* glabrous, or pubescent at apex or all over.

30. Helictotrichon Besser ex Schult. & Schult. fil.

Avenula (Dumort.) Dumort.; *Avenochloa* Holub; *Amphibromus* Nees; *Trisetum* section *Avenula* Dumort.

Tufted *perennial herbs* with or without short rhizomes.

Culms erect, stout. *Leaves* flat or channelled. *Inflorescence* a rather sparsely branched, more or less diffuse panicle. *Glumes* unequal, the lower 1- to 3-veined, the upper 3- to 5-veined. *Lemma* 5- to 7-veined, variously shortly toothed at apex, with a long, bent, dorsal awn. *Rhachilla* segments pubescent, the hairs in a longer tuft at the apex of each segment around the lemma. *Caryopsis* glabrous or pubescent at apex.

About 100 species in temperate Eurasia, extending across the tropical mountains to temperate regions throughout the world.

Dixon, J. M. (1991). *Avenula pratensis* (L.) Dumort. in Biological flora of the British Isles. *Jour. Ecol.* **79**: 829–846.
Dixon, J. M. (1991). *Avenula pubescens* (Hudson) Dumort. in Biological flora of the British Isles. *Jour. Ecol.* **79**: 846–865.

1. Ovary and caryopsis glabrous; lower glume 4–5 mm; upper glume 5–7 mm; lemmas 5–8 mm (excluding awn), papillose   **3. neesii**
1. Ovary and caryopsis with pubescent apex; lower glume 7–15 mm; upper glume 10–20 mm; lemmas 9–17 mm(excluding awn), smooth to scabrid   **2.**
2. Lower culm-sheaths softly pubescent; spikelet with 2–3(–4) flowers; rhachilla hair-tuft 3–6(–7) mm; palea with smooth keels   **1. pubescens**
2. Culm-sheaths glabrous; spikelets with 3–6(–8) flowers; rhachilla hair-tuft 1–3 mm; palea with scabrid keels   **2. pratense**

1. H. pubescens (Huds.) Besser ex Pilger
Downy Oat Grass

Avena pubescens Huds.; *Avenula pubescens* (Huds.) Dumort.; *Avenochloa pubescens* (Huds.) Holub.

Loosely tufted *perennial herb*, with short rhizomes. *Culms* 30–100 cm, erect or bent at the base, slender to somewhat stout, smooth, with 2–3 nodes. *Leaves* 2–30 cm × 2–6 mm, green, long-linear, pointed or obtuse at apex, folded when young, becoming flat, softly hairy or glabrescent; sheaths green or purple, tubular at first, but soon splitting, the lower loosely hairy with spreading or deflexed hairs, the upper nearly smooth; ligules up to 8 mm, acute at apex, membranous. *Inflorescence* a panicle 6–20 × up to 6 cm, green or purplish, erect or nodding, loose, lanceolate to oblong, glistening; branches clustered, fine, flexuous or straight, slightly rough, with 1–3 spikelets; pedicels 6–20 mm. *Spikelets* 10–17 mm, oblong, 2- to 3(–4)-flowered, breaking up at maturity beneath each lemma; rhachilla with hairs up to 3–6(–7) mm. *Glumes* persistent, finely pointed at apex, thin, 1- to 3-veined, the lower 7–13 mm and narrowly lanceolate, the upper 10–18 mm and wider. *Lemmas* 9–14 mm, narrowly oblong-lanceolate in side view, rounded on the back, toothed at the tips, tough except for the thin margins, 5-veined, bearded at the base with hairs 2–6(–7) mm, rough upwards, awned from the back at about the middle, the awn bent and twisted in the lower part and 12–22 mm. *Paleas* nearly as long as the lemmas, narrowly elliptical, with smooth keels. *Anthers* 5–7 mm. *Caryopsis* linear, hairy at the top, enclosed by the hardened lemma and palea. *Flowers* 5–7. $2n = 14$.

Native. Lowland grassland and the lower slopes of

hills, usually on base-rich soils. Throughout most of the British Isles and common on chalk and limestone. Europe from Arctic Norway to north Portugal and Bulgaria; Siberia; introduced in North America. Our plant is subsp. **pubescens**, which occurs throughout the range of the species.

2. H. pratense (L.) Besser   Meadow Oat Grass
Avena pratensis L.; *Avenula pratensis* (L.) Dumort.;
Avenochloa pratensis (L.) Holub

Densely tufted *perennial herb*. *Culms* 30–80 cm, erect, slender, stiff, with 1–2 nodes in the lower part. *Leaves* 4–30 × 1–5 mm, bluish-green above, glabrous, long-linear, obtuse at apex, stiff to rigid, folded when young, opening out, minutely rough on the margins, smooth beneath; sheaths rounded on the back or keeled upwards, glabrous, smooth or minutely rough; ligules membranous, the upper 2–5 mm, the lower shorter. *Panicles* 4–18 cm, green or purplish, glistening, erect, narrow, contracted; branches rough, paired or solitary, with 1–2 spikelets, the lower up to 3.5 cm, the upper shorter. *Spikelets* 11–28 mm, narrowly oblong to oblong, 3- to 6(–8)-flowered, breaking up at maturity beneath each lemma; rhachis shortly hairy. *Glumes* persistent, lanceolate to oblong-lanceolate, finely pointed at apex, firm except for the translucent margins, 3-veined, the lower 10–15 mm, the upper 12–20 mm. *Lemmas* 10–17 mm, narrowly oblong-lanceolate in side view, rounded on the back, toothed at the tip, tough except for the thin upper part and margins, 5-veined, minutely rough upwards, bearded at the base with hairs 1–3 mm, awned from just above the middle, with the awn 12–22(–27) mm and bent and twisted in the lower part. *Paleas* narrowly elliptical with minutely hairy keels. *Anthers* 5–8 mm. *Caryopsis* ellipsoid, hairy at the tip, enclosed by the hardened lemma and palea. *Flowers* 6–7. $2n = 126$.

Native. Natural grasslands, mainly on the chalk and limestone, usually in shorter turf than *H. pubescens* and commoner in mountains. Found over most of the British Isles except the Outer Isles of Scotland, central Wales and the extreme south-west of England. West and central Europe, extending southwards to north-east Spain and the Apennines; Caucasus. Our plant is subsp. **pratense** which occurs throughout the range of the species.

× **pubescens**
This hybrid has been recorded from Lincolnshire but never confirmed. It is not known elsewhere.

3. H. neesii (Steud.) Stace   Swamp Wallaby Grass
Amphibromus neesii Steud.

Tufted *perennial herb* without or with short rhizomes. *Culms* up to 1 m, erect, striate, glabrous. *Leaves* up to 70 cm × 1–2 mm, pale bluish-green, narrowly long-linear, acute at apex, glabrous; sheaths glabrous; ligules up to 9 mm, membranous, lanceolate, long-acute at apex. *Inflorescence* a sparsely branched, more or less diffuse panicle; rhachilla segments pubescent. *Spikelets* 9–15 mm, with (2–)4–7 flowers. *Glumes* unequal, the lower

4–5 mm, membranous, lanceolate, acute at apex, 1- to 3-veined and glabrous, the upper 5–7 mm, membranous, lanceolate, acute at apex, 3- to 5-veined and glabrous. *Lemmas* 5–8 mm excluding awn, lanceolate, gradually narrowed to a toothed apex 5- to 7-veined, papillose; awn dorsal, up to 18 mm, bent, minutely ascending-hairy; tuft at base 1–2 mm. *Caryopsis* ellipsoid, glabrous at top, enclosed by the hardened lemma and palea. *Flowers* 8–10.

Introduced. Wool alien found on tips and waste ground and in fields. Recorded in scattered localities in England. Native of Australia.

31. Arrhenatherum P. Beauv.

Loosely tufted *perennial herbs*. *Culms* erect, up to 150 cm. *Leaf sheath* not fused or rarely fused to a tube. *Inflorescence* a fairly well branched, more or less diffuse panicle. *Spikelets* with 2(–5) flowers, the lower (lowest) male, the upper bisexual, rarely both lower male. *Glumes* unequal, the lower 1-veined, the upper 3-veined. *Lemmas* 7-veined, bifid at apex, the lemma of the male flower with a long, bent dorsal awn, that of the bisexual flowers awnless or with a short terminal awn or rarely with a dorsal bent awn. *Rhachilla* segments with an apical hair-tuft 1–2 mm. *Caryopsis* pubescent at apex.

Six species in Europe, Mediterranean region and Middle East.

Pfitzenmeyer, C. D. C. (1962). *Arrhenatherum elatius* (L.) Beauv. ex J. & C. Presl in Biological flora of the British Isles. *Jour. Ecol.* **50**: 235–245.

1. Basal internodes not swollen; nodes usually glabrous
                                        1(a). elatius subsp. **elatius**
1. Culms with (1–)2–6(–8) swollen, globose basal
   internodes 6–10 mm in diameter; nodes often hairy
                                        1(b). elatius subsp. **bulbosum**

1. A. elatius (L.) P. Beauv. ex J. & C. Presl
                                                    False Oat-Grass
A. avenaceum P. Beauv.; *Avena elatior* L.

Loosely tufted *perennial herb*, with yellowish roots. *Culms* 50–180 cm, erect or slightly spreading, stout, smooth or hairy at the nodes, with 3–5 nodes, sometimes with the basal internodes swollen. *Leaves* 10–40 cm × 4–10 mm, green, long-linear, finely pointed at apex, flat, loosely to sparsely hairy above or glabrous, rough; sheaths rounded on the back, smooth, rarely rough or loosely hairy; ligules 1–3 mm, membranous, rounded at apex. *Inflorescence* a panicle 10–30 cm, green or purplish, shining, lanceolate to oblong, erect or nodding, loose or rather dense; branches clustered, rough; pedicels 1–10 mm. *Spikelets* 7–11 mm, oblong, 2- (rarely 3–5)-flowered, the lowest or lower usually male, rarely bisexual like the upper, the flowers falling together at maturity. *Glumes* persistent, finely pointed at apex, membranous, minutely rough, the lower 4–6 mm, lanceolate and 1-veined, shorter than the upper, the upper 7–10 mm, narrowly ovate and 3-veined, as long as or shorter than the spikelet. *Lemmas* 8–10 mm, rounded on the back, narrowly ovate or oblong-ovate, pointed at apex, 7-veined, firm except for the thin tips, short-bearded at the base, the upper or both loosely hairy or glabrous on the back, minutely rough upwards, the lower awned from the back in the lower third and the awn 10–20 mm, the upper awnless or with a fine, short bristle from near the tip or an awn from the back above the middle. *Paleas* lanceolate, acute at apex, with minutely hairy keels. *Anthers* 4–5 mm. *Caryopsis* ellipsoid, hairy, enveloped by the hardened lemma. *Flowers* 6–9. $2n = 28$.

(a) Subsp. elatius

Basal internodes not swollen. *Nodes* usually glabrous. $2n = 28$.

(b) Subsp. bulbosum (Willd.) Hyl.

Onion couch

A. elatius var. *bulbosum* (Willd.) St Amans;
A. tuberosum (Gilib.) F. W. Schultz

Culms with (1–)2–6(–8) swollen, globose basal internodes 6–10 mm in diameter. *Nodes* often hairy. $2n = 28$.

Native. Subsp. *elatius* is very common in rough grasslands, hedgerows, waysides, field margins, shingle and gravel banks and waste places. Throughout the British Isles. Most of Europe but only on mountains in the south and rare in the south-west. Subsp. *bulbosum* is of scattered occurrence and is probably under-recorded. It used to be a troublesome weed of agriculture, where the swollen, corm-like basal structures serve as effective propagules. South and west Europe.

32. Avena L.

Annual herbs. Culms erect, usually tall. *Leaves* flat. *Inflorescence* a diffuse panicle. *Spikelets* with 2–3 flowers, all bisexual, or the distal 1 or 2 reduced and male or sterile. *Glumes* subequal, 7- to 11-veined. *Lemmas* 7- to 9-veined, bifid or with 2 bristles at the apex, with or without a long, bent, dorsal awn. *Rhachilla segments* with or without a hair-tuft. *Stamens* 3. *Lodicules* 2, lanceolate. *Ovary* pubescent at apex or all over.

Chater, A. O. (1993). *Avena strigosa*, Bristle Oat, and other cereals as crops and casuals in Cardiganshire v.c. 46. *Welsh Bull. B.S.B.I.* 55: 7–9.

About 25 species, mainly in the Mediterranean region and Middle East, extending to northern Europe and widely introduced in other temperate regions with 2 species in Ethiopia.

1. Lemma bifid, the 2 apical points (1–)3–9 mm and each with 1 or more veins entering from the main body of the lemma and reaching the apex   2.
1. Lemma bifid, the 2 apical points 0.5–2.0 mm and without veins or veins not reaching the apex   3.
2. Rhachilla disarticulating between the flowers at maturity, releasing 1-fruited disseminules each with an elliptical basal scar; lemma with dense long hairs on the lower half   **1. barbata**
2. Rhachilla not disarticulating at maturity, the whole spikelets acting as disseminules or the flowers breaking away irregularly without a basal scar; lemma glabrous or sparsely pubescent in the lower half   **2. strigosa**

3. Rhachilla disarticulating at maturity at least above the glumes, often also between the flowers, hence at least the lowest flower with a basal scar; lemmas with a long, strongly bent awn, usually pubescent   4.
3. Rhachilla not disarticulating at maturity, the whole spikelets acting as dissseminules, or the flowers breaking away irregularly without a basal scar; lemmas usually without awns, if awned then the awn nearly straight, usually glabrous   5.
4. Rhachilla disarticulating at maturity between the flowers, releasing 1-fruited disseminules each with an ovate basal scar   **3. fatua**
4. Rhachilla disarticulating at maturity only above the glumes, releasing 2- to 3-fruited disseminules, hence only the lowest flower with an ovate basal scar   **4. sterilis**
5. Awn (when present) with a distinct column; rhachilla segments breaking at their apex and falling attached to the lower flower   **5. sativa**
5. Awn without a distinct column; rhachilla segments breaking at their base and falling attached to the upper flower   **6. byzantina**

1. A. barbata Pott ex Link   Slender Oat

Annual herb. Culms up to 100 cm, solitary or fascicled, erect or geniculately ascending, smooth and glabrous even at the nodes. *Leaves* up to 30 cm × 7 mm, long-linear, tapering to a long acuminate tip, scabrid on the upper surface and on the margins, glabrous or sparsely covered with hairs; sheaths clasping the stems, striate, glabrous or sparsely hairy, more or less rough towards the top; ligule 1–3 mm, membranous, truncate at apex, scabrid on the outer surface. *Inflorescence* a panicle up to 30(–50) cm, effuse, with spreading branches or somewhat one-sided; branches fascicled; rhachilla disarticulating between the flowers at maturity, releasing 1-fruited disseminules each with an elliptical basal scar. *Spikelets* 18–30 mm, 2- to 3-flowered. *Glumes* 20–30 mm, more or less equal, the lower 3- to 5-veined, the upper 5- to 7-veined, lanceolate or elliptical, acute at apex. *Lemmas* articulated to the rhachilla and caducous, the lowest up to 25 mm, cleft at the apex into 2 aristate lobes, very hairy below the awn insertion on the dorsal surface, the aristae not exceeding the glumes, awned on the back, the awns 30–60 mm, twisted below the knee. *Paleas* shorter than lemmas, narrowly elliptical, hairy on the keels. *Anthers* about 2.5 mm. *Caryopsis* hairy, oblong, adherent to the lemma and palea. *Flowers* 6. $2n = 14, 28$.

Introduced. Rare grain alien, naturalised in Guernsey since 1970. Mediterranean region eastwards to India; Macaronesia; introduced into South Africa, Australia and North America. Our plant is subsp. **barbata** which occurs throughout most of the range of the species.

2. A. strigosa Schreb.   Bristle Oat

Annual herb. Culms 60–120 cm, tufted or solitary, erect or bent at the base, stout, smooth, with 3–5 nodes. *Leaves* 8–25 cm × 5–10 mm, green, long-linear, finely pointed at apex, flat, firm, rough; sheaths rounded on the back, the lower loosely hairy, the upper smooth;

ligules 2–5 mm, obtuse at apex, membranous. *Inflorescence* a panicle 8–30 × up to 10 cm, green, erect, narrowly ovate, nodding; branches clustered, spreading, fine, loosely divided, rough; pedicels unequal; rhachilla not disarticulating at maturity, the whole spikelets acting as disseminules or the flowers breaking away irregularly without a basal scar. *Spikelets* 15–26 mm, loosely scattered, pendulous, narrowly oblong, 2-flowered. *Glumes* persistent, as long as the spikelet, equal or slightly unequal, lanceolate, finely pointed at apex, rounded on the back, smooth, 7- to 9-veined, becoming papery except for the narrow, membranous margins. *Lemmas* 10–18 mm, lanceolate or oblong-lanceolate in side view, finely 2-toothed at tip with each tooth becoming a fine bristle 3–6(–9) mm, becoming tough and rigid, stiffly hairy all over or only in the upper part or glabrous, rough upwards, 7-veined, awned from about the middle of the back, with the stout awn 17–28(–35) mm, rough, bent, dark brown and twisted in the lower part. *Paleas* narrowly elliptical, shorter than the lemmas, minutely hairy on the 2 keels. *Anthers* 2.5–4.0 mm. *Caryopsis* hairy, narrowly ellipsoid, tightly enclosed by the hard lemma and palea. *Flowers* 7–8. $2n = 42$.

Introduced. Formerly, and rarely still, grown as a minor crop in Wales, Scotland and Ireland, and there a frequent cornfield weed, sometimes naturalised. Also an infrequent grain casual. North, west and central Europe.

3. A. fatua L. Wild Oat

Annual herb. Culms 30–150 cm, tufted or solitary, erect or bent at the base, stout, smooth, with 3–5 nodes. *Leaves* 10–45 cm × 3–15 mm, green, narrowly linear-lanceolate, finely pointed at apex, flat, rough; sheaths rounded on the back, the basal usually loosely hairy, the rest smooth; ligules up to 6 mm, obtuse at apex, membranous. *Panicles* 10–40 × up to 20 cm, green, nodding, narrowly to broadly pyramidal, loose; branches widely spreading, mostly clustered, fine, loosely divided, rough; pedicels unequal; rhachilla disarticulating at maturity between the flowers releasing 1-fruited disseminules each with an ovate basal scar. *Spikelets* 18–25(–30) mm, loosely scattered, pendulous, narrowly oblong, 2- to 3-flowered, with all lemmas awned. *Glumes* 20–30 mm, persistent, lanceolate, finely pointed at apex, as long as the spikelet, equal or slightly unequal, smooth, 7- to 11-veined, becoming papery except for the thinner margins. *Lemmas* 14–20 mm, narrowly oblong-lanceolate in side view, shortly 2- to 4-toothed at the apex, rounded on the back, becoming tough and rigid, stiffly hairy in the lower half, rough above, 7- to 9-veined, finally brown, with a dense beard 1.5–4.0 mm around the ovate basal scar, awned from the middle of the back, with the stout awn 2.5–4.0 mm, bent, twisted and dark brown in the lower part. *Paleas* shorter than the lemmas, lanceolate, densely minutely hairy on the 2 keels. *Anthers* about 3 mm. *Caryopsis* oblong, hairy or glabrous, tightly enclosed by the hard lemma and palea. *Flowers* 6–9. $2n = 42$.

Introduced. Weed of arable, waste and rough ground.

Common in most of England, but scattered elsewhere in the British Isles. Most of Europe, but native only in the south.

× sativa

This hybrid resembles *A. sativa* in appearance, but has longer awns and tardily disarticulating spikelets which are *A. fatua*-influenced. The lemma has 2 minute teeth at the apex and the awn is inserted halfway up the back and is twisted in the lower half. The caryopsis has no tuft of hairs at the base and no basal scar, but up to 5 hairs about 3 mm are found at its base. The caryopsis is paler in colour than *A. sativa*. It also has a lower fertility. $2n = 42$.

Occurs rarely in Britain where fields of *A. sativa* are grown and are infested with *A. fatua*. The hybrids may be more frequent than they are recorded. They also occur sporadically over Europe and throughout much of Canada and the U.S.A. They are rarely persistent in one place.

4. A. sterilis L. Winter Wild Oat

Annual herb. Culms 60–180 cm, tufted or solitary, erect or bent at the base, stout, smooth, with 2–4 nodes. *Leaves* up to 60 cm × 6–14 mm, green, narrowly linear-lanceolate, finely pointed at apex, firm, rough, glabrous; sheaths rounded on the back, the basal slightly hairy, the upper smooth; ligules up to 8 mm, obtuse at apex, membranous. *Inflorescence* a panicle 15–45 × 8–25 cm, green, nodding, pyramidal, very loose; branches clustered, spreading, rough; pedicels 5–35 mm; rhachilla disarticulating at maturity only above the glumes, releasing 2- to 3-fruited disseminules, hence only the lowest flower with an ovate basal scar. *Spikelets* 23–32 mm, scattered, pendulous, lanceolate, 2-awned, 2- to 3-flowered. *Glumes* 20–32 mm, persistent, lanceolate, finely pointed at apex, rounded on the back, as long as the spikelet, 9- to 11-veined, smooth, finally papery except for the thinner shining margins. *Lemmas* 15–25 mm, narrowly lanceolate, rounded on the back, becoming tough except for the 2-toothed membranous tip, mostly stiffly hairy except for the rough upper third, bearded at the base with hairs up to 5 mm, finally light to dark brown, 7-veined, the lowest (largest) with an ovate-shaped thickening at the base, awned from the middle of the back, the awn 30–55 mm, stout in the lowest part, bent, twisted, dark brown and minutely hairy, where a third flower is present it has no awn. *Paleas* lanceolate, with 2 minutely hairy keels. *Anthers* 2.5–3.0 mm, hairy. *Caryopsis* oblong, tightly enclosed by the hard lemma and palea. *Flowers* 7–8. $2n = 42$.

Introduced. Weed of arable, waste and rough ground, particularly on heavy soils. Common in most of England, very scattered elsewhere in the British Isles. Native of south Europe, North Africa and south-west Asia. Our plant is mostly subsp. **ludoviciana** (Durieu) Gillet & Magne (*A. ludoviciana* Durieu) which occurs throughout most of the range of the species. Subsp. **sterilis** differs in its spikelets having 3–5 flowers, the glumes 32–45 mm and the lemmas 25–33 mm. It is a rare grain alien. It also occurs throughout most of the range of the species.

5. A. sativa L.   Oat

Annual herb. Culm up to 150(–200) cm, solitary or fascicled, erect, smooth, glabrous. *Leaves* up to 45 cm × 4–15 mm, bluish-green, linear, acuminate at apex, flat, rough on both surfaces and the margins, glabrous, rounded at the base; sheaths clasping or rather loose, glabrous; ligules short, membranous, toothed. *Inflorescence* a panicle up to 25 cm, pyramidal, nodding, more or less unilateral; rhachilla below and between the flowers tough, breaking with a horizontal irregular fracture at their apex and falling attached to the lower flower; none with a basal scar. *Spikelets* 17–30 mm, lanceolate, pendulous, 2- to 3-flowered. *Glumes* 17–30 mm, more or less equal, broadly elliptical, acuminate at apex, 7- to 9-veined, papery, membranous on the margins. *Lemmas* 12–25 mm, chartaceous, elliptic-oblong, acute at apex, shining, smooth, glabrous, markedly 7-veined in the upper half, unawned or awned, the awn if present on the lowest lemma only, rather slender, not twisted and with a distinct column. *Palea* enclosed by the margins of the lemma, narrowly lanceolate, 2-keeled, scabrid on the keels. *Anthers* 2–3 mm. *Caryopsis* hairy, oblong, adherent to the lemma and palea. *Flowers* 7–9. 2n = 42.

Varieties of this species occur in which the lemma and palea are mostly of the same delicate texture as the glumes and they do not adhere to the ripe seed. They are called 'Naked Oats' (cf. Chater, 1993).

Introduced. Formerly a common crop, now much less so but some 130,000 hectares are still grown; frequent as a casual on tips and waysides. In the Middle Ages oaten cakes and cheese formed an important part of the peasant diet in Britain. With the development of potatoes, oats became less popular. Oats were also an important food of the horse, which caused a further decline when the farm-horse became redundant after the Second World War. Throughout the British Isles. Origin probably the Mediterranean region. This genus first appears in archaeological records from the Middle East as a weed in wheat fields, but as agriculture spread northwards its adaptability in the climate of north-west Europe led to its domestication there in about 2000 BC. Other forms continued to be evolved and *A. fatua* is now one of the worst weeds in the world (Sampson in *Bot. Mus. Leafl. Harvard* 16: 265–303 (1954)).

6. A. byzantina K. Koch   Algerian Oat

Annual herb. Culms 60–100(–180) cm, glabrous. *Leaves* up to 30 cm × 1.5–6.0(–12) mm, green, long-linear, acute at apex, scabridulous; ligules 6–7 mm. *Inflorescence* a panicle up to 25 cm, more or less patent; rhachilla not disarticulating at maturity, but breaking at the base and falling attached to the upper flower. *Spikelets* 25–35 mm with 3(–4) flowers. *Glumes* subequal, 7- to 9-veined. *Lemma* 15–20 mm, narrowly lanceolate, glabrous except for a basal tuft of long hairs, apex with teeth 0.5–1.5 mm; awn 25–35 mm, without a distinct column, flexuous but never geniculate, the lowest

1(–2) lemmas awned, the others without awns. *Palea* three-quarters as long as lemma, lanceolate. *Anthers* 2–3 mm. *Caryopsis* hairy, oblong, adherent to the lemma and palea. *Flowers* 7–9.

Probably derived from *A. sativa* in cultivation.

Introduced. Occurring occasionally as a grain contaminant. Cultivated as a cereal in southern Europe, mostly on dry or saline soils.

33. Gaudinia P. Beauv.

Annual herbs sometimes lasting for a few years. *Stems* erect. *Leaves* pubescent. *Inflorescence* a spike whose axis breaks into 1-spikeleted segments when fruiting. *Spikelets* with 3–11 flowers, all except the most apical bisexual. Lower glume 3- to 8-veined, about half as long as upper, the upper 5- to 11-veined. *Lemmas* 5- to 9-veined, minutely bifid at apex, with a long, bent, dorsal awn. *Rhachilla* segments more or less glabrous. *Ovary* with a distinct hairy apex remaining conspicuous as a projection when in fruit.

Four species in the Mediterranean area.

1. G. fragilis (L.) P. Beauv.   French Oat grass
Avena fragilis L.

Annual herb. Culms several, 15–120 cm, erect or ascending, cylindrical, smooth, shiny, glabrous. *Leaves* 1.0–6.5 cm × 0.6–4.0 mm, long-linear, acute at apex, flat, more or less villous; sheaths more or less villous; ligule very short, truncate at apex. *Inflorescence* up to 35 cm, a distichous spike. *Spikelets* 7–20 mm, laterally compressed, sessile, more or less appressed to the concave rhachis, with 3–11 flowers. *Glumes* shorter than the spikelets, glabrous or scabrid on the veins or sometimes villous, the lower 3–6 mm, lanceolate and acute at apex, the upper (7–)8–10(–12) mm, oblong and obtuse at apex. *Lemmas* 7–11 mm, lanceolate, glabrous or sometimes villous, with a geniculate awn 5–13 mm. *Palea* 3.5–7.5 mm, hairy on keel. *Anthers* 4–5 mm. *Caryopsis* 2.5 × 0.5 mm, oblong. *Flowers* 5–7. 2n = 14 + 0–2B.

Introduced. Naturalised in grassy fields, rough ground and waysides on a wide range of soils. Local in south-central England, south-west Ireland and the Channel Islands; an infrequent grain alien casual elsewhere. Native of the Mediterranean region.

34. Trisetum Pers. nom. conserv.

Perennial herb without rhizomes or stolons. *Culms* loosely tufted, erect or spreading. *Leaves* and sheaths pubescent. *Inflorescence* a well-branched panicle. *Spikelets* with 2–4 flowers, all except the most apical bisexual. *Glumes* unequal, the lower 1-veined, the upper 3-veined. *Lemmas* 5-veined, bifid with 2, short, bristle-like points at apex, with a long, bent dorsal awn. *Rhachilla* segments pubescent, the hairs less than 1 mm at apex. *Ovary* glabrous.

About 70 species in all temperate regions except Africa.

1. Culms 25–50 cm; leaves 2–4 mm wide
                              1(a). flavescens subsp. **flavescens**
1. Culms 50–100 cm; leaves 5–10 mm wide
                              1(b). flavescens subsp. **purpurascens**

1. T. flavescens (L.) P. Beauv.   Yellow Oat Grass
Avena flavescens L.; *T. pratense* Pers.

Loosely tufted *perennial. Culms* 20–100 cm, erect or spreading, slender, stiff to weak, unbranched, hairy near the nodes, or quite smooth, with 2–5 nodes. *Leaves* up to 35 cm × 2–10 mm, green, softly hairy or glabrous, narrowed to a fine point at apex, flat, firm, often hairy above, mostly smooth beneath; sheaths rounded on the back, the lower often hairy; ligules 0.5–2.0 mm, rounded at apex, membranous. *Inflorescence* a panicle 5–20 × 1.5–7.0 cm, erect or nodding, loose to rather dense, usually yellowish, less often greenish, purplish or variegated yellow and purple, glistening; branches clustered, fine, loosely divided, rough; pedicels 1–4 mm. *Spikelets* 5–7(–8) mm, oblong or finally wedge-shaped, compressed, 2- to 4-flowered, breaking up at maturity beneath each lemma; rhachis shortly hairy. *Glumes* persistent, keeled, finely pointed at apex, membranous, rough on the keels, shining, unequal, the lower 3.0–4.5 mm, narrowly lanceolate and 1-veined, the upper 4–6 mm, elliptical and 3-veined. *Lemmas* 4.0–5.7 mm, loose, narrowly lanceolate or narrowly oblong in side view, narrowed upwards and tipped with 2 fine, short teeth or bristle-points, firm except for the thinly membranous tips and margins, finely 5-veined, minutely rough upwards, awned from or near the middle of the back, the awn 2.5–9.0 mm, bent at and twisted below the middle when dry. *Paleas* nearly as long as the lemmas, lanceolate, whitish and very thin. *Anthers* 2–3 mm. *Caryopsis* about 3 mm, oblong, enclosed by the firm back of the lemma. *Flowers* 5–7.

Subsp. *purpurascens* can easily be overlooked as at first glance it looks like *Arrhenatherum* with which it flowers, earlier than subsp. *flavescens*.

(a) Subsp. flavescens
Culms 25–50 cm. *Leaves* 2–4 mm wide. *Panicle* rather short and ovoid, rather lax. *Flowers* 6–7. 2n = 28.

(b) Subsp. purpurascens (DC.) Arcangeli
Avena purpurascens DC.
Culms 50–100 cm. *Leaves* 5–10 mm wide. *Panicle* long and wide, dense. *Flowers* 5–7. 2n = 14.

Native or introduced. Meadows, pastures, hillsides and waysides. Throughout most of lowland British Isles, common in England and south-east Scotland, rather scattered elsewhere. Most of Europe; Caucasus; North Africa; introduced in North America. Subsp. *flavescens* is the native plant. It occurs throughout most of the range of the species. Subsp. *purpurascens* has been sown over a large area of the Gog Magog Hills, near Cambridge. This grass has apparently been much sold, especially to local Councils, for covering roadside verges and should be looked for elsewhere.

35. Koeleria Pers.

Tufted *perennial herbs* without rhizomes or stolons. *Culms* erect or procumbent, variously hairy. *Leaves* glabrous or hairy. *Inflorescence* a spike-like panicle with very short branches. *Spikelets* with 2–3(–5) flowers, all except the most apical bisexual. *Glumes* unequal, the lower 1-veined, the upper 3-veined. *Lemmas* 3-veined. *Rhachilla* segments shortly pubescent. *Ovary* glabrous.

About 35 species in temperate regions throughout the world.

Domin, K. (1907). Monographic der Gattung *Koeleria. Biblioth. Bot. (Stuttgart)* **65**: 1–354.
Grime, J. P. et al. (1988). *Koeleria macrantha* (Ledeb.) Schultes in *Comparative plant ecology.* London.
Ujhelyi, J. (1972) Evolutionary problems of the European *Koelerias. Symp. Biol. Hung.* **12**: 163–176.

1. Basal leaf-sheaths splitting into fibres, which persist and form a dense, thickened fibrous network at the base of the plant   **1. vallesiana**
1. Basal leaf-sheath not splitting into fibres,when old loosely imbricate and usually corrugate   **2.**
2. Leaves of non-flowering shoots usually flat or plicate, usually green, sometimes glaucous, slightly scabrid or smooth above; lemmas acuminate at apex, glabrous
          2(a). macrantha subsp. **macrantha**
2. Leaves of non-flowering shoots usually convolute, glaucous and silvery, scabrid; lemmas obtuse and minutely puberulent at least below
          2(b). macrantha subsp. **glauca**

1. K. vallesiana (Honck.) Gaudin   Somerset Hair Grass
Poa vallesiana Honck.

Densely tufted *perennial herb*, thickened at the base. *Culms* 10–40 cm, erect, slender, stiff, downy with very fine short hairs especially towards the panicle, with 1–3 nodes below the middle. *Leaves* 3–12 cm × up to 3 mm, greyish-green, very narrowly long-linear, straight or curved, obtuse at apex, rolled and bristle-like or opening out, stiff, closely ribbed above, smooth except for the rough margins; sheaths smooth, rounded on the back, the basal splitting into fibres, which persist and form a dense, thickened, fibrous network at the base of the plant; ligules up to 0.5 mm, membranous. *Inflorescence* a panicle 15–70 × 6–12 mm, spike-like, dense, oblong to ovate-oblong, obtuse at apex, silvery-green or tinged with purple; pedicels extremely short. *Spikelets* 4–6 mm, oblong or wedge-shaped, flattened, densely overlapping, 2- to 3-flowered, breaking up at maturity between the lemmas. *Glumes* persistent, pointed at apex, equal or slightly unequal, rough or minutely hairy on the keels above the middle, firm and green about the veins, the rest very thin and whitish, the lower 3.5–4.5 mm, narrow and 1-veined, the upper 4.0–5.5 mm, elliptical or obovate and 3-veined. *Lemmas* 4–5 mm, overlapping, their tips shortly exceeding those of the glumes, keeled upwards, elliptical, pointed at apex, sometimes with a very short awn from the tip, similar in texture to the glumes, 3-veined, minutely rough or with short hairs. *Paleas* elliptical, slightly shorter than the lemmas, 2-keeled, thin. *Anthers* 2.0–2.5 mm. *Caryopsis* oblong, enclosed by the slightly hardened lemma and palea. *Flowers* 6–8. 2n = 42.

Native. Short limestone grassland at seven sites in the Mendip Hills, Somerset. West Europe; north Apennines; North Africa. Our plant is subsp. **vallesiana**.

2. K. macrantha (Ledeb.) Schult.   Crested Hair Grass
Aira macrantha Ledeb.; *K. albescens* auct.

Compactly tufted *perennial*, sometimes with slender wiry rhizomes. *Culms* 10–60 cm, erect or slightly curved at the base, slender, stiff, downy especially towards the panicle or glabrous, with 1–3 nodes. *Leaves* up to 20 cm × 1.0–2.5 mm, green or glaucous-green, long-linear, with a fine blunt tip, rolled and bristle-like or opening out, finely hairy or glabrous and smooth; sheath rounded on the back, at first entire, densely to loosely hairy, especially the lower, or the upper glabrous; ligules up to 1 mm, membranous. *Inflorescence* a panicle 1–10 cm × 5–20 mm, silvery-green or purplish, glistening, spike-like, erect, very dense, often lobed or interrupted in the lower part, narrowly oblong or tapering upwards; branches very short, hairy. *Spikelets* 4–6 mm, densely clustered, on very short pedicels, oblong or wedge-shaped, compressed, 2- to 3-flowered, breaking up at maturity beneath the lemmas, glabrous or downy hairy. *Glumes* 4.0–5.5 mm, persistent, pointed or obtuse at apex, with thin membranous margins, the lower three-quarters the length of the upper, narrowly oblong and 1-veined, the upper oblong or elliptic-oblong and 3-veined. *Lemmas* 3.5–5.5 mm, as long as the upper glume or with their tips exserted, oblong, pointed at apex, keeled upwards, firm except for the thin margins, 3-veined. *Paleas* about as long as the lemmas, lanceolate, thin, 2-keeled. *Anthers* about 2 mm. *Caryopsis* 2.5–3.0 mm, oblong, enclosed by the hardened lemma. *Flowers* 6–7.

Very variable in stature, leaf rigidity and inrolling, pubescence and colour. There are probably more taxa than the following two subspecies.

(a) Subsp. **macrantha**
K. gracilis Pers. nom. illegit.; *K. britannica* (Domin ex Druce) Ujh.
Leaves of non-flowering shoots usually flat or plicate, usually green, sometimes glaucous, slightly scabrid or smooth above. *Glumes* acuminate at apex. *Lemmas* acuminate, glabrous. 2n = 14, 28 + 0–6B, 42, 70.

(b) Subsp. **glauca** (Schrad.) P. D. Sell
Aira glauca Schrad.; *K. glauca* (Schrad.) DC.
Leaves of non-flowering shoots usually convolute, glaucous and silvery-scabrid. *Glumes* obtuse at apex. *Lemmas* obtuse or mucronulate, minutely puberulent at least below. 2n = 14 (28, 42, 70).

Native. Limestone or sandy, base-rich grassland and dunes. Throughout the British Isles, mostly on calcareous soils in the south, mostly coastal in the north. Most of Europe and north temperate Asia. Both subspecies occur throughout much of Europe and subsp. *glauca* seems to be the plant of the more sandy soils.

× vallesiana
This hybrid is intermediate in leaf sheath characters and is sterile. 2n = 35.
Native. Occurs in most *K. vallesiana* populations. Endemic.

36. Rostraria Trin.

Annual herbs. Culms erect. *Leaves* with glabrous to pubescent sheaths. *Inflorescence* a spike-like panicle with very short branches. *Spikelets* with 3–5(–11) flowers, all except the apical one to few bisexual. *Glumes* unequal, the lower 1-veined, the upper 3-veined. *Lemmas* 5-veined, shortly bifid with a short subterminal awn. *Rhachilla segments* pubescent. *Caryopsis* glabrous.

About 10 species in the Mediterranean region and Middle East.

1. R. cristata (L.) Tzvelev   Mediterranean Hair Grass
Lophochloa cristata (L.) Hyl.; *Koeleria phleoides* (Villars) Pers.; *Festuca cristata* L.

Annual herb. Culms up to 50(–60) cm, fasciculate, rarely solitary, simple, smooth, glabrous, leafy almost to the panicle. *Leaves* up to 18 cm × 3–8 mm, linear, acuminate at apex, flat, flaccid, loosely hairy on both surfaces, scabrid on the margins; sheaths clasping, covered with loose, spreading hairs, densely hairy in the throat; ligules about 2 mm, truncate at apex, toothed. *Inflorescence* a panicle 5–60(–120) × 4–15(–20) mm, cylindrical, sometimes lobed. *Spikelets* 3.0–7.5 mm, 3- to 7-flowered, compressed, elliptical; rhachilla disarticulating above the glumes and underneath the flowers. *Glumes* unequal in length and breadth, the lower 2.5–4.0 mm, 1-veined, oblong and acute or acuminate at apex, the upper 3.5–5.5 mm, 3-veined, broader, elliptical and acute at apex, both silvery-hyaline on the margins and scaberulous on the keels. *Lowest lemma* 4–5 mm, elliptical, acute at apex, prominently 5-veined, reticulate between the veins, bidentate at the apex, awned in the sinus, the awn 1–3 mm and straight. *Palea* shorter than the lemma, lanceolate, 2-veined. *Anthers* 0.2–0.6 mm. *Caryopsis* narrowly oblong. *Flowers* 5–6. 2n = 26.

Introduced. Characteristic wool alien of cultivated and waste places. Scattered localities in Britain. Native of south Europe and the Mediterranean region, eastwards to central Asia.

37. Deschampsia P. Beauv.

Densely tufted *perennial herbs* usually without rhizomes or stolons. *Stems* erect. *Leaves* flat or setaceous. *Inflorescence* a very diffuse panicle with fine branches. *Spikelets* with 2 flowers, both bisexual or sometimes proliferating. *Glumes* unequal, the lower 1-veined, the upper (1–)3-veined. *Lemmas* 4- to 5-veined, rounded, obtuse or jaggedly toothed at apex, with a dorsal, rarely subterminal, straight or bent awn. *Rhachilla segments* pubescent with a longer hair-tuft at the base of each. *Caryopsis* glabrous.

About 40 species in temperate regions throughout the world.

Davy, A. J. (1980) *Deschampsia cespitosa* (L.) Beauv. in Biological flora of the British Isles. *Jour. Ecol.* **68**:1075–1096.
Grime. J. P. et al. (1988). *Deschampsia cespitosa* (L.) Beauv. and *D. flexuosa* (L.) Trin. in *Comparative plant ecology*. London.
Scurfield, G. (1954). *Deschampsia flexuosa* (L.) Trin. in Biological flora of the British Isles. *Jour. Ecol.* **42**: 225–233.

Stewart, A., Pearman, D. A. & Preston, C. D. (1994). *Scarce plants in Britain*. Peterborough. [*D. cespitosa* subsp. *alpina* and *D. setacea*.]

1. Spikelets proliferating   2.
1. Spikelets not proliferating   3.
2. Leaves distinctly hooded at apex; panicle branches and pedicels smooth, the main branches usually reflexed; awn arising from the middle of the upper half of the lemma   **1(a). cespitosa** subsp. **alpina**
2. Leaves scarcely or not hooded at apex; panicle branches and pedicels with minute, sometimes very sparse pricklets, the main branches rarely reflexed; awn arising from the lower half of the lemma   **1(c). cespitosa** subsp. **cespitosa**
3. Leaves more than 1 mm wide even if rolled up; awns not or scarcely exceeding the glumes   4.
3. Leaves less than 1 mm even if opened out; awns conspicuously exceeding the glumes   5.
4. Spikelets 2–3(–3.5) mm; hair-tuft at the base of the lower lemma not reaching the apex of the rhachilla segments above; lowland woodland   **1(b). cespitosa** subsp. **parviflora**
4. Spikelets (3–)3.5–5.0(–6) mm; hair-tuft at the base of the lower lemma reaching the apex of the rhachilla segments above; lowland meadows and uplands   **1(c). cespitosa** subsp. **cespitosa**
5. Ligules 2–8 mm, very acute; lemmas 2.3–3.0 mm, toothed at apex, with marginal teeth the longest; palea bifid   **2. setacea**
5. Ligules 0.5–3.0 mm, obtuse at apex; lemmas 3.0–5.5 mm, subacute to minutely toothed at apex with the marginal teeth not longer than the inner ones; palea entire at apex   **3. flexuosa**

1. D. cespitosa (L.) P. Beauv.   Tufted Hair Grass
Aira cespitosa L.

Densely tufted *perennial herb*, often forming large tussocks. *Culms* 10–200 cm, erect or bent near the base, slender to stout, smooth, with 1–3 nodes. *Leaves* 2–60 cm × 2–5 mm, green, glabrous, long-linear, sharply pointed to obtuse at apex, sometimes hooded, flat or rolled, ribbed above, smooth beneath, rough on the margins; sheaths rounded on the back or somewhat keeled, smooth or rough upwards; ligules up to 15 mm, membranous. *Panicles* 5–50 × up to 20 cm, open and loose to rather dense, erect or nodding, ovate to oblong, green, silvery, golden, purple or variegated, sexual or proliferating; branches very slender, spreading, rough or smooth, bare below; pedicels 1–6 mm. *Spikelets* 2–6 mm, loosely scattered or clustered, lanceolate to oblong, breaking up at maturity beneath each lemma, the upper flower sometimes replaced by a proliferating plantlet, rhachis hairy. *Glumes* persistent, as long as the spikelet or slightly shorter, keeled, pointed at apex, membranous, shining, equal or nearly so, the lower lanceolate and 1-veined, the upper wider and 3-veined. *Lemmas* 3–8 mm, enclosed in the glumes or with their tips protruding, rounded on the back, ovate or oblong, toothed or lobed at the apex, membranous, 5-veined, bearded at the base, with an awn up to 4 mm from near the base or towards and above the middle. *Paleas* shorter than the

lemmas, lanceolate. *Anthers* 1.5–2 mm. *Caryopsis* ellipsoid, enclosed by the thin firm lemma and palea. *Flowers* 6–8.

(a) Subsp. **alpina** (L.) Hook. fil.
Aira alpina L.; *D. alpina* (L.) Roem. & Schult.
Leaves distinctly hooded at apex. *Panicle* branches and pedicels smooth. *Spikelets* 4.0–5.5 mm, at least some proliferating. *Lemma* with awn arising from middle or above; hair-tuft of lower not reaching the apex of the rhachilla segment above. Reproduction appears to be entirely by propagules. $2n = 39, 52$.

(b) Subsp. **parviflora** (Thuill.) Dumort.
Leaves not or scarcely hooded at apex. *Panicle* branches sometimes rough with small protrusions. *Spikelets* 2–3(–3.5) mm, not proliferating. *Lemma* with the awn arising from lower part; hair-tuft of lower not reaching the apex of the rhachilla segments above. $2n = 26$.

(c) Subsp. **cespitosa**
Leaves not or scarcely hooded at apex. *Panicle* branches and pedicels rough with small protrusions. *Spikelets* (3–)3.5–5.0(–6) mm, rarely proliferating. *Lemma* with the awn arising from the lower half; hair-tuft of lower reaching apex of the rhachilla segment above. $2n = 26, 39, 52$.

Native. Subsp. *cespitosa* occurs in lowland damp meadows, waysides and ditches. These plants seem to be diploids. In similar habitats in hilly country to the north they appear to be diploid, triploid or tetraploid. Subsp. *cespitosa* is common throughout the British Isles. Subsp. *parviflora* occurs in woods and shady places in lowland Britain, especially on heavy soils, where it is common. It has not been confirmed from Ireland. Subsp. *alpina* occurs in damp rocky places in mountains, often where the snow lies late. It is frequent between 800 and 1200 m in the west and central Highlands of Scotland, and very local in Caernarvonshire, the Lakes, Co. Kerry and Mayo. The species occurs throughout most of Europe, but only on mountains in the south; circumboreal; mountains of North Africa.

2. D. setacea (Huds.) Hack.   Bog Hair Grass
Aira uliginosa Weihe & Boenn.; *Aira setacea* Huds.

Densely tufted *perennial herb* with numerous closely packed vegetative shoots. *Culms* 20–60 cm, erect, slender, stiff, smooth, with 2–3 nodes. *Leaves* 5–20 cm × 0.2–0.4 mm in diameter, green, glabrous, bristle-like, very fine, sharply pointed at apex, inrolled, or opening out and up to 1 mm wide, rough; sheaths tight, smooth; ligules 2–8 mm, narrowly lanceolate, finely pointed at apex, membranous. *Inflorescence* a panicle 6–15 × up to 7 cm, loose, lanceolate to ovate; rhachis rough; branches hair-like, rough, divided towards the tips; pedicels 1–4 mm. *Spikelets* 4–5 mm, in clusters towards and at the tips of the branches, narrowly oblong, becoming wedge-shaped, 2-flowered, breaking up at maturity above the glumes, variegated with purple and pale yellow, the rhachis hairy. *Glumes* persistent, obtuse or pointed at apex, membranous, shining, the lower 3.3–3.6 mm, slightly shorter than the upper, narrowly oblong and 1-veined, the upper 3.7–3.8 mm as long as the spikelet, elliptic-oblong and 3-veined. *Lemmas* 2.3–3.0 mm, rounded on the back, oblong, unequally 4-toothed with the outer teeth longest, shortly bearded at the base, membranous, finely 4-veined, minutely rough upwards, awned from near the base, with the bent awn brown and twisted in the lower half and up to 6 mm. *Paleas* nearly as long as the lemmas, linear-elliptical, 2-keeled, the keels minutely hairy upwards. *Anthers* 1.5–2.0 mm. *Caryopsis* about 1.5 mm, brown, ellipsoid, enclosed between the slightly hardened lemma and palea. *Flowers* 7–8. $2n = 14$.

Native. In dry or wet areas on peat or sand in *Calluna* heathland, often in small depressions below heath level, rarely in bogs or on the margins of rock-pools. Very local and scattered in Britain and Ireland, mostly near the coasts in west and north Scotland and in south-central England. West Europe from north Spain to south-west Norway and west Poland.

3. D. flexuosa (L.) Trin.   Wavy Hair Grass
Aira flexuosa L.

Loosely to densely tufted *perennial* herb, sometimes with slender rhizomes. *Culms* 20–100 cm, erect or bent at the base, slender, wiry, smooth, with 1–3 nodes. *Leaves* up to 20 cm × 0.3–0.8 mm, green, glabrous, bristle-like, pointed or obtuse at apex, tightly inrolled, rather stiff, rough towards the tip; sheaths rounded on the back, often slightly rough upwards; ligules 0.5–3.0 mm, obtuse at apex, membranous. *Inflorescence* a panicle 4–15 × up to 8 cm, open and very loose; rhachis rough upwards; branches hair-like, rough, flexuous, divided in the upper part, spreading; pedicels 3–10 mm. *Spikelets* 4.0–6.0 mm, oblong or slightly wedge-shaped, loosely scattered, usually 2-flowered, breaking up at maturity above the glumes, purplish, brownish or silvery. *Glumes* persistent, keeled upwards, very thin, pointed at apex, minutely rough or smooth, the lower slightly shorter than the upper, ovate and 1-veined, the upper elliptic-ovate, as long as the spikelet and 1- to 3-veined. *Lemmas* 3.0–5.5 mm, rounded on the back, elliptic-oblong, subacute and minutely toothed at the tip, membranous, minutely rough, short-bearded at the base, finely 4-veined, awned from near the base, the awn brown and twisted in the lower half and 4–7 mm. *Paleas* about as long as the lemmas, narrowly elliptical, entire at the apex, 2-keeled, rough on the keels. *Anthers* 2–3 mm. *Caryopsis* 2.0–2.5 mm, ellipsoid, enclosed by the slightly hardened lemma and palea. *Flowers* 6–7. $2n = 26, 28, 56$.

Native. On sandy and peaty soils, usually in dry places on acid heaths, moors, open woods and the drier parts of bogs. Throughout the British Isles, but absent from much of central England and Ireland where soils are not suitable. Most of Europe; temperate Asia; Caucasus; Japan; North America; only in the mountains in the south.

38. Holcus L. nom. conserv.

Densely tufted or rhizomatous *perennial herbs*. *Culms* erect. *Leaves* pubescent. *Inflorescence* a rather compact panicle. *Spikelet* with 2 flowers, the lower bisexual, the upper male. *Glumes* subequal in length but unequal in width, the lower 1-veined, the upper 3-veined. *Lemmas* 5-veined, more or less rounded at apex, the lower awnless, the upper with a dorsal awn arising from the upper half. *Rhachilla segments* more or less glabrous, but lemmas with a basal tuft of hairs. *Ovary* glabrous.

Six species in Europe, North Africa and the Middle East.

Beddows, A. R. (1961). Holcus lanatus L. in Biological flora of the British Isles. *Jour. Ecol.* **49**: 421–430.
Beddows, A. R. (1971). The inter- and intra-specific relationships of *Holcus lanatus* L. and *H. mollis* L. sensu lato (Gramineae). *Bot. Jour. Linn. Soc.* **64**: 183–198.
Carroll, C. P. & Jones, K. (1962). Cytotaxonomic studies in *Holcus*. 3. A morphological study of the triploid F_1 hybrid between *Holcus lanatus* L. and *H. mollis* L. *New Phytol.* **61**: 72–84.
Jones, K. (1958). Cytotaxonomic studies in *Holcus*. 1. The chromosome complex *Holcus mollis* L. *New Phytol.* **57**: 191–210.
Jones, K. & Carroll, C. P. (1962). Cytotaxonomic studies in *Holcus*. 2. Morphological relationships in *Holcus mollis* L. *New Phytol.* **61**: 63–71.
Ovington, J. D. & Scurfield, G. (1956). *Holcus mollis* L. in Biological flora of the British Isles. *Jour. Ecol.* **44**: 272–280.

1. Nodes not bearded; awn hooked, not exserted   **1. lanatus**
1. Nodes bearded; awn not hooked, exserted   **2. mollis**

1. H. lanatus L.   Yorkshire Fog

Loosely to compactly tufted, softly hairy *perennial herb*. *Culms* 20–100 cm, erect or ascending from a bent base, slender to somewhat stout, downy at the nodes, rarely almost glabrous, with 2–5 nodes. *Leaves* 4–20 cm × 3–10 mm, greyish-green or green, softly hairy, rarely nearly glabrous, long-linear, narrowed to a fine point at apex; sheaths usually with reflexed hairs, rounded on the back; ligules 1–4 mm, membranous. *Inflorescence* a panicle 3–20 × 1–8 cm, whitish, pale green, pinkish or purple, lanceolate to oblong or ovate, very dense to rather loose, erect or nodding; branches hairy, closely divided; pedicels 1–4 mm. *Spikelets* 4–6 mm, falling entire at maturity, 2-flowered, with the lower flower bisexual and the upper usually male. *Glumes* equal, or with the upper longer and broader, as long as the spikelet, stiffly hairy on the keels and nerves, minutely rough or hairy on the sides, thinly papery, the lower narrowly lanceolate or oblong and 1-veined, the upper ovate to elliptical, usually tipped with an awn up to 1 mm and 3-veined. *Lemmas* 2.0–2.5 mm, enclosed by the glumes, keeled upwards, obscurely 3- to 5-veined, firm and shining, the lower boat-shaped, obtuse, awnless and with an equally long elliptical palea, the upper narrow, awned on the back near the tip, with the awn up to 2 mm, becoming recurved like a fish-hook when dry, the palea shorter than the lemma. *Anthers* 2.0–2.5 mm. *Caryopsis* narrowly ellipsoid, enclosed by the hardened lemma and palea. *Flowers* 5–9. $2n = 14$.

Native. Meadows, pastures, rough grassland, waste ground, waysides and open woodland on a wide range of soils, under dry or wet conditions. Common throughout the British Isles. Europe except the Arctic; temperate Asia; North Africa; introduced in North and South America.

× mollis = H. × hybridum Wein

This hybrid resembles *H. mollis*, but has more obtuse glumes, less exserted awns and more hairy culms and has hairy sheaths of various lengths. $2n = 21$.

The records for Britain and Ireland are scattered and it may be under-recorded. Elsewhere recorded only from Germany.

2. H. mollis L.   Creeping Soft Grass

Perennial herb, with tough, creeping rhizomes, forming compact tufts or loose mats. *Culms* 20–100 cm, erect or more often spreading, slender, loosely to densely bearded at the nodes, otherwise smooth, with 4–7 nodes. *Leaves* 4–20 cm × 3–12 mm, greyish-green, narrowly linear-lanceolate, pointed at apex, shortly hairy or glabrous, rough or nearly smooth. *Inflorescence* a panicle 4–12 cm, narrowly oblong to ovate, compact to somewhat loose, whitish, pale grey or purplish; branches hairy; pedicels 1–4 mm. *Spikelets* 4–7 mm, elliptical or oblong, flattened, falling entire at maturity, 2-flowered, with the lowest flower bisexual and the upper functionally male. *Glumes* slightly unequal, the upper as long as the spikelet, pointed at apex, thinly papery, with short, stiff hairs on the keels and veins, minutely rough on the sides, the lower narrowly lanceolate and 1-veined, the upper elliptical or ovate and 3-veined. *Lemmas* 2.5–3.0 mm, obliquely lanceolate in side view, enclosed by the glumes, obscurely 5-veined, bearded at the base, smooth or minutely hairy above, firm, shining, the lower awnless, the upper awned on the back just below the apex, with the awn 3.5–5.0 mm, slightly bent and protruding beyond the glumes. *Paleas* about as long as the lemmas, oblong, minutely hairy on the nerves. *Anthers* about 2 mm long. *Caryopsis* narrowly ellipsoid, enclosed between the lemma and palea. *Flowers* 6–8. $2n = 28, 35, 42, 49$.

Native. Woods, hedgerows, heaths and poor grassland, and sometimes a weed of arable land, mostly on acid soils. Common throughout much of the British Isles, but absent from areas of base-rich soils. Scattered throughout Europe; introduced in North America.

39. Corynephorus P. Beauv. nom. conserv.

Densely tufted *perennial herbs* without rhizomes. *Culms* erect or spreading. *Leaves* bristle-like. *Inflorescence* a rather compact panicle. *Spikelets* with 2 flowers, both bisexual. *Glumes* subequal, 1-veined or the upper also with very short laterals. *Lemmas* with one central and often 2 pairs of very short lateral veins, obtuse to very shortly bifid at apex, with a bent dorsal awn with a club-shaped apex. *Rhachilla segments* pubescent and with a tuft of hairs at the base of each lemma. *Ovary* glabrous.

Five species in Europe and the Mediterranean area east to Iran.

Marshall, J. E. (1967). *Corynephorus canescens* (L.) Beauv. in Biological flora of the British Isles. *Jour. Ecol.* 55: 207–220.

1. C. canescens (L.) P. Beauv.   Grey Hair Grass
Aira canescens L.; *Weingaertneria canescens* (L.) Bernh.

Densely tufted *perennial herb*. *Culms* 10–35 cm, erect or spreading, very slender, smooth or slightly rough, with 2–7 nodes below the middle. *Leaves* up to 6 cm × 0.3–0.5 mm, greyish, glabrous, bristle-like, stiff, sharply pointed at apex, tightly inrolled, minutely and densely rough; sheaths usually purplish, minutely rough; ligules 2–4 mm, pointed at the apex, membranous. *Inflorescence* a panicle 1.5–8.0 × 0.5–1.5 cm, purple or variegated with pale green, narrow, lanceolate to narrowly oblong, loose when in flower, afterwards rather dense; branches short; pedicels 1–3 mm. *Spikelets* 3–4 mm, lanceolate to narrowly oblong, compressed, breaking up at maturity below each lemma, 2-flowered. *Glumes* persistent, narrowly lanceolate and pointed at apex in side view, equal or nearly so, shining, membranous with thinner white tips and margins, 1-veined. *Lemmas* 1.5–2.0 mm, enclosed by the glumes, ovate, obtuse at apex, thin, obscurely veined, with a tuft of minute hairs at the base, the awn with its lower half orange or brown and twisted when dry, bearing a ring of minute hairs at the middle, and with the terminal part club-shaped and enclosed by the glumes. *Paleas* nearly as long as the lemmas, narrowly elliptical. *Anthers* 1.0–1.5 mm. *Caryopsis* ellipsoid, enclosed by the lemmas and palea. *Flowers* 6–7. $2n = 14$.

Native. Open sand on leached, fixed dunes and inland sandy heaths on acid soils. On and near the coasts of Suffolk and Norfolk, rare inland in Suffolk, Worcestershire, Inverness-shire and East Lothian. Probably introduced on the coasts of Moray and formerly in Lancashire and Glamorgan. Naturalised in Staffordshire. South Norway and the Baltic States, southwards to north Italy and north-west Africa; introduced in North America.

40. Aira L.

Annual herbs. Culms erect to decumbent. *Leaves* very narrow. *Inflorescence* a compact to diffuse panicle. *Spikelets* with 2 flowers, both bisexual. *Glumes* subequal, 1- to 3-veined. *Lemmas* with 5 short veins, shortly bifid, with dorsal awn slightly bent. *Rhachilla segments* extremely short. *Lemma* with short hair-tuft at base. *Ovary* glabrous.

Eight species in Europe and the Mediterranean regions east to Iran.

1. Sheaths smooth; pedicels mostly shorter than the spikelets   **1. praecox**
1. Sheaths scabrid; pedicels mostly longer than the spikelets   **2.**
2. Spikelets 2.2–2.6 mm; seed 1.1–1.5 mm   **2(a). caryophyllea** subsp. **multiculmis**
2. Spikelets 2.5–3.5 mm; seed 1.2–1.9 mm   **3.**
3. Spikelets 2.5–3.0 mm; pedicels up to 10 mm   **2(b). caryophyllea** subsp. **caryophyllea**
3. Spikelets 3.0–3.5 mm; pedicels up to 5 mm   **2(c). caryophyllea** subsp. **armoricana**

1. A. praecox L.   Early Hair Grass

Annual herb. Culms 2–20 cm, few to many in small tufts or solitary, erect, spreading or prostrate, very fine, smooth or very minutely hairy, with 2–3 nodes. *Leaves* up to 5 cm × 0.3–0.5 mm, green, glabrous, narrowly linear, obtuse at apex, involute, smooth or minutely rough; sheaths smooth; ligules up to 3 mm, obtuse at apex, membranous. *Inflorescence* a panicle 0.5–5.0 cm × 2–8 mm, silvery, purplish or pale green, spike-like, narrowly oblong; branches erect, very short; pedicels 1–3 mm. *Spikelets* 2.5–4.0 mm, crowded, ovate or oblong, 2-flowered. *Glumes* persistent, 2.5–4.0 mm, as long as the spikelet, similar, obliquely lanceolate in side view, pointed at apex, keeled upwards, minutely rough on the keels, shining, with thin membranous tips and margins, 1- to 3-veined. *Lemmas* similar, slightly shorter than and enclosed by the glumes, narrowly lanceolate, rounded on the back, finely 2-toothed at the narrowed tip, minutely rough upwards, with a tuft of very short hairs at the base, finely 5-veined, firm, bearing an awn on the back one-third of the way from the base, the awn yellowish-brown, twisted and bent below the middle and projecting from the tips of the glumes. *Paleas* shorter than the lemmas, narrowly elliptical. *Anthers* about 0.3 mm. *Caryopsis* narrowly ellipsoid, tightly enclosed by the firm brown lemma and palea. *Flowers* 4–6. $2n = 14$.

Native. Well-drained, sandy, gravelly and rocky ground, walls, heaths and dunes, preferring acid soils. Common throughout the British Isles. West Europe, extending eastwards to south-west Finland, west Czechoslovakia and north-west Italy; introduced in North America.

2. A. caryophyllea L.   Silver Hair Grass

Annual herb. Culms 3–50 cm, few to many in tufts or solitary, erect or spreading, very slender, smooth, with 2–3 nodes below the middle. *Leaves* 0.5–5.0 cm × about 0.3 mm, greyish-green, glabrous, thread-like, obtuse at apex, inrolled, minutely rough on the veins; sheaths minutely rough upwards; ligules up to 5 mm, toothed, membranous. *Inflorescence* a panicle 1–12 × 1–12 cm, very loose to fairly dense, with widely spreading branches; rhachis often wavy; branches bare at the base, usually loosely divided into threes at intervals, hair-like; pedicels 1–10 mm. *Spikelets* 2.2–3.5 mm, silvery or tinged purple, in small, loose clusters at the tips of the branches, ovate to oblong, 2-flowered. *Glumes* 2.2–3.5 mm, persistent, similar, obliquely lanceolate and pointed in side view, minutely rough on the keel, shining, thinly membranous, 1- to 3-veined. *Lemmas* slightly shorter than and enclosed by the glumes, narrowly ovate, finely 2-toothed, minutely rough above, with a tuft of short hairs at the base, firm, awned from the back, one-third above the base, the awns twisted and bent below the middle and projecting from the tips of the glumes. *Paleas* elliptical, shorter than the lemmas. *Anthers* 0.3–0.6 mm. *Caryopsis* 1.1–1.9 mm, ellipsoid, tightly enclosed by the firm brown lemma and palea. *Flowers* 5–7.

(a) Subsp. **multiculmis** (Dumort.) Bonnier & Layens

A. multiculmis Dumort.
Spikelets 2.2–2.6 mm. *Pedicels* less than 5 mm. *Anthers* 0.3–0.5 mm. *Caryopsis* 1.1–1.5 mm. $2n = 28$.

(b) Subsp. **caryophyllea**
Spikelets 2.5–3.0 mm. *Pedicels* up to 10 mm. *Anthers* 0.3–0.4 mm. *Caryopsis* 1.2–1.6 mm. $2n = 28$.

(c) Subsp. **armoricana** (Albers) Kerguélen
A. armoricana Albers
Spikelets 3.0–3.5 mm. *Pedicels* less than 5 mm. *Anthers* 0.3–0.5 mm. *Caryopsis* 1.5–1.9 mm.

Native. The species occurs in dry sandy or gravelly ground, among rocks, and on walls, heaths and dunes. It is frequent throughout the British Isles, and in south, west and central Europe, North Africa and Macaronesia, and introduced in North America. Subsp. *caryophyllea* occurs throughout the range of the species. Subsp. *multicaulmis* is mainly in the south-western part of the species range, especially in bulb fields in the Isles of Scilly. Subsp. *armoricana* has been found in the British Isles only in west Cornwall. The variation is partly geographical, but may also concern chromosome number, as subsp. *caryophyllea* is sometimes considered to be $2n = 14$ and sometimes $2n = 28$.

Tribe **10. Phalarideae** Rchb.

Annual or *perennial herbs*, with or without rhizomes, without stolons. *Leaves* with a membranous ligule; sheaths not fused. *Inflorescence* a panicle, usually contracted. *Spikelets* with (2–)3 florets, the lower (1–)2 male or sterile and often much reduced, the upper one bisexual, rarely the spikelets in groups consisting of one central bisexual and 4–6 surrounding sterile or male spikelets. *Glumes* 2, equal or unequal, one or both nearly as long as to longer than the rest of the spikelet. *Lemmas* of bisexual florets 3- to 7-veined, keeled or rounded on the back, awnless. *Palea* with one vein. Lower 2 lemmas minute to longer than upper, without or with 4–5 veins, awnless or with a dorsal, bent awn. *Stamens* 2 or 3. *Stigmas* 2. *Lodicules* 2 or absent. *Ovary* glabrous.

41. Hierochloe R. Br. nom. cons.

Perennial herb with rhizomes. *Culms* erect. *Leaves* sparsely hairy or glabrous. *Inflorescence* a diffuse panicle. Lower 2 flowers with 3 stamens, and a 5-veined lemma slightly longer than lemma of the bisexual flower; terminal flower with 2 stamens, a 5-veined lemma and 2 lodicules. *Glumes* subequal, keeled, slightly shorter than the rest of the spikelet, 1- to 3-veined. Crushed or dried plants smell strongly of coumarin and continue to do so in the herbarium for many years.

About 30 species in temperate and Arctic regions generally, except Africa.

1. H. odorata (L.) P. Beauv. Holy Grass
Holcus odoratus L.; *Anthoxanthemum nitens*
(G. Weber) Schouten & Veldk.

Aromatic *perennial herb,* with slender, extensive, strong, creeping rhizomes, forming compact tufts or patches. *Culms* 20–50 cm, erect, slender, slightly rough near the panicle or smooth, with few nodes. *Leaves* 3–30 cm × 1.5–5.0(–10) mm, glossy green beneath, long-linear, finely pointed at apex, sparsely and minutely hairy above or glabrous, rough on the margins; sheaths rounded on the back, minutely rough; ligules 1.5–2.0(–4) mm, membranous. *Inflorescence* a panicle 4–10 × up to 8 cm, ovate, loose; branches spreading, naked below; pedicels up to 4 mm, smooth. *Spikelets* 3.5–5.0 mm, green or purplish at the base, golden-brown upwards, 3-flowered, the lower 2 flowers male, the uppermost bisexual, breaking up above the glumes at maturity, the flowers falling together. *Glumes* persistent, broad, ovate, obtuse at apex, slightly shorter than to exceeding the flowers, keeled, with wide, membranous, shining margins, 1- to 3-veined. *Lower 2 lemmas* 3.5–4.5 mm, broadly elliptical, very obtuse at apex, rough with minute hairs, shortly hairy on the margins, firm except for the thin tip, 5-veined; paleas lanceolate, with 2 minutely rough keels; anthers 3, up to 3 mm. *Terminal lemma* slightly shorter than the others, ovate, becoming hardened, shortly hairy at the tip, 3- to 5-veined; palea ovate, 1-veined; anthers 2, about 2.5 mm. *Flowers* 3–5. Apomictic. $2n = 28, 42.$

Native. Banks of rivers and lakes, wet meadows and cliff-bases near the sea. Caithness, Kirkcudbrightshire, Roxburghshire and Renfrewshire in Scotland and Lough Neagh in Ireland, very local. North and central Europe, southwards to the Alps. Circumboreal.

42. Anthoxanthemum L.

Annual or tufted *perennial herbs. Culms* erect. *Leaves* flat. *Inflorescence* a contracted panicle; lower 2 flowers sterile, with the 4- to 5-veined lemma slightly longer than the lemma of the bisexual flowers and with a long dorsal awn; several flowers with 2 stamens, a 5-veined, awnless lemma and a 1-veined palea. *Lodicules* absent. *Glumes* very unequal, the lower 1-veined, the upper 3-veined and longer than the rest of the spikelets. Has a strong scent of courmarin which is retained in hay and in pressed specimens.

About 18 species in temperate Eurasia and Africa including the tropical mountains, and in central America; introduced in other temperate regions.

1. Perennials with non-flowering shoots; awn about as
   long as the upper glume **1. odoratum**
1. Annual; awn distinctly longer than the upper glume
   2. aristatum

1. A. odoratum L. Sweet Vernal Grass
A. alpinum auct.

Tufted *perennial herb* smelling strongly of coumarin. *Culms* 10–100 cm, erect or spreading, slender to relatively stout, unbranched, rather stiff, smooth, with 1–3 nodes. *Leaves* 1–12(–30) cm × 1.5–5.0(–9) mm, green,

linear, finely pointed at apex, variable in size, flat, loosely hairy or glabrous, rough or smooth; sheaths rounded on the back, loosely to densely bearded at the apex, otherwise smooth or loosely hairy; ligules 1–5 mm, obtuse at apex, membranous. *Inflorescence* a panicle 1–12 cm × 6–15 mm, green or purplish, spike-like, very dense to somewhat loose, ovate to narrowly oblong; branches short; pedicels up to 1 mm, hairy. *Spikelets* 6–10 mm, lanceolate, compressed, with 3 flowers, the two lower sterile, the third bisexual, the three falling together at maturity. *Glumes* persistent, keeled, finely pointed at apex, loosely to sparingly hairy, thinly membranous, the lower ovate, about half the length of the upper and 1-veined, the upper as long as the spikelet, enclosing the flowers, ovate to elliptical and 3-veined. *Sterile lemmas* 3.0–3.5 mm, narrowly oblong, obtusely 2- lobed at the tip, firm, brown and hairy except for the white membranous tip, 4- to 5-veined, the lower awned from above the middle with a straight, 2–4 mm awn, the upper with a stouter bent awn from the base which is 6–9 mm; dark brown and strongly twisted below. *Fertile lemma* about 2 mm, shining brown, subrotund, smooth. *Palea* as long as the lemma, ovate, 1-veined. *Anthers* 2, 3.0–4.5 mm. *Caryopsis* about 2 mm, ellipsoid, enclosed between the lemma and palea. *Flowers* 4–7. Protogynous. $2n = 20.$

A very variable grass in size, leafiness and hairiness.

Native. Grassy places on acid or calcareous, heavy or light soils, lowland or montane. Throughout the British Isles. Throughout Europe; North Africa; Macaronesia; circumboreal, but only on mountains in the south; introduced in North and South America, Australia and Tasmania.

2. A. aristatum Boiss. Annual Vernal Grass
A. puelii Lecoq & Lamotte

Annual herb. Culms 10–40 cm, loosely tufted or solitary, erect or spreading, slender, branched especially in the lower part, smooth, with 4–5 nodes. *Leaves* 0.8–6 cm × 1–5 mm, green, linear, finely pointed, thin, glabrous or sparsely hairy above; sheaths rounded on the backs, with a few spreading hairs near the ligules, smooth; ligules up to 2 mm, obtuse at apex, membranous. *Inflorescence* a panicle 1–4 cm × up to 12 mm, spike-like, moderately dense to loose, lanceolate to ovate or oblong, pale green; branches short; pedicels up to 0.3 mm. *Spikelets* 5.0–7.5 mm, lanceolate to oblong, compressed, with 3 flowers, the lower 2 barren, the uppermost bisexual, the flowers falling together at maturity. *Glumes* persistent, glabrous, finely pointed at apex, sometimes tipped with a short mucro, rough on the keels, thinly membranous, the lower ovate, about half the length of the upper and 1-veined, the upper as long as the spikelet and enclosing the flowers, ovate or elliptical and 3-veined. *Sterile lemmas* 3–4 mm, narrowly oblong, 2-lobed or entire and toothed at the tip, keeled, brown, firm and hairy except for the whitish membranous tips, finely 4- to 5-veined, the lower with a fine awn 4–5 mm from just above the middle, the upper with a stouter awn 7–10 mm from near the base, the awns bent and the upper twisted and dark brown below the middle;

paleas absent. *Fertile lemma* about 2 mm, subrotund, firm, smooth and shiny; palea lanceolate, 1-veined. *Anthers* 2, 2.5–3.5 mm. *Caryopsis* ellipsoid, tightly enclosed by the hardened lemma and palea. *Flowers* 6–10. $2n = 10$.

Introduced. Formerly naturalised as an arable weed in rough or cultivated, sandy ground in Surrey, Suffolk and Cambridgeshire, but not seen since 1970. Now a rare casual in England, south Wales and the Channel Islands. Native of south Europe and North Africa.

43. Phalaris L.

Annual or *perennial herbs* with rhizomes. *Stems* erect. *Leaves* flat. *Inflorescence* a contracted, often spike-like panicle; lower 2 flowers reduced to scales or rarely only one present; terminal flower with 3 stamens and a 5-veined, awnless lemma. *Lodicules* 2. *Glumes* equal, sharply keeled, longer than the rest of the spikelet.

Fifteen species in the north temperate zone, but mainly Mediterranean with a secondary centre in California; also in South America.

Grenfell, A. L. (1985). More on *Phalaris paradoxa* and related species. *B.S.B.I. News* **39**: 8.

1. Perennial with very short to long rhizomes and tillers at
   flowering time   2.
1. Annual without rhizomes or tillers   3.
2. Panicle at least distinctly lobed, usually with
   conspicuous branches; glumes strongly keeled, but not
   winged   **1. arundinacea**
2. Panicle not lobed, oblong to lanceolate in outline,
   without visible branches; glumes with a distinct wing
   on the keel   **2. aquatica**
3. Spikelets in groups of 3–7, 1 bisexual, the rest sterile,
   falling as a group when the seeds are ripe   **6. paradoxa**
3. Spikelets all bisexual, the (2–)3 flowers of each spikelet
   falling separately at maturity leaving their glumes on
   the panicle   4.
4. At least 1 glume on at least some of the spikelets with
   the wing on its keel minutely toothed   **5. minor**
4. Wings on keels of glume entire   5.
5. Sterile flowers 2, half as long as the fertile flower
                                                  3. canariensis
5. Sterile flowers 1–2, one-third as long as the fertile
   flower   **4. brachystachys**

1. P. arundinacea L.   Reed Canary Grass

Robust *perennial herb*, spreading extensively by creeping rhizomes. *Culms* 60–200 cm, stout, erect or bent at the base, smooth, with 4–6 nodes. *Leaves* 10–35 cm × 6–18 mm, green or whitish-green, sometimes striped cream and green (var. *picta* L.), linear, finely pointed, flat, firm, rough in the upper part, with cross-veins between the veins; sheaths smooth, rounded on the backs; ligules 2.5–16.0 mm, obtuse at apex, becoming torn, membranous. *Inflorescence* a panicle 5–25 × 1–4 cm, lanceolate to oblong, dense or somewhat loose below, lobed; branches spreading when in flower, closely divided, very rough; pedicels very short. *Spikelets* 5.0–6.5 mm, greenish, purplish or whitish-green, densely crowded, oblong,

flattened, 1-flowered, breaking up at maturity above the glumes. *Glumes* equal or the upper slightly longer, 5.0–6.5 mm, as long as the spikelet, sharply keeled, narrowly lanceolate in side view, pointed at apex, firm, minutely rough, 3-veined. *Sterile lemmas* 1.0–1.5 mm, narrow, one on each side of the fertile lemma, with short hairs. *Fertile lemma* 3–4 mm, enclosed by the glumes, broadly lanceolate in side view, pointed at apex, keeled, firm, 5-veined, with appressed hairs, becoming smooth and glossy below. *Palea* as long as the lemma, elliptical, 2-veined. *Anthers* 2.5–3.0 mm. *Caryopsis* narrowly ellipsoid, tightly enclosed by the hardened lemma and palea. *Flowers* 6–8. $2n = 27$–$31, 35, 42$.

Native. In ditches, wet meadows and marshes, by lakes and rivers and on rough and waste ground. Common throughout most of the British Isles. Most of Europe; circumboreal; Japan; Macaronesia.

2. P. aquatica L.   Bulbous Canary Grass
P. tuberosa L.

Perennial herb with short rhizomes. *Culms* up to 1.5 m, loosely tufted, the lateral geniculately ascending, prominently swollen at the base, smooth, glabrous. *Leaves* up to 35 cm × 2–15 mm, green or greyish, linear, acuminate at apex, rounded-truncate at the base, scaberulous to almost smooth on both surfaces, scaberulous on the margins; sheaths clasping, except the uppermost which is somewhat inflated, smooth, glabrous; ligule 4–6 mm, entire. *Inflorescence* a panicle 3–12 × 1–3 cm, dense, lanceolate or oblong in outline; branches usually very short or exceptionally up to 25 mm, with closely packed spikelets. *Spikelets* about 5 mm, firmly compressed, elliptical, acute at apex, pale with green veins. *Glumes* scarious on the margins, keeled, as long as the spikelets, 4–5 mm, winged along the whole length of the keel, the wing with an entire margin. *Sterile lemmas* 1, sometimes 2, minute or up to 2 mm. *Fertile lemma* 3.0–4.5 mm, ovate, acute at apex, densely pubescent. *Palea* ovate, similar to fertile lemma. *Anthers* 3.0–3.5 mm. *Caryopsis* compressed, ellipsoid. *Flowers* 6–7. $2n = 28$.

Introduced. Grown as game cover and food, sometimes for grazing or silage; naturalised in fields and waste ground and also a wool-alien casual. Widely scattered in Britain and increasingly frequent in central and south England. Native of the Mediterranean region and eastwards to Iraq; Macaronesia. Introduced and cultivated in India, Africa, Australia and North America.

3. P. canariensis L.   Canary Grass

Tufted *annual herb*. *Culms* 20–120 cm, erect or bent at the base, slender to rather stout, stiff, smooth. *Leaves* 5–25 cm × 4–12 mm, green, glabrous, linear, narrowed to a fine point at apex, flat, rough; sheaths smooth or rather rough, rounded on the back, the upper somewhat inflated; ligules 3–8 mm, membranous. *Inflorescence* a panicle 15–50 × 12–22 mm, whitish except for the green veins, ovate to ovate-oblong, erect, very dense, spike-like, stiff; branches very short. *Spikelets* 6–10 × up to 6 mm, much flattened, closely packed, obovate, sterile flowers 2, half as long as the fertile flower, breaking up above the glumes at maturity. *Glumes* persistent, 6–10 mm, as long as the spikelet, equal, similar, oblanceolate, abruptly pointed at apex, keeled, with the green keel broadly winged above the middle, minutely rough, firm, 3- to 5-veined, slightly hairy. *Sterile lemmas* 2, 3.0–4.5 mm. *Fertile lemma* 5–6 mm, broadly elliptical when opened out, keeled, tough, hairy, 5-veined, becoming smooth and glossy. *Palea* lanceolate, 2-veined. *Anthers* 3–4 mm. *Caryopsis* narrowly ellipsoid, tightly enclosed by the hardened lemma and palea. *Flowers* 6–9. 2*n* = 12.

Introduced. Usually a bird-seed casual on tips and waste ground. Sometimes more or less naturalised. Sometimes grown as a crop for bird-seed in south and central England. Frequent throughout the British Isles. Native of north-west Africa and the Canary Islands.

4. P. brachystachys Link   Confused Canary Grass

Annual herb. Culms up to 60 cm, loosely tufted, rarely solitary, erect or geniculately ascending, smooth, glabrous. *Leaves* up to 30 cm × 3–12 mm, dark green, flat, linear, acuminate at apex, flaccid, scaberulous on both surfaces and the margins, glabrous; sheaths rather loose, the uppermost somewhat inflated, smooth, glabrous; ligules up to 3 mm, hyaline or milky-white. *Inflorescence* a panicle 15–40 × 10–15 mm, densely cylindrical or ellipsoid, rounded at apex; branches and pedicels very short. *Spikelets* 5.0–8.5 mm, firmly compressed, rather glaucous; sterile flowers 1–2, one-third as long as the fertile flower. *Glumes* equal, as long as the spikelet, 5.0–8.5 mm, hyaline on the margins, keeled, 3-veined, winged in the upper half or two-thirds, the wings entire. *Sterile lemmas* 2, about 1 mm, ovate, acute at apex. *Fertile lemma* about 5 mm, compressed, keeled, with short appressed hairs. *Palea* lanceolate, similar to fertile lemma. *Anthers* about 3 mm. *Caryopsis* compressed, ellipsoid. *Flowers* 6–7. 2*n* = 12.

Introduced. Frequent wool alien on tips and waste ground. Scattered records in England. Native of the Mediterranean eastwards to Iran; Atlantic islands.

5. P. minor Retz.   Lesser Canary Grass

Annual herb. Culms up to 100 cm, solitary or more often loosely tufted, erect or geniculately ascending, smooth, glabrous, slender or stout. *Leaves* up to 25 cm × 10 mm, dark green, rounded at the base, linear, tapering to a long acuminate point, more or less scaberulous on both surfaces and on the margins, glabrous; sheaths somewhat loose, the uppermost inflated, minutely scabrid, glabrous; ligule 6–8 mm, membranous. *Inflorescence* a panicle 2–5 × 1.0–1.5 cm, very dense and cylindrical, ovate- or lanceolate-cylindrical, spikelets closely packed on short branches and pedicels; rachis scaberulous. *Spikelets* all bisexual, the 2–3 flowers of each falling separately at maturity leaving the glumes on the panicles, elliptical, acute at apex and firmly compressed. *Glumes* equal, 4.0–6.5 mm, as long as the spikelet, rather pale, with 3 green veins, keeled, smooth, glabrous, winged in the upper two-thirds, the wings undulate-denticulate. *Sterile lemma* 1, about 1 mm, subulate, curved. *Fertile lemma* about 3 mm, with appressed hairs but finally smooth and

glabrous, indurated, shining, broadly ovate, acute at apex. *Palea* ovate, similar to fertile lemma. *Anthers* about 2 mm. *Caryopsis* compressed, ellipsoid. *Flowers* 6–7. 2*n* = 28.

Introduced. Rather frequent wool-alien casual on tips and in waste places in Britain; naturalised in sandy places in Guernsey since at least 1791. Native of south Europe, Mediterranean region east to central Asia; Macaronesia; an introduced weed throughout the temperate regions and the tropics.

6. P. paradoxa L.   Awned Canary Grass

Annual herb. Culms up to 65(–85) cm, fascicled, those in the centre of the tuft erect, the outer geniculately ascending, very leafy, smooth, glabrous. *Leaves* up to 30 cm × 2–6 mm, linear, acuminate at apex, glabrous, scaberulous on the upper surface and the margins, smooth or scaberulous beneath; sheaths somewhat loose, the uppermost inflated and often partially clasping the base of the panicle, glabrous, scaberulous; ligules 3–4 mm. *Inflorescence* a panicle 2–7 × 1.0–2.5 cm, very dense, cylindrical to obovoid-cylindrical, or oblanceolate, partially enclosed at the base or free. *Spikelets* of 3 kinds; those at the base of the panicle which are neuter, deformed and consisting of empty, shapeless, club-like scales; those which resemble fertile spikelets, but which are sterile; and the fertile spikelets, so arranged that in a group of spikelets the central is fertile with a scabrid pedicel or sessile and is surrounded by sterile spikelets. *Sterile spikelets* consist of 2 empty, club-shaped glumes on a smooth, glabrous pedicel, the glumes about 4 mm, narrow, keeled. *Fertile spikelets* 6–8 mm, on a coarsely scabrid pedicel and firmly compressed, the glumes with an awn about 3 mm, winged on the keel, with an erect tooth-like wing, 7-veined, smooth and glabrous. *Sterile lemmas* 2, about 0.3 mm, very minute, filiform or absent. *Fertile lemma* about 3 mm, ovate, acute at apex, indurated, with a few long soft hairs towards the tip, finally smooth, glabrous and shining. *Palea* ovate, similar to fertile lemma. *Anthers* about 2 mm. *Caryopsis* ellipsoid. *Flowers* 6–8. 2*n* = 14.

Introduced. Widespread casual from wool and grain, grown with game-bird seed mixture and becoming naturalised in arable fields. Scattered in Britain and becoming increasingly common in central and south England and south Wales. Native of the Mediterranean region east to Afghanistan; Macaronesia; Ethiopia; South Africa; introduced and naturalised in many temperate regions of the world.

Tribe 11. Agrostideae Griseb.

Annual or *perennial herbs*, with or without rhizomes and/or stolons. *Leaves* with a membranous ligule; sheaths not fused. *Inflorescence* a panicle, contracted or diffuse, often spike-like, rarely a raceme or spike. *Spikelets* with one bisexual plus one male or sterile flower. *Glumes* 2, more or less equal, usually at least as long as the rest of the spikelet. *Lemmas* 3- to 7-veined, rounded on back, awnless or with a terminal, subterminal or dorsal awn. *Stamens* 3. *Stigmas* 2. *Lodicules* 2 or absent. *Ovary* glabrous.

44. Agrostis L.

Vilfa Adans.; *Trichodium* Michx, *Decandolia* Bast.;
Agraulus P. Beauv.

Annual or *perennial herbs* with or without rhizomes
and/or stolons. *Culms* usually more or less erect. *Leaves*
flat or setaceous. *Inflorescence* a slightly contracted to
very diffuse panicle with obvious branches. *Glumes* 2,
equal or nearly so, 1- to 3-veined, longer than the rest of
the spikelets. *Lemma* 3- to 5-veined, awnless or with a
subterminal or dorsal awn, with or without a hair-tuft
on the callus. *Palea* less than three-quarters as long as
the lemma, sometimes vestigial, usually weakly 2-veined
or veinless, disarticulating at maturity at the base of the
lemma.

A very plastic genus showing much genetic variation
and hybridisation. Palea length and presence of hair-
tufts on the lemma-callus are important though minute
characters. Presence of rhizomes as opposed to stolons
is very reliable, but their absence is not, as they are often
not developed until the plant is in fruit, and in some
habitats never.

About 220 species in temperate regions throughout
the world and on tropical mountains.

Bradshaw, A. D. (1958). Natural hybridisation of *Agrostis
tenuis* Sibth. and *A. stolonifera* L. *New Phytol.* **57**: 66–84.
Grime, J. P. et al. (1988). *A. canina* L.; *A. capillaris* L.; *A.
stolonifera* L.; *A. vinealis* Schreb. in *Comparative plant
ecology*. London.
Ivimey-Cook, R. B. (1958). *Agrostis setacea* Curt. in Biological
flora of the British Isles. *Jour. Ecol.* **47**: 697–706.
Philipson, W. R. (1937). A revision of the British species of the
genus *Agrostis* Linn. *Jour. Linn. Soc. London (Bot.)* **51**:
73–151.

1. Palea minute, less than two-fifths as long as lemma,
    or absent   2.
1. Palea more than two-fifths as long as lemma, about
    half as long to nearly as long   6.
2. Main panicle branches bare of spikelets for the
    proximal two-thirds of their length; anthers
    0.2–0.6 mm   3.
2. Main panicle branches bare of spikelets for less than
    the proximal half of their length; anthers 1–2 mm   4.
3. Leaves 1–5 mm wide; spikelets more than 2 mm;
    lemma 1.5–1.7 mm, distinctly exceeding the caryopsis
                                                              10. scabra
3. Leaves less than 1 mm wide; spikelets less than 2 mm;
    lemma 1.0–1.3(–1.5) mm, not or scarcely exceeding
    the caryopsis   **11. hyemalis**
4. Rhizomes and stolons absent; tiller leaves less than
    0.3 mm wide, bristle-like; panicle always with more
    or less erect branches   **7. curtisii**
4. Rhizomes or stolons usually present; tiller leaves
    more than (0.6–)1.0 mm wide even when inrolled, flat
    or inrolled; panicle often with patent to erecto-patent
    branches at or after flowering   5.
5. Rhizomes absent, stolons usually present and bearing
    tufts of leaves or shoots at the nodes; ligule usually
    more than 1.5 times as long as wide, acute to
    acuminate at apex   **8. canina**

5. Stolons absent, rhizomes usually present; ligules
    usually less than 1.5 times as long as wide, acute to
    subobtuse at apex   **9. vinealis**
6. Lemma callus with a tuft of hairs more than 0.3 mm;
    anthers 0.2–0.8 mm   7.
6. Lemma callus glabrous or very shortly hairy; anthers
    1.0–1.5 mm   8.
7. Panicle branches bare of spikelets for less than the
    proximal half of their length, erect to erecto-patent
    after flowering; rhachilla not extended above lemma-
    base; lemmas with awn absent or up to 0.5 mm, not
    exserted from glumes   **5. lachnantha**
7. Panicle branches bare of spikelets for the proximal
    two-thirds of their length, patent to erecto-patent
    after flowering; rhachilla extended above the lemma
    base reaching more than halfway up the lemma;
    lemmas with awn more than 2 mm, well-exserted from
    glumes   **6. avenacea**
8. Lemma callus with a tuft of hairs more than 0.3 mm;
    awn absent or, if present, arising from the basal one-
    third of the lemma and often exceeding the glumes   9.
8. Lemma callus glabrous or with hairs less than 0.2
    mm; awn absent, or if present, arising from the apical
    half of the lemma, rarely exceeding the glumes   11.
9. All spikelets awned; lemmas 5-veined, the outer
    excurrent at apex, hairy
                                       3(III). castellana var. **castellana**
9. Some or all spikelets without awns; lemmas 3-veined,
    the outer not excurrent at apex, glabrous   10.
10. All spikelets awnless   **3(i). castellana** var. **mutica**
10. Only terminal spikelets awned
                                         3(ii). castellana var. **mixta**
11. Rhizomes usually present with more than 3 scale-
    leaves, stolons usually absent or poorly developed;
    ligules of culm-leaves truncate; panicle-branches
    patent or nearly so after flowering   12.
11. Rhizomes absent, or short with fewer than 3 scale-
    leaves; stolons usually well developed; panicle
    contracted after flowering   13.
12. Leaves rarely more than 5 mm wide; ligules of tillers
    shorter than wide; fruiting panicle branches with
    spikelets all well separated   **1. capillaris**
12. Leaves often more than 5 mm wide; ligules of tillers
    longer than wide; fruiting panicle branches bearing
    spikelets in small, more or less dense clusters at their
    tips   **2. gigantea**
13. Plants forming a dense short tuft; leaves glaucous or
    greyish-green   14.
13. Plants not forming a dense, short tuft; leaves green   15.
14. Leaves narrow and flat; panicle often lax and not
    lobed; salt-marshes   **4(i). stolonifera** var. **marina**
14. Leaves often folded; panicles narrow and meagre;
    chalk grassland   **4(ii). stolonifera** var. **calcicola**
15. Plant not tufted; panicle long, lax and narrowly
    pyramidal; wet places   **4(v). stolonifera** var. **palustris**
15. Plant tufted; panicle short, dense and lobed; usually
    in dry places   16.
16. Culms usually much inclined; leaves short and
    folded; loose sand   **4(iii). stolonifera** var. **maritima**
16. Culms becoming erect; leaves flat; grassy places
                                   4(iv). stolonifera var. **stolonifera**

1. A. capillaris L. Common Bent

A. tenuis Sibth.; *A. vulgaris* With; *A. pumila* L.; *A. polymorpha* Huds. nom. illegit.; *Decandolia vulgaris* (With.) Bast.; *Vilfa vulgaris* (With.) P. Beauv.; *Vilfa divaricata* var. *pumila* (L.) Gray; *A. laxa* Gray; *A. alba* var. *pumila* (L.) Plues

Tufted *perennial herb*, spreading by short rhizomes and sometimes by stolons, forming a loose or dense tuft. *Culms* 10–70 cm, erect or spreading, slender, usually smooth, with 2–5 nodes. *Leaves* 1–15 cm × 1–5 mm, green, glabrous, linear, finely pointed at apex, flat or inrolled, soft to stiff, rough or nearly smooth; sheaths rounded on the back, smooth; ligules 0.5–2.0 mm, mostly shorter than broad, membranous. *Panicles* 1–20 × 1–12 cm, green or purplish, oblong to ovate or pyramidal, open and very loose, rarely somewhat dense, erect or slightly nodding; rhachis smooth or rough above; branches clustered, spreading, hair-like, bare in the lower part, divided above, smooth or rough; pedicels 1–3 mm. *Spikelets* 2.0–3.5 mm, lanceolate to narrowly oblong, 1-flowered, breaking up at maturity above the glumes. *Glumes* persistent, equal or slightly unequal, 2.0–3.5 mm, as long as the spikelet, lanceolate, finely pointed, membranous except for the thinner margins, 1-veined, the lower slightly rough on the keel above the middle. *Lemma* two-thirds to three-quarters the length of the spikelet, ovate or elliptical, obtuse at apex, 3- to 5-veined, callus minutely hairy, awnless, or rarely with a short awn from the middle of the back. *Palea* elliptical, half to two-thirds the length of the lemma. *Anthers* 1.0–1.5 mm. *Caryopsis* ellipsoid, enclosed by the thin lemma and palea. *Flowers* 6–8. 2n = 28.

Native. Heaths, moorlands, pastures and waste ground, usually on acid soils. Large areas, particularly on the poorer soils on hills and mountains, are dominated by this grass. Abundant throughout the British Isles. Europe, north Asia and North Africa. Introduced in North America, Australia, New Zealand and Tasmania.

× **castellana** = **A.** × **fouilladei** P. Fourn.

This hybrid is partially fertile and variably intermediate between the parents, and is thus difficult to determine.

There are scattered records in England arising from introduced seed-mixtures and perhaps *in situ*. It is likely to be found to be more frequent.

× **gigantea** = **A.** × **bjoerkmannii** Widén

This hybrid is intermediate in leaf width, and in the size of the ligule and panicle between those of the parents. The rhizomes are strong and 1.5–2.8 mm wide, greater than either parent, the panicle is ovate-oblong as in *A. capillaris* with the pedicels clustered, not spreading. It is a vigorous, highly sterile pentaploid. 2n = 35.

Native. There are a few scattered records in Britain, but it is probably under-recorded. It is widespread in Fennoscandia on sandy shores of lakes and rivers and on roadsides, but its distribution elsewhere in Europe in not clear.

× **stolonifera** = **A.** × **murbeckii** Fouill. ex P. Fourn.

This hybrid is intermediate between the parents in its ligule size (under 1.0 mm) and panicle-shape and usually has both rhizomes and stolons. It is a vigorous, highly sterile tetraploid with abundant, indehiscent anthers. 2n = 28.

Native. Recorded from scattered localities in Britain, but is probably common throughout the British Isles. It is widespread in northern Europe especially on coasts in the spray zone.

× **vinealis** = **A.** × **sanionis** Asch. & Graebn.

This hybrid has the awns absent to long and basal, the palea about one-third as long as the lemma and the ligule of the tillers about as long as wide. It is a highly sterile tetraploid. 2n = 28.

Native. There are scattered records in Britain on poor sandy soils. It also occurs in north-west Europe.

2. A. gigantea Roth Black Bent

A. nigra With.; *A. repens* Curt.; *A. alba* auct.; *A. seminuda* Knapp; *Vilfa nigra* (With.) Gray; *Vilfa divaricata* Gray

Loosely tufted *perennial herb*, with tough creeping rhizomes 1.0–1.8 in diameter. *Culms* 40–120 cm, erect or ascending from a curved or procumbent base, rooting and branching from the lower nodes with up to 6 decumbent tillers, slender to stout, smooth, with 3–6 nodes. *Leaves* 5–20 cm × 2–8 mm, dull green, glabrous, linear, finely pointed at apex, rolled when young, afterwards flat, firm, rough; sheaths rounded on the back, smooth or rather rough; ligules 1.5–12.0 mm, very obtuse, irregularly toothed, membranous, often curled on the margins. *Panicles* 8–25 × 3–15 cm, green or purplish, erect, oblong to ovate, usually open and very loose, much branched; branches up to 12, clustered, spreading, of various lengths, divided above the naked base, rough; pedicels 0.5–3.0 mm. *Spikelets* 2–3 mm, very numerous, lanceolate to oblong, 1-flowered, scabridulous, breaking up above the glumes at maturity. *Glumes* persistent, lanceolate in side view, finely pointed at apex, 2– 3 mm, as long as the spikelets, equal or slightly unequal with the upper shorter, membranous, 1-veined, rough on the keels. *Lemma* two-thirds to three-quarters the length of the glumes, ovate-oblong or oblong, very obtuse at apex, 3- to 5-veined, minutely hairy at the base, thin, awnless or rarely with a short awn from or near the tip. *Palea* elliptical, half to two-thirds the length of the lemma. *Anthers* 1.0–1.5 mm. *Caryopsis* ellipsoid, enclosed by the delicate lemma and palea. *Flowers* 6–8. 2n = 42 + 0–4B.

Native. Grassy places, cultivated headlands and rough waste ground, mostly on disturbed soils. Throughout the British Isles, common in central and southern England, of scattered occurrence elsewhere. Most of Europe; North Africa; south-west Asia; North America. Our plant is subsp. **gigantea.**

× **stolonifera**

This hybrid has intermediate ligule size (3.5–6.0 mm) and panicle shape between the two parents. It often has strong rhizomes and stolons. The hybrid is a vigorous, highly sterile pentaploid. $2n = 35$.

Native. There are a few, very scattered records in Britain north to Orkney. There are also records from Finland and Sweden.

3. A. castellana Boiss. & Reut.   Highland Bent

Perennial herbs with short rhizomes and sometimes stolons. *Culms* 20–80 cm, usually smooth. *Leaves* 10–25 cm × 1–2 mm, greyish-green, long-filiform, subobtuse at apex, flat or involute, nearly always aculeolate on the ribs; sheaths glabrous; ligule 1–3 mm. *Inflorescence* a panicle (6–)8–16(–25) cm; branches patent at anthesis, rarely afterwards, aculeolate, very unequal, branched 2 or 3 times. *Spikelets* 2.5–4.0 mm, greenish-yellow to purplish. *Glumes* 2.5–3.0 mm, lanceolate, long-acute at apex, aculeolate on the keel at least distally, sometimes with appressed hairs. *Lemma* 1.6–2.0 mm, usually with few to many, usually long hairs, the lateral veins raised in the distal half to three-quarters and excurrent for 0.2–0.5 mm, but glabrous and without excurrent veins when unawned; awn up to 5 mm, arising near the base; callus with hairs up to 0.5 mm. *Palea* elliptical, half to two-thirds as long as lemma. *Anthers* 1.0–1.5 mm. *Caryopsis* ellipsoid, enclosed by the lemma and palea. *Flowers* 7 8. $2n - 28 + 0$ 4B, 42 + 0–3B.

(i) Var. mutica Hack.
A. olivetorum Godr.
All spikelets awnless. *Lemmas* 3-veined, the outer not excurrent at apex, glabrous.

(ii) Var. mixta Hack.
Only terminal spikelets awned. *Lemma* 3-veined, the outer not excurrent, glabrous.

(iii) Var. castellana
A. hispanica Boiss. & Reut.
All spikelets awned. *Lemmas* 5-veined, the outer excurrent at apex, hairy.

Introduced. Lawns, waysides and amenity and sports areas where it has been sown and soon escapes. Throughout Britain, probably under-recorded, and seems to be getting more common. Native of south Europe and North Africa. All three varieties seem to occur in Britain. Their native habitats and distributions are unknown.

4. A. stolonifera L.   Creeping Bent
A. alba auct.

Tufted *perennial herb*, spreading by leafy stolons and forming a close turf. *Culms* 8–40 cm, erect or ascending from a bent or prostrate base, rooting from the lower nodes, slender, smooth, with 2–5 nodes. *Leaves* 1–10 cm × 0.5–5.0 mm, green, greyish-green or bluish-green, glabrous, linear, finely pointed at apex, rolled when young, afterwards flat, closely veined, minutely rough; sheaths rounded on the back, mostly smooth; ligules 1–6 mm, obtuse at apex, membranous. *Panicles* 1–13 × 0.4–2.5 cm, green, whitish or purplish, linear to lanceolate or oblong, open in flower, afterwards contracted

and often dense, or only loose below, frequently lobed; branches clustered, closely divided, rough; pedicels 0.5–2.0 mm. *Spikelets* 2–3 mm, densely clustered, lanceolate to narrowly oblong, 1-flowered, breaking up above the glumes at maturity. *Glumes* persistent, 2–3 mm, as long as the spikelet, equal or slightly unequal, narrowly lanceolate to oblong-lanceolate in side view, pointed at apex, membranous, 1-veined, rough upwards on the keels. *Lemma* up to three-quarters the length of the glumes, ovate or oblong, very obtuse at apex, finely 5-veined, thin, usually awnless, rarely with a short awn from near the tip. *Palea* lanceolate, up to two-thirds the length of the lemma. *Anthers* 1.0–1.5 mm. *Caryopsis* ellipsoid, enclosed by the delicate lemma and palea. *Flowers* 7–8. $2n = 28, 30, 32, 35, 42, 44, 46$.

(i) Var. marina (Gray) P. D. Sell
Vilfa stolonifera var. *marina* Gray; *A. glaucescens* Don ex Hook.; *A. stolonifera* subvar. *salina* Jansen & Wachter
Stolons few and usually short; plant forming a close turf. *Culms* becoming erect. *Leaves* narrow, flat, frequently glaucous. *Panicle* often rather lax, usually not lobed.

(ii) Var. calcicola (Philipson) P. D. Sell
A. stolonifera ecad. *calcicola* Philipson
Stolons either numerous and short, or absent and the plant tufted, forming a close turf. *Culms* erect or geniculate. *Leaves* greyish-green, flat or folded. *Panicle* narrow and meagre.

(iii) Var. maritima (Lam.) Koch
A. maritima Lam.; *Milium maritimum* (Lam.) Clem.; *Vilfa maritima* (Lam.) P. Beauv.; *A. straminea* Hartm.; *Vilfa stolonifera* var. *maritima* (Lam.) Gray; *A. lobata* Sinclair; *A. bryoides* Dumort.; *A. alba* var. *minor* With.; *A. stolonifera* var. *compacta* Hartm.; *A. alba* var. *maritima* (Lam.) Meyer; *A. maritima* var. *pseudopungens* Lange; *A. alba* var. *subrepens* Bab.; *A. salina* Dumort.; *A. alba* var. *subjungens* Hack.; *A. alba* subsp. *salina* (Dumort.) K. Richter; *A. alba* var. *pseudopungens* (Lange) Asch. & Graebn.; *A. stolonifera* subvar. *arenaria* Jansen & Wachter
Growing as isolated plants in loose sand with numerous widely creeping stolons. *Culms* usually much inclined. *Leaves* short, folded; sheaths usually purple. *Panicles* usually dense and lobed. *Glumes* broad and short.

(iv) Var. stolonifera
A. coarctata Ehrh. ex Hoffm.; *A. brevis* Knapp; *Decandolia stolonifera* var. *coarctata* (Ehrh. ex Hoffm.) Bast.; *Vilfa coarctata* (Ehrh. ex Hoffm.) P. Beauv.; *Vilfa stolonifera* var. *brevis* (Knapp) Gray; *A. stolonifera* var. *angustifolia* Sinclair; *A. alba* var. *coarctata* (Ehrh. ex Hoffm.) Rchb.; *A. signata* var. *coarctata* (Ehrh. ex Hoffm.) Schur; *A. stolonifera* var. *coarctata* (Ehrh. ex Hoffm.) Čelak.; *A. depressa* Vasey; *A. alba* var. *condensata* Hack. ex Druce; *A. reptans* Rydb.
Growing as isolated plants usually with numerous, widely creeping stolons. *Culms* becoming erect. *Leaves* about 3–4 mm wide, usually pale green, flat; sheaths often a reddish-purple. *Panicles* usually dense and lobed.

(v) Var. palustris (Huds.) Farw.

A. palustris Huds.; *A. polymorpha* var. *palustris* (Huds.) Huds.; *A. mutalis* Knapp; *Apera palustris* (Huds.) Gray; *A. stolonifera* var. *latifolia* Sinclair; *A. alba* var. *pallens* Gaudin; *A. stolonifera* var. *flagellare* Neilr.; *A. densissima* Druce nom. nud.

Stolons long, often numerous, with no indication of a tuft. *Culms* tall and usually few. *Leaves* long and distant. *Panicle* long and usually narrowly pyramidal.

Native. Damp meadows, ditches, marshes, by lakes, ponds, canals and rivers, damp arable land and rough ground, shingle and dune-slacks, chalk downs and salt-marshes from sea level to 2000 m. Abundant throughout the British Isles. Europe; temperate Asia; North Africa; North America. Introduced in Australia, New Zealand, South Africa and South America. The varieties are ecological. Var. *marina* forms a close turf in salt-marshes and areas within tidal spray, and has been used for the formation of lawns. Var. *calcicola* forms a close turf on chalk downs. Var. *maritima* is a plant of loose sand, mainly coastal. Var. *stolonifera* is a plant of a wide variety of grassland and waste places. Var. *palustris* is a plant of wet places in the lowlands.

× vinealis

This hybrid often has stolons and rhizomes, it has long basal awns or no awns, the palea is about half as long as the lemma and the ligules of the tiller leaves are longer than wide.

Native. It has been recorded only from Cornwall, but is probably overlooked.

5. A. lachnantha Nees   African Bent

Loosely tufted *annual* or *perennial herb*. *Culms* 40–70 cm, pale greyish-green, erect, glabrous. *Leaves* 4–20 cm × 2–4 mm, greyish-green, long-linear, gradually narrowed to an acute apex, flat, minutely scabrid; sheath minutely scabrid; ligule less than 7 mm, acute to obtuse, those of the tillers longer than wide. *Inflorescence* a narrow, lax panicle 10–30 × 1–5 cm, contracted at fruiting; branches erecto-patent to erect, minutely scabrid. *Spikelets* 2–3 mm, 1-flowered, breaking up above the glumes at maturity. *Glumes* subequal, 2.0–2.5 mm, with a green midrib and membranous margins, lanceolate, gradually narrowed to an acute apex, minutely ascending-hairy on the keel, 1-veined. *Lemma* 1.9–2.1 mm, membranous, narrowly lanceolate, acute at apex, finely 5-veined, hairy, usually awnless. *Palea* nearly as long as lemma, membranous. *Anthers* about 0.6 mm. *Caryopsis* enclosed in lemma and palea. *Flowers* 8–9.

Introduced. Wool alien in cultivated and waste places. Rather infrequent in scattered localities in England. Native of central and South Africa.

6. A. avenacea J. F. Gmel.   Blown Grass

Calamagrostis avenacea (J. F. Gmel.) Becherer; *Avena filiformis* G. Forster, non *Agrostis filiformis* Vill.; *A. filiformis* (G. Forster) Spreng.; *Lachnagrostis filiformis* (G. Forster) Trin.; *Calamagrostis filiformis* (G. Forster) Cockayne, non Griseb.; *Deyeuxia filiformis* (G. Forster)

Petrie; *A. retrofacta* Willd.; *Vilfa retrofacta* (Willd.) P. Beauv.; *Lachnagrostis retrofacta* (Willd.) Trin.; *Deyeuxia retrofacta* (Willd.) Kunth; *Calamagrostis retrofacta* (Willd.) Link ex Steud.; *A. debilis* Poir.; *Vilfa debilis* (Poir.) P. Beauv.; *A. forsteri* Rich. ex Roem. & Schult.) Trin.; *Deyeuxia forsteri* (Rich. ex Roem. & Schult.) Steud.; *Lachnagrostis willdenowii* Trin.; *Calamagrostis willdenowii* (Trin.) Steud.; *A. ligulata* Steud.

Tufted *annual* or *perennial herb*. *Culms* 15–75 cm, more or less erect, striate, glabrous. *Leaves* 8–25 cm × 1–3 mm, greyish-green, long-linear, acute to acuminate at apex, minutely scabrid on the veins; sheaths pale green, striate, glabrous; ligules 3–8 mm, acute to rounded at apex, minutely scabrous, those of the tillers longer than wide. *Inflorescence* a large, lax panicle; branches divaricately spreading or drooping at maturity, minutely scabrous. *Spikelets* solitary and narrow. *Glumes* subequal, 2.0–3.5 mm, greyish-green with a narrow membranous margin, narrowly lanceolate, long-acute at apex, scabrous along the keel and occasionally the sides, 1-veined. *Lemma* 1.2–2.0 mm, lanceolate, truncate and very shortly 4-toothed at apex, 4-veined, hairy, awned from about the middle; awn 2.5–6.0 mm, geniculate, minutely scabrous. *Palea* slightly shorter than or subequal to the lemma. *Anthers* 0.2–0.8 mm. *Caryopsis* oblong, more or less compressed, grooved near apex. *Flowers* 8–9. 2*n* = 56.

A very variable species as is shown by the number of names which it has been given.

Introduced. Frequent wool alien naturalised in waste and rough ground and by roads and railways. There are scattered records in Britain. Native of Australia and New Zealand.

7. A. curtisii Kerguélen   Bristle Bent

A. setacea Curtis, non Vill.; *Vilfa setacea* P. Beauv.; *Trichodium setaceum* Roem. & Schult.; *Agraulus setaceus* Gray

Densely tufted *perennial herb*, with numerous slender vegetative shoots growing in isolated tufts or forming a close turf. *Culms* 10–60 cm, erect or sometimes bent below, very slender, rough near the nodes and panicles, with 2–3 nodes. *Leaves* up to 20 cm × 0.2–0.3 mm, green or greyish-green, very fine and bristle-like, finely pointed at apex, erect or flexuous, grooved above, glabrous, minutely rough; sheaths rounded on the back, slightly rough, the basal straw-coloured; ligules 2–4 mm, narrow, membranous. *Panicles* 3–10 cm × 5–15 mm, purple or green, erect, narrowly oblong or lanceolate, dense and spike-like before and after flowering; branches clustered, divided, minutely rough; pedicels 1–4 mm. *Spikelets* 3–4 mm, lanceolate or narrowly oblong, 1-flowered, breaking up at maturity above the glumes. *Glumes* persistent, 2.5–3.5 mm, lanceolate, finely pointed at apex, rounded on the back below, keeled above, firmly membranous, minutely rough, unequal, the upper three-quarters of the length of to nearly as long as the lower, 1-veined or the upper 3-veined. *Lemma* about 2.5 mm, about two-thirds the length of the glumes, ovate-oblong, very obtuse at apex, thin, minutely bearded at the base, finely 5-veined, with the outer veins produced at the tip into minute points, awned from near the base, with the very fine awn up to 5 mm and projecting from the glumes. *Palea* oblong, minute. *Anthers* 1.5–2.0 mm. *Caryopsis* about 1.5 mm, narrowly ellipsoid, covered by the thin lemma. *Flowers* 6–7. $2n = 14$.

Native. Dry sandy or peaty heaths. Locally common in south-west England, extending to south Wales and Surrey; formerly in Sussex. South-west Europe; North Africa.

8. A. canina L.   Velvet Bent
Agraulus caninus (L.) P. Beauv.; *A. fascicularis* Curt. ex Sinclair; *A. tenuifolia* Curt.; *Trichodium caninum* (L.) Schreb.

Tufted *perennial herb* with stolons but without rhizomes, rooting at the nodes and producing there tufts of fine leafy shoots, eventually forming a rather close turf. *Culms* 15–75 cm, erect or more often ascending from a bent or prostrate base, sometimes branching and rooting in the lower half, slender, 2- to 4-noded, smooth. *Leaves* 2–15 cm × 1–3 mm, bright green or greyish-green, long-linear, gradually narrowed to a pointed apex, finely ribbed above, glabrous; sheath rounded on the back, smooth to rough upwards; ligules less than 4 mm, those of tillers longer than wide, membranous, acute to acuminate at apex. *Inflorescence* usually a loose and open panicle 3–16 × up to 7 cm, becoming contracted at fruiting, erect or nodding, purplish, reddish or green; branches clustered, fine, bare in the lower part, divided above, minutely rough. *Spikelets* 1.6–3.0 mm, lanceolate or narrowly oblong, 1-flowered, breaking up above the glume at maturity. *Glumes* 1.5–4.0 mm, pale green, sometimes tinged with purple, lanceolate, acute at apex, 1-veined, the lower as long as the spikelet and minutely rough on the keel, the upper 1.6–2.0 mm and rough only near the tip. *Lemma* about two-thirds the length of the glumes, ovate-oblong, obtuse at apex, very thin, 4- to 5-veined, minutely hairy at the base; awn absent, or present from near the base and bent. *Palea* minute. *Anthers* 1.0–1.5 mm. *Caryopsis* ellipsoid, about 1.1 × 0.3 mm, enclosed by the thin lemma. *Flowers* 6–8. $2n = 14$.

Native. Wet meadows, low places on heaths, common land, marshes, ditches, and pond-sides on acid soils. Common throughout the British Isles, but it has been much confused with *A. vinealis*. Throughout Europe and in temperate Asia and north-east America.

9. A. vinealis Schreb.   Brown Bent
A. canina subsp. *montana* (Hartm.) Hartm.; *A. canina* var. *arida* Schlechtend.

Densely tufted *perennial herb*, usually with slender, creeping, scaly rhizomes. *Culms* 10–60 cm, erect or bent at the base, slender, smooth, with 1–2 nodes. *Leaves* 2–15 cm × (0.6–)1–5 mm, green or greyish-green, glabrous, narrowly linear, finely pointed at apex, flat or inrolled, firm, sometimes bristle-like, closely veined

above, rough on both sides or smooth beneath; sheaths rounded on the back, smooth; ligules 1–5 mm, less than 1.5 times as long as wide, the uppermost pointed. *Panicles* 2–20 cm, green, purplish or brownish, usually contracted and somewhat dense before and after flowering, lanceolate to oblong or narrowly ovate; branches clustered, hair-like, mostly erect, rough, divided above the naked base; pedicels 0.5–2.0 mm. *Spikelets* 2.0–3.3 mm, rather closely clustered, lanceolate to narrowly oblong, 1-flowered, breaking up at maturity above the glumes. *Glumes* persistent, 2–3 mm, as long as the spikelet, equal or slightly unequal with the upper shorter, membranous, 1-veined, or the upper 3-veined, the lower rough on the keel, the upper smooth or nearly so. *Lemma* about three-quarters the length of the glumes, ovate or ovate-oblong, very obtuse at apex, thin, minutely rough, 4- to 5-veined, minutely hairy at the base, with a fine bent awn 2.0–4.5 mm on the back near the base or in the lower third, or awnless. *Palea* minute. *Anthers* 1.0–1.8 mm. *Caryopsis* ellipsoid, enclosed by the thin lemma. *Flowers* 6–8. $2n = 28$.

Native. Dry and sandy or peaty heaths, moors and hillsides. Common throughout much of the British Isles. North, west and central Europe, but distribution imperfectly known.

10. A. scabra Willd.   Rough Bent
A. hyemalis auct.

Annual or *perennial* tufted herb. *Culms* 30–85 cm, erect, pale green, often becoming suffused brownish-purple, striate. *Leaves* 8–20 cm × 1–5 mm, long-linear or filiform, acute at apex, flat, minutely scabrous; sheaths scabrous, striate and glabrous; ligules 1.5–5.0 mm, those of the tillers longer than wide, hyaline, acute to obtuse at apex. *Inflorescence* a panicle 15–25 cm; branches ascending and becoming widely spreading in fruit, sometimes drooping, brittle, scabrous. *Spikelets* 2.0–3.5 mm, lanceolate, loosely arranged at ends of branchlets. *Glumes* unequal, scarious, lanceolate, acute at apex, scabrous on the keels, the lower about 2.0 mm, the upper about 2.5 mm. *Lemma* 1.5–1.7 mm, distinctly exceeding the caryopsis, lanceolate, acute at apex, the callus sparsely hairy; awn absent or very short. *Palea* minute, less than two-fifths as long as lemma. *Anthers* 0.4–0.8 mm. *Caryopsis* ellipsoid, enclosed by the lemma. *Flowers* 7–10.

Introduced. Frequent grain alien now naturalised in waste places, rough ground and by roads and railways. Scattered localities in Britain. Native of North America.

11. A. hyemalis (Walt.) Britton, Sterns & Poggenb.
                                                      Small Bent
Cornucopiae hyemalis Walt.

Tufted, *annual* to *perennial* herb. *Culms* 30–60 cm, very slender, erect, striate, glabrous. *Leaves* 3–9 cm × up to 1 mm, erect, filiform, acute at apex, flat or rolled; sheaths glabrous; ligule 0.5–4.0 mm, acute to obtuse at apex, those of the tillers longer than wide, lacerate. *Inflorescence* a usually purple panicle 4–30 cm, very loose and open at maturity; branches erect or ascending to spreading or finally deflexed, filiform, scabrous, simple or forking only near the tip. *Spikelets* 1.5–1.7 mm, crowded at the end of branchlets. *Glumes* subequal, 1.5–2.1 mm, lanceolate, acute at apex, scabrous on the keel. *Lemma* 1.0–1.3(–1.5) mm, not or scarcely exceeding the caryopsis, lanceolate, obtuse at apex, the callus glabrous. *Palea* absent. *Anthers* about 0.2 mm. *Caryopsis* ellipsoid, embraced by the lemma.

Introduced. Wool and grain alien perhaps naturalised in rough ground and by roads and railways. Scattered records from Britain. Perhaps confused with *A. scabra* and more frequent than has been recorded. Native of North America.

44 × 51. × Agropogon P. Fourn.

Agrostis × Polypogon

Variously intermediate between *Agrostis stolonifera* and *Polypogon* and sterile. It differs from *A. stolonifera* in its more compact panicle with shorter pedicels and disarticulation at maturity, if any, near the base of the pedicel. From *Polypogon* it differs in having persistent spikelets, acute not emarginate glumes and an apical awn.

44/4 × 51/1. × A. littoralis (Sm.) C. E. Hubb.
Perennial Beard Grass
= A. stolonifera × P. monspeliensis
Polypogon littoralis Sm.; *A. lutosus* auct.; *Agrostis littoralis* auct.

Perennial herb, loosely tufted or creeping. *Culms* 8–60 cm, bent at the base and then erect, or ascending from a prostrate base which has many nodes, from which stolons develop, slender, smooth. *Leaves* 3–20 cm × 2–11 mm, greyish-green, glabrous, linear, finely pointed at apex, flat, rough towards the tips or all over; sheaths smooth; ligules 3–7 mm, obtuse at apex, becoming toothed, membranous. *Inflorescence* a panicle 2–18 × 0.6–7.0 cm, green or purplish, erect, moderately to very dense, lanceolate to narrowly ovate or oblong, more or less lobed; branches closely divided, rough; pedicels very short. *Spikelets* 2–3 mm, persistent, narrowly oblong, laterally compressed, 1-flowered. *Glumes* similar, 2.0–2.5 mm, narrowly oblong or elliptical, narrowed upwards, minutely notched at the tip, rounded on the back below, strongly keeled above, membranous, 1-veined, rough, especially in the lower half, awned from the tip, with the fine straight awn 2.0–2.5(–3.0) mm. *Lemma* 0.8–2.8 mm, between half and two-thirds the length of the glumes, elliptical-oblong, with a very blunt, minutely toothed tip, smooth, very thin, obscurely 5-veined, awned from the back just below the tip, the awn 1–2 mm. *Palea* about three-quarters the length of the lemma, lanceolate, finely 2-veined. *Anthers* about 1 mm. *Flowers* 6–9. 2n = 28.

Native. Maritime sand or mud where it is sporadic with its parents. Coasts of England from Dorset to Norfolk. Elsewhere in Britain a bird-seed casual on rubbish tips. South and west Europe; Canary Islands.

44/4 × 51/2. × A. robinsonii (Druce) Melderis & D. C. McClint.
= A. stolonifera × P. viridis
Agrostis × robinsonii Druce

This hybrid is a short-lived perennial which differs from *A. stolonifera* in its bifid glumes with a long apical awn, and from *P. viridis* in its sparsely scabrid glumes and the palea being less than three-quarters as long as the lemma.

Native. Has occurred with the parents in Guernsey in 1924 and 1953.

45. Calamagrostis Adans.

Perennial herbs with rhizomes. *Culms* erect. *Leaves* flat or convolute. *Inflorescence* a more or less diffuse to slightly contracted panicle. *Glumes* 2, equal or nearly so, 1- to 3-veined, longer than the rest of the spikelet. *Lemmas* 3- to 5-veined, with apical or dorsal awn, with a conspicuous basal hair-tuft more than half as long as the lemma. *Palea* about two-thirds as long as the lemma, 2-veined; disarticulating at maturity at the base of the lemma.

About 270 species in temperate regions throughout the world and on tropical mountains.

Stewart, A., Pearman, D. A. & Preston, C. D. (1994). *Scarce plants in Britain*. Peterborough. [*C. stricta*.]

1. Hairs at base of lemma not reaching lemma apex; lemma minutely rough, with an awn arising from the middle or lower half   2.
1. Hairs at base of lemma reaching at least to the apex of lemma; lemma smooth, with awn arising from middle, apex or upper half   3.
2. Spikelets (2.5–)3–4(–4.6) mm; lower glume acute at apex   **4. stricta**
2. Spikelets (4–)4.5–6.0 mm; lower glume acuminate at apex   **5. scotica**
3. Culms mostly with 5–8 nodes; ligules 7–10(–14) mm; pollen absent; anthers indehiscent
3. purpurea subsp. **phragmitoides**
3. Culms mostly with 2–5 nodes; ligules (1–)2–9(–12) mm; pollen present; anthers dehiscent   4.
4. Upper side of leaf scabrid, not hairy; ligules 4–9(–12) mm; lemmas 3-veined, with basal hairs more than 1.5 times as long as lemma   **1. epigejos**
4. Upper side of leaf hairy, often sparsely so; ligules (1–)2–6 mm; lemmas 3- to 5-veined, with basal hairs 1.1–1.5 times as long as lemma   **2. canescens**

1. C. epigejos (L.) Roth   Wood Smallreed
Arundo epigejos L.

Perennial herbs, forming tufts or tussocks, with strong, creeping rhizomes. *Culms* 60–200 cm, erect or slightly spreading, slender to moderately stout, rough near the panicle, otherwise smooth, with 2–3 nodes. *Leaves* up to 70 cm × 4–10 mm, dull green, glabrous, long-linear, finely pointed at apex, flat, rather coarse and scabrid, strongly veined above; sheaths smooth; ligules 4–9(–12) mm, membranous, becoming torn. *Inflorescence* a panicle 15–30 × 3–6 cm, purplish, brown and/or green,

erect, lanceolate to oblong, dense before and after flowering; rhachis rough upwards; branches retrorsely
scabrid, up to 8 cm; pedicels 0.3–1.0 mm. *Spikelets*
4.2–7.5 mm, densely clustered, narrowly lanceolate.
Glumes persistent, equal or nearly so, 4.2–7.5 mm, very
narrowly lanceolate and finely pointed in side view,
firmly membranous, 1-veined or the upper 3-veined,
rough on the keels. *Lemma* 2.5–3.0 mm, about half the
length of the glumes, lanceolate-oblong, thinly membranous, smooth, finely 3-veined, with a fine awn 1.4–1.7
mm from the toothed tip or from the back, surrounded
by white hairs 2.8–4.6(5.5) mm from the base which
much exceed the lemma. *Palea* up to two-thirds the
length of the lemma, 2-veined. *Anthers* 1.6–2.0 mm.
Caryopsis narrowly ellipsoid, enclosed by the thin
lemma and palea. *Flowers* 6–8. $2n = 28, 42, 56$.

Native. Open places in damp woods, thickets, ditches,
fens and dune-slacks, usually on light sands or heavy
clays. Throughout much of Britain and common in
parts of south and east England, rare in most of
Scotland, Wales and Ireland. Most of Europe; temperate Asia; Japan; introduced in North America.

2. C. canescens (Wigg.) Roth   Purple Smallreed
C. lanceolata Roth; *Arundo canescens* Wigg.

Perennial herb, forming loose tufts or patches, with
slender rhizomes. *Culms* 60–120 cm, erect or slightly
spreading, moderately slender, smooth, with 3–5 nodes.
Leaves up to 40 cm × 3–6 mm, green, long-linear, finely
pointed at apex, flat, with short hairs above, rough on
both sides, closely veined; sheaths smooth; ligules
(1–)2–6 mm, membranous. *Inflorescence* a panicle 10–25
× 2.5–6.0 cm, purplish or greenish, rarely yellowish,
lanceolate to oblong, moderately loose, becoming flexuous or finally nodding; rhachis rough; branches slightly
spreading, rough, very slender, the lower up to 8 cm;
pedicels 0.5–3.0 mm. *Spikelets* 4.5–7.0 mm, clustered,
lanceolate. *Glumes* persistent, 4.5–5.5(–6.0) mm, narrowly lanceolate in side view, finely pointed at apex,
slightly unequal, membranous, 1-veined, minutely
rough. *Lemmas* 2.0–2.5(–2.8) mm, half to nearly two-
thirds the length of the glumes, ovate, thinly membranous, smooth, finely 3- to 5-veined, with a very short,
fine awn from the narrowly two-toothed tip, surrounded
and shortly exceeded by fine white hairs 3–4(–4.5) mm
from the base. *Palea* about two-thirds the length of the
lemma, lanceolate, finely 2-veined. *Anthers* about 1.5
mm. *Caryopsis* ellipsoid, enclosed by the delicate lemma
and palea. *Flowers* 6–7. $2n = 28$.

Native. Marshes, fens and wet, open woodland.
Scattered in England, in Kirkcudbrightshire and
Selkirkshire in Scotland. Central and north Europe,
southwards to north Spain, north Italy and Bulgaria;
west Siberia.

× **stricta** = **C.** × **gracilescens** (Blytt) Blytt
C. halleriana var. *gracilescens* Blytt; *C.* × *conwentzii*
Ulbr.

This hybrid is intermediate between the parents in
bract position, glume shape, length of callus hairs in

relation to flower length, awn length and point of origin
of awn on the lemma.

Native. In the British Isles it is known for certain only
by a canal in Yorkshire where partially fertile octoploids
($2n = 56$) intermediate between the parents and sterile
tetraploids ($2n = 28$) closer to *C. stricta* occur along with
C. stricta. In other populations of *C. stricta* in England
and Scotland past introgression with *C. canescens* has
probably taken place.

3. C. purpurea (Trin.) Trin.   Scandinavian Smallreed
Arundo purpurea Trin.

Laxly caespitose, long rhizomatous *perennial herb*.
Culms usually 90–200 cm, rather slender to stout, with
5–8 nodes. *Leaves* 10–60 × 5–10(–15) mm, green, linear,
acute at apex, flat or sometimes convolute, the upper
surface glabrous, hairy or scabrid, with rather prominent veins; sheaths smooth; ligules 7–10(–14) mm.
Inflorescence a panicle, 5–35 cm, nodding at maturity,
often rather dense, the branches scabrid. *Spikelets* 4–7
mm, brownish-green. *Glumes* 4.5–6.0 mm, subequal,
ovate-lanceolate to narrowly lanceolate, acuminate at
apex, sparsely scabridulous. *Lemma* 2.6–3.5 mm, 5-
veined, notched at apex, minutely hairy, awn about 0.5
mm, arising from apical third and scarcely exceeding the
lemma; basal hairs 4–5 mm, 1.1–1.5 times as long as
lemma. *Rhachilla* scarcely prolonged. *Anthers* about 1.7
mm, yellow, without pollen, thecae not separating at
maturity. *Flowers* 6–8. Apomictic. $2n = 56$.

Native. Fens, marshes, ditches and lakesides. About
six localities in Argyllshire, Perthshire, Aberdeenshire
and Angus, and one locality in Westmorland. North-
central east Europe, southwards to northern
Switzerland and south-central Russia. Our plant, not
recognised until 1980, is subsp. **phragmitoides** (Hartm.)
Tzvelev (*C. phragmitoides* Hartm.). It occurs throughout the range of the species. This species is said to be
very variable and if the variants are treated as apomic-
tics with binomials, it is probable that many taxa would
be recognised.

4. C. stricta (Timm) Koeler   Narrow Smallreed
C. neglecta auct.; *Arundo stricta* Timm; *C. neglecta*
subsp. *stricta* (Timm) Tzvelev

Perennial herb, forming compact tufts, with slender,
creeping rhizomes. *Culms* 30–100 cm, erect or slightly
spreading, slender, smooth or rough near the panicle,
with 2–3 nodes. *Leaves* up to 60 cm × 1.5–5.0 mm,
green, narrowly linear, finely pointed at apex, inrolled or
flat, shortly hairy and closely veined above, smooth
beneath, rough on the margins; sheaths smooth; ligules
1–3 mm, membranous. *Inflorescence* a panicle 7–20 ×
1–3 cm, yellowish-brown, purple or greenish, erect,
lanceolate to narrowly oblong, dense to somewhat
loose; rhachis rough; branches erect or nearly so, rough,
the lower up to 6 cm; pedicels 0.5–2.0 mm. *Spikelets*
(2.5–)3–4(–4.6) mm, densely clustered, lanceolate.
Glumes persistent, 3–4 mm, lanceolate and more or less
pointed in side view, equal or nearly so, firmly membranous, rough on the keel and minutely so on the sides, 1-

veined, or the upper 3-veined. *Lemma* 2.5–3.2 mm, three-quarters to four-fifths the length of the glumes, ovate-oblong, with a broadly toothed tip, membranous, minutely rough, 5-veined, with a ring of white hairs at the base up to three-quarters as long as the lemma, bearing a fine straight awn on the back, one-third to halfway above the base, the awn reaching or slightly exceeding the tip of the lemma. *Palea* elliptical, about two-thirds the length of the lemma, 2-keeled. *Anthers* 2.0–2.5 mm. *Caryopsis* narrowly elliptical. *Flowers* 6–8. $2n = 28$.

Native. In neutral marshes and fens and by shallow lakesides, ascending to 340 m. Very scattered in the north of the British Isles south to Cheshire and Co. Antrim, extinct in Suffolk. North and central Europe, north and central Asia; Japan; North and South America.

5. C. scotica (Druce) Druce   Scottish Smallreed
C. strigosa auct.; *Deyeuxia neglecta* Kunth nom. illegit.; *Deyeuxia neglecta* var. *scotica* Druce; *C. neglecta* auct.

Perennial herb forming compact tufts, with slender, creeping rhizomes. *Culms* up to 90 cm, erect or slightly spreading, slender, smooth, with 2–3 nodes. *Leaves* up to 30 cm × 3–5 mm, green, long-linear, finely pointed at apex, inrolled or flat, shortly hairy and prominently veined above, smooth beneath; sheaths smooth; ligules up to 3 mm, obtuse at apex, membranous. *Inflorescence* a panicle 7–16 × 1–3 cm, brownish or tinged with purple, lanceolate to narrowly oblong, erect, rather dense; rhachis rough in the upper part; branches erect after flowering, closely divided, very rough, the lower up to 6 cm; pedicels 1–3 cm. *Spikelets* (4.0–)4.5–6.0 mm, densely clustered, lanceolate, 1-flowered. *Glumes* persistent, 4.5–5.5 mm, narrowly lanceolate in side view, finely pointed at apex, similar, equal or nearly so, firmly membranous, rough on the keels and minutely so on the sides, 3-veined or the lower 1-veined. *Lemma* three-quarters to four-fifths the length of the glumes, ovate, obtuse at apex, with a minutely toothed tip, membranous, 5-veined, minutely rough, ringed at the base with white hairs up to three-quarters as long as the lemma, bearing a fine straight awn on the back one-third of the way from the base, the awn reaching or slightly exceeding the tip of the lemma. *Palea* about two-thirds the length of the lemma, elliptical, membranous. *Anthers* about 2 mm. *Caryopsis* about 2 mm, ellipsoid, enclosed by the papery lemma and palea. *Flowers* 6–8.

Native. Marshes and fens. One locality, formerly three in Caithness, and possibly one in Roxburghshire. Endemic.

45 × 46. × Calammophila Brand
Calamagrostis × Ammophila

This hybrid is sterile and it can be intermediate between the parents or nearer to one or the other. It is usually nearer to the *Ammophila* parent with long rhizomes and tall culms, and has dense panicles which are often purple-tinged.

1. × C. baltica (Flügge ex Schrad.) Brand
Purple Marram
= **A. arenaria × C. epigejos**
× *Ammocalamagrostis baltica* (Flügge ex Schrad.) P. Fourn.; *Arundo baltica* Flügge ex Schrad.; *Ammophila baltica* (Flügge ex Schrad.) Dumort.

Sterile *perennial herb*, spreading extensively by stout rhizomes. *Culms* up to 150 cm, erect or spreading, stout, smooth or rough near the panicle, with few nodes. *Leaves* up to 100 cm × up to 7 mm, green or purplish beneath, long-linear, finely pointed at apex, flat and spreading on the sand with the lower surface uppermost, or becoming inrolled, tough, prominently ribbed above, with the ribs rough or minutely and stiffly hairy, smooth beneath; sheaths rigid, rounded on the back, smooth; ligules 1–25 mm, lanceolate. *Inflorescence* a panicle 13–25 × 1.7–3.0 cm, pale green or purplish, linear-lanceolate to lanceolate, tapering upwards, very dense; rough branches, the lower up to 7 cm; pedicels 1–4 mm, rough. *Spikelets* 9–12 mm, narrow, closely overlapping, compressed, 1-flowered. *Glumes* 9–12 mm, narrowly lanceolate, finely pointed at apex, equal or slightly unequal, as long as the spikelet, firm, keeled upwards, rough on the keels, the lower 1- to 3-veined, the upper 3-veined. *Lemma* 7–9 mm, lanceolate and pointed in side view, firm, minutely rough, 3- to 7-veined, with a straight awn 0.8–2.0 mm from just below the minutely 2-toothed tip, densely bearded at the base with fine silky hairs 5–7 mm. *Palea* slightly shorter than the lemma, lanceolate, 2- to 4-veined. *Anthers* 4.0–4.5 mm. *Flowers* 6–8. $2n = 28, 42$.

Native. Sand-dunes on the coasts of Norfolk, Suffolk, the Cheviots and Sutherland. Planted for binding the sand in Norfolk, Suffolk and Hampshire. Coasts of the Baltic, North Sea and north France, where other variants more exactly intermediate between the two parents and others nearer *C. epigejos* occur. The British plants which are nearer to *A. arenaria* are referable to var. **baltica.**

46. Ammophila Host

Perennial herb with strong rhizomes. *Leaves* tightly inrolled; ligules 1–30 mm, narrow. *Inflorescence* a compact, linear-ellipsoid panicle. *Glumes* 2, subequal, the lower 1- to 3-veined, the upper 3-veined, slightly longer than the rest of the spikelet. *Lemmas* 5- to 7-veined, awnless or with a minute subapical awn less than 1 mm, with the basal hair-tuft less than half as long as the lemma. *Palea* nearly as long as the lemma, 2- to 4-veined. *Spikelets* disarticulating at maturity at the base of the lemma.

Two species in Europe and North Africa, and on the east coast of North America.

Huiskes, A. H. L. (1979). *Ammophila arenaria* (L.) Link in Biological flora of the British Isles. *Jour. Ecol.* **67**: 363–382.

1. Ligule 10–30 mm   **1. arenaria**
1. Ligule 1–3 mm   **2. breviligulata**

1. A. arenaria (L.) Link   Marram

A. arundinacea Host; *Psamma arenaria* (L.) Roem. &
Schult.; *Arundo arenaria* L.; *Calamagrostis arenaria*
(L.) Roth; *Phalaris ammophila* Link; *Arundo littoralis*
P. Beauv. ex Steud.

Perennial herb, spreading extensively by stout,
branched rhizomes and forming compact tufts. *Culms*
50–150 cm, erect or spreading, moderately stout, rigid,
smooth, with few nodes. *Leaves* up to 60 cm × up to 6
mm, greyish-green, long-linear, sharply pointed at
apex, tightly inrolled, rigid, closely ribbed above, with
the ribs minutely and densely hairy, smooth beneath;
sheaths overlapping, smooth; ligules 10–30 mm,
narrow, firm, acute at apex. *Inflorescence* a panicle 7–22
× 1.0–2.5 cm, pale, spike-like, narrowly oblong to
lanceolate-oblong, cylindrical, tapering upwards;
branches erect; pedicels 1–4 mm. *Spikelets* 10–16 mm,
narrowly oblong, compressed, closely overlapping, 1-
flowered, breaking up at maturity above the glumes.
Glumes 10–16 mm, persistent, narrow, pointed or
obtuse at apex, equal or slightly unequal, minutely
rough on the sides and on the keels, firm, the lower
mostly 1-veined, the upper 3-veined. *Lemma* 8–12 mm,
lanceolate, obtuse at apex, keeled, 5- to 7-veined,
minutely rough, with a projecting point 0.2–0.8 mm
near the apex, surrounded at the base with fine white
hairs 3–5 mm. *Palea* nearly as long as the lemma,
linear-lanceolate, 2- to 4-veined. *Anthers* 4–7 mm.
Caryopsis ellipsoid, enclosed by the hooded lemma and
palea. *Flowers* 6–8. $2n = 28, 56$.

Native. Common and often abundant and dominant
on sand-dunes on the coast. All round the coasts of the
British Isles; a rare casual inland; sometimes planted as
a sand-binder. Coasts of west Europe to about 62° N in
Norway; introduced in North America. The British
plant is subsp. **arenaria** which occurs in north and west
Europe southwards to north-west Spain.

2. A. breviligulata Fern.   American Marram

Perennial herb spreading extensively by stout, branched
rhizomes and forming compact tufts. *Culms* 50–100 cm,
erect, coarse. *Leaves* up to 60 cm × up to 6 mm, greyish-
green, long-linear, sharply pointed at apex, tightly
inrolled, rigid, closely ribbed above, the ribs minutely
hairy, smooth beneath; sheaths overlapping, smooth;
ligules 1–3 mm, narrow, chartaceous, rounded at apex.
Inflorescence a panicle 13–40 cm, whitish-brown or
slightly purplish-tinged, linear-cylindrical, the rhachis
puberulous. *Spiklets* 8–12 mm, oblong, compressed,
closely overlapping, breaking up at maturity below the
glumes. *Glumes* 8–12 mm, unequal, chartaceous,
pointed or obtuse at apex, keeled and minutely rough on
the keels. *Lemma* 8–12 mm, lanceolate, obtuse at apex,
crowded at the base by hairs. *Palea* nearly as long as the
lemma, linear-lanceolate. *Anthers* 6–8 mm. *Caryopsis*
3.0–3.6 mm, ellipsoid. *Flowers* 6–8. $2n = 28$.

Introduced. Planted on sand-dunes at Newborough,
Anglesey in the late 1950s and now well naturalised in
one small area. Native of eastern North America.

47. Gastridium P. Beauv.

Annual herbs. Leaves flat; ligules membranous.
Inflorescence a compact, linear-ellipsoid panicle. *Glumes*
2, unequal, 1-veined, much longer than rest of spikelet,
linear-lanceolate, with a swollen, more or less hemispherical base. *Lemmas* 5-veined, with a dorsal, usually
bent awn, without a basal hair-tuft. *Palea* more or less as
long as the lemma, 2-veined. *Spikelets* disarticulating at
maturity at the base of the lemma.

Two species in Europe and North Africa, east to Iran.

Trist, P. J. O.(1983). The past and present status of *Gastridium
    ventricosum* (Gouan) Schinz & Thell. as an arable colonist in
    Britain. *Watsonia* **14**: 257–261.
Trist, P .J. O. (1986). The distribution, ecology, history and
    status of *Gastridium ventricosum* (Gouan) Schinz & Thell. in
    the British Isles. *Watsonia* **16**: 43–54.

1. Culms procumbent to erect; spikelets (2–)3–5 mm;
   lemmas nearly glabrous to sparsely hairy, with an awn
   0–4 mm, rarely exceeding the glumes   **1. ventricosum**
1. Culms usually erect; spikelets (4–)5–8 mm; lemmas
   hairy to densely so, with an awn 4–7(–8) mm, often
   exceeding the glumes   **2. phleoides**

1. G. ventricosum (Gouan) Schinz & Thell.   Nit Grass
Agrostis ventricosa Gouan; *G. lendigerum* (L.) Desv.; *G.
australe* P. Beauv.

Annual herb. Culms 2–70 cm, loosely tufted or solitary,
slender, erect, or bent near the base and procumbent,
smooth. *Leaves* 2–10 cm × 2–4 mm, green, glabrous,
linear, finely pointed at apex, flat, rough, or smooth
except on the margins; sheaths rounded on the back,
minutely rough upwards or smooth; ligules 1–3 mm,
membranous. *Inflorescence* a panicle (1–)5–15 cm ×
5–12 mm, narrowly lanceolate to narrowly oblong,
cylindrical or tapering upwards, open in flower but
afterwards rather dense, pale green, shining; branches
closely divided, rough; pedicels 0.5–2.0 mm. *Spikelets*
(2–)3–5 mm, densely overlapping, narrowly oblong, 1-
flowered, breaking up above the glumes at maturity.
Glumes 3–5 mm, persistent, narrowly lanceolate,
rounded on the back, hardened, shining and swollen
near the base, then slightly constricted, keeled, narrowed above and pointed, rough on the keels, 1-veined,
unequal, the lower as long as the spikelet, the upper
three-quarters or more the length of the lower. *Lemma*
rounded on the back, broadly elliptical, with a blunt,
minutely toothed tip, 0.8–1.0 mm, thin, finely 5-veined,
minutely hairy at the base, sparingly short-hairy on the
sides or glabrous, awned from the back near the apex, or
awnless, the awn 0–4 mm, very slender, bent at and
twisted below the middle. *Palea* as long as the lemma,
elliptical, 2-veined. *Anthers* 0.6–0.8 mm. *Caryopsis* ellipsoid, tightly enclosed between the lemma and palea.
Flowers 6–8. $2n = 14$.

Native. Formerly a frequent weed of arable land. On
rather bare or sparsely grassed, calcareous, well-drained
soil; also a frequent wool and grain alien. As a native
very local near the coast in south-west Britain from
south Devon and the Isle of Wight to Glamorgan, and
perhaps in the Channel Islands. As an alien scattered in

England and south Wales. South and west Europe; North Africa; Macaronesia.

2. G. phleoides (Nees & Meyen) C. E. Hubb.

Eastern Nit Grass

Lachnagrostis phleoides Nees & Meyen; *G. ventricosum* subsp. *phleoides* (Nees & Meyen) Tzvelev

Annual herb. Culms up to 60 cm, erect when solitary, but usually more or less densely fascicled and then the outermost stems prominently decumbent at the base, smooth, glabrous, with few nodes. *Leaves* up to 10 cm × 1–3 mm, linear, long-acuminate at apex, rounded at base, scabrid on both surfaces and on the margins, often flaccid; sheaths glabrous, scaberulous, striate; ligules 2–6 mm, pointed at apex, membranous. *Inflorescence* a panicle 5–18 cm × 5–15 mm, erect, dense, silvery-green, lanceolate or cylindrical and then slightly tapering at tip and base, with short, densely spiculate branches. *Spikelets* (4–)5–8 mm, lanceolate, shortly pedicellate, erect. *Glumes* unequal, with a prominent bulge at the base, the lower as long as the spikelet, acuminate at apex, keeled, 1-veined, hyaline on the margins and scabrid on the keel, the upper 1.0–1.5 mm, lanceolate, acuminate at apex, keeled above, scabrid on the keel and 1-veined. *Lemma* 1.0–1.5 mm, broadly elliptical, truncate at apex, 5-veined, hyaline, loosely hairy on the dorsal surface, the central vein produced from the dorsal surface above the middle as an awn, the awn 4–7(–8) mm with the lower half twisted. *Palea* elliptical, hyaline, 2-veined, much narrower than the lemma. *Anthers* about 0.5 mm. *Caryopsis* ellipsoid, tightly enclosed between the lemma and palea. *Flowers* 5–6. 2n = 28.

Introduced. A rather infrequent wool alien. In scattered localities in England. Native of the east Mediterranean region eastwards to Iran; Macaronesia. Introduced into Australia, South Africa, West Indies, North and temperate South America.

48. Lagurus L.

Annual herb. Leaves flat; ligule membranous. *Inflorescence* a very compact, ovoid, densely silky-hairy panicle. *Glumes* 2, equal, 1-veined, linear-lanceolate, tapered to an apical awn, longer than the rest of the spikelet. *Lemmas* 5-veined, with 2 apical bristles reaching about as far as the glume awns, with a dorsally bent awn well exserted from the glumes, hairy but without a basal hair-tuft. *Palea* shorter than the body of the lemma, 2-veined. *Spikelets* disarticulating at maturity at the base of the lemma. *Caryopsis* enclosed by the hardened lemma and palea.

1. L. ovatus L.

Hare's-tail

Softly hairy *annual herb. Culms* 5–60 cm, solitary or tufted, erect or ascending from a bent base, slender, with few nodes, loosely to densely hairy. *Leaves* 1–20 cm × 2–14 mm, greyish-green, hairy, linear to narrowly lanceolate, pointed at apex, flat, velvety; sheaths loose, rounded on the back, the upper somewhat inflated; ligules up to 3 mm, obtuse at apex, membranous, hairy. *Inflorescence* a panicle 1–7 cm × 6–20 mm, spike-like,

dense, globose to ovoid or oblong-cylindrical, very softly hairy, bristly, erect or at length nodding, pale, rarely tinged with purple; branches very short; pedicels 0.5–2.0 mm. *Spikelets* 8–10 mm, densely overlapping, narrow, 1-flowered, breaking up at maturity above the glumes. *Glumes* persistent, equal, 8–10 mm, narrowly lanceolate, tapering into a fine bristle as long as the spikelet, thinly membranous, 1-veined, closely hairy with fine spreading hairs. *Lemma* 4–5 mm, rounded on the back, elliptical, narrowed into 2 teeth, with each tooth terminated by a fine straight bristle 3–5 mm, membranous, 5-veined, hairy near the base, awned from the back in the upper third, the awn 8–18 mm and bent and twisted below the middle. *Palea* narrow, shorter than the lemma, 2-keeled. *Anthers* 1.5–2.0 mm. *Caryopsis* 2.5–3.0 mm, narrowly ellipsoid, enclosed by the hardened lemma and palea. *Flowers* 6–8. 2n = 14.

Introduced. Planted and now often abundant on mobile sand-dunes in Jersey and Guernsey; planted but still rare at Dawlish in Devon; a rare casual elsewhere in England and south Wales. Native in the Mediterranean region, south Europe, Madeira and the Canary Islands. Introduced in North and South America, Australia and South Africa.

49. Apera Adans.

Annual herbs. Leaves flat or rolled; ligule membranous. *Inflorescence* a diffuse to more or less contracted panicle. *Glumes* 2, unequal, the lower 1-veined, the upper 3-veined, more or less as long as the rest of the spikelets. *Lemmas* 5-veined, with a subterminal, more or less straight awn much longer than the body, minutely hairy at the base. *Palea* shorter than to more or less as long as the body of the lemma, 2-veined. *Spikelets* disarticulating at maturity at the base of the lemma.

Three species, from Europe to Afghanistan.

Easy, G. (1992). 'Breckland bent' in Cambridgeshire. *Nat. Cambridgeshire* **34**: 43–45.

Stewart, A., Pearman, D. A. & Preston, C. D. (1994). *Scarce plants in Britain.* Peterborough. [*A. interrupta* and *A. spica-venti.*]

1. Ligules 3–10 mm; panicle very diffuse; spikelet
   2.4–3.0 mm; anthers 1–2 mm **1. spica-venti**
1. Ligules 2–5 mm; panicle loosely contracted; spikelet
   1.8–2.5 mm; anthers 0.3–0.4 mm **2. interrupta**

1. A. spica-venti (L.) P. Beauv.   Loose Silky Bent

Agrostis spica-venti L.

Annual herb. Culms 20–100 cm, tufted or solitary, erect or bent near the base, slender to relatively stout, with 3–5 nodes below the middle, smooth. *Leaves* 7–25 cm × 3–10 mm, green, glabrous, linear, pointed at apex, flat, rough all over or smooth beneath; sheaths smooth or rough near the apex, green or purple; ligules 3–10 mm, oblong, membranous. *Inflorescence* a panicle 10–25 × 3–15 cm, ovate or oblong, usually open and very diffuse, or sometimes appearing congested, much branched, green or purplish; branches usually numerous, spreading, very fine, rough; pedicels 1–3 mm. *Spikelets* 2.4–3.0 mm, scattered, oblong, 1-flowered, breaking up at maturity. *Glumes* slightly to markedly unequal, finely pointed, rough on the keel above the middle, membranous, the lower three-quarters or more the length of the upper, narrowly lanceolate and 1-veined, the upper oblong-lanceolate and 3-veined. *Lemmas* as long as the upper glumes or slightly shorter, rounded on the back, lanceolate-oblong, pointed in side view, minutely rough above the middle, minutely bearded at the base, firmly membranous, finely 5-veined, awned from near the tip, with a fine straight or flexuous awn 5–10 mm. *Palea* as long as the lemma or slightly shorter, narrowly elliptical, 2-veined. *Anthers* 1–2 mm. *Caryopsis* ellipsoid, tightly enclosed between the hardened lemma and palea. *Flowers* 6–8. $2n = 14$.

Native or of ancient introduction. Locally frequent in dry, sandy arable fields, where it sometimes forms dense swards, and adjacent waste ground. Locally frequent in east and south-east England, elsewhere scattered and casual. Most of Europe; Siberia; introduced in North America.

2. A. interrupta (L.) P. Beauv.   Dense Silky Bent
Agrostis interrupta L.

Annual herb. Culms 10–90 cm, tufted or solitary, erect or bent near the base, with 2–4 nodes below the middle, smooth. *Leaves* up to 12 cm × 1–4 mm, green, linear, pointed at apex, flat, or rolled when dry, minutely hairy or rough on the veins above, smooth beneath except towards the tip; sheaths green or purple, smooth; ligules 2–5 mm, oblong, membranous. *Inflorescence* a panicle 3–20 cm × 4–15 mm, green or purplish, erect, contracted, more or less interrupted below, continuous and moderately dense above; branches erect or slightly spreading, bearing spikelets close to the base, rough; pedicels 0.5–2 mm. *Spikelets* 1.8–2.5 mm, narrowly oblong, 1-flowered, breaking up at maturity. *Glumes* persistent, unequal, pointed at apex, membranous, rough on the keel above the middle, the lower 1.5–2.0 mm, lanceolate and 1-veined, the upper 1.7–2.0 mm and broader, about as long as the spikelet. *Lemma* 1.8–2.5 mm, from slightly shorter to slightly longer than the upper glume, rounded on the back, lanceolate-oblong and pointed in side view, firmly membranous, finely 5-veined, minutely rough above the middle, minutely hairy at the base, awned from near the tip, the very fine, straight awn 4–10 mm and scabrid in the upper half. *Palea* about three-quarters the length of the lemma, narrowly elliptical. *Anthers* 0.3–0.4 mm. *Caryopsis* ellipsoid, enclosed by the lemma and palea. *Flowers* 6–7. $2n = 14$.

Usually regarded as introduced but Easy (1992) considers it may have been native on the sands and chalklands of eastern England. Dry sandy fields and rough ground, road and rail verges and chalk and gravel excavations. Locally frequent in East Anglia, a rare casual elsewhere. West and central Europe and in scattered localities eastwards to south-east Russia and central Asia; North Africa; introduced in North America.

50. Mibora Adans.

Annual herb. Leaves flat or with inrolled margins; ligules membranous. *Inflorescence* a slender, 1-sided raceme with pedicels less than 0.5 mm. *Glumes* 2, equal, 1-veined, longer than the rest of the spikelet. *Lemmas* 5-veined, awnless, hairy but without basal hair-tuft. *Palea* as long as lemma, 2-veined. *Spikelet* disarticulating at maturity at the base of the lemma.

Two species in western Europe and North Africa.

1. M. minima (L.) Desv.   Early Sand Grass
Agrostis minima L.; *M. verna* P. Beauv.; *Chamagrostis minima* (L.) Borkh.

Small, compactly tufted *annual herb. Culms* 2–15 cm, erect or spreading, numerous, very slender, unbranched, smooth, closely sheathed at the base, the upper part green or purplish. *Leaves* mostly at the base of the plant, up to 20 × 0.5 mm, smooth, glabrous, very narrowly linear, obtuse at apex, flat or with inrolled margins; sheaths round, overlapping, delicate; ligules up to 1 mm, very obtuse at apex, thinly membranous. *Inflorescence* spike-like, very slender, secund, erect, 5–20 × about 1 mm, reddish-purple or green; rhachis very slender, smooth. *Spikelets* almost stalkless, appressed to and loosely to closely overlapping in 2 rows on one side of the rhachis, 1.8–3.0 mm, oblong, obtuse at apex, 1-flowered, breaking at maturity above the glumes. *Glumes* persistent, as long as the spikelet, equal or nearly so, oblong or elliptic-oblong, very obtuse at apex, rounded on the back, thinly membranous, 1-veined, smooth. *Lemma* about two-thirds the length of the glumes, very broad, truncate or minutely toothed at the apex, shortly and densely hairy, thinner than the glumes, finely 5-veined. *Palea* as long as the lemma, elliptical, 2-veined, shortly hairy. *Anthers* 1.0–1.7 mm. *Caryopsis* ellipsoid, loosely enclosed between the delicate lemma and palea. *Flowers* 4–5, sometimes again 8–9. $2n = 14$.

Native in moist sandy places by the sea in Anglesey and the Channel Islands and possibly Glamorgan. Naturalised in a few places on the south and east coasts of Britain north to East Lothian; rare casual elsewhere.

51. Polypogon Desf.

Annual or *perennial herbs* with stolons. *Leaves* flat; ligule membranous. *Inflorescence* a contracted or semi-diffuse panicle. *Glumes* 2, equal, 1-veined or the upper 3-veined, much longer than the rest of the spikelet, awned from the apex or unawned. *Lemmas* 5-veined, truncate and finely toothed at the apex, without a basal hair-tuft, awnless or with a short terminal awn. *Palea* nearly as long as lemma, 2-veined. *Spikelets* disarticulating at maturity near the base or apex of the pedicel.

Eighteen species in warm, temperate regions of the world and on tops of tropical mountains.

Stewart, A., Pearman, D. A. & Preston, C. D. (1994). *Scarce plants in Britain.* Peterborough. [*P. monspeliensis.*]

1. Annual; glumes and lemma awned; glumes hairy
                                                   1. monspeliensis

1. Perennial or annual; glumes and lemma unawned; glumes scabrid   **2. viridis**

1. P. monspeliensis (L.) Desf.   Annual Beard Grass
P. paniceus (L.) Lag.; *Alopecurus monspeliensis* L.

Annual herb. Culms 6–80 cm, in small tufts or solitary, erect or bent towards the base, slender to rather stout, with 3–6 nodes, usually branched near the base, rough beneath the panicle or smooth. *Leaves* 5–15 cm × 2–8 mm, green, glabrous, long-linear, pointed at apex, flat, rough on the veins; sheaths smooth, or rough upwards, the upper somewhat inflated; ligules 3–15 mm, oblong, toothed, membranous. *Inflorescence* a panicle 1.5–16.0 cm × 10–35 mm, very dense, covered with fine bristles, narrowly ovate to narrowly oblong, cylindrical or sometimes lobed, pale green or yellowish; branches closely divided; pedicels very short, jointed. *Spikelets* 2–3 mm, narrowly oblong, 1-flowered, falling entire at maturity with a minute piece of the pedicel. *Glumes* similar, much exceeding the lemma, rounded on the back below, keeled above, narrowly oblong when opened out, obtuse and slightly notched at the apex, rough with minute projections especially in the lower part, minutely hairy on the margins, 1-veined, awned from the tip, with the fine, straight awn 3.5–7.0 mm. *Lemma* about half the length of the glumes, very smooth and shining, broadly elliptical, with a broad, obtuse, minutely toothed tip, very thin, obscurely 5-veined, awnless, or with an awn up to 2 mm long from the tip. *Palea* oblanceolate, slightly shorter than the lemma. *Anthers* 0.4–0.7 mm. *Caryopsis* ellipsoid, enclosed by the lemma and palea. *Flowers* 6–8. $2n = 28$.

Native. Damp brackish marshes and salt-marshes and edges of pools, creeks and ditches near the sea. Coasts of south and south-east England from Dorset to Norfolk, and in Guernsey. Elsewhere in Britain north to central Scotland as a casual of tips and waste ground, sometimes becoming naturalised. South and west Europe; North Africa; south-west Asia; Macaronesia; introduced in North America.

2. P. viridis (Gouan) Breistr.   Water Bent
P. semiverticillata (Forskål) Hyl.; *Agrostis semiverticillata* (Forskål) C. Christ.; *Agrostis verticillata* Vill.; *Agrostis viridis* Gouan

Perennial or *annual herb*, loosely tufted, with long tailing stolons, rooting at the nodes. *Culms* 10–60 cm, ascending from a bent or prostrate base, slender, often branching in the lower part, smooth. *Leaves* 3–18 cm × 2–10 mm, greyish-green or green, glabrous, linear, pointed, flat, closely veined, rough; sheaths rounded on the back, rather loose, smooth; ligules 1.5–6.0 mm, obtuse at apex, membranous. *Inflorescence* a panicle 2–15 cm × 6–40 mm, pale green or purplish, erect, ovate or oblong, dense, lobed, sometimes interrupted below; branches clustered, closely divided, crowded with spikelets mostly right to the base, rough; pedicels very short, jointed on the branches. *Spikelets* 1.7–2.2 mm, very numerous, oblong, awnless, 1-flowered, falling at maturity with the pedicel attached. *Glumes* equal, as long as the spikelet, oblong or elliptical, obtuse at apex when opened out, rounded on the back below, keeled above the middle, minutely rough, 1- to 4-veined, or the upper 3-veined,

firmly membranous. *Lemma* about half the length of the glumes, rounded on the back, broadly elliptical, with a very blunt, minutely toothed, broad tip, smooth, finely 5-veined, thin. *Palea* as long as the lemma and similar in texture, elliptical, 2-veined. *Anthers* 0.5–0.7 mm. *Caryopsis* pale brown, about 1 mm, ellipsoid, enclosed by the thin lemma and palea. *Flowers* 6–8. $2n = 28$.

Introduced. Naturalised on roadsides, rough ground and by pools in Guernsey and Jersey. Elsewhere in Britain an occasional casual on tips and waste land. Native in south Europe; south-west Asia; north India; North Africa; Macaronesia. Introduced in North and South America, South Africa and South Australia.

52. Alopecurus L.

Annual or *perennial herbs* without rhizomes, sometimes with stolons. *Leaves* flat or rolled; ligule membranous. *Inflorescence* a very contracted, spike-like panicle. *Glumes* 2, equal, 3-veined, sometimes with their margins fused proximally round the spikelet, slightly shorter to slightly longer than the rest of the spikelet, keeled, rounded, acute or apiculate at apex. *Lemmas* 4-veined, obtuse to truncate or notched at apex, without a basal hair-tuft, sometimes with margins fused proximally round the carpel and stamens, with a dorsal, long or short awn from the lower half. *Palea* absent. *Spikelets* at maturity disarticulating near the base of the pedicel.

Thirty-six species in the north temperate zone and in South America.

Naylor, R. E. L. (1972). *Alopecurus myosuroides* Huds. in Biological flora of the British Isles. *Jour. Ecol.* **60**: 611–622.
Stewart, A., Pearman, D. A. & Preston, C. D. (1994). *Scarce plants in Britain*. Peterborough. [*A. aequalis*, *A. borealis* and *A. bulbosus*.]
Trist, P. J. O. (1981). The survival of *Alopecurus bulbosus* Gouan in former sea-flooded marshes in East Suffolk. *Watsonia* **13**: 313–316.
Trist, P .J. O. & Wilkinson, M. J. (1988). *Alopecurus × plettkei* Mattfeld in Britain. *Watsonia* **17**: 301–308.

1. Lemmas unawned or with an awn shorter than the body and not exserted from the glumes or exserted by less than 0.5 mm   2.
1. Lemmas with awn longer than the body and exserted from the glumes by more than 1.0 mm   3.
2. Panicles more than 3 times as long as wide; glumes with hairs less than 0.5 mm; lemma with margins fused proximally for one-third to half their length; lowlands   **4. aequalis**
2. Panicles less than 3 times as long as wide; glumes with hairs more than 0.5 mm; lemma with margins fused proximally for less than one-quarter of their length; mountains   **5. borealis**
3. The 2 glumes fused only at their extreme base; margins of lemma free or fused proximally for less than one-quarter of their length   4.
3. The 2 glumes fused proximally for about one-quarter to half of their length   5.
4. Basal culm internode 0–1(–1.5) mm wider than the normal culm width; glumes obtuse at apex   **2. geniculatus**

4. Basal culm internode swollen (1–)2.0–4.5(–6) mm
   wider than the normal culm width; glumes acute at
   apex   **3. bulbosus**

5. Perennial; glumes fused proximally for about one-
   quarter of their length, conspicuously hairy, the hairs
   more than 0.5 mm, keel unwinged   **1. pratensis**

5. Annual; glumes fused proximally for one-third to half
   of their length, subglabrous or with hairs less than
   0.5 mm on keel, margins and at base, with a winged
   keel   **6. myosuroides**

1. A. pratensis L.   Meadow Foxtail

Loosely or compactly tufted *perennial herb*. *Culms*
30–120 cm, erect or bent at the base, slender to rather
stout, with few nodes, smooth, green or whitish-green in
the upper part. *Leaves* up to 40 cm × 3–10 mm, green,
glabrous, linear, finely pointed at apex, finally flat, rough
or nearly smooth; sheaths smooth, cylindrical, split, the
basal turning dark brown, the upper green or whitish-
green and somewhat inflated; ligules 1.0–2.5 mm, membranous. *Inflorescence* a very dense, spike-like panicle
2–13 cm × 5–10 mm, cylindrical, obtuse at apex, soft,
green or purplish; branches short, erect; pedicels very
short. *Spikelets* 4–6 mm, lanceolate-oblong or elliptical,
flattened, 1-flowered, falling entire at maturity. *Glumes*
more or less as long as spikelet, narrowly lanceolate and
pointed in side view, with their margins united towards
the base for about a quarter of their length, 3-veined,
firm, fringed with fine hairs on the keels, hairy on the
sides. *Lemma* as long as or slightly longer than the
glumes, ovate or elliptical, rather blunt at apex, keeled,
with the margins united below the middle, 4-veined,
membranous, smooth, awned from the lower third of
the back, with the awn exceeding the glumes by 3–5 mm.
Palea absent. *Anthers* 2.0–3.5 mm, yellow or purple.
Caryopsis more or less ellipsoid, enclosed in the lemma
between the glumes. *Flowers* 4–7. Protogynous. 2n = 28.

Very variable as many strains have been introduced to
improve grassland. Some yield succulent leafy herbage
which is nutritious and very palatable to stock. Forms
with yellow or striped yellow leaves are sometimes
grown in gardens.

Native. Grassy places, mostly on damp, rich soils,
especially in river valleys and low-lying areas. Common
in the lowlands throughout the British Isles. Most of
Europe; west Siberia; Caucasus.

2. A. geniculatus L.   Marsh Foxtail

Perennial herb. *Culms* spreading, grooved, narrow and
usually ascending from a bent or prostrate base and
rooting at the nodes, sometimes extensively creeping,
occasionally floating in water, slender, with 5–7 nodes,
smooth, whitish-green in upper part, rarely slightly
tuberous at the basal node. *Leaves* 2–12 cm × 2–7 mm,
green or greyish-green, glabrous, linear, pointed at apex,
flat, spreading, rough on the veins or smooth beneath;
sheaths smooth, whitish-green, the upper somewhat
inflated; ligules 2–5 mm, membranous, obtuse at apex.
Inflorescence a panicle 15–70 × 3–7 mm, very dense,
spike-like, narrowly cylindrical, obtuse at apex, green or
tinged with blue or purplish; pedicels very short.

Spikelets 2.0–3.3 mm, oblong, 1-flowered, flattened,
falling entire at maturity. *Glumes* more or less equalling
spikelets, narrowly oblong, obtuse at apex, keeled, with
the margins free nearly to the base, thinly membranous,
3-veined, fringed with silky hairs on the keel and with
appressed hairs on the sides. *Lemma* slightly shorter
than or as long as the glumes, broadly oblong or ovate,
very obtuse at apex, keeled at apex, with the margins
united near the base, thinly membranous, smooth, 4-
veined, awned just above the base, with the awn exceeding the glumes by 1.5–3.0 mm. *Palea* absent. *Anthers*
1.2–2.0 mm, yellow or purple. *Caryopsis* narrowly ellipsoid, enclosed in the lemma between the thin glumes.
Flowers 6–8. 2n = 28.

Native. Low places in wet meadows, ditches and
marshes, and sides of ponds. Frequent to common
throughout the British Isles. Most of Europe; west
Siberia; introduced in North and South America.

× **pratensis** = **A.** × **brachystylus** Peterm.
A. × *hybridus* Wimm.
This highly sterile hybrid is intermediate in habit and
fusion of lemma and glumes, spikelet length
(3.5–4.2(–4.8) mm) and anther length (1.8–2.5 mm).
   It is scattered throughout Britain in low areas by
ditches and small streams. It is widespread in
Continental Europe.

3. A. bulbosus Gouan   Bulbous Foxtail

Perennial herb forming small compact or spreading
tufts. *Culms* 11–29 cm, erect, or ascending from a bent
base, slender, unbranched, with 1–4(–5) nodes, smooth,
the basal one or two joints swollen or bulbous. *Leaves*
1.5–12.0 cm × 1.0–3.5 mm, greyish-green or green,
glabrous, narrowly linear, pointed at apex, flat or rolled,
closely veined, smooth except for the rough margins;
sheaths smooth, the uppermost somewhat inflated;
ligules 2–6 mm, membranous, obtuse at apex.
Inflorescence a spike-like panicle 15–70 × 3–5 mm, very
dense, narrowly cylindrical, green tinged with bluish-
grey or purple; pedicels extremely short. *Spikelets* 3–4
mm, oblong, flattened, 1-flowered, falling entire at
maturity. *Glumes* similar, free to the base, as long as the
spikelet, keeled, narrowly oblong or lanceolate, sharply
acute at apex, 3-veined, firmly membranous, short hairy
on the keels and on the sides below the middle. *Lemma*
up to four-fifths the length of the glumes, keeled,
broadly oblong, very obtuse at apex, with the margins
free to the base, membranous, 4-veined, minutely hairy
at the tip, otherwise glabrous, awned on the back from
near the base, the awn 3.5–6.5 mm and protruding
beyond the glumes for 3–4 mm. *Palea* absent. *Anthers*
1.2–2.3 mm. *Caryopsis* about 1.6 mm, compressed, elliptic-oblong in side view, enclosed in the lemma between
the glumes and falling with them at maturity. *Flowers*
5–8. 2n = 14.

Native. Tidal salt-marshes and brackish seaside
meadows in winter-wet hollows, ditch edges and tram-
pled tracks, generally uncommon, but sometimes locally
abundant. Coasts of south and east Britain from

Cornwall and Carmarthenshire northwards to Lincolnshire, and in Guernsey. Coasts of north and west Europe, eastwards to north-west Yugoslavia.

× geniculatus = A. × plettkei Mattf.

This highly sterile hybrid is intermediate in culm shape and swelling and forms dense mats with long straggling radiating stolons.

It has been found with the parents in south and east England north to Lincolnshire. It is also recorded in central Europe.

4. A. aequalis Sobol.   Orange Foxtail
A. fulvus Sm.

Annual, biennial or short-lived *perennial herb. Culms* 10–35 cm, usually ascending from a strongly bent or prostrate base, sometimes rooting at the nodes, slender, with few to many nodes, smooth, whitish-green towards the panicle. *Leaves* 2–10 cm × 2–5 mm, green, glabrous, linear, long-pointed at apex, flat, finely veined, with the veins rough; sheaths smooth, the upper whitish-green and somewhat inflated; ligules up to 5.5 mm, membranous, obtuse at apex. *Inflorescence* a panicle 1–5 cm × 3–6 mm, pale or greyish-green, sometimes tinged with blue, very dense, spike-like, cylindrical; pedicels very short. *Spikelets* 2.0–2.5 mm, elliptical or oblong, very obtuse at apex, flattened, 1-flowered, falling entire at maturity. *Glumes* more or less equalling spikelets, similar, keeled, with margins free nearly to the base, narrowly oblong, blunt at apex, thinly membranous, 3-veined, with a fringe of silky hairs on the keels and shorter ones on the sides. *Lemma* as long as or very slightly longer than the glumes, keeled, broadly elliptical, very obtuse at apex, thinly membranous, 4-veined, smooth, with the margins united for up to half their length, awned on the back just below the middle, with the awn included in the glumes or protruding 0.5 mm from them. *Palea* absent. *Anthers* 0.7–1.0 mm. *Caryopsis* ellipsoid, enclosed in the lemma between the thin glumes. *Flowers* 6–9. $2n = 14$.

Native. Meadows, marshes, ditches and pools, in shallow water and especially on drying mud where it sometimes forms dense carpets. Scattered in central and south Britain to north Yorkshire, and in east Inverness. Absent from Ireland. Most of Europe, but absent from many islands; Siberia; North Africa; North America.

× geniculatus = A. × haussknechtianus Asch. & Graebn.

This highly sterile hybrid is intermediate between the parents in spikelet and awn length and fusion of lemma margin.

It is recorded from Norfolk, Huntingdonshire, Bedfordshire and Merionethshire. It is also known from central and north Europe.

5. A. borealis Trin.   Alpine Foxtail
A. alpinum Sm., non Vill.

Perennial herb forming loose tufts; rhizomes slender, creeping. *Culms* 10–45 cm, erect or bent at the base, slender, with 2–3 nodes, smooth. *Leaves* 6–10(–22) cm ×

1.5–6.0 mm, green, glabrous, narrowly linear-lanceolate, pointed at apex, flat, closely veined and slightly rough above, smooth beneath, the upper somewhat inflated; ligules 1–2 mm, membranous. *Inflorescence* a spike-like panicle, 10–30 × 7–12 mm, greyish-green or tinged with purple, very dense, erect, broadly cylindrical or ovoid, silky-hairy; pedicels extremely short. *Spikelets* 3.0–4.5 mm, elliptical or ovate, obtuse at apex, 1-flowered, flattened, falling entire at maturity. *Glumes* more or less equalling spikelets, with their margins united near the base, semi-elliptical and pointed in side view, 3-veined, with the veins green, membranous, clothed with fine spreading hairs, especially on the keels. *Lemma* as long as the glumes or slightly longer, keeled, broadly ovate, very obtuse at apex, with the margins united towards the base, 4-veined, membranous, slightly hairy near the margins or glabrous, awnless or with awns up to 2.5 mm on the back from about one-third above the base, with the awn sometimes projecting beyond the tip of the spikelet. *Palea* absent. *Anthers* 2.0–2.5 mm. *Flowers* 7–8. $2n = 112$–130.

Native. Wet grassy slopes, moist rocks and mountain springs and flushes between 600 and 1200 m. Very local in northern Britain from Westmorland, Cumberland and Durham to Inverness-shire and Ross-shire. Arctic Russia, Spitsbergen, Novaya Zemlya; Urals; Greenland.

6. A. myosuroides Huds.   Black Grass
A. agrestis L.

Annual herb. Culms 20–80 cm, loosely to compactly tufted, or sometimes solitary, erect or bent at base, slender, smooth, with few nodes. *Leaves* 3–16 cm × 2–8 mm, green, glabrous, linear, pointed at apex, flat, rough on both sides or smooth beneath; sheaths green or purplish, smooth, the uppermost somewhat inflated; ligules 2–5 mm, obtuse at apex, membranous. *Inflorescence* a spike-like panicle, 2–12 cm × 3–6 mm, dense, narrowly cylindrical, tapering upwards, yellowish-green, pale green or purplish; pedicels very short. *Spikelets* 4.5–7.0 mm, narrowly oblong or oblong-lanceolate, flattened, 1-flowered, falling entire at maturity. *Glumes* more or less equalling spikelets, united by their margins for one-third to half their length, narrowly oblong to lanceolate, pointed at apex, 3-veined, firm, narrowly winged on the keels, minutely hairy on the keels and veins near the base. *Lemma* as long as or slightly longer than the glumes, ovate, obtuse at apex, keeled, with the margins united for one-third to half above the base, membranous, 4-veined, smooth, awned on the back from near the base, with the awn exceeding the tip of the lemma by 4–8 mm. *Palea* absent. *Anthers* 3–4 mm. *Caryopsis* ellipsoid, tightly enclosed in the lemma between the hardened glumes. *Flowers* 5–8. $2n = 14$.

Native. A common weed of arable land and waste places on both heavy and light soils, and despite the use of herbicides still a serious weed in some areas. Common in south, central and eastern England, very scattered in the south-west, Wales and north to central Scotland, and there mostly a casual as it is in Ireland. South and west Europe; North Africa; western Asia; introduced in North America and New Zealand.

53. Beckmannia Host

Annual or *perennial herbs* without stolons, rarely with rhizomes. *Inflorescence* a long, narrow raceme of closely packed, appressed spikes. *Spikelets* with 1 or 2 flowers, the first bisexual, the second bisexual, male or sterile. *Glumes* 2, equal, 3-veined, with connecting veins in between, enclosing the rest of the spikelet except for the apiculate tip of the lemmas, strongly keeled and hooded. *Lemmas* 5-veined, apiculate but not awned, without basal hair-tuft. *Palea* nearly as long as lemma, 2-veined. Disarticulating at maturity below the glumes. *Stamens* 3. *Lodicules* 2.

Two species in the north temperate zone.

1. B. syzigachne (Steud.) Fernald
American Slough Grass

Panicum syzigachne Steud.; *B. eruciformis* subsp. *syzigachne* (Steud.) Grett.; *B. eruciformis* var. *baicalensis* Kuzn.; *B. baicalensis* (Kuzn.) Hultén; *B. eruciformis* subsp. *baicalensis* (Kuzn.) Hultén; *B. syzigachne* subsp. *baicalensis* (Kuzn.) Koyana & Kawano; *B. eruciformis* auct.

Robust *annual* or short-lived *perennial herb. Culms* up to 100 cm, erect, hollow. *Leaves* up to 35 cm × 5–10 mm, linear, tapering to a pointed apex, flat; sheaths glabrous; ligule 6–11 mm, acuminate at apex, entire or commonly lacerate and usually folded back in growth, shortly hairy. *Inflorescence* a long, narrow raceme of closely packed, appressed spikes up to 30 cm; spikes 1–2 cm. *Spikelets* 2.2–3.2 mm, with one bisexual flower or sometimes a second male, bisexual or sterile one, disarticulating below the glumes at maturity. *Glumes* 2, equal, about 3 mm, strongly compressed and inflated, semicircular, abruptly acute at apex, 3-veined with connecting veins in between, glabrous. *Lemmas* about 3 mm, lanceolate, much narrower than glumes, acuminate at apex, 5-veined, glabrous. *Palea* nearly as long as lemma, 2-veined. *Stamens* 3; anthers 0.7–1.2 mm. *Lodicules* 2. *Flowers* 7–8. 2n = 14.

This species has been mistaken for *B. eruciformis* (L.) Host, which differs in having 2 regularly bisexual flowers, a usually hairy lemma and anthers 1.5–2.0 mm. It is a long-lived perennial and is often swollen at the base.

Introduced. Casual of waste ground and tips mainly from grain or bird-seed. Scattered records in southern England and south Wales and naturalised in a few sites in the Bristol Channel area. It is a native of North America. *B. eruciformis*, which should be looked for, is native of Eurasia.

54. Phleum L.

Annual or *perennial herbs,* sometimes with rhizomes or stolons. *Leaves* flat, rarely rolled; ligules membranous. *Inflorescence* a very contracted spike-like panicle. *Glumes* 2, equal, 3-veined, strongly keeled, with stiff hairs on the keel, apiculate to shortly awned at apex. *Lemmas* 3- to 7-veined, irregularly truncate to rounded at apex, without a basal hair-tuft, unawned. *Palea* as long or nearly as long as lemma, 2-veined; disarticulating at maturity below the lemma.

Fifteen species in the north temperate zone and South America.

Stewart, A., Pearman, D. A. & Preston, C. D. (1994). *Scarce plants in Britain.* Peterborough. [*P. alpinum.*]

1. Annual without tillers at flowering time; glumes acute to subacute at apex, gradually narrowed to an awn; anthers less than 1 mm   **5. arenarium**
1. Perennial with tillers at flowering time; glumes obtuse or truncate at apex, abruptly or very abruptly narrowed to an awn; anthers more than 1 mm   **2.**
2. Panicles (1–)2–3(–5) times as long as wide; glumes 5.0–8.5 mm including awns which are 2–4 mm; mountains above 600 m in northern Britain   **3. alpinum**
2. Panicles (2–)3–20(–30) times as long as wide; widespread   **3.**
3. Culms not swollen at base; ligules 0.5–2.0 mm; glumes obtuse and shortly awned at apex; central and east England   **4. phleoides**
3. Culms usually swollen at base; ligules 1–9 mm; glumes truncate and shortly awned at apex   **4.**
4. Leaves 3–9 mm wide; ligule usually obtuse; panicle 6–10 mm wide; spikelets (3.5–)4–5 mm including awns which are (0.8–)1–2 mm   **1. pratense**
4. Leaves 2–6 mm wide; ligules usually acute at apex; panicle 3.0–7.5 mm wide; spikelets 2.0–3.5 mm including awns which are 0.2–1.2 mm   **2. bertolonii**

1. P. pratense L.
Timothy

Loosely to densely tufted *perennial herb. Culms* 40–150 cm, erect or ascending from a bent base, mostly stout, smooth or nearly so, with 3–6 nodes, the basal 1–3 internodes very short and usually swollen or bulbous. *Leaves* up to 45 cm × 3–9 mm, linear, narrowed at apex to a fine tip, flat, firm, tough all over or only on the upper part and margins; sheaths smooth, rounded on the back, the basal becoming dark brown; ligules up to 6 mm, obtuse at apex, membranous. *Inflorescence* a panicle 6–15(–30) cm × 6–10 mm, green, greyish-green or purplish, dense, spike-like, cylindrical; pedicels extremely short. *Spikelets* (3.5–)4–5 mm, oblong, flattened, tightly packed, 1-flowered, breaking up at maturity above the glumes. *Glumes* 3.5–5.5 mm, persistent, narrowly oblong, truncate at apex, keeled, minutely rough on the membranous sides, 3-veined, the keels fringed with stiff, spreading, white hairs and produced at the tip into a rigid, rough awn (0.8–)1.0–2.0 mm, the lower glumes softly hairy on the margins. *Lemma* two-thirds to three-quarters the length of the glumes, broad and very blunt at apex, 5- to 7-veined, membranous, minutely hairy. *Palea* lanceolate, about as long as the lemma. *Anthers* about 2 mm. *Caryopsis* ellipsoid, enclosed between the thin lemma and palea. *Flowers* 6–8. 2n = 21, 35, 36, 42, 49, 56, 63, 70, 84.

P. pratense and *P. bertolonii* sometimes grow in large colonies with every degree of intermediate, but hybrids appear to be recorded only in Sweden.

Native. Meadows, rough grassland, waste places, field margins and waysides. Common throughout the British Isles. North temperate regions.

2. P. bertolonii DC.   Smaller Cat's-tail
P. nodosum auct.; *P. hubbardii* D. Kovats.; *P. pratense*
subsp. *bertolonii* (DC.) Bornm.; *P. pratense* subsp.
serotinum (Jordan) Berher

Loosely to compactly tufted *perennial herb*, sometimes
with leafy stolons. *Culms* 10–50 cm, erect or ascending
from a bent or prostrate base, slender, stiff, smooth, with
2–6 nodes, the basal 1–2 internodes short and usually
swollen or bulbous. *Leaves* 3–12 cm × 2–6 mm, green to
greyish-green, glabrous, linear, finely pointed at apex,
firm, minutely rough on the veins and margins, or only
on the latter; sheaths rounded on the back, smooth, the
basal becoming dark brown, the upper slightly inflated;
ligules 1–4 mm, usually acute at apex, membranous.
Inflorescence a panicle 1–6 cm × 3.0–7.5 mm, dense,
spike-like, pale or whitish-green, narrowly cylindrical;
pedicels extremely short. Spikelets 2.0–3.5 mm, oblong,
tightly packed, flattened, 1-flowered, breaking up at
maturity above the glumes. *Glumes* persistent, narrowly
oblong, truncate at apex, keeled, with a rough awn
0.2–1.2 mm, minutely rough or minutely hairy on the
membranous sides, 3-veined, the keels fringed with stiff,
white, spreading hairs and the margins of the lower
glumes softly hairy. *Lemma* two-thirds to three-quarters
the length of the glumes, very broad and obtuse at apex,
membranous, 5- to 7-veined, minutely hairy on the
veins. *Palea* elliptical, nearly as long as the lemma.
Anthers 1–2 mm. *Caryopsis* ellipsoid, enclosed by the
thin lemma and palea. *Flowers* 6–8. $2n = 14$.

Native. Old pastures, grassland on hills and downs,
waysides and waste places. Probably throughout the
British Isles. North temperate regions.

3. P. alpinum L.   Alpine Cat's-tail
P. commutatum Gaudin

Loosely tufted *perennial herb*; rhizomes short, creeping.
Culms 10–50 cm, erect, or ascending from a bent or
curved base, slender, with 2–4 nodes, smooth. *Leaves* up
to 15 cm × up to 6 mm, green, glabrous, linear, narrowed
to a fine blunt tip, flat, minutely rough on the margins,
otherwise smooth; sheaths rounded on the back,
smooth, the back becoming dark brown, the upper
somewhat inflated; ligules up to 2 mm, very obtuse at
apex, membranous. *Inflorescence* a panicle 1–5 cm ×
6–12 mm, spike-like, very dense, erect, short-bristled,
oblong, broadly cylindrical, usually purplish; pedicels
very short. *Spikelets* 3.0–3.8 mm, oblong, densely
packed, compressed, 1-flowered, breaking up at maturity above the glumes. *Glumes* 5.0–8.5 mm including
awns, narrowly oblong, truncate at apex, keeled, with
broad, minutely rough, membranous sides, 3-veined, the
keels fringed with stiff, white, spreading hairs, with a
rough, rigid, straight or slightly curved awn 2–4 mm at
the tip, the lower glumes with softly hairy margins.
Lemma about two-thirds the length of the glumes,
broadly elliptic-oblong, very obtuse at apex, membranous, finely 3- to 5-veined, minutely hairy on the veins.
Palea lanceolate, slightly shorter than the lemma, membranous. *Anthers* 1.0–1.5 mm. *Flowers* 7–8. $2n = 28$.

Native. Wet slopes, rock-ledges and other wet places

between 600 and 1200 m. Very local in Westmorland,
Cumberland, Dunbartonshire, Perthshire, Angus, Aber-
deenshire, Banffshire, Inverness-shire and Ross-shire.
Arctic regions and high mountains of Europe and Asia;
North America; Andes of South America, South Georgia.

4. P. phleoides (L.) H. Karst.
                               Purple-stemmed Cat's-tail
P. boehmeri Wibel; *Phalaris phleoides* L.

Densely tufted *perennial herb*. *Culms* 10–70 cm, erect or
nearly so, slender and very narrow below the panicle
with 2–3 nodes below the middle, smooth, often purplish. *Leaves* up to 12 cm × 1.0–3.5 mm, greyish-green,
glabrous, linear, narrowed to a fine, obtuse apex, flat or
rolled, rough especially on the margins; sheaths smooth,
rounded on the back, the lower purple-tinged; ligules
0.5–2 mm, very blunt, membranous. *Inflorescence* a
panicle 1.5–10.0 cm × 4–6 mm, erect, very dense, spike-
like, narrowly cylindrical, green or purplish; pedicels
very short. *Spikelets* 2.5–3.0 mm, oblong, densely overlapping, compressed, 1-flowered, breaking up above the
glume at maturity. *Glumes* persistent, similar, equal, as
long as the spikelet, narrowly oblong, abruptly narrowed into a rough point up to 0.5 mm, keeled, rigid
about the keels, the margin membranous, 3-veined,
rough or with short, stiff, hairs on the keels. *Lemma* two-
thirds to three-quarters the length of the glumes,
rounded on the back, ovate, obtuse, firm, finely 5-
veined, minutely hairy or smooth. *Palea* elliptical, about
as long as the lemma, oblong, 2-veined. *Anthers* about
1.5 mm. *Caryopsis* about 1.3 mm, ellipsoid, enclosed in
the hardened lemma and palea. *Flowers* 6–8. $2n = 14, 28$.

Native. Dry sandy and chalky grassland and adjacent
roadsides and rough ground. Norfolk, Suffolk,
Cambridgeshire, Bedfordshire, Hertfordshire and
Essex. Much of Europe; temperate Asia; North Africa.

5. P. arenarium L.   Sand Cat's-tail
Annual herb. *Culms* 1–15(–30) cm, tufted or solitary,
erect or bent at the base, very slender, stiff, with 1–4
nodes below the middle, smooth. *Leaves* 5–60 × up to 4
mm, whitish-green or pale green, glabrous, linear-lanceolate, narrowed to a fine apex, flat, closely veined above,
minutely rough on the margins; sheaths rounded on the
back, smooth, the uppermost somewhat inflated; ligules
up to 7 mm, membranous. *Inflorescence* a panicle 5–50
× 3–7 mm, spike-like, very dense, narrowly cylindrical
or obovoid, usually narrowed at the base, rounded at the
apex, pale green or whitish-green, sometimes tinged
with purple; pedicels very short. *Spikelets* 3–4 mm,
lanceolate to oblong, densely overlapping, flattened, 1-
flowered, breaking up at maturity above the glumes.
Glumes persistent, equal, as long as the spikelet, gradually narrowed into a sharply pointed rough tip, keeled,
firmly membranous, 2- to 3-veined, with short, stiff,
spreading hairs on the keel above the middle. *Lemma*
about one-third the length of the glumes, rounded on
the back, broad, very obtuse at apex, finely 5-veined,
membranous, minutely hairy. *Palea* obovate, as long as
the lemma. *Anthers* 0.3–1.0 mm. *Caryopsis* broadly

ellipsoid, enclosed between the thin lemma and palea. *Flowers* 5–7. 2*n* = 14.

Native. Sand-dunes and shingle by the sea, and heaths inland. Frequent on most coasts of the British Isles except north and west Scotland; inland in East Anglia. South and west Europe north-eastwards to Gotland; mountains of North Africa; south-west Asia.

Tribe 12. Bromeae Dumort.

Annual or *perennial herbs* with or without rhizomes and without stolons. *Leaves* with a membranous ligule; sheaths usually fused when young but soon splitting. *Inflorescence* a panicle, rarely slightly contracted. *Spikelets* with several to many bisexual flowers, the apical one or 2, sometimes more often reduced and male or sterile. *Glumes* 2, unequal, 1- to 9-veined, much shorter than the rest of the spikelet, unawned. *Lemmas* 5- to 11-veined, rounded or keeled on the back, usually minutely bifid at apex, usually with a long, subterminal awn. *Stamens* 2 or 3. *Stigmas* 2. *Lodicules* 2. *Caryopsis* with a pubescent terminal appendage, the styles arising below it.

55. Bromus L.

Annual herbs. Leaves usually flat; ligules membranous. *Inflorescence* a panicle, but in dry conditions sometimes reduced to a single spikelet. *Spikelets* ovoid to narrowly so, terete or slightly compressed. *Lower glumes* 3- to 5-veined, upper glumes 5- to 7(–9)-veined. *Lemmas* 7- to 9(–11)-veined, rounded on the back, subacute to obtuse or rounded-obtuse and minutely bifid at apex. *Stamens* 3. *Stigmas* 2. *Caryopsis* with a hairy terminal appendage, the styles arising below it, tightly enclosed by the hardened lemma and palea.

About 50 species, mainly in the temperate regions.

Smith, P. (1968). Serological distincness of *Bromus pseudosecalinus* P. Smith sp. nov. *Feddes Repert. Spec. Nov. Regni Veg.* **77**: 61–64.

Smith, P. (1968). The *Bromus mollis* aggregate in Britain. *Watsonia* **6**: 327–344.

Smith, P. (1970). Taxonomy and nomenclature of the Bromegrasses. (*Bromus* L. s.l.) *Notes Roy. Bot. Gard. Edinburgh.* **30**: 361–375.

Smith, P. M. (1973).Observations of some critical Brome grasses. *Watsonia* **9**: 319–332.

1. Lemma with the margins wrapped around the caryopsis when mature, hence the lemma margins not overlapping the next higher lemma, but rhachilla more or less revealed between the flowers; rhachilla disarticulating tardily; caryopsis thick, with inrolled margins   2.
1. Lemma margins not wrapped around the caryopsis, overlapping the next higher lemma and obscuring the rhachilla; rhachilla disarticulating readily; caryopsis thin, flat or with weakly inrolled margins   3.
2. Sheaths usually glabrous or sparingly hairy; spikelets 12–24 mm, glabrous or hairy; lemma 6.5–9.0(–10) mm; palea equalling the lemma; caryopsis 6–9 mm
                                             8. secalinus
2. Sheaths hairy; spikelets 8–12 mm, glabrous; lemma 3–6 mm; palea shorter than lemma; caryopsis 4.0–4.5 mm   **9. pseudosecalinus**

3. Panicle with mostly subsessile spikelets densely clustered in groups of three; extinct   **7. interruptus**
3. Panicle varying, but spikelets not subsessile in groups of 3   4.
4. Panicle branches long, forming a very open panicle; anthers 3–5 mm, half as long as lemma   **1. arvensis**
4. Panicles varying, often not very open; anthers 0.2–3.0 mm, less than half as long as lemma   5.
5. Awns curved or bent outwards at maturity, their apices widely diverging   6.
5. Awns more or less straight to slightly curved or curved at maturity, the apices of the more apical lemmas more or less parallel or even convergent (if more or less curved outwards the culms procumbent to ascending)   9.
6. Panicle branches and pedicels much shorter than spikelets; spikelets less than 18 mm; pedicels less than 10 mm   7.
6. At least some panicle branches and pedicels on well-grown plants longer than spikelets; spikelets more than 18 mm; pedicels often more than 15 mm   8.
7. Culms 2–12(–20) cm; usually with less than 10 spikelets; lemma 5.5–8.5 mm; maritime and native
                           4(b). hordeaceus subsp. **ferronii**
7. Culms usually more than 15 cm, with usually more than 10 spikelets; lemmas 8–11 mm; alien
                           4(c). hordeaceus subsp. **divaricatus**
8. Panicle usually lax, with patent, pendulous branches; lemmas 8–10 mm   **10. japonicus**
8. Panicles usually rather stiffly erect; lemmas 11–20 mm   **11. lanceolatus**
9. Lemmas (4.5–)5.5–6.5 mm; caryopsis longer than palea   **6. lepidus**
9. Lemmas 6.5–11 mm; caryopsis shorter than to as long as palea   10.
10. Panicle more or less lax with at least some pedicels longer than spikelets; lemmas rather coriaceous, with rather obscure veins; anthers 1–3 mm   11.
10. Panicle more or less dense, usually with all pedicels shorter than spikelets; lemmas papery, with prominent veins; anthers 0.5–1.5 mm   12.
11. Spikelets 15–28 mm; lowest rhachilla segments mostly 1.3–1.7 mm; lemmas 8–11 mm; anthers mostly 1.0–1.5 mm   **2. commutatus**
11. Spikelets 10–16 mm; lowest rhachilla segments mostly 0.7–1.0 mm lemmas 6.0–8.5 mm; anthers mostly 2–3 mm   **3. racemosus**
12. Lemmas 8–11 mm, usually hairy
                           4(d). hordeaceus subsp. **hordeaceus**
12. Lemmas 6.5–8.0 mm, usually glabrous   13.
13. Culms (2–)4–7(–12) cm, procumbent to ascending; caryopsis shorter than palea; maritime dunes
                           4(a). hordeaceus subsp. **thominei**
13. Culms usually more than 12 cm, usually erect; caryopsis about as long as palea; widespread
                           5. × pseudothominei

1. B. arvensis L.   Field Brome
Serrafalcus arvensis (L.) Parl.

Annual herb. Culms 25–90 cm, loosely tufted or solitary, erect or spreading, slender to somewhat stout, unbranched, with 2–5 nodes, smooth. *Leaves* 5–20 cm × 2–5 mm, green, loosely hairy, rough, linear, finely pointed at apex; sheaths tubular, soon splitting, rounded on the back, the lower softly hairy, the upper glabrous; ligules 2–4 mm, toothed, membranous. *Inflorescence* a panicle 8–25 × up to 20 cm, erect or sometimes nodding, open and very loose, broadly ovate to elliptical, green or purplish; branches clustered, eventually widely spreading, fine, unequal, rough, the longer bearing up to 8 spikelets; pedicels up to 30 mm long. *Spikelets* 10–20 × 2.5–4.0 mm, loosely scattered, lanceolate to oblong, slightly compressed, 4- to 10-flowered, breaking up at maturity beneath each lemma. *Glumes* persistent, unequal, pointed at apex, with membranous margins and tips, the lower lanceolate, 3.5–6.5 mm, 3-veined, the upper 4.7–5.5(–8.0) mm, narrowly ovate to elliptical, 5- to 7-veined. *Lemmas* 6.0–7.5(–9.0) mm, overlapping, rounded on the back, elliptical or slightly obovate when opened out, pointed at apex and sometimes with 2 teeth, firm except for the membranous tips and margins, 7-veined, minutely rough, with a fine, straight, rough awn 4.5–7.0(–10) mm long from near the tip. *Paleas* narrowly elliptical, as long as the lemmas, the keels loosely fringed with short hairs. *Anthers* 3–5 mm. *Caryopsis* narrowly ellipsoid, hairy at the tip, tightly enclosed by the hardened lemma and palea. *Flowers* 6–8. 2*n* = 14.

Introduced. Naturalised in east Gloucestershire. Mostly a casual weed of arable land and grassy and waste places. Scattered in central and south Britain and decreasing, sporadic elsewhere. South and south-central Europe; west Asia.

2. B. commutatus Schrad.   Meadow Brome
B. pratensis Ehrh. ex Hoffm., non Lam.; *Serrafalcus commutatus* (Schrad.) Bab.

Annual or *biennial herb. Culms* 40–120 cm, loosely tufted or solitary, erect or spreading, slender to moderately stout, unbranched, with 3–5 nodes, smooth. *Leaves* up to 30 cm × 3–9 mm, green, linear, finely pointed at apex, loosely hairy, rough; sheaths tubular, splitting, the lower softly hairy, the upper thinly hairy or glabrous; ligules 1–4 mm, membranous, becoming torn. *Inflorescence* a panicle 6–25 × up to 12 cm, green or purplish, loose, open or somewhat contracted, eventually drooping to one side; branches clustered, fine, unequal, rough or minutely hairy, bearing 1–4 spikelets; pedicels longer than the spikelets. *Spikelets* 15–28 × 4.5–6.0 mm, eventually nodding, lanceolate to oblong, slightly compressed, 4- to 10-flowered, breaking up at maturity beneath each lemma. *Glumes* persistent, slightly unequal, acute at apex, the lower 5–7 mm, oblong-lanceolate and 3- to 5-veined, the upper 6–9 mm, elliptical and 5- to 9-veined. *Lemmas* 8–11 × 5–6 mm, overlapping, rounded on the back, broadly elliptical or slightly obovate, obtuse or slightly toothed at apex, firm except for the membranous margins, minutely rough, finely 7- to 11-veined, with a fine, straight, rough awn 4–10 mm long from just below the tip. *Paleas* narrowly elliptical, shorter than the lemmas, the keels loosely fringed with short, stiff hairs. *Anthers* 1.0–1.5 mm. *Caryopsis*

narrowly ellipsoid, hairy at the tip, tightly enclosed by the hardened lemma and palea. *Flowers* 5–7. 2*n* = 14, 28, 56.

Native. Grassy places, waysides, waste places, old pastures, and margins of arable land. Locally frequent in central and south Britain, but decreasing; very scattered and casual elsewhere. Nearly the whole of Europe except Fennoscandia and north-east Russia; North Africa; south-west Asia; introduced in North America.

3. B. racemosus L.   Smooth Brome
Serrafalcus racemosus (L.) Parl.

Annual or *biennial herb. Culms* 25–110 cm, loosely tufted or solitary, erect or slightly spreading, slender to relatively stout, unbranched, with 2–5 nodes, smooth or minutely hairy. *Leaves* 5–20 cm × 2–5 mm, green, linear, finely pointed at apex, loosely and softly hairy; sheaths tubular, soon splitting, the lower loosely and softly retrorse hairy, the upper thinly hairy or glabrous; ligules 1–3 mm, membranous, toothed. *Inflorescence* a panicle 4–14 cm × 15–40 mm, green or purplish, erect, rather stiff and open, finally somewhat contracted and nodding; branches clustered, fine, rough, ascending, unequal, up to 60 mm, bearing 1–4 spikelets; pedicels up to 25 mm. *Spikelets* 10–16 × 3–5 mm, narrowly ovate to oblong, slightly compressed, 4- to 8-flowered, breaking up at maturity beneath each lemma. *Glumes* persistent, unequal, pointed at apex, the lower 4–6 mm, lanceolate-oblong and 3-veined, the upper 4.5–7.0 mm, ovate to elliptical and 5- to 7-veined. *Lemmas* 6.0–8.5 × 4–5 mm, overlapping, with an uninterrupted curve on the back, hooded, broadly elliptical or slightly obovate, obtuse, minutely and obscurely rough, stiff, finely 7- to 9-veined, with a fine, straight, rough awn 5–9 mm, from near the tip of the lemma. *Paleas* oblanceolate, shorter than the lemmas, the keels fringed with short, stiff hairs. *Anthers* mostly 2–3 mm. *Caryopsis* narrowly ellipsoid, hairy at the tip, tightly enclosed by the hardened lemmas and paleas. *Flowers* 6–7. 2*n* = 28.

Native. Grassy places, roadsides, waste ground, and damp old pasture. Locally frequent in central and south Britain, but decreasing; a scattered casual elsewhere. Most of Europe except for the greater part of Russia.

4. B. hordeaceus L.   Soft Brome

Annual or *biennial herbs. Culms* up to 100 cm, loosely tufted or solitary, erect, prostrate or ascending, slender to relatively stout, shortly hairy at the nodes, with 2–5 nodes. *Leaves* up to 20 cm × 2–7 mm, linear, finely pointed at apex, greyish-green, flat, flaccid, softly short-hairy; sheaths tubular, soon splitting, rounded on the back, softly hairy, or the upper glabrous; ligules up to 2.5 mm, hairy, toothed. *Inflorescence* a panicle 1.5–16.0 × up to 6 cm, at first erect and loose, afterwards contracted and nodding, greyish-green or purplish; branches clustered, minutely hairy, with 1–5 spikelets; pedicels 2–10 mm. *Spikelets* 12–22 × 3.5–6.0 mm, narrowly ovate to oblong, slightly compressed, 6- to 12-flowered, softly and shortly hairy, breaking up at maturity beneath each lemma. *Glumes* persistent, pointed at apex, the lower 5–8 mm, ovate to oblong and 3- to 7-veined; the upper 6–9 mm, elliptical and 5-to 7-veined. *Lemmas* 8–11 × 4.5–5.5 mm, closely overlapping, rounded on the back, obovate to elliptical, blunt at apex, with narrow membranous margins, prominently 7- to 9-veined, with a fine, rough awn 5–10 mm from just below the apex. *Paleas* narrowly elliptical, slightly shorter than the lemmas, the keels fringed with short, stiff hairs. *Anthers* 0.2–2.0 mm. *Caryopsis* narrowly ellipsoid, hairy at the top, slightly enclosed by the hardened lemma and palea. *Flowers* 5–7. 2n = 28.

(a) Subsp. **thominei** (Hardouin) Braun-Blanquet
B. thominei Hardouin
Culms (2–)4–7(–12) cm, procumbent to ascending. *Panicle* up to 3 cm, with few spikelets, erect. *Lemmas* 6.5–7.5 mm, with a hyaline margin, bluntly angled, usually glabrous, rarely pubescent, with a straight or slightly curved awn 3–7 mm weakly divaricate at maturity.

(b) Subsp. **ferronii** (Mabille) P. M. Sm.
B. ferronii Mabille
Culms 2–12(–20) cm, prostrate or erect to ascending. *Panicle* up to 5 cm, ovate and dense with few spikelets, stiffly erect. *Lemmas* 5.5–8.5 mm, densely pubescent, with an awn 2–5(–7) mm curved outwards at maturity.

(c) Subsp. **divaricatus** (Bonnier &Layens) Kerguélen
B. hordeaceus subsp. *molliformis* (Lloyd) Maire & Weiller; *B. molliformis* Lloyd
Culms to 60 cm, erect. *Panicle* up to 10 cm, with many spikelets, stiffly erect, with many short branches and pedicels. *Lemmas* 8–11 mm, pubescent, with an awn 5–10 mm curved outwards at maturity.

(d) Subsp. **hordeaceus**
B. mollis L.; *Serrafalcus mollis* (L.) Parl.
Culms up to 100 cm, erect. *Panicle* up to 10(–16) cm, with few to many spikelets, often drooping to one side at maturity. *Lemmas* 8–11 mm, with hyaline margin, bluntly angled, pubescent, with a more or less straight awn 4–11 mm.

Native. Frequent throughout the lowlands of the British Isles. Almost all Europe; Caucasus; North Africa; introduced in North America. The subspecies are ecotypic. Subsp. *thominei* occurs in sandy places by the sea, from the Channel Islands through Britain to Scotland. It also occurs elsewhere on the coasts of western Europe. Subsp. *ferronii* occurs on grassy cliff-tops and sandy or shingly areas by the sea in the Channel Islands and south and south-west Britain, and scattered north to Kirkcudbrightshire and Angus. It occurs elsewhere in north-west Europe. Subsp. *divaricatus* is introduced in grassy and waste places in scattered localities in England. It is native of south Europe. Subsp. *hordeaceus* is frequent in grassy and waste places and waysides throughout the British Isles. It occurs through the range of the species.

5. B. × pseudothominei P. M. Sm.   Lesser Soft Brome
   B. hordaceus × lepidus
B. thominei auct.; *B. mollis* var. *leiostachys* Hartm.; *B. gracilis* var. *micromollis* Krösche; *B. lepidus* forma *lasiolepis* Holmb.

Annual or *biennial herb. Culms* 20–70 cm, erect, slender. *Leaves* 2–20 cm × 1–5 mm, linear, more or less acute at apex, glabrous or more or less hairy; sheaths hairy; ligule membranous, toothed. *Inflorescence* an erect, lax or dense panicle 1–10 cm, sometimes reduced to a single spikelet, the branches shorter than the spikelets. *Spikelets* 10–15 × 3–4 mm, lanceolate, hairy or glabrous; with a short pedicel. *Glumes* unequal, the lower lanceolate, the upper ovate-lanceolate, both acute at apex and with prominent veins. *Lemmas* 6.5–8.0 mm, ovate-lanceolate, usually bluntly angled with a narrow, hyaline margin, obovate, rounded on the back, acute at apex, with a straight awn 3–7 mm. *Palea* elliptical, shorter than lemmas, prominently veined, ciliate on the veins. *Anthers* 0.2–1.5 mm. *Caryopsis* narrowly ellipsoid, usually equalling the palea. *Flowers* 5–7. 2n = 28.

This hybrid is most variable; some plants are nearer to *B. hordeaceus* and others nearer to *B. lepidus*. It occurs sometimes with one parent, sometimes with the other, sometimes with both and sometimes with neither.

Probably native. Grassland, waysides and rough ground. Scattered throughout the British Isles and frequent in central and south Britain. It is also widespread in Europe and probably elsewhere.

6. B. lepidus O. Holmb.   Slender Soft Brome

Annual or *biennial herb. Culms* 10–90 cm, loosely tufted or solitary, erect or bent at the base, slender, minutely hairy or glabrous, with 2–6 nodes. *Leaves* 5–20 cm × 2–5 mm, green, linear, finely pointed at apex, flat, softly hairy; sheaths tubular, soon splitting, the lower softly hairy, the upper thinly hairy to glabrous except for the hairy nodes; ligules up to 1 mm, membranous, toothed. *Inflorescence* a panicle 2–10 × 1–4 cm, bright green or tinged with purple, erect and open when in flower, rather dense and nodding in fruit; branches clustered, minutely hairy, up to 4 cm, with 1–3 spikelets. *Spikelets* 5–15 × 2–4 mm, lanceolate to ovate or oblong, slightly compressed, 3- to 11-flowered, breaking up at maturity beneath each lemma. *Glumes* persistent, pointed at apex, the lower 4–5 mm, ovate to oblong and 3- to 7-veined, the upper 5–6 mm, ovate-elliptical to broadly elliptical and 5- to 7-veined. *Lemmas* (4.5–)5.5–6.5 × up to 3.5 mm, the hyaline margin in young spikelets giving the appearence of green and white variegation, closely overlapping at first, later with the lower margins incurved, rounded on the back, obovate, angular above the middle, 2-toothed, pointed at apex, 7-veined, minutely and obscurely rough, firm except for the broad thin tips and margins, with a straight rough awn 2.0–5.5 mm from near the tip. *Paleas* elliptical, shorter than the lemmas and caryopsis, the 2 keels fringed with short hairs. *Anthers* 0.5–2.0 mm. *Caryopsis* narrowly ellipsoid, with its hairy top visible at the tip of the lemma. *Flowers* 5–7. 2n = 28.

Probably introduced. Grassland, waysides and rough ground. Scattered throughout the British Isles. North-west and north-central Europe.

7. B. interruptus (Hack.) Druce   Interrupted Brome

Annual or *biennial herb. Culms* 20–100 cm, loosely tufted, or solitary, erect, slender to relatively stout, unbranched, with 2–4 nodes, minutely hairy. *Leaves* 6–20 cm × 2–6 mm, green, long-linear, pointed at apex, flat, softly hairy, sheaths tubular, soon splitting, rounded on the back, the lower softly hairy, the upper with shorter hairs; ligules 1–2 mm, membranous, toothed. *Inflorescence* a panicle 2–9 cm × up to 20 mm, greyish-green, stiffly erect, dense, oblong, usually interrupted, sometimes reduced to a single spikelet; branches up to 15 mm. *Spikelets* 10–15 × 5–8 mm, sessile or nearly so, in dense clusters, plump, broadly ovate to broadly oblong, 5- to 11-flowered, softly hairy, slowly breaking up at maturity beneath each lemma. *Glumes* persistent, unequal, obtuse or abruptly pointed, the lower 5–7 mm, oblong to elliptical and 3- to 7-veined, the upper 6–9 mm, ovate to broadly elliptical and 5- to 9-veined. *Lemmas* 7.5–9.0 × 5.0–5.5 mm, overlapping, rounded on the back, obovate or elliptic-obovate, minutely 2- toothed, moderately firm except for the narrow membranous margins, prominently 7- to 9-veined; with a fine, rough, straight or flexuous awn 4–8 mm from near the tip. *Paleas* narrowly elliptical, shorter than the lemmas, split to the base between the slightly hairy keels. *Anthers* 1.0–1.5 mm. *Caryopsis* narrowly ellipsoid, tightly enclosed by the lemma and palea. *Flowers* 5–8. 2n = 28.

Endemic. Arable and waste land, and road and track sides. Formerly scattered in south-central and south-east England from Kent to Somerset and Lincolnshire; last seen in Cambridgeshire in 1972; retained in cultivation.

8. B. secalinus L.   Rye Brome
Serrafalcus secalinus (L.) Bab.

Annual or *biennial herb. Culms* 62–88(–120) cm, loosely tufted or axillary, erect, stiff, slender to somewhat stout, unbranched, with 5–7 nodes, smooth. *Leaves* 10–25 cm × 4–12 mm, green, linear, pointed at apex, loosely hairy, rough; sheaths tubular, soon splitting, rounded on the back, glabrous or the lower obscurely hairy; ligules 1–2 mm, membranous, toothed. *Inflorescence* a panicle 5–20 cm, green or purplish, erect and finally nodding, loose, open or contracted; branches clustered, fine, rough or minutely hairy, unequal, up to 8 cm, bearing 1–4 spikelets; pedicels up to 3 cm long. *Spikelets* 12–24 × 4–7 mm, ovate to oblong, slightly compressed, 4- to 11-flowered, very slowly breaking up at maturity beneath each lemma. *Glumes* persistent, obtuse, unequal, firm, the lower 4–6 mm, ovate to oblong and 3- to 5-veined, the upper 5–8 mm, ovate to elliptical and 5- to 7-veined. *Lemmas* 6.5–9.0(–10) × 4.5–5.5 mm, at first overlapping, finally with tightly incurved margins, rounded on the back, broadly elliptical, obtuse, 7-veined, becoming tough and rigid except for the membranous margins,

nearly smooth, with a fine, straight awn 4.5–9.0 mm from near the tip, or awnless. *Paleas* narrowly elliptical, as long as the lemmas, the keels fringed with short, stiff hairs. *Anthers* 1–2 mm. *Caryopsis* 6–9 mm, oblanceolate, hairy at the tip, inrolled, tightly enclosed by the inrolled hardened lemma and palea. *Flowers* 6–7. 2n = 28.

Introduced. Weed of cereals, marginal and waste ground, much decreased, and now an infrequent casual. Formerly frequent in central and south England, now scattered throughout most of the British Isles except north and west Scotland. South and south-central Europe; west Asia; introduced in North Africa and North America.

9. B. pseudosecalinus P. M. Sm.   Smith's Brome
B. secalinus var. *hirsutus* auct.

Annual herb. Culms 30–60 cm, erect, slender. *Leaves* 10–25 × 4–10 mm, linear, acute at apex, hairy; sheaths hairy; ligule about 0.5 mm, membranous. *Inflorescence* a panicle 3–7(–10) cm, lax at flowering, contracted and somewhat nodding later. *Spikelets* 8–12 mm, lanceolate at first, becoming ovate, glabrous, disarticulating tardily at maturity; flowers imbricate at first, erecto-patent after anthesis. *Glumes* unequal, the lower 3.8–4.5 mm and lanceolate, the upper 4.5–6.3 mm, ovate or lanceolate, obscurely veined. *Lemmas* 3–6(–7) mm, the bluntly angled margins becoming inrolled revealing the rhachilla, with a straight or slightly flexuous awn 2–6 mm. *Palea* elliptical, shorter than lemma, ciliate on the veins. *Anthers* 1.2–1.7 mm. *Caryopsis* ellipsoid, 4.0–4.5 mm, longitudinally incurved at maturity. *Flowers* 6–7. 2n = 14.

Introduced. Grassy fields and waysides, probably a grass-seed contaminant. Scattered in Britain and Ireland. Origin uncertain and distribution not well known.

10. B. japonicus Thunb. ex Murray   Thunberg's Brome
B. patulus Mert. & Koch; *Serrafalcus patulus* (L.) Parl.

Annual herb. Culms up to 80 cm, erect or ascending, glabrous. *Leaves* 10–20 cm × 3–7 mm, linear, pointed at apex, flat, softly hairy; sheaths hairy; ligules membranous. *Inflorescence* a panicle, usually large and effuse and very pendulous at maturity; branches mostly longer than the spikelets, with 2–5 at the lower nodes, slender and flexuous, with 1–3 spikelets. *Spikelets* 20–40 mm, with 6–12 flowers, glabrous to hairy. *Lower glume* 4–5 mm, the upper 6–7 mm. *Lemmas* 8–10 mm, ovate, with prominent membranous margins, shortly 2-toothed at the apex, with awns 4–14 mm, flattened at base and widely divergent at maturity. *Palea* elliptical, much shorter than the lemma. *Caryopsis* ellipsoid, tightly enclosed by the hardened lemma and palea. *Flowers* 6–7. 2n = 14.

Introduced. A sporadic casual in waste places from bird-seed, wool and other sources. Scattered records in England. Native of southern Europe.

11. B. lanceolatus Roth   Large-headed Brome
B. macrostachys Desf.

Annual herb. Culms up to 70 cm, erect or geniculately ascending, glabrous or rarely puberulent near the panicle, sometimes hairy at the nodes. *Leaves* up to 30 cm × 3–5 mm, linear, acute or acuminate at apex, flat, flaccid, glabrous or softly hairy; sheaths hairy, sometimes densely so, with soft, white hairs. *Inflorescence* a panicle 5–10(–15) cm, erect, dense when immature or depauperate, becoming laxer with age; branches and pedicels scabrid or hairy, rather rigid and tough, mostly shorter than the spikelets. *Spikelets* 20–50 × 6–10 mm, lanceolate to oblong-lanceolate, with 8–20 flowers, usually with dense, long hairs. *Lower glume* 5–9 mm, the upper 8–12 mm. *Lemmas* 11–20 mm, oblanceolate, often deeply bifid at apex, with a bluntly angled margin, with an awn 6–12 mm, which is flat, often twisted at the base and patent at maturity. *Palea* elliptical, much shorter than the lemma. *Anthers* about 1 mm. *Caryopsis* ellipsoid. *Flowers* 6–7. *2n = 28.*

Introduced. Sporadic casual in waste places from wool and bird-seed and as a garden escape. Scattered records in central and southern Britain. Native of southern Europe.

56. Bromopsis (Dumort.) Fourr.
Zerna auct.

Perennial herbs with long to very short rhizomes. *Leaves* flat or inrolled; ligules membranous. *Inflorescence* a loose to dense panicle. *Spikelets* narrowly oblong, tapered to apex, terete or slightly compressed. *Lower glumes* 1(–3)-veined, the upper 3(–5)-veined. *Lemmas* 5- to 7-veined, rounded to slightly keeled on the back, acute to shortly acuminate and minutely bifid at apex. *Stamens* 3. *Stigmas* 2. *Caryopsis* enclosed by the lemma and palea.

About 25 species in temperate regions of the northern hemisphere.

Stewart, A., Pearman, D. A. & Preston, C. D. (1994). *Scarce plants in Britain.* Peterborough. [*B. benekenii.*]

1. Leaf sheaths with distinct pointed auricles at apex; inflorescence very lax, the branches pendulous or all moved to one side   2.
1. Leaf sheaths without or with short, rounded auricles at apex; inflorescence dense to fairly lax, the branches erect to erecto-patent   3.
2. Panicle branches pendulous; lowest panicle node usually with 2 branches, both long with more than 3 spikelets and a small, pubescent scale   **1. ramosa**
2. Panicle branches swept to one side; lowest panicle node usually with more than 2 branches, some with 1 or very few spikelets and a small, more or less glabrous scale   **2. benekenii**
3. Plant densely tufted with a short rhizome; leaves with tillers usually folded or rolled along the long axis; lemmas with awns (2–)3–8 mm   **3. erecta**
3. Plant not densely tufted, with long rhizomes; leaves of tillers usually flat; lemmas awnless or frequently with awns up to 3(–6) mm   4.
4. Culm nodes glabrous or with short hairs just below; leaf-sheaths usually glabrous; lemmas glabrous to scabrid or with sparse hairs on the margin, with awn absent or rarely up to 3 mm   **4(a). inermis** subsp. **inermis**
4. Culm nodes hairy; leaf-sheaths usually hairy; lemmas appressed-hairy on the margin, with an awn 0–6 mm
4(b). inermis subsp. **pumpelliana**

1. B. ramosa (Huds.) Holub   Hairy Brome
Bromus ramosus Huds.; *Zerna ramosa* (Huds.) Lindm.; *Bromus asper* Murray

Loosely tufted *perennial herb.* Culms 45–190 cm, erect, slender to stout, unbranched, with 3–5 nodes, hairy. *Leaves* up to 60 cm × 6–16 mm, dark green, linear, gradually narrowed to a pointed apex, flat, drooping, rough; sheaths tubular, splitting, rounded on the back, hairy with stiff, reflexed hairs, with narrow auricles at the apex; ligules up to 6 mm, firm, jagged. *Inflorescence* a panicle 15–45 cm, loose, open, nodding, green or purplish; branches usually in pairs, spreading and drooping, divided, rough, bearing up to 9 spikelets, the lowest branches with a minute ciliate scale at the base; pedicels 6–30 mm. *Spikelets* 20–40 × 4–6 mm, pendulous, narrowly lanceolate to narrowly oblong, compressed, loosely 4- to 11-flowered, breaking up beneath each lemma. *Glumes* persistent, keeled, the lower 6–8 mm, subulate and 1-veined, the upper 9–11 mm, oblong-lanceolate, abruptly pointed at apex and 3- to 5-veined. *Lemmas* 10–14 mm, at first overlapping, later with incurved margins, narrowly oblong-lanceolate, pointed at apex, keeled or becoming rounded on the back, firm except for the narrow, membranous margins, 7-veined, with short hairs near the margins and on the backs, and a fine, straight awn 4–8 mm from near the tip. *Paleas* narrowly elliptical, shorter than the lemmas, with rough keels. *Anthers* 3–4 mm. *Caryopsis* narrowly ellipsoid, hairy at the tip, enclosed by the lemma and palea. *Flowers* 7–8. *2n = 14, 42.*

Native. Woods, wood-margins and hedgerows, sometimes persisting on roadsides and ditchbanks in areas formerly wooded. Frequent throughout most of lowland British Isles except Jersey and Outer Isles of Scotland. West, central and south Europe, eastwards to Gotland and eastern Romania; North Africa.

2. B. benekenii (Lange) Holub   Lesser Hairy Brome
Bromus benekenii (Lange) Trimen; *Zerna benekenii* (Lange) Lindm.

Perennial herb, forming small tufts. *Culms* 45–120 cm, erect or ascending, slender to somewhat stout, minutely and obscurely hairy, with few nodes. *Leaves* 10–25 cm × 4–12 mm, dull green, linear, finely pointed at apex, flat, drooping, shortly hairy, or almost glabrous, somewhat rough; sheaths tubular, splitting, with short pointed auricles at the tip, the lower densely to loosely hairy with short, reflexed soft hairs, the upper minutely hairy or glabrous; ligules 1–3 mm, obtuse at apex, membranous. *Inflorescence* a panicle 12–20 mm, green, rather narrow, contracted, nodding in the upper part, rather loose to fairly dense; branches 1–3 together, more or less appressed, not spreading and swept to one side, very slender, rough, bearing 1–3 spikelets, the lower up to 75 mm, the lowest with a hairless scale at the base; pedicels 3–6 mm, very slender, rough. *Spikelets* 15–25 mm, lanceolate to oblong, compressed, 3- to 5-flowered, breaking up beneath each lemma. *Glumes,* persistent, keeled, unequal, membranous, the lower 7–9 mm, linear-lanceolate or narrowly lanceolate, finely pointed at apex and

1-veined, the upper oblong-lanceolate, pointed at apex and 3-veined. *Lemmas* 11–14 mm, narrowly lanceolate-oblong or oblong, pointed or blunt at apex, slightly keeled or rounded on the back, becoming firm, 5-veined, with appressed hairs near the margin, rough on the back, tipped with a fine rough awn 5–8 mm. *Paleas* narrowly elliptical, shorter than the lemmas, with minutely hairy or rough keels. *Anthers* 2.8–3.0 mm. *Caryopsis* narrowly ellipsoid, hairy at the tip, enclosed by the lemma and palea. *Flowers* 6–8. $2n = 28$.

Native. Woods, wood-margins and hedgerows, where it usually flowers a month earlier than *B. ramosus*. Scattered in mainland Britain, perhaps much overlooked. Scattered through Europe and temperate Asia; North Africa.

3. B. erecta (Huds.) Fourr.   Upright Brome
Bromus erectus Huds.; *Zerna erecta* (Huds.) Gray

Densely tufted *perennial herb* with a short rhizome. *Culms* erect or slightly spreading, slender to stout, stiff, with 3–4 nodes, smooth, rarely hairy. *Leaves* up to 30 cm × 2–3 mm, green, long-linear, finely pointed, loosely hairy or glabrous, tough, the basal inrolled or flat, the upper broader, flat and up to 6 mm wide; sheaths tubular, rounded on the back, the lower sparsely hairy with spreading hairs or glabrous, the upper usually smooth; ligules 1.5–3.0 mm, membranous. *Inflorescence* a panicle 10–25 cm, purplish, reddish or green, erect or rarely nodding, loose or rather dense; branches clustered, erect or spreading, mostly short, rough, bearing 1–4 spikelets. *Spikelets* 15–40 mm, narrowly lanceolate to narrowly oblong, slightly compressed, breaking up at maturity beneath each lemma. *Glumes* persistent, finely pointed at apex, the lower 7–12 mm, subulate and 1- to 3-veined, the upper 8–14 mm, narrowly lanceolate and 3-veined. *Lemmas* 8–15 mm, overlapping, later with the margins incurved, narrowly lanceolate or oblong-lanceolate in side view, pointed or slightly 2-toothed at apex, keeled on the back, firm except for the membranous margins and apex, 7-veined, sparsely hairy with a fine, straight or flexuous awn (2–)3–8 mm long from the tip. *Paleas* narrowly elliptical, shorter than the lemmas with 2 rough keels. *Anthers* 5–7 mm, orange or reddish-orange. *Caryopsis* narrowly ellipsoid, hairy at the top, tightly enclosed by the hardened lemma and palea. *Flowers* 6–7. $2n = 42, 56$.

Native. Dry grassland and grassy slopes, roadside banks and sometimes waste land, especially on calcareous soils. Common on base-rich soil in central, south and east England; scattered north to Fifeshire and west to Cornwall, Pembrokeshire and Caernarvonshire. Local in Ireland and the Channel Islands. South, west and central Europe; North Africa; west Asia.

4. B. inermis (Leysser) Holub   Hungarian Brome
Bromus inermis Leysser; *Zerna inermis* (Leysser) Lindm.

Strongly rhizomatous *perennial herb*. *Culms* 30–150 cm, erect, glabrous or hairy at the nodes. *Leaves* up to 30 cm × 9 mm, dark green, linear, gradually narrowed to a pointed apex, flat, usually glabrous; sheaths split, rounded on the back, glabrous or hairy, with narrow rounded auricles at apex; ligules 1–2 mm, membranous, obtuse at apex, jagged. *Inflorescence* a panicle 10–25 cm, erect, spreading, often 1-sided; branches 3–7 at a node; pedicels 6–20 mm. *Spikelets* 15–30 × 3–5 mm, lanceolate or linear-oblong, erect, with 5–10 flowers, breaking up beneath each lemma. *Glumes* unequal, 8–10 mm, linear-lanceolate or lanceolate, with an obtuse hyaline apex, glabrous. *Lemma* 8–10 mm, linear-lanceolate, obtuse or shallowly emarginate at apex, glabrous or appressed-hairy on margins, unawned or with an awn up to 6 mm. *Palea* shorter than lemma. *Anthers* 4–5 mm. *Caryopsis* narrowly ellipsoid, hairy at tip, enclosed by the lemma and palea. *Flowers* 7–8.

(a) Subsp. inermis
Culm-nodes glabrous or with short hairs just below them. *Leaf-sheaths* usually glabrous. *Lemmas* glabrous to scabrid or with sparse hairs on the margin; awn absent or rarely up to 3 mm. $2n = 56$.

(b) Subsp. pumpelliana (Scribn.) W. A. Weber
B. pumpelliana (Scribn.) Holub; *Bromus pumpellianus* Scribn.; *B. inermis* subsp. *pumpellianus* (Scribn.) Wagnon
Culm-nodes hairy. *Leaf-sheaths* usually hairy. *Lemmas* appressed-hairy on the margin, with an awn 0–6 mm. $2n = 42, 56$.

Intermediates occur between the two subspecies.

Introduced. Subsp. *inermis* was formerly sown for fodder and is now mostly a seed contaminant naturalised and casual in grassy places, waysides and field margins. Scattered in Britian. Native or naturalised in much of Europe. Subsp. *pumpelliana* is naturalised in a single locality in south Essex. It is native of North America.

57. Anisantha K. Koch

Annual herbs. Leaves flat; ligules membranous, toothed or jagged. *Inflorescence* an open to condensed panicle. *Spikelets* more or less parallel-sided or widening at apex, slightly compressed. *Lower glume* 1(–3)-veined, upper glume 3(–5)-veined. *Lemma* 7-veined, rounded on the back, acute to acuminate and minutely bifid at apex. *Stamens* 2 or 3.

About 11 species in both temperate zones.

1. Lemma 20–36 mm   2.
1. Lemma 9–20 mm   3.
2. Panicle lax, with branches spreading laterally; callus scar at base of lemma nearly circular   **1. diandra**
2. Panicle dense, with erect branches; callus scar at base of lemma elliptical   **2. rigida**
3. Panicle lax, with branches spreading laterally or pendulous   4.
3. Panicle dense, with stiffly erect branches   5.
4. Inflorescence simple or the larger branches slightly branched and with up to 3(–5) spikelets; spikelets with 1–2 sterile apical flowers; lemmas 13–20 mm   **3. sterilis**

4. Inflorescence compound, the larger branches with 3–8
   spikelets (except in depauperate plants); spikelets with
   more than 3 sterile apical flowers; lemmas
   9–13 mm   **4. tectorum**
5. Spikelets with 1–2 sterile apical flowers; lemmas
   12–20 mm   **5. madritensis**
5. Spikelets with more than 3 sterile apical flowers;
   lemmas 9–13(–15) mm   **6. rubens**

1. A. diandra (Roth) Tutin ex Tzvelev   Great Brome
A. gussonei (Parl.) Nevski; *Bromus gussonei* Parl.;
Bromus diandrus Roth; *Bromus maximus* auct.

Annual herb. Culms 35–80 cm, loosely tufted or solitary, erect or usually spreading, slender to relatively
stout, unbranched, with 3–6 nodes, usually hairy on
the panicles. *Leaves* 10–25 cm × 4–8 mm, green, linear,
finely pointed at apex, flat, thinly to loosely hairy,
rough; sheaths tubular, soon splitting, rounded on the
back, loosely hairy with spreading hairs; ligules 3–6
mm, membranous, jagged. *Inflorescence* a panicle up to
25 cm × 25 cm, very loosely, nodding, very variable
in size, bearing few to many spikelets, green or purplish; branches in clusters of 2–4, up to 10 cm, spreading, unequal, scabrid, bearing 1 spikelet, or the longest
with 2. *Spikelets* 7–9 cm (including awns), finally
drooping, loosely scattered, oblong, becoming wedge-
shaped and gaping, compressed, loosely 5- to 8-flow-
ered, breaking up at maturity beneath each lemma.
Glumes persistent, unequal, narrow, finely pointed at
apex, the lower 15–25 mm, subulate and 1- to 3-veined,
the upper 18–35 mm, narrowly lanceolate and 3- to 5-
veined. *Lemmas* 20–36 mm, overlapping, narrowly
lanceolate in side view, finely pointed at apex with 2
teeth 4–7 mm, broadly rounded on the back, with the
middle vein projecting, very rough, 7-veined, firm
except for the narrow, membranous margins, with a
straight, stout, rough awn 3.5–6.5 cm from just below
the tip, the callus scar at the base nearly circular. *Paleas*
narrowly elliptical, shorter than the lemmas, with the 2
keels fringed with short stiff hairs. *Anthers* 0.8–1.5 mm.
Caryopsis narrowly ellipsoid, hairy at the tip, tightly
enclosed by the hardened lemma and palea. *Flowers*
5–7. $2n = 56$.

Introduced. Naturalised in waste ground, waysides,
and open grassland and heathland on sandy soils.
Frequent in the Channel Islands and East Anglia.
Elsewhere in Britain north to central Scotland scattered
and usually casual. Native of south and west Europe;
North Africa; south-west Asia; Macaronesia.

2. A. rigida (Roth) Hyl.   Ripgut Brome
Bromus rigidus Roth; *Bromus maximus* Desf.

Annual herb. Culms 20–60 cm, loosely tufted, erect or
decumbent at the base, stout, smooth, glabrous, but
crisply tomentose to retrorsely hairy below the panicle.
Leaves up to 25 cm × 8 mm, flat, linear, acuminate at
apex, sparsely villous on both surfaces, scaberulous on
the margins; sheaths with patent hairs; ligules membranous. *Inflorescence* a panicle 6–12 cm, open or compact,
the lower branches up to 2 cm, straight, fasciculate
and often hairy. *Spikelets* 25–35 mm, cuneate, 4- to 9-

flowered, erect, glabrous to hairy. *Glumes* unequal,
smooth, glabrous, hyaline except for the veins, the lower
15–20 mm, 1-veined, lanceolate and acuminate at apex,
the upper 25–30 mm, 3-veined and broader. *Lemmas*
22–30 mm, narrowly oblong-lanceolate, broadly silvery-
hyaline on the margins, bifid at the apex, with the lobes
up to 4 mm, conspicuously 7-veined, callus-scar at base
elliptical, awned, the awn 30–50 mm, very stout and
rough. *Palea* narrowly elliptical, much shorter than the
lemma. *Anthers* up to 5 mm. *Caryopsis* oblong, flattened, adherent to the lemma and palea. *Flowers* 5–6.
Cleistogamous. $2n = 42$.

Introduced. Waste places, waysides, and open grassland on sandy soils. Infrequently naturalised in the
Channel Islands and southern Britain. A rare casual
elsewhere. South and west Europe; Caucasus; North
Africa; Macaronesia.

3. A. sterilis (L.) Nevski   Barren Brome
Bromus sterilis L.

Annual or *biennial herb.* Culms 15–100 cm, loosely
tufted or solitary, erect or spreading, slender to rather
stout, unbranched, with 3–5 nodes, smooth. *Leaves*
5–25 cm × 2–7 mm, green or purplish, linear, finely
pointed, flaccid to firm, softly and shortly hairy, rough,
sheaths tubular, soon splitting, rounded on back,
shortly and softly hairy, the upper often glabrous; ligules
2–4 mm, membranous, toothed. *Inflorescence* a panicle
up to 25 × 25 cm, very loose and open, nodding, variable in size and sometimes reduced to a single spikelet,
green or purplish; branches widely spreading, up to 12
cm, flexuous, unequal, slender, rough, each bearing 1
spikelet, or the longest with up to 3; pedicels 2–7 cm.
Spikelets (3–)4–6 cm (including awns), loosely scattered
and drooping, oblong, becoming wedge-shaped and
gaping, compressed, 4- to 10-flowered, 1–2 at apex
sterile, breaking up at maturity beneath each lemma.
Glumes persistent, unequal, narrow, finely pointed at
apex, the lower 6–14 mm, subulate and 1-veined, the
upper narrowly oblong-lanceolate, 10–20 mm and 3-
veined. *Lemmas* 13–20 mm, at first overlapping, later
separate, each with an obtuse, hard base, linear-lanceolate in side view, finely pointed at apex, slightly keeled or
becoming rounded on the back, finely 2-toothed at apex,
minutely rough, 7-veined, firm except for the narrow,
shining, membranous margins, with a fine, rough,
straight awn 15–30 mm from just below the apex. *Paleas*
narrowly elliptical, shorter than the lemmas, the 2 keels
fringed with very short stiff hairs. *Anthers* 1.0–1.8 mm.
Caryopsis narrowly ellipsoid, hairy at the tip, tightly
enclosed by the hardened lemma and palea. *Flowers* 5–7.
$2n = 14$.

Native. Roadsides, waste ground, field margins,
gardens and open grassland; increasing in some areas.
Throughout lowland British Isles, common in the south-
ern and central area, scattered in Scotland and Ireland.
South, west and central Europe, north to south Sweden;
North Africa; south-west Asia; introduced in North
America.

4. A. tectorum (L.) Nevski   Drooping Brome
Bromus tectorum L.

Annual herb. Culms 10–60 cm, loosely tufted or solitary, slender, unbranched, with 2–5 nodes, minutely hairy or smooth. *Leaves* 3–16 cm × 2–4 mm, green, linear, finely pointed at apex, flat, softly hairy; sheaths tubular, soon splitting, rounded on the back, softly hairy, or the upper glabrous; ligules up to 5 mm, membranous, jagged. *Inflorescence* a panicle 4–18 cm, loose, or contracted and rather dense, drooping to one side, green or purplish, glistening; branches clustered, fine, flexuous, spreading, shortly hairy or rough, the longer bearing up to 8 spikelets. *Spikelets* 25–35 mm (including awns), nodding, narrowly oblong, becoming wedge-shaped, gaping, compressed, 4- to 8-flowered more than 3 of which are sterile, breaking up at maturity beneath each lemma. *Glumes* persistent, finely pointed at apex, thinly membranous, the lower 5–8 mm, subulate and 1-veined, the upper 7–11 mm, narrowly oblong-lanceolate and 3-veined. *Lemmas* 9–13 mm, overlapping, linear-lanceolate in side view, finely pointed and with 2 teeth 1–2 mm long at the apex, becoming rounded on the back, with thinly membranous margins, rough, 7-veined, with a fine, straight, rough awn 10–18 mm from just below the tip. *Paleas* narrowly elliptical, shorter than the lemmas, with 2 keels loosely fringed with short, stiff hairs. *Anthers* 0.5–1.0 mm. *Caryopsis* narrowly ellipsoid, hairy at tip, tightly enclosed by the lemmas and palea. *Flowers* 5–7. $2n = 14$.

Introduced. Waste ground, roadsides and open grassland on sandy soils. Naturalised in Norfolk and Suffolk; a rare casual elsewhere. Most of Europe; Asia; North Africa; Canary Islands.

5. A. madritensis (L.) Nevski   Compact Brome
Bromus madritensis L.

Annual herb. Culms 10–60 cm, loosely tufted or solitary, erect or spreading, slender, unbranched, with 2–4 nodes, smooth or rarely minutely hairy beneath the panicle. *Leaves* 3–20 cm × 2–5 mm, green, linear, finely pointed at apex, short-haired or glabrous; sheaths tubular, soon splitting, the lower softly hairy, the upper glabrous; ligules 1.5–4.0 mm, membranous, jagged. *Inflorescence* a panicle 4–15 × 1.5–6.0 cm, erect, or slightly inclined, contracted and rather dense or somewhat loose, purple or green; branches clustered, fine, unequal, erect or slightly spreading, minutely rough, 5–35 mm, bearing 1–2 spikelets. *Spikelets* 35–60 mm (including awns), oblong, becoming wedge-shaped and gaping, compressed, loosely 6- to 13-flowered, the upper 1–2 sterile, breaking up at maturity beneath each lemma. *Glumes* persistent, narrow, unequal, finely pointed at apex, the lower 6–11 mm, subulate and 1-veined, the upper linear-lanceolate, 10–16 mm and 3-veined. *Lemmas* 12–20 mm, at first overlapping, later with tightly incurved margins, linear-lanceolate in side view, finely pointed at apex with 2 teeth at the tip 1–2 mm, finely rounded on the back, rough, with narrow, shining, membranous margins, 7-veined, awned from just below the tip, with the fine, rough awn slightly diverging and 12–20 mm. *Paleas* very

narrowly elliptical, shorter than the lemmas, with the 2 keels loosely fringed with short, stiff hairs. *Anthers* 0.5–1.0 mm. *Caryopsis* narrowly ellipsoid, hairy at the tip, tightly enclosed by the hardened lemma and palea. *Flowers* 5–7. $2n = 28$.

Introduced. Naturalised in waste ground, roadsides and open grassland. Local in south-west England, south Wales, Suffolk and the Channel Islands. Occasional casual elsewhere in Britain north to central Scotland and in southern Ireland. Native of south and west Europe; North Africa; south-west Asia; Macaronesia; introduced in North and South America, South Africa and Australia.

6. A. rubens (L.) Nevski   Foxtail Brome
Bromus rubens L.

Annual herb. Culms up to 45 cm, loosely tufted or solitary, erect or decumbent at the base, smooth, glabrous except for just below the panicle where it is densely but shortly hairy, simple. *Leaves* up to 12 cm × 5 mm, linear, acuminate at apex, loosely hairy on both surfaces, scaberulous and ciliate on the margins; sheaths closely clasping, the upper slightly keeled and sometimes glabrous, the lower densely covered with white, short, retrorse hairs; ligule up to 5 mm, lacerate. *Inflorescence* a panicle up to 10 cm, very dense, erect, more or less long exserted, obovate or obovate-oblong, cuneate at base; branches short, the lower in fascicles, hairy. *Spikelets* 15–27 mm, somewhat sterile, wedge-shaped, 4- to 10-flowered, at least the 3 upper ones sterile and often reddish. *Glumes* unequal, hyaline on the margins, hairy or glabrous, the lower 5–7 mm, subulate and 1-veined, the upper 8–10 mm, narrowly lanceolate, acuminate at apex and 3-veined. *Lemmas* 9–13(–15) × 2.0–2.5 mm, oblong-elliptical, 5- to 7-veined, rounded on the back, bifid at the acute tip; glabrous or hairy, awned below the tip, the awn 8–12 mm. *Palea* elliptical, shorter than the lemma. *Anthers* 2–3, 0.3–1.0 mm. *Caryopsis* oblong, flattened, adherent to the lemma and palea. *Flowers* 5–6. $2n = 28$.

Introduced. A rather infrequent wool alien, casual in waste places. Sporadic and very scattered in Britain. Native in the Mediterranean region east to central Asia; Macaronesia; introduced in North America.

58. Ceratochloa DC. & P. Beauv.

Bromus section *Ceratochloa* (DC. & P. Beauv.) Griseb.

Perennial, sometimes *annual* or *biennial herbs,* usually without rhizomes. *Leaves* flat. *Inflorescence* a panicle. *Spikelets* ovoid to narrowly so, compressed. *Glumes* unequal, the lower 3- to 5-veined, the upper 5- to 7-veined. *Lemmas* 7- to 11(–13)-veined, acute to acuminate and minutely bifid at the apex, strongly keeled on the back. *Stamens* 3.

1. Leaves less than 3(–4) mm wide; leaves and sheaths densely hairy with long, patent hairs; lemmas 8–13 mm   2.

1. Leaves 4–14 mm wide; leaves and sheaths glabrous to conspicuously long-patent hairy   3.

2. Lemmas sparsely to densely hairy, with an awn
(3–)4–8(–12) mm   **4. staminea**

2. Lemmas glabrous to sparsely hairy, with an awn up to
1(–2) mm or absent   **5. brevis**

3. Lemmas awnless or with an awn up to 3.5(–5) mm,
with 9–11(–13) veins; palea half to two-thirds as long
as the body of the lemma   **3. cathartica**

3. Lemmas with an awn (3–)4–10(–12) mm, with 7–9
veins; palea three-quarters as long to as long as the
body of the lemma   4.

3. Leaves and sheaths glabrous to sparsely hairy; lemmas
glabrous to sparsely hairy with awns
(4–)6–10(–12) mm   **1. carinata**

3. Leaves and sheaths more or less hairy; lemmas
conspicuously hairy with awns
(3–) 4–6 (–7) mm   **2. marginata**

1. C. carinata (Hook. & Arn.) Tutin California Brome
Bromus carinatus Hook. & Arn.

Short-lived *perennial herb*. *Culms* 30–80 cm, stout, erect,
glabrous. *Leaves* 20–30(–45) cm × 4–14 mm, linear, narrowed to an acuminate apex, tough, flat, glabrous to
sparsely hairy; sheaths glabrous but shortly hairy at the
mouth and 2–3 cm below; ligule about 2 mm, truncate at
apex and torn. *Inflorescence* a large, loose panicle
15–30(–40) cm, with long patent or nodding branches.
Spikelets 22–45 mm, with 5–9(–12), lax flowers, oblong
or ovate-oblong, strongly compressed. *Glumes* unequal,
lanceolate, the lower 6–9 mm, hyaline at the margins,
3–5 veined and keeled on the back, the upper 10–13 mm,
acuminate at apex and 5- to 7-veined. *Lemmas* 10–18
mm, oblong, acuminate at apex, with 7–9 veins, glabrous
or sparsely hairy, strongly keeled on back, awned at the
tip, the awn (4–)6–10(–12) mm. *Palea* elliptical, three-
quarters as long as to nearly as long as the lemma.
Anthers 4–5 mm, shorter in cleistogamous flowers.
Flowers 6–8. Flowers cleistogamous or chasmogamous.
$2n = 56$.

Introduced. Seed contaminant and rarely grown for
fodder. Naturalised in rough ground, field borders, waysides and river-banks. Scattered records in Britain,
mostly the south and especially by the River Thames;
Guernsey; Co. Dublin. Native of western North
America.

2. C. marginata (Nees ex Steud.) B. D. Jackson
                                        Western Brome
Bromus marginatus Nees ex Steud.

Tufted *perennial herb*. *Culms* up to 1 m, very shortly
hairy. *Leaves* 6–12 mm wide, linear, usually hairy;
sheaths nearly glabrous to hairy. *Inflorescence* an erect,
rather narrow panicle 10–30 cm, the lower branches
erect or more or less spreading. *Spikelets* 25–40 mm,
oblong-lanceolate, strongly compressed, with 5–9
flowers. *Glumes* unequal, scabrous or scabrous-hairy,
the lower 7–9 mm, 3- to 5-veined and subacute at apex,
the upper 9–11 mm, 5- to 7-veined and obtuse at apex.
Lemmas 11–14 mm, lanceolate, subcoriaceous, conspicuously hairy, with 2, subacute, hyaline apical teeth and a
stout awn (3–)4–6(–7) mm between them, with 7–9 veins.
Palea elliptical, about three-quarters as long as to about

equalling the lemma. *Caryopsis* ellipsoid. *Flowers* 8–9.
$2n = 42$.

Introduced. Casual and sometimes naturalised on
rough and waste ground. A few localities in south-east
England. Native of western North America.

3. C. cathartica (Vahl) Herter   Rescue Brome
Bromus catharticus Vahl; *C. unioloides* (Willd.) P.
Beauv.; *Bromus unioloides* Willd.; *Bromus willldenowii*
Kunth

Tufted, often robust *annual* or *perennial herb*. *Culms* up
to 1 m, erect or spreading. *Leaves* up to 30 cm × 4–10
mm, linear, attenuate at apex, flat, very shortly hairy on
upper surface, minutely scabrous along the veins and
margins; sheaths shortly to long-hairy; ligule 4–5 mm,
obtuse at apex, often laciniate, shortly hairy.
Inflorescence a large, loose panicle up to 30 cm, the
primary branches whorled. *Spikelets* 15–35 mm, with
6–12 flowers, ovoid, compressed. *Glumes* unequal,
hooded at apex, the lower 6–13 mm and 3- to 7-veined;
the upper 7–16 mm and 5- to 9-veined, both strongly
keeled and scabrous along the keels. *Lemmas* 8–19 mm,
strongly keeled, overlapping when young, later more or
less spreading, 9–11(–13)-veined, unawned or with an
awn 0.5–3.5(–5) mm. *Palea* elliptical, half to two-thirds
as long as the lemma. *Caryopsis* linear to linear-obloid,
dorsally compressed, usually adherent to the palea and
tightly enclosed in the lemma. *Flowers* 6–9. Usually
cleistogamus. $2n = 42$.

Introduced. A grain and wool alien, rarely grown for
fodder. Casual or naturalised on rough ground, roadsides and field borders. Scattered records in central and
southern Britain and the Channel Islands. Native of
Central and South America.

4. C. staminea (Desv.) Stace   Southern Brome
Bromus stamineus Desv.; *Bromus valdivianus* Philippi

Annual or *perennial herb*. *Culms* up to 1 m. *Leaves*
2–3(–4) mm wide, linear, acute at apex, more or less
long-hairy; sheaths long-hairy. *Panicle* 10–20 cm;
branches stiff and ascending. *Spikelets* 18–24 mm,
greenish, with 4–6 flowers. *Glumes* unequal, ovate-lanceolate, acute at apex, hairy, the lower 7–8 mm and 5- to 7-
veined, the upper longer and 9-veined. *Lemmas* 8–13
mm, obovate, keeled, weakly 9-veined, sparsely to
densely hairy, with a prominent white or slightly purplish hyaline margin; awn (3–)4–8(–12) mm. Flowers
7–9. $2n = 42$.

Introduced. A casual sometimes naturalised for a
while on rough ground and waysides. Recorded from a
few places in southern England. Native of South
America.

5. C. brevis (Nees ex Steud.) B. D. Jackson
                                        Patagonian Brome
Bromus brevis Nees ex Steud.

Annual or *perennial herb*. *Culms* up to 1 m. *Leaves* 2–4
mm wide, linear, acute at apex, more or less long-hairy.
Inflorescence a narrow, erect panicle with short
branches. *Spikelets* ovate-oblong. *Glumes* unequal,
ovate-lanceolate, acute at apex, the lower 6.5–7.0 mm

and 7-veined, the upper 8–9 mm and 9-veined. *Lemmas* with curved keels, closely overlapping, usually with awnless tips, the awn when present up to 1(–2) mm, glabrous or sparsely pubescent. *Flowers* 7–9. $2n = 42$.

Introduced. Wool alien on tips and waste ground. In a very few scattered localities in Britain. Native of South America.

Tribe **13. Brachypodieae** Harz

Annual or *perennial herbs* with rhizomes. *Leaves* with a membranous ligule; sheaths not fused. *Inflorescence* a raceme with usually one spikelet at each node and pedicels less than 2.8 mm. *Spikelets* with many bisexual florets, the apical 1 or 2 reduced and male or sterile. *Glumes* 2, unequal, 3- to 9-veined, much shorter than the rest of the spikelet, sometimes shortly awned. *Lemmas* mostly 7-veined, rounded on back, acute or acuminate at apex and usually with a short to long awn. *Stamens* 3. *Stigmas* 2. *Caryopsis* with a pubescent terminal appendage, the styles arising from below it.

59. Brachypodium P. Beauv.

Annual or rhizomatous *perennial herbs*. *Leaves* rolled or flat; ligule membranous; sheaths not fused. *Inflorescence* a raceme usually with one spikelet at each node and pedicels less than 2(–2.8) mm. *Spikelets* with many bisexual flowers, the apical 1 or 2 reduced and male or sterile. *Glumes* 2, unequal, 3- to 9-veined, much shorter than the rest of the spikelets, sometimes shortly awned. *Lemmas* mostly 7-veined, rounded on back, acute to acuminate at apex and usually with a short to long awn. *Stamens* 3. *Stigmas* 2. *Lodicules* 2. *Caryopsis* with a terminal, shortly hairy appendange, the styles arising from below it.

Sixteen species in temperate Eurasia, extending southwards on tropical mountains from Mexico to Bolivia.

1. Annual; spikelets distinctly compressed; anthers 0.3–1.0 mm **3. distachyon**
1. Perennials with rhizomes and tillers; spikelets subterete; anthers 3.5 –4.5 mm **2.**
2. Plant strongly rhizomatous, usually scarcely tufted; leaves usually less than 6 mm wide; culms with 3–4(–5) nodes, raceme usually erect; lemmas with awns 1–5 mm **1. pinnatum**
2. Plant weakly rhizomatous, usually densely tufted; culms with 4–8 nodes; raceme usually pendulous at apex; leaves usually more than 6 mm wide; lemmas with awns 7–15 mm **2. sylvaticum**

1. B. pinnatum (L.) P. Beauv.   Tor Grass
Bromus pinnatus L.

Perennial herb forming loose to compact tufts; rhizomes wiry, scaly, spreading. *Culms* 30–120 cm, usually erect, slender to somewhat stout, stiff, unbranched, with 2–3(–4) nodes, smooth, usually glabrous. *Leaves* up to 45 cm × 2–6(–10) mm, green or yellowish-green, linear, finely pointed at apex, rolled or flat, stiff to flaccid, erect, sparsely hairy or glabrous, rough; sheaths rounded on the back, glabrous and smooth or the lower shortly

hairy; ligules up to 2 mm, obtuse at apex, membranous. *Inflorescence* a raceme 4–25 cm, spike-like, erect or sometimes nodding, bearing 3–15 spikelets, green or yellowish; rhachis slender; pedicels 1–2 mm. *Spikelets* 20–40 mm, cylindrical, lanceolate to narrowly oblong, straight or curved, alternating in 2 rows on opposite sides of the rhachis, overlapping, 8- to 22-flowered, usually solitary, very rarely in clusters of 2–3, breaking up at maturity beneath each lemma. *Glumes* persistent, lanceolate to narrowly ovate, pointed at apex, rounded on the back, firm, glabrous, the lower 3–5 mm and 3- to 6-veined, the upper 5–9 mm and 5- to 7-veined. *Lemmas* 6–10 mm, overlapping, rounded on the back, lanceolate-oblong, pointed at apex, firm, 7-veined, glabrous and smooth, rarely shortly hairy, tipped with a fine awn 1–5 mm. *Paleas* as long as the lemmas, narrowly oblong, the 2 keels fringed with minute hairs. *Anthers* 3.5–4.5 mm. *Caryopsis* narrowly ellipsoid, hairy at the apex, tightly enclosed by the hardened lemma and palea. *Flowers* 6–8. $2n = 28$.

Native. Grassland, mainly on chalk and limestone, sometimes spreading rapidly and dominating large areas. Common in central, south and eastern England, scattered in north-west and south-west England, Wales and Ireland, casual in Scotland. Europe to about 62° N; North Africa; Siberia.

× **sylvaticum** = **B. cugnacii** A. Camus
This hybrid is probably the identity of some variably sterile plants found growing with the parents. They are intermediate in awn length, hairyness of culms, leaves, glumes and general habit.

2. B. sylvaticus (Huds.) P. Beauv.   False Brome
Festuca sylvatica Huds.

Compactly tufted *perennial herb* with weak rhizomes. *Culms* 30–90 cm, erect or spreading, slender to moderately stout, unbranched, with 4–8 nodes, hairy at the nodes and often also towards them, otherwise smooth and glabrous. *Leaves* up to 35 cm × (4–)6–12 mm, green, narrowed to the sheath, linear, pointed at apex, erect or finally drooping, soft, mostly loosely hairy, rarely glabrous and rough; sheaths rounded on the back or keeled upwards, loosely hairy with spreading or reflexed hairs or the upper smooth, rarely all glabrous; ligules 1–6 mm, membranous. *Inflorescence* a raceme 6–20 cm, spike-like, loose, erect, or more often nodding at apex, bearing 4–12 spikelets, green; rhachis slender; pedicels 0.5–2.0 mm. *Spikelets* 20–40 mm, 8- to 16-flowered, breaking up at maturity beneath each lemma. *Glumes* persistent, unequal, rounded on the back, sharply pointed, firm, usually hairy; the lower 6–8 mm, lanceolate and 5- to 7-veined, the upper 8–11 mm, lanceolate to narrowly oblong and 7- to 9-veined. *Lemmas* 7–12 mm, overlapping, rounded on the back, oblong-lanceolate, pointed at apex, tipped with a fine, rough awn 7–15 mm, tough, 7-veined, shortly and stiffly hairy, rarely only rough or quite smooth. *Paleas* as long as or nearly as long as the lemmas, narrowly oblong, obtuse at apex, with 2, short-haired keels. *Anthers* 3.5–4.0 mm.

Caryopsis narrowly ellipsoid, hairy at the tip, tightly enclosed by the hardened lemma and palea. *Flowers* 7–8. $2n = 18$.

Native. Woods, scrub, shaded hedgerows, ditches and open grassland. Common throughout the British Isles except in north Scotland. Europe; North Africa; Macaronesia; west Asia to north-west Himalaya; Japan.

3. B. distachyon (L.) P. Beauv.   Stiff Brome
Bromus distachos L.; *Trachynia distachya* (L.) Link

Annual herb. Culms up to 45 cm, erect or geniculately ascending, fasciculate and spreading, rarely solitary, smooth or sometimes retrorsely scabrid under the panicle, shortly pubescent or glabrous. *Leaves* 1–6 cm × 2–5 mm, dark green or glaucescent, flat, ending in a rigid, acute tip, with thickened, scabrid margins, strongly veined on the upper surface, sparsely hairy or glabrescent on both surfaces; sheaths striate, glabrous; ligules about 1 mm or vestigical. *Inflorescence* consisting of 1–3(–7) spikelets often closely aggregated at the tip of the peduncle. *Spikelets* 10–30 mm, somewhat laterally compressed, lanceolate or oblong, 8- to 12-flowered. *Glumes* unequal, coriaceous, the lower 5–6 mm and 5-veined, the upper 7–9 mm, rather broader, 7-veined and mucronate. *Lowest lemma* 8–9 mm, oblong-elliptical, acute at apex, mucronate or shortly awned, glabrous or covered on the back with minute scabridities, 7-veined, the veins confluent at the apex; upper lemmas with an awn about 9 mm. *Paleas* as long as the lemma, oblong, obtuse at apex, bristly pectinate on the keels at the tip. *Anthers* 0.3–1.0 mm. *Caryopsis* adnate to the palea, free from the lemma, terete or oblong-ellipsoid, somewhat compressed. *Flowers* 4–5. $2n = 10, 28, ?30$.

Introduced. Casual in waste places from wool or rarely grain. Very scattered in England. Native of the Mediterranean region east to central Asia; Macaronesia; Ethiopia; South Africa.

Tribe 14. Triticeae Dumort.

Tribe *Hordaeae* Spenner

Annual or *perennial herbs* with or without rhizomes and without stolons. *Leaves* with a membranous ligule; sheaths not fused. *Inflorescence* a spike with 1–3 spikelets at each node. *Spikelets* with one to many flowers; often some flowers, or some spikelets if spikelets more than one per node, male or sterile. *Glumes* 2, more or less equal, 1- to 11-veined, often with a long terminal awn. *Lemmas* mostly 5- to 7-veined, rounded on back, acute to obtuse at apex, awnless or with a terminal awn. *Stamens* 3. *Stigmas* 2. *Lodicules* 2. *Caryopsis* usually pubescent or pubescent at apex, sometimes with a pubescent terminal appendange, the styles arising from below it.

60. Elymus L.

Agropyron auct.; *Roegneria* K. Koch

Perennial herbs without rhizomes. *Leaves* mostly flat; ligules membranous. *Inflorescence* a spike, with one spikelet at each node. *Spikelets* with several to many

flowers all but the apical 1 or 2 bisexual, flattened broadside on to the rhachis. *Glumes* equal, acute to narrowly acute, often awned, 2- to 5-veined. *Lemmas* 5-veined, usually long-awned, sometimes awnless or short-awned. *Spikelets* breaking up below each lemma at maturity.

Numerous species through temperate latitudes of both southern and northern hemisphere and best represented in Asia.

Grime, J. P. et al. (1988). *Elymus caninus* (L.) L. in *Comparative plant ecology*. London.

1. Glumes 15–35 mm   **2. canadensis**
1. Glumes 5–14 mm   2.
2. Spikelets distant at least towards the base of the spike, not reaching the spikelet above   **3. scabrus**
2. Spikelets condensed and contiguous   3.
3. Lemmas with awns 3–20 mm   **1(a). caninus** subsp. **caninus**
3. Lemmas without awns or with awns up to 3 mm
         1(b). caninus subsp. **donianus**

1. E. caninus (L.) L.   Bearded Couch
Triticum caninum L.; *Roegneria canina* (L.) Nevski; *Agropyron caninum* (L.) P. Beauv.

Loosely tufted *perennial herb* without rhizomes. *Culms* 30–110 cm, erect or bent in the lower part, slender, unbranched, with 2–5 nodes, minutely hairy towards and at the nodes or smooth. *Leaves* 10–30 cm × 4–13 mm, bright green, linear, finely pointed at apex, flat, rather thin, rough, finely veined, loosely hairy above or glabrous; rounded on the back, glabrous or the lower shortly hairy; ligule up to 1.5 mm, membranous. *Inflorescence* a spike 5–20 cm, green or tinged with purple, erect, curved or nodding, slender; rhachis rough or minutely hairy along the angles. *Spikelets* 10–20 mm, sessile, alternating in 2 rows on opposite sides of the rhachis, with their broader sides pressed to it, lanceolate to oblong, scabrid, 2- to 6-flowered, breaking up at maturity beneath each lemma. *Glumes* 7–11 mm, persistent, equal or slightly unequal, lanceolate to narrowly oblong, rounded on the back, sharply pointed at apex, sometimes with a short awn 1.7–4.0(–6.0) mm from the top, rigid, prominently 2- to 6-veined, the veins rough. *Lemma* 8–13 mm, overlapping, rounded on the back, lanceolate-oblong, rigid, 5-veined, minutely hairy at the base, often with minute scattered hairs in the upper part or smooth, awnless or narrowed at the tip into a straight or flexuous awn up to 20 mm. *Paleas* narrowly elliptical, about as long as the lemmas, with 2 rough keels. *Anthers* 2.0–3.5 mm. *Caryopsis* narrowly ellipsoid, hairy at the top, enclosed by the hardened lemma and palea. *Flowers* 6–8.

(a) Subsp. **caninus**
Lemmas with awn 3–20 mm. $2n = 28$.

(b) Subsp. **donianus** (F. B. White) P. D. Sell
Agropyron donianum F. B. White; *Elymus donianus* (F. B. White) Å. & D. Löve; *Elymus trachycaulus* subsp. *donianus* (F. B. White) Å. & D. Löve; *Roegneria donianus* (F. B. White) Melderis
Lemmas without awns or with awns up to 3 mm.

Native. The species is scattered throughout Britain and Ireland, but is absent from large parts of central and north Scotland and Ireland. It occurs throughout Europe, but is rare in the south and in temperate Asia. The subspecies are ecotypic. Subsp. *caninus* occurs in woods and along hedgerows and other shady places, throughout the range of the species. Subsp. *donianus* occurs in rock-crevices, gullies and on cliff-ledges, usually near water in the mountains of central and north Scotland, and is late-flowering. Similar taxa are found elsewhere in north and central Europe.

2. E. canadensis L.   Canadian Couch

Coarse, tufted *perennial herb. Culms* 70–200 cm, often glaucous, erect. *Leaves* 10–30 cm × 10–20 mm, green to glaucous, linear, gradually narrowed to a sharp point at apex, thin, flat, glabrous to sparsely rigid-hairy on the upper surface, with 20–40(–70) veins; sheaths glabrous or rarely shortly hairy; ligules membranous. *Inflorescence* a thick, bristly, nodding or drooping spike, often interrupted below, 10–25 cm; rhachis mostly hidden. *Spikelets* usually slightly spreading, crowded, 2- to 5-flowered. *Glumes* 15–35 × 0.5–2.0 mm, linear-lanceolate, narrowed at apex, 2- to 4-veined, scabrous and sometimes stiff-hairy. *Lemmas* lanceolate, narrowed at apex, usually stiffly hairy, rarely glabrous, strongly 5-veined above; awn 20–30 mm, curved. *Paleas* elliptical. *Caryopsis* oblong. *Spikelets* breaking up below each lemma at maturity. *Flowers* 7–8. 2*n* = 28.

Introduced. Formerly naturalised in Kent. Native of North America.

3. E. scabrus (R. Br.) Å. Löve   Australian Couch
Triticum scabrum R. Br.; *Agropyrum scabrum* (R. Br.) P. Beauv.; *Festuca scabra* auct.

Loosely tufted *perennial herb. Culms* up to 120 cm. *Leaves* 3–25 cm × 2–6 mm, linear, attenuate at apex, flat or convolute, often sparse to moderately rigid-hairy; ligules up to 0.5 mm, membranous, truncate at apex. *Inflorescence* a spike (5–)15–25 cm, elongated. *Spikelets* distant at least at base, not reaching as far as the one above, 15–25 mm excluding awns. *Glumes* 5–14 mm, narrowly ovate, obtuse, mucronate, acute or acuminate at apex, with awns as long as the body, 3- to 8-veined. *Lemmas* 7–13 mm, narrowly ovate, smooth or with minute hairs towards the apex, 5-veined; awns 15–33 mm, straight or curved. *Paleas* elliptical, 2-keeled. *Caryopsis* oblong, flat or convex on inner side, enclosed by the hardened lemma and palea. *Spikelets* breaking up below each lemma at maturity. *Flowers* 7–8. 2*n* = 28, 41–43, 57, 63, 64–66, 94.

Introduced. Occasional wool alien of fields and waste ground. Scattered records in England. Native of Australia.

61. **Elytrigia** Desv.

Agropyron auct.; *Thinopyrum* Å. Löve; *Elymus* auct.

Perennial herbs with long rhizomes. *Leaves* flat or inrolled; ligules less than 1 mm. *Inflorescence* a spike, with one spikelet at each node. *Spikelets* with several to many flowers, all but the apical 1 or 2 bisexual, flattened broadside on to the rhachis. *Glumes* 3- to 11-veined, acute to very obtuse at apex, rarely awned. *Lemmas* 5-veined, unawned or with a short or rarely long awn. *Spikelets* not breaking up at maturity, usually falling whole or rhachis breaking up.

Numerous species, best represented in Asia, but extending through temperate latitudes of the northern and southern hemispheres.

Grime J. P. et al. (1988). *Elymus repens* (L.) Gould in *Comparative plant ecology*. London.
Palmer, J. H. & Sager, G. R. (1963). *Agropyron repens* (L.) P. Beauv. in Biological flora of the British Isles. *Jour. Ecol.* **51**: 783–794.

1. Veins on upper side of leaf densely and minutely hairy; rhachis of inflorescence breaking up between each spikelet at maturity   **3. juncea** subsp. **boreoatlantica**
1. Veins on upper side of leaves glabrous to scabrid, sometimes with sparse long hairs; rhachis of inflorescence not breaking up between each spikelet at maturity   2.
2. Leaves with their upper side veins more or less flat-topped, at least their middle and lower sheaths with a minute, often sparse, fringe of hairs on the exposed free margin   **2. atherica**
2. Leaves with their upperside veins with rounded tops, their sheaths with a glabrous margin   3.
3. Leaves inrolled, 1.5–6.0(–9.5) mm wide; with thick, round-topped ribs on the upper surface; spikes mostly under 8 cm; glumes (4.4–)6–9 mm, mostly with 3 veins; lemmas 6–10 mm, awnless or with awns up to 2.8(–4.6) mm   **1(b). repens** subsp. **arenosa**
3. Leaves mostly flat, 3–10 mm wide; with fine veins on the upper surface; spikes up to 20(–30) mm; glumes 7–12 mm, with (3–)5–7 veins; lemmas 8–12 mm, sometimes with awns up to 15 mm   4.
4. Lemmas without awns
            1(a,i). repens subsp. **repens** var. **repens**
4. Lemmas with awns up to 15 mm
            1(a,ii). repens subsp. **repens** var. **aristata**

1. E. repens (L.) Desv. ex Nevski   Common Couch
Agropyron repens (L.) P. Beauv.; *Elymus repens* (L.) Gould

Extremely variable *perennial*, forming tufts or large patches, spreading extensively by creeping, wiry rhizomes. *Culms* 30–150 cm, erect or bent below, slender to rather stout, smooth, with 3–5 nodes. *Leaves* 6–30 cm × 1.5–10 mm, dull green, sometimes bluish or greyish-green, linear, finely pointed at apex, flat or inrolled, soft to rather stiff, smooth or rough beneath, usually loosely to sparsely hairy above, but sometimes glabrous; sheaths rounded on the back, with short, spreading auricles at the apex, glabrous, the lower loosely to closely hairy, smooth; ligules less than 1 mm, membranous. *Inflorescence* a spike 5–20(–30) cm, green or sometimes bluish-green, erect, straight, slender, loose to compact; rhachis tough, rough on the margins, glabrous or rarely softly hairy. *Spikelets* 10–20 mm, oblong or wedge-shaped, 3- to 8-flowered, falling entire at maturity, sessile, alternating in 2 rows on opposite sides of the rhachis, with the broader sides appressed to it, one-third to half their length apart. *Glumes* 5–12 mm, similar, equal or nearly so, lanceolate-oblong, blunt or pointed at apex, tough, 3- to 7-veined, rough upwards on the keels. *Lemmas* 6–12 mm, overlapping, lanceolate-oblong, obtuse or pointed, sometimes sharply, at apex, often with an awn up to 15 mm, keeled upwards, tough, 5-veined. *Paleas* nearly as long as the lemmas, with 2 rough keels. *Anthers* 3.5–6.0 mm. *Caryopsis* hairy at the top, tightly enclosed by the hard lemma and palea. *Flowers* 6–8.

(a) Subsp. repens
Leaves mostly flat, 3–10 mm wide, with fine, well-spaced veins on the upper surface. *Spikes* up to 20(–30) cm. *Spikelets* with 3–8 flowers. *Glumes* 7–12 mm, with (3–)5–7 veins. *Lemmas* 8–12 mm, sometimes with awns to 15 mm. $2n = 28, 42, 56$.

(i) Var. repens
Lemmas without awns.

(ii) Var. aristata (Döll) P. D. Sell
Lemmas with awns up to 15 mm.

(b) Subsp. arenosa (Spenner) Å. Löve
Elymus repens subsp. *arenosus* (Spenner) Melderis; *Agropyrum maritimum* Jansen & Wachter, non (L.) P. Beauv.
Leaves inrolled, 1.5–6.0(–9.5) mm wide, with thick, round-topped ribs on the upper surface. *Spikes* mostly under 8 cm. *Spikelets* with (2–)3–6 flowers. *Glumes* (5.0–)6–9 mm, mostly with 3 veins. *Lemmas* 6–10 mm, blunt or mucronate and sometimes with awns (0.3–)1.8–2.8(–4.6) mm.

Native. Throughout the British Isles. Europe; North Africa; Macaronesia; Siberia; North America. The subspecies are ecotypic. Subsp. *repens* var. *repens* is an abundant weed of cultivated fields and waste places, field margins, roadsides and rough grassland including grassy places by the sea. It occurs throughout the range of the species. Var. *aristata* also seems to occur throughout the range of the species. Subsp. *arenosa* occurs on maritime sand-dunes in the south and east of England and perhaps elsewhere. It also occurs on the coast of north-west Europe.

2. E. atherica (Link) Kerguélen ex Carreras Mart.
Sea Couch
Triticum athericum Link; *Elymus athericus* (Link) Kerguélen; *E. pycnanthus* (Godr.) Melderis; *Agropyron pycnanthum* (Godr.) Godr.; *Agropyron pungens* auct.; *Triticum pungens* auct.; *Agropyron repens* subsp. *pungens* auct.; *E. pungens* auct.; *Triticum repens* subsp. *pungens* auct.

Perennial herb, forming tufts or large patches, spreading extensively by wiry rhizomes. *Culms* 20–120 cm, erect or bent at the base, slender to stout, rigid, unbranched, smooth, with 3–4 nodes. *Leaves* 8–35 cm × 2–6 mm, bluish-grey or greyish-green, glabrous, linear, with sharply pointed, hard tips, flat or tightly inrolled, stiff,

smooth beneath, closely and prominently veined above, the veins flat-topped, rough on the margins and the veins or the latter smooth; sheaths rounded on the back, smooth, with short, narrow, prominent auricles at the top, at least the middle and lower with a minute, often sparse, fringe of hairs on the exposed free margin; ligules less than 1 mm, membranous. *Inflorescence* a spike 4–20 cm, erect, stiff, compact, slender to stout; rhachis tough, rough. *Spikelets* 10–20 mm, closely overlapping, one-fifth to half their length apart, singly and alternating in 2 rows on opposite sides of the rhachis, with their broader sides pressed to the flattened rhachis, falling entire or breaking up at maturity, oblong or elliptic-oblong, compressed, 3- to 10-flowered. *Glumes* 8–10 mm, similar, equal or nearly so, lanceolate-oblong, pointed at apex, keeled, tough and rigid, rough on the keels, prominently 4- to 7-veined. *Lemmas* 7–11 mm, closely overlapping, lanceolate-oblong, blunt or pointed at apex, sometimes with awns up to 10 mm, keeled upwards, tough, 5-veined. *Paleas* about as long as the lemmas, narrowly elliptical, 2-keeled, with the keels rough. *Anthers* 5–7 mm. *Caryopsis* narrowly ellipsoid, minutely hairy at the top, enclosed by the hard lemma and palea. *Flowers* 6–8. $2n = 42$.

Native. Frequent on the margins of salt-marshes and brackish creeks, sandy and gravelly muds, and shingle and consolidated sand-dunes, often covering large areas. Frequent round the coast of the British Isles north to Co. Louth, Dumfriesshire and Yorkshire. West and south Europe; North Africa.

× **juncea = E. × obtusiuscula** (Lange) D. C. McClint.
Elymus × obtusiusculus (Lange) Melderis & D. C. McClint.; *Agropyron × obtusiusculum* Lange; *Agropyron hackelii* Druce; *Agropyron acutum* auct.
This hybrid closely resembles *E. × laxa*, but has some minute hairs on the free margin of the leaf-sheaths. It is intermediate between the parents. $2n = 35$.

Occurs frequently with the parents on the coasts of Britain north to Durham and Cumberland, and in south-west Ireland and the Channel Islands. It also occurs on the coasts of Continental Europe and in Sweden, Poland and Corsica.

× **repens = E. × oliveri** (Druce) Kerguélen ex Carreras Mart.
Agropyron × oliveri Druce; *Elymus × oliveri* (Druce) Melderis & D. C. McClint.
This hybrid is intermediate between the parents in its leaf veins and the hairiness of the leaf-sheath margin. It is sterile with indehiscent anthers and empty pollen.

Occurs with the parents in scattered places round the coasts of Britain and south-west Ireland and is probably not uncommon. It also occurs on the coasts of western Europe and the Baltic.

3. E. juncea (L.) Nevski
Sand Couch
Perennial herb, forming loose tufts or mats, spreading extensively by long, slender, wiry rhizomes. *Culms* 20–60 cm, erect, spreading or drooping, slender to rather stout, rather brittle, unbranched, smooth, with a few nodes

above the base. *Leaves* 10–35 cm × 2–6 mm, bluish-grey, linear, finely pointed at apex, spreading or drooping, flat or rolled, stiff to rather soft, smooth beneath, prominently veined above, with the veins minutely and densely hairy, without auricles; sheaths overlapping, rounded on the back, smooth; ligules 0.5–1.0 mm, truncate at apex, membranous. *Inflorescence* a spike 4–20 cm, stout, straight or curved; rhachis smooth, fragile, readily breaking just above each spikelet. *Spikelets* 15–28 mm, sessile, alternating in 2 rows on opposite sides of the rhachis, with their broad sides appressed to it, breaking up at maturity beneath each lemma, oblong, elliptical or wedge-shaped, 3- to 8-flowered, their own length or less apart. *Glumes* 9–20 mm, similar, equal or slightly unequal, narrowly oblong, obtuse at apex, keeled or rounded on the back, very tough and rigid, prominently 7- to 11-veined, smooth. *Lemmas* 11–20 mm, overlapping, oblong or oblong-lanceolate, obtuse or emarginate at apex, with a very short, hard apical mucro, rounded on the back below, keeled above, thick and rigid, 5-veined, smooth. *Paleas* narrowly elliptical, shorter than the lemmas, with 2 keels minutely hairy. *Anthers* 6–8 mm. *Caryopsis* narrowly ellipsoid, minutely hairy at the top, enclosed by the hard lemma and palea. *Flowers* 6–8. $2n = 28$.

All the plants of the British Isles are referable to subsp. **boreoatlantica** (Simonet & Guin.) N. Hyl. (*Agropyron junceiforme* (Å. & D. Löve) Å. & D. Löve; *Elymus farctus* subsp. *boreoatlanticus* (Simonet & Guin.) Melderis.

Native. Maritime sand-dunes and beaches. Common all round the coasts of the British Isles. The species is found on all the coasts of Europe. Subsp. *boreoatlantica* is confined to the coasts of north and west Europe.

× **repens** = E. × **laxa** (Fr.) Kerguélen

Triticum × laxum Fr.; *Elymus × laxus* (Fr.) Melderis & D. C. McClint.; *Agropyron × laxum* (Fr.) Tutin

This hybrid is sterile and has a tardily breaking, smooth, or slightly scabrid rhachis, distant lower spikelets and conspicuously scabrid or more or less shortly hairy leaf-veins. It is distinguished from *E. × obtusiuscula* by the glabrous free margin of the sheaths. $2n = 35, 49$.

It occurs with the parents on the coasts of Britain north to Merionethshire and Yorkshire, and on the coasts of Ireland and the Channel Islands. It also occurs on the coasts of western Europe and Sweden and Poland.

61 × 64. × Elytrordeum Hyl.

Elytrigia × Hordeum

1. × E. langei (K. Richt.) Hyl.

E. repens × H. secalinum

Agropyron × langei K. Richt.; × *Agrohordeum langei* (K. Richt.) E. G. & A. Camus; × *Elyhordeum langei* (K. Richt.) Melderis

This hybrid exists in two forms, one clearly intermediate, the other differing from *E. repens* in that the rhachis disarticulates and the glumes and lemmas are awned, and

from *H. secalinum* in having rhizomes and mostly one spikelet with 3–4 flowers per node. $2n = 49$.

Wet meadows, fields and roadsides. In a few scattered localities in England from the Isles of Scilly to Northumberland. Also in Denmark.

62. Leymus Hochst.

Elymus auct.

Perennial herbs with long rhizomes. *Leaves* flat with a convolute margin, glaucous, scabrid; ligule short. *Inflorescence* a spike; rhachis tough. *Spikelets* 2–3 at each node, with 3–6 flowers, subsessile, usually imbricate, all but the most apical bisexual, flattened broadside on to the rhachis. *Glumes* narrowly lanceolate, 3- to 5-veined, finely pointed at apex, but not awned. *Lemmas* mostly 7-veined, not or shortly awned. *Palea* nearly as long as the lemma, 2-keeled. Spikelets breaking at maturity below each lemma.

About 40 species in the north temperate zone, one species in Argentina.

Bond, T. E. T. (1952). *Elymus arenarius* L. in Biological flora of the British Isles. *Jour. Ecol.* **40**: 217–227.

1. L. arenarius (L.) Hochst.   Lyme Grass
Elymus arenarius L.

Robust *perennial herb* forming large tufts or masses; rhizome long and stout. *Culms* erect or spreading, stout, unbranched, smooth. *Leaves* up to 60 cm × 8–20 mm, bluish-grey, sharply pointed at apex, flat or inrolled, linear, rigid, minutely rough above on the prominent veins, smooth beneath; sheaths smooth, with 2 narrow, spreading auricles at the apex; ligules up to 1 mm, firmly membranous, minutely hairy. *Inflorescence* a spike 15–35 cm × 12–25 mm, stout, compact, stiff. *Spikelets* stalkless, usually in pairs, the pairs alternating on opposite sides of the rhachis, 20–32 mm, oblong or wedge-shaped, 3- to 6-flowered, breaking up at maturity beneath each lemma. *Glumes* persistent, similar, narrowly lanceolate, firmly pointed in side view, as long as or nearly as long as the spikelet, keeled, usually rigid-hairy especially on the keel, 3- to 5-veined. *Lemmas* decreasing in size upwards, the lowest 15–25 mm, lanceolate, pointed, tough, 7-veined, densely hairy with short, soft hairs. *Paleas* as long as the lemmas, very narrowly elliptical, 2-keeled. *Anthers* 7–8 mm. *Caryopsis* narrowly ellipsoid, tightly enclosed by the lemma and palea, about 10 mm, hairy at the top. *Flowers* 7–8. $2n = 56$.

Native. Mobile sand on maritime dunes; rare casual or garden escape inland. Frequent round the coasts of most of the British Isles, but absent from large parts of southern England and Ireland, and from the Channel Islands. North and west Europe, from the Arctic to north-west Spain. It is useful as a sand-binder.

63. Hordelymus (Jessen) Jessen

Perennial herbs with very short rhizomes. *Leaves* flat; ligules small, membranous. *Inflorescence* a compressed, linear to oblong spike. *Spikelets* (2–)3 at each node of its rhachis, dispersed together at maturity, the triplets arranged in 2 longitudinal rows, each triplet with a central, usually bisexual spikelet and lateral bisexual spikelets. *Glumes* linear-lanceolate, connate at base, 1- to 3-veined, with a long awn. *Lemmas* 5-veined, with a very long awn. *Spikelets* breaking up at maturity above the glumes.

A single species in Europe and North Africa eastwards to the Caucasus.

Stewart, A., Pearman, D. A. & Preston, C. D. (1994). *Scarce plants in Britain*. Peterborough.

1. Hordelymus europaeus (L.) Jessen   Wood Barley
Hordeum europaeum (L.) All.; *Hordeum sylvaticum* Huds.; *Elymus europaeus* L.

Short-lived *perennial herb*; tufts loose. *Culms* 40–120 cm, erect or bent below, slender to stout, 3- to 4-noded, hairy at the nodes, smooth. *Leaves* 10–30 cm × 5–14 mm, green, flat, linear, pointed at apex, loosely to sparsely hairy, rough above and on the margins; sheaths beset with spreading or reflexed hairs, or the upper glabrous and smooth, with short, spreading auricles at the apex; ligules less than 1 mm, membranous. *Inflorescence* a spike 5–10 cm × 7–12 mm, green, rhachis persistent. *Spikelets* at each node of the rhachis, rarely in pairs, usually triplets alternating on opposite sides of the rhachis, usually 1- rarely 2-flowered, narrow, breaking up at maturity above the glumes. *Glumes* including awns 14–17 mm, persistent, erect, side by side in front of the lemma, linear-lanceolate, flat, rigid, narrowed into a fine, straight, rough awn. *Lemmas* 8–10 mm, lanceolate, broadly rounded on the back, becoming tough, 5-veined, rough upwards, narrowed into a fine, rough awn 15–25 mm. *Paleas* narrowly elliptical, as long as the lemmas, 2-keeled. *Anthers* 3–4 mm. *Caryopsis* about 7 mm, narrowly ellipsoid, tightly enclosed by the hardened lemma and palea, hairy at the top. *Flowers* 6–7. $2n = 28$.

Native. Local in woods and copses, usually on calcareous soils. Britain north to Northumberland, formerly in Co. Antrim and Berwickshire. In the Chilterns it is sometimes locally abundant, especially in beech woods with undergrowth. Europe, from central Spain and the Crimea north to south Sweden; mountains of North Africa.

64. Hordeum L.

Critesium Raf.

Annual or less often *perennial herbs* without rhizomes. *Leaves* flat. *Inflorescence* a compressed, linear to oblong spike. *Spikelets* 3 at each node of the rhachis, dispersed together at maturity, the triplets arranged in 2 longitudinal rows, the fertile flowers in 2, 4 or 6 longitudinal rows, each triplet with a middle bisexual spikelet and 2 lateral bisexual, male or sterile spikelets. *Flowers* 1, rarely 2. *Glumes* linear-subulate to lanceolate, 1- to 3-veined, with a long awn. *Lemmas* of bisexual flowers 5-veined, ovate, with a very long awn. *Palea* narrowly ovate, keeled. *Rhachilla* usually prolonged in spikelets. *Rhachis* breaking up at each node at maturity or below each bisexual lemma.

About 40 species in temperate regions throughout the world.

Most of the species are superficially similar. The spikelets, flowers and lemmas are referred to as middle or lateral according to their position in the triplet.

Bor, N. L. in C. C. Townsend & E. Guest (1968). *Flora of Iraq* 9: 244–257. Baghdad.

1. Rhachis not breaking up at maturity, the caryopsis containing flowers breaking away from the rest of the spikelet, which remains on the rhachis; awns of the lemma of the middle spikelet usually more than 10 cm, rarely very short   2.
1. Rhachis breaking up at maturity, the triplet of spikelets forming the dispersal unit; all awns less than 10 cm   3.
2. All 3 flowers in each triplet producing a caryopsis and with a long-awned lemma   **1. vulgare**
2. Only the middle flower of each triplet producing a caryopsis; lateral lemmas awnless or more or less so   **2. distichon**
3. Lateral flowers extremely reduced, usually simply an awn-like outgrowth; glumes of lateral spikelets more than 30 mm, awn-like from base to apex; awn of lemma of middle spikelet usually more than 5 cm   **6. jubatum**
3. Lateral flowers male or sterile but with obvious flower construction; glumes of lateral spikelets less than 30 mm; if more than 2 cm then at least one pair distinctly widened at base; awns of lemma of middle spikelet less than 5 cm   4.
4. Perennials with tillers at time of flowering   5.
4. Annuals without tillers   6.
5. Upper leaves 1.5–2.0(–3.0) mm wide; proximal part of glumes with very short, soft hairs more than 0.1 mm; awns strongly divergent at maturity; anthers 1–2 mm   **7. pubiflorum**
5. Upper leaves 2–6 mm wide; proximal part of glumes with minute rough projections; awns stiffly erect at maturity; anthers 3–4 mm   **8. secalinum**
6. Leaf sheaths usually with well-developed pointed auricles; glumes of the middle spikelet with conspicuous marginal hairs more than 0.5 mm   7.
6. Leaf sheaths usually without or with small rounded auricles; glumes of middle spikelet with only minute projections   9.
7. Middle spikelet with a stalk less than 0.6 mm; palea of middle spikelet longer than those of lateral spikelets   **3(c). murinum** subsp. **murinum**
7. Middle spikelet with a stalk 0.6–1.5 mm; palea of middle spikelet shorter than those of lateral spikelets   8.
8. Leaves usually glaucous; anthers of middle spikelets usually blackish and less than 0.6 mm, less than one-third as long as those of the lateral spikelets   **3(a). murinum** subsp. **glaucum**
8. Leaves not glaucous; anthers of middle spikelets usually yellowish, more than 0.6 mm, half as long as to equalling those of the lateral spikelets   **3(b). murinum** subsp. **leporinum**
9. Lateral spikelets distinctly stalked, the stalk about 1 mm and about as long as the stalk of the lateral

spikelets; the longest awn of the triplet of spikelets usually less than 10 mm   10.

9. Lateral spikelets sessile or nearly so, the stalk up to 0.5 mm and about as long as stalk of the lateral spikelets; the longest awn of the triplet of spikelets usually more than 10 mm   11.

10. Lemmas of lateral spikelets markedly acuminate at apex or with the awn less than 2 mm, the whole including the awn 2.8–6.0 mm   **4. pusillum**

10. Lemmas of lateral spikelets obtuse to acute at apex, 1.7–3.3 mm   **5. euclaston**

11. Lower leaf-sheaths glabrous to hairy, the hairs less than 0.2 mm; glumes of lateral spikelets very unequal, the inner with a more or less winged basal part 0.7–1.2 mm wide   **9. marinum**

11. Lower leaf-sheaths hairy with the hairs less than 0.5 mm; glumes of lateral spikelets slightly unequal, the inner with flattened basal part 0.3–0.7 mm wide   **10. geniculatum**

1. H. vulgare L.   Six-rowed Barley

Annual herb. Culms up to 100 cm, solitary or loosely fasciculate, erect, smooth, glabrous. *Leaves* up to 60 cm × 10–15 mm, flat, linear, acute at apex, glabrous, scaberulous on both surfaces, scabrid on the margins; sheaths falcate-auriculate, rather loose, striate, smooth, glabrous, the lower scarious, very loose; ligule very short, membranous. *Inflorescence* a spike 4- to 6-sided, 6–10 cm (excluding awns); rhachis tough, not articulated, flattened, ciliate on the margins. *Spikelets* in threes at each node of the rhachis, all bisexual, 1-flowered, with the rhachillas produced as a glabrous or hairy filament. *Glumes* and awn 20–30 mm, very narrow, hairy on the dorsal surface, attenuate into a scabrid awn. *Lemma* papyraceous or chartaceous, broadly elliptical, acute at apex, distinctly 5-veined towards the tip, drawn out above into a scabrid awn up to 18 cm. *Palea* narrowly ovate, as long as the lemma, of the same texture, 2-keeled. *Caryopsis* elliptical, hairy at apex. *Flowers* 5–6. $2n = 14$.

Introduced. A barley now rarely cultivated, but a grain-alien casual. A rare relic in waste places, waysides and field borders. Scattered throughout the British Isles. Cultivated throughout Europe and in most temperate regions of the world, or on mountains in the tropics.

2. H. distichon L.   Two-rowed Barley

Annual herb. Culms up to 90 cm, solitary or loosely fasciculate, smooth, glabrous, erect, green. *Leaves* up to 60 cm × 14–15 mm, acuminate at apex, flat, scaberulous on both surfaces, glabrous, very scabrid on the margins; upper leaf-sheaths somewhat loose, falcately auriculate at the mouth, smooth, glabrous, the lower very loose, scarious, disintergrating, slipping from the stems; ligule membranous, very short, truncate at apex. *Inflorescence* a spike, erect or somewhat curved, 6–12 cm excluding the awns, laterally compressed; rhachis tough, briefly ciliate on the margins. *Spikelets* in threes at the nodes, the middle fertile and sessile, the lateral pedicillate and sterile. *Glumes of fertile spikelets* with awns 9–10 mm, equal, narrowly elliptical, glabrous or slightly hairy,

prolonged into an awn; lemma about 10 mm, coriaceous, elliptical or oblong-elliptical, acute at apex, 5-veined, glabrous, scaberulous on the veins towards the tip, awned, the awns up to 12 cm, flattened, scabrid, caryopsis falling easily from the glumes. *Glumes of sterile spikelets* similar to those of fertile spikelets; lemma 8–9 mm, narrowly oblong, obtuse at apex, glabrous, flat, more or less awnless; palea shorter than the lemma. *Caryopsis* elliptical, hairy at apex. *Flowers* 5–6. $2n = 14$.

Introduced. The common cultivated barley. Common relic in waste places, fields and waysides where it sometimes persists. Throughout the British Isles. Cultivated in most temperate regions of the world, and has been grown since prehistoric times.

3. H. murinum L.   Wall Barley

Annual herb. Culms 6–60 cm, loosely tufted or solitary, erect or spreading, slender to somewhat stout, smooth, with 3–5 nodes. *Leaves* 2–20 cm × 2–8 mm, light green or glaucous, finely pointed at apex, with narrow, spreading, long-pointed, overlapping auricles at the base, rather weak, loosely hairy or glabrous; sheaths rounded on the back, the lower usually hairy, the upper smooth, slightly inflated; ligules up to 1 mm, membranous. *Inflorescence* a spike 4–12 cm × 10–30 mm, erect or inclined, dense, compressed, bristly, with the awns erect or slightly spreading, breaking up at maturity beneath each cluster of 3 spikelets, green or tinged with purple. *Spikelets* 1-flowered, in threes, the middle one bisexual and sessile, the two lateral male or sterile, on short stalks, the 3 falling together. *Lateral spikelets* with glumes bristle-like and long-awned, including the fine, stiff awn 16–30 mm, slightly dissimilar, the upper very slightly wider near the base where it is fringed with short hairs; lemma 7–11 mm, lanceolate, terminated by an awn 10–40(–50) mm. *Middle spikelet* with glumes in front of the lemma, bristle-like and long-awned, including the awn up to 26 mm, fringed with hairs in the lower part; lemma 7–12 mm, lanceolate, broadly rounded on the back, 5-veined, rough towards the tip, tipped with a stiff awn 18–50 mm; palea as long as the lemma. *Anthers* 0.2–1.5 mm. *Caryopsis* narrowly ellipsoid, hairy at the top, tightly enclosed by the hardened lemma and palea. *Flowers* 5–8.

(a) Subsp. glaucum (Steud.) Tzvelev
H. glaucum Steud.
Leaves usually glaucous. Prolongation of the rhachilla of the lateral spikelets stout and orange-brown. *Middle spikelet* with a stalk 0.6–1.5 mm. *Palea* of middle spikelets 0.7–0.8 times as long as palea of lateral spikelets. *Anthers* of middle spikelet 0.2–0.6 mm. $2n = 14$.

(b) Subsp. leporinum (Link) Arcangeli
H. leporinum Link
Leaves green. Prolongation of the rhachilla of the lateral spikelets slender and green. *Middle spikelet* with a stalk 0.6–1.5 mm. *Palea* of middle spikelet 0.7–0.9 times as long as palea of lateral spikelet. *Anthers* of middle spikelet 0.7–1.4 mm. $2n = 28$.

(c) Subsp. **murinum**

Leaves green. Prolongation of rhachilla of lateral spikelets slender and green. *Middle spikelet* with a stalk less than 0.6 mm. *Palea* of middle spikelet 1.0–1.4 times as long as palea of lateral spikelet. *Anthers* of middle spikelet 0.7–1.4 mm. $2n = 28$.

The species occurs in much of Europe, but is absent from most of Russia; North Africa. Subsp. *murinum* is native in waste and rough ground and open areas in rough grassland. It is common in central, south and east England and the Channel Islands, becoming more scattered to the north-east and south-west of Scotland and north Wales. In Ireland it is local, mainly in the south and east. It occurs almost throughout the range of the species. Subsp. *glaucum* is introduced and is a frequent casual of waste ground, sometimes becoming naturalised for a short time. It is native of the Mediterranean region and south-west Europe. Subsp. *leporinum* is found in similar places to subsp. *glaucum* and is also introduced. It is native of southern Europe.

4. H. pusillum Nutt.   Little Barley

Annual herb. Culms up to 45 cm, erect or geniculate, striate, glabrous. *Leaves* up to 6 cm × 1.5–2.5 mm, greyish-green; erect, long-linear, narrowed to a more or less acute apex, with minute teeth on the veins and margins; sheaths shortly hairy, without auricles; ligules very short, membranous. *Inflorescence* an erect spike 2–7 × 1.0–1.5 cm, linear-oblong, rhachis hairy. *Spikelets* in threes at each node, the middle one bisexual and sessile, the lateral male or sterile, considerably reduced and stalked, the 3 falling together. *Lateral spikelets* with glumes slightly unequal, the lower very narrow and attenuate into an awn which together are 10–16 mm, the inner of similar length but narrowly elliptical and dilated above the base; lemmas including awns 2.8–6.0 mm, lanceolate, acuminate at apex, the awns 1.0–1.5 mm; surface, margins and awns with minute teeth. *Middle spikelet* with both glumes dilated near the base, narrowly elliptical and narrowed into a slender awn as in the latter; lemmas with awns 5–10 mm; margins, surface and awns with minute teeth. *Anthers* about 0.5 mm. *Caryopsis* glabrous at the top, tightly enclosed by the hardened lemma and palea. *Flowers* 8–10. $2n = 14, 28$.

Introduced. Rather frequent wool alien on tips and waste ground in fields. Scattered localities in England. Native of North America.

5. H. euclaston Steud.   Argentine Barley

Annual herb. Culms up to 45 cm, ascending, erect or geniculate, striate, glabrous. *Leaves* up to 45 cm × 1.5–2.0 mm, greyish-green, narrowly linear, narrowed to a more or less acute apex, rigid, with minute teeth on the veins and margin; sheaths hairy, without auricles; ligule short, rounded at apex, membranous. *Inflorescence* a spike up to 3.8 cm, linear-oblong, compressed, rhachis hairy. *Spikelets* in threes at each node, the middle one bisexual and sessile, the lateral male or sterile, considerably reduced, the three falling together. *Lateral spikelets* with glumes slightly unequal, the lower very narrow and

attenuate into an awn which together are 10–16 mm, the inner of similar length but narrowly elliptical and dilated above the base; lemmas including awns if present 1.7–3.3 mm, lanceolate, obtuse to acute at apex; margins, surface and awns with minute teeth. *Anthers* about 0.5 mm. *Caryopsis* glabrous at top, tightly enclosed by the hardened lemma and palea. *Flowers* 8–10. $2n = 14, 28$.

Introduced. Wool alien on tips and waste ground and in fields. Scattered localities in England, but relative abundance not known owing to confusion with *H. pusillum*. Native of South America.

6. H. jubatum L.   Foxtail Barley
Critesium jubatum (L.) Nevski

Tufted, short-lived, fragile *perennial herb. Culms* 20–60 cm, erect or geniculate-ascending, striate, glabrous. *Leaves* up to 15 cm × 2–5 mm, long-linear, gradually narrowed to a point at apex, scabridulous above, hairy beneath; sheaths more or less patent-hairy; ligule up to 1 mm, membranous. *Inflorescence* a dense, linear, nodding spike 3–8(–11) cm; rhachis shortly hairy; spikelets disarticulating readily at maturity. *Spikelets* 1-flowered, in threes alternating on opposite sides of the rhachis, the middle one bisexual and sessile, the 2 lateral ones sterile, on short curved stalks, the three falling together. *Lateral spikelets* with glumes more than 30 mm, awn-like from base to apex, the spikelets extremely reduced and usually just an awnlike outgrowth. *Middle spikelet* with glumes awnlike and (5–)20–60 mm, the lemma (5–)6–8 mm, elliptical, with an awn (2–)5–10 cm nearly as long as the glumes, ultimately more or less divaricate and 5-veined. *Anthers* 1.0–1.5 mm. *Caryopsis* elliptical, hairy at the tip, tightly enclosed in the hardened lemma and palea. *Flowers* 5–8. $2n = 14, 28, 42$.

Introduced. An alien from wool, birdseed, and grass-seed, casual in waste places and naturalised especially along main roads, particularly those salted in winter. Frequent in eastern Britain from Kent to central Scotland, a scattered casual or garden outcast elsewhere. Native of North America.

7. H. pubiflorum Hook. fil.   Antarctic Barley

Tufted *perennial herb* with tillers at time of flowering. *Culms* up to 50 cm, erect, glabrous, striate. *Leaves* 4–20 × 1.5–2.0(–3.0) mm, long-linear, gradually narrowed to an acute apex, minutely hairy; sheaths glabrous or shortly hairy; ligule up to 1 mm, membranous. *Inflorescence* a linear to ovoid spike, 1–5 cm; rhachis shortly hairy; spikelets disarticulating readily at maturity. *Spikelets* 1-flowered, in threes alternating on opposite sides of the rhachis, the 2 lateral spikelets sterile and much reduced, often to a single awn-like lemma, with short pedicels, the middle spikelet sessile, fertile and bisexual, the 3 spikelets falling together. *Lateral spikelets* with glumes up to 30 mm, awn-like from base to apex, with short, soft hairs in the upper part. *Middle spikelet* with glumes 30–80 mm, awn-like from base to apex and hairy in the upper part; lemmas with an awn 10–16 mm, strongly divergent at maturity. *Anthers*

1.0–2.0 mm. *Caryopsis* ellipsoid, enclosed in the hardened lemma and palea. *Flowers* 8–10. $2n = 14$.

Introduced. Wool alien found on tips, waste ground and in fields. Scattered records in England; possibly overlooked for *H. jubatum*. Native of South America.

8. H. secalinum Schreb.   Meadow Barley
H. pratense Huds.; *H. nodosum* auct.

Tufted *perennial herb*. *Culms* 20–80 cm, erect or bent at the base, slender, 3- to 5-noded, smooth. *Leaves* up to 15 cm × 2–6 mm, green or greyish-green, linear, finely pointed at apex, with very short, spreading auricles at the base, loosely short hairy or glabrous, rough or smooth beneath; sheaths rounded on the back, the lower softly hairy, the upper glabrous and smooth; ligules less than 1.0 mm, membranous, not embracing the stem. *Inflorescence* a spike, erect or inclined, dense, 20–80 × 7–15 mm, with awns erect or slightly spreading, breaking up at maturity beneath each cluster of spikelets. *Spikelets* 1-flowered, in threes alternating on opposite sides of the rhachis, the central one bisexual and stalkless, the 2 lateral male or sterile and often much reduced in size, on very short stalks, the 3 falling together. *Lateral spikelets* with *glumes* similar, bristle-like and long-awned, with bristles in the upper part, including awn up to 14 mm; *lemma* 4–6 mm, very narrow to lanceolate, tipped with a fine awn up to 3 mm which is minutely hairy or rough in the upper part. *Middle spikelet* with *glumes* like those of the lateral spikelets, placed in front of the lemma, with minute bristles in the upper part; *lemma* 6–9 mm, broadly lanceolate, 5-veined, smooth on the back, tipped with a fine, stiff, erect awn 6–12 mm; *palea* narrowly elliptical, as long as the lemma. *Anthers* 3–4 mm. *Caryopsis* oblong, hairy at the tip, tightly enclosed in the hardened lemma and palea. *Flowers* 6–7. $2n = 14, 28$.

Native. Coastal and inland meadows, pastures and waysides, especially on moist heavy soils. Frequent and often locally abundant in the southern part of England, less common to rare in the north and west, rare and scattered in the Channel Islands and Ireland. South Sweden to central Spain, Sicily and Bulgaria; North Africa.

9. H. marinum Huds.   Sea Barley
H. maritimum Stokes

Annual herb. *Culms* 10–40 cm, loosely tufted or solitary, erect or spreading from a bent base, slender, stiff, unbranched, 3- to 4-noded, smooth. *Leaves* 15–80 × 1.0–3.5 mm, bluish-green, linear, tapering to a fine point at apex, often with obscure auricles at the base, flat, minutely hairy or glabrous; sheaths rounded on the back, smooth; ligules up to 1 mm, membranous. *Inflorescence* a spike, 20–60 × 15–30 mm, green or purplish, stiff, dense, oblong to ovate, bristly, with the awns at first erect, later spreading, breaking up at maturity beneath each cluster of spikelets. *Spikelets* 1-flowered, in threes alternating on opposite sides of the rhachis, with the middle one bisexual and stalkless, the 2 lateral sterile and on very short stalks, the 3 falling together. *Lateral spikelets* with *glumes* dissimilar, rough, becoming very

rigid, the lower bristle-like and long-awned, together 8–26 mm, the upper 4–6 mm, broadly winged on one side with an awn 10–22 mm; *lemma* 3–5 mm, tipped with a straight awn 3–5 mm. *Middle spikelet* with *glumes* 10–24 mm including the fine straight awn; bristle-like throughout; *lemma* 6–8 mm, narrowly ovate, rounded on the back, smooth, 5-veined, tipped with awn up to 24 mm; *palea* narrowly lanceolate, as long as the lemma. *Anthers* 1.3–1.5 mm. *Caryopsis* narrowly ellipsoid, hairy at the tip, tightly enclosed by the hardened lemma and palea. *Flowers* 6. $2n = 14$.

Native. Coastal and tidal river marshes and ditchbanks, bases of tidal defences and on disturbed ground on the landward side of low walls by estuaries. Locally common in east and south Britain; north to the Cheviots; a rare casual elsewhere. Southern Europe and the coasts of western Europe, north to England.

10. H. geniculatum All.   Mediterranean Barley
H. hystrix Roth; *H. marinum* subsp. *gussoneanum* (Parl.) Thell.; *H. gussoneanum* Parl.; *H. winkleri* Hack.

Loosely tufted *annual herb*. *Culms* up to 40 cm, solitary or several, erect or geniculate at the base, rather slender, simple, smooth, glabrous. *Leaves* up to 6 cm × 2–3 mm, long-linear, acuminate at apex, flat, hairy or glabrous, but the margins scabrid; upper sheaths slightly inflated, the uppermost sometimes embracing the inflorescence and smooth and glabrous, the lower more or less scarious and glabrous or more or less covered with minute hairs; ligules less than 1 mm, glabrous. *Inflorescence* a dense spike 20–40 × 10–15 mm, green, stiff, cylindrical, the awns erect or slightly spreading, breaking up at maturity beneath each cluster of spikelets. *Spikelets* 1-flowered, in threes alternating on opposite sides of the rhachis, with the middle one bisexual and stalkless, the 3 falling together. *Lateral spikelets* often reduced to a series of slightly unequal, awn-like scales, the inner with a flattened basal part 0.3–0.7 mm wide, as wide as the basal part of the outer ones. *Middle spikelet* with setaceous, scabrid glumes, with awns up to 20 mm; *lemma* 6–7 mm, elliptical, acute at apex, 5-veined, papyraceous, sessile, with an awn 15–20 mm; *palea* as long as lemma, 2-veined, 2-keeled. *Anthers* about 1 mm. *Caryopsis* elliptical, hairy at apex. *Flowers* 5–6. $2n = 14, 28$.

Introduced. Wool alien on tips and waste ground and in fields; naturalised in Guernsey during the nineteenth century. Of scattered occurrence in Britain. Central and south Europe, eastern Mediterranean region eastwards to central Asia; tropical Africa.

65. Secale L.

Annual tufted herbs. *Leaves* flat; ligule short and truncate. *Inflorescence* usually a simple, laterally compressed, dense, distichous spike. *Spikelets* solitary at each node of the rhachis, sessile, with 2, rarely 3 bisexual flowers. *Glumes* linear to narrowly oblong-elliptical, acute at apex, 1-veined, awnless or shortly awned, keeled, scabrid. *Lemmas* lanceolate, acuminate at apex, keeled, 5-veined, usually very long-awned. *Palea* hyaline, membranous, 2-keeled. *Spikelets* disarticulating at maturity below each caryopsis, leaving the glumes, lemma and palea on the rhachis.

Four species. Eastern Europe to central Asia; also Spain and South Africa.

1. S. cereale L.   Rye

Annual herb. Culms up to 150 cm, solitary or loosely fasciculate, erect or very shortly geniculate at the base, very shortly hairy below the spike, otherwise smooth and glabrous. *Leaves* up to 30 cm × 2–10 mm, glaucous, flat, acuminate at apex, flaccid, scaberulous on the upper surface, smooth beneath, sometimes sparsely hairy on the upper surface, scabrid on the margins; sheaths closely appressed, smooth and glabrous, the lower becoming scarious; ligule very short, truncate at apex. *Spikes* up to 20 cm, laterally compressed; rhachis tough, flat, densely ciliate on the margins and sometimes on both surfaces. *Spikelets* compressed, 2-flowered. *Glumes* 6–13 mm, linear or narrowly oblong-elliptical, acute at apex, awnless or gradually passing into a short awn, 1-veined, keeled, scabrid. *Lemmas* 7–15 mm, firmly compressed, 5-veined, lanceolate, pectinate-spinulose on the keel, otherwise smooth and glabrous, passing gradually into a scabrid awn 2–5 cm. *Palea* rather shorter than the lemma, hyaline, 2-keeled, scabrid on the keels. *Caryopsis* free between lemma and palea, oblong, hairy at apex. *Flowers* 5–7. 2n = 14, 28.

Introduced. Formerly widely cultivated, now on a much smaller scale. Relic of cultivation, or on roadsides, waste places and tips. Frequent throughout much of the British Isles. Cultivated throughout much of Europe. It probably originated in Turkey as a weedy derivative of wild *S. montanum* Guss., infesting wheat and barley fields and eventually being itself taken into cultivation when agriculture spread northwards into colder climates.

65 × 66. × Triticosecale Wittons ex A. Camus

Secale cereale × Triticum

Gregory, R. S. (1987). *Triticale* breeding in F. G. H. Lupton, *Wheat breeding: its scientific basis*. London & New York.

This intergeneric hybrid resembles a vigorous Rye plant, but has a pendulous spike and long awns. Chromosomes from 28 to 70 have been recorded. It is a Man-made hybrid which is now being grown on a small scale, but may become more widespread.

66. Triticum L.

Annual herbs. Leaves flat; ligules membranous. *Inflorescence* a distichous spike. *Spikelets* 1 per node of the rhachis, each with 3–7(–9) flowers, the apical 2 or more sterile and reduced. *Glumes* equal, chartaceous, keeled, truncate to bifid at apex, apiculate to shortly awned. *Lemmas* 3-veined, keeled, truncate to bifid at apex, apiculate to very long-awned, coriaceous, scabrid. *Palea* membranous with 2 keels.

Triticum contains a polyploid series in which there are diploid, tetraploid and hexaploid representations, with chromosome numbers of 14, 28 and 42. Genome analysis of the karyotypes has shown that the diploids have one *Triticum* genome, the tetraploids have this genome and another derived from *Aegilops speltoides* Tausch, and the hexaploids have both these and a third derived from *Aegilops squarrosa* L.

Ten to twenty species in the eastern Mediterranean region to Iran.

Bor, N. L. in C. C. Townsend & E. Guest (1968). *Flora of Iraq.* 9: 194–208. Baghdad.
Percival, J. (1921). *The wheat plant.* London.
Peterson, R. R. (1965). *Wheat, botany, cultivation and utilization.* London & New York.

1. Glumes keeled in the upper three-quarters only
                                                        1. aestivum
1. Glumes strongly keeled throughout   **2.**
2. Spike slender, laterally compressed; glumes nearly as long as the lowest flower; endosperm flinty   **2. durum**
2. Spike stout, nearly square in section; glumes about two-thirds as long as the lowest flower; endosperm mealy   **3. turgidum**

1. T. aestivum L.   Bread Wheat
T. hybernum L.; *T. sativum* Lam.; *T. vulgare* Vill.

Annual herb. Culms 40–150 cm, thin-walled and hollow, rarely partially filled with pith, smooth and glabrous, sometimes pubescent at the nodes. *Leaves* up to 60 cm × 6–16 mm, linear, acuminate at apex, flat, scaberulous on the margins; sheaths membranous, usually glabrous, sometimes hairy; ligules truncate at apex, membranous. *Spike* 6–18 cm, quadrangular in section, lax to dense; rhachis tough, continuous, not bearded at the nodes, glabrous. *Spikelets* 10–15 mm × 9–18 mm, 3- to 6-flowered. *Glumes* about 10 mm, glabrous, hairy or villous, reddish, yellowish or pale, oblique, usually keeled in the upper three-quarters only, toothed at the tip. *Lemmas* rounded on the dorsal surface in the lower quarter, keeled above, awnless or with awns up to 16 cm. *Palea* 2-veined, 2-keeled, ciliate on the keels. *Caryopsis* free between the lemma and palea, oblong-elliptical, hairy at the apex; endosperm mealy to flinty. *Flowers* 5–7. 2n = 42.

Introduced. The common cultivated wheat. Common and often persisting as a relic in fields and on waysides, waste ground and tips. Throughout the British Isles. Cultivated for making bread throughout the temperate regions of the world. Perhaps originated in south-west Asia. It is an allohexaploid with one set of 14 chromosomes derived from *Aegilops speltoides* Tausch, one from *Aegilops tauschii* Cosson, and one from *T. monococcum* L.

2. T. durum Desf.   Pasta Wheat
Annual herb. Culms 70–140 cm, thick-walled and hollow, glabrous, smooth or rather scaberulous. *Leaves* up to 60 cm × 7–16 mm, linear, acuminate at apex, glabrous on the upper surface, puberulous on the lower, flat; sheaths not fused; ligules membranous. *Spike* 4–11 cm × 12–17 mm, quadrangular in section, dense; rhachis narrow, tough, sometimes brittle, ciliate on the margins, bearded below the insertion of the spikelets.

Spikelets 10–15 × 8–15 mm, 5- to 7-flowered, with 2–4 fertile flowers. *Glumes* 8–12 mm, coriaceous, firmly keeled from base to apex, toothed at the apex, hairy or glabrous, nearly as long as the lowest flower. *Lemmas* thin, pale, 9- to 15-veined, the lowest long-awned, the awns up to 20 cm, white, red or black. Palea 2-veined, 2-keeled, ciliate on the keels. *Caryopsis* free between lemma and palea, oblong-elliptical, hairy at the apex; endosperm flinty. *Flowers* 6–7. $2n = 28$.

Introduced. A rare casual, perhaps under-recorded. Widely distributed as a cultivated plant in the Mediterranean region eastwards to central Asia. Also cultivated in North, Central and South America. It is the wheat principally used for making pasta.

3. T. turgidum L.   Rivet Wheat

Annual herb. Stems 120–170 cm, thick-walled and more or less solid, smooth and glabrous. *Leaves* up to 45 cm × 10–20 mm, when young covered on both surfaces with soft white hairs which are velvety to the touch, flat; sheaths not fused; ligules membranous. *Spikes* 7–12 cm × 10–15 mm, stout, quadrangular or oblong in section, often compound and branched below, dense; rhachis tough, ciliate on the margins, bearded on the joints below the spikelets. *Spikelets* 10–13 × 8–15 mm, 5- to 7-flowered of which 3–5 are fertile. *Glumes* 8–11 mm, about two-thirds as long as the lowest flower, white, yellow, red or blackish, sometimes glaucous, glabrous or pubescent, broadly ovate, coriaceous, with a prominent 1-veined keel, terminating in an acute, curved, apical tooth. *Lemmas* rounded on the dorsal surface, with an awn 8–16 cm. *Palea* 2-veined, 2-keeled, ciliate on the keels. *Caryopsis* free between lemma and palea, oblong-elliptical, hairy at the apex; endosperm mealy. *Flowers* 6–7. $2n = 28$.

Introduced. Formerly a cereal crop, especially in north and west Britain and Ireland. Now rarely grown, and in such cases for animal feed. Occasional casual. Scattered in Britain and Ireland. Widely cultivated in southern Europe, possibly native in south-west Asia. As a crop it is said to have a high yield and be free from rust disease.

Subfamily 3. Arundinaoideae Tateoka

Annual or *perennial herbs.* Leaves without a false petiole or cross-veins; ligule a fringe of hairs; sheaths not fused. *Inflorescence* a panicle. *Spikelets* with (1–)2 to many flowers, with all flowers except often the distal ones bisexual or the lowest male or sterile. *Glumes* 2. *Lemmas* firm, 1- to 9-veined, sometimes bifid at apex, awnless or with a terminal awn. *Stamens* 3. *Stigmas* 2. *Lodicules* 2. *Embryo* arundinoid; photosynthesis C3-type only, with non-Kranz leaf anatomy; fusoid cells absent, microhairs present. *Chromosome base-number* 9 or 12.

Tribe 15. Arundineae Dumort.

Tribe *Cortaderieae* Zotov; Tribe *Danthonieae* Zotov; Tribe *Molinieae* Jirasek
As subfamily.

67. Danthonia DC. nom. conserv.

Densely tufted *perennial herb. Leaves* flat or inrolled; ligule a ring of hairs. *Inflorescence* a small panicle with rarely more than 12 spikelets. *Spikelets* with 2–6 flowers, all or all except the most apical bisexual. *Glumes* subequal, ovate, chartaceous, about equalling the spikelet, 5- to 7-veined. *Lemmas* coriaceous, 7- to 9-veined, minutely 3-toothed at apex, awnless, with a tuft of short hairs at the base and a fringe up to halfway on each side. *Palea* obtuse or 2-toothed at apex, with 2 keels near the margin. *Rhachilla* prolonged.

Twenty species in Europe and North and South America.

1. D. decumbens (L.) DC.   Heath Grass
Triodia decumbens (L.) P. Beauv.; *Sieglingia decumbens* (L.) Bernh.

Densely tufted *perennial herb.* Culms 10–60 cm, erect, spreading to almost prostrate, slender, stiff, 5- to 3-noded, smooth. *Leaves* 5–25 × 2–4 mm, green, linear, obtuse or abruptly pointed at apex, flat or inrolled, stiff, sparsely hairy or glabrous, rough upwards; sheaths rounded on the back, usually with fine patent hairs, bearded at the apex; ligules a fringe of dense, short hairs. *Panicles* narrow, compact or loose, 2–7 cm, containing 3–12 spikelets; branches erect or spreading, rough, with 1–3 spikelets. *Spikelets* 6–12 mm, elliptical or oblong, 2- to 6-flowered, breaking up at maturity above the glumes, purplish or green. *Glumes* persistent, rounded at base, keeled above, as long as the spikelet or nearly so, equal, or the lower slightly longer, lanceolate to narrowly ovate, obtuse to acute at apex, 5- to 7-veined, with smooth, thin, transparent sides. *Lemmas* closely overlapping, 5–7 mm, rounded and smooth on the back, broadly elliptical, shortly 3-toothed at the tip, becoming tough and rigid, densely short-bearded at the base, usually short-haired on the margins up to the middle, 7- to 9-veined. *Paleas* elliptical, with 2 short-haired keels. *Anthers* 0.2–0.4 mm, or about 2 mm. *Caryopsis* ellipsoid, tightly enclosed between the lemma and the palea. *Flowers* 6–8. Flowers often cleistogamous and sometimes solitary spikelets hidden in the basal leaf-sheaths. $2n = 24, 36$.

Native. Sandy and peaty soils, usually acid, but also mountain limestone, on moorland, heaths, mountains and the poorer acid grassland. Throughout the British Isles. Most of Europe; Caucasus; North Africa; North America; Madeira. Only on mountains in the south.

68. Rytidosperma Steud.

Densely tufted *perennial herbs. Leaves* closely inrolled; ligules of very short hairs. *Inflorescence* a rather compact to elongated panicle. *Spikelets* with 6–10 flowers, the lower ones bisexual, the upper male or sterile and reduced. *Glumes* lanceolate to narrowly ovate, with wide hyaline margins, slightly longer to slightly shorter than the rest of the spikelet (excluding awns). *Lemmas* 7-veined, with 2 long, acuminate lobes at apex tipped with straight awns, the one from the sinus between the lobes long and bent, with dense, white silky hairs at the base and middle reaching or nearly reaching the apex of the body of the lemma.

About 90 species, mainly in Australia, New Zealand and southern Africa, but also in Indonesia, Himalayas, Arabia, Ethiopia, Madagascar and Argentina.

1. R. racemosum (R. Br.) Connor & Edgar
Wallaby Grass
Danthonia racemosa R. Br.; *Danthonia racemosa* var. *biaristata* Benth.; *Danthonia penicillata* F. Muell.; *Danthonia penicillata* var. *racemosa* (R. Br.) Maiden & Betsche; *Notodanthonia racemosa* (R. Br.) Zotov

Densely tufted *perennial herb*. *Culms* up to 60 cm, erect, slender. *Leaves* 5–15 cm, linear, closely inrolled, attenuate towards the apex, sparsely to rarely densely hairy on the outer surface, the hairs with tubercular bases, minutely scabrous on the inner surface; ligules of hairs about 0.2 mm. *Inflorescence* a compact to elongated panicle 3–15 cm, linear to slightly branched in the lower half. *Spikelets* 7–16 mm excluding awns, with 6–10 flowers, the flowers cohering in a semi-cylindrical mass, the lower ones bisexual, the upper ones male or sterile and reduced. *Glumes* 7–15 mm, subequal to irregularly unequal, lanceolate to narrowly ovate, acute to erose-truncate at apex, with a wide hyaline margin, slightly shorter to slightly longer than the rest of the spikelet (excluding awns), 5- to 7-veined, glabrous or scabrid. *Lemmas* 5–11 mm, 7-veined, aristate or truncate or shortly, abruptly and obliquely acuminate at the apex and bilobed, the lobes long and tipped with a straight awn 2–9 mm at apex, and with a bent terminal awn 5–15 mm from the sinus between the lobes; with dense, white silky hairs at the base and middle reaching or nearly reaching the apex of the body of the lemma. *Stamens* 3. *Styles* 2, distinct; stigmas plumose. *Caryopsis* glabrous, loosely enclosed by, but free from lemma and palea. *Flowers* 6–8. $2n = 24$.

Two varieties are known. Var. **racemosum** has the lobes of the lemma long-aristate and 2 mm or more long. Var. **obtusatum** (F. Muell. ex Benth.) Connor (*Danthonia racemosa* var. *multiflora* Benth.; *Danthonia racemosa* var. *obtusata* F. Muell. ex Benth.) has the lemma lobes truncate or very shortly and obliquely pointed, the point less than 1 mm. These varieties do not seeem to have been considered when British plants are recorded.

Introduced. An occasional wool alien of fields and waste places and on tips. In scattered localities in England where it does not overwinter. Native of Australia and New Zealand.

69. Schismus P. Beauv.

Tufted *annual herbs*. *Leaves* flat or involute; ligule a row of hairs. *Inflorescence* a rather compact panicle. *Spikelets* with 5–10 flowers, the lower ones bisexual, the upper ones male or sterile and reduced, laterally compressed. *Glumes* subequal, lanceolate, with a wide hyaline margin, slightly longer than or slightly shorter than the rest of the spikelet, 5- to 7-veined. *Lemmas* 7- to 9-veined, deeply and acutely 2-lobed at apex, awnless or

more or less so, with short hairs at the base and long silky hairs on the back, not reaching the apex. *Palea* shorter than to as long as the lemma.

Five species in South Africa, Mediterranean region and Middle East.

Bor, N. L. in C. C. Townsend & E. Guest (1968). *Flora of Iraq* 9: 377–382. Baghdad.

1. Lowest lemma with obtuse to acute lobes one-sixth to one-quarter as long as the rest of the lemma; palea longer than the entire part of the lemma   **1. barbatus**
1. Lowest lemma with acuminate lobes one-third to half as long as the rest of the lemma; palea about as long as the entire part of the lemma   **2. arabicus**

1. S. barbatus (L.) Thell.   Kelch Grass
S. calycinus (L.) Duval-Jouve; *Festuca calycina* L.

Annual herb. *Culms* up to 15 cm, fascicled, erect or geniculately ascending, smooth, glabrous. *Leaves* 1.5–4.0 cm × 1–2 mm, acuminate at apex, flat or convolute, filiform, flexuous, hairy above towards the base, with a tuft of long hairs at the junction of the sheath and lamina; sheaths rather loose, broadly hyaline, particularly the lower on the margins, markedly striate on the rounded dorsal surface, smooth, glabrous; ligule a fringe of hairs. *Panicle* 1–4 cm, rather dense, elliptic-ovate or oblong in outline, shortly pedicellate; rhachis erect, smooth below, scabrid above; branches short, slender, scaberulous. *Spikelets* 5–7 mm, cuneate, 5- to 10-flowered, rhachilla disarticulating below each flower. *Glumes* usually less than 5 mm, lanceolate, acute at apex, glabrous. *Lemma* 2–3 mm, hairy near the margins, the lowest with acute to obtuse lobes one-sixth to one-quarter as long as the rest of the lemma; 9-veined. *Palea* longer than the entire part of the lemma. *Anthers* minute. *Caryopsis* elliptical, glabrous. *Flowers* 5–6. $2n = 12$.

Introduced. Characteristic wool alien in fields and waste land, and on tips. Of scattered occurrence in England. Native of the Mediterranean region eastwards to central Asia.

2. S. arabicus Nees   Arabian Kelch Grass

Annual herb. *Culms* up to 25 cm, very slender, densely fasciculate, erect, sometimes geniculately ascending or spreading, smooth, glabrous, shining, usually glaucous. *Leaves* 1.5–4.0 cm × 1.0–2.0 mm, usually glaucous, flat or convolute, filiform, flexuous, acuminate at apex, loosely covered with long, white hairs, rarely glabrous, scaberulous on the surfaces and scabrid on the margins; sheaths rather loose, broadly hyaline, particularly the lower on the margins, markedly striate on the rounded dorsal surface, smooth, glabrous; ligule a fringe of hairs. *Panicle* 1–5 × 1.0–1.5 cm, rather dense, elliptic-ovate or oblong in outline, of shortly pedicellate spikelets; rhachis erect, smooth below, scabrid above; branches short, slender, ascending, scaberulous. *Spikelets* 5–7 mm, cuneate, 5- to 10-flowered; rhachilla disarticulating below each flower. *Glumes* as long as the spikelets, persistent, green, lanceolate or oblong, acute at apex, more or less equal in length, with broad, scarious margins, glabrous, 3- to 7-veined. *Lower lemmas* 2–3 mm, elliptical or obovate when flattened, cleft at the apex into 2 acuminate lobes, 7- to 9-veined, hairy in the lower half, sometimes with an awn in the sinus. *Palea* shorter than or about as long as the lemma, just reaching the base of the cleft. *Anthers* minute, 0.3–0.4 mm. *Caryopsis* elliptical, glabrous. *Flowers* 5–6. 2n= 12.

Introduced. Wool alien on tips, fields and waste land. Rather rare in England. Native of east Mediterranean and west Asia.

70. Cortaderia Stapf nom. cons.

Densely tufted dioecious or gymnodioecious or monoecious *perennial herbs. Leaves* flat, green or glaucous, with sharply serrated margins; ligules a line of hairs. *Inflorescence* a very large, spreading panicle. *Spikelets* with 2–7 flowers, silvery or sometimes pinkish. *Glumes* slightly unequal, lanceolate, hyaline, 1-veined, at least the upper more or less as long as to longer than the rest of the spikelet and with a long terminal awn. *Lemmas* 3- to 5-veined, acuminate or bifid and long-awned at apex, with a tuft of long, fine hairs at the base reaching more or less to the apex of the body of the lemmas. *Rhachilla* villous, disarticulating above the glumes and just above each flower. *Lodicules* hairy. *Stamens* 3. Styles free to their bases; stigmas 2, brown.

Twenty-four species, mainly in South America, but four species in New Zealand and one in New Guinea.

1. Plants dioecious or gymnodioecious; spikelets all
   unisexual; lemma acuminate   **1. selloana**
1. Plants monoecious or female; spikelets bisexual or
   unisexual, lemma bifid, the lobes bristle-like   **2. richardii**

1. C. selloana (Schult. & Schult. fil.) Asch. & Graeb.
                                    Pampas Grass
Arundo selloana Schult. & Schult. fil.

Densely tufted dioecious or gymnodioecious, *perennial herb* forming tussocks often more than one metre across. *Culms* up to 3 m, erect. *Leaves* 1–3 m, glaucous, flat, hard, long-linear, acute at apex, with fiercely cutting serrated edges; ligule a line of hairs. *Inflorescence* a very large spreading panicle (30–)50–100 cm; branches erecto-patent in the male, patent in female plants, a silky-hairy mass when fruiting. *Spikelets* 12–16 mm, including awns, laterally compressed, silvery and sometimes pinkish, with 2–7 flowers. *Rhachilla* villous. *Glumes* hyaline, lanceolate, slightly unequal, 1-veined, at least the upper more or less as long as or longer than the rest of the spikelet and with a long, terminal awn. *Lemmas* 3- to 5-veined, membranous, acuminate and long-awned at apex, with a tuft of long, fine hairs at the base reaching more or less to the apex of the body of the lemma. *Palea* hairy. *Caryopsis* glabrous, free from both lemma and palea. *Flowers* 9–10.

Introduced. Grown for ornament and becoming naturalised where thrown out or planted in rough ground, waysides, old gardens and maritime cliffs and dunes. Scattered in south-west England and South Wales the Channel Islands and Co. Wexford. Casual on dumps elsewhere. Native of South America.

2. C. richardii (Endl.) Zotov   Toetoe
Arundo richardii Endl.

Densely tufted, monoecious or female *perennial herb. Culms* up to 3(–6) m, rigid, erect. Leaves numerous, 0.6–1.2 m, green, flat, hard, long-linear, curved, with cutting, serrated edges; ligule a line of hairs. *Inflorescence* a panicle, male and female or female, 30–60 cm, dense, much branched, silvery, white or yellowish-white, the branches drooping, a silky mass when fruiting. *Spikelets* laterally compressed, with 2–7 flowers. *Rhachilla* villous. *Glumes* hyaline, slightly unequal, lanceolate, 1-veined. *Lemmas* 3- to 5-veined, bifid at apex, the lobes bristle-like, with a tuft of hairs at the base. *Caryopsis* glabrous, free from both lemma and palea. *Flowers* 9–10.

Introduced. Naturalised near Aberystwyth. Native of New Zealand.

71. Molinia Schrank

Densely tufted *perennial herb. Leaves* flat; ligule a dense fringe of short hairs. *Inflorescence* a more or less diffuse to a more or less contracted panicle. *Spikelets* with 1–4 flowers, all except the apical bisexual, laterally compressed. *Glumes* slightly unequal, ovate, membranous, 1- to 3-veined, much shorter than the rest of the spikelet. *Lemma* membranous, 3- to 5-veined, acute to obtuse at apex, awnless, glabrous.

Two to four species in Europe, western Russia, Turkey, China and Japan.

Trist, P. J. O. & Sell, P. D. (1988). Two subspecies of *Molinia caerulea* (L.) Moench. *Watsonia* 17: 153–157.

1. Culms up to 65 cm; panicles up to 30 cm, narrow, with
   branches up to 5 cm; spikelets 3.0–5.5 mm; lemmas
   3–4 mm   **1(a). caerulea** subsp. **caerulea**
1. Culms usually 65–162 cm; panicles usually 30–65 cm;
   with branches of very uneven length and up to 15 cm;
   spikelets (3.0–) 4.0–7.5 mm; lemmas 3–6 mm
                             1(b). caerulea subsp. **arundinacea**

1. M. caerulea (L.) Moench   Purple Moor Grass
Aira caerulea L.

Densely tufted, tough-rooted *perennial herb*, often forming large tussocks. *Culms* 15–130(–162) cm, erect, slender to somewhat stout, stiff, smooth, with one node towards the base, the basal internode about 5 cm and club-shaped. *Leaves* 10–45 cm × 3–10 mm, green, linear, long-tapering to a fine point at apex, flat, slightly hairy on the upper surface or glabrous, minutely rough on the margins, falling from the sheaths in winter; sheaths rounded on the backs, smooth, hairy at the top; ligule a dense fringe of short hairs. *Panicles* 5–65 × 1–10 cm, very variable, dark to light purple, brownish, yellowish or green, ranging from very dense and spike-like to open and very loose; branches slender, smooth or minutely rough; pedicels variable in length. *Spikelets* 3.0–7.5 mm, lanceolate to oblong, loosely 1- to 4-flowered, breaking up at maturity beneath each lemma, with a slender, rough rhachis. *Glumes* persistent, unequal to nearly equal, lanceolate to ovate or oblong, shorter than the lemmas, membranous, the lower 1.5–3.0 mm and 0- to 1-veined, the upper 2.5–4.0 mm and 1- to 3-veined. *Lemmas* 3–6 mm, narrowly lanceolate to narrowly oblong in side view, pointed or obtuse at apex, membranous, rounded on the back, 3- to 5-veined, firm, smooth. *Paleas* elliptic-oblong, with minutely rough keels. *Anthers* 1.5–3.0 mm, purple. *Stigmas* purple. *Caryopsis* ellipsoid, enclosed by the hardened lemma and palea. *Flowers* 7–9.

(a) Subsp. caerulea
Culms up to 65 cm. *Panicle* up to 30 cm, narrow, with branches up to 5 cm. *Spikelets* 3.0–5.5 mm. *Lemmas* 3–4 mm. 2*n* = 36.

(b) Subsp. arundinacea (Schrank) K. Rich.
M. arundinacea Schrank; *M. litoralis* Host; *M. altissima* Link; *M. caerulea* subsp. *altissima* (Link) Domin
Culms usually 65–162 cm. *Panicles* usually 30–65 cm, with branches of very uneven length and up to 15 cm. *Spikelets* (3.0–)4.0–7.5 mm. *Lemmas* 3–6 mm. 2*n* = 18.

Native. Subsp. *caerulea* occurs on heaths, moors, bogs, fens, mountain grassland, cliffs and lake shores, where it is at least seasonally wet. Common throughout the British Isles. Subsp. *arundinacea* occurs in fens, fen-scrub and by rivers and canals. Scattered in central and southern Britain north to south and west Scotland and in southern Ireland. Both subspecies occur in most of Europe, but subsp. *arundinacea* is less common. The species also occurs in North Africa, Caucasus and Siberia and is introduced in North America.

72. Phragmites Adans.

Tichoon Roth; *Xenochloa* Roem. & Schult.

Extensively rhizomatous *perennial herb*. *Leaves* flat; ligule a line of hairs. *Inflorescence* a large, spreading panicle. *Spikelets* with 2–6 or more flowers, the lowest male or sterile, the rest bisexual. *Glumes* unequal, narrowly elliptic-ovate, 3- to 5-veined, much shorter than the rest of the spikelet. *Lemma* lanceolate, acute to acuminate at apex, with 1–3 veins, awnless. *Rachilla* segments with long, white, silky hairs becoming very conspicuous in fruit.

Three of four species, cosmopolitan.

Clayton, W. D. (1968). The correct name of the common reed. *Taxon* **17**: 168–169.
Haslam, S. M. (1972). *Phragmites communis* Trin. in Biological flora of the British Isles. *Jour. Ecol.* **60**: 585–610.

1. P. australis (Cav.) Trin. ex Steud.   Common Reed
Arundo australis Cav.; *P. communis* Trin.; *Arundo phragmites* L.

Robust *perennial herb*, spreading by stout, creeping rhizomes and stolons. *Culms* 1.5–3.0 m, erect, rigid, stout, closely sheathed, usually unbranched, smooth, with many nodes. *Leaves* 20–60 cm × 10–30 mm, greyish-green, flat, linear, long-tapering to a very fine curved or flexuous tip, contracted at the base, smooth, tough, closely veined, ultimately falling from the sheaths;

sheaths rounded on the back, overlapping; ligule a dense fringe of short hairs. *Inflorescence* a panicle 15–40 cm, purplish or brownish, erect or finally nodding, loose to dense, soft, much branched, with smooth or nearly smooth branches, hairy at intervals; pedicels short. *Spikelets* 10–16 mm, lanceolate, 2- to 6-flowered, with the lowest flower male and the others bisexual, the hairy rhachis breaking up at maturity beneath each fertile lemma. *Glumes* persistent, membranous, smooth, 3- to 5-veined, the lower half to two-thirds the length of the upper, the upper about half the length of the lowest lemma. *Lowest lemma* 9–13 mm, narrowly lanceolate, pointed at apex, membranous, mostly 3-veined, smooth. *Fertile lemma* narrower, more finely pointed, thinner, 1- to 3-veined, surrounded by white, silky hairs up to 9 mm from the rhachilla segments. *Paleas* 3–4 mm, narrowly elliptical. *Anthers* 1.5–2.0 mm. *Caryopsis* ellipsoid, enclosed by the thin lemma and palea. *Flowers* 8–10. 2*n* = 48, 96.

Native. Covering large areas of swamp and fen and frequent in the shallow water of ditches, rivers and lakes. Common in suitable places throughout the British Isles. Cosmopolitan, except in a few tropical regions. The culms are used for thatching buildings.

Subfamily 4. Chloridoideae Rouy

Annual to *perennial herbs*. *Leaves* without false petiole or cross-veins; ligules usually a fringe of hairs, sometimes membranous or a membrane fringed with hairs; sheaths not fused. *Inflorescence* usually an umbel or a raceme of spikes or racemes, sometimes a panicle. *Spikelets* with 1 to many flowers, with all flowers except often the distal ones bisexual. *Glumes* 2. *Lemmas* membranous to firm, 1- to 3-veined, awnless or often awned. *Stamens* 3. *Stigmas* 2. *Lodicules* absent or 2. *Embryo* chloridoid or rarely arundinoid; photosynthesis C3-plus C4-type, with Kranz leaf anatomy; fusoid cells absent; microhairs present. *Chromosome base-number* more than 7.

Tribe 16. Eragrostideae Stapf

Tribe *Sporoboleae* Stapf

Annual or *perennial herbs* with or without rhizomes. *Leaves* with ligules membranous or a fringe of hairs. *Inflorescence* a diffuse to contracted panicle, an umbel of spikes or a raceme of racemes. *Spikelets* with 3 to many flowers, or with one flower, all except 1–2 apical bisexual. *Glumes* 2, much shorter than the rest of the spikelet. *Lemmas* (1-) to 3-veined, sometimes with extra veins close to the midrib, usually keeled, awnless or rarely shortly awned.

73. Leptochloa P. Beauv.

Diplachne P. Beauv.

Perennial herbs with rhizomes, but often behaving as annuals. *Leaves* usually inrolled. Inflorescence a loose, long panicle of racemes, the spikelets well spaced out, the rhachis ending in a spikelet. *Spikelets* with 6–10(–14) flowers. *Glumes* unequal, 1-veined. *Lemmas* 3-veined, slightly keeled, with a prominent midrib and submarginal laterals, with long, silky hairs at the base and on the margins and at the base of the dorsal midline, bifid or shouldered or with 1 to few teeth on either side at the apex, with a very short apical awn. *Spikelets* disarticulating between the flowers.

About 40 species found throughout the tropics and in warm parts of America and Australia.

1. Lemmas with the apical awn shorter than the lateral teeth **1. muelleri**
1. Lemmas with the apical awn or mucro longer than the lateral teeth **2.**
2. Lemma with 2 acute subapical teeth; anthers more than 1 mm **2. fusca**
2. Lemma with 1–2 minute subapical teeth or merely a shoulder; anthers less than 0.5 mm **3. uninervia**

1. L. muelleri (Benth.) Stace   Mueller's Beetle Grass
Diplachne muelleri Benth.

Perennial, sometimes *annual herb* with or without rhizomes. *Culms* up to 1.5 m. Leaves 25–55 cm × 1–8 mm, linear, narrowed to an acute apex, usually inrolled, rarely flat, minutely scabrous; sheaths loose, glabrous; ligule 3–8 mm, membranous, acute or laciniate at apex. *Inflorescence* an open panicle of racemes 10–40 cm; branches erect, initially compact, later spreading with unbranched racemes, the rhachis ending in a spikelet. *Spikelets* pale green, narrowly elliptical, with 8–12 flowers. *Glumes* unequal, the lower narrowly ovate and acute at apex, the upper narrowly oblong, obtuse to acute at apex and 1-veined. *Lemmas* with 2 rounded subapical teeth and a minute awn between and shorter than them, 3-veined, keeled, hairy on the veins. *Palea* hairy on the veins. *Caryopsis* narrowly obovoid, flattened. *Flowers* 8–9.

Introduced. Recorded as an occasional casual in waste places. Native of Australia.

2. L. fusca (L.) Kunth.   Brown Beetle Grass
Festuca fusca L.; *Diplachne fusca* (L.) P. Beauv. ex Roem. & Schult.; *Diplachne repatrix* (L.) Druce

Perennial or annual herb with or without rhizomes, often rooting and branching from the lower nodes. *Culms* up to 1.5 m. *Leaves* 25–55 cm × 1–8 mm, linear, narrowed to an acute apex, usually inrolled, rarely flat, minutely scabrous; sheaths loose, glabrous; ligules 3–8 mm, membranous, acute or laciniate at apex. *Inflorescence* an open panicle of racemes 10–40 cm; branches erect, initially compact, later spreading with unbranched racemes, the rhachis ending in a spikelet. *Spikelets* 8–15 mm, greyish-green or olive green, slightly overlapping, narrowly elliptical, 6- to 14-flowered. *Glumes* unequal, the lower 3.5–4.5 mm, narrowly ovate and acute at apex, the upper 4.5–5.0 mm, narrowly oblong, acute to obtuse at apex, mucronulate and 1-veined. *Lemma* 3–6 mm, hairy on the veins; awn 0.5–1.5 mm, between 2 shorter teeth at apex. *Palea* hairy on the veins. *Anthers* more than 1 mm. *Caryopsis* obovoid-oblong. *Flowers* 8–10. $2n = 20$.

Introduced. Wool alien which appears as an occa-sional casual in fields, waste places and on tips. Scattered records in England. Native from tropical Africa to Australia.

3. L. uninervia (C. Presl) Hitchc. & Chase
                           One-nerved Beetle Grass
Diplachne uninervia (C. Presl) Parodi; *Megastachya uninervia* C. Presl

Tufted *perennial* or *annual herb.* Culms 20–90 cm, erect or geniculately ascending. *Leaves* 5–35 cm × 2–10 mm, long-linear, flat or soon inrolled, narrowed at apex; ligule 2.5–6.0 mm, hyaline, usually lacerate into several strap-shaped parts. *Inflorescence* a panicle of well-spaced racemes, the racemes 20–40 and 2–8 cm; rhachis ending in a spikelet. Spikelets 4–9 mm, scarcely laterally compressed, 6- to 12-flowered. *Glumes* unequal, 1.5–3.5 mm, subacute, 1-veined. *Lemmas* 1.8–3.0 mm, obovate, obtuse or mucronate at apex with 1–2 minute teeth or only a shoulder on each side, 3-veined. *Anthers* less than 0.5 mm. *Flowers* 8–9. $2n = 20$.

Introduced. Recorded as an occasional casual in waste places. Native of North and South America.

74. **Eragrostis** Wolf

Annual or tufted *perennial herbs. Leaves* flat or inrolled; ligule a fringe of hairs. *Inflorescence* a usually diffuse panicle. *Spikelets* often narrow and parallel-sided, with 3 to many flowers. *Glumes* subequal to unequal, (0–)1(–3)-veined. *Lemma* 3-veined, keeled, acute to obtuse, rounded or emarginate at apex; awnless or shortly apiculate-awned. *Spikelets* articulating between the flowers or the caryopsis falling free leaving a persistent rhachilla. *Pericarp* adherent to seed.

Distinguished from *Poa,* which it superficially resembles, by its 3-veined lemma and its ligule being a ring of hairs. Over 50 species have been recorded from the British Isles of which 14 of the commonest are included here. Most are rare casuals. Ryves (1980) gives a key to 51 species.

Ryves, T. B. (1974). An interim list of the wool alien grasses from Blackmoor, North Hants 1969–1972. *Watsonia* **10**: 35–48.
Ryves, T. B. (1980). Alien species of *Eragrostis* P. Beauv. in the British Isles. *Watsonia* **13**: 111–117.

1. Annual; leaf margins with prominent crateriform glands (not bulbous- based hairs); panicle less than 30 cm; caryopsis without a dorsal pit **2.**
1. Annual or perennial; leaf margins without prominent glands (except *E. neomexicana* occasionally) **3.**
2. Leaves glabrous; pedicel without a prominent gland; spikelets 2–4 mm wide, often olive or grey; lemmas 2.0–2.8 mm **1. cilianensis**
2. Leaves often with sparse, coarse hairs; pedicel with a prominent gland; spikelets 1.3–2.0 mm wide, often purplish; lemmas 1.2–2.0 mm **2. minor**
3. Culm nodes with a ring of glandular tissue beneath; pedicels with a gland **3. barrelieri**
3. Culm nodes without or sometimes with a ring of glandular tissue beneath; branch axes with glandular tissue occasionally; pedicels without a gland **4.**

4.   Sheaths with many prominent circular glands, with
     or without short hairs; spikelets 5–8 × 1.5–3.0 mm
                                                 4. neomexicana
4.   Sheaths without prominent glands   5.
5.   Spikelets short, less than 5 mm, with 3–5(–6) flowers
                                                 5. trachycarpa
5.   Spikelets usually more than 5 mm, some with 5–20
     flowers   6.
6.   Basal sheath strongly compressed, glabrous,
     spreading like a fan; spikelets 6–10 × about 2 mm,
     appressed, olive green; glumes very short, unequal;
     caryopsis 1.4–1.5 mm, compressed, lumpy   **6. plana**
6.   Basal sheaths not strongly compressed   7.
7.   Mature spikelets more than 1.5 mm wide, often 4
     times as long as wide   **7. brownii**
7.   Mature spikelets less than 1.5 mm wide, often more
     than 5 times as long as wide   8.
8.   Perennial   9.
8.   Annual   11.
9.   Top of sheath very hairy; leaves more or less hairy,
     flat; panicle diffuse, about 15 × 15 cm; spikelets
     lanceolate, 3–4 mm; lemma acute at apex, closely
     overlapping   **8. lugens**
9.   Top of sheath usually not very hairy; spikelets linear
     to lanceolate, more than 4 mm; lemmas rather loose   10.
10. Culms branched, geniculate, 30–60 cm; lower leaf-
     sheath papery with rounded, well-separated veins;
     spikelets 1.0(–1.5) mm wide; lemmas about 1.5 mm
                                                 9. lehmanniana
10. Culms unbranched, erect or geniculate; 30–120 cm;
     lower leaf-sheath tough with flattened close-set veins;
     spikelets 1.5(–2) mm wide; lemmas 2.0–2.5 mm
                                                 10. curvula
11. Throat of sheath without a tuft of long hairs
                                                 11. parviflora
11. Throat of sheath when young with a conspicuous
     tuft of long (2 mm), white, stiff hairs   12.
12. Immature spikelets with the upper lemma equalling
     lower lemma, about 1.5 mm   **12. pectinacea**
12. Immature spikelets with the upper lemma shorter
     than the lower lemma   13.
13. Branch axils glabrous; spikelets often yellowish-
     green; lower lemma 2–3 mm; caryopsis ovoid,
     1.0–1.5 mm   **13. tef**
13. Branch axils usually with long, white hairs; spikelets
     purplish-grey; lower lemma about 1.5 mm; caryopsis
     oblong, 0.5–1.0 mm   **14. pilosa**

1. E. cilianensis (All.) Vign. ex Janchen   Stink Grass
E. megastachya (Koerler) Link; *Poa cilianensis* All.; *E. major* (L.) Host

Annual herb. Culms up to 75 cm, erect or geniculate at the base, smooth, glabrous, simple or branched towards the base, with purple nodes which are often decorated with a row of depressed glands on the lower margin. *Leaves* up to 20 cm × 5–8 mm, acuminate at apex, flat, rigid, minutely rough on both surfaces, glabrous, margins with raised crateriform glands, veins on the lower surface often with depressed glands; sheaths clasping or some-what inflated, often with depressed oval or circular glan-

dular patches; ligule 1–2 mm, membranous, ciliate. *Panicle* up to 20 × 12 cm, lax or contracted, ovate or elliptical in outline. *Spikelets* 2–4 mm wide, oblong or elliptical in outline, greenish-grey, compressed, 6- to 40-flowered, on pedicels without a crateriform gland. *Glumes* 1.5–2.0 mm, unequal, keeled, lanceolate or ovate-lanceolate, the lower 1-, the upper 3-veined, with one to several glands on the keels. *Lemma* 2.0–2.8 mm, broadly elliptical, acute or retuse at apex, 3-veined, with lateral veins close to the margin, whitish, greyish or tinged with purple, sometimes with glandular keels. *Palea* 2-veined. *Anthers* minute. *Caryopsis* globose, ellipsoid or obovoid, without a dorsal pit. Flowers 6–8. $2n = 20$.

Introduced. Frequent casual on tips and waste ground. Scattered in England and Wales. Native of the Mediterranean region and Macaronesia, eastwards to the warmer parts of Asia; north-east tropical Africa; introduced in South Africa, Australia and North and South America.

2. E. minor Host   Small Love Grass
E. pooides P. Beauv.

Loosely tufted *annual herb. Culms* up to 60 cm, ascending or erect, smooth, with sparse, long hairs. *Leaves* up to 15 cm × 2–5 mm, triangular-linear, acute at apex, flat, glabrous or sprinkled with tubercular-based hairs, usually with a row of crateriform glands along the margin or the mid-vein; sheaths often with various glandular hairs and tubercular-based hairs; ligules reduced to a ciliate rim. *Inflorescence* a fairly dense, more or less ovate panicle 4–20 × up to 8 cm, stiffly branched, the lower branches solitary or in pairs, often with glands on the pedicels and branchlets. *Spikelets* 3–9(–15) × 1.3–2.0 mm, yellowish-green, leaden grey or purplish, linear to narrowly ovate-elliptical, with 6–10(–40) flowers. *Glumes* subequal, 1.0–1.7 mm, ovate, acute at apex, usually 1-veined, often glandular-hairy on the keel. *Lemmas* 1.2–2.0 mm, greyish with a purple apex, broadly ovate, obtuse to acute at apex, 3-veined, often glandular-hairy on the keel. *Palea* a little shorter than the lemma, 2-keeled. *Caryopsis* 0.7–0.8 mm, ellipsoid, without a dorsal pit. *Flowers* 7–9. $2n = 40$.

Introduced. Fairly frequent wool and grain alien casual on tips and waste ground. Scattered records in central and southern Britain. Native of temperate and subtropical regions of the Old World.

3. E. barrelieri Daveau   Mediterranean Love Grass
Annual herb forming lax tufts. *Culms* up to 60 cm, geniculately ascending to erect, with a ring of glands below the nodes. *Leaves* linear, narrowed to an acute apex, flat or loosely rolled, glabrous; sheaths striate, glabrous; ligule reduced to a ciliate rim. *Inflorescence* a rigid panicle about 12 × 4 cm; branches and spikelets stiffly spreading. *Spikelets* 5–15 mm, green to olive green, greyish or purplish, with 10–20 flowers; pedicels with a gland. *Glumes* unequal, acute at apex. *Lemmas* 2.0–2.5 mm, subobtuse at apex, veins conspicuous. *Anthers* about 0.4 mm. *Caryopsis* about 1 mm, elliptic-oblong. Flowers 8–9. $2n = 60$.

Introduced. Rare casual on tips and waste places. Few records in England. Native of the Mediterranean region.

4. E. neomexicana Vasey ex L. Dewey
New Mexico Love Grass

Loosely tufted *annual herb*. *Culms* 25–100 cm, rather stout, geniculate in lower part, sometimes with a ring of glands just below the nodes. *Leaves* 2–10 mm wide, long-linear, flat; sheaths sparingly papillose-hairy or glabrous and usually with rows of circular glandular depressions on the keel and near the veins; ligule reduced to a ciliate rim. *Inflorescence* 9–40 cm; branches ascending or inclined. *Spikelets* 5–8 × 1.5–3.0 mm, dark greyish-green or brownish-purple, appressed, with 6–16 flowers; pedicels without a gland. *Glumes* unequal, ovate, pointed at apex. *Lemmas* 1.6–2.5 mm, ovate, obtuse at apex. *Flowers* 8–9. $2n = 56$.

Introduced. Rare casual on tips and in waste places. A few records in England. Native of North America.

5. E. trachycarpa (Benth.) Domin
Rough-fruited Love Grass

E. nigra var. *trachycarpa* Benth.

Glabrous tufted *annual herb*. *Culms* up to 60 cm, erect. *Leaves* in a basal tuft, up to 20 cm × 3 mm, linear or inrolled and filiform, narrowed at apex, minutely hispid; sheaths striate; ligule reduced to a ciliate rim. *Inflorescence* a diffuse, spreading, ovate panicle up to 40 cm; branches long and capillary. *Spikelets* solitary at the apex of branches, 2–5 × 0.7–1.2 mm, greenish-purple to purplish, narrowly ovate, with 3–5(–6) flowers; pedicels without a gland. *Glumes* 1.0–1.8 mm, subequal or the lower shorter, ovate, blunt at apex, 1-veined, minutely hispid along the keel. *Lemmas* 1.5–2.0 mm, broadly ovate, obtuse at apex, obscurely 3-veined, sometimes minutely hispid towards the keel apex. *Flowers* 8–9.

Introduced. A rather rare casual found in a few localities in Britain. Native of Australia

6. E. plana Nees   Compressed Love Grass

Perennial herb forming dense, very strong tufts. *Culms* 30–100 cm, more or less erect, strongly compressed, glabrous, smooth. *Leaves* long-linear, acute at apex, rigid; sheaths strongly compressed, often spreading and fan-shaped, pale green, striate to smooth and shiny, glabrous; ligule reduced to a ciliate rim. *Inflorescence* a long, narrow panicle; branches appressed or spreading with shining patches of glandular hairs at their bases and on the central axis below the lowermost branches which are single or clustered. *Spikelets* 6–10 × about 2 mm, shiny olive grey, appressed to the branches, with 7–14 flowers; pedicels without a gland. *Glumes* very unequal, often reduced to 2 small scales. *Lemmas* about 2.5 mm, tapering to an obtuse apex; veins pale with conspicuous pit-like glands. *Anthers* about 1.6 mm. *Caryopsis* 1.4–1.5 mm, dark brown, compressed. *Flowers* 8–9. $2n = 20$.

Introduced. Rare casual on tips and waste places. A few localities in England. Native of Africa.

7. E. brownii (Kunth) Nees ex Steud.
Brown's Love Grass

Poa brownii Kunth; *Poa polymorpha* R. Br.; *E. urvillei* Steud.

Variable, tufted *perennial herb*. *Culms* up to 40 cm, erect or ascending. *Leaves* up to 22 cm × 3 mm, setaceous or narrow and flat, minutely hispid; sheaths often ciliate at the orifice; ligule reduced to a ciliate rim. *Inflorescence* a loose and spreading or dense and contracted panicle 2–16 cm; branches narrow, not pendulous. *Spikelets* 4–9 × 1.5–2.5 mm, yellowish-green to purplish-green, narrowly ovate to narrowly elliptical, with 6–21 flowers, pedicels without a gland. *Glumes* subequal, 1.0–2.5 mm, ovate, acute at apex, 1-veined, minutely scabrid along the keel. *Lemmas* 1.5–2.0 mm, broadly ovate, acute to obtuse at apex, 3-veined, minutely scabrid on the keel and on the back towards the apex. *Flowers* 8–9. $2n = 40$.

Introduced. A rather rare casual found in a few localities in Britain. Native of Australia.

8. E. lugens Nees   Mourning Love Grass

Tufted *perennial herb*. *Culms* 12–80 cm, wiry, geniculate below, erect, ascending or spreading. *Leaves* often crowded towards the lower part, 10–25 cm × 1.5–3.0 mm, long-linear, acute at apex, flat, folded or involute, hairy above, rarely beneath; sheath very hairy at top; ligule reduced to a ciliate rim. *Inflorescence* an erect, open panicle about 15 × 15 cm; branches numerous, spreading or ascending, the axils of the lower long-hairy. *Spikelets* 3–4 × less than 1.5 mm, pedicels without a gland. *Glumes* 0.7–1.2 mm, thin, falling early. *Lemmas* 1.3–2.0 mm, only with slightly hyaline margins above, ovate, obtuse or abruptly acute at apex, closely overlapping; veins very obscure. *Caryopsis* about 0.7 mm. *Flowers* 8–9. $2n = 40, 80$.

Introduced. Rare casual of tips and waste places. Few records in England. Native of North America.

9. E. lehmanniana Nees   Lehmann's Love Grass

Tufted *perennial herb*. *Culms* 30–60 cm, usually repeatedly geniculate, occasionally erect, branched, glabrous, or hairy with bulbous-based hairs. *Leaves* yellowish-green, narrowly linear, tapering to a rigid point, rolled or often flat at the base, some veins finer than others; sheaths papery, pale or yellowish-green, with rounded, well separated veins, more or less rounded at apex, appressed silky-hairy at the base; ligule reduced to a ciliate rim. *Inflorescence* a lax and open panicle 6–20 cm, lower branches single or opposite, central axis flattened or grooved opposite the branches, the branches and rhachis rather stiff. *Spikelets* 4–8 × 1.0–1.5 mm, greyish-green to dark olive grey, linear, often much flattened, with 4–13 flowers; pedicels without a gland. *Glumes* subequal to rather unequal, acute at apex. *Lemmas* about 1.5 mm, obtuse at apex, membranous. *Anthers* about 0.8 mm. *Caryopsis* about 0.6 mm, obovoid or oblong. *Flowers* 8–9. $2n = 40, 60$.

Introduced. Rare casual of tips and waste places. Recorded from a few localities in England. Native of southern Africa.

10. E. curvula (Schrad.) Nees   African Love Grass
Poa curvula Schrad.; *E. chloromelas* Steud.

Tufted *annual* or *perennial herb.* Densely tufted *perennial herb.* Culms up to 1.2 m, slender or robust, usually erect or geniculate, unbranched. *Leaves* up to 30 cm × 2(–3) mm, linear, usually rolled or filiform, narrowed at apex, minutely hispid; sheath with flattened, close-set veins, apressed silky-hairy below; ligule reduced to a ciliate rim. *Inflorescence* a variable, loose and spreading to narrow and contracted panicle 6–30 × up to 15 cm, usually hairy in the branched axils. *Spikelets* (3–)4–16 × 1.0–1.5(–2.0) mm, greyish-green, linear to linear-lanceolate, with 4–13 flowers; pedicels without a gland. *Glumes* lanceolate, acute at apex, 1-veined, the lower 1–2 mm, the upper 1.5–2.2 mm. *Lemmas* 2.0–2.5 mm, ovate-elliptical, obtuse at apex, obscurely 3-veined, loose. *Caryopsis* up to 0.8 mm. *Flowers* 8–9. 2*n* = 20, 40, 60.

Introduced. Occasional wool alien casual on tips and waste ground. Scattered localities in England and Wales, more or less naturalised in Southampton, Hampshire. Native of tropical Africa.

11. E. parviflora (R. Br.) Trin.   Weeping Love Grass
Poa parviflora R. Br.; *E. pellucida* (R. Br.) Steud.; *E. multicaulis* Steud.; *E. pilosa* auct.

Tufted *annual herb.* Culms up to 90 cm, erect or ascending, usually slender, glabrous. *Leaves* up to 35 cm × 4 mm, linear, narrowed at apex, flat or inrolled, minutely hispid; sheaths striate, throat without hairs; ligule reduced to a ciliate rim. *Inflorescence* an open, spreading, often slightly nodding, elliptical or ovate panicle 15–45 cm; branchlets capillary, the lowermost often more or less whorled. *Spikelets* 3–11 × 0.8–1.5 mm, greenish to purplish, linear to lanceolate, with 5–19 flowers; pedicels without a gland. *Glumes* subequal, ovate, obtuse at apex, the lower 0.7–1.0 mm, the upper 1.0–1.5 mm and obscurely 1-veined. *Lemmas* 1.2–1.7 mm, broadly ovate, obtuse at apex, obscurely 3-veined. *Flowers* 8–9. 2*n* = 40.

Introduced. Fairly frequent wool alien and grain casual on tips and waste ground. Scattered localities in England. Native of Australia.

12. E. pectinacea (Michx) Nees   Pectinate Love Grass
Poa pectinacea Michx

Loosely tufted, diffuse *annual herb.* Culms numerous, 15–60 cm, ascending, spreading and geniculate in the lower part. *Leaves* 2–5 mm wide, linear, acute at apex, mostly flat; sheaths usually folded, softly keeled, conspicuously hairy at the corner, with long white hairs; ligules reduced to a ciliate rim. *Inflorescence* an obovoid, usually open and diffuse, inclined or nodding panicle 5–40 cm; branches numerous and ascending, with flexuous branchlets. *Spikelets* 3–10 × up to 1.5 mm, greyish, more or less appressed, with 10–15 flowers; pedicels without a gland. *Glumes* unequal, 1.0–1.5 mm, ovate, acute at apex, scabrous on the keel. *Lemmas* 1.5–1.6 mm, greyish, obtuse at apex, when immature the upper equalling the lower. *Caryopsis* 0.7–0.8 mm. *Flowers* 8–9. 2*n* = 40, 60.

Introduced. Rare casual on tips and in waste places. A few records in England. Native of North America.

13. E. tef (Zucc.) Trotter   Teff
Poa tef Zucc.

Annual herb forming sparse tufts. *Culms* up to 1.2 m, erect, finely striate, glabrous. *Leaves* 2–4 mm wide, loosely rolled or flat, linear, with a setaceous tip; sheaths somewhat keeled, glabrous; ligule reduced to a ciliate rim. *Inflorescence* a large diffuse or contracted panicle up to 13 × 15 cm; branches long, filiform, glabrous and flexible, the lower in a pseudo-whorl or clustered, very rarely single. *Spikelets* straw-coloured or green, often tinged reddish, linear or linear-lanceolate, on long flexible pedicels; pedicels without a gland. *Glumes* unequal, the lower little more than half the length of the upper, acute at apex. *Lemmas* 2–3 mm, with prominent green veins. *Anthers* about 0.3 mm. *Caryopsis* 1.0–1.5 mm. *Flowers* 8–9. 2*n* = 40.

Introduced. Wool or grain alien casual on tips and waste ground. Scattered localities in England. Native of tropical Africa where it is a minor grain crop.

14. E. pilosa (L.) P. Beauv.   Jersey Love Grass
Poa pilosa L.

Loosely tufted *annual herb.* Culms up to 70 cm, erect or ascending, smooth, glabrous. *Leaves* up to 20 cm × 1.0–3.5 mm, linear, narrowed at apex, flat, glabrous, scabridulous above and on the margin; sheaths glabrous; ligules reduced to a ciliate rim. *Inflorescence* an open, elliptical to ovate panicle 4–25 × 15 cm; lowest branches whorled except in the smallest panicles, capillary, usually with a few, long, white hairs in the axils. *Spikelets* 3–7 × 0.7–1.5 mm, purplish-green, linear to linear-lanceolate, with 4–14 flowers. *Glumes* unequal, the lower about 0.5 mm, narrowly ovate and veinless, the upper 1.0–1.5 mm, ovate, acute at apex and 1-veined. *Lemmas* 1.2–1.5 mm, greyish with a purple apex, broadly ovate to oblong, obtuse to subacute at apex, obscurely 3-veined, minutely scabrid along the keel. *Caryopsis* 0.5–1.0 mm, oblong. *Flowers* 8–9. 2*n* = 40, 160.

Introduced. Naturalised in Jersey in the Channel Islands since 1961. Native of tropical and warm temperate regions of the world.

75. Eleusine Gaertn.

Annuals or tufted *perennial herbs.* Leaves usually folded; ligules membranous, with a sparse to dense fringe of hairs. *Inflorescence* an umbel or very short raceme of spikes, with very crowded spikelets, the rhachis ending in a spikelet. *Spikelets* with 3–15 flowers. *Glumes* subequal to unequal, 1- to 7-veined. *Lemmas* with 3 main veins and 2–4 extra veins close to the mid-vein, keeled, acute to apiculate at apex, not awned. *Spikelets* disarticulating between the flowers; pericarp not adhered to the seed, which eventually falls out separately.

Nine species, mostly in east and north-east tropical Africa.

Phillips, S. M. (1972). A survey of the genus *Eleusine* Gaertn. (Gramineae) in Africa. *Kew Bull.* 27: 251–270.

1. Lemma and glumes obtuse at apex, with a hooded tip
2. tristachya

1. Lemma and glumes pointed to obtuse and apiculate at apex, without a hooded tip 2.

2. Spikes 1–3 cm, in a very short terminal raceme; lemmas about 1.5 mm wide from keel to edge **3. multiflora**

2. Spikes (3.5–)5–15 cm, all or most in a terminal umbel; lemmas less than 1 mm wide from keel to edge 3.

3. Ligules sparsely and minutely hairy at apex; lower glumes 1.1–2.3 mm, 1-veined; upper glume 1.8–3.0 mm; lemmas 2.4–4.0 mm; seeds 1.0–1.3 mm
1(a). indica subsp. **indica**

3. Ligules with a strong fringe of hairs at apex; lower glumes 2.0–3.5 mm, (1–)2- to 3-veined; upper glumes 3.0–4.7 mm; lemmas 3.7–5.0 mm; seeds 1.2–1.6 mm
1(b). indica subsp. **africana**

1. E. indica (L.) Gaertn. Yard Grass
Cynosurus indica L.; *E. indica* var. *monostachya* F. M. Bailey

Coarse, tufted *annual herb*. Culms 30–90 cm, ascending or prostrate. *Leaves* 3–35 cm × 3–8 mm, linear, obtuse at apex, glabrous or minutely scabrous on the margin; sheaths keeled; ligules 0.5–1.0 mm, membranous, with few to numerous cilia round the margin. *Inflorescence* of (1–)2–15 spikes mostly in an umbel, (3.5–)5–15 × under 1 cm. *Spikelets* 3.5–7.8 mm, laterally compressed, with 3–9 bisexual flowers. *Glumes* subequal, shorter than lemmas, membranous, the lower 1.0–3.5 mm, lanceolate, more or less acute at apex and 1- to 3-veined, the upper 1.8–4.7 mm and 3- to 5-veined. *Lemmas* 2.4–5.0 mm, acute at apex and keeled, 3- to 5-veined. *Palea* slightly shorter than lemma with 2 winged keels. *Anthers* 0.6–0.7 mm. *Styles* distinct, slender with a thickened base; stigmas plumose. *Seeds* 1.0–3.5 mm, enclosed within a free hyaline pericarp. *Flowers* 8–9.

(a) Subsp. indica
Ligule sparsely and minutely hairy at apex. *Lower glume* 1.1–2.3 mm, 1-veined; upper glume 1.8–3.0 mm. *Lemmas* 2.4–4.0 mm. *Seeds* 1.0–1.3 mm, with very fine, close striations between and at right-angles to the main ridges. $2n = 18$.

(b) Subsp. africana (Kenn.-O'Byrne) S. M. Phillips
E. africana Kenn.-O'Byrne
Ligules with a strong fringe of hairs at apex. *Lower glume* 2.0–3.5 mm, (1–)2- to 3-veined; upper glume 3.0–4.7 mm. *Lemmas* 3.7–5.0 mm. *Seeds* 1.2–1.6 mm, with granulations between the main ridges. $2n = 36$.
Introduced. Casual from bird-seed, grain, wool, cotton or pulse aliens. On tips and waste ground. Scattered records in England. Subsp. *indica* is native of Asia, but is now a world-wide weed. Subsp. *africana* is a native of Africa, but is now more widespread. Distribution of the subspecies in England is unknown, but subsp. *indica* seems to be commoner in grain, pulse and cotton aliens, whereas subsp. *africana* is much the commoner and perhaps the only wool alien.

2. E. tristachya (Lam.) Lam. American Yard Grass
Cynosurus tristachyus Lam.

Tufted *annual herb*. Culms up to 40 cm, oblique or ascending. *Leaves* 2–20 cm × 2–4 mm, linear, obtuse at apex, minutely scabrous on the margin, particularly towards the apex; sheaths keeled; ligules about 0.5 mm, ciliolate. *Inflorescence* an umbel of 2–5 digitate or subdigitate spikes 1–3 cm, the axes flattened, with crowded spikelets. *Spikelets* 3–7 mm, with 3–11, bisexual flowers. *Glumes* unequal, the lower 1.5–2.5 mm, ovate, acute to obtuse at apex with a hooked tip and 1-veined, the upper 2.5–3.0 mm, elliptic-ovate, obtuse at apex and more or less 3-veined. *Lemmas* 3.5–4.0 mm, 3- to 5-veined, obtuse at apex and hooded, keeled. *Palea* slightly shorter than lemma, with 2 winged keels. *Styles* distinct, slender from a thickened base; stigmas plumose. *Seeds* enclosed with free hyaline pericarp. *Flowers* 8–9. $2n = 18$.
Introduced. Fairly frequent wool alien on tips and waste ground. Scattered records in England. Native of South America.

3. E. multiflora Hochst. ex Rich.
Fat-spiked Yard Grass
Eragrostis kwaiensis Peter

Tufted *annual herb*. Culms up to 45 cm, fairly slender, ascending. *Leaves* 6–26 cm × 3–6 mm, linear, abruptly pointed at apex, flat or loosely folded, sparsely hairy; ligules about 1 mm, membranous, truncate at apex. *Inflorescence* of 2–8, short, broad spikes alternating on a short axis towards the top of the culm and forming a short terminal raceme; spikes 1–3 cm × 8–16 mm, oblong to ovate. *Spikelets* 7–11 mm, 5- to 15 flowered, disarticulating above the glumes and between the flowers. *Glumes* subequal, 2.8–4.2 mm, winged on the keel, scabrid on the upper part of the wing, the lower narrowly oblong, subacute at apex and 1-veined, the upper oblong, obtuse to acute at apex and 3-veined. *Lemmas* 3.3–5.2 mm, about 1.5 mm wide from keel to edge, narrowly ovate, acute or obtuse and mucronate at apex. *Palea* narrowly winged on the keels, scabrid along the wing margins. *Anthers* 0.8–1.0 mm. *Seeds* 1.0–1.2 mm, broadly oblong, enclosed within a free hyaline pericarp. *Flowers* 8–9. $2n = 18$.
Introduced. Fairly frequent wool alien on tips and waste ground. Scattered localities in England. Native of tropical Africa.

76. Dactyloctenium Willd.

Annual herbs, but sometimes rooting at the lower nodes. *Leaves* flat; ligule membranous, sometimes slightly fringed at apex. *Inflorescence* an umbel of spikes with very crowded spikelets, the rhachis ending in a short projection. *Spikelets* with 2–5 flowers. *Glumes* unequal, 1-veined, the upper with a long, awn-like point. *Lemmas* 3-veined, keeled, acuminate to acute-apiculate at apex. *Spikelets* disarticulating above the glumes, not between the flowers, the pericarp not adherent to the seed, which eventually falls out separately.
Thirteen species native round the hinterland of the

Indian Ocean from Natal to northern India and one species native in Australia.

1. D. radulans (R. Br.) P. Beauv.   Button Grass
Eleusine radulans R. Br.; *D. australiense* Scribn.;
D. aegyptia auct.

Short, slender *annual herb. Culms* up to 40 cm, decumbent, ascending or rarely erect, sometimes rooted at the lower nodes, branched. *Leaves* 2.5–12.0 cm × 2–6 mm, linear, attenuate to the apex, flat, usually with long tuberculate-based hairs along the margin of the lower part; ligule about 0.5 mm, membranous, sometimes slightly fringed at apex. *Inflorescence* an umbel of 3–10 spikes 5–10(–15) cm of very crowded spikelets, the rhachis ending in a short projection. *Spikelets* about 5 mm, with 2–5 flowers. *Glumes* unequal, 1-veined, the lower 1–2 mm, lanceolate, acute at apex and with a thick, scabrid keel, the upper 1.5–3.0 mm, oblong-elliptical, obtuse at apex and the keel extended into a stout awn 1.0–2.5 mm. *Lemmas* 3–4 mm, keel minutely scabrid and extended into an awn about 0.5 mm. *Stamens* 3. *Styles* distinct. *Seed* subglobose, the pericarp not adherent, the seed eventually falling out separately. *Flowers* 8–9.

Introduced. Fairly frequent wool-alien on tips and waste ground. Scattered records in England. Native of Australia.

77. Sporobolus R. Br.

Tufted *perennial herbs. Leaves* flat or convolute; ligule a fringe of hairs. *Inflorescence* a narrow, spiciform panicle, with short to long, closely appressed, erect branches, the ultimate ones with closely borne, small spikelets resembling at a casual glance a spikelet with many flowers. *Spikelets* with 1 flower. *Glumes* unequal, 0- to 1-veined, much shorter than the rest of the spikelet. *Lemmas* 1- to 3-veined, rounded on the back, acute to acuminate at apex, awnless, inrolled and more or less cylindrical, but tapering at apex. *Spikelet* disarticulating below the lemma; pericarp not coherent to seed, which eventually falls out separately.

About 100 species in the tropics and subtropics.

Clayton, W. D. (1965). Studies in Gramineae VI: The *Sporobolus indicus* complex. *Kew Bull.* **19**: 287–296.

1. Lemma 2.1–2.5 mm   **1. africanus**
1. Lemma 1.3–1.9 mm   2.
2. Inflorescence densely spicate, rarely interrupted at base; stamens 3   **2. indicus**
2. Inflorescence interrupted in the lower half, spikelets in dense clusters on appressed primary branches; stamens 2   **3. elongatus**

1. S. africanus (Poir.) Robyns & Tournay
African Dropseed
Agrostis spicata Thunb., non Vahl; *Agrostis capensis* Willd., non Lam.; *Agrostis africana* Poir.; *Vilfa capensis* P. Beauv.; *S. capensis* (P. Beauv.) Kunth; *S. batesii* A. Chev.

Densely tufted *perennial herb. Culms* up to 1 m, erect to ascending, greyish-green, striate, glabrous. *Leaves* 5–30

cm × 1.5–3.0 mm, greyish-green, long-linear, very narrowly tapered to an acute apex, flat, glabrous; sheaths compressed, striate, glabrous; ligule a fringe of hairs. *Inflorescence* a narrow, spike-like panicle 10–35 cm; branches not exceeding 2 cm. *Spikelets* 2.0–2.5 mm. *Glumes* unequal, 1.5–2.0 mm, dull greyish-green, lanceolate, the lower obtuse, the upper acute at apex. *Lemmas* 2.1–2.5 mm, lanceolate, acute at apex. *Anthers* about 0.5 mm. *Seed* half to two-thirds as long as lemma. Flowers 8–9. 2*n* = 36.

Introduced. Rather frequent wool alien on tips and rough ground. Scattered localities in England. Native of Africa.

2. S. indicus (L.) R. Br.   Ratstail Dropseed
Agrostis indica L.; *Agrostis elongata* Lam.; *Agrostis compressa* Poir.; *Vilfa elongata* (Lam.) P. Beauv.; *Axonopus poiretii* Roem. & Schult.; *S. lamarkii* Desv.; *Agrostis tenuissima* Spreng.; *Vilfa exilis* Trin.; *Vilfa berteroana* Trin.; *Vilfa indica* (L.) Steud.; *S. berteroanus* (Trin.) Hitchc. & Chase; *S. angustus* Buckl.; *S. poiretii* (Roem. & Schult.) Hitchc.; *S. indicus* var. *exilis* (Trin.) Koyama

Tufted *perennial herb. Culms* up to 80 cm, slender or stout. *Leaves* up to 50 cm × 7 mm, rolled in bud, becoming flat or convolute, linear, attenuate to apex, minutely scabrid; sheaths often more or less papery; ligules up to 0.5 mm, a fringe of hairs. *Inflorescence* a narrow panicle 10–35 cm, with short branches to densely spiciform-linear, rarely interrupted. *Spikelets* 1.5–2.5 mm, lanceolate. *Glumes* unequal, the lower 0.5–1.0 mm, broadly oblong to broadly ovate and obtuse at apex, the upper elliptical and bluntly acuminate at apex. *Lemma* 1.5–1.9 mm, ovate-elliptical, acute at apex, more or less smooth, 1- to 3-veined. *Stamens* 3. *Styles* free, short; stigmas plumose. *Mature caryopsis* with a loose pericarp, almost as long as lemma and palea. *Flowers* 8–9. 2*n* = 18, 36.

Introduced. A rare casual of waste places, but may have been overlooked. Scattered records for England. Native of tropical America.

3. S. elongatus R. Br.   Slender Dropseed
S. indicus var. *elongatus* (R. Br.) F. M. Bailey

Tufted *perennial herb. Culms* up to 100 cm, usually slender, smooth. *Leaves* up to 50 cm × 4 mm, linear, flat or convolute, flexuous, attenuate towards apex, smooth or minutely striate; sheaths striate; ligule up to 0.5 mm. *Inflorescence* a spiciform panicle 10–40 cm, interrupted in lower half, branches appressed to the rhachis, densely spiculate to base. *Spikelets* 1.5–2.0 mm, narrowly ovoid. *Glumes* unequal, the lower 0.5–1.0 mm, oblong and obtuse at apex, the upper 0.7–1.2 mm, ovate and obtuse at apex. *Lemma* 1.3–1.9 mm, ovate-elliptical, more or less smooth. *Stamens* 2. *Styles* free, short; stigmas plumose. *Mature caryopsis* with a loose pericarp, almost as long as lemma and palea. *Flowers* 8–9. 2*n* = 36.

Introduced. A rare casual of waste places, but may have been overlooked. Scattered records for England. Native of Australia.

Tribe **17. Cynodonteae** Dumort.

Tribe *Spartineae* Gren. & Godr.; Tribe *Chlorideae* Rchb.; Tribe *Zoysiae* Benth.

Annual or *perennial herbs*, with or without rhizomes and stolons. *Leaves* with ligules membranous or a fringe of hairs. *Inflorescence* an umbel or raceme of spikes, or a spike-like panicle. *Spikelets* with one bisexual flower, sometimes with 1–2(–3) extra sterile or male flowers distal to it. *Glumes* (1–)2, shorter to longer than the rest of the spikelet. *Lemmas* 1- to 3(–9)-veined, usually keeled, awned or not.

78. Chloris Sw.

Annual or *perennial herbs*, sometimes with stolons. *Leaves* flat or folded; ligule membranous, with a well-marked fringe of hairs. *Inflorescence* an umbel of (4–)6–many long, slender spikes. *Spikelets* with 2–3 flowers, the lowest bisexual, the others reduced and male or sterile. *Glumes* unequal, 1-veined, narrowly acute at apex, shorter to slightly longer than the rest of the spikelet. *Lowest lemma* bisexual, 3-veined, keeled, minutely to deeply bifid at apex with a long terminal or subterminal, straight awn; upper lemmas variously reduced, but the second of similar shape to the upper and only that one long-awned, thus the spikelets are 2-awned. *Spikelets* disarticulating above the glumes.

About 55 species in the tropical and temperate regions of both hemispheres.

1. Lemma with low rounded to transversely or obliquely truncate lobes either side of the awn, forming a very shallow notch   **1. truncata**
1. Lemma with sharply acute lobes or teeth either side of the awn, forming a deep notch   2.
2. Lowest lemma without an apical tuft of hairs, producing a more or less glabrous spike   **2. divaricata**
2. Lowest lemma with dense tufts of silky hairs at apex producing feathery spikes   **3. virgata**

1. C. truncata R. Br.   Windmill Grass
C. truncata forma *abbreviata* Thell.

Compact *perennial herb*, often stoloniferous. *Culms* up to 45 cm, erect or creeping. *Leaves* 1.5–15(–22) cm × 15–25 mm, linear-oblong, obtuse-mucronate or subacute at apex, minutely scabrous, rarely hairy; ligules about 0.5 mm, membranous, truncate at apex, ciliate on margin. *Inflorescence* an umbel of 5–12 spikes, the spike 4–19 cm, rigid or rather flaccid, spreading or divaricate and hairy at the base. *Spikelets* 2.5–3.5 mm, with 2 or 3 flowers; pedicels about 0.5 mm. *Glumes* unequal, lanceolate, acuminate or mucronate at apex, the lower 1–2 mm, the upper 2.5–4.0 mm, both minutely scabrous at least on the keel. *Lower lemma* 3–4 mm, obovate with 2 minute, obtuse apical lobes, inflated, hairy on the upper margin, otherwise glabrous, smooth or sometimes minutely scabrous, its awn 10–15 cm; upper lemma about 1.5 mm, obovate, cuneate or truncate at apex and shallowly notched, the lobes rounded or truncate, inflated, appressed to the lower lemma, its awn 6–9 mm

and filiform. *Stamens* 3. *Styles* short and distinct; stigmas plumose. *Caryopsis* ellipsoid to linear, usually trigonous, smooth and shiny, enclosed in the lemma and palea. *Flowers* 8–9. $2n = 40$.

Introduced. Fairly frequent wool alien on tips and waste ground and in fields. Scattered records in England. Native of Australia.

2. C. divaricata R. Br.   Australian Rhodes Grass
C. divaricata var. *cynodontoides* (Balansa) Lazarides

Slender, more or less glabrous, compact *perennial herb*, sometimes stoloniferous. *Culms* up to 60 cm, erect or decumbent. *Leaves* 2–12 cm × 1–2 mm, linear, obtuse to acute at apex, usually minutely scabrous; ligules about 0.5 mm, truncate at apex, ciliate. *Inflorescence* an umbel of 3–8 spikes; spikes 4–20 cm, slender, stiff or flaccid, somewhat erect or spreading. *Spikelets* 3.0–4.5 mm, with 2 flowers, bisexual; pedicels about 1 mm. *Glumes* unequal, membranous, lanceolate, acuminate at apex, the lower 1–2 mm, the upper 2.0–2.5 mm, both minutely scabrous or smooth. *Lower lemma* with bilobed apex, the lobes acute, awned from the sinus, 2.0–3.5 mm, minutely scabrous, the awn 3–16 mm and capillary; upper lemma 1.0–1.3 mm, divided into 2, acute lobes about half as long as lemma, the awn 5–9 mm and filiform. *Stamens* 3. *Styles* short and distinct; stigmas plumose. *Caryopsis* ellipsoid to linear, usually trigonous, smooth and shiny, enclosed in lemma and palea. *Flowers* 8–9.

Introduced. Fairly frequent wool alien on tips and waste ground and in fields. Scattered records in England. Native of Australia.

3. C. virgata Sw.   Feathery Rhodes Grass
C. gabreilae Domin; *C. decora* Nees ex Steud.; *C. barbata* var. *decora* (Nees es Steud.) Benth.

More or less glabrous *annual herb*. *Culms* up to 1 m, erect or decumbent. *Leaves* 5–25 cm × 3–6 mm, linear, gradually narrowed to the apex, smooth or minutely scabrous, sometimes with a few hairs; ligule 0.5–1.0 mm, truncate at apex, ciliolate on margin. *Inflorescence* in an umbel of 7–19 spikes, the spikes 2–9 cm, erect, stiff, appressed. *Spikelets* 3–4 mm, with 2 flowers. *Glumes* usually hyaline, very narrowly ovate, minutely scabrid, the lower 1.5–2.0 mm, acuminate or mucronate, the upper 3.5–4.5 mm, aristate at apex. *Lower lemma* 3.0–3.5 mm, entire or notched at apex, the lobes acute, usually grooved between the midrib and the margin, the lower margin hairy, the upper margin bearded with stiff, white hairs 2–3 mm, smooth and sometimes bearded on the midrib in the lower part, the awn 6–100 mm, filiform and stiff; upper lemma about 2 mm, truncate at apex, the awn slightly shorter than the lower lemma awn. *Stamens* 3. *Styles* short and distinct; stigma plumose. *Caryopsis* ellipsoid to linear, mostly trigonous, smooth and shiny, enclosed in the lemma and palea. *Flowers* 8–9. $2n = 14$, 20, 26, 30.

Introduced. Fairly frequent wool alien on tips and waste ground and in fields. Scattered records in England. Native of tropical Africa.

79. **Cynodon** Rich. nom. conserv.

Capriola Adans. nom. rejic.; *Dactilon* Vill. nom. rejic.; *Fibichia* Koel. nom. rejic.

Perennial herbs with rhizomes and/or stolons. *Leaves* flat; ligule membranous or a fringe of hairs. *Inflorescence* an umbel of 3–6, slender spikes. *Spikelets* with one bisexual flower, laterally compressed. *Glumes* subequal, shorter than the rest of spikelet, membranous, keeled, 1-veined, narrowly acute at apex. *Lemma* 3-veined, keeled, subacute at apex, unawned. *Palea* 2-keeled. *Stamens* 3. *Spikelets* disarticulating above glumes.

About 8 species in the Old World tropics extending into warm temperate regions.

1. Ligules a fringe of short hairs less than 0.5 mm with a longer tuft at each edge; rhachilla a spikelet continued beyond the base of the flower as a fine projection between the upper glumes and the flower, more than half as long as the flower   **1. dactylon**
1. Ligule membranous, 0.4–1.0 mm, with sparse hairs at apex and a longer tuft at the base of each edge; rhachilla of spikelet not extended beyond the base of the flower   **2. incompletus**

1. **C. dactylon** (L.) Pers.   Bermuda Grass

Capriola dactylon (L.) Kuntze; *Panicum dactylon* L.

Mat-forming *perennial herb*, spreading by prostrate stolons and scaly rhizomes, the stolons branching profusely, rooting at the many nodes and there developing short leafy shoots and flowering culms. *Culms* 8–30 cm, erect or bent at the base, very slender, smooth. *Leaves* mostly crowded at the base, 2–15 cm × 2–4 mm, greyish-green or green, with sparse short hairs or nearly glabrous, flat, spreading, minutely rough, linear, narrowed to an obtuse tip; sheaths short, rounded; ligules a dense row of short hairs less than 0.5 mm. *Spikes* in a cluster of 3–6 at the end of the culm, 20–50 × 1.0–1.5 mm, very slender, soon spreading, straight or curved, rhachis very slender, 3-angled. *Spikelets* without stalks, borne in 2 rows on one side of and appressed to the rhachis, overlapping, 2.0–2.8 mm, purple or green, ovate-oblong, much compressed, 1-flowered, breaking up at maturity above the glumes; rhachilla continued beyond the base of the flower as a fine projection between the upper glume and the flower. *Glumes* more or less persistent, equal or slightly unequal, 1.5–2.3 mm, narrow, acute at apex, keeled, 1-veined, membranous. *Lemma* as long as the spikelet, boat-shaped, obtuse at apex, keeled, 3-veined, firm, densely and minutely hairy on the keel and often near the margins. *Palea* about as long as the lemma, 2-keeled. *Anthers* up to 1.5 mm. *Caryopsis* ellipsoid, tightly enclosed between the hardened lemma and the palea. *Flowers* 8–9. $2n = 36$.

Probably introduced. Locally naturalised in rough, sandy ground and short grassland by the sea in southwest England, south Wales and the Channel Islands. Elsewhere scattered in England and Wales as a casual from wool and other sources, and naturalised for a short period. World-wide in warm areas.

2. **C. incompletus** Nees   African Bermuda Grass

Slender, stoloniferous *perennial herb* without rhizomes. *Culms* up to 30 cm, but generally less than 15 cm. *Leaves* 1.5–10 cm × 10–30 mm, narrowly triangular to linear-triangular, attenuate at apex, minutely scabrous and often hairy with long tubercular-based hairs; ligule 0.4–1.0 mm, membranous, lacerate, with sparse hairs at the apex and a longer tuft of hairs at base of each edge. *Inflorescence* an umbel of long, slender spikes; spikes 3–5 in a whorl, each 2–5 cm, long and slender. *Spikelets* with one bisexual flower; rhachilla not extended beyond the base of the flower. *Glumes* unequal, 1–3 mm, shorter than the rest of the spikelet, 1-veined, acute to acuminate at apex, smooth or minutely scabrid. *Lemma* 2–3 mm, 3-veined, keeled, obtuse or very shortly mucronate at apex, ciliate along the keel. *Stamens* 3. *Styles* distinct; stigmas plumose. *Caryopsis* oblong, tightly enclosed between the hardened lemma and palea. *Flowers* 8–9.

Introduced. A fairly frequent wool alien on tips and rough ground and in fields. Casual or sometimes more or less naturalised in scattered localities in England. Native of South Africa.

80. **Spartina** Schreb.

Strongly rhizomatous *perennial herbs*. *Leaves* flat or convolute; ligule a dense fringe of hairs. *Inflorescence* of (1–)2–12(–30) spikes arranged in a raceme. *Spikelets* with one bisexual flower, strongly compressed laterally, in 2 rows, closely appressed to one face of the triangular rhachis. *Glumes* unequal, chartaceous, the upper as long as or longer than the rest of the spikelet, 1- to 9-veined, narrowly acute at apex and awned or not, the lower 1-veined. *Lemma* 1- to 5-veined, keeled, acute or minutely notched at apex, unawned, coriaceous, with a wide membranous margin. *Palea* slightly shorter than lemma. *Spikelets* falling entire at maturity.

About 15 species on the coasts of the Americas and the Atlantic coasts of Europe and Africa.

Goodman, P. J. (1969). *Spartina* Schreb. in Biological flora of the British Isles. *Jour. Ecol.* **57**: 285–287.

Goodman, P .J., Braybrooks, E. M., Marchant, C. J. & Lambert, J. M. (1969a). *Spartina alterniflora* × *S. maritima* in Biological flora of the British Isles. *Jour. Ecol.* **57**: 302–313.

Goodman, P. J., Braybrooks, E. M., Marchant, C. J. & Lambert, J. M. (1969b). *Spartina* × *townsendii* H. & J. Groves sensu lato in Biological flora of the British Isles. *Jour. Ecol.* **57**: 298–301.

Marchant, C. J. & Goodman P. J. (1969a). *Spartina alterniflora* Loisel. in Biological flora of the British Isles. *Jour. Ecol.* **57**: 291–295.

Marchant, C. J. & Goodman P. J. (1969b). *Spartina glabra* Muhl. in Biological flora of the British Isles. *Jour. Ecol.* **57**: 295–297.

Marchant, C. J. & Goodman P. J. (1969c). *Spartina maritima* (Curtis) Fernald in Biological flora of the British Isles. *Jour. Ecol.* **57**: 287–291.

Stewart, A., Pearman, D. A. & Preston, C. D. (1994). *Scarce plants in Britain*. Peterborough. [*S. maritima*.]

1. Spikes with 2 rows of crowded spikelets; upper glumes scabrid on the keel with rigid pricklets more than 0.3 mm, with an awn 3–8 mm   **5. pectinata**
1. Spikes with 2 rows of spikelets each spaced out 1–3 cm; upper glume glabrous or with soft hairs less than 0.3 mm on the keel, awnless   2.

2. Glumes glabrous or with hairs on keel only, sometimes very sparse on body also   **4. alterniflora**
2. Glumes softly pubescent on keel and body   3.

3. Ligules 1.8–3.0 mm at longest point; anthers (5–)7–10(–13) mm, with full pollen more than 45 μm across   **3. anglica**
3. Ligules 0.2–1.8 mm at longest point; anthers 4–8(–10) mm, if more than 7 mm then indehiscent with empty pollen; pollen less than 45 μm across   4.

4. Ligules 0.2–0.6 mm; anthers 4.0–6.5 mm, dehiscent with full pollen   **1. maritima**
4. Ligules 1.0–1.8 mm; anthers 5–7(–10) mm, indehiscent with empty pollen   **2. × townsendii**

1. S. maritima (Curtis) Fernald   Small Cord Grass
Dactylis maritima Curtis; *Dactylis stricta* Aiton; *S. stricta* (Aiton) Roth

Stiff *perennial herb* with tough, extensively creeping rhizomes, forming small tufts or loose patches. *Culms* 15–50 cm, erect, moderately stout, closely sheathed below, smooth, with many nodes. *Leaves* 2–18 cm × 3.5–4.5(–6) mm, green or purplish, erect and close to culm or spreading slightly, narrowly linear-lanceolate, with a fine, hard tip, stiff, flat or inrolled, closely ribbed above, smooth, finally disarticulating from the sheaths; sheaths persistent, smooth, rounded on the back; ligules 0.2–0.6 mm, as a dense fringe of hairs. *Panicles* 4.0–7.5(–10) cm, very narrow, exceeding the leaves, erect, of 1–5 spikes. *Spikes* 3–8 cm, spaced, rigid, one-sided, green or purplish; rhachis 3-angled, smooth, elongated at the tip into a bristle up to 14 mm. *Spikelets* 11–17 mm, closely overlapping, in 2 rows on one side of and appressed to the rhachis, flattened, 1-flowered, narrowly oblong, falling entire at maturity. *Glumes* softly and shortly hairy, keeled, the lower narrow, up to four-fifths the length of the upper, 1-veined, and firmly membranous, the upper linear-lanceolate, as long as the spikelet, firm except for the membranous margins and 3-veined. *Lemmas* a little shorter than the palea, narrowly oblong-lanceolate, minutely hairy upwards, 1-veined, firm except for the wide membranous margins. *Palea* lanceolate, slightly shorter than the spikelet, 2-veined, smooth. *Anthers* 4.0–6.5 mm. *Caryopsis* enclosed by the lemma and palea between the firm glumes. *Flowers* 7–9. $2n = 60$.

Native. Bare tidal sand or mud by the sea or in estuaries. Formerly from Devon, now local from the Isle of Wight north to Lincolnshire. Introduced in Co. Dublin. Coasts of south and west Europe from the Netherlands to Yugoslavia; North Africa.

2. S. × townsendii H. & J. Groves
Townsend's Cord Grass
= **S. alterniflora × maritima**

Deep-rooting *perennial herb*, with relatively numerous, vegetative shoots, forming close tussocks or patches and finally meadows, spreading by extensively creeping, scaly rhizomes. *Culms* 30–130 cm, erect, slender to stout, closely sheathed, with many nodes. *Leaves* 6–30 cm × 4–12 mm, narrowly linear-lanceolate, tapering to a fine, hard point at apex, the upper spreading at an angle of 20–45°, flat or rolled, firm, smooth, closely flat-ribbed above; sheaths overlapping, rounded on the back, smooth; ligules a dense fringe of silky hairs 1.0–1.8 mm. *Panicles* up to 25 cm, erect, contracted, linear to narrowly oblong, of 2–9 spikes, yellow or yellowish-green. *Spikes* 4–15 cm × 2–3 mm, erect, stiff; rhachis 3-angled, smooth, terminating in a bristle 10–20 mm. *Spikelets* 12–18 × 2.0–2.5 mm, closely overlapping, in 2 rows on one side of and appressed to the rhachis, narrowly oblong, flattened, mostly 1-flowered, minutely and sparsely hairy. *Glumes* keeled, pointed at apex, the lower up to two-thirds the length of the upper and 1-veined, the upper as long as the spikelet, oblong-lanceolate, tough except for the membranous margin, and 1- or closely 3-veined, both softly pubescent on keel and body. *Lemma* oblong-lanceolate, shorter than the upper glumes, keeled, rough on the keel, minutely hairy, 1- to 3-veined, with membranous margins. *Palea* lanceolate, slightly longer than the lemma, 2-veined. *Anthers* 5–7(–10) mm, very narrow, not opening; pollen sterile. *Flowers* 7–9. $2n = 62$.

This sterile hybrid arose prior to 1870 in Southampton Water, Hampshire, where it still occurs with its parents, and has spread to Dorset and elsewhere in Hampshire. It is scattered on the coasts of south Britain and east Ireland where it has either been introduced with *S. anglica* or been derived from it. It occurs in the absence of one or both parents, but not *S. anglica* from which it originates by a halving process of the chromosomes, thus reversing the doubling. It is frequently planted as a mud-binder and has been introduced to west Europe, North and South America, Australia and New Zealand.

3. S. anglica C. E. Hubb.   Common Cord Grass

Deep rooted *perennial herb*, spreading by soft, stout, fleshy rhizomes, forming large clumps and extensive meadows. *Culms* 30–130 cm, erect, stout, smooth, with many nodes. *Leaves* 10–45 cm × 6–15 mm, green or greyish-green, narrowly linear-lanceolate, with a fine, hard point at apex, flat or inrolled upwards, firm, closely flat-ribbed above, smooth, the upper widely spreading; sheaths overlapping, rounded on the back, smooth; ligules of dense, silky hairs 1.8–3.0 mm. *Panicles* 12–40 cm, erect, finally contracted and dense, of 2–12 spikes, overtopping the leaves. *Spikes* up to 25 cm, erect or slightly spreading, stiff; rhachis 3-angled, smooth, terminating in a bristle up to 50 mm. *Spikelets* 14–21 mm, closely overlapping, in 2 rows on one side of and appressed to the rhachis, narrowly oblong, flattened, 1- rarely 2-flowered, loosely to closely pubescent, falling entire at maturity. *Glumes* keeled, pointed at apex, the lower two-thirds to four-fifths the length of the upper and 1-veined, the upper as long as the spikelet, lanceolate-oblong, tough except for the membranous margins and 3- to 6-veined, both softly pubescent on keel and body. *Lemma* shorter than the upper glume, lanceolate-oblong, 1- to 3-veined, with broad, membranous margins, shortly hairy. *Palea* a little longer than the lemma, narrowly lanceolate, 2-veined. *Anthers* (5–)7–10(–13) mm; pollen fertile, more than 45 μm across. *Caryopsis* of a long green embryo enclosed between the lemma, palea and glumes. *Flowers* 7–11. $2n = 122$.

Native. An amphidiploid which arose in Southampton Water, Hampshire about 1890, by the doubling of chromosomes of *S.* × *townsendii* (*S. alterniflora* × *maritima*). It has spread naturally to Dorset and Sussex and has been extensively planted elsewhere in the British Isles north to central Scotland, and has become dominant over large areas of tidal mud-flats.

4. S. alterniflora Loisel.   Smooth Cord Grass
S. glabra Muhlenb. ex Bigelow

Robust *perennial herb*, with fleshy rhizomes forming clumps or beds. *Culms* 40–100 cm, erect, stout, closely sheathed, smooth. *Leaves* 10–40 cm × 5–12 mm, green, narrowly linear-lanceolate, with a fine, hard point at apex, flat, firm, equalling or overtopping the spikes, smooth, closely ribbed above, persistent; sheaths smooth, rounded on the back; ligule a dense fringe of hairs 1.0–1.8 mm. *Panicle* 10–25 × 3–5 cm, erect, contracted, of 3–13 scattered spikes. *Spikes* erect or slightly spreading, slender; rhachis 3-angled, smooth, tipped by a bristle 15–27 mm. *Spikelets* 10–18 mm, loosely overlapping, in 2 rows on one side of and appressed to the rhachis, lanceolate or narrowly oblong, flattened, 1-flowered, falling at maturity. *Glumes* keeled, pointed at apex, minutely hairy or rough on the keels, otherwise smooth, the lower very narrow, half to four-fifths the length of the spikelet, 1- to 3-veined and membranous, the upper broader, as long as the spikelet, 5- to 9-veined and tough except for the membranous margins. *Lemma* slightly shorter than the palea, lanceolate-oblong to narrowly ovate, 1- to 5-veined, thin, smooth. *Palea* a little shorter than the upper glume, lanceolate, 2-veined, membranous. *Anthers* 5–7 mm. *Caryopsis* enclosed by the lemma, paleas and glumes. *Flowers* 7–11. $2n = 62$.

Introduced. Coastal mud-flats. Planted in three sites in Hampshire from the early 1800s, but now extinct in all but one of them, despite early spread. Also planted in Ross-shire from 1920 and Dorset in 1963. Native of North America.

5. S. pectinata Bosc ex Link   Prairie Cord Grass
Perennial herb with firm, creeping rhizomes 3–8 mm thick. *Culms* 75–200 cm, erect, unbranched, leafy. *Leaves* 20–60 cm × 5–10 mm, broad and flat at the base, linear-lanceolate, involute towards the tip; ligules 1–3 mm. *Inflorescence* a panicle 10–30 × 2–6 cm. *Spikes* 5–20(–30), 2–15 cm × 3–7 mm, appressed or diverging, overlapping, with 2 rows of crowded spikelets. *Spikelets* 40–80 per spike, 8–12 mm. *Glumes* with a strongly scabrid keel, glabrous to sparsely hairy elsewhere, the lower 5–10 mm including an awn-like tip, the upper

15–25 mm including an awn-like tip about one-third to half its entire length. *Lemma* much shorter than the upper glume, apically narrowed and bidentate, pectinate on the upper half of the dorsal keel. *Palea* longer than the lemma but shorter than the upper glume. *Anthers* 5–7 mm with full pollen. Flowers 7–11. $2n = 40$.

Introduced. Grown for ornament. Persistent and spreading where it has been neglected by a freshwater lake in Galway since 1967 and by a canal in Hampshire since 1986. Native of North America.

81. Tragus Haller nom. conserv.
Annual herbs. Leaves flat; ligule a dense ring of hairs. *Inflorescence* a spike or a spike-like panicle, with 2–5 spikelets on an extremely short branch at each node. *Spikelets* with one bisexual flower. *Glumes* very unequal, the lower absent or vestigial, the upper at least as long as the flower, 5- to 7-veined, each vein with a line of long, hooked spines, acute at apex, unawned. *Lemma* 3-veined, acute at apex. Each *nodal group* of spikelets falling as a bur at maturity, the spikelets facing the centre of the bur with the glume hooks spines outermost.

Seven species throughout the tropics.

1. Upper glume 2–3 mm   **3. berteronianus**
1. Upper glume 3.5–4.5 mm   2.
2. Spikelets 3–5 at all or almost all nodes; upper glume
   with 7 veins and rows of spines, one pair of veins
   sometimes thinner and with smaller spines   **1. racemosus**
2. Spikelets 2 at all or almost all nodes; upper glume with
   5 veins and rows of spines   **2. australianus**

1. T. racemosus (L.) All.   European Bur Grass
Cenchrus racemosus L.; *Phalaris muricata* Forssk.; *Lappago racemosa* (L.) Honck.; *T. muricatus* Moench nom. illegit; *Lappago biflora* Roxb. nom. illegit.; *T. biflorus* Schult.; *Lappago decipiens* Fig. & De Not.; *T. decipiens* (Fig. & De Not.) Boiss.; *Nazia racemosa* (L.) Kuntze; *T. paucispina* Hack.; *T. arenarius* Bremek. et Oberm.

Annual herb. Culms up to 40 cm, decumbent or procumbent, rooting at the lower nodes, more or less branched, smooth. *Leaves* 1–3 cm × 2–3 mm, linear, flat, rigid, acuminate at apex, margins with tubercle-based setae; sheath somewhat inflated; ligule a line of stout hairs. *Inflorescence* a spike-like panicle 2–10 cm × 6–10 mm, with 2–5 spikelets on very short branches at each node. *Spikelets* 3.5–4.5 mm, with one bisexual flower. *Glumes* very unequal, the lower minute and hyaline, the upper 3.5–4.5 mm, lanceolate, acuminate at apex, 7-veined, one vein sometimes thinner than the others, longer than the flowers, with smaller spines, chartaceous, with hooked spines on the veins. *Lemma* 3-veined, hyaline, lanceolate, acute at apex, setulose on the back. *Palea* as long as the lemma. *Stamens* 3. *Styles* 2. *Caryopsis* enclosed by the lemma and palea. *Flowers* 8–10. $2n = 40$.

Introduced. Fairly frequent as a wool alien on tips and rough ground and in fields. Scattered records in southern England. Native of southern Europe.

2. T. australianus S. T. Blake   Australian Bur Grass
T. racemosus auct.

Tufted *annual herb. Culms* up to 60 cm, erect, oblique or geniculately ascending. *Leaves* (1–)2.5–6.0 cm × 3–5 mm, lanceolate, flat, acute to acuminate at apex, more or less undulate on the margin, subcordate at base; spinulosely ciliate on the margin; sheaths striate; ligules minute, ciliolate. *Inflorescence* of dense, stiff, cylindrical, spike-like panicles 5.0–7.5 cm; branches about 1 mm. *Spikelets* 2 on each branch, slightly unequal, 3.4–5.0 mm long, each spikelet with one bisexual flower. *Glumes* unequal, the lower suppressed or minute and hyaline, the upper 3.5–4.5 mm, lanceolate, acute to acuminate at apex, subcordate at base, as long as spikelets, 5-veined, densely spiny on the veins with thick uncinate spines. *Lemma* 2.5–3.0 mm, almost as long as the glume, membranous, oblong, acute-mucronate at apex, 3-veined. *Palea* as long as lemma, 2-veined. *Stamens* 3. *Styles* distinct; stigmas plumose. *Caryopsis* oblong to ellipsoid, slightly dorsally compressed, enclosed by the lemma and palea. *Flowers* 8–10.

Introduced. Fairly frequent wool alien on tips and rough ground and in fields. Scattered records in England. Native of Australia.

3. T. berteronianus Schult.   African Bur Grass
T. occidentalis Nees; *Lappago racemosa* var. *erecta* Kunth; *T. ciliatus* Leprieur ex Kunth; *Lappago phleoides* De Not.; *Lappago berteronianus* (Schult.) Steud.; *T. racemosus* var. *brevispiculus* Doel.; *Nazia occidentalis* (Nees) Scribn.; *T. mongolorum* Ohwi

Annual herb. Culms up to 60 cm. *Leaves* 2–10 cm × 1.5–2.0 mm, linear, graduating to a fine point at the apex, narrowed below, hairy along the margin; sheaths glabrous; ligule a dense ring of hairs. *Inflorescence* a spike-like panicle 2–15 × 0.4–0.6 cm, with 2 spikelets at almost all the nodes, the upper usually sterile. *Spikelets* (1.8–)2–3(–3.4) mm, each with one flower. *Glumes* very unequal, the lower minute and hyaline, the upper 2–3 mm, with 5 veins each bearing a row of hooked spines. *Lemma* 3-veined, acute at apex. *Palea* as long as lemma. *Stamens* 3. *Style* 2. *Caryopsis* enclosed by the lemma and palea. *Flowers* 8–10. 2n = 20.

Introduced. Fairly frequent wool alien on tips and rough ground and in fields. Scattered records in England. Native of southern Europe.

Subfamily 5. Panicoideae A. Br.

Annual to *perennial herbs. Leaves* without false petiole or cross-veins; ligule absent or a fringe of hairs or a membrane fringed with hairs, rarely membranous; sheaths not fused. *Inflorescence* a panicle, a spike, or an umbel or raceme of racemes or spikes. *Spikelets* with 2 flowers, the upper bisexual, the lower male or sterile, sometimes with paired spikelets, one bisexual the other male or sterile. *Glumes* 2. *Lemmas* firm to thick, 5- to 11-veined, awnless or with a terminal awn. *Stamens* 3. *Stigmas* 2. *Lodicules* usually 2, sometimes absent in the female. *Embryo* panicoid; photosynthesis C3-type alone or C3 plus C4-type; with or without Kranz leaf anatomy; fusoid cells absent; microhairs present. *Chromsome base-number* 5, 9 or 10.

Tribe 18. Paniceae R. Br.

Annual or less often *perennial herbs*, with or without rhizomes or stolons. *Leaves* with ligule absent, membranous, a fringe of hairs or a membrane fringed with hairs. *Inflorescence* a panicle, a spike or an umbel or raceme of racemes or spikes. *Spikelets* all the same, all bisexual.

82. Panicum L.

Annual herbs. Leaves flat or folded; ligule a dense fringe of hairs or membranous with a fringe of hairs round the top. *Inflorescence* a diffuse panicle. *Spikelets* with 2 flowers, the lower male or sterile with the lemma more or less as long as the spikelet, the upper bisexual, smaller, concealed between the upper glume and the lower glume. *Glumes* unequal, the lower much shorter than the spikelet, the upper more or less as long as the spikelet and closely resembling the lower lemma. *Lower lemma* 5- to 11-veined, awnless. *Spikelets* falling whole at maturity.

About 470 species, pantropical and extending to the temperate regions of South America.

1. Leaf-sheaths glabrous; lower glume less than one-third as long as the spikelet   2.
1. Leaf-sheaths with long patent hairs; lower glume more than one-third as long as the spikelet   3.
2. Spikelets 2.0–2.8 mm, subacute to obtuse at apex; lower flower usually male, with a well-developed palea
                                                                 1. schinzii
2. Spikelets 2.7–3.5 mm, acute to acuminate at apex; lower flower sterile, without or with a much reduced palea   **2. dichotomiflorum**
3. Spikelets 2.0–3.5 mm   **3. capillare**
3. Spikelets 4–6 mm   **4. miliaceum**

1. P. schinzii Hack.   Transvaal Millet
P. laevifolium Hack.

Annual herb. Culms up to 1 m, erect or ascending, glabrous, compressed to subterete, branched from most of the nodes. *Leaves* 7–20 × 5–8.5 cm, pale green, linear, shortly tapering to a very acute point, suberect, flat or folded, glabrous; sheaths lax, mostly shorter than the internodes, pale, striate, glabrous; ligules about 1 mm, ciliate, membranous. *Inflorescence* a panicle 15–30 cm, very lax and open, ovoid, branches solitary and remote, branchlets finely filiform; rhachis smooth. *Spikelets* 2.0–2.8 mm, pale green tinged with purple, oblong, subacute to obtuse at apex, not compressed, glabrous; flowers 2, the lower usually male, the upper bisexual. *Glumes* unequal, the lower much broader than long, obtuse to subacute at apex and less than one-third as long as the spikelet, the upper thin, green, oblong, obtuse at apex and 7- to 9-veined. *Lower flower* with palea well-developed. *Spikelets* falling whole at maturity. *Flowers* 8–10. 2n = 18, 48.

Introduced. Frequent casual from wool and bird seed on tips and waste ground. Scattered records from England. Native of tropical and southern Africa.

2. P. dichotomiflorum Michx   Autumn Millet

A somewhat succulent *annual herb*. *Culms* 50–200 cm, ascending or spreading from a geniculate base, branching. *Leaves* 10–50 cm × 2–20 mm, linear, attenuate at apex, flat, scaberulous or sometimes sparsely hairy on the upper surface, the white midrib usually prominent; sheaths glabrous; ligule a dense ring of white hairs 1–2 mm. *Inflorescence* a many-flowered, terminal and axillary panicle up to 40 cm, mostly included in the upper sheath, the main branches rather stiff and ascending, the branches rather short and appressed against the main branches. *Spikelets* shortly pedicelled, 2.7–3.5 mm, narrowly oblong-ovate, acute at apex, 7-veined, glabrous; flowers 2, the lower sterile, the upper bisexual. *Glumes* unequal, only about a quarter as long as the spikelet. *Lemma* of fertile flower smooth and shining, lemma of sterile flowers with a very reduced palea. *Spikelets* falling whole at maturity. *Flowers* 8–10. $2n = 36, 54$.

Introduced. Constant casual from soya-bean waste and occasionally from wool and other sources. Scattered records in southern England. Native of North America.

3. P. capillare L.   Witch Grass

Annual herb. *Culms* 20–80 cm, branching freely from the base, usually somewhat spreading from the base, papillose-hispid to rarely nearly glabrous. *Leaves* 10–25 cm × 5–15 mm, linear, gradually narrowed to a pointed apex, shortly hairy on midrib beneath; sheaths hispid; ligule 1–3 mm. *Inflorescence* a many-flowered panicle, often making up half the total length of the plant, the branches divaricate and spreading at maturity, the whole panicle breaking away and rolling before the wind. *Spikelets* 2.0–3.5 mm, ovate, attenuate and acuminate at apex, 7- to 9-veined, glabrous; 2-flowered, the lower sterile with a much reduced palea, the upper bisexual. *Glumes* unequal, the lower large and clasping, nearly half as long as the upper glume and acute at apex, the upper glume strongly ribbed. *Lower lemma* 5- to 11-veined, awnless; upper lemma awnless. *Spikelets* falling whole at maturity. *Caryopsis* tightly enclosed in the lemma and palea. *Flowers* 8–10.

Introduced. Frequent casual from bird seed and sometimes other sources on tips and waste ground. Scattered records in Britain and the Channel Islands. Native of North America.

4. P. miliaceum L.   Common Millet

Tufted *annual herb*. *Culms* 20–120 cm, erect or ascending, simple or sparingly branched; nodes slightly to densely short hairy. *Leaves* 15–30 cm × 8–25 mm, linear, long-tapering towards the slender, pointed apex, the base more or less as wide as the sheath, rounded and subcordate, the margin flat and often ciliate towards the base, otherwise glabrous except, sometimes, for being minutely hispid on the midrib beneath, but sometimes loosely hairy with tuberculate-based hairs; sheaths hirsute with spreading tubercular-based hairs; ligule a ciliate rim, the hairs about 1.2 mm. *Inflorescence* a diffuse panicle up to 30 cm, primary branches up to

three-quarters the length of the panicle; pedicels about 2 mm. *Spikelets* 4–6 mm, 2-flowered, the lower male or sterile, the upper bisexual. *Glumes* unequal, the lower half to three-quarters as long as the spikelet and 5- to 7-veined, the upper as long as the spikelet and 11- to 13-veined. *Lower flower* sterile, its lemma similar to the upper glume and more or less as long, the palea hyaline and up to one-third as long as the lemma; upper flower smooth and shining, its lemma 3.0–3.5 mm, crustaceous and its margin embracing the palea. *Palea* crustaceous, about as long as the lemma. *Stamens* 3. *Styles* distinct. *Caryopsis* tightly enclosed in the lemma and palea. *Flowers* 8–10. $2n = 36, 54, 72$.

Before flowering, plants may be mistaken for *Zea mays* or *Sorghum bicolor*, but these have the membranous part of the ligule longer, not shorter, than the distal fringe and the leaf-sheaths do not have long, patent hairs over the whole surface.

Introduced. A common casual from bird seed and sometimes other sources on tips and waste ground. Scattered in Britain and the Channel Islands. Native of Asia. It is grown as a cereal crop in some countries.

83. Echinochloa P. Beauv. nom. conserv.

Annual herbs. *Culms* erect. *Leaves* flat; ligules absent or represented by a ring of hairs. *Inflorescence* a panicle of more or less dense spikes or racemes, or the secondary racemes again racemosely branched, often with long, stiff hairs especially in tufts at the branch points. *Spikelets* with 2 flowers, the lower male or sterile with the lemma more or less as long as the spikelet, the upper bisexual, smaller and crowded between the upper glume and lower lemma. *Glumes* unequal, the lower much shorter than the upper and more or less as long as the spikelet, the upper closely resembling the lower lemma. *Lower lemma* 5- to 7-veined, awned or awnless; upper lemma awnless. *Spikelets* falling whole at maturity. *Caryopsis* enclosed in lemma and palea.

1. Inflorescence with lateral spikes or racemes all or mostly separate, obviously branched; lower flower male or sterile, the lemma awned or not   2.

1. Inflorescence with fat lateral spikes or racemes close together and forming entirely or for the most part a single, lobed, elongate head; lower flower sterile, its lemma not awned   3.

2. Leaves usually more than 10 mm wide; lower primary branches of inflorescences usually branched again; spikelets 3–4 mm; lower flower usually sterile; lower lemma often awned   **1. crus-galli**

2. Leaves usually less than 8 mm wide; primary branches of inflorescence simple; spikelets 1.5–3.0 mm; lower flower usually male; lower lemma acute to apiculate at apex with an awn 0–2 mm   **3. colona**

3. Spikelets 3–4 mm; glumes and lower lemma bright green, usually tinged purplish, sometimes completely purplish; lower lemma acuminate at apex   **2. utilis**

3. Spikelets 2.5–3.5 mm; glumes and lower lemma yellowish-green to straw-coloured; lower lemma acute to subacute, sometimes minutely apiculate at apex

4. frumentacea

1. E. crus-galli (L.) P. Beauv. Cockspur
Panicum crus-galli L.

Tufted *annual herb. Culms* 30–120 cm, erect or spreading, rather stout, usually branched, smooth. *Leaves* 8–35 cm × (8–)10–20 mm, green, glabrous, linear, finely pointed at apex, flat, soft, rough on the thickened margins, smooth beneath; sheaths smooth, keeled; ligules absent. *Inflorescence* a panicle 6–20 × up to 8 cm, green or purplish, lanceolate to ovate, erect or nodding, formed of few to many, scattered or clustered, spike-like racemes, often branched again; racemes up to 6 cm, very dense, their rhachis rough and usually with bristly hairs; pedicels 0.2–2.0 mm. *Spikelets* 3–4 mm, crowded, in pairs or clusters on one side of the rhachis, falling entire at maturity, ovate-elliptical in back view, semi-elliptical in side view, pointed at apex, or awned, with 2 flowers, the lower male or sterile and the upper bisexual. *Glumes* membranous, the lower broad, about one-third the length of the spikelet and 3- to 5-veined, the upper covering the rounded back of the spikelet, 5-veined, the veins bearing short, stiff hairs. *Lower lemma* as long as the spikelet, similar to the upper glume, but flat or depressed on the back, pointed at apex or abruptly narrowed into a short cusp or an awn up to 5 cm, 5- to 7-veined; palea ovate, shorter than the lemma. *Upper lemma* as long as the spikelet, rounded on the back, tough, white or yellowish, smooth; palea as long as the lemma. *Anthers* about 1 mm. *Caryopsis* broadly ellipsoid, pale brownish, tightly enclosed by the hard lemma and palea. *Flowers* 8–10. $2n = 36, 54$.

Introduced. Sometimes naturalised as a weed in cultivated ground. Most often a casual on tips, waysides and waste ground, mostly from bird seed. Scattered throughout most of the British Isles, but particularly in south Britain. Throughout the warm regions of the world.

2. E. utilis Ohwi & Yabuno Japanese Millet
E. frumentacea auct.; *E. frumentacea* subsp. *utilis* (Ohwi & Yabuno) Tzvelev

Annual herb. Culms up to about 1 m, erect, simple or with few branches; nodes glabrous. *Leaves* 15–35 cm × 4–25 mm, linear, tapering upwards; sheaths glabrous; ligules absent, the area in which they would occur glabrous. *Inflorescence* an erect panicle, 7–20 cm; branches 1–5 cm, fat, the whole forming a single lobed head. *Spikelets* crowded, 3–4 mm without awns, bright green and often purplish. *Lower glume* one-quarter to one-third as long as the spikelet, 5-veined, the veins scabrid, usually scabrid-pubescent between the veins; upper glume slightly shorter than spikelets, 5-veined, the veins scabrid, minutely scabrid-pubescent between the veins. *Lower flower* with the lemma 7-veined, the palea shorter than the lemma, and acuminate at the apex, the upper flowers sterile with the lemma 2.5–3.0 mm and withered at the tip. *Stamens* 3. *Styles* distinct. *Caryopsis* brownish, enclosed in the lemma and palea. *Flowers* 6–8. $2n = 54$.

Introduced. Casual bird seed alien on tips and waste ground. Scattered in Britain, mainly the south. A culti-

vated derivative of *E. crus-galli*, which originated in Japan.

3. E. colona (L.) Link Shama Millet
Panicum colonum L.

Annual herb. Culms up to 60 cm, loosely tufted, usually decumbent at the base and rooting at the nodes, finally erect, smooth, glabrous or rarely sparsely hairy at the lower nodes. *Leaves* up to 10 cm × 3–8 mm, linear, acuminate at apex, rounded at base, glabrous on both surfaces, flat, flaccid, thin, scaberulous on the margins; lower sheaths very lax, the upper clasping, smooth, glabrous and striate; ligule absent. *Inflorescence* a raceme up to 3 cm, ascending, with closely packed spikelets spaced along a central rhachis, each raceme of shortly pedicellate spikelets which are secund on a scabrid rhachis. *Spikelets* in 4 rows, 1.5–3.0 mm, ovate, acute at apex, green or tinged with purple. *Glumes* very unequal, the lower about 1 mm, broadly ovate, cuspidate at apex, 3-veined and roughly hairy on the veins, the upper as long as the spikelet, elliptical, obtuse at apex, 5- to 7-veined and roughly hairy or scabrid on the veins. *Lower flower* male, its lemma similar to the upper glumes, but flat and acute to apiculate at apex and its *palea* oblong, acute at apex, hyaline, the upper flower bisexual, its lemma 1.5–2.0 mm, ovate, acute at apex, more or less cuspidate, whitish and its palea of the same texture. *Anthers* about 1 mm. *Caryopsis* broadly elliptical. *Flowers* 6–8. $2n = 36, 48$.

Introduced. Casual on tips and waste ground. Occasional in south Britain. A weed throughout the warmer regions of the world.

4. E. frumentacea Link White Millet
E. crus-galli var. *frumentacea* (Link) W. F. Wight; *E. crus-galli* var. *edulis* A. S. Hitchc.; *E. colona* var. *frumentacea* (Link) Ridley; *Panicum crus-galli* var. *edule* (A. S. Hitchc.) Thell. ex Bouly de Lesd.; *Panicum frumentaceum* auct.

Annual herb. Culms up to 1.5 m, erect, branching from the lower nodes; nodes puberulous to glabrous. *Leaves* 5–30 cm × 3–20 mm, linear, long-tapering upwards, glabrous on both surfaces, scabrid on the margins; sheaths glabrous; ligules absent but sometimes the area where the ligule would be, shortly hairy. *Inflorescence* a more or less erect panicle with branches 1–3 cm. *Spikelets* crowded, 2.5–3.5 mm, occasionally bearing 2 fertile flowers as well as a sterile flower. *Lower glumes* one-third to a half as long as the spikelet, 3- or rarely 5-veined, scabrid on the veins, scabrid-pubescent between the veins; upper glume slightly shorter than spikelet, 5- to 7-veined, the veins hispid, scabrid-pubescent between the veins. *Lower flower* with the lemma yellowish-green to straw-coloured and 7-veined, the palea more or less shorter than the lemma and acute to apiculate at apex; the upper flower with the lemma about 2–3 mm and the tip withered. *Palea* subequal to lemma. *Stamens* 3. *Styles* distinct. *Caryopsis* whitish, enclosed in the lemma and palea. *Flowers* 6–8. $2n = 54$.

Introduced. Casual bird seed alien on tips and waste

ground. Scattered localities in Britain. A cultivated derivative of *E. crus-galli* which originated in Japan.

84. Brachiaria (Trin.) Griseb.

Panicum subtaxon *Brachiaria* Trin.

Annual herbs. Leaves flat; ligule a dense fringe of hairs. *Inflorescence* a panicle of racemes with the spikelets in 2 rows on one side of the rhachis. *Spikelets* with 2 flowers, the lower male or sterile with the lemmas more or less as long as the spikelet, the upper glume and the lower lemma. *Glumes* unequal, the lower less than half as long as the upper, the upper closely resembling the lower lemma. *Lower lemma* 5- to 7-veined, obtuse at apex, awnless, the upper obtuse to rounded at apex, the lower obtuse to rounded at apex and awnless. *Spikelets* falling whole at maturity.

About 100 species in the tropics, mainly the Old World.

1. Spikelets 1.7–3.7 mm, pubescent; racemes 3–14
   1. eruciformis
1. Spikelets 3.5–4.5 mm, glabrous; racemes 2–6
   2. platyphylla

1. B. eruciformis (Sm.) Griseb.

Small-flowered Signal Grass

Panicum eruciforme Sm.

Annual herb. Culms up to 60 cm, loosely tufted from a decumbent and geniculately ascending base, rooting at the nodes, smooth, glabrous, very shortly hairy at the nodes or glabrous. *Leaves* up to 9 cm × 3–7 mm, linear, somewhat rigid, more or less hairy on both surfaces or more or less glabrous above, scabrid on both surfaces and on the margins, greyish-green. *Inflorescence* of 3–14 racemes, 5–25 mm, spike-like, curved, ascending and appressed to the central rhachis which is 1.6 cm, very shortly pedicellate. *Spikelets* 1.7–3.7 mm, dorsally compressed, elliptical in outline, more or less hairy. *Lower glumes* 0.2 mm, a hyaline veinless scale which is truncate at apex, the upper glume as long as the spikelet, elliptical, obtuse at apex and 5-veined. *Lemma* of the lower flower similar to the upper glumes, 5-veined, hairy, neuter or enclosing a male flower, its palea hyaline; lemma of the upper flower 1.2–1.7 mm, elliptic-oblong, obtuse at apex, very firm, shining white and with a long palea of the same texture. *Anthers* about 1 mm. *Caryopsis* oblong-elliptical, dorsally compressed. *Flowers* 6–9. $2n = 18$.

Introduced. Rare bird seed alien of tips and waste places. South Europe and the east Mediterranean area eastwards to India; tropical and South Africa.

2. B. platyphylla (Griseb.) Nash

Broad-leaved Signal Grass

Panicum platyphyllum Griseb.

Tufted *annual herb. Culms* 25–90 cm, 1.0–2.5 mm thick, mostly ascending, sometimes decumbent with a stoloniferous base. *Leaves* 5–12 cm × 6–12 mm, linear, flat and glabrous; sheaths glabrous; ligule about 0.5 mm, a fringe of hairs. *Inflorescence* a spike-like panicle of racemes which are 8–30 cm, subsecund on a flattened,

twisted rhachis with 2 rows of spikelets on the abaxial side; racemes 2–6, many-flowered. *Spikelets* 3.5–4.5 mm, compressed and appearing flattish, glabrous, with 2 flowers, the lower male or sterile, the upper bisexual. *Lower glume* about one-third the length of the spikelet, the upper glume and the sterile lemma about as long as the spikelet. *Lemma* of sterile flowers 5- to 7-veined, as large as the spikelet, the lemma of the fertile flowers cartilaginous-indurate, obtuse at apex, marginally revolute, awnless, and enclosing a palea of the same texture, superficially, irregularly, microscopically and transversely rugose. *Flowers* 8–9. $2n = 18, 36$.

Resembles *Echinochloa colona*, but differs from it in being more pubescent, the spikelets borne in only 2 rows, the lower lemma obtuse at apex, the well-developed ligule and often the shorter hair of the leaf sheath.

Introduced. Casual from bird seed and soya-bean waste, occasionally on tips and waste ground. Scattered records in southern Britain. Native of North America.

85. Urochloa P. Beauv.

Annual herbs. Leaves flat. *Inflorescence* of racemes along an axis, the rhachis narrowly membranously winged. *Spikelets* single or paired, in 2 rows, subtended by a long bristle; flowers 2, the lower male or sterile, the upper bisexual. *Glumes* usually unequal, the upper shortly acuminate and 7-veined. *Lower lemma* shortly acuminate, the upper lemma with a distinct terminal awn. *Palea* as long as lemma. *Stamens* 3. *Styles* distinct. *Caryopsis* tightly enclosed by the hardened lemma and palea.

Twelve species in the Old World tropics, mainly Africa.

1. U. panicoides P. Beauv.

Sharp-flowered Signal Grass

Annual herb. Culms 10–100 cm, tufted, often ascending from a prostrate, rooting base. *Leaves* 2–5 cm × 5–15 mm, semi-amplexicaul, coarse, glabrous or pubescent, the margins tuberculate-ciliate at least near the base. *Inflorescence* of 2–7(–10) racemes on a rhachis 1–6 cm, each raceme 1–6 cm, bearing single or sometimes paired spikelets on a narrowly winged rhachis; pedicels clothed with white hairs. *Spikelets* (2.5–)3.5–5.0 mm, elliptical and acute at apex, subtended by long bristles. *Lower glume* one-quarter to half as long as the spikelets, ovate, 3- to 5-veined and obtuse to subacute at apex, the upper glume glabrous or pubescent and often with cross-veins. *Lower lemma* sometimes has a stiffly hairy fringe, the upper with a mucro 0.3–1.0 mm. *Flowers* 8–9. $2n = 36, 48$.

Introduced. Casual on tips and waste ground from bird-seed. Of scattered occurrence in central and south England. Native of Africa and west Asia.

86. Eriochloa Kunth

Annual or *perennial herb. Leaves* flat; ligule reduced to a ciliate rim. *Inflorescence* a rather regular raceme of racemes, the main branches more or less appressed to the main axis and often slightly branched again, the spikelets in barely recognisable rows. *Spikelets* with 2 flowers, the lower sterile with the lemma almost as long as the spikelet and the palea absent, the upper bisexual, smaller, concealed between the upper glume and lower lemma, and the palea with a small bud-like swelling at the apex of the pedicel. *Lower glume* more or less absent; the upper glume as long as the spikelet. *Lower lemma* 5-veined, acuminate at apex and awned to 2 mm, the upper lemma obtuse to rounded at apex with an awn 0.3–1.0 mm. *Spikelets* falling whole, disarticulating immediately below the bead-like swelling.

Thirty species in the tropics.

The following species may be an aggregate of closely allied species which occur in Australia and which have not been looked for in the English plants.

1. E. pseudoacrotricha (Stapf. ex Thell.) S. T. Blake
Cup Grass

E. ramosa var. *pseudoacrotricha* Stapf ex Thell.

Perennial herb. Culms up to 1 m, erect or ascending, branching, pubescent at the nodes. *Leaves* 3–30 cm × 1.5–4.0 mm, linear, flat, attenuate at apex, shortly hairy near the ligule and sometimes on the lower surface, otherwise glabrous; sheaths glabrous or shortly hairy at the base; ligule reduced to a ciliate rim. *Inflorescence* a panicle up to 18 × 1 cm, with 2–10 appressed or slightly spreading racemes each 2–10 cm. *Spikelets* 4–6 mm including awn, solitary or in pairs, with a distinct bead-like swelling at its base; flowers 2, the lower male or sterile, the upper bisexual. *Glumes* unequal, the upper tapering to a bristle, the lower reduced to a minute cupular sheath clasping the bead-like swelling. *Lemma* of lower flowers more or less as long as the spikelet; lemma of upper flower two-fifths to three-fifths as long as spikelet, 5-veined. *Palea* more or less as long as the lemma. *Stamens* 3. *Styles* distinct. *Flowers* 8–9.

Introduced. A rather infrequent wool alien on tips and in fields and waste places. Scattered records in England. Native of Australia.

87. Paspalum L.

Perennial herbs with widely spreading stolons. *Leaves* flat; ligule membranous. *Inflorescence* a raceme of 2(–4) racemes, with spikelets in 2 rows on one side of the rhachis. *Spikelets* with 2 flowers, the lower sterile with the lemma as long as the spikelet and the palea very small, the upper bisexual, smaller, concealed between the upper glume and the lower lemma and with a palea. *Lower glume* more or less absent, the upper glume very similar to the lower lemma. *Lower lemma* 3- to 5-veined and acute or slightly apiculate at apex, the upper lemmas subacute at apex and apiculate. *Spikelets* falling whole.

About 330 species in the tropics, predominantly in the New World.

1. P. distichum L.
Finger Grass

P. paspalodes (Michx) Scribn.; *Digitaria paspalodes* Michx

Perennial herb with widely spreading stolons. *Culms* up to 50 cm, decumbent at the base, rooting at the nodes, finally erect, smooth, glabrous except for the stiffly hairy

nodes. *Leaves* up to 12 cm × 2–4 mm, flat, usually more or less erect, linear, acute at apex, glabrous on the surfaces but ciliate towards the base of the scaberulous margins; lower sheaths scarious and loose, the upper green, keeled and ciliate on the margin but otherwise smooth and glabrous; ligules rarely up to 1.5 mm, very narrow, membranous. *Inflorescence* a raceme 2–7 cm, in pairs, spike-like, each consisting of a flattened rhachis on the under surface of which are situated 2 rows of shortly pedicellate, closely packed spikelets. *Spikelets* 2.5–3.5 mm, with 2 flowers, oblong or elliptic-oblong, acute at apex, flattened on one surface, convex on the other. *Lower glume* absent or represented by a small triangular scale, the upper similar to the lower lemma, as long as the spikelets, 3- to 5-veined and appressed-pubescent. *Lemma* of the lower neuter, flower flat, the shape of and as long as the spikelet and 3- to 5-veined, the lemma of the upper bisexual spikelet indurated, smooth, glabrous and apiculate at the tip. *Flowers* 6–9. $2n = 40, 48, 60$.

Introduced. Naturalised in damp ground by the sea at Mousehole, Cornwall since 1971, and by a canal in Middlesex since 1984. Native of America, widespread as an introduction in the warmer parts of the world.

88. Setaria P. Beauv. nom. conserv.

Annual or *perennial herbs* with rhizomes. *Leaves* flat; ligule a dense fringe of hairs. *Inflorescence* a dense spike-like panicle with very or more or less vestigial, crowded branches, sometimes interrupted in the lower part. *Spikelets* with 2 flowers, the lower male or sterile, the upper bisexual and slightly smaller, oblong to ovate, subtended by bristles which persist on the panicle after the spikelets have fallen. *Glumes* unequal, the lower absent or up to two-thirds as long as the spikelet, the upper about two-thirds as long as the spikelet. *Lower lemma* 5-veined, obtuse at apex, awnless; upper lemma obtuse to rounded at apex, awnless.

About 100 species in the tropics and subtropics.

1. Bristles (4–)5–12 below each spikelet; upper glume scarcely longer than lower glume, about half to two-thirds as long as the spikelet; lower sterile flower with the palea almost as long as the lemma $\qquad$ 2.

1. Bristles 1–3 below each spikelet; upper glume much longer than lower glume, about two-thirds to as long as the lemma $\qquad$ 3.

2. Perennial; spikes less than 5 mm wide when mature (excluding bristles); spikelets 2.0–2.5(–3.0) mm
1. parviflora

2. Annual; spikes more than 6 mm wide when mature (excluding bristles); spikelets (2.5–)3.0–3.3 mm   **2. pumila**

3. Main rachis (often rather sparsely) hispid, with pricklets less than 0.2 mm; bristles usually with backwardly-directed (rarely forwardly directed) barbs $\qquad$ 4.

3. Main rachis densely pubescent with hairs more than (0.2–)0.5 mm; bristles always with forwardly directed barbs $\qquad$ 6.

4. Leaf-sheaths glabrous; spikelets 1.5–2.0 mm   **4. adhaerens**

4. Leaf-sheaths pubescent on margin; spikelets 2.0–2.3 mm $\qquad$ 5.

5. Bristles with backwardly directed barbs
                                    3(i). verticillata var. **verticillata**
5. Bristles with forwardly directed barbs
                                    3(ii). verticillata var. **ambigua**
6. Panicle often more than 15 cm × 15 mm, the bristles
   often not or scarcely longer than the spikelet-clusters;
   spikelets disarticulating below the upper lemma
   leaving the glumes and lower lemma on the rhachis;
   upper lemma smooth   **7. italica**
6. Panicle rarely as much as 15 cm × 15 mm, the bristles
   always much longer than the spikelets; spikelets falling
   whole, leaving only pedicels and bristles on the
   rhachis; upper lemma finely transversely rugose   7.
7. Leaves glabrous; upper glume as long as or almost as
   long as the spikelets; spikelets (1.8–)2.0–2.5(–2.7) mm
                                                5. viridis
7. Leaves hairy, often sparsely so; upper glume two-thirds
   to three-quarters as long as spikelets; spikelets
   (2.5–)2.7–3.0 mm   **6. faberi**

1. S. parviflora (Poir.) Kerguélen
                              Knotroot Bristle Grass
S. geniculata P. Beauv.; *Panicum geniculatum* Willd.,
non Lam.

Perennial herb with a short, knotty rhizome. *Culms*
20–100 cm, geniculate at the lower nodes, otherwise
erect, slender, compressed. *Leaves* 6–30 × 3–8 mm,
linear, flat, long- pointed at apex, strictly erect; ligule a
fringe of hairs. *Inflorescence* a yellowish panicle, 1–10
cm × 3–4 mm (excluding bristles), cylindrical, dense,
not interrupted; rhachis with dense, short hairs; bristles
8–12, mostly 7–10 mm, numerous, stiff, with forwardly
directed barbs. *Spikelets* 2.0–2.5(–3.0) mm, with 2
flowers, the lower male or sterile with the lemma as long
as the spikelet, the upper bisexual, slightly smaller and
concealed between or protruding from within the upper
glume and lower lemma. *Glumes* unequal, the lower
about one-third the length of the spikelet, the upper
longer and with an excurrent midrib. *Lemma* of the
sterile flower equalling that of the fertile flower, 5-
veined, obtuse at apex and awnless, the lemma of the
fertile flower elliptic-ovate, acute at apex, striate and
transversely rugose. *Palea* of sterile flower well-developed. *Caryopsis* tightly enclosed by the hard lemma and
palea. *Flowers* 8–10. $2n = 72$.

Introduced. Casual from grain and bird seed, on tips
and waste ground. Scattered localities in southern
England and south Wales. Native of North America.

2. S. pumila (Poir.) Schult.   Yellow Bristle Grass
S. lutescens F. T. Hubb.; *S. glauca* auct.

Annual herb. Culms 6–75 cm, loosely tufted or solitary,
erect or bent and ascending, slender, with 2–4 nodes,
rough under the panicle. *Leaves* 6–30 cm × 4–10 mm,
green, linear, finely pointed at apex, flat, hairy towards
the base or glabrous, minutely rough on the margins;
sheaths smooth, the lower compressed, keeled; ligule a
dense fringe of fine hairs. *Inflorescence* a panicle 1–15
cm × (4–)6–10 mm, spike-like, dense, very bristly, erect,
cylindrical; branches up to 1 mm, usually bearing a
single spikelet and beneath it 5–10 bristles, the rhachis

shortly hairy; bristles up to 10 mm, yellowish to reddish-
yellow, fine, minutely rough with forward-directed
barbs, straight or wavy. *Spikelets* (2.5–)3.0–3.3, broadly
elliptical in back view, obtuse at apex, broadly semi-
elliptical in side view, falling entire at maturity, 1- to 2-
flowered, with the lower flower male or barren and the
upper bisexual. *Glumes* broadly ovate, thinly membranous, the lower up to half the length of the spikelet and
3-veined, the upper up to two-thirds the length of the
spikelet and 5-veined. *Lower lemma* as long as the
spikelet, broad, flat or slightly depressed on the back, 5-
veined; palea flat, broadly elliptical. *Upper lemma* as
long as the spikelet, broadly boat-shaped, rounded on
the back, prominently transversely wrinkled, tough,
rigid, becoming yellow or brown; palea broadly elliptical, as long as the lemma. *Anthers* up to 1.5 mm.
Caryopsis broadly ellipsoid, tightly enclosed by the hard
lemma and palea. *Flowers* 7–10. $2n = 36, 72$.

Introduced. Casual in cultivated and waste ground
and on tips. Occasional in south and central Britain and
the Channel Islands, very scattered in northern Britain.
Warm temperate regions of the Old World.

3. S. verticillata (L.) P. Beauv.   Rough Bristle-grass
Panicum verticillatum L.

Annual herb. Culms up to 50 cm, loosely tufted, erect or
somewhat decumbent at the base and genicately
ascending, smooth, glabrous but scaberulous below the
panicle. *Leaves* up to 30 cm × 4–16 mm, linear, acuminate at apex, flat, scaberulous on the upper surface and
on the margins, flaccid, sparsely hairy on both surfaces
with bulbous-based hairs or almost glabrous; sheaths
somewhat compressed and loose, ciliate on the margins;
ligule very short, densely ciliate. *Inflorescence* a *panicle*
up to 10 cm × 10–15 mm, narrow, erect, cylindrical,
rarely very dense, usually somewhat open and often
interrupted towards the base, very bristly, the bristles
3–18 mm; rhachis angled, scabrid with barbs; branches
short. *Spikelets* 2.0–2.3 mm, elliptical in outline,
rounded on the back, flat on the face, supported by 1–3
scabrid bristles. *Glumes* very unequal, the lower 0.6–0.7
mm, ovate, rounded or acute at apex and 1- to 3-veined,
the upper as long as the spikelet, two-thirds as long as
the lemma, convex, membranous, 5- to 7-veined.
Lemma of the neuter lower flower flat and as long as
the spikelet, the palea short; lemma of the bisexual
flower indurated, a little shorter than the spikelet and
finally granular, the palea of the same texture and flat.
Anthers about 0.3 mm. *Caryopsis* tightly enclosed in the
hardened lemma and palea. *Flowers* 7–9. $2n = 18, 36$,
c. 54.

(i) Var. verticillata
Bristles subtending spikelets with backwardly directed
barbs.

(ii) Var. ambigua (Guss.) Park.
S. verticillatiformis Dumort.; *S. ambigua* (Guss.) Guss.,
non Schrad.; *Panicum verticillatum* var. *ambigua* Guss.
Bristles subtending spikelets with forwardly directed
barbs.

Introduced. Wool and bird-seed casual. Occasional in south and central Britain and the Channel Islands. Central and south Europe and the Mediterranean region eastwards to Indonesia; Atlantic Islands; introduced and naturalised in most warmer parts of the world.

4. S. adhaerens (Forssk.) Chiov. Adherent Bristle Grass
S. verticillata subsp. *aparine* (Steud.) T. Durand & Schinz; *Panicum adhearens* Forssk.

Tufted *annual herb*. *Culms* 15–60 cm, usually decumbent basally, the lower nodes occasionally rooting and geniculate, sparsely to moderately branched, ascending above. *Leaves* up to 30 cm × 5–11 mm, linear, acute at apex, shortly and sparsely papillose-hairy on both surfaces; sheaths glabrous with narrow hyaline margins towards the summit; ligule absent. *Inflorescence* a cylindrical panicle 2–6(–8) cm, the axis retrorsely scabrid-hispid on the angles, but mostly hidden in the dense mass of flowers, the bristles 1–3, short and with retrorse barbs. *Spikelets* 1.5–2.0 mm, 2-flowered, the lower male or reduced, the upper bisexual. *Lower glumes* much shorter than the spikelet, several veined and membranous, the upper glume nearly as long as the spikelet, several veined and membranous. *Lower lemma* several-veined, membranous and usually not as long as the upper lemma, the upper lemma indurate, strongly convex and the margin revolute and clasping the palea of the same texture. *Caryopsis* tightly enclosed in the lemma and palea. *Flowers* 7–9. 2n = 18.

Introduced. Casual from bird seed and perhaps other sources on tips and waste and cultivated ground. Scattered in southern England and the Channel Islands. Native of the tropics of the Old World.

5. S. viridis (L.) P. Beauv.   Green Bristle Grass
Panicum viride L.

Loosely tufted *annual herb*. *Culms* 10–60 cm, erect or more usually bent at the base, slender, with 3–5 nodes, sometimes branched in the lower part, rough towards the panicle. *Leaves* 3–30 cm × 4–10 mm, green, narrowly linear-lanceolate, finely pointed at the apex, flat, glabrous, minutely rough; sheaths hairy on the margins, round; ligule a dense fringe of silky hairs. *Inflorescence* a panicle 1–10 × 4–10 cm, spike-like, very bristly, mostly erect, very dense, cylindrical or tapering upwards, greenish or purplish; branched up to 3 mm; bristles up to 10 mm, straight or wavy, 1–3 beneath each spikelet, minutely rough. *Spikelets* (1.8–)2.0–2.5(–2.7) mm, elliptic-oblong in back view, very obtuse at apex, semi-elliptic-oblong in side view, 2-flowered, falling entire at maturity. *Glumes* thinly membranous, the lower up to one-third the length of the spikelet, ovate and 1- to 3-veined, the upper covering the whole of the back of the spikelet and 5-veined. *Lower lemma* resembles the upper glume, 5- to 7-veined; palea lanceolate, up to half the length of the lemma, very thin. *Upper lemma* as long as the spikelet, elliptic-oblong, obtuse at apex, pale, rounded on the back, becoming tough and rigid, very finely wrinkled; palea lanceolate, flat on the back.

Anthers about 0.8 mm. *Caryopsis* ellipsoid, tightly enclosed by the hardened lemma and palea. *Flowers* 8–10. 2n = 18.

Introduced. Weed of cultivated and waste ground and on tips, often only casual, sometimes naturalised. Frequent in south and central Britain and the Channel Isles, sporadic elsewhere. Warm temperate regions of the.Old World.

6. S. faberi Herrm.   Nodding Bristle Grass
S. autumnalis Ohwi

Annual herb. *Culms* 40–100(–180) cm, slightly scabrous above, branching below. *Leaves* 10–30 cm × 8–20 mm, linear, flat, with a long drawn out tip, strigose-hairy on the upper surface with shorter hairs beneath; sheath ciliate; ligule a fringe of hairs. *Inflorescence* a panicle 5–10 cm × up to 6 mm (excluding bristles), cylindrical, green or slightly purplish, nodding; bristles numerous. *Spikelets* dense, (2.5–)2.7–3.0 mm, with 2 flowers, the lower male or sterile, the upper bisexual, narrowly to broadly ovate, obtuse at apex. *Glumes* unequal, the lower about one-third as long as the spikelet, the upper two-thirds to three-quarters as long as the spikelet and broadly ovate. *Lemma* of lower flower as long as the spikelet; lemma of upper flower ovate, subobtuse at apex, punctate and minutely and transversely rugulose. *Stamens* 3. *Styles* free. *Caryopsis* elliptical, enclosed between the hardened lemma and palea. *Flowers* 8–10. 2n = 36.

Introduced. Characteristic casual from soya-bean waste and grain. Scattered records in southern England. Native in east Asia. Naturalised in North America from where it came to England.

7. S. italica (L.) P. Beauv.   Foxtail Bristle Grass
Panicum italicum L.

Annual herb. *Culms* up to 1 m, erect or geniculately ascending, simple or more or less branched from the base, smooth, glabrous, but puberulous below the panicle. *Leaves* up to 45 cm × 6–20 mm, linear, tapering to an acuminate tip, flat, scabrid on the upper surface and on the margins; sheaths rather loose, smooth, glabrous on the surface, ciliate on the margin, striate; ligule a densely ciliate rim. *Inflorescence* a panicle up to 30 cm × 15–30 mm, spike-like, erect or nodding, often interrupted below, cylindrical and continuous or more or less lobed; rhachis stout, angled, hairy. *Spikelets* 2.5 × 3.5 mm, persistent, closely packed in small groups, each supported by 2–5, antrorsely barbed bristles up to 10 mm, disarticulating below the upper lemma leaving the glumes and lower lemma on the rhachis. *Lower glume* about 1 mm, broadly ovate, acute at apex and 1- to 3-veined, the upper 1.5–2.0 mm, elliptical, acute or obtuse at apex, 5- to 7-veined and about two-thirds as long as the lemma. *Lemma* of the neuter lower flower similar to the upper glume, flat on the back, 5-veined; palea if present a minute hyaline scale. *Lemma* of the bisexual upper flower as long as the spikelet, yellowish or reddish, crustaceous, smooth or nearly so; palea of the same texture. *Anthers* about 1 mm. *Caryopsis* tightly enclosed between the hardened lemma and palea. *Flowers* 7–8. $2n = 18$.

Introduced. Common casual from bird seed on tips and waste ground. It is the well-known cage birds' Millet-spray. Of scattered occurrence in Britain and the Channel Islands. Derived from *S. viridis* as a crop, probably in China. Cultivated in the Mediterranean and throughout tropical and subtropical regions.

89. Digitaria Haller nom. conserv.

Annual herbs sometimes rooting at the lower nodes. *Leaves* flat; ligule membranous. *Inflorescence* of unilateral racemes arranged subdigitately, or upon a short central axis and bearing the spikelets in appressed groups of 1–5 or more. *Spikelets* dorsally compressed, with 2 flowers, the lower sterile and represented by a lemma usually as long as the spikelet, the upper bisexual, as long as or slightly shorter than the lower and concealed between or protruding from within the upper glume and lower lemma. *Glumes* very unequal, the lower very short or absent, the upper about one-third as long as the spikelet. *Lemma* of upper flower chartaceous or cartilaginous, more or less acute at apex and awnless, with its hyaline margins enfolding and concealing most of the palea, the lemma of the lower flower 5- to 7-veined, obtuse at apex and awnless. *Spikelet* falling whole.

About 200 species in tropical and warm temperate regions.

1. Leaf-sheaths glabrous except at mouth; apex of pedicel slightly cup-shaped; spikelets 2.0–2.3(–2.5) mm, minutely pubescent; upper glume and lower lemma more or less the same length   **1. ischaemum**
1. Leaf-sheaths usually pubescent; apex of pedicel truncate; spikelets 2.5–3.5(–3.8) mm, sparsely appressed-pubescent; upper glume less than three-quarters as long as lemma   2.
2. Upper glume one-third to half (to two-thirds) as long as spikelet, rather abruptly acute at apex; lower lemma with minutely rough veins, rarely ciliate   **2. sanguinalis**
2. Upper glume (half–) two-thirds to three-quarters as long as spikelet, tapering to an acute apex; lower lemma with smooth veins, often ciliate   **3. ciliaris**

1. D. ischaemum (Schreb. & Schweigg.) Muhl.

Smooth Finger Grass
D. humifusa Pers.; *Panicum glabrum* Gaudin; *Panicum ischaemum* Schreb. & Schweigg.

Loosely or compactly tufted *annual herb*. Culms 10–35 cm, ascending or spreading from a bent base or prostrate, very slender, usually branched, smooth, with 2–4 nodes. *Leaves* 2–12 cm × 2–7 mm, green or tinged with purple, narrowly lanceolate, more or less rounded at the base, linear, finely pointed at apex, flat, glabrous, minutely rough on the margins; sheaths glabrous or with a few hairs at the apex, smooth; ligules 1–2 mm, membranous. *Inflorescence* of 2–8 racemes borne at or near the apex of the stem, 1.5–7.0 cm, purplish, spike-like, very slender, finely spreading; rhachis narrowly winged, minutely rough on the margins. *Spikelets* 2.0–2.3(–2.5)

mm, in pairs or threes, on one side of the rhachis only, contiguous, shortly and unequally pedicelled, the pedicel slightly cup-shaped at apex, falling entire at maturity, somewhat flattened on the back, elliptical or oblong-elliptical, rather obtuse at apex, 2-flowered, minutely pubescent. *Lower glume* outermost and very minute or suppressed; upper glume elliptical, as long as the spikelet or nearly so, 3-veined, minutely hairy and thin. *Lower lemma* the same length as the upper glume, flat on the back, similar in outline and length to the spikelet, 5-veined, thin, hairy like the upper glume and with a minute palea, but no flower. *Upper lemma* as long as the spikelet, brown or purplish-brown, rigid except for the the thin margins which fold over the palea, smooth. *Anthers* 0.5–0.7 mm. *Caryopsis* ovoid or ellipsoid, tightly enclosed by the hardened lemma and palea. *Flowers* 8–9. $2n = 36$.

Introduced. Weed of cultivated ground, waste places and tips. Very scattered in southern England, formerly more common. Native of south Europe.

2. D. sanguinalis (L.) Scop.   Hairy Finger Grass
Panicum sanguinale L.

Loose-growing *annual herb*. Culms 10–30(–60) cm, mostly ascending from a bent or prostrate base, rooting from the lower nodes, slender, sometimes branched, with 3–8 nodes, loosely bearded at the nodes or glabrous and smooth. *Leaves* 3–10(–20) cm × 3–8(–14) mm, narrowly linear-lanceolate, rounded at the base, finely pointed at apex, flat, hairy or glabrous, minutely rough on the margins; sheaths closely to sparsely hairy, the hairs with tuberculate bases, or glabrous, loose; ligules 1–2 mm, membranous. *Inflorescence* of racemes 4–18 cm, spike-like, 4–10 clustered at or near the apex of the stem, very slender, finely spreading; rhachis 3-angled, rough on the angles. *Spikelets* 2.5–3.3 mm, in pairs along one side of the racemes, touching, unequally short-pedicelled, falling entire at maturity, somewhat flattened, ovate-elliptical to oblong-elliptical, pointed at apex, 2-flowered. *Lower glume* outermost and minute, the upper glume up to half the length of the spikelet, lanceolate to narrowly ovate, abruptly pointed at apex, 3-veined, thin and minutely hairy. *Lower lemma* flat or nearly so on the back, corresponding in length and outline to the spikelet, 7-veined, minutely rough on the veins, finely and obscurely hairy, with a minute palea. *Upper lemma* as long as the spikelet, pointed at apex, firm except for the broad, thin margins folding over the back of the palea. *Anthers* about 0.6 mm. *Caryopsis* oblong-ellipsoid, tightly enclosed by the hardened lemma and palea. *Flowers* 8–10. $2n = 36$.

Introduced. Weed of cultivated and waste places and on tips. Scattered over southern Britain and the Channel Islands, becoming more common. Native of south Europe.

3. D. ciliaris (Retz.) Koeler   Tropical Finger Grass
Panicum ciliare Retz.; *D. adscendens* (Kunth) Henrard

Annual stoloniferous *herb*. Culms up to 1 m, decumbent to ascending, rooting at the nodes, branched or unbranched, nodes glabrous or with few hairs. *Leaves* 5–15 cm × 4–8 mm, linear or linear-lanceolate, narrowed at apex, flat, glabrous or sparsely hairy at the base; sheaths more or less glabrous or usually with sparse tuberculate-based hairs; ligules 1–2 mm, membranous. *Inflorescence* of 4–9 racemes, digitate or subdigitate; racemes uusally 3–15 cm, sometimes longer; pedicel truncate at apex. *Spikelets* 2.6–3.5(–3.8) mm, in pairs, falling whole. *Lower glume* less than 0.5 mm, the upper glume half to three-quarters the length of the spikelet, 3-veined, tapering to an acute apex and usually with at least a few hairs. *Lemma* of lower flower as long as spikelet, 7-veined, silky-pubescent at least along the margin and sometimes also with tubercular-based hairs; lemma of upper flower about as long as spikelet and glabrous. *Palea* more or less as long as lemma. *Stamens* 3. *Styles* distinct. *Caryopsis* tightly enclosed by the hardened lemma and palea. *Flowers* 8–9. $2n = 54$.

Introduced. Bird seed, wool or soya-bean casual, sometimes naturalised and occurring in cultivated ground. Scattered records in southern Britain. Widespread in tropical areas of the Old World and introduced elsewhere.

90. Cenchrus L.

Annual to tufted *perennial herbs. Leaves* flat; ligule a fringe of hairs. *Inflorescence* spike-like; rhachis angular, bearing groups of 1–few spikelets on a very short stalk, the group surrounded and enclosed by a spiny bur composed of spikes and bristles, which are connate below to form a disc or cup and fall with the spikelets. *Spikelets* lanceolate or ovate, unawned, with 2 flowers, the lower sterile, the upper bisexual. *Glumes* very unequal, the lower up to half as long as the spikelet or absent, the upper as long as the spikelet. *Lemma* of the lower flower as long as the spikelet of the lower flower, the lemma of the upper flower firmly membranous to coriaceous.

Twenty-two species occurring throughout the tropics.

1. United whorl of flattened spines forming the bur subtended by smaller, finer bristles **1. echinatus**
1. United flattened spines forming the bur in more than one whorl, without subtending bristles 2.
2. United flattened spines forming the bur cleft on one side only; spines mostly more than 50; spikelets 6 mm or more **2. longispinus**
2. United flattened spines forming the bur cleft on 2 sides; spines usually less than 45; spikelets less than 6 mm **3. incertus**

1. C. echinatus L.   Spiny Sandbur

Annual herb. Culms up to 60(–90) cm, branching, glabrous. *Leaves* 5–25 cm × 3–12 mm, linear-lanceolate, tapering to a sharp point, glabrous to shortly hairy beneath, sparsely hairy above; sheaths glabrous, hairy on the margin, or rarely hairy all over; ligules a dense ciliate rim. *Inflorescence* a spike-like panicle up to 10 cm, with the rhachis bearing groups of one to few spikelets on a very short stalk, the group surrounded and enclosed by a spiny bur composed of fused spines and smaller, finer bristles 4–10 mm, the spines rigid, unequal

in length and retrorsly barbed, the inner ones flattened and fused for about half their length, shortly pubescent and ciliate on the margin. *Spikelets* in clusters of 2 or 3, rarely more, 4.5–7.0 mm; flowers 2, the lower male or sterile and reduced to a lemma, the upper bisexual. *Glumes* subequal or the lower small or suppressed, the lower 1- to 3-veined, the upper as long as the spikelet and 1- to 7-veined. *Lemma* of lower flower about as long as the spikelet, 3- to 7-veined. *Palea* as long as lemma. *Stamens* 3. *Styles* free or connate near the base. *Flowers* 8–9. $2n = 34, 68$.

Introduced. Infrequent casual on tips and waste ground from wool, bird seed and soya-bean waste. Scattered in southern England. Originally native of America, now pan-tropical.

2. C. longispinus (Hack.) Fernald
                            Long-spined Sandbur

C. pauciflorus auct.

Annual herb often forming spreading clumps. *Culms* up to 90 cm, decumbent to erect, branched, nodes glabrous. *Leaves* 3–19 cm × 3–7 mm, more or less linear, flat or occasionally folded or involute, attenuate at apex, scabrous to sparingly hairy, lower surface sometimes smooth and glabrous; sheaths glabrous except for the margin and throat which are hairy; ligules 0.7–1.7 mm, ciliate. *Inflorescence* a spike-like panicle up to 10 cm, with the rhachis bearing groups of one to few spikelets on a very short stalk, the group surrounded and enclosed by a spiny bur composed of fused spines 8–12 mm, the spines usually (40–)50–70, retrorsely barbed, united in the lower part but cleft on one side, shortly hairy in the lower part and on stalks 1–5 mm. *Spikelets* in clusters of 2 or 3, 6–8 mm; flowers 2, the lower male or sterile and reduced to a lemma, the upper bisexual. *Glumes* subequal or the lower small and suppressed, the lower 1- to 3-veined, the upper as long as the spikelet and 1- to 7-veined. *Lemma* of the lower flower about as long as the spikelet and 3- to 7-veined, the palea up to as long as the lemma or suppressed; lemma of upper flower as long as the spikelet, 5- to 7-veined, the palea as long as the lemma. *Stamens* 3. *Styles* free or connate near the base. *Flowers* 8–9. $2n = 34$.

Introduced. Reported as a casual. Some records of *C. echinatus* may be it. Native of North and Central America.

3. C. incertus M. A. Curtis   Few-spined Sandbur
C. pauciflorus auct.

Annual or occasionally *biennial herb* forming clumps. *Culms* up to 80 cm, decumbent to erect, branched, nodes glabrous. *Leaves* 2–18 cm × 0.2–0.6 mm, more or less linear, flat or folded, attenuate at apex, scabrous on the margin and upper surface; sheaths glabrous or sparingly hairy early, the margin and throat glabrous to hairy; ligules 5–15 mm, ciliate. *Inflorescence* a spike-like panicle up to 10 cm, with the rhachis bearing groups of one to few spikelets on a very short stalk, the group surrounded and enclosed by a spiny bur composed of fused spines 5.5–10.0 mm, the spines usually 8–45, retrorsly barbed, united in lower part but cleft on 2 sides, hairy in the lower part and on a stalk 0.5–2.0 mm. *Spikelets* in clusters of 2–4, 3.5–5.5 mm; flowers 2, the lower male or sterile and reduced to a lemma, the upper bisexual. *Glumes* subequal or the lower small and suppressed, the lower 1- to 3-veined, the upper as long as the spikelet and 1- to 7-veined. *Lemma* of lower flower about as long as the spikelet, 5- to 7-veined. *Palea* as long as lemma. *Stamens* 3. *Styles* free or connate near the base. *Flowers* 8–9. $2n = 34$.

Introduced. Reported as a casual. Some records of *C. echinatus* may be it. Native of tropical America.

Tribe 19. **Andropogoneae** Dumort.

Maydeae Dumort. nom. inval.

Annual or *perennial herbs* with rhizomes. *Leaves* linear, flat; ligule membranous, breaking up into an apical fringe of hairs. *Inflorescence* a compact to diffuse panicle or umbel of racemes. *Spikelets* in pairs, one bisexual and the other male or sterile, separated into male and female panicles in *Zea* and only the male paired.

91. **Sorghum** Moench nom. cons.

Usually robust *annual* or *perennial herbs*. *Leaves* flat. *Inflorescence* a large, terminal panicle with persistent branches bearing short fragile racemes. *Spikelets* in dissimilar pairs. *Sessile spikelets* with 2 flowers, dorsally compressed; lower flower reduced to a hyaline lemma; upper flower bisexual, the lemmas 2-dentate with a glabrous, geniculate awn. *Pedicellate spikelets* male or sterile, smaller than the sessile, unawned ones. *Spikelets* falling whole or persistent.

About 20 species in the tropics and subtropics of the Old World and one species endemic to Mexico.

1. Perennial; leaves usually less than 20 mm wide;
   spikelets deciduous at maturity **1. halapense**
1. Annual; leaves usually more than 20 mm wide;
   spikelets persistent **2. bicolor**

1. S. halapense (L.) Pers.   Johnson Grass
Holcus halapensis L.; *Andropogon halapensis* (L.) Brot.

Perennial herbs with long, spreading rhizomes. *Culms* up to 250 cm, erect or somewhat decumbent at the base and rooting at the nodes, smooth, glabrous but with finely silky-pubescent nodes. *Leaves* up to 100 cm × up to 20 mm, long-linear, tapering to a fine acuminate tip, rounded at the base, glabrous, smooth, very scabrid on the margins; sheaths striate, glabrous; ligules short, membranous, erose, hairy on the back. *Inflorescence* a panicle, 15–30 cm, made up of many pairs of spikelets, one sessile, one pedicellate, deciduous at maturity; branches terminating in few-noded false racemes; pedicels similar, nearly as long as the spikelets. *Sessile spikelet* 4.0–5.5 mm, ovate or elliptical, acute at apex; lower glume flat on the back, keeled on the margins above and more or less hairy on the dorsal surface, the upper glume 3-veined and somewhat keeled towards the apex; *lower lemma* sterile, hyaline, as long as the glumes, with the palea absent; fertile lemma short, cleft at the

apex, hyaline, awned in the cleft, with the palea absent or a minute, hyaline scale; awn perfect, 10–16 mm long of which 5 mm comes from a chestnut, twisted column; *anthers* about 3 mm. *Pedicellate spikelets* 4.5–6.0 mm, male, as long as the sessile spikelets, but narrower, purplish, the upper navicular; lemmas hyaline; paleas usually absent; anthers about 3 mm. *Caryopsis* obovoid or ellipsoid. *Flowers* 6–10. $2n = 40, 41, 42, 43$.

Introduced. Casual on tips and in waste ground, occasionally naturalised for a short period. Scattered records in southern Britain. Mediterranean region eastwards to central Asia; Macaronesia; cultivated in most of the warmer regions of the world.

2. S. bicolor (L.) Moench   Great Millet
Holcus bicolor L.; *S. vulgare* Pers.

Annual herb. *Culms* up to 250 cm, erect, solitary or fasciculate, solid, sometimes pubescent at the nodes. *Leaves* up to 60 × 6 cm, linear, acuminate at apex, rounded at base, hairy on the upper surface towards the base, otherwise glabrous, scabrid on the margins; sheaths rather loose or the upper tightly clasping; ligule very short, scarious. *Inflorescence* a panicle up to 25 cm, very variable, dense or loose, erect or the peduncle curved; branches short; racemes stout, compact, 2- to 4-noded; spikelets in pairs, one sessile, one pedicellate. *Sessile spikelet* 4–6 × 3–4 mm, elliptical or obovate; rhachilla not disarticulating above the glumes; glumes more or less equal, the shape of the spikelet, variously coloured from pale yellow to chestnut-brown or black; lower lemma empty, hyaline and palea absent; upper lemma about 3 mm, broadly ovate, 2-lobed at the apex, awned in the notch, the palea a minute, hyaline scale or absent; awn up to 10 mm, feebly twisted at the base; anthers about 3 mm. *Caryopsis* broadly obovoid or globular. *Pedicellate spikelets* neuter, 3–4 mm, lanceolate, persistent. *Flowers* 7–9. $2n = 20$.

Introduced. Casual on tips and in waste places. Scattered records in southern Britain and the Channel Islands. Native of tropical Africa, but widely cultivated elsewhere including the Mediterranean region.

92. **Zea** L.

Mays Mill. nom. illegit.

Annual herb. *Leaves* flat. Male and female inflorescences separate. *Male inflorescence* a large terminal panicle of false spike-like racemes. *Female inflorescence* axillary, of numerous spikelets arranged in longitudinal rows on a thickened axis, the whole enclosed in leaf-sheaths. *Male spikelets* in pairs, one subsessile, the other stalked; flowers 2; glumes equal, membranous; lemma and palea hyaline; stamens 3; lodicules 2. *Female spikelets* with 2 flowers, the lower sterile; glumes wider than long, fleshy below, hyaline above; lemma short, hyaline; lodicules absent; style long, shortly bifid at apex, hairy throughout its length.

Four species in central America.

Mangelsdorf P. C. (1974). *Corn: its origin, evolution and improvement.* Cambridge, Mass.

1. Z. mays L.   Maize

Robust *annual herb. Culms* up to 300 cm, smooth, glabrous, shining, solid, rooting from the lower nodes, erect. *Leaves* up to 75 × 2.5–12.0 cm, broadly linear, acuminate at apex, flat, hairy or glabrous on the upper surface, ciliate on the margins, smooth and glossy beneath, drooping; sheaths rather loose, rounded on the back, striate, often overlapping, smooth, glabrous; ligules 1–2 mm, membranous. *Male panicle* up to 45 × 20 cm, of spreading or drooping false racemes on a stout, terete, smooth, glabrous rhachis; racemes up to 30 cm; spikelets all alike, 2-flowered; glumes 6–10 mm, membranous, elliptical, obtuse at apex, many-veined, smooth, glabrous or stiffly hairy; lemmas and palea hyaline. *Female panicle* consisting of a thick, fleshy rhachis seated in a number of papery spathe-like sheaths; spikelet sessile, half sunk in the fleshy or fibrous rhachis, conical in shape, arranged in longitudinal rows; lower glume emarginate, ciliate, the upper acute or 2-lobed and ciliate; lemma of the lower flower empty and hyaline, the palea shorter or absent; lemma of the upper flower containing a female flower, membranous, the palea hyaline, the stigmas very long, drooping, often tinged with purple and 2-lobed at the tip. *Caryopsis* usually cuneate and flattened above. *Flowers* 9–10. $2n = 20 + 1$–7B.

Introduced. Grown on a large scale in southern England as fodder and on a small scale as grain. Casual on tips and waste ground in southern and central Britain and the Channel Islands. Maize was cultivated by the American Indians long before the time of Columbus, and reached Europe about 500 years ago. It is now cultivated world-wide and is a staple cereal for tropical regions.

Order 5. TYPHALES Lindl.

Perennial marsh or aquatic *herbs* with rhizomes. *Leaves* linear, sheathing at base. *Flowers* unisexual, hypogynous, small, densely crowded in spikes or heads. *Perianth* small, sepaloid, often of scales or threads. *Stamens* 2 or more. *Ovary* 1-celled; ovule solitary, pendulous. *Fruit* dry. *Seeds* with endosperm.

The order Typhales consists of two small, closely related families, each with a single genus and about 22 species in all.

168. SPARGANIACEAE Rudolphi nom. conserv.

Aquatic or semi-aquatic *perennial herbs* with rhizomes, and rooted in mud. *Stems* simple or branched, leafy. *Leaves* alternate, simple, linear, entire, sessile with a sheathing base, exstipulate; on underground parts scale-like. *Flowers* in a terminal spike, raceme or panicle of globose, unisexual heads, the upper male, the lower female, hypogynous, actinomorphic. *Perianth* of 1–6, inconspicuous scales often not easy to distinguish from bracts. *Stamens* 3–8; filaments sometimes partially united. *Stigmas* 1–2(–3), linear to subcapitate. *Ovary* 1- to 2-celled, with 1–2 ovules. *Fruit* a small, dry, spongy drupe with 1–2(–3) seeds.

One genus with about 14 species. Generally distributed in Arctic to warm regions, but absent from Africa and South America.

1. Sparganium L.

As family.

Cook, C. D. K. (1961). *Sparganium* L. in Britain. *Watsonia* **5**: 1–10.

Cook, C. D. K. (1962). *Sparganium* L. in Biological flora of the British Isles. *Jour. Ecol.* **50**: 247–255.

Cook, C. D. K. & Nicholls, M. S. (1986). A monographic study of the genus *Sparganium* (Sparganiaceae). Part I. Subgenus *Xanthosparganium* Holmberg. *Bot. Helv.* **96**: 213–267.

Cook, C. D. K. & Nicholls, M. S. (1987). A monographic study of the genus *Sparganium* (Sparganiaceae). Part II. Subgenus *Sparganium. Bot. Helv.* **97**: 1–44.

1. Inflorescence branched, the male heads borne at the apex of branches as well as on the main axis; perianth segments dark-tipped   2.
1. Inflorescence not branched, the heads all at the apex of the main axis; perianth segments more or less translucent, not dark-tipped   5.
2. Fruits distinctly shouldered below the beak   3.
2. Fruits gradually rounded below the beak   4.
3. Fruits (3–)4–6(–7) mm across, with a flat top (excluding beak)   **1(a). erectum** subsp. **erectum**
3. Fruits 2.5–4.5 mm across, with a rounded top (excluding beak)   **1(d). erectum** subsp. **microcarpum**
4. Fruits 4–7 mm across, subglobose, abruptly contracted to the beak   **1(b). erectum** subsp. **oocarpum**
4. Fruits 2.0–4.5 mm across, ellipsoid, gradually tapered to the beak   **1(c). erectum** subsp. **neglectum**
5. Stem leaves not inflated, but strongly keeled at base; male heads 3–10, clearly separated   **2. emersum**
5. Stem leaves sometimes inflated, but not keeled at base; male heads 1–3(–4), if more than 1 then close together and appearing more or less as one elongated head   6.
6. Bract of lowest female head more than 10 cm, more than twice as long as the inflorescence; male heads mostly 2–3   **3. angustifolium**
6. Bract of lowest female head less than 10 cm, barely longer than the inflorescence; male head usually 1   **4. natans**

Subgenus 1. Sparganium

Perianth segments thick, with a dark brown to black apex. *Seeds* with 6–10, longitudinal ridges.

1. S. erectum L.   Branched Bur-reed

Glabrous, emergent or rarely floating or submerged *perennial herb*, with rhizomes up to 60 cm × 6–9 mm in diameter; corms 1.5–4.5 cm in diameter, when mature hard and woody, each bearing up to 9 rhizomes. *Stems* 20–100(–150) cm, erect and nearly always branched. *Leaves* (30–)50–150(–350) cm × (6–)10–20(–28) mm, yellowish-green, often pink at base, long-linear, narrowed to an obtuse apex, entire, keeled throughout, obliquely erect and fern-like, usually emergent but plants in very deep or swiftly flowing water may develop submerged or partly floating leaves, but usually remain

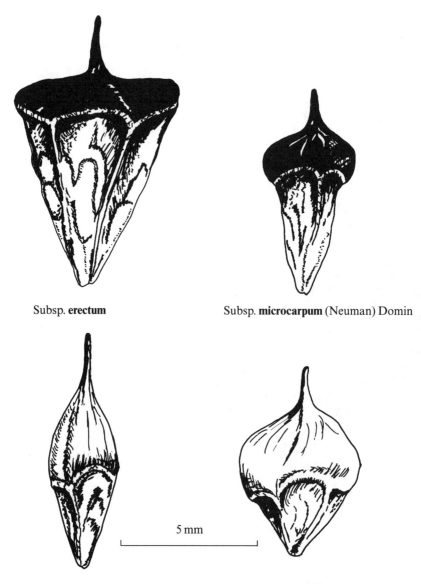

Subsp. **erectum**   Subsp. **microcarpum** (Neuman) Domin

5 mm

Subsp. **neglectum** (Beeby) K. Richt.   Subsp. **oocarpum** (Čelak.) Domin

Fruits of the subspecies of **Sparganium erectum** L. (After Cook, 1961.)

sterile. *Inflorescence* with (0–)2–5(–9) branches, with (0–)1–3(–6) female heads at the proximal end and (0–)6–9(–16) male heads at the distal end, the upper 1 or rarely 2 branches are often without female heads, the lowest branch often bears a solitary female head without any males, occasionally the solitary head is topped by some males or rarely the branch has male heads only, the terminal part of the inflorescence above the last branch is without female heads and with (3–)5–15(–22) male heads; bracts leaf-like below, erect and keeled from base to apex, scale-like above, the lowest bract usually exceeding the inflorescence. *Female*

heads borne on axillary branches 12–20 mm in diameter at anthesis, 15–32 mm in diameter in fruit. *Male heads* borne above female heads or branch or terminating the main axis, usually remote and distinct at anthesis, before anthesis green with black flecks and about 5 mm in diameter, at anthesis 10–12 mm in diameter. *Female flowers* with perianth segments somewhat thickened and dark brown to black at apex, not translucent, almost entire at apex; ovaries with 1, 2 or rarely 3 carpels, but a least half the ovaries are unicarpellate; more than half the styles undivided; stigmas (1.5–)2–4 mm, somewhat irregularly bent or coiled. *Male flowers* with perianth segments thickened and dark brown or black at apex; filaments about 5 mm; anthers 1.2–1.5 mm. *Fruits* very variable in form, 4–12 × 2.0–7.5 mm. *Seeds* with 6–10, longitudinal ridges. *Flowers* 6–8.

(a) Subsp. **erectum**

S. erectum var. *angustifolium* Warnst.; *S. ramosum* forma *platycarpum* Čelak.; *S. ramosum* subsp. *polyedrum* Graebn.; *S. ramosum* forma *conocarpum* Čelak.; *S. ramosum* subvar. *dolichocarpum* Graebn.; ?*S. draco* F. Herm.; *S. ramosum* Huds.

Fruits (4–)6–10(–12) × (3–)4–6(–7) mm, cuneate-obpyramidal, with a distinct shoulder between the upper and lower parts; lower part (4–)5–8(–10) mm, obpyramidal, distinctly 3- to 6-angled in transverse section, pale brown; upper part flattened, pale to dark brown or blackish, matt, smooth, very abruptly contracted into a beak; beak not more than 2 mm; bilocular ovaries frequent, but less than half the total ovaries; perianth segments not visible between the fruits on mature fruiting heads. $2n = 30$.

(b) Subsp. **oocarpum** (Čelak.) Domin

S. neglectum var. *oocarpum* Čelak.; *S. neglectum* subsp. *oocarpum* (Čelak.) Ostenf.; *S. oocarpum* (Čelak.) Fritsch; *S. erectum* subsp. *neglectum* var. *oocarpum* (Čelak.) Hayek; *S.* × *tardivum* Topa

Fruits 5–8 × 4–7 mm, uniform pale to medium brown and shiny, widely ovoid to almost spherical, shoulder between upper and lower parts indistinct; lower part 2.5–5.0 mm, almost circular in transverse section; upper part abruptly contracted into a beak and sometimes wrinkled or creased at base of the beak but otherwise smooth; beak up to 2 mm; bilocular ovaries very rare; perianth segments visible between the fruits on mature fruiting heads.

(c) Subsp. **neglectum** (Beeby) K. Richt.

S. neglectum Beeby; *S. ramosum* subsp. *neglectum* (Beeby) Neuman

Fruits (6–)7–9(–10) × 2.0–3.5(–4.5) mm, uniform pale brown to straw-coloured, upper and lower parts essentially alike in form, colour and texture, without a distinct shoulder; lower part 3–5 mm, about half the fruit length, hardly angled and almost ellipsoid in transverse section; upper part smooth without wrinkles, gradually tapered into a beak; beak (2–)2.5–3.5 mm, sometimes dark brown; bilocular ovaries extremely rare; perianth segments visible between the fruits on mature fruiting heads.

(d) Subsp. **microcarpum** (Neuman) Domin

S. ramosum forma *microcarpum* Neuman; *S. microcarpum* (Neuman) Čelak.; *S. erectum* subsp. *neglectum* var. *microcarpum* (Neuman) Hayek

Fruits (5.5–)6–7(–9) × (2–)2.5–3.5(–4.5) mm, acutely obpyramidal below and domed above, upper and lower parts different in form and texture with a distinct shoulder between them; lower part 4–6 mm, acutely obpyramidal, 3- to 6-angled in transverse section, the flattened faces sometimes bellied, pale to reddish-brown; upper part domed and wider than the lower part with a slight

constriction below the shoulder, brown to black, matt or somewhat shining near the shoulder, with irregular longitudinal creases or wrinkles at base of beak, tapering rather abruptly into a beak; beak up to 2 mm, dark brown to black; bilocular ovaries rare; perianth segments visible on mature fruiting heads.

Native. By ponds, lakes, slow-flowing rivers and canals and in marshy fields and ditches. Common throughout most of the British Isles. Europe from the Arctic Circle southwards to North Africa, extending through temperate Asia southward to the western Himalayas and eastwards through Japan into north-western North America. Also in south-eastern Australia where it may have been introduced. No ecological differences seem to exist between the four subspecies, and all four have been recorded growing together. Subsp. *erectum* is mainly in central and south Britain and the Isle of Man, and rare in Ireland. It is common through much of central and southern Europe, but largely absent from Scandinavia. It extends eastwards to the Caucasus and south to Turkey. Subsp. *neglectum* occurs throughout the British Isles. It is found in much of Europe, extending northwards to about 58° N in Sweden and southwards to Morocco. It is recorded from European Russia and south into Greece and Turkey. Subsp. *oocarpum* is intermediate in morphology between subsp. *erectum* and subsp. *neglectum* and may have arisen through hybridisation. It is found in Ireland, south and central Britain and the Channel Islands. There are isolated records through much of Europe and in Turkey and Iraq. Subsp. *microcarpum* occurs throughout the British Isles. It is also found throughout Europe and is particularly common in Scandinavia. It extends eastwards to European Russia and south around the Sea of Azov and in Caucasus and Turkey.

Subgenus 2. **Xanthosparganium** Holmb.

Perianth segments thin, uniformly pale brown. *Seeds* smooth.

2. S. emersum Rehmann   Unbranched Bur-reed

S. angustifolium subsp. *emersum* (Rehmann) Brayshaw; *S. simplex* var. *gracilis* Meinsh.; *S. splendens* Meinsh.; *S. simile* Meinsh.; *S. subvaginatum* Meinsh.; *S. diversifolium* Graebn.; *S. chlorocarpum* Rydb.; *S. simplex* auct.; *S. simplex* var. *longissimum* E. M. Fr.

Glabrous, submerged, floating or erect *perennial herb* with rhizomes up to 50 cm × (1–)2–3(–4) mm. *Stems* simple, erect, 20–60(–80) cm or when floating to 180 cm. *Leaves* of fertile plants (10–)20–50(–80) cm × (1.5–)4–10(–12) mm, long-linear, usually distinctly keeled throughout their length, narrowed to an obtuse apex, entire, not inflated at base, usually erect and partially emergent; plants in deep or swiftly flowing water usually remain sterile and the submerged or floating leaves are up to 220 cm × 18 mm, and flat to somewhat keeled. *Inflorescence* simple, the heads sessile or stalked; bracts usually erect and keeled, with hyaline margins. *Female heads* (1–)3–4(–6), appearing white to yellowish-green at anthesis, in fruit 1.6–2.5(–3.5) cm in diameter, the lower usually axillary and with peduncles 0–4(–8) cm, the upper sessile and some usually supra-axillary. *Male heads* (3–)4–7(–10), before anthesis pale yellowish-green, remote and distinct at anthesis, separated from the uppermost female head by a 3–20(–40) mm long internode. *Female flowers* with perianth segments pale brown, thin, spathulate, translucent, with erose tips and scarcely clawed at base, half to two-thirds as long as the fruit, united at base, usually with a pedicel 1–2 (–3) mm; stigmas (1.0–)1.5–2.0(–2.5) mm, obliquely attached to a 2–3 mm style. *Male flowers* with filaments 5–7 mm; anthers 1.0–1.5(–2.0) mm. *Fruits* 3.5–5.5 × 1.8–2.5 mm, brown and shiny, fusiform, sometimes slightly constricted around the middle, tapering above to a beak 2.0–4.5(–6.0) mm, tapering below to a stalk up to 4 mm. *Seeds* smooth. *Flowers* 6–7. $2n = 30$.

Native. By ponds, lakes, slow-flowing rivers and canals and in marshy fields and ditches, but rarely out of water. Frequent throughout the British Isles. Throughout the north temperate regions from the Arctic Circle south to 40° N. All material from the British Isles is referable to subsp. **emersum**.

3. S. angustifolium Michx   Floating Bur-reed
S. simplex var. *angustifolium* (Michx) Engelm.; *S. affine* A. Schnizl.; *S. oligocarpon* Ångstr.; *S. vaginatum* Larss.; *S. borderi* Focke; *S. simplex* var. *multipedunculatum* Morong; *S. affine* var. *zosteraefolium* Neuman; *S. affine* var. *deminutum* Neuman; *S. affine* var. *microcephalum* Neuman; *S. kawakamii* Hara

Glabrous, usually submerged or floating, rarely emergent *perennial herb* with rhizomes up to 35 cm × about 3 mm, with scale leaves up to 16 × 3 mm. *Stems* (7–)30–100(–175) cm × 1–3 mm, decumbent or ascending, usually floating, flexuous, or rarely erect and emergent. *Leaves* (20–)30–80(–250) cm × (1.5–)2–4(–10) mm, usually submerged below and floating distally, long-linear, narrowed to an obtuse apex, entire, flat or rounded abaxially, without a distinct midrib or a keel even at the sheathing and inflated base. *Inflorescence* 1–9 cm, simple; bracts inflated at base, with hyaline margin, the lowermost more than 10 cm and (1.5–)2.5–4.0(–4.5) times as long as the inflorescence. *Female heads* (1–)2–4 (–5), in fruit (8–)10–20(–24) mm in diameter, remote, the lowest axillary or supra-axillary and stalked, the others axillary or supra-axillary and usually sessile. *Male heads* (1–)2–3(–4), usually crowded and appearing as one elongated head terminating the main axis, usually remote from the uppermost female head with an internode (0–)10–20(–50) mm. *Female flowers* with perianth segments pale brown, spathulate, translucent, with erose tips, scarcely clawed below, about two-thirds as long as the fruit, attached at the base of the fruit to a short pedicel; stigmas (0.6–)0.8–1.0(–1.2) mm, ovate-lanceolate, obliquely attached to a style 1.5–2.0 mm. *Male flowers* with filaments 3–5 mm; anthers (0.8–)0.9–1.0 (–1.2) mm, narrowly oblong. *Fruits* 3.0–5.5 × 2.0–2.5 mm, brownish, ellipsoid to fusiform, usually somewhat constricted near the centre, tapering below to an obconic

base with a pedicel (0.8–)1.0–1.5 mm, narrowed upwards to an acute apex with a conical-based beak 1.5–2.0(–2.2) mm. *Seeds* smooth. *Flowers* 8–9. $2n = 30$.

Native. In acid or alkaline lakes, pools or ditches with high organic content, mainly in mountainous districts. Local in west Wales, north-west England, western and northern Scotland, and Ireland except for much of the centre. Iceland to north Portugal and the Alps and Macedonia, east in Asia to Kamchatka; North America.

4. S. natans L.   Least Bur-reed
S. minimum Wallr.; *S. rostratum* Larss.; *S. ratis* Meinsh.; *S. septentrionale* Meinsh.; *S. flaccidum* Meinsh.; *S. perpusillum* Meinsh.

Glabrous, usually submerged or floating *perennial herb* with rhizomes up to 20 cm × 1–2 mm, with scale leaves up to 12 × 2 mm. *Stems* (6–) 8–40(–100) cm × 1–3 mm, decumbent or ascending, usually floating and flexuous, or rarely erect and emergent. *Leaves* 6–40(–60) cm × (1.5–)2–6(–10) mm (or when terrestrial 7–20 cm × 2.5–4.0 mm), long-linear, obtuse at apex, deep green, flat to rather concave, but without a distinct midrib or a keel even at the base; sheaths not inflated. *Inflorescence* 1.5–8.0 cm, simple; bracts with usually somewhat inflated bases with hyaline margins, the lowermost less than 10 cm and scarcely exceeding the inflorescence. *Female heads* 1–3(–4), in fruit (5–)7–10(–15) mm in diameter, usually remote, axillary or rarely supra-axillary, the uppermost sessile, the lowermost sessile or with a more or less straight peduncle up to 28 mm. *Male heads* terminal, solitary, remote from the uppermost female head. *Female flowers* with perianth segments attached at the base of the fruit or along a short pedicel, thin, pale brown, elliptical to cuneate-spathulate with erose tips, scarcely clawed below, half to two-thirds as long as the fruit; stigmas (0.3–)0.5–0.8 mm, ovate, obliquely attached to a style 0.5–0.1 mm. *Male flowers* with filaments 3–5 mm; anthers (0.3–)0.5–0.8 mm, oblong. *Fruits* (1.5–)2–4 (–6) × 1.0–2.5 mm, dull greenish or brownish, tapered below, sessile or with a pedicel not exceeding 1 mm, rather abruptly tapered at apex to an acute summit, with a conical-based, slender beak 0.5–1.0(–1.5) mm. *Seeds* smooth. *Flowers* 6–7. $2n = 30$.

Native. In acid or alkaline lakes, pools or ditches with high organic content. Scattered throughout most of the British Isles, but absent from much of central and south England and south Wales. Throughout Europe south to the Pyrenees and southern Bulgaria; North America.

169. TYPHACEAE Juss. nom. conserv.

Stout, glabrous, aquatic or semi-aquatic *perennial herbs* with rhizomes. *Stems* erect, simple. *Leaves* alternate, simple, linear, entire, thick, sessile with a sheathing base; exstipulate. *Flowers* very numerous, in a dense, cylindrical spike, unisexual, the male at the top, the female lower down, hypogynous, actinomorphic. *Perianth* of few to many bristles and/or narrow scales. *Stamens* 1–5; filaments usually fused. *Style* 1; stigmas clavate to linear.

Ovary 1-celled, stalked, with 1 ovule. *Fruit* a 1-seeded capsule, dispersed by wind. Wind-pollinated.

One genus containing about 8 species. Throughout the world from the Arctic Circle to 30° S.

1. Typha L.

As family.

T. laxmannii Lepech. has been introduced in a lake near Hextable, Kent.

Smith Galen ? (1987). *Typha*: its taxonomy and the ecological significance of hybrids. *Arch. Hydrobiol. Beih. Ergebn. Limnol.* **27**: 129–138.
Grime, J. P. et al. (1988). *Typha latifolia* L. in *Comparative plant ecology.* London.

1. Leaves 8–25 mm wide; female flowers without scales
   1. latifolia
1. Leaves 3–6(–10) mm wide; female flowers with scales
   2. angustifolia

1. T. latifolia L.   Bulrush

Glabrous, aquatic or semi-aquatic *perennial herb* with fibrous roots and creeping rhizomes. *Stems* up to 3 m, bluish-green, erect, robust, smooth, simple. *Leaves* up to 3.5 m × 8–25 mm, some usually overtopping the stems, bluish-green, long-linear, gradually tapering to an obtuse apex, entire, flat, thick, smooth, sheathing at base with the leaf margins free; scale-like leaves present on underground parts. *Inflorescence* a dense, dark brown, cylindrical spike 18–30 mm wide, the male and female parts usually contiguous or separated by not more than 2.5 cm; male part (3–)6–14(–21) cm; female part (5–)8–15(–21) cm. *Male flowers* with simple hairs; pollen grains in tetrads. *Female flowers* without scales; hairs 30–50, borne on a zone (0.7–)1.0–1.6(–3.5) mm at the base of the gynophore. *Stigmas* lanceolate to rhombic, fleshy, persistent, brown or dark brown, longer than hairs. *Seeds* (0.9–)1.2–1.5(–1.6) mm, cylindrical. *Flowers* 6–7. *Seeds* shed 2–3. $2n = 30$.

Native. Reed-swamps and margins of lakes, ponds and slow-flowing rivers and ditches. Frequent throughout most of the British Isles but absent from much of north and west Scotland. From the Arctic Circle to 30° S, except in South Africa, southern Asia, Australia and Polynesia.

2. T. angustifolia L.   Lesser Bulrush

Glabrous, aquatic or semi-aquatic *perennial herb* with fibrous roots and creeping rhizome. *Stems* up to 3 m, bluish-green, erect, robust, smooth, simple. *Leaves* up to 3.5 m × 3–6(–10) mm, some usually overtopping the stems, dark green, long-linear, gradually tapering to a very narrow, obtuse apex, entire, flat, smooth, sheathing at base, the margin free but parallel, usually closed at the throat and auriculate above; scale-like leaves present on underground parts. *Inflorescence* a dense, reddish-brown, cylindrical spike 13–25 mm in width, the male and female parts separated by (0.5–)3–8(–12) cm; female part 8–20 cm. *Male flowers* with hairs or linear scales which are entire or forked at their apex; pollen grains single. *Female flowers* with spathulate, dark brown,

opaque, firm scales. *Stigmas* broad, flat-topped, the same colour as the scales, exceeding the scales and hairs. *Seeds* 1.0–1.3 mm, cylindrical. *Flowers* 6–7. $2n = 30$.

Native. Reed-swamps and margins of lakes, ponds and slow-flowing rivers and ditches. Scattered throughout most of the British Isles, but absent from central and north Scotland, Isle of Man and most of Ireland. Widely distributed, but absent from America south of Louisiana and California and from Africa.

× **latifolia** = **T. × glauca** Godr.
In this hybrid the leaf width and the degree of separation of the male and female parts of the inflorescence are more or less intermediate between the parents. The stigma, perianth hairs, and pollen grain characters vary from one parent to the other. The hybrid is highly sterile, but can form large clones by vegetative means.

Scattered throughout England and probably much overlooked. Occurs widely in mainland Europe and in North America.

Subclass 4. ZINGIBERIDAE Cronq.

Leaves narrow and parallel-veined. *Outer perianth segments* well differentiated from the inner, but still petaloid in colour.

Order 1. BROMELIALES Lindl.

Consists of one family Bromeliaceae.

170. BROMELIACEAE Juss. nom. conserv.

Glabrous, glaucous, dome-shaped, evergreen, *shrubby plants*. *Leaves* closely spiralled, simple, linear, with strong spines on the margin, sessile with a sheathing base; exstipulate. *Flowers* more or less sessile in dense, more or less globose terminal heads about 5 cm in diameter, bisexual, epigynous, more or less actinomorphic. *Perianth segments* 6, 3 outer and 3 inner, free, coloured. *Stamens* 6. *Style* 1; stigmas 3, linear. *Ovary* inferior, 3-celled, each cell with numerous ovules on axile placenta. *Fruit* a berry.

A large family of 46 genera and about 2,100 species. Mostly tropical, all but one species which is found in west Africa, are natives of the New World.

McClintock, D. (1975). A note on *Fascicularia, Ochagavia* and *Rhodostachys*. *Watsonia* **10**: 289–290.

1. Leaves flat distally, with marginal spines all directed forward; inflorescences more or less sessile in terminal leaf-rosettes; inner perianth segments blue, with basal, scale-like nectaries   **1. Fascicularia**
1. Leaves concave throughout, with the lower marginal spines patent or recurved; inflorescences arising from terminal leaf-rosettes on distinct stalks more than 10 cm; inner perianth segments pink, without scale-like nectaries   **2. Ochagavia**

1. Fascicularia Mez

Robust, cushion-like *perennial herb*. *Stem* branched. *Leaves* in dense, crowded rosettes. *Inflorescence* a sessile, terminal head. *Inner perianth segments* blue, with basal, scale-like nectaries. *Fruit* a berry.

Five species confined to Chile.

1. F. bicolor (Ruiz Lopez & Pavon) Mez

Rhodostachys

Bromelia bicolor Ruiz Lopez & Pavon; *F. pitcairniifolia* auct.

Long-lived cushion-like *perennial herb* up to 75 cm high and 2 m across. *Stem* short and thick, bearing several short branches. *Leaves* up to 35 × 2.5 cm, numerous and crowded, dark, somewhat glaucous green, but the innermost, before and during anthesis, bright red on the proximal part of the upper surface, coriacous, flat distally, ensiform, tapered from a wide ovate-triangular, sheathing base to an acute apex, sparsely covered with white scales when young, with sharp, forwardly directed spines on the margin, the lower spines patent with forwardly directed tips. *Inflorescence* a dense, sessile terminal head about 5 cm in diameter. *Outer perianth segments* sepaloid, about 20 mm, oblong-lanceolate, erect, at first with dense, whitish scales. *Inner perianth segments* a little longer than the outer, blue, erect but with out-turned apex, oblong, obtuse at apex, with basal, scale-like nectaries. *Hypanthial tube* about 5 mm. *Stamens* 6, slightly shorter than the petals; filaments dilated below; anthers yellow, linear. *Style* reaching to the middle of the anthers, thickened below; stigmas 3, short. *Ovary* oblong, plano-convex, hairy in the upper part. *Fruit* a berry. *Seeds* not ripening in British Isles. *Flowers* 7–9.

Introduced. Well established where planted on maritime dunes and shingle. Isles of Scilly, Cornwall and Guernsey in the Channel Islands. Native of Chile.

2. Ochagavia Philippi

Robust, cushion-like *perennial herb*. *Stem* branched. *Leaves* in dense, crowded rosettes. *Inflorescence* a stalked, terminal head. *Inner perianth segments* pink, without scale-like nectaries. *Fruit* a berry.

About five species in South America.

1. O. carnea (Beer) L. B. Sm. & Looser

Tresco Rhodostachys

Bromelia carnea Beer; *Rhodostachys carnea* (Beer) Mez; *O. lindleyana* (Lem.) Mez; *Fascicularia littoralis* auct.

Long-lived, cushion-like *perennial herb* up to 60 cm high and 2 m across. *Stem* short and thick, bearing several short branches. *Leaves* 20–50 × 2.5–3.0 cm, numerous and crowded, dark, somewhat glaucous green, not or scarcely turning red on the proximal part of the upper surface, concave throughout, coriaceous, linear, gradually narrowed from a wide, sheathing base to an acute apex, with persistent whitish scales on the lower surface, with long, stout spines on the margin of which the lower are patent or recurved. *Inflorescence* a dense, stalked, terminal head 5.0–7.5 cm in diameter. *Outer perianth segments* 12–27 mm, narrowly elliptical, acuminate at apex, with whitish scales on the outer surface. *Inner perianth segments* 20–30 mm, pink to lavender, with a small

mucro at apex, without scale-like nectaries. *Hypanthial tube* about 5 mm. *Stamens* 6, longer than the petals; anthers bright yellow, linear-oblong. *Style* overtopping the anthers; stigmas 3, short. *Ovary* glabrous. *Fruit* a berry, 13–18 mm, ellipsoid. *Seeds* not ripening in the British Isles. *Flowers* 7–9.

Introduced. Well established where planted on dunes at Appletree Banks on Tresco and near Normandy, St Mary's in the Isles of Scilly. Native of coastal Chile.

Subclass 5. LILIIDAE Takht.

Leaves typically narrow and paralled-veined. *Outer perianth segments* usually petaloid in form and texture, sometimes differentiated from the inner perianth segments but still petaloid in appearance, only rarely green and herbaceous.

Order 1. LILIALES Lindl.

Perennial herbs often with corms, bulbs or rhizomes, rarely shrubs. *Leaves* mostly linear. *Flowers* usually bisexual, sometimes unisexual, hypogynous to epigynous, actinomorphic to zygomorphic. *Perianth* of 2 whorls, usually both petaloid, rarely both sepaloid, very rarely unlike each other. *Stamens* in 1 or 2 whorls, often 3 or 6. *Ovary* syncarpous, usually 3-celled; ovules 1–many in each cell; placentation axile or parietal. *Seeds* with endosperm.

This order contains 15 families and nearly 8,000 species.

171. **PONTEDERIACEAE** Kunth nom. conserv.

Aquatic, glabrous *perennial herbs* with rhizomes. *Stems* creeping or floating. *Leaves* emergent, alternate, simple, entire, petiolate with a sheathing base; exstipulate. *Flowers* in a simple, terminal spike with a leaf-sheath below it, sessile, bisexual, hypogynous, slightly zygomorphic. *Perianth segments* 6, fused at base, bluish-violet. *Stamens* 6, unequal, in 2 groups of 3. *Style* 1; stigma capitate. *Ovary* 3-celled, with 2 cells empty, or 1-celled; ovule 1. *Fruit* a capsule, with 1 seed.

Nine genera and 34 species. Aquatic habitats in tropical and subtropical regions.

1. **Pontederia** L.

As family.

About 5 species in the New World.

Clements, E. J. (1980). *Pontederia cordata* L. *B.S.B.I. News* **26**: 18.

Lowden, R. M. (1973). Revision of the genus *Pontederia* L. *Rhodora* **75**: 426–487.

1. P. cordata L.   Pickerelweed

Aquatic, glabrous *perennial herb* with rhizomes. *Stem* 90–130 cm, simple or branching at the base, with short internodes. *Leaves* mostly basal, 5–25 × 0.4–21 cm, soft, linear-lanceolate to broadly ovate, acute at apex, entire, cordate or sagittate at base, rarely auricled; petiole 4–60 cm, the base sheathed and clasping. *Inflorescence* 2–16 cm, a simple, dense, terminal spike with a spathe-like sheath below it; peduncle 5–33 cm. *Perianth segments* 6, 10–15 mm, blue to bluish-violet, fused in the basal part, the lobes slightly unequal, oblong to linear, the largest with 2 yellow spots near the base, glandular-hairy. *Stamens* 6; anthers blue. *Style* 1; stigma 3-lobed. *Capsule* 6–10 mm, reniform, with 6 toothed ridges. *Seeds* 3.5–4.5 × 2.0–2.5 mm, reniform. *Flowers* 6–9.

Introduced. Grown for ornament and naturalised at edges of ponds and gravel pits. Scattered in southern England, probably increasing. Native of North and South America.

172. **LILIACEAE** Juss. nom. conserv.
ALLIACEAE; ALSTROEMERIACEAE; AMARYLLIDACEAE; ASPARAGACEAE; ASPHODELACEAE; COLCHICACEAE; CONVALLARIACEAE; HEMEROCALLIDACEAE; HYACINTHACEAE; MELANTHIACEAE; RUSCACEAE; TRILLIACEAE

Mostly glaorous, usually erect *perennial herbs*, with rhizomes, corms, bulbs or tuberous roots, occasionally small, evergreen *shrubs*. *Leaves* all basal or alternate or occasionally whorled, simple, entire, sessile or petiolate, with or without a sheathing base, exstipulate, sometimes reduced to scales and functionally replaced by leaf-like lateral stems (*cladodes*). *Flowers* solitary and terminal or axillary, or in racemes or umbels, less often in cymes or panicles, bisexual or rarely functionally dioecious, hypogynous or epigynous, actinomorphic or slightly zygomorphic. *Perianth segments* 6, free or fused, the 3 outer not or obviously different from the 3 inner, rarely 4 in 2 whorls of 4–6 each; sometimes with a tubular or collar-like *corona* within the rows of segments. *Stamens* usually 6, rarely 3, 4 or 7–12. *Styles* 3–4(–5) with a capitate to linear stigma, or 1 with 1 or 3–4(–5) stigmas. *Fruit* a capsule or berry, the capsule sometimes fleshy.

One of the largest families of flowering plants with about 325 genera and 4,600 species. Cosmopolitan.

The Liliaceae is here regarded in its broadest sense, some other classifications dividing it into up to 13 families. The subfamilial and generic classification follows that of Stace (1991). It is one of the most important families for horticulture and contains many outstandingly beautiful plants. Many species have large swollen storage organs such as bulbs, corms, rhizomes and thick fleshy roots.

1. Leaves with 2 identical surfaces borne on 2 opposite sides of the stem, each with a leaf-base sheathing that of the next higher leaf | 2.
1. Leaves not as above | 3.
2. Perianth segments greenish-white; filaments glabrous; styles 3 | **1. Tofieldia**
2. Perianth segments yellow; filaments densely hairy; style 1 | **2. Narthecium**
3. Leaves (3–)4(–8) in a whorl on the stem below a single flower | **19. Paris**
3. Leaves not all in 1 whorl on the stem below the flower; often with more than 1 flower | 4.

4. Flowers *Crocus*-like, appearing from the soil in autumn without leaves or stems; ovary subterranean, emerging with the leaves to fruit in the spring | **8. Colchicum**
4. Flowers with stems and leaves; ovary above ground at flowering time | 5.
5. Leaves all reduced to small scarious scales, with green, more or less leaf-like stems (cladodes) arising from the axils on the main stem | 6.
5. At least some leaves green and photosynthetic | 7.
6. Perennial herb; cladodes needle-like or slightly flattened and linear, borne in clusters | **36. Asparagus**
6. Evergreen shrubs; cladodes ovate, obovate or obovate-lanceolate, borne singly | **37. Ruscus**
7. Flowers in an umbel with 1 to few spathe-like bracts at their base, or solitary with a spathe-like bract at the base; sometimes the flowers replaced by bulbils | 8.
7. Flowers in a cyme or raceme with or without bracts, but without spathe-like bract(s), or rarely solitary or in an umbel without spathe-like bracts at its base | 21.
8. Flowers entirely replaced by bulbils | 9.
8. At least some flowers present | 10.
9. Stems less than 4 cm; leaves less than 10 cm | **10. Gagea**
9. Stem more than 5 cm; at least some leaves more than 15 cm | **26. Allium**
10. Funnel-like or collar-like corona present inside the perianth | **35. Narcissus**
10. Corona absent | 11.
11. Ovary more or less inferior | 12.
11. Ovary superior | 16.
12. Perianth golden yellow | **32. Sternbergia**
12. Perianth white, pink or red, often tinged green | 13.
13. Perianth pink or red, often tinged green | 14.
13. Perianth white with green or yellow patches | 15.
14. Perianth segments free, whitish tinged with green and often pale pink or dull red outside; ovary semi-inferior | **27. Nectaroscordum**
14. Perianth segments fused at base into a short tube, purplish-pink or pink, often whitish towards the base, or rarely all white; ovary inferior | **31. Amaryllis**
15. All 6 perianth segments similar | **33. Leucojum**
15. Three inner perianth segments shorter and blunter than the 3 outer | **34. Galanthus**
16. Perianth yellow | 17.
16. Perianth white or various shades of red or blue | 18.
17. Leaves less than 12(–15) mm wide; bracts (spathe) at base of inflorescence linear, more than 15 mm; flowers 1–5(–7) | **10. Gagea**
17. Leaves more than 12 mm wide; bracts (spathe) at base of inflorescence ovate, less than 15 mm; flowers more than 5 | **26. Allium**
18. Perianth segments fused for more than 10 mm, usually some shade of blue | 19.
18. Perianth segments free or fused for less than 5 mm, rarely blue or bluish | 20.
19. Flowers in an umbel, held horizontally, slightly zygomorphic, 30–50 mm | **29. Agapanthus**
19. Flowers solitary, erect, actinomorphic, 25–35 mm | **30. Tristagma**

20. Plant with an onion-like smell when fresh; perianth segments free; style arising from the base of the ovary   **26. Allium**

20. Plant without onion-like smell when fresh; perianth segments fused at base; style arising from the top of the ovary   **28. Nothoscordum**

21. Ovary inferior   **38. Alstroemeria**

21. Ovary superior   22.

22. Leaves strongly cordate at base; perianth segments and stamens 4   **16. Maianthemum**

22. Leaves cuneate to rounded at base   23.

23. Perianth segments united in a tube for more than one-fifth of their length   24.

23. Perianth segments free or united just at the base   31.

24. Perianth yellow to orange or red, more than 35 mm   25.

24. Perianth white to blue, pink or purple, very rarely pale yellow, less than 35 mm   26.

25. Flowers fewer than 20, perianth more than 50 mm, funnel-shaped   **6. Hemerocallis**

25. Flowers very numerous; perianth less than 50 mm, tubular to narrowly campanulate   **7. Kniphofia**

26. Flowers borne in groups of 1–5 in the axils of the main foliage leaves   **17. Polygonatum**

26. Flowers terminal, or in a terminal inflorescence, bracts absent or much reduced from leaves   27.

27. Plant rhizomatous; leaves linear to narrowly elliptical, narrowed at base   28.

27. Plant with a bulb; leaves linear, not narrowed at base 29.

28. Leaves with a distinct petiole, up to 30 × 10 cm including petiole; flowers pendulous, stalked, usually white, with perianth tube longer than lobes
                                        15. Convallaria

28. Leaves only slightly narrowed at base, up to 40 × 2 cm; flowers erect to patent, sessile, pink, with the perianth tube shorter than the lobes   **18. Reineckia**

29. Corolla contracted at mouth; perianth tube more than twice as long as lobes   **25. Muscari**

29. Corolla spread open at mouth; perianth tube less than twice as long as lobes   30.

30. Perianth tube much shorter than lobe, the lobes bent outwards at the junction of the tube   **24. Chionodoxa**
(See also × *Chionoscilla allenii*)

30. Perianth tube about as long as lobes, the lobes gradually curved outwards   **23. Hyacinthus**

31. Leaves all basal; inflorescence bractless or with bracts much reduced from leaves   32.

31. Stems bearing at least 1 leaf, or leaves all basal but at least the lowest bract more or less leaf-like   37.

32. Inflorescence a panicle; filaments densely hairy   **5. Simethis**

32. Inflorescence a raceme or a solitary flower; filaments glabrous   33.

33. Flowers solitary   **11. Erythronium**

33. Flowers in a raceme   34.

34. Bracts 2 per flower; perianth segments fused at the extreme base   **22. Hyacinthoides**

34. Bract absent or 1 per flower; perianth segments free   35.

35. Perianth segments usually blue, sometimes pink or pure white   **21. Scilla**

35. Perianth segments white with a green to reddish-brow stripe on the outside   36.

36. Plant without a bulb, with swollen roots; bracts brown   **3. Asphodelus**

36. Plant with a bulb; bracts whitish   **20. Ornithogalum**

37. Perianth segments less than 20 mm   38.

37. Perianth segments more than 20 mm   40.

38. Perianth segments white with purplish veins **9. Lloydia**

38. Perianth segments yellow or greenish-yellow and sometimes with a green stripe   39.

39. Styles 3–4   **4. Veratrum**

39. Style 1   **10. Gagea**

40. Stigma sessile   **12. Tulipa**

40. Stigmas on obvious styles   41.

41. Filaments more or less rigidly fixed to the base of the anthers; stigmas linear   **13. Fritillaria**

41. Filaments loosely fixed to the middle of the anther; stigmas not or scarcely longer than wide   **14. Lilium**

Subfamily 1. Melanthioideae Engl.

Plant rhizomatous. *Leaves* all or mostly basal, vertical, flat with 2 identical faces. *Inflorescence* a terminal raceme. *Perianth segments* yellow to greenish-white, free. *Styles* 1–3. *Ovary* superior. *Fruit* a capsule.

1. Tofieldia Huds.

Perennial herbs with rhizomes. *Leaves* numerous, mostly basal, densely tufted in 2 opposite ranks, narrowly sword-shaped. *Flowers* in racemes. *Perianth* segments 6, greenish-white, free or almost so. *Stamens* 6, free; filaments glabrous; anthers ovate, introrse, about as long as wide. *Styles* 3, short, persistent. *Ovary* trigonous, superior; carpels free above. *Capsule* splitting where ovary-cells meet. *Seeds* oblong, curved.

About 18 species in north temperate and Arctic regions and in the Andes.

Stearn, W. T. (1947). The nomenclature and synonymy of *Tofieldia calyculata* and *T. pusilla*. *Jour. Linn. Soc. London (Bot.)* 53: 194–204.

Stewart, A., Pearman, D. A. & Preston, C. D. (1994). *Scarce plants in Britain*. Peterborough.

1. T. pusilla (Michx) Pers.   Scottish Asphodel
Anthericum calyculatum auct.; *T. palustris* auct.; *Helonias borealis* auct.; *Narthecium pusillum* Michx; *Narthecium borealis* Wahlenb.; *T. borealis* (Wahlenb.) Wahlenb.; *T. minima* Druce nom. illegit.

Glabrous *perennial herb* with short rhizomes and fibrous roots. *Stems* 1.5–12.0(–25.0) cm, pale green, smooth, erect, leafless. *Leaves* 1–8 cm × 1–2 mm, dark green, tufted, narrowly sword-shaped, acute at apex, with 3–7 veins. *Inflorescence* a terminal raceme with 5–10 flowers; pedicels short, with a three-lobed, scarious bract at the base. *Perianth segments* 6, free, 1.5–2.5 × 0.9–1.0 mm, greenish-white, lanceolate, oblanceolate or ovate, obtuse at apex. *Stamens* 6; filaments about 1 mm, greenish-white, glabrous; anthers greenish-yellow. *Styles* 3, short and persistent. *Capsule* 2.5–3.0 mm, broadly ellipsoid. *Seeds* numerous, very small, narrowly oblong. *Flowers* 6–8. Pollinated by insects or selfed. $2n$ = 30.

Native. By streams and in calcareous flushes in mountains dropping to sea level in Sutherland. Very local in northern England (Upper Teesdale), locally frequent in central and north Scotland. North Europe eastwards to the central Urals; Alps; west Carpathians; eastern Siberia; Arctic North America. British plants are all referable to subsp. **pusilla** which occurs throughout the range of the species.

2. **Narthecium** Huds. nom. conserv.
Abama Adans.

Perennial herbs with rhizomes and fibrous roots. *Leaves* numerous, basal and cauline, in 2 opposite ranks. *Flowers* in racemes. *Perianth segments* 6, yellow, free or almost so. *Stamens* 6, free; filaments densely hairy; anthers linear, more than twice as long as wide. *Style* 1, simple; stigma slightly 3-lobed. *Ovary* trigonous, superior; carpels completely united. *Capsule* splitting along the centre of the ovary-cells. *Seeds* with long, fine projections at each end.

About 8 species in north temperate regions.

Summerfield, R. J. (1974). *Narthecium ossifragum* (L.) Hudson in Biological flora of the British Isles. *Jour. Ecol.* **62**: 325–339.

1. **N. ossifragum** (L.) Huds.   Bog Asphodel
Anthericum ossifragum L.; *Abama ossifraga* (L.) DC.

Perennial herb with a white, jointed, creeping rhizome with copious, fibrous roots. *Stem* 5–45 cm, yellowish-brown, terete, striate, hollow, glabrous. *Leaves* medium matt yellowish-green, the basal 40–70 × 3–5 mm, linear-ensiform, acute at apex, white and sheathing at the base, 3–6(–8)-veined, the cauline 4–25 × 2–5 mm, orange-brown, scale-like, alternate, distant, membranous, strongly ribbed, inflated, sheathing at the base, keeled and folded together. *Inflorescence* a terminal raceme with 6–20 flowers; pedicels 3–10 mm, with 2 linear-lanceolate, coloured bracts. *Flowers* 12–15 mm in diameter, without scent. *Perianth segments* free, 6–9 × 1–2 mm, pale golden yellow, tinged orange at apex beneath, linear-oblong, obtuse at apex. *Stamens* 6; filaments yellow, 3–4 mm, densely hairy; anthers orange, linear. *Style* 1, simple; stigma 3-lobed, pale yellowish-green. *Capsule* 8–12 mm, orange, narrowly ellipsoid. *Seeds* numerous, almost filiform, with projections at each end. *Flowers* 7–9. Pollinated by insects. $2n = 26$.

Dwarf forms are recorded from the Faeroes and similar plants occur in Shetland, Orkney, Fair Isle and Hebrides.

Native. Bogs and other wet, peaty acid places on heaths, moors and mountains. Common in Ireland and west and north Britain, absent from most of central and east Britain. North and west Europe eastwards to southeast Sweden and southwards in the mountains to north Portugal.

Subfamily 2. **Asphodeloideae** (Vent.) Engl.

Plants rhizomatous or with swollen roots. *Leaves* all or nearly all basal, linear. *Inflorescence* a raceme or terminal, compound cyme. *Perianth segments* of various colours, but not blue. *Style* 1. *Ovary* superior. *Fruit* a capsule.

3. **Asphodelus** L.

Perennial herbs with swollen roots. *Leaves* all or nearly all basal, linear, with a membranous, sheathing base. *Flowers* in racemes, actinomorphic. *Perianth segments* white with a greenish or reddish stripe on the outside, erecto-patent, free. *Stamens* 6, free; filaments glabrous, expanded at base; anthers dorsifixed, introrse. *Style* 1. *Ovary* superior; ovules 2 per cell. *Fruit* a capsule, splitting along the centre of the ovary cells.

About 12 species from the Mediterranean to the Himalayas.

A. fistulosus L. is a grain and wool casual and possibly also a garden escape.

1. **A. albus** Mill.   White Asphodel

Glabrous *perennial herb* with swollen, oblong tubers and fibrous roots. *Stems* 30–100(–150) cm, solid, smooth, more or less glaucous, usually simple or sometimes with a few, short branches at the base of the inflorescence. *Leaves* 15–60 × (0.5–)1–2(–3) cm, more or less glaucous, long-linear, gradually narrowed to an acute apex, keeled, entire. *Inflorescence* a dense, branched raceme; pedicels articulated; bracts membranous, dark brown, the larger longer than the pedicels. *Perianth segments* 6, 15–20 mm, white with a greenish- to reddish-purple stripe on the outside, narrowly elliptical, rounded-obtuse at apex, erecto-patent. *Stamens* 6, free; filaments white, dilated at base; anthers brownish-orange. *Capsule* 8–15 × 6–13 mm, ovoid-ellipsoid, trigonous, truncate and emarginate at apex. *Seeds* 6, black. *Flowers* 5–6. $2n = 28, 56$.

Introduced. Naturalised on a grassy bank at the foot of Mont Rossignot in Jersey since the early 1970s and on a railway bank near Tunbridge Wells West station in Kent. Native of south and central Europe.

4. **Veratrum** L.

Perennial herbs with rhizomes. *Leaves* basal and on the stem, elliptical to ovate. *Flowers* in dense, terminal racemes or panicles. *Perianth segments* pale to bright yellowish-green, free, patent. *Stamens* 6, free; filaments glabrous. *Styles* 3–4, free. *Ovary* superior; ovules numerous per cell. *Fruit* a capsule splitting along the centre of the ovary cells.

About 45 species, widely distributed in the northern hemisphere.

Mathew, B. (1989). A review of *Veratrum. The Plantsman* **11**: 34–61.

1. **V. viride** Aiton   Green False Helleborine

Tall, robust *perennial herb* with a stout rhizome. *Stems* 60–200 cm, erect, stout, leafy to the top. *Leaves* 15–30 × 6–14 cm, medium green, ovate to broadly elliptical, acute to acuminate at apex, entire, with a sessile, sheathing base, glabrous. *Inflorescence* 20–50 cm, a freely branched panicle, the branches ascending and subtended by leafy bracts; pedicels 2–4 mm. *Perianth segments* 10–15 × 3–5 mm, yellowish-green, ovate or elliptical, acute to acuminate at apex, narrowed to the base, hairy outside at least on the midrib and margin. *Stamens* 6, free; filaments erect, somewhat shorter than the perianth and free from it at the base. *Styles* 3–4. *Capsule* 18–25 × 9–12 mm, ovoid. *Seeds* yellowish to brown, lanceolate, tapering to an acute apex, winged. *Flowers* 6–7. 2*n* = 32.

Introduced. Sometimes persisting for a short period when thrown out of gardens. A clump has survived in woodland at Douglaston Estate, Milngavie in Dunbartonshire for many years. Native of North America.

5. Simethis Kunth nom. conserv.
Pubilaria Raf.

Perennial herbs with short rhizomes. *Leaves* all basal, linear, with sheathing bases. *Flowers* in a cymose panicle, actinomorphic. *Perianth segments* purplish outside, white inside, more or less patent, free. *Stamens* 6; filaments densely hairy. *Style* 1. *Ovary* superior; ovules 2 per cell. *Fruit* a capsule, splitting along the centre of the ovary cells.

One species mainly in the west Mediterranean region.

1. S. planifolia (L.) Gren.   Kerry Lily
Anthericum planifolium L.; *Anthericum bicolor* Desf.;
Pubilaria planifolia (L.) Druce

Glabrous *perennial herb* with short, erect rhizomes covered with the brown, fibrous remains of the leaf bases; roots tufted, fusiform, fleshy. *Leaves* usually all basal, 15–60 cm × 2.5–7.5 mm, exceeding the stem, bluish-green, long-linear, acute at apex, entire, with a sheathing, scarious, fibrous base. *Inflorescence* terminal, lax, cymose with several erecto-patent branches, each with 3–5 flowers; pedicels with short bracts. *Perianth segments* 8–11 × 3–4 mm, white inside, purplish with white margins beneath, elliptical, obtuse at apex, free, patent, 5- to 7-veined. *Stamens* 6; filaments 2–3 mm, white, villous; anthers yellow. *Style* 1, straight. *Capsule* 5–6 mm, broadly ellipsoid. *Seeds* 3–6, about 5 mm, shining black, with a whitish appendage, ovoid. *Flowers* 5–7. 2*n* = 48.

Native. Rocky heathland near the sea, growing with *Ulex* over about 30 km, near Derrynane, Co. Kerry and formerly naturalised on heathland near the sea in Dorset and Hampshire. South-west Europe, Morocco, Algeria, Tunisia.

(**Hosta elata** Hyl., **H. fortunei** (Baker) L. H. Bailey, **H. lancifolia** (Thunb.) Engl., **H. sieboldiana** (Hook.) Engl. and **H. ventricosa** (Salisb.) Stearn have been recorded as garden escapes.)

6. Hemerocallis L.
Perennial herbs with short rhizomes and more or less fleshy roots. *Leaves* linear, in 2 ranks. *Flowers* in a cymose panicle, slightly zygomorphic. *Perianth segments* yellow to orange, fused to form a tube. *Stamens* 6, deflexed downwards; filaments glabrous. *Style* 1. *Ovary* superior, trigonous; ovules many per cell. *Fruit* a capsule splitting along the centre of the ovary-cells.

About 15 species from eastern Asia. Cultivated in China and Japan for several centuries and the U.S.A. more recently where large numbers of varieties of hybrid origin and unknown parentage have been created.

H. citrina Baroni has been recorded as a garden escape.

Bailey, L. H. (1930). *Hemerocallis* : the day lilies. *Gentes Herbarium* **2**: 143–156.
Kitchingman, R. M. K. (1985). Some species and cultivars of *Hemerocallis*. *The Plantsman*. **7**: 68–69.
Stout, A. B. (1934). *Day lilies*. New York.

1. Flowers orange, more or less scentless; perianth segments with anastomosing veins, dull orange, undulate on margins   **1. fulva**
1. Flowers bright yellow, sweet-scented; perianth segments without anastomosing veins, lemon-yellow, flat   **2. lilioasphodelus**

1. H. fulva (L.) L.   Orange Day-lily
H. lilioasphodelus var. *fulva* L.

Perennial herb growing in clumps, with spreading rhizomes and fusiform, tuberous roots. *Stems* up to 1 m, exceeding the leaves, erect, forked and slightly branched above. *Leaves* 30–90 × 1.0–2.5(–3.5) cm, green, in 2 ranks, linear, acute at apex, keeled. *Inflorescence* terminal, cymose, with 6–12(–20), scentless flowers. *Perianth segments* 7–10 × 2–3 cm, dull orange, usually with zones or stripes of darker colour, fused at base from one-quarter to half their length, the tube yellow, linear-lanceolate or elliptic-oblong, obtuse at apex, margins undulate, recurved; veins anastomosing. *Stamens* 6, deflexed downwards; filaments glabrous; anthers linear-oblong. *Style* 1, deflexed downwards. *Capsule* never forms. *Flowers* 6–8. It seems possible that all plants are sterile triploids with 2*n* = 33.

Introduced. Forming clumps on rough ground, banks, cliffs, dunes, quarries and grassy places. Scattered over much of Britain and the Channel Islands. Origin uncertain, but perhaps China.

2. H. lilioasphodelus L.   Yellow Day-lily
H. lilioasphodelus var. *flava* L. nom. illegit.; *H. flava* L. nom. illegit.

Perennial herb growing in clumps, with a short rhizome; root partly fibrous and partly tuberous. *Stem* up to 80 cm, exceeding the leaves, but often bent over with the weight of the flowers. *Leaves* 50–65 × 1.0–1.4 cm, green, in 2 ranks, linear, acute at apex, entire, keeled, recurved. *Inflorescence* terminal, cymose, with 8–12 fragrant flowers. *Perianth segments* 6–10 × 1.5–2.5 cm, lemon yellow, fused at base for one-quarter of their length, linear-lanceolate, acute at apex, with flat margins; veins not anastomosing. *Stamens* 6; deflexed downwards. *Style* 1, deflexed downwards. *Capsule* 3–4 × about 2 cm, ellipsoid or obovate-oblong, rarely found. *Seeds* about 6 mm, obovoid, shining black. *Flowers* 5–6. 2*n* = 22.

Said to be self-sterile and capsules rarely found, but can produce abundant seed and spreads extensively by clonal growth.

Introduced. Naturalised on rough ground, roadsides and railway banks and grassy places. Scattered through Britain, though less common than *H. fulva* in England and Wales, but more common in Scotland. Native of eastern Asia and perhaps Italy and Yugoslavia.

7. Kniphofia Moench nom. conserv.

Tufted *perennial herbs* with short, thick rhizomes. *Leaves* usually basal, in several ranks, linear. *Flowers* in a dense raceme, slightly zygomorphic. *Perianth segments* red to yellow, fused to form a proximal tube cylindrical to campanulate. *Stamens* 6, often becoming spirally twisted and withdrawn after shedding pollen; filaments glabrous. *Style* 1. *Ovary* superior; ovules many per cell. Fruit a *capsule* splitting along the centre of the ovary cells.

About 70 species mainly in eastern and southern Africa, with one species in Madagascar and one in southern Arabia.

Many plants in cultivation may be hybrid cultivars of which *K. uvaria* is often one of the parents. **K. rufa** Leichtlin ex Baker is a garden escape on dunes at Point of Ayr, Flintshire.

Codd, L. E. (1963). The South African species of *Kniphofia* (Liliaceae). *Bothalia* **9**: 363–511.

Taylor, J. (1985). *Kniphofia* – a survey. *The Plantsman* **7**: 129–160.

1. Leaves up to 80.0 × 1.8 cm; bracts rounded to subacute at apex; perianth segments 28–40 mm; stamens included or just exserted   **1. uvaria**
1. Leaves up to 200 × 4 cm; bracts acute to acuminate at apex; perianth segments 24–35 mm; stamens exserted 4–15 mm   **2. praecox**

1. K. uvaria (L.) Oken   Red-hot-poker

Aloe uvaria L.; *Aletris uvaria* (L.) L.; *K. alooides* Moench nom. illegit.; *K. burchellii* (Lindl.) Kunth; *K. odorata* Heyh.; *K. bachmannii* Baker; *K. occidentalis* Berger; *Veltheimia uvaria* (L.) Willd.; *Tritoma uvaria* (L.) Link; *Tritoma burchellii* Herb. ex Lindl.; *Tritomanthe uvaria* (L.) Link; *Triclissa uvaria* (L.) Salisb.

Densely tufted *perennial herbs* with short rhizomes. *Stems* 50–120 cm. *Leaves* 10–20 per stalk, 35–80 × 0.6–1.8 cm, dull green to glaucous, linear, stiffly erect to arcuate-spreading, keeled, V-shaped in section, usually tough and often drying with a hard, fibrous texture, margin and keel smooth to sparingly serrulate, with a few scattered teeth mainly towards the apex. *Peduncle* overtopping or equal to the leaves. *Inflorescence* 4.5–11.0 × about 5.8 cm, oblong to globose, dense to subdense or more or less lax at apex; buds spreading, brilliant red to greenish tinged with red; flowers orange-yellow to greenish-yellow, becoming pendulous. *Bracts* 3–9 mm, broadly ovate to oblong-ovate, rounded or obtuse to subacute at apex, the margin entire to erose-denticulate. *Pedicels* (1.5–)3–5 mm, elongating to 8 mm

in fruit. *Perianth* 28–40 mm long, about 3 mm broad above the ovary, widening to 4–5 mm broad at the throat, subcylindrical, almost straight; lobes about 2 mm, ovate, obtuse at apex, slightly spreading. *Stamens* included or just exserted at anthesis, later withdrawn. *Style* subequal to the anthers at anthesis, finely exserted by 3–5 mm. *Capsule* 7–14 mm, ovoid-triquetrous. *Flowers* 7–9. $2n = 12$.

Introduced. Much grown in gardens and very persistent when thrown out or planted on dunes, beaches, railway embankments and waste ground, usually near the sea. Scattered in south and west Britain from Kent to Flintshire and in the Channel Islands. Native of South Africa.

2. K. praecox Baker   Greater Red-hot-poker

Densely tufted *perennial* with a short rhizome. *Stems* up to 200 cm. *Leaves* more or less in 4 ranks, 12 or more per stalk, 90–200 × 2–4 cm, dark green, linear, erect to more or less spreading, recurved, fairly rigid, deeply keeled below and channelled above, keel and margin serrulate to obscurely serrulate, rarely almost smooth. *Peduncle* 120–200 cm, well overtopping the leaves, with several triangular sterile bracts below the inflorescence. *Inflorescence* 12–30 × 6–7 cm, tapering somewhat towards the apex, subcylindrical, dense to very dense; buds spreading, orange to reddish-orange or red. *Flowers* pale yellow, often brown-tipped, especially when dry. *Pedicels* 4–5(–8) mm, elongating to 8–10 mm in fruit. *Bracts* 8–12 × 1.5–2.0 mm, lanceolate to linear-lanceolate, tapering gradually to the small, rounded apex; margin finely erose-denticulate. *Perianth* 2.4–3.4 cm, subcylindrical to narrowly funnel-shaped. *Stamens* exserted by 4–15 mm at anthesis. *Style* subequal to the stamens at anthesis, eventually exserted by about 16 mm. *Capsule* 5–7 mm, subglobose, erect. *Flowers* 7–9. $2n = 12$.

Introduced. Grown and naturalised with *K. uvaria* and much confused with it. Certainly in Cornwall and Channel Islands and probably elsewhere. Native of South Africa.

Subfamily 3. Wurmbaeoidae

Plants with a corm. *Leaves* appearing in spring with the fruits, more or less all on stems, linear-oblong. *Flowers* appearing in autumn without the leaves, 1–few, each arising from ground. *Perianth segments* pinkish to pale purple, proximally united into a long tube. *Styles* 3. *Ovary* superior, but subterranean at flowering, emerging above the ground at the stem apex when fruiting. *Fruit* a capsule.

8. Colchicum L.

Perennial herbs with a corm. *Leaves* on stem, linear-oblong, not present when flowering. *Flowers* 1–few, emerging directly from the ground in autumn. *Perianth segments* pinkish to pale purple, fused to form a tube. *Stamens* 6. *Styles* 3. *Ovary* superior, subterranean. *Capsule* splitting where the ovary-cells meet.

About 80 species from Europe, west and central Asia and North Africa.

Butcher, R. W. (1954). *Colchicum autumnale* L. in Biological flora of the British Isles. *Jour. Ecol.* **42**: 249–257.

1. C. autumnale L.   Meadow Saffron

Glabrous *perennial herb,* with an ovoid to almost spherical corm 2.5–6.0 × 2–4 cm, dark brown, membranous or subcoriaceous and with a neck 2–4 cm. *Leaves* developing after anthesis, (3–)4(–5), 15–35 × 2–5(–7) cm, glossy green, broadly lanceolate, subobtuse at apex, entire. *Flowers* 1–3, 2.5–3.5 cm in diameter, appearing out of the ground in autumn, narrowly bell-shaped. *Perianth* tube 5–20 cm, narrow; segments 40–60 × 10–15 mm, pale purple, rarely white, oblong or elliptical-oblong, subobtuse at apex. *Stamens* 6; filaments 10–16 mm, pink; anthers 5–8 mm, yellow. *Styles* 3, curved at apex; stigmas long-decurrent. *Ovary* underground at anthesis, the stem elongating in fruit. *Capsule* 3–5 cm, obovoid, deeply 3-lobed, transversely rugose, splitting at the apex along the septa. *Seeds* numerous, globose, brown. *Flowers* 5–10. Pollinated by insects and selfed, protogynous. $2n = 38$.

Native. Damp meadows and open woodland on rich soils. Local in central and southern Britain, common around the River Severn estuary, very local in southern Ireland, naturalised in Jersey and Co. Armagh. South, west and central Europe eastwards to White Russia and north-west Ukraine.

Subfamily 4. Lilioideae Engl.

Plant with a rhizome or a bulb. *Leaves* all basal, all on stem, or both, linear to ovate or elliptical. *Inflorescence* a single flower or flowers in a spike or raceme. *Perianth segments* of various colours, but not blue, free or fused. *Style* one or absent. *Ovary* superior. *Fruit* a berry or capsule.

9. Lloydia Salisb. ex Rchb. nom. conserv.

Perennial herbs with a bulb. *Leaves* at base usually 2, with 2–4(–5) on the stem, more or less linear. *Flowers* 1–2(–3) at the stem apex. *Perianth segments* white with purplish veins, free, erecto-patent. *Stamens* 6, inserted at base of perianth segments. *Style* 1. *Capsule* splitting along the centre of the ovary cells.

About 12 species in the temperate northern hemisphere.

Woodhead, N. (1951). *Lloydia serotina* (L.) Reichenb. in Biological flora of the British Isles. *Jour. Ecol.* **39**: 198–203.

1. L. serotina (L.) Rchb.   Snowdon Lily
Bulbocodium serotinum L., *Bulbocodium autumnale* L.; *Anthericum serotinum* (L.) L.; *Phalangium serotinun* (L.) Poir.; *L. alpina* Salisb. nom. illegit.

Glabrous *perennial herb. Bulb* brown, ovoid, elongate; tunic persistent, membranous, enclosed in the remains of old leaves; with filiform stolons. *Stem* 5–15(–40) cm, compressed, bluntly angled, slender, with 2–4(–5) leaves. *Leaves* at base 2, 7–20(–30) cm × 1–2 mm, medium green, more or less linear or filiform, curved, usually exceeding the stem; the cauline linear to linear-lanceolate, the upper subtending the pedicels, bract-like,

shorter and wider. *Flowers* 10–15 mm in diameter, usually solitary or in pairs, rarely 3, terminal, campanulate-infundibuliform; peduncles 20–25 mm. *Perianth segments* 9–15 × 2.5–5.0 mm, free, erecto-patent, white with reddish or purplish veins, obovate or elliptical, obtuse or rounded at apex, narrowed at base. *Stamens* 6, inserted at the base of the perianth segments; filaments 3–4 mm; anthers yellow. *Style* 1, shorter than stamens; stigma entire or 3-lobed. *Capsule* 5–7 mm in diameter, globose, 3-ribbed. *Seeds* 2.0–2.5 mm, numerous, brown, ellipsoid, with one side straight. *Flowers* 6. Pollinated by flies etc.; protandrous. $2n = 24$.

Native. Cracks in basic mountain rock from 460 to 760 m. Confined to about five sites in the Snowdon range, Caernarvonshire. Arctic Russia, Alps, Carpathians, Bulgaria, Urals, Caucasus; Soviet Asia, Japan, Himalayas, western China; western North America from Alaska to Oregon and New Mexico.

10. Gagea Salisb.

Perennial herbs with 1–2 bulbs. *Leaves* basal and on stem, more or less linear. *Flowers* 1–5 at stem apex. *Perianth segments* yellow with a green band, free, more or less patent. *Stamens* 6. *Style* 1. *Capsule* splitting along the centre of the ovary-cells.

About 50 species, mostly in temperate central Asia, but also in Europe.

Rix, E. M. & Woods, R. G. (1981). *Gagea bohemica* (Zauschner) J. A. & J. H. Schultes in the British Isles, and a general view of the *G. bohemica* species complex. *Watsonia* **13**: 268–270.

Slater, F. M. (1990). *Gagea bohemca* (Zauschner) J. A. & J. H. Schultes (*G. saxatilis* Koch) in Biological flora of the British Isles. *Jour. Ecol.* **78**: 535–546.

Stewart, A., Pearman, D. A. & Preston, C. D. (1994). *Scarce plants in Britian.* [*Gagea lutea.*]

Uphof, J. C. T. A review of the genus *Gagea* Salisb. *Plant Life* **14**: 124–132 (1958); **15**: 151–161 (1959); **16**: 163–176 (1960).

1. Bulb 1; basal leaf 1, linear-lanceolate   **1. lutea**
1. Bulbs 2, in a common tunic; basal leaves 2–4, filiform
                                                            2. bohemica

1. G. lutea (L.) Ker. Gawl.   Yellow Star-of-Bethlehem
Ornithogalum luteum L.; *G. fascicularis* Salisb. nom. illegit.; *G. silvatica* (Pers.) Loudon

Perennial herb. Bulbs solitary, without a collar. *Stem* 8–25 cm, glabrous, with a small leaf-like bract 2–3 cm subtending the inflorescence and another much larger leaf or bract a short distance below, occasionally a third bract and two inflorescences are present. *Basal leaf* solitary (7–)15–45 cm × 7–15 mm, deep green, linear-lanceolate or linear, suddenly contracted to a hooded, acuminate apex, often curled, flat; the cauline 2, similar to basal. *Inflorescence* subumbellate. *Flowers* 1–7(–10); cylindrical-campanulate; pedicels 15–50 mm, glabrous or slightly hairy, unequal. *Perianth segments* 6, free, 10–18 × 2.0–2.5 mm, yellow with a green band on the outside, linear-lanceolate or linear-oblong, obtuse at apex. *Stamens* 6, inserted at the base of the perianth; filaments 5–7 mm, linear-lanceolate; anthers deep yellow to orange-yellow. *Style* trigonous; stigmas obscurely lobed. *Capsule* globose. *Seeds* 2.0–2.5 mm, elliptical or semi-lunar, with a strophiole. *Flowers* 3–5. *Pollinated* by insects, weakly protogynous. Sometimes persists for many years without flowering. $2n = 72$.

Non-flowering plants can be easily confused with *Hyacinthoides non-scriptus*.

Native. Damp, base-rich woods, woodland borders, wooded limestone pavements, hedgerows, shady river banks and rough fields. Scattered through Britain north to central Scotland, rare except in central and northern England. Most of Europe; temperate Asia to Kamchatka and Japan.

2. G. bohemica (Zauschn.) Schult. & Schult. fil.

Early Star-of-Bethlehem
Ornithogalum bohemicum Zauschn.; *G. saxatilis* Koch

Perennial herb. *Bulbs* 2 in a common tunic; tunic pale chestnut-brown, papery. *Stem* 12–37 mm, almost glabrous, with 4 leaves. *Basal leaves* 2–4 on a flowering bulb, 4–185 × about 1 mm, filiform, D-shaped in section, glabrous or with short, crisp hairs, the cauline 15–40 × 2–4 mm, narrowly lanceolate and long-ciliate. *Inflorescence* usually 1-flowered, rarely up to 4-flowered; bracts 1–2 per flower; flowering stem often replaced by a group of about 15–25 bulbils, white at flowering time, later becoming dark brown, each covered with a reticulate tunic. *Perianth segments* usually 6, sometimes 5–10, 11–18 × 2–4 mm, bright and shining yellow inside, greenish-veined outside, narrowly oblong-lanceolate, obtuse at apex, occasionally transitional to a bract. *Stamens* 6, inserted at the base of the perianth segments; filaments about 7–8 mm; anthers about 1.0–2.5 mm; sometimes staminodal. *Style* 5–6 mm, filiform, glabrous. *Ovary* obovate. *Capsule* 6–7 mm, obovate and emarginate, but has only once been seen on British plants. *Flowers* 1–3. $2n = 60$.

Over its total range it is a very variable species which is not easily studied on herbarium sheets and perhaps ought to be divided into several infraspecific taxa.

Native. Cracks and ledges of sunny basic rock in very short vegetation. Stannor Rocks in Radnorshire where it was first found in 1965 and is confined to an area of about one hectare. West France, central and south Europe eastwards to western Turkey, Syria and Israel.

11. Erythronium L.

Perennial herbs with a bulb. *Leaves* usually 2, basal or almost so, oblong, spotted. *Flowers* solitary, terminal. *Perianth segments* rose-pink to purple, sometimes white, free, inner with basal appendage, usually recurved. *Stamens* 6. *Style* 1, deeply 3-fid. *Ovary* superior. *Capsule* splitting along the centre of the ovary cells.

About 25 species from temperate North America, with one species in Eurasia.

1. E. dens-canis L.   Dog's-tooth Violet

Glabrous *perennial herb.* *Bulb* ovoid-cylindrical. *Stem* 10–30 cm. *Leaves* usually 2, 5–9 × 1.5–4.0 cm, opposite, ovate-elliptical to oblong, reddish-spotted; petiole 2–4

cm, amplexicaul. *Flowers* solitary, nodding, terminal. *Perianth segments* 6, free, 18–30 × 4–10 mm, bright purplish-pink, sometimes spotted at the base, oblong-lanceolate, subobtuse at apex, strongly deflexed from near the base, the inner with basal appendages. *Stamens* 6, free; filaments fusiform; anthers bluish, linear-oblong. *Style* deeply 3-fid; stigma deeply lobed. *Capsule* ovoid-trigonous. *Seeds* few. *Flowers* 4–5. $2n = 24$.

Introduced. Sometimes planted in woodland gardens where it is very persistent, and a persistent garden escape elsewhere. Native of southern Europe and south-west Asia.

12. Tulipa L.

Perennial herbs with bulbs. *Bulb-tunics* usually lined inside with hairs. *Leaves* cauline, few, alternate, somewhat fleshy, decreasing in size up the stem. *Flowers* 1–2(–4) at the stem-apex, cup-shaped, erect when mature. *Perianth segments* 6, free, petaloid, caducous. *Nectaries* absent. *Anthers* basifixed, introrse; filaments dilated towards the base. *Style* absent; stigma sessile. *Fruit* a globose or ellipsoid capsule. *Seeds* flat, numerous.

About 100 species in central and south Europe, temperate Asia and North Africa.

Cronk, Q. C. B. (1977). *Tulipa saxatilis* Sieb. ex Sprengel. *B.S.B.I. News* **16**: 19.

Hall, A. D. (1940). *The Genus Tulipa*. London.

1. Flower rounded at base; filaments glabrous; buds erect
                                                    3. gesneriana
1. Flower narrowed at base; filaments hairy near the base; buds nodding   2.
2. Perianth segments yellow, the outer sometimes tinged green, pink or crimson   **1. sylvestris**
2. Perianth segments pink to lilac-purple with a white-edged yellow basal blotch   **2. saxatilis**

1. T. sylvestris L.   Wild Tulip

Bulb 18–45 × 8–22 mm, often stoloniferous; tunic tough, with a few, straight hairs inside towards the apex, occasionally glabrous. *Stem* (5–)8–45 cm, stout, flexuous, glabrous or downy, bluish-green or green. *Leaves* 2–3(–4), up to 30 cm × 18 mm, bluish-green, rather fleshy, linear-lanceolate or linear, acute at apex, channelled, glabrous. *Buds* nodding. *Flowers* 1(–2), 40–90 mm in diameter, often fragrant. *Perianth segments* yellow, rarely cream, usually ciliate at base, the outer 20–63 × 4.5–27.0 mm, narrowly elliptical, acute at apex, sometimes tinged outside with green, pink or crimson, the inner 21–70 × 6–26 mm, elliptic-oblanceolate, shortly acuminate at apex. *Stamens* 5; filaments 5–14 mm, yellow, with a hairy boss near the base; anthers 2.5–9.0 mm, yellow. *Stigma* 3-lobed. *Ovary* green. *Capsule* 17–30 × 14–16 mm, globose to oblong-ellipsoid. *Flowers* 4–5. Pollinated by small insects and selfed; homogamous. $2n = 48$.

Introduced. Naturalised in woods, meadows and neglected estates. Scattered and rare in England and south and central Scotland, formerly much more frequent. Native of south and south-east Europe, extending northwards to north-west France and to about 55°N in east-central Russia; widely naturalised in north and central Europe.

2. T. saxatilis Sieber ex Spreng.   Rock Tulip
T. bakeri A.D. Hall

Bulb 20–35 × 15–30 mm, markedly stoloniferous; tunic coriaceous, with a few, straight hairs inside towards the apex. *Stem* 15–45 cm, yellowish-green, slightly compressed, glabrous. *Leaves* 2–4, 10–38 × 1.2–4.5 cm, shiny yellowish-green above, oblong-lanceolate, linear-lanceolate or oblong-elliptical, fleshy, broadly channelled, acute and hooded at apex, entire, glabrous. *Buds* nodding. *Flowers* 1–2(–4), 70–90 mm in diameter, scentless or fragrant. *Perianth segments* pink to lilac-purple with a distinct, white-edged, yellow basal blotch inside, the outer 38–53 × 9–18 mm, elliptical to elliptic-oblong, acute to subacute at apex, narrowed at base, the inner 38–55 × 16–30 mm, oblong-elliptical to oblong-obovate, obtuse to subacute at apex, often shortly apiculate, narrowed at base. *Stamens* 5; filaments 8–20 mm, deep yellow, hairy at base; anthers 4.5–8.0 mm, dark brown. *Stigma* 3-lobed, pale yellow, shorter than anthers. *Ovary* green. *Capsule* rarely formed. *Flowers* 4 5. 2*n* − 24, 36.

Usually triploid. Diploid plants, which appear to have rather smaller, deeper coloured flowers have been called *T. bakeri*.

Introduced. Sometimes grown in gardens. Naturalised on rough, stony ground between Tresco Abbey Gardens and Appletree Banks in Tresco in the Isles of Scilly since 1976 where it spreads rapidly by stolons. Native of Crete.

3. T. gesneriana L.   Garden Tulip

Bulb 40–70 × 40–50 mm; tunics papery or rather coriaceous, glabrous or with a few, straight hairs inside towards the apex. *Stems* 30–60 cm, green or glaucous, glabrous or finely pubescent. *Leaves* 2–5(–7), up to 35 × 8 cm, green or glaucous, elliptic-oblong to lanceolate, the uppermost linear-lanceolate or linear, obtuse or acute at apex, surfaces glabrous, margin entire and ciliate towards the apex. *Buds* erect. *Flowers* solitary, 40–90 mm in diameter, often scented, rounded at base. *Perianth segments* scarlet, orange, yellow, white, pink or purplish or streaked, with or without a dark, basal blotch inside, the outer 45–75 × 18–32 mm, lanceolate to elliptical, more or less acute at apex, the inner 38–82 × 21–41 mm, elliptic-oblanceolate to oblanceolate or obovate, obtuse at apex, often shortly apiculate. *Stamens* 5; filaments 6–14 mm, yellow or purple, glabrous; anthers 8.5–15.0 mm, yellow or purple. *Stigmas* 3-lobed, shorter than anthers. *Capsule* ellipsoid. *Flowers* 4–5.

A complex species from which the majority of garden tulip cultivars have been derived. The popularity of the garden tulip led to a great influx of importations into Europe from south-west and south-central Asia from the fifteenth century onwards, from which numerous variants were selected and persisted and reproduced

vegetatively. Plants in which the normal flower colour is variegated with another colour and are called 'broken' are infected by a virus.

Introduced. Persistent on rubbish tips, waysides and rough ground where they have been thrown out from gardens. Scattered in central and south England and the Channel Islands. No certainly wild counterpart has yet been identified, although similar plants occur in south-west and south-central Asia.

13. Fritillaria L.

Perennial herbs with a bulb. *Leaves* all or most on the stem, linear, elliptical or oblong-elliptical. *Flowers* 1–5, at stem apex, pendulous, cup-shaped. *Perianth segments* pink to purple or orange to yellow, sometimes white, free. *Filaments* more or less rigidly fused to the base of the anthers. *Style* 1; stigmas 3, linear. *Capsule* splitting along the centre of the ovary-cells.

About 100 species found throughout the temperate regions of the northern hemisphere excluding North America, with most species in the Mediterranean area, the mountains of south-west Asia and in California.

King, N. & Wells, D. (1993). *Fritillaria meleagris* L. Fritillary in Gillam, B. *The Wiltshire Flora*. Newbury. 95–97.
Oswald, P. (1992). The Fritillary in Britain – a historical perspective. *British Wildlife* **3**: 200–210.
Stewart, A., Pearman, D. A. & Preston, C. D. (1994). *Scarce plants in Britian*. Peterborough.

1. Flowers 3–5 in an umbel, orange or yellow   **2. imperialis**
1. Flowers solitary or rarely 2, chequered purple and cream, chequered dirty purple or rarely white   2.
2. Flowers chequered dirty purple or chequered purple and cream   **1(1). meleagris** forma **meleagris**
2. Flowers white   **1(2). meleagris** forma **albiflora**

1. F. meleagris L.   Fritillary

Bulb up to 25 mm in diameter, of 2 large scales. *Stem* 12–50 cm, bluish-green, erect. *Leaves* mostly cauline, 3–6(–8), 5–20 cm × 3–10 mm, bluish-green, alternate, narrowly linear, gradually narrowed to a slender point at the apex, entire, rounded at base, glabrous. *Flowers* terminal, usually solitary, rarely 2, 28–40 mm in diameter, pendulous even when mature, campanulate. *Perianth segments* 30–50 × 10–19 mm, dull pink, chequered with dirty purple, sometimes purple and cream, sometimes white with a greenish-yellow mid-vein outside, and a little purple at base, oblong, more or less acute at apex, each with a green nectary 7–10 mm near the base. *Stamens* 6; filaments 10–16 mm, whitish or pale green, papillose; anthers about 6 mm, yellow. *Style* 1, pale green; stigma branches 3, green with a yellow tip, papillose, protruding slightly beyond the anthers. *Capsules* 6-angled, not winged, flat-topped, erect. *Seeds* many, flat. *Flowers* 4–5. 2*n* = 24.

(1) Forma meleagris
Perianth segments dull pink, chequered with dirty purple or sometimes purple and cream.

(2) Forma albiflora Zahar.
Perianth segments white.

Native. Damp meadows and pastures where it thrives best if it is cut for hay then grazed. Survival is mainly by vegetative reproduction, but seed is freely produced. Local in England north to Leicestershire and Staffordshire, but much less common than formerly. Its main stations are now in Suffolk and the Thames Valley. The two forms grow in mixed populations. Much planted in gardens and sometimes escaping and sometimes naturalised in England and Wales. From central Russia southwards to the southern Alps and central Yugoslavia; naturalised elsewhere in Europe.

2. F. imperialis L.   Crown Imperial

Bulb up to 10 cm in diameter, a compressed sphere, with a pungent, foxy smell. *Stem* 50–150 cm, erect, thick, angled, yellowish-green, often suffused and mottled brownish-purple, glabrous. *Leaves* in 3 or 4 whorls of 4–8, 12–15 × 2–9 cm, medium green and rather waxy, ascending, the lower broadly oblong-elliptical, the upper narrowly elliptical, drawn out to a slender, acute tip, entire, rounded at base, slightly folded, glabrous. *Bracts* 10–20, in a close group above the flowers, 11–13 × 1.5–2.5 cm, linear-lanceolate, drawn out at apex to a point. *Flowers* 3–5 in an umbel, broadly campanulate, 70–80 mm in diameter, pendulous. *Perianth segments* 40–55 × 25–30 mm, orange or yellow, the outer narrower than the inner, elliptical, narrowed to a blunt apex, entire, narrowed to a rounded base, each with a large circular nectary about 5 mm in diameter near its base. *Stamens* 6; filaments 35–45 mm, cream; papillose towards the apex, 3-fid, the branches 1–4 mm, exceeding the anthers. *Capsule* 6-angled, winged. *Seeds* many, flat. *Flowers* 3–5. 2*n* = 24.

Introduced. Commonly grown in gardens and sometimes thrown out and persisting for a while in waste and grassy places and on rubbish tips. Native of south Turkey eastwards to north-west India.

14. Lilium L.

Perennial herbs with a bulb composed of fleshy, overlapping scales. *Stem* erect, unbranched, leafy. *Leaves* basal and cauline or all cauline, linear to elliptical. *Flowers* few to many, in terminal racemes, funnel-shaped, cup-shaped or with perianth segments diverging and strongly rolled back to form a turk's-cap. *Perianth segments* free, each with a nectar-bearing furrow or gland at their base. *Stamens* 6, borne at the base of the perianth segments; filaments loosely fixed to the middle of the anthers. *Style* 1, with a 3-lobed stigma. *Fruit* a capsule, with 3 cells. *Seeds* many, flat.

About 100 species in the temperate northern hemisphere.

Synge, P. M. (1980). *Lilies*. London.
Woodcock, H. B. D. & Stearn, W. T. (1950). *Lilies of the world*. London.

1. Outside of perianth segments white   2.
1. Outside of perianth segments pinkish-purple,
    maroon, red, orange or yellow   3.
2. Perianth segments 5–8 cm, white, inside with a
    yellow base   **1. candidum**

2. Perianth segments 3.0–4.5 cm, white inside as well as
    outside   **4(i). martagon** var. **album**
3. Perianth segments white inside   **2. regale**
3. Perianth segments not white inside   4.
4. Perianth segments pink or dark purplish or maroon   5.
4. Perianth segments yellow, orange or red   9.
5. Stem glabrous or nearly so   6.
5. Stem hairy   7.
6. Perianth segments dull purplish-pink
                             4(iv). martagon var. **martagon**
6. Perianth segments maroon
                    4(vi). martagon var. **sanguineopurpureum**
7. Perianth segments maroon, unspotted
                         4(ii). martagon var. **cattaniae**
7. Perianth segments purplish-pink to wine red   8.
8. Buds glabrous; leaves downy on the veins of the
    undersurface   **4(iii). martagon** var. **hirsutum**
8. Buds hairy; leaves glabrous on the veins of the
    undersurface   **4(v). martagon** var. **pilosiusculum**
9. Perianth segments ascending and curved outwards
    forming a cup   **3. × hollandicum**
9. Perianth segments diverging and very strongly rolled
    back forming a turk's-cap   10.
10. Leaf-veins glabrous on the lower surface   11.
10. Leaf-veins with at least some hairs on the lower
    surface   13.
11. Perianth segments unspotted
        5(b,ii). pyrenaicum subsp. **carniolicum** var. **albanicum**
11. Perianth segments spotted   12.
12. Perianth segments greenish-yellow to yellow
        5(a,i,1). pyrenaicum subsp. **pyrenaicum** forma **pyrenaicum**
12. Perianth segments orange-red
        5(a,i,2). pyrenaicum subsp. **pyrenaicum** forma **rubrum**
13. Perianth segments with warts towards the base
    inside; filaments smooth   14.
13. Perianth segments without warts; filaments papillose   15.
14. Perianth segments red or orange; spotted
        5(c,i). pyrenaicum subsp. **carniolicum** var. **carniolicum**
14. Perianth segments yellow, spotted or unspotted
        5(b,iii). pyrenaicum subsp. **carniolicum** var. **jankae**
15. Perianth segments deep yellow
        5(c,i). pyrenaicum subsp. **ponticum** var. **ponticum**
15. Perianth segments deep orange
        5(c,ii). pyrenaicum subsp. **ponticum** var. **artvinense**

1. L. candidum L.   Madonna Lily
L. album Houtt.

Perennial herb with the rootstock a bulb composed of overlapping white or yellowish scales. *Stem roots* absent. *Stem* 90–200 cm, dark reddish-purple, glabrous. *Basal leaves* produced in autumn and overwintering, up to 19.5 × 2.5 cm, oblanceolate, 3- to 5-veined, glabrous; cauline leaves up to 7.5 × 1.0 cm, lanceolate, scattered, glabrous. *Inflorescence* a raceme. *Flowers* 5–20, 7–8 cm in diameter, broadly funnel-shaped, fragrant; pedicels erect or spreading; sometimes bracteolate. *Perianth segments* 6, free, 5–8 × 1–4 cm, white with a yellow base inside, obovate, recurved slightly at the rounded apex.

Stamens 6; filaments white; anthers 5–18 mm, yellow. *Style* 1, longer than stamens, green; stigma 3-lobed. *Capsule* obovoid, more or less 6-angled. *Seeds* numerous, flat. *Flowers* 6–7. 2*n* = 24.

Introduced. A garden plant which persists for a short time when thrown out. Native of Greece, Yugoslavia and south-west Asia; long cultivated all round the Mediterranean.

2. L. regale E. E. Wilson   Royal Lily

Perennial herb with the rootstock an ovoid bulb composed of lanceolate or ovate-lanceolate, yellowish-brown scales turning vinous-purple. *Stem roots* present. *Stem* 50–200 cm, greyish-green, spotted and flushed with reddish-purple, strong and wiry. *Cauline leaves* numerous, but scattered, 12 cm × 5 mm, dark green, 1-veined, acute at apex, entire, scabrid on the margin and veins beneath. *Inflorescence* a raceme. *Flowers* 1–30, funnel-shaped, fragrant; pedicels 18–130 mm, spreading or somewhat ascending. *Perianth segments* 6, free, 12–15 cm, the outer about 2 cm wide, the inner up to 3.7 cm wide, pinkish-purple outside, white inside, recurved at the rounded apex; perianth tube flushed with yellow inside. *Stamens* 6; filaments white; anthers yellow. *Style* and stigma green. *Capsule* obovoid, more or less 6-angled. *Seeds* numerous, flat. *Flowers* 6–7. 2*n* = 24.

Introduced. A garden plant which persists for a short time when thrown out. Native of western China. L. × **imperiale** E. E. Wilson (*L. regale* × *L. sargentiae* E. E. Wilson) is recorded as a casual garden escape.

3. L. × hollandicum Woodcock & Stearn   Orange Lily
L. bulbiferum L. × **maculatum** Thunb.
L. umbellatum auct.

Perennial herb with the rootstock a bulb composed of overlapping scales. *Stem roots* present. *Stem* 36–70 (–100) cm, pale green, erect, robust, angled, glabrous. *Stem leaves* numerous, dark green above, paler beneath, narrowly elliptical, acute at apex, 3-veined, glabrous. *Inflorescence* a terminal umbel or short raceme. *Flowers* 110–170 mm in diameter, scentless, upright, cup-shaped. *Perianth segments* 6, 60–90 × 25–37 mm, red, orange or yellow, with brownish-purple spots and dashes and raised tramlines down the middle inside, ovate or lanceolate, obtuse at apex. Stamens 6; filaments 25–40 mm, yellow; anthers 12–15 mm, brownish; pollen orange. Style 1, greenish-yellow; stigma brown. *Capsule* obovoid, more or less 6-angled. *Flowers* 6–7. 2*n* = 24.

Introduced. A garden plant which persists for a short time when thrown out. Also grown as a crop in the East Anglian Fens where it occurs as a relic. A hybrid of garden origin.

4. L. martagon L.   Martagon Lily

Perennial herb with the rootstock a yellow, ovoid bulb 25–75 mm, composed of overlapping, narrowly oblong or lanceolate scales. *Stem* 90–200 cm, stout, purplish or green, hairy or glabrous, cylindrical. *Stem roots* present. *Leaves* up to 16 × 6.5 cm, usually in whorls, dark green, spathulate or oblanceolate, acute at apex, 7- to 9-veined, glabrous or pubescent on the veins beneath.

Inflorescence a terminal raceme. *Flowers* 3–50, about 40 cm in diameter, turk's-cap-like, scented, nodding; pedicels recurved, ebracteolate. *Perianth segments* 3.0–4.5 cm × 6–10 mm, pink to dark muddy purplish-red or maroon, rarely white, often spotted, oblong-lanceolate, obtuse at apex, clawed, reflexed; nectary furrows with warty margins. *Stamens* 6; filaments pinkish; anthers orange-yellow or reddish-purple. *Style* and stigma purplish or green, stigma 3-lobed. *Capsule* obovoid, more or less 6-angled. *Seeds* numerous, flat. *Flowers* 8–9. Pollinated by Lepidoptera or selfed. Homogamous. 2*n* = 24 + 0–2B.

(i) Var. album Weston
Stem green and glabrous. *Leaves* glabrous. *Perianth segments* 3.0–4.5 cm, white inside as well as outside.

(ii) Var. cattaniae Vis.
L. martagon var. *dalmaticum* Elwes
Stem purplish, hairy. *Perianth segments* maroon, unspotted.

(iii) Var. hirsutum Weston
Stem purplish, hairy. *Leaves* downy beneath. *Perianth segments* purplish-pink, spotted.

(iv) Var. martagon
Stem purplish, glabrous or nearly so. *Leaves* glabrous or nearly so. *Perianth segments* pale purplish-pink.

(v) Var. pilosiusculum Freyn
Stem purplish, arachnoid hairy. *Leaves* hairy only on the margin. *Perianth segments* wine-red with sparse spots.

(vi) Var. sanguineopurpureum Beck
Stem purplish, glabrous. *Flowers* dark maroon with heavy spotting.

Introduced. A garden escape well naturalised in woods and grassy places where it is sometimes abundant. Scattered in Britain north to central Scotland. Native of most of Europe from north-east France, Estonia and central Urals, southwards to central Spain, central Greece and Caucasus; naturalised in Scandinavia, Siberia and northern Mongolia. All the varieties are grown in gardens and could escape.

5. L. pyrenaicum Gouan   Pyrenean Lily

Perennial herb with the rootstock a broadly ovoid to almost globose bulb about 70 mm, composed of yellowish-white, oblong-lanceolate scales which become pinkish on exposure. *Stem* 25–90(–135) cm, green, sometimes with sparse purple spots, glabrous. *Stem roots* often present. *Leaves* all cauline, 3–15 cm × 3–20 mm, very numerous, ascending or spreading, dark green, linear to lanceolate or narrowly elliptical, acute at apex, spirally arranged, 5- to 7-veined, minutely ciliate on the margin, the veins glabrous to downy on the lower surface. *Inflorescence* a terminal raceme. *Flowers* 1–12, about 35 mm in diameter, turk's-cap-like, faintly to strongly and unpleasantly scented, nodding; pedicels recurved, bracteolate. *Perianth segments* 3–7 cm × 4–13 mm, greenish-yellow, yellow, orange and red with dark purplish or reddish-brown spots and lines inside, especially towards the base, and usually with purplish warts. *Stamens* 6; filaments greenish or yellowish, smooth or papillose; anthers 4–10 mm, with orange-red or brown pollen. *Style* long; stigma 3-lobed. *Capsule* obovoid, more or less 6-angled. *Seeds* numerous, flat. *Flowers* 5–6. $2n = 24$.

(a) Subsp. pyrenaicum
Stem 30–135 cm, green, with sparse purple spots. *Leaves* 7–15 cm × 3–20 mm, with 3–15 veins which are glabrous beneath. *Flowers* up to 12. *Perianth segments* 4.0–6.5 cm, yellow to greenish-yellow with purple spots and lines inside and purple warts. *Filaments* smooth.

(1) Forma pyrenaicum
Perianth segments greenish-yellow to yellow.

(2) Forma rubrum Stoker
Perianth segments orange-red.

(b) Subsp. carniolicum (Bernh. ex Koch) V. A. Matthews
L. carniolicum Bernh. ex Koch
Stem up to 120 cm, green. *Leaves* 3–11 cm × 4–17 mm, with 3–9 veins which are glabrous or downy beneath. *Flowers* up to 6(–12). *Perianth segments*, 3–7 cm, yellow, orange or red, often spotted with brownish-purple and with purplish, rarely yellow warts. *Filaments* smooth.

(i) Var. carniolicum (Bernh. ex Koch) V. A. Matthews
Stem up to 120 cm. *Leaf-veins* densly downy beneath. *Flowers* red or orange, spotted.

(ii) Var. albanicum (Griseb.) V. A. Matthews
L. albanicum Griseb.; *L. carniolicum* subsp. *albanicum* (Griseb.) Hayek
Stem up to 40 cm. *Leaf-veins* glabrous beneath. *Flowers* yellow, unspotted.

(iii) Var. jankae (A. Kern.) V. A. Matthews
L. jankae A. Kern.; *L. carniolicum* subsp. *jankae* (A. Kern.) Aschers. & Grabn.
Stem up to 80 cm. *Leaf-veins* downy beneath. *Flowers* yellow, spotted or unspotted.

(c) Subsp. ponticum (K. Koch) V. A. Matthews
L. ponticum K. Koch
Stem 15–90 cm, green. *Leaves* 3–8 cm × 8–20 mm, with 7–15 veins which are downy beneath. *Flowers* up to 5, occasionally to 12. *Perianth segments* 3.5–5.5 cm, deep yellow or deep orange with dense red-brown or purplish stripes and spots towards the base inside, but lacking warts. *Filaments* papillose.

(i) Var. ponticum (K. Koch) V. A. Matthews
L. georgicum Manden.; *L. carniolicum* subsp. *ponticum* (K. Koch) P. H. Davis & D. M. Hend.
Stem usually stout. *Perianth segments* deep yellow, 7–10 mm wide.

(ii) Var. artvinense (Miscz.) V. A. Matthews
L. artvinense Miscz.; *L. ponticum* var. *artvinense* (Miscz.) P. H. Davis & D. M. Hend.

Stem more slender than in var. *ponticum* and with fewer, less crowded leaves. *Perianth segments* deep orange, 5–6 mm wide.

Introduced. All the above variants are grown in gardens and may be the ones which escape. Well naturalised in woods, hedgerows, roadsides and field margins. Scattered in Britain, mostly in the west and north and in Jersey. As a native plant the species occurs in northern Spain and the Pyrenees, south-east Alps; Balkans and north-eastern Turkey. Subsp. *pyrenaicum* occurs in northern Spain and south-west France. Subsp. *carniolicum* occurs in south-east Europe. Var. *albanicum* occurs in south-west Yugoslavia, Albania and north-west Greece. Var. *jankae* occurs in north-west Italy, north-west Yugoslavia, western Bulgaria and central Romania. Subsp. *ponticum* occurs in north-east Turkey and Georgia. Var. *ponticum* occurs in north-east Turkey. Var. *artvinense* occurs in north-east Turkey and Georgia. British populations need to be looked at.

15. **Convallaria** L.

Glabrous *perennial herb* with creeping rhizomes. *Stems* solitary erect. *Leaves* all basal or more or less at base of stem, narrowly to broadly elliptical, petiolate. *Inflorescence* a terminal, one-sided raceme with 15–20 pendulous flowers. *Perianth segments* white, fused into a tube longer than the lobes, forming a campanulate flower. *Stamens* 6. *Style* simple; stigma capitate. *Fruit* a red berry.

One very variable species distributed throughout the north temperate regions.

Ponert, J. (1968). *Biosistematiceskaja Monografija Roda Convallaria* L. s. str. Leningrad.

1. C. majalis L.   Lily-of-the-valley
Glabrous *perennial herb,* with creeping rhizomes and fibrous roots. *Stems* 10–35 cm, solitary, erect, pale green, with several green or violet scales below and 1–4 leaves above. *Lamina of leaves* 3–22 × 0.5–10.5 cm, all more or less convolute at base, bluish-green above, paler beneath, narrowly to broadly elliptical, acute to acuminate at apex, entire, cuneate to attenuate at base; petioles 1–24 cm, the sheathing bases simulating an elongation of the stem. *Inflorescence* a terminal, one-sided raceme with 5–20 flowers, ellipsoid in bud; peduncles 5–15 mm, curving downwards; bracts ovate-lanceolate, shorter than peduncles. *Flowers* 5–10 mm in diameter, more or less campanulate, sweetly scented, pedicellate. *Perianth segments* 6, usually fused for about two-thirds of their length, creamy white or rarely pink, 5–10 mm, obovate-lanceolate, reflexed at apex. *Stamens* 6; anthers attached to near the base of the perianth. *Style* 1, simple; stigma capitate. *Berry* 5–7 mm in diameter, red, more or less globose. *Seeds* 2–6. *Flowers* 5–6. $2n = 38$.

Native. Dry woods, scrub and hedgebanks usually on base-rich soil. Scattered throughout Britain north to central Scotland. Most of Europe except the extreme north and south; north-east Asia. Much grown in gardens where it can soon become a weed and frequently escapes and becomes naturalised. The cultivated plants are often much more robust than the wild ones and variants occur with pink flowers, with purple spots at the base of the filaments and double flowers. It can spread by division as well as seed.

16. Maianthemum G. H. Weber nom. conserv.
Unifolium Ludw.

Perennial herbs with slender, creeping rhizomes, bearing solitary, long-stalked leaves. *Stems* erect, with scales at base and 2–3 leaves above. *Leaves* ovate, cordate. *Flowers* in terminal racemes, numerous, white. *Perianth segments* 4, free, spreading or reflexed. *Stamens* 4, shorter than perianth. *Style* 1, simple, short; stigma slightly 2-lobed. *Fruit* a red berry.

Three species extending around the whole north temperate region.

M. kamtschaticum (Cham.) Nakai occurs as a garden escape in woodland at Porlock, Somerset.

1. M. bifolium (L.) F. W. Schmidt   May Lily
Convallaria bifolia L.; *Smilacina bifolia* (L.) Desf.; *M. convallaria* G. H. Weber nom. illegit.; *Unifolium bifolium* (L.) Greene

Perennial herbs with slender, creeping rhizomes. *Stems* 5–20(–25) cm, erect, pale green, with short, stiff, white hairs above, with 2 scale leaves at the base and 2 foliage leaves near the apex. *Leaves* with lamina 4–6(–8) × 2.5–5.0 cm, dark green above, paler beneath, alternate, ovate, acute at apex, entire, cordate below with rounded lobes and a wide sinus, with short, white hairs on the veins; petiole 3–20 mm, pale green, sheathed at base; transverse veins present. *Inflorescence* a terminal raceme 1–4 cm. *Flowers* 8–20, 2.5–6.0 mm in diameter; pedicels slender, articulated. *Perianth segments* 4, free, patent, 2–3 × 0.8–1.0 mm, white, oblong, obtuse at apex. *Stamens* 4, inserted at base of perianth; filaments 1.0–1.5 mm, pale; anthers cream. *Style* 1, simple, short; stigma slightly 2-lobed. *Berry* 5–6 mm in diameter, pale green and spotted at first, turning red. *Seeds* 1–3. *Flowers* 5–6. Pollinated by insects and selfed, protogynous. $2n = (30, 32) 36 (38, 42)$.

Native. Woods on acid soils. Very local in Durham, Yorkshire, Lincolnshire and Norfolk; formerly elsewhere in north and east England and Midlothian. Europe southwards from 69° 50' N to north Spain, north Apennines, south Carpathians and Crimea; northern Asia to Kamchatka and Korea.

17. Polygonatum Mill.

Perennial herbs with horizontal, creeping rhizomes. *Stems* solitary, unbranched, erect, leafy. *Leaves* all cauline, alternate, opposite or whorled, linear-elliptical to elliptical, sessile. *Flowers* 1–6, in axillary, stalked, pendulous clusters. *Perianth segments* fused into a more or less cylindrical tube longer than the lobes, white or cream with green markings. *Stamens* 6, borne on the perianth tube. *Style* 1, with a 3-lobed stigma. *Fruit* a purple or bluish-black berry.

About 50 species found in the north temperate region.

P. biflorum (Walter) Elliot is recorded as a garden relic at Virginia Water, Surrey.

Stewart, A., Pearman, D. A. & Preston, C. D. (1994). *Scarce plants in Britain*. Peterborough. [*P. odoratum.*]

1. Leaves linear, narrowly lanceolate or normally oblong, mostly in whorls of 3–8   **4. verticillatum**
1. Leaves elliptical, lanceolate-oblong or ovate, all alternate   2.
2. Perianth tube not contracted in the middle; flowers 1 or 2 per leaf axil; filaments glabrous   **3. odoratum**
2. Perianth tube slightly contracted in the middle; flowers 1–6 per leaf axil; filaments sparsely pubescent   3.
3. Stems terete; perianth 9–15(–20) × 2–4 mm   **1. multiflorum**
3. Stem ridged to slightly angled; perianth 15–22(–25) × 3–6 mm   **2. × hybridum**

1. P. multiflorum (L.) All.   Solomon's Seal
Convallaria multiflora L.

Perennial herb with long, thick, creeping rhizome. *Stems* 30–80 cm, terete, glabrous, erect then arching. *Leaves* all on stem, with lamina 5–15 cm × 20–75 mm, bright green above, glaucous beneath, alternate, subdistichous and ascending in two directions at an obtuse angle with each other, elliptical, lanceolate, oblong or ovate, subacute at apex, entire, sessile or shortly petiolate, the lower semi-amplexicaul, glabrous. *Inflorescence* of axillary racemes. *Flowers* 1–6 per raceme, not scented, usually bisexual, pendulous; peduncles 10–20 mm, slender; pedicels short. *Perianth segments* fused into a more or less cylindrical tube longer than the lobes, 9–20 × 2–4 mm, somewhat contracted in the middle; greenish-white, green and bearded on the inside of the lobes; lobes 6, ovate or lanceolate, obtuse at apex. *Stamens* 6, inserted inside and adnate to the perianth tube; filaments 4–7 mm, minutely hairy; anthers yellow. *Style* 1, about as long as the stamens, linear. *Berry* 7–12 mm in diameter, bluish-black with a white bloom, more or less globose. *Seeds* 6 or fewer, pale, enclosed in a dark green pulp, globose. *Flowers* 5–6. $2n = 18, 32$.

Native. Woods, mostly on basic soils. Locally frequent in southern Britain and scattered north to northern England. Much of Europe, but absent from parts of the south-west and many islands; temperate Asia to Japan.

2. P. × hybridum Brügger   Garden Solomon's Seal
P. multiflorum × odoratum
P. intermedium Boreau, non Dumort.

Perennial herb with long, creeping rhizomes. *Stems* up to 1 m, ridged to slightly angled, erect, becoming arching, glabrous. *Leaves* all on stem, with lamina 3–11 × 1–4 cm, bright green above, glaucous beneath, alternate, subdistichous and ascending in two directions at an obtuse angle with each other, elliptical, lanceolate or ovate, subacute at apex, entire, sessile or shortly petiolate, the lower semi-amplexicaul, glabrous. *Inflorescence* of axillary racemes. *Flowers* 1–6 per raceme, not scented, usually bisexual, pendulous; peduncles 10–15 mm, slender; pedicels short. *Perianth segments* fused into a more or less cylindrical tube longer than the lobes,

15–22(–25) mm, slightly contracted in the middle, greenish-white, green and bearded on the inside of the lobes; lobes 6, ovate or lanceolate, obtuse at apex. *Stamens* 6; inserted inside of and adnate to the perianth tube; filaments 4–7 mm, minutely hairy; anthers yellow. *Style* 1, about as long as the stamens, linear. *Berry* rarely produced. *Flowers* 5–6. $2n = 19$.

Hybrid of garden origin often recorded as *P. multiflorum*. Much grown in gardens and naturalised in woods, scrub and rough ground. Throughout the Channel Islands and Britain, very scattered in Ireland.

3. P. odoratum (Mill.) Druce   Angular Solomon's Seal
Convallaria polygonatum L.; *Convallaria odorata* Mill.; *Convallaria angulosa* Lam. nom. illegit. *P. officinale* All.; *P. anceps* Moench nom. illegit.; *P. polygonatum* (L.) Voss nom. illegit.; *P. sigillum* Druce nom. illegit.

Perennial herb with long, creeping rhizomes. *Stems* 15–40 cm, distinctly angled, glabrous, erect or erect then arching. *Leaves* all on stem, with lamina 5–10 × 1–5 cm, bright yellowish-green above, glaucous beneath, alternate, subdistichous and ascending in two directions at an obtuse angle with one another, ovate, elliptic or elliptic-oblong, acute at apex, entire, sessile or shortly petiolate, the lower semi-amplexicaul, glabrous. *Inflorescence* of axillary racemes. *Flowers* 1–2 per raceme, scented, usually bisexual, pendulous; peduncles 5–10 mm; pedicels short. *Perianth segments* fused into a more or less cylindrical tube longer than the lobes, 15–30 × 4–9 mm, not contracted in the middle, creamy-white, green and bearded on the inside of the lobes, ovate or lanceolate, subacute at apex. *Stamens* 6, inserted inside and adnate to the perianth tube; filaments 4–7 mm, green, glabrous; anthers yellow. *Style* 1, about as long as the stamens, green. *Berry* 8–12 mm in diameter, bluish-black, more or less globose. *Seeds* 6 or fewer, 4–5 mm, broadly ellipsoid. *Flowers* 6–7. Pollinated by bumblebees or selfed; homogamous. Spread is mainly vegetatively. $2n = 20$.

Native. Well-drained steep banks and lesser cliffs in woods on Carboniferous and oolitic limestone. Very local in north-west England, the Peak District and around the River Severn estuary, very scattered in Wales, rarely naturalised elsewhere in Britain. Throughout most of Europe southwards from 66° N in Finland to the mountains of Spain and Portugal, central Italy, Greece and the Caucasus; Morocco; Siberia; western Himalaya; China.

4. P. verticillatum (L.) All.   Whorled Soloman's Seal
Convallaria verticillata L.

Perennial herb with a long, creeping rhizome. *Stems* 30–80 cm, erect, distinctly angled, yellowish-green, faintly spotted with red, glabrous. *Leaves* all on stem, the middle and upper in whorls of 3–8, with lamina 4–15 cm × 3–25 mm, bright green above, glaucous beneath, linear, narrowly elliptical, narrowly lanceolate or narrowly oblong, acute at apex, entire, sessile, glabrous. *Inflorescence* of axillary racemes. *Flowers* 1–4 per raceme, not scented, usually bisexual, pendulous;

peduncles 5–10 mm; pedicels very short. *Perianth segments* fused into a more or less cylindrical tube longer than the lobes, 5–10 × 1.5–3.0 mm, slightly contracted in middle, greenish-cream, the lobes green on the inside; lobes 6, lanceolate or elliptical, subacute at apex. *Stamens* 6, inserted inside and adnate to the perianth tube; filaments 4–7 mm, glabrous; anthers pale yellow. *Style* 1, about as long as the stamens, linear. *Berry* 7–8 mm in diameter, turning red then purple, more or less globose. *Seeds* 6 or fewer, about 3 mm, broadly ellipsoid. *Flowers* 6–7. $2n = 28$.

Native. Mountain woods. Very rare in Perthshire, formerly in Angus and Northumberland. Arctic Norway southwards to the mountains of north Spain, central Italy and Bulgaria and eastwards to Latvia and Romania; Caucasus; Turkey; Himalayas.

18. Reineckea Kunth nom. conserv.

Perennial herbs with prostrate, stem-like rhizomes. *Stems* erect. *Leaves* all basal, in a rosette, linear, slightly narrowed at base. *Flowers* sessile, numerous, erect to patent, in a short terminal spike, pale to deep pink. *Perianth segments* fused into a tube slightly shorter than the patent to reflexed lobes. *Stamens* 6. *Style* 1, equalling stamens. *Fruit* a red berry.

A single species found in China and Japan.

1. R. carnea (Andrews) Kunth   Reineckea
Sansevieria carnea Andrews

Perennial herb with a thick, fleshy, creeping rhizome. *Stem* up to 15 cm, arising from a bud on the rhizome, simple, interruptedly angular. *Leaves* all basal, sessile, with lamina 15–40 × up to 3 cm, dark green above, paler beneath, linear to linear-lanceolate, pointed at apex, entire, gradually narrowed at base, slightly channelled and keeled with the sides deflected, glabrous. *Inflorescence* a terminal spike; bracts ovate, membranous. *Flowers* numerous, erect to patent, sessile. *Perianth segments* 8–12 mm, pale to deep pink, fused into a tube 3–5 mm which is slightly shorter than the lobes; lobes patent to reflexed, linear-oblong, flat but becoming reflexed, subacute at apex. *Stamens* 6; filaments filiform, arising from the mouth of the tube, yellow; anthers short, yellow. *Style* 1, equalling stamens, clavate; stigma 3-lobed. *Berry* red, seldom formed in cultivation. *Flowers* in early summer. $2n = 38, 42$.

Introduced. Planted for ground cover and naturalised in woodland borders, near Helford, Lizard, Cornwall. Native of China and Japan.

19. Paris L.

Perennial herbs with a creeping rhizome. *Stem* erect, unbranched. *Leaves* (3–)4(–8) in a whorl in the upper half of the stem, broadly obovate to subrotund. *Inflorescence* a single, terminal, erect, long-stalked flower. *Perianth segments* 8(–12), 4(–6) lanceolate or linear-lanceolate outer, 4(–6) linear inner, all green, free and patent. *Stamens* 8(–12). Style 1, short; stigmas 4(–5), linear. *Fruit* a dehiscent black berry.

About 20 species in temperate Eurasia.

1. P. quadrifolia L.   Herb Paris

Glabrous, foetid *perennial herb* with a creeping rhizome and slender, pale brown roots. *Stems* 15–40 cm, erect, pale yellowish-green, slightly purplish towards the base. *Leaves* (3–)4(–8), 6–15 × 5–8 cm, dull rather dark green on upper surface, paler beneath, broadly obovate or subrotund, rounded to a shortly acuminate apex, entire, broadly cuneate at base, more or less sessile, grouped in a whorl in the upper half of the stem. *Flowers* 4(–6)–numerous, solitary; pedicels 20–80 mm. *Outer perianth segments* 4(–6), 20–35 × 4–7 mm, pale green, lanceolate or linear-lanceolate, gradually narrowed to a long-acuminate apex, reflexed and curved back and inwards; inner segments 4(–6), 15–25 × about 1 mm, pale green, linear. *Stamens* 8(–12); filaments 2.5–3.0 mm, green, flat; anthers yellowish-green. *Style* 1, purple, curved downwards and shorter than the stamens. *Ovary* purple, 4- to 5-celled. *Fruit* a berry-like capsule, up to 20 mm in diameter, black, suborbicular, finally dehiscent. *Flowers* 4–8. Pollinated by flies and selfed; strongly protogynous. $2n = 20$.

Native. Damp woods on calcareous soils; ascending to 360 m. Rather local in Britain and absent from most of Wales, south-west England and north and west Scotland. Most of Europe, but rare in the Mediterranean region; Caucasus; Siberia.

Subfamily 5. Scilloideae Engl.
Tribe *Scilleae* Rchb.

Perennial herbs with a bulb. *Leaves* all basal, linear or nearly so. *Inflorescence* a terminal raceme. *Perianth segments* usually white or blue, sometimes pink or brownish, very rarely pale yellow, free or fused. *Style* 1, with 1, often 3-lobed, stigma. *Ovary* superior. *Fruit* a capsule splitting along the centre of the ovary cells.

20. Ornithogalum L.

Perennial herbs with bulbs which are usually subterranean with whitish or brownish papery tunics. *Stems* erect. *Leaves* in a basal rosette, linear. *Flowers* in a corymbose raceme, each with a bract. *Perianth segments* 6, free, equal, white with a green stripe on the outside. *Stamens* 6, inserted on a receptacle; filaments flat. *Style* 1. *Fruit* a capsule with many seeds.

About 80 species, with two main areas of distribution, the Mediterranean area and southern Africa.

O. arabicum L. is recorded as a persistent garden escape on a sandy bank, near Vale Pond, Guernsey.

Czapik, R. (1968). Chromosome number of *Ornithogalum umbellatum* L. from three localities in England. *Watsonia* 6: 345–349.

Green, D. (1993). *Ornithogalum pyrenaicum* L. in Gillam, B. *The flora of Wiltshire*. Newbury.

Stewart, A., Pearman, D. A. & Preston, C. D. (1994). *Scarce plants in Britain*. Peterborough. [*O. pyrenaicum.*]

1. Bracts longer than pedicels; filaments with 1 acute lobe at the apex on either side of the anther   **3. natans**
1. Bracts shorter than pedicels at least on the lower flowers; filaments without apical lobes   2.

2. Inflorescence an elongated raceme; perianth segments less than 14 mm   **1. pyrenaicum**
2. Inflorescence corymbose; perianth segments more than 14 mm   3.
3. Bulblets subspherical; flowers up to 20; perianth segments up to 30 mm   **2(a). umbellatum** subsp. **umbellatum**
3. Bulblets elongated; flowers 4–12; perianth segments 15–20 mm   **2(b). umbellatum** subsp. **angustifolium**

Subgenus 1. Beryllis (Salisb.) Baker

Bulb progressively renewed over 2–3 years, of free scales, the outer sometimes tunicate. *Leaves* without a white stripe. *Inflorescence* many-flowered, elongated-cylindrical. *Pedicels* patent in flower, more or less erect and appressed to the axis of the inflorescence when in fruit. *Filaments* entire. *Ovary* and capsule with 3 obtuse angles. *Germination* epigeal.

1. O. pyrenaicum L.   Spiked Star-of-Bethlehem

Glabrous *perennial herb* with an ovoid bulb with free scales, the outer of which is sometimes tunicate, the bulb being progressively renewed over 2–3 years. *Stem* 30–100 cm, pale green, stout. *Leaves* (4–)5–8, appearing at the end of winter and often withered at the time of anthesis, 30–60 cm × 3–12 mm, glaucous, long-linear, narrowed to an obtuse apex, entire. *Inflorescence* a narrow, elongated raceme with more than 20 flowers. *Flowers* 15–22 mm in diameter, faintly scented; pedicels 10–20 mm, slender, patent in flower, more or less erect and appressed to the axis of the inflorescence in fruit; all more or less equal; bracts 10–12 mm, shorter than the pedicels, whitish, lanceolate, acuminate at apex, thin. *Perianth segments* with pale cream margins and a yellowish-green centre wider than the margins and with green veins, linear-oblong; the outer 8–13 × 2–3 mm, flat at first, becoming inrolled at the margins. *Stamens* about three-quarters as long as the perianth; filaments yellowish-cream, oblanceolate, acuminate at apex, entire; anthers cream. *Style* 2.5–3.3 mm, slightly longer to slightly shorter than the ovary, very pale green; stigma green. *Capsule* 12–15 × 10–11 mm, ellipsoid, 3-grooved; seeds about 2 mm, pyramidal, grey. *Flowers* 6–7. Vegetative reproduction occurs on lateral buds and seeds are abundant. $2n = 16, 18$.

Flower colour is variously recorded. All British plants seem to be as above. They need to be compared with fresh material from Continental Europe.

Native. Woods and scrub, hedges, roadsides and ditch-banks, very local, but often abundant, usually on oolitic limestone and clays. The young inflorescences were formerly sold in Bath markets as 'Bath Asparagus'. Very local in the Bath area and formerly more widespread in that region of south-central England; local in Bedfordshire and Huntingdonshire; rarely naturalised elsewhere in England. South-west and south-central Europe, northwards to Belgium; Turkey; mountains of Morocco.

Subgenus 2. Ornithogalum

Bulb completely renewed each year by one series of concrescent fleshy scales. *Leaves* with a white stripe on the upper surface, flat to channelled. *Inflorescence* more or less corymbiform, elongating in fruit. *Pedicels* ascending in flower, ascending to deflexed in fruit. *Perianth segments* patent, appressed to the capsule after anthesis. *Filaments* entire. *Ovary* and capsule with 6 more or less obtuse angles. *Germination* epigeal.

2. O. umbellatum L.   Star-of-Bethlehem

Glabrous *perennial herb*; bulb about 25 mm, surrounded by tufts of leaves arising from subspherical or elongated offsets, without dormant bulbs. *Stem* 15–30(–40) cm, pale green. *Leaves* (4–)6–9, 15–30 cm × 2–5(–8) mm, green with a white stripe on the upper surface, linear, tapering to an acute apex, entire, flat to grooved. *Inflorescence* more or less corymbiform, elongating in fruit. *Flowers* (4–)8–20, 30–40 mm in diameter; pedicels up to 10 cm, ascending, becoming patent in fruit; bracts shorter than or equalling pedicels. *Perianth segments* 15–30 mm, white with a green stripe on the back, lanceolate to oblong-lanceolate, subacute at apex. *Stamens* about half as long as the perianth; filaments lanceolate, acuminate, white; anthers yellowish-white. *Style* shorter than ovary. *Capsule* 10–15 mm, obovoid to ovoid, angles equally spaced, rounded and thickened, separated by shallow furrows. *Seeds* numerous. *Flowers* 4–5. *Pollinated* by insects and selfed.

(a) Subsp. umbellatum

Bulblets subspherical. *Flowers* up to 20. *Perianth segments* up to 30 mm. $2n = 54$.

(b) Subsp. angustifolium (Boreau) P. D. Sell

Bulblets elongated. *Flowers* 4–12. *Perianth segments* 15–22 mm. $2n = 27$.

Native. Grassy places, rough ground and open woods. Scattered throughout Britain and the Channel Islands, but perhaps native only in East Anglia. South and south-central Europe, extending north to Sweden, Denmark and south-west Russia, but very doubtfully native in the northern part of its range. Our native plants appear to be referable to subsp. *angustifolium*. The larger, more handsome subsp. *umbellatum* is grown in gardens and is sometimes found amongst the escapes.

Subgenus 3. Myogalum (Link) Baker

Bulb progressively renewed over 3–4 years, of free scales, the outer tunicate at the base. *Leaves* with a white stripe on the upper surface, flat or concave. *Inflorescence* a more or less unilateral raceme. *Pedicels* recurved in fruit. *Perianth* more or less widely campanulate. At least the inner 3 filaments tricuspidate at apex. *Ovary* and *capsule* with 6, equidistant, rounded-obtuse angles. *Germination* epigeal.

3. O. nutans L.   Drooping Star-of-Bethlehem

Glabrous *perennial herb*. *Bulb* about 5 cm, with numerous offsets. *Stem* 25–60 cm. *Leaves* 4–6, 25–60 cm × (6–)10–15 mm, green with a broad whitish stripe, long-linear, acute at apex, entire, channelled. *Inflorescence* a unilateral raceme. *Flowers* 3–12(–15), 40–60 mm in diameter, drooping; pedicels 5–12 mm, more or less equal, curved; bracts 20–30 mm, whitish, ovate-lanceolate, acuminate at the apex, thin, longer than pedicels. *Perianth segments* 15–32 mm, white with a green band covering most of the back, oblong-lanceolate, obtuse at apex. *Stamens* shorter than the perianth; filaments white, broad, deeply divided at the apex with the yellow anther in the sinus. *Style* slender. *Capsule* broadly ovoid, pendulous, 6-grooved. *Flowers* 3–5. Pollinated by insects and selfed; protandrous. $2n = 16, 30, 42$.

Introduced. Grown in gardens and naturalised in grassy places, usually in small quantity. Scattered in central and south Britain. Native in central Europe; naturalised elsewhere in Europe.

21. Scilla L.

Glabrous *perennial herbs* with ovoid to spherical bulbs composed of numerous free scales which are progressively renewed annually; roots annual or perennial. *Stems* erect, 1–4 per bulb. *Leaves* all basal, few to several, linear, appearing before, with or after the flowers. *Flowers* few to many, in terminal racemes, each without or with a solitary bract. *Perianth segments* 6, free, usually blue, rarely white or pink. *Stamens* 6, inserted on the base of the perianth; filaments narrow or flattened; anthers dorsifixed. *Style* 1, straight; stigma small, truncate. *Ovary* superior, 3-celled. *Fruit* a capsule. *Seeds* spherical or oblong.

About 90 species found in Europe, Asia and southern Africa.

Ainsworth, C. C., Parker, J. S. & Horton, D. M. (1983). Chromosome variation and evolution in *Scilla autumnalis*. *Kew chromosome conference* II. ed. by P. D. Brandham & M. D. Bennett. London. pp. 261–268.

Meikle, R. D. (1962). Lily group discussion on Scillas and Chinodoxas. *Lily Year Book* **25**: 116–133.

Speta, F. (1980). Die frühjahrsblühanden Scilla-Arten des östlichen Mittelmerraumes. *Naturk.-Jahrb. Stadt Linz* **25**: 19–198.

Stewart, A., Pearman, D. A. & Preston, C. D. (1994). *Scarce plants in Britain*. Peterborough. [*S. autumnalis*.]

1. Bracts more than 4 mm   **2.**
1. Bracts absent or less than 4 mm   **4.**
2. Lower bracts more than 3 cm; flowers usually more than 20   **7. peruviana**
2. Lower bracts less than 3 cm; flowers less than 15   **3.**
3. Leaves 2–5 mm wide; perianth segments 5–8 mm **6. verna**
3. Leaves 10–30 mm wide; perianth segments 8–12 mm
                                                        5. liliohyacinthus
4. Flowers drooping; perianth segments 12–16 mm
                                                            4. siberica
4. Flowers erect to patent; perianth segments 2–10 mm.   **5.**
5. Flowering from July to September without the leaves; perianth segments 3–6 mm   **8. autumnalis**
5. Flowering from February to April with the leaves; perianth segments 5–10 mm   **6.**
6. Leaves 2(–5), sheathing the stem to about halfway; inflorescence 1- to 5(–10)-flowered; bracts usually absent   **1. bifolia**

6. Leaves 3–7, not or scarcely sheathing the stem;
   inflorescence 7- to 15(–20)-flowered; bracts always
   present   7.
7. Bracts 2–3 mm, auriculate at base; perianth segments
   9–10 mm   **2. bithynica**
7. Bracts about 1 mm, more or less truncate at apex;
   perianth segments 6–7 mm   **3. messeniaca**

Subgenus 1. Scilla

Roots annual, simple. *Flowering* in spring with or after
the leaves. *Inflorescence* 1- to 15-flowered. *Seeds* globose,
carunculate. *Germination* epigeal.

1. S. bifolia L.   Alpine Squill

Glabrous *perennial herb. Bulb* 5–25 mm in diameter,
pinkish beneath the brown tunic. Stems solitary, 5–30
cm, terete, glabrous. *Leaves* usually 2, occasionally 3–5,
5–20 cm × 3–15 mm, green, appearing with the flowers,
broadly linear or wider near the tip, slightly hooded at
apex, glabrous, partially sheathing the stem near the
base. *Inflorescence* 1- to 10-flowered, slightly one-sided,
in a loose raceme; bracts absent or if present about 1
mm, ovate-lanceolate; lower pedicels 10–40 mm. *Flower*
1, 10–20 mm in diameter, erect, cup-shaped, slightly
scented. *Perianth segments* 5–10 × 1–3 mm, blue to purplish-blue, sometimes white or pink, ovate to elliptical,
slightly hooded at apex, more or less spreading. *Stamens*
6; filaments pale blue; anthers dark blue. *Style* blue;
stigmas dark blue. *Capsule* subglobose. *Seeds* 2–3 mm,
brownish, with an irregular white appendage; testa
smooth. *Flowers* 3–4. $2n = 18, 36, 54$.

This plant has been divided into several species and
subspecies by Speta (1980) and it is not known if more
than one taxon escapes from cultivation.

Introduced. Grown in gardens and naturalised and
spreading where planted in churchyards and on banks.
Scattered records in southern England. Central and
southern Europe and western Asia.

2. S. bithynica Boiss.   Bithynian Squill
S. amoena var. *bithynica* (Boiss.) Baker

Glabrous *perennial herb. Bulb* about 10 × 15–20 mm,
rounded-ovoid, dark brown to greyish-brown on the
outside, first row of inner scales sometimes flushed purplish. *Stems* usually 2, 8–30 cm, slender, glabrous.
Leaves usually 3, sometimes 2, 4 or 5, 7–38 cm × 4–13
mm, dark green, long-linear, shortly attenuate to an
obtuse apex, entire, narrowed to the base. *Inflorescence* a
7–10(–15)-flowered raceme; bracts 2–3 mm, elongate-
triangular, cordate, biauriculate at base; pedicels 3–12
mm, slender, suberect, purplish. *Flowers* 18–21 mm in
diameter, more or less erect. *Perianth segments* 9–10 ×
about 3.5 mm, mauvish-blue, elliptical, obtuse at apex,
1-veined, spreading. *Stamens* 6; filaments 4.5 × 0.7 mm,
linear, narrowed at apex, mauvish-blue; anthers about
2.5 mm, dark blue. *Style* 2.0–3.5 mm; stigma dark
mauvy-blue. *Capsule* about 7 × 5 mm. *Seeds* with a
scarcely developed appendage. *Flowers* 4. $2n = 12$.

Introduced. Garden escape established by the River
Cray at St Mary Cray, Kent, reported from Milford,
Salisbury, Wiltshire, a long-persistent relic on the site of

Warley Place Gardens, Essex and spreading in grassland
in the Cambridge Botanic Garden. Native of Bulgaria
and north-west Turkey.

3. S. messeniaca Boiss.   Greek Squill

Glabrous *perennial herb. Bulb* 20–30 mm in diameter,
ovoid; tunic pale brown. *Stems* 5–15 cm, shorter than
the leaves, pale green, angular. *Leaves* 5–7, 15–25 ×
1.2–2.5 cm, appearing with the flowers and sheathing
the stem at their bases, dark glossy green with a pale
midrib, broadly long-linear, abruptly acuminate at apex,
entire. *Inflorescence* a raceme, ovate-oblong, dense, with
(7–)10–20 flowers; bracts solitary, very minute, truncate
or sometimes deeply bifid at apex; lower pedicels 4–8
mm, ascending or somewhat spreading. *Flowers* 14–18
mm in diameter, erect or slightly spreading. *Perianth
segments* 6–7 × 3–4 mm, pale blue to lilac-blue, linear to
lanceolate, obtuse at apex. *Stamens* 6; filaments 4–6 mm,
blue; anthers dark violet-blue. *Style* slightly longer than
the ovary, pale blue. *Capsule* about 7 mm, nearly spherical, obtusely 3-angled. *Flowers* 3–4.

Introduced. Grown in gardens and naturalised at the
edge of Smallcombe Wood, Bath, Somerset. Native of
southern Greece.

4. S. siberica Haw.   Siberian Squill

Glabrous *perennial herb. Bulb* 15–20 mm in diameter,
ovoid, whitish under the dark purple outer scales. *Stems*
1–4, 10–20 cm, angled, finely ribbed, glabrous. *Leaves*
2–4, 10–15 cm × 10–20 mm, bright shiny green, appearing with the flowers, broadly linear, slightly hooded at
apex, glabrous, sheathing the stem for about one-third
of its length. *Inflorescence* a 1–2(–5)-flowered raceme;
bracts 1–2 mm, white, ovate, truncate or bifid at apex;
pedicels 8–12 mm, purple, equalling or shorter than the
perianth. *Flowers* 25–30 mm in diameter, drooping, cup-
to funnel-shaped. *Perianth segments* 12–16 × 4–6 mm,
deep blue with a darker central stripe which has a
whitish apex, sometimes white, elliptic-oblong, obtuse at
apex. *Stamens* 6; filaments 5–8 mm, white below, blue
above; anthers greyish-blue. *Style* 4–5 mm; white;
stigma pale. *Capsule* 8–10 mm in diameter, globose.
Seeds about 2 mm, pale brown, ovoid; appendage cylindrical, peg-like. *Flowers* 3–4. $2n = 12$.

Introduced. Much grown in gardens and more or less
naturalised where planted and neglected. Scattered in
south-east England. Native of Russia, Turkey and Iran.

5. S. liliohyacinthus L.   Pyrenean Squill

Glabrous *perennial herb. Bulb* 30–50 mm in diameter,
ovoid, with loose, overlapping yellow scales, without a
tunic. *Stems* 2 or 3, 15–40 cm, robust. *Leaves* appearing
with the flowers, 6–10, 15–30 cm × 10–30 mm, glossy
yellowish-green, linear or linear-oblanceolate, obtuse at
apex, entire. *Inflorescence* a 5- to 15-flowered raceme;
bracts 10–25 mm, ovate, papery, whitish; pedicels 15–30
mm, ascending. *Flowers* 18–22 mm in diameter, curved
upwards. *Perianth segments* 8–12 mm, bright blue, ovate
to elliptical, rounded-obtuse at apex. *Stamens* 6; filaments 7–9 mm, pale blue; anthers dark blue. *Capsule*
subglobose. *Flowers* 4–6. $2n = 22$.

Introduced. Naturalised where planted or neglected in open woodland. Berkshire, Roxburghshire and Berwickshire. Native of central and south France and northern Spain.

6. S. verna Huds.   Spring Squill

Glabrous *perennial herb* with an ovoid bulb 15–30 mm across, with numerous bulblets. *Stems* 5–30 cm, terete, pale green flushed brownish-red. *Leaves* 2–7, 3–20 cm × 2–5 mm, glossy green, long-linear, obtuse at apex, entire, shorter than to about as long as the stem. *Inflorescence* a dense, corymbose cluster; bracts 5–15 mm, linear-subulate, scarious, longer than the pedicels; lower pedicels 5–12 mm, ascending. *Flowers* 2–12, 10–16 mm in diameter, scentless. *Perianth segments* 5–8 × 2–3 mm, violet-blue, ovate-lanceolate, acute at apex, more or less ascending. *Stamens* 6; filaments 2–4 mm, lanceolate, white; anthers purplish-blue. *Capsule* 4–6 mm, subglobose, trigonous. *Flowers* 4–5. Pollinated by insects. $2n = 22$.

Native. Dry short grassland near the sea, especially on cliff-tops. Locally common on cliffs of western Britain from Devon to Shetland and down the east coast south to the Cheviots; east coast of Ireland. Western Europe from northern Portugal to the Faeroes.

7. S. peruviana L.   Portuguese Squill

Glabrous *perennial herb*. Bulb 60–80 mm in diameter, ovoid, with a brown tunic, the outer scales of which are woolly. *Stem* usually one, 15–50 cm, stout, green with whitish bases, glabrous. *Leaves* usually 9–15 in a dense rosette, 40–60 cm × 10–40(–60) mm, linear to lanceolate, acute at apex, entire, sometimes ciliate. *Inflorescence* 5–20 cm, densely corymbose, subhemispherical, with 5–20(–100) flowers; bracts 30–80 mm, papery, subulate, narrowed to a slender, acute apex; pedicels 30–50 mm, patent to ascending. *Flowers* 10–30 mm in diameter, spreading or erect. *Perianth segments* 5–18 mm, deep violet-blue or white, elliptical, more or less obtuse at apex. *Stamens* 6; filaments about 2 mm wide, lanceolate-elliptical, coloured like the flowers violet-blue or white; anthers yellow. *Style* and stigma blue. *Capsule* ovoid, acuminate at apex. *Flowers* 5–7. $2n = 16$.

An extremely variable species which has been split into many specific and infraspecific taxa.

Introduced. Naturalised where planted and neglected; known for over 60 years on a railway bank near Pontac, Jersey. Very scattered in Britain and the Channel Islands. Native of south-west Europe and north-west Africa.

Subgenus 2. Prospero (Salisb.) Chouard.

Roots perennial, branched. *Flowering* in late summer or autumn before the leaves. *Inflorescence* 4- to 20-flowered. *Seeds* ellipsoid. *Germination* epigeal.

8. S. autumnalis L.   Autumn Squill

Glabrous *perennial herb*. Bulb 13–50 × 10–20 mm, ovoid or subglobose; tunic membranous, the outermost scales dull brown, the inner whitish. *Stems* 2 or more, 7–25 cm,

elongating after anthesis, erect, slender, purplish. *Leaves* 8–10 or more, appearing after the flowers, 7–12 cm × 1–2 mm, linear-filiform, more or less acute at apex, rather prominently veined, glabrous. *Inflorescence* 1.5–10 × 1–2 cm, lax or dense, racemose, with 4–20 flowers; pedicels 5–10 mm, filiform, spreading and becoming arcuate-recurved in fruit. *Flowers* 6–12 mm in diameter, suberect, widely campanulate; bracts absent. *Perianth segments* 3–6 × 1.0–1.5 mm, mauvish-blue or purplish, oblong, obtuse or subacute at apex, more or less patent, with a prominent median vein. *Stamens* 6, free; filaments about 2 mm, purplish, filiform, somewhat dilated below; anthers about 1.0 × 0.4 mm, bluish-purple, oblong. *Style* about 2 mm, straight, slender; stigma truncate. *Capsule* 3–4 mm in diameter, subglobose, with an indented apex, papery, pale brown, usually enveloped by the persistent perianth. *Seeds* about 3.0 × 1.8 mm, shining brown or blackish, pyriform, irregularly compressed, minutely rugulose. *Flowers* 7–9. $2n = 28, 42$.

Two cytological races occur in Britain; a tetraploid from east Cornwall eastwards and in Jersey; a hexaploid in the south of Cornwall and in Guernsey (Ainsworth, Parker & Horton, 1983).

Native. Dry, short grassland, usually near the sea; in the Thames valley on acidic grass-heath on golf courses. Local in the Channel Islands and south-west England, Isle of Wight and Thames valley; formerly in Glamorgan. South and west Europe to north France and Hungary; north-west Africa; Asia east to Iran.

22. Hyacinthoides Heist. ex Fabr.

Endymion Dumort.

Perennial herbs with bulbs, with coalescent scales completely renewed each year. *Stems* erect. *Leaves* all basal, appearing with the flowers. *Flowers* in a raceme; bracts 2, subtending each flower; bracteoles absent. *Perianth segments* erect or patent, usually bluish, pink or rarely white. *Stamens* 6, inserted between the base and just above the middle of the perianth. *Capsule* subglobose or broadly ovoid. *Seeds* globose or broadly ovoid, unwinged.

This genus is hardly distinct from *Scilla*, *H. italica* being equally applicable to both genera. *H. non-scripta* and *H. hispanica* are really geographically replacing subspecies which hybridise where they meet. Rather than change the nomenclature yet again current opinion is followed, and the hybrid of the two species, which is by far the commonest plant of gardens and thus garden escapes, is given a binomial.

Has from 3 to 5 species in west Europe and North Africa.

Blackman, G. E. & Ritter, A. J. (1954). *Endymion non-scriptus* (L.) Garcke in Biological flora of the British Isles. *Jour. Ecol.* **42**: 629–638.

Grime, J. P. et al. (1988). *Hyacinthoides non-scripta* (L.) Chouard ex Rothm. in *Comparative plant ecology*. London.

1. Perianth segments 5–8 mm, patent; stamens inserted at the base of the perianth   **1. italica**

1. Perianth segments more than 14 mm, erect to
   erecto-patent   2.
2. Leaves 7–15 mm wide; perianth tubular, segments
   strongly recurved at apex   **2. non-scripta**
2. At least some leaves more than 15 mm wide; perianth
   not tubular, erect or spreading at apex but not
   recurved   3.
3. Flowers 12–17 mm long, 10–20 mm in diameter,
   variable in colour; stamens variable in length
                                                   3. × variabilis
3. Flowers 15–22 mm long, 15–25 mm in diameter; stamens
   subequal in length   **4. hispanica**

1. H. italica (L.) Rothm.   Italian Bluebell
Scilla italica L.

Plant glabrous. *Bulb* 15–35 × 15–30 mm, subglobose or
ellipsoid; tunic pale yellowish-brown. *Leaves* 2–6, 10–25
cm × 3–6(–12) mm, yellowish-green, linear to linear-
lanceolate, obtuse to acute at apex, entire, shorter than
to longer than scape, broadly channelled. *Scape* 10–40
cm, yellowish-green, terete. *Raceme* 10–50 cm, dense,
conical, 6- to 20-flowered; bracts paired, 2–20 mm,
linear or narrowly linear-lanceolate, acute at apex.
Flowers 8–12 mm in diameter, erect in bud, scentless;
pedicels 7–20 mm, slender. *Perianth segments* 5–8 ×
1.5–2.5 mm, bluish-violet, rarely white, lanceolate to
oblong-lanceolate, more or less obtuse at apex, patent.
Stamens inserted at base of perianth; filaments 2.5–4.0
mm, whitish; anthers blue. *Style* slightly longer than
anthers. *Capsule* subglobose-trigonous. *Seed* black,
globose. *Flowers* 3–5. 2*n* = 16.

Introduced. Widely grown in gardens; naturalised in a
neglected estate at Warley Place in south Essex; casual
elsewhere. Native of south-east France, north-west Italy
and south-west Portugal.

2. H. non-scripta (L.) Chouard ex Rothm.   Bluebell
Hyacinthus non-scriptus L.;*Endymion non-scriptus* (L.)
Garcke; *Scilla non-scripta* (L.) Hoffmanns. & Link;
Scilla nutans Sm.

Plant glabrous. *Bulb* 15–30 × 10–22 mm, ovoid or subglobose; tunic pale yellowish-brown. *Leaves* 3–6, 20–45
cm × 7–15 mm, at first erect, but spreading and flopping
after flowering, bright green on both sides, smooth,
shining, succulent, linear, obtuse to rounded or more or
less pointed at apex, entire, tapered below, canaliculate,
bluntly keeled. *Scape* 20–50 cm, pale green, purplish
above, erect, terete, solid, brittle, angular from the lowermost flower upwards. *Raceme* 4- to 16-flowered,
drooping at tip, unilateral; bracts paired, bluish, the
lower linear-lanceolate, longer than pedicel, the upper
smaller; whole inflorescence sometimes bracteate.
Flowers 10–28 mm in diameter, erect in bud, nodding
when fully open, bell-shaped, scented; pedicels 3–10
mm, afterwards elongating to about 30 mm and becoming erect. *Perianth segments* 14–20 × 3.0–4.5 mm, violet-
blue with a darker midrib, rarely pink or white, oblong,
more or less parallel and erect so that the lower part of
the flower appears cylindrical, rounded-obtuse and
recurved at apex. *Stamens* unequal, the outer inserted

just above the middle of the perianth, the inner lower;
filaments 8–9 mm, pale violet; anthers white or cream.
Style blue, shorter than anthers; stigma cream. *Capsule*
12–15 × 8–10 mm, more or less ovoid. *Seed* 2.5–3.0 mm
in diameter, black, globose or broadly ovoid. *Flowers*
4–6. *Fruits* 6–7. Pollinated by insects. 2*n* = 16.

Native. Often dominant over large areas in the drier
parts of woods, hedgerows, shady banks and sometimes
grassland; in upland areas on heaths, spoil heaps and
cliffs. Frequent to abundant throughout the British Isles.
Western Europe from central Spain to the Netherlands;
locally naturalised in central Europe.

3. H. × variabilis P. D. Sell   Hybrid Bluebell
    H. hispanica × non-scripta

Plant glabrous. *Bulb* 15–30 × 10–20 mm, ovoid or subglobose; tunic pale yellowish-brown. *Leaves* 3–8, 20–50
cm × 10–30 mm, at first erect but spreading and flopping after flowering, bright green on both sides, smooth,
shining, succulent, linear or oblong, obtuse to rounded
or more or less pointed at apex, entire, tapered below,
canaliculate, bluntly keeled. *Scape* 20–50 cm, pale yellowish-green, terete, solid, angular at the top. *Raceme*
5–15 cm, 4- to 15-flowered, erect or drooping, lax or
rather dense, not too markedly unilateral; bracts paired,
the larger 10–20 mm, the smaller 6–10 mm, narrowly
linear-lanceolate, bluish. *Flowers* 10–20 mm in diameter,
sweet-scented, erect in bud, nodding or erect when open;
pedicels 8–20 mm, elongating after flowering and
becoming erect. *Perianth segments* 12–17 mm, mauve,
pink or white, with a bluer or darker midrib outside,
paler inside, spreading or recurved, oblong, rounded or
obtuse at apex. *Stamens* variable in length and insertion
but mainly above the middle; filaments 6–10 mm,
whitish or bluish; anthers whitish or bluish. *Style* bluish,
equalling or exceeding the anthers. *Capsule* 15–18 ×
7–10 mm, more or less ovoid. *Seed* 2–3 mm in diameter,
broadly ovoid, black.

Introduced or derived in gardens. Naturalised in
woods, copses, banks, waysides and waste places. Widely
dispersed in Britain and the Channel Islands, where it is
fertile and forms a complete spectrum between the
parents and is often found in the absence of both. Arises
naturally in the Iberian Peninsula where the ranges of
both parents meet.

4. H. hispanica (Mill.) Rothm.   Spanish Bluebell
Scilla hispanica Mill.; *Endymion hispanicus* (Mill.)
Chouard; *Endymion patulus* Dumort.; *Scilla
campanulata* Ait.; *Endymion campanulatus* (Ait.) Willk.

Plant glabrous. *Bulb* 15–30 × 10–20 mm, ovoid or subglobose; tunic pale yellowish-brown. *Leaves* 4–8, 20–50
cm × 10–35 mm, at first erect, but spreading and flopping after flowering, bright green on both sides, smooth,
shining, succulent, linear or oblong, obtuse to rounded
or more or less pointed at apex, entire, tapered below,
canaliculate, bluntly keeled. *Scape* 30–60 cm, pale yellowish-green, terete, solid, angular at the top. *Raceme*
10–20 cm, 4- to 20-flowered, erect or nearly so, more or
less lax, not unilateral or only slightly so; bracts paired,

the larger 15–25 mm, the smaller 7–10 mm, narrowly linear-lanceolate, bluish. *Flowers* 15–25 mm in diameter, sweet-scented, erect in bud, patent when open; pedicels 10–20 mm, elongating after flowering and becoming erect. *Perianth segments* 15–22 mm, mauve with a blue mid-vein outside, paler inside, spreading but not recurved, oblong, rounded-obtuse at apex. *Stamens* all equal, the outer inserted below the middle of the perianth; filaments 8–10 mm, whitish or tinted blue; anthers bluish. *Style* bluish, equalling anthers. *Capsule* 18–22 × 15–18 mm, more or less ovoid. *Seed* 3.0–4.0 mm in diameter, black, broadly ovoid.

Introduced. Grown in gardens and becoming naturalised in woods, copses and waste places, but almost certainly over-recorded for *H. hispanica* × *non-scripta*. Scattered in Britain and the Channel Islands. Native of Spain and Portugal.

23. Hyacinthus L.

Perennial herbs with bulbs. *Leaves* present in spring, 2 or more on each flowering bulb. *Flowers* in a raceme arising from between the leaves, bisexual; bracts 1, minute, 2-lobed. *Perianth segments* fused at base into a tube about as long as the lobes, the lobes gradually curved outwards, blue, pink or white, very rarely pale yellow. *Stamens* inserted on the perianth tube, the filaments almost wholly fused to it. *Ovary* superior, 3-celled. *Capsule* conical-spherical. *Seeds* black, wrinkled, with an appendage.

Three species from west and central Asia.

Bentzer, B. & von Bothmer, R. (1974). Cytology and morphology of the genus *Hyacinthus* L. s. str. (Liliaceae). *Bot. Not.* **127**: 297–301.

1. H. orientalis L. Hyacinth

Glabrous *perennial herb* with a large bulb and a membranous tunic of free scales. *Stem* up to 30(–50) cm, thick, hollow, collapsing in fruit. *Leaves* present in spring, 2–4 on each bulb, 15–35 cm × 5–40 mm, bright green, linear to linear-lanceolate, rounded at apex, entire, flat, channelled, suberect at first. *Inflorescence* a loose or dense raceme, oblong; bracts minute. *Flowers* 2–40, 20–25 mm in diameter, very fragrant; pedicels 4–8 mm. *Perianth* 10–35 mm, the tube slightly constricted near the middle, inflated near the base; segments subequal, pale to deep violet-blue, pale to deep pink, white or yellowish, oblong to ovate, patent or recurved, as long as or shorter than the tube. *Stamens* attached to the perianth at or below the middle of the tube; anthers longer than the filaments, dark blue. *Style* much shorter than the perianth tube. *Capsule* 10–15 mm, conical-spherical, fleshy. *Seeds* black, wrinkled, with an appendage. *Flowers* 2–4.

Introduced. Much grown in gardens and long persistent where thrown out or neglected. Scattered over southern England. Native of central and southern Turkey, north-west Syria and Lebanon.

24. Chionodoxa Boiss.

Perennial herbs with small bulbs; tunics brown. *Leaves* usually 2. *Flowers* in a loose raceme; bracts absent or with one rudimentary one. *Perianth segments* fused into a tube much shorter than the lobes, blue or pinkish, often with a white central area. *Stamens* inserted at the apex of the perianth tube, fully exserted; filaments white, flattened, of unequal length. *Ovary* superior. *Capsule* about spherical. *Seeds* broadly ovoid, with an appendage.

Six species from western Turkey, Crete and Cyprus.

Speta, F. (1976). Über *Chionodoxa* Boiss. ihre Gliederung und Zugehörigkeit zu *Scilla* L. *Naturk. Jahrb. Stadt Linz* **21**: 9–79.

1. Flowers bright blue, without a pale central zone
                                                1. sardensis
1. Flowers blue or pink with a white central zone   **2.**
2. Scape with 4–15 slightly drooping flowers; perianth tube 3–5 mm   **2. forbesii**
2. Scape with 1–3 erect flowers; perianth tube 2–4 mm   **3. luciliae**

1. C. sardensis Whittall ex Barr
Lesser Glory-of-the-Snow
Scilla sardensis (Whittall ex Barr) Speta

Bulb 10–20 mm, ovoid; tunic brown. *Leaves* usually 2, 3.5–20 cm × 5–14 mm, dark green, sometimes tinged bronze, linear, margins inturned and entire, shortly narrowed to an obtuse apex, glabrous. *Scapes* usually solitary, 7–20(–40) cm, dark green tinged bronze, glabrous. *Raceme* 4–12(–15)-flowered; pedicels 10–70 mm, erecto-patent, green tinged bronze, glabrous; bracts inconspicuous, caducous, subulate, membranous. *Flowers* 15–22 mm, slightly nodding. *Perianth tube* 3–5 mm in diameter, subglobose; segments (5–)8–10 × 2–4 mm, deep blue throughout, at first erecto-patent, recurving with age, narrowly elliptical or ovate, rounded at apex. *Stamens* 6; filaments alternately 3 and 4 mm, white; anthers 2–3 mm, yellow or bluish. *Capsule* subglobose. *Seeds* broadly ovoid, with a pale appendange. *Flowers* 3–4. $2n = 18$.

Some plants, particularly when the above species grows together with *C. forbesii*, as at Cambridge, are like *C. sardensis*, but have some faint whiteness at the base of the perianth segments, and may be hybrids between the two species.

Introduced. Grown in gardens where it spreads by seed. Recorded as naturalised in Surrey, Essex and Kirkcudbright and the lawns of the Botanic Garden, Cambridge, and probably elsewhere. Native of Turkey.

2. C. forbesii Baker Glory-of-the-Snow
C. luciliae auct.; *C. siehei* Stapf; *C. tmolusi* Whittall

Bulb 10–28 mm, subglobose; tunic dark brown. *Leaves* usually 2, 7–28 cm × 3–15 mm, erect or spreading, green or slightly glaucous above, sometimes tinged brown, linear, the margins inturned, entire, narrowed to a blunt point at apex, glabrous. *Scapes* mostly solitary, rarely 2–3, up to 30 cm, green, suffused brown, glabrous. *Raceme* 2- to 15-flowered; pedicels 8–40 mm, green flushed brownish, erecto-patent or patent, ebracteate or with caducous, subulate or filiform, membranous bracts up to 3(–10) mm. *Flowers* 20–27 mm in diameter, slightly nodding. *Perianth tube* 3–5 mm in diameter, subglobose; segments 6, 10–15 × 4–6 mm, deep rich blue with a white basal area, at first erecto-patent, recurving with age, elliptical, ovate or oblong, rounded at apex. *Stamens* 6; filaments alternately 2 and 3 mm, white; anthers 3–4 mm, pale yellow. *Capsule* subglobose. *Seeds* broadly ovoid, with a pale appendage. *Flowers* 3–4. $2n = 18$.

Introduced. Much grown in gardens where it spreads by seed and becomes well naturalised where it is neglected. Scattered over Britain, but mainly in the south. Native of Turkey.

3. C. luciliae Boiss.   Pale Glory-of-the-Snow
C. gigantea Whittall

Perennial herb. Bulb 10–28 mm, subglobose; tunic dark brown. *Leaves* usually 2, 7–20 cm × 3–15 mm, erect or spreading, green or slightly glaucous, linear, the margins inturned, entire, narrowed to a blunt point at apex, glabrous, often recurved. *Scapes* mostly solitary, up to 14 cm, green, glabrous. *Raceme* 1–3-flowered; pedicels 8–40 mm, green, erect. *Flowers* 20–27 mm in diameter, erect. *Perianth tube* 2–4 mm, subglobose; segments 12–20 × 3–8 mm, pale lavender-blue with a white central zone. *Stamens* 6; filaments 2.5–3.0 mm, white; anthers pale yellow. *Capsule* subglobose. *Seeds* broadly ovoid, with a pale appendage. *Flowers* 3–4. $2n = 18$.

Introduced. Grown in gardens where it spreads by seed and becomes well naturalised where it is neglected. Lawns of Cambridge Botanic Garden; should be looked for elsewhere. Native of western Turkey.

24 × 21. × Chionoscilla Nicholson
Chionodoxa × Scilla

1. × C. allenii Nicholson
C. forbesii × S. bifolia

A hybrid intermediate between the parents with very short perianth tube, a small pale central zone to the flower and pale blue filaments.

This garden hybrid seems to be fertile and spreads by seed in the Cambridge Botanic Garden and may be found elsewhere.

25. Muscari Mill.

Perennial herbs with bulbs, with or without offsets. *Leaves* 2–7, basal, present in spring. *Flowers* in racemes, contracted at mouth, the apical group sterile, the lower ones fertile and often of a different colour; bract 0–1, when present, minute. *Perianth segments* fused for most of their length, blue to blackish-blue with white lobes, or brownish. *Stamens* inserted about halfway up the perianth-tube, included, with narrow filaments. *Ovary* superior, 3-celled. *Capsule* strongly angled. *Seeds* black often shiny.

About 30 species from the Mediterranean area and south-west Asia.

M. azureum Fenzl is an established garden escape in rough grass on the site of old parkland at Curry Rivel, Somerset.

Speta, F. (1982). Über die Abgrenzung und Ghederung der Gattung *Muscari*, und über ihre Beziehungen zu anderen Vertreten de Hyacinthaceae. *Bot. Jahrb. Syst.* **103**: 247–291.
Stuart, D. C. (1966). *Muscari* and allied genera. *RHS Lily Year Book* **29**: 125–138.

1. Fertile flowers brownish-buff, on pedicels mostly more than 5 mm; apical sterile flowers bright bluish-violet, some on pedicels more than 5 mm   **4. comosum**
1. All flowers blue to blackish-blue, on a pedicel less than 5 mm   2.
2. Leaves linear, with a more or less spathulate apex; corolla more or less spherical with strongly recurved lobes   **3. botryoides**
2. Leaves linear to oblanceolate; corolla ellipsoid-ovoid, distinctly longer than wide with erecto-patent lobes   3.
3. Perianth of fertile flowers blackish-blue   **1. neglectum**
3. Perianth of fertile flowers bright blue   **2. armeniacum**

1. M. neglectum Guss. ex Ten.   Grape Hyacinth
M. atlanticum Boiss. & Reut.; *M. racemosum* L. nom. rej.

Perennial herb. Bulbs 10–25 mm, with or without offsets; tunics dark to reddish-brown. *Stems* 4–30 cm, often as long as or longer than leaves, glabrous. *Leaves* 3–6, 6–40 cm × 1.5–8.0 mm, bright green, sometimes reddish at base, linear to linear-oblanceolate, acute at apex, entire, channelled to involute, glabrous. *Inflorescence* a dense, ovoid raceme, bearing fertile flowers below and sterile above. *Fertile flowers* 3.5–7.5 mm, very dark to blackish-blue, ovoid to oblong-urceolate or ellipsoid-ovoid, strongly constricted, the lobes 0.3–1.0 mm, white, patent; pedicels 0.5–5.0 mm, patent or deflexed, shorter than perianth. *Sterile flowers* up to 20, smaller and paler than fertile. *Fruiting raceme* lax. *Capsule* 8–10 × 7–10 mm, ovoid to obovoid, emarginate to apiculate at apex. *Flowers* 4–5. Pollinated by insects and selfed; protogynous. $2n = 36, 45, 54$.

Native. Dry grassland, hedgebanks and field borders. Very local in Suffolk and Cambridgeshire, formerly in Norfolk; rarely naturalised elsewhere. Much of Europe, north to northern France and south-central Russia; North Africa.

2. M. armeniacum Leichtlin ex Baker
Garden Grape Hyacinth

Perennial herb. Bulb with or without offsets; tunic dark brown. *Stem* 10–40 cm, usually exceeding leaves, pale green becoming bluish above, glabrous. *Leaves* (2–)3–5(–7), 10–30 cm × 1–10 mm, shining green below, glaucous above, linear or oblanceolate, rarely linear-elliptical, acute at apex, entire, channelled, glabrous. *Inflorescence* a raceme, 1.5–5.0 cm, ovoid to cylindrical, dense to very dense, becoming lax in fruit, bearing fertile flowers below and sterile above; bracts minute. *Fertile flowers* 3.5–5.5 × 2.5–3.5 mm, bright blue, sometimes with a purplish tinge, rarely white, obovoid to oblong-urceolate, lobes pale or white; pedicels 1–5 mm, patent to deflexed, usually shorter than the flowers. *Sterile* flowers few, smaller, paler or concolorous. *Capsule* 8–12 × 6.5–8.0 mm, obovoid, emarginate at apex. *Flowers* 4–5. $2n = 18$.

Introduced. Common in gardens from which it escapes on to rough ground, banks and grassy places. Spreads both vegetatively and by seed. Scattered in Britain, mainly the south and in the Channel Islands. Native of south-east Europe to south-west Asia from the Balkans to the Caucasus.

3. M. botryoides (L.) Mill. Compact Grape Hyacinth
Hyacinthus botryoides L.

Perennial herb. Bulb with slender annual roots, not producing offsets; tunic pale or greyish-brown. *Stems* 7–30 cm, almost always exceeding the leaves, pale green, glabrous. *Leaves* 2–3(–4), 5–25 cm × 5–12 mm, glaucous, paler above, erect, linear-oblanceolate or rarely linear, abruptly contracted into a hooded or shortly acuminate apex, entire, often prominently ribbed, channelled, glabrous. *Inflorescence* a raceme, 1–7 cm, dense at first becoming laxly cylindrical, bearing fertile flowers below and a few sterile flowers above; bracts minute. *Fertile flowers* 2.5–5.0 × 2.5–4.0 mm, bright blue with white, recurved lobes, globose, strongly constricted; pedicels (0.5–)2–5 mm, as long as or shorter than the flower, patent or deflexed; stamens inserted in 2 series in the perianth tube. *Sterile flowers* few, smaller and paler; pedicels very short. *Capsule* 4–6 × 4–6 mm, globose. *Flowers* 4–6. 2n = 18, 36.

Somewhat variable in leaf shape and density of raceme.

Introduced. Grown in gardens where it remains when neglected, or thrown out on dumps; naturalised in sand or gravel pits and on roadsides. Scattered in southern England. Native of south Europe, extending locally westwards to north-west France.

4. M. comosum (L.) Mill. Tassel Hyacinth
Hyacinthus comosus L.; *Bellevalia comosa* (L.) Kunth; *Leopoldia comosa* (L.) Parl.

Perennial herb. Bulb 4–5 × 3–4 cm, ovoid; tunic papery, fuscous externally, pinkish internally. *Stem* solitary, 15–60 cm. *Leaves* (1–)2–4, 15–40 cm × 8–20 mm, glaucous-green, linear, acuminate at apex, hyaline-membranous, scabridulous or shortly ciliate on the margins, long-sheathing at base. *Inflorescence* 8–30 × 2–4 cm, bearing 10–50 fertile flowers, and a conspicuous terminal tuft or coma of sterile flowers; bracts less than 1.5 mm, whitish, membranous. *Fertile flowers* brownish, or yellowish at apex and base; perianth tube 5–6 × about 4 mm in diameter, shortly cylindrical, distinctly constricted at apex, the lobes very short, recurved, pallid, ovate-deltoid, about 1.5 mm and almost as wide at base; stamens inserted in 2 series near the middle of the perianth tube, free part of filaments about 1.5 mm; anthers 1–2 × 1.0–1.2 mm, violet-blue, shortly oblong; style about 2 mm, stigma truncate; ovary about 2 × 2 mm, sessile, ovoid-triquetrous. *Sterile flowers* with subglobose or oblong-globose perianth, violet-blue; pedicels up to 20 mm. *Capsule* about 8 × 3 mm, broadly ovoid-triquetous, valves pale brown, distinctly veined, rounded or slightly emarginate at apex. *Seeds* about 2.5 mm in diameter, black, subglobose, rugulose-reticulate, minutely pitted. *Flowers* 4–7. 2n = 18.

Introduced. Persistent weed of cultivated and rough ground, sometimes also on dunes and in open grassland. Local in south-west England, south Wales and the Channel Islands, rare and casual elsewhere. Much of Europe, north to north France and east to Transcaucasia and Iran.

Subfamily 6. Allioideae Engl.
Tribes *Allieae* Kunth, *Agapantheae* Engl.

Perennial herbs with rhizomes or bulbs, mostly smelling of garlic or onion when fresh. *Leaves* usually all basal, sometimes on stem, linear to more or less cylindrical, linear-oblong or elliptical. *Inflorescence* a terminal umbel, usually with a scarious spathe at its base, sometimes reduced to one flower or some or all flowers replaced by bulbils, but still with a spathe. *Perianth segments* of various colours, free or fused at base. *Style* 1, with capitate to 3-lobed stigma. *Ovary* superior or semi-inferior. *Fruit* a capsule, splitting along the centre of the ovary-cells.

26. Allium L.

Perennial herbs with well-formed bulbs, in some species forming a short rhizome, usually smelling of onion or garlic when fresh. *Leaves* linear to more or less cylindrical, or elliptical, basal or sheathing the stem, solid or hollow. *Flowers* in an umbel, some or all often replaced by bulbils, enclosed at first within a spathe consisting of one or more bracts. *Perianth segments* free or more or less so, white to greenish, pink, purple or yellow. *Stamens* 6; filaments free or united at base. *Style* arising from the base of the ovary. *Ovary* superior, 3-celled; ovules usually 2 per cell. *Fruit* a 3-celled capsule.

About 690 species in Europe, Asia and America and most abundant in central Asia, but extending into tropical Africa, Sri Lanka and Mexico. Our species are here given conventional status, but W. T. Stearn considers all the British species of *Allium* except *A. schoenoprasum* and *A. ursinum* to be long-established aliens.

This account is taken from W. T. Stearn (1978 and 1980). Several species are important for human consumption. In Ancient Greece garlic was known as *skorodon*, σκόροδον, or *skordon*, σκόρδον, the onion as *krommuon*, κρόμμυον (earlier *kromuon* κρόμυον), and the leek as *prason*, πράσον, hence the use of names in *Allium* formed from scorodon, scordon, scordum, crommyum and prasum. Garlic in Latin was *allium* or *alium*, from which comes the Italian *aglio*, Spanish *ajo*, Portuguese *alho* and French *ail*. The Latin for onion is *caepa, cepa* or *caepe*, and for leek *porrum* or *porrus*.

A. cyrillii Ten. and **A. tuberosum** Rottler ex Spreng. have been recorded as casuals.

Barling, D. M. (1971). Studies on Gloucestershire populations of *Allium paradoxum* (Bieb.) G. Don. *Watsonia* **8**: 379–384.
Davies, D. (1992). *Alliums; the Ornamental Onions*. London.
Frost, L. C., Houston, L., Lovatt, C. M. & Beckett, A. (1991). *Allium sphaerocephalon* L. and introduced *A. carinatum* L., *A. roseum* L. and *Nectaroscordum siculum* (Ucria) Lindley on St Vincent's Rocks, Avon Gorge, Bristol. *Watsonia* **18**: 381–385.

Grime, J. P. et al. (1988). *Allium ursinum* L. in *Comparative plant ecology*. London.

Jones, H. A. & Mann, L. K. (1963). *Onions and their allies*. London.

Oswald, P. (1993). Native and naturalised garlics in the Cambridge University Botanic Garden. *Nat. Cambridgeshire* **35**: 67–75.

Richens, R. H. (1947). *Allium vineale* L., in Biological flora of the British Isles. *Jour. Ecol.* **34**: 209–226.

Stearn, W. T. (1978). European species of *Allium* and allied genera of Alliaceae: a synonimic enumeration. *Ann. Mus. Goulandris* **4**: 83–198.

Stearn, W. T. (1980). *Allium* L. in Tutin et al. *Flora Europaea* **5**: 49–69. Cambridge.

Stewart, A., Pearman, D. A. & Preston, C. D. (1994). *Scarce plants in Britain*. Peterborough. [*A. oleraceum, schoenoprasum* and *scorodoprasum*.]

1. Inflorescence consisting entirely of bulbils   2.
1. Inflorescence with at least 1 flower   10.
2. Leaves terete, circular or semicircular in section   3.
2. Leaves flat to strongly keeled   5.
3. Stem hollow, inflated and bulging just below the middle; leaves usually more than 4 mm wide   **2. cepa**
3. Stem solid or nearly so, not inflated; leaves less than 4 mm wide   4.
4. Spathe of 2 persistent valves each with the apical attenuate part much longer than the basal part
                                      13(i). oleraceum var. **oleraceum**
4. Spathe of 1, more or less deciduous valve with the apical attenuate part about as long as the basal part
                                      20(iii). vineale var. **compactum**
5. Stems triangular in section; leaves all basal   **11. paradoxum**
5. Stems terete, more or less circular in section, at least some leaves borne on the stem   6.
6. Leaves less than 4 mm wide   7.
6. Leaves more than 5 mm wide   8.
7. Perianth segments 1.5–2.0 mm wide
                                      14(a). carinatum subsp. **carinatum**
7. Perianth segments 2.5–3.0 mm wide
                                      13(ii). oleraceum var. **complanatum**
8. Leaves 2–5; main bulb single, often with numerous small bulblets outside its covering   **18. scorodoprasum**
8. Leaves 4–10, main bulb composed of several more or less equal bulblets within a common cover   9.
9. Stem erect and straight   **15(i). sativum** var. **sativum**
9. Flower stem before anthesis coiled at the top into 1 or 2 loops   **15(ii). sativum** var. **ophioscorodon**
10. Inflorescence with bulbils and flowers   11.
10. Inflorescence with flowers only   25.
11. Leaves terete, more or less circular or semicircular in section   12.
11. Leaves flat to strongly keeled   16.
12. Stem hollow, inflated and bulging just below the middle; leaves usually more than 4 mm wide   **2. cepa**
12. Stem solid or more or less so, not inflated; leaves less than 4 mm wide   13.
13. Stamens shorter than perianth segments; filaments simple   14.

13. Stamens longer than perianth segments; inner 3 filaments divided distally into 3 points, the middle one anther-bearing   15.
14. Leaves 2–3 mm wide, semi-terete, usually hollow at least below   **13(i). oleraceum** var. **oleraceum**
14. Leaves 3–4 mm wide, flat
                                      13(ii). oleraceum var. **complanatum**
15. Spathe 2-valved; lateral points of the inner 3 filaments less than twice as long as the central point   **19. sphaerocephalon**
15. Spathe 1-valved; lateral points of the inner 3 filaments more than twice as long as the central point
                                      20(ii). vineale var. **vineale**
16. Stems triangular in section   **11. paradoxum**
16. Stems more or less circular in section   17.
17. Perianth segments yellow   **8(ii). moly** var. **bulbilliferum**
17. Perianth segments pink to white, greenish or purplish   18.
18. Leaves often less than 5 mm wide; filaments simple   19.
18. Leaves more than 5 mm wide; inner 3 filaments divided at the top into 3 points, the middle one anther-bearing   20.
19. Spathe shorter than pedicels; stamens shorter than perianth segments   **5(ii). roseum** var. **bulbiferum**
19. Spathe longer than pedicels; stamens longer than perianth segments   **14(a). carinatum** subsp. **carinatum**
20. Stamens longer than perianth segments   21.
20. Stamens shorter than perianth segments   23.
21. Bulb scarcely swollen at base, without bulblets; spathe persistent at least until flowering; style shorter than perianth segments   **17. porrum**
21. Bulb swollen at base with bulbils around it within a common cover; spathe usually deciduous before flowering; style longer than perianth segments   22.
22. Flowers and bulbils 6–8 mm in a rather compact umbel   **16(ii). ampeloprasum** var. **bulbiferum**
22. Flowers and bulbils 8–15 mm with a rather loose umbel, often with pedicels bearing secondary heads
                                      16(iii). ampeloprasum var. **babingtonii**
23. Main bulb single, often with numerous small bulblets outside its cover; leaves 2–5; outer part of inner 3 filaments 2–3 times as long as the central anther-bearing point   **18. scorodoprasum**
23. Main bulb composed of several more or less equal bulblets within their common cover; leaves 4–10; outer part of inner 3 filaments about as long as the central anther-bearing point   24.
24. Stem erect and straight   **15(i). sativum** var. **sativum**
24. Flower stem before anthesis curled at the top into 1 or 2 loops   **15(ii). sativum** var. **ophioscorodon**
25. Leaves more or less circular in section   26.
25. Leaves flat to strongly keeled   30.
26. Stem hollow, inflated and bulging just below the middle; leaves usually more than 4 mm wide   27.
26. Stem not inflated; leaves usually less than 4 mm wide   28.
27. Perianth segments 3.0–4.5 mm   **2. cepa**
27. Perianth segments 7–9 mm   **3. fistulosum**
28. Stamens shorter than perianth segments; filaments simple   **1. schoenoprasum**

28. Stamens at least as long as perianth segments; inner 3 filaments divided into 3 points, the middle one anther-bearing   29.

29. Spathe 2-valved; lateral points of the inner 3 filaments less than twice as long as the central point   **19. sphaerocephalon**

29. Spathe 1-valved; lateral points of the inner 3 filaments more than twice as long as the central point   **20(i). vineale** var. **capsuliferum**

30. Perianth segments yellow   **8(i). moly** var. **moly**

30. Perianth segments white to pink, greenish or purplish 31.

31. Leaves with a distinct petiole, the lamina elliptical to narrowly so   **12. ursinum**

31. Leaves without a petiole, linear to filiform   32.

32. Stem triangular in section   33.

32. Stem more or less circular in section   35.

33. Spathe 1-valved; stigma simple   **6. neapolitanum**

33. Spathe 2-valved; stigma 3-lobed   34.

34. Umbel 1-sided, with pendulous flowers; perianth never opening more than 45°   **9. triquetrum**

34. Umbel not 1-sided, with erect and pendulous flowers; perianth opening more than 45° at first, but less so later   **10. pendulinum**

35. Inner 3 filaments divided into 3 points, the middle one anther-bearing   36.

35. Filaments simple   37.

36. Bulb scarcely swollen at base, without bulblets; spathe persistent at least until flowering; style shorter than perianth segments   **17. porrum**

36. Bulb swollen at base with bulblets around it within a common cover; spathe usually deciduous before flowering; style longer than perianth segments   **16(i). ampeloprasum** var. **ampeloprasum**

37. Leaves conspicuously ciliate   **7. subhirsutum**

37. Leaves glabrous   38.

38. Leaves less than 3 mm wide; spathe with valves much longer than pedicels; stamens longer than perianth segments   **14(a). carinatum** subsp. **pulchellum**

38. Most of leaves more than 4 mm wide; spathe with valves rarely as long as pedicels; stamens shorter than perianth segments   39.

39. Leave more than 20 mm wide   **21. nigrum**

39. Leaves less than 15 mm wide   40.

40. Covering of bulb with undulating or net-like markings; spathe 2-valved   **4. unifolium**

40. Covering of bulb minutely pitted; spathe with one primary valve often deeply 2-lobed   **5(i). roseum** var. **roseum**

Section 1. Schoenoprasum Dumort.

Bulbs cylindrical or very narrowly conical, on a short rhizome. *Stems* terete. *Leaves* almost basal or sheathing the lower quarter to one-third of the stem, cylindrical, fistular. *Spathe* equalling or shorter than the pedicels, which are without bracteoles at the base. *Perianth* campanulate. *Stamens* simple. *Ovary* with deep, rounded nectariferous pits; ovules 2 in each loculus. *Stigma* entire. *Seeds* angular.

1.A. schoenoprasum L.   Chives

A. sibiricum auct.; *A. montanum* Schrank, non F.W. Schmidt; *A. sibthorpianum* Schult. & Schult. fil.; *A. smithii* Nyman; *A. foliosum* Clarion ex DC.; *Cepa schoenoprasa* (L.) Moench; *Cepa tenuifolia* Gray nom. illegit.

Tufted *perennial herb*, smelling of onions. *Bulbs* 5–10 mm in diameter, very narrowly conical, clustered on a short rhizome; outer tunics membranous, sometimes splitting into coriaceous strips. *Stems* 5–50 cm, hollow. *Leaves* 1–2, up to 35 cm × 1–6 mm, bright green, sheathing up to the lower third of the stem, sometimes almost basal, cylindrical, fistular. *Spathe* up to 15 mm, with the 2–3, ovate lobes mucronate, equalling or shorter than the umbel, persistent. *Inflorescence* an umbel 1.5–5.0 cm in diameter, hemispherical or ovoid, dense, with 8–30 flowers; pedicels 2–15 mm. *Perianth segments* 7–15 × 2.5–4.0 mm, lilac or pale purple, occasionally white, usually lanceolate, acute or acuminate at apex. *Stamens* included; filaments 3–6 mm, about half as long as the perianth segments; anthers yellow. *Capsule* about 4 mm. *Seeds* angular. *Flowers* 6–7. Reproduction by both seed and bulbs. $2n = 16$.

Native. Thin soil in rocky ground, sometimes in crevices, usually on limestone, sometimes near the sea, inland sometimes in deeper soil. Local in south-west and northern England and south Wales; formerly in Berwickshire and East Mayo. The mild-flavoured leaves, which are usually chopped are used for garnishing soups, omelettes, salads and sandwiches. It is primarily home-grown, where it is also planted for ornament. From gardens it occurs as a relic or throw-out elsewhere in Britain. Northern Europe and Asia from Scandinavia and arctic Russia to Japan, and extending south in the mountains to north-west Portugal, Corsica, central Apennines, Greece and Turkey; North America from Newfoundland and Alaska to New York and Washington.

Section 2. Cepa (Mill.) Borkh.
Cepa Mill.

Bulbs cylindrical to subglobose, usually clustered on a short rhizome. *Leaves* sheathing the lower part of the stem, distichous, fistular. *Stem* terete, fistular. *Spathe* shorter than or almost equalling pedicels. *Perianth* stellate to campanulate. *Stamens* simple or with small teeth at the base of the inner filaments. *Ovary* with distinct, nectariferous pores; ovules 2 in each loculus. *Stigma* entire. *Seeds* angular.

2. A. cepa L.   Onion
Cepa esculenta Gray; *Cepa vulgaris* Renault

Biennial or *perennial herb*, smelling of onions. *Bulbs* varying in size and shape from cultivar to cultivar, often depressed-globose and up to 10 cm in diameter; outer tunics membranous. *Stem* up to 100 cm and up to 30 mm in diameter, tapering from the inflated lower part. *Leaves* up to 10, up to 40 cm × 20 mm in diameter, glaucous, usually almost semicircular in section and slightly flattened on upper side, basal in the first year, in second year their bases sheathing for the lower one-sixth of the stem. *Spathe* often 3-valved, persistent, shorter than the umbel. *Inflorescence* an umbel 4–9 cm in diameter, subglobose or hemispherical, dense, with many flowers, bulbils or bulbils and flowers; pedicels up to 40 mm, almost equal. *Perianth* stellate; segments 3.0–4.5 × 2.0–2.5 mm, white with a green stripe, slightly unequal, the outer ovate, the inner oblong, obtuse or acute at apex. *Stamens* exserted; filaments 4–5 mm, the outer subulate, the inner with an expanded base up to 2 mm wide and shortly toothed on each side. *Ovary* whitish. *Capsule* about 5 mm. *Seeds* black, angular. *Flowers* 7–9. $2n = 16$.

There are numerous cultivars of the common Onion resulting from some 3,000 years of cultivation. The culinary uses of onions are exceedingly numerous. They are eaten raw, or cooked in soups, sauces, stews, curries and a great variety of other savoury dishes and they are a main ingredient of many pickles and chutneys. The edible part of the onion is the bulb which is composed of fleshy, enlarged leaf-bases. In shape onions of different varieties may be flattened-globose, globose or ovoid and silvery-white to pale or dark brown and occasionally purplish-red. *Spanish Onions* are characteristically large and mild in flavour. *Spring Onions*, which are used in salads, are grown from seed and harvested when young. *Shallots* were long thought to be a distinct species, wrongly called *A. ascalonicum*, but they are now considered to be a variety of *A. cepa*. The *Shallot* group, which includes the *Potato Onion* or *Multiplier Onion*, multiply freely by producing lateral bulbs and can be called var. **proliferum** (Moench) Targ.-Toz. *Tree Onion* produces mostly or only bulbils in the inflorescence.

Introduced. Widely cultivated in household gardens and also grown as a crop. About 6,000 hectares are grown in the United Kingdom yearly, mainly in Lincolnshire, Norfolk, Isle of Ely, Suffolk, Kent and the Vale of Evesham. Sometimes occurs as a throw-out from gardens in waste places or by tracks where bulbs have fallen off the waggons which cart them from the fields. Unknown as a wild plant, though probably derived from the Asian *A. oschaninii* B. Fedtsch.

3. A. fistulosum L.   Welsh Onion
Cepa fistulosa (L.) Gray

Perennial herb smelling of onions. *Bulb* 10–25 mm in diameter, cylindrical, adnate to a short rhizome; outer tunics membranous. *Stem* 12–70 cm, 10–20 mm in diameter, tapering from the inflated middle part. *Leaves* 2–6, 6–30 cm × 5–15 mm, their bases sheathing in the lower quarter to one-third of the stem, fistular, circular in section, terete. *Spathe* 1- to 2-valved, valves up to 2 cm, ovate, acute at apex, almost equalling the umbel. *Inflorescence* an umbel 1.5–5.0 cm in diameter, subglobose or broadly ovoid, dense, many-flowered; pedicels 3–20 mm, unequal. *Perianth* conically campanulate; segments 6–7 × about 2 mm, yellowish-white, unequal, the outer cymbiform, lanceolate, the inner 7–9 × 3 mm, narrowly ovate, acuminate at apex. *Stamens* long-exserted; filaments 8–12 mm, simple; anthers yellow. *Capsule*

about 4 mm. *Seeds* black, angular. *Flowers* 7–9. $2n = 16$.

Introduced. Welsh Onion is not a native of Wales nor has it ever been cultivated there in any quantity. The word may be a corruption of the German word *welsche* (foreign), applied to this onion when it was introduced into Europe towards the end of the Middle Ages. It is a home-garden crop used as a substitute for Spring Onions and for seasoning and occurs from time to time as a casual garden throw-out. It is not known in the wild state, though it has been the principal onion of China and Japan since prehistoric times, but has never become popular in the West.

4. A. unifolium Kellogg   American Onion

Perennial herb smelling of onions. *Bulb* 10–15 mm, ovoid, arising terminally on a stout lateral rootstock, the old one not persisting; outer tunics pale, with obscure, narrow, horizontal, undulating reticulations. *Stem* 20–60 cm. *Leaves* 2–4, 10–40 cm × 2–8 mm, long-linear, flat, scarcely keeled. *Spathe* 20–30 mm, lanceolate-ovate, acuminate at apex. *Inflorescence* an umbel, loose and with many flowers; pedicels 20–35 mm, rather stout. *Perianth segments* 10–17 mm, rose-pink to lilac, sometimes white, oblong-ovate, more or less acuminate. *Stamens* scarcely two-thirds as long as perianth, wide at base; anthers yellow or purplish. *Capsules* 4–5 mm. *Seeds* about 3.0 × 1.5 mm. *Flowers* 4–6. $2n = 14$.

Introduced. A garden plant, naturalised in damp woodland behind Cardross Park in Dunbartonshire. Native of western North America.

Section **3. Molium** G. Don ex Koch

Bulbs ovoid or subglobose; not rhizomatous. *Leaves* almost basal, arranged spirally with short above-ground sheath, flat or slightly keeled. *Stem* terete or angled. *Spathe* shorter than or equalling pedicels. *Perianth* stellate to campanulate or cylindrical. *Ovary* with distinct nectariferous pores; ovules 2 in each loculus. *Stigma* entire. *Seeds* angular.

5. A. roseum L.   Rosy Garlic

Perennial herb smelling of garlic. *Bulb* 10–15 mm in diameter, broadly ovoid or subglobose; exterior tunic crustaceous, greyish-brown, deeply and conspicuously packed with circular or hexagonal pits, interior smooth, whitish; bulblets usually numerous, ovoid. *Stem* 10–65 cm, terete. *Leaves* 2–6, 10–35 cm × 5–14 mm, matt above, glossy below, linear, slender and acuminate at apex, flat or slightly keeled, glabrous, distinctly scabrid, base long-sheathing, smooth or papillose on the margins. *Spathe* 10–15 mm, papery, 1-valved, deeply 3- to 4-lobed at anthesis. *Inflorescence* an umbel up to 7 cm in diameter, with or without bulbils, usually many-flowered, often rather lax; pedicels 10–40 mm, slender, glabrous. *Perianth* campanulate or broadly cup-shaped with a rounded base; segments 7–12 × 3.0–5.5 mm, usually pink, sometimes white, free almost to base, oblong, narrowly elliptical or narrowly obovate, rounded or obtuse at apex, translucent, the midrib often prominent and dark pink. *Stamens* included; filaments 5.0–5.5 mm, subulate, tapering from a broad base;

anthers about 1.5×1.0 mm, yellow, oblong. *Style* about 5 mm; stigma clavate. *Ovary* about 3.5 mm in diameter, trigonous-subglobose, glabrous. *Capsule* 4.0–4.5 mm. *Seeds* about 3.0×2.0–2.5 mm, black, angular, bluntly papillose. *Flowers* 5–6.

(i) Var. roseum
A. illyricum Jacq.; *A. roseum* var. *grandiflorum* Briq. Umbel without bulbils, with numerous flowers. $2n = 32$.

(ii) Var. bulbiferum DC.
A. carneum Targ.-Tozz.; *A. ambiguum* Sm., non DC.; *A. incarnatum* Hornem.; *A. tenorii* Spreng.; *A. amoenum* G. Don; *A. roseum* var. *bulbilliferum* Vis.; *A. roseum* var. *carneum* Rchb.; *A. roseum* subsp. *bulbiferum* (DC.) E. F. Warb.
Umbel with flowers and bulbils. $2n = 32$.

A very variable species which has been divided into numerous named taxa based on the width of leaf, length of pedicels, presence or absence of bulbils in the umbels and colour of the flowers. Variants with bulbils replacing flowers in the umbel have probably arisen independently in different populations and, owing to their ease of multiplication, formed clones.

Introduced. Grown in gardens, escaping and well naturalised in rough or cultivated ground, hedgerows, waysides and old dunes. Frequent in the south-west of England, south Wales and the Channel Islands. Native of south Europe and the Mediterranean regions; Azores. Var. *roseum* is rarely naturalised, var. *bulbiferum* more commonly so.

6. A. neapolitanum Cirillo   Neapolitan Garlic
Perennial herb smelling of garlic. *Bulb* 10–20 mm in diameter; outer tunic crustaceous and greyish-brown, the inner white and smooth. *Stem* 25–50(–70) cm, triquetrous, the angles sometimes narrowly winged, glabrous. *Leaves* 2–4, 6–50 cm $\times$ 15–40 mm, linear, slenderly acuminate at apex, flat, glabrous, somewhat keeled, the base sheathing the stem for 6–10 cm, minutely papillose on the margins. *Spathe* 1, 15–20 $\times$ 10–15 mm, membranous, shortly cuspidate at apex. *Inflorescence* an umbel, hemispherical, many-flowered, rather lax; pedicels 15–35 mm, angular, glabrous. *Perianth* cup-shaped; segments 10–12 $\times$ 3–4 mm, spreading, white, translucent, ovate, obtuse at apex. *Stamens* included; filaments 5–6 mm, subulate, expanded towards the base, glabrous; anthers about 1.5–0.5 mm, greyish-green, oblong. *Style* about 4 mm, tapering to apex; stigma truncate. *Ovary* about 2.5 mm in diameter, trigonous-subglobose, pale. *Capsule* about 6 mm in diameter, membranous, pale brown. *Seeds* about 2.5×2.5 mm, black, irregularly angular, closely papillose with rows of blunt papillae. *Flowers* 3–5. $2n = 14, 21, 28, 31, 32, 33, 34, 35, 36$.

Introduced. Grown in gardens, escaping and naturalised in rough and cultivated ground, hedgebanks and waysides. Frequent in south-west England and the Channel Islands, occasionally elsewhere. Native of the eastern Mediterranean region.

7. A. subhirsutum L.   Hairy Garlic
A. hirsutum Lam. nom. illegit.; *A. ciliatum* Cirillo; *A. pulchrum* E. D. Clarke; *A. clusianum* Retz.

Perennial herb smelling of garlic. *Bulbs* up to 15 mm in diameter, subglobose; tunic membranous, not pitted. *Stem* 7–45 cm, terete. *Leaves* 2–3, 6–45 cm $\times$ 2–10 mm, almost basal, linear, flat, scarcely keeled, ciliate. *Spathe* up to 13 mm, 1-valved, persistent, shorter than the pedicels. *Inflorescence* an umbel 2.5–7.0 cm in diameter, hemispherical and lax, or fastigiate; pedicels up to 40 mm, 3–5 times as long as the perianth segments, irregularly patent. *Perianth* stellate; segments 7–9 mm, white, lanceolate to oblanceolate, obtuse to acute at apex. *Stamens* half to one-third as long as perianth; filaments 4–6 mm, subulate, entire; anthers usually brown, occasionally yellow. *Capsule* about 3 mm.

Introduced. Garden plant, escaping and naturalised in rough and cultivated ground, hedgebanks and waysides. Frequent in south-west England and the Channel Islands. Native of the Mediterranean region.

8. A. moly L.   Yellow Garlic
A. aureum Lam. nom. illegit.; *A. flavum* Salisb., non L.; *Cepa moly* (L.) Moench

Perennial herb smelling of garlic. *Bulb* up to 2.5 cm in diameter, subglobose; outer tunic chartaceous. *Stem* 12–35 cm, terete. *Leaves* (1–)2(–3), 20–30 cm $\times$ 15–35 mm, almost basal, linear-lanceolate to lanceolate, glabrous, glaucous, keeled below. *Spathe* 2-valved, persistent, shorter than pedicels. *Inflorescence* an umbel 4–7 cm in diameter, fastigiate or hemispherical, many-flowered, without bulbils or rarely with bulbils and a few flowers; pedicels 15–35 mm, unequal, ascending. *Perianth* stellate; segments 9–12 $\times$ 4–5 mm, yellow, with a greenish line outside, the outer elliptical, the inner oblanceolate, acute at apex. *Stamens* shorter than perianth segments; filaments 5–6 mm, yellow, simple. *Ovary* not prominently angled. *Capsule* covered by persistent, connivent perianth segments. *Flowers* 6–8. $2n = 14$.

(i) Var. moly
Inflorescence with flowers, without bulbils.

(ii) Var. bulbilliferum Rouy
Inflorescence with bulbils and a few flowers.

Introduced. Garden plant, escaping and naturalised on warm banks and hedgerows and in fields. Scattered in southern England and Jersey, rarely elsewhere. Native of France and Spain. Var. *moly* is more frequent than var. *bulbilliferum*.

Section 4. Briseis (Salisb.) Stearn

Bulbs subglobose, not rhizomatous. *Stem* triquetrous, flaccid after anthesis. *Leaves* almost basal, with short above-ground sheath, so strongly keeled as to be triquetrous. *Spathe* shorter than pedicels, 2-valved, persistent. *Ovary* with minute nectariferous pores; ovules 2 in each loculus. *Stigma* 3-lobed. *Seeds* angular, with a white elaiosome.

9. A. triquetrum L.   Three-cornered Garlic

Perennial herb smelling of garlic. *Bulb* 15 mm in diameter, subglobose; outer tunic membranous. *Stem* 10–45 cm, sharply triquetrous. *Leaves* usually 2–5, almost basal, 12–42 cm × 5–17 mm, linear, flat, scarcely keeled, with a short above-ground sheath. *Spathe* up to 2.5 cm, 2-valved, scarious, the valves lanceolate. *Inflorescence* an umbel with 3–15 flowers, usually one-sided, lax, without bulbils; pedicels up to 25 mm, longer than perianth. *Perianth* drooping, campanulate, never opening more than 45°; segments 10–18 × 2–5 mm, white, with a distinct longitudinal green stripe, lanceolate, acute at apex. *Stamens* 6–7 mm, included, shorter than perianth; filaments simple. *Capsules* 6–7 mm. *Seeds* angular, with a white elaiosome. *Flowers* 4–6. *Seeds* dispersed by ants. $2n = 18$.

Introduced. Weed of rough, waste and cultivated land, copses, hedgerows and roadsides. Common in south-west England and the Channel Islands; scattered elsewhere in England, Wales and Ireland north to Ayrshire. Native of the western Mediterranean region.

10. A. pendulinum Ten.   Italian Garlic
A. triquetrum var. *pendulinum* (Ten.) Regel

Perennial herb smelling of garlic. *Bulb* about 10 mm in diameter, subglobose; tunic membranous. *Stem* 15–30 cm, triquetrous, flaccid after anthesis. *Leaf* 1, up to about 30 cm × 3–8 mm, soon withering, almost basal, with short, above-ground sheath, so strongly keeled as to be triquetrous. *Spathe* shorter than pedicels, 2-valved, persistent. *Inflorescence* an umbel with 5–9 flowers, not secund, without bulbils; pedicels up to 40 mm, ascending at anthesis, later drooping. *Perianth* stellate, campanulate, opening more than 45° at first; segments 3–5 × 1.1–1.4 mm, white, with a longitudinal green stripe, lanceolate, acute at apex. *Stamens* simple, included; filaments about 4.5 mm. *Capsule* 4–6 mm. *Seeds* angular, with a white appendage. *Flowers* 4–6. $2n = 14$.

Introduced. Naturalised in neglected estates at Warley Place Gardens in Essex and at Enfield in Middlesex. Native of central Mediterranean region.

11. A. paradoxum (M. Bieb.) G. Don
                                        Few-flowered Garlic
Scilla paradoxa M. Bieb.

Perennial herb smelling of garlic. *Bulb* 5–10 mm in diameter, subglobose, not rhizomatous; tunic membranous. *Stem* 15–30 cm, triquetrous, triangular in section, flaccid after anthesis. *Leaf* 1, up to 30 cm × 5–15 mm, almost basal, linear, obtuse at apex, flat, scarcely keeled, with short, above-ground sheath. *Spathe* shorter than pedicels, 2-valved, persistent. *Inflorescence* and umbel often reduced to one flower in Europe, usually with small, green, subglobose bulbils and often then without flowers; pedicels 20–45 mm. *Perianth* campanulate; segments 10–12 × about 6 mm, white, with a faint longitudinal green stripe, oblong, obtuse at apex. *Stamens* simple, included; filaments about 5 mm. *Stigma* 3-lobed. *Capsule* rarely produced in Europe. *Seeds* angular, with a white appendage. *Flowers* 4–5. $2n = 16$.

The flowers of this species are often abnormal and

rarely produce seeds. D. M. Barling (1971) found 180 abnormal plants in 250 studied. It propagates, as a naturalised species, almost entirely by bulbils.

Introduced. Naturalised and locally abundant in woods, grassy places, rough ground, river banks and roadsides. Scattered throughout much of England and Scotland. Native of the Caucasus.

Section **5. Ophioscorodon** (Wallr.) Bubani

Bulbs subcylindrical, not rhizomatous. *Stem* sharply angled. *Leaves* basal, petiolate, with a broad resupinate lamina. *Spathe* shorter than pedicels, 2-valved, persistent. *Perianth* stellate. *Stamens* simple. *Ovary* with 2 ovules in each loculus. *Seeds* subglobose.

12. A. ursinum L.   Ramsons
Moly latifolium Gray; *Cepa ursina* (L.) Berh.

Perennial herb smelling of onions. *Bulb* about 4 × 1 cm, narrow, subcylindrical; outer tunic papery, with a few parallel fibres at base. *Stem* 10–50(–75) cm, yellowish-green, usually 2-angled and semi-cylindrical, sometimes 3-angled. *Leaves* 2–3, all basal; lamina 6–24 cm × 15–80 mm, dark green above, paler beneath, narrowly elliptical to narrowly ovate, acute at apex, entire, attenuate to the petiole and rounded or subcordate at the base; petiole up to 30 cm, yellowish-green. *Spathe* shorter than pedicels, 2-valved, persistent. *Inflorescence* an umbel 2.5–6.0 cm in diameter, with 6–28 flowers, lax; pedicels 10–25 mm, ascending. *Perianth* stellate; segments 7.0–12.0 × 2.0–2.5(–4.0) mm, white, lanceolate, acute at apex. *Stamens* simple; filaments 7–8 mm, white; anthers cream. *Stigma* simple. *Capsule* 3–4 mm. *Seeds* subglobose. *Flowers* 4–6. Pollinated by insects and selfed; protandrous. $2n = 14$.

Native. Woods and other damp shady places, sometimes forming large populations almost devoid of other species. Frequent, often abundant, over most of the British Isles, but only one site in the Channel Islands. Europe, from about 61° N in Norway, and central Russia, south to central Spain, Corsica, Sicily and the Balkan Peninsula; Caucasus; Turkey. Our plant is subsp. **ursinum** which is found in west and central Europe and north and central Italy. It is replaced further east by subsp. *ucrainicum* Kleopow & Oxner.

Section **6. Codonoprasum** Rchb.

Bulbs ovoid, not rhizomatous. *Stem* terete. *Leaves* sheathing up to two-thirds of the stem. *Spathe* 2-valved, the valves unequal, each with an ovate or lanceolate, usually strongly veined base, narrowed above into a tail-like appendage, longer than the pedicels. *Perianth* campanulate or cup-shaped, never stellate. *Stamens* simple. *Ovary* with or without minute, inconspicuous nectariferous pores; ovules 2 in each loculus. *Stigma* entire. *Seeds* angular.

13. A. oleraceum L.   Field Garlic
A. virens Lam.; *A. virescens* DC.; *Cepa oleracea* (L.) Bernh.

Perennial herb smelling of garlic. *Bulb* 10–15 mm in diameter, ovoid, not rhizomatous; outer tunics membra-

nous. *Stems* 25–100 cm, terete. *Leaves* 2–4, 15–30 cm ×
2.0–4.0 mm, linear and flat to filiform and terete, fistular
in lower part, channelled above, prominently ribbed
beneath, the veins usually scabrid with minute teeth;
sheathing the lower half or more of the stem. *Spathe*, 2-
valved; valves unequal, persistent, lanceolate at base,
contracted above into a long slender appendage longer
than the pedicels, the longer up to 20 cm. *Inflorescence* a
diffuse, lax umbel, 5- to 40-flowered, with few to many
bulbils, sometimes with bulbils only, the outermost
pedicels curving downwards at anthesis; pedicels 15–60
mm, unequal, slightly flattened or winged. *Perianth*
campanulate; segments 5–7 × 2.5–3.0 mm, whitish, variously tinged with green, pink or brown, the outer narrowly obovate, the inner oblong-elliptical,
obtuse-mucronate at apex. *Stamens* included, simple; filaments 4.5–5.5 mm, simple, connate at base into an
annulus about 1.5 mm high; *anthers* yellow or reddish.
Ovary at anthesis narrowly obovoid, about 4 times as
long as wide, rounded-truncate at apex. *Stigma* entire.
Capsule rarely produced. *Seeds* angular. *Flowers* 7–8.
Reproduced from bulb fragments, offsets and bulbils. 2*n*
= 32, 140.

(i) Var. **oleraceum**
Leaves 2–3 mm wide, semi-terete, usually hollow at least
below.

(ii) Var. **complanatum** Fr.
A. complanatum (Fr.) Boreau
Leaves 3–4 mm wide, flat.

Native. Dry grassy, often south-facing places on steep
slopes over chalk and oolite and Carboniferous limestone. It is most frequent on the banks of floodplain
meadows or open sandy banks in the middle reaches of
river systems. Var. *oleraceum* is scattered throughout
England and very scattered in Wales, Scotland and
Ireland. Var. *complanatum* is confined to the north. The
main centres of distribution are based on the river com-
plexes of the Vale of York, Trent, Ure, Severn and Avon.
The species occurs in Europe from Scandinavia and
northern Russia to north Spain, Corsica, central Italy,
Yugoslavia, Bulgaria and the Caucasus.

14. A. carinatum L.   Keeled Garlic
A. flexum Waldst. & Kit.; *A. violaceum* Willd.; *Cepa
carinata* (L.) Gray
Perennial herb smelling of garlic. *Bulb* about 10 mm in
diameter, ovoid, not rhizomatous; outer tunics membranous, sometimes breaking into longitudinal strips.
Stems 30–60 cm, terete, faintly ridged. *Leaves* 2–4,
10–20 cm × 1.0–2.5 mm, linear, slightly channelled
above, ribbed below with 3–5 prominent veins; sheathing the lower half to two-fifths of the stem. *Spathe* 2-
valved; valves unequal, persistent, lanceolate at base,
contracted above into a long slender appendage, the
longer up to 12 cm longer than the pedicels.
Inflorescence a diffuse umbel, the outer pedicels curving
downwards; pedicels 10–25 mm, unequal. *Perianth* cup-
shaped; segments 4–6 × 1.5–2.0 mm, purple, cymbiform, oblong-elliptical, obtuse at apex. *Stamens* simple,

long-exserted; filaments 6.5–9.0 mm; anthers purple;
pollen yellow. *Ovary* narrowly obovoid, narrowed at the
base, rounded at apex. *Stigma* entire. *Capsule* about 5
mm. *Seeds* angular. *Flowers* 8.

(a) Subsp. **carinatum**
Umbel with bulbils. *Flowers* 0–30. *Capsules* rarely produced. 2*n* = 16, 24, 26.

(b) Subsp. **pulchellum** Bonnier & Layens
A. cirrhosum Vandelli; *A. coloratum* Spreng.; *A.
pulchellum* G. Don nom. illegit.; *A. flavum* var.
purpurascens Mertens & Koch; *A. flavum* var.
capsuliferum Koch; *A. carinatum* var. *capsuliferum*
(Koch) Koch; *A. flavum* var. *pulchellum* Regel; *A.
carinatum* var. *pulchellum* Fiori
Flowers often numerous. *Capsule* with fertile seed abundantly produced. 2*n* = 16.

Introduced. Garden escape naturalised in rough
ground, grassy places and roadsides. Scattered throughout England, very scattered in Wales, Scotland and
Ireland. Scandinavia and Börnholm to east France,
Switzerland and the Balkan Peninsula. Both subspecies
seem to occur in the British Isles, but their distribution is
not known. Subsp. *carinatum* is native throughout the
range of the species except for the southern part of the
Balkan Peninsula. Subsp. *pulchellum* is native in southern Europe, westwards to south-east France.

Section 7. **Allium**

Bulbs ovoid or subglobose; not rhizomatous. *Stems*
usually terete. *Leaves* linear, flat or fistular, sheathing the
lower quarter or more of the stem. Spathe 1- or 2-
valved, usually beaked and caducous. *Perianth* cylindrical, campanulate or ovoid, with permanently connivent
segments. *Stamens* usually dimorphic, the outer 3
usually simple, rarely toothed or tricuspidate, the inner 3
always tricuspidate with a broad, flat, basal lamina
topped by a central anther-bearing cusp and 2 (rarely 4)
sterile, usually much elongated cusps. *Ovary* with distinct, nectariferous pores; ovules 2 in each loculus.
Stigma entire. *Seeds* angular.

15. A. sativum L.   Garlic
Perennial herb smelling of garlic. *Bulb* 30–60 mm in
diameter, depressed-ovoid, composed of 5–15(–60) bulblets; outer tunics membranous. *Stem* 25–100(–200) cm,
terete. *Leaves* 6–12, up to 60 cm × 5–25(–30) mm, linear,
flat, keeled, obtuse at apex, sheathing the lower half of
the stem. *Spathe* up to 25 cm, 1-valved, with a long beak,
caducous. *Inflorescence* an umbel 2.5–5.0 cm in diameter, usually with few flowers, which often abort and
wither in bud, and many bulbils; pedicels 10–20 mm,
unequal. *Perianth* cup-shaped; segments 3–5 mm, greenish-white, pink, or rarely white or purple, smooth, the
outer lanceolate and acute at apex, the inner ovate-
lanceolate. *Stamens* included or equalling the perianth,
dimorphic, the outer 3 filaments 6–8 mm, simple or tricuspidate, the inner 3 filaments with the basal lamina
broadly oblong, 1.5–2.0 mm wide and one-third as long
to about as long as the central cusp, lateral cusps 2 or 4,

much longer than the central cusp. *Ovary* with distinct nectariferous pores. *Stigma* entire. *Seeds* black, angular. *Flowers* 8–9. $2n = 14$.

(i) Var. sativum

Stem erect and straight.

(ii) Var. ophioscorodon (Link) Döll

A. controversum Schrad.; *A. ophioscordon* Link; *A. sativum* subsp. *ophioscordon* (Link) Holub
Stem before anthesis coiled at the top into one or two loops.

A. sativum is very variable and was probably derived from *A. longicuspis* Regel from central Asia. It is cultivated widely in southern Europe for its bulbs (garlic) and for flavouring.

Introduced. About 50 hectares are grown for food flavouring, mainly in the Isle of Wight. Naturalised on the shore at Port Dinllaen, Caernarvonshire and in a salt-marsh beside the River Lune at Lancaster. Also grown in gardens and casual where thrown out. In scattered localities in England and Wales. Origin unknown.

16. A. ampeloprasum L.　　　　　　Wild Leek

Perennial herb smelling of onions. *Bulbs* 20–60 mm in diameter, broadly ovoid or subglobose; outer tunic membranous; *bulblets* usually numerous, yellowish. *Stem* 45–180 cm, stout, terete. *Leaves* 4–10, up to 50 cm × 5–40 mm, linear, flat, channelled, with a scabrid margin, sheathing for the lower half or one-third of the stem. *Spathe* 1-valved, caducous. *Inflorescence* a dense, globose umbel 5–9 cm in diameter with up to 500 flowers, but with as few as 30 flowers in the variety with bulbils; pedicels 15–50 mm, unequal. *Perianth* cup-shaped or campanulate; segments 4.0–5.5 × 1.3–2.4 mm, white, pink or dark red, the outer mostly oblong-lanceolate, concave, subacute and mucronate at apex, the inner mostly narrowly ovate or spathulate, obtuse or rounded at apex, equalling or a little shorter than outer, both with large papillae especially on the keel. *Stamens* more or less exserted; outer 3 with filaments 4–6 mm, usually simple, the lower one-third oblong and 0.5–1.0 mm wide, contracted above, the inner 3 with the basal lamina oblong-elliptical and 1.5–2.5 mm wide, usually at least as wide as the perianth segments, nearly twice as long as the central cusps. *Stigma* entire. *Capsule* about 4 mm. *Seeds* angular. *Flowers* 7–8. $2n = 16, 32, 40, 48, 56, 80$.

(i) Var. ampeloprasum

A. ampeloprasum var. *holmense* Asch. & Graebn.
Umbel of dense flowers, without bulbils.

(ii) Var. bulbiferum Syme

Flowers and bulbils 6–8 mm, in a rather compact umbel.

(iii) Var. babingtonii (Borrer) Syme

A. babingtonii Borrer
Flowers and bulbils 8–15 mm, with a rather loose umbel, often with some pedicels bearing secondary heads.

Native. Rocky or sandy places and rough ground near the sea. The species occurs very locally in south-west

England, south and north-west Wales, north and central-west Ireland and in the Channel Islands. South and west Europe; Macaronesia. Var. *ampeloprasum* occurs in south-west England and Wales. It occurs in Europe throughout the range of the species. Var. *bulbiferum* occurs in the Channel Islands. Var. *babingtonii* is endemic to south-west England, Wales and west Ireland.

17. A. porrum L.　　　　　　　　　　Leek

A. ampeloprasum var. *porrum* (L.) Gray

Biennial herb smelling of onions, often grown as an annual. *Bulbs* simple, in the first year little wider than the thick neck, whitish. *Stem* up to 1 m, thick, solid, leafy in lower third. *Leaves* up to 1 m × 100 mm, bluish-green, long-linear, obtuse at apex, flat, keeled. *Spathe* 1-valved, persistent until flowering. *Inflorescence* a dense, many-flowered umbel 6–10 cm in diameter; pedicels 15–50 mm, unequal. *Perianth* cup-shaped; segments about 4 mm, pinkish, ovate or lanceolate, obtuse at apex. *Stamens* more or less exserted; outer 3 filaments with 2 outer cusps longer than the central one. *Style* shorter than perianth segments. *Capsule* about 4 mm. *Seeds* angular. *Flowers* 7–8. $2n = 32$.

Introduced. It was probably the Romans who first introduced the leek to these islands. Much grown as a vegetable. Important areas of production are the Thames Valley, Isle of Ely, Lancashire, Lincolnshire, West Norfolk and the Vale of Evesham. Casual where thrown out or a relic. Said to be a cultigen of *A. ampeloprasum*. The edible part is the elongated blanched bases of the leaves which are usually cooked and used in combination with other vegetables in soups and stews. The custom in Wales of wearing a leek on St David's Day is said to date back to AD 640 when King Cadwallader fought a victorious battle against the Saxons and the Welsh wore leeks in their hats to distinguish them from the enemy.

18. A. scorodoprasum L.　　　　　　Sand Leek

Perennial herb smelling of onions. *Bulbs* 10–20 mm in diameter, ovoid; outer tunics membranous, sometimes breaking into fibrous strips; bulblets reddish-black. *Stem* 20–90 cm, terete. *Leaves* 2–5, up to 27 cm × 20 mm, linear, narrowed at base, flat or channelled, solid, sheathing one-third to half of the stem. *Spathe* about 15 mm, shortly beaked, deciduous. *Inflorescence* an umbel 1–5 cm in diameter; pedicels up to 20 mm, unequal. *Perianth* ovoid; segments lilac to dark purple, the outer 4.0–6.5 × 1.5–2.5 mm, lanceolate or narrowly ovate, the inner 4–7 × 1.5–3.5 mm, narrowly oblong or ovate, subacute to obtuse at apex, papillose along the keel. *Stamens* included, usually papillose; outer 3 filaments 2.5–4.5 mm, simple, very narrowly triangular, the inner 3 with the basal lamina 3 or more times a long as the central cusp; anthers yellow. *Stigma* entire. *Capsule* about 5 mm. *Seeds* angular. *Flowers* 5–8. Reproduces by bulb offsets and by bulbils. $2n = 16, 24$.

Native or long-established alien. Dry grassland, scrub and open woodland on sandy soils. Local in Britain from Derbyshire and Lincolnshire north to

Aberdeenshire. Naturalised in south-west Ireland and rarely elsewhere. Europe from Scandinavia, Finland and central Russia to south-east France, central Italy and the Balkan peninsula. Our plant is subsp. **scorodoprasum,** which is mainly in north and central Europe, but extending locally southwards to Bulgaria and the Crimea. At least part of its distribution may be due to its former cultivation as a culinary plant.

19. A. sphaerocephalon L.   Round-headed Leek
A. descendens L.

Perennial herb smelling of onions. *Bulb* 20–30 × 20–25 mm, broadly ovoid or subglobose; tunic whitish, membranous, or the outermost subcoriaceous, yellowish-brown, coarsely lacerate at apex and base; bulblets usually present, about 10 × 5–8 mm, compressed, acuminate at apex, stipitate, concealed under the base of the leaf sheaths, pale or yellowish-brown and lustrous. *Stems* 20–90 cm, slender, terete, finally ridged. *Leaves* 2–6, sheathing the lower half or quarter of the stem up to 30 cm × 1–4 mm, fistulose, compressed, caniculate, glabrous, narrowly linear, tapering at the apex. *Spathe* 10–20 × 10–15 mm, persistent, 2- to 4-valved, shortly beaked, papyraceous, reflexed at anthesis, much shorter than the inflorescence. *Umbel* 2–4(–6) cm in diameter, dense, many-flowered, globose or broadly ovoid, without bulbils; pedicels 5–20(–30) mm. *Perianth* 4–6 × 2.5–3.0 mm, reddish-purple, ovoid, constricted towards the apex; segments subequal, 4.0–5.5 × 2.0–2.5 mm, ovate, concave, somewhat carinate, obtuse or mucronate-subacute at apex, smooth or minutely papillose externally especially along the keel. *Stamens* conspicuously exserted; outer filaments 5–6 × about 1 mm, simple, subulate, thinly papillose, the inner 3-cuspidate, with an oblong papillose basal part about 2.5 × about 1 mm, and a median antheriferous cusp about 2.5 mm, flanked by 2 filiform sterile cusps about 4–5 mm; anthers 1.0–1.2 mm, purplish, oblong. *Style* about 4 mm; stigma truncate. *Capsule* about 3.5 mm in diameter, pale yellowish-brown, broadly ovoid or subglobose. *Seeds* about 2.5 × 0.5 mm, black, angled, minutely and closely papillose-verruculose. *Flowers* 6–8. Pollinated by insects and selfed; protandrous. $2n = 16$.
    Native. Known only from two localities. First found on limestone rocks, St Vincent's Rocks, Bristol in 1847, and on sandy, waste ground by the sea at St Aubin's Bay, Jersey in 1836. Mediterranean region north to Belgium, central-west Germany and south-central Russia. Our plant is subsp. **sphaerocephalon,** which occurs throughout the range of the species.

20. A. vineale L.   Wild Onion
Perennial herb smelling of onions. *Bulbs* 10–20 mm, ovoid; outer tunics splitting into strips with parallel fibres; bulblets yellowish. *Stem* 30–120 cm, terete. *Leaves* 2–4, 15–60 cm × 1.4–4.0 mm, subcylindrical, fistular, sheathing the lower one-third to two-thirds of the stem. *Spathe* usually 3 cm or more, 1-valved, caducous, the beak as long as or a little longer than the base. *Inflorescence* an umbel 2–5 cm in diameter, subglobose,

ovoid or hemispherical, many-flowered and with no bulbils, with no flowers and many bulbils, or few-flowered and with several bulbils; pedicels 5–30 mm, unequal. *Perianth* campanulate; segments 2.0–4.5 × 1.2–1.5 mm, pink to dark red or greenish-white, the outer narrowly oblong-ovate, subacute or obtuse at apex, very concave, the inner narrowly oblong or almost narrowly obovate, rounded at apex, rarely narrowly oblong-ovate and subacute at apex, smooth. *Stamens* almost included to distinctly exserted; outer 3 filaments 3.5–4.0 mm, simple, the inner 3 with the basal lamina slightly or distinctly longer than the central cusp, the lateral cusps much longer than the central cusp; anthers yellow. *Capsule* 3.0–3.5 mm. *Flowers* 6–7. Pollinated by insects; strongly protandrous. $2n = 32, 40$.

(i) Var. **capsuliferum** Koch
A. rifoense Panov; *A. vineale* subsp. *capsuliferum* (Koch) Ceschon.
Umbels with flowers only.

(ii) Var. **vineale**
Umbels with flowers and bulbils.

(iii) Var. **compactum** (Thuill.) Boreau
A. compactum Thuill.
Umbels with bulbils only, compact.

    Native. Grassy places, rough ground, banks, waysides and as a field weed. Common in southern England, frequent to scattered in the rest of the British Isles, but absent from north Scotland. Most of Europe except the extreme north and central and east Russia; North Africa; Caucasus, Lebanon. Distribution of varieties unknown.

Section 8. Melanocrommyum Webb & Berth.
Bulbs ovoid; not rhizomatous. *Leaves* basal, with no above-ground sheath. *Stem* usually longer than leaves, terete. *Spathe* shorter than pedicels, becoming 2- to 4-fid, persistent. *Perianth* usually stellate, with segments ultimately deflexed. *Stamens* simple. *Ovary* with 4–8 ovules in each loculus. *Stigma* entire. *Seeds* angular.

21. A. nigrum L.   Broad-leaved Onion
Moly speciosum Moench nom. illegit.

Perennial herb, smelling of onions. *Bulb* 20–30 × 20–30 mm, ovoid; tunic membranous, the outermost greyish-brown, smooth or rugulose; bulblets sometimes numerous or represented by a subglobose, fleshy gemma enveloped in a cucullate, gemmiferous leaf. *Leaves* (2–)3–6, all basal, up to 50 × 9 cm, broadly linear or linear-lanceolate, fleshy, acuminate at apex, glabrous, with smooth or papillose, denticulate margins. *Stems* 60–90 cm × 5–10 mm, terete, striate, robust. *Spathe* up to 30 × 25 mm, persistent, papery, 2- to 4-valved, ultimately reflexed, the valves ovate-concave, shortly acute at apex. *Umbel* 5–10 cm in diameter, hemispherical, dense, many-flowered, usually without bulbils; pedicels 2–5 cm, glabrous, bluntly angular. *Perianth* 15–25 mm in diameter, widely cup-shaped; segments 10–13 × 3–4 mm, whitish or pinkish with a greenish or purplish, papillose midrib, oblong-elliptical, obtuse at apex. *Stamens* included; filaments 3–4(–5) mm, subulate, glabrous, fleshy, connate at the base into a narrow annulus; anthers about 3.5 × 1.4 mm, yellowish, oblong. *Style* about 1.5 mm; stigma very shortly and obscurely 3-lobed. *Ovary* about 4 mm in diameter, trigonous, subglobose, greenish or blackish. *Capsule* about 10 × 8 mm, coriaceous, trigonous, subglobose-ovoid. *Seeds* about 4 × 3 mm, angular, black, minutely verruculose. *Flowers* 4–6. $2n = 16$.

Introduced. Grown in gardens, escaping and long naturalised in rough ground. In a few places in southern England. Native of south Europe and the Mediterranean region.

27. Nectaroscordum Lindl.

Perennial herbs with bulbs, smelling of garlic when fresh. *Leaves* 3 or 4, basal, linear, strongly keeled, with a sheathing base. *Flowers* up to 30 in a loose umbel, sweetly scented, subtended by a deciduous spathe. *Perianth segments* free, whitish tinged with green and often pale pink or dull red outside. *Stamens* 6, free. *Ovary* semi-inferior, 3-celled with numerous ovules per cell. *Fruit* a 3-celled capsule. *Seeds* black, angled.

Two species in southern Europe and western Asia.

Stearn, W. T. (1955). *Allium bulgaricum* in *Bot. Mag.* **170**: t. 257.

1. N. siculum (Ucria) Lindl.   Honey Garlic
Allium dioscoridis auct.

Perennial herb smelling strongly of garlic. *Bulb* 15–30 mm in diameter, solitary, subglobose; outer tunic membranous. *Stem* 50–125 cm, terete, covered in lower one-third by an erect, sheathing leaf. *Leaves* 3–4, all basal, 30–60 × 1–5 cm, linear, strongly keeled, with a sheathing base. Spathe 6.0–7.5 cm, beaked, quickly deciduous. *Umbel* terminal, at first completely enclosed in a spathe, many-flowered, lax; pedicels 2–7 cm, unequal, curved downwards at anthesis, erect in fruit. *Perianth* broadly cup-shaped; segments whitish or cream, tinged pale pink or dull greenish-red, persistent, free, 3- to 7-veined, coriaceous, the outer 3 about 12 × 8 mm and oblong, the inner about 17 × 11 mm and pandurate, abruptly contracted into a cuneate base, slightly apiculate at apex. *Stamens* 6, free, included; filaments subulate; anthers greenish-yellow. *Ovary* semi-inferior, turbinate, 3-locular; ovules numerous. *Capsule* 5–7 mm, ovoid. *Seeds* numerous, black, compressed, angled. *Flowers* 4–6. $2n = 18$.

(a) Subsp. siculum
Allium siculum Ucria; *Trigonea sicula* (Ucria) Parl.
Perianth segments dull greenish-red.

(b) Subsp. bulgaricum (Janka) Stearn
N. bulgaricum Janka; *Allium bulgaricum* (Janka) Prodan; *Allium meliophilum* Janka
Perianth segments dull greenish-white tinged with pale pink outside, with green midvein, red inside near the base.

Introduced. Probably both subspecies are amongst those which escape from gardens and become natu-

ralised in rough ground. Of scattered occurrence in southern England. Subsp. *siculum* is native of the west Mediterranean region, and subsp. *bulgaricum* of southeast Europe, from Turkey to the Crimea.

28. Nothoscordum Kunth nom. conserv.

Perennial herbs with bulbs, not smelling of garlic or onion. *Leaves* basal, linear, not or scarcely keeled, with a sheathing base. *Flowers* 6–15, in a loose umbel, sweetly scented, subtended by a spathe of 2 papery bracts. *Perianth segments* shortly fused at base, greenish-white with a pinkish midrib. *Stamens* 6, free. *Style* arising from top of ovary. *Ovary* superior, 3-celled, with numerous ovules in each cell. *Fruit* a membranous capsule. *Seeds* black, angular.

About 35 species, all American.

Ravena, P. (1991). *Nothoscordum gracile* and *N. borbonicum* (Alliaceae). *Taxon* **40**: 485–487.

1. N. borbonicum Kunth   Honey-bells
N. inodorum auct.; *N. gracile* auct.; *N. fragrans* auct.; *Allium gracile* auct.; *Allium fragrans* auct.

Perennial herb not smelling of onions. *Bulb* about 20 × 15 mm, ovoid; tunics pale brown, papery; bulblets less than 5 mm, usually numerous, ovoid. *Stems* 30–60 cm, erect, slender, striate, glabrous. *Leaves* 4–6, 20–40 cm × 2–15 mm, all basal, linear, glabrous, striate. *Spathe* membranous, pallid, 2-valved, the valves 8–10 × 5–7 mm, acute or acuminate at apex. *Umbel* 1–4 cm, lax, 6- to 15-flowered; pedicels 10–40 mm, unequal, erect, glabrous, striate. *Perianth* 10–15 × 10–15 mm, cup-shaped, base shortly tubular-infundibuliform, fragrant and not smelling of onions; segments 5–8 × 2.5–3.0 mm, whitish with a brownish or pinkish midrib, oblong, obtuse or subacute at apex. *Stamens* included; filaments 5–10 mm, about 1.5 mm wide at base, linear-lanceolate, contiguous in the lower half or less; anthers about 1 mm, greenish-yellow, oblong. *Style* about 6 mm; stigma subcapitate. *Ovary* about 3.0 × 2.5 mm, trigonous-obovoid. *Capsule* 6–7 × about 5 mm, obovoid, valves separating widely on dehiscence, papery, pale brown. *Seeds* about 3 × 2 mm, subovoid, bluntly angular and shortly rostrate, shining black, rugulose-cerebriform. *Flowers* 4–6, opening mid-afternoon to late evening. $2n = 18$.

Introduced. Grown in gardens from which it escapes and becomes naturalised in rough ground. Well established about Mont Cambrai and Rue da Haut in Jersey. Scattered over south and south-west England and the Channel Islands. Native of South America, naturalised in many parts of the Old World.

29. Agapanthus L'Hér. nom. conserv.

Perennial herbs with a short tuber-like rhizome, not smelling of garlic or onions, forming a dense clump. *Leaves* basal, oblong-linear, scarcely keeled. *Flowers* few to many in an umbel, slightly zygomorphic. *Perianth segments* fused in lower half, usually bright blue, very rarely white. *Stamens* 6, inserted on the perianth tube; anthers dorsifixed, introrse. *Ovary* superior, 3-celled, with numerous ovules in each cell. *Fruit* a loculicidal capsule. *Seeds* numerous, black, flat and winged.

Ten species were recognised in Leighton's monograph, but this is a very conservative approach to wild populations where many more taxa could be recognised. In cultivation they all appear to be interfertile. Native of South Africa.

Leighton, F. M. (1965). The Genus *Agapanthus* L'Héritier. *Jour. South African Bot.* suppl. volume **4**.

1. A. praecox Willd.   African Lily

Perennial herb forming a dense clump, with a short tuber-like rhizome. *Stems* up to 1 m, terete, pale green. *Leaves* up to 20 on each shoot, 20–70 × 1.5–5.5 cm, bright green, long-linear, gradually narrowed to an obtuse or more or less acute apex, entire, evergreen, soft and arcuate. *Inflorescence* an umbel with many flowers; pedicels 4–12 cm. *Flowers* 35–40 mm in diameter. *Perianth* 30–50 mm, pale to medium blue, rarely white in cultivation, the tube about half the length of the perianth, the segments spreading and often crisped on the margins, the outer 6–8 mm, wide and lanceolate, the inner 9–11 mm wide, oblanceolate or ovate-lanceolate and often emarginate at apex. *Stamens* 6, about as long as the perianth; filaments mauve, anthers dark mauve; exserted. *Style* mauve, exserted. *Capsule* 22–25 mm, ellipsoid but parallel-sided for much of its length. *Seeds* black, flat and winged. *Flowers* 12–2. 2*n* = 32.

Introduced. Cultivated for ornament and naturalised on sandy soil by the sea on Tresco in the Isles of Scilly; also in the Channel Islands. Native of South Africa. Our plant is subsp. **orientalis** (F. M. Leight.) F. M. Leight. (*A. orientalis* F. M. Leight.; *A. umbellatus* var. *maximus* Edwards) which is native to the Cape Province and Natal.

30. Tristagma Poepp.

Ipheion Raf.

Perennial herbs with a bulb, smelling of garlic if bruised when fresh. *Leaves* basal, linear, slightly keeled. *Flowers* solitary, erect, actinomorphic, terminal, sweetly scented, subtended by a spathe of 2 partly united bracts. *Perianth segments* fused in lower half, usually pale bluish-violet with a dark midrib outside. *Stamens* 6, within the perianth tube. *Ovary* superior, with numerous ovules in each cell. *Fruit* a many-seeded capsule.

Ten or more species in temperate South America.

Turrill, W. B. *Ipheion uniflorum* in *Bot. Mag.* **169**: t. 185.

1. T. uniflorum (Lindl.) Traub   Spring Starflower

Triteleia uniflora Lindl.; *Milla uniflora* Graham; *Brodiaea uniflora* (Lindl.) Engl.; *Ipheion uniflorum* (Graham) Raf.; *Leucocoryne uniflora* (Lindl.) Greene; *Beauverdia uniflora* (Lindl.) Herter

Perennial herb with a bulb and fleshy roots, smelling of garlic if bruised when fresh. *Stems* up to 35 cm, pale green, terete. *Leaves* 20–25 cm × 4–8 mm, bright yellowish-green, linear, obtuse at apex, entire, somewhat keeled and channelled, flaccid. *Spathe* 25–30 mm, scarious, some distance below the flower. *Flowers* solitary, termi-

nal, with a sweet smell, 30–43 mm in diameter; pedicels 30–50 mm, not articulated. *Perianth* 25–35 mm; tube 10–16 mm, dull purple tinged with green, narrow; segments 14–20 × 7–12 mm, white or violet-blue, the abaxial surface with a brownish-purple stripe which in the outer segments is tinged with green, patent, imbricate, lanceolate to more or less ovate, rounded-obtuse to acute at apex. *Stamens* 6; 3 inserted near the base of the perianth, 3 higher up; filaments 3–4 mm, green; anthers orange. *Stigma* pale green, capitate, obscurely 3-lobed, just visible at the mouth of the tube. *Capsule* with many seeds. *Flowers* 4–5. 2*n* = 12.

Introduced. Cultivated for ornament and increasingly becoming a weed of cultivated and waste ground in some areas. Naturalised in Cornwall, Isles of Scilly and the Channel Islands. Native of Uruguay and warm temperate Argentina.

Subfamily 7. Amaryllidoideae Pax

Perennial herbs with bulbs. *Leaves* all basal, linear to narrowly elliptical or linear-oblong. *Inflorescence* of one flower or a terminal umbel, usually with a scarious spathe at base. *Perianth segments* free or fused at the base, white to yellow or orange, rarely pink, sometimes with a funnel- or collar-shaped corona within the rows of perianth segments. *Style* 1, with simple or slightly 3-lobed stigma. *Ovary* inferior. *Fruit* a capsule dehiscing irregularly or along the centre of the ovary cells, often slightly succulent.

31. Amaryllis L. nom. conserv.

Perennial herbs with a bulb. *Leaves* appearing after the flowers, all basal, linear-oblong. *Flowers* in an umbel, erecto-patent, trumpet-shaped, slightly zygomorphic, with a corona. *Perianth segments* fused at base into a short tube, pink, or purplish-pink, often white towards the base, rarely all white, all more or less similar. *Stamens* 6, deflexed then curving upwards towards the apex. *Ovary* inferior, 3-celled, with few ovules in each cell. *Fruit* a loculicidal capsule, dehiscing irregularly.

A single species in South Africa.

Baker, J. G. (1888). *Handbook of the Amaryllidae*. London.
Goldblatt, P. (1984). Proposal to conserve 1176 *Amaryllis* and typification of *A. belladonna* (Amaryllidaceae). *Taxon* **33**: 511–516.

1. A. belladonna L.   Jersey Lily

Glabrous *perennial herb* with a bulb about 10 cm in diameter. *Stems* up to 60 cm, stout, solid, often purplish. *Leaves* appearing after the flowers; 30–45 × 1.5–3.0 cm, linear-oblong, rounded at apex, narrowed to the base, entire. *Spathe* up to 8 cm, 2-valved. *Inflorescence* an umbel of 4–8(–12) flowers; each flower subtended by a linear bracteole; pedicels 2–4 cm. *Flowers* trumpet-shaped, fragrant, erecto-patent. *Perianth* 5–10 cm, purplish-pink or pink, often whitish towards the base or rarely all white; tube 10–15 mm; segments 6, oblanceolate, acute at apex, slightly clawed, the 3 outer each with a small hairy inwardly pointing appendage. *Stamens* 6, shorter than the perianth; filaments much longer than anthers. *Style* deflexed and then curving upwards; stigma capitate. *Capsule* subglobose. *Seeds* rather few. *Flowers* 8–10. $2n = 22$.

A. belladona can be confused with *Nerine sarniensis* (L.) Herb. (Guernsey Lily), also from South Africa, and grown in gardens in the Channel Islands, but *Nerine* has much more open flowers with well-exposed stamens and perianth tube less than 5 mm.

Introduced. Grown for ornament in the Isles of Scilly and the Channel Islands, especially Jersey, where it is a frequent relic in old fields, hedges, rough ground and sandy places; recently recorded from Poldhu, Cornwall. Native of South Africa.

32. Sternbergia Waldst. & Kit.

Perennial herbs with a bulb, the tunics membranous. *Leaves* basal, narrowly lanceolate, appearing with or just before the flowers in autumn. *Flowers* solitary, erect, actinomorphic, without a corona. *Perianth segments* fused at base into a narrow tube, yellow. *Stamens* 6, in two unequal whorls, borne at the top of the perianth tube. *Ovary* inferior, 3-celled. *Fruit* subglobose, intermediate between a capsule and a berry.

About 8 species centred on Turkey, extending west to Spain, east to Kashmir and south to Israel.

1. S. lutea (L.) Ker Gawler ex Spreng.   Winter Daffodil
Amaryllis lutea L.; *Oporanthus lutea* (L.) Herb.; *S. aurantiaca* Dinsm.

Glabrous *perennial herb*; bulb 2–4 cm in diameter, the tunic membranous. *Stems* 2.5–10.0(–20) cm. *Leaves* 4–6, appearing with or just before the flowers, 4–15 cm × 2–15 mm, usually bright shining green, long-linear, obtuse at apex, entire to minutely crenulate, slightly channelled above, keeled beneath. *Spathe* 3–6 cm. *Flowers* golden yellow, crocus-like. *Perianth tube* 5–20 mm; lobes 30–55 × 4–20 mm, oblanceolate to obovate, rounded at apex. *Stamens* 6; filaments 15–35 mm, yellow; anthers yellow. *Style* 1, yellow; stigma capitate. *Fruit* subglobose. *Flowers* 8–10. $2n = 22$.

Introduced. Naturalised on grassy slopes by the sea at Gory Castle in Jersey since before 1919; formerly in Guernsey. Native in south Europe and west Asia from Spain to Iran and Russia.

(**Crinum × powellii** Baker is recorded as a garden escape in Jersey and has persisted for 30 years in a derelict garden at Farnham, near Saxmundham in Suffolk.)

33. Leucojum L.

Perennial herbs with bulbs. *Leaves* basal, linear and strap-like. *Spathe* of 1 or 2 bracts. *Flowers* solitary or few in an umbel, white, campanulate, pendulous, actinomorphic, without a corona or tube. *Perianth segments* free, all more or less alike. *Stamens* 6; anthers blunt at tip, opening by pores. *Ovary* inferior, 3-celled. *Capsule* erect, more or less pyriform. *Seeds* numerous.

About 10 species in south, west and central Europe, North Africa and south-west Asia.

Farrell, L. (1979). The distribution of *Leucojum aestivum* L. in the British Isles. *Watsonia* **12**: 325–332.
Gilham, B. (1993). *The Wiltshire flora*. Newbury.
Knowles, M. C. & Phillips, R. A. (1910). On the claim of the snowflake (*Leucojum aestivum*) to be native in Ireland. *Proc. Roy. Iris Acad.* **28** Sect. B. **8**: 387–399.
Stern, F. C. (1956). *Snowdrops and snowflakes*. London.

1. Flowers 1(–2); perianth segments 15–25 mm   **1. vernum**
1. Flower (1–)2–7; perianth segments 10–22 mm   2.
2. Leaves 7–20 mm wide; umbel of (2–)3–5(–7) flowers; spathe 7–11 mm wide; perianth segments 13–22 mm
                    2(a). aestivum subsp. **aestivum**
2. Leaves 5–15 mm wide; umbel of (1–)2–4 flowers; spathe 4–7 mm wide; perianth segments 10–17 mm
                    2(b). aestivum subsp. **pulchellum**

1. L. vernum L.   Spring Snowflake
Perennial herb. Bulb 15–30 mm in diameter, subglobose; tunic brown, thin. *Leaves* 3–5, 8–30 cm × 3–16 mm, deep green, linear, obtuse at apex, glabrous. *Scape* 15–30 cm, slightly hollow, narrowly 2-winged. *Sheath* up to 7 cm, membranous. *Spathe* 27–40 mm, single, membranous, convolute at the base. *Flowers* usually solitary, rarely 2, up to 40 mm in diameter; pedicels 25-40 mm. *Perianth segments* 15–25 × 8–12 mm, white, with a green or yellow blotch near the tip on the outside of each segment, broadly oblong to obovate, bluntly acuminate at apex. *Stamens* 6; filaments about 2 mm, white; anthers pale yellow. *Ovary* green, subglobose. Style white, the apex tipped with green markings; stigma rostrate. *Capsule* pyriform. *Seeds* about 7 mm, whitish, with a yellowish-white, wrinkled appendage. *Flowers* 2–4. Pollinated by bees and Lepidoptera; homogamous. *Seeds* distributed by ants. $2n = 22$.

Possibly native. Damp scrub and stream banks. Single sites in Somerset and Dorset. Elsewhere grown in gardens and occasionally escaping. Europe, from Belgium and north France to north Spain, north Italy and central Yugoslavia.

2. L. aestivum L.   Summer Snowflake
Nivaria monodelpha Medicus; *Nivaria aestivalis* (L.) Moench

Plant foetid. *Bulb* 25–40 mm in diameter, ovoid, in clumps; tunic brown. *Leaves* 30–50 cm × 5–15 mm, bright green, linear, obtuse at apex, glabrous. *Scape* 30–60 cm, deep green, triangular, stout, hollow. *Sheath* 30–60 mm, membranous. *Spathe* 30–65 × 4–11 mm, single, lanceolate, obtuse at apex, membranous. *Flowers* (1–)2–5(–7), up to 34 mm in diameter; pedicels up to 60 mm, slender. *Perianth segments* 10–22 × 5–8 mm, white with a green patch just below the apex on the outside, obovate or oblong-obovate, obtuse or bluntly acuminate at apex. *Stamens* 6; filaments 4–5 mm, white; anthers yellow. *Ovary* green. *Style* white with a green stigma overtopping the stamens. *Capsule* pyriform. *Seeds* black, without an appendage. *Flowers* 4–5. Pollinated by bees; homogamous. *Seeds* distributed by water.

(a) Subsp. **aestivum**
Robust. Leaves (7–)12–20 mm wide. *Wings of scape* with hyaline, remotely denticulate margins. *Umbel* of

(2–)3–6(–7) flowers. *Spathe* 7–11 mm wide. *Perianth segments* 13–22 × 8–10 mm. Flowering in late spring. $2n =$ 22, 24.

(b) Subsp. **pulchellum** (Salisb.) Briq.
L. pulchellum Salisb.
Less robust. Leaves 5–15 mm wide. *Wings* of scape entire. *Umbel* of (1–)2–4(–5) flowers. *Spathe* 4–7 mm wide. *Perianth segments* 10–17 × 6–7 mm. Flowering much earlier than subsp. *aestivum.* $2n = 22$.

Native and introduced. Subsp. *aestivum* is native in wet meadows and willow thickets in over 30 localities in Wiltshire and Dorset, and in the valley of the River Thames in Berkshire, Oxfordshire and Buckinghamshire. It was formerly also in Devon, Hampshire, Kent and Middlesex. In Ireland it is scattered from Wexford and Cork to Antrim and Fermanagh. It is widely grown in gardens where it spreads by seeds, and also escapes and becomes naturalised. West, central and south Europe, eastwards to Turkey and the Caucasus. Subsp. *pulchellum* is widely grown in gardens from which it often escapes. It is native of the west Mediterranean region.

34. Galanthus L.

Perennial herbs with bulbs. *Leaves* 2, basal; linear and strap-like or oblanceolate, enclosed in a tubular, membranous sheath at the base. *Spathe* of 2 fused bracts. *Flowers* solitary, nodding, actinomorphic, often double. *Perianth segments* in 2 whorls of 3, free, the inner whorl smaller and blunter; without a corona. *Stamens* 6, inserted at the base of the perianth, shorter than the inner segments; anthers basifixed, opening by terminal pores. *Style* slender, exceeding the anthers; stigma capitate. *Ovary* inferior, 3-celled. *Capsule* ellipsoid, or almost spherical, opening by 3 flaps; seeds with an appendage.

About 12 species from Europe, Turkey, Iran and Caucasus.

The naturalised British plants are much underrecorded and should particularly be looked for in cemeteries where they are planted on graves, and escape into the surrounding grassland and often adjacent areas. They are in almost all Cambridgeshire cemeteries where several taxa are involved. They also occur in most Cardiganshire cemeteries, but there they are nearly all *G. nivalis.*

Stern, F. C. (1956). *Snowdrops and snowflakes.* London.
Webb, D. A. (1978). The European species of *Galanthus* L. *Bot. Jour. Linn. Soc.* **76**: 307–313.

1. Flower double with numerous perianth segments which grade from the outer to the inner or are unequal
1(2). nivalis forma **pleniflorus**
1. Flowers single, with 3 outer perianth segments and 3 inner ones   2.
2. Leaves with margins folded back or under especially when young   3.
2. Leaves flat, or with margins convolute, especially when young   5.

3. Inner perianth segments with green patches at both base and apex   **3(b). plicatus** subsp. **byzantinus**
3. Inner perianth segments with a green patch only at apex   4.
4. Leaves glaucous
3(a,i). plicatus subsp. **plicatus** var. **plicatus**
4. Leaves bright yellowish-green sometimes with a dull glaucous stripe down the centre
3(a,ii). plicatus subsp. **plicatus** var. **viridifolius**
5. Leaves flat, more or less linear, less than 10 mm wide
1(1). nivalis forma **nivalis**
5. Leaves inrolled at least when young, oblanceolate, at least one more than 10 mm wide   6.
6. Inner perianth segments with green patches at both apex and base   **2(i). elwesii** var. **elwesii**
6. Inner perianth segments with a green patch at the apex only   7.
7. Leaves glaucous   **2(ii). elwesii** var. **monostictus**
7. Leaves bright shining green   **4. woronowii**

1. G. nivalis L.   Snowdrop

Perennial herb. Bulb 15–27 × 10–24 mm, ovoid to shortly oblong. *Leaves* 3–30 × 4–10 mm, glaucous, linear, flat, obtuse at apex, entire, glabrous. *Scape* 8–25 cm, compressed, glaucous, glabrous. *Sheath* up to 6 cm. *Spathe* bifid, green in the middle with broad scarious margins, glabrous. *Flowers* up to 40 mm in diameter, solitary, often double, nodding, faintly sweet-scented; pedicel slender. *Outer perianth segments* 12–28 × 6–10 mm, white, elliptical, oblanceolate, or narrowly obovate, concave, rounded at apex, shortly clawed. *Inner perianth segments* 4–11 × 4–7 mm, white with a horseshoe-shaped, green patch near the apex on the outer surface and 6–7 green lines on the inner surface, oblanceolate or obovate, emarginate at apex with rounded lobes, attenuate at base. *Stamens* 6; filaments 1–2 mm, white; anthers orange or yellow. *Ovary* dark green. *Style* about 9 mm, slender. *Capsule* 12–14 mm, spherical or almost so. *Seeds* few, white, with an appendage. *Flowers* 2–3. Pollinated by bees; homogamous. *Fruits* 5–6. $2n = 24$.

(1) Forma **nivalis**
Flowers single, with 3 outer perianth segments and 3 inner ones.

(2) Forma **pleniflorus** P. D. Sell
Flowers double, with numerous perianth segments which grade from the outer to the inner or are of various lengths.

Probably usually introduced. Local in damp woods and by streams and very commonly cultivated and planted in cemeteries, often escaping and becoming naturalised. From Cornwall and Kent to Dunbarton and Moray; a rare escape in Ireland. From north-central France and White Russia southwards to the Pyrenees, Sicily and south Greece; north Syria, Asia Minor, Caucasus and south-east Russia; naturalised further north. Forma *pleniflorus* is widely grown in gardens and cemeteries and often escapes.

× plicatus

This hybrid has narrower leaves and shows partial folding back of the leaf margin, but it is usually smaller in all its parts than *G. plicatus*.

It is frequently found naturalised in England and Wales with or without one or both of its parents. It is widespread in gardens.

2. G. elwesii Hook. fil.   Greater Snowdrop

Perennial herb. Bulb up to about 27 × 22 mm, more or less globose, fleshy. *Leaves* 6–15(–25) cm × 6–30 mm, glaucous, broadly linear or oblanceolate, convolute, rounded and hooded at the front of the apex, glabrous. *Scape* up to 30 cm, glaucous, glabrous. *Sheath* up to 80 mm, membranous. *Spathe* 25–40 × 5–6 mm, convolute, green in the centre of each bract with a membranous border. *Flowers* 40–50 mm in diameter, solitary, nodding, more or less sweet-smelling; pedicels up to 20 mm, slender. *Outer perianth segments* 15–30 × 10–14 mm, white, broadly obovate, saucer-like, obtuse at apex, with a short claw at base. *Inner perianth segments* 8–16 × 4–7 mm, white with green patches at the base and apex or only at the apex on the outside, the two areas sometimes joined by a narrow green line or coalescing, inside with green stripes and a white margin, oblong, emarginate at apex with rounded lobes, cuneate at base. *Stamens* 6; filaments 1–2 mm, white; anthers yellow, apiculate at apex. *Ovary* green. *Style* about 9 mm, slender. *Capsule* 10–20 × 10–20 mm, broadly ellipsoid to spherical. *Seeds* with an appendage. *Flowers* 12–3, earlier than the other species in Britain. 2n = 24, 48.

(i) Var. elwesii

Inner perianth segments white with green patches at the base and apex outside, the two patches sometimes coalesced. 2n = 24, 48.

(ii) Var. monostictus P. D. Sell

G. caucasicus auct.
Inner perianth segments white with one green blotch outside at the apex.

Introduced. Both varieties are grown in gardens and naturalised in woods, damp grassland and cemeteries. Local in southern and eastern England. Native of south-east Europe and the Caucasus.

× nivalis

This hybrid has glaucous leaves, and is intermediate in leaf convolution and amount of green on the outer side of the inner perianth segments.

Rarely occurs where both parents are naturalised in south-east and east England.

× plicatus

This hybrid looks like *G. elwesii* subsp. *elwesii* but has the leaf margins curved under, slightly hooded leaf apices at the front and variable amounts of green on the inner perianth segments.

Little Abington cemetery, Cambridgeshire, mixed with subsp. *elwesii* and subsp. *plicatus*.

3. G. plicatus M. Bieb.   Pleated Snowdrop

Perennial herb. Bulb 20–25 × 20–29 mm, fleshy, more or less globose. *Leaves* 5–15(–30) cm × 4–25 mm, glaucous or bright yellowish-green sometimes with a glaucous midrib, the edges folded back and hooded at the back of the leaf, linear to slightly oblanceolate or narrowly elliptical, obtuse at apex, glabrous. *Scapes* up to 25 cm, slightly glaucous, glabrous. *Sheath* up to 66 mm, membranous. *Spathe* up to 50 × 8 mm, convolute, membranous with 2 green lines. *Flowers* 30–50 mm in diameter, solitary, nodding, sweet-smelling; pedicels up to 30 mm, slender. *Outer perianth segments* 15–30 × 5–14 mm, white, oblong or narrowly elliptical, rounded at apex, with a short claw at base. *Inner perianth segments* 7–12 × 5–7 mm, white outside with a green blotch at the apex, and sometimes also at the base, inside with green stripes and a white margin, oblanceolate, emarginate at apex with rounded lobes, cuneate at base. *Stamens* 6; filaments 1–2 mm, white; anthers yellow, apiculate at apex. *Ovary* green. *Style* about 9 mm, slender. *Capsule* up to 16 × 12 mm, ellipsoid to almost spherical. *Seeds* with an appendage. *Flowers* 2–3.

(a) Subsp. plicatus

G. latifolius Salisb. nom. illegit.
Inner perianth segments with a green patch on the outer side at the apex only. 2n = 24.

(i) Var. plicatus

Leaves glaucous.

(ii) Var. viridifolius P. D. Sell

Leaves bright yellowish-green, sometimes with a glaucous midrib.

(b) Subsp. byzantinus (Baker) D. A. Webb

Leaves glaucous. *Inner perianth segments* with green patches on the outer side at both the apex and the base. 2n = 24.

Introduced. All three taxa are grown in gardens and are naturalised in a number of places in woods and cemeteries in southern and eastern England. Subsp. *plicatus* is native of Romania and the Crimea, and subsp. *byzantinus* occurs in Turkey. Both varieties are said to occur in the Crimea.

4. G. woronowii Losinsk.   Woronow's Snowdrop

G. latifolius Rupr., non Salisb.; *G. ikariae* auct.
Perennial herb. Bulb 18–35 × 16–28 mm, ovoid. *Leaves* 5–16 cm × 5–30 mm, bright or deep green, linear to narrowly oblanceolate, convolute, obtuse and recurved at apex, glabrous. *Scapes* 11–25 cm. *Flowers* 30–50 mm in diameter, solitary. *Outer perianth segments* 17–26 × 6–11 mm, white, oblong-elliptical, rounded at apex. *Inner perianth segments* about 9 × 5 mm, the outside white with a green patch at the apex which often extends beyond the middle, the inside white with green stripes, narrowly obovate, not flared at apex, cuneate at base. *Stamens* 6; filaments 0.5–1.5 mm; anthers 5–6 mm. *Ovary* green. *Capsule* spherical or almost so. *Seeds* with an appendage. *Flowers* 2–4.

Introduced. Grown in gardens and cemeteries and sometimes escaping onto grassland and waste places. Several Cambridgeshire cemeteries, a beach in Swanage, Dorset, a churchyard at Liss, Hampshire, woodland on Reigate Heath, Surrey and Eaglefield Park, Berkshire. Native of the western Transcaucasus.

35. Narcissus L.

Perennial herbs with bulbs. *Leaves* several with each bulb, the basal flat and broad to almost cylindrical. *Spathe* a single scarious bract. *Flowers* solitary or few in an umbel, pendulous to erecto-patent, actinomorphic, with a corona; perianth segments and corona fused to form a hypanthial tube between the base of the perianth segments and the apex of the ovary. *Perianth segments* 6, white to yellow. *Corona* more or less conspicuous, white to yellow or orange, free from the stamens. *Stamens* 6, in 2 whorls. *Ovary* inferior, 3-celled. *Capsule* ellipsoid to almost spherical. *Seeds* many.

About 50 species with a vast number of subspecies, varieties, cultivars and hybrids. Most originate in Europe, especially France, Spain and Portugal, though a few are natives of North Africa and the eastern Mediterranean area. They are extremely popular garden plants and are grown as crops for both flowers and bulbs. A thorough study of the genus in the British Isles would make a much commoner distribution of many of the species.

This genus, perhaps more than any other, exhibits the taxonomic difficulties which arise from long-established cultivation, hybridisation and selection. They are grown for ornament, for naturalisation on banks and slopes and as important cut-flower and bulb crops. In the last 30 years they have appeared more and more on road-sides, in waste places, by tracks and in fields. Some have clearly been planted. Others have appeared where people have thrown away garden rubbish and yet others remain as scattered plants in fields where they have been grown as crops. When the bulb crops are carted from the fields bulbs tend to slip off the vehicle and become estab-lished by tracks, margins of fields and waste land. Plants cultivated rarely match up to wild plants, but often have the same main characters. Many of the species have cul-tivated types anyway. If precise naming is required one can only run them down to one of the species or hybrids which follow and then match them up with a cultivar name which can be appended after the scientific name. This will not tell you their origin, but will enable you to apply names to those plants which look alike and they can be placed next to a cultivar they resemble. No point can been seen in putting them in the Divisions of the *International Daffodil Check-list* as the cultivars are not attached to any scientific name.

Blanchard, J. W. (1990). *Narcissus: A guide to wild daffodils.* Woking.
Bowles, E. A. (1934). *A hand-book of Narcissus.* London.
Burbidge, F. W. (1875). *The Narcissus: its history and culture.* London.
Fernandes, A. (1951). Sur la phylogenie des espèces du genre *Narcissus* L. *Bol. Soc. Brot.* ser. 2, **25**: 113–190.
Fernandes, A. (1968). Key to the identification of native and naturalised taxa of the genus *Narcissus* L. *Daffodil and tulip year book* 37–66.
Jefferson-Brown, M. J. (1969). *Daffodils and Narcissi.* London.
McClintock, D. (1975). *The wild flowers of Guernsey.* London.
Meyer, F. G. (1966). Narcissus species and wild hybrids. *Amer. Hort. Mag.* **45**: 47–76.
Pugsley, H. W. (1915). *Narcissus poeticus* and its allies. *Jour. Bot. (London)* **53** Suppl. 2.
Pugsley, H. W. (1933). A monograph of *Narcissus*, subgenus *Ajax. Jour. Hort. Soc.* **58**: 17–93
Pugsley, H. W. (1939). Notes on *Narcissi. Jour. Bot. (London)* **77**: 333–337.

1. Perianth-tube obconical, that is with a narrow base widening evenly to the insertion of the segments and funnel-shaped, narrowly so in some hybrids; stamens in a single whorl, the filaments as long as or longer than the anthers   2.
1. Perianth-tube cylindrical, parallel-sided for most of its length or widening a little towards the insertion of the segments; stamens in 2 whorls, filaments of lower shorter than the anthers   24.
2. Flowers bilaterally symmetrical; stamens deflected downwards, the filaments then curving upwards towards their apices; corona funnel-shaped   **6. bulbocodium**
2. Flowers radially symmetrical; stamens not as above; corona flattish, bell-shaped or cylindrical   3.
3. Perianth tube narrowly funnel-shaped, 3–4 times as long as wide   4.
3. Perianth tube broadly funnel-shaped, up to 1.5 times as long as wide   7.
4. Perianth segments white   **22. × boutigyanus**
4. Perianth segments yellow   5.
5. Corona yellow   6.
5. Corona deep golden or orange-yellow   **21. × incomparabilis**
6. Flowers 1–4; perianth segments 15–25 mm   **19. × odorus**
6. Flowers solitary; perianth segments 25–30 mm   **23. × bernardii**
7. Perianth and corona white or cream   **7. moschatus**
7. At least the corona yellow or orange   8.
8. Perianth segments reflexed   9.
8. Perianth segments spreading or erect-spreading   11.
9. Perianth segments white or cream   **15. × dichromus**
9. Perianth segments yellow   10.
10. Perianth segments 25–35 mm   **16. × monochromus**
10. Perianth segments up to 20 mm   **17. cyclamineus**
11. Perianth tube 8–15 mm   12.
11. Perianth tube 15–25 mm   13.
12. Corona 25–40 mm   **13. bicolor**
12. Corona 16–25 mm   **14. minor**
13. Corona white to pale yellow   14.
13. Corona deep golden yellow   18.
14. Pedicel 10–25 mm; corona white   **7. moschatus**
14. Pedicel 3–10 mm; corona pale yellow   15.
15. Corona 3–4 cm, distinctly expanded and usually lobed on the margin   **8. macrolobus**

15. Corona 2.0–3.5 cm; scarcely expanded at the margin
    and usually only slightly lobed   16.
16. Plant dwarf; flowers small, 35–40 mm long
    10(ii). pseudonarcissus var. **humilis**
16. Plant medium to tall; flowers medium to large   17.
17. Flowers single
    10(i,1). pseudonarcissus var. **pseudonarcissus** forma
    pseudonarcissus
17. Flowers double   **10(i,2). pseudonarcissus** var.
    pseudonarcissus forma **pleniflorus**
18. Perianth segments up to 20 mm   **11. hispanicus**
18. Perianth segments 20 mm or more   19.
19. Perianth segments pale yellow or cream, paler than
    the corona   20.
19. Perianth segments deep golden yellow, the same
    colour as the corona   23.
20. Pedicel 8–25 mm; corona conspicuously expanded at
    margin; flowers horizontal or ascending   **9. nobilis**
20. Pedicel 3–12 mm; corona scarcely expanded at
    margin; flowers horizontal or drooping   21.
21. Plant dwarf; flowers small 35–40 mm long
    10(ii). pseudonarcissus var. **humilis**
21. Plant medium to tall; flowers medium to large   22.
22. Flowers single   **10(i,1). pseudonarcissus** var.
    pseudonarcissus forma **pseudonarcissus**
22. Flowers double   **10(i,2). pseudonarcissus** var.
    pseudonarcissus forma **pleniflorus**
23. Leaves 20–50 cm × 5–15 mm   **11. hispanicus**
23. Leaves 20–30 cm × 6–10 mm   **12. obvallaris**
24. Leaves cylindrical or semicylindrical, usually not
    channelled down the inner face, 1–4 mm wide; scape
    terete   **5. jonquilla**
24. Leaves mostly flat and channelled down the inner
    face, 3–25 mm wide; scape usually compressed   25.
25. Umbel with 4–20 flowers, rarely fewer; corona without
    a red or scarious margin   26.
25. Umbel with 1–3 flowers; corona with a red or
    scarious margin   32.
26. Corona yellow or orange   27.
26. Corona white   30.
27. Flowers up to 4 cm in diameter   28.
27. Flowers 4–5 cm in diameter   29.
28. Perianth segments white   **1(a). tazetta** subsp. **tazetta**
28. Perianth segments pale to deep yellow **20. × intermedius**
29. Perianth segments cream-coloured or pale yellow;
    corona medium to bright yellow
    1(b). tazetta subsp. **italicus**
29. Perianth segments bright or golden yellow; corona
    deep yellow to orange   **1(c). tazetta** subsp. **aureus**
30. Flowers 2.0–2.5 cm in diameter
    2(a). papyraceus subsp. **panizzianus**
30. Flowers 2.5–4.0 cm in diameter   31.
31. Leaves glaucous; scape very compressed, 2-edged;
    corona usually finely scalloped
    2(b). papyraceus subsp. **papyraceus**
31. Leaves green; scape less flattened, not 2-edged;
    corona entire   **2(c). papyraceus** subsp. **polyanthus**

32. Flowers in umbels of 2–3; pollen sterile
    18. × medioluteus
32. Flowers solitary; pollen fertile   33.
33. Stamens of lower whorl not projecting from the
    perianth tube; perianth segments overlapping   34.
33. Stamens of lower whorl all projecting from the
    perianth tube; perianth segments not overlapping   38.
34. Flowers 35–50 mm in diameter   35.
34. Flowers 50–70 mm in diameter   36.
35. Leaves 3–7 mm wide   **3(a). poeticus** subsp. **verbanensis**
35. Leaves 10–13 mm wide   **3(b). poeticus** subsp. **hellenicus**
36. Corona disc-like   **3(c). poeticus** subsp. **poeticus**
36. Corona 3.0–3.5 mm deep   37.
37. Leaves 10–13 mm wide; corona deep yellow with a
    red margin   **3(d). poeticus** subsp. **recurvus**
37. Leaves 7–9 mm wide; corona deep yellow with a red
    margin and a narrow white zone beneath
    3(e). poeticus subsp. **majalis**
38. Corona flat and disc-like   39.
38. Corona cup-shaped   40.
39. Corona bright yellow with a red margin
    4(a). radiiflorus subsp. **exertus**
39. Corona wholly red   **4(b). radiiflorus** subsp. **poetarum**
40. Corona bright yellow edged with red
    4(c). radiiflorus subsp. **radiiflorus**
40. Corona bright yellow with a red margin and a white zone
    in between   **4(d). radiiflorus** subsp. **stellaris**

Section **1. Tazettae** DC.
Section *Hermione* (Salisb.) Spreng.

Leaves flat or channelled. *Flowers* in umbels, concolorous or bicoloured. *Perianth tube* cylindrical. *Corona* fairly short. *Filaments* straight, much shorter than anthers; anthers dorsifixed, included, or the upper whorl slightly exserted. *Flowering* in spring, rarely in autumn.

1. N. tazetta L.   Bunch-flowered Daffodil

Perennial herb. Bulb 30–50 × 20–35 mm, ovoid; covered with thin, papery, dark brown, dull or shining scales. *Leaves* 13–30(–50) cm × 4–15(–25) mm, glaucescent, rather fleshy, linear or lorate, obtuse at apex, entire, flat, veins obscure, glabrous. *Scape* 12–45 cm, robust, fistulose or solid, often distinctly compressed. *Spathe* 30–35 × 6–10 mm, papery, pale brownish, usually tubular towards the base, unilaterally split above, acute at apex. *Flowers* (2–)3–8(–17), 20–35(–40) mm in diameter, sweet-scented, forming a loose, subsecund umbel; pedicels 15–35 mm, angular, glabrous, lengthening to 50–75 mm in fruit. *Perianth tube* about 20 mm × 5 mm at apex, narrow, greenish. *Perianth segments* 8–22 × 6–10 mm, white, spreading or reflexed, ovate-oblong, obtuse or subacute with a mucronate or cuspidate apex. *Corona* 3–6 mm high by 6–11 mm in diameter, yellow or tinged orange, cup-shaped, entire or irregularly lobulate, sometimes shallowly plicate. *Stamens* inserted in 2 series near the apex of the perianth tube; filaments up to 2 mm; anthers included, medifixed, 7–8 × about 1.5 mm. *Style* about 20 mm, included in the corona, apex shortly 3-lobed. *Capsule* 10–15 × about 9 mm, oblong-trigonous, pale brown, transversely rugulose. *Seeds* about 3 × 2 mm, irregularly angular and compressed, black, minutely rugulose. *Flowers* 1–4.

A very polymorphic species which has been cultivated for centuries and much of the variation may have been derived from horticultural selection.

(a) Subsp. **tazetta**
Perianth segments white. *Corona* bright to deep yellow. $2n = 22$.

(b) Subsp. **italicus** (Ker Gawl.) Baker
Perianth segments cream-coloured or very pale yellow. *Corona* medium to bright yellow. $2n = 22$.

(c) Subsp. **aureus** (Loisel.) Baker
Perianth segments bright or golden yellow. *Corona* deep yellow to orange. $2n = 22$.

Introduced. All subspecies are grown in gardens, from which they escape and sometimes persist for a while, or remain as a relic, particularly in south-west England, the Isles of Scilly and the Channel Islands. The species is native of the west and central Mediterranean region. As native plants the subspecies are geographical. Subsp. *tazetta* is in the west Mediterranean region, subsp. *italicus* in the north and east parts of the Mediterranean, and subsp. *aureus* in south-east France, north-west Italy and Sardinia.

2. N. papyraceus Ker Gawl.   Paper-white Daffodil
Perennial herb. Bulb 40–70 × 25–35 mm, globose; covered with thin, papery, dark brown, dull or shining scales. *Leaves* 10–45 cm × 4–12 mm, glaucous or green, long-linear, obtuse at apex, entire, usually keeled, erect, glabrous. *Scape* 14–55 cm, solid, striate, more or less compressed, sharply keeled. *Spathe* 30–50 mm, papery. *Flowers* up to 20 in an umbel, 25–50 mm in diameter, fragrant; pedicels 30–50 mm, elongating in fruit. *Perianth tube* 15–25 × 3–5 mm, green below to white above. *Perianth segments* 8–18 × 8–11 mm, white, ovate or lanceolate, obtuse-apiculate at apex, spreading or incurving, overlapping at base. *Corona* 3–4 mm high by 7–8 mm in diameter, white, cup-shaped, entire or shortly notched or undulating. *Stamens* inserted in 2 series near the apex of the perianth tube; filaments up to 2 mm; anthers biserriate, 3 in the corona, and 3 in the perianth tube. *Style* equal to upper anthers, overtopped by anthers or shorter than anthers. *Capsule* oblong-trigonous, pale brown, transversely rugulose. *Seeds* about 3 × 2 mm, irregularly angular and compressed, black, minutely rugulose. *Flowers* 12–4.

(a) Subsp. **panizzianus** (Parl.) Arcang.
N. panizzianus Parl.
Leaves glaucous. *Scape* compressed, 2-edged. *Flowers* 20–25 mm in diameter. *Corona* entire, but undulating. $2n = 22$.

(b) Subsp. **papyraceus**
Leaves glaucous. *Scape* compressed, 2-edged. *Flowers* 25–40 mm in diameter. *Corona* usually crenulate. $2n = 22$.

(c) Subsp. **polyanthus** (Loisel.) Asch. & Graebn.
N. polyanthus Loisel.
Leaves green. *Scape* almost terete. *Flowers* 25–40 mm in diameter. *Corona* entire. $2n = 22$.

Introduced. All subspecies are grown in gardens. A rather rare relic of cultivation in the Isles of Scilly and the Channel Islands. The species is native of the Mediterranean region and south-west Europe. Subsp. *papyraceus* occurs throughout the range of the species. Subsp. *polyanthus* is native of southern France and is naturalised in Spain and Italy. Subsp. *panizzianus* is native of south-east France and northern Italy, Portugal and south-west Spain.

Section 2. Narcissus

Leaves flat, usually fairly wide. *Flowers* solitary, bicoloured. *Perianth tube* cylindrical, slender. *Perianth segments* patent or slightly deflexed. *Corona* small. *Filaments* much shorter than anthers; anthers dorsifixed, partly exserted, or the lower whorl included. *Flowering* in spring or early summer.

3. N. poeticus L.   Pheasant's-eye Daffodil
Perennial herbs. Bulb 12–35 mm in diameter, ovoid or ovoid-elongate; outer scales pale brown with darker veins. *Leaves* 20–40 cm × 3–13 mm, dark green to glaucous, erect or drooping, more or less keeled and channelled, linear, rounded-obtuse at apex, glabrous. *Scape* 20–45 cm, more or less compressed and 2-edged, slender to stout, striate to ribbed. *Spathe* 30–50 mm, more or less membranous. *Flowers* solitary, 35–70 mm in diameter, scented; pedicel 10–45 mm, slender to stout. *Perianth tube* 20–30 mm, cylindrical, pale to deep green. *Perianth segments* 15–25 × 6–20 mm, white or pale cream, obovate-suborbicular, imbricate, without a distinct claw, patent or slightly deflexed. *Corona* 8–15 mm in diameter, 2.0–3.5 mm deep, yellow, with a green centre with a red or orange rim below which is sometimes a more or less white band at maturity, discoid or cup-shaped, with a plicate-crenulate-dentate margin. *Stamens* unequal, 3 anthers exserted, 3 included within the perianth tube. *Style* included in the perianth tube, or equalling or exceeding the stamens. *Capsules* 12–20 mm, ellipsoid, subglobose or triangular-obovoid, sometimes trigonous. *Flowers* 4–5. $2n = 14$.

(a) Subsp. **verbanensis** (Herb.) P. D. Sell
N. verbanensis (Herb.) Pugsley; *N. poeticus* var. *verbenensis* Herb.
Scape 20–30 cm. *Leaves* 3–7 mm wide. *Flowers* 35–50 mm in diameter. *Perianth segments* markedly mucronate at apex. *Corona* cup-shaped, 8–9 mm in diameter, about 2 mm high, deep yellow, edged with red.

(b) Subsp. **hellenicus** (Pugsley) Hayek
N. hellenicus Pugsley
Scape 30–45 cm. *Leaves* 10–13 mm wide. *Flowers* 35–50 mm in diameter. *Perianth segments* obtuse-mucronate. *Corona* cup-shaped, 12–14 mm in diameter, about 3 mm high, pale yellow, edged crimson below which a faint and narrow white zone sometimes develops.

(c) Subsp. **poeticus**

N. ornatus Haw.

Scape 30–40 cm. *Leaves* 6–9 mm wide. *Flowers* 55–70 mm in diameter. *Perianth segments* subacute or mucronulate. *Corona* discoid, 13–15 mm in diameter, yellow with a red or orange-red rim.

(d) Subsp. **recurvus** (Haw.) P. D. Sell

N. recurvus Haw.; *N. poeticus* var. *recurvus* (Haw.) A. Fern.

Scape 30–45 cm. *Leaves* 10–13 mm wide. *Flowers* 50–70 mm in diameter. *Perianth segments* obtuse-mucronate. *Corona* cup-shaped, 12–14 mm in diameter, 3.0–3.5 mm deep, yellow with a red margin.

(e) Subsp. **majalis** (Curtis) P. D. Sell

N. majalis Curtis; *N. poeticus* var. *majalis* (Curtis) A. Fern.

Scape 25–40 cm. *Leaves* 7–9 mm wide. *Flowers* 50–70 mm in diameter. *Perianth segments* obtuse-mucronate to subacute at apex. *Corona* cup-shaped, 12–14 mm in diameter, about 3 mm deep, yellow with a narrow red margin and a narrow white zone beneath.

Introduced. All the subspecies are grown in British gardens and some at least escape and are naturalised, especially in the south of England. Native of Europe from east-central France, southwards to central Spain, south Italy and north-west Greece. The subspecies are geographically discrete, subsp. *verbanensis* in Italy, subsp. *hellenicus* in Greece, subsp. *poeticus* widespread, subsp. *recurvus* in Switzerland and subsp. *majalis* in France.

4. N. radiiflorus Salisb.   Small Pheasant's-eye Daffodil

Perennial herb. Bulb 20–30 mm in diameter, ovoid; outer scales pale or whitish-brown with well-marked darker veins. *Leaves* 20–40 cm × 5–11 mm, green to glaucous, mostly erect, sometimes drooping at the tips, more or less keeled and channelled, linear, obtuse at apex, glabrous. *Scape* 30–50 cm, more or less compressed and 2-edged, striate, slender. *Spathe* 30–50 mm, more or less membranous. *Flowers* solitary, 50–70 mm in diameter, scented; pedicels 10–45 mm, slender. *Perianth* tube 20–30 mm, slender, pale green. Perianth segments 22–30 × 6–22 mm, white to greenish-white, sometimes tinged yellow at base, narrowly obovate, not imbricate, rounded to a mucronate apex, cuneate at base to an obvious claw, spreading or slightly reflexed, sometimes twisted. *Corona* almost discoid to cup-shaped, 8–13 mm in diameter, up to 2.5 mm high, deep yellow with a bright red or orange-red rim, sometimes with a narrow white band below it, sometimes with the yellow suffused red throughout. *Stamens* subequal, with all the anthers more or less exserted. *Style* exserted, a little exceeding the anthers. *Capsule* 15–18 mm, narrowly ellipsoid, ellipsoid or obovoid, sometimes more or less trigonous. *Flowers* 4–5. 2*n* = 14.

(a) Subsp. **exertus** (Haw.) P. D. Sell

N. majalis var. *exertus* Haw.; *N. exertus* (Haw.) Pugsl. *Corona* flat and disc-like, bright yellow with a red margin.

(b) Subsp. **poetarum** (Haw.) P. D. Sell

N. poetarum Haw.; *N. poeticus* var. *poetarum* (Haw.) Burb. & Baker

Corona flat and disc-like, wholly red.

(c) Subsp. **radiiflorus**

N. angustifolius Haw.; *N. poeticus* subsp. *radiiflorus* (Salisb.) Baker; *N. poeticus* subsp. *angustifolius* (Haw.) Asch. & Graebn.

Corona cup-shaped, 2.0–2.5 mm high, bright yellow edged with red.

(d) Subsp. **stellaris** (Haw.) P. D. Sell

N. stellaris Haw.; *N. seriorflorens* Schur; *N. stelliflorus* Schur.

Corona cup-shaped, 2.0–2.5 mm high, bright yellow with a red margin and a white zone in between.

Introduced. A frequent relic of cultivation in the Channel Islands and probably elsewhere. All the subspecies are grown in gardens. As native plants they are geographically separated with the exception of subsp. *poetarum* whose origin in unknown. Subsp. *exertus* occurs in Switzerland and adjacent France, subsp. *radiiflorus* in Switzerland, Austria and Yugoslavia, and subsp. *stellaris* from south France to Romania and perhaps further east.

Section **3. Jonquillae** DC.

Leaves narrow. *Flowers* solitary or in umbels of 2–5(–8), more or less concolorous, yellow or rarely green. *Perianth tube* cylindrical to narrowly infundibuliform, slender. *Perianth segments* patent or deflexed, usually wide, often imbricate. *Corona* small to fairly large, usually wider than long. *Filaments* much shorter than the anthers; anthers dorsifixed, included, or the upper whorl partly exserted. *Flowering* in spring.

5. N. jonquilla L.   Jonquil

Perennial herb. Bulb 10–25 mm in diameter, ovoid; outer scales dark brown. *Leaves* up to 20 cm × 1–4 mm, dark green, erect and arching, rolled so that they are almost terete, very narrowly linear, shortly acute at apex, channelled on the inner surface, slightly striate on the outer, glabrous. *Scape* up to 24 cm, terete, slender, medium to dark green, smooth and glabrous, striate. *Spathe* 25–40 × up to 2.5 mm, more or less membranous. *Flowers* 1–5, in umbels, 20–35 mm in diameter, strongly scented; pedicel up to 50 mm; slender. *Perianth tube* 18–30 mm, greenish-yellow to yellow, cylindrical. *Perianth segments* 10–15 × 7–10 mm, yellow, ovate, elliptical or obovate, rounded-mucronate at apex, cuneate at base, spreading. *Corona* 9–15 mm in diameter, 2–4 mm high, same colour as the perianth segments, margin finely scalloped. *Stamens* included in the corona, 3 exserted, 3 in the perianth tube; filaments and anthers yellow. *Style* between the upper and lower anthers, greenish. *Flowers* 4–5. 2*n* = 14.

Introduced. Sometimes escapes from gardens and may persist for a short time. Native of south and central Spain and east Portugal; widely cultivated for its perfume and naturalised elsewhere in southern Europe.

Section **4. Bulbocodii** DC.

Leaves narrow, semi-cylindrical. *Flowers* solitary, concolorous. *Perianth tube* broadly obconical. *Perianth segments* narrow, erecto-patent. *Corona* large, obconical or broadly infundibuliform. *Filaments* curved, ascending distally, much longer than anthers; anthers dorsifixed, widely exserted from the perianth tube and sometimes from the corona. *Flowering* in winter or spring.

6. N. bulbocodium L.   Hoop-petticoat Daffodil

Perennial herb. Bulb 15–20 mm in diameter, ovoid; outer scales whitish or pale brown. *Leaves* 2 to several to each bulb, up to 25(–45) cm × 1–2 mm, dark green, usually erect, flat above with a slight channel, rounded beneath, very narrowly linear, rounded to acute at apex, glabrous. *Scape* up to 25(–45) cm, dark, slightly flattened, slender, smooth and glabrous. *Spathe* up to 50 × 5–7 mm, brownish. *Flowers* solitary, 25–35 mm in diameter, slightly scented; pedicel 5–30 mm, slender. *Perianth tube* 4–25 mm, slightly widened at apex, green. *Perianth segments* 6–22 × 1.5–7.0 mm, linear to narrowly triangular, acute or mucronate at apex, yellow, ascending. *Corona* 7–25 × 9–35 mm, hardly narrowed at the margin, same colour as the perianth segments. *Stamens* included in the corona, and lying along its bottom, all the same length or slightly variable, filaments and anthers yellow. *Style* as long as or slightly longer than the corona, yellow; stigma pinkish. *Flowers* 3–4. $2n = 14, 21, 28, 35, 42, 49, 56$.

Introduced. Long naturalised on Jethou in the Channel Islands, at Hextable in Kent and by seed at Virginia Water in Berkhire. Native of south-west Europe.

Section **5. Pseudonarcissi** DC.
Section *Ajax* (Salisb. ex Haw.) Dumort.

Leaves flat, usually fairly wide. *Flowers* usually solitary, rarely in umbels of 2–4. *Perianth tube* broadly obconical. *Perianth segments* usually patent to erecto-patent, rarely deflexed, fairly narrow. *Corona* large, more or less cylindrical, at least as long as wide. *Filaments* straight, as long as anthers or up to twice as long; anthers subbasifixed (rarely dorsifixed), exserted from the perianth tube but usually concealed within the corona. *Flowering* in spring or early summer.

7. N. moschatus L.   White Daffodil
N. albus Haw.; *N. candidissimus* DC.; *Ajax patulus* Salisb.; *Ajax albus* (Haw.) Haw.; *Ajax moschatus* (L.) Haw.; *Ajax moschatus* var. *candidissimus* (DC.) Herb.; *N. pseudonarcissus* subsp. *moschatus* (Haw.) Baker; *N. alpestris* Pugsley; *N. tortuosus* Haw.; *N. albescens* Pugsley; *N. albicans* Herb., non Spreng.; *N. moschatus* var. *albicans* Herb.

Perennial herb. Bulb 20–50 mm, ovoid or ovoid-attenuate, with pale brown or brownish-white scales. *Leaves* 10–40 cm × 5–12 mm, more or less glaucous, long-linear, narrowed to an obtuse apex, entire, nearly flat to more or less channelled, sometimes keeled and twisted. *Scape* 10–40 cm, slender to stout, more or less erect,

more or less compressed, 2-edged, more or less striate. *Spathe* scarious. *Flowers* solitary, 70–120 mm in diameter, nearly horizontal or more or less drooping, more or less scented; pedicel 10–25 mm, slender, straight below, arcuate-recurved above; pedicels 12–20 mm, rather narrow, slender. *Perianth tube* 12–20 mm, green at base, shaded yellow above, or with bright green stripes. *Perianth segments* 20–60 mm, pure white to sulphur-white, oblong, lanceolate-elliptical, oblong or ovate-lanceolate, acute or obtuse-mucronate at apex, slightly imbricate, erecto-patent, incurved or drooping, undulate and often spirally twisted, equalling or shorter than the corona. *Corona* 25–40 mm, the colour as the segments, straight, more or less spirally dilated, longitudinally plicate, more or less 6-lobed, the lobes rounded and crenate or entire, but little crisped or plicate, transversely rugose within. *Stamens* inserted 3–4 mm above the base of the perianth tube; anthers straw-coloured, yellow or buffish-yellow. *Style* rather long; stigmas small, more or less 3-lobed. *Capsule* 12–25 mm, narrowly or oblong-ellipsoid and tapered at both ends, subterete or bluntly trigonous. *Flowers* 3–4. $2n = 14, 28, 29$.

Introduced. Grown in gardens and fields and naturalised on roadsides, by tracks and in waste places. Native of the Pyrenees and northern Spanish mountains.

8. N. macrolobus (Jord.) Pugsley
                                Pale-flowered Daffodil
Ajax macrolobus Jord.; *N. pallidiflorus* Pugsley; *N. pallidiflorus* forma *asturica* Pugsley; *N. pallidiflorus* var. *intermedius* Pugsley; *N. macrolobus* var. *pallescens* Pugsley; *N. pseudonarcissus* subsp. *pallidiflorus* (Pugsley) A. Fern.

Perennial herb. Bulb 20–30 mm, ovoid or subrotund, with pale brown scales. *Leaves* 15–30 cm × 5–12 mm, more or less glaucous, more or less flat, attenuate to an obtuse apex, erect. *Scape* 15–30 cm, erect, subequalling or a little shorter than the leaves, rather stout, not much attenuate above, little compressed, obscurely 2-edged, coarsely striate or furrowed. *Spathe* scarious. *Flowers* solitary, 90–120 mm in diameter, nearly scentless, drooping or horizontal; pedicels 5–10 mm, not slender, strongly deflexed. *Perianth tube* 15–20 mm, scarcely half as long as the corona, suffused below with soft yellow. *Perianth segments* 40–60 mm, creamy white to primrose yellow or straw-coloured, broadly elliptical or ovate-lanceolate, obtuse-mucronate to subacute or acuminate at apex, erect-spreading and more or less twisted, equalling or slightly shorter than the corona. *Corona* 3–4 cm, about 30 mm across, pale yellow abruptly dilated, spreading and recurved at its mouth, 6-lobed, sometimes more or less obscurely so; lobes irregularly and sparingly incised and plicate with a crenate, undulate margin. *Stamens* inserted 3–4 mm above the base of the tube; anthers without a dark apical spot. *Capsule* 12–25 mm, oblong or more or less subrotund, not trigonous, sometimes with obscure, broad and shallow furrows, not rugose. *Flowers* 2–6. $2n = 14$.

Introduced. Heathy roadsides near Aldeburgh, Suffolk, fields at Meldreth, Cambridgeshire and probably elsewhere. Native of the Pyrenees and Cordillera Cantabrica.

9. N. nobilis (Haw.) Schult. fil. Large-flowered Daffodil
Ajax nobilis Haw.; *N. leonensis* Pugsley

Perennial herb. Bulb 30–35 mm, subrotund, with brownish scales. *Leaves* 15–50 cm × 8–14(–17) mm, erect, green or glaucous, attenuate above and obtuse at apex, sometimes twisted. *Scape* 15–50 cm, erect, rather stout, more or less compressed, 2-edged, more or less striate. *Spathe* up to 10 cm, often tinted green. *Flowers* solitary, 70–90 mm in diameter, horizontal or ascending, strongly scented; pedicels 8–25 mm, more or less erect. *Perianth tube* 20–25 mm, yellow or greenish tinted. *Perianth segments* 30–35 mm, pale yellow or cream, elliptical, elliptic-oblong, elliptic-lanceolate or ovate-lanceolate, obtuse-mucronate at apex, imbricate below, spreading or erect-spreading, sometimes undulate and twisted, equalling or longer than the corona. *Corona* 30–45 mm, golden yellow, straight, expanded above with a spreading margin, deeply dentate or irregularly cut into shallow lobes, more or less rugulose within. *Stamens* 6, inserted 4–7 mm above the base of the perianth tube. *Style* exceeding stamens. *Capsule* 20–25 mm, broadly ellipsoid, nearly terete, not furrowed. *Flowers* 3–4. $2n = 28$.

Introduced. This is said not to be much grown in gardens, but many naturalised daffodils in East Anglia seem to fit its description. Native of Spain.

10. N. pseudonarcissus L.   Daffodil
N. festalis Salisb. nom. illegit.; *Ajax festalis* Salisb. nom. illegit. *Ajax pseudonarcissus* (L.) Haw.

Perennial herb. Bulb 20–30 mm, ovoid, with brownish scales. *Leaves* 12–35 cm × 6–12 mm, erect, glaucous, somewhat channelled, attenuate above, obtuse at apex. *Scape* 20–35 cm, erect, equalling or slightly exceeding the leaves, generally rather slender, moderately compressed, 2-edged, usually distinctly striate. *Spathe* 2–6 cm, scarious. *Flowers* solitary, 50–75 mm in diameter, drooping or nearly horizontal, strongly scented. *Perianth tube* 15–22 mm, rather narrow. *Perianth segments* 35–60 × 17–25 mm, yellow or cream, oblong-lanceolate to ovate-lanceolate or elliptical, obtuse-mucronulate, acute or more rarely acuminate at apex, more or less imbricate below, ascending over the corona, waved and in well-grown garden plants often more or less spirally twisted, usually about as long as the corona. *Corona* 2.0–3.5 cm, yellow, straight, scarcely expanded or spreading at the margin, without distinct lobes but cut irregularly into numerous, short, dentate, crenate or serrate and imbricate lobules, strongly plicate above and finely transversely rugulose within. *Stamens* straight and subequal, uniseriate; filaments inserted 3–4 mm above base of perianth tube; anthers without dark apical spot. *Style* shortly exceeding stamens. *Capsule* 12–25 mm, obovoid or subrotund or rarely oval-ellipsoid, very obtuse or subtruncate, roundly trigonous or nearly terete, often furrowed and generally rugose. *Flowers* 2–4. Pollinated by bumble-bees and other insects; homogamous. $2n = 14$.

(i) Var. **pseudonarcissus**
Plant medium to tall. *Flowers* medium to large.

(1) Forma **pseudonarcissus**
Flower single.

(2) Forma **pleniflorus** P. D. Sell
Flowers double.

(ii) Var. **humilis** Pugsley
Plant dwarf. *Flower* small, 35–40 mm long.

Native. Damp woods and grassland. Local, but often abundant in England, Wales and Jersey; commonly naturalised in parks, grassy places and roadsides throughout the British Isles. Forma *pleniflorus* is common. Var. *humilis* occurs in Wales, northern England and Scotland and is perhaps the truly wild daffodil. Belgium, France, north and central Spain, Portugal, north Italy, western Germany, and perhaps Switzerland. Naturalised elsewhere in Europe.

11. N. hispanicus Gouan   Spanish Daffodil
N. major Curtis; *N. grandiflorus* Salisb.; *N. pseudonarcissus* subsp. *major* (Curtis) Baker

Perennial herb. Bulb 40–50 mm, ovoid; with brown scales. *Leaves* 40–50 cm × 5–15 mm, erect, glaucous, flat and more or less spirally twisted, obtuse at apex. *Scape* 40–60(–90) cm, erect, stout below but attenuate above, much compressed and acutely 2-edged, finely striate. *Spathe* up to 6 cm, scarious. *Flower* solitary, 50–70 cm in diameter, suberect or nearly horizontal. *Perianth tube* about 18 mm. *Perianth segments* 18–40 mm, deep golden yellow, oblong-lanceolate, subacute at apex, very slightly imbricate below, spreading-incurved, regularly spirally twisted, as long as the corona. *Corona* 40–45 mm in diameter, golden yellow, with an abruptly dilated and widely spreading margin, which is obscurely lobed, deeply crenate-dentate and plicate-recurved. *Stamens* straight and subequal, uniseriate, fairly short; filaments inserted 3–4 mm above the base of the perianth-tube; anthers with a minute dark apical spot. *Capsule* 20–30 mm, oblong-ellipsoid, bluntly trigonous with shallow furrows. *Flowers* 3–5. $2n = 14$.

Introduced. Widely grown in gardens and naturalised over the whole of the British Isles. Native of south-west Europe.

12. N. obvallaris Salisb.   Tenby Daffodil
N. sibthorpii Haw.; *N. pseudonarcissus* var. *bromfieldii* Syme; *N. pseudonarcissus* subsp. *obvallaris* (Salisb.) A. Fern.; *Ajax lobularis* Haw.; *N. lobularis* (Haw.) Schult. & Schult. fil.

Perennial herb. Bulb 25–35 mm, subrotund-ovoid; with pale brown scales. *Leaves* 20–30 cm × 6–11 mm, erect, glaucous, flat and not twisted, obscurely keeled, slightly attenuate above, obtuse at apex. *Scape* 20–30 cm, erect, more or less stout, 2-edged but not much compressed, coarsely striate. *Spathe* thick, scarious. *Flowers* solitary, 50–60 mm in diameter, ascending or nearly horizontal, slightly scented; pedicels 10–15 mm. *Perianth tube* 12–15 mm, green. *Perianth segments* 25–30 mm, deep golden yellow, ovate, obtuse and mucronate at apex, imbricate, spreading, somewhat incurved and undulate but not twisted, shorter than the corona. *Corona* 30–35(–44) mm long, 25–30 mm in diameter, deep golden yellow, more or less longitudinally plicate, dilated above with spreading or slightly reflexed margin, with 6 well-marked, rounded lobes which are sometimes irregularly or sparingly undulate-plicate-crenate and sometimes entire. *Stamens* straight and subequal, uniseriate; filaments inserted about 3 mm above the base of the perianth tube; anthers with a minute dark apical spot. *Style* longer than in *N. hispanicus*. *Capsule* 20–25 mm, narrowly oblong or oblong-obovoid, subtruncate, obscurely trigonous with flattish sides and scarcely furrowed. *Flowers* 2–4. $2n = 14$.

Origin unknown, but may have been derived from *N. hispanicus* in cultivation. Long naturalised in Pembrokeshire and Carmarthenshire and probably elsewhere.

13. N. bicolor L.   Two-coloured Daffodil
N. tubaeflorus Salisb.; *Ajax lorifolius* Salisb.; *Ajax bicolor* (L.) Salisb.; *N. bicolor* var. *lorifolius* (Salisb.) Haw.; *N. abscissus* Schult. fil.; *Ajax abscissus* (Schult. fil.) Haw.; *Oileus abscissus* (Schult. fil.) Haw.; *Ajax muticus* Gay; *N. pseudonarcissus* subsp. *bicolor* (L.) Baker; *N. abscissus* var. *serotinus* (Jord.) Pugsley; *Ajax serotinus* Jord.; *Ajax tubulosus* Jord.; *N. abscissus* var. *tubulosus* (Jord.) Pugsley

Perennial herb. Bulb 25–60 mm, ovoid, with brown scales. *Leaves* 30–40 cm × 12–20 mm, green or glaucous, long-linear, not attenuate but obtuse at apex, entire, erect or recurved, nearly flat, not twisted. *Scape* 30–40 cm, erect, slender to robust, moderately compressed, acutely 2-edged, finely striate. *Spathe* scarious. *Flowers* 80–100 mm in diameter, ascending or horizontal, almost scentless; pedicels 15–35 mm, rather slender, nearly erect, but slightly curved above. *Perianth tube* 8–12 mm, obconical, orange-yellow, sometimes with a greenish tinge. *Perianth segments* 40–50 mm, pale or sulphur yellow, lanceolate or ovate-lanceolate, obtuse or mucronate, acute or acuminate at apex, rounded and imbricate below, subequal to or longer than the corona. *Corona* 25–40 mm in diameter, golden yellow, slightly ventricose below, somewhat dilated above with more or less spreading margin, more or less 6-lobed, the lobes shortly rounded-obtuse or subtruncate. *Stamens* long, 1–4 mm from the base of the perianth tube; anthers without apical spot. *Style* long. *Capsule* 15–25 mm, broadly ellipsoid or obovoid, trigonous but scarcely furrowed. *Seed* obtuse. *Flowers* 3–5. $2n = 14$.

Introduced. Cultivated in gardens and escaping onto roadsides and waste places. Native of the Pyrenees and Corbières.

14. N. minor L.   Small-flowered Daffodil
N. exiguus Salisb.; *Ajax pygmaeus* Salisb.; *Ajax minor* (L.) Haw.; *N. pseudonarcissus* subsp. *minor* (L.) Baker; *Ajax minor* var. *minimus* Haw.; *Ajax minimus* (Haw.) Haw.; *Ajax minor* var. *humilior* Herb.; *N. pumilus* Salisb.; *Ajax pumilus* (Salisb.) Haw.; *Ajax cuneiflorus* Salisb.; *N. nanus* Spach; *Ajax nanus* (Spach.) Haw.; *N.*

parviflorus (Jord.) Pugsley; *Ajax parviflorus* Jord.; *N. provincialis* Pugsley

Perennial herb. Bulb 20–30 mm, subrotund-ovoid or ovoid, with thin, whitish-brown scales. *Leaves* 8–25 cm × 4–14 mm, more or less glaucous, long-linear, more or less attenuate to an obtuse apex, flat, slightly channelled, undulate, erect-spreading. *Scape* 12–22 cm, nearly erect or inclined, slender to stout, 2-edged but little compressed, striate. *Spathe* thinly membranous. *Flowers* 40–80 mm in diameter, horizontal or nodding, faintly scented; pedicels 10–20 mm, porrect or slightly curved, slender, usually nearly as long a the capsule. *Perianth tube* 9–15(–18) mm, narrowly to broadly obconical. *Perianth segments* 16–22 mm, pale to deep yellow, ovate, ovate-lanceolate or oblong, obtuse-mucronate, acute or acuminate at apex; erect-spreading or nearly flat, twisted, imbricate. *Corona* 16–25 mm, a little longer than the perianth segments, straight, golden yellow, subcylindrical, not or slightly expanded at apex, obscurely crenate-lobate to distinctly plicate and irregularly incised-crenate. *Stamens* short, inserted 3–5 mm above the base of the perianth tube. *Stigma* more or less 3-lobed. *Capsule* 12–25 mm, rotund-ellipsoid to oblong-obovoid, terete, more or less furrowed. *Flowers* 3–5. $2n = 14$.

Introduced. Grown in gardens and escaping onto roadsides and waste places. Naturalised in the Ballaugh Curraghs on the Isle of Man and known from 1885 to 1945 at Charles, near South Molton, Devon. Native of the Pyrenees and northern Spain.

15. N. × dichromus P. D. Sell
                White-and-Yellow Daffodil
N. cyclamineus × moschatus

Perennial herb. Bulb about 20–30 mm, ovoid, with brownish scales. *Leaves* up to 35 cm × 5–12 mm, slightly glaucous, long-linear, obtuse at apex, channelled above and keeled below, erect. *Scape* up to 35 cm, exceeding the leaves, rather slender, compressed, erect, finely striate. *Spathe* 5.0–5.5 cm, brownish-scarious. *Flowers* solitary, 70–80 mm in diameter, horizontal, almost scentless. *Perianth tube* 13–15 mm, yellowish-green. *Perianth segments* 35–37 × 20–22 mm, white, ovate, more or less obtuse at apex, reflexed backwards. *Corona* 23–25, shorter than the perianth segments, parallel-sided but slightly spreading towards the apex, pale yellow, irregularly dentate. *Stamens* 6, inserted near the base of the tube; anthers orange. *Style* pale greenish; stigma just above anthers. *Flowers* 2–3.

Introduced. Grown in gardens and fields and escaping onto roadsides and tracks and in waste places. Garden origin.

16. N. × monochromus P. D. Sell   Reflexed Daffodil
N. cyclamineus × pseudonarcissus

Perennial herb. Bulb about 20–30 mm, ovoid, with brownish scales. *Leaves* up to 35 cm × 5–12 mm, green, long-linear, obtuse at apex, channelled above and keeled below, erect. Scape up to 35 cm, usually slightly exceeding the leaves, rather slender, compressed, erect, finely striate. *Spathe* 4.5–5.0 cm, greenish turning scarious-brown. *Flowers* solitary, 65–75 mm in diameter, horizontal, scented. *Perianth tube* 10–15 mm, yellowish-green. *Perianth-segments* 30–35 × 20–22 mm, pale yellow, ovate, more or less obtuse at apex, reflexed backwards. *Corona* 22–25 mm, shorter than the perianth segments, parallel-sided but slightly spreading towards the apex, slightly deeper yellow than the perianth segments, irregularly dentate. *Stamens* 6, inserted near the base of the tube; anthers orange. *Style* greenish-white; stigma just above anthers. *Flowers* 2–3. $2n = 21$.

Introduced. Grown in gardens and fields and escaping onto roadsides and tracks and in waste places. Garden origin.

17. N. cyclamineus DC.   Cyclamen-flowered Daffodil

Perennial herb. Bulb about 16 × 12 mm, ovoid; with whitish scales. *Leaves* 15–25 cm × 4–5 mm, bright green, ascending, thick, channelled above and broadly keeled below, obtuse at apex. *Scape* 15–20(–30) cm, erect, nearly terete, obscurely 2-edged, finely striate. *Spathe* about 20 mm, green at flowering, later scarious. *Flower* solitary, very narrow; pedicels 15–25 mm, rather slender, not compressed, nearly straight below and strongly recurved above. *Perianth tube* 2–3 mm, short and broad. *Perianth segments* up to 20 mm, reflexed backwards at almost 180° to the corona, deep yellow, linear-oblong, subacute at apex. *Corona* up to 20 mm, deep yellow, cylindrical, but widened slightly towards the apex, its margin toothed or scalloped. *Stamens* straight and subequal, uniseriate, inserted close to the base of the perianth tube. *Style* slightly exceeding stamens, with a rather large stigma. *Capsule* 12–18 mm, narrowly obovoid, obscurely trigonous, not furrowed. *Flowers* 2–3. $2n = 14$.

Introduced. Naturalised on a Wisley roadside and at Henley Park in Surrey. Native of Spain and Portugal.

Intersectional hybrids

18. N. × medioluteus Mill.   Primrose-peerless
     N. poeticus × tazetta
N. × biflorus Curtis

Perennial herb. Bulb 25–60 × about 45 mm, subglobose. *Leaves* 23–75 cm × 7–14 mm, scarcely glaucous, linear, obtuse at apex, entire, flat. *Scape* 30–70 cm, compressed, 2-edged. *Spathe* 15–45 mm, scarious. *Flowers* (1–)2(–3), 30–50 mm in diameter, fragrant, erect, finely striate; pedicels 25–35 mm. *Perianth tube* 20–25 mm, cylindrical, expanded at the mouth. *Perianth segments* 18–22 mm, white or cream, subrotund or obovate, obtusemucronate, imbricate. *Corona* 9–12 × 3–5 mm, bright yellow, usually with crenate, whitish margins. *Anthers* aborted. *Capsule* not formed. *Flowers* 4–5. $2n = 17, 24$.

Introduced. A rather frequent relic of cultivation in most of the British Isles, but more frequent in the south. Of garden origin or perhaps France.

19. N. × odorus L.   Hybrid Jonquil
     N. jonquilla × pseudonarcissus aggr.
N. × infundibulus Poir.

Perennial herb. Bulb 25–30 mm. *Leaves* 35–50 × 6–8 mm, bright green, long-linear, obtuse at apex, entire,

strongly keeled, semi-terete. *Scape* 25–40 cm, subcylindrical. *Spathe* 50–70 mm, scarious. *Flowers* 1–4, usually fragrant; pedicels 25–35 mm. *Perianth tube* 15–25 mm, narrowly infundibular. *Perianth segments* 15–25 mm, yellow, broadly elliptical, rounded at apex. *Corona* 13–18 mm, 17–20 mm in diameter, yellow, funnel-shaped, lobes subentire, with erect, nearly truncate margin. $2n = 14$.

Introduced. Grown in gardens and naturalised in the Isles of Scilly and the Channel Islands. Known for over 100 years in a field about 2 miles south of St Austell, Cornwall. Of garden origin.

20. N. × intermedius Loisel.   Intermediate Jonquil
     N. jonquilla × tazetta

Perennial herb. Bulb 28–30 mm. *Leaves* 30–45 cm × 5–8 mm, bright green, deeply grooved, long-linear or almost cylindrical, obtuse at apex, entire. *Scape* 25–40 cm, more or less terete. *Spathe* 30–40 mm, scarious. *Flowers* in umbels of 3–6, fragrant; pedicels 20–40 mm. *Perianth tube* 14–20 mm, cylindrical-trigonous, slender. *Perianth segments* 10–18 mm, bright yellow. *Corona* 3–4 × 6–7 mm, slightly deeper in colour than the perianth, its margin slightly lobed. $2n = 17$.

Introduced. Grown in gardens and occasionally found in waste places. Native of southern France, Spain and Italy.

21. N. × incomparabilis Mill.   Nonesuch Daffodil
     N. poeticus × pseudonarcissus
N. barrii Baker

Perennial herb. Bulb 25–30 mm. *Leaves* 17–35 cm × 8–12 mm, somewhat glaucous, long-linear, obtuse at apex, entire, channelled, nearly flat, twisted. *Scape* 17–40 cm, compressed, more or less glaucous. *Spathe* about 35 mm, scarious. *Flowers* solitary, slightly fragrant. *Perianth tube* 20–25 mm, narrowly obconical or narrowly infundibuliform. *Perianth segments* 25–30 × 12–16 mm, pale or medium yellow, narrowly obovate, patent. *Corona* 13–22 × 17–20 mm, about two-thirds as long as the perianth segments, deep golden or orange-yellow, undulate, lobed and frilled. *Anthers* yellow. $2n = 14, 21, 28$.

Introduced. Widely cultivated for ornament and grown as a crop in bulb fields. Naturalised in fields, waste land and by tracks and roads. Of garden origin.

22. N. × boutigyanus Philippe
                          White-and-Orange Daffodil
     N. moschatus × poeticus

Perennial herb. Bulb 25–30 mm. *Leaves* 17–35 cm × 8–12 mm, glaucous, long-linear, obtuse at apex, entire, channelled, twisted. *Scape* up to 40 cm, compressed, twisted. *Spathe* a little longer than perianth tube. *Flowers* solitary, slightly fragrant. *Perianth tube* 20–25 mm, narrowly obconical. *Perianth segments* 25–40 × 10–40 mm, white, narrowly obovate, spreading. *Corona* 12–15 mm, about half to two-thirds as long as perianth segments, medium yellow or orange, much lobed and frilled. *Anthers* medium yellow. *Flowers* 3–4. $2n = 17$.

Introduced. Widely cultivated for ornament and

grown as a crop in bulb fields. Naturalised in fields, waste land and by tracks and roads. Of garden origin.

23. N. × bernardi DC.　　　　Bernard's Daffodil
N. hispanicus × poeticus

Perennial herb. Bulb 25–30 mm. *Leaves* 15–35 cm × 8–12 mm, glaucous, long-linear, obtuse at apex, entire, channelled, twisted. *Scape* up to 40 cm, compressed, twisted. *Spathe* much longer than perianth tube. *Flowers* solitary, sweet-scented. *Perianth tube* 20–25 mm, narrowly obconical. *Perianth segments* 25–30 × 10–15 mm, pale yellow, spreading, flat to slightly cupped. *Corona* about as long as perianth segments and the same pale colour, frilled, margins sometimes reflexed. *Anthers* pale yellow. *Flowers* 3–4. 2n = 14, 21.

Introduced. Cultivated for ornament and grown as a crop in bulb fields. Naturalised in fields, waste land and by tracks and roads. Of garden origin.

Subfamily 8. Asparagoideae Vent
Tribe *Asparageae* DC.

Perennial herbs or shrubs with rhizomes. *Leaves* all cauline, reduced to small scales and replaced functionally by cladodes which are stems arising from their axils. *Inflorescence* an inconspicuous cluster of 1 to few flowers borne in the scale-like leaf axil on the main stem or on a cladode. *Flowers* unisexual, rarely bisexual. *Perianth segments* greenish to yellowish-white, free or fused just at base, without a corona. *Stamens* 6. *Style* 1, or more or less absent; stigma capitate or 3-lobed. *Ovary* superior, 3-celled. *Fruit* a spherical red berry.

36. Asparagus L.

Perennial herbs with rhizomes. *Leaves* reduced to small scarious scales at the nodes. *Cladodes* leaf-like, borne in clusters of 4–15(–25) in the axils of the leaves, needle-like or slightly flattened. *Flowers* unisexual by abortion, rarely bisexual, borne in the axils of scale leaves on the main stems or on a cladode. *Perianth segments* 6, fused at base, greenish to yellowish-white. *Stamens* 6, inserted at the base of the perianth segments. *Ovary* superior, 3-celled. *Fruit* a globose berry. *Seeds* 1–6.

About 50–60 species in the Old World, but absent from Australasia.

A. aethiopicus L. and **A. setaceus** (Kunth) Jessop have been recorded as casual garden escapes.

1. Stems up to 1.5(–2) m, erect; pedicels 6–10(–15) mm; seeds usually 5–6　　**1(a). officinalis** subsp. **officinalis**
1. Stems up to 30(–60) cm, procumbent to decumbent; pedicels 2–6(–8) mm; seeds usually less than 5
　　　　　　　　1(b). officinalis subsp. **prostratus**

1. A. officinalis L.

Glabrous *perennial herb* with a short rhizome. *Stems* up to 1.5(–2.0) m, green, procumbent to erect, smooth, much branched. *Leaves* reduced to small, scarious scales at the nodes with conical, hardened bases, those of the main stem up to 5 mm, triangular-lanceolate, acute at apex, those of the branches less than 1.5 mm, ovate,

acute at apex, sagittate. *Cladodes* 4–25 × 0.3–0.4 mm, in fascicles of 4–15(–25) in the axils of the reduced leaves, needle-like to slightly flattened, stiff. *Flowers* 1–2(–3) in the axils of the scale leaves on the main stem, sometimes among the cladodes, unisexual by abortion, rarely bisexual, male larger than female; pedicels 2–10(–15) mm, arching downwards, jointed about the middle or just below it. *Perianth segments* fused at base, (4–)4.5–6.5(–10) × 1.5–2.5 mm, greenish to yellowish-white, narrowly oblanceolate or oblong, obtuse at apex. *Stamens* 6, inserted at the base of the perianth segments; filaments 1.0–2.5 mm, pale yellowish-green; anthers yellow, about as long as filaments. *Style* 1; stigma capitate or 3-lobed, green. *Berry* 5–8(–10) × (6–)8–10 mm, green turning red, globose, but flattened at base and apex. *Seeds* (1–)2–6, 3.5–4.5 × 2–3 mm, dark blackish-brown, ellipsoid. *Flowers* 7–9.

(a) Subsp. officinalis　　　　Garden Asparagus
A. officinalis var. *altilis* L.; *A. altilis* (L.) Asch.
Stems up to 1.5 (–2.0) m, erect. *Cladodes* (5–) 10–20 (–25) mm, on main lateral branches, flexible, usually green. *Pedicels* 6–10 (–15) mm. *Seeds* usually 5–6. 2n = 20.

(b) Subsp. prostratus (Dumort.) Corb.
　　　　　　　　　　Wild Asparagus
A. prostratus Dumort.; *A. maritimus* auct.; *A. officinalis* var. *maritimus* auct.
Stems up to 30(–60) cm, procumbent to decumbent. *Cladodes* 4–10(–15) mm, rigid, usually glaucous. *Pedicels* 2–6(–8) mm. *Seeds* usually less than 5. 2n = 40.

The species occurs throughout Britain and Ireland and most of Europe. Subsp. *officinalis* is grown for its young shoots and is usually considered to be a luxury vegetable. It has been known since the time of the ancient Greeks and, as well as being grown in gardens, about 500 hectares are produced commercially in Britain, particularly in Suffolk, Norfolk, Essex, Worcestershire and Lancashire. It was certainly grown by the seventeenth century, perhaps earlier, and is a long-standing crop which can remain on a site for twenty years. It escapes and is well naturalised in dry sandy soils amongst sparse grassland especially on maritime dunes and East Anglian heathland. It spreads as a weed in gardens by seed. It occurs almost throughout the range of the species. Subsp. *prostratus* is native on grassy sea-cliffs and is very local in Cornwall, Pembrokeshire, south-east Ireland and the Channel Islands. It formerly occurred at other places in south-west England and Wales north to Anglesey. It occurs elsewhere on the coasts of western Europe from north-west Germany to north Spain.

37. Ruscus L.

Evergreen *shrubs*, with short rhizomes and simple or branched stems. *Leaves* reduced to small, scarious scales, bearing in the axils a broad, green, flattened, leaf-like *cladode*. *Flowers* unisexual by abortion, solitary or in clusters on the surface of the cladode and subtended by a bract. *Perianth segments* 6, free, the inner smaller than the outer, green, sometimes tinted purple. *Stamens*

6, the filaments united into a fleshy tube which is present in the flowers of both sexes; in the male flowers the tube is topped by 3 sessile anthers, which are represented in the female flower by minute papery flanges. *Stigma* more or less entire, protruding from the neck of the staminal tube. *Ovary* filling the space within the staminal tube, vestigial in male flowers. *Fruit* a red berry, with thin flesh and 1–4 large seeds.

Seven species in the Mediterranean region, west Europe, Madeira and Azores.

Yeo, P. F. (1968). A contribution to the taxonomy of the genus *Ruscus. Notes Roy. Bot. Gard. Edinb.* **28**: 237–264.

1. Stem with 1 or more whorls of branches; bract of inflorescence very small   **1. aculeatus**
1. Stem unbranched or occasionally with 1 branch; bract of inflorescence 10–33 mm   **2. hypoglossum**

Series **1. Ramosae** Yeo

Stems branched. *Cladodes* less than 3(–4) cm, spine-tipped.

1. R. aculeatus L.   Butcher's Broom

Evergreen, glabrous *shrub*; rhizome creeping, thick, fibrous. *Stems* 25–80 cm, dull dark green, stiff, much branched, markedly striate. *Leaves* about 3 × 2 mm, narrowly triangular, with a long, pointed tip, entire, brown-scarious. *Cladodes* 10–30(–40) × 6–16 mm, dark dull green, narrowly to broadly ovate, entire, thick, rigid, twisted at base, spine-tipped. *Flowers* 1–2, unisexual by abortion, on the upper surface of the cladode in the axil of a small scarious bract. *Male flowers* with 3 outer and 3 inner perianth segments; the outer about 2.5 × 1.5 mm, rounded, the inner about 2 × 1 mm, narrower and more pointed; filaments united into a purplish-brown tube about 2 mm; anthers 3; *ovary* vestigial. *Female flowers* with similar perianth segments to male; style 1, protruding beyond the staminal tube, green; stigma slightly paler; ovary filling the space in the staminal tube. *Fruit* 8–13 × 10–17 mm, bright red, more or less globose, but often broader than long and slightly flattened. *Flowers* 1–4. *Fruits* 10–5. 2*n* = 40.

Native. Dry woods, hedgerows and amongst rocks, local. Widespread in southern England extending north to Caernarvonshire and Norfolk; commonly cultivated and escaping to become naturalised and then extending into northern England and Scotland. Mediterranean region north to Transylvania, south and west Switzerland and north France; Azores.

Series **2. Simplices** Yeo

Stems unbranched or with a single branch. *Cladodes* normally more than 3 cm.

2. R. hypoglossum L.   Spineless Butcher's Broom
R. alexandrinus Gars. nom. illegit.; *R. humilis* Salisb. nom. illegit.; *R. troadensis* E. D. Clarke; *R. hypophyllum* var. *hypoglossum* (L.) Baker; *R. hypophyllum* subsp. *hypoglossum* (L.) Domin

Evergreen, glabrous *shrub*; rhizome short. *Stems* 20–40 cm, oblique, sometimes arching above, unbranched or

rarely with 1 branch. *Leaves* 4–12 × 2–6 mm, ovate or triangular-ovate, sometimes auricled at base, 3- to 5-veined. *Cladodes* up to 22, 3–10(–11) × 1.0–3.3(–5.2) cm, matt yellowish-green, usually obovate to obovate-lanceolate, sometimes more or less ovate, more or less acuminate with a more or less acute point at the apex, entire, narrowed, channelled and twisted at base, sessile. *Flowers* unisexual, on the upper surface of the cladode in the axil of a bract; bract 10–33 × 3–13 mm, green, ovate or lanceolate, 5- to 11-veined. *Male flowers* with 3 outer and 3 inner perianth segments, the outer 3–4 × 1.0–1.7 mm, patent, violet-tinged, linear, with the margins recurved, the inner 1.5–3.0 × about 0.5 mm, violet-tinged, linear and the margins revolute; staminal column 3.0–3.5 × 1.0 mm, subcylindrical, slightly tapered from near the apex to the base; anthers 3, violet; ovary vestigial. *Female flowers* with similar perianth segments to male; style 1, protruding beyond the staminal tube; stigma spherical, violet; ovary unilocular; ovules 2, superposed; staminal column 2.5 × 1.2 mm, narrowly ovoid, dark violet, tapering towards the apex, with 6 obscure or whitish veins or with faint greenish striations. *Fruit* 8–13 × 8–13 mm, more or less globose, bright red. *Flowers* 1–4. 2*n* = 40.

Introduced. Grown for ornament in gardens and naturalised in a few shady places in western England and in Midlothian. South-east and east-central Europe, extending westwards to north-west Italy and eastwards to northern Turkey.

Subfamily **9. Alstroemerioideae** Pax

Perennial herbs with tuberous roots. *Leaves* all cauline, linear-lanceolate. *Inflorescence* a terminal, simple or compound umbel, without a spathe. *Flowers* slightly zygomorphic. *Perianth segments* 6, the outer shorter and broader than the inner, orange with darker markings, free, without a corona. *Stamens* 6, curved. *Style* 1; stigma 3-lobed. *Ovary* inferior, 3-celled. *Fruit* a capsule, splitting along the centre of the ovary cells.

38. Alstroemeria L.

Perennial herb with tuberous roots. *Leaves* all cauline, linear-lanceolate. *Umbel* with 3–7 rays, each ray with 1–3 flowers. *Perianth segments* orange, the outer tipped green, the inner spotted and streaked red. *Stamens* 6, shorter than perianth segments. *Ovary* inferior, 3-celled. *Fruit* a capsule.

About 50 species, native of South America.

1. A. aurea Graham   Peruvian Lily
A. aurantiaca D. Don ex Sweet

Glabrous *perennial herb* with tuberous roots. *Stems* up to 1 m, erect. *Leaves* 7–10 × 0.5–2.0 cm, all cauline, alternate, linear-lanceolate, obtuse at apex, entire, sessile. *Inflorescence* a terminal simple or compound umbel, without a spathe. *Umbel* with 3–7 rays, each ray bearing 1–3 flowers. *Perianth segments* 6, almost equal, 4–6 × 1–2 cm, the outer orange or yellow, broadly obovate, obtuse at apex with a green tip, narrowed to a slender claw, the inner narrower, orange or yellow with red spots and streaks or the lower without marks, lance-olate, with an acute projection at the apex. *Stamens* 6, shorter than the perianth segments; filaments recurved; anthers orange. *Style* slender and downward curving; stigmas 3-fid. *Capsule* ellipsoid, acuminate at apex, 6-ribbed. *Seeds* subglobose, tuberculate. *Flowers* 6–8. $2n = 16$.

Introduced. Much grown in gardens and naturalised in grassy places and the sites of old gardens. Scattered over Britain north to central Scotland. Native of Chile.

(**Smilex aspera** L. and **S. excelsa** L. are recorded as garden relics.)

173. IRIDACEAE Juss. nom. conserv.

Perennial herbs, with rhizomes, corms or rarely bulbs or swollen roots. *Stems* usually erect. *Leaves* all or mostly basal, those of the stem alternate and usually smaller and few, not distinguished into a lamina and petiole, entire, sessile and usually with a sheathing base; exstipulate. *Flowers* solitary or in terminal spikes or panicles usually with sheathing, often spathe-like bracts, bisexual, epigynous, actinomorphic or more rarely zygomorphic. *Perianth* in 2 series, the 6 segments usually connate at base into a longer or shorter tube, sometimes free, the 3 outer not or obviously different from the 3 inner. *Stamens* 3, free or partially connate. *Style* usually with 3 branches, the branches divided or not with the stigmas at the tip or on the underside of branches. *Ovary* inferior, 3-celled, with numerous ovules on axil placentas, rarely 1-celled with 3 parietal placentas. *Capsule* loculicidal, opening by valves, usually with a conspicuous circular scar at the top marking the point of attachment of the perianth.

About 60 genera and 800 species, widely distributed throughout the world.

Baker, J. G. (1892). *Handbook of the Irideae.* London.
Innes, C. (1985). *The World of Iridaceae.* Ashington.

1. Style branches broad and petaloid; flowers like those of an *Iris* — 2.
1. Style branches not petaloid, narrow; flowers not like those of an *Iris* — 3.
2. Plant without rhizome or bulb; roots tuberous; ovary 1-celled — **5. Hermodactylus**
2. Plant with rhizomes or bulb; roots not tuberous; ovary 3-celled — **6. Iris**
3. Flowers 1–few, erect, arising direct from the ground or on very short stems like a *Crocus* — 4.
3. Flowers few to many, erect or laterally directed, arising from aerial green stems in spikes or panicles, not like a *Crocus* — 5.
4. Leaves subterete, without a white line; perianth tube less than 10 mm, sheathed by a green bract — **8. Romulea**
4. Leaves flat, channelled and with a central whitish line on the upper side; perianth tube more than 15 mm, sheathed by a white or brown bract — **9. Crocus**
5. Perianth actinomorphic, with radial symmetrical lobes and a straight tube; plant with or without a corm — 6.
5. Perianth zygomorphic, often with bilaterally asymmetrical lobes but sometimes only due to the curved tube; plant with a corm — 12.
6. Perianth tube fused proximally into a lobe more than 5 mm — 7.
6. Perianth lobes completely free or fused proximally into a tube less than 5 mm — 8.
7. Perianth tube more than 30 mm, the lobes shorter; bracts less than 20 mm, 3-toothed at apex, without dark streaks — **11. Ixia**
7. Perianth tube less than 20 mm, the lobes more than 20 mm; bracts more than 20 mm, deeply and jaggedly toothed at apex, with irregular dark longitudinal streaks — **12. Sparaxis**
8. Inner perianth segments about twice as long as outer — **1. Libertia**
8. Inner and outer perianth segments more or less the same length — 9.
9. Stem terete, arising from a corm; flowers sessile — **11. Ixia**
9. Stem flattened and narrowly winged, arising from a rhizome or fibrous roots; flowers stalked — 10.
10. Perianth segments twisting spirally after flowering; filaments free, arising from the top of a short perianth tube — **4. Aristea**
10. Perianth segments not twisting after flowering; filaments fused either just at the base or for most of its length, arising from the base of the perianth — 11.
11. Plants with a corm; outer perianth segments more than 25 mm — **2. Homeria**
11. Plants with fibrous roots and short or no rhizomes; outer perianth segments less than 20 mm — **3. Sisyrinchium**
12. Style 3-branched, each branch bifid — 13.
12. Style with 3 simple branches, or unbranched with a 3-lobed stigma — 14.
13. Leaves tough; bracts more than 20 mm; spike erect, with flowers on 2 sides of the axis; seeds winged — **7. Watsonia**
13. Leaves soft; bracts less than 15 mm; spike bent horizontally near the lowest flower, with flowers on 1 side; seeds not winged — **13. Freesia**
14. Uppermost perianth lobe more than twice as long as the rest — **15. Chasmanthe**
14. Uppermost perianth lobe slightly longer to slightly shorter than the rest — 15.
15. Perianth lobes much narrowed at the base; style branches much widened distally — **10. Gladiolus**
15. Perianth lobe not narrowed at the base; style branches filiform with minutely capitate stigmas — **14. Crocosmia**

1. Libertia Spreng. nom. conserv.

Perennial herbs with short, creeping rhizomes and fibrous roots. *Stems* erect, with few reduced leaves. *Leaves* numerous, mostly basal, linear, flat with 2 identical faces. *Inflorescence* a small, terminal panicle. *Flowers* actinomorphic. *Perianth* segments free, spreading, the inner usually longer than the outer but of a similar shape and white. *Stamens* 3; filaments slightly fused at base. *Style* with 3, entire, linear, spreading branches. *Fruit* a many-seeded, 3-celled capsule.

About 20 species from Australasia and the temperate or montane regions of South America.

L. caerulescens Kunth has escaped from a garden on to the gravelly bed of the River Flesk, near Killarney, Kerry.

1. Pedicels shorter than and obscured by bracts; inner perianth segments 12–18 mm   **1. formosa**
1. Pedicels exceeding bracts by 5–10 mm; inner perianth segments 6–9 mm   **2. elegans**

1. L. formosa Graham   Chilean Iris
L. chilensis (Molina) Klotzsch ex Baker nom. illegit.; *L. ixioides* Klatt; *L. grandiflora* auct.

Densely tufted, glabrous *perennial herb* with short rhizomes. *Stem* up to 1.2 m, pale green, erect, unbranched. *Leaves* dark green, rigid; the basal 15–45(–75) cm × 6–12 mm, linear, acute at apex, entire, flat with 2 identical faces, with 1 or 2 reduced sheathing leaves on the stem just below the inflorescence. *Flowers* about 25 mm in diameter, in many, dense, terminal umbels; outer bract of each umbel, large, obovate and membranous; inner bract smaller and oblong; pedicel shorter than bracts. *Perianth segments* 6, the outer 6–9 mm, small, greenish-brown at apex, whitish at base, ovate or oblong and keeled, the inner 12–18 mm, white or pale yellow, entire, slightly crisped, retuse at apex, cordate at base, with a distinct midrib and very faint laterals. *Stamens* 3, 6–10 mm, inserted into the base of the corolla, opposite to the outer perianth segments; filaments pure white, erect, connate for about one-quarter of their length; anthers yellow, incumbent, oblong. *Style* white, shorter than the stamens, divided into 3 where the filaments cohere. *Capsule* spherical, about 6 mm in diameter. *Flowers* 5–6. 2*n* = 114.

Introduced. A garden plant naturalised on rough ground, waysides and rocky lake shores and coasts. Isles of Scilly, Cornwall, Isle of Man, Kintyre and Co. Kerry. Native of Chile.

2. L. elegans Poepp.   Lesser Chilean Iris

Densely tufted, glabrous *perennial herb* with short rhizomes. *Stem* rather slender, pale green, erect, branched above. *Leaves* 15–45 cm × 6–12 mm, dark green, rigid, linear, acute at apex, entire, flat with 2 identical faces, with 1 or 2 reduced sheathing leaves on the stem just below. *Flowers* about 13 mm in diameter, large, in terminal umbels; outer bracts ovate, membranous, the inner short, 6–9 mm; pedicel white, exceeding bracts by 5–10 mm. *Perianth segments* 6; the outer lanceolate-elliptical, membranous, the inner 6–9 mm, white, ovate-subrotund, obtuse at apex. *Stamens* 3, inserted at the base of the corolla, opposite the outer perianth segments; filaments white, erect, connate for about one-quarter of their length; anthers yellow, incumbent, oblong. *Style* shorter than the stamens, divided into 3 where the filaments cohere; stigma obtuse. *Capsule* cylindrical. *Flowers* 5–6.

Introduced. A rare garden plant formerly well naturalised on a railway embankment at Helensburgh in Dunbartonshire.

2. Homeria Vent.

Perennial herb with a corm covered by a netted tunic. *Stem* erect, glabrous, usually branched. *Leaf* 1, trailing. *Inflorescence* with several flowers, each flower subtended by 2 leafy bracts. *Flowers* pale golden yellow or pink, actinomorphic. *Perianth segments* 6, free, almost equal or the inner slightly smaller, with a basal claw, their lamina with a basal nectary. *Stamens* 3; filaments united. *Style* with 3, entire, flattened branches. *Fruit* a cylindrical capsule.

Contains 31 species from southern Africa.

1. H. collina (Thunb.) Salisb.   Cape Tulip
Moraea collina Thunb.; *H. breyniana* G. Lewis

Perennial herb with a corm about 10 mm in diameter, covered by a brownish, netted tunic. *Stems* 16–38 cm, leafy, erect, simple or branched. *Leaf* one, usually trailing, up to 60 cm × 4–10 mm, linear to strap-like, entire; leaves and bracts on the stem green, lanceolate, long-acuminate at apex. *Inflorescence* of several flowers. *Flowers* scented, each with 2 bracts. *Perianth segments* up to 35 mm, free, subequal, pale golden yellow, pink or peach, with or without darker veins outside, obovate or oblong, rounded at apex, the claw papillose at base, with deep golden, green-edged nectaries. *Stamens* 3; filaments up to 6 mm, united, slightly downy at base; anthers up to 5 mm. *Style* with 3, flattened branches 5–6 mm. *Capsule* cylindrical, flat-topped. *Flowers* 6–8.

Introduced. Self-sown in gardens in Tresco in the Isles of Scilly and in Guernsey. Native of South Africa.

3. Sisyrinchium L.

Perennial herbs with fleshy or fibrous roots, rhizomes very short or absent. *Stems* erect, simple or branched. *Leaves* mostly basal, 2-ranked, linear to strap-shaped, often blackening on drying. *Inflorescence* a terminal cyme or panicle of terminal and lateral cymes; spathe bracts usually in pairs. *Flowers* solitary or in clusters, actinomorphic, white, yellow, blue or purple. *Perianth segments* very shortly fused at base, spreading, more or less equal in size. *Stamens* 3, arising from the base of the perianth; filaments slightly fused at base to fused for most of their length. *Style* with 3, entire, linear branches. *Fruit* a globose- or ellipsoid-trigonous capsule.

Species delimitation in this genus has caused many problems; the genus has been variously estimated to contain between 70 and 200 species in the New World. Some authors have also considered that one or two species may be native to western Europe.

Ingram, R. (1967). On the identity of the Irish populations of *Sisyrinchium*. *Watsonia* **6**: 283–289.
Parent, G. H. (1980). Le Genre *Sisyrinchium* L. (Iridaceae) en Europe: un Bilan Provisoire. *Lejeunia* **99**: 1–40.

1. Perianth segments blue   2.
1. Perianth segments mostly cream to yellow   3.
2. At least some stems branched, each branch with 1 terminal inflorescence; flowers 15–20 mm in diameter, pale blue; pedicels arched to pendulous in fruit
        1. bermudiana

2. Stem unbranched, with 1 terminal inflorescence; flowers 25–35 mm in diameter, violet-blue; pedicels erect in fruit   **2. montanum**

3. Stem unbranched, with terminal and several lateral clusters; leaves more than 10 mm wide   **5. striatum**

3. Stem branched or unbranched, with one terminal cluster 4.

4. Stem unbranched; perianth segments bright yellow   **3. californicum**

4. Stem branched; perianth segments cream to pale yellow   **4. laxum**

1. S. bermudiana L.   Blue-eyed Grass

S. gramineum auct.; *S. anceps* auct.; *S. graminoides* Bickn.; *S. hibernicum* Å. & D. Löve; *S. angustifolium* Mill.

Glabrous, green or glaucous *perennial herb* drying black, with fibrous roots and a short rhizome, or rhizome absent. *Stems* 15–50 cm, usually branched, ascending to suberect, winged, with the wing margins slightly broadened upwards. *Leaves* up to 5 mm wide, from half to three-quarters as long as the stem, or rarely equalling it, thin and grass-like, minutely serrulate or denticulate on the margin. *Branches* each with one terminal inflorescence, winged, mostly suberect and unequal. *Bracts* of spathe usually green, sometimes purplish, usually serrate-scabrous on the keel, the outer one with scarious margins. *Flowers* 15–20 mm in diameter. *Perianth segments* 6–10 mm, pale blue, sparsely pubescent on the outer surface, connate, obovate, retuse at apex, long-mucronate. *Stamens* 3, inserted in the throat of the short perianth tube; filaments connate almost to top. *Capsule* about 5 mm, globose-trigonous, spreading or even recurved on a slender pedicel. *Seeds* black, globose, more or less pitted. *Flowers* 6–7. 2n = 64, 88.

Probably native. Wet meadows and stony ground by lakes. Western Ireland from Cork to Donegal where it was first found in 1845. Eastern North America. There has been much discussion as to whether our plant is a native endemic, *S. hibernicum*, or introduced from America, *S. bermudiana*.

2. S. montanum Greene   American Blue-eyed Grass

S. bermudiana auct.; *S. gramineum* Lam. nom. illegit.; *S. anceps* Cav. nom. illegit.; *Ferraria pulchella* Salisb. nom. illegit.; *S. montanum* var. *crebrum* Fernald

Glabrous, green, *perennial herb* often drying black, with fibrous roots and a short rhizome. *Stems* up to 60 cm, simple or occasionally forked, stiffly erect, flattened with winged margins. *Leaves* 1–3 mm wide, about half the length of the stem, but sometimes equalling it, smooth or minutely serrulate. *Inflorescence* usually solitary. *Bracts* of spathe deep green, often purple-tinged, very unequal, the outer often twice the length of the inner. *Flowers* 25–35 mm in diameter; fruiting pedicels erect. *Perianth segments* 10–18 mm, violet-blue, sparsely pubescent on outer surface, obovate-cuneate, apiculate at apex. *Stamens* 3, inserted in the throat of the short perianth tube; filament tube always equal to style. *Capsules* 1–9, dark when ripe, often purplish tinged, 3–6 mm, globose. *Seeds* mostly obliquely obovate-oblong,

often angled, brownish, smooth or with a coarse shallow pitting. *Fruiting pedicels* erect, scarcely exceeding the inner spathe bract. *Flowers* 6–7. 2n = 96.

Introduced. Naturalised in grassy places, rough ground and waysides. Scattered in Britain north to Inverness-shire. Has been much confused with *S. bermudiana*. Native of North America.

3. S. californicum (Ker Gawl.) W. T. Aiton   Yellow-eyed Grass

Marica californica Ker Gawl.; *Olsynium luteum* Raf.; *S. brachypus* (Bickn.) J. N. Henry; *Hydrastylus brachypus* Bickn.; *Hydrastylus borealis* Bickn.; *S. boreale* (Bickn.) J. N. Henry; *Hydrastylus californicus* (Ker Gawl.) Salisb.; *Echthronema californica* (Ker Gawl.) Herb.; *Bermudiana californica* (Ker Gawl.) Greene; *S. convolutum* Klatt; *S. flavidum* Kellogg; *S. flavum* Hoffm. ex Steud.; *S. lineatum* Torr.

Glabrous, tufted *perennial herb*. *Stem* up to 60 cm, unbranched, erect, leafless, broadly winged. *Leaves* in grass-like tufts, 12–30 × 3–6 mm, about half as long as the stem, bluish-green, linear, acute at apex, entire. *Inflorescence* of 2–7 flowers, in a terminal cyme. *Bracts* of spathe 15–50 mm, equal or unequal, green. *Flowers* about 40 mm in diameter; pedicels erect when fruiting. *Perianth segments* 8–18 mm, bright yellow with brownish veins, oblanceolate to obovate or elliptical-oblong, rounded at apex. *Stamens* 3; filaments connate for less than half their length; anthers 2.5–4.0 mm, yellow. *Style* with slender, spreading branches. *Capsule* 8–12 mm, blackish-purple when ripe, oblong-ovoid or ellipsoid-trigonous. *Seeds* about 1.2 mm, black, subglobose, strongly pitted. *Flowers* 6–7. 2n = 34.

This taxon is sometimes divided into two species: those which have the perianth segments 11–18 mm and come from California are *S. californicum* sensu stricto and those which have the perianth segments 8–12 mm from Washington Co. are *S. boreale* (Bickn.) J. N. Henry. Those from Oregon are dwarf and constitute another taxon, *S. brachypus* (Bickn.) J. N. Henry. The above description covers all of them.

Introduced. Grown in gardens and naturalised in damp grassy places near the sea. Monmouthshire, Pembrokeshire, Co. Wexford and Galway. Native of western North America.

4. S. laxum Otto ex Sims   Veined Yellow-eyed Grass

S. iridifolium Kunth subsp. *valdivianum* (Phil.) Ravenna

Glabrous *perennial herb*. *Stems* up to 45 cm, branched, flattened and winged. *Leaves* much shorter than stem, up to 10 mm wide, green, ensiform, acute at apex, entire. *Inflorescence* terminal on stem branches with 2–3 flowers. *Bracts* green with a pale margin, linear-lanceolate, acute at apex. *Flowers* up to 15 mm in diameter. *Perianth segments* 12–15 mm, subequal, whitish to pale yellow with purple veins, broadly elliptical, rounded-mucronate at apex. *Stamens* 3; filaments connate; anthers yellow. *Style* with slender, spreading branches. *Capsule* ovoid. *Seeds* black. *Flowers* 6–7.

Introduced. Jersey. A persistent weed in a garden at St

Ouen, Jersey, from which, however, it has not spread; and spreading onto gravel paths of St Brelade's churchyard, Jersey, where it was planted many years ago. Native of South America.

5. S. striatum Sm.   Pale Yellow-eyed Grass
Marica striata (Sm.) Ker Gawl.; *Phaiophleps nigricans* (Phil.) R. C. Foster

Glabrous, evergreen *perennial herb* with slender root fibres and often short-lived. *Stems* 45–75 cm, unbranched, erect, strongly compressed, narrowly winged. *Leaves* in tufts of 8–10, 20–35 × 12–20 mm, greyish-green sometimes cream-striped, long-linear to narrowly linear-lanceolate, acute at apex, entire. *Bracts* of spathe ovate, slightly chaffy. *Inflorescence* of several sessile clusters spread along the stem. *Flowers* 10–20 in a cluster; pedicels equalling bracts. *Perianth segments* 15–18 mm, pale yellow and strongly veined with purplish-brown, oblanceolate, rounded at apex. *Stamens* 3; filaments united only at base; anthers yellow. *Capsule* about 6 mm in diameter, spherical. *Flowers* 6–7. 2n = 18.

Introduced. Grown in gardens and naturalised on tips, waste ground, banks and waysides. Scattered in southern England from Surrey to the Isles of Scilly. Native of South America.

4. Aristea Ait.

Perennial herbs with creeping rhizomes bearing fibrous roots. *Stems* erect. *Leaves* mostly basal, 2-ranked, linear. *Inflorescence* a loose terminal panicle of few-flowered clusters. *Flowers* actinomorphic, blue. *Perianth segments* free except for a small area at the base, more or less equal, twisting spirally after flowering. *Stamens* 3; filaments free, inserted at the top of the short perianth tube. *Style* very slender, with a 3-lobed stigma. *Fruit* cylindrical.

About 50 species, almost 40 of which are in southern Africa, the rest further north in tropical Africa.

1. A. ecklonii Baker   Blue Corn Lily
A. dichotoma auct.; *A. caerulea* auct.

Perennial herb with creeping rhizome bearing fibrous roots. *Stems* 30–60 cm, erect, flattened, bearing several much reduced leaves. *Leaves* 15–60 × 0.6–1.2 cm, dark green, red at base, linear, acute at apex, entire, not rigid. *Inflorescence* a loose terminal panicle of few-flowered clusters. *Flowers* actinomorphic, around 2 cm in diameter; pedicels 6–12 mm. *Spathe* with outer bracts up to 6 mm, green with a membranous margin, lanceolate. *Perianth segments* more or less equal, 8–15 mm, blue, the outer oblong, the inner ovate, fused for less than 5 mm at base. *Stamens* 3; filaments free, inserted at the top of the perianth tube. *Style* very slender, with 3-lobed stigma. *Capsule* 12–18 mm, cylindrical. *Flowers* 5–7. 2n = 64.

Introduced. A garden escape naturalised in rough ground on Tresco, Isles of Scilly. Native of South Africa.

5. Hermodactylus Mill.

Perennial herbs with a rootstock of 2–4, palmately branched tubers. *Stems* erect. *Leaves* in two ranks, narrowly linear, subterete, 4-angled, longer than stem.

Flowers terminal, solitary, actinomorphic, partly enclosed in a spathe, yellowish-green with purplish-brown marks. *Perianth segments* shortly fused at base into a tube. *Stamens* 3, opposite the outer perianth segments. *Style* with 3, linear branches each with two lobes at the apex beyond the stigma. *Ovary* one-celled, with a short, slender, sterile beak at the apex.

A single species found in the central and east Mediterranean region.

1. H. tuberosus (L.) Mill.   Snake's-head Iris
Iris tuberosa L.

Glabrous *perennial herb* with a rootstock of 2–4, palmately branched tubers, without rhizomes, bulbs or corms. *Stems* 20–40 cm, slender, erect, covered with sheathing leaves. *Leaves* in 2 ranks, greyish-green, up to 50 cm and longer than stems, 1.5–3.0(–5.0) mm wide, narrowly linear, acute at apex, entire, 4-angled. *Flowers* 4–5 cm in diameter, scented, terminal and solitary. *Spathes* 1–2, equalling or exceeding flowers, green. *Pedicel* 10–50 mm. *Perianth segments* fused at the base into a tube up to 5 mm; the outer 40–50 mm, yellowish-green with purplish-brown to blackish on the limb, ovate-subrotund to obovate-oblong, obtuse at apex, patent; the inner 20–25 mm, yellowish-green, linear-oblanceolate, mucronate or cuspidate at apex, erect. *Stamens* 3, opposite the outer perianth segments; anthers yellow. Style with 3 linear branches each with 2 lobes at apex beyond the green stigma. *Ovary* with a very slender sterile upper portion 6–8 mm. *Capsule* obovoid. *Flowers* 4–5. 2n = 20.

Introduced. A garden plant naturalised in grassy or waste places, sandhills and hedgerows. Somerset, Devon and Cornwall. Native of the Mediterranean region.

6. Iris L.

Perennial herbs with rhizomes or rarely bulbs. *Stems* erect. *Leaves* ensiform or linear, flat with two identical faces, subterete or 4-angled. *Flowers* 1 to several, borne within 2 spathes, terminal, cymose, actinomorphic. *Perianth segments* united at the base into a hypanthial tube; outer (sometimes called *falls*) usually deflexed and larger than inner, narrowed towards the base into a claw (*haft*), the blade sometimes bearded; inner (sometimes called *standards*) usually erect or arching, sometimes horizontal or deflexed, narrowed towards the base into a claw, rarely bearded or very reduced. *Stamens* 3; filaments free, borne at base of outer perianth segments. *Style* with 3, long, broad branches each with 2 lobes at the apex protruding beyond the stigma, each covering a stamen. *Capsule* cylindrical to ellipsoid, more or less round to triangular in cross-section, often with 3–6 ribs. *Seeds* numerous, sometimes bearing a fleshy appendage (*aril*).

Contains about 250 species distributed throughout the northern hemisphere. Numerous species and varieties are cultivated in gardens.

Dykes, W. R. (1915). *The Genus Iris*. Cambridge.
Lynch, R. Irwin (1904). *The book of the Iris*. London & New York.
Mathew, B. (1981). *The Iris*. London. Reprinted with revision 1989. Portland, Oregon.

I. danfordiae (Baker) Boiss., **I. reticulata** M. Bieb. and **I. × unguicularis** Poir. have been recorded as garden escapes.

1. Plant with a bulb; leaves subterete or slightly flattened, angled or channelled   2.
1. Plant with a rhizome; leaves flat, not channelled or angled, vertical with 2 identical faces   4.
2. Perianth tube more than 10 mm   **12. × hollandica**
2. Perianth tube less than 10 mm   3.
3. Leaves dying down in winter; haft of outer perianth segments more than 20 mm wide, no longer than blade   **10. latifolia**
3. Leaves evergreen; haft of outer perianth segments less than 20 mm wide, 1.5–2.0 times as long as blade   **11. xiphium**
4. Outer perianth segments bearded on inner face   **1. germanica**
4. Outer perianth segments not bearded, but sometimes softly hairy   5.
5. Perianth segments predominantly yellow or yellow and white, without blue, purple, mauve or violet or only the veins or spots with it   6.
5. Perianth segments predominantly of some shade of blue, purple, mauve or violet   8.
6. Stems distinctly compressed; leaves evergreen, dark green, with a stinking smell when crushed; seeds bright orange to orange-red   **9. foetidissima**
6. Stems subterete; leaves dying down in winter, mid- to pale green, not stinking; seeds brownish   7.
7. Perianth segments yellow, the outer often with brownish or purple spots on the veins; style lobes petaloid, deeply serrate   **3. pseudacorus**
7. Inner perianth segments white with a large yellow patch on the blade; style lobes petaloid, subentire   **8. orientalis**
8. Leaves evergreen, dark green, with stinking smell when crushed; seeds bright orange   **9. foetidissima**
8. Leaves dying in winter, mid- to pale green, not stinking; seeds brownish   9.
9. Stems hollow; bracts brown and papery when flowering; perianth tube usually 4–7 mm   **2. sibirica**
9. Stems solid; bracts at least partly green when flowering; perianth tube 7–20 mm   10.
10. Ovary without sterile apical part; capsule without or with a short beak less than 5 mm   11.
10. Upper part of ovary sterile, narrower than ovary below it and perianth tube above, forming an acuminate beak on the capsule more than 5 mm   12.
11. Outer perianth segments glabrous on central patch; capsules setting many seeds   **4. versicolor**
11. Outer perianth segments hairy on central patch; capsules not setting seed or setting a few seeds   **5. × robusta**
12. Leaves mostly less than 10 mm wide; flowers mostly more than 8 cm across; capsule with a beak 5–8 mm, with 1 rib where the 2 ovary cells meet   **6. ensata**
12. Leaves mostly more than 10 mm wide; flowers mostly less than 8 cm across; capsule with a beak 8–16 mm, with 2 ridges where the 2 ovary cells meet   **7. spuria**

Subgenus 1. Iris

Stock a rhizome. *Stem* usually solid. *Leaves* isobilateral.

Section 1. Iris

Sect. *Pogiris* Tausch; Sect. *Pogoniris* (Spach) Baker

Rhizomes stout, of uniform diameter. *Stems* simple or branched. *Outer perianth segments* with a beard of multicellular hairs. *Seeds* without an aril.

1. I. germanica L.   Bearded Iris

Robust rhizomatous *perennial herb*, with a stout, horizontal rhizome of uniform diameter and tough roots. *Stems* up to 1.2 m, glabrous, usually branched, the branches spreading, much exceeding the bracts or cauline leaves. *Leaves* glaucous-grey, flat; basal up to 57 cm × 25–45 mm, equitant, isobilaterally folded, ensiform, entire, with numerous longitudinal, parallel veins; cauline usually much reduced, not tubular or closely investing the stems; bracts about 40 × 20 mm, oblong-navicular, acute at apex, herbaceous and greenish towards the base with a brownish scarious apex and margins. *Inflorescence* rather open, with a 2- to 3-flowered terminal head of flowers and 2–4 lateral, 2(–3)-flowered heads. *Flowers* 9–10 cm in diameter, fragrant, shortly pedicellate. *Perianth tube* 20–30 mm, greenish with purplish streaks extending from the bases of the inner perianth segments; outer perianth segments 90–100 × 40–50 mm, broadly obovate, reddish-lilac or violet, sometimes white or yellow, rounded at apex, obscurely clawed at base, the beard about 40 mm, conspicuous, its hairs long, white and tipped orange; inner perianth segments up to 70 mm, usually broader than outer, with plicate-undulate margins. *Stamens* 3; filaments 10–15 mm, whitish, glabrous; anthers about 20 × 2 mm, creamy, linear, with a shortly sagittate base. *Style* branches about 50 × 15 mm, paler than the perianth segments, deeply 2-lobed at apex, the lobes irregularly dentate-lacerate and acute. *Capsule* 38–50 mm, trigonous, tapering to a point at the apex. *Seeds* rarely formed in the British Isles, pyriform or ellipsoid, not compressed, dark reddish-brown, wrinkled, without an aril. *Flowers* 5–6. $2n = 36, 44, 48$.

Introduced. Much grown in gardens and naturalised on rough and waste ground, banks, waysides and derelict gardens. Frequent in central and south Britain and the Channel Islands. Of unknown and probably garden origin, but perhaps native of the east Mediterranean region.

Section 2. Spathula Tausch

Rhizome usually slender, of uniform diameter. *Stems* simple or branched. *Outer perianth segments* glabrous or puberulent with unicellular hairs. *Seeds* with an aril.

2. I. siberica L.   Siberian Iris

I. maritima Mill.; *I. flexuosa* Murray; *I. pratensis* Lam.; *I. angustifolia* Mill.; *I. stricta* Moench; *I. acuta* Willd.; *I. trigonocarpa* A. Braun; *I. cirrhiza* Posp.; *Xiphium sibericum* (L.) Schrank; *Xiphium flexuosum* (Murray) Alef.; *Xyridion flexuosum* (Murray) Klatt; *Xyridion sibericum* (L.) Klatt; *I. orientalis* Thunb., non Mill.

Glabrous *perennial herb* with a slender, shortly creeping, closely tufted rhizome clothed with the fibrous remains of old leaves. *Stems* 50–120 cm, much overtopping the leaves, somewhat compressed, usually branched. *Leaves* dying in autumn, 25–80 × 2–10 mm, green or slightly glaucous and sometimes tinged pink near the base, flat, linear, acute at apex, entire, not rigid; bracts 30–50 mm, lanceolate, acute at apex, brown and papery. *Inflorescence* rather open, with up to 12 flowers. *Flowers* 6–7 cm in diameter; pedicels 12–75 mm. *Perianth tube* 4–7(–12) mm, with many indistinct veins, dark violet; outer perianth segments 30–60 mm, the blade obovate, subrotund or oblong, bluish-violet with a white, strongly violet-veined area in the centre, narrowed to a paler haft which has prominent dark veining; inner perianth segments smaller and more uniform violet, broadly lanceolate, with slightly channelled haft; there are nearly white-flowered cultivars and others from all shades of blue to deep violet-blue. *Stamens* 3; filaments free, purplish or white; anthers deep blue or cream. *Style* shorter than inner perianth segments. *Capsule* 30–40 mm, twice as long as wide, ellipsoid or oblong, obtusely trigonous. *Seeds* large, flat, thin and D-shaped, with an aril. *Flowers* 5–7. 2n = 28.

Introduced. A common garden plant which escapes and becomes naturalised in rough, often wet or shaded ground. Scattered through Britain north to Inverness-shire. Native from central Europe to southern Asia.

3. I. pseudacorus L.   Yellow Iris
I. palustris L. ex Moench; *I. acoriformis* Boreau; *I. bastardii* Boreau; *I. sativa* Mill.; *I. lutea* Lam.; *I. longifolia* Lam. & DC.; *I. paludosa* Pers.; *I. curtopetala* Delavay; *I. acoroides* Spach; *Xiphium pseudacorum* (L.) Pavlov; *Xiphion acoroides* Alef.; *Limniris pseudacorus* (L.) Fuss; *Xyridion pseudacorus* (L.) Klatt; *Xyridion acoroideum* (Spach) Klatt; *I. flava* Tornab.

Glabrous *perennial herb* with a stout, horizontal, creeping rhizome of uniform diameter, clothed with the fibrous remains of old leaves. *Stems* 60–160 cm, usually branched, slightly compressed. *Leaves* dying down in winter, 50–90 cm × 10–30 mm, more or less glaucous, flat, long-linear to ensiform, acute at apex, entire, with a prominent midrib; bracts 4–10 cm, green with a narrow, membranous margin. *Inflorescence* with 4–12 flowers, the lower on long, erect peduncles. *Flowers* 7–10 cm in diameter; pedicels 20–50 mm. *Perianth tube* 10–15 mm, narrowly infundibuliform; outer perianth segments 50–75 × 20–30 mm, reflexed, the blade pale to medium yellow and longer than the haft which usually has a brighter spot at the throat and radiating brownish veins, without a beard; inner perianth segments 20–30 × 4–8 mm, yellow, oblong, erect. *Stamens* 3; filaments cream; anthers cream to orange, edged with dark purple or wholly purplish-brown. *Styles* broad, keeled, oblong or almost triangular; stigma prominent and tongue-like. *Capsule* 32–56 mm, oblong and bluntly trigonous with 6 faint furrows and a short beak. *Seeds* flattened, D-shaped or nearly circular, with a smooth, brown testa and aril. *Flowers* 6–8. 2n = 24, 30, 32, 34, 40.

This plant shows much variation in the size of its parts, amount of glaucousness in its leaves and stem, depth of colour in the yellow of the perianth segments from pale (var. *bastardii* (Boreau) Lynch) to deep yellow, presence or absence of a deeper colour in the throat of the perianth which varies from deep yellow to orange and various degrees of brown or violet veining in the perianth segments. In cultivation dwarf, double-flowered, almost white and variegated forms occur.

Native. Wet meadows, fens and ditches and by lakes, streams and rivers. Common throughout the British Isles. Most of Europe; North Africa; Caucasus and western Asia. Medicinally the plant has been used as a strong purgative.

4. I. versicolor L.   Purple Iris
Xiphion versicolor (L.) Alef.; *I. caroliniana* S. Watson; *I. caurina* Herb.; *I. pulchella* Regel; *I. flaccida* Spach

Glabrous *perennial herb* with a stout, creeping rhizome bearing the fibrous remains of old leaves. *Stems* 20–80(–100) cm, branched, solid. *Leaves* dying in winter, 35–60 cm × 8–25 mm, equalling or slightly shorter than the stems, rather glaucous, often tinged with purple near the base, flat, ensiform, acute at apex, entire, thickened along the centre but not bearing a raised midrib; bracts 30–80 mm, green. *Inflorescence* with 2–9 flowers, the lower on long ascending peduncles. *Flowers* 60–80 mm in diameter; pedicels 25–55 mm. *Perianth tube* 7–10 mm, stout; outer perianth segments 40–60 × 15–27 mm, some shade of violet, bluish-purple, reddish-purple, lavender or dull slaty-purple, rarely white, the blade often having a greenish-yellow blotch in its centre surrounded by a white area variegated with purple veins which continue down the haft, ovate or subrotund, spreading; inner perianth segments 25–40 × 8–12 mm, coloured like the general colour of the outer, oblanceolate, obtuse at apex, erect. *Stamens* 3; filaments purplish; anthers dark purple or violet. *Styles* narrow with a low, rounded keel which is deeper in colour than the sides; stigma broad, triangular and tongue-like. *Capsule* 35–50 mm, oblong, obtusely trigonous with a short beak. *Seeds* numerous, flat, pale brown, with an aril. *Flowers* 5–7. 2n = 72, 84, 108, 112.

Introduced. Grown in gardens and naturalised by lakes, ponds and rivers and in reed-swamps. Scattered in Britain north to central Scotland. Native of eastern North America from Hudson Bay to the Gulf of Mexico.

5. I. × robusta E. S. Anderson   Windermere Iris
    I. versicolor × virginica L.

Perennial herb with a rhizome. *Stems* up to 80(–100) cm, solid. *Leaves* dying in winter, mid- to pale green, with 2 identical faces, flat, not channelled or angled; bracts at least partly green at flowering time. *Perianth tube* 7–20 mm; perianth segments violet to purple or greyish-purple, the outer with the central, greenish-yellow patch pubescent, surrounded by a whitish, purple-veined area. *Capsule* without or with a very short beak. *Seeds* brownish, with an aril. *Flowers* 6–7.

Introduced. Naturalised in reed swamp and rough pasture by Lake Windermere, Westmorland without either parent. Native of eastern North America and of garden origin.

6. I. ensata Thunb.   Japanese Iris
I. kaempferi Siebold ex Lem.

Glabrous *perennial herb* with a fairly stout, shortly creeping rhizome. *Stems* 60–90 cm, rarely branched, solid. *Leaves* dying in winter, 20–60 cm × 4–12 mm, shorter than the stems, slightly glaucous, ensiform, acute at apex, entire, flat, with a prominent midrib; bracts up to 75 mm, linear-lanceolate, acute at apex, green, covered with minute papillae. *Inflorescence* with 3 or 4 flowers. *Flowers* 80–150 mm in diameter; pedicels 12–50 mm. *Perianth tube* 12–18 mm, pale green; outer perianth segments 75–100 mm, the blade purple or reddish-purple, elliptic or obovate, obtuse at apex, the haft yellowish, the yellow spreading to the base of the blade; inner perianth segments about two-thirds as long as the outer, reddish-purple, erect, narrowly oblanceolate, obtuse at apex. *Stamens* 3; filaments short, purple; anthers pale yellow, more or less flushed with mauve, twice as long as the filaments. *Styles* deep violet; stigma entire, broad, with an irregularly dentate edge. *Capsule* up to 25 mm, tapering at both ends, obtusely acuminate-beaked, the sides concave, the angles grooved. *Seeds* brown, flat and more or less subrotund, with an aril. *Flowers* 6–7. 2*n* = 24.

Introduced. Naturalised in a swamp at Dartford Heath in Kent. Native of Japan, northern China and Russia in the Sakhalin, Amur and Ussuri regions.

7. I. spuria L.   Blue Iris

Glabrous *perennial herb* with a rather slender but hard rhizome which remains clothed with the bases of old leaves. *Stems* 30–90 cm, terete, usually unbranched, solid. *Leaves* dying in winter, 25–90 × 6–20 mm, about equalling the stem, dark green or slightly glaucous, linear-ensiform, tapering gradually to an acute apex, entire, flat, stiff; bracts 60–80 mm, green with a semi-transparent apex, lanceolate, acute at apex, sometimes slightly keeled, somewhat foetid. *Inflorescence* with 2–4 flowers. *Flowers* 60–80 mm in diameter; pedicels 15–25 mm. *Perianth tube* 7–10 mm, broad; outer perianth segments 30–80 × 10–30 mm, lilac with violet veins and a yellow median signal streak, the blade elliptical to subrotund, obtuse at apex, the haft separated from the blade by an obvious constriction; inner perianth segments 30–60 × 8–20 mm, deep violet blue edged with yellow in the lower part, narrowly obovate, obtuse at apex, erect. *Stamens* 3; filaments broad, dark purple; anthers purple edged with yellow. *Styles* narrowly oblong; stigma bifid, with 2 distinct points. *Capsule* 25–40 mm, oblong, with a double ridge at each angle, acuminately beaked from 8–16 mm. *Seeds* brown, smooth, angular, with a loose membranous testa and an aril. *Flowers* 6–7. 2*n* = 22.

This is really an aggregate species whose segregates have been treated in several ways. The British plant is thought to be subsp. *spuria*.

Introduced. Naturalised in fen ditches in Lincolnshire since 1835 and also in Somerset; formerly in Dorset. Native from Europe to central Asia. Subsp. *spuria* is mainly a central European plant.

8. I. orientalis Mill.   Turkish Iris
I. ochroleuca L.; *I. spuria* subsp. *ochroleuca* (L.) Dykes; *I. gigantea* Carrière

Nearly glabrous *perennial herb* with a hard, compact rhizome. *Stems* 40–120 cm, usually with one branch, sometimes more, solid, flattened. *Leaves* dying in winter, 60–90 × 1–2 cm, mid- to pale green to slightly glaucous, ensiform with a characteristic spiral twist, acute at apex, entire; bracts green, papery, lanceolate, acuminate at apex. *Inflorescence* with several flowers from each bract. *Flowers* 8–10 cm in diameter; pedicels 25–75 mm. *Perianth tube* about 12 mm, funnel-shaped; outer perianth segments 36–50 mm, white flushed with yellow at the centre, the blade subrotund, deeply and widely emarginate at the apex, reflexed at a right angle and not spreading, the haft narrow; inner perianth segments white, erect, narrow and slender and sometimes shortly hairy. *Stamens* 3; filaments slightly shorter than the anthers, pale yellow; anthers pale buff. *Styles* about 48 mm, with parallel sides; stigma bilobed, the lobes subentire. *Capsule* about 50 mm, more or less beaked, with 3 conspicuous, double, longitudinal ridges. *Seeds* flattened or wedge-shaped, with loose, white, semi-transparent wrinkled testa and an aril. *Flowers* 6–7. 2*n* = 28.

Introduced. Naturalised in limestone scrub at Sand Point, Kewstoke in Somerset since at least 1950, on banks about Northfleet and Swanscombe in Kent since 1984 and for many years in a field at Abbotsbury, Dorset. Native of the eastern Mediterranean region.

9. I. foetidissima L.   Stinking Iris
I. foetida Thunb.

Glabrous *perennial herb* with a rather slender, slow-growing rhizome and giving off a strong unpleasant smell when bruised. *Stems* 30–90 cm, branched, solid, angled on one side, slightly compressed, bearing 3–4 leaves. *Leaves* 30–70 cm × 10–25 mm, dark green, evergreen, broadly linear-ensiform, acute at apex, flat, entire; bracts 5–10 cm, green with narrow, scarious margins, linear-lanceolate, acute at apex. *Inflorescence* of 1–3 flowers. *Flowers* 5–7 cm in diameter; pedicels 2–10 cm. *Perianth tube* about 10 mm, stout; outer perianth segments 30–50 × 10–20 mm, dull violet more or less tinged with dull yellow, and with darker veins, rarely clear yellow, oblong-obovate to oblanceolate, obtuse at apex and sometimes emarginate, curved downwards; inner perianth segments 25–40 × 5–9 mm, oblanceolate, obtuse at apex and emarginate, with a short haft. *Stamens* 3; filaments short; anthers long. *Styles* thick, rather short, trigonous; stigmas 3, brownish, petaloid, with a central midrib, 2-lobed at apex. *Capsule* 4–7 cm, oblong-ellipsoid. *Seeds* 5–7 mm, deep orange or orange-red, with an aril. *Flowers* 5–7. 2*n* = 40.

Plants with the flowers uniform pale lemon-yellow without purple lines have been named var. *citrina*

Bromf. Their ecology and distribution needs further study.

Native. Dry places in woods, hedges, banks and cliffs near the sea, mostly on calcareous soils. Locally frequent in the Channel Islands, Britain north to north Wales and Norfolk; naturalised in scattered localities in Britain and Ireland elsewhere. South and west Europe from France southwards, east to north-east Italy; North Africa, Macaronesia.

Subgenus 2. **Xiphion** (Mill.) Spach

Stock a bulb. *Stem* unbranched, hollow. *Leaves* not isobilateral.

Section 3. **Xiphion** (Mill.) Parl.

Roots fibrous, rarely with a few fleshy ones. *Stems* unbranched, hollow. *Leaves* linear, channelled. *Outer perianth segments* suberect, at least half as long as the inner. *Seeds* without an aril.

10. I. latifolia (Mill.) Voss   English Iris
I. xiphioides Ehrh.; *Xiphion latifolia* Mill.; *I. pyrenaica* Bubani; *I. cepifolia* Stokes; *I. anglica* hort. ex Steud.; *I. argentea* hort.; *Xiphium jacquinii* Schoute

Glabrous *perennial herb* with a large, ovate bulb which has a thin, membranous, dark brown tunic, splitting into fibres at the apex; leafless in winter. *Stems* 25–65 cm, unbranched, hollow. *Leaves* 25–65 cm × 5–8 mm, more or less equalling the stem, the outer surface glaucous-green, the inner silvery-grey, long-linear, tapering to an acute point, entire, grooved, the upper becoming smaller and gradually merging into the bracts; bracts 7.5–10.0 mm, green, inflated and sharply keeled. *Inflorescence* terminal, of 2–3 flowers. *Flowers* 8–12 cm in diameter; pedicels 25–70 mm. *Perianth segments* 60–75 × 30–35 mm, usually violet with a yellow or orange centre, suberect, the blade subrotund or obovate-oblong, gradually narrowing to the haft which is 25–30 mm wide and obovate; inner perianth segments 45–65 × 15–20 mm, oblanceolate. *Stamens* 3; filaments white, stained and spotted with purple; anthers white, edged with blue. *Style* broad, widening in the upper part, very sharply keeled; stigmas with 2 points. *Capsule* up to 10 cm, tapering at both ends. *Seeds* dark reddish-brown, globose, wrinkled, without an aril. *Flowers* 6–7. 2n = 42.

The English Iris received its confusing name because Matthias de l'Obel saw plants of it in England in the sixteenth century and passed it on to other authors. Although it was soon known that it was not native, it was still called the English Iris.

Introduced. Grown in gardens and naturalised in a hayfield at Fawkham, Kent and in several grassy places in Shetland. Native of the Pyrenees and Cordillera Cantabrica.

11. I. xiphium L.   Spanish Iris
I. variabilis Jacq.; *Xiphium vulgare* Mill.; *Xiphium verum* Schrank; *I. coronaria* Salisb.; *I. hispanica* Steud.; *I. serotina* Willk.; *I. lusitanica* Ker Gawl.; *Xiphium sordidum* Salisb.; *I. spectabilis* Spach; *I. taitii* Foster

Glabrous *perennial herb* with an ovate bulb, covered with thin, membranous tunics not splitting into fibres at the apex and producing bulblets in pairs on opposite sides of the base. *Stems* 30–60 cm, not usually branched, hollow. *Leaves* present during the winter, 20–70 cm × (1–)3–5 mm, glaucous, long-linear, acute at apex, entire, channelled, the upper shorter and wider and merging into the bracts; bracts 10–12 cm, green. *Inflorescence* of 1–3 flowers. *Flowers* with pedicels up to 12 cm. *Perianth tube* 1–3 mm, sometimes almost absent; outer perianth segments 45–65 × 18–25 mm, usually violet with an orange or yellow centre, rarely mostly yellow or white, suberect, the blade subrotund, the haft 7–10 mm wide, narrowly oblong; inner perianth segments 45–65 × 15–20 mm, oblanceolate. *Stamens* 3; filaments and anthers varying in colour. *Capsule* 50–75 mm, long and narrow, with a hollow running down each face. *Seeds* small, yellowish-brown, subrotund and compressed or D-shaped, without an aril. *Flowers* 4–5. 2n = 34.

Introduced. Grown in gardens and cultivated in bulb fields; naturalised in old fields, rough ground and waste places. Isles of Scilly and the Channel Islands. Native of south-west Europe.

12. I. × hollandica hort.   Dutch Iris

Perennial herb with an ovate bulb and fibrous roots. *Stem* up to 50 cm, hollow, unbranched. *Leaves* present during the winter, 20–40 cm × 0.5–3.0 mm, subterete or slightly flattened, angled, channelled, long-linear, acute at apex, entire; bracts green. *Inflorescence* of several flowers. *Flowers* 8–12 cm in diameter, with rather long pedicels. *Perianth tube* more than 10 mm; segments white, yellow or bronze and blue to purple, the inner suberect and at least half as long as the outer. *Stamens* 3; filaments varying in colour. *Capsule* long and narrow. *Seeds* without an aril. *Flowers* 4–6.

Introduced. Grown in gardens and bulb fields and naturalised as a relic in old fields, rough ground and waste places. Frequent in the Channel Islands. Of garden origin consisting of a complex of hybrids involving *I. filifolia* Boiss., *I. latifolia*, *I. tingitana* Boiss. & Reut. and *I. xiphium*.

7. **Watsonia** Mill. nom. conserv.

Perennial herbs with more or less spherical corms. *Stems* erect. *Leaves* basal and cauline, 2-ranked, ensiform, usually with a prominent main vein. *Flowers* in 2 ranks in a spike, slightly zygomorphic. *Perianth* forming a curved tube at the base longer than the segments, segments more or less equal. *Stamens* 3, arising below the throat of the perianth tube; filaments free. *Style* very slender, with 3 bifid stigmas. *Fruit* an oblong capsule.

Contains about 60 species from South Africa.

W. beatricis Mathew & L. Bolus is recorded as a garden escape on the foreshore of Rhu, Dunbartonshire.

Goldblatt, P. (1989). *The Genus* Watsonia. National Botanic Gardens.
Roux, J. P. (1980). Studies in the genus *Watsonia* Miller. *Jour. South African Bot.* **46**: 365–378.

1. W. borbonica (Pourr.) Goldblatt   Bugle Lily
Lomenia borbonica Pourr.; *Gladiolus pyramidatus*
Andrews; *W. pyramidata* (Andrews) Klatt; *W. rosea*
Banks ex Ker Gawl.; *Neuberia rosea* (Klatt) Eckl.; *W.
striata* Klatt

Perennial herb. Corms 3–4 cm in diameter; tunic greyish-
brown, coarsely reticulate, inner layers often unbroken.
Stems 1.2–2.0 m, bearing two or more, imbricate, green,
sheathing bract leaves in the upper part, usually with
several branches, the lower sometimes branched again;
bracts subtending the branches dry and brown.
Inflorescence a spike with the main axis up to 20-flow-
ered, lateral spikes up to 10-flowered. *Flowers* slightly
zygomorphic. *Perianth tube* 12–15(–20) × 1.5–2.0 mm in
the lower part, emerging 3–8 mm from the bracts, upper
part 8–20 mm × 7–10 mm, flared and funnel-shaped,
nearly horizontal and slightly pendulous, widening at
the mouth; segments (26–)30–36 × 13–18 mm, spreading outwards to almost right angles to the tube, obovate
to oblanceolate, often acuminate at apex. *Stamens* 3,
arising below the throat of the perianth tube; filaments
13–18(–25) mm, unilateral, either arcuate-horizontal
and lying under the outer perianth segments or decumbent and lying above the outer perianth
segments; anthers 10–13 mm, unilateral, either arcuate-
horizontal and sometimes widely separated with the
central above the two lateral anthers or decumbent and
violet. *Style* dividing at or beyond the apex of the
anthers, arching above or below the stamens; stigma
branches up to 7 mm. *Ovary* 3–4 mm, oblong. *Capsules*
ovoid to oblong-truncate 20–25(–30) × 7–10 mm, sometimes widest at apex. *Seeds* 8–12 × up to 2.5 mm,
2–winged, the distal wing 2–3 mm, the proximal smaller
and sometimes vestigial. *Flowers* 10–12. $2n = 18$.

Introduced. Grown in gardens and naturalised in
rough ground. Well established on Appletree banks and
perhaps elsewhere on Tresco, Isles of Scilly. Native of
South Africa. Our plant is the white-flowered variant of
subsp. **ardernei** (Sander.) Goldblatt (*W. ardernei* Sander.).

8. Romulea Maratti nom. conserv.
Trichonema Ker Gawl.

Perennial herbs with corms which have hard brown
tunics, usually asymmetrical at the base. *Stem* short.
Leaves basal and cauline, subterete, 4-grooved. *Flowers*
upright, surrounded by an outer bract and an inner
bracteole. *Perianth* united into a short tube below, segments all similar. *Stamens* 3; filaments free, borne in the
perianth tube. *Style* very slender with 3 bifid stigmas, the
branches thread-like. *Capsule* borne on an elongated
stem. *Seeds* spherical.

Contains about 10 species in the Mediterrranean
region extending northwards to south-west England,
and about 70 species in South Africa, mainly in the Cape
Province.

1. Basal leaves 2; perianth 9–19 mm, the tube 2.5–5.5 mm
               1. columnae subsp. **columnae**
1. Basal leaves several; perianth 15–45 mm; the tube
  2–8 mm   **2. rosea** var. **australis**

1. R. columnae Sebast. & Mauri   Sand Crocus
Ixia parviflora Salisb., non *Romulea parviflora* Bubani;
Trichonema parviflora (Salisb.) Gray; *Trichonema
columnae* (Sebast. & Mauri) Rchb.; *R. parviflora*
Bubani nom. illegit.

Glabrous *perennial herb* with a small, ovoid, asymmetrical corm producing offsets freely. *Corm* obliquely flattened with a crescentic basal ridge on one side; tunic
brown, hard and smooth. *Sheathing leaves* 1–2, tubular
below and sheathing the aerial shoot. *Leaves* 3–8(–20)
cm, appearing before the flowers, bright dark green; the
basal 2, distichous, linear, simple, subterete, 4-grooved,
erect, twisted or curved; the cauline 1–6, 0.6–1.0 mm in
diameter, fairly short, erect or appressed or long and
slender. *Stem* usually underground at anthesis, 1- to 3-
flowered. *Flowers* each in the axil of a cauline leaf;
pedicels short, semi-terete, recurved after flowering.
Bract 6–13 mm, green with a narrow, scarious margin,
tinged purple or spotted with reddish-brown. *Bracteoles*
almost entirely scarious. *Perianth* 9–19 mm; tube 2.5–5.5
mm; segments equal, tinged green on the outer surface,
usually mauve with darker veins, sometimes white or
pale yellow inside at the base, lanceolate or oblanceolate, acute at apex. *Stamens* 3; filaments 3–5 mm, yellow,
often glabrous; anthers yellow, reaching about halfway
up the perianth. *Style* filiform with 2- to 3-fid branches;
stigmas below the top of the anthers. *Fruiting stem*
about 5 cm above ground. *Capsule* 5–11 × about 4 mm,
obovoid-trigonous to subglobose-trigonous, on a scape
up to 50 mm. *Seeds* numerous, globose, brownish,
without a strophiole. *Flowers* 3–5. $2n = c. 60$.

Native. In short, sandy turf near the sea. Very local
near Dawlish in Devon, common in all the Channel
Islands, formerly in Cornwall. Western Europe from
northern France southwards; Mediterranean region;
Macaronesia. Our plant is subsp. **columnae** which
occurs throughout the range of the species. Many varieties have been named and our plant has been referred to
var. *occidentalis* Bég.

2. R. rosea (L.) Eckl.   Onion Grass
Trichonema purpurascens Sweet; *Ixia rosea* L.

Glabrous *perennial herb* with a subglobose corm
rounded at the base. *Tunic* with basal teeth bent to one
side. *Sheathing leaves* 1–2, tubular below and sheathing
the aerial shoot. *Leaves* appearing before the flowers;
the basal several, 15–25 × 1.0–2.5 mm, erect or more or
less so, linear, entire, subterete, 4-grooved; the cauline
5–6. *Stem* usually underground at anthesis, short, 1- to
3-flowered. *Flowers* each in the axil of a cauline leaf;
pedicels 30–80 mm. *Bract* 1.5–2.5 mm, green with a
narrow scarious margin. *Bracteoles* with a wide scarious
margin, brown or spotted. *Perianth* 15–45 mm, tube
2.5–8.0 mm, segments white with yellow inside at the
base, lanceolate, obtuse at apex. *Stamens* 3; filaments
hairy at base; anthers yellow, reaching about halfway up
the perianth. *Styles* filiform, with two 3-fid branches;
stigmas below the top of the anthers. *Capsule* ovate-subrotund, membranous. *Seeds* numerous, globose. *Flowers*
4–5. $2n = 18$.

Introduced. Naturalised at the base of a wall and a sparsely grassed area on gravel. Cobo in Guernsey, first found in 1969. Native of South Africa. Our plant is the albino variant of var. **australis** (Ewart) De Vos. It is almost unknown in gardens of the northern hemisphere.

9. Crocus L.

Perennial herbs with symmetrical corms. *Sheathing leaves* 3–5, papery, tubular and whitish, enclosing the whole of the newly developing aerial shoot. *Leaves* linear, more or less flat, green with a central whitish channel on the upper surface, in some species present at the time of flowering, in others not until much later. *Spathe* subtending the base of the scape so that it is produced adjacent to the newly developing corm. *Bract* and sometimes a bracteole at the apex of the scape where the ovary is attached, subtending or sheathing the ovary and perianth tube until well above soil level. *Perianth* arises from the top of the ovary in the form of a long tube which serves instead of a stem to carry the showy part of the flower above the surface and is divided into 6 segments, which are normally in 2 subequal whorls. *Stamens* 3, joined to the outer perianth segments. *Style* 1, with 3 to many branches. *Capsule* pushed above the soil as it developes from the fertilised ovary. *Seeds* very variable. About 80 species in Europe and west and central Asia.

C. longiflorus Raf., **C. serotinus** Salisb. and **C. versicolor** Ker Gawl. have been recorded as garden escapes.

Mathew, B. (1982). *The Crocus.* London.
Wurzell, B. (1992). Spring flowering Crocuses. *B.S.B.I. News* **60**: 36–38.

1. Flowers appearing in autumn, usually without the leaves, never predominantly yellow   2.
1. Flowers appearing in spring, with or immediately before the leaves, often predominantly yellow   6.
2. Filaments densely pubescent; throat uniformly deep yellow; anthers creamy white   **12. pulchellus**
2. Filaments glabrous or papillose; throat not yellow or very pale yellow or with yellow blotches only; anthers often yellow   3.
3. · Plant stoloniferous; perianth segments not conspicuously veined, throat white or lilac   **3. nudiflorus**
3. Plant non-stoloniferous; perianth segments conspicuously darker veined outside, throat whitish, lilac-purple, pale yellow or with yellow blotches   4.
4. Style usually with many slender branches   **11. speciosus**
4. Styles with 3 main branches   5.
5. Anthers creamy white   **4. kotschyanus**
5. Anthers yellow   **5. sativus**
6. Perianth segments predominantly pale to deep yellow, sometimes tinged or striped dark purple   7.
6. Perianth segments predominantly white or pale mauve or lilac-purple to dark purple, sometimes yellow on throat   9.
7. Leaves 2.5–4.0 mm wide   **10. flavus**
7. Leaves 0.5–2.5 mm wide   8.
8. Corms with reticulated fibres; leaves 0.5–1.0 mm

wide; perianth segments bright yellow or orange, sometimes stained purplish at base   **7. ancyrensis**
8. Corms membranous, splitting into rings at the base; leaves 0.5–2.5 mm wide; perianth segments pale yellow to orange-yellow sometimes striped or suffused bronze or purple   **8. chrysanthus**
9. Perianth segments 15–30 × 4–12 mm, throat yellow; spathe absent   10.
9. Perianth segments 24–55 × 7–15 mm, throat white, mauve or purple; spathe present   12.
10. Tunic not forming rings at base   **6. sieberi**
10. Tunic forming rings at base   11.
11. Plant and leaves tall; perianth segments with conspicuously pencilled dark purple stripes below
                                   9(a). biflorus subsp. **biflorus**
11. Plant and leaves shorter and more compact; perianth segments pale lilac with 3–5 purple stripes, broader than in subsp. *biflorus*   **9(b). biflorus** subsp. **adamii**
12. Leaves 2–8 mm wide; perianth tube usually purple, only white if the rest of the perianth is white   **1. vernus**
12. Leaves 1–3 mm wide; perianth tube white, segments pale lilac to purple   **2. tommasinianus**

Section 1. Crocus

Spathe present.

1. C. vernus (L.) Hill   Spring Crocus
C. purpureus Weston; *C. napolitanus* Mord. Laun. & Loisel.

Corm 8–20 mm in diameter, depressed-globose; tunic finally fibrous, the fibres parallel or slightly reticulated, especially at the apex of the corm. *Sheathing leaves* 3–4, white or greenish. *Leaves* 2–4, present at flowering time, 90–150 × 2–6(–8) mm, shorter than or equalling the flowers, medium to bright green with a whitish median stripe above, slightly paler beneath, linear, shortly narrowed to a truncate apex, entire, glabrous, ciliate or pubescent. *Spathe* present, whitish. Bract white or sometimes veined greenish or purplish. *Bracteole* absent. *Flowers* in spring, 1(–2), 40–85 mm in diameter. *Perianth tube* 2.5–15 cm, purple or white, but never white if the rest of the flower is purple or lavender; segments 6, 15–55 × 4–20 mm, purple or lavender, white, or striped purple and white on the outside, white or purplish and pubescent or glabrous in the throat, oblanceolate or obovate, obtuse to rounded or sometimes emarginate at apex. *Stamens* 3; filaments 5–14 mm, white, glabrous; anthers 5–18 mm, yellow. *Style* yellow, shortly divided into 3 deep yellow to orange, rarely creamy or whitish branches, each much expanded and frilled at the apex, usually exceeding, sometimes equalling the anthers. *Capsule* 15–20 × 6–10 mm, ellipsoid, carried well above ground level at maturity. *Seeds* 2–3 mm in diameter, reddish-brown, subglobose; raphe and caruncle poorly developed. *Flowers* 2–6.

(a) Subsp. vernus
Corm 8–20 mm. *Perianth tube* 5–15 cm; segments (25–)30–55 × (20–)40–60(–80) mm, purple or lavender, sometimes white, sometimes purple and white. *Anthers* 10–18 mm. $2n = 8, 10, 12, 16, 18, 20, 22, 23.$

(b) Subsp. **albiflorus** (Kit. ex Schult.) Asch. & Graebn.

Corm 5–8 (–10) mm. *Perianth tube* 2.5–6.0(–9) cm; segments 15–30 × 4–10 mm, often white, but can be purple or striped. $2n = 8$.

Introduced. This very variable species is the most commonly grown *Crocus* in gardens, and is often naturalised in meadows, on banks, and in churchyards and waste places. The species is scattered over the British Isles and both subspecies may grow intermixed. In their native areas subsp. *vernus* occurs from Italy and Austria to west Russia and subsp. *albiflorus* from France and Spain to Yugoslavia and Albania.

2. C. tommasinianus Herb.   Early Crocus

Corm 6–14 mm in diameter, depressed-globose; tunic finely fibrous, the fibres parallel or slightly reticulated, especially at the corm apex. *Sheathing leaves* 3–4, white, membranous. *Leaves* 3–4, present at flowering time, 40–70 × 1–3 mm, equalling or exceeding the flowers, dark green with a very distinct silvery-white median stripe above, paler beneath, very narrowly linear, truncate at apex, entire, glabrous. *Spathe* present. *Bract* membranous, densely spotted and veined greyish and greenish. *Bracteole* absent. *Flowers* in spring, 1–2, 45–55 mm in diameter. *Perianth tube* 3.5–10 cm, white; segments 6, 24–45 × 8–20 mm, pale lilac to purple, often silvery or buff on the outside, sometimes with darker purple tips, white and sparsely pubescent in the throat inside, obovate, subacute or obtuse at apex. *Stamens* 3; filaments 5–7 mm, white, filiform, glabrous; anthers 10–15 mm, yellow. *Style* orange; divided into 3 orange branches, each much expanded and fimbricate at the apex, slightly shorter than or just exceeding the anthers. *Capsule* about 20 × 8 mm, purplish-tinted, oblong. *Seeds* 2–3 mm in diameter, reddish-brown, subglobose; raphe and caruncle prominent. *Flowers* 2–4. $2n = 16$.

Introduced. Commonly grown in gardens and naturalised in cemeteries and other grassy places. Of scattered occurrence in Britain north to Yorkshire, but probably much under-recorded. Native of south-east Europe.

× **vernus**
The perianth of this hybrid is intermediate in shape and size between the parents and paler than normal *C. vernus* with a delicately bluish-lilac tube.

Occurs in bulb mixtures and is naturalised in London cemeteries and probably elsewhere.

3. C. nudiflorus Sm.   Autumn Crocus
C. multifidus Ramond; *C. aphyllus* Ker Gawl.; *C. fimbriatus* Lapeyr.; *C. pyrenaeus* Herb.

Corm 8–15 mm in diameter, depressed-globose, usually producing stolons which are clothed with leaf-sheaths, and eventually producing cormlets; tunic papery, interspersed with strong parallel fibres, the older tunics becoming more or less wholly fibrous. *Sheathing leaves* 3–4, membranous. Leaves 3–4, absent at flowering and remaining below ground for a long period afterwards,

up to 17 cm × 2–4 mm, green with a whitish medium stripe above, narrowly linear, acute at apex, entire, glabrous. *Spathe* present. *Bract* membranous, white suffused green at apex. *Bracteole* absent. *Flowers* in autumn, solitary, 50–100 mm in diameter. *Perianth tube* 10–22 cm, white suffused lilac or purple at the apex; segments 30–60 × 9–20 mm, deep purple or rarely lilacpurple, not prominently veined, white or lilac and glabrous or papillose-pubescent in the throat, equal, elliptical to oblanceolate, obtuse at apex. *Stamens* 3; filaments 9–12 mm, white, glabrous or papillose; anthers 12–19 mm, yellow. *Style* orange, much dissected into many slender branches. *Capsule* 13–20 mm, ellipsoid. *Seeds* 3–5 mm, brown, with a prominent caruncle and inconspicuous raphe. *Flowers* 9–10. $2n = 48$.

Introduced. Widely grown in gardens and naturalised in fields, parks and meadows, and on grassy banks. Scattered through England and Wales, especially northwest England. Native of south-west Europe. Formerly grown commercially as a substitute for saffron.

4. C. kotschyanus K. Koch   Kotschy's Crocus
C. zonatus Gay

Corm 10–30 mm in diameter, somewhat flattened or depressed-globose, upright, but sometimes misshapen, usually forming many offsets, occasionally producing stolons; tunic thinly membranous with the fibres parallel near the base and weakly reticulate at the apex. *Sheathing leaves* 3–4, white. *Leaves* 4–6, absent at flowering time, up to 38 cm × 1.5–4.0 mm, green with a whitish median line above, the keel nearly as wide as the lamina, glabrous. *Spathe* present. *Bract* membranous, white, not or only slightly exserted from the sheathing leaves. *Bracteole* smaller than and completely concealed in the bract. *Flowers* in autumn, 1(–2), 40–70 mm in diameter, fragrant. *Perianth tube* 3–13 cm, white; segments (24–)30–45 × 5–18 mm, pale to mid-bluish-lilac, usually with darker, nearly parallel longitudinal veins, whitish with 2 yellow blotches at the base of each segment which sometimes coalesce into a large V-shaped blotch in the throat, obovate or oblanceolate, roundedobtuse, subacute or bluntly and abruptly acuminate at apex. *Stamens* 3; filaments 3.0–8.5 mm, white or pale yellow, glabrous or papillose; anthers (7–)9–20 mm, creamy white. *Style* divided into 3 creamy yellow or yellow branches, each of which is sometimes subdivided into several shorter branches, equalling or overtopping the anthers. *Capsule* 11–25 × 6–7 mm, oblong or ellipsoid, carried well above ground level at maturity. *Seeds* 2–3 mm, subglobose or ellipsoid, with an indistinct raphe and caruncle. *Flowers* 9–10. $2n = 8, 10$.

Introduced. Naturalised and on a grassy trackside at Wisley Common in Surrey and in a small meadow at Shrubland Park and in rough grass in Stowmarket, both in Suffolk, since 1981. Native of Turkey, Syria and Lebanon.

5. C. sativus L.   Saffron Crocus
Corm 10–15 mm in diameter, depressed-globose; tunic fibrous, the fibres finely netted. *Sheathing leaves* 3–5,

white, membranous. *Leaves* 7–12, present at flowering times, up to 15 cm × 0.5–1.5 mm, equalling or exceeding the flowers, greyish-green with a narrow, whitish median stripe, linear, obtuse at apex, entire, glabrous or ciliate. *Spathe* present. *Bract* and bracteole very unequal, white, membranous. *Flowers* in autumn, 1–5, fragrant, 50–60 mm in diameter. *Perianth tube* 3–5(–7) cm, lilac-purple; segments 35–50 × 10–15 mm, lilac-purple, conspicuously veined darker purple and stained darker purple towards the base, lilac-purple and hairy in the throat, oblanceolate or obovate, obtuse at apex. *Stamens* 3; filaments 3–7 mm, white or purplish, glabrous or slightly papillose at base; anthers 10–15 mm, yellow. *Style* divided into 3 red branches 25–32 mm, equalling or exceeding the anthers. Sterile. *Flowers* 9–11. 2*n* = 24.

Introduced. Formerly cultivated in the Middle Ages as the source of the spice, Saffron, which was prepared from the dried style branches. A sterile triploid of unknown origin, formerly widely cultivated in southern Europe and Asia, and still grown as a crop in a few Mediterranean countries, and elsewhere in Europe as an ornament.

Section **2. Nudiscapus** (Herb.) B. Mathew

Spathe absent.

6. C. sieberi Gay   Sieber's Crocus
C. sieberianus Herb.

Corm 7–15 mm in diameter, ovoid or depressed-globose; tunic fibrous, the fibres finely reticulated; corm with no persistent brown neck. *Sheathing leaves* 3–5, membranous, white, sometimes greenish at the apex. *Leaves* 4–7, present with the flowers, 2–7(–10) cm × up to 2 mm, about equal to the flower, green with a whitish median line above, linear, narrowed into an obtuse apex, entire, glabrous. *Spathe* absent. *Bract* and bracteole subequal, or the bracteole much narrower than the bract, membranous, white, often stained greenish. *Flowers* in spring, 1–3, 50–70 mm in diameter, fragrant. *Perianth tube* 2.5–5.0 cm, white or purplish, often stained yellow at the apex; segments (15–)20–30 × 7–11 mm, white, usually stained purple on the exterior of the outer segments, these external markings in the form of a longitudinal stripe, or broad horizontal bands or an overall stain, deep yellow or orange and glabrous in the throat, equal, oblanceolate or obovate, obtuse or rounded at apex. *Stamens* 3; filaments 3–7 mm, yellow, glabrous; anthers 8–12 mm, yellow. *Style* 1, rather obscurely divided into 3, yellow to orange-red branches, each of which is much expanded and frilled or lobed, slightly shorter than, equalling or slightly exceeding the anthers. *Capsule* 20–25 × 6–8 mm, raised above the ground on a short pedicel, ellipsoid, acute at apex. *Seeds* 3–4 mm, reddish-brown, ellipsoid, with a prominent caruncle and fleshy raphe. *Flowers* 3–6. 2*n* = 22.

Introduced. Naturalised by a trackside on Wisley Common, Surrey. Native of the Balkans.

7. C. ancyrensis (Herb.) Maw   Ankara Crocus

Corm 10–15 mm in diameter, subglobose; tunic strongly fibrous, the fibres coarsely reticulated. *Sheathing leaves* 3,

white or green-tinged at apex. *Leaves* (2–)3–6, present at time of flowering, up to 28 cm × 0.5–1.0(–1.5) mm, shorter than to just exceeding the flower, slightly greyish-green, with a whitish median line above, linear, entire. *Spathe* absent. *Bract* and bracteole membranous, white, subequal or the bracteole slightly smaller. *Flowers* in spring, 1–2(–3), scented, 30–40 mm in diameter. *Perianth tube* 4–6 cm, yellow or purplish; segments more or less equal, (13–)15–30 × 9–13 mm, bright yellow or orange, sometimes stained purplish at base when associated with a purple tube, the throat yellow and glabrous, obovate, obtuse or rounded at apex. *Stamens* 3; filaments 2–4 mm, yellow, glabrous; anthers (6–)8–13 mm, yellow. *Style* divided into 3 slender orange or reddish-orange branches, sometimes each somewhat expanded at the apex, slightly shorter than to slightly exceeding the anthers. *Capsule* 10–15 × about 7 mm, often purplish-tinged, ellipsoid-oblong, carried up to 30 mm above ground level at maturity. *Seeds* 3–4 mm, brown, ellipsoid, with a prominent raphe and caruncle. *Flowers* 2–4. 2*n* = 10.

Introduced. Widely grown in gardens and naturalised on a wide grassy strip by Old Church Lane in Tottenham, London. Native of central and north Turkey.

8. C. chrysanthus (Herb.) Herb.   Golden Crocus
C. annulatus var. *chrysanthus* Herb.

Corm 6–15 mm in diameter, depressed-globose or ovoid; tunics membranous or coriaceous, splitting into rings at the base, the rings entire or with tooth-like projections. *Sheathing leaves* 3–5, the upper green or brown-stained. *Leaves* 3–7, present at time of flowering, shorter than to exceeding the flower, up to 15 cm × 0.5–2.5 mm, usually greyish-green, with a whitish median line above, linear, narrowed to a subacute apex, entire. *Spathe* absent. *Bract* and bracteole unequal, the bracteole much narrower, membranous white or brownish. *Flowers* in spring, 1–4, 30–60 mm in diameter, fragrant. *Perianth tube* 3–7 cm, yellow, brown, purple or cream; segments (13–)15–35(–40) × 5–15 mm, pale yellow to orange-yellow, sometimes striped or suffused bronze or purple on the exterior, rarely creamy white, yellow and glabrous in the throat, subequal, obovate or oblanceolate, obtuse to subacute at apex. *Stamens* 3; filaments 3–6 mm, yellow, glabrous or papillose; anthers 6–14 mm, yellow, sometimes with blackish basal lobes, rarely becoming wholly blackish. *Style* 1, dividing into 3, slender or expanded, yellow to orange-red branches, shorter than to exceeding the anthers. *Capsule* 10–20 × 5–7 mm, ellipsoid, often purple-tinged, just above the ground level at maturity. *Seeds* 2–3(–4) mm, pale brown to deep reddish-brown, broadly ellipsoid, with an indistinct raphe, but very prominent caruncle; testa densely papillose. *Flowers* 2–7. 2*n* = 8, 10, 12, 13, 14, 16, 18, 19, 20.

Introduced. Naturalised on grassy pathsides in Surrey, Cambridgeshire and London cemeteries. Native of the Balkans and Turkey.

9. C. biflorus Mill.   Silvery Crocus

Corm 7–15 mm in diameter, flattened-globose; tunic usually coriaceous, occasionally splitting lengthways, forming rings at the base. *Sheathing leaves* 3–5, papery, usually white, yellowish or brownish, often distinctly speckled brownish-red. *Leaves* usually 3–5, present at flowering time, up to 15 cm × 0.5–2.0 mm, shorter than, equalling or exceeding the flowers, green or greyish-green, with narrow whitish median stripe, linear, curved, obtuse at apex, entire, glabrous. *Spathe* absent. *Bract* and bracteole present, silvery-white or brownish speckled. *Flowers* in spring, 1–4, fragrant, 45–55 mm in diameter. *Perianth tube* 3–6(–10) cm, whitish; segments 18–30 × 4–12 mm, white or lilac-blue outside with 3–5 purple or brownish-purple longitudinal bands on the outer ones, often also some lateral feathering, the throat inside yellow and glabrous, oblanceolate or obovate, obtuse or subacute at apex. *Stamens* 3; filaments 2–5 mm, yellow, glabrous or sparsely papillose-pubescent; anthers 6–11 mm, yellow, sometimes with blackish basal lobes. *Style* yellow, divided into 3 orange or reddish, apically expanded branches which equal or exceed the anthers. *Capsule* 10–15 mm, ellipsoid, rising to just above ground level at maturity. *Seeds* 2–3 mm in diameter, brown or reddish; raphe indistinct; caruncle flattened. *Flowers* 1–4.

(a) Subsp. **biflorus**
Plant and *leaves* tall. *Perianth segments* white above, conspicuously pencilled dark purple below. $2n = 8, 10$.

(b) Subsp. **adamii** (Gay) B. Mathew
C. adamii Gay; *C. biflorus* var. *violaceus* Boiss.; *C. biflorus* var. *tauricus* Trautv.; *C. tauricus* (Trautv.) Pur.; *C. geghartii* Sosn.
Plant and *leaves* shorter and more compact. *Perianth segments* uniformly pale lilac with 3–5 purple stripes and broader than in subsp. *biflorus*. $2n = 12, 16, 18, 20, 22$.

Introduced. Subsp. *biflorus* was formerly naturalised between about 1830 and 1850 in Barton Park, near Bury St Edmunds in Suffolk. Now both subspecies are naturalised in London cemeteries and perhaps elsewhere. The species is native of southern Europe and Turkey. Subsp. *biflorus* occurs in Italy, Sicily, Rhodes and north-west Turkey. Subsp. *adamii* occurs from south Yugoslavia and Bulgaria eastwards to the Crimea, Caucasus and north Iran.

× chrysanthus
The most obvious hybrids are those which combine the delicate shades of yellow of *C. chrysanthus* with the delicate shades of lilac of *C. biflorus*, but hybrid swarms seem to occur in cultivation.
Commonly naturalised in London cemeteries and perhaps elsewhere. Occasionally occurs naturally in Turkey, where the parent species are usually ecologically separated.

10. C. flavus Weston   Yellow Crocus
C. maesiacus Ker Gawl.; *C. luteus* Lam.; *C. aureus* Sm.; *C. lageniflorus* Salisb.; *C. lacteus* Sabine
Corm (8–)10–15 mm in diameter, depressed-globose; tunics membranous, with fibrous points at the apex,

splitting at the base into coarse parallel fibres. *Sheathing leaves* 4–5, rather tough and persisting as a long brown neck at the apex of the corm. *Leaves* 4–8, present at the time of flowering, 90–140 cm × 2.5–4.0 mm, shorter than to equalling the flower, green with a whitish median line above, papillose to ciliate on the margins and keel, occasionally glabrous, linear, gradually narrowed above to an acute apex. *Spathe* absent. *Bracts* and bracteole membranous, white, often suffused or speckled brown, very unequal, the broad bract sheathing the narrowly linear bracteole. *Flowers* in spring, 1–4, 45–85 mm in diameter, fragrant. *Perianth tube* 5–15 cm, yellow or brownish; segments 20–35 × 4–12 mm, pale yellow to deep orange-yellow, sometimes striped or suffused brownish on the perianth tube and base of segments, yellow and glabrous or pubescent in the throat, more or less equal, oblanceolate to obovate, obtuse to subacute at apex. *Stamens* 3; filaments 3–7 mm, yellow, glabrous or pubescent; anthers 8–15 mm, yellow, markedly divergent, distinctly tapered to the apex. *Style* 1, rather obscurely divided into about 3 short, pale yellow to orange branches which are entire or sometimes shortly lobed and usually much shorter than the anthers. *Capsule* 12–30 × 8–10 mm, raised well above ground level at maturity, ellipsoid. *Seeds* 2–3 mm, subglobose, with a prominent caruncle and wing-like raphe; testa brown, rugose on the main body and sparsely covered with papillae. *Flowers* 2–4. $2n = 8$.
Introduced. Has been grown in gardens for at least 400 years where many varieties were developed, most of which have now been lost. Naturalised in churchyards and other grassy places. Scattered localities in England. Native of south-east Europe and Turkey.

× angustifolius Weston = **C. × stellaris** Haw.
With larger golden flowers. *Flowers* 2–4.
Widely grown in gardens for several centuries and often escaping. Many of the plants called *C. flavus* are probably this. In nature the parents do not grow together. The hybrid is a sterile triploid, wholly maintained by vegetative propagation.

11. C. speciosus M. Bieb. Bieberstein's Crocus
Corm 8–12 mm in diameter, depressed-globose; tunics coriaceous, splitting into entire, horizontal rings at the base, usually with a brown neck of old sheathing leaves at the apex. *Sheathing leaves* 3–4, often reddish-spotted and sometimes green-veined. *Leaves* (3–)4(–5), emerging long after flowering is finished, up to 30 cm × (2–)3–5 mm, deep green, with a whitish median line above, linear, entire, glabrous or with a ciliate or scabrid margin, or pubescent on the upper surface. *Spathe* absent. *Bract* and bracteole subequal or the bracteole smaller, membranous, white. *Flowers* in autumn, 1(–2), 40–90 mm in diameter, fragrant. *Perianth tube* 5–20 cm, white or purplish; segments (30–)37–60(–75) × 10–22 mm, lilac-blue, conspicuously darker veined and often speckled darker on the exterior, sometimes with silvery or white exterior, whitish or occasionally faintly yellow and glabrous in the throat, equal or subequal, oblanceo-

late or obovate, subacute or obtuse, sometimes shortly acuminate at apex. *Stamens* 3; filaments 4–11 mm, white or pale yellow, glabrous or minutely papillose-pubescent; anthers 10–24 mm, yellow. *Style* divided into many, slender, yellow to deep orange branches, usually much exceeding the anthers. *Capsule* 12–25 × 6–9 mm, ellipsoid, carried just above ground level at maturity. *Seeds* about 3 mm, subglobose, reddish-brown, raphe and caruncle poorly developed. *Flowers* 9–11. $2n = 8, 10, 12, 14, 18$.

Introduced. Abundantly naturalised in churchyards at Chiddingfold in Surrey and Cold Ash, near Newbury, Berkshire. Occurs as a garden escape in a few other places in south-east England. Native of south-west Asia.

12. C. pulchellus Herb.   Hairy Crocus

Corm 7–12 mm in diameter, depressed-globose; tunics coriaceous or membranous, splitting into horizontal rings at the base. *Sheathing leaves* 3–4, papery, white, often tinged or spotted brownish or purplish at apex. *Leaves* (3–)4(–5), emerging long after flowering, up to 17 cm × 4–5 mm, green with a whitish median line on the upper surface, long-linear, entire, glabrous. *Spathe* absent. *Bract* and bracteole subequal or the bracteole smaller, membranous, white or reddish-brown, speckled at the apex, usually slightly exserted from the sheathing leaves. *Flowers* in autumn, 1(–2), 30–40 mm in diameter, honey-scented. *Perianth tube* 4–15 cm, white or purplish; segments 18–40(–50) × 8–20 mm, clear pale to mid-bluish-lilac with darker longitudinal veins, deep yellow and glabrous to slightly pubescent in the throat, equal, obovate, rounded or obtuse at apex. *Stamens* 3; filaments 3–6 mm, yellow, densely pubescent; anthers 7–13 mm, creamy white. *Capsule* 10–20 mm, ellipsoid, carried on a short pedicel just above ground level at maturity. *Seeds* subglobose, reddish-brown, raphe indistinct. *Flowers* 9–11. $2n = 12$.

Introduced. Naturalised in a churchyard at Bildeston, Suffolk since 1983. Native of the Balkans and west Turkey.

10. Gladiolus L.

Perennial herbs usually with a corm, often producing cormlets at base or on stolons. *Stems* erect. *Leaves* 2-ranked, linear to narrowly lanceolate, the lowest reduced to sheaths, the upper much shorter and clasping the stem. *Inflorescence* a spike. *Flowers* bisexual, zygomorphic, sessile, each with a bract and bracteole. *Perianth segments* fused at base into a tube, segments unequal, cream, white, yellow, yellowish-orange and red to magenta. *Stamens* 3; filaments free, arising from the perianth tube. *Style* slender, with 3 short terminal lobes greatly widened distally. *Capsule* subglobose, ellipsoid or oblong-trigonous.

About 180 species in Africa, Madagascar, western Asia and the Mediterranean region extending to northern Europe.

Hamilton, A. P. (1976). A history of the garden gladiolus. *The Garden* **101**: 424–428.

1. Corm more than 4 cm in diameter; widest leaves at least 2 cm wide; flowers 12–20, crowded   **4. × hortulanus**
1. Corm and leaves narrower; flowers fewer; if these characters do not apply, the flowers are not crowded   2.
2. Anthers longer than filaments   **1. italicus**
2. Anthers equalling or shorter than filaments   3.
3. Plant usually more than 50 cm; inflorescence often branched, with 10 or more flowers   **2. communis** subsp. **byzantinus**
3. Plant usually less than 50 cm; inflorescence unbranched; flowers 10 or fewer   **3. illyricus**

1. G. italicus Mill.   Italian Gladiolus
G. segetum Ker Gawl.

Perennial herb with a corm usually less than 4 cm. *Stems* 50–100 cm, rounded but flattened, bluish-green, erect, glabrous. *Leaves* 3–5, 18–65 cm × 5–17 mm, bluish-green, long-linear, acute at apex, entire, sheathing at base, the sheaths usually pale to dark red, often spotted white or pale green, the cauline 5–10 mm wide and irregularly veined; bracts always exceeding the bracteoles, the lower bracts frequently resembling and grading into the cauline leaves. *Inflorescence* usually a weakly distichous, lax spike with 6–16 flowers. *Flowers* 30–40 mm in diameter, loosely spaced. *Perianth tube* shorter than the segments, straight; perianth segments unequal, 40–50 mm, bright purplish-red to pale pink, oblong to elliptic-oblong or oblanceolate, subacute at apex, scarcely overlapping. *Stamens* 3; filaments 5–8 mm, brownish-purple; anthers longer than filaments, yellow. *Capsule* broadly ellipsoid-trigonous, about 8 mm. *Seeds* about 2.5 mm, brown, globose-pyriform or D-shaped, not winged. *Flowers* 6–8. $2n = 120, 171 + 2$.

Introduced. Grown in gardens and occasionally found on tips where it does not persist. Native in south Europe to Afghanistan, extending into North Africa and the Canary Islands.

2. G. communis L.   Eastern Gladiolus
Perennial herb with a corm usually less than 4 cm. *Stems* 50–100 cm, bluish-green, erect, rounded but flattened, glabrous. *Leaves* 4–5, 30–70 cm × 8–20 mm, bluish-green, long-linear, acute at apex, entire, sheathing at the base, the sheaths green or red-veined. *Inflorescence* a lax spike, often with 1–3 axillary branches, with 10–20 flowers. *Perianth tube* much shorter than segments; perianth segments unequal, pink, red or magenta, 30–45 × 15–25 mm, the upper lateral segments oblanceolate to rhombic, the lower segments frequently blotched or with white and dark red central lines. *Stamens* 3; filaments 20–25 mm, reddish; anthers brownish, shorter than filaments. *Styles* as long as the stamens. *Capsule* oblong-trigonous. *Seeds* dark, with a pale wing. *Flowers* 6–8. $2n = 90, 120$.

Introduced. A persistent relic of cultivation in old bulb-fields, field margins, roadsides and rough ground. Scattered in southern England and frequent in the Channel Islands and Isles of Scilly. Our plant is subsp. **byzantinus** (Mill.) A. P. Ham. (*G. byzantinus* Mill.). The species occurs in southern Europe and North Africa, the subspecies almost throughout the range of the species.

3. G. illyricus Koch   Wild Gladiolus

Perennial herb with a corm about 10 mm in diameter, producing numerous offsets. *Stem* 25–50(–90) cm, bluish-green, erect, rounded but flattened, very slender, glabrous. Leaves 4–5, 8–25(–40) cm × 3–10 mm, bluish-green with brown tips and very narrow pale margins, long-linear, gradually narrowed at apex to a sharp point, entire, sheathing at the base and forming a wing on the stem. *Inflorescence* a lax, secund spike with 3–8(–10) flowers, not usually branched. *Flowers* 35–40 mm in diameter, sweet-scented. *Perianth tube* shorter than segments; perianth segments unequal, the middle upper segment 25–40 × 15–17 mm, reddish-purple, obovate to broadly elliptical, the outer upper segments 20–25 × 10- –14 mm, oblanceolate, subobtuse at apex and reddish-purple, the lower segments 30–40 × 9–12 mm, reddish-purple with a paler or yellowish margin with a reddish central area, obovate to oblanceolate, subobtuse at apex. *Stamens* 3; filaments 20–23 mm, bluish-red; anthers 12–13 mm, shorter than the filaments, brownish. *Styles* 30–32 mm, bluish-red; stigma deep bluish-red and curved over the stamens. *Capsule* about 10 × 5 mm, broadly ellipsoid, emarginate at apex. *Seeds* about 5 × 2 mm, dark with a pale wing. *Flowers* 6–8. 2n = 90.

Native. Among bracken in scrub. New Forest, Hampshire. Formerly in the Isle of Wight. South and west Europe. The species shows considerable geographical variation over its total range and our plant may be a distinct subspecies.

4. G. × hortulanus L. H. Bailey   Garden Gladiolus

Perennial herb with a corm usually more than 4 cm in diameter. *Stems* up to 120 cm, bluish-green, erect, glabrous. *Leaves* bluish-green, the largest usually more than 2 cm wide, long-linear, acute at apex, entire, sheathing at base. *Inflorescence* a dense spike, with 12–20 flowers. *Flowers* 6–15 cm in diameter. *Perianth* variable in colour, white, yellow and pink to red, in the smaller cultivars more often with definite blotches on the segments, in the larger more often uniform or bicoloured; perianth segments broad and strongly overlapping. *Anthers* longer than filaments. *Flowers* 6–8.

G. × hortulanus covers the commonly grown garden and florists' hybrids. Although selection has been for large size, smaller cultivars have also been raised and make it difficult to distinguish their parent species. *G. cardinalus* Curtis, *G. saundersii* Hook., *G. natalensis* (Eckl.) Hook. fil.; *G. papilio* Hook. fil.; *G. callianthus* Marsis; *G. tristis* L. and *G. carneus* Delaroche are among the species used to form this group of hybrids.

Hybrid of garden origin. Sometimes found thrown out on tips where it does not usually persist for long.

11. Ixia L. nom. conserv.

Perennial herbs with corms surrounded by papery or fibrous tunics. *Stems* simple or little branched. *Leaves* few, mostly basal, 2-ranked, linear or lanceolate, usually with a false midrib. *Inflorescence* a spike or raceme of spikes, bracts and bracteoles present. *Flowers* actinomorphic, few to many, in two ranks or spirally arranged. *Perianth segments* united at base into a long or short tube, white, yellow or red, more or less equal. *Stamens* 3; filaments free, arising from the top of or within the perianth tube. *Style* thread-like, with a 3-lobed stigma. *Fruit* a thin-walled capsule opening by 3 flaps. *Seeds* numerous, angular.

Contains 45 species all from the Cape Province of South Africa.

I. maculata L. var. **nigroalbida** (Klatt) Baker escapes from cultivation in the Isles of Scilly and may persist.

Lewis, G. J. (1962). South African Iridaceae: the genus *Ixia* L. *Jour. South African Bot.* **28**: 45–195.

1. Leaves 2–5 mm wide; perianth tube 2–3 mm, much shorter than the segments   **1. campanulata**
1. Leaves 3–12 mm wide; perianth tube 40–70 mm, much longer than the segments   **2. paniculata**

1. I. campanulata Houtt.   Red Corn Lily
I. speciosa Andrews; *I. crateroides* Ker Gawl.

Perennial herb. Corm 1.0–1.8 cm in diameter, ovoid or irregularly subglobose; tunic of fine reticulate fibres extending up in a neck. *Stem* 10–35 cm, simple, rarely with a long branch arising near the base, straight or flexuose. *Leaves* 5–10, 5–20 cm × 2–5 mm, linear or subulate, erect or suberect, half as long as the stem or more, entire, 3–6-veined, the midrib and margins slightly prominent. *Inflorescence* a spike, with 1–9 flowers; bracts 4–6 × 5–7 mm, whitish or pink, subrotund, tricuspidate, usually conspicuously 3-nerved; bracteoles bicarinate and bicuspidate. *Flowers* rotate-campanulate or crateriform. *Perianth tube* 2–3 mm, much shorter than the segments; *segments* 12–25 × 8–12 mm, deep crimson to pure white, usually red with a longitudinal white band on the outside, obovate or obovate-oblong, obtuse at apex. *Stamens* 3, arising at the throat of the perianth-tube; filaments 3–5 mm; anthers 6–8 mm, linear. *Style* reaching to the middle or top of the anthers or just above them, the branches of the stigma 3–4 mm, linear and minutely ciliate. *Ovary* 2.5–4.0 mm, subglobose. *Flowers* 10–11. 2n = 20.

Introduced. Grown in gardens and bulb fields and naturalised in old fields or rough ground. St Martin's and St Mary's in the Isles of Scilly. Native of South Africa.

2. I. paniculata Delaroche   Tubular Corn Lily
Morphixia paniculata (Delaroche) Baker; *Tritonia paniculata* (Delaroche) Klatt; *I. longiflora* Bergius

Perennial herb. Corm 10–15 mm in diameter, globose; tunic pale brown, soft and rather membranous. *Stem* 30–100 cm, the lower half sometimes fairly stout, usually with one or two patent or suberect branches. *Leaves* 15–60 × 0.3–1.2 cm, erect or suberect, linear to lanceolate. *Inflorescence* a spike, 5- to 18-flowered; bract 8–15 mm, pale brown, oblong or lanceolate, acute or shortly cuspidate and sometimes lacerate at apex;

bracteoles bicarinate, bicuspidate or bidentate. *Perianth tube* 40–70 mm, slightly and gradually expanded from base to throat. Segments 15–25 × 3–8 mm, pale to deep cream or yellowish tinged with red, less than half as long as the tube, oblong or linear-oblong, obtuse or subemarginate at apex. *Stamens* 3, arising in the perianth tube; filaments 5–6 mm, shortly decurrent, not exserted; anthers 6–8 mm, usually purplish, linear, sagittate. *Style* reaching to base or middle or top of anthers, or sometimes extended well above them, the branches of the stigma 1.5–2.0 mm, spathulate, recurved at apex. *Ovary* 3–4 mm, rotund. *Flowers* 9–11.

Introduced. Grown in gardens and bulb fields and naturalised in old fields or rough ground. Tresco in the Isles of Scilly. Native of South Africa.

12. Sparaxis Ker Gawl.

Perennial herbs with corms covered with a papery or fibrous tunic, the stems sometimes bearing small corms after flowering. *Stems* erect, rigid. *Leaves* mostly basal, in a fan, with several veins and a conspicuous false midrib. *Inflorescence* a spike with few flowers; bracts large, jaggedly toothed. *Flowers* more or less actinomorphic. *Perianth segments* united into a straight tube shorter than the free, more or less equal segments. *Stamens* 3; filaments free, borne in the perianth tube, slightly asymmetrically arranged. *Style* very slender, with 3 linear branches. *Capsule* spherical, firm, containing several large, smooth seeds.

Contains 6 species, all from the Cape Province of South Africa.

Goldblatt, P. (1969). The genus *Sparaxis. Jour. South African Bot.* **35**: 219–252.

1. S. grandiflora (Delaroche) Ker Gawl.

Plain Harlequin-flower

Ixia grandiflora Delaroche

Perennial herb. Corm 6–15 mm in diameter, globose; tunic of white, fine fibres. *Stems* 1–3, 8–45 cm, firm, erect, simple. *Leaves* mostly basal, 6–10, 3–30 cm × 4–13 mm, distichous, ensiform, lanceolate or falcate, acute, acuminate or obtuse-acuminate at apex, entire, with several veins and a prominent false midrib. *Inflorescence* a lax spike with 1–6 flowers; bracts dry, firm, scarious, usually marked with brown streaks especially on the margin, rarely uniformly white, deeply lacerated, the apices with 1–3 long cusps, which are straight or slightly twisted and entire; bracteoles similar but smaller, 2–cusped. *Flowers* more or less actinomorphic. *Perianth tube* 10–14 mm, yellow, purple or black; segments 24–30 × 12–16 mm, uniformly cream-coloured, or marked with black blotches at the base, or marked externally with broad purple stripes, or uniformly reddish-purple, violet-purple or yellow with or without dark spots and streaks, lanceolate to ovate or spathulate, subacute to obtuse at apex. *Stamens* white or yellow; filaments 7–9 mm, curved; anthers facing upwards and slightly curved, rather longer than the filaments. *Style* white or yellow, usually reaching almost to the apex of the anther, recurved and lying behind the stamens

against the posterior perianth lobe; stigma branches filiform, curved, 6–10 mm. *Ovary* ovoid. *Capsule* small globose, firmly membranous. *Seeds* fairly numerous, smooth and suborbicular. *Flowers* 8–9. 2n = 20.

Introduced. Grown in bulb fields and naturalised in old fields. Isles of Scilly. Native of South Africa.

13. Freesia Eckl. ex Klatt nom. conserv.

Perennial herbs with corms covered with fibrous tunics. *Stems* erect. *Leaves* in 2 ranks, mostly basal, with a few prominent veins and usually a false midrib. *Inflorescence* a spike, its axis bent sharply at an angle of about 90° at the level of the lowest flower, the flowers borne on the upper side of the axis; bracts 2, membranous or green. *Flowers* slightly zygomorphic. *Perianth segments* forming a very narrow tube at their base which gradually widens into a broader tube, ending in 6 free segments, the tube curved. *Stamens* 3; filaments free, not or scarcely projecting from the perianth tube. *Style* very slender, with 3 bifid stigmas. *Fruit* a few-seeded capsule.

Contains 19 species, all from the Cape Province of South Africa.

Smith, D. (1979). *Freesias*. London.

1. F. × hybrida L. H. Bailey   Freesia

F. refracta auct.

Glabrous *perennial herb* with more or less conical corms covered with fibrous tunics. *Stems* 10–40 cm, sparsely branched, neat and tufted or tall and graceful, erect, flexuous, terete. *Leaves* 15–30 × 0.5–1.0 cm, linear-ensiform, acute at apex, entire; bracts 15–25 mm, ovate, subacute at apex, green with a narrow, scarious margin. *Inflorescence* a spike, short and erect to sparsely branched and horizontal, with numerous flowers. *Flowers* 40–50 mm in diameter, single or double, scarcely scented to sweetly scented. *Perianth tube* 1.5–3.5 cm, slender below and broader above, yellow, often with green veins; perianth segments unequal or subequal, 8–15 mm, white, yellow, blue, lilac, pink, silvery white or bronze, the 3 upper 15–18 mm wide, broadly elliptical, rounded at apex, the 2 lateral similar, the lower 20–25 mm wide, very broadly elliptical with a slightly hooded apex. *Stamens* 3; filaments 12–15 mm, white or pale yellow; anthers white, cream or bluish. *Style* white, cream, blue or pink; stigma white, yellow, bluish or pinkish, with 3 deeply forked branches. *Capsule* few-seeded. *Flowers* 4–5. 2n = 22, 33, 44.

The florists' Freesias are a complex of hybrids involving *F. lactea* Klatt (*F. alba* Foster), *F. corymbosa* (Burm. fil.) N. E. Br., *F. refracta* (Jacq.) Eckl. ex Klatt, *F. armstrongii* Watson, *F. sparrmanii* (Thunb.) N. E. Br., *F. zanthospila* Klatt and *F. leichtlinii* Klatt and have been grouped under the hybrid name *F. × hybrida*. Although species of *Freesia* were first grown in Europe in the mid-eighteenth century, selective hybridization did not start until the late nineteeth century between the yellow-flowered *F. leichtlinii* and the white *F. lactea*.

Introduced. Grown in bulb fields and naturalised in and by old fields. Isles of Scilly and the Channel Islands. The parents are all native of South Africa.

14. **Crocosmia** Planch.

Perennial herbs with small, flattened corms and creeping stolons which readily produce new corms. *Stems* erect. *Leaves* mostly basal, 2-ranked, conspicuously ribbed, sometimes folded. *Inflorescence* usually a panicle of spikes, more rarely a simple spike, overtopping the leaves; bracts small, scarcely exceeding the ovaries. *Flowers* zygomorphic. *Perianth segments* fused at base into a tube longer or shorter than the free segments, slender throughout or slender at the base and then abruptly expanded, the free parts rather unequal and spreading. *Stamens* 3, free, borne on the perianth tube, arching to the upper side of the flower, the anthers either just projecting from the mouth of the perianth tube, or with very long filaments, the anthers borne at the level of the tips of the perianth segments or beyond them. *Style* slender, with 3 short branches. *Fruit* a capsule opening by 3 splits and containing many seeds.

Contains about 7 species in southern Africa.

De Vos, M. P. (1984). The African genus *Crocosmia* Planchon. *Jour. South African Bot.* **50**: 463–502.

Kostelijk, P. (1984). *Crocosmia* in gardens. *The Plantsman* **5**: 246–253.

1. Leaves 3–7 cm wide, conspicuously plicate; perianth 4–6 cm — 2.
1. Leaves up to 3 cm wide, ribbed but not plicate; perianth 2–4(–5) cm — 3.
2. Ribs of leaves bearing fine hairs; axis of the spike conspicuously zig–zag; perianth tube 3–4 cm — **1. paniculata**
2. Ribs of leaves glabrous; axis of the spike not zig–zag; perianth tube less than 3 cm — **2. masoniorum**
3. Perianth lobes about half as long as tube, more or less erect; perianth tube very narrow at base, abruptly widened distally — **3. pottsii**
3. Perianth lobes about as long as tube, more or less spreading; perianth tube gradually expanded distally — **4. × crocosmiiflora**

1. **C. paniculata** (Klatt) Goldblatt   Aunt Eliza
Curtonus paniculatus (Klatt) N. E. Br.; *Antholyza paniculata* Klatt

Perennial herb. Corms 25–40(–50) mm in diameter, depressed-globose, single or in superimposed groups of 2–3; tunics brown, membranous, later fibrous towards the top. *Stems* 100–150(–180) cm, often reddish-brown, strong, cylindrical. *Basal leaves* several, 40–60(–90) × (1.5–)2.5–5.0(–8) cm, plicate, elliptic-ensiform, tapering to an acute to acuminate apex, entire, almost stalk-like at base, with numerous prominent veins, often with minute hair-stumps on the veins; the cauline 15–20 cm, the lower ones sometimes plicate. *Inflorescence* a large panicle 10–20(–35) cm, on a rigid scape up to 9 mm in diameter, with 2–5, distichous, almost horizontal primary and sometimes secondary branches, each with a straight, thin, zig-zag axis up to 15 cm, with numerous, densely arranged flowers which are at first distichous and later become somewhat secund on the upper side of the branches; bracts 7–10 mm, reddish-brown, membranous, obovate. *Flowers* strongly zygomorphic, narrowly

funnel-shaped. *Perianth tube* (25–)30–40(–45) mm, more than twice as long as the segments, 5–7 mm in diameter at the oblique mouth; segments (10–)15–18 × 6–9 mm, very unequal, elliptical to lanceolate, obtuse at apex. *Stamens* 3; filaments 25–40 mm, curved towards the upper segment; anthers 6–8 mm. *Style* (35–)50–65 mm; stigma branches 4–6 mm, slightly widened or sometimes shortly bifid at apex, reaching and over-topping the anthers. *Capsules* globose, 3-lobed. *Seeds* about 4 mm, globose or angled. *Flowers* 9–11. $2n = 22$.

Introduced. Grown in gardens and naturalised on roadsides, in quarries and in rough and waste ground. Local in west Scotland, west Ireland and south Britain. Native of South Africa.

2. **C. masoniorum** (L. Bolus) N. E. Br.
Giant Montbretia
Tritonia masoniorum L. Bolus

Perennial herbs. Corms 20–25 mm in diameter, several, subglobose, superimposed one upon another; tunics soon disintegrating. *Stem* 50–75 cm, robust, cylindrical, shining. *Basal leaves* up to 60 × 2–5 cm, several, elongated-elliptical, plicate, tapering gradually to an acuminate tip and a narrow stalk-like base, entire, with a strong middle vein only in the lower part and 6–12 more glabrous veins; the cauline up to 5, the lower imbricate, the upper distant and smaller. *Inflorescence* a simple or 1- to 2-branched spike, with ascending branches, the main axis curved, not zig-zag, with numerous flowers secund on the upper side of the curve, sometimes in approximating pairs; bracts 4–10 mm, oblong to ovate, membranous. *Flowers* slightly zygomorphic, funnel-shaped, the later somewhat salver-shaped. *Perianth tube* 18–25 mm, funnel-shaped, slightly curved; narrow in the lower half, widening gradually to 5–10 mm in diameter at the throat; segments 20–30 × 3–7 mm, slightly unequal, scarlet-orange to orange-yellow, drying to brownish-yellow, elliptical, obtuse at apex, spreading, at length somewhat recurved. *Stamens* 3, well exserted; filaments 30–35 mm; anthers 7–9 mm, usually overtop-ping the perianth segments. *Style* 42–45 mm, well exserted; stigma branches 2–4 mm, shortly bifid or widened, fimbriate. *Capsule* 7–9 mm, depressed-globose, 3-lobed. *Seeds* about 4 mm in diameter, wrinkled. *Flowers* 9–11. $2n = 22$.

Introduced. Grown in gardens and naturalised in rough and waste ground. Found in three sites in Dunbartonshire where it may now be extinct. Native of South Africa.

3. **C. pottsii** (Macnab ex Baker) N. E. Br.
Potts's Montbretia
Tritonia pottsii (Macnab ex Baker) Baker; *Montbretia pottsii* Macnab ex Baker

Perennial herb. Corms 20–25 mm in diameter, depressed-globose, often 2 or more in a vertical row or connected by long, slender stolons; tunic brown, papery or some-times somewhat fibrous. *Stems* 70–100(–120) cm, erect, strong and cylindrical. *Basal leaves* several, (10–)50–80 × (0.5–)0.8(–1.8) cm, plane, linear-lanceolate, acuminate at apex, with a prominent middle vein and several more closely spaced glabrous veins, firm, entire; the cauline 2–3, up to 20 cm. *Inflorescence* a lax panicle 15–35 cm, with 1–3(–5) ascending, more or less straight or later faintly zig-zag branches 15–20 cm, each with numerous, fairly laxly arranged, distichous flowers, or sometimes a simple spike up to 32 cm; bracts 4–8 mm, reddish-brown, ovate, membranous. *Flowers* zygomorphic, funnel-shaped. *Perianth tube* 14–20 mm, funnel-shaped, slightly curved; segments 8–15 × 5–7 mm, unequal, bright orange, orange-yellow or flushed brick-red on outside, the outer lanceolate, the inner oblanceolate or elliptical, obtuse at apex. *Stamens* 3; filaments 8–12 mm, slightly curved; anthers 6–7 mm, scarcely exserted from the perianth tube. *Style* 20–30 mm, orange; stigma branches 2–3 mm, overtopping the anthers. *Capsule* up to 8 mm in diameter, shortly obovoid to subglobose, deeply 3-lobed. *Flowers* 9–11. 2*n* = 22.

Introduced. Rarely grown in gardens, but naturalised by roads and rivers in Kirkcudbrightshire, Wigtownshire, Argyll and the Clyde Islands. Native of South Africa.

4. C. × crocosmiiflora (Lemoine) N. E. Br. Montbretia
= **C. aurea** (Hook.) Planchon × **pottsii**
Tritonia × crocosmiiflora (Lemoine) Nicholson; *Montbretia crocosmiiflora* Lemoine

Glabrous *perennial herb*. *Corms* up to 2 cm in diameter, often several in a row, freely stoloniferous. *Stem* 30–90 cm, slender, simple or with 1–2(–4) branches. *Leaves* 4–8, (7–)10–30(–40) cm × 5–20 mm, shorter than stem, green, ensiform, acute at apex, entire; bracts small, oblong, reddish. *Inflorescence* a 10- to 20-flowered spike, distichous, secund. *Flowers* 2.5–5.0 cm in diameter, zygomorphic, funnel-shaped. *Perianth tube* infundibuliform, gradually widened, slightly bent at apex; perianth segments 2.5–4.0(–5.0) cm, longer or shorter than the tube, oblong, spreading, the dorsal segments slightly larger than the rest, deep orange suffused with red. *Stamens* 3, extending nearly to apex of perianth; anthers yellow. Capable of producing some fertile seed. *Flowers* 9–11. 2*n* = 22–24, 33.

Originated in cultivation. Much grown in gardens and naturalised in hedgerows, woods, by lakes and rivers and on waste ground. Scattered throughout the British Isles, common in Ireland, west Britain and the Channel Islands. Both parents are native of South Africa.

(**Antholyza ringens** L. is recorded as a garden escape on Tresco in the Isles of Scilly.)

15. Chasmanthe N. E. Br.

Perennial herbs with large, flattened-spherical corms which have fibrous necks. *Stems* erect. *Leaves* mostly basal, 2-ranked. *Inflorescence* an erect spike; bracts longer than the ovaries. *Flowers* strongly zygomorphic. *Perianth* conspicuously curved, tubular below and

broadening into the upper part; free segments 6, the upper much longer than the others and directed forwards, the two upper lateral parallel to the uppermost, the 3 lower deflexed. *Stamens* 3, conspicuously exserted; filaments free, attached to the perianth tube where it widens. *Style* long exserted, very narrow, with 3 terminal, thin lobes. *Capsule* spherical, opening by 3 splits and containing many brightly coloured seeds.

Contains 9 species in South and South-west Africa.

1. C. bicolor (Gasp. ex Ten.) N. E. Br. Chasmanthe
Antholyza bicolor Gasp. ex Ten.; *Antholyza aethiopica* var. *bicolor* (Gasp. ex Ten.) Baker

Perennial herb with corms 20–25 mm in diameter covered by a fibrous tunic. *Stem* 70–130 cm, more or less terete, erect, longer than leaves. *Leaves* 6–7, 40–80 × 2.5–3.5 cm, green, often with a silky sheen, sword-shaped, acute at apex, entire, slightly narrowed at base, with a strong midrib. *Inflorescence* a spike, 25–30 cm, sometimes with 1–2 basal branches, rather laxly 12- to 28-flowered, more or less secund. *Flowers* strongly zygomorphic, up to 8 cm in diameter. *Perianth tube* 3.0–3.5 cm, the lower part 6–10 mm, sometimes twisted, yellow on the lower side, orange-red on the upper side; uppermost segment 30 40 × 6 7 mm, dark red, pink or green, linear-spathulate, longitudinally folded, other lobes 5–8 mm, the 3 outer erect, the inner 2 spreading or recurved. *Stamens* 3, attached to the perianth tube where it widens; filaments free. *Style* 5–8 cm, the three terminal lobes 3.0–3.5 mm. *Capsule* spherical. *Seeds* many, ovoid. *Flowers* 4–6.

Introduced. Grown for ornament and naturalised in damp, shady places nearby. In a few places on Tresco in the Isles of Scilly. Native of South Africa.

174. AGAVACEAE Endl. nom. conserv.

Perennials, often with thick, woody, sparsely branched stems with leaves tufted at the branch-ends, or with stemless, giant rosettes of leaves. *Leaves* in rosettes, tough, often succulent, simple, sessile, entire or distinctly toothed, the base more or less sheathing; exstipulate. *Flowers* in large terminal panicles, bisexual, hypogynous or epigynous, actinomorphic or slightly zygomorphic. *Perianth segments* 6, proximally fused into a tube. *Stamens* 6, borne on the perianth tube.

Contains 20 genera and about 700 species. Mainly arid regions of the tropics and subtropics.

1. Leaf-rosettes at the ends of woody branches; perianth whitish   2.
1. Leaf-rosettes sessile on the ground or more or less so; perianth reddish or greenish- or brownish-yellow   3.
2. Leaves spine-tipped, strongly recurved; perianth more than 4 cm   **1. Yucca**
2. Leaves often sharply pointed, but not spine-tipped, not recurved; perianth less than 1 cm   **3. Cordyline**
3. Leaves with extremely strong spines at margin and apex   **2. Agave**
3. Leaves spineless   **4. Phormium**

1. Yucca L.

Perennials. Stems woody, usually branched. *Leaf-rosettes* at ends of branches, the leaves stiff, fleshy, entire or with a few inconspicuous teeth, with a sharp spine at apex. *Flowers* in large, terminal panicles, bisexual, hypogynous, actinomorphic. *Perianth* campanulate, with lobes longer than tube. *Stamens* 6. *Ovary* superior. *Fruit* an indehiscent capsule.

About 30 species from arid regions of the United States, Mexico, Guatemala and the West Indies.

Y. gloriosa L. and **Y. rupicola** Scheele have been recorded as casuals.

1. Y. recurvifolia Salisb. Curved-leaved Spanish-dagger
Y. recurva Haw.; *Y. obliqua* Regel; *Y. pendula* Groenl.; *Y. gloriosa* var. *recurvifolia* (Salisb.) Engelm.

Perennial up to 2 m, forming large leaf-rosettes topped by a stalked, erect panicle. *Stems* woody, usually branched. *Leaf-rosettes* at the ends of branches. *Leaves* 70–100 × 3–5 cm, more or less glaucous, with narrow yellow or brown margins, long-lanceolate, tapering to a sharp spine at apex, entire or with a few, inconspicuous teeth, stiff, strongly recurved and often folded above. *Inflorescence* 60–120 cm, a stalked, erect panicle. *Flowers* loosely arranged, 5–7 cm in diameter, creamy white or greenish-white. *Perianth* 5–8 cm; lobes ovate or ovate-lanceolate, acute at apex. *Stamens* 6; filaments 20–25 mm. *Style* 1, short and thick; stigma 3-lobed. *Capsule* 5–6 cm, oblong, erect, with 6 winged ribs. *Seeds* 6–7 × 7–8 mm, blackish, not glossy. *Flowers* 9–10. 2n = 60.

Introduced. Naturalised on sand-dunes at Crymlyn Burrows in Glamorgan since 1982, at Dawlish in Devon and in a gravel pit at Broadway, Worcestershire, where it was presumably planted. Native of the south-eastern United States.

2. Agave L.

More or less stemless *perennials* 2–3 m in diameter. *Leaf-rosettes* sessile on the ground, leaves with an extremely sharp spine at apex and many more along the margins. *Inflorescence* a panicle on a tall scape, the flowers borne in clusters at the ends of the branches, the axis of the inflorescence as broad as the crown of the rosette. *Flowers* epigynous, bisexual, actinomorphic. *Perianth* tubular, the 6 lobes much shorter than the tube. *Stamens* 6. *Ovary* inferior. *Fruit* a dehiscent capsule. *Seeds* black, flattened.

Over 100 species, mostly from Central America and the West Indies, but extending north and south from this centre.

Gentry, H. S. (1982). *Agaves of Continental North America.* Tucson.

1. A. americana L.   Century Plant

Perennial forming large leaf-rosettes and rarely in our area tall scapes with large inflorescences, freely suckering and almost stemless. *Leaf-rosettes* sessile on the ground, 1–2 m tall, 2.0–3.7 m in diameter. *Leaves* 20–60, 1–2 m × 5–25 cm, greyish-glaucous to pale green, some-times variegated, oblanceolate, usually acuminate with a very sharp spine at the apex 3–5 cm, sinuous-dentate on the margin with spines about 8 mm, narrowed towards the base, smooth, leathery, rigidly spreading or recurved towards the apex. *Inflorescence* a panicle 5–9 m, the shaft slender and straight, with scarious, rather small triangular bracts, the panicles usually long-ellipsoid in outline, rather open, with 15–35 spreading branches in the upper one-third to half of the shaft. *Flowers* 7–10 cm, opening yellow over a greenish ovary, pedicellate. *Perianth* with tube 8–20 × 16–20 mm, funnel-shaped; lobes unequal, the outer 25–35 mm, sometimes red-tipped, linear-lanceolate, the apex cucullate, the inner shorter. *Stamens* 6; filaments 60–90 mm, rather flattened; anthers 30–36 mm, yellow. *Stigma* trilobate. *Capsules* 4–5 cm, oblong, with a short beak. *Seeds* 7–8 × 5–6 mm, lunate, shiny black.

Introduced. Very persistent where planted, surviving from suckers when the main rosette dies after a single flowering. A few places in the Channel Islands and Tresco on the Isles of Scilly. Native of Mexico; widely planted and naturalised in the Mediterranean region.

3. Cordyline Comm. ex A. L. Juss.

Tree-like plant with a well-developed, simple or branched stem or trunk. *Leaf-rosettes* at ends of branches; leaves entire, sharply pointed at apex but without a spine. *Inflorescence* a dense, terminal panicle, the branches more or less at right angles, the axes almost hidden by flowers. *Flowers* subsessile, sweetly scented, hypogynous, actinomorphic, bisexual. *Perianth* with a short tube and wider, longer, spreading lobes. *Stamens* 6, attached to the top of the perianth tube. *Ovary* superior, 3-celled. *Fruit* a berry, becoming dry with age. *Seeds* black, curved.

About 20 species in south-east Asia, Polynesia and Hawaii.

1. C. australis (G. Forst.) Endl.   Cabbage Palm
Dracaena australis G. Forst.; *Dracaenopsis australis* (G. Forst.) Planch.

Tree-like plant. Stem up to 20 m, well-developed and in mature plants trunk-like, simple or branched. *Leaf-rosettes* of young plants scattered along the stem, in older plants forming a dense round head at the top of the stem or branches. *Leaves* 30–100 × 3–6 cm, pale green and very similar on both surfaces, long-linear, narrowed to a sharply acute apex, entire, little inclined to droop at tip but bending from the base when old; midrib indistinct; veins fine, equal and more or less parallel. *Inflorescence* a dense terminal panicle 60–150 × 30–60 cm, up to 3- to 4-branched, the branches coming off at right angles and well spaced; lower bracts green and foliaceous; ultimate racemes 10–20 × about 2 cm, axes almost hidden by flowers. *Flowers* white and sweet-smelling. *Perianth* 5–6 mm; lobes free nearly to base, oblanceolate or obovate, rounded at apex, spreading. *Stamens* more or less equalling perianth; filaments more or less flattened. *Stigma* more or less 3-lobed. *Berry* about 4 mm, globose, whitish. *Seeds* about 2.5 mm, shiny black, more or less comma-shaped. *Flowers* 6–8. $2n = 38$.

Introduced. Much planted in western Britain, Ireland and the Channel Islands; persistent in south-west England, southern Ireland and the Channel Islands, and producing seedlings in the Channel Islands. Native of New Zealand.

4. **Phormium** Forst. & G. Forst.

Perennial herbs with rhizomes. *Stems* more or less absent. *Leaf-rosette* sessile on the ground; leaves entire, not spiny, rough and coriaceous, folded proximally, nearly flat distally. *Inflorescence* long, the axis bearing alternate, deciduous bracts some of which are scarious, the upper bracts subtending and entirely enclosing the short, alternately branched flowering branches. *Flowers* bisexual, hypogynous, slightly zygomorphic. *Perianth* with short tube and longer lobes, more or less tubular. *Stamens* 6, projecting from the perianth. *Ovary* superior, 3-celled. *Fruit* a dehiscent capsule. *Seeds* black, flattened.

Two species in New Zealand.

1. Leaves up to 3 m, stiff; flowers predominantly reddish; inner perianth lobes not or only slightly recurved at apex
 1. tenax
1. Leaves usually less than 2 m, more flexible; flowers greenish with tones of orange or yellow; inner perianth lobes usually markedly recurved at apex **2. cookianum**

### 1. **P. tenax** Forst. & G. Forst.	New Zealand Flax

Tall, tufted *perennial herb,* brightly coloured at base, increasing by budding from a short, stout, branched rhizome. *Stems* more or less absent. *Leaf-rosettes* 1–2 m, sessile on the ground. *Leaves* 1–3 m × 5–12 cm, dark green above, glaucous beneath, sometimes variegated with white or creamy yellow, or tinted bronze or purplish, linear-ensiform, long-acute at apex, entire, stiff, without spines, strongly keeled, equitant, marked by fine, close, longitudinal striations, folded proximally, nearly flat distally, fibrous and extremely tough. *Inflorescence* terminal, flowering stem up to 4(–6) m, terete, sheathed with a series of alternate, deciduous bracts of progressively smaller size, the lower ones empty, the upper ones each subtending and entirely enclosing a comparatively short, alternately branched, flowering lateral, the smaller bracts quite scarious. *Flowers* numerous, predominantly dull red. *Perianth* 2.5–5.0 cm; tubular, curved; lobes unequal, the 3 outer lanceolate, erect and acute at apex, the 3 inner longer with erect, spreading or slightly recurved tips. *Stamens* 6, inserted at the base of the perianth segments; filaments filiform; anthers linear-oblong. *Style* 1, slender, equalling or exceeding the stamens; stigmas small, capitate. *Capsule* up to 10 cm, erect, trigonous, abruptly contracted at the apex, not twisted. *Seeds* 9–10 × 4–5 mm, shining black, more or less elliptical, plate-like but more or less twisted. *Flower* 6–8. $2n = 32$.

Introduced. Very persistent where planted on cliffs or rocky places by the sea. Naturalised in Cornwall, Isles of Scilly, Cork, Isle of Man and the Channel Islands. Self-

sown mainly in the Isles of Scilly. Native of New Zealand and Norfolk Island.

### 2. **P. cookianum** Le Jolis	Lesser New Zealand Flax
P. colensoi Hook. fil.; *P. hookeri* Gunn ex Hook. fil.

Tall, tufted *perennial herb,* pale at base, budding from a short, stout, branched rhizome. *Stems* more or less absent. *Leaf-rosettes* 1–2 m in diameter, sessile on the ground. *Leaves* 1–2 m × 3.5–6.5 cm, linear-ensiform, long-acute at apex, entire, less stiff than *P. tenax*, rather pale green above, more or less glaucous beneath, not usually marked, without spines, strongly keeled, folded proximally, nearly flat distally, fibrous. *Inflorescence* terminal, flowering up to 2 m, terete, sheathed with a series of alternate, deciduous bracts of progressively smaller size, the lower one empty, the upper ones each subtending and entirely enclosing a comparatively short, alternatively branched, flowering lateral, the smaller bracts scarious. *Flowers* numerous, greenish, with tones of orange or yellow. *Perianth* 2.5–4.0 cm, tubular, curved; lobes unequal, the 3 outer lanceolate, erect and acute at apex, the 3 inner longer with strongly recurved tips, one usually more than the other two. *Stamens* 6, inserted at the base of the perianth segments; filaments filiform; anthers linear-oblong. *Style* 1, slender; equalling or exceeding the stamens; stigma small, capitate. *Capsule* often more than 10 cm, sometimes to 20 cm, cylindrical, pendulous, gradually narrowed to its apex, twisted and spirally curled in age. *Seeds* 8–10 mm, shining black, more or less elliptical, plate-like. *Flowers* 6–8. $2n = 32$.

Introduced. Planted on cliffs and rocky places near the sea and very persistent. St Martin's in the Isles of Scilly, where it is often self-sown. Native of New Zealand.

175. **DIOSCOREACEAE** R. Br. nom. conserv.

Usually dioecious, slender, herbaceous or woody *twiners* with tuberous rhizomes or stocks. *Leaves* spirally arranged, simple, entire, with palmate main veins and a network of smaller ones, petiolate, stipulate. *Flowers* in axillary, simple or branched racemes, inconspicuous, epigynous, actinomorphic. *Perianth* campanulate, of 3 + 3 more or less equal segments united below into a short tube. *Stamens* in male flowers borne on the base of the perianth tube, 3 + 3 or 3 with or without staminodes replacing the missing set, in the female flower rudimentary or absent. *Ovary* in the female flower inferior, 3-celled, with 2 superimposed anatropous ovules on axile placentation in each cell; style 1, divided above into 3 stigmatic lobes; ovary in male flower rudimentary or absent. *Fruit* a berry with up to 6 seeds.

Includes 8–9 genera, containing about 750 species. Widespread in tropical and warm temperate regions.

1. **Tamus** L.

Perennial, dioecious herbs with large hypogeal stem tubers. *Stems* annual, twining. *Leaves* entire, cordate. *Flowers* small, in axillary racemes. *Stamens* 6, rudimentary in female flowers. *Style* 1, with three 2-lobed stigmas. *Fruit* a berry, incompletely 3-celled.

One species in south and west Europe, North Africa, Caucasus and west Asia.

1. T. communis L. Black Bryony

Perennial herb, with a large, irregular, dark-coloured, subterranean tuber. *Stems* up to 20(–60) m, scandent, twisting to the left, glabrous. *Leaves* with lamina 3–15 × 2–10 cm, broadly ovate, dark glossy green above, paler beneath, finely acuminate at apex, entire, cordate at base, glabrous; veins 3–9, curving palmately from base; petiole up to 10 cm, slender, glabrous; stipules reduced to two short, horn-like prominences. *Inflorescence* a lax, axillary raceme, the male ones often up to 15 cm, the female ones usually much shorter and sometimes reduced to clusters of flowers; bracts minute, about 1.5 × 0.3 mm, subulate, scarious; pedicels up to 3 mm, slender. *Male flowers* about 5 mm in diameter, with a campanulate, pale green, glabrous perianth, divided more than halfway into 6, oblong lobes 2–3 × 0.8–1.0 mm; stamens 6; filaments about 0.6 mm, glabrous; anthers about 0.3 mm, pale yellow. *Female flowers* about 4 mm in diameter, with a conspicuous ovoid or ellipsoid, glabrous ovary 2–3 × 1.5–2.0 mm, the perianth divided almost to the base into 6 narrow, oblong, recurved lobes; styles united for about 0.3 mm, forming a short column; stigmas about 0.5 mm, recurved. *Fruit* 10–13 mm in diameter, pale red when ripe, subglobose or ovoid-ellipsoid, glabrous. *Seeds* about 3 mm in diameter, globose, pale yellow or brownish, obscurely rugose. *Flowers* 5–7. Visited for nectar by many insects, including small bees. $2n = 48$.

Native. Wood-margins, scrub and hedgerows scrambling over trees and bushes. Common in Britain, north to Cumberland and Durham, local in Channel Islands, rarely introduced further north and in Ireland and the Isle of Man. South and west Europe; North Africa; Israel and Syria; coastal regions of Turkey; Caucasus.

Order 2. ORCHIDALES Bromhead

Perennial herbs without bulbs, but often with tubers, often epiphytes or saprophytes. *Leaves* simple, often rather thick. *Flowers* mostly bisexual, epigynous, zygomorphic. *Perianth* of 2 whorls, usually both petaloid, sometimes the outer sepaloid. *Stamens* 2 or 1; pollen usually stuck together in masses (*pollinia*). *Ovary* usually 1-celled, often twisted through 180°; ovules numerous; placentation parietal. *Fruit* usually a capsule. *Seeds* minute, without endosperm and with undifferentiated embryo.

This order consists of 4 families including some 15,000 species most of which are in the Orchidaceae.

176. ORCHIDACEAE Juss. nom. cons.

Perennial herbs with rhizomes, a vertical stock or tuberous roots; terrestrial, sometimes heterotrophic, usually with associated mycorrhizal fungi. *Stems* sometimes swollen at the base to form pseudobulbs. *Leaves* entire, spirally arranged or distichous, rarely subopposite, reduced to scales or sheaths in heterotrophic species. *Inflorescence* a spike or raceme. *Flowers* zygomorphic, epigynous, usually bisexual. *Perianth segments* 6, in two whorls, all petaloid or sepaloid and petaloid; median inner segment (*labellum*) usually larger and of a different shape from the others, usually directed downwards owing to the ovary or the pedicel twisting through 180°, often with a basal spur. *Anthers* and *stigma* borne on a column formed from the fused filaments and style. *Stamens* usually 1, rarely 2, with a sessile or subsessile, 2-locular anther behind or at the summit of the column; pollen grains single or in tetrads, bound by elastic threads into packets (*pollinia*) which may be narrowed into a sterile, stalk-like caudicle; pollinia sometimes furnished with 1 or 2 viscid bodies (*viscidia*); viscidia sometimes enclosed in 1(–2), simple or 2-lobed, membranous, pocket-like outgrowths of the rostellum (*bursicles*). *Stigmas* 3, rarely all fertile, usually with the median sterile and often consisting of a beak-like process (*rostellum*) between the anther and fertile stigmas; ovary inferior, 1-locular with parietal placentation or rarely 3-locular. *Fruit* a capsule, dehiscing by 6 longitudinal slits; seeds numerous, minute, with undifferentiated embryo and no endosperm.

A very large family consisting of about 750 genera and 18,000 species, distributed throughout the world except for a few isolated islands and Antarctica. The ecological range is equally wide, with species occurring in all but the most extreme environments. Most temperate zone species are terrestrial, but in the tropics, where they form a important part of the vegetation, they live chiefly as epiphytes.

The orchid flower is exceedingly complicated. The perianth is composed of 6 parts, 3 of which are outside the others. In some cases these appear as 3 sepals and 3 petals, but in others some or all of the outer parts are coloured. All descriptions will therefore refer to them as *outer* and *inner perianth segments* and describe their coloration, which may vary considerably within a species. The central segment of the inner 3 segments is usually larger and more intricately coloured than the other parts and is called the *labellum*. The labellum is usually at the front or bottom of the flower, but is occasionally at the top or back. It is sometimes drawn out into a *spur* at the base. The male and female organs instead of being separate are joined together to form a special structure called the *column*. In some species this does actually look like a column, but in others it is much diversified in shape with all manner of lumps and projections. The stamen, placed near the top or towards the base of the column, has a 2-locular anther, which, when ripe resembles two bag-like containers placed side by side or more rarely diverging from one another. The pollen grains (except in one species) are joined together in various ways to form more or less solid masses termed *pollinia*. Each of these pollinia contains thousands or even millions of individual pollen grains. Part of the pollinium itself may be sticky or it may be furnished with a special adhesive

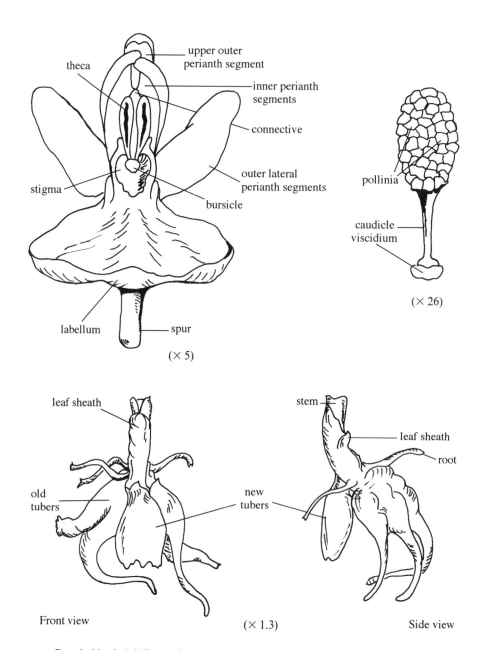

theca

upper outer
perianth segment

inner perianth
segments

connective

outer lateral
perianth segments

stigma

bursicle

pollinia

caudicle
viscidium

(× 26)

labellum   spur

(× 5)

leaf sheath

stem

leaf sheath

root

old
tubers

new
tubers

Front view   (× 1.3)   Side view

Dactylorhiza fuchsii (Druce) Soó (Common Spotted Orchid) (After Summerhayes, 1951)

attachment, the *viscidium*. The base of the pollinia may be narrowed into a sterile stalk termed the *caudicle*. The viscidia are sometimes enclosed in a simple or 2-lobed, membranous, pocket-like outgrowth of the rostellum, the *bursicle*. The stigmas are three in number and are borne on the column, usually below or in front of the stamen containing the pollinia. In most species only two of the stigmas are receptive, the third (showing great diversity in structure) lying between the receptive stigmas and the stamen, being non- functional and termed the *rostellum*. Abnormalities in flower structure

are widespread in the orchids. They may take the form of partial or complete doubling (which appears grotesque in the zygomorphic orchid flower), in having more or fewer perianth segments, in having 3 labella and no lateral inner perianth segments, or in having 3 similar inner perianth segments and no labellum.

The orchid flowers hold a high position of achievement in the wide range and complexity of the devices they have developed to ensure that cross-pollination is effective. In many species the structure of the flower and of the visiting insect have become so delicately adjusted to one another as to reach maximum efficiency in this process. The adaptation is often so precise that only one species of insect is effectively able to pollinate the flower, although other species may visit it. The flowers of some of our *Ophrys* species bear a strong resemblance to the animals after which they have been named. There are some foreign orchids which are visited only by the males of a particular species, the flowers resembling the female. Such a system would depend on the exact synchronisation of the flowering of the orchid with the appearance of the males of a particular species of insect, which in the course of time may break down owing to changes in climate or other causes. This may be the reason for the scarcity of insect visitors to some of our species of orchid. In such cases a self-pollination mechanism has often been developed for when cross-pollination has not been effected. A few species seem to be cleistogamous and are either automatically self-pollinated or possibly apomictic. The insects visit the flower not only to get pollen, but to obtain nectar or other liquids, which are frequently, but not always, in some part of the labellum, and particularly in the hollow spur which is an obvious feature of the flower.

The ovules in most orchid flowers are very numerous, so it might be thought that the fertilisation of such a large number would be an unlikely event. The special nature of the pollen, however, provides against this. Even a small part of a single pollinium contains thousands or tens of thousands of pollen grains so that the danger of the ovules remaining unfertilised is unlikely if any pollen at all is brought to the flower.

When the capsules are ripe they split lengthways into 6 parts which, however, remain joined together at both ends. Three of the valves bear the seeds, the others are narrow strips of the capsule wall. In dry weather the capsule contracts, thus causing the slits to open and allowing the seeds to escape. In damp weather the dry fruit wall absorbs moisture and lengthens, and the splits are drawn together again. Although the seeds are so small and rather like dust, they are normally longer than broad and more or less flattened. There is often a thickened portion in the centre containing the embryo, which is surrounded by a flattened, net-like testa. The seeds are very light and easily carried for long distances by the wind. The dried capsules on the top of the stem move to and fro in the wind shedding the seeds a few at a time so that they are distributed far and wide. Such large numbers of seeds need to be produced because of the difficulty of establishment of new plants.

The minuteness of orchid seeds means that there is little room for food storage, consequently the beginning of growth is slow and nothing resembling a normal-looking seedling is visible for months or even years. Leaves may not be produced until two or more years have passed and even after their appearance development is slow, and several more years may elapse before flowers are formed. Owing to the great difficulty of raising orchids from seed, most of the available data are based on a careful examination of seedlings in the field. The lack of food in the minute seeds is made up for by

their association with fungi, a phenomenon usually termed endomycorrhizal. These fungi occur as hyphal coils within the host cells. They obtain their food from the abundance of dead plant remains occurring in the soil and by secretion of enzymes obtain energy-producing foods. This food is, in due course, transferred to the orchid itself, and serves to nourish it. In all orchids which have been examined, infection of the germinating seed takes place at a very early stage and it is thought that the seeds of some species will not germinate unless the proper fungus is present in the soil. The seed of the orchid on germination gives rise to a special structure known as a protocorm or mycorrhizome which is always heavily infected by the fungus. At first appearance it is rather like a small peg-top with a minute bud-like projection at the broader end. Subsequent growth varies considerably in different species, depending amongst other things on the type of growth finally adopted by the adult. There is a great difference between British orchid species as regards their dependence on the mycorrhizal fungus in the early stages, some manufacturing part of the food required at an early stage while others do not contribute anything until many years have passed. These differences extend into the adult life of the plant, some plants being virtually independent of the fungus as an adult, while others always obtain a substantial proportion of their food from the fungal associate. Some species of orchid can carry on life by means of rhizomes heavily infected with mycorrhizal fungus when deep shade does not allow them to produce either leaves or flowers. With partial clearing of trees and undergrowth and the return of normal light, flowering can commence again. Some species may persist for a long period without flowering and there are records of them doing so for over twenty years when in thick forest.

The orchids frequently spread vegetatively by means of rhizomes, tubers, a combination of rhizomes and tubers, root buds, or in one species by producing buds along the margins of the leaf. Plants that have arisen vegetatively can often be recognised because they occur in clusters or form lines along the rhizome and are exactly alike in colour and morphology. As many as 130 individuals have been found round the parent plant.

Extensive natural hybridisation in both temperate and tropical species of orchid, together with the vast number (over 45,000) of artificially produced hybrids involving, on occasions, up to five genera and 20 species in one plant, makes classification difficult to deal with. Records of hybrids between British native orchids are scattered through the literature, but not a great deal of attention has been paid to them. Wherever species occur adjacent to one another or in a mixed population and are known to hybridise, a careful examination of the populations should be made. Hybrids are not always easy to recognise and show all stages of intermediacy between the parents. In some cases, although the two species may cross elsewhere in their geographical range, they may not have overlapping flowering times in the British Isles.

The identification of the parents of British orchid

hybrids can rarely be more than tentative and depends on a sound knowledge of their possible morphological variation. Study of the details of the reproductive parts is not always helpful. In a case where bursicles were present in one parent and not in the other, the hybrids had them present, absent or imperfectly formed. Only in *Dactylorhiza* have any extensive experimental or biometrical studies been carried out on orchid hybrids.

Many species of orchid can be recognised by the shape and colouration of the labellum. Most orchids should not be collected and even if they are they make poor herbarium specimens, losing their colour and becoming difficult to interpret. A single flower preserved in alcohol and a colour photograph showing the complete front of the labellum are more useful. A side view of the individual flower will show the presence or absence of a spur. As there is much variation in labellum colour, photographs of individual flowers on several plants in a population are even more helpful.

Buttler, K. P. (1991). *Field guide to orchids of Britain and Europe*. Swindon. [Very good colour photographs.]

Camus, A. A., Bergman, P. & Camus E. G. (1908). *Monographie des orchidées*. Paris.

Camus, A. A. & Camus E. G. (1921–1929). *Iconographie des orchidées*. Paris.

Danesch, O. & E. (1968–1969). *Orchideen Europas*. 1 & 2. Bern.

Davies, P. & J., & Huxley, A. (1988). *Wild Orchids of Britain and Europe*. London.

Ettlingers, D. M. T. (1976). *British & Irish orchids: a field guide*. London & Basingstoke.

Godfery, H. M. (1933). *Monograph & iconograph of native British Orchidaceae*. Cambridge. [The best descriptions ever made of British orchids.]

Lang, D. (1989). *A guide to the wild orchids of Great Britain and Ireland*. Ed. 2. Oxford.

Summerhayes, V. S. (1951). *Wild orchids of Britain*. London.

Williams, J. G., Williams, A. E. & Arlott, N. (1978) *A field guide to the orchids of Britain and Europe*. London.

1.	Plant without green leaves	2.
1.	Plant with green leaves or with green bract-like leaves on a green stem	4.
2.	Stem swollen above the base; labellum directed upwards, pink; spur fairly long, directed upwards	**4. Epipogium**
2.	Stem not swollen; labellum directed downwards; spur absent or very short and adnate to the ovary	3.
3.	Stem with numerous brownish scales; labellum brown, about twice as long as other perianth segments which are connivent into a open hood	**5. Neottia**
3.	Stem with 2–4, long, brown-veined, sheathing scales; labellum whitish, with crimson lines or spots, about as long as the perianth segments which are curved downwards	**11. Corallorhiza**
4.	Flowers without a spur	5.
4.	Flowers with a spur	15.
5.	Labellum large, inflated, slipper-shaped; perianth segments 6–9 cm	**1. Cypripedium**
5.	Labellum neither inflated nor slipper-shaped; perianth segments less than 6 cm	6.

6.	Labellum with a distinctly coloured, insect-like, velvety central area (*speculum*)	**23. Ophrys**
6.	Labellum without an insect-like, velvety speculum	7.
7.	Labellum on upper side of flower and directed upwards in some or all of the yellowish-green flowers	8.
7.	Labellum on lower side of flower and directed downwards	9.
8.	Leaves 2.5–8.0 cm, oblong-elliptical or ovate-elliptical; labellum variously orientated, about as long as the outer perianth segments	**9. Liparis**
8.	Leaves 0.5–1.0 cm, obovate, the margin fringed with tiny green bulbils; labellum always on the upper side of the flower, shorter than the outer perianth segments	**10. Hammarbya**
9.	Labellum with a concave basal part (*hypochile*) separated by a constriction or fold from the tongue-like distal part (*epichile*)	10.
9.	Labellum not clearly divided by a constriction into a hypochile and an epichile	12.
10.	Outer perianth segments not more than 4 mm	**8. Goodyera**
10.	Outer perianth segments at least 7 mm	11.
11.	Flowers suberect, sessile or subsessile, not in a secund spike; column longer than wide	**2. Cephalanthera**
11.	Flowers patent or pendulous, pedicellate, in a more or less secund spike; column not longer than wide	**3. Epipactis**
12.	Flowers white, arranged in 1–3, spiral rows or in a secund spike	**7. Spiranthes**
12.	Flowers yellowish or greenish, neither arranged in spiral row nor in a secund spike	13.
13.	Plant with rhizomes; labellum 2-lobed	**6. Listera**
13.	Plant with 1 or 2 tubers; labellum 3-lobed, the median lobe sometimes bifid	14.
14.	Plant with 1 tuber at anthesis; median lobe of labellum entire	**12. Herminium**
14.	Plant with 2 tubers at anthesis; median lobe of labellum deeply bifid	**21. Aceras**
15.	Labellum with the median lobe much exceeding the lateral and spirally twisted	**22. Himantoglossum**
15.	Labellum entire or with the median lobes not much exceeding the lateral and not spirally twisted	16.
16.	Labellum divided by a constriction into a concave basal part (*hypochile*) and a tongue-like distal part (*epichile*)	**2. Cephalanthera**
16.	Labellum not clearly divided by a constriction into a hypochile and an epichile	17.
17.	Spur less than 3 mm; perianth segments connivent into a hood	18.
17.	Spur exceeding 5 mm; perianth segments connivent or patent	21.
18.	Flowers green, sometimes red-tinged	**17. Coeloglossum**
18.	Flowers white or pink or marked purplish	19.
19.	Flowers in a short, ovoid spike	**20. Orchis**
19.	Flowers in a narrow, cylindrical spike	20.
20.	Perianth segments obtuse at apex; viscidia not enclosed in a bursicle	**15. Pseudorchis**
20.	Perianth segments acute to acuminate at apex; viscidia enclosed in a 2-lobed bursicle	**19. Neotinea**

21. Spur less than 11 mm 22.
21. Spur more than 11 mm 23.
22. Emerging flower-spike not enclosed in spathe-like leaves; bracts leaf-like; spur descending **18. Dactylorhiza**
22. Emerging flower-spike enclosed by thin spathe-like leaves; bracts membranous; spur more or less ascending **20. Orchis**
23. Spur more than 15 mm; flowers greenish-white, in lax spikes **13. Platanthera**
23. Spur less than 15 mm; flowers usually pink (rarely white), in dense spikes 24.
24. Labellum deeply 3-lobed, with 2 longitudinal ridges at the base; tubers entire **14. Anacamptis**
24. Labellum shallowly 3-lobed, without ridges at the base; tubers palmately lobed **16. Gymnadenia**

Subfamily **1. Cypripedioideae** Melch.

Fertile stamens 2; pollen not united into pollinia. *Fertile stigmas* 3.

1. Cypripedium L.

Perennial herbs with creeping rhizomes. *Stems* leafy. *Flowers* 1(–3), large. *Perianth segments* spreading, the outer median ovate to elliptical, erect, the lateral connate in the lower half; the inner linear-lanceolate, twisted. *Labellum* large, inflated. *Spur* absent. *Column* stout, surmounted by a large, petaloid, sterile anther partly closing the mouth of the labellum. *Stigmas* triangular-peltate. *Rostellum* absent. *Viscidia* absent. *Bursicles* absent.

1. **C. calceolus** L. Lady's Slipper

Perennial herb; tubers 0; rhizome thick, short-jointed, creeping, often branching and forming new shoots; roots numerous, cylindrical, sinuous. *Stems* 20–50 cm, erect, flexuous, terete, downy, with 3–4, short, broad, obtuse, brown or green, leafless sheaths at the base. *Leaves* 3–5, (7–)10–17 × 3.5–7.0(–10) cm, rather greyish-green, elliptical to ovate-oblong, acute to acuminate at apex, wavy and ciliate on the margins, somewhat folded and narrowed to a clasping base, with more or less 11 principal veins, with short, scattered hairs on the lower surface and sometimes above, mainly on the veins. *Bracts* broadly lanceolate, more or less resembling the leaves. *Flowers* 1(–3), hanging from the tip of the curved ovary, very large, with a bright yellow, pouch-shaped, inflated labellum and brownish-purple, very rarely yellow, white or greenish perianth, and a sweet, orange-like smell. *Outer perianth segments* 60–90 mm, brownish-purple, rarely yellow, lanceolate to ovate-lanceolate, acute at apex, with wavy, sometimes reflexed edges, several-veined, downy on the inner face, hairy at the base, the median broader and erect, the 2 lateral joined together except at the 2-toothed tip and pointed downwards. *Inner perianth segments* 40–60 mm, asymmetrical, brownish-purple, linear-lanceolate, tapering to an acute apex, twisted, with a downy midrib and long hairs in front of the base. *Labellum* about 30 mm, yellow with reddish spots inside, inflated and bag-shaped, curved forwards and slightly compressed from back to

front, with a rounded opening in front with inrolled margins, a small auriculate aperture on each side between the base of the labellum and the column, and several concentric veins. *Spur* absent. *Column* yellowish-green, 3-lobed, with a short, thick base, the side-lobes short, curved and horn-like, each bearing an anther on its lower surface; middle lobe long, curved obliquely downwards, expanding into a large, thick body whose undersurface forms the stigma; back of column prolonged into a stalk bearing a greatly enlarged staminode, which is petaloid, oblong, subcordate at base, obtuse at apex, trough-like with upturned sides, strongly keeled below and yellowish or white with crimson spots. *Anthers* cup-shaped, open, 2-celled. *Stigma* close to and facing the hairy base of the labellum, triangular-peltate with rounded angles, somewhat convex but depressed in the centre, dry, non-viscous, covered with papillae pointing obliquely downwards. *Rostellum* absent. *Ovary* long, slender, curved, stalked, 6-ribbed, with short, erect hairs. *Capsule* large, oblong, 6-ribbed. *Seeds* numerous, oblong. *Flowers* 5–6. Cross-pollinated by insects which crawl from the cavity of the labellum past the stigmatic disc and the fertile anthers. $2n = 22$.

Native. North-facing grassy slopes on limestone. Now known in only one locality in Yorkshire. Formerly widespread in open woods on limestone in northern England. Central and north Europe northwards to about 70° N in Scandinavia and southwards to north Greece, but not in the Mediterranean lowlands; northern Asia.

Subfamily **2. Orchidoideae** Melch.

Fertile stamen 1; pollen united in pollinia. *Fertile stigmas* 2, often confluent.

Tribe **1. Neottieae** (Lindl.) Camus

Labellum of various shapes and often without well-marked lobes, often constricted near the middle so delimiting proximal and distal parts. *Fertile stamen* 1, borne at the back of the column; pollen dispersed on 2, often rather friable pollinia which are sessile or with an apical stalk, with or without a sticky pad. *Receptive stigmas* 2, a third stigma absent or as a sterile bulge.

2. Cephalanthera Rich.

Perennial herb with a short, creeping rhizome. *Stem* leafy. *Flowers* few, large, suberect, more or less sessile, in a lax spike, scentless. *Perianth segments* similar, usually connivent to make campanulate flowers. *Labellum* constricted between the suberect, concave basal part (*hypochile*) and the forwardly directed distal part (*epichile*) which has a recurved apex and 3–9 interrupted longitudinal ridges above. *Spur* absent, or very short. *Column* long, erect. *Anther* hinged to the summit of the column; pollinia 2, clavate, each more or less completely divided into longitudinal halves; caudicles absent; pollen grains single, in powdery masses. *Stigma* large, elliptical. *Rostellum* absent. *Bursicles* absent. *Capsule* erect.

Stewart, A., Pearman, D. A. & Preston, C. D. (1994). *Scarce plants in Britain*. Peterborough. [*C. longifolia*.]

1. Flowers purplish-pink; ovaries with glandular hairs; labellum acute   **3. rubra**
1. Flowers white with yellow or orange marks on the labellum; ovaries glabrous; labellum obtuse   **2.**
2. Lower leaves ovate to rather narrowly so; bracts longer than ovaries; outer perianth segments obtuse at apex   **1. damasonium**
2. Lower leaves lanceolate to narrowly elliptic-oblong; bracts shorter than ovaries; outer perianth segments acute at apex   **2. longifolia**

1. C. damasonium (Mill.) Druce   White Helleborine
Serapias grandiflora L. nom. illegit.; *Serapias damasonium* Mill.; *Serapias latifolia* Mill., non Huds.; *Epipactis alba* Crantz; *Serapias lonchophyllum* L. fil.; *Serapias lancifolia* Murray nom. illegit.; *Epipactis pallens* Swartz nom. illegit.; *C. grandiflora* Gray nom. illegit.; *C. alba* (Crantz) Simonk.; *C. latifolia* Janchen nom. illegit.

Perennial herb with a hard, short, woody rhizome; roots many, short, stiff, corky. *Stem* 15–60 cm, nearly terete, erect, yellowish-green, rigid, solid, often flexuous, angled, glabrous, but rough above with translucent ridges, leafy throughout, with 2 3, loose-fitting, brown, membranous, ribbed sheaths at base, the upper sometimes green-tipped. *Leaves* 5–10 cm × 5–30 mm, dull or greyish-green, the lowest shortly ovate-lanceolate, the middle ones ovate-oblong, the upper lanceolate, obtuse or acute at apex, entire but often wavy at the margin, with prominent parallel veins, glabrous. *Inflorescence* a spike 4–10 × 2–5 cm, with 3–12 flowers, the lower distant, the upper closer together. *Bracts* 7–30 × 1–7 mm, the lower ovate-lanceolate to lanceolate, acute at apex, the lowermost often leaf-like so that the flower appears solitary and axillary, the uppermost linear. *Flowers* large, erect, sessile, creamy white, scentless, tubular at the base and often closed, but for a short time the perianth segments spread and the lip turns down at the apex. *Outer perianth segments* 15–20 mm, creamy white, oblong, obtuse at apex, entire or minutely toothed, 5-veined. *Inner perianth segments* shorter, oblong-lanceolate, rounded at apex, slightly clawed, 5-veined. *Labellum* shorter than the other perianth segments, the basal half saccate with a deep yellow or orange patch within and a rounded lobe on each side clasping the column, making the flower tubular, the distal part heart-shaped, obtuse, finely crenate, often mucronate, narrowed to a hinge, broader than long, trough-like above, curved downwards at the tip and with 3 or 5, parallel, orange-yellow, interrupted, longitudinal crests. *Spur* absent. *Column* up to 10 mm, curved, semi-cylindrical, flat in front, whitish, ending behind in a point like the nib of a pen to which the anther is fixed. *Anther* semi-ovoid, obtuse at apex, hood-like, rough outside with minute papillae, with a staminode on each side; pollinia 2, cylindrical, curved; caudicles absent. *Stigma* on the front of the column immediately below the anther, elliptical, broader than long, concave, glis-

tening greenish-white. *Rostellum* absent. *Ovary* 15–30 mm, erect, spindle-shaped, rough, twisted, 6-ribbed. *Capsule* 20–30 mm, erect, slightly curved, hexagonal. *Seeds* oblong; testa reticulate, transparent. *Flowers* 5–6. Usually self-pollinated, the pollen falling on the stigma. $2n = 32, 36$.

Native. Shady woods and banks, particularly under *Fagus sylvatica* where there is little ground cover, on calcareous soils. Locally frequent in southern England, north to Northamptonshire and Herefordshire. Much of Europe northwards to Denmark and Gotland; North Africa; Caucasus; Asia Minor.

× **longifolia** = **C.** × **schulzei** Camus, Bergon & A. Camus
Intermediate in leaf and flower characters.
Found in Hampshire in 1974 and 1975 with both parents.

2. C. longifolia (L.) Fritsch   Narrow-leaved Helleborine
Serapias helleborine var. *longifolia* L.; *Serapias longifolia* (L.) Huds.; *Serapias grandiflora* auct.; *Serapias xiphophyllum* L. fil. nom. illegit.; *Serapias ensifolia* Murray nom. illegit.; *Epipactis ensifolia* Schmidt; *Cephalanthera ensifolia* (Schmidt) Rich.

Perennial herb with a knotted rhizome; roots stiff, wiry. *Stem* 20–60 cm, pale green, leafy throughout with short internodes, slightly ridged above, with 2–4, whitish or green-tipped sheaths at the base. *Leaves* 7–20 cm × 3–25 mm, alternate and mostly in two opposite ranks, lanceolate-linear or oblong, tapering to an acute apex the lower obtuse, entire, rounded to a clasping base, obliquely erect or horizontally spreading, often folded, stiff, with 3–5 prominent veins and numerous fainter ones, sometimes exceeding the spike, glabrous. *Inflorescence* a spike 3–13 × 2–5 cm, with 3–15 flowers, lax. *Bracts* mostly 5–7 × 0.5–1.0 mm, but lowest usually large and up to 10 cm, linear, acute at apex, 1- to 5-veined. *Flowers* large, obliquely erect, pure white, opening wider than *C. damasonium*, scentless. *Outer perianth segments* 10–16 mm, white, lanceolate, acute at apex, keeled, 1- to 5-veined. *Inner perianth segments* shorter and broader, white, elliptical, obtuse at apex, 3- to 5-veined. *Labellum* shorter than other perianth segments, jointed; the basal half somewhat sack-shaped, with 2, rounded, auriculate flaps embracing the column and white with an orange blotch at the base, the distal half somewhat heart-shaped, broader than long, curved, trough-like, white with 3–5, orange-yellow, parallel crests and a turned down, round orange tip covered with dense papillae. *Spur* absent. *Column* long, erect, flat in front, rounded behind, with a nib-like filament at the back from the tip of which the anther is suspended to below the middle of its back. *Anther* whitish, ovoid, rounded, flat in front, tilted forwards and firmly pressed against the back of the stigma, with an obtuse, white staminode on each side at the base, tending to keep it in position; pollinia 2, long, curved, yellowish-white, sometimes split in two or nearly so, their convex centres projecting slightly from the cells of the anther, where they remain until removed by insects. *Stigma* on the front of the column just below the anther, transversely elliptical or reniform, concave, whitish, glistening, with thickened margins. *Rostellum* absent. *Ovary* 10–15 mm, sessile, cylindrical, slender, glabrous, twisted, 6-ridged. *Capsules* erect, slightly curved, hexagonal. *Seeds* oblong, transparent. *Flowers* 5–7. Cross-pollinated by small bees. Long-lived with a poor reproduction capacity. $2n = 32$.

Native. Woods, particularly of beech and oak/ash, and shady places on calcareous soils, less common, but much more widespread than *C. damasonium*. Scattered in Britain and Ireland north to west Sutherland, but has shown a massive decline this century. Europe northwards to about 63° N in Norway; North Africa; west Asia to the Himalayas.

3. C. rubra (L.) Rich.   Red Helleborine
Serapias rubra L.

Perennial herb with short roots and slender, horizontal rhizomes from which buds can arise. *Stem* solitary, 15–45(–60) cm, slender, solid, striate, glabrous below, with numerous, very short glandular hairs in the upper part. *Leaves* 5–8, rather flaccid, glabrous; the lower 2–7 cm × 3–10 mm, brownish and sometimes with a green tip, linear-oblanceolate or narrowly elliptical, obtuse at apex, cuneate at base with the lowest sheathing the stem, 7- to 9-veined; the middle and upper dark green, 5–15 cm × 5–22 mm, narrowly lanceolate to oblong-lanceolate, acute at apex, narrowed to a semi-amplexicaul base, with 3–5 prominent and several intermediate veins. *Inflorescence* a spike up to 12 cm, lax, with 2–7(–15) flowers. *Bracts* 10–45 × 1–4 mm, linear-lanceolate, acute at apex, longer or shorter than the flowers, 3- to 7-veined, with numerous, very short glandular hairs towards the base. *Flowers* large, scentless, rose, opening fairly wide. *Outer perianth segments* 15–25 × 2–5 mm, rose, linear-lanceolate, subacute at apex, subpatent, with very short glandular hairs on the outer surface. *Inner perianth segments* 14–20 × 2–5 mm, rose, lanceolate, acute at apex, connivent. *Labellum* 15–17 mm, erect, more or less parallel with the column at the base (making the flower tubular), then curving outwards, the basal part forming a bag with an erect-rounded lobe on each side, white with several irregular yellowish ridges diverging fan-wise to the edges, the distal part white with reddish-violet margins and apex, lanceolate, acute at apex and with 7–9 narrow, yellow, crested ridges on its inner surface. *Spur* absent. *Column* violet-rose, erect, slender, slightly curved, half as long as the labellum, flat in front, rounded behind, with a triangular tooth to which the anther is fixed on its back. *Anther* leaning forwards at the top of the column against the back of the stigma, broad, hood-like, covered with papillae outside; pollinia 2, long and narrow, more or less crescent-shaped, whitish, greyish-blue or violet-rose. *Stigma* broad, hood-like, glistening, on the front of the column immediately below the anther. *Rostellum* absent. *Ovary* 10–15 mm, sessile or shortly stalked, green with violet ribs, with numerous, very short glandular hairs. *Capsule*

about 20 mm, erect, slightly trigonous. *Seeds* minute; testa reticulate, transparent. *Flowers* 6–7. Cross-pollinated by small bees. $2n = 48$.

Native. Beech woods or scrub on chalk or limestone. This species appears and disappears in varying numbers from period to period. Mycorrhizal activity is nearly always at a comparatively high level, and the plant depends on it for much of its food. Should its habitat become overgrown, the plant may become much reduced in size and even disappear, but the mycorrhiza can keep the rhizomes alive for a very long period. Very rare in Hampshire, Buckinghamshire and Gloucestershire; formerly elsewhere in southern England. Europe northwards to about 60° N in Scandinavia and in western Asia.

3. Epipactis Zinn nom. conserv.
Helleborine Mill.

Perennial herbs with horizontal or vertical rhizomes and pale brown, fleshy roots. *Stems* leafy. *Flowers* pedicellate, patent or pendulous, borne spirally in racemes or more or less on one side of the stem. *Perianth segments* spreading or connivent, dull reddish-brown or greenish, all similar but inner smaller. *Labellum* usually in 2 parts separated by a narrow joint or fold, the basal part (*hypochile*) forming a nectar-containing cup, the distal part (*epichile*) a more or less cordate or triangular, forwardly-directed terminal lobe. *Spur* absent. *Column* short, with a shallow cup (*clinandrium*) at its apex. *Anther* free, hinged at the back of the summit of the column, behind the stigma and rostellum; pollinia 2, tapering towards their apices near which they are attached to the rostellum, each more or less divided longitudinally into halves; caudicles absent; pollen grains globose, cohering in tetrads bound loosely together by weak elastic threads, so that the masses are very friable. *Stigma* prominent, more or less transversely oblong. *Rostellum* placed centrally above the stigma, large and globose, persistent, evanescent and absent at flower-opening or absent. *Viscidia* absent or as a white sticky cap. *Bursicles* absent. *Ovary* straight, often resembling a twisted stalk. Cross-pollinated by insects or self-pollinated; perhaps sometimes apomictic.

A number of species in the north temperate zone are imperfectly understood and indeed it is a very difficult genus to understand. The first four species are cross-pollinating and some of them interfertile. The remainder are mainly self-pollinating and there is considerable disagreement about their taxonomy. The ecology, distribution and biology may be better understood if all the taxa and not just the species are considered.

Richards, A. J. (1988). Cross-pollination by wasps in *Epipactis leptochila* (Godf.) Godf. s.l. *Watsonia* 16: 180–182.
Richards, A. J. & Porter, A. F. (1982). On the identity of a Northumberland *Epipactis*. *Watsonia* 14: 121–128.
Stewart, A., Pearman, D. A. & Preston, C. D. (1994). *Scarce plants in Britain*. Peterborough. [*E. atrorubens, E. leptochila* and *E. phyllanthes*.]

Young, D. P. (1949). Studies in British *Epipactis*. *Epipactis dunensis* and *E. pendula*. Watsonia 1: 102–108.

Young, D. P. (1949). Studies in British *Epipactis*. The differentiation of *E. pendula* from *E. vectensis*. *Watsonia* 1: 108–113.

Young, D. P. (1952). Studies in British *Epipactis*. *Epipactis phyllanthes* G.E. Sm., an overlooked species. *Watsonia* 2: 253–259.

Young, D. P. (1952). Studies in British *Epipactis*. A revision of the *phyllanthes–vectensis–pendula* group. *Watsonia* 2: 259–276.

Young, D. P. (1962). Studies in British *Epipactis*. *Epipactis leptochila*, with some notes on *E. dunensis* and *E. muelleri*. *Watsonia* 5: 127–135.

Young, D. P. (1962). Studies in the British *Epipactis*. Some further notes on *E. phyllanthes*. *Watsonia* 5: 136–139.

Young, D. P. (1962). Studies in the British *Epipactis*. Seed dimensions and root diameters. *Watsonia* 5: 140–142.

Young, D. P. (1970). Bestimmung und Verbreitung der autogamen *Epipactis* -Arten. *Jahresb. Naturw. Ver. Wuppertal* 23: 43–52.

1. Rhizome long and creeping; labellum strongly constricted between the basal and distal portions, the basal with an upright auricle on each side ...... 2.

1. Rhizome short or nearly absent; labellum not or only slightly constricted between the basal and distal portions, the basal without or with obscure auricles on each side ...... 3.

2. Stem up to 50(70) cm; raceme many flowered
    1(i). palustris var. **palustris**

2. Stem up to 15 cm; raceme few-flowered
    1(ii). palustris var. **ericetorum**

3. Inflorescence rhachis glabrous or nearly so; flowers hanging as soon as they open; leaves often shorter than internodes ...... 4.

3. Inflorescence rhachis more or less pubescent; at least the younger flowers usually patent to erecto-patent; leaves usually longer than internodes ...... 8.

4. Labellum not clearly divided into 2 parts ...... 5.

4. Labellum clearly divided into basal and distal parts ...... 6.

5. Labellum like the other inner perianth segments
    7(i). phyllanthes var. **phyllanthes**

5. Labellum with the basal part formed by a shallow, ventricose depresssion   **7(iv). phyllanthes var. degenera**

6. Labellum with the distal part about equalling the basal part and reflexed at apex
    7(v). phyllanthes var. **pendula**

6. Labellum with the distal part longer than the basal part and porrect ...... 7.

7. Plant tall; raceme with numerous flowers, the distal part of the labellum often whitish or pinkish
    7(ii). phyllanthes var. **vectensis**

7. Plant much shorter; raceme with few, greenish flowers   **7(iii). phyllanthes** var. **cambrensis**

8. Ovary hairy to densely hairy; perianth segments dark wine red to deep rose all over   **2. atrorubens**

8. Ovary glabrous to sparsely hairy; perianth segments usually greenish, often marked or tinged with pink, purple or violet ...... 9.

9. Upper leaves spirally arranged; rostellum secreting obvious, white, persistent viscidium; pollinia becoming detached as integral units ...... 10.

9. Upper leaves usually obviously 2-ranked; rostellum

without or with sparse, soon disappearing viscidium; pollinia crumbling apart ...... 11.

10. Leaves greyish-green, often tinged violet, the lowest considerably longer than wide; distal part of the labellum at least as long as wide, with 2, smoothly pleated, pinkish bosses near the base   **3. purpurata**

10. Leaves dark green, the lowest often wider than long; distal part of labellum wider than long, with 2, usually rough, brownish bosses near the base
    4. helleborine

11. Stigma with 2 basal bosses, with the rostellum appearing tricornate; ovary glabrous or nearly so; inner perianth segments rose; distal part of labellum wider than long   **5. youngiana**

11. Stigma without marked basal bosses hence not appearing tricornate; ovary usually hairy; inner perianth segments pale green, sometimes tinted pink; distal part of labellum longer than wide or wider than long ...... 12.

12. Outer perianth segments acute at apex; labellum with the distal half as long as or longer than wide, yellowish-green or green ...... 13.

12. Outer perianth segments obtuse at apex; labellum with the distal half as long as wide or wider than long, greenish, often tinged with pink ...... 14.

13. Flowers opening; labellum with the distal half yellowish-green with a white margin
    6(i). muelleri var. **leptochila**

13. Flowers cleistogamus; labellum with the distal half green   **6(ii). muelleri** var. **cleistogama**

14. Anther stipitate; ovary hairy
    6(iii). muelleri var. **dunensis**

14. Anther sessile; ovary glabrous or nearly so
    6(iv). muelleri var. **muelleri**

1. E. palustris (L.) Crantz   Marsh Helleborine
Serapias helleborine var. *palustris* L.; *Serapias longifolia* L., non Huds.; *Serapias palustris* (L.) Mill.; *E. longifolia* All. nom. illegit.; *Helleborine palustris* (L.) Schrank; *Helleborine longifolia* Britton & Rendle nom. illegit.

Perennial herb with cylindrical or irregularly thickened rhizome which often creeps horizontally, giving off one or more stolons from the ends of which new buds arise; roots pale brown, slender, rather long. *Stem* 10–50(–70) cm, erect, solid, green, sometimes streaked with violet-red below, angled and with short, whitish hairs above, leafy below the middle, with one or more, acute or rarely obtuse, close-fitting to rather loose leafless sheaths at the base. *Leaves* 4–8, spirally arranged, 5–15 × 2–4 cm, yellowish-green or greyish-green, matt, erect or slightly spreading, ovate-oblong to linear-lanceolate, acute and sometimes hooded at the stiff tip, entire, more or less folded, with rather long sheaths, with 3–5 thickened veins beneath and several intermediate translucent ones. *Inflorescence* a raceme 6–15(–20) × 3–5 cm, 7- to 14-flowered, turned to one side, at first drooping, lax. *Bracts* 5–25 × 1–4 mm, lanceolate, acute at apex, tapering, the lowest about equalling or slightly exceeding the ovary, the upper much shorter, with 5–7 veins. *Flowers* rather large, campanulate but finally opening wider, at first drooping, then horizontal, finally pendulous, creamy white or rarely pure white, more or less tinged or streaked with violet-red or brown, scentless. *Outer perianth segments* 8–12 mm, the lateral slightly longer than the median and rounded on one side, brownish, whitish or dull green on the outside, violet-red within, lanceolate, obtuse to acute at apex, keeled, concave, with 3–5 prominent veins, with short hairs on the outside, glabrous inside. *Inner perianth segments* slightly shorter than outer, white flushed violet-red at base, lanceolate, obtuse at apex, glabrous, with 3–7, sometimes branched veins. *Labellum* 10–12 mm, equal to outer perianth segments, basal half cup-like, shallow, cut low in front with an upright auricle on each side, white with rosy parallel veins and many bright yellow raised spots down the middle, the distal half joined on by a narrow elastic hinge, broadly elliptical, rounded or truncate at apex, flat or slightly folded, with wavy, frilled, upturned edges, white with rose-red veins, with a quadrangular, furrowed, yellow plate, with 3–4 teeth in front projecting over its base. *Spur* absent. *Column* short, yellowish-green, expanding upwards, at an obtuse angle with the ovary. *Anther* relatively large, elliptical or subquadrangular, slightly 2-lobed; pollinia 2, oval, yellowish-white, friable; caudicles absent; pollen grains cohering in fours but not compressed, tetrads tied together by fine elastic threads, which join to form a brown line down the front of the pollinium. *Stigma* transversely elliptical or subquadrangular, slightly 2-lobed. *Rostellum* nearly glandular, projecting from the middle of the upper edge of the stigma; viscidium obvious, white and sticky, remaining in the open flowers. *Ovary* 10–16 mm, narrowly pyriform, shortly hairy, with 6 ridges, brownish or dull green, finally pendulous, stalk twisted. *Seeds* ovoid; testa transparent. *Flowers* 6–8. Cross-pollinated by hive bees and other insects. $2n = 40, 44, 46, 48$.

(i) Var. **palustris**
Stems up to 50(–70) cm. *Raceme* many-flowered.

(ii) Var. **ericetorum** Asch. & Graebn.
Stems up to 15 cm. *Raceme* few-flowered.

Native. A decreasing, but still locally frequent, plant of fens and dune-slacks. Locally frequent in the British Isles north to central Scotland, extinct in many inland sites. Europe to about 60° N in Scandinavia; temperate Asia; North Africa. The varieties appear to be ecotypic. Var. *ericetorum* occurs in moist sandy ground between dunes on the south coast of Wales and England and in Wexford.

2. E. atrorubens (Hoffm.) Besser
Dark-red Helleborine
Serapias latifolia auct.; *Serapias longifolia* auct.; *Serapias atrorubens* Hoffm.; *E. atropurpurea* Raf.; *E. media* Fr.; *E. ovalis* Bab.; *Helleborine media* (Fr.) Druce; *E. latifolia* subsp. *ovalis* (Bab.) Moore & More; *Helleborine atropurpurea* (Raf.) Schinz & Thell.; *E. helleborine* subsp. *atrorubens* (Hoffm.) Syme; *E. helleborine* var. *rubiginosa* Crantz; *E. rubiginosa*

(Crantz) Koch; *E. latifolia* subsp. *rubiginosa* (Crantz) Hook. fil.; *E. latifolia* subsp. *atrorubens* (Hoffm.) Hook. fil.

Perennial herb with a thick, hard, dark brown rhizome; roots many, long, slender, sinuous, pale brown, sometimes forming a swelling from which fresh rootlets spring; younger roots have dense, short hairs. *Stem* (15–)20–60(–106) cm, usually solitary, stiff, erect, solid, flexuous, violet-red below, with rather dense, short, whitish hairs especially above, and 2–3, loose-ribbed, somewhat funnel-shaped basal sheaths, the uppermost often with a green tip. *Leaves* 5–10, in 2 rows, 4–10 × 1.5–4.5 cm, each clasping the base of the next, often reddish beneath, the lowest short and broad, the rest oblong-ovate to ovate-lanceolate, sharply pointed at apex, keeled, folded, stiff, many-veined. *Inflorescence* a raceme 7–25 × 3–5 cm, 8- to 18-flowered, stiff and spike-like. *Bracts* 10–35 × 1–4 mm, often reddish at the base, lanceolate, acute at apex, 3- to 7-veined. *Flowers* rather small, wine red, rarely dull rose or greenish, at first campanulate, later wider open, said to smell of vanilla. *Outer perianth segments* 6–7 mm, dark wine-red or rarely dull rose or greenish outside, greenish with red veins inside, ovate, concave, hooded, keeled. *Inner perianth segments* slightly shorter than outer, but of similar colour, broadly ovate, keeled, minutely and irregularly toothed, 3- to 5-veined. *Labellum* 5.5–6.5 mm, basal half cup-shaped, green, its red edges folded down in front leaving a triangular entrance to the cup, which is spotted inside with violet-red or brown and glistening with nectar, the distal half brighter and often darker in colour, transversely elliptical, minutely toothed with a small, acute, reflexed tip, its basal bosses forming a raised triangular central area often brighter and darker in colour than the lip, strongly wrinkled, sometimes almost tubercled or with tooth-like triangular folds. *Spur* absent. *Column* very short, white, convex behind, green and flat in front below the stigma with a shallow cup at the apex below the anther, its low walls ending in a white translucent staminode on each side. *Anther* yellow, hood-like, obtuse, hinged at the back, minutely papillose outside; pollinia 2, creamy white, without caudicles, deposited in the cup of the column. *Stigma* transversely oblong with a tooth at each upper corner and a small projection in the centre. *Rostellum* whitish, elliptical, resting on the central projection of the stigma; viscidia well developed, white and sticky, remaining in the open flower. *Ovary* 6–7 mm, pyriform, 6-ridged, rough with short hairs, shortly stalked, olive green often tinged with red. *Capsule* small, ovoid, pendulous. *Seeds* short and broad, transparent. *Flowers* 6–7. Cross-pollinated by bees and wasps. $2n = 40$.

Native. Limestone rocks and screes, in woods or in the open up to 610 m in Scotland. Locally frequent in central-west Ireland, north Scotland, north Wales and north England south to Derbyshire; formerly in Breconshire. Europe northwards to Arctic Scandinavia; Caucasus; north Iran.

× **helleborine** = E. × **schmalhausenii** K. Richt.
Helleborine schmalhausenii (K. Richt.) Vollm.;
Helleborine crowtheri Druce nom. nud.
Plants apparently intermediate in morphology have been recorded growing with both parents in north Wales, north England and north Scotland, but wide variation in the parents makes it difficult to be sure if these are really hybrids.

3. E. purpurata Sm.   Violet Helleborine
E. sessilifolia Peterm.; *E. latifolia* var. *violacea* Dur.-Duq.; *E. violacea* (Dur.-Duq.) Boreau; *Helleborine sessilifolia* (Peterm.) Druce; *Helleborine purpurata* (Sm.) Druce; *E. latifolia* subsp. *purpurata* (Sm.) Hook. fil.

Perennial herb with a descending, deeply buried rhizome often with enlarged knots and blackish scales; roots worm-like, fleshy, thickening downwards, springing from various nodes at different depths. *Stems* often clustered, 20–70(–80) cm, greyish-green flushed with violet, erect, rigid, solid, rather slender, with dense, short, whitish hairs above, and 2–3, rather loose, brown, leafless sheaths at the base, the upper often green-tipped. *Leaves* spirally arranged, 5–10, 6–10 × 2–5 cm, dull greyish-green, often flushed with violet, distant, ovate-lanceolate to lanceolate, long-acute at apex, entire, with long sheaths, with 3–7 main and numerous fainter veins, with short translucent papillae on the margins, glabrous. *Inflorescence* a raceme 15–25 × 2–5 cm, many-flowered, rather dense, one-sided, at first nodding but finally erect. *Bracts* 5–60 × 1–10 mm, often flushed with violet, lanceolate to linear, acute at apex, tapering, spreading or pointed downwards, margins minutely toothed, usually with 3 main veins and several subsidiary ones. *Flowers* rather large, greenish-white, horizontal, opening wide, faintly scented. *Outer perianth segments* 10–12 mm, green outside, smooth and whitish-green within, lanceolate, obtuse to subacute at apex, hooded or drawn together at the tip, concave, obscurely 5-veined with a green keel. *Inner perianth segments* shorter than outer, whitish, sometimes faintly tinged with rose, or greenish-white, ovate-lanceolate, obtuse or acute at apex, green-keeled in the lower part only, obscurely 5-veined. *Labellum* 8–10 mm, the basal half cup-shaped, elliptical, glossy pale green slightly tinged with violet outside, olive-green mottled with violet within and obscurely 5-veined, the distal half broader than long, dull white with 2–3, more or less confluent, smooth bosses and a cord-like central boss faintly tinged with violet or rose, cordate, acute at apex, finely toothed and reflexed at the tip. *Spur* absent. *Column* short, with a shallow cup at the top enclosed by a low, semicircular, wavy-edged wall, ending on each side in a rounded staminode almost touching the back of the stigma. *Anther* sessile, narrow, laterally compressed, obtuse at apex, granular outside, pale yellowish-cream, leaning forward over the cup-shaped basal half of the labellum; pollinia 2, yellowish-white, short, thick, ovoid, adhering to the rostellum just below their apex. *Stigma* in front of column, oblong, quadrangular, with rounded corners, its lower edge

turned up into a lip, with a keel at the back running up to the rostellum. *Rostellum* whitish, globular, smaller than in *E. helleborine*; viscidium obvious, white, sticky and remaining in the open flower. *Ovary* 6–8 mm, clavate, straight, tapering gradually to the twisted stalk, dark green tinged with violet, with 6 ridges with few short, stiff hairs. *Capsule* 15–17 mm, trigonous, sides nearly flat, rough with the bases of worn-off hairs, horizontal, not usually pendulous. *Seeds* long and narrow, straight or curved, obtuse at apex. *Flowers* 8–9. Cross-pollinated by wasps. $2n = 40$.

Native. Woods, especially of beech, usually on calcareous or sandy soils or clay-with-flints. Frequent in south-east and south-central England, north to Shropshire and Leicestershire. North-west and central Europe northwards to Denmark and south-eastwards to Bulgaria; west Siberia.

4. E. helleborine (L.) Crantz Broad-leaved Helleborine
Serapias helleborine L.; *Serapias helleborine* var. *latifolia* L.; *Serapias latifolia* (L.) Huds.; *E. latifolia* (L.) All.; *Helleborine latifolia* (L.) Moench; *Helleborine helleborine* (L.) Druce nom. illegit.; *E. media* auct.; *E. helleborine* subsp. *media* Syme; *E. helleborine* subsp. *latifolia* (L.) Syme; *E. helleborine* var. *viridans* Crantz; *E. latifolia* subsp. *viridans* (Crantz) Hook. fil.; *E. atroviridis* W.R. Linton

Perennial herb with numerous, moderately thick, pale brown, cylindrical roots; rhizome if present small and woody. *Stem* 25–80(–100) cm, pale green often tinged with violet in the lower part, solid, terete, leafy, with short, whitish hairs in the upper part, with 2 or more ribbed, leafless sheaths at the base. *Leaves* 4–10, spirally arranged, 5–17 × 2.5–10.0 cm, dull green, matt, broadly ovate to ovate-lanceolate, sometimes almost subrotund, acute or acuminate at apex, entire, their ribbed sheaths enclosing the stem, decreasing in size upwards, the transition gradual or abrupt, and sometimes crowded near the middle, the uppermost bract-like with a long-tapering apex, all spreading, with about 5 main veins strongly raised beneath the rather weak and flaccid lamina. *Inflorescence* a raceme (7–)10–40 × 2.5–3.5 cm, 15- to 50-flowered, loose or dense, one-sided and spike-like, nodding at first, but finally erect. *Bracts* 10–40(–70) × 1–8(–13) mm, spreading, lanceolate, acute at apex, the lowest often exceeding the flowers, 9- to 19-veined. *Flowers* rather large, campanulate at first, but opening wide later, drooping, usually scentless, rarely faintly sweet-smelling, greenish to dull reddish-violet. *Outer perianth segments* 10–13 mm, green or dull reddish-violet, obliquely ovate to lanceolate, acute at apex, concave, keeled, 3- to 5-veined. *Inner perianth segments* slightly shorter than outer, green tinged with rose or reddish-violet, ovate, keeled, semi-transparent, 5- to 7-veined. *Labellum* 9–11 mm, the basal half cup-shaped, green outside, dark reddish-brown and glistening with nectar inside, 5-veined, the front edges turned down in a fold connecting it with the distal half, which is rose to violet-rose, cordate to triangular, with an acute, reflexed tip, and with 2 wart-like, smooth or wrinkled bosses at the base. *Spur* absent. *Column* short and squat, the thin green walls forming a shallow cup on the summit and ending on each side in a whitish, rounded staminode almost touching the back of the stigma. *Anther* yellowish, sessile or nearly so, in back view ovoid, in profile rhomboidal, leaning forwards over the cup of the column; pollinia 2, yellow, clavate, not friable, firmly attached to the rostellum. *Stigma* transversely oblong, angles rounded, lipped below and facing forwards. *Rostellum* large, milky white, subglobose, seated on a square, green projection; viscidium obvious, white and sticky, remaining in the open flower. *Ovary* 4–10 mm, green, somewhat pyriform, 6-ridged, glabrous or with a few scattered hairs, tapering into a short, twisted stalk. *Capsule* about 11 mm, oblong-ovate, shortly stalked, hanging. *Seeds* long and narrow, transparent. *Flowers* 7–10. Said to be cross-pollinated by wasps or self-pollinated. $2n = 36, 38, 40, 44$.

Variable in leaf shape and flower colour, shape of labellum and the rugosity of its basal bosses and whether a central boss is present or absent.

Native. Woods, hedgerows, wood-margins and clearings, shady banks and scree slopes. Frequent in Britain and Ireland, rare in north Scotland. Most of Europe to about 68° N in Norway; North Africa; temperate Asia east to Japan; introduced in North America.

$\times$ **purpurata = E. $\times$ schulzei** P. Fourn.
Fertile and difficult to distinguish because of the variation within the parents. Plants with long leaves, purplish colouration and dull purple flowers are the most convincing.
Widely recorded where the parents grow together.

5. E. youngiana A. J. Richards & A. F. Porter
Young's Helleborine
Rhizomes and *roots* unknown, but presumably a perennial herb. *Stem* usually solitary, sometimes in pairs or groups, 30–58 cm, pale green, with numerous, short hairs from the uppermost leaf upwards, glabrous in the lower part, with 1–2 leafless sheaths below. *Leaves* 4–7, in two ranks, 2–8 × 1–5 cm, yellowish-green, lanceolate to ovate-lanceolate, acute to subacuminate at apex, the margin somewhat wavy, erecto-patent, with subequal papillae on the margin, glabrous. *Inflorescence* a raceme about 4.0 × 1.5 cm, usually one-sided. *Bracts* 5–40 × 2–15 mm, linear-lanceolate, shorter than flowers. *Flowers* rather large, more or less campanulate at first but opening later, green and pink, patent to more or less nodding. *Outer perianth segments* 8–11 × 5–6 mm, green or with rose margins, ovate-lanceolate, acute or acuminate at apex, cucullate, keeled. *Inner perianth segments* 6–8 × 5–6 mm, rose with a paler central zone, ovate, more or less acuminate at apex. *Labellum* small, the basal half about 4 mm, hemispherical and usually purple-spotted inside, the distal half 3–4 × 4–5 mm, rose with a green central zone and 2 purple basal bosses, cordate, acuminate at apex and markedly reflexed. *Spur* absent. *Column* inclined. *Anthers* 2–3 mm, sessile or with a short stalk; pollinia 2, soon fragmenting and falling,

but remaining for a short while in the cup of the column. *Stigma* with 2 basal bosses, with the rostellum appearing tricornate. *Rostellum* bearing a very small glandular viscidium which usually appears ineffective in preventing self-pollination and usually disappears soon after the opening of the flower, or before, the other 2 projections on the lower angle of the stigma. *Ovary* 7–8 × 3–4 mm, green, pyriform, glabrous or slightly pubescent, with 6 longitudinal ribs, not swelling until after flowering. *Capsule* 10–13 × 6–7 mm, obovoid, with persistent, shrivelled calyx. *Seeds* 1.0–1.1 × about 0.2 mm. *Flowers* 7–8. Self-pollinated by the pollen falling on the stigma from the disintegrating pollinia.

Native. Woodland on heavy, often metal-polluted soils. Rare in Northumberland and Lanarkshire. Endemic.

6. E. muelleri Godfery   Narrow-lipped Helleborine
Perennial herb with a decending, deep, knotted rhizome with dark brown scales; roots numerous, long, fleshy, sinuous, pale brown. *Stems* 20–70 cm, erect, rigid, terete, solid, sometimes rather woody at the base, pale green or whitish tinged with violet below, hairy, sometimes densely so above, with 2 or 3, ribbed, violet-tinged, leafless sheaths at the base which later turn brown. *Leaves* 4–10, in two ranks, 4–10 × 1.5–5.0 cm, yellowish-green to dark green, broadly lanceolate to broadly ovate, acute and often tapering at apex, with undulate margins, often with short internodes, rigid, rather rough to the touch, with 3 raised veins on either side of the thick midrib and 4–8 translucent veins between each pair. *Inflorescence* a raceme 5–25 × 2–4 cm, 10- to 25(–40)-flowered, drooping but finally erect, narrow, rather lax, one-sided. *Bracts* 5–20 × 1–5 mm, yellowish-green to dark green, linear-lanceolate, tapering to an acute apex, up to twice as long as the flowers, 3- to 11-veined. *Flowers* rather small, yellowish-green, scentless, erecto-patent to slightly drooping, opening or cleistogamous. *Outer perianth segments* 7–15 mm, pale green with transparent edges, lanceolate, tapering to an obtuse or acute apex, concave, spreading, 3-veined, with a rough keel. *Inner perianth segments* shorter and broader than outer, pale green sometimes tinted pale pink, ovate-lanceolate, acute at apex, with a bright, thick keel and faint veins on each side. *Labellum* 4–10 mm, the basal half cupuliform, pale greenish outside with a raised keel, mottled reddish inside, the distal half cordate-acuminate, yellowish-green or pale pink. *Spur* absent. *Column* short, divided into two parts at the apex by a V-shaped incision on each side, with a curved, nib-shaped stalk at the back supporting the anther, the front wall on each side ending in a staminode. *Anther* cream-coloured or yellowish-green, stipitate or sessile, ovate with an acute apex, projecting over the upper edge of the stigma, convex behind, more or less flat in front, minutely papillose outside; pollinia 2, creamy white, joined at the apex, thicker, rounded and slightly divergent at the base, split for most of their length. *Stigma* transversely oblong, leaning slightly backwards. *Rostellum* globose; viscidium minute or absent. *Ovary* 5–10 mm, green, spindle-shaped, with 6

ridges, hairy to nearly glabrous. *Capsule* horizontal or drooping, yellowish-green, ellipsoid, with a stalk up to 6 mm. *Seeds* long, narrow, obtuse at apex. *Flowers* 6–8. Automatically self-pollinated by the loosened pollen masses falling from the pollinia onto the stigma; rarely cross-pollinated by wasps.

(i) Var. **leptochila** (Godfery) P. D. Sell
E. viridiflora var. *leptochila* Godfery; *Helleborine leptochila* (Godfery) Druce; *E. leptochila* (Godfery) Godfery
Flowers opening. *Outer perianth segments* acute at apex. *Labellum* with the distal half as long as or longer than wide, yellowish-green with a white margin, flat at apex. *Anther* stipitate. *Ovary* hairy. $2n = 36$.

(ii) Var. **cleistogama** (C. Thomas) P. D. Sell
E. cleistogama C. Thomas; *E. leptochila* var. *cleistogama* (C. Thomas) D. P. Young
Flowers cleistogamous. *Outer perianth segments* acute at apex. *Labellum* with the distal half as long as or longer than wide, green, flat at apex. *Anther* stipitate. *Ovary* hairy.

(iii) Var. **dunensis** (T. & T. A. Stephenson) P. D. Sell
E. viridiflora forma *dunensis* T. & T. A. Stephenson; *E. viridiflora* var. *dunensis* (T. & T. A. Stephenson) Wilmott; *E. leptochila* var. *dunensis* (T. & T. A. Stephenson) T. & T. A. Stephenson
Flowers opening. *Outer perianth segments* obtuse at apex. *Labellum* with the distal half as long as wide, greenish, often tinted pale pink, apex recurved. *Anther* stipitate. *Ovary* hairy.

(iv) Var. **muelleri**
Flowers opening. *Outer perianth segments* obtuse at apex. *Labellum* with the distal half wider than long, greenish or tinted pale pink. *Anther* sessile. *Ovary* almost glabrous.

Native. Woods mostly on calcareous or heavy-metal polluted soils, river-gravels and dunes. Scattered in Britain north to Lanarkshire. North, west and central Europe. Var. *leptochila* occurs mainly in the deep shade of beech, ash–hazel coppice or birch scrub. It is mainly in southern England from Kent to south Devon northwards to the Chiltern and Cotswold Hills, with outliers further north and west, sometimes on substrata rich in heavy metals, where intermediates with var. *dunense* can be found. In Europe it occurs in the north-west and central regions, south-eastwards to Yugoslavia and north Greece. Var. *cleistogama* is an endemic, only recorded under beech trees on the steep western slope of the Cotswolds near Wotton-under-Edge in Gloucestershire. Var. *dunensis* is found in somewhat peaty, but not very moist, hollows in coastal dunes, chiefly amongst *Salix repens*, but sometimes beneath planted pines. Probably endemic, in Anglesey and north-west England from Southport to south Cumberland, north Lincolnshire and Northumberland (Holy Island). Var. *muelleri* has been recorded from Sussex. It is also found in open woods and clearings in west and central Europe.

7. E. phyllanthes G. E. Sm. Green-flowered Helleborine
Perennial herb with a short, horizontal or ascending rhizome; roots numerous, thick, fleshy, emerging from the rhizome and sometimes the buried portion of the stem. *Stems* solitary or more rarely 2–3 together, (8–)20–45(–65) cm, plus 5–20 cm below ground level, stout, glabrous or with very sparse, short hairs, with 1–3 sheaths at the base. *Leaves* 3–6, in two ranks, 3.5–6.0 (–7.0) × 3–5 cm, mid- to dark green, subrotund, ovate or lanceolate, acuminate at apex, often undulate and interruptedly ciliolate on the margins, the lower with rather long sheaths, smooth and thick in texture, main veins few and not prominent. *Inflorescence* a raceme 5–15 × 3–4 cm, with up to 35 flowers which are sometimes aggregated. *Bracts* 7–50 × 0.5–7.0 mm, linear to linear-lanceolate, long-acute at apex, entire, 5- to 7-veined. *Flowers* rather small, opening slightly, widely, or not at all, usually hanging more or less vertically downwards. *Outer perianth segments* 8–10 mm, yellowish-green to green, lanceolate to ovate-lanceolate, acuminate at apex. *Inner perianth segments* green, sometimes with a violet tinge, lanceolate to ovate-lanceolate, acuminate at apex. *Labellum* variable, usually more or less green, the basal half cup-shaped to shallowly concave, greenish or whitish and sometimes small, the apical half sometimes not clearly differentiated, ovate-lanceolate to cordate, acute to acuminate at apex, usually longer than wide and greenish-white to pinkish. *Spur* absent. *Column* rapidly decaying after anthesis. *Anther* sessile or stipitate, cuneiform or cylindrical; pollinia 2, disintegrating in bud. *Stigma* base with 2 small, blunt bosses. *Rostellum* and viscidium minute and shrivelling at an early stage, *Ovary* 9–13 mm, pyriform, 6-ribbed, more or less glabrous. *Seeds* 1.0–1.5 mm, tapered at each end. *Flowers* 7–8. Self-pollinated by the pollen falling from the disintegrating pollinia onto the stigma. $2n = 36$.

This species is represented by an aggregate of local variants differing mainly in details of floral structure. There is a continuous range from open flowers with labellum clearly divided into 2 parts to cleistogamous flowers with the labellum not clearly divided into 2 parts. Intermediates occur between the recognised varieties.

(i) Var. **phyllanthes**
Flowers rarely opening widely, usually cleistogamous. *Labellum* not divided into 2 parts, in form, colour and texture like the lateral inner perianth segments, ovate to lanceolate. *Anthers* sessile or stipitate.

(ii) Var. **vectensis** (T. & T. A. Stephenson) D. P. Young
Helleborine viridiflora forma *vectensis* T. & T. A. Stephenson; *E. leptochila* var. *vectensis* (T. & T. A. Stephenson) T. & T. A. Stephenson; *E. viridiflora* var. *vectensis* (T. & T. A. Stephenson) Wilmott; *E. vectensis* (T. & T. A. Stephenson) Brooke & Rose
Flowers often cleistogamous. *Labellum* more or less closely embracing the stigma, the basal part 2.5–3.5 mm, hemispherical and entirely green, the distal part longer than the basal, cordate-deltoid, usually elongate

and acuminate, often whitish or pinkish, usually with 2 lateral bosses, the joint between the basal and distal parts perfectly formed and with a central sinus. *Anther* sessile or subsessile.

(iii) Var. cambrensis (C. Thomas) P. D. Sell
E. cambrensis C. Thomas
Similar to var. *vectensis*, but shorter and with fewer, pale, uncoloured flowers and a long, slender ovary.

(iv) Var. degenera D. P. Young
Flowers rarely opening widely and often cleistogamous. *Labellum* not clearly divided into a basal and a distal part, the basal part represented by a shallow or ventricose depression, the distal part variable in width and often with basal lateral bosses. *Anthers* cuneiform or ovate-cylindrical, sessile to long-stipitate.

(v) Var. pendula D. P. Young
E. pendula C. Thomas, non A. A. Eaton
Flowers rarely cleistogamous. *Labellum* with clearly defined basal and distal parts, the basal part about 4 mm, the distal part as long as or slightly longer, cordate, rugose at the base with 2 bosses, acuminate at apex and usually strongly reflexed. *Anther* cuneiform, sessile.

Native. Woods on calcareous or sandy soils, sometimes heavy-metal polluted, and on dunes. Scattered in Britain north to the Cheviots, Westmorland and Lanarkshire; very scarce in Ireland. It also occurs on the Atlantic coast of France, in the Pyrenees, and in Denmark. Var. *phyllanthes* and var. *degenera* are mainly in southern England. Var. *pendula* is often on sand-dunes and mainly in northern England and North Wales, but southwards to Herefordshire, Oxfordshire and Hertfordshire, where intermediates with var. *vectensis* are not infrequent. It also occurs in a few localities in Ireland. Var. *vectensis* is mainly in southern England and the south Midlands, but north to south-east Yorkshire. Var. *cambrensis* is known only from sand-dunes in south Wales.

4. Epipogium Emelin ex Borkh.
Pinkish *perennial saprophytic herbs* with coralloid rhizomes; roots absent. *Stem* with few, sheathing scales. *Green leaves* absent. *Flowers* in racemes, drooping. *Labellum* uppermost, with 2, short, rounded, lateral lobes at the base. *Spur* directed vertically upwards, about as long as the labellum. *Column* pointing downwards. *Anther* helmet-like, sessile in the concave summit of the column; pollinia 2, pyriform, with caudicles attached to their base, ending in a viscidium; pollen in packets of tetrads. *Stigma* horseshoe-shaped, on the overhanging base of the upper side of the column. *Rostellum* large, cordate, at the apex of the column. *Ovary* not twisted. Bursicles absent.
Two species in north and central Europe; northern Asia and the Himalayas.

1. E. aphyllum Swartz   Ghost Orchid
Satyrium epipogium L.; *Orchis aphylla* Schmidt, non Forssk.; *Limodorum epipogium* (L.) Swartz; *E. gmelinii*

Rich. nom. illegit.; *E. epipogium* (L.) Karsten nom. illegit.

Perennial saprophytic herb; rhizome whitish, much branched, resembling coral, the branches very short and forked or trilobed, the lobes rounded at apex, often spreading like a fan; roots absent; often sending out 1 or 2, threadlike, sinuous, whitish stolons, with buds at intervals (protected by semitransparent scales) which give rise to new rhizomes. *Stem* 5–25 cm, white tinged with dull rose or pinkish-brown, with numerous, short, pale rose dashes, erect, more or less translucent, weak, slightly hollow, much swollen at the base then tapering suddenly to its very weak attachment to the rhizome. *Leaves* represented only by 2–3, brownish, basal-sheathing scales, and 1–2, long, close-fitting, usually dark-edged sheaths higher up the stem. *Flowers* 1–4(–7), distant, pendulous, on slender stalks, the perianth segments directed downwards, the labellum and spur upwards, yellowish, more or less tinged with rose, smelling of bananas. *Bracts* 6–7 × 0.5–2.0 mm, membranous, semitransparent, oblong or rhomboidal, obtuse at apex, sometimes 3-veined. *Outer perianth segments* 10–11 mm, linear, yellowish, sometimes tinged red, margins incurved. *Inner perianth segments* equalling outer, yellowish with a few, short, violet lines, lanceolate, obtuse at apex, semitransparent, incurved on the margins, curving downwards. *Labellum* bent sharply back in the middle so that the back of the distal half nearly touches the spur, both pointing obliquely upwards, the basal half with 2, short, rounded lobes directed forwards, the distal half cordate, rather pointed, very concave, with eroded or entire edges, white with violet spots and with irregularly tubercled, violet-tinged crests, making it deeply channelled down the middle. *Spur* about 8 × 4 mm, erect, thick, sack-shaped, white tinged with yellow, sometimes lilac or reddish, with lines of violet spots inside, glabrous and shining within. *Column* with a broad, protruding, flat-topped base above which it expands into a deep cup at the apex. *Anther* helmet-like, rounded, sessile in the concave summit of the column with a slightly protruding point just above the rostellum, fastened to the back of the column by a narrow band; pollinia 2, pyriform, pale yellow, granular; caudicles long, elastic, ribbon-like, attached to the base of the pollinia and running up nearly their whole length, each fastened to the rostellum. *Stigma* on the base and part of the face of the column. *Rostellum* large, white, cordate, in a fork at the apex of the column. *Ovary* about 8 mm, ovoid, short and thick, yellowish, streaked or spotted with violet, glabrous, stalk short and curved. *Capsule* almost globose, pendulous, opening by short slits. *Flowers* 6–8. Cross-pollinated by bumble-bees and other insects.
Native. In the deep shade of beech or oak woods on leaf litter or rotten stumps. Very rare and known only in one site each in Herefordshire, Oxfordshire and Buckinghamshire; formerly in Shropshire. North and central Europe southwards to the Pyrenees, north Greece and the Crimea; Caucasus; Siberia; Himalayas.

5. Neottia Guett. nom. conserv.

Yellowish-brown, *saprophytic perennial herb;* rhizomes short, creeping; roots thick, fleshy, forming a nest-like mass. *Stems* densely covered with brownish scales. Green *leaves* absent. *Flowers* numerous, in a spike-like raceme, pale brownish. *Perianth segments* subequal or the inner somewhat shorter, connivent into an open hood. *Labellum* saccate at the base, with 2 distal lobes. *Spur* absent. *Column* long and slender. *Anther* hinged to the back of the column, directed forwards; pollinia 2, more or less cylindrical; pollen tetrads loosely united by a few, weak threads, so that the pollen is friable. *Stigma* large, prominent. *Rostellum* broad, flat, arching over the stigma. *Viscidia* absent. *Bursicles* absent.

Nine species in temperate Europe and Asia.

1. N. nidus-avis (L.) Rich.   Bird's-nest Orchid
Ophrys nidus-avis L.; *Epipactis nidus-avis* (L.) Crantz; *N. abortiva* Gray, non Clairv.; *Listera nidus-avis* (L.) Hook.

Whole plant, including the flowers, yellowish-brown, without chlorophyll. *Perennial saprophytic herb;* rhizomes short, creeping; roots short, thick, obtuse, fleshy, in a dense mass resembling a bird's nest. *Stem* 20–50 cm, erect, stiff, cylindrical, usually stout, slightly viscid in the upper part with glandular hairs, covered below with brown, oblong or lanceolate, close-fitting or loose, brown, sheathing scales, the lower short and acute at apex, the upper longer, lanceolate and obtuse at apex. *Green leaves* absent. *Inflorescence* a raceme 5–20 × 2–4 cm, spike-like, cylindrical, long, dense above, lax lower down. *Bracts* 5–10 × 0.5–1.0 mm, lanceolate, tapering to an acute apex, 1-veined. *Flowers* moderately large, yellowish-brown, honey-scented. *Outer perianth segments* 4–6 mm, spreading in an erect fan, obovate, minutely toothed at apex, 1-veined. *Inner perianth segments* similar to outer, but sometimes slightly smaller. *Labellum* twice as long as the outer perianth segments, 8–12 mm, darker brown than the other segments, directed obliquely forwards, at right angles to the column, hollowed out into an oval cup glistening with honey, with 2, small, tooth-like side-lobes at the base, oblong, then dividing into 2, broad, widely spreading, curved, oval or strap-like, often crenulate lobes, rounded at the apex and glandular-hairy beneath. *Spur* absent. *Column* nearly in line with the ovary, but sloping slightly backwards, cylindrical in section with a long-waisted body, pale brownish-white, ending at the back in a short, rounded tooth, projecting in front in a spout-like, slightly notched lip on the upper surface of which is the stigma. *Anther* oblong, slightly cordate, papillose outside, hinged to the back of the column, projecting forwards over the stigma nearly at right angles to the column; pollinia 2, bipartite, pale yellow, linear-oblong, friable. *Stigma* narrowly kidney-shaped, appearing V-shaped from the front. *Rostellum* projecting forwards immediately above the stigma and extending some distance beyond it, strap-shaped, trough-like, grooved above, curving downwards at the tip. *Ovary* not twisted, on a twisted stalk half its length or more, ovate, with sparse to numerous glandular hairs, 6-ridged. *Capsule* about 12 mm, erect, ovate, appearing 3-sided, half the ridges being more prominent than the rest. *Seeds* oblong, with a transparent testa. *Flowers* 6–7. Cross-pollinated by small, crawling insects which touch the sensitive rostellum, or self-pollinated. $2n = 36$.

Native. On leaf litter in shady woods, especially under beech on calcareous soils. Scattered throughout the British Isles and locally frequent in southern England, Europe, northwards to about 64° N in Norway and in south Sweden and south Finland; Caucasus; Siberia.

6. Listera R. Br. ex. W. T. Aiton nom. conserv.
Diphryllum Raf.

Perennial herbs; rhizomes short; roots numerous, slender. *Stems* erect. *Leaves* usually 2, subopposite, sessile, borne just below the middle of the stem. *Inflorescence* a rather lax, spike-like raceme. *Flowers* inconspicuous, green or reddish-green. *Perianth segments* subequal, patent or somewhat convergent. *Labellum* long and narrow, forking distally into 2 narrow segments and sometimes with 2 lateral lobes near the base. *Spur* absent. *Column* short, erect. *Anther* hinged to the back of the column; pollinia 2, clavate, each more or less divided into halves; caudicles absent; pollen friable in loosely bound tetrads. *Stigmas* transversely elongated, prominent. *Rostellum* broad and flat, arching over the stigma and expelling its viscid contents explosively in a terminal drop when touched. *Viscidia* absent. *Bursicles* absent.

About 30 species in north temperate and Subarctic zones.

1. Plant 20–60(–75) cm; leaves 5–20 cm; raceme 7–28 cm;
   flowers green   **1. ovata**
1. Plant 6–10(–25) cm; leaves 1.0–2.5 cm; raceme
   1.5–6.0 cm; flowers reddish-green   **2. cordata**

1. L. ovata (L.) R. Br.   Common Twayblade
Ophrys ovata L.; *Epipactis ovata* (L.) Crantz; *Epipactis ovalifolia* Stokes nom. illegit.; *Diphryllum ovatum* (L.) Druce

Perennial herb; rhizome short, creeping, deeply buried; roots many, long, sinuous, fairly thick. *Stem* 20–60(–75) cm, pale green, whitish and thicker below, erect, solitary, pubescent above with 1–2, small, green, triangular, bract-like leaves, glabrous below with 2–3, membranous, scale-like sheaths at the base. *Leaves* 2, spreading horizontally, just below the middle of the stem and almost opposite, 5–20 × 4–12 cm, green, broadly ovate-elliptical, obtuse-mucronate at apex, entire, sessile and sheathing at the base, rather thick, with 5 or more principal veins. *Inflorescence* a raceme 7–28 × 1–3 cm, with numerous flowers, rather lax. *Bracts* very small, green, ovate-lanceolate, tapering at apex, glabrous or nearly so. *Flowers* inconspicuous, yellowish-green, spreading, shortly stalked. *Outer perianth segments* 5–6 mm, green- or violet-edged, ovate, obtuse at apex, concave, loosely connivent. *Inner perianth segments* more or less equalling outer, yellowish-green, narrowly oblong, subacute at apex. *Labellum* 7–15 mm, green, cuneate at base, with 2, small, tooth-like, erect side-lobes, deeply divided at the apex into 2, linear, obtuse, yellowish-green lobes, rarely with an intermediate tooth, projecting forwards at the base and then turning sharply downwards, a nectar-secreting furrow running down its centre. *Spur* absent. *Column* short, rising at the back into a pale green, white-edged, notched hood, arching over the anther; in front the white rostellum curves forward, in profile like the spout of a jug, on which lie the 2 pollinia. *Anther* ovate, wide open in front, with a shallow partition between the cells; pollinia 2, pale yellow, oblong, each more or less divided into 2, sessile. *Stigma* on the front of the column immediately below the rostellum, transversely elliptical, convex, prominent. *Rostellum* rather broadly tongue-shaped, very finely striate, with 2, longitudinal, very shallow furrows in which the pollinia lie. *Ovary* almost globose, not twisted, green or tinged violet, glabrous or hairy, with 6 slight ridges, shorter than its curved, erect, twisted stalk. *Capsule* globose, with 3 flat ridges, on an upwards curved stalk. *Seed* oblong, reticulate. *Flowers* 5–7. Cross-pollinated by small insects which crawl up the centre furrow of the labellum. $2n = 32, 34, 35, 38, 42$.

Native. Moist woods, hedgerows, pastures, dunes and sometimes *Calluna* moors. Frequent, and sometimes locally abundant throughout the British Isles. Most of Europe northwards to about 70° N in Norway; Caucasus; Siberia.

2. L. cordata (L.) R. Br.   Lesser Twayblade
Ophrys cordata L.; *Epipactis cordata* (L.) All.; *Diphryllum cordatum* (L.) Druce

Perennial herb; rhizome creeping, very slender; roots few, whitish, filiform. *Stem* 6–10(–25) cm, pale green or reddish-tinted, erect, slender, angled and fluted above, glabrous or slightly pubescent, with 1–2, brownish, scale-like, lanceolate, close-fitting, leafless sheaths at the base. *Leaves* 2, at about the middle of the stem, appearing opposite, spreading horizontally, $1.0–2.5 \times 1.0–2.5$ cm, green and shining above, paler and rather greyish-green beneath, ovate-deltate, rounded-mucronate at apex, entire, with incurved or wavy margins, glabrous, with numerous reticulate veins. *Inflorescence* a raceme $1.5–6.0 \times 0.7–1.5$ cm, short, lax, with 4–12 flowers. *Bracts* minute, triangular-ovate, acute at apex. *Flowers* very small, green, more or less suffused with red. *Outer perianth segments* 2.0-2.5 mm, green, oblong, rounded at apex, spreading, persisting in fruit. *Inner perianth segments* similar to outer, green outside, reddish-purple within. *Labellum* 3.5–4.5 mm, linear, brownish-red, horizontal or pendulous, with an erect, lanceolate, acute tooth on each side at the base, and divided about the middle into 2, linear, tapering, divergent terminal lobes. *Spur* absent. *Column* thick, very short, with a hood at the apex protecting the anther. *Anther* oblong, with a small, obtuse tooth at the apex; pollinia 2, bipartite, clavate, without caudicles. *Stigma* reniform. *Rostellum* oblong, 2- to 3-toothed at apex. *Ovary* nearly globose, angular, pale green, glabrous, with 6 reddish ridges, shorter than its twisted stalk. *Capsule* more or less

globose. *Seeds* oblong. *Flowers* 6–9. Cross-pollinated by minute flies and Hymenoptera. $2n = 38, 40, 42$.

Native. Upland woods, especially of *Pinus sylvestris*, and moors in usually wet, acid places, often amongst *Sphagnum* and under *Calluna* or other moorland shrubs. Frequent in Scotland up to 825 m; scattered in Ireland; in Britain south to Derbyshire and Devon; absent from central, south-central and south-east England except for one site in Sussex. Europe from the Pyrenees and Apennines to Iceland, northernmost Scandinavia and the Kola Peninsula; Transcaucasia; northern Asia; North America.

7. Spiranthes Rich. nom. conserv.

Gyrostachys Pers. ex Dumort.; *Ibidium* Salisb. ex Druce nom. illegit.

Small, *perennial* herbs; roots fleshy, more or less tuberous. *Leaves* several, at base and on stem. *Flowers* small, fragrant, in spikes, the axis twisted so that the flowers are arranged in one or more spiral rows. *Perianth segments* subequal, free or variously connate or connivent. *Labellum* entire, frilled and more or less recurved distally, furrowed below and embracing the base of the column to form the lower part of the perianth tube, its edges being overlapped by the inner perianth segments. *Spur* absent. *Column* horizontal, with the circular stigma on its underside facing the labellum. *Anther* hinged to the back of the column, resting on the rostellum; pollinia 2, each of two plates of coherent pollen tetrads attached to the viscidium near their summits; caudicle absent. *Rostellum* narrow, projecting beyond the stigma and consisting of a narrowly elliptical viscidium supported between 2, long, narrow teeth which are left behind like the prongs of a fork when the viscidium is removed. *Bursicles* absent.

About 25 species in the north temperate zone and South America.

Stewart, A., Pearman, D. A. & Preston, C. D. (1994). *Scarce plants in Britain.* Peterborough. [*S. romanzoffiana*.]
Wilmott, A. J. (1927). The Irish '*Spiranthes Romanzoffiana*'. *Jour. Bot. (London)* **65**: 145–149.

1. Flowering stem bearing only bract-like, appressed scales, the true leaves of the current season being in a lateral basal rosette   **1. spiralis**
1. Flowering stems bearing true leaves   2.
2. Flowers in a single spirally twisted row; bracts 6–9 mm; flowers 6–8 mm excluding ovary   **2. aestivalis**
2. Flowers in 3 spirally twisted rows; bracts 10–20(–30) mm; flowers 10–14 mm excluding ovary   **3. romanzoffiana**

1. S. spiralis (L.) Chev.   Autumn Lady's-tresses
Ophrys spiralis L.; *Neottia spiralis* (L.) Swartz; *Ibidium spirale* (L.) Salisb. nom. illegit.; *S. autumnalis* Rich. nom. illegit.; *Gyrostachys autumnalis* Dumort. nom. illegit.

Perennial herb; roots 2(–5), pale brown, carrot-shaped, thick, tapering to an obtuse point, hard, smooth, with short transparent hairs. *Stem* 6–20(–35) cm, with or without the remains of last year's leaves at the base, terete, solid, pale green, with transparent glandular hairs, with several, bract-like, sheathing, lanceolate, close-fitting leaves with membranous edges and 3–5 veins. *Leaves* of current season 4–5, appearing with or after the flowers in a lateral basal rosette which will flower next season, 2.0–3.5 × 0.5–1.0 cm, bluish-green with transparent margins, ovate-elliptical, acute at apex, entire, broadly sheathing, with a thick keel and 3–5 veins. *Inflorescence* a spike 3–12 cm, 6- to 20-flowered, slender. *Bracts* 6–7 × 1–3 mm, lanceolate, tapering at apex, incurved, white-edged, 3-veined. *Flowers* very small, white, sweet-scented by day, in a single rank, twisting spirally round the axis or more rarely all turned to one side. *Outer perianth segments* 6–7 mm, white and crystalline with a faint green nerve, oblong, slightly tapering to an obtuse apex, ciliate or minutely toothed, slightly glandular-hairy outside. *Inner perianth segments* white, strap-shaped, obtuse at apex, 1-veined, adhering to the slightly longer upper outer perianth segment and together forming an upper lip to the flower, with recurved tips. *Labellum* 6–7 mm, oblong, trough-shaped, enlarged, rounded and turning down at the tips, forming a trumpet-like tube with the united upper outer and inner perianth segments, pale green with a broad crystalline jagged margin, closely embracing the column at the base, where there are 2, white, glistening, rounded, honey-secreting glands, each with a ring of papillae round the base and a shallow receptacle below. *Spur* absent. *Column* green, obconical, horizontal, tapering to an acute point. *Anther* sessile, resting face downwards on the upper side of the column, ovate, acute at apex, already brown and shrunken when the flower opens; pollinia 2, cream-coloured, narrow, tapering from a rounded base to an acute apex, and attached to the back of the viscidium. *Stigma* on undersurface of column, slanting gently upwards, scutate, rounded and fringed with white hairs below, with a short point at each upper corner, and in the middle of the upper edge of the labellum a brown or greyish, linear, viscid gland supported between 2, rather long, very narrow teeth, left behind like the prongs of a fork when the viscidium is removed, a membrane extending backwards from the edge of the stigma over the back of the anther, forming a clinandrium to shelter the pollinia. *Rostellum* deeply bifid, with narrow, acute lobes. *Ovary* 3–4 mm, narrow, bent at apex, green, usually not twisted, with 3 rounded ridges, with glandular hairs. *Capsule* about 6 mm, obovoid. *Flowers* 8–9. Pollinated by bumble-bees. 2n = 30.

Native. Hilly pastures, downs, moist meadows and grassy coastal dunes, usually on calcareous substratum. Locally frequent in the British Isles north to Yorkshire, Isle of Man and Co. Sligo; extinct in many inland sites. Europe northwards to Denmark and central Russia; North Africa; Asia Minor.

2. S. aestivalis (Poir.) Rich.	Summer Lady's-tresses
Ophrys aestivalis Poir.; *Gyrostachys aestivalis* (Poir.) Dumort.; *Ibidium aestivalis* (Poir.) Druce

Perennial herb; tubers 2–6, thick at base, tapering downwards; roots few, short. *Stems* 10–40 cm, erect, glandu-

lar-hairy in upper part with the remains of a few withered leaves at the base. *Leaves* 5–12 cm × 4–9 mm, linear-lanceolate, obtuse at apex, entire, slightly trough-like, bright green, glossy on both sides, with one or more veins each side of the keel and numerous cross-veins, mostly basal, but 1–2 on the lower part of the stem passing into lanceolate scales in the upper part and present at anthesis. *Inflorescence* a spike 3–10 cm, with 6–20 flowers in one spiral row, slightly twisted, glandular-hairy. Bracts 6–9 × 1–2 mm, lanceolate, tapering, 3- to 5-veined, clasping. *Flowers* small, 6–8 mm excluding ovary, pure white, tubular, slightly scented at night. *Outer perianth segments* 6–7 mm, forming a tube with the labellum, white with a greenish keel, linear-lanceolate, obtuse at apex, the lateral often curving outwards, glandular-hairy. *Inner perianth segments* shorter than outer, linear, obtuse at apex, white, 1-veined. *Labellum* 6–7 mm, pure white, oblong, trough-like, expanded, turned down and irregularly toothed at the tip, forming a slight sack or receptacle with the ovary at the base, with 2 nectar-secreting nipples. *Column* green, horizontal, slender, acute at apex, shorter than sepals. *Anther* at apex of upper surface of the column, nib-shaped, brown, acute at apex, with a short, thick, curved filament or stalk attached to the grooved middle of its back; a delicate whitish, transparent membrane forms a tooth lying flat on the anther on each side of this filament, and then runs along the edge of the column to the corner of the stigma, forming a clinandrium for the protection of the pollinia; pollinia 2, rather long, yellowish-white or sulphur-yellow, parallel with and attached to the upper surface of the viscidium. *Stigma* at the apex of its lower surface scutate, green, rounded below, truncate above, with 2 short, central, acute teeth supporting the short, brown, linear viscidium. *Rostellum* deeply bifid, with narrow, acute lobes. *Ovary* about 9 mm, sessile, 6-ridged, slightly twisted, usually glandular-hairy. *Capsule* oblong. *Flowers* 7–8. Probably pollinated by moths. 2n = 30.

Native. Marshy ground with sedges and rushes. Marshy ground in the New Forest, Hampshire until 1959; bog in Guernsey until 1914; by a pond in Jersey until 1926; now extinct. Central and south Europe northwards to Belgium and Germany; North Africa; Asia Minor.

3. S. romanzoffiana Cham.	Irish Lady's-tresses
Neottia gemmipara Sm.; *S. gemmipara* (Sm.) Lindl.; *S. cernua* auct.; *Gyrostachys gemmipara* (Sm.) Kuntze.; *Gyrostachys romanzoffiana* (Cham.) MacMill.; *Ibidium romanzoffianum* (Cham.) Druce; *Gyrostachys stricta* Rydb.; *S. stricta* (Rydb.) A. Nelson; *S. romanzoffiana* subsp. *gemmipara* (Sm.) A. R. Clapham; *S. romanzoffiana* subsp. *stricta* (Rydb.) A. R. Clapham.

Perennial herb; roots 2–6, fusiform-cylindrical, fleshy, thick. Stems 12–25 cm, pale green, erect, triangular and more or less glandular-hairy in the upper part. *Leaves* basal and cauline, 5–10(–15) cm × 5–10 mm, brown, erect on the flowering shoot of the current season, linear-oblanceolate to bract-like, tapering to an acute apex, entire, glabrous, 3-veined, loosely sheathing the stem. *Inflorescence* a spike 2.5–5.0(–8.0) × about 2 cm, with 12–35 flowers in 3 spiral rows, slightly twisted, stout, dense, with dense glandular hairs. *Bracts* 10–20(–30) × 1–3 mm, lanceolate or ovate, acute to acuminate at apex, concave, 3-veined, sometimes with intermediate veins, sheathing the ovary. *Flowers* rather large, 10–14 mm excluding ovary, white, scented. *Outer perianth segments* about 12 mm, adherent with the inner ones, their tips free and turned up, the lateral ones joined together beneath the labellum, lanceolate, acuminate at apex, white, greenish at the base with 3 green veins, glandular-hairy on the outer surface. *Inner perianth segments* linear, obtuse at apex, 3-veined, adherent to the outer segments. *Labellum* white with green veins, tongue-shaped, trough-like at base, curving downwards to the broader, rounded, frilled, finely toothed apex. *Column* horizontal, ending in a sharp, dark brown beak. *Anther* cordate, brown, resting face downwards on the back of the rostellum; pollinia 2, wax-like, cream-coloured, split at the apex, without caudicles, lying horizontally on the upper surface of the viscidium and attached to it by their centres. *Stigma* crescent-shaped, glistening, slanting gently upwards. *Rostellum* green, nib-shaped, ending in a linear, viscid gland, supported by green, tapering sides, left behind like the prongs of a fork when the viscidium is removed. *Ovary* 4–6 mm, cylindrical, very shortly stalked, not twisted, turned to one side, flat in front, convex behind, pale green, 3-ridged, bent at tip so that the flowers stand out nearly at right angles, glandular-hairy. *Capsule* oblong. *Flowers* 7–8. Cross-pollinated by insects. $2n = 60$.

The south-west Irish plant (subsp. *gemmipara*) has broad leaves and denser flower-spikes than those in the north or in Scotland and the Hebrides (subsp. *stricta*), while the Devon plant seems to be intermediate. Both variants occur in North America. *S. romanzoffiana* was originally described from Unalaska, which may constitute a third taxon (subsp. *romanzoffiana*). The problem cannot be satisfactorily solved without examining a large number of living populations.

Native. Wet peaty pastures and meadows, especially where flooded in winter, species-rich upland bogs and wet, stony slopes of banks and rivers. Restricted to a few localities in west Scotland and Ireland and one locality on Dartmoor in Devon. Not elsewhere in Europe, but in North America.

8. Goodyera R. Br.

Peramium Salisb. nom. illegit.

Small *perennial herbs* with creeping rhizomes. *Stems* with scales. *Leaves* ovate, stalked, in a basal rosette. *Spike* twisted, one-sided. *Flowers* small, more or less spirally arranged or not, sessile. *Labellum* in 2 parts, a basal, pocket-like hypochile and a distal narrow, spout-like epichile. *Spur* absent. *Column* horizontal, short. *Anther* hinged to the back of the column, resting on the rostellum; pollinia 2, partly divided lengthways, attached to the rostellum just beneath their summits,

usually without caudicles; pollen in tetrads, cohering in packets. *Stigma* roundish, on its lower side facing the labellum. *Rostellum* projecting beyond the stigma and consisting of 2 short, curved horns enclosing the more or less circular viscidium. *Bursicles* absent.

About 40 species in the north temperate zone, tropical Asia, Madagascar and New Caledonia.

1. 'G. repens (L.) R. Br.   Creeping Lady's-tresses
Satyrium repens L.; *Neottia repens* (L.) Swartz; *Paramium repens* (L.) Salisb. nom. illegit.

Perennial herb with creeping, short-jointed, white, slender rhizomes often running through the fallen leaf layer and ending in a rosette of leaves so that a plant may give rise to a number of others, all connected; roots few, short, brown, thickly clothed with hairs. *Stem* 10–25 cm, pale green, stiffly erect, cylindrical, ridged above, glandular-hairy, with a leafless sheath at the base which sometimes has a leaf-like tip, and appressed scales above. *Leaves* in a basal rosette, 1.5–3.0 × 1–2 cm, dark green often marbled with lighter green, ovate or ovate-lanceolate, obtuse or acute at apex, entire, narrowing into a winged stalk, firm, keeled, conspicuously net-veined, persisting through the winter, glabrous. *Inflorescence* a spike 3–7 × 1–2 cm, slender, lax, one-sided, rather dense, often with a slight spiral twist. *Bracts* 10–15 × 1–2 mm, pale green, whitish and glossy within, linear-lanceolate, tapering to an acute apex, 1-veined, ciliate, sometimes glandular-hairy. *Flowers* small, white, sweet-scented, rather tubular, glandular-hairy, nearly at right-angles with the ovary. *Outer perianth segments* 3–4 mm, white or tinged green, ovate, obtuse at apex, concave, glandular-hairy outside, the lateral slightly spreading, the upper longer, horizontal and 1- to 2-veined. *Inner perianth segments* white, lanceolate, obtuse at apex, 1-veined, lying close to the upper outer segment and about the same length. *Labellum* shorter than outer segments, undivided, the basal half forming a deep rounded bag-shaped pouch, the distal half narrow, tongue-shaped, obtuse at apex, slightly, sometimes considerably, shorter than the basal pouch, furrowed, and curved sharply down in front. *Spur* absent. *Column* short, broad, projecting forwards with a triangular nib-shaped filament at the back, to the summit of which the anther is fixed by its back. *Anther* stalked, brownish, hood-like, 3-toothed in front, resting on the upper surface of the rostellum; pollinia 2, short, yellow, ovoid, without caudicles, attached to the viscidium. *Stigma* nearly circular, not fringed with hairs at the base, prolonged at the apex into a rostellum. *Rostellum* projecting beyond the stigma and consisting of 2, short, acute, curved horns enclosing and supporting the viscidium, which is nearly circular but truncate at the apex. *Ovary* about 9 mm, shortly stalked or sessile, turbinate, slightly bent forwards at the apex, compressed laterally, later somewhat triangular, pale green, glandular-hairy, 3-ridged, slightly twisted if sessile, if not the stalk is twisted. *Capsule* pyriform, distinctly stalked. *Flowers* 7–8. Cross-pollinated by bumble-bees. $2n = 30$.

Native. Local in pine woods, rarely under birch or on

moist, fixed dunes. From Cumberland, Yorkshire (extinct) and Northumberland to Sutherland and Orkney; introduced under pines in Norfolk and Suffolk. Central and north Europe from the Pyrenees and Balkans to northern Scandinavia, and in Russia; Turkey, Afghanistan, Himalayas, Siberia and Japan; North America.

Tribe 2. Epidendrae Melch.

Labellum rather small, simple or with 2, short, lateral lobes. *Anther* borne at the apex of the column; pollinia 2, each divided into 2, sessile, with minute viscidia. *Stigmas* 2, the third a minute sterile bulge, the rostellum.

9. Liparis Rich. nom. cons.

Pseudorchis Gray; *Sturmia* Rchb.

Small *perennial herbs*, with 2, ellipsoid pseudo-bulbs (parent and daughter) side by side. *Stem* angled. *Leaves* 2, cauline. *Inflorescence* a raceme of small, greenish-yellow flowers. *Perianth segments* free, narrow, spreading. *Labellum* pointing in various directions, most commonly upwards, broad, entire. *Spur* absent. *Column* long and slender. *Anther* on top of the column, deciduous; pollinia 2, each of two, flat plates of waxy pollen and each attached to one of the viscidia; caudicles absent. *Stigma* transversely oblong, depressed, flanked by lateral wings of the column. *Rostellum* minute, with 2 evanescent viscidia. *Bursicles* absent.

About 250 species, some epiphytic, widely distributed in temperate and tropical regions.

1. Leaves oblong-elliptical, acute at apex
                  1(i). loeselii var. loeselii
1. Leaves broadly elliptical, obtuse at apex
                  1(ii). loeselii var. ovata

1. L. loeselii (L.) Rich.   Fen Orchid

Ophrys loeselii L.; *Ophrys lilifolia* auct.; *Cymbidium loeselii* (L.) Swartz; *Malaxis loeselii* (L.) Swartz; *Pseudorchis loeselii* (L.) Gray; *Sturmia loeselii* (L.) Rchb.

Perennial herb with 2 pseudo-bulbs above ground, side by side, the older enveloped in the reticulate remains of last year's leaves; in young plants there is a short rhizome emitting thread-like roots, in older plants the roots are thicker and very hairy. *Stem* 6–20 cm, smooth, glabrous, with usually 3, rarely 4 or 5, almost winged angles above, and 2–3 greenish or whitish basal leafless sheaths; between the leaves at the base it swells into an elliptical, green, shining bulge, with one (or more) new bulbs enclosed in soft pale scales. *Leaves* 2, nearly opposite, erect, rather greasy looking green, oblong-elliptical to broadly elliptical, acute or obtuse at apex, keeled, many-veined, with elongated sheathing bases, glabrous. *Inflorescence* a loose spike, 1- to 10(–15)-flowered. *Bracts* 3–20 × 0.5–2.0 mm, lanceolate, keeled, 1-veined. *Flowers* rather small, greenish-yellow. *Outer perianth segments* 4–5 mm, yellowish or yellowish-green, linear, obtuse or acute at apex, spreading, slightly inrolled. *Inner perianth segments* similar to outer, but narrower

and often shorter. *Labellum* usually pointing upwards, about as long as the perianth segments, oblong or oblong-ovate, obtuse at apex, folded, trough-like, curved, usually crenate, sometimes wavy at margin, undivided, much broader than, nearly as long as and of a deeper yellow than the outer segments. *Spur* absent. *Column* erect, flat in front above, but rounded out at the base, narrowed in the middle, with rounded, crenate, forwardly directed side-wings at the apex protecting the anther and stigma, with a shallow cup at the apex in which the pollinia are deposited, and a furrow down the front. *Anther* sessile at the apex of the column, ending in a deciduous, membranous appendage; pollinia 2, each of two flat plates of waxy pollen and each attached to one of the viscidia; caudicles absent. *Stigma* small, quadrangular, transversely oblong, depressed, with rather prominent margins. *Rostellum* minute, horizontal, toothed. *Ovary* spindle-shaped, slightly 3-angled, 6-ribbed, straight or slightly twisted at the base, erect in fruit, the stalk rather long, twisted, 3-angled and furrowed. *Capsules* rather large. *Flowers* 6–7. Probably cross-pollinated by insects. $2n = 32$.

(i) Var. loeselii
Leaves oblong-elliptical, acute at apex.

(ii) Var. ovata Ridd. ex Godfery
Leaves broadly elliptical, obtuse at apex.

Native. Rare and decreasing in wet fen peat, the edges of lakes and pools and in dune-slacks on the coast. Norfolk, Devon, Glamorgan and Carmarthenshire; formerly in Suffolk, Cambridgeshire, Huntingdonshire and Surrey. Europe, from about 61° N in Scandinavia, southwards to south-west France, Romania and south Russia; North America. The varieties are both geographical and ecological. Var. *loeselii* is in fens in eastern England. Var. *ovata* is in dune-slacks in Devon and Wales.

10. Hammarbya Kuntze

Small, green *perennial herb* with a pseudo-bulb covered by pale, sheathing scales and connected to the daughter pseudo-bulb above it by a short, vertical stolon. *Leaves* 2(–4), cauline, short and broad. *Inflorescence* a short raceme, of up to 20, minute, yellowish-green flowers which are turned through 360° so that the labellum is directed upwards. *Perianth segments* spreading. *Labellum* short, entire. *Spur* absent. *Column* very short. *Anther* hinged to the top of the column behind the rostellum; pollinia 2, each of two thin, flat plates of waxy pollen, broad below and tapering upwards, standing in a hollow formed by 2 lateral, membranous lobes of the column; pollen in tetrads which cohere firmly; caudicles absent. *Stigma* in a deep fold in front of the column. *Rostellum* a membrane above the stigma and in front of the anther, surmounted by a small, viscid mass. *Bursicles* absent.

One species in central and north temperate Europe, northern Asia and North America.

Stewart, A., Pearman, D. A. & Preston, C. D. (1994). *Scarce plants in Britain*. Peterborough.

1. H. paludosa (L.) Kuntze   Bog Orchid
Ophrys paludosa L.; *Ophrys palustris* Huds.; *Malaxis paludosa* (L.) Swartz

Small, glabrous *herb* with the old, ovoid, more or less angled pseudo-bulb buried in moss or peat, and tapering below into a slender root, the daughter pseudo-bulb clothed with brownish, acute scales and connected to the old pseudo-bulb by a rhizome 10–20 mm. *Stem* 3–12 cm, erect, slender, 3- to 5-angled above, with 1–2 sheaths below. *Leaves* 2(–4), 5–10(–30) × 2–5(–10) mm, pale yellowish-green, ovate or oblong, more or less acute at apex, concave, rather thick, entire, 3- to 7-veined, often with little gland-like bulbils fringing the apex. *Inflorescence* a raceme 1.5–5.0 × 0.5–0.8 cm, spike-like, rather dense, becoming lax, with up to 20 flowers. *Bracts* 2–3 × 0.5–1.0 mm, lanceolate, acute at apex. *Flowers* very small, yellowish-green, labellum pointing upwards. Lateral *outer perianth segments* 2.5–3.0 mm, yellowish-green, more or less ovate, obtuse at apex, with the tips recurved, 1-veined, the median slightly longer and broader, obtuse at apex and pointing downwards. Lateral *inner perianth segments* yellowish-green, linear-lanceolate, 1-veined, spreading with the tips curved back. *Labellum* yellowish-green, lanceolate, acute at apex, shorter than the outer perianth segments and sometimes darker, erect, concave, 3-veined, its base clasping the column. *Spur* absent. *Column* very short, with a green, obtuse, membranous lobe, concave within on each side of the apex forming a hollow to shelter the pollinia, making the column appear toothed at apex. *Anther* seen from behind in the bud ovate-cordate, obtuse at apex, united to the column by its broad base; pollinia 2, waxy, each composed of 2 plates of pollen. *Stigma* on the front of the column in a horseshoe-like space between the sides of the hollow and continued downwards as a pocket-like cavity behind the oblong projection of the front of the column which is covered with a thin layer of viscid material. *Rostellum* a minute tongue-shaped mass of viscid matter on the apex of the stigma which, by the time the flower opens, has become attached to the thin upper ends of the pollinia. *Ovary* small, turbinate, not twisted; stalk twisted. *Flowers* 6–9. Cross-pollinated by small insects. Reproduces by bulbils on the leaf-tips as well as by seed. 2*n* = 28.

Native. On wet *Sphagnum* or other mosses in bogs, up to 500 m in Scotland. Formerly scattered throughout the British Isles except central England, now very rare except in central Wales, northern England, north-west Scotland and locally in Hampshire. Central and northern Europe, from France, Switzerland and Yugoslavia to just north of the Arctic Circle in Sweden, Finland and north-west Russia; Siberia; North America.

11. Corallorrhiza Ruppius ex Gagnebin

Brown, saprophytic, rootless *perennial herbs* with coral-like, much-branched, fleshy rhizomes. *Stems* with sheathing scales, but no green leaves. *Flowers* small, in spikes, pendulous. *Perianth segments* more or less spreading, free, the outer median and inner lateral some-

what convergent. *Labellum* short, entire or 3-lobed, the lateral lobes very small or absent. *Spur* very short or absent, when present more or less adnate to the ovary. *Column* long, erect. *Anther* terminal on the column, lid-like, deciduous; pollinia 4, subglobose; pollen waxy or powdery. *Stigma* discoid or triangular. *Rostellum* small, globose. *Viscidia* 2, distinct. *Bursicles* absent.

About 15 species in Europe, temperate Asia, North America and Mexico.

Stewart, A., Pearman, D. A. & Preston, C. D. (1994). *Scarce plants in Britain.* Peterborough.

1. C. trifida Chätel.   Coralroot Orchid
Ophrys corallorhiza L.; *C. neotia* Scop. nom. illegit.; *C. inata* R. Br. nom. illegit.; *C. corallorhiza* (L.) Karsten nom. illegit.

Perennial herb; rhizome coral-like, horizontal, fleshy, cream-coloured, with knob-like, rounded branches. *Stem* arising from the branch of a rhizome, erect, slender, solid, glabrous, pale yellowish-green, with 2–4, long, membranous, brown, whitish or green, obtuse or acute, mucronate, brown-veined leafless sheaths often up to the middle of the stem, loose and slightly open at the tip. Green *leaves* absent. *Inflorescence* a lax spike, 4- to 12- flowered. *Bracts* 1.5–3.0 × 0.2–0.7 mm, membranous, sometimes shorter than the stalk of the ovary, triangular, acute or truncate at apex, 1-veined, sometimes with 1–2 short teeth. *Flowers* small, inconspicuous, greenish-yellow with a white labellum and reddish markings at the base. *Outer lateral perianth segments* about 5 mm, curving forwards on each side of the labellum, yellow or yellowish-green, linear-lanceolate with incurved margins, the upper outer segment concave, leaning forwards over the inner segments and 1-veined. *Inner perianth segments* yellowish, spotted with reddish-brown within, sometimes streaked with violet, nearly flat with a little, spur-like depression at the base, elliptic-oblong, obtusely pointed at apex, 1-veined. *Labellum* about 5 mm, white, crystalline, with crimson blotches, lines or dots and two broad, distant, longitudinally slightly raised ridges at the base, oblong, tongue-like, directed upwards and then sharply downwards, as long as but broader than the outer perianth segments, side lobes small, rounded or tooth-like at the base, sometimes absent, middle lobe obtusely pointed or notched. *Column* long, slightly curved forwards, convex behind, flat and streaked with violet in front, truncate at apex, not winged. *Anther* hinged to the back of the column and soon deciduous, nearly flat, yellowish turning brown, without a beak; pollinia 4, ovoid to globular. *Stigma* discoid or triangular, in front of the column just below the anther. *Rostellum* small, globose. *Ovary* about 7 mm, spindle-shaped, not twisted, flattened in front with 3 prominent and 3 lesser obtuse ridges; stalk short and twisted. *Capsules* pendulous. *Flowers* 5–8. Normally self-pollinated but some cross-pollinated by insects. Most flowers set seed. 2*n* = 42.

This species is unusual in that it obtains its nutrients by photosynthesis, parasitism and saprophytism.

Native. Damp peaty or mossy woods, especially of

birch, pine or alder, and in moist dune-slacks. Scattered in the north of Britain south to Yorkshire. Europe, from the Pyrenees, Apennines, north Greece and Crimea to northernmost Scandinavia and the Kola Peninsula; Siberia; North America.

Tribe 3. Orchideae Melch.

Labellum often large and conspicuous, variably lobed. *Spur* absent or short to long. *Anther* borne in front of the column; pollinia 2, usually on long stalks, each with a sticky pad or the two sharing a sticky pad. Two of the stigmas receptive with a third (rostellum) as a small to large sterile bulge.

12. Herminium L.

Small herbs, with one fully developed, entire root-tuber at anthesis, daughter tubers at the tips of slender stolons. *Leaves* usually 2. *Spike* slender, dense. *Flowers* small, green, more or less campanulate. *Perianth segments* incurved, connivent. *Labellum* 3-lobed. *Spur* absent. *Column* very short. *Anther* adnate to the top of the column; pollinia 2, ovoid, diverging downwards and each attached basally by a very short caudicle to a viscidium which almost equals it in size; pollen in tetrads bound together by elastic thread. *Stigma* 2-lobed. *Rostellum* represented by 2 large viscidia each covered by a delicate skin, but not enclosed in a pouch.

Contains about 30 species in Europe and Asia.

Stewart, A.; Pearman, D. A. & Preston, C. D. (1994). *Scarce plants in Britain*. Peterborough.

1. H. monorchis (L.) R. Br.   Musk Orchid
Ophrys monorchis L.

Small, glabrous herb with root-tubers, one fully developed at anthesis and sessile, the rest, usually 2–3, sometimes 4–5, on long slender rhizomes up to 10 cm and semitransparent; roots few, rather slender. *Stem* 7–15 (–30) cm, yellowish-green to dark green, stiff, erect, slender, rounded below, angled above. *Leaves* 2(–4), 2–7 × 0.3–2.0 cm, yellowish- or bluish-green, oblong or elliptic-oblong, obtuse to acute at apex, entire, keeled, flat or slightly folded, spreading, with one principal and fainter intermediate veins on each side; with 1(–3) sessile, acute bracts above. *Inflorescence* a spike 1–5(–10) × 0.5–1.0 cm, slender, cylindrical, erect, with usually many, more or less dense flowers, often one-sided. *Bracts* usually shorter than ovary, but the lowest sometimes as long as the flowers, lanceolate, tapering, 1-veined. *Flowers* very small, yellowish-green or green, more or less campanulate, sweet-scented. *Outer perianth segments* 2.5–3.0 mm, connivent, making the flower almost tubular, the lateral lanceolate, obtuse at apex, 3-veined, the upper broader, ovate-oblong, rounded at apex. *Inner perianth segments* about 3.5 mm, longer and narrower than the outer with a rounded angle or tooth on each side, then suddenly narrowed and strap-shaped with an obtuse apex. *Labellum* 3.5–4.0 mm, greenish, 3-lobed, with a small, cup-like hollow at the base; side lobes short, widely divergent or curved downwards; middle lobe longer and broader, linear, obtuse at apex. *Spur* absent. *Column* small, short

and broad. *Anther* rounded, with a large, plate-like, quadrilateral staminode on each side, with rounded angles; pollinia 2, white, ovoid, relatively large; caudicles short, thick, elastic; viscidium large, obscurely triangular, with the edges turned up. *Stigma* 2-lobed. *Ovary* pale green, sessile, erect, cylindrical, slightly twisted, bent downwards at apex. *Capsule* oblong, twisted, 6-ridged. *Seeds* very small, linear. *Flowers* 6–7. Pollinated by minute flies, beetles and Hymenoptera, the viscidia becoming attached to their legs. $2n = 40$.

Native. Chalk and limestone grassland and the floors of old quarries. Local in southern Britain north to Gloucestershire and Bedfordshire; formerly in East Anglia and Glamorgan. Has become much rarer this century. Much of Europe from the Pyrenees, Apennines, northern Greece, Crimea and Caucasus to northern Scandinavia and the Kola Peninsula; northern and central Asia eastwards to China, Korea and Japan.

13. Platanthera Rich. nom. cons.

Lysias Salisb. nom. nud.

Perennial herbs with entire, ellipsoid, tapering root-tubers. *Stems* with 2–3 leaves near the base plus some reduced ones higher up. *Flowers* white in a usually fairly dense spike, very fragrant. *Outer lateral perianth segments* spreading, outer median and inner lateral pair connivent in an ovoid, more or less erect hood. *Labellum* linear-oblong, entire. *Spur* usually long and slender. *Column* rather short. *Anther* adnate to the top of the column; pollinia 2, narrowed downwards into slender caudicles, each attached laterally, close to its base, either directly to one of the 2 naked viscidia, or indirectly through a short, connecting stalk; pollen in packets of tetrads more or less firmly tied by elastic threads. *Stigma* more or less oblong, depressed. *Rostellum* represented only by the viscidia. *Bursicles* absent. *Capsule* spindle-shaped.

About 200 species occur in the north temperate and tropical zones.

1. Spike often more or less pyramidal; flowers about 18–23 mm across, greenish; labellum 10–16 mm; pollinia 3–4 mm, divergent downwards; viscidia about 3–4 mm apart, circular   **1. chlorantha**
1. Spike more or less cylindrical; flowers 11–18 mm across, whitish; labellum 6–12 mm; pollinia about 2 mm, parallel; viscidia about 1 mm apart, elliptical
                                                    2. bifolia

1. P. chlorantha (Custer) Rchb.
                          Greater Butterfly Orchid
Orchis bifolia auct.; *Orchis chlorantha* Custer; *Orchis virescens* Zollick. ex Custer nom. in syn.; *Orchis virescens* Zollick. ex Gaudin, non Muhl. ex Willd.; *Habenaria chlorantha* (Custer) Bab., non Spreng.; *Habenaria chloroleuca* Ridl. nom. illegit.; *Habenaria virescens* Druce, non Spreng.; *Habenaria montana* auct.; *Habenaria bifolia* subsp. *chlorantha* (Custer) Syme

Perennial glabrous herb with 2 ovoid root tubers which are narrowed downwards; roots brown, few, stout. *Stem* 20–60 cm, pale yellowish-green, erect, rigid, robust, solid, angled or ridged above, with 1–3, brown, oblong,

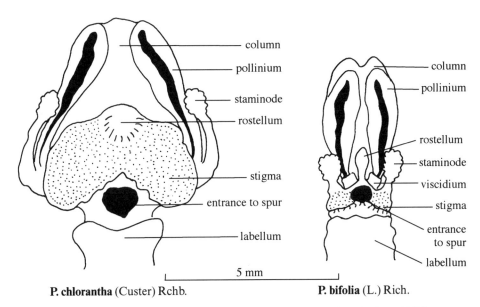

P. chlorantha (Custer) Rchb.   P. bifolia (L.) Rich.

Columns of the flowers of **Platanthera** Rich. (After Clapham, Tutin & Warburg, 1962.)

ribbed, leafless sheaths at base. *Lower leaves* usually 2, rarely 3, near the base, 5–15(–20) × 1–5(–7) cm, pale, usually yellowish-green, glossy beneath and sometimes above, elliptical or elliptical-oblanceolate, obtuse at apex, entire, narrrowed towards the base, strongly keeled, with about 7 transparent parallel veins on each side of the thick midrib and numerous translucent marks between them; upper leaves 2–6, linear-lanceolate or bract-like. *Inflorescence* a spike 5–20 × 3–7 cm, more or less pyramidal, lax. *Bracts* 5–20 × 2–5 mm, green, lanceolate, acuminate at apex, 1-veined. *Flowers* 18–23 mm across, greenish, becoming greenish-white, heavily scented especially at night. *Lateral outer perianth segments* 10–11(–13) × 5–6 mm, lanceolate or ovate-lanceolate, obtuse at apex, obscurely 3-veined, the upper shorter and broader, triangular with a rounded apex and slightly inclined forwards. *Inner perianth segments* 7–9 × about 2 mm, linear-lanceolate, curved, rounded out in the front at the base with their tips nearly touching, faintly 3-veined. *Labellum* 10–16 × 2.0–2.5 mm, slightly narrowed towards the rounded apex, green below, greenish-white distally. *Spur* 19–28 × about 1 mm, yellowish-white more or less suffused with green, compressed from above at the base so that the entrance is transversely elliptical, often flattened and somewhat clavate at the obtuse apex, horizontal or directed downwards. *Column* with the basal wall curving round in a white-tipped wing overlapping the stigma on each side. *Anther* flattened and truncate above, the green, wavy connective furrowed down the back separating the 2 cells, which slope backwards and diverge downwards; pollinia 2, pale yellow, narrowed downwards into slender caudicles; viscidia 2, facing each other, one on each side of the stigma, 3–4 mm apart, thin, flat and circular. *Stigma* concave, green-edged above, situated above the spur entrance. *Rostellum* represented only by the viscidia. *Ovary* sessile,

about 15 mm, slender, cylindrical, curved, twisted, with 6 obtuse ridges. *Capsule* about 20 mm, spindle-shaped, brown, with 3 acute angles, erect. *Flowers* 5–7. Cross-pollinated mainly by noctuid moths to whose eyes the pollinia become attached. $2n = 42$.

Native. Woods and open grassland, usually on calcareous soils. Throughout the British Isles except Orkney and Shetland; much commoner than *P. bifolia* in southern Britain. Europe; Caucasus; Siberia.

2. P. bifolia (L.) Rich.   Lesser Butterfly Orchid
Orchis bifolia L.; *Lysias bifolia* (L.) Salisb. nom. illegit.; *Habenaria bifolia* (L.) R. Br.

Perennial glabrous herbs with 2, ovoid tubers which are narrowed downwards; roots few and short. *Stems* 15–30(–45) cm, erect, slender, angled above, solid, with 2–3, brownish or whitish, tapering, membranous, ribbed sheaths at the base. *Lower leaves* 2, 3–9(–15) × 0.7–3.0(–4) cm, bright or greyish-green, glossy, elliptical or oblong, obtuse to rather acute at apex, sometimes wavy on the entire margin, with numerous veins, tapering below to a flat, whitish, winged stalk pressed against the stem; upper leaves small and bract-like, erect, with 3–5 obscure veins, slightly decurrent. *Inflorescence* a spike 2.5–20.0 × 2–4 cm, more or less cylindrical, usually lax. *Bracts* 10–20 × 2–5 mm, green, lanceolate to ovate-lanceolate, obtuse at apex, somewhat keeled, slightly decurrent, with 3–5, very obscure veins. *Flowers* 11–18 mm across, whitish, less strongly night-scented than *P. chlorantha*. *Lateral outer perianth segments* 8–10 mm, white, lanceolate, obtuse at apex, spreading, the upper broader and slightly shorter, erect, ovate and slightly stalked. *Inner perianth segments* slightly shorter than the outer, white or greenish-white, linear-lanceolate, curved, erect and more or less arched over the column. *Labellum* 6–12 mm, white with a greenish tip, linear-oblong, obtuse at apex, entire, directed obliquely downwards. *Spur* 15–20(–27) mm, greenish at least at the tip, very slender, acute at apex and sometimes slightly clavate, usually horizontal. *Column* small, the basal part whitish, curving round on each side with a long white staminode at the base. *Anther* an upright green membrane curved forwards like a horseshoe, convex behind and ending on each side with a white cell; pollinia 2, about 2 mm, cream-coloured, vertical, parallel; viscidia 2, about 1 mm apart, orange-yellow, elliptical. *Stigma* on the lower front of the column, not extending below the entrance to the spur. *Rostellum* represented only by the viscidia. *Ovary* linear, twisted, curved, tapering. *Capsule* spindle-shaped. *Flowers* 5–7. Pollinated mainly by sphingid moths, to whose proboscides the pollinia become attached. $2n = 42$.

Native. Heaths, moorlands, boggy or calcareous flushes and open woodland. Widespread in southern England though nowhere abundant, still sparser in central and eastern England, but becoming more frequent in the west and north and there in more open habitats; frequent in much of Ireland, but rarer in the east and north. Most of Europe to 70° N in Norway; northern Asia; North Africa.

× **chlorantha = P.** × **hybrida** Brügger
Habenaria hybrida Druce
The parents are interfertile when artificially crossed. In nature the two species sometimes grow together, but have different pollinators. Hybrids have been recorded from several places but have never been confirmed. They are most likely to be aberrant plants of one or the other species.

13 × 15. × Pseudanthera McKean

Platanthera × Pseudorchis

13/1 × 15/1. × **P. breadalbanensis** McKean
Platanthera chlorantha × Pseudorchis albida
This supposed hybrid was described from four plants found with the parents in Perthshire in 1980. They had the tall robust habit of *P. chlorantha* with the flowers about 14 mm, but the labellum was 3-lobed and the spur entrance not visible (*Watsonia* **14**: 129–131 (1982)).

14. Anacamptis Rich.

Perennial herb with 2 more or less globose, entire tubers. *Stems* leafy. *Inflorescence* a conical spike. *Flowers* small. *Outer lateral perianth segments* spreading. *Labellum* deeply 3-lobed, with 2 obliquely erect guide-plates decurrent on its base from the lateral lobes of the column. *Spur* long and slender. *Column* short. *Anther* adnate to the top of the column; pollinia 2, narrowed downwards into caudicles which are attached by their bases to a single transversely elongated, narrowly saddle-shaped viscidium enclosed in the rostellum pouch; pollen in packets of tetrads united by elastic threads. *Stigmas* 2, on the rounded lateral lobes of the column which are continued downwards as guide-plates. *Rostellum* a centrally placed pouch between the bases of

the stigmas and partially closing the entrance to the spur.

One species in Europe and North Africa.

1. Flowers white   **1(1).** pyramidalis forma **albiflora**
1. Flowers rose to deep red   **2.**
2. Flowers pale to bright rose, rose-red or violet-rose
     1(2). pyramidalis forma **pyramidalis**
2. Flowers blood red   **1(3).** pyramidalis forma **sanguinea**

1. A. pyramidalis (L.) Rich.   Pyramidal Orchid
Orchis pyramidalis L.; *Orchis appendiculata* Stokes nom. illegit.

Perennial glabrous herb with 2, more or less globose tubers; roots few and short. *Stem* 20–60(–75) cm, green, erect, solid, often rather flexuous, slightly angled above, often slender, with 2–3, brown, truncate or tapering leafless sheaths at the base. *Leaves* green, gradually decreasing in size, the lower 8–25 × 0.3–1.5 cm, linear to oblong-lanceolate, acute at apex, keeled, with numerous parallel veins and numerous cross-veins, the upper loosely embracing each other, the uppermost bract-like and sometimes membranous. *Inflorescence* a spike 2–8 × 2–4 cm, conical at first, often lengthening later and becoming ovate or oblong, many-flowered, dense. *Bracts* 5–13 × 1–2 mm, linear-lanceolate, tapering to a slender point, green or coloured, slightly exceeding the ovary. *Flowers* small, usually pale or bright rose, rose-red or violet-red, rarely bright blood red or white, with a foxy smell. *Outer perianth segments* 4–6 mm, free, ovate-lanceolate to lanceolate, 1-veined, keeled, the lateral curved and spreading, but not reflexed, the upper connivent with the inner segments to form a helmet. *Inner perianth segments* slightly shorter, ovate or lanceolate, somewhat hooded, 1-veined. *Labellum* 6–9 mm, broadly wedge-shaped, deeply 3-lobed, flat, the lobes nearly equal, oblong, truncate or rounded at apex, usually entire, sometimes crenate or dentate, sidelobes divergent; the side wings of the column run down the labellum in the form of 2 slightly divergent plates with a prolonged rounded apex, which act as guide-plates and converge towards the entrance of the spur. *Spur* long and slender, often exceeding the ovary. *Column* short, obtuse at apex, white, more or less tinged with colour, with a concave wing on each side. *Anther* ovate, granular outside, white tinged with pale rose; pollinia 2, narrowed downwards into the caudicles; viscidium 1, transversely elongated, narrowly saddle-shaped, enclosed in the rostellum pouch. *Stigmas* 2, white, oval, on the concave side wings of the column, with a whitish or pinkish rugose staminode between each and the anther. *Rostellum* embraced by the anther, low down, separating the stigmas and partly blocking the entrance to the spur. *Ovary* sessile, cylindrical, green, twisted, often flushed reddish-violet. *Seeds* oblong. *Flowers* 6–8. Pollinated by day- and night-flying Lepidoptera, the viscidium coiling tightly round their proboscides after withdrawal.

(1) Forma **albiflora** Fors.-Major
Flowers white.

(2) Forma **pyramidalis**

Flowers pale to bright rose, rose-red or violet-rose.

(3) Forma **sanguinea** (Druce) P.D. Sell

Flowers blood red.

Native. Chalk and limestone grassland and calcareous dunes up to 244 m. Locally frequent throughout the British Isles north to North Ebudes and Fife; Channel Islands. Europe northwards to southern Scandinavia and central Russia; West Asia; North Africa. Forma *pyramidalis* occurs throughout the range of the species, forma *albiflora* is scattered in mixed populations with forma *pyramidalis*. Forma *sanguinea* is recorded only from Galway.

14 × 16. × Gymnanacamptis Asch. & Graebn.

Anacamptis × Gymnadenia

14/1 × 16/1. × **G. anacamptis** (Wilms) Asch. & Graebn.

A. pyramidalis × G. conopsea

G. aschersonii E.G. Camus, Bergon & A. Camus nom. illegit.; *Habenaria anacamptis* (Wilms) Druce

This hybrid has the plates on the labellum of *A. pyramidalis* and the scent and cylindrical spike of *G. conopsea*.

It has been recorded from Hampshire, Gloucestershire and Co. Durham.

15. Pseudorchis Seg.

Leucorchis E. Mey.; *Biochia* Parl.; *Entaticus* Gray nom. illegit.

Perennial herbs with deeply palmate root tubers, so that the more or less cylindrical segments may appear separate. *Stem* with large lower leaves and small upper ones. *Inflorescence* a narrowly cylindrical, dense spike. *Flowers* small, greenish-white. *Spur* short. *Outer lateral perianth segments* connivent with the outer median and inner perianth segments forming a hood, and the labellum more or less connivent with the hood so that the flower becomes almost campanulate. *Column* short and erect. *Anther* broad, obtuse; pollinia 2, very small; caudicles very short; viscidia elliptical, naked. *Stigma* more or less reniform; staminodes large. *Rostellum* long. *Bursicles* absent.

Three species in Europe and west Asia.

1. P. albida (L.) Å. & D. Löve   Small-white Orchid

Satyrium albidum L.; *Orchis albida* (L.) Scop.; *Habenaria albida* (L.) R. Br.; *Gymnadenia albida* (L.) Rich.; *Entaticus albidus* (L.) Gray; *Platanthera albida* (L.) Lindl.; *Peristylus albidus* (L.) Lindl.; *Leucorchis albida* (L.) E. Mey.; *Biochia albida* (L.) Parl.

Perennial glabrous herb with deeply palmate root tubers, the segments widely divergent, long, thickened at base and gradually tapering to apex; roots long and fleshy and much thinner than the tubers. *Stem* 10–30(–40) cm, erect, stiff, solid, cylindrical, slightly angled above, pale green, with 2–3 brownish or whitish leafless sheaths at the base. *Lower leaves* 2.5–8.0 × 1.0–1.7 cm, up to 4,

glossy green above, paler beneath, oblong-ovate to oblong-lanceolate, shortly pointed, entire, narrowed at base and sheathing, rather thick, keeled, with 4–6 veins on each side of the midrib; upper leaves small, bract-like, acute at apex. *Inflorescence* a spike 3–7 × 0.7–2.0 cm, cylindrical, dense, rather one-sided. *Bracts* 4–7 × 1–2 mm, green, lanceolate, tapering to an obtuse apex, sessile, 1- to 3-veined. *Flowers* 2.0–2.5 mm, almost tubular, white, faintly scented, turned to one side and half-drooping. *Outer perianth segments* white, sometimes greenish- or yellowish-white, ovate, obtuse at apex, keeled, 1-veined, loosely connivent with the inner segments to form a more or less campanulate flower. *Inner perianth segments* ovate, obtuse at the hooded apex, shortly clawed, 2-veined. *Labellum* directed forwards, white, yellowish or greenish-white, usually deeply 3-lobed; lateral lobes linear-lanceolate, acuminate at apex; middle lobe larger and rather obtuse at apex. *Spur* 2.0–2.5 mm, yellowish, compressed from back to front, pendulous. *Column* short, erect. *Anther* broad, obtuse or with a little point at apex, greenish-white, clavate; pollinia 2, very small, pale cream; caudicles very short; viscidia elliptical, naked. *Stigma* more or less reniform, with lateral lobes, or oblong with rounded angles; staminodes large, short and rather flat. *Rostellum* long. *Ovary* about 7 mm, slightly twisted, with 3 rather broad ridges. *Capsules* 6–7 cm. *Seeds* faintly striate. *Flowers* 6–7. Probably cross-pollinated by tiny insects. $2n = 42$.

Native. Usually in rough, well-drained pasture, especially in hilly districts and tolerant of both calcareous and non-calcareous soils. Frequent in central, west and north Scotland, very scattered and decreasing in northern Britain, Wales and Ireland; formerly south to Derbyshire and in Sussex. Much of west and central Europe from northernmost Norway southwards to the Pyrenees, southern Apennines and southern Bulgaria. The plants of the British Isles are subsp. **albida**, which occurs throughout most of the range of the species.

16. Gymnadenia R. Br.

Perennial herb with laterally compressed, palmately lobed tubers. *Stems* leafy. *Flowers* in dense spikes. *Outer lateral perianth segments* spreading or curved downwards; outer median and inner lateral segments connivent into a hood. *Labellum* shortly 3-lobed, directed downwards, without raised plates at base. *Spur* long and slender. *Column* short, erect. *Anther* wholly adnate to the column; pollinia 2, convergent and narrowed below into caudicles, each of which is attached basally to a long, linear viscidium; viscidia close together and naked; pollen in pockets of tetrads bound together by elastic threads. *Stigmas* 2, on lateral lobes of the column. *Rostellum* elongated, projecting between the viscidia. *Bursicles* absent.

Ten species in Europe, North America and northern Asia.

1. Lateral outer perianth segments mostly 4–5 × about
   2 mm; labellum (3–)3.5–4.0(–5.0) mm wide, obscurely
   lobed   **1(a). conopsea** subsp. **borealis**

1. Lateral outer perianth segments mostly 5–7 × about 1 mm; labellum (4.5–)5.5–7.0(–8.0) mm, conspicuously lobed
2. Lateral outer perianth segments about 5–6 mm; labellum usually 5–6 × 5.5–6.5 mm, scarcely wider than long; spur usually 12–14 mm

1(b). conopsea subsp. conopsea

2. Lateral outer perianth segments 6–7 mm; labellum usually (3.0–)3.5–4.0(–4.5) × (5.5–)6.5–7.0(–8.0) mm, much wider than long; spur normally 14–16 mm

1(c). conopsea subsp. densiflora

1. G. conopsea (L.) R. Br.   Fragrant Orchid
Orchis conopsea L.; *Habenaria conopsea* (L.) Benth., non Rchb. fil.; *Habenaria gymnadenia* Druce

Perennial glabrous herb with palmately lobed tubers, the segments rather thick, tapering and obtuse; roots few, short and rather thick. *Stem* 15–40(–75) cm, green, sometimes purplish above, erect, terete, or angled above, leafy, with 2–3, brown, membranous sheaths at the base. *Leaves* green, the lower 3–5, 6–15 × 0.5–3.0 cm, erect to slightly spreading, more or less narrowly oblong-lanceolate or linear-lanceolate, obtuse to subacute and slightly hooded at apex, entire, keeled and folded and with 1–2, sometimes more veins on each side of the midrib, the upper 2–3 smaller, lanceolate or bractlike, tapering to a fine point. *Inflorescence* a spike 2–10(–16) × 1.5–3.0 cm, more or less cylindrical, rather dense. *Bracts* 3–20 × 0.2–2.0 mm, green, often violet-tinged, lanceolate, tapering to a fine point at apex, 1- to 3-veined. *Flowers* 7.0–14.5 mm across, rose-pink to rose-purple, rarely white or bright magenta, strongly scented. *Outer lateral perianth segments* 4–7 × 1–2 mm, oblong-ovate, obtuse to acute at apex, spreading horizontally or downwardly curved, their margins revolute, the upper forming a hood with the inner segments. *Inner perianth segments* broader and shorter, slightly hooded at apex, connivent. *Labellum* (3–)4–7(–8) mm, longer than broad or broader than long, 3-lobed, the lobes more or less equal and rounded, sometimes obscurely crenate or dentate. Spur (8–)11–16(–17) mm, slender, tapering, curved downwards. *Column* very short. *Anther* pyriform, adnate to the column; pollinia 2, pale green or greenish-yellow, attached basally to the long linear viscidium, narrowed to the caudicles. *Stigmas* 2, on the base of the anther. *Rostellum* elongated, projecting between the viscida. *Ovary* green, tinged violet, slender, cylindrical, twisted and curved, with 3 longitudinal ridges. *Flowers* 6–8. Pollinated by moths, the viscidia becoming attached to the proboscides. 2n = 40, 80, c. 97, c. 117, c. 119.

(a) Subsp. **borealis** (Druce) F. Rose
Habenaria gymnadenia var. *borealis* Druce
Flowers (7–)8–10(–12) mm across, dark pink to purple, smelling of cloves. *Lateral outer perianth segments* 4–5 × about 2 mm, bent downwards. *Labellum* (3.5–)4.0–4.5(–5.0) × (3.0–)3.5–4.0(–5.0) mm, obscurely lobed. *Spur* (8–)11–14(–15) mm.

(b) Subsp. **conopsea**
Flowers (7–)10–11(–13) mm across, rose-pink, sweet-scented. *Lateral outer perianth segments* 5–6 × about

1 mm, bent downwards. *Labellum* (4.0–)5.0–6.0(–6.5) × (4.5–)5.5–6.5(–7.0) mm, conspicuously lobed. *Spur* (11–)12–14(–17) mm.

(c) Subsp. **densiflora** (Wahlenb.) E. G. Camus, Bergon & A. Camus
Orchis conopsea var. *densiflora* Wahlenb.
Flowers (10–)11–13(–14) mm across, rose-pink, smelling of cloves. *Lateral outer perianth segments* 6–7 × about 1 mm, held horizontally. *Labellum* (3–)3.5–4.0(–4.5) × (5.5–)6.5–7.0(–8.0) mm, conspicuously lobed. *Spur* (13–)14–16(–17) mm.

Native. Scattered throughout the British Isles. Most of Europe to northernmost Scandinavia; north and west Asia. The subspecies are ecological. Subsp. *borealis* is in base-rich to poor hilly grassland in Scotland, Cardiganshire in Wales, north and south-west England and in bogs in Hampshire and Sussex. Subsp. *conopsea* is on dry chalk and limestone grassland in Britain north to Durham and in Northern Ireland. It is widespread in Europe. Subsp. *densiflora* is scattered and local in base-rich fens and usually north-facing chalk grassland in Britain north to Westmorland, and scattered throughout Ireland where it is the commonest subspecies. It also occurs in Continental Europe.

[Old records of the European **G. odoratissima** (L.) Rich., which is distinguished by its small flowers, lateral outer perianth segments 2.5–3.0 mm, labellum 2.3–3.0 mm and spur 4–5 mm, have never been confirmed.]

16 × 15. × Pseudadenia P. F. Hunt
Gymnadenia × Pseudorchis

16/1 × 15/1. × P. schweinfurthii (Hegelm. ex A. Kern.) P. F. Hunt
G. conopsea × P. albida
× *Gymleucorchis schweinfurthii* (Hegelm. ex A. Kern.) T. & T. A. Stephenson

Intermediate between the parents in size and perianth shape, especially the spur, with pale pink flowers.

Recorded from several places in northern Britain and still frequent in north-west Scotland with both parents. It is also recorded from Austria, Germany, Norway, Sweden and Switzerland.

17. Coeloglossum Hartm. nom. conserv.
Satyrium L., non Sw.

Small perennial herbs with palmate root tubers. *Leaves* decreasing in size up the stem. *Flowers* small, green and brown. Outer and inner lateral *perianth segments* connivent into a hood. *Labellum* narrowly oblong, shallowly 3-lobed. *Spur* very short, saccate. *Column* short, erect. *Anther* adnate to the top of the column; pollinia 2, clavate, converging above, but narrowing downwards into widely separated caudicles, each attached basally to one of the 2 oblong viscidia which are covered by a membrane, but only partially enclosed in small pouches; pollen in packets of tetrads bound by elastic threads. *Stigma* central, reniform, depressed. *Rostellum* of 2 widely separated protuberances, one on each side of the upper edge of the stigma.

Two species in the north temperate and Arctic zones.

1. C. viride (L.) Hartm.   Frog Orchid

Satyrium viride L.; *Satyrium fuscum* Huds.; *Orchis viridis* (L.) Crantz; *Habenaria viridis* (L.) R. Br.; *Entaticus viridis* (L.) Gray; *Platanthera viridis* (L.) Lindl.; *Peristylus viridis* (L.) Lindl.

Perennial glabrous herb with 2 tubers, each with 2–4 tapering segments; roots few, short. *Stem* 6–35(–40) cm, erect, cylindrical, angled above and often with reddish ridges, with 1–2, brown, obtuse, leafless sheaths at base. *Leaves* 2–5, bluish-green, the lower 3–8(–11) × 1–3 cm and broadly oblong, the lowest sometimes more or less rotund, obtuse at apex, entire, with a winged petiole and with 3–5 veins each side of the midrib, the upper smaller, lanceolate and acute at apex. *Inflorescence* a spike 2–10(–15) × 1.0–2.5 cm, cylindrical, rather lax. *Bracts* 5–8 × 1–8 mm, green or suffused with reddish-brown, lanceolate to oblong, more or less obtuse at apex, 3-veined. *Flowers* small, green and reddish-brown, faintly scented. *Outer perianth segments* 4.5–6.0(–7.0) mm, ovate, more or less obtuse at apex, forming a semi-globular hood, 3- to 5-veined. *Inner perianth segments* linear or linear-lanceolate, more or less acute at apex, 1-veined, often protruding from the hood. *Labellum* 3.5–9.0 × 1.5–3.0 mm, green, often edged red, narrowly oblong, straight and more or less parallel-sided, hanging almost vertically, 3-lobed near its tip, the outer lobes narrowly oblong and parallel, the middle usually much shorter, rounded or tooth-like, flat. *Spur* about 2 mm, greenish-white, semitransparent, flattened from back to front, sometimes grooved at the apex, which is slightly curved forwards, with a green band running from the column down the middle of the back and up the front. *Column* short, broad. *Anther* broader than long, brownish, clavate, adnate to the top of the column; pollinia 2, pale greenish-yellow, clavate, narrowed downwards into widely separated caudicles; viscidia 2, oblong, each enclosed in a membrane and separated by a thickened, arched ridge. *Stigma* reniform, on the back of the chamber at the base of the labellum, bordered by dark side-lines. *Rostellum* of 2 widely separated protuberances, one each side of the upper edge of the stigma. *Ovary* greenish-brown, cylindrical or spindle-shaped, twisted, with 6 slight ridges. *Flowers* 6–8. Cross-pollinated by insects. $2n = 40$.

Very variable in height and size of parts. Large plants with long bracts exceeding the flowers have been called var. *bracteatum* K. Richt., but many intermediates occur. The variation does not seem to fit into any ecological or geographical patterns.

Native. Pastures and grassy hillsides, especially on calcareous soils, occasionally on dunes and rock-ledges up to 1000 m in Scotland. Formerly locally frequent throughout the British Isles, now extinct in many places in southern Scotland and central and eastern England. Europe southwards to central Spain, south Italy, Bulgaria and Crimea, but only on mountains in the south; western Asia; North America.

17 × 18. × Dactyloglossum P. F. Hunt & Summerh.
Coeloglossum × Dactylorhiza

Grose, J. D. (1950). *Coeloglossum viride* × *Orchis fuchsii* on the Wiltshire downs. *Watsonia* **2**: 207–208.

17/1 × 18/1. × D. mixtum (Asch. & Graebn.) Rauschert
    C. viride × D. fuchsii

Orchis mixta Domin, non Retz.; *Orchicoeloglossum mixtum* Asch. & Graebn.; *Orchicoeloglossum dominianum* E. G. Camus, Bergon & A. Camus nom. illegit.; *Habenaria mixta* (Asch. & Graebn.) Druce

Most plants of this hybrid are smaller than *D. fuchsii*, but the size and arrangement of the leaves are as in *C. viride*, though they can be spotted. The flower is the colour of *D. fuchsii*, but overlaid with pale green giving a distinct mottled effect. The labellum is intermediate between those of the parents but variable.

Widely scattered in Britain and in Co. Down, wherever the parents grow together. Also found in Continental Europe.

17/1 × 18/2. × D. drucei (A. Camus) Soó
    C. viride × D. maculata

Orchicoeloglossum drucei A. Camus; *Habenaria websteri* Druce

The overall resemblance of this hybrid is with *D. maculata*, but the flower is suffused green and the labellum narrower than *D. maculata*.

A few scattered records in Britain. Also reported from Continental Europe.

17/1 × 18/5. × D. viridella (Hesl.-Harr.) Soó
    C. viride × D. purpurella

Orchis × viridella Hesl.-Harr.

This hybrid is usually small, with narrow yellowish-green leaves which are spotted or not, the labellum is intermediate between that of the parents and the flowers are rich rosy purple with a green tinge and darker purple spots and loops.

Recorded from Co. Durham, the Ebudes and Outer Hebrides. Endemic.

17 × 16. × Gymnaglossum Rolfe
Coeloglossum × Gymnadenia

17/1 × 16/1. × G. jacksonii (Quirk) Rolfe
    C. viride × G. conopsea

× *Gymplatanthera jacksonii* Quirk; *Habenaria jacksonii* (Quirk) Druce; *Coeloglossogymnadenia jacksonii* (Quirk) A. Camus; *Coeloglossogymnadenia quirkii* A. Camus; *Coeloglossohabenaria jacksonii* (Quirk) Druce

This hybrid usually resembles *Gymnadenia*, but the pink flowers are tinged green. The labellum is intermediate in shape between those of the parents.

Recorded occasionally through much of the British Isles. Also occurs over much of Continental Europe.

18. Dactylorhiza Nevski

Orchis subgenus *Dactylorchis* Klinge; *Orchis* subgenus *Dactylorhiza* Nevski nom. illegit.; *Dactylorchis* (Klinge) Vermuel.

Perennial herbs with 2–3 root tubers, which are usually palmately and deeply 2- to 5-lobed, the lobes narrowed and prolonged distally. *Leaves* several, the lower sheathing the stem, the upper transitional to the bracts and not sheathing, though sometimes clasping. Emerging *spike* not enclosed in spathe-like leaves. *Bracts* always herbaceous. *Outer lateral perianth segments* more or less erect or spreading, not contributing to the helmet over the column. *Inner perianth segments* connivent and forming a helmet over the column. *Labellum* not lobed, or more or less 3-lobed, porrect or slightly decurved. *Spur* usually less than 10 mm, down-pointed, mostly rather wide. *Column* erect. *Anther* adnate to the top of the column; pollinia 2, narrowed below to the caudicle and attached by a basal disc to one of the viscidia; pollen in packets of tetrads united by elastic threads. *Stigma* more or less 2-lobed, roofing the entrance to the spur. *Rostellum* 3-lobed, the middle lobe short, lamelliform. *Capsule* erect. *Bursicle* present.

There is much controversy about what is a species in the genus, but there are probably about 30 or more. Europe, temperate Asia, North Africa and the Canary Islands. Many localities are being lost through drainage.

Dactylorhiza is a very difficult genus to classify, not only because there is great variation, but because ready hybridisation seems to be possible between all taxa. Hybrids within a ploidy level can be highly fertile; those between ploidy levels are highly, but not completely, sterile, even in the case of triploid hybrids. Some classifications treat all the Marsh Orchids, which have the same chromosome number, as one species, and divide the Spotted Orchids into two species which usually have different chromosome numbers. This treatment accepts the usual division of the Spotted Orchids, but divides up the Marsh Orchids into species on the basic shape of the labellum. Ecological and geographical populations are treated as subspecies or varieties.

. Individual plants are often difficult to identify. The best method of approach is to examine, as far as possible, a total population in the field, before making any identifications. This may enable you to know which taxa are present before dealing with the hybrids. Sometimes, however, one of the parents may be absent. In other cases, when one taxon only seems to be present, often in numbers, there may be a few aberrant plants which may defy any attempt at identification. If a statistical approach to identification is taken, it is very important to eliminate hybrids first.

The first major classification of the genus in this country was made by H. W. Pugsley (1935, 1939, 1940), it was much improved upon by J. Heslop-Harrison (1948, 1949, 1951, 1954, 1968), and an excellent account of the hybrids was made by R. H. Roberts (1975). No serious attempt has been made to define nothotaxa below binomial rank, as although one often knows which infraspecific taxa are present it is difficult to define the variants morphologically.

Adcock, E. M., Gorton, E. & Morries, G. F. (1983). A study of some *Dactylorhiza* populations in Greater Manchester. *Watsonia* **14**: 377–389.

Bateman, R. M. & Denholm, I. (1983). A reappraisal of the British and Irish dactylorchids. I. The tetraploid marsh-orchids. *Watsonia* **14**: 347–376.

Bateman, R. M. & Denholm, I. (1985). A reappraisal of the British and Irish dactylorchids. 2. The diploid marsh-orchids. *Watsonia* **15**: 321–355.

Bateman, R. M. & Denholm, I. (1988). A reappraisal of the British and Irish dactylorchids. 3. The spotted-orchids. *Watsonia* **17**: 319–349.

Foley, M. J. Y. (1990). An assessment of populations of *Dactylorhiza traunsteineri* (Sauter) Soó in the British Isles and a comparison with others from Continental Europe. *Watsonia* **18**: 153–172.

Heslop-Harrison, J. (1948). Field studies on *Orchis* L. I. The structure of dactylorchid populations on certain islands in the Inner and Outer Hebrides. *Trans. Proc. Bot. Soc. Edinb.* **35**: 26–66.

Heslop-Harrison, J. (1949). Notes on the distribution of the Irish dactylorchids. *Veröff. Geobot. Inst. Rübel, Zürich* **25**: 100–113.

Heslop-Harrison, J. (1951). A comparison of some Swedish and British forms of *Orchis maculata* L. sens. lat. *Svensk. Bot. Tidsskr.* **45**: 608–635.

Heslop-Harrison, J. (1954). A synopsis of the dactylorchids of the British Isles. *Ber. Geobot. Forsch. Inst. Rübel, Zurich* **1953**: 53–82.

Heslop-Harrison, J. (1968). Genetic system and ecological habit as factors in dactylorchid variation. *Jber. Naturw. Ver. Wuppertal* **21–22**: 20–27.

Hunt, P. F. & Summerhayes, V. S. (1965). *Dactylorhiza* Nevski, the correct generic name of the Dactylorchids. *Watsonia* **6**: 128–133.

Kenneth, A. G. & Tennant, D. J. (1984). *Dactylorhiza incarnata* (L.) Soó subsp. *cruenta* (O. F. Mueller) P. D. Sell in Scotland. *Watsonia* **15**: 11–14.

Kenneth, A. G. & Tennant, D. J. (1988). Further notes on *Dactylorhiza incarnata* subsp. *cruenta* in Scotland. *Watsonia* **16**: 332–334.

Kenneth, A. G., Lowe, M. R. & Tennant, D. J. (1988). *Dactylorhiza lapponica* (Laest. ex Hartman) Soó in Scotland. *Watsonia* **17**: 37–41.

Løjtnant, B. *Dactylorhiza purpurella* ssp. *majaliformis* Nelson ex Løjtnant. *Bot. Tidsskr.* **74**: 175–176.

Lord, R. M. & Richards, A. J. (1977). A hybrid swarm between the diploid *Dactylorhiza fuchsii* (Druce) Soó and the tetraploid *D. purpurella* (T. & T. A. Steph.) Soó in Durham. *Watsonia* **11**: 205–210.

Lowe, M. R., Tennant, D. J. & Kenneth, A. G. (1986). The status of *Orchis francis-drucei* Wilmott. *Watsonia* **16**: 178–180.

Nelson, E. (1976). *Monographie und Ikongraphie der Orchidaceen* - Gattung, III. *Dactylorhiza*. Zurich.

Pugsley, H. W. (1935). On some marsh orchids. *Jour. Linn. Soc. London (Bot.)* **49**: 553–592.

Pugsley, H. W. (1939). Recent work on dactylorchids. *Jour. Bot. (London)* **77**: 50–56.

Pugsley, H. W. (1940). Further notes on British dactylorchids. *Jour. Bot. (London)* **78**: 177–181.

Roberts, R. H. (1961). Studies on Welsh orchids. I. The variation of *Dactylorchis purpurella* (Steph.) Vermeul. in North Wales. *Watsonia* **5**: 23–36.

Roberts, R. H. (1961). Studies on Welsh orchids. II. The occurrence of *Dactylorchis majalis* (Reichb.) Vermeul. in Wales. *Watsonia* **5**: 37–42.

Roberts, R. H. (1966). Studies on Welsh orchids. III. The coexistence of some of the tetraploid species of marsh orchids. *Watsonia* **6**: 260–267.

Roberts, R. H. (1975). *Dactylorhiza* Nevski (*Dactylorchis* (Klinge) Vermuel.) in Stace, C. A. *Hybridization and the flora of the British Isles*. 495–506. London, New York, San Francisco.

Roberts, R. H. (1988). The occurrence of *Dactylorhiza traunsteineri* (Sauter) Soó in Britain and Ireland. *Watsonia* **17**: 43–47.

Roberts, R. H. (1989). Errors and misconceptions in the study of marsh-orchids. *Watsonia* **17**: 455–462.

Stewart, A., Pearman, D. A. & Preston, C. D. (1994). *Scarce plants in Britain*. Peterborough. [*D. traunsteineri* subsp. *traunsteineri*.]

Tennant, D. J. & Kenneth, A. G. (1983). The Scottish record of *Dactylorhiza traunsteineri* (Sauter) Soó. *Watsonia* **14**: 415–417.

1. Stem solid for much of its length; leaves nearly always spotted; upper non-sheathing, bract-like leaves which are transitional to bracts (1–)2–6 or more; lower bracts up to 3 mm wide; outer lateral perianth segments patent or drooping; spur usually under 2 mm wide at the middle   2.

1. Stem usually hollow for much of its length; leaves often not spotted; upper non-sheathing, bract-like leaves which are transitional to bracts 0–1(–3); lower bracts usually more than 3 mm wide; outer lateral perianth segments more or less erect; spur usually over 2 mm wide at the middle   10.

2. Green leaves (excluding bracts) 7–12 (or more), the basal widest and obovate-oblong with a broad, rounded apex, the next 2 longest, the rest progressively shorter, narrower and more acute at apex, all usually with solid and more or less transversely elongated spots, sometimes unspotted; labellum with 3 subequal lobes, the central usually longer than the lateral; spur 1–2 mm in diameter   3.

2. Green leaves (excluding bracts) 4–8 (or more), all more or less narrowly oblong and more or less acute at apex, usually with small, solid circular spots, sometimes unspotted; labellum with central lobe much smaller and usually shorter than the lateral lobes; spur 1 mm or less in diameter   9.

3. Flowers white or cream, sometimes with faint markings   4.

3. Flowers from very pale pink to bright reddish-purple with deeper markings   5.

4. Leaves often 20–40 mm wide; spike often pyramidal; flowers not or slightly scented; labellum 5–9(–11) mm wide
    white-flowered forms of **1(b). fuchsii** subsp. **fuchsii** and **1(c). subsp. hebridensis**

4. Leaves rarely above 15 mm wide; spike often narrow and cylindrical; flowers strongly scented; labellum 8–12 mm wide   **1(a). fuchsii** subsp. **okellyi**

5. Stems (8–)20–50(–70) cm, little suffused purple; leaf-spotting not heavy; flowers pale to deep pink or lilac-pink; labellum 8–12 mm wide; spur 5–7 mm   6.

5. Stems 15–30 cm, often suffused purple; leaf spotting tending to be heavy; flowers often deeply coloured

rose-pink to reddish-purple; labellum 8–15 mm wide; spur 7.0–8.5 mm   7.

6. Stems tall and very leafy; spike long; labellum with central lobe distinctly longer than laterals
    1(b,i). fuchsii subsp. **fuchsii** var. **fuchsii**

6. Stems shorter and less leafy; spike short; labellum with lobes subequal in length
    1(b,ii). fuchsii subsp. **fuchsii** var. **trilobata**

7. Leaves narrow, not recurved; labellum usually less than 6.7 mm long and 9.5 mm wide
    1(c,i). fuchsii subsp. **hebridensis** var. **alpina**

7. Basal leaves elliptical, often recurved; labellum usually more than 6.7 mm long and 9.5 mm wide   8.

8. Floral bracts usually less than 8 mm; spur at middle less than 1.8 mm wide
    1(c,ii). fuchsii subsp. **hebridensis** var. **hebridensis**

8. Floral bracts often more than 8 mm; spur up to 2 mm wide at middle
    1(c,iii). fuchsii subsp. **hebridensis** var. **cornubiensis**

9. Labellum broader than long, middle lobe much smaller than lateral lobes, from white to deep lilac-pink, but predominantly pale; spur hardly tapering; flowers May to July   **2(a). maculata** subsp. **ericetorum**

9. Labellum about as broad as long, with the middle lobe more prominent than last, pale purple; spur slender and pointed; flowers late July and August
    2(b). maculata subsp. **rhoumensis**

10. Stem cavity usually more than half its total diameter; leaves erect, widest near to base, usually distinctly narrowed and hooded at apex, often yellowish-green; sides of labellum usually reflexed and with 2 distinct, dark, loop-shaped marks   11.

10. Stem cavity often less than half its total diameter; leaves often spreading, widest well above the base, not or broadly hooded at apex, usually dark green; labellum usually with sides not or little reflexed, usually without 2 distinct, dark loops   17.

11. At least some plants with leaves which have spots on both surfaces   **3(f). incarnata** subsp. **cruenta**

11. Plants with unmarked leaves or rarely with some spots on upper surface only   12.

12. Perianth white, pale yellow or cream   13.

12. Perianth pink to purple   14.

13. Labellum yellow or cream, usually more than 6.5 × 8.0 mm, with well-marked lobes, the lateral lobes usually shallowly toothed
    3(a). incarnata subsp. **ochroleuca**

13. Labellum white, usually less than 6.5 × 8.0 mm, the lobes less well marked and the lateral lobes entire
    albino variants of other incarnata subspecies

14. Perianth pale salmon pink; bracts usually lacking anthocyanin   15.

14. Perianth reddish or purplish; bracts usually strongly suffused with anthocyanin   16.

15. Plant up to 80 cm; bracts often much longer than flowers; labellum up to 8 × 10 mm
    3(b). incarnata subsp. **gemmana**

15. Plant up to 40 cm; bracts not so obviously longer than flowers; labellum up to 8 × 7 mm
    3(c). incarnata subsp. **incarnata**

16. Plants typically short and stout, mostly less than 20 cm; leaves lesw than 9 cm and typically curved; labellum brick red, crimson-red or reddish-purple
                           3(d). incarnata subsp. **coccinea**

16. Plant 20–40 cm, often slender; leaves often more than 9 cm and erect; labellum purple or reddish-purple
                           3(e). incarnata subsp. **pulchella**

17. Total number of leaves often less than 5, the widest usually less than 1.5(–2.0) mm; labellum usually distinctly 3-lobed, the central lobe distinctly exceeding the 2 laterals and usually more than half as long as the unlobed basal part   18.

17. Total number of leaves usually 6 or more, the widest usually more than (1.5–)2.0 mm wide; labellum entire or indistinctly 3-lobed, the central shorter than the laterals and usually less than one-third as long as the unlobed basal part   19.

18. Stems up to 30(–45) cm; leaves without marks or with small faint spots, transverse bars, or broken lines formed into rings on the upper surface; bracts strongly suffused reddish-purple all over without spots or rings; labellum reddish- or lilac-purple with many dark markings more or less to the margin
                       4(a). traunsteineri subsp. **traunsteineri**

18. Stems up to 21 cm; leaves with strong dark spots and rings on the upper surface; bracts suffused reddish-purple at the margins and apex, with dark spots and rings on the upper side; labellum reddish-purple, usually with heavy dark spots and lines
                       4(b). traunsteineri subsp. **lapponica**

19. Labellum with angled outer lobes making them deltoid   20.

19. Labellum with broad, rounded outer lobes   21.

20. Leaves unmarked or with small spots less than 1 mm in diameter, often only in the upper part; labellum 5–6 × 6–8 mm   **5(a). purpurella** subsp. **purpurella**

20. Leaves with dull violet spots more than 2 mm in diameter; labellum 6–8 × 9–11 mm
                 5(b). purpurella subsp. **majaliformis**

21. Leaves not marked with spots or bars   22.

21. Leaves heavily marked with spots and/or bars   23.

22. Labellum more or less 3-lobed, more or less flat or the edges of the outer lobes reflexed, central lobe well-marked, usually one-third to half as long as the unlobed part
     6(c,2). comosa subsp. **occidentalis** forma **kerryensis**

22. Labellum obscurely lobed, flat but more or less turned up at the edges, the middle lobe short with a rounded tip   **7(1). praetermissa** forma **praetermissa**

23. Labellum obscurely lobed, flat but more or less turned up at the edges, the middle lobe short with a rounded tip   **7(2). praetermissa** forma **junialis**

23. Labellum more or less 3-lobed, flat or more or less reflexed at the edges, the middle lobe usually one-third to half as long as the unlobed part   24.

24. Plant 5–9 cm; sheathing leaves 2–3, 12–16 mm wide; spur often not more than 3.5 mm wide
                 6(a). comosa subsp. **scotica**

24. Plant 9–50 cm; sheathing leaves usually 4 or more, 12–20 mm wide   25.

25. Plant 15–32 cm; sheathing leaves usually 4 or more, 12–20 mm wide; spur usually less than 3.0 mm wide
                 6(b). comosa subsp. **cambrensis**

25. Plant 20–50 cm; spur not usually more than 3.5 mm wide
    6(c,1). comosa subsp. **occidentalis** forma **occidentalis**

1. D. fuchsii (Druce) Soó   Common Spotted Orchid
Orchis fuchsii Druce; *Orchis maculata* subsp. *fuchsii* (Druce) Druce; *Dactylorchis fuchsii* (Druce) Vermuel.; *D. maculata* subsp. *fuchsii* (Druce) Hyl.

Perennial glabrous herb; tubers palmate, flattened, thick at the base, with 2–4, tapering, more or less divergent segments; roots rather short, thick, cylindrical. *Stems* (5–)7–50(–70) cm × 1.5–7.5(–11) mm in diameter, pale green, sometimes flushed with violet above, erect, cylindrical, solid for much of their length. *Leaves* usually bright to dark green, usually sparsely to densely marked on the upper surface with more or less evenly distributed, more or less transversely elongated dark spots or blotches, sometimes unmarked, keeled; basal broadly elliptical to obovate-oblong; sheathing (2–)3–6(–7), 4.5–16.5 × 0.5–5.5 cm, usually crowded towards the base of the stem, narrowly to broadly lanceolate or oblong-lanceolate, subacute and rarely hooded at apex, entire; non-sheathing (1–)2–6(–9), similar to sheathing, but gradually decreasing in size upwards. *Inflorescence* 1.5–10(–22) × 2–4 cm, fairly lax to dense, conical or more or less cylindrical. *Bracts* 5–23(–38) × 1–3 mm, green, sometimes tinged violet, linear-lanceolate, gradually tapering to a fine point, 1- to 3-veined, usually more or less equalling the ovaries. *Flowers* medium-sized, pale pink or lilac-pink to deep reddish-purple, or white, sometimes more or less scented. *Outer perianth segments* 6–9 mm, lanceolate, obtuse to acute at apex, 1- to 3-veined, the lateral segments spotted and erect or spreading, the upper unspotted. *Inner perianth segments* shorter than outer, ovate-lanceolate to linear-lanceolate, more or less obtuse at apex, 1- to 3-veined, connivent, unspotted, often edged with violet. *Labellum* 4.5–8.5(–9.5) × 7–12(–13.5) mm, white to pale pink, lilac-pink or reddish-purple, with pale to bold dashes and/or loops, or rarely with them absent or with a solid blotch, deeply 3-lobed, the middle lobe at least as long as and about as wide as the lateral lobes, the lateral lobes rounded and divergent with more or less parallel sides, entire or slightly toothed, the middle lobe more or less triangular and obtuse to acute at apex. *Spur* 3.5–10.0 × 1.0–2.5 mm, straight to rather decurved, usually slightly tapering. *Column* erect. *Anther* erect, purplish, pyriform; pollinia 2, dark green; caudicles yellow, transparent; viscidia colourless, in a whitish pouch. *Stigma* with a violet line on its roof. *Rostellum* 3-lobed, the middle lobe short and lamelliform. *Ovary* often flushed with violet, sessile, cylindrical, twisted, with 6 rather acute, longitudinal ridges. *Flowers* 6–8. Pollinated by bees. 2n = 40.

The infraspecific taxa are both geographical and ecological.

(a) Subsp. **okellyi** (Druce) Soó
Orchis maculata var. *okellyi* Druce; *Orchis okellyi* (Druce) Druce; *Orchis maculata* subsp. *okellyi* (Druce) Druce; *Dactylorchis fuchsii* subsp. *okellyi* (Druce) Vermuel.
Stems 15–20 cm, often without, sometimes with a little purplish colouring. *Leaves* without spots or with some faint spotting, narrow. *Spike* often cylindrical, not usually dense. *Flowers* white or cream and unmarked or with faint pinkish markings, strongly scented. *Labellum* 5–9(–11) mm wide. *Spur* 5.5–6.5 mm. *Flowers* June and early July.

(b) Subsp. **fuchsii**
Stems (8–)20–50(–70) cm, little suffused purple. *Leaf spotting* usually not heavy. *Spike* often pyramidal, but only rather dense. *Flowers* pale to deep pink or lilac-pink, rarely white. *Labellum* 8–12 mm wide. *Spur* 5–7 mm.

(i) Var. **fuchsii**
Stems tall and very leafy. *Spike* long. *Labellum* with the middle lobe distinctly longer than laterals.

(ii) Var. **trilobata** (Bréb.) P. D. Sell
Orchis maculata var. *trilobata* Bréb.; *D. fuchsii* var. *rhodochilla* Turner Ettl.
Plant shorter and less leafy. *Spike* short. *Labellum* with lobes subequal in length.

(c) Subsp. **hebridensis** (Wilmott) Soó
Orchis hebridensis Wilmott; *Dactylorchis maculata* subsp. *hebridensis* (Wilmott) Vermeul.; *Orchis fuchsii* subsp. *hebridensis* (Wilmott) A. R. Clapham; *Dactylorchis fuchsii* subsp. *hebridensis* (Wilmott) Hesl.-Harr. fil; *D. maculata* subsp. *hebridensis* (Wilmott) E. Nelson
Stems 15–30 cm, often suffused purplish. *Leaf spotting* tending to be heavy. *Spike* often pyramidal. *Flowers* often deeply coloured rose-pink to reddish-purple, rarely white. *Labellum* 8–15 mm wide. *Spur* 7.0–8.5 mm.

(i) Var. **alpina** (Landwehr) R. M. Bateman & Denholm
Dactylorchis fuchsii forma *alpina* Landwehr
Leaves narrow, not recurved. *Spike* rarely dense. *Floral bracts* usually under 8 mm. *Labellum* usually less than 6.7 mm long and 9.5 mm wide. *Spur* at middle less than 1.8 mm wide. *Flowers* late June and July.

(ii) Var. **hebridensis** (Wilmott) R. M. Bateman & Denholm
Basal leaves elliptical, often recurved. *Spike* often dense. *Floral bracts* usually less than 8 mm. *Labellum* usually more than 6.7 mm long and 9.5 mm wide. *Spur* at middle less than 1.8 mm wide. *Flowers* late June and July.

(iii) Var. **cornubiensis** (Pugsl.) Soó
Orchis maculata var. *cornubiensis* Pugsl.; *Orchis fuchsii* var. *cornubiensis* (Pugsl.) A. R. Clapham; *D. fuchsii* subsp. *cornubiensis* (Pugsl.) Soó
Basal leaves elliptical, often recurved. *Spike* not dense. *Floral bracts* often more than 8 mm. *Labellum* usually

more than 6.7 mm long and 9.5 mm wide. *Spur* at middle up to 2 mm wide. *Flowers* June.

Native. Damp woods, banks, meadows, marshes, dune-slacks and fens, usually on base-rich soils. More or less throughout the British Isles. Much of Europe, but absent from much of the south. Subsp. *fuchsii* occurs throughout much of the British Isles, but is less common in areas of the south-west peninsula, Wales and central and north Scotland where base rich soils are scarce, and where it is replaced by *D. maculata*. It is absent from much of the Western Isles and absent or infrequent in western Ireland. Var. *fuchsii* is the plant of woods and scrub. Var. *trilobata* is the plant of open grassland. Subsp. *okellyi* occurs on the limestone pavement of the Burren Hills, Co. Clare and a few other places in western Ireland, in the Isle of Man and on the western coasts of Argyll and Sutherland. Subsp. *hebridensis* occurs in inland and coastal localities in central and west Scotland, northern England, Wales, Devon and Cornwall and Ireland. Var. *alpina* is found in the mountains and on the coasts of Scotland, Wales and Ireland. Var. *hebridensis* occurs in the Western Isles and in western Ireland. Most populations occupy coastal machair. Var. *cornubiensis* is confined to the northern coasts of Devon and Cornwall.

× **incarnata** = **D.** × **kernerorum** (Soó) Soó
Orchis ambigua A. Kern., non Martrin.-Donos; *Orchis kernerorum* Soó; *Orchis curtisiana* Druce
This hybrid is usually taller than *D. incarnata*, with a thinner, less hollow stem and yellowish-green leaves which are broadest near the middle, narrowly hooded at the tip and unspotted or with rather pale spots. The flowers vary from pale salmon pink to a bluish-purple with a flat, 3-lobed labellum marked with a pattern of broken loops and dots. They are highly sterile.
Scattered in Britain and Ireland where they are usually found as single plants in mixed populations of the parents. They are also found in central and west Europe. The common form would appear to be nothosubsp. **kernerorum**, which seems to be *D. fuchsii* subsp. *fuchsii* × *D. incarnata* subsp. *incarnata*. Nothosubsp. **variabilis** (Hesl.-Harr.) P. D. Sell (*Orchis variabilis* Hesl.-Harr.; *D. variabilis* (Hesl.-Harr.) Soó) occurs in the Hebrides and is *D. fuchsii* subsp. *hebridensis* × *D. incarnata* subsp. *incarnata*.

× **maculata** = **D.** × **transiens** (Druce) Soó
Orchis transiens Druce
This hybrid usually has the robust habit of *D. fuchsii*, but its leaves are narrower and the lowest are not so bluntly rounded at the tip. The flowers have a heavily marked labellum with broader, more crenate lateral lobes than in *D. fuchsii*, and a long, straight cylindrical spur which is not so slender as in *D. maculata*. They are highly sterile. $2n = 60$.
Scattered throughout the British Isles, but probably over-recorded. They occur, in small numbers, where the soil varies from acid to base-rich. The hybrid occurs also in Continental Europe. The common plant is

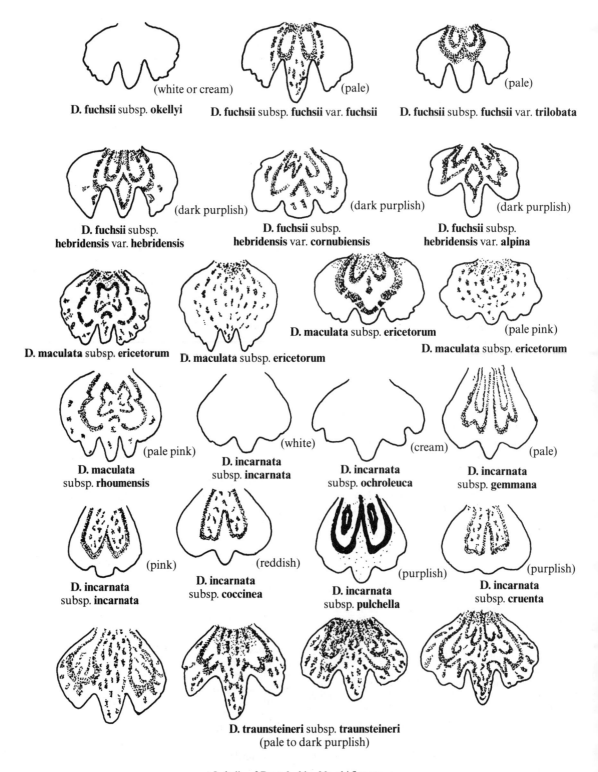

(white or cream)

D. fuchsii subsp. **okellyi**

(pale)

D. fuchsii subsp. **fuchsii** var. **fuchsii**

(pale)

D. fuchsii subsp. **fuchsii** var. **trilobata**

(dark purplish)

D. fuchsii subsp.
hebridensis var. **hebridensis**

(dark purplish)

D. fuchsii subsp.
hebridensis var. **cornubiensis**

(dark purplish)

D. fuchsii subsp.
hebridensis var. **alpina**

D. maculata subsp. **ericetorum**

(pale pink)

D. maculata subsp. **ericetorum**

D. maculata subsp. **ericetorum**

D. maculata subsp. **ericetorum**

(pale pink)

D. maculata
subsp. **rhoumensis**

(white)

D. incarnata
subsp. **incarnata**

(cream)

D. incarnata
subsp. **ochroleuca**

(pale)

D. incarnata
subsp. **gemmana**

(pink)

D. incarnata
subsp. **incarnata**

(reddish)

D. incarnata
subsp. **coccinea**

(purplish)

D. incarnata
subsp. **pulchella**

(purplish)

D. incarnata
subsp. **cruenta**

D. traunsteineri subsp. **traunsteineri**
(pale to dark purplish)

Labella of **Dactylorhiza** Nevski flowers

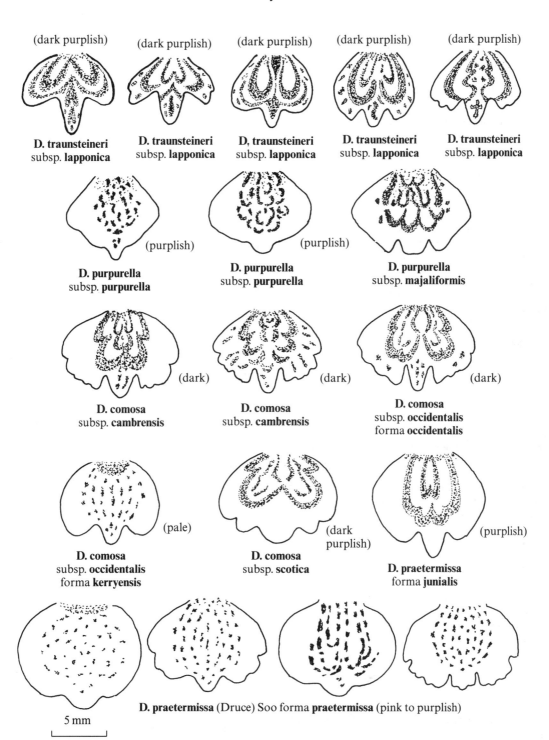

(dark purplish)

D. traunsteineri
subsp. **lapponica**

(dark purplish)

D. traunsteineri
subsp. **lapponica**

(dark purplish)

D, traunsteineri
subsp. **lapponica**

(dark purplish)

D. traunsteineri
subsp. **lapponica**

(dark purplish)

D. traunsteineri
subsp. **lapponica**

(purplish)

D. purpurella
subsp. **purpurella**

(purplish)

D. purpurella
subsp. **purpurella**

D. purpurella
subsp. **majaliformis**

(dark)

D. comosa
subsp. **cambrensis**

(dark)

D. comosa
subsp. **cambrensis**

(dark)

D. comosa
subsp. **occidentalis**
forma **occidentalis**

(pale)

D. comosa
subsp. **occidentalis**
forma **kerryensis**

(dark
purplish)

D. comosa
subsp. **scotica**

(purplish)

D. praetermissa
forma **junialis**

D. praetermissa (Druce) Soo forma **praetermissa** (pink to purplish)

5 mm

Labella of **Dactylorhiza** Nevski flowers

nothosubsp. **transiens**, which is between *D. fuchsii* subsp. *fuchsii* and *D. maculata* subsp. *ericetorum*. In the Hebrides occurs nothosubsp. **corylensis** (Hesl.-Harr.) P. D. Sell (*Orchis corylensis* Hesl.-Harr.; *D. corylensis* (Hesl.-Harr.) Soó), which is *D. fuchsii* subsp. *hebridensis* × *D. maculata* subsp. *ericetorum*.

× **praetermissa** = **D.** × **grandis** (Druce) P. F. Hunt
Orchis grandis Druce; *Orchis mortonii* Druce; *Orchis mortonensis* A. Camus; *D. mortonii* (Druce) Soó comb. inval.; *Orchis braunii* var. *townsendiana* Rouy; *D.* × *townsendiana* (Rouy) Soó
This hybrid is usually tall and robust and the stems have a narrow cavity. The leaves are whitish beneath, and usually marked with faint, transverse spots, the upper are narrower and more acute and the lower shorter and blunter than those of *D. praetermissa*. The spike is conical, many-flowered and dense, the flowers intermediate between those of the parents. The pollen is highly sterile and seed production very low. 2*n* = 60.

Often abundant with the parents in marshes, wet meadows, fens and occasionally chalk downs. Throughout the distribution of *D. praetermisa*.

× **purpurella** = **D.** × **venusta** (T. & T. A. Stephenson) Soó
Orchis × *venusta* T. & T. A. Stephenson
This hybrid is very variable. The stems can be solid or with a narrow hollow, the leaves are whitish beneath and marked with pale to deep purple, transversely elongated, large spots, or may be unspotted, the flowers are in a dense, pointed spike and intermediate between those of the parents. The pollen is highly sterile and seed production is very low. 2*n* = 60.

Wet meadows, marshes and dune-slacks. Throughout the range of *D. purpurella* and often abundant where the parents grow together. The most widespread plant is northosubsp. **venusta** which is between *D. fuchsii* subsp. *fuchsii* and *D. purpurella* subsp. *purpurella*. Nothosubsp. **hebridella** (Wilmott) P. D. Sell (*Orchis hebridella* Wilmott; *D. hebridella* (Wilmott) Soó) is found in the Hebrides and is between *D. fuchsii* subsp. *hebridense* and *D. purpurella* subsp. *purpurella*.

· × **traunsteineri** = **D.** × **kelleriana** P. F. Hunt nom. inval.
This hybrid has the dwarf habit and narrow leaves of *D. traunsteineri*, which are often marked with dark purple, narrow transverse bars or transversely elongated spots or rings. The flowers have a trilobed labellum which is nearer *D. fuchsii* in its markings. The pollen is highly sterile. 2*n* = 60.

Has been recorded in north-west Wales, Yorkshire and Ireland where there are populations of *D. traunsteineri*.

2. D. maculata (L.) Soó   Heath Spotted Orchid
Orchis maculata L.; *Dactylorchis maculata* (L.) Vermuel.

Perennial glabrous herb; tubers palmate, with 2–4, ovoid, divergent segments; roots rather short and thick. *Stems* 15–50 cm × 1.3–5.5(–8.5) mm in diameter, pale green, sometimes purplish above, solid, erect. *Leaves* usually bright to dark green, usually sparsely to densely marked on the upper surface with usually solid, round to slightly elongated spots; basal 0–1, elliptical or oblanceolate-oblong; sheathing (1–)2–4(–5), 4–16(–19) × 0.5–2.0 (–2.5) cm, usually much crowded towards the base of the stem, narrowly lanceolate, often distinctly hooded at apex, usually recurved; non-sheathing (1–)2–5(–7), narrowly lanceolate or linear, more or less acute at apex. *Inflorescence* 1–8(–10) × 1.5–3.0 cm, lax to dense, rather few-flowered and tending to remain pyramidal. *Bracts* 4–20(–25) × 1–3 mm, green, often suffused purplish, linear-lanceolate, finely pointed at apex, 3- to 5-veined, equalling to longer than the ovaries. *Flowers* medium sized, from white to deep lilac-pink, but usually pale, with faint to bright markings, faintly scented. *Outer perianth segments* 7–11 mm, the lateral spreading horizontally or drooping, narrowly lanceolate, reflexed at edges, sometimes slightly twisted, the upper connivent with the inner segments. *Inner perianth segments* shorter than the outer, white or pale mauve, edged violet, 1- or obscurely 3-veined. *Labellum* (4–)5–9(–11) × (5–)6–13(–15) mm, usually shallowly lobed, the lateral lobes broadly rounded and entire or crenulate, often more or less deflexed, the middle lobe much smaller, triangular and usually shorter than the laterals, the whole marked with irregular double loops, often broken into bars or dots. *Spur* (2.5–)3–8(–9.5) × 0.6–2.2 mm, straight to slightly decurved, cylindrical to slightly tapering. *Column* erect. *Anther* erect, reddish-purple; pollinia 2, olive green; caudicles yellow, transparent; viscidia colourless, in a whitish pouch. *Stigma* on the roof of the mouth of the spur, edged with a violet line. *Rostellum* 3-lobed, the middle lobe short and lamelliform. *Ovary* cylindrical, with 6 longitudinal ridges. *Flowers* 5–7(–8). Pollinated by bees and flies.

D. maculata shows a great range of variability in size and flower, particularly in the size, shape, colour and patterning of the labellum, but it does not tend to form local races. Subsp. *maculata* is a plant of Continental Europe which does not occur in the British Isles. The two races given below need further consideration, but are kept for the time being, particularly because they have different chromosome numbers. The continental subsp. *elodes* is often regarded as distinct, but is here considered to be a synonym of subsp. *ericetorum*.

(a) Subsp. **ericetorum** (E. F. Linton) P. F. Hunt & Summerh.
Orchis elodes Griseb.; *Orchis maculata* var. *praecox* Webster; *Orchis maculata* subsp. *ericetorum* E. F. Linton; *Orchis maculata* subsp. *elodes* (Griseb.) A. Camus; *Orchis ericetorum* (E. F. Linton) E. S. Marshall; *Dactylorchis maculata* subsp. *ericetorum* (E. F. Linton) Vermuel.; *Dactylorchis elodes* (Griseb.) Vermuel.; *Dactylorchis maculata* subsp. *elodes* (Griseb.) Vermuel.; *D. maculata* subsp. *elodes* (Griseb.) Soó
Labellum broader than long, middle lobe much smaller than lateral lobes, from white to deep lilac-pink, but predominantly pale. *Spur* hardly tapering. *Flowers* May to July. 2*n* = 80.

(b) Subsp. **rhoumensis** (Hesl.-Harr. fil.) Soó
Orchis fuchsii subsp. *rhoumensis* Hesl.-Harr. fil.;
Dactylorchis maculata subsp. *rhoumensis* (Hesl.-Harr. fil.) Soó
Labellum about as broad as long, with the middle lobe longer than in subsp. *ericetorum* and thus approaching *D. fuchsii*, pale purple. *Spur* slender and pointed. *Flowers* late July and August. $2n = 40$.

Native. The common spotted orchid of peaty areas from quite dry heaths to damp moorland and often in open woodland. Throughout the British Isles but commonest in the north and west where there are more suitable habitats. Much of Europe except the south-east. Almost all the plants in the British Isles are subsp. *ericetorum*, which also occurs in northern France, Belgium, Holland, Germany and in high altitudes in Sweden. Subsp. *rhoumensis* is endemic to the island of Rhum, where it occurs on mineral soil influenced by blown sand with a pH of 4.9–5.5, and is absent from the main areas of machair.

× **praetermissa = D. × hallii** (Druce) Soó
Orchis hallii Druce; *Orchis scotica* Druce nom. nud.
This hybrid is similar to *D. × grandis*, but has longer, narrower leaves, a cylindrical spike and lilac-purple flowers with a broad, shallowly trilobed labellum with a pattern of lines and dots, and the spur is long and straight. The hybrids show normal pollen and seed fertility. $2n = 80$.

These hybrids probably occur throughout the range of *D. praetermissa*, but are rarer than *D. × grandis* because of habitat preferences. They are also recorded from Holland.

× **purpurella = D. × formosa** (T. & T. A. Stephenson) Soó
Orchis formosa T. & T. A. Stephenson
This hybrid is very variable, the long, narrow leaves are spotted or unspotted, the flowers are in a small, oblong spike varying from pale lilac to dull purple, the labellum with rounded, crenate lateral lobes and a small middle lobe, with a pattern of purple spots and broken lines. It appears to have normal fertility. $2n = 80$.

Occurs throughout the range of *D. purpurella* and is probably the commonest hybrid orchid in northern Britain and Ireland. It is also recorded from Norway.

× **traunsteineri = D. × jenensis** (Brand) Soó
Orchis schulzei K. Richt., non Hausskn.; *Orchis jenensis* Brand
This hybrid has a slender, flexuous stem with narrow, subacute leaves more or less faintly marked with round spots, a short spike with the bracts as long as the flowers, and lilac flowers with a large, trilobed labellum, the middle lobe slightly longer than the lateral and lightly marked with reddish flecks and dots. The pollen appears to be sterile and no seed set has been observed.

This hybrid has been found in populations of *D. traunsteineri* in north-west Wales, Yorkshire and Ireland. It is also recorded from north and central Europe.

3. D. incarnata (L.) Soó   Early Marsh Orchid
Orchis latifolia L. nom. rej.; *Orchis incarnata* L.; *Orchis impudica* Crantz; *Orchis divaricata* Rich.; *Orchis strictifolia* Opiz; *Orchis angustifolia* Wimm. & Grab.; *Orchis fistulosa* Stokes nom. illegit.; *Dactylorchis incarnata* (L.) Vermuel.

Perennial, glabrous herb; tubers palmate, thick, flattened, the 2–4 segments long and tapering; roots numerous, horizontal, sometimes very long. *Stem* (5–)10–60(–90) cm × 3–9(–12) mm in diameter, stout, stiff, hollow, the cavity often more than half its diameter, angled, pale green, sometimes purplish above. *Leaves* yellowish-green to bright green, unmarked, or marked on both surfaces, or marked only on the upper surface, strongly keeled; basal (0–)1, elliptical or lanceolate-oblong; sheathing (2–)3–4(–5), 4–20(–30) × 1.0–3.5 cm, more or less evenly distributed along the stem or crowded towards the base, oblong-lanceolate, acute and more or less hooded at apex, erect to erect-spreading or recurved; non-sheathing (0–)1(–2), similar to sheathing but smaller. *Inflorescence* 2.5–8.0 × 2–3 cm, fairly lax to dense, ovoid to cylindrical. *Bracts* 7–20 × 1–4 mm, green, often suffused purplish, the lower and sometimes all exceeding the flowers, lanceolate, tapering to a fine point, keeled, usually 3-veined. *Flowers* rather small, white or cream and unmarked or, pale to deep salmon pink, brick-red to crimson-red or bright purple with darker markings. *Outer perianth segments* 5–6(–9) mm, the lateral erect or reflexed, often spotted, lanceolate to ovate-lanceolate, more or less obtuse at apex and with 3 veins, the upper narrower, obtuse and hooded at apex, with one vein, and arched forwards or connivent with the inner segments. *Inner perianth segments* shorter, oblong-lanceolate, connivent, with 1–3 obscure veins. *Labellum* 4.5–9.0 × 5–9 mm, ovate-rhomboid to more or less subrotund, entire to 3-lobed, with the sides often strongly reflexed so that it appears to be quite narrow, white, cream, pale yellow, brick-red, crimson-red, or bright purple, usually variously marked with spots, lines and loops. *Spur* 5.5–8.0 × 2.5–5.5 mm, more or less decurved, slightly tapering. *Column* short, small. *Anther* ovate, coloured like inner perianth segments; pollinia 2, greyish-green; viscidia in a pouch. *Stigma* longer than broad, with a purple edge, very sticky and glistening. *Rostellum* rose or violet. *Ovary* sessile, cylindrical, with 3 broad and 3 narrow ridges. *Flowers* 5–8. Visited especially by bumble-bees. $2n = 40$.

Variation ties up to a considerable extent with ecology. Where two of its habitats are adjacent two subspecies may occur in close proximity. White-flowered forms may occur in any population. When two subspecies grow together they tend to have different flowering times and mostly do not intergrade.

(a) Subsp. **ochroleuca** (Wüstnei ex Boll) P. F. Hunt & Summerh.
Orchis incarnata var. *ochroleuca* Wüstnei ex Boll; *Orchis ochroleuca* (Wüstnei ex Boll) Schur; *Orchis latifolia* var. *ochroleuca* (Wüstnei ex Boll) Pugsl.; *Orchis strictifolia* var. *ochroleuca* (Wüstnei ex Boll)

Hyl.; *Orchis incarnata* subsp. *ochroleuca* (Wüstnei ex Boll) Schwarz; *Dactylorchis incarnata* subsp. *ochroleuca* (Wüstnei ex Boll) Hesl.-Harr. fil.; *Orchis incarnata* var. *straminea* Rchb. fil.; *D. incarnata* var. *straminea* (Rchb. fil.) Soó

Stems 20–60 cm, robust. *Leaves* unspotted, the longest often more than 9 × 2 cm. *Basal bracts* usually more than 25 mm, the floral more than 18 mm. *Labellum* usually less than 6.5 × 8.0 mm, cream or very pale yellow, deepening in colour towards the entrance to the spur, markings absent, deeply 3-lobed, the outer lobes often notched.

Albino variants of other subspecies are usually less robust and lack the characteristic, large, deeply 3-lobed labellum with the outer lobes notched. Flowers up to 2 weeks later than subsp. *incarnata* when growing with it.

(b) Subsp. **gemmana** (Pugsl.) P. D. Sell
Orchis latifolia var. *gemmana* Pugsl.; *Dactylorchis incarnata* subsp. *gemmana* (Pugsl.) Hesl.-Harr. fil.
Stems 30–80 cm, robust. *Leaves* unspotted, 15–30 cm. *Basal bracts* much longer than flowers. *Labellum* up to 8 × 10 mm, very faint salmon-pink to almost white, with markings, 3-lobed, the outer lobes entire.

(c) Subsp. **incarnata**
Orchis lanceata Dietrich; *Dactylorchis incarnata* subsp. *lanceata* (Dietrich) Vermuel.; *Orchis palmata* subsp. *incarnata* (L.) Syme; *Orchis latifolia* subsp. *incarnata* (L.) Hook. fil.; *D. incarnata* forma *punctata* (Vermuel.) R. M. Bateman & Denholm
Stems 20–40 cm, rather slender. *Leaves* unspotted, 8–15 cm. *Basal bracts* usually exceeding flowers, but not so obvious as in subsp. *gemmana*. *Labellum* up to 8 × 7 mm, pale to rather deep salmon pink, with markings, shallowly 3-lobed. Flowering begins in May, reaching a peak in early June in southern England and extending to July in the north.

(d) Subsp. **coccinea** (Pugsl.) Soó
Orchis latifolia var. *coccinea* Pugsl.; *Orchis strictifolia* subsp. *coccinea* (Pugsl.) A. R. Clapham; *Dactylorchis incarnata* subsp. *coccinea* (Pugsl.) Hesl.-Harr. fil.; *Orchis incarnata* var. *dunensis* Druce; *Dactylorchis incarnata* var. *dunensis* (Druce) Vermuel.; *D. incarnata* var. *dunensis* (Druce) Landwehr; *Orchis incarnata* forma *atriruba* Godfery
Stems 5–20 cm, typically short and stout. *Leaves* unspotted, often less than 9 cm, typically curved. *Basal bracts* rather short in a dense inflorescence. *Labellum* up to 6.5 × 8.0 mm, brick red, crimson-red or reddish-purple, with markings, shallowly 3-lobed. Flowers up to 2 weeks later where it grows with subsp. *incarnata*.

(e) Subsp. **pulchella** (Druce) Soó
Orchis incarnata var. *pulchella* Druce; *Orchis latifolia* var. *pulchella* (Druce) Pugsl.; *Orchis strictifolia* var. *pulchella* (Druce) A. R. Clapham; *Dactylorchis incarnata* subsp. *pulchella* (Druce) Hesl.-Harr. fil.; *Orchis traunsteineri* var. *serotina* Hausskn.; *Orchis incarnata* var. *serotina* (Hausskn.) Hausskn.; *Orchis serotina* (Hausskn.) Schwarz; *Dactylorchis incarnata* var. *serotina* (Hausskn.) Vermuel.; *D. incarnata* subsp. *serotina* (Hausskn.) D. M. Moore & Soó; *Orchis angustifolia* var. *haussknechtii* Klinge; *Orchis incarnata* var. *borealis* Neuman; *Orchis incarnata* var. *pulchriora* Druce; *Orchis latifolia* var. *cambrica* Pugsl.
Stems 20–40 cm, often slender. *Leaves* unspotted, often more than 9 cm. *Basal bracts* rather short, often purplish. *Labellum* up to 6.5 × 8.0 mm, purple or purplish-violet with markings, entire or shallowly 3-lobed.

Variable in stature and shape, dimensions and marking of the labellum.

(f) Subsp. **cruenta** (O. F. Müll.) P. D. Sell
Orchis cruenta O. F. Müll.; *Orchis latifolia* var. *cruenta* (O. F. Müll.) Lindl.; *Orchis incarnata* var. *cruenta* (O. F. Müll.) Blytt & Dahl; *Orchis incarnata* subsp. *cruenta* (O. F. Müll.) Asch. & Graebn.; *Dactylorchis cruenta* (O. F. Müll.) Vermuel.; *Dactylorchis incarnata* subsp. *cruenta* (O. F. Müll.) Hesl.-Harr. fil.; *D. cruenta* (O. F. Müll.) Soó; *Orchis haematodes* Rchb.; *Orchis cruenta* var. *haematodes* (Rchb.) Neuman; *D. incarnata* var. *haematodes* (Rchb.) Soó; *Orchis latifolia* var. *brevifolia* Rchb. fil.; *Orchis cruenta* var. *brevifolia* (Rchb. fil) Neuman; *Orchis incarnata* var. *haematodes* Schulze; *Dactylorchis incarnata* var. *haematodes* (Schulze) Vermuel.; *Orchis cruentiformis* Neuman; *Orchis cruenta* var. *lanceolata* Neuman; *D. incarnata* subsp. *cruenta* var. *lanceolata* (Neuman) Landwehr; *Orchis incarnata* var. *hyphaematodes* Neuman; *D. incarnata* var. *hyphaematodes* (Neuman) Landwehr
Stems up to 20 cm, often purplish. *Leaves* usually spotted on one or both surfaces, up to 9 cm. *Basal bracts* usually rather short, often purplish. *Labellum* up to 6.5 × 8.0 mm, purple or purplish-violet, with markings, entire or shallowly 3-lobed.

Native. Marshes, fens, bogs, dune-slacks and damp pastures. Throughout the British Isles. In most of Europe, but rare in the Mediterranean region. Subsp. *ochroleuca* occurs in reed-swamp on calcareous fen peat in a few localities in East Anglia and possibly in western Ireland and south Wales. It is widely scattered through Scandinavia, the Baltic States, Germany and the Alps and perhaps further east. Albino forms of the other subspecies are sometimes mistaken for it. Subsp. *gemmana* occurs in wet meadows and swamp in Norfolk and perhaps elsewhere in south-east England and Wicklow and Galway in Ireland, but large plants of subsp. *incarnata* are sometimes mistaken for it. It is also recorded from the Netherlands and Germany. Subsp. *incarnata* is characteristic of more or less waterlogged soil of high mineral content and of a pH of 6.0–7.5 or higher. It is the most frequent subspecies in south and south-east England, but is more scarce in the north and west of the British Isles as far north as Argyllshire and the Inner Hebrides. It occurs throughout most of the range of the species. Typical subsp. *coccinea* favours moist grassland in coastal dunes, especially hollows or slacks and particularly along the west coast of Britain and Ireland, but is also found locally inland as near the shores of the central Irish loughs. Taller, slender plants with the deep flower colour are probably best placed in subsp. *incarnata*. It seems to be endemic. Subsp. *pulchella* occurs in fen or marsh with *Schoenus nigricans* and a pH substratum of 6.2–7.5, poor-fen with neither *Schoenus* nor *Sphagnum* species and a pH of 5.2–6.8, and soligenous bog with a *Sphagnum* carpet and pH of 5.0–6.0. It has been recorded in suitable habitats throughout the British Isles, but particularly in the north and west where it replaces subsp. *incarnata* to a considerable extent. It is scattered in continental Europe. Subsp. *cruenta* occurs in a few highly calcareous fens in Co. Mayo, Galway and Clare in western Ireland and in neutral flushes in Ross-shire in Scotland. In continental Europe it occurs in the Alps, Scandinavia and Russia.

× **maculata** = D. × **carnea** (A. Camus) Soó
Orchis × *carnea* A. Camus
This hybrid has the strict habit of *D. incarnata*, the yellowish-green leaves are usually marked with large round spots, the labellum is nearly flat and broadly rounded and its colour varies from pale salmon-pink to bright reddish-purple with a pattern of dots arranged in parallel loops, and the spur has a wide throat and is slender and distinctly curved. The pollen is sterile and no seed seems to be set after selfing. $2n = 60$.

Scattered through Britain and Ireland, but rare as the parents do not usually grow near one another. There are scattered records in Continental Europe.

× **praetermissa** = D. × **wintonii** (A. Camus) P. F. Hunt
Orchis × *wintonii* A. Camus
This hybrid is most like *D. praetermissa* with a hollow stem, erect yellowish-green, more or less hooded leaves broadest near the middle, an oblong spike with long, narrow, often incurved bracts and pale lilac to bright violet flowers (depending on the subspecies of *D. incarnata*) which are smaller than in *D. praetermissa*. It is sterile.

Scattered through south and central Britain north to Lancashire, but generally rare and usually as single plants with the parents in base-rich meadows and fens.

× **purpurella** = D. × **latirella** (P. M. Hall) Soó
Orchis × *latirella* P. M. Hall
This hybrid is intermediate in habit between the parents, its leaves are narrower and more yellowish-green than in *D. purpurella* and are sometimes indistinctly spotted, the purplish flowers are in a cylindrical spike and the sides of the labellum are often somewhat reflexed.

Scattered in north and west Britain, usually as single plants with the parents in base-rich marshes and dune-slacks. Endemic.

× **traunsteineri** = D. × **dufftii** (Hausskn.) Peitz
Orchis × *dufftii* Hausskn.; *Orchis* × *lehmannii* Klinge; *Dactylorchis lehmannii* (Klinge) Soó; *D.* × *lehmannii* (Klinge) Soó; *D.* × *thellungiana* (Braun-Blanq.) Soó
Hybrids of this presumed parentage have the habit of *D. traunsteineri*, with narrow yellowish-green leaves with a few small spots, long purplish bracts and the flowers intermediate between the parents.

These hybrids are rare in areas of rich-fen in north-west Wales, Yorkshire and Co. Wicklow. They have been widely recorded in Continental Europe.

4. D. traunsteineri (Sauter ex Rchb.) Soó
Narrow-leaved Marsh Orchid
Orchis traunsteineri Sauter ex Rchb.; *Dactylorchis traunsteineri* (Sauter ex Rchb.) Vermeul.

Perennial glabrous herb; tubers palmate, short and thick; roots numerous, thick. *Stem* 6–30(–45) cm, rarely exceeding 5 mm in diameter, slender, slightly hollow, flexuous, pale green, often suffused purplish in the upper part. *Leaves* pale, dull or medium green, often with dark purplish-brown markings, occasionally heavily marbled; basal one, broadest at the middle; sheathing 2–4(–5), 3–12 × 0.8–1.8 cm, linear-oblong, oblong-lanceolate, oblong-oblanceolate or narrowly elliptical, obtuse to acute at apex, entire, more or less spreading, sometimes undulate, sometimes recurved; non-sheathing 0–2, similar to sheathing but smaller. *Inflorescence* 2.5–7.0 × 2.0–3.5 cm, lax and often secund or subsecund, with 3–20 flowers. *Bracts* 10–25 × 1–3 mm, greenish, frequently tinged purplish on the margin, occasionally over the whole bract, sometimes spotted on both surfaces, linear-lanceolate, more or less acute at apex. *Flowers* pale to deep lilac or reddish-purple with darker markings. *Outer perianth segments* 7–8 mm, lilac-purple, oblong-lanceolate, obtuse at apex, erect or obliquely spreading. *Inner perianth segments* lilac- or reddish-purple, obliquely ovate. *Labellum* 5.5–9.5 × 7–11 mm, pale to deep lilac- or reddish-purple with deeper markings, deltate to obcordate or transversely elliptical, usually distinctly 3-lobed with the middle lobe longer than the laterals, or at least with a projecting point. *Spur* 6.5–10.5 × 2.0–3.5 mm, usually more or less cylindrical and straight, sometimes curved. *Column* short, small. *Anther* elliptical, rose-violet; pollinia greyish-green; caudicles yellow; viscidia in a pouch. *Stigma* on the roof of the mouth of the spur, bordered by a purple line. *Rostellum* 3-lobed, the middle lobe short. *Ovary* 9–15 mm, 6-ridged. *Flowers* 5–6. $2n = 80$.

(a) Subsp. traunsteineri
Orchis majalis subsp. *traunsteineroides* Pugsl.; *Orchis traunsteinerioides* (Pugsl.) Pugsl.; *Dactylorchis traunsteinerioides* (Pugsl.) Vermeul.; *D. traunsteineri* subsp. *traunsteinerioides* (Pugsl.) Soó; *D. traunsteinerioides* (Pugsl.) Landwehr; *D. traunsteineri* subsp. *hibernica* Landwehr; *Orchis latifolia* var. *eboracensis* Godfery; *Orchis majalis* subsp. *traunsteinerioides* var. *eboracensis* (Godfery) Pugsl.; *D. majalis* subsp. *traunsteinerioides* var. *eboracensis* (Godfery) R. M. Bateman & Denholm; *Orchis francisdrucei* Wilmott; *D. traunsteineri* subsp. *francis-drucei* (Wilmott) Soó; *D. majalis* subsp. *traunsteinerioides* var. *francis-drucei* (Wilmott) R. M. Bateman & Denholm
Stems up to 30(–45) cm. *Leaves* unspotted or with a few, small faint spots or transverse bars, slightly hooded at apex. *Bracts* strongly suffused reddish-purple all over, mostly without spots or rings. *Labellum* reddish- or

lilac-purple, usually with many dark markings more or less to the margin.

(b) Subsp. **lapponica** (Laest. ex Hartm.) Soó
Orchis pseudocordigera Neuman; *Dactylorchis lapponica* subsp. *pseudocordigera* (Neuman) Vermeul.; *D. pseudo-cordigera* (Neuman) Soó; *Orchis angustifolia* var. *lapponica* Laest. ex Hartm.; *Orchis traunsteineri* var. *lapponica* (Laest. ex Hartm.) Hartm.; *Orchis lapponica* (Laest. ex Hartm.) Rchb. fil.; *Orchis latifolia* var. *lapponica* (Laest. ex Hartm.) Rchb. fil.; *Dactylorchis lapponica* (Laest. ex Hartm.) Vermuel.; *D. cruenta* subsp. *lapponica* (Laest. ex Hartm.) E. Nelson; *D. majalis* subsp. *lapponica* (Laest. ex Hartm.) Sünd.
Stems up to 21 cm. *Leaves* heavily marked with spots, rings or blotches, not or scarcely hooded at apex. *Bracts* suffused reddish-purple, usually with heavy dark spots and lines. *Labellum* reddish-purple, usually with heavy dark spots and lines.

Native. The species is local in Ireland, western Scotland, north-west Wales, northern England and East Anglia, and very scattered in central and southern England. Also in northern and central Europe eastwards to Poland and Russia. Subsp. *traunsteineri* occurs throughout much of the range of the species in wet, basc-rich habitats, but is rare in the north. Subsp. *lapponica* is local in Westerness, Kintyre, North Ebudes and Outer Hebrides. It also occurs in northern Fennoscandia.

5. D. purpurella (T. & T. A. Stephenson) Soó
                              Northern Marsh Orchid
Orchis purpurella T. & T. A. Stephenson; *Dactylorchis purpurella* (T. & T. A. Stephenson) Vermuel.

Perennial glabrous herb; tubers palmate, with 2–4 slender segments; roots numerous, thick. *Stem* (5–)10–30(–45) cm often exceeding 5 mm in diameter, erect, slightly hollow, fluted above. *Leaves* dull or bluish-green, unspotted, with a few small spots near the apex, or with large violet spots all over; basal 1, broadest more or less at the middle; sheathing usually 4 or more, 3–15 × 1.5–3.0 cm, often slightly more crowded towards the base of the stem, more or less lanceolate, obtuse to acute at apex, entire, slightly spreading; non-sheathing usually 1–2, like sheathing but smaller. *Inflorescence* 2–7 × 2.0–3.5 cm, usually rather dense, ovoid to cylindrical, often square-topped, with up to 20(–30) flowers. *Bracts* 5–20 × 2–6 mm, often edged or flushed with purple, linear-lanceolate, more or less acute at apex. *Flowers* rich reddish-purple, magenta or pinkish. *Outer perianth segments* 6–8 mm, reddish-purple with irregular darker rings and lines, lanceolate to ovate, slightly hooded, keeled, 3-veined, the lateral often nearer vertical than horizontal. *Inner perianth segments* reddish-purple or magenta, ovate, obtuse at apex, sometimes minutely crenate, concave, 1-veined, connivent with the upper outer segments. *Labellum* 5.0–9.0 × 6–11 mm, mostly vivid reddish-purple, but sometimes magenta or pinkish, heavily marked with dark lines, loops and blotches, broadly deltate with the angles more or less

straight, undivided, with a short protruding middle lobe or shallowly 3-lobed, more or less flat. *Spur* 6–9 mm, conical, obtuse at apex, purplish. *Column* short, nearly white, forming a chamber over the spur entrance. *Anther* hood-like, purplish with darker spots; pollinia 2, pyriform, green; caudicles pale yellow, transparent; viscidia small, in a pouch. *Stigma* scutellate, glistening, edged with purple. *Rostellum* elliptical, pale purplish. *Ovary* up to 10 mm, cylindrical, twisted, erect, with 6, often purple-spotted ridges. *Capsule* up to 15 mm, oblong, with 3 prominent ridges. *Flowers* from the end of 5 to the end of 7. $2n = 80$.

The two subspecies seem to be ecologically separated. Subsp. *majaliformis*, as its name implies tends towards *D. comosa* (*D. majalis*), but although its labellum is broad it tends to be angled and thus rhomboid rather than rounded, and is often not clearly 3-lobed. Both subspecies are variable.

(a) Subsp. **purpurella**
Orchis praetermissa var. *pulchella* Druce; *Orchis purpurella* var. *pulchella* (Druce) Pugsl.; *D. purpurella* var. *pulchella* (Druce) Soó; *Orchis purpurella* var. *maculosa* T. Stephenson; *D. purpurella* var. *maculosa* (T. Stephenson) R. M. Bateman & Denholm; *Orchis purpurella* var. *crassifolia* T. Stephenson; *D. purpurella* var. *crassifolia* (T. Stephenson) Landwehr; *D. majalis* subsp. *purpurella* var. *atrata* A. J. Richards
Leaves unmarked or with small spots often only in the upper part. *Bracts* unmarked. *Labellum* 5–8 × 6–10 mm, side lobes usually not separated by sinuses.

(b) Subsp. **majaliformis** E. Nelson ex Løjtnant
Leaves usually with dull violet spots larger than in subsp. *purpurella*. *Bracts* spotted. *Labellum* 6–8 × 9–11 mm, side lobes sometimes separated from middle lobe by shallow sinuses.

Native. Bogs, marshes and wet areas where there is some calcareous water and the soil is not over-acid, in Scotland in damp meadows up to 450 m. Subsp. *purpurella* is frequent in northern Britain south to Yorkshire, Derbyshire and Pembrokeshire, is frequent in northern Ireland and scattered in southern Ireland; formerly in the New Forest. It also occurs in north-west Continental Europe. Subsp. *majaliformis* is restricted to damp coastal localities within 100 m of the sea along the north and north-west coasts of Scotland and at two sites in Cardiganshire in Wales. It also occurs in Denmark.

6. D. comosa (Scop.) P. D. Sell Western Marsh Orchid
Orchis comosa Scop.

Perennial glabrous herb; tubers deeply palmate, with 2–4 slender segments; roots numerous, thick. *Stem* (6–)10–30(–35) cm, only occasionally exceeding 5 mm in diameter, erect, slightly hollow, angular above. *Leaves* glaucous or bright green, often heavily marked with spots and blotches, sometimes ring-spotted, sometimes unmarked; basal 1, broadest more or less at the middle; sheathing leaves usually 4, rarely more, 5–12 × 1.5–2.5 cm, often more or less crowded towards the base of the stem, more or less lanceolate, obtuse to acute at apex, entire, slightly spreading and sometimes curved; non-sheathing often 2, like sheathing but smaller. *Inflorescence* 4–7 × 2.5–4.0 cm, usually dense, more or less ovoid, with up to 24 flowers. *Bracts* 10–20(–30) × (2–)4–6 mm, green, often flushed purple, linear-lanceolate, more or less acute at apex. *Flowers* usually deep reddish-purple, sometimes paler. *Outer perianth segments* 7–10 mm, reddish-purple, often marked, linear-lanceolate, obtuse at apex, the lateral often more nearly horizontal than vertical. *Inner perianth segments* reddish-purple, oblong, pointed at apex, connivent with the outer upper segment. *Labellum* (6–)7–8(–9) × (8–)9–11(–12) mm, deep reddish-purple, usually heavily marked with dark loops often broken into lines or dots, sometimes only dots on a pale background, usually broader than long and broadest at the middle, 3-lobed, the outer lobes broad and rounded, often shallowly toothed and more or less flat or reflexed, the middle lobe shorter than or exceeding the lateral lobes. *Spur* (5–)7–9 × 2.5–4.5 mm, conical, obtuse at apex, purplish. *Column* short, erect. *Anther* rose-violet; pollinia greyish-green; caudicles yellow. *Stigma* edged with purple. *Rostellum* rose or violet. *Ovary* slender, twisted, 6-ridged, often tinted purple. *Capsule* oblong, with 3 prominent ridges. *Flowers* 5–6(–7). 2n = 80.

The 3 subspecies are geographically separated.

(a) Subsp. **scotica** (E. Nelson) P. D. Sell
D. majalis subsp. *scotica* E. Nelson; *D. majalis* subsp. *occidentalis* var. *scotica* (E. Nelson) R. M. Bateman & Denholm
Plant 5–9 cm. *Sheathing leaves* 2–3, 12–16 mm wide; non-sheathing leaves usually 1. *Labellum* usually less than 7.5 × 9.5 mm; lateral lobes usually reflexed. *Spur* usually not more than 3.0 mm wide.

(b) Subsp. **cambrensis** (R. H. Roberts) R. H. Roberts
Dactylorchis majalis subsp. *cambrensis* R. H. Roberts; *D. latifolia* subsp. *cambrensis* (R. H. Roberts) Soó; *D. majalis* subsp. *cambrensis* (R. H. Roberts) R. H. Roberts; *D. majalis* subsp. *occidentalis* var. *cambrensis* (R. H. Roberts) R. M. Bateman & Denholm
Plant 15–32 cm. *Sheathing leaves* usually 4 or more, 12–20 mm wide; non-sheathing leaves usually 2. *Labellum* usually more than 7.5 × 9.5 mm; lateral lobes often more or less flat. *Spur* usually less than 3.0 mm wide.

(c) Subsp. **occidentalis** (Pugsl.) P. D. Sell
Orchis majalis var. *occidentalis* Pugsl.; *Orchis majalis* subsp. *occidentalis* (Pugsl.) Pugsl.; *Orchis occidentalis* (Pugsl.) Wilmott; *Dactylorchis occidentalis* (Pugsl.) Vermuel; *Dactylorchis majalis* subsp. *occidentalis* (Pugsl.) Hesl.-Harr. fil.; *D. latifolia* subsp. *occidentalis* (Pugsl.) Soó; *D. majalis* subsp. *occidentalis* (Pugsl.) P. D. Sell
Plant 20–50 cm. *Sheathing leaves* usually 4 or more, 12–20 mm wide; non-sheathing leaves usually 2. *Labellum* usually more than 7.5 × 9.5 mm; lateral lobes more or less flat. *Spur* not usually less than 3.0 mm wide.

(1) Forma **occidentalis** (Pugsl.) P. D. Sell
Leaves heavily marked with spots and blotches. *Flowers* dark; labellum with a pattern of heavy lines and dashes.

(2) Forma **kerryensis** (Wilmott) P. D. Sell
Orchis kerryensis Wilmott; *Orchis occidentalis* subsp. *kerryensis* (Wilmott) A. R. Clapham; *Dactylorchis kerryensis* (Wilmott) Vermuel.; *D. latifolia* subsp. *occidentalis* var. *kerryensis* (Wilmott) Soó; *D. kerryensis* (Wilmott) P. F. Hunt & Summerh.; *D. majalis* subsp. *kerryensis* (Wilmott) Senghas; *D. majalis* subsp. *occidentalis* var. *kerryensis* (Wilmott) R. M. Bateman & Denholm
Leaf markings absent. *Flowers* pale with labellum markings of dots or dots and dashes.

Native. Damp, boggy pastures, wet dune-slacks, marshes and fens. Locally frequent in western Ireland, rare in Wales and western Scotland. West and central Europe, Baltic region and northern Russia. *D. comosa* subsp. *comosa* (*D. majalis* subsp. *alpestris* (Pugsl.) Senghas) and *D. comosa* subsp. *majalis* (Rchb.) P. D. Sell (*D. majalis* subsp. *majalis*) do not occur in the British Isles. Subsp. *scotica* is local and endemic to north-west Scotland. Subsp. *cambrensis* occurs in marshes and dune-slacks in Merionethshire, Cardiganshire, Caernarvonshire and Anglesey, where it is endemic. Subsp. *occidentalis* is endemic to south-west and west Ireland where it grows in boggy pastures and dune-slacks as well as the limestone country of the Burren. The two forms of subsp. *occidentalis* sometimes grow in pure populations, but are often mixed.

× **fuchsii** = D. × **braunii** (Halácsy) Borsos & Soó
Orchis braunii Halácsy
This hybrid is robust, the leaves are usually heavily marked with transversely elongated, dark purple spots, and the flowers are in a large, pointed spike and intermediate between the parents. The pollen is highly sterile and the seed-set very low. 2n = 60.
Recorded from western Scotland and Anglesey where the parents grow together in base-rich marshes. They also occur in central Europe.

× **incarnata** = D. × **aschersoniana** (Hausskn.) Soó
Orchis × *aschersoniana* Hausskn.
This hybrid has a more slender stem than *D. incarnata*, the leaves are rather broad and yellowish-green, marked with faint spots or unspotted, the flowers vary from pale pink to reddish-purple, but are longer and more vividly coloured than in *D. incarnata* with the labellum showing various stages of intermediacy between the parents.
Recorded from Cardiganshire, Outer Hebrides and Co. Limerick. Also recorded from north and central Europe.

× **maculata** = D. × **dinglensis** (Wilmott) Soó
Orchis × *dinglensis* Wilmott; *D.* × *townsendiana* auct.
This hybrid is usually taller than *D. comosa* and has narrower, more acute, spotted leaves, the flowers have a broad, flat labellum and are pale to dark reddish-purple with broken lines and dots and the spur is intermediate

between that of the parents. It shows a high degree of fertility. $2n = 80$.

Nothosubsp. **dinglesis** is *D. comosa* subsp. *occidentalis* × *D. maculata* subsp. *ericetorum*. Nothosubsp. **robertsii** F. Horsman is *D. comosa* subsp. *cambrensis* × *D. maculata* subsp. *ericetorum*. The hybrid has been found in Cardiganshire and in Irish populations of *D. comosa*. It also occurs in north and central Europe.

× **purpurella**

These presumed hybrids are intermediate between the parents in the characters of the leaves and flowers. They are fully fertile. Variation in *D. purpurella* sometimes suggests there is more hybridisation than actually exists, only a few hybrids probably occurring.

Recorded in Anglesey, Cardiganshire and western Scotland. Endemic.

7. D. praetermissa (Druce) Soó
Southern Marsh Orchid
Orchid praetermissa Druce; *Dactylorchis praetermissa* (Druce) Vermuel.; *D. majalis* subsp. *praetermissa* (Druce) D. M. Moore & Soó

Perennial glabrous herb; tubers 2, palmate, with 3–4 short to long segments; roots long and rather thick. *Stem* 15–50(–70) cm, usually exceeding 5 mm in diameter, usually robust, hollow, but with thick walls, green. *Leaves* greyish-green or dark green, unmarked or rarely with large, elongated, ringed spots; basal one, broadest more or less at the middle; sheathing leaves 10–14 × (1.5–)2.0–3.5 cm, usually 4 or more, often crowded towards the base of the stem, oblong-lanceolate, more or less acute at the sometimes slightly hooded apex, entire, erect or more or less spreading; non-sheathing often 2, like sheathing but smaller. *Inflorescence* 5–17 × 2–4 cm, dense, cylindrical, with numerous flowers. *Bracts* 7–25 × 1–5 mm, linear-lanceolate to lanceolate, more or less acute at apex, often suffused purplish. *Flowers* dark pinkish-purple or magenta, sometimes paler. *Outer perianth segments* 8–19 mm, pinkish-purple or magenta, narrowly lanceolate, obtuse to acute at apex, 1- to 3-veined, the lateral spreading horizontally or suberect. *Inner perianth segments* shorter than outer, pinkish-purple or magenta, ovate-lanceolate, obtuse at apex, concave, connivent with the upper outer segment. *Labellum* 9–12 × 9–14 mm, pale to deep pinkish-purple or magenta usually marked with dots and fine lines, but sometimes heavier, broadly transversely elliptical, scarcely or shallowly 3-lobed, the lateral lobes rounded, flat but turned up at the edges, the middle lobe usually short with a rounded tip. *Spur* 6–9 mm, slightly tapering, obtuse at apex. *Column* short, erect. *Anther* purplish; pollinia greyish-green; caudicles yellow. *Stigma* edged with purple. *Rostellum* purplish. *Ovary* spindle-shaped, twisted, 6-ridged. *Capsule* oblong, with 3 prominent ridges. *Flowers* 6–7. $2n = 80$.

The following two forms can be found growing together. They seem to differ only in colouring.

(1) Forma **praetermissa**
Leaves without markings. *Labellum* marked with fine lines and dots.

(2) Forma **junialis** (Vermuel.) P. D. Sell
Orchis latifolia var. *junialis* Vermuel.; *Orchis pardalina* Pugsl.; *Dactylorchis praetermissa* var. *junialis* (Vermuel.) Vermuel.; *D. praetermissa* subsp. *junialis* (Vermuel.) Soó; *D. majalis* subsp. *pardalina* (Pugsl.) E. Nelson; *D. majalis* subsp. *praetermissa* var. *junialis* (Vermuel.) Senghas
Leaves with large ring markings. *Labellum* marked with heavy loops and dashes.

Native. Slightly acid to calcareous damp places in fens, marshes, bogs, meadows, gravel pits and waste tips. Frequent in Britain north to Northumberland and Lancashire; Channel Islands. Also on the north-west European mainland. Forma *junialis* is by far the less common of the two forms.

× **purpurella** = D. insignis (T. & T. A. Stephenson) Soó
Orchis insignis T. & T. A. Stephenson; *Orchis salteri* T. Stephenson
These putitive hybrids are intermediate between the parents in habit and vegetative characters, but the flower is nearer to *D. purpurella*. $2n = 80$.

Has been recorded only from Cardiganshire and Merionethshire. Endemic.

× **traunsteineri**
Plants with narrower leaves and less dense inflorescences, and intermediate labellum have been considered to be this hybrid. They are fully fertile and appear to backcross and form a complete range of intermediates linking the species. $2n = 80$.

Recorded from base-rich fens in Norfolk and Suffolk. Endemic.

18 × 16. × **Dactylodenia** Garay & H. R. Sweet
Dactylorhiza × Gymnadenia
× *Dactylogymnadenia* Soó
Hybrids between these two genera often resemble *Dactylorhiza* in general appearance, but have scented flowers with a longer spur. Parentage is sometimes difficult to determine when there is more than one species of *Dactylorhiza* nearby.

18/1 × 16/1. × D. st-quintinii (Godfery) J. Duvign.
D. fuchsii × G. conopsea
?*Orchis heinzeliana* Reichardt; ?*Orchigymnadenia heinzeliana* (Reichardt) A. Camus; *Orchis hibernica* Druce nom. nud.; ?× *D. heinzeliana* (Reichardt) Garay & H. R. Sweet; × *Orchigymnadenia st-quintinii* Godfery
In this hybrid the characteristic short, rounded basal leaf of *D. fuchsii* is usually absent, the leaf spots are usually present, and the flowers usually resemble *D. fuchsii*, but have the long, slender spur and characteristic sweet scent of *G. conopsea*.

Occurs in scattered localities throughout most of the British Isles. There is a complicated array of nothotaxa, the only one of which appears to have a name is *D. fuchsii* subsp. *hebridensis* × *G. conopsea* ?subsp. *conopsea* as *Orchigymnadenia cookei* Hesl.-Harr. (× *Dactylogymnadenia cookei* (Hesl.-Harr.) Soó).

18/3 × 16/1. × D. vollmannii (M. Schulze) Peitz
   D. incarnata × G. conopsea
Orchigymnadenia vollmannii M. Schulze-Menz
This hybrid has twice been recorded for Cornwall, once doubtfully. It has also occurred in Sweden.

18/2 × 16/1. × D. legrandiana (A. Camus) Peitz
   D. maculata × G. conopsea
Orchis evansii Druce; *Orchigymnadenia evansii* (Druce) T. & T. A. Stephenson; *Habenaria evansii* (Druce) Druce; *Orchigymnadenia legrandiana* A. Camus
This hybrid has the leaves variable in shape but usually spotted, the labellum resembles *D. maculata* being pale lilac with fine spots and occasional loops, but the spur is long and slender and the flowers fragrant as in *G. conopsea*.
   Recorded in scattered localities through much of the British Isles. Also recorded from Continental Europe.

18/4 × 16/1. × D. wintonii (Druce) Peitz
   D. praetermissa × G. conopsea
Habenaria wintonii Druce; *Orchigymnadenia wintonii* (Druce) Tahourdin; *Habenaria quirkiana* Druce nom. illegit.; *Dactylogymnadenia wintonii* (Druce) Soó
This hybrid is robust, with the leaves variable and usually faintly spotted, and the flowers intermediate between the parents but with the faint scent of *G. conopsea*.
   The two species are usually separated ecologically, but there are records of this hybrid from Devon and Hampshire.

18/5 × 16/1. × D. varia (T. & T. A. Stephenson) Aver
   D. purpurella × G. conopsea
Orchigymnadenia varia T. & T. A. Stephenson; *Habenaria varia* (T. & T. A. Stephenson) Druce; *Dactylogymnadenia varia* (T. & T. A. Stephenson) Soó
The general appearance of this hybrid is of a fairly robust *D. purpurella* with finely spotted leaves, with a long slender spike of pale flowers and the long slender spur of *G. conopsea*.
   Recorded for Cumberland, various parts of Scotland, especially the western islands, and Co. Down in Ireland. Endemic.

19. Neotinea Rchb. fil.

Perennial herbs with 2 small root tubers. *Leaves* 2–3(–4), near the base of the stem, usually spotted. *Inflorescence* a short, dense, one-sided spike. *Flowers* whitish or pink. *Perianth segments* all connivent. *Labellum* directed forwards, 3-lobed. *Spur* very short. *Column* very short and small. *Anther* adnate to the top of the column; pollinia 2, clavate, each narrowed downwards into a short caudicle attached loosely to one of the naked viscidia which are enclosed in a bursicle; pollen in tetrads bound by elastic threads. *Stigmas* 2, large, more or less crescentic, joined below, borne on lateral wings of the column. *Rostellum* a broad, flat plate between the stigma lobes, its apex curving over the viscidia. *Capsule* spindle-shaped.
   One species in south and west Europe; Macaronesia; North Africa; Cyprus; Turkey.

Stearn, W. T. (1974). Multum pro parvo: the nomenclatural history and synonymy of *Neotinea maculata* (Orchidaceae). *Ann. Mus. Goulandris (Kifissia)* **2**: 69–81.

1. N. maculata (Desf.) Stearn   Dense-flowered Orchid
Satyrium maculatum Desf.; *Orchis intacta* Link; *N. intacta* (Link) Rchb. fil.; *Habenaria intacta* (Link) Lindl. ex Benth.; *Aceras densiflorum* (Brot.) Boiss.

Perennial glabrous herb; tubers 2, about 20 × 12 mm, ovoid; roots few, short. *Stem* 8–30(–40) cm, erect, often flexuous, cylindrical, with brownish, acute, membranous basal sheaths. *Leaves* 3–6, the basal 3–12 × 1–3 cm, yellowish-green, with small, purplish-brown spots in longitudinal lines, or unspotted, elliptical-oblong, obtuse-mucronate at apex, entire, with 3 principal veins on each side of the midrib, the cauline narrower, erect and clasping the stem, acute at apex, the uppermost bract-like and adpressed. *Inflorescence* a spike 2–6 × 0.8–1.2 cm, narrow, cylindrical, dense, often one-sided. *Bracts* 3–6 × 1–2 mm, whitish or reddish towards the tip, ovate-lanceolate or lanceolate, acute at apex, sometimes with a tooth at the side, 1- to 3-veined. *Flowers* 15–20(–30), very small, white or rarely pink, never opening fully, said to smell of vanilla. *Outer perianth segments* 3–4 mm, connivent, forming a nearly closed helmet, coherent at base, free at tip, greenish-white or pale rose, lanceolate to ovate-lanceolate, acute at apex. *Inner perianth segments* greenish, very narrow, linear, acute at apex, 1-veined. *Labellum* very small, equal to or exceeding the outer segments, 3-lobed, white or rose, with 2 or 3 pale rose or violet markings at the deeply channelled base, side lobes very narrow, linear, acute at apex, usually shorter than the middle lobe which is linear or tongue-shaped and notched at the apex. *Spur* very short, conical, obtuse at apex, flattened laterally. *Column* very short, notched at apex. *Anther* adnate to the top of the column; pollinia 2, very small, pale green, clavate; caudicle very short; viscidia globular, enclosed in a 2-lobed bursicle. *Stigmas* 2, large, more or less crescentic. *Rostellum* a broad, flat plate between the stigma lobes. *Ovary* short, pale green, spindle-shaped, twisted, with 3 slightly raised ridges. *Capsule* spindle-shaped, tapering at both ends, 3-ridged. *Flowers* 4–6. Cross- or self-pollinated. $2n = 40$.
   Native. Mainly on rocky or grassy hillsides, in pastures or on grassy banks and roadsides in the limestone country near the coast, and on calcareous sand-dunes and rarely peat soil over acid rocks. West of Ireland from Co. Clare to Co. Mayo, including the Burren Hills, inland to Co. Offaly and a single locality on the north coast of the Isle of Man. Mediterranean region and Portugal eastwards to Greece; Madeira; Canary Islands; North Africa; Cyprus; Turkey.

20. Orchis L.

Strateuma Salisb. nom. nud.

Perennial herbs with 2–3, rounded, unforked root tubers. *Basal leaves* in a rosette; sheathing leaves present. *Flowers* in a spike which is usually enclosed by thin, spathe-like leaves during emergence. *Bracts* membranous. *Perianth segments*, except the labellum, connivent into a helmet over the column, or the outer lateral segments spreading or turned upwards. *Labellum* usually 3-lobed, usually directed downwards. *Spur* descending or ascending. *Column* erect. *Anther* adnate to the top of the column; pollinia 2, each narrowed below to a caudicle and attached by a basal disc to one of the viscidia, both of which are enclosed in a bursicle; pollen in packets of tetrads united by elastic threads. *Stigma* more or less 2-lobed, roofing the entrance to the spur. *Rostellum* 3-lobed, the middle lobe short. *Capsule* erect.

About 35 species found in Europe, temperate Asia, North Africa and the Canary Islands.

Farrell, L. (1985). *Orchis militaris* L. in Biological flora of the British Isles. *Jour. Ecol.* **73**: 1041–1053.

Foley, M. J. Y. (1992). The current distribution and abundance of *Orchis ustulata* L. (Orchidaceae) in the British Isles – an updated summary. *Watsonia* **19**: 121–126.

Rose, F. (1948). *Orchis purpurea* Huds. in Biological flora of the British Isles. *Jour. Ecol.* **36**: 366–377.

Stewart, A., Pearman, D. A. & Preston, C. D. (1994). *Scarce plants in Britain*. Peterborough. [*O. morio, O. purpurea* and *O. ustulata*.]

1. Outer lateral perianth segments erect or spreading   2.
1. All perianth segments except the labellum connivent to form a helmet over the column   3.
2. Leaves never spotted; bracts 3-veined or the lowest sometimes 5-veined; labellum with the central lobe shorter than the laterals or absent; spur shorter than ovary   **1. laxiflora**
2. Leaves usually dark-spotted; bracts 1-veined or the lower sometimes 3-veined; labellum with the central lobe exceeding the laterals; spur at least as long as ovary   **2. mascula**
3. Outer perianth segments green-veined, not coherent; labellum shallowly 3-lobed, not shaped like a man   **3. morio**
3. Outer perianth segments not green-veined, more or less coherent; labellum shaped somewhat like a man, the central lobe much longer than the more or less slender lateral lobes (arms) and forking distally into 2 branches (legs), often with a short tooth between them (penis)   4.
4. Outer perianth segments free to the base; helmet dark purplish-or brownish-red; labellum 4–8 mm, white, sometimes tinged rose; spur about one-quarter the length of the ovary   **4. ustulata**
4. Outer perianth segments coherent below; helmet not dark purplish- or reddish-brown; labellum at least 10 mm; spur about half the length of the ovary   5.
5. Helmet heavily blotched with dark reddish-purple; labellum usually suffused with violet or pale rose and with darker spots, terminal lobelets wider than long   **5. purpurea**
5. Helmet whitish, greyish or pale violet, flushed, veined and spotted with violet or purple; labellum with terminal lobelets longer than wide.   6.
6. Two main lobelets of terminal lobe of labellum linear, about as wide as lateral lobes   **6. simia**
6 Two main lobelets of terminal lobe of labellum oblong, more than twice as wide as lateral lobes   7.

7. Lowest leaves (1.5–) 2.5–5.0 cm wide; spike with 11–42 flowers   **7(i). militaris** var. **militaris**
7. Lowest leaves 1.2–3.0 cm wide; spike with 2–26 flowers   **7(ii). militaris** var. **tenuifrons**

Section **1. Morio** Dumort.

Plant not hay-scented on drying. *Bracts* about equalling the ovary, membranous. *Outer perianth segments* not coherent. *Labellum* 3-lobed, not shaped like a man, the middle lobe truncate or emarginate, or labellum 2-lobed.

1. O. laxiflora Lam.   Loose-flowered Orchid
O. ensifolia Vill.

Perennial glabrous herb; tubers 10–30 × 10–25 mm, brownish, oblong to nearly globose, sessile to shortly stalked; roots pale brown, rather short. *Stem* 20–80 cm, green, purplish above, erect, sometimes slightly flexuous, terete, angled, rough, with rather loose, brownish basal sheaths. *Leaves* 12–17 × 1–2 cm, shining green, never spotted, spreading, linear or lanceolate, acute at apex, entire, keeled and more or less folded, with prominent parallel veins, the upper small and bract-like. *Inflorescence* a spike 6–20 × 4–6 cm, ovoid or cylindrical, lax, with 7–16 flowers. *Bracts* 10–20 × 2–4 mm, shorter than to longer than the ovary, linear-lanceolate, acute at apex, somewhat membranous, tinged with purple, 3- to 5-veined. *Flowers* large, usually dark purple, rarely rose or white, with the centre of the labellum white, scentless. *Outer perianth segments* 7–10 mm, free, the lateral dark purple, ovate, obtuse at apex, 3- to 6-veined, erect and back to back, the upper elliptical, obtuse at apex, concave, 5-veined and curving upwards. *Inner perianth segments* about two-thirds as long as outer, obliquely oblong, rather obtuse at apex, 3- to 6-veined, forming a helmet. *Labellum* 6–10 × 8–10 mm, dark purple or rose with a keeled white centre and a channel leading into the spur, rarely white, the strongly reflexed sides almost touching behind, when flattened out transversely elliptical, broader than long, with rounded, more or less crenate or dentate lateral lobes separated by a broad, truncate, sinuous middle lobe much shorter than the lateral and sometimes absent, making the labellum 2-lobed. *Spur* 10–15 mm, dark purple, cylindrical, but expanding into a chamber at the mouth, flattened, obtuse and squarish or notched at the apex. *Column* about 4 mm, white. *Anther* pyriform, obtuse at apex, violet, with a white fold at the base; pollinia 2, greenish; caudicles hyaline, whitish, flattened; viscidia oval, transparent, colourless, enclosed in a violet bursicle. *Stigma* 2-lobed, on roof of spur, violet-edged. *Rostellum* 3-lobed. *Ovary* 10–16 mm, tinged purple, long-linear, curved, twisted, sessile. *Capsule* spindle-shaped. *Seeds* oblong, rounded at apex; testa transparent. Flowers 5–6. $2n = 36$.

The Channel Island plants are referable to subsp. **laxiflora**, which is recognised by the spur widening at the apex and the short, broad middle lobe of the labellum.

Native. Wet meadows and bogs, where it may be open to threat from modern farming practices. Locally common in Guernsey and Jersey. The species occurs in west, central and south Europe, extending northwards to Gotland and eastwards to south Russia; west Asia; North Africa. Subsp. *laxiflora* occurs in west and south Europe.

× **morio** = O. × **alata** Fleury

Usually nearer *O. morio*, but with longer, more acute leaves, and larger flowers with the outer segments not connivent. Sometimes like a dwarf *O. laxiflora* with deeper coloured flowers.

Occurs sporadically with the parents in Jersey, and occurred in Guernsey in 1949.

2. O. mascula (L.) L.   Early Purple Orchid

O. morio var. *mascula* L.; *O. compressiflora* Stokes nom. illegit.

Perennial herb; tubers 2, 15–35 × 10–20 mm, subglobose or ovoid (testicle-like, hence its Latin name); roots few, rather slender. *Stem* 15–60 cm, erect, stout, cylindrical, pale green, often purplish above, angled above, sometimes hollow at the base, with 3–5 leaves in the lower half and 2–3 sheaths above. *Leaves* 5–20 × 0.5–3.5 cm, bright or greyish-green, very glossy beneath, less so above, unspotted or with large, dark, irregular, blackish-purple blotches or spots, oblong-lanceolate, acute or obtuse at apex, entire, keeled and often folded, with about 3 transparent veins on each side of the midrib and fainter veins in between, the surface sometimes wavy giving a crimped appearance, the lower close together and spreading, the upper erect and loosely clasping the stem, the uppermost thin, membranous, acute at apex and often purplish. *Inflorescence* a spike 4–15 × 3–5 cm, ovoid or cylindrical, rather lax, especially below, with 10–45 flowers. *Bracts* 12–20 × 1.0–1.5 mm, usually about as long as the ovary, narrowly purple-tinged, when flowers are white transparent with a green vein, narrowly linear-lanceolate, long-acute at apex, membranous, 1-veined or the lower 3-veined. *Flowers* rather large, reddish-violet, magenta, lilac, rose, pale pink or very rarely white, with yellowish, whitish or greenish throat, scentless or with an unpleasant smell of cats. *Outer perianth segments* 6–8 mm, the 2 lateral obliquely ovate-lanceolate, obtuse or acute at apex, 1-veined, shorter and paler than outer segments. *Inner perianth segments* more or less ovate, obtuse or subacute at apex, 1-veined, shorter and paler than outer segments. *Labellum* 8–15 × 8–15 mm, deep reddish-violet to pale rose or magenta, yellowish or white in the centre, velvety with erect papillae at the base, with a few, interrupted lines or spots formed of dense tufts of bright reddish-violet hairs, pointing downwards and forwards, wedge-shaped at base, convex, the reflex sides making it look longer than wide, 3-lobed, the side-lobes more or less rhomboidal, more or less dentate or crenate towards the tip, the middle lobe slightly longer, with 2 entire or crenate lobelets at the apex. *Spur* 10–15 mm, at least as long as ovary, cylindrical, slightly curved upwards, enlarged and clavate at apex. *Column* short, with a little point at the apex. *Anthers* ovate, purplish or greenish-grey; pollinia 2, dark green, or yellow when the flowers

are white; caudicles yellow and transparent, enclosed in a rose-violet bursicle. *Stigmas* 2, confluent on the roof and sides of the chamber, edged with a purplish line. *Rostellum* 3-lobed. *Capsule* erect. *Seeds* rounded at apex; testa transparent. *Flowers* 4–6. Visited by bumble-bees. $2n = 42$.

All our plants are referable to subsp. **mascula**. They show considerable variation in size, flower colour, spotting and width of leaves and shape of perianth segments, but do not seem to form distinct populations.

Native. Moist meadows and pastures, open woods, copses, thickets or on rock-ledges, usually on calcareous or neutral soils. In the Burren it grows in crevices in almost bare limestone. Frequent to common throughout the British Isles up to 900 m. Mainly west and west-central Europe, reaching 70° N in Norway; North Africa; north and west Asia. Subsp. *mascula* is mainly in west and west-central Europe.

× **morio** = O. × **morioides** Brand

This hybrid has the appearance of *O. mascula*, but has green-veined outer perianth segments and a broad truncate lobe to the labellum.

There are scattered records from England and Wales. Scattered in Continental Europe.

3. O. morio L.   Green-winged Orchid

Perennial herb; tubers 10–15 × 5–15 mm, subglobose or ovoid; roots few, short. *Stem* 10–30(–40) cm, yellowish-green tinged with violet above, erect, terete, more or less hollow, angled, with 2–3 whitish, membranous sheaths at base. *Leaves* 10–80 × 5–12 mm, green or bluish-green, elliptic-oblong, lanceolate or oblong-lanceolate, acute or obtuse at apex, entire, keeled, broadly sheathed at base, the lower spreading or recurved, the upper 2–3 erect, rather broader and loosely clasping the stem, the uppermost bract-like, membranous, acute at apex and often purplish; veins 1–2 on each side of the midrib with fainter ones in between and numerous cross-veins. *Inflorescence* a spike 1.5–7.0 cm, ovoid or cylindrical, lax, with 4–14 flowers. *Bracts* 8–12 × 1.2–3.0 mm, membranous, glossy, more or less tinged with purple, lanceolate, more or less acute at apex, the lower 3- to 5-veined, the upper shorter, 1-veined, all sheathing the ovary, which they equal or slightly exceed. *Flowers* rather large, deep purple, reddish-purple, lilac, rarely salmon pink, more rarely white, the outer perianth segments with conspicuous green veins, the middle of the labellum paler and spotted, very faintly scented. *Outer perianth segments* 6–10 mm, forming a short, almost globular helmet, the lateral ovate-oblong, concave, keeled, rounded at the apex and with 3–7, conspicuous green or bronze-purple veins, the upper narrower, oblong and 3- to 5-veined. *Inner perianth segments* paler, shorter and narrower, linear-oblong, rounded and concave at the tip, 1- to 3-veined, forming an interior hood beneath the helmet. *Labellum* up to 10 × 10 mm, transversely oblong, broader than long, the side lobes broadly rounded and folded back, the middle lobe short, broad, truncate, sometimes notched, usually shorter than but sometimes equalling or slightly exceeding the side lobes, the edges toothed or slightly crenate and the surface densely covered with very short minute papillae. *Spur* 8–10 mm, shorter than the ovary, cylindrical, thick, horizontal or slightly curved upwards, flattened, clavate, truncate or often notched at the tip, densely covered inside with minute papillae. *Column* short, purplish. *Anthers* purple; pollinia 2, greenish or, when flowers are white, yellow; viscidia disc-like, enclosed in a whitish, tinted purple bursicle. *Stigma* on the roof of the throat of the flower, bordered with a purple line, depressed and greenish in the middle. *Rostellum* 3-lobed. *Capsule* erect. *Seeds* rather narrow, hyaline. *Flowers* 5–6. Pollinated by bees, especially bumble-bees. $2n = 36$.

A very variable species. Sometimes all the plants in a population have the same flower colour, while other populations may contain all the known flower colours. Var. *bartlettii* Hesl.-Harr. is a small plant with small flowers, while var. *churchillii* Druce is a tall plant at the other end of the size range. All the plants of Britain and Ireland are referable to subsp. **morio**, which occurs throughout the range of the species.

Native. Moist meadows and pastures, field borders, open woods and grassy slopes on base-rich to neutral soils. Formerly frequent over much of England and Wales, Ireland and the Channel Islands up to 300 m, but its number of localities has been much reduced and it is now very local. Europe southwards from south Norway, Gotland, Estonia and central Russia; Caucasus; Turkey; Siberia; everywhere decreasing.

Section 2. Orchis

Plants not hay-scented when drying. *Bracts* of spike usually short, membranous. *Perianth segments* connivent to form a helmet over the column, the outer more or less coherent below. *Labellum* shaped like a man, its middle lobe longer than the lateral lobes (arms) and forked distally (legs), often with a short tooth in the sinus (penis).

4. O. ustulata L.   Burnt Orchid

Perennial glabrous herb; tubers 8–20 × 7–15 mm, sessile, globose or ovoid. *Stem* 8–20(–30) cm, pale yellowish-green, slender, cylindrical, solid, angled above, somewhat striate, channelled, with 2–3, white, membranous sheaths at the base. *Leaves* 2.5–10.0 × 0.5–1.0 cm, rather bluish-green, unspotted, oblong, elliptic-oblong or lanceolate, acute at apex, entire, erect or spreading, keeled, folded, with numerous parallel veins, the upper more or less loosely embracing the stem, the uppermost bract-like, membranous, 3-veined. *Inflorescence* a spike 2–5 × 1.2–2.0 cm, ovoid at first, gradually lengthening, dense, many-flowered. *Bracts* up to 10 mm, purplish-red, lanceolate, acute or obtuse at apex, membranous, the lower keeled, with 1–3 greenish or reddish veins. *Flowers* very small, honey-scented, with a nearly globular, dark brownish-red helmet, later becoming much paler, and a pure white, crimson-spotted labellum. *Outer perianth segments* 3.0–3.5 mm, free to the base, dark purplish- or brownish-red, greenish within, ovate,

obtuse at apex, keeled, 3-veined. *Inner perianth segments* paler, shorter and narrower than outer, linear-spathulate, obtuse at apex, which is sometimes notched or slightly toothed. *Labellum* 4–8 mm, longer than wide, directed forwards and downwards, 3-lobed, somewhat concave, with a groove at the base leading into the spur, white rarely tinged rose, with a few, bright crimson spots, the side-lobes divergent, rather broad, oblong, rounded or squarish at apex and often crenate, the middle lobe longer and often broader, widening downwards, and ending in 2 short, more or less divergent, crenulate lobules, sometimes with a tooth between them. *Spur* about 2 mm, conical, rounded at apex, compressed from back to front, directed downwards and forward. *Column* very short, whitish. *Anther* ovate, pale yellow; pollinia 2, pale lemon yellow, very short; caudicles short, brighter yellow; viscidia 2, in a bursicle. *Stigma* partly concealed by rostellum. *Rostellum* 3-lobed. *Ovary* sessile, cylindrical, twisted, green, with 6 scarcely prominent, longitudinal ridges. *Capsule* about 10 mm, cylindrical, with 3 obtuse ridges. *Seeds* oblong, slightly narrowed above. *Flowers* 5–6. $2n = 42$.

Native. Short grassland over chalk and limestone. Formerly frequent over much of England, now greatly reduced and very local. Central Europe from Denmark, Gotland and Estonia southwards to northern Spain, southern Italy, north Balkan Peninsula and Ukraine; Caucasus; Siberia.

5. O. purpurea Huds.   Lady Orchid

O. fusca Jacq.; *Strateuma grandis* Salisb. nom. illegit.

Perennial glabrous herb; tubers ellipsoid or subglobose; roots thick, short, numerous. *Stem* 20–50(–100) cm, pale green below, often dark dull purple above, sometimes with green lines, terete, solid, angular, channelled. *Leaves* 3–5(–7), 7–15 × 3–5 cm, bright or greyish-green, shining on upper surface, paler, glossy and greyer green beneath, elliptic-oblong or oblong-lanceolate, the lower obtuse at apex, the upper acute at apex and clasping at the base. *Inflorescence* a spike 5–10(–15) × 3–4 cm, ovoid to oblong, lax or dense. *Bracts* 4–7 mm, purplish, narrow, ovate or triangular, thin, membranous, usually 1-veined. *Flowers* many, large, not scented, with a dark red helmet and pale labellum, rarely white. *Outer perianth segments* 12–14 mm, heavily blotched with dark reddish-purple outside, green or whitish mottled with purple inside, ovate, obtuse or acute at apex, 3-veined, coherent towards the base, but easily separated, the tips slightly spreading. *Inner perianth segments* whitish or pale violet with violet spots, narrow, linear, acute at apex. *Labellum* (8–)10–15 mm, pendulous, white above, the edges more or less suffused with violet or pale rose, plentifully spotted with tiny tufts of rather long violet papillae, and white beneath sometimes edged violet or rose, broad and flat, 3-lobed, the side-lobes narrow, linear, curved, rounded, pointed, truncate or spathulate at tip and spotted with violet hair tufts, the middle lobe broad, widening gradually from the base downwards, ending in 2 short, divergent, rhomboidal or rounded, often truncate, irregularly toothed or crenate lobules,

with a short tooth between them; the base of the labellum curves up on each side to form a heart-shaped
chamber, on the roof of which is the stigma, bordered by
a purple line. *Spur* shortly cylindrical, curved forwards,
compressed from back to front, truncate, notched or
slightly bilobed. *Column* very short, white or rose, nearly
as broad as long. *Anther* purple-eyed, ovoid; pollinia 2,
pale green; caudicles pale yellow, flat, ribbon-like; viscidia oval, hyaline, enclosed in a white bursicle. *Stigma*
cordate, glistening. *Rostellum* 3-lobed. *Ovary* 10–18
mm, green or tinged with purple, sessile, twisted, linear,
often curved, with 6, sometimes purple-spotted ridges.
Seeds transparent. *Flowers* 5. Cross-pollinated by
insects. $2n = 40, 42$.

This species has much variation in flower colour. The
commonest variant has a dark reddish-violet or brownish-purple helmet and pale-coloured labellum.
Sometimes the helmet is green flecked with purple;
sometimes the helmet is dull rose outside and green
within with pale rose markings, the labellum white with
very pale spots; sometimes the helmet is greenish-white
with pink veins, the labellum pure white with extremely
faint spots; sometimes the helmet is dark reddish-
purple, the lip broadly mottled at the edge with bright
purple. Plants in the open are often darker-flowered
than those in the shade. Sometimes the whole flower is
white, var. *albiflora* A. Camus. The shape of the flower
parts are also variable. Var. *pseudomilitaris* Druce has
smaller flowers with the lobes of the labellum narrower
and is said to resemble *O. militaris*.

Native. Woods and scrub, rarely open grassland, on
chalk. Locally frequent on the North Downs in Kent
and one site in Oxfordshire. Formerly in a few other
places in south-east England. Europe, from Denmark
southwards to north Spain, north Greece and Crimea;
Caucasus; Turkey.

6. O. simia Lam.   Monkey Orchid

O. tephrosanthos Vill.; *O. macra* Lindl.; *O. militaris*
subsp. *simia* (Lam.) Hook. fil.

Perennial glabrous herb; tubers 10–25 × 10–20 mm,
ovoid or subglobose; roots few, short, rather thick. *Stem*
15–30(–45) cm, greyish-green, often purplish above,
terete, solid, angled, with 2–3 close-fitting, acute sheaths
at base. *Leaves* 3–5, 4–10 × 1.5–2.5 cm, greyish-green,
unspotted, glossy, oblong-ovate or oblong-lanceolate,
more or less obtuse and mucronate or acute at apex,
folded, keeled, slightly hooded, with parallel veins
depressed beneath, the uppermost often membranous
and sometimes bractlike. *Inflorescence* a spike 2–5 × 3–4
cm, ovoid or broadly cylindrical, fairly dense and untidy.
Bracts 4–6 × 1–2 mm, greenish, rose or whitish, membranous and semitransparent, tapering to a long-acute
point. *Flowers* of medium size, faintly scented, helmet
white or violet, with variable labellum, rarely pure white.
Outer perianth segments about 10 mm, forming a rather
long, open helmet, white or pale violet with minute
violet dots and 2–3 raised, white veins, lanceolate, acute
at apex, partly adherent but with the tips free. *Inner perianth segments* shorter, white, edged violet, adherent to

outer segments for half their length, linear, acute at
apex, minutely denticulate. *Labellum* 14–16 mm, longer
than wide, white, rose or deep crimson, paler below, with
2 long, slender, upturned, linear, pale violet or purple
arm-like segments on each side, the basal pair with a
short, tail-like, violet appendix between them, the
central area dotted with small tufts of violet and white
or red papillae. *Spur* pinkish-white, half as long as the
ovary, pointing downwards, narrow at the neck, flattened from back to front, wide at the truncate, sometimes notched apex. *Column* short, white at the back,
with 2 dark purple, eye-like spots at apex, forming a
chamber over the mouth of the spur. *Anther* oblong,
truncate at apex, sometimes with a short beak; pollinia
2, each narrowed into a caudicle; viscidia 2, in a bursicle.
Stigma on the back of the chamber above the spur, bordered by a purple line. *Rostellum* projecting downwards
making the stigma appear cordate. *Ovary* about 10 mm,
cylindrical, twisted, curved, with 6 violet-tinged ridges.
Flowers 5–6. Probably cross-pollinated by bees, flies and
butterflies. $2n = 42$.

Variable in flower colour. Three variants have been
described from Britain. Some plants from Kent have a
whitish helmet and a very narrow deep crimson central
lobe of the labellum. Some Oxford plants differ in their
broader, white, red-dotted central labellum lobe. The
third more frequent variant, has a violet helmet with a
pink, crimson-dotted centre, the central labellum lobe
with purple arms (var. *macra* (Lindley) Godfery).

Native. Chalk grassland, open scrub and field
borders. Very rare, two sites in Kent and two in
Oxfordshire. Formerly very scattered elsewhere in
south-east England and one site in Yorkshire. Central
and south Europe from Belgium to Spain, Italy, northern part of Balkan Peninsula and Crimea; Caucasus;
North Africa; western Asia.

7. O. militaris L.   Military Orchid

O. galeata Poir.; *O. rivinii* Gouan; *O. tephrosanthes*
auct.; *Strateuma militaris* (L.) Salisb. nom. invalid.

Perennial herb; tubers 2, 10–25 × 7–20 mm, brownish,
ovoid, ellipsoid or subglobose; roots rather few, brownish, thin. *Stem* (10–)20–60 cm, pale green, often tinged
with violet in the upper part, erect, terete or slightly
angled, without leaves or sheaths in the upper part.
Leaves 2–5 near the base and 2–3 sheaths higher up the
stem; basal 3–19 × 1.2-5.0 cm, pale, glossy yellowish-
green on upper surface, paler and slightly glaucous
beneath, rather thick, oblong-lanceolate, oblong-
oblanceolate or oblong-ovate, obtuse to acute at apex,
entire, flat, long-attenuate at base; sheathing leaves membranous, acute at apex. *Inflorescence* a spike 3.5–14.0 ×
3.5–8.0 cm, ovoid or conical, later cylindrical, dense to
rather lax, with 2–42 flowers. *Bracts* 4–6, 1.5–7.0(–11) ×
1.5–3.0 mm, usually much shorter than the ovary, membranous, often tinged rose or violet, triangular-lanceolate or triangular-ovate, acute or obtuse at apex, entire.
Flowers rather large, smelling of coumarin, rose or pale
reddish-violet. *Outer perianth segments* 10–15 × 5–7
mm, whitish- or greyish-pink outside, whitish with longi-

tudinal purplish lines inside, connivent into a helmet, ovate or ovate-lanceolate, more or less acute at apex. *Inner perianth segments* 10–12 × 2.0–2.5 mm, coloured as outer, linear-lanceolate, acute at apex. *Labellum* 12–18 × 7–14 mm, 3-lobed, the lateral lobes 5–8 × 1.5–2.0 mm, linear, curved, obtuse, pale to dark purple and with an elongated, central, whitish area with spots, the middle lobe 16–17 mm, variable, but narrow and usually longer than the lateral lobes, nearly white in the centre with 2 rows of tufts of purple papillae which come together at the base and 2-lobed at apex, the lobules purple, ovate, oblong or linear, obtuse, entire or minutely toothed and with a short, narrow, acute tooth between the lobules. *Spur* 5.0–8.5 × 1.5–2.5 mm, narrowly cylindrical, directed downwards, about half as long as the ovary. *Column* obtuse. *Anther* violet-purple, ovoid; pollinia 2, dark bluish-green; viscidia yellowish-white, enclosed in a bursicle. *Stigma* cordate. *Rostellum* 3-lobed. *Ovary* green, often violet-tinged, slender, linear, much twisted, sessile. *Capsule* 15–20 mm, erect, oblong with 6, prominent ridges. *Seeds* oblong, rounded at apex, reticulate. *Flowers* 5–6. Cross-pollinated by insects. $2n = 42$.

The two varieties are geographically isolated.

(i) Var. **militaris**
Stem 20–60 cm. *Lowest leaves* 6.5–19.0 × (1.5–)2.5–5.0 cm. *Spike* 5–14 × 3.5–8.0 cm, with 11–42 flowers.

(ii) Var. **tenuifrons** P. D. Sell
Stem 10–35 cm. *Lowest leaves* 3–15 × 1.2–3.0 cm. *Spike* 3–7 × 3–4 cm, with 2–26 flowers.

Native. Var. *militaris* occurs in an old chalk pit with invading trees and shrubs, near Mildenhall in Suffolk. It also occurs in central Europe from southern Sweden, Gotland and Estonia southwards to central Spain, central Italy, Bulgaria and southern Russia; central Asia; Turkey. Var. *tenuifrons* is endemic to one site in Buckinghamshire and sporadically at two sites in Oxfordshire. It was formerly more widespread in the Mid Thames valley.

× **simia = O.** × **beyrichii** A. Kern.
O. grenieri A. Camus
This hybrid is sometimes nearer one parent and sometimes the other. When near *O. simia* it is more robust with a broader undivided part of the labellum and the lobes are often curled upwards and more spathulate. When near *O. militaris* it has longer, narrower terminal loblets to the labellum and longer side lobes.

Occurred up to the mid nineteenth century in the Mid Thames valley where the parents used to grow together.

(**Serapias lingua** L. was found in a meadow with native *Orchis laxiflora* at Vagon in Guernsey in 1992. Its status is uncertain. Two plants of **S. parviflora** Parl. were reported from east Cornwall in 1989, but were probably planted.)

21. Aceras R. Br.

Perennial herbs with entire root tubers, hay-scented on drying. *Stems* leafy. *Spikes* long and narrow. *Flowers* greenish-yellow, often tinged reddish-brown. All *perianth segments*, except the labellum, connivent to form a helmet over the column. *Labellum* shaped like a man (hence its name), with long slender lateral lobes (arms) near its base, and a long, narrow middle lobe which forks distally into 2 slender segments (legs), which are rather shorter than the lateral lobes, sometimes with a small tooth in the sinus (penis). *Spur* absent, nectar secreted from 2 tiny depressions at the base of the labellum. *Column* very short. *Anther* adnate to the top of the column; pollinia 2, narrowed downwards into caudicles whose bases are attached to the 2 contiguous, more or less globose viscidia which are enclosed in a common bursicle, the viscidia (often coherent) and the pollinia are usually removed together, but occasionally withdrawn separately; pollen in packets of tetrads, united by elastic threads. *Rostellum* minute. *Capsule* erect.

One species in Europe, western Asia and North Africa.

Differs from *Orchis* Section *Orchis* only by the absence of a spur.

Stewart, A., Pearman, D. A. & Preston, C. D. (1994). *Scarce plants in Britain*. Peterborough.

1. A. anthropophorum (L.) Aiton fil.   Man Orchid
Ophrys anthropophora L.; *Orchis anthropophora* (L.) All.; *Loroglossum anthropophorum* (L.) Rich.; *Himantoglossum anthropophorum* (L.) Spreng.

Perennial glabrous herb; tubers 15–25 × 5–20 mm, thick, ovoid or subglobose; roots several, moderately thick. *Stem* 20–40(–60) cm, pale green, erect, solid, cylindrical, slightly ridged above, with obtuse, membranous, leafless sheaths at base. *Lower leaves* several, crowded, 6–12 × 1.5–2.5 cm, dark green or greyish-green, glossy on both sides, with numerous white dots beneath, suberect or spreading, oblong to oblong-lanceolate, more or less acute at apex, entire, sheathing at base, keeled, with 3–4 veins on each side of the midrib and fainter ones between each pair, the upper leaves smaller and grading into bracts. *Inflorescence* a spike 4–14 × 1.5–3.0 cm, narrowly cylindrical, many-flowered, becoming lax. *Bracts* 2.5–8.0 × 1–2 mm, membranous, 1-veined, tapering to an acute apex. *Flowers* green, edged with red and with a green, yellow or dull red labellum. *Outer perianth segments* 6–7 mm, adherent for about half of their length, free at the tip, forming with the inner segments a short, globose or ovoid helmet, pale green, edged violet- or brownish-red, ovate to ovate-lanceolate, obtuse to rather acute at apex, 1-veined. *Inner perianth segments* slightly shorter and much narrower than outer, greenish, linear, more or less obtuse at apex, 1-veined. *Labellum* 12–15 mm, greenish or yellowish, often edged with brownish-red, sometimes pure yellow or red, pendulous, shaped like the body of a man, with 2 long linear, very narrow side lobes and 2 shorter, equally narrow terminal lobules, sometimes with a small tooth between; the base of the lip is divided into 2 whitish, shining thickened folds which curve round and join the base of the column, enclosing a shallow cup on the floor of which are 2 small depressions which secrete nectar. *Spur* absent. *Column* very short. *Anther* yellowish-green, short, ovate, obtuse at apex; pollinia 2, small, sulphur-yellow; caudicles transparent, darker yellow; viscidia 2, colourless, nearly globular, touching or coherent, enclosed in a bursicle. *Stigma* on the roof and sides of the stigmatic chamber formed by the base of the labellum and the column. *Rostellum* minute. *Ovary* green, sessile, erect, slightly spreading, cylindrical or slightly triangular, twisted, 6-ridged. *Capsule* 10–15 mm, erect. *Flowers* 6–7. Cross-pollinated by small insects. Most reproduction is by tubers, despite plentiful seed production. $2n = 42$.

Native. Chalk and limestone grassland and scrub, field margins, roadsides, old chalk pits and limestone quarries. Local and decreasing, from Somerset and Oxfordshire to Kent, and in scattered localities north to Lincolnshire and Derbyshire; frequent only in Kent, Mediterranean region and western Europe northwards to central England and the Netherlands; Cyprus; North Africa.

21 × 20. Orchiaceras A. Camus
Aceras × Orchis

Bateman, R. M. & Farrington, O. S. (1987). A morphometric study of × *Orchiaceras bergonii* (Nanteuil) Camus and its parents (*Aceras anthropophorum* (L.) Aiton f. and *Orchis simia* Lamarck) in Kent. *Watsonia* **16**: 397–407.

21/1 × 20/7. × O. bergonii (Nanteuil) A. Camus
A. anthropophorum × O. simia
This hybrid is intermediate in flower colour and perianth shape, but the labellum is similar to that of *O. simia*.

Found at an *O. simia* locality near Faversham in Kent in 1985.

22. Himantoglossum Koch nom. cons.

Perennial herbs with entire tubers. *Stems* tall, stout and leafy. *Spikes* long, cylindrical, bracts shorter than the large flowers. All *perianth segments* except the labellum connivent into a helmet. *Labellum* 3-lobed, the lateral lobe short, the central lobe very long and narrow, spirally coiled in bud and twisted when mature. *Spur* very short. *Column* rather short, erect. *Anther* adnate to the top of the column; pollinia 2, narrowed below into caudicles both of which are attached to a single, more or less 4-angled viscidium enclosed in a bursicle; pollen in packets of tetrads united by elastic threads. *Stigma* 1, large. *Rostellum* beak-like, projecting above the stigma.

Four species in Europe and the Mediterranean region.

1. H. hircinum (L.) Spreng.   Lizard Orchid
Satyrium hircinum L.; *Orchis coriophora* auct.; *Orchis hircina* (L.) Crantz; *Loroglossum hircinum* (L.) Rich.

Perennial glabrous herb; tubers 2, ovoid or subglobose, entire; roots short, rather thick. *Stem* 20–40(–90) cm, pale green, faintly marbled with purple, solid, cylindrical, smooth, obscurely angled above. *Leaves* pale to dark green, the lower 4–6, 6–15 × 3–5 cm, elliptical-oblong to oblong-lanceolate, obtuse to acute at apex,

entire, keeled and erect or spreading, the upper smaller, tapering to an acute apex, clasping the stem and many-veined. *Inflorescence* a spike 1–25(–50) × 4–12 cm, rather lax, cylindrical. *Bracts* 5–50 × 1–2 mm, pale green or whitish, often tinged rose, narrow linear, tapering to an acute apex, more or less membranous, often with inrolled edges, with a green central vein and about 3 others. *Flowers* greenish, with very long, ribbon-like, twisted labella hanging obliquely downwards, with a tangled and untidy appearance and a goat-like smell. *Outer perianth segments* 7–10 mm, forming a helmet, cohering at base, free or not at the tip, whitish-green, sometimes flushed violet, paler inside with numerous violet-red lines and spots, ovate, rounded at apex, concave, 3- or 4-veined, the upper arched forward and rather boat-shaped. *Inner perianth segments* slightly shorter than outer, very narrow, linear, spotted, 1-veined. *Labellum* 3–50 mm, linear, wedge-shaped and broader at base, with green, strongly crimped edges continued on the linear, short, narrow, acute, curly, green or reddish side lobes, the middle lobe ribbon-like, trough-like at the base, white, spotted reddish-violet and with a dense fur of minute white papillae as far as the junction with the side lobes, the rest of the labellum pale green, bifid, notched or 2- to 4-toothed at apex. *Spur* short, conical, sock-like, rounded at apex, directed downwards. *Column* rather short, erect. *Anther* greenish-white, pyriform; pollinia 2, short, olive green, pyriform; caudicles thick, yellow, longer than the pollinia, bent at apex; viscidium elliptical or quadrangular, enclosed in a bursicle. *Stigma* obtusely 4-cornered or ovate-cordate, bordered by a dark line. *Rostellum* beak-like, projecting beyond the stigma. *Ovary* about 10 mm, pale green, rather spindle-shaped, twisted, pedicellate, with 3 slight, longitudinal ridges. *Capsule* long, tapering at base, with marked ridges. *Flowers* 5–7. Cross-pollinated by Hymenoptera. $2n = 24$.

Native. Wood margins, by woodland paths, amongst bushes, on field borders and in grassland, chiefly over chalk and limestone. Scattered places, often sporadic, in south and east England west to Somerset and north to Suffolk, formerly north to Yorkshire and Lancashire and west to Devon; formerly in Jersey. Now more widespread than at the beginning of the century. South, south-central and west Europe, northwards to England and the Netherlands and eastwards to the Balkan Peninsula and Crimea; North Africa. All British material is referable to subsp. **hircinum** which is widespread in Continental Europe except for the south-east.

23. Ophrys L.

Perennial herbs with entire, ovoid or subglobose tubers. *Stems* leafy. *Inflorescence* a lax-flowered spike. *Perianth segments* more or less spreading, the inner lateral segments usually shorter than the outer. *Labellum* large, entire to 3-lobed, often convex, velvety, usually dark-coloured and conspicuously marked, usually with a glabrous central area called the speculum. *Spur* absent. *Column* long, erect. *Anther* adnate to the top of the column; pollinia 2, narrowed downwards into long caudicles which are attached basally to separate, more or less globose viscidia enclosed in two distinct bursicles; pollen in packets of tetrads more or less firmly united by elastic threads. *Stigma* 1, large, central, depressed. *Rostellum* minute. *Ovary* not twisted.

About 30 species in Europe, western Asia and North Africa.

An easily recognisable genus by the large, spurless labellum, which often resembles an insect, and the 2 distinct bursicles. **O. bertolonii** Moretti from the Mediterranean region, was found in Dorset in 1976 (*Nature* 263: 186 (1976)), but was almost certainly planted. It has the labellum curved upwards anteriorly and the speculum scutelliform.

Lang, D. C. (1991). A new variant of *Ophrys apifera* Hudson in Britain. *Watsonia* 18: 408–410.

1. Outer perianth segments yellowish-green, the inner lateral dark violet and filiform; labellum distinctly 3-lobed and longer than wide, the lateral lobes spreading, the long central lobe deeply emarginate or 2-lobed
      1. insectifera
1. Inner lateral perianth segments oblong to linear-oblong, never filiform; labellum entire or with often indistinct basal lateral lobes, not or only just longer than wide 2.
2. Outer perianth segments yellowish- to brownish-green; inner segments yellowish-green, at least half as long as the outer; labellum lacking a terminal glabrous appendage **2. sphegodes**
2. Outer perianth segments pink or greenish-pink; inner segments pink or greenish-pink, less than half as long as outer; labellum with a glabrous terminal appendage 3.
3. Labellum only slightly convex with the basal lobes usually represented only by bosses, ending in a wide and often 3-toothed, upcurving, yellowish appendage
      3. fuciflora
3. Labellum strongly convex with 2 small, hairy basal lobes and with the central lobe ending in a long, narrow, tooth-like appendage curved downwards and backwards so as to be more or less invisible from above 4.
4. Labellum broad, with a purplish or dark brown area in the central lobe 5.
4. Labellum narrow, without a dark brown area in the central lobe 6.
5. Labellum mostly purplish or dark brown
      4(1). apifera forma **apifera**
5. Labellum greenish at base, the rest brown
      4(2). apifera forma **bicolor**
6. Labellum with 2 confluent yellow areas
      4(5). apifera forma **botteronii**
6. Labellum greenish-yellow or pale yellow with irregular brown markings 7.
7. Labellum greenish-yellow **4(3). apifera** forma **flavescens**
7. Labellum pale yellow with irregular brown markings.
      4(4). apifera forma **trolli**

1. O. insectifera L.   Fly Orchid
O. muscifera Huds.; *O. myodes* Jacq.

Perennial glabrous herb; tubers 2, globose or ovoid, the younger often stalked; roots short, rather thick. *Stem* 15–60 cm, slender, terete, solid, with whitish or brown-

ish, rarely green-tipped, obtuse, ribbed basal sheaths. *Leaves* few, 4.0–12.5 × 0.3–8.0 cm, bluish-green, the lower oblong to elliptical, more or less acute at apex, entire, keeled, folded and with about 13 veins, the upper narrower, tapering and clasping the stem. *Inflorescence* a spike 2–14 × 1.0–2.5 cm, lax, with 4–12 flowers. *Bracts* 3–25 × 1–2 mm, erect, lanceolate, folded, with inrolled edges, with about 6 veins. *Flowers* small, green with purple or reddish-brown labellum, scentless. *Outer perianth segments* 6–8 mm, yellowish-green, narrowly lanceolate, obtuse at apex, spreading, with rolled back edges, 3-veined. *Inner perianth segments* shorter, dark violet to purplish-brown, the edges rolled back to make them filiform, velutinous. *Labellum* 9–10 × 6–7 mm, rich mahogany brown, paler at apex, deflexed, papillose, 3-lobed, the lateral lobes patent, short, oblong, slightly convex and densely velvety, the middle lobe broadening downwards, bilobed or deeply notched, rarely with a short tooth in the middle; speculum reniform or square, shiny, pale bluish-violet. *Spur* absent. *Column* shorter than inner segments, with a short, blunt beak. *Anther* adnate to the top of the column; pollinia 2, bright yellow; caudicles transparent yellow; viscidia in 2 orange bursicles. *Stigma* on the inner surface of the chamber at the base of the column. *Rostellum* minute. *Ovary* pale green, sessile, erect, 6-ridged. *Flowers* 5–7. Sparingly visited by a small wasp and setting seed only if visited. 2n = 36.

Native. Woods, scrub, grassland, spoil heaps, fens and lakesides on calcareous soil. Britain north to Westmorland and Yorkshire, scattered and local but frequent in south-east England. Europe northwards to about 67° N in Norway.

2. O. sphegodes Mill.   Early Spider Orchid
O. aranifera Huds.; *O. fucifera* Curtis nom. illegit.

Perennial almost glabrous herb; tubers 2, globose or ovoid, the younger often stalked; roots short, rather thick. *Stem* 10–45 cm, yellowish-green, erect, terete, often flexuous. *Leaves* 4–10 × 0.5–1.5 cm, the lower green or greyish-green, elliptical-oblong to ovate-lanceolate, rather obtuse at apex, often mucronate, entire, spreading or recurved, with 1–14 veins, the upper narrower and more acute at apex. *Inflorescence* a spike 4–10 × 1.5–4.0 cm, lax, with 2–10 flowers. *Bracts* 10–30 × 1–4 mm, pale green, lanceolate, more or less obtuse at apex, concave, with 7–9 veins. *Flowers* medium sized, scentless, finally turning a yellowish-brown. *Outer perianth segments* 6–10(–12) mm, yellowish-green, oblong-ovate to oblong-lanceolate, obtuse at apex, edges rolled back, the lateral spreading, the upper erect and slightly arched forwards. *Inner perianth segments* 4–8 mm, green or brownish, spreading, strap-shaped, rounded or squarish at apex, wavy at margin, 1-veined. *Labellum* (8–)10–12 × 8–12 mm, subrotund to ovate-oblong, strongly convex, subentire or 3-lobed at base, the lateral lobes small and with basal protuberances, the middle lobe entire or emarginate, usually with no terminal appendage; central lobe velvety, dull purplish-brown, later turning yellowish and with various, glabrous, bluish markings often H- or horseshoe-shaped. *Spur* absent. *Column* nearly at right angles to the labellum, slightly curved forwards, forming a small, arched chamber at base. *Anther* like the head of a bird with yellowish eyes and a short, obtuse beak; pollinia 2, yellow; caudicles yellow; viscidia enclosed in bursicles. *Stigma* glistening, on the inner surface of the chamber at the base of the column. *Rostellum* minute. *Ovary* curved, 6-ridged. *Flowers* 4–6. Visited by bees and setting seed only if visited. $2n = 36$.

There is considerable variation in labellum shape, rounded and long pointed ones sometimes occurring on the same plant. Colonies of very small plants and others with dark brown outer perianth segments occur. Abnormalities in labellum and perianth segments are fairly frequent. Var. *fucifera* Rchb. fil., the Drone Orchid, was defined as having the inner perianth segments rough or downy in front and the labellum without lobes, but these characters do not often go together.

Native. Short, open grassland or on spoil heaps on chalk or limestone. Very local from Kent to Dorset and Gloucestershire; introduced and naturalised in Hertfordshire; formerly extended to Cornwall and Northamptonshire, and in Denbighshire and Jersey. West, central and south Europe northwards to England and central Germany, and eastwards to the Crimea; North Africa. All British plants are referable to subsp. **sphegodes** which is widespread in Continental Europe.

3. O. fuciflora (Crantz) Moench   Late Spider Orchid

O. insectifera var. *arachnites* L.; *Orchis fuciflora* Crantz; *Orchis arachnites* (L.) Scop.; *O. arachnites* (L.) Reichard; *O. apifera* subsp. *arachnites* (L.) Hook. fil.; *O. holosericea* auct.

Perennial nearly glabrous herb; tubers 2, entire, ovoid or globose; roots short, rather thick, sinuous, tapering to a narrow, flattened point. *Stem* 15–35(–55) cm, erect, solid, terete, sinuous. *Leaves* 4–10 × 0.5–2.5 cm, greyish-green, elliptical-oblong, obtuse to subacute at apex, entire or wavy-edged, spreading or erect, the basal sometimes withered at anthesis, the upper loosely clasping the stem, with about 15 veins. *Inflorescence* a spike 3–9 × 2–4 cm, 2- tò 6(–14)-flowered, lax. *Bracts* 5–35 × 1–5 mm, green, lanceolate, acute or obtuse at apex, concave. *Flowers* rather large, rose or white. *Outer perianth segments* 9–13 mm, usually pale rose, rarely white, ovate-oblong, obtuse at apex, margins rolled back, keeled, with 1–3 green veins. *Inner perianth segments* short, rose, rarely white, triangular or linear-lanceolate, with 2 rounded auricles at the base, faintly 3-veined, densely velvety-hairy. *Labellum* 9–13(–16) × 11–15(–17) mm, broader than long, obovate-subrotund or obscurely angled, slightly convex, 3-lobed, the lateral lobes broad but very short, usually reduced to bosses, sometimes absent, the middle lobe large, with a pair of small, spreading or reflexed triangular teeth at its distal corners and terminating in a thick, wide and often 3-toothed, flat or upcurved appendage, velvety, maroon or dark brown and marked with a bold symmetrical pattern in greenish-yellow. *Spur* absent. *Column* as long

as the inner perianth segments, forming an arched chamber. *Anther* on top of the arched chamber of the column; pollinia 2, bright yellow; caudicles bright yellow, short, translucent, ribbon-shaped; viscidia colourless, in yellowish bursicles. *Stigma* on the inside surface of the chamber of the column. *Rostellum* minute. *Ovary* sessile, linear, curved, slightly twisted, 6-ridged. *Capsule* slightly enlarged towards the apex, with 3 ˙more or less prominent ridges. *Seeds* marked with nearly parallel transverse striae. *Flowers* 6–7. Cross-pollinated by bees and other insects. $2n = 36$.

The outer perianth segments vary from dark pink to nearly white, and there is much variation in the shape and markings of the labellum. Distinguished from the closely related *O. apifera* by its more obovate, less convex labellum with its spreading or upturned, never reflexed, appendage.

Native. Short grassland on chalk. Very local in Kent. Central and south Europe and the Near East. British plants are referable to subsp. **fuciflora**, which occurs throughout the range of the species.

4. O. apifera Huds.   Bee Orchid

Perennial nearly glabrous herb; root tubers 2, entire, sub-globose; roots short, thick. *Stem* 15–45(–60) cm, solid, terete, somewhat flexuous, often stout. *Leaves* 3–8 × 0.5–2.0 cm, yellowish-green, paler and slightly glossy beneath, elliptical-oblong, obtuse to acute at apex, entire, diminishing rapidly up the stem and grading into the bracts, keeled, many-veined, the upper more or less embracing the stem. *Inflorescence* a spike 3–12 × 2–5 cm, lax, with 2–7(–11) flowers. *Bracts* 10–40 × 1–7 mm, lanceolate, acute at apex. *Flowers* rather large. *Outer perianth segments* (8–)10–15 mm, pale rose to bright violet-rose, rarely white, spreading, finally much reflexed, oblong, obtuse at apex and more or less hooded, with 3–5 green veins. *Inner perianth segments* about half as long as the outer or less, green or purplish-brown, linear with rolled back margins, covered in front with whitish hairs. *Labellum* 10–15 mm, resembling a bumble-bee, strongly convex, semi-globose, 3-lobed, the lateral lobes small, more or less bordered with yellow, more or less rounded, with basal protuberances and a blunt, forwardly and upwardly directed apex, the middle lobe much longer, curved downwards distally and terminating in 2 truncate terminal segments and a long, narrowly triangular central appendage, curled up behind so as to be invisible from the front, velvety, purplish to dark brown, marked with glabrous yellow spots distally and greenish-yellow horseshoe or H patches basally, sometimes greenish-yellow and very narrow, sometimes pale yellow with irregular brown markings and narrower. *Spur* absent. *Column* long, green, at right angles to the labellum, its base forming a hemispherical chamber. *Anther* with a long flexuous beak; pollinia 2, yellow, pyriform; caudicles yellow, long, very slender and thread-like, flexible; viscidia in bursicles. *Stigma* inside the chamber at the base of the column. *Rostellum* minute. *Ovary* green, sessile, curved forwards, linear, not twisted, slightly 3-sided. *Capsule* oblong with prominent ridges. *Seeds* with transverse striae. *Flowers* 6–7. The pollinia fall forwards and downwards on to the stigma while remaining attached to their long slender caudicles, so that the flowers habitually, though not invariably, self-pollinate in this country. $2n = 36$.

There is considerable variation in the colour and shape of the parts of the flower. A semi-peloric form has been recorded in Sussex. Five variants are recorded and require recognition.

(1) Forma **apifera**
Labellum broad with a purplish or dark brown area in the middle lobe.

(2) Forma **bicolor** (Naegeli) P. D. Sell
Labellum greenish at base lacking any pattern, the rest dark brown.

(3) Forma **flavescens** (Rost) P. D. Sell
O. apifera var. *flavescens* Rost
Labellum very narrow and greenish-yellow.

(4) Forma **trollii** (Hegetschw.) P. D. Sell
                                             Wasp Orchid
O. trollii Hegetschw.; *O. apifera* var. *trollii* (Hegetschw.) Druce
Labellum narrow and pointed, pale yellow with irregular brown markings.

(5) Forma **botteronii** (Chodat) P. D. Sell
O. apifera subsp. *jurana* Ruppert
Labellum with two confluent yellow areas.

Native. Grassland, scrub, spoil heaps and sand-dunes on calcareous or base-rich soils, especially in recently disturbed sites. Locally frequent in Britain north to Cumberland and Durham and in the Channel Islands; scattered in Ireland. South, west and central Europe northwards to Northern Ireland; North Africa. Always uncertain and fluctuating in its appearance. Most plants from the British Isles are referable to subsp. **apifera**, which occurs throughout the range of the species. Forma *bicolor* has been recorded from Anglesey. Forma *flavescens* is widespread, especially in Suffolk and Sussex. Forma *trollii* occurs in several places and has long been known in Gloucestershire and Dorset. Forma *botteronii* occurs in Wiltshire.

$\times$ **fuciflora** = **O.** $\times$ **albertiana** A. Camus
This hybrid shows the strongly convex labellum, long, narrow, deflexed outer perianth segments and long, curved anther-beak of *O. apifera*, and the short, triangular inner perianth segments and undivided labellum with a short, apical tooth of *O. fuciflora*. $2n = 36$.

It is known only from Kent where there is a substantial colony of *O. fuciflora*. It is also recorded from France, Switzerland and Italy.

$\times$ **insectifera** = **O.** $\times$ **pietzschii** Kuempel ex Rauschert nom. inval.
This hybrid is intermediate in the colour of the perianth segments and the shape of the labellum.

It has occurred in woodland in Somerset since 1968. It was originally described from Germany in 1971.

Juncus alpinoarticulatus Chaix, Fl. Vap. 74 (1785).
var. **marshallii** (Pugsley) P. D. Sell comb. nov.
J. marshallii Pugsley in *Jour. Bot. (London)* **69**: 282 (1931).
J. nodulosus var. *marshallii* (Pugsley) P. W. Richards in A. R. Clapham, Tutin & E. F. Warb., *Fl. Brit. Isles* 1251 (1952).
Holotype: Shore of Loch Ussie, near Conan, E. Ross, 8 Aug. 1892, E. S. Marshall in Herb. Marshall (**CGE**) labelled typus by H. W. Pugsley.

Schoenoplectus × carinatus (Sm.) Palla in *Bot. Jahrb. Syst.* 10: 299 (1888).
nothosubsp. **kuekenthalianus** (Junge) P. D. Sell stat. nov.
Scirpus kuekenthalianus Junge in *Jahrb. Hamburg. Wiss. Anst.* **22**, **Beih.** **3**: 73 (1905). = *S. lacustris* subsp. **tabernaemontani** (C. C. Gmel.) Å. & D. Löve × **S. triqueter** (L.) Palla

Carex flava L., *Sp. Pl.* 975 (1753).
subsp. **jemtlandica** (Palmgr.) P. D. Sell comb. nov.
C. lepidocarpa subsp. *jemtlandica* Palmgr. in Lindm., *Sv. Fan.-fl.* ed. 2, 153 (1926).
subsp. **scotica** (E. W. Davies) P. D. Sell comb. nov.
C. lepidocarpa subsp. *scotica* E. W. Davies in *Watsonia* **3**: 71 (1953).
subsp. **bergrothii** (Palmgr.) P. D. Sell comb. & stat. nov.
C. bergrothii Palmgr., *Comment. Biol.* **20**(3): 4 (1958).
subsp. **oedocarpa** (Andersson) P. D. Sell comb. nov.
C. oederi* oedocarpa Andersson, *Cyper. Scand.* 25 (1849).
subsp. **serotina** (Mérat) P. D. Sell comb. nov.
C. serotina Mérat, *Nouv. Fl. Env. Paris* ed. 2, **2**: 54 (1821).
subsp. **pulchella** (Lönnr.) P. D. Sell comb. nov.
C. oederi* pulchella Lönnr., *Obs. Pl. Suec.* 24 (1854).

Puccinellia × hybrida O. Holmb. in *Bot. Not.* **1920**: 106 (1920).
nothosubsp. **mixta** (O. Holmb.) P. D. Sell stat. nov.
P. × *mixta* O. Holmb. in *Bot. Not.* **1920**: 106 (1920).

Puccinellia fasciculata (Torr.) E. P. Bicknell in *Bull. Torrey Bot. Club* **35**: 197 (1908).
var. **pseudodistans** (Crép.) P. D. Sell comb. nov.
Glyceria pseudodistans Crép., *Nouv. Rem. Glyceria Gr. Heleochloa* 15 (1865).

Koeleria macrantha (Ledeb.) Schultes in Schultes & Schultes fil., *Mant.* **2**: 345 (1824).
subsp. **glauca** (Schrad.) P. D. Sell stat. & comb. nov.
Poa glauca Schkuhr, *Cat. Hort. Wittenb.* 49 (1799), non Vahl in Oeder, *Fl. Dan.* **6**(17): 3 (1790); *Aira glauca* Schrad., *Fl. Germ.* **1**: 256 (1806).

Agrostis stolonifera L., *Sp. Pl.* 62 (1753)
var. **marina** (Gray) P. D. Sell comb. nov.
Vilfa stolonifera var. *marina* Gray, *Nat. Arrang. Brit. Pl.* **2**: 146 (1821).
var. **calcicola** (Philipson) P. D. Sell stat. nov.
A. stolonifera ecas *calcicola* Philipson in *Jour. Linn. Soc. London (Bot.)* **51**: 98 (1937).

Elymus caninus (L.) L., *Fl. Suec.* ed. 2, 39 (1755).
subsp. **donianus** (F. B. White) P. D. Sell comb. nov.
Agropyron donianum F. B. White in *Proc. Perthshire Soc. Nat. Sci.* **1**: 41 (1889).

Elytrigia repens (L.) Desv. ex Nevski in *Trudy Bot. Inst. Akad. Nauk SSSR* ser. 1, *Fl. Sist. Vyss. Rast.* **1**: 14 (1933).
var. **aristata** (Döll) P. D. Sell comb. nov.
Agropyron repens var. *aristatum* Döll, *Fl. Bad.* 128 (1857).

Ornithogalum umbellatum L., *Sp. Pl.* 307 (1753).
subsp. **angustifolium** (Boreau) P. D. Sell stat. nov.
O. angustifolium Boreau, *Notes Pl. Franç.* **4**: 14 (1847).

Galanthus nivalis L., *Sp. Pl.* 288 (1753).
forma **pleniflorus** P. D. Sell forma nova
Holotype: Abundant in old copse, by Shepreth churchyard, Cambs, v.c. 29, 52/393475, 19 Feb. 1994, P. D. Sell 94/26 (**CGE**).
A forma *nivali* modo floribus plenis differt.

Galanthus elwesii Hook. fil. in *Bot. Mag.* **101**: t. 6166 (1875).
var. **monostictus** P. D. Sell var. nov.
Holotype: Grassland, Botanic Garden, Cambridge, 52/452572, 17 Feb. 1994, P. D. Sell 94/14 (**CGE**).
Perianthii segmenta interiora maculis singulis extra ad apices signata.

Galanthus plicatus M. Bieb., *Fl. Taur.-Caucas.* **3**: 255 (1819).
var. **viridifolius** P. D. Sell var. nov.
Holotype: Sawston churchyard, Cambs. v.c. 29, 52/487493, 16 Feb. 1995, P. D. Sell 95/5, and J. G. Murrell (**CGE**).
Folia clare flaviusculoviridia, interdum costa glauca.

Narcissus poeticus L., *Sp. Pl.* 289 (1753).
subsp. **verbanensis** (Herbert) P. D. Sell stat. nov.
N. poeticus var. *verbanensis* Herbert, *Amaryll.* 317 (1837).
subsp. **recurvus** (Haw.) P. D. Sell stat. nov.
N. recurvus Haw., *Syn. Pl. Succ. App.* 331 (1812).
subsp. **majalis** (Curtis) P. D. Sell stat. nov.
N. majalis Curtis in *Bot. Mag.* sub. t. 193 (1793).

Narcissus radiiflorus Salisb., *Prodr. Stirp. Chap. Allerton* 225 (1796).
  subsp. **exertus** (Haw.) P. D. Sell stat. nov.
N. majalis var. *exertus* Haw., *Narciss. Rev.* 150 (1819).
  subsp. **poetarum** (Haw.) P. D. Sell stat. nov.
N. poetarum Haw., *Mon. Narciss.* 14 (1831).
  subsp. **stellaris** (Haw.) P. D. Sell stat. nov.
N. stellaris Haw., *Mon. Narciss.* 15 (1831).

Narcissus pseudonarcissus L., *Sp. Pl.* 289 (1753).
  forma **pleniflorus** P. D. Sell forma nova
Holotype: Roadside bank of B1125, west of Dunwich, E. Suffolk, v.c. 25, 62/449716, 23 March 1995, P. D. Sell no. 65/20 and J. G. Murrell **(CGE)**.
A forma *pseudonarcisso* modo floribus plenis differt.

Narcissus × dichromus P. D. Sell nothospecies nova
N. cyclamineus × moschatus
Holotype: Cultivated, Botanic Garden, Cambridge, v.c. 29, 52/454572 22 March 1995, P. D. Sell 95/16 **(CGE)**.
Herba perennis. *Bulbus* circiter 20–30 mm longus, ovoideus, squamis infuscatis. *Folia* usque ad 35 cm longa, 5–12 mm lata, aliquantum glauca, longe linearia, ad apicem obtusa, supra canaliculata, subtus carinata, erecta. *Scapus* usque ad 35 cm altus, folia excedens, aliquantum gracilis, compressus, erectus, subtiliter striatus. *Spatha* 5.0–5.5 cm longa, infuscatoscariosa. *Flos* solitarius, 70–80 mm diametro, horizontalis, fere inodorus. *Perianthii tubus* 13–15 mm longus, flaviusculo-viridis. *Perianthii segmenta* 35–37 mm longa, 20–22 mm lata, alba, ovata, ad apicem magis minusve obtusa, reflexa. *Corona* 23–25 mm longa, quam perianthii segmenta brevior, lateribus parallelis, sed ad apicem leviter expansa, pallide flava, irregulariter dentata. *Stamina* 6, perianthii tubi basin versus inserta; antherae aurantiacae. *Stylus* pallide viridiusculus; stigma antheras parum excedens.

Narcissus × monochromus P. D. Sell nothospecies nova
N. cyclamineus × pseudonarcissus
Holotype: Cultivated, Botanic Garden, Cambridge, v.c. 29, 52/454572 as Cv. February Gold, 8 Feb. 1995, P. D. Sell 95/4 **(CGE)**.
Herba perennis. *Bulbus* circiter 20–30 mm longus, ovoideus, squamis infuscatis. *Folia* usque ad 35 cm longa, 5–12 mm lata, viridia, longe linearia, ad apicem obtusa, supra canaliculata, subtus carinata, erecta. *Scapus* usque ad 35 cm altus, folia plerumque paulo excedens, aliquantum gracilis, compressus, erectus, subtiliter striatus. *Spatha* 4.5–5.0 cm longa, viridula brunneolescens. *Flos* solitarius, 65–75 mm diametro, horizontalis, odorus. *Perianthii tubus* 10–15 mm longus, flaviusculoviridis. *Perianthii segmenta* 30–35 mm longa, 20–22 mm lata, pallide flava, ovata, ad apicem magis minusve obtusa, reflexa. *Corona* 22–25 mm longa, quam perianthii segmenta brevior, lateribus parallelis, sed ad apicem leviter expansa, quam perianthii segmenta aliquantum flavior, irregulariter dentata. *Stamina* 6, perianthii tubi basin versus inserta; antherae aurantiacae. *Stylus* viridiusculo-albus; stigma antheras parum excedens.

Epipactis muelleri Godfery in *Jour. Bot. (London)* **59**: 106 (April, 1921).
  var. **leptochila** (Godfery) P. D. Sell comb. nov.
E. viridiflora var. *leptochila* Godfery in *Jour. Bot. (London)* **57**: 38 (1919); *E. leptochila* (Godfery) Godfery in *Jour. Bot. (London)* **59**: 146 (May, 1921).
  var. **cleistogama** (C. Thomas) P. D. Sell comb. nov.
E. cleistogama C. Thomas in Ridd., Hedley & W.R. Price, *Fl. Gloucestershire* 612 (1948).
  var. **dunensis** (T. & T. A. Stephenson) P. D. Sell comb. nov.
E. viridiflora forma *dunensis* T. & T. A. Stephenson in *Jour. Bot. (London)* **56**: 2 (1918).

Epipactis phyllanthes G. E. Sm. in *Gard. Chron.* **1852**: 660 (1852).
  var. **cambrensis** (C. Thomas) P. D. Sell stat. nov.
E. cambrensis C. Thomas in *Watsonia* **1**: 284 (1950).

Anacamptis pyramidalis (L.) Rich., *De Orchid. Eur.* 33 (1817).
  forma **sanguinea** (Druce) P. D. Sell stat. nov.
Orchis pyramidalis var. or subvar. *sanguinea* Druce in *Rep. Bot. Soc. Exch. Club Brit. Isles* **8**: 539 (1929).

Dactylorhiza fuchsii (Druce) Soó
  var. **trilobata** (Bréb.) P. D. Sell comb. nov.
Orchis maculata var. *trilobata* Bréb., *Fl. Normand.* cd. 5, 387 (1879).

Dactylorhiza × kernerorum (Soó) Soó, *Nom. Nov. Gen. Dactylorhiza* 9 (1962).
  nothosubsp. **variabilis** (Hesl.-Harr.) P. D. Sell stat. nov.
Orchis variabilis Hesl.-Harr. in *Proc. Univ. Durham Phil. Soc.* **10**: 308 (1941).
= **D. fuchsii** subsp. **hebridensis × incarnata** subsp. **incarnata**

Dactylorhiza × transiens (Druce) Soó, *Nom. Nov. Gen. Dactylorhiza* 9 (1962).
  nothosubsp. **corylensis** (Hesl.-Harr.) P. D. Sell stat. nov.
Orchis corylensis Hesl.-Harr. in *Proc. Univ. Durham Phil. Soc.* **10**: 308 (1941).
= **D. fuchsii** subsp. **hebridensis × maculata** subsp. **ericetorum**

Dactylorhiza venusta (T. Stephenson & T. A. Stephenson) Soó, *Nom. Nov. Gen. Dactylorhiza* 9 (1962).
  nothosubsp. **hebridella** (Wilmott) P. D. Sell stat. nov.
Orchis hebridella Wilmott in *Jour. Bot. (London)* **77**: 193 (1939).
= **Orchis fuchsii** subsp. **hebridensis × purpurella** subsp. **purpurella**

Dactylorhiza comosa (Scop.) P. D. Sell comb. nov.
Orchis comosa Scop., *Fl. Carn.* ed. 2, **2**: 198 (1772).
  subsp. **scotica** (E. Nelson) P. D. Sell comb. nov.
D. majalis subsp. *scotica* E. Nelson in *Taxon* **28**: 593 (1979).
  subsp. **occidentalis** (Pugsley) P. D. Sell comb. nov.
Orchis majalis var. *occidentalis* Pugsley in *Jour. Linn. Soc. London (Bot.)* **49**: 586 (1935).

subsp. **occidentalis** forma **occidentalis** (Pugsley) P. D. Sell stat. nov.
Orchis majalis var. *occidentalis* Pugsley in *Jour. Linn. Soc. London (Bot.)* **49**: 586 (1935).
  subsp. **occidentalis** forma **kerryensis** (Wilmott) P. D. Sell stat. nov.
Orchis kerryensis Wilmott in *Proc. Linn. Soc. London Sess.* 148, 126 (1936).
  subsp. **cambrensis** (R. H. Roberts) R. H. Roberts comb. nov.
Dactylorchis majalis subsp. *cambrensis* R. H. Roberts in *Watsonia* **5**: 41 (1961).
  subsp. **majalis** (Rchb.) P. D. Sell comb. nov.
Orchis majalis Rchb., *Iconograph Bot. Pl. Crit.* **6**: 7 (1828).

Dactylorhiza praetermissa (Druce) Soó, *Nom. Nov. Gen. Dactylorhiza* 5 (1962).
  forma **junialis** (Verm.) P. D. Sell stat. & comb. nov.
Orchis latifolia var. *junialis* Verm. in *Ned. Kruidk. Arch.* **37**: 151 (1930).

Orchis militaris L., *Sp. Pl.* 941 (1753).
  var. **tenuifrons** P. D. Sell var. nov.
Holotype: Hills between Streetly and Whitchurch, Oxfordshire, W. Borrer **(CGE)**.
Caulis 10–35 cm altus. *Folia* infima 3–15 cm longa, 1.2–3.0 cm lata. *Spica* 3–7 cm longa, 3–4 cm lata, floribus 2–26.

Ophrys apifera Huds., *Fl. Angl.* 340 (1762).
  forma **bicolor** (Nägeli) P. D. Sell stat. nov.
O. bicolor Nägeli in *Ber. Schweiz. Bot. Ges.* **23**: 64 (1914).
  forma **flavescens** (Rosbach) P. D. Sell stat. nov.
O. apifera var. *flavescens* Rosbach, *Fl. Trier* **1**: 182 (1880).
  forma **trollii** (Hegetschw.) P. D. Sell stat. nov.
O. trollii Hegetschw., *Fl. Schweiz* 874 (1840).
  forma **botteronii** (Chodat) P. D. Sell stat. nov.
O. botteronii Chodat in *Bull. Soc. Bot. Genév.* **5**: 187 (1889).

ABBREVIATIONS

Authors of taxa are consistent with the abbreviations in:
Brummitt, R. & Powell, C. E. (1992). *Authors of plant names*. Kew.

Abbreviations of journals in the references follow:
Lawrence, H. M. et al. (1968).
Botanico–Periodicum–Huntianum. Pittsburgh.

Metric measurements follow standard abbreviations:

mm	=	millimetre
cm	=	centimetre
m	=	metre
km	=	kilometre

Infraspecific taxa are abbreviated:

subsp.	=	subspecies
var.	=	variety
cv.	=	cultivar

GLOSSARY

abaxial Of a lateral organ, the side away from the axis, normally the lower side.

accrescent Becoming larger after flowering, usually applied to the calyx.

achene A dry, indehiscent, one-seeded fruit, more or less hard, with a papery to leathery wall.

achene pit see **receptacular alveole**.

acicle A slender prickle with a scarcely widened base.

actinomorphic Radially symmetrical, having more than one plane of symmetry.

acumen The tip of an acuminate point.

acuminate Curved inwards on both sides to a point. Often wrongly used for gradually narrowed to a point (see **acute**).

acute With a point. Gradually narrowed to a point is *long-acute*, but is often called *acuminate*.

adaxial Of a lateral organ, the side towards the axis, normally the upper side.

adherent Joined or fused.

adnate Joined to another organ of a different kind.

aerial Above ground or above water.

alien Not native. Believed on good evidence to have been introduced by Man and now more or less naturalised.

allopolyploid A polyploid derived by hybridisation between two different species with doubling of the chromosome number.

alternate Lateral organs on an axis, one per node, successive ones on opposite sides. Commonly used also to include spiral arrangements.

amphimixis Reproducing by seed resulting from normal sexual fusion (adjective **amphimictic**).

amplexicaul Clasping the stem.

anastomosing Joining up to form loops, usually refering to veins.

anatropous (of an ovule) Bent over against the stalk.

androecium The male parts of the flower, the stamens.

andromonoecious Having male and bisexual flowers on the same plant.

anemophilus Wind-pollinated.

angustiseptate A fruit with the septum across the narrowest diameter.

annual Completing its life cycle in under 12 months, but often not within one calendar year.

annular Ring-shaped.

annulus Special thick-walled cells forming part of the opening mechanism of a fern sporangium, often forming a ring.

anther The pollen-bearing part of a stamen, usually terminal on a filament.

anthesis At the time of flowering.

aphyllopodous Without basal leaves.

apiculate With an apiculus.

apiculus With a small, broad point at the apex.

apomictic Reproducing by seed not formed by sexual fusion, or spreading vegetatively over wide areas.

appendage A small extra protrusion or extension such as on a petal, sepal or seed.

appressed Pressed close to another organ, but not united with it.

arachnoid Appearing as if covered with cobwebs.

archegonium The structure containing the female sexual cell in many land plants.

arcuate Curved so as to form a quarter of a circle or more.

aril The succulent covering around a seed, outside the testa, but not the pericarp.

aristate Extended into a long bristle.

ascending Sloping or curving upwards.

asperous Rough to the touch.

attenuate Gradually tapering.

auricles Small ear-like projections at the base of a leaf, especially in grasses.

autopolyploid A polyploid derived from one diploid species, by mutiplication of its chromosome sets.

autotrophic Neither parasitic nor saprophytic.

awn A stiff, bristle-like projection from the tip or back of the glumes and/or lemma in grasses, or from a fruit, usually the indurated style, e.g. *Erodium,* or less frequently the tip of a leaf.

axil The angle between the main and lateral axis.

axile Of a placenta formed by the central axis of an ovary, that is connected by septa to the wall.

axillary Arising in the axil of a leaf or bract.

base-rich Soils containing a relatively large amount of free basic ions, e.g. calcium, magnesium, etc.

basic number see **chromosomes**.

basifixed (of anthers) Joined by the base to the filament and not capable of independent movement.

beak A narrow, usually apical projection.

berry A fleshy fruit, usually several-seeded, without a stony layer surrounding the seeds.

biennial Completing its life cycle in more than one year, but less than two years, not flowering in the first year.

bifid Divided into two, usually deeply at the apex.

bifurcate Dividing into two branches.

biotype A genetically fixed variant of a taxon, particularly to some condition.

bird-seed alien An alien introduced as a contaminant of bird-seed.

bisexual Of a flower, bearing both sexes.

blade The main part of a flat organ, e.g. petal, leaf.

bloom A delicate, waxy, easily removed covering to fruit or leaves.

bog A community on wet, very acid peat.

bract Modified, often scale-like leaf subtending a flower, or less often a branch.

bracteate With a bract or bracts.

bracteole A supplementary or secondary bract, or a bract once removed.

bud-scales Scales enclosing a bud before it expands.

bulb A swollen, underground organ consisting of a short stem bearing a number of swollen, fleshy leaf-bases or scales with or without a tunic, the whole enclosing the next year's bud.

bulbil A small bulb or tuber arising in the axil of a leaf or in an inflorescence on the aerial part of the plant.

bullate With the surface raised into blister-like swellings.

caducous Falling off at an early stage.

caespitose Tufted.

calcicole More frequently found upon or confined to soils containing free calcium carbonate.

calcifuge Not normally found on soils containing free calcium carbonate.

callus A horny region at the base of the lemma in grasses.

calyx The sepals as a whole, the outer whorl of the perianth if different from the inner, including the calyx tube and calyx lobes.

campanulate Bell-shaped, widest at the mouth.

campylotropous When the ovule is bent so that the stalk appears to be attached to the side midway between the micropyle and chalaza.

capillary Hair-like.

capitate Head-like, as in a tight inflorescence, a knob-like stigma or style, or a stalked gland.

capitulum An aggregate head of sessile flowers in the Dipsaceae and Asteraceae.

capsule A dry, many-seeded dehiscent fruit formed from more than one carpel.

carpel One of the units of which the gynoecium is composed; the basic reproductive unit of the Magnoliophyta. One to many per flower, if more than one then separate or fused.

carpophore A stalk-like sterile part of a flower between the receptacle and the carpels, as in the Apiaceae and Caryophyllaceae.

cartilaginous Hard, not green, and easily cut with knife.

caryopsis A fruit in the Poaceae with the ovary wall and seed coat united.

casual An alien plant not naturalised.

catkin A dense spike of reduced flowers on a long axis, often wind-pollinated.

cauline Pertaining to the stem.

cell (of an ovary) The chambers into which the ovary may be divided (often each one corresponding to a carpel).

chartaceus Of papery texture.

chasmogamous Of flowers which open normally.

chlorophyll The green colouring matter of leaves, etc.

chromosomes Small deeply staining bodies, found in all nuclei, which determine most or all of the inheritable characters of organisms. Two similar sets of these are normally present in all vegetative cells, the number (diploid number, $2n$) usually being constant for a single species. The sexual reproductive cells normally contain half this number (haploid number, n).

Polypoids are multiples of more than 2 of the haploid number.

cilia Small whip-like structures by means of which some sexual reproductive cells swim.

ciliate With hairs projecting from the margin.

circumscissile Dehiscing transversely, the top of the capsule coming off like a lid.

cladode A green leaf-like lateral shoot in the Liliaceae functionally replacing the leaves, which are reduced to scales.

clavate Club-shaped, slender and thickened towards the apex.

claw The narrow part of a flat organ of which the broader part is the blade or lamina.

cleistogamous Of flowers which never open and are self-pollinated in the bud stage. Opposite of chasmogamous.

clone An area of plant from vegetative spread; or from seeds from an apomictic plant.

column A stout stalk formed by fusion of various floral parts in the Orchidaceae, Geraniaceae, etc.

columnar Column-like.

commissure The faces by which the two carpels are joined together in the Apiaceae.

compound Of an inflorescence, with the axis branched; of a leaf, made up of several distinct leaflets.

compressed Flattened.

concolorous Of approximately the same colour throughout.

cone A compact body composed of an axis with lateral organs bearing spores or seeds, in the Lycopodiophyta and Pinophyta.

cone-scales The lateral organs of a cone.

connate Organs of the same kind growing together and becoming joined, though distinct in origin.

connective Part of an anther connecting its two halves.

connivent Of two or more organs with their bases wide apart, but their apices approaching one another.

contiguous Touching at the edge with no gap between.

contorted With each lobe of the perianth overlapping the next with the same edge and appearing twisted.

convergent Of more than two organs with their apices closer together than their bases.

convolute Rolled together, coiled.

cordate A flat object with incurved base on both sides.

coriaceous Of a leathery texture.

corm A short, usually erect and tunicated, swollen underground stem of one year's duration, that of the next year arising at the top of the old one and close to it.

corolla The inner whorls of the perianth, if different from the outer; all the petals, corolla-tube and corolla-lobes.

corona A long or short tube within the perianth segments in the Liliaceae.

corymb A raceme with the pedicels becoming short towards the top, so that all the flowers are at approximately the same level.

corymbose Corymb-like, strictly not a corymb but a flat-topped cyme.

cotyledon The first leaf or leaves of a plant, already present in the seed and usually differing in shape from the other leaves. Cotyledons may remain within the testa or may be raised above the ground and become green during germination.

crenate With round teeth on the margins of a flat organ.

crenulate The diminutive of crenate.

crisped Curled.

culm The stem of a grass.

cuneate Wedge-shaped, i.e. flat and narrowed at the base of an organ.

cuneiform Wedge-shaped with the thin end at the base of a solid object.

cupule A fruit which is a nut surrounded by a husk formed of fused scales in the Fagaceae and Corylaceae.

cuspidate Abruptly narrowed to a point.

cyme An inflorescence in which each flower terminates the growth of a branch, the more distal flowers being produced by longer branches lateral to it.

cymose In the form of a cyme.

decaploid see **polyploid**.

deciduous Not persistent, the leaves falling in autumn or petals falling after anthesis.

decumbent Lying on the ground but turning up at the ends.

decurrent Having the base prolonged down the axis, as in leaves where the blade continues down the petiole or stem as a wing.

decussate Opposite, with successive pairs at right angles to each other.

deflexed Bent sharply downwards.

dehiscent Opening naturally to shed seeds or pores.

deltate Shaped like the Greek letter Δ.

dentate With patent teeth at the margin of a flat organ.

denticulate With small teeth.

depressed-globose Like globose but wider than long.

diadelphous With one stamen free, and the rest connate, in the Fabaceae.

dichasium A cyme in which the branches are opposite and more or less equal.

didymus Formed of two similar parts attached to each other by a small portion of their surface.

digitate see **palmate**.

dimorphic Occurring in two forms.

dioecious Having the sexes on different plants.

diploid Having two matching sets of chromosomes.

disc (disk) Anything disc-shaped, such as the fleshy, sometimes nectar-secreting, portion of the receptacle, surrounding or surmounting the ovary.

disc flower The central, eligulate flowers on a receptacle in the Asteracea.

dissected Deeply divided into segments.

distal At the end away from the point of attachment.

distichous Arranged in two diametrically opposed rows.

divaricate Divided into widely divergent branches.

diverging, divergent With organs having their apices further apart than their bases.

dominant The chief constituents of a particular plant community, e.g. oaks in an oak wood or heather on a moor.

dorsifixed Of anthers, attached by their back.

dorsiventral With a distinct upper side and lower side.

drupe A more or less fleshy fruit with one or more seeds each surrounded by a stony layer.

dry Not succulent.

e- Without, e.g. eglandular, ebracteate.

ebracteate Without bracts.

ectomycorrhizal An association of roots with a fungus which may form a layer outside the root.

effuse Spreading widely.

eglandular Without glands.

elaiasome An oily appendage to seeds offering food-bodies to ants.

ellipsoid A solid, elliptical in side view; broadly and narrowly can be used as in elliptical.

elliptical A flat shape widest in the middle and 1.2–3.0 times as long as wide; if less, broadly so, if more, narrowly so. By some authors this shape is called lanceolate.

emarginate Shallowly notched at the apex.

embryo-sac see **gametophyte**.

endemic Confined to one particular area, i.e. in this book, to the British Isles.

endomycorrhizal An association of roots with a fungus which may form a layer inside the root.

endosperm In the Magnoliophyta, the nutritive tissue for the embryo in the developing seed; it may or may not remain as the food-store in the mature seed.

ensiform A flat, narrow shape broadest in the middle, both ends with straight sides but gradually narrowed.

entire The margin of a flat shape not toothed or lobed.

entomophilus Insect-pollinated.

epicalyx Organs on the outside of a flower, calyx-like but outside and additional to the calyx.

epicormic Of new shoots borne direct from the trunk of a tree.

epigeal Above ground; in epigeal germination, the cotyledons are raised above the ground.

epigynous Of a flower with an inferior ovary.

epipetalous Inserted upon the corolla.

equitant Of distichous leaves folded longitudinally and overlapping in their parts.

erect Upright.

erecto-patent Between erect and patent.

erose Appearing as if gnawed.

escape A plant growing outside a garden, but having spread vegetatively or by seed from one.

evergreen Retaining its leaves throughout the year.

exceeding Longer than.

exserted Protruding from.

exstipulate Without stipules.

extravaginal Of branches which break through the sheaths of the subtending leaves.

extrorse (of anthers) Opening towards the outside of the flower.

falcate Sickle-shaped.

false fruit An apparent fruit, actually formed from tissue in addition to the real fruit.

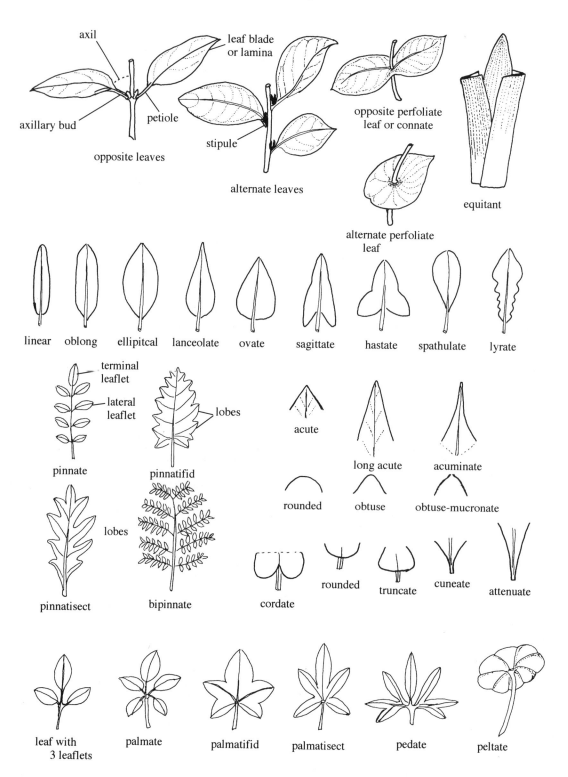

axil

leaf blade
or lamina

opposite perfoliate
leaf or connate

axillary bud

petiole

equitant

opposite leaves

stipule

alternate leaves

alternate perfoliate
leaf

linear   oblong   ellipitcal   lanceolate   ovate   sagittate   hastate   spathulate   lyrate

terminal
leaflet

lateral
leaflet

lobes

acute

long acute   acuminate

pinnate

pinnatifid

rounded   obtuse   obtuse-mucronate

lobes

pinnatisect

bipinnate

cordate   rounded   truncate   cuneate   attenuate

leaf with
3 leaflets

palmate   palmatifid   palmatisect   pedate   peltate

Leaves

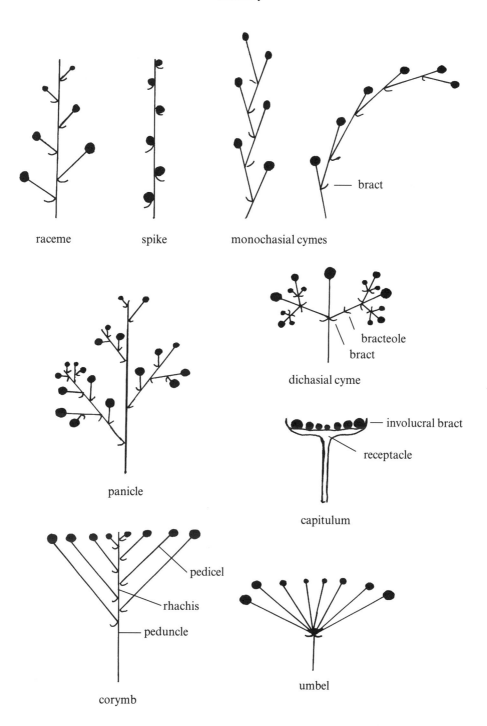

raceme   spike   monochasial cymes

— bract

panicle

dichasial cyme

bracteole
bract

— involucral bract

receptacle

capitulum

pedicel

rhachis

peduncle

corymb

umbel

Inflorescences

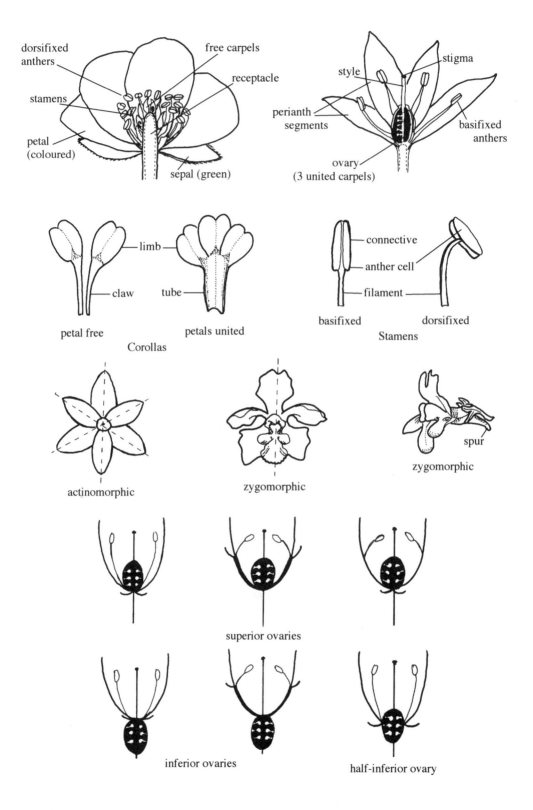

dorsifixed anthers

free carpels

receptacle

stamens

petal (coloured)

sepal (green)

style

stigma

perianth segments

basifixed anthers

ovary (3 united carpels)

limb

claw

tube

petal free

petals united

Corollas

connective

anther cell

filament

basifixed

dorsifixed

Stamens

actinomorphic

zygomorphic

zygomorphic

spur

superior ovaries

inferior ovaries

half-inferior ovary

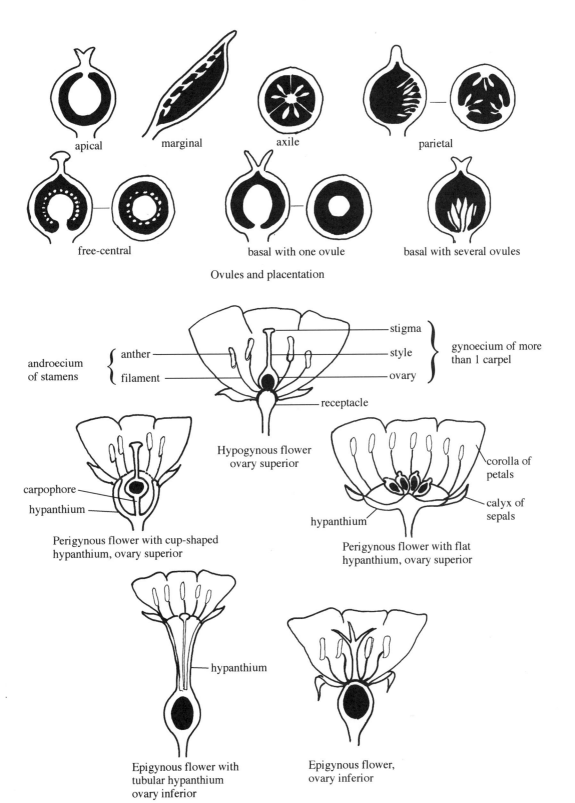

apical marginal axile parietal

free-central basal with one ovule basal with several ovules

Ovules and placentation

androecium
of stamens
{ anther
  filament

stigma
style } gynoecium of more
              than 1 carpel
ovary

receptacle

Hypogynous flower
ovary superior

carpophore
hypanthium

Perigynous flower with cup-shaped
hypanthium, ovary superior

corolla of
petals

calyx of
sepals

hypanthium

Perigynous flower with flat
hypanthium, ovary superior

hypanthium

Epigynous flower with
tubular hypanthium
ovary inferior

Epigynous flower,
ovary inferior

fasciculate In tight bundles.

fastigiate A plant with upright branches forming a narrow outline.

fen A community on alkaline, neutral or slightly acid wet peat.

fertile Producing seed capable of germination or (of anthers) containing viable pollen.

fibrous roots A root system where there is no main axis.

filament The stalk of an anther, the two together forming the stamen.

filiform Thread-like.

fimbriate With the margins divided into a fringe.

fistular Hollow and cylindrical; tube-like.

flexuous Wavy, of a stem or hair.

floccose A short, whitish indumentum often of stellate hairs.

flore pleno A double flower with many more petals than normal, usually due to conversion of the stamens to petals; in the Asteraceae, a double capitulum with all or many disc flowers converted to ray flowers.

floret A small flower.

flush Wet ground, often on hillsides, where water flows though not in a definite channel.

foliaceaus Leaf-like, of an organ not usually a leaf.

follicle A dry, usually many-seeded fruit dehiscent along its side, formed from one carpel.

free Separate, not fused to another organ or to one another except at the point of origin.

free-central Of a placenta formed by the central axis of any ovary which is not connected by septa to the wall.

fruit The ripe seeds and structure surrounding them, whether fleshy or dry; strictly the ovary and seeds, but often used to include other associated parts such as the fleshy receptacle, as in the rose and strawberry.

fugacious Withering or falling off very rapidly.

funicle The stalk of the ovary.

fusiform Spindle-shaped.

gametophyte The haploid generations of a plant that bears the true sex organs; in the Pteridophyta the prothallus, in the Pinophyta and Magnoliophyta the pollen grains (male) and embryo-sac (female).

gamopetalous Having the petals joined into a tube, at least at the base.

geniculate Bent abruptly to make a knee.

gibbous With a rounded swelling on one side, as on the base of the calyx in *Acinos*.

glabrescent Becoming glabrous. Sometimes wrongly used for sparsely hairy.

glabrous Hair-less.

gland A small globose or oblong vesicle containing oil, resin or other liquid, sunk in on the surface of or protruding from any part of a plant. When furnished with stalks they are called glandular hairs.

glandular Furnished with glands.

glaucous Bluish-white or bluish-green.

globose Spherical.

glumaceous Resembling a glume.

glume A flat bract-like organ(s) subtending a flower in the Cyperaceae and Poaceae.

grain alien Alien introduced as a contaminant of grain.

granulose With a fine, sand-like surface texture.

gymnodioecious Having female and bisexual flowers on the same plant.

gynoecium The group of female parts of a flower made up of one or more ovaries, with their styles and stigmas.

hairs Generally, hairs can be described in terms of *pubescent, tomentose, lanate, hispid, strigose,* etc. Usually it is indicated if the hairs are *eglandular, glandular, stellate* or *plumose.* If precise comparisons are made of the indumentum it is often diagnostic of a species, but very difficult to put into a few words. In some critical genera, where the indumentum is a very important character, a more detailed account of the hairs is given. There are two main types of hair, *branched* and *simple.* Branched hairs are either more or less *stellate,* or *plumose* or *subplumose,* i.e. pinnately branched with the branches projecting longer than the diameter of the hair. *Simple* hairs with a capitate tip are *glandular,* or when more or less the same throughout, *eglandular.* Eglandular hairs usually include those hairs with minute side projections not longer than the diameter of the hairs, although these are sometimes called *denticulate* hairs. Two rarer types of hair are *medifixed* hairs in the form of a T and *glochidiate* hairs in the form of an anchor. The abundance of hairs is indicated thus: *few* or *sparse* when the hairs in question form only a small proportion of the total indumentum or are scattered; *numerous* when the hairs are abundant but separate enough to be individually distinct; *dense* when they form a continuous indumentum. The length of the hairs is referred to as *very short* up to 0.3 mm, *short* 0.3–0.7 mm, *medium* 0.7–1.5 mm, and *long* 1.5–4.0 mm. If more than 4 mm the length is given. *Colour, rigidity, angle* and *waviness* can be added. In those cases when an attempt has been made to be very precise in a particular group of plants it is so stated after the generic description.

half-epigynous Of a flower with a semi-inferior ovary.

haploid Having only one set of chromosomes, as in gametophytic tissue.

hastate A leaf with two spreading basal lobes and a much longer erect central lobe.

heath A lowland community dominated by heath or ling, usually on sandy soils with a shallow layer of peat.

hemiparasite see **parasite.**

heptaploid see **polyploid.**

herb A plant dying down to ground level each year.

herbaceous Not woody, dying down each year; leaf-like as opposed to woody, horny, scarious or spongy.

hermaphrodite Bisexual.

heterochlamydeous Having the perianth segments in two distinct series which differ from one another.

heterophyllus Having leaves of more than two distinct forms.

heterosporus Having spores of two sorts, megaspores female and microspores male, as in all Pinophyta and Magnoliophyta and a few Pteridophyta.

heterostylous Having two forms, not sexes, of flowers on different plants, the two forms with different styles and pollen.

hexaploid Plants having six sets of chromosomes (see **polyploid**).

hilum The scar on a seed where it left its point of attachment.

hirsute Clothed with long stiff hairs.

hispid Coarsely and stiffly hairy.

homochlamydeous or **homolochlamydeous** Having all the perianth segments similar.

homogamous The anthers and stigmas maturing simultaneously.

homosporous Having spores of approximately the same size, as in most Pteridophyta.

hyaline Thin and translucent.

hybrid A plant originating by the fertilisation of one species by another.

hybrid swarm A series of plants originating by hybridisation between two or more species and subsequently recrossing with the parents and between themselves, so that a continuous series of forms arises.

hypanthium The extension of the receptacle above the base of the ovary in perigynous and epigynous flowers.

hypogeal Below ground; in hypogeal germination, the cotyledons remain below ground.

hypogynous Of a flower with a superior ovary, the calyx, corolla and stamens inserted at the base of the ovary.

imbricate Overlapping at the edges.

imparipinnate Pinnate with an unpaired terminal leaflet.

impressed Sunk below the surface.

included Not exserted.

incurved Curved inwards.

indehiscent Not dehiscent.

indumentum The hairy coverings as a whole.

indurated Hardened and toughened.

indusium A small flap or pocket of tissue covering a group of sporangia in many Pteridophyta.

inferior An ovary which is borne below the point of origin of the petals, sepals and stamens and is fused with the receptacle (hypanthium) surrounding it.

inflated Of an organ which is dilated, leaving a gap between it and its contents.

inflexed Bent inwards.

inflorescence A group of flowers with their branching system and associated bracts and bracteoles.

insertion The position and form of the point of attachment of an organ.

intercalary leaf The leaves on the main stem between the topmost branches and the lowest flowers of the terminal spike in *Rhinanthus*.

internode The stem between adjacent nodes.

interpetiolar Between the petioles.

interrupted Not continuous.

intrapetiolar Between the petiole and the stem.

intravaginal Within the sheath.

introduced A plant which owes its existence, in the area covered by this Flora, to deliberate or accidental importation by Man.

introgression The acquiring of characteristics by one species from another by hybridisation followed by back-crossing.

introrse Anthers opening towards the middle of the flower.

involucel An involucre at the base of a flower, formed by the united bracteoles, in the Dipsaceae.

involucral Forming an involucre.

involucre Bracts forming a more or less calyx-like structure round or just below the base of a usually condensed inflorescence, e.g. *Anthyllis*, Asteraceae.

involute With the margins rolled upwards.

isodiametric Of any shape or organ more or less the same distance across in any plane.

isomerous Having the same number of parts in two or more different floral whorls.

jaculator The indurated funicle in Acanthaceae which acts as an ejector of its seeds.

keel A longitudinal ridge on an organ, like the keel of a boat; the lower petal or petals when shaped like the keel of a boat, as in *Fumaria* and Fabaceae.

labellum The central inner perianth segment, which appears to be the lower one, but is actually the uppermost due to the flower twisting through 180° and usually different from all the other perianth segments.

lacerate Deeply and irregularly cut, appearing as if torn.

laciniate Irregularly and deeply toothed.

lamina The blade of a leaf.

laminar In the form of a flat leaf.

lammas growth Extra, usually abnormal growth, put on in summer by some trees.

lanate With matted hairs, as in sheeps wool,

lanceolate Very narrowly ovate.

lanuginose With woolly indumentum.

latex Milky juice.

latiseptate Fruit with the septum across the widest diameter.

lax Loose or diffuse, not dense.

leaflet A division of a compound leaf.

leaf-opposed A lateral organ borne on the stem on the opposite side from a leaf, not in a leaf-axil as usual.

leaf rosette A radiating cluster of leaves often at the base of the stem at soil level.

legume A usually dry, mostly many-seeded fruit dehiscent along two sides, formed from one carpel.

lemma A bract borne on the abaxial side of the flower, in the Poaceae.

lenticular Convex on both faces and more or less circular in outline.

ligulate Strap-shaped.

ligule A minute membranous flap at the base of the leaves of *Isoetes* and *Selaginella*; the strap-shaped part of a ligulate flower in the Asteraceae; the short projection in the axil of a leaf in the Cyperaceae and Poaceae.

limb The flattened, expanded part of a calyx or corolla the base of which is tubular.

linear Long and narrow with more or less parallel margins.

lip Part of the distal region of a calyx or corolla sharply differentiated from the rest due to fusion or close association of its parts.

lobe Divided substantially, but not into separate leaflets.

loculicidal Splitting down the middle of each cell of the ovary.

lodicule Two minute scales at the base of the ovary in the Poaceae.

long-shoot Stem of potentially unlimited growth, especially in trees or shrubs.

lower side The under surface of a flat organ.

lunate Crescent moon-shaped.

lyrate More or less lyre-shaped.

marsh A community on wet or periodically wet, but not peaty soil.

meadow A grassy field cut for hay.

mealy With a floury texture.

megasporangium In a heterosporous plant, the sporangium bearing megaspores.

megaspore In a heterosporous plant, the female spore which gives rise to a female gametophyte.

meiosis Special form of cell division (in sporangia, pollen-sacs or ovules) in which the chromosome number is halved, producing haploid spores.

membranous Like a membrane in consistency.

mericarp A one-seeded portion split off from a syncarpus ovary at maturity.

-merous E.g. 5-merous, having the parts in fives.

micron A micrometre, i.e. one-thousanth of a millimetre or one millionth of a metre.

microsporangium In a heterosporus plant, the sporangium bearing microspores.

microspores In a heterosporus plant, the male spores which give rise to the male gametophytes.

midrib The central main vein.

monadelphus (of stamens) United into a single bundle by the fusion of the filaments.

monocarpic Living for one year, flowering, fruiting and then dying.

monochasium A cyme with one lateral branch at each node.

monochlamydeous Having only one series of perianth segments.

monoecious Having male and female organs on the same plant.

monomorphic Occuring in one form, not dimorphic or trimorphic.

monopodial Of a stem in which growth is continued from year to year by the same apical growing point.

moor Upland communities, often dominated by heather, on dry or damp but not wet peat.

mucro The tip of a mucronate object.

mucronate Having a very short, bristle-like tip.

mucronulate The diminutive of mucronate.

mull soil A fertile woodland soil with no raw humus layer.

muricate Rough with short firm projections.

muticous Without an awn or mucro.

mycorrhiza An association of roots with a fungus which may form a layer outside the roots (ectomycorrhizal) or within the outer tissues (endomycorrhizal).

naked Devoid of hair or scales, or not enclosed.

native A plant growing in an area where it was not put by the hand of Man.

naturalised An alien plant which has become self-perpetuating in the British Isles, or a native plant which is transferred to a new locality by Man and is self-perpetuating.

nec Nor, nor of.

nectariferous Nectar-bearing.

nectar-pit A nectariferous pit.

nectary Any nectariferous organ, usually a small knob or a modified petal or stamen.

nerve see **vein**.

nodding Bent over and pendulous at the tip.

node The position of a stem where leaves, flowers or lateral stems arise.

nodule A small, more or less globose swelling.

non Not, not of.

nonaploid see **polyploid**.

nothomorph One of more than two variants of a particular hybrid.

nucellus The tissue between the embryo and the integument in an ovule.

nut A dry, indehiscent, one-seeded fruit with a hard, woody wall.

nutlet A small nut, or a woody-walled mericarp.

ob- The other way up from normal, usually flattened or widened at the distal rather than the proximal end.

obdiplostemonous The stamens in two whorls, the outer opposite the petals, the inner opposite the sepals.

oblong A flat shape with the middle part parallel-sided, 1.2–3.0 times as long as wide; if less, broadly so, if more, narrowly so.

obtuse Blunt.

ochrea Sheathing stipules in the Polygonaceae.

octoploid see **polyploid**.

opposite Of two organs arising laterally at one node on opposite sides of the stem.

orbicular Like an orb.

orthotropous (ovule) Straight and with the axis of the ovule in the same line as that of the funicle.

oval Elliptical.

ovary The basal part of the gynoecium containing the ovules.

ovate A flat shape widest nearer the base 1.2–3.0 times as long as wide; if less, broadly so, if more, narrowly so.

ovoid A solid shape, ovate in side view.

ovule An organ inside the ovary of the Magnoliophyta, naked in the Pinophyta, that contains the embryo-sac, which in turn contains the egg developed into a seed after fertilization.

palate A projecting part of the lip closing the mouth of the corolla in *Antirrhinum*.

palea A bract on the adaxial side of a flower in the Poaceae.

palmate Consisting of more than three leaflets arising from the same point.

panduriform Lanceolate in outline with a dip in each side below the middle and with an obtuse apex.

panicle A compound or much-branched inflorescence, either racemose or cymose.

papilla A small, nipple-like projection.

pappus The hairs, scales or scarious margin at the top of a fruit (achene or cypsela) in the Asteraceae.

parasite Plant which gets all or part of its nourishment by attachment (often under the ground) to other plants.

parietal Of a placenta formed by a central axis of an ovary that is connected by septa to the wall.

paripinnate Pinnate without an unpaired terminal leaflet.

partial septum A septum which is incomplete.

pasture A grassy field grazed during summer.

patent Projecting at more or less right angles, spreading.

pectinate Lobed with the lobes resembling and arranged like the teeth of a comb.

pedate With five leaflets of a leaf arising from the same point, the leaflets obovate.

pedicel Flower stalk.

peduncle Stalk of a group of flowers.

peltate Of a flat organ with its stalk inserted in the lower surface.

pendant Pendulous.

pentaploid See **polyploid**.

perennial Living for more than two years and usually flowering each year.

perianth The floral leaves including petals and sepals.

perianth segments, lobes or **tube** Petals and sepals, usually used when they are not or little differentiated.

pericarp The wall of a fruit, originally the ovary wall.

perigynous A flower with a superior ovary, but with the calyx, corolla and stamens inserted above the base of the ovary on an extension of the receptacle (hypanthium) that is not fused with the ovary.

perigynous zone The annular region between the gynoecium and the other floral parts in perigynous or epigynous flowers.

perisperm The nutritive tissue derived from the nucellus in some seeds.

perispore A membrane surrounding a spore.

persistent Remaining attached longer than normal.

petal One of the segments of the inner whorl(s) of the perianth.

petaloid Brightly coloured and resembling petals.

petiole The stalk of a leaf.

petiolate With a petiole.

phyllary One of the involucral bracts surrounding the capitulum in the Asteraceae.

phyllode A green, flattened petiole resembling a leaf.

pilose Hairy.

pinna The primary division of a more than 2-pinnate leaf.

pinnate A compound leaf, with more than 3 leaflets arising in opposite pairs along the rhachis; 2-pinnate, with the pinnae themselves pinnate.

pinnatifid Pinnately cut, but not into separate portions, the lobes connected by the lamina as well as the midrib or stalk.

pinnatisect Like pinnatifid but with some of the lower divisions reaching very nearly or quite to the midrib.

pinnule The ultimate division of a more than 2-pinnate leaf, usually applied only in ferns.

placenta The part of the ovary to which the ovules are attached.

placentation The position of the placentae in the ovary. The chief types of placentation are: *apical,* at the apex of the ovary; *axile,* in the axils formed by the meeting of the septa in the middle of the ovary; *basal,* at the base of the ovary; *free-central,* on a column or projection arising from the base in the middle of the ovary, not connected with the wall by septa; *parietal,* on the wall of the ovary or on an intrusion from it; *superficial,* when the ovules are scattered uniformly all over the inner surface of the wall of the ovary.

plastic Varying in form according to environmental conditions, not according to genetic characteristics.

plumose Feather-like.

pollen The microspores of Pinophyta and Magnoliophyta.

pollen-sac The microsporangium of a species of Pinophyta or Magnoliophyta; one of the chambers in an anther in which the pollen is formed.

pollinia Regularly shaped masses of pollen formed by a large number of pollen grains cohering, as in the Orchidaceae.

polygamous Having male, female and bisexual flowers on the same or different plants.

polyploid Having more than two sets of chromosomes e.g. 3, triploid; 4, tetraploid; 5, pentaploid; 6, hexaploid; 7, heptaploid; 8, octoploid; 9, nonaploid; 10, decaploid.

pome A fruit in which the seeds are surrounded by tough but not woody or stony layers, derived from the inner part of the fruit wall, and the whole fused with the deeply cup-shaped, fleshy receptacle, e.g. apple.

porrect Directed outwards and forwards.

premorse Ending abruptly and appearing as if bitten off at the lower end.

prickle Spiny outgrowth with a broadened base.

pricklet A small spiny outgrowth without a broadened base.

procumbent Trailing along or loosely lying on the ground.

proliferating Inflorescences bearing plantlets instead of flowers and fruits.

pro parte Partly; in part.

prostrate Lying closely along the surface of the ground.

protandrous Stamens maturing before the ovary.

prothallus The small gametophyte generation of a plant bearing the true sex organs, mostly applied to the free-living gametophytes of Pteridophyta.

protogynous Ovary maturing before the stamens.

proximal At the end near the point of attachment.

pruinose With a bloom.

pteridophytes Ferns and fern allies, i.e. Lycopodiophyta and Pteridophyta.

puberulous With very short hairs.

pubescent With short, soft hairs, but sometimes used as a general word for hairy.

punctate Marked with dots or transparent spots.

punctiform A small more or less circular dot.

pungent Sharply and stiffly pointed so as to prick.

raceme An unbranched racemose inflorescence in which the flowers are borne on pedicels.

racemose An inflorescence, usually conical in outline, whose growing points commonly continue to add to the inflorescence and in which there is usually no terminal flower. A consequence of this mode of growth is that the youngest and smallest branches or flowers are normally nearest the apex.

radical (of leaves) Arising from the base of the stem or a rhizome.

radiate A central region of tubular flowers and an outer region of ligulate flowers in the Asteraceae.

rank A vertical file of lateral organs; 2-ranked, etc., with two ranks of lateral organs.

raphe The united portions of the funicle and outer integument in a conotropous ovule.

ray Anything which radiates outwards, e.g. branches of an umbel; stigma ridges in *Papaver* or *Nuphar*.

ray flowers The outer ligulate flowers of a capitulum in the Asteraceae.

receptacle The flat, concave or convex part of the stem or peduncle from which the parts of the flower arise; often used to include the perigynous zone.

receptacular alveole The pits in which the achenes fit in the receptacle of a capitulum in the Asteraceae.

recurved Curved down or back.

reflexed Bent down or back.

regular Actinomorphic.

reniform Kidney-shaped.

replum The adjacent wall tissue to the placentae.

resilient Springing sharply back when bent out of position.

resiniferous Producing resin.

reticulate Marked with a network, usually of veins.

retuse Notched at the apex.

revolute Rolled downwards.

rhachilla The short, slender axis of the flower in the Poaceae.

rhachis The axis of an inflorescence or pinnate leaf.

rhizomatous Bearing a rhizome.

rhizome An underground stem lasting more than one growing season.

rhombic Having the shape of a diamond in a pack of playing cards.

rigid Stiff.

rostellum A beak-like process formed by the sterile stigma in the flower of the Orchidaceae.

rounded Without a point or angle.

rugose With a wrinkled surface.

rugulose Finely rugose.

ruminate Looking as though chewed.

runcinate Pinnately lobed with the lobes directed backwards towards the base of the leaf.

saccate Pouched.

sagittate The base of a leaf which is cut by straight lines upwards on either side from the margin to the petiole to leave an inverted ∧.

salt-marsh The series of communities growing on intertidal mud or sandy mud in sheltered places on coasts and in estuaries.

samara a dry indehiscent fruit, part of the wall of which forms a flattened wing.

saprophyte A plant deriving its nourishment from decaying organisms.

scabrid Rough to the touch.

scaridulous The diminutive of scabrid.

scale-leaf A leaf reduced to a small scale.

scape A flowering stem of a plant in which all the leaves are basal with none on the stem.

scarious Of thin papery texture and not green.

schizocarp A fruit which breaks into one-seeded portions or mericarps.

sclerenchyma Woody tissue in a partly or mostly non-woody organ.

scorpioid A monochasial cyme that is coiled up like a scorpion's tail when young.

scrambler A plant sprawling over other plants, fences, etc.

scrub A community dominated by shrubs.

secund All directed towards one side.

seed A fertilised ovule.

self-compatible Self-fertile, able to self-fertilise.

self-incompatible Self-sterile, not able to self-fertilise.

semi-inferior Of an ovary of which the lower part is inferior but the upper part is free and projects above the sepals, etc.

sensu lato In the broad sense.

sensu stricto In the narrow sense.

sepal One of the segments of the outer whorls of the perianth.

sepaloid Resembling sepals.

septicical Dehiscing along the septa of the ovary.

septum A wall of membrane dividing the ovary into cells.

sericeous With silky, appressed, straight hairs.

serrate Toothed with the teeth pointing towards the apex.

serration With serrate teeth.

sessile Not stalked.

setaceous Shaped like a bristle, but not necessarily rigid.

sheath Long stem, sheathing and often cylindrical round the lower part of a leaf in the Poaceae.

short-shoot A short stem of strictly limited growth usually lateral on a long-shoot, especially on trees and shrubs.

shrub A woody plant branching abundantly from the base and not reaching a very large size.

silicula A dehiscent, 2-valved, 2-celled capsule less than 3 times as long as wide in the Brassicaceae.

siliqua A dehiscent, 2-valved, 2-celled capsule more than 3 times as long as wide.

simple Not compound.

sinuate Having a wavy outline.

sinus The space of indentation between a lobe or teeth;

the space at the base of a leaf both sides of the petiole.

solitary Borne singly.

sorus A group of sporangia in the Pteridophyta.

spadix Sterile axis on which the flowers of an araceous inflorescence are packed, which often extends distinctly as a succulent appendix.

spathe An ensheathing bract in the Lemnaceae, Araceae and Hydrocharitaceae.

spathulate Paddle- or spoon-shaped.

spermatophyte A seed plant belonging to the Pinophyta or Magnoliophyta.

spermatozoid A male reproductive cell capable of moving by means of cilia.

spike A racemose inflorescence in which the flowers (or spikelets in Poaceae) have no stalks.

spikelet One to many flowers in a discrete group in *Limonium,* Cyperaceae and Poaceae.

spine A sharp, stiff, straight, woody outgrowth, usually not greatly widened at the base.

spinose Spine-like.

spiny With spines.

spiral Lateral organs on the axis, one per node, successive ones not at 180° to each other.

sporangiophore A structure, not leaf-like, bearing sporangia.

sporangium A structure containing spores.

spore The haploid product of meiotic division produced on the sporophyte and developing into the gametophyte.

sporophyll A leaf-like structure or one regarded as homologous with a leaf, bearing sporangia.

spreading Growing out divergently, not straight or erect.

spur A protrusion or tubular or pouch-like outgrowth of any part of the flower.

stamen The basic male reproductive unit of the Magnoliophyta, one to many per flower, sometimes fused.

staminode A sterile stamen, sometimes modified to perform some other function.

standard The large, often erect adaxial petal of the zygomorphic flowers of the Fabaceae.

stellate Star-shaped with radiating arms.

stem-leaves Leaves borne on the stem as opposed to basally.

sterile Not producing seed capable of germination; or anthers not viable of pollen.

stigma The receptive surface of the gynoecium to which the pollen grains adhere.

stipel A structure similar to a stipule but at the base of the leaflets of a compound leaf.

stipitate Having a short stalk or stalk-like base.

stipule A scale-like or leaf-like appendage usually at the base of a petiole, sometimes adnate to it.

stipulate With stipules.

stolon An aerial or procumbent stem, usually not swollen.

stoloniferous With stolons.

stoma; stomata Pores in the epidermis which can be closed by changes in shape of the surrounding cells.

stomium The part of the sporangium wall in the ferns, which ruptures during dehiscence.

striate Marked with long narrow depressions or ridges.

strict Growing up at a small angle to the vertical.

strigose With stiff, appressed hairs.

strophiole A small, hard appendage outside the testa of a seed.

style The part of the gynoecium connecting the ovary with the stigma.

stylopodium The enlarged base of the style in the Apiaceae.

sub- Almost, as in subacute, subglabrous, subglobose, subentire, subequal.

subshrub A perennial with a short woody surface stem producing aerial herbaceous stems.

subtended Of a lateral organ, to have another organ in its axil.

subulate Awl-shaped, narrow, pointed and more or less flattened.

succulent Fleshy and juicy or pulpy.

sucker A shoot arising adventitiously from the root of a tree or shrub often at some distance from the main stem.

suffruticose A dwarf shrub or undershrub.

superior Of any ovary that is borne above the calyx, corolla and stamens, or if below or partly below them then not fused laterally to the receptacle.

suture The line of junction of two carpels.

sympodial Of a stem in which the growing point either terminates in an inflorescence or dies each year, growth being continued by a new lateral growing point.

syncarpous (ovary) Having the carpels united to one another.

tap-root A main descending root bearing laterals.

taxon Any taxonomic grouping such as family, genus or species.

tendril A spirally coiled, thread-like outgrowth from a stem or leaf, used by the plant to climb and support.

tepal One of the segments of the perianth, sometimes used when sepals and petals are not differentiated.

terete Round, not ridged, grooved or angled.

terminal Borne at the end of a stem and limiting its growth.

ternate A compound leaf with 3 leaflets, which may be similarly divided again, 2-ternate, etc.

testa The outer coat of a seed.

tetrad A group of 4 spores cohering in a tetrahedral shape or as a flat plate and originating from a single spore mother cell.

tetraploid see **polyploid**.

tetraquetrous Square in section.

thallus The plant body when not differentialted into a stem, leaf, etc.

thorn A woody, sharp-pointed structure formed from a modifed branch.

throat The opening where the tube joins the limb of the corolla or calyx.

tiller Leafy shoot.

tomentose With a dense covering of short cottony hairs.

tooth A shallow division of a leaf, calyx or corolla or the apex of a capsule.

transverse Lying crossways.

tree A woody plant usually more than 5 m with a single trunk.

triangular A flat shape with three sides, widest at the base and 1.2–3.0 times as long as wide with the two sides gradually narrowing to a point.

trifid Split into three, but not to the base.

trifoliolate A term used in the Fabaceae for ternate.

trigonous Triangular in section, with obtuse to rounded angles.

trimorphic Occurring in three forms.

tripartite Divided into three parts.

triploid see **polyploid**.

triquetrous A solid body triangular in section and acutely angled.

trullate A flat shape, widest nearer the base and more or less angled (not rounded) there, 1.2–3.0 times as long as wide, if less broadly so, if more narrowly so.

truncate Of the base or apex of a flat organ, straight or flat.

tube The fused part of a corolla or calyx, or a hollow, cylindrical, empty prolongation of an anther.

tuber Swollen roots or subterraneous stems.

tuberous Tubercle-like.

tubercle A small more or less spherical or elliptical swelling.

tuberculate With a surface texture covered in minute tubercles.

tubular In the form of a hollow cylinder.

tufted Of elongated organs or stems which are clustered together.

tunic A dry, usually brown and more or less papery covering round a bulb or corm.

turbinate Top-shaped.

turion A detachable winter-bud, by means of which many water plants perennate.

twig Ultimate branch of a woody stem.

umbel An inflorescence in which all the pedicels arise from one point.

unarmed Devoid of thorns, spines or pricklets.

undulate Wavy at the edge in the plane at right angles to the surface.

unifacial With only one surface, not with a lower side and underside.

unilocular Having a single cavity.

unisexual Of a flower bearing organs of only one sex.

upperside The upper surface of a flat organ.

urceolate More or less globular to cylindrical but strongly contracted at the mouth.

valve A deep division or lobe, or a lobe of a capsule apex.

valvate Of perianth segments with their edges in contact with, but not overlapping in bud.

vein A strand of vascular tissue consisting of more than one vascular bundle.

verrucose Covered in small wart-like outgrowths.

villous Shaggy.

viscid Sticky.

viscidium Two viscid bodies to which the pollinia are attached in the Orchidaceae.

vitta(e) Resin canal(s) on the fruits of Apiaceae.

viviparus With flowers proliferating vegetatively and not forming seed.

waste place Uncultivated more or less open habitat much influenced by Man.

whorl More than two organs of the same kind arising at the same level.

wing Extension of an organ.

woody Hard and wood-like.

wool alien An alien introduced as a contaminate of raw wool imports.

woolly Clothed with shaggy hairs.

zygomorphic Having only one plane of symmetry.

INDEX

Accepted latin names and the page number on which the account of them occurs are in **bold** type. Synonym latin names are in *italic*. Vernacular names are in roman. Where a vernacular name is the same as the latin genus, the latin genus and species are placed first.